Historical summary of operations management

Approximate Date	Contribution/Concept	Originator
1776	Division of labor	Adam Smith
1790	Interchangeable parts	Eli Whitney
1911	Principles of scientific management	Frederick W. Taylor
1911	Motion study; use of industrial psychology	Frank and Lillian Gilbreth
1912	Chart for scheduling activities	Henry Gantt
1913	Moving assembly line	Henry Ford
1915	Mathematical model for inventory management	F. W. Harris
1930	Hawthorne studies on worker motivation	Elton Mayo
1935	Statistical procedures for sampling and quality control	H. F. Dodge, H. G. Romig, W. Shewhart, L. H. C. Tippett
1940	Operations research applications in warfare	Operations research groups
1947	Linear programming	George Dantzig
1951	Commercial digital computers	Sperry Univac
1960	Extensive development of quantitative tools	Numerous
1975	Emphasis on manufacturing strategy	W. Skinner
1980s	Emphasis on quality, flexibility, time-based competition	Japanese manufacturers
1990s	Information highway (Internet)	Numerous

INSTRUCTOR'S EDITION

PRODUCTION/OPERATIONS MANAGEMENT

SIXTH EDITION

INSTRUCTOR'S EDITION

PRODUCTION/OPERATIONS MANAGEMENT

SIXTH EDITION

WILLIAM J. STEVENSON

Rochester Institute of Technology

Boston Burr Ridge, IL Dubuque, IA Madison, WI New York San Francisco St. Louis
Bangkok Bogotá Caracas Lisbon London Madrid
Mexico City Milan New Delhi Seoul Singapore Sydney Taipei Toronto

NOTE TO THE INSTRUCTOR

Production/operations management is a dynamic discipline. It involves the integration of numerous activities and processes to produce products and services in a highly competitive global environment. In the chapters in this book, students are faced with the actual issues practicing operations managers deal with every day. There is a considerable amount of material for the student to cover and to understand. To help students learn the material, this new edition provides a number of learning tools, as well as many that have worked well for students using the earlier editions. This section describes the key features of the book, the chapter elements, the supplements that may be used in teaching the course, highlights of the sixth edition, and suggested applications for classroom instruction. By providing this support, it is our hope that you and your students will have the tools you need to make this learning experience a rewarding one.

MAJOR PEDAGOGICAL FEATURES

A number of key features in this text have been specifically designed to help students learn operations concepts and problem-solving techniques.

Examples. Throughout the book, whenever a quantitative technique is described, a student-level example is included that illustrates the technique. These are designed in a boxed "example/solution" format that is easy to find and easy to read and study.

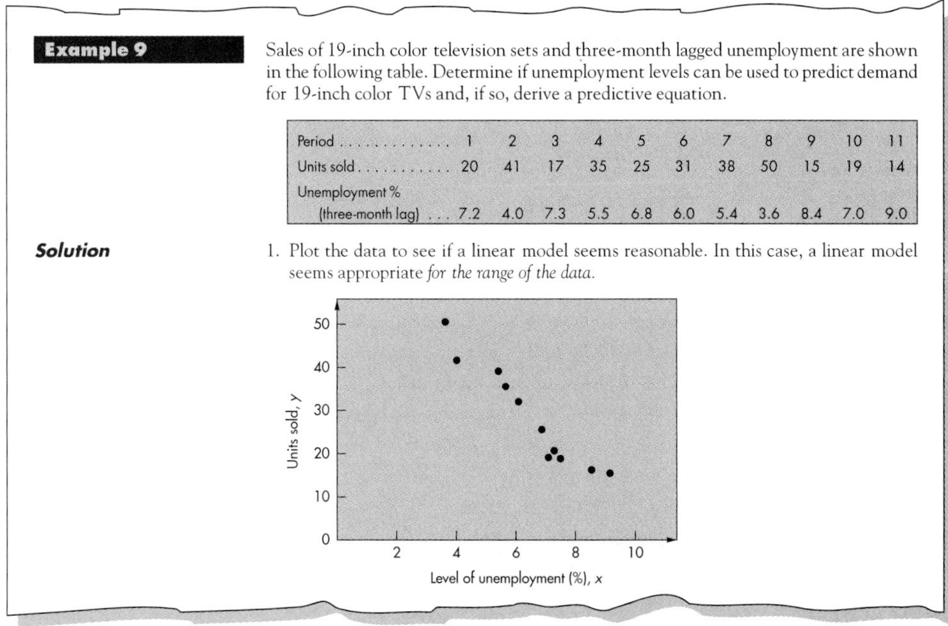

Example 9

Sales of 19-inch color television sets and three-month lagged unemployment are shown in the following table. Determine if unemployment levels can be used to predict demand for 19-inch color TVs and, if so, derive a predictive equation.

Period	1	2	3	4	5	6	7	8	9	10	11
Units sold	20	41	17	35	25	31	38	50	15	19	14
Unemployment % (three-month lag)	7.2	4.0	7.3	5.5	6.8	6.0	5.4	3.6	8.4	7.0	9.0

Solution

1. Plot the data to see if a linear model seems reasonable. In this case, a linear model seems appropriate *for the range of the data.*

Solved Problems. At the end of chapters and chapter supplements, "solved problems" are provided to illustrate problem solving and the core concepts in the chapter. These have been carefully prepared to enhance and support student understanding as well as to serve as examples and confidence builders for students prior to their own work on chapter assignment problems.

Excel Spreadsheet Solutions. Where applicable, the solved problems include 'screen shots' of a spreadsheet solution. Many of these were taken from the Excel templates, which are included on the CD-ROM in the back of the text.

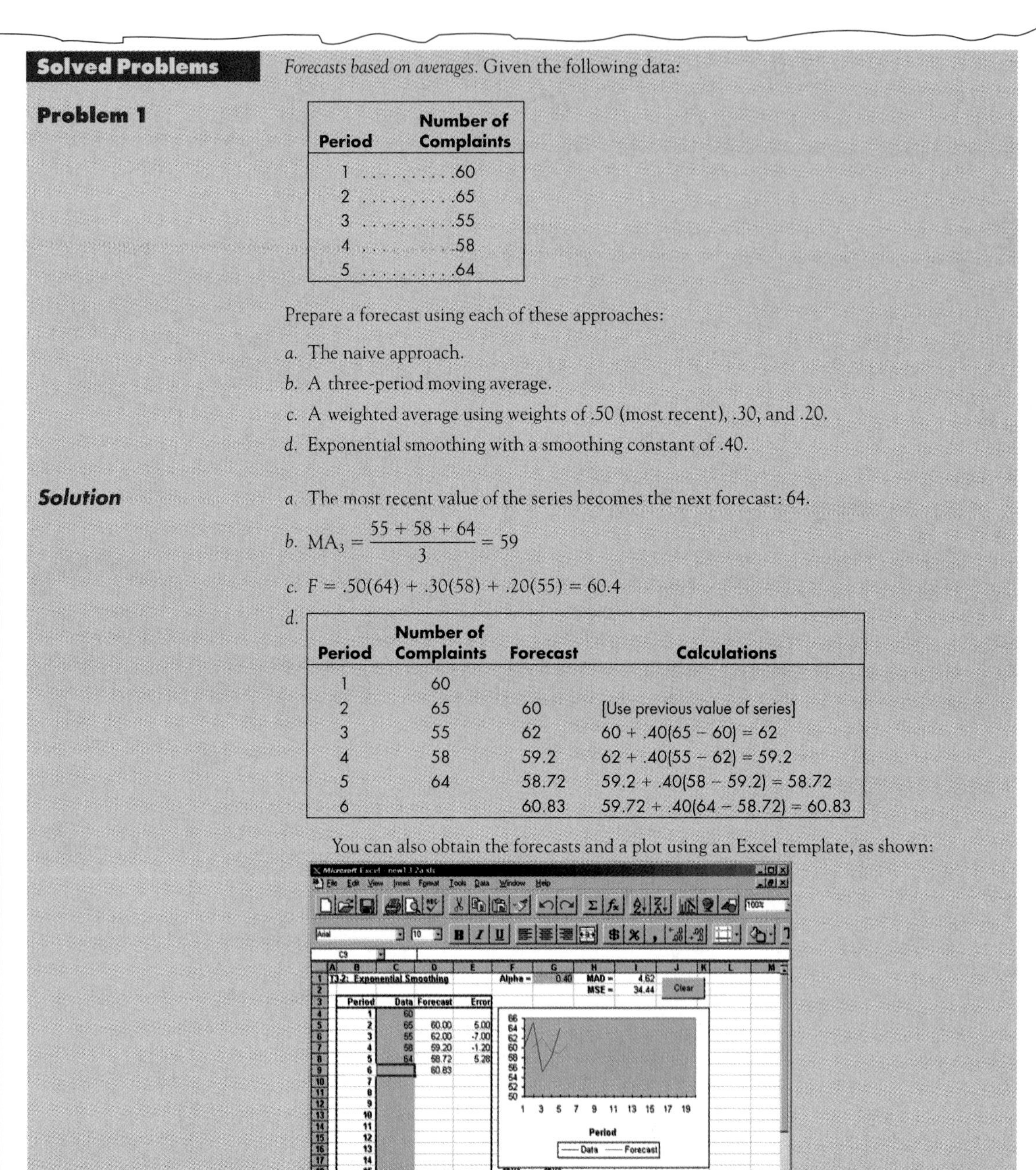

Solved Problems

Problem 1

Forecasts based on averages. Given the following data:

Period	Number of Complaints
1	60
2	65
3	55
4	58
5	64

Prepare a forecast using each of these approaches:

a. The naive approach.

b. A three-period moving average.

c. A weighted average using weights of .50 (most recent), .30, and .20.

d. Exponential smoothing with a smoothing constant of .40.

Solution

a. The most recent value of the series becomes the next forecast: 64.

b. $MA_3 = \dfrac{55 + 58 + 64}{3} = 59$

c. $F = .50(64) + .30(58) + .20(55) = 60.4$

d.

Period	Number of Complaints	Forecast	Calculations
1	60		
2	65	60	[Use previous value of series]
3	55	62	60 + .40(65 − 60) = 62
4	58	59.2	62 + .40(55 − 62) = 59.2
5	64	58.72	59.2 + .40(58 − 59.2) = 58.72
6		60.83	59.72 + .40(64 − 58.72) = 60.83

You can also obtain the forecasts and a plot using an Excel template, as shown:

Readings and Newsclips. Throughout the text, and in the assignment sections of some chapters, the text includes short newsclips as well as larger, more complete readings. These highlight important real-world applications, provide examples of production/operations issues and offer further elaboration of the text material. They also provide a basis for classroom discussion and generate interest in the subject matter. Many of the end of chapter readings include assignment questions.

READING

CUSTOM-MADE, DIRECT FROM THE PLANT

By Otis Port in New York

The newest manufacturing combines state-of-the-art fabrication and product-delivery technologies with the old-world artisan's aim to please.

It is 2010, and you need a new suit—fast—for an overseas business trip. No problem. You head to the department store at the mall. You step into a kiosk-like contraption, and an optical scanner automatically measures your body. Seconds after you record your choice of style and fabric, the information is relayed to a plant where lasers cut the cloth precisely to your size. A few days later, the suit is ready.

You may not have to wait 15 years for this kind of high-tech couture. Manufacturing is fast entering a new age of industrial excellence: mass customization. Not only clothes, but a huge variety of goods, from autos to computers, will be manufactured to match each customer's taste, specifications, and budget. Mass customization will mark the culmination and synthesis of agile manufacturing, virtual companies, and total quality management. If ever the term "paradigm shift" was

THE EVOLUTION OF THE FACTORY

Factories will soon rival craftshops in their ability to customize

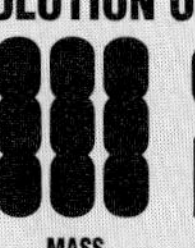

FACTORY FOCUS	MASS PRODUCTION	FLEXIBLE PRODUCTION	MASS CUSTOMIZATION
	1900 TO 1970	1971 TO 2000	2001 TO 2020
TYPICAL NUMBER OF:			
Machine tools	150	50 TO 30	25 TO 20
Products made	10 TO 15	100 TO 1,000	UNLIMITED
PRODUCTS REWORKED DUE TO POOR QUALITY	25% OR MORE	0.02% OR LESS	UNDER 0.0005%

NEWSCLIP

PRODUCTIVITY GAINS CURB INFLATION

www.burgerking.com

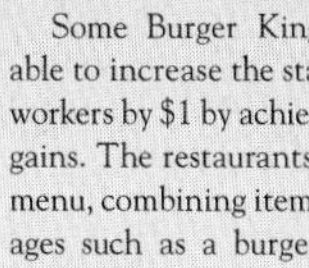

Wage increases can lead to inflationary pressure—they can cause the prices consumers pay for products and services to rise. Unless, that is, they are offset by gains in productivity, which lead to an increase in profits. If that happens, a portion of the resulting profits can be used to cover the wage increases without having to raise prices.

Some Burger Kings were recently able to increase the starting pay of new workers by $1 by achieving productivity gains. The restaurants restructured the menu, combining items into meal packages such as a burger, fries, and soft drink. This enabled the counter staff to enter orders with a single key stroke instead of multiple key strokes on their point-of-sale machines, reducing the time needed to take an order. That, in turn, enabled them to take orders more quickly, increasing productivity and, consequently, reducing labor requirements, which produced higher profits.

Source: Based on "Despite Pay Increases, Gains in Productivity, Profits Curb Inflation," *The Wall Street Journal*, May 22, 1997, p. A1.

CHAPTER ELEMENTS

Within each chapter, you will find the following elements which are designed to facilitate learning. Most all of these have been carefully developed over six editions and have proven to be successful.

Learning Objectives. Every chapter and supplement lists the learning objectives as a short guide to studying the chapter.

Chapter Outlines. Every chapter and supplement includes this quick overview of topics covered with page number references.

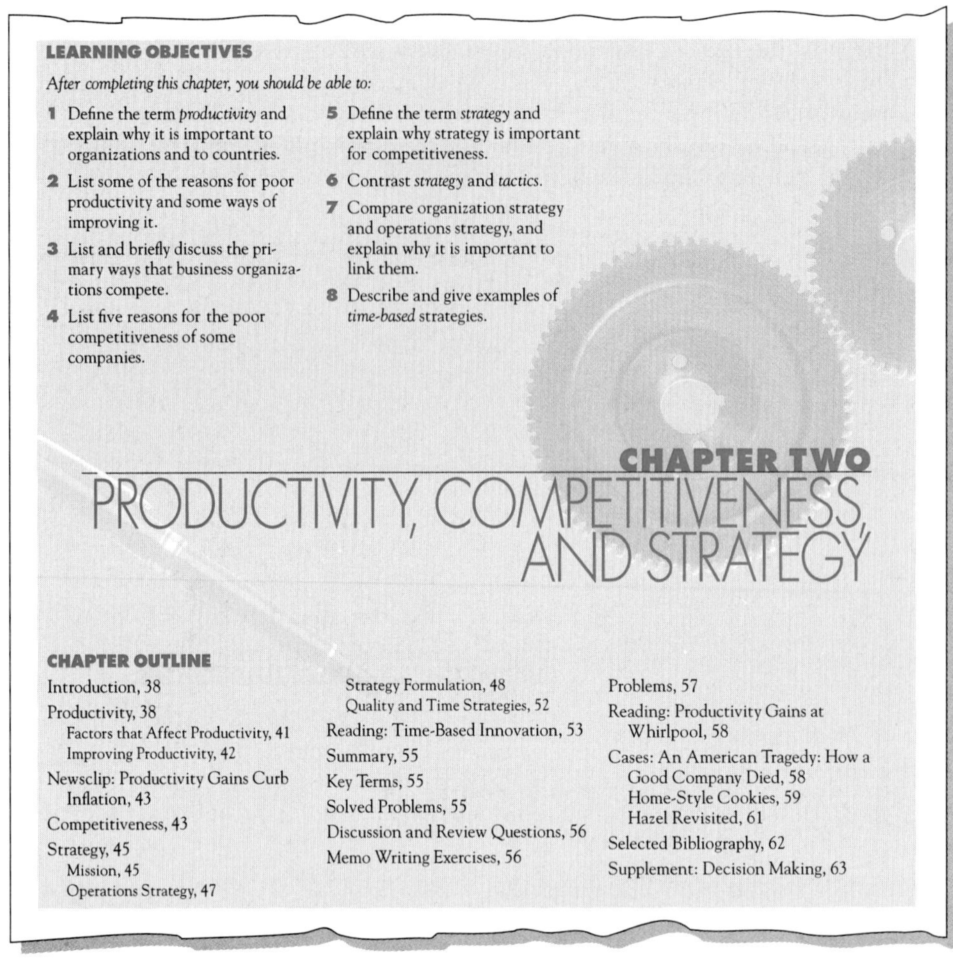

LEARNING OBJECTIVES

After completing this chapter, you should be able to:

1 Define the term *productivity* and explain why it is important to organizations and to countries.
2 List some of the reasons for poor productivity and some ways of improving it.
3 List and briefly discuss the primary ways that business organizations compete.
4 List five reasons for the poor competitiveness of some companies.
5 Define the term *strategy* and explain why strategy is important for competitiveness.
6 Contrast *strategy* and *tactics*.
7 Compare organization strategy and operations strategy, and explain why it is important to link them.
8 Describe and give examples of *time-based* strategies.

CHAPTER TWO

PRODUCTIVITY, COMPETITIVENESS, AND STRATEGY

CHAPTER OUTLINE

Introduction, 38
Productivity, 38
Factors that Affect Productivity, 41
Improving Productivity, 42
Newsclip: Productivity Gains Curb Inflation, 43
Competitiveness, 43
Strategy, 45
Mission, 45
Operations Strategy, 47
Strategy Formulation, 48
Quality and Time Strategies, 52
Reading: Time-Based Innovation, 53
Summary, 55
Key Terms, 55
Solved Problems, 55
Discussion and Review Questions, 56
Memo Writing Exercises, 56
Problems, 57
Reading: Productivity Gains at Whirlpool, 58
Cases: An American Tragedy: How a Good Company Died, 58
Home-Style Cookies, 59
Hazel Revisited, 61
Selected Bibliography, 62
Supplement: Decision Making, 63

Figures and Photos. Throughout the text, color photographs have been included to show operations, in both manufacturing and service settings. These are intended to help create interest and provide students with images of the real world of operations.

All of the more than 300 graphic illustrations in the text have been color coded consistently to help students learn while stimulating interest.

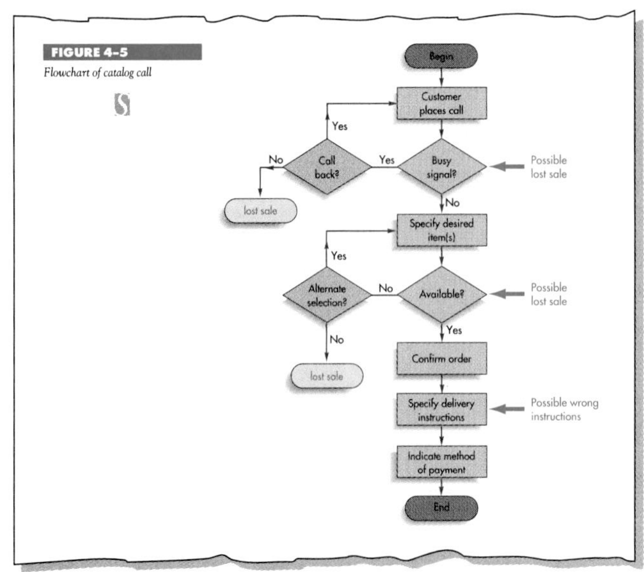

Chrysler's 1.3 mile automated durability road simulates bad roads and is used for testing the integrity of their automobiles and trucks. Chrysler vehicles are "guided" by computer-controlled robots.

A milestone crash test performed at Ford Motor Corporation to plan for a new side-impact head and chest air bag.

Operations Strategy. Throughout the text, whenever the concepts being presented have significant strategic implications for firms, an icon and head is placed to signify that text will relate these strategic issues.

Operations Strategy

Process selection often requires engineering expertise. However, many managers charged with selecting a process may have little technical knowledge; their education and experience lie in marketing, finance, and the like. Consequently, they may tend to delegate technical decisions to engineers. Engineering white elephants are not uncommon, and neither are systems based on narrow viewpoints of problems and solutions. In the long run, the solution may be to hire and promote managers who have both managerial and technical skills and expertise. In the short run, managers must work with technical experts, asking questions and increasing their understanding of the benefits and limitations of sophisticated processing equipment and technology, and ultimately make decisions themselves. Thus, there is a real need for *management of technology.*

Memo Writing Exercises. This feature, which was suggested by a reviewer, helps students to develop their communications skills by completing assignments that require analysis and organization, and communicating clearly in writing. In order to help students 'get started,' a new detailed example is included in Chapter 1.

Problems. All chapters and supplements have numerous problems, ranging from simple practice types which apply techniques, to more difficult conceptual problems that provide a challenge and require students to integrate concepts. Many new problems have been added in this edition.

Operations Tours. This feature gives students a descriptive look at operations in action at manufacturers or service organizations. These "real life" illustrations show direct application of the text information and reinforce the importance of the concepts students are studying.

OPERATIONS TOUR
IN THE CHIPS AT JAYS **Neil Steinberg**

A potato chip is a delicate thing. Fragile. A pound of pressure will crush it. So when you're moving 250 tons of chips through your plant, as they do every day at Jays Foods, you need to have a system.

"You don't buy potato crumbs, you buy potato chips," said Tom Howe, CEO and co-owner of the Chicago company, at 99th and Cottage Grove. Jays makes 125 different types and brands of chips and several hundred varieties of popcorn, puffs, twists, pretzels and assorted bagged munchies.

Jays combats the tendency of potato chips to crush into flinders with a variety of conveyor belts, radial filling chutes and gently vibrating slides, where masses of chips, a yard deep, are gradually massaged forward, the outer layer of chips shearing away like the face of a glacier.

The raw material is far easier to handle. An entire semi-trailer of sturdy North Dakota "chipping" potatoes can be emptied in a matter of minutes, by backing the trailer onto a hydraulic lift, tilting it 45 degrees and letting the potatoes—grown for their thin skins and low moisture—tumble out.

Cases. Many chapters include short cases. The cases were selected to provide a broader, more integrated thinking opportunity for students without taking a full "case" approach.

Instructor's Edition Annotations. Most problems have answers and key intermediate results printed in color in the margins in the Instructor's Edition. These are intended to provide a quick reference for the instructor in class and to help in choosing problems for assignment; or for providing hints or helps to students in class discussion.

Icons. Icons are included throughout the text, in the margins, inside readings, newsclips, or next to photos to point out relevant applications in a discussion or concept. These include: service icons to alert students to examples that are service oriented, global icons to point out the international aspects of a particular application, and internet icons to direct students to a web address where they can seek additional updated information on a company discussed in the text. These icons are intended to help students integrate and have opportunities to expand upon what they learn from the text and in class.

For the instructor, two additional icons are provided in the instructor's edition—a video icon to recommend a particular video segment that applies to the topics being covered and a Powerpoint icon that points to a portion of the text, a figure or a graphic that can be accessed in the Powerpoint portion of the Presentation CD-ROM described below.

END OF CHAPTER MATERIAL

Summaries. Chapters and supplements contain summaries that provide an overview of the material covered.

Key Terms. Key terms are highlighted in the text, and then repeated in the margin with brief definitions for emphasis, and they are listed at the end of each chapter (along with page references) to aid in reviewing.

Manufacturing Design

The term **design for manufacturing (DFM)** is also used to indicate the designing of products that are compatible with an organization's capabilities. A related concept in manufacturing is **design for assembly (DFA).** A good design must take into account not only how a product will be fabricated, but also how it will be assembled. Design for assembly focuses on reducing the number of parts in an assembly, as well as the assembly methods and sequence that will be employed.

Environmental regulations and recycling have given rise to another concern for designers, **design for recycling (DFR).** Here the focus is on designing products to allow for dis-assembly of used products for the purpose of recovering components and materials for reuse.

design for manufacturing (DFM) Designers take into account the organization's capabilities when designing a product.

design for assembly (DFA) Design focuses on reducing the number of parts in a product and on assembly methods and sequence.

Discussion and Review Questions. Each chapter and each supplement has a list of discussion and review questions. These precede the problem sets and are intended to serve two main purposes. They enable students to review the chapter concepts before trying to apply any quantitative techniques, and they provide a basis for classroom dialogue and discussion.

TEACHING SUPPORT

This text is accompanied by teaching supplements that you and your students may use in the course.

For the Student

STUDENT CD-ROM. The Student CD-ROM, packaged with each copy of the text and the instructor's edition, includes Powerpoint slides, Excel Templates and data files, self quizzes, study outlines, a tutorial on the simplex method, video clips, and a web reference file.

- *Study Guide.* Co-authored by Paul Van Ness, the study guide contains key ideas, a glossary, problems and solutions, and questions and answers.

For the Instructor

PRESENTATION CD-ROM. An all-in-one lecture aid, the Presentation CD-ROM is offered to adopters of the text. It contains the Instructor's Manual, Student Study Guide, Test Bank/Computest, Powerpoint slides, Excel Templates and data files, HOM Lecture Screens, Study Outlines, Web links, and video clips from the Irwin/McGraw-Hill video series. The Instructor's Manual and Test Bank are also available in print formats, and Computest is available on disk in Windows and Macintosh platforms. Each of these key components is described below:

- *Instructor's Manual.* Prepared by Ceyhun Ozgur, this manual includes a brief guide to using the Irwin/McGraw-Hill video series, "notes" for each chapter and supplement, complete solutions to all text problems, and teaching hints for the readings, cases, and memo exercises. Also included are several enrichment modules that provide coverage of such topics as Simplex, Vogel's approximation method (VAM), and an expanded discussion of models.
- *Test Bank and Computest.* Prepared by Seung Lae Kim, the written manual and Computest package each include over 2,000 questions and problems for exams. All of these items have been class-tested by the author or contributors.
- *Powerpoint Lecture Slides.* Prepared by Ralph Butler, the powerpoint slides draw on the highlights of each chapter and provide an opportunity for the instructor to emphasize the most relevant visuals in their class discussions.
- *Excel Templates.* Templates created by Lee Tangedahl allow the instructor and students to solve a number of problems in the text or additional problems you may assign using pre-formatted templates.

IRWIN/McGRAW-HILL WEB SITE

The OM Center edited by Byron Finch provides additional operations management resources for both students and instructors. Please consider this as your site for pedagogical support or reference and for students seeking current OM information. Among its features, the site covers OM resources by topic, contains over fifty company tours, offers internet published articles and business news, and lists OM publications and organizations. To explore, visit http://www.mhhe.com/pom/

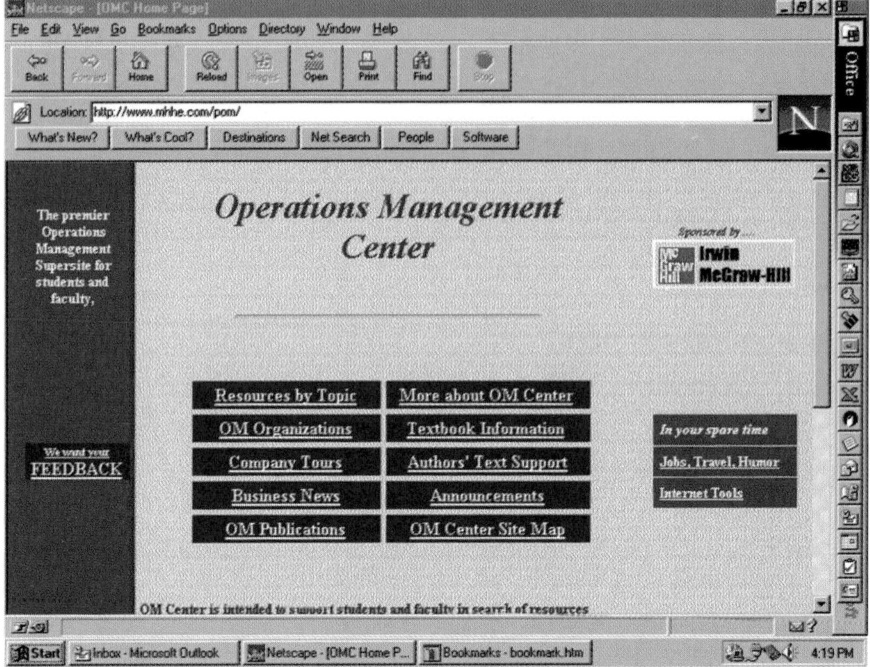

Irwin/McGraw-Hill Operations Management Video Series. This series of professionally developed video segments is being provided free to adopters. The series currently contains 21 segments which are *briefly* described below. Each segment was developed with three primary intentions—to provide some insight into a particular operation concept, such as inventory, etc.; second, to take students on-site to a real plant or operation; and, third, to hear from real managers and operators of the critical importance of operations concepts to business today.

VOLUME I

Tape 1

"Lean Production" (13 minutes) This segment shows and discusses how Lean Production is used at Caterpillar, Cummins Engine, and Navistar.

"Quality" (13 minutes) George Bush at the Baldrige Award ceremony opens this segment, which goes on to list the seven basic tools. Then Zytec, Motorola, and Hewlett-Packard are used as examples.

"Manufacturing Processes" (10 minutes) This segment presents the Hayes-Wheelright continuum, from customized, low-volume to mechanized, high-volume manufacturing.

Tape 2

"Computer-Integrated Manufacturing" (12 minutes) This segment presents an on-site tour of the Nucor Steel minimill, focusing on the automation system.

"Inventory Management" (11 minutes) This segment contrasts Navistar's high turnover-low inventory heavy truck manufacturing system with the high inventory-service parts business at Caterpillar.

"Services" (11 minutes) This segment features First National Bank of Chicago and particularly the operations aspect of its check-clearing system. Interviews are included.

VOLUME II

"Service Systems and the Service-System Design Matrix" (12 minutes) First National Bank of Chicago is again highlighted with a survey of their customer services.

"Improving Operations Methods" (12 minutes) This segment illustrates how Bernard Welding Equipment Company reduced set-up time and changed from a push to pull system to better serve their customers.

"Layout Improvements and Equipment Strategies" (10 minutes) This segment features Bernard Welding, this time illustrating a reconfigured layout based on process flow.

"Supplier Development Outreach Program" (16 minutes) From Toyota, this tape describes changes and improvements made at Flex-n-Gate, a manufacturer in Danville, Illinois, which supplies Toyota with pickup truck bumpers.

VOLUME III

"Reengineering at Caterpillar" (10 minutes) This segment shows how a team and a specialist work together to reengineer the blueprint drawing system at the small engine division at CAT.

"Washburn Guitars" (12 minutes) This segment provides a tour of the Chicago plant which produces 150,000 top line electric guitars annually. The video focuses on the flexibility and scheduling rules.

"Shouldice Hospital, Canada" (25 minutes) This video describes the specialty service of Shouldice—hernia surgery.

VOLUME IV

Tape 1

"Value-Driven Production at Trek" (9 minutes) This segment describes the distinctive approach Trek uses to manufacture high-quality, mass-customized bikes for their customers. On-site footage features interviews and examples of unique uses of materials, testing, and assembly.

"Scheduling Services—The United Solution" (9 minutes) This segment gives an overview of the scheduling system used by United Airlines in planning flights, maintaining its equipment, and delivering efficient service to its customers.

"Quality Product and Process Design at Detroit Diesel" (11 minutes) This segment features Detroit Diesel and its focus on improving quality from design of products through manufacturing of engines.

Tape 2

"Production Tour of the Vision Light System at Federal Signal" (12 minutes) This segment follows the production and assembly process for vision lights at Federal Signal Corporation, beginning with the original bill of materials schedule, through parts inspections and assembly processes.

"JIT at Federal Signal" (12 minutes) JIT concepts at Federal Signal are featured that reduce waste, improve quality, and closely meet customer demand in manufacturing a variety of products.

VOLUME V

"TriState-Converting to JIT—Assessment" Part 1 (10 minutes) This video describes and shows the assessment and planning steps taken by TriState Industries, a supplier of metal products for the construction industry, as it prepares to deliver a new product and double its output based on a new contract.

"TriState-Converting to JIT—Implementation" Part 2 (9:50 minutes) This segment shows how TriState implemented the new cells, eliminated a large percentage of WIP and instituted a Kanban control system.

"International Logistics Featuring American President Lines" (11 minutes) This segment describes the state of the art American President Line operations at LA harbor, including electronic tracking from point of origin, electronic loading maps and schedules, and services provided to global manufacturers and suppliers.

HIGHLIGHTS OF THE 6TH EDITION

This edition has a number of additions and improvements. Among them are the following:

- **Spreadsheet applications** have been added in many chapters.
- A new chapter on **Supply Chain Management** has been added.
- A new chapter on **Quality** has been added, and the quality chapters are now together in a sequence, Chapters 9, 10, 11.
- **Forecasting,** formerly the subject of Chapter 10, has been moved up to Chapter 3.
- The **Maintenance** supplement has been expanded, and now includes quantitative material.
- More real **company examples** have been added throughout the text.
- Several new **Operations Tours** have been added, including Boeing, Jays Chips, and PSC (bar code scanners).
- The **video series** has been expanded to include Volumes IV and V.
- Many new **Readings** and **Newsclips** have been added.
- Several **new cases** have been added, and some existing cases have been revised.
- More than two-thirds of the photos have been replaced.
- Increased emphasis has been given to **services** throughout the text.
- Global applications of operations and the effect of globalization on operations has been given greater emphasis.
- The book is printed in a **new format,** with a **larger and darker font,** more distinct **page numbers,** and **restyled tables.**
- Extensive **error checking** of Answers to Problems and Instructor Manual solutions has been done.

HOM OPERATIONS MANAGEMENT SOFTWARE FOR WINDOWS, BY MOSES, SESHADRI, AND YAKIREVICH

This new software package developed at NYU offers powerful Windows-based programs for solving real-world type operating problems, such as forecasting, process analysis, waiting line design and analysis, project management, MRP and inventory management, and capacity planning and management. All programs include standard edit, file, and help icons and constant "run," "results," and "plot" options. HOM imports and exports files to and from Excel and each module has detailed, step-by-step "how to solve" dialog boxes for quick reference or for beginning students.

The HOM package also includes data sets drawn from classic Operations cases as well as two integrative cases which utilize all the modules, one from a service company (United Branch Bank) and one from a manufacturer (Ice Queen).

HOM is available as an independent text/software package or at a discount, shrink-wrapped with the text.

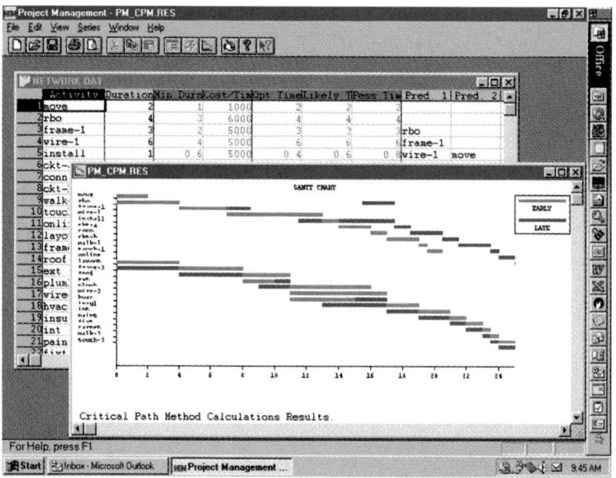

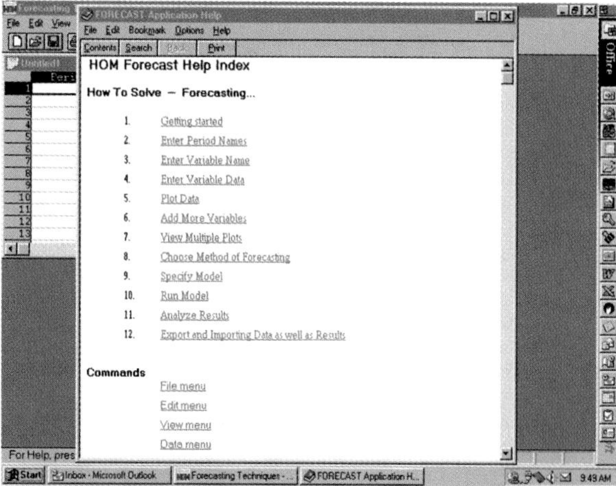

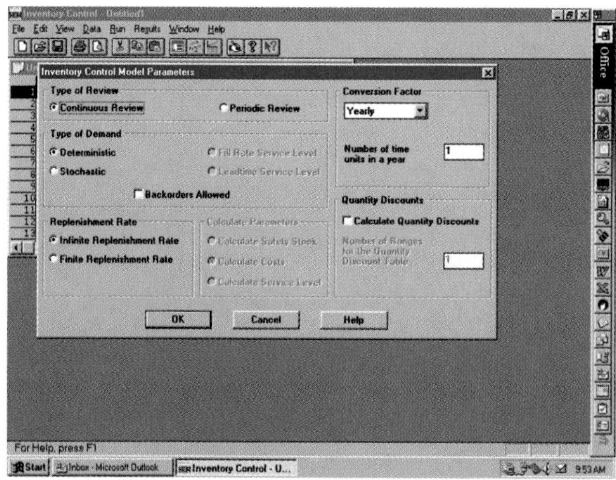

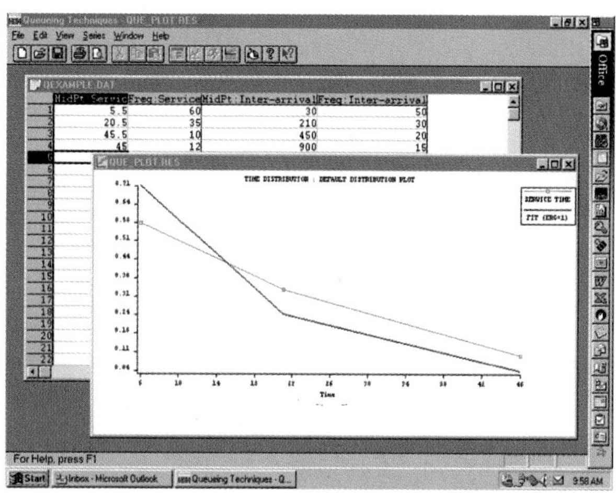

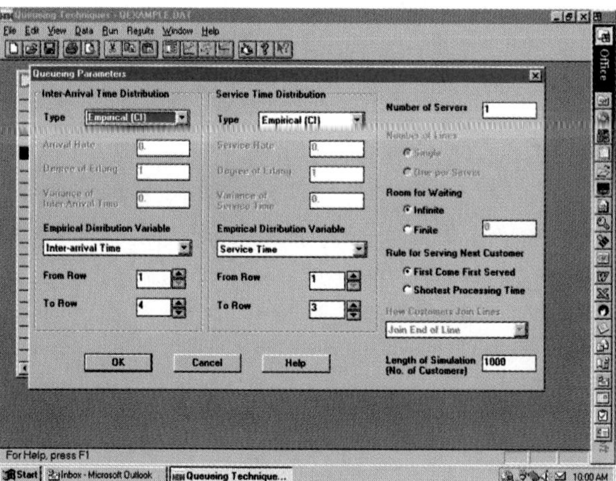

PRODUCTION/OPERATIONS MANAGEMENT

SIXTH EDITION

PRODUCTION/OPERATIONS MANAGEMENT

SIXTH EDITION

WILLIAM J. STEVENSON

Rochester Institute of Technology

Boston Burr Ridge, IL Dubuque, IA Madison, WI New York San Francisco St. Louis
Bangkok Bogotá Caracas Lisbon London Madrid
Mexico City Milan New Delhi Seoul Singapore Sydney Taipei Toronto

This Book Is Dedicated to You

Irwin/McGraw-Hill

A Division of The **McGraw-Hill** *Companies*

PRODUCTION/OPERATIONS MANAGEMENT

This book is printed on acid-free paper.

international 1 2 3 4 5 6 7 8 9 0 VNH/VNH 9 3 2 1 0 9 8
domestic 1 2 3 4 5 6 7 8 9 0 VNH/VNH 9 3 2 1 0 9 8

ISBN 0-07-366112-0 (student ed.)
ISBN 0-07-366113-9 (instructor's ed.)
ISBN 0-07-115856-1 (international edition)

Vice president and editorial director: *Michael W. Junior*
Publisher: *Jeffrey J. Shelstad*
Executive editor: *Richard T. Hercher, Jr.*
Developmental editor: *Wanda J. Zeman*
Senior marketing manager: *Colleen J. Suljic*
Senior project manager: *Jean Lou Hess*
Senior production supervisor: *Heather D. Burbridge*
Senior designer: *Laurie J. Entringer*
Cover and interior designer: *Ellen Pettengel Design*
Supplement coordinator: *Cathy L. Tepper*
Compositor: GAC *Shepard Poorman Communications*
Typeface: *10.5/12 Goudy*
Printer: *Von Hoffmann Press, Inc.*

Library of Congress Cataloging-in-Publication Data

Stevenson, William J.
Production/operations management / William J. Stevenson. — 6th ed.
p. cm.
Includes index.
ISBN 0-07-366112-0
1. Production management. I. Title.
TS155.S7824 1999
658.5—dc21 98-24796

When ordering the title, use ISBN 0-07-115856-1

http://www.mhhe.com

THE IRWIN/MCGRAW-HILL SERIES OPERATIONS AND DECISION SCIENCES

Operations Management

Bowersox and Closs
Logistical Management: The Integrated Supply Chain Process
First Edition

Chase, Aquilano, and Jacobs
Production and Operations Management
Eighth Edition

Chu, Hottenstein, and Greenlaw
PROSIM for Windows
Third Edition

Cohen and Apte
Manufacturing Automation
First Edition

Davis, Aquilano, and Chase
Fundamentals of Operations Management
Third Edition

Dilworth
Operations Management
Second Edition

Dobler
Purchasing and Supply Management
Sixth Edition

Flaherty
Global Operations Management
First Edition

Fitzsimmons and Fitzsimmons
Service Management: Operations, Strategy, Information Technology
Second Edition

Hill
Manufacturing Strategy: Text and Cases
Second Edition

Hopp and Spearman
Factory Physics
First Edition

Lambert and Stock
Strategic Logistics Management
Third Edition

Leenders and Fearon
Purchasing and Supply Chain Management
Eleventh Edition

Lotfi and Pegels
Decision Support Systems for Operations and Management Science
Third Edition

Melnyk and Denzler
Operations Management
First Edition

Moses and Seshadri
HOM Operations Management Software for Windows
First Edition

Nahmias
Production and Operations Analysis
Third Edition

Nicholas
Competitive Manufacturing Management
First Edition

Pinedo, Chao
Operations Scheduling
First Edition

Sanderson and Uzumeri
Managing Product Families
First Edition

Schroeder
Operations Management
Fourth Edition

Schonberger and Knod
Operations Management: Customer-Focused Principles
Sixth Edition

Stevenson
Production/Operations Management
Sixth Edition

Vollmann, Berry, and Whybark
Manufacturing Planning and Control Systems
Fourth Edition

Zipkin
Foundations of Inventory Management
First Edition

Quantitative Methods and Management Science

Bodily, Carraway, Frey, and Pfeifer
Quantitative Business Analysis: Casebook
First Edition

Bodily, Carraway, Frey, and Pfeifer
Quantitative Business Analysis: Text and Cases
First Edition

Bonini, Hausman, and Bierman
Quantitative Analysis for Business Decisions
Ninth Edition

Hesse
Managerial Spreadsheet Modeling and Analysis
First Edition

Lotfi and Pegels
Decision Support Systems for Operations and Management Science
Third Edition

Stevenson
Introduction to Management Science
Second Edition

PREFACE

The material in this book is intended as an introduction to the field of production and operations management. It is suitable for both undergraduate and graduate students. The field of production and operations management is dynamic, and very much a part of many of the good things that are happening in business organizations. The book is intended to be interesting and informative. Much of what you learn will have practical application.

The subject matter represents a blend of concepts from industrial engineering, cost accounting, general management, quantitative methods, and statistics. Production and operations activities, such as forecasting, choosing a location for an office or plant, allocating resources, designing products and services, scheduling activities, and assuring and improving quality are core activities and often strategic issues in business organizations. Some of you are or will be employed directly in these areas, while others will have jobs that are indirectly related to this area. So whether this is your field of study or not, knowledge of this field will certainly benefit you and the organization you work for.

The text contains more material than one could normally hope to cover in a one-semester course. Rather than rely on the author's personal bias, each instructor can choose those topics most suited to his or her own proclivities. Those who prefer an analytic quantitative emphasis, for example, will be quite comfortable with the abundance of examples and student problems. Those who prefer a more qualitative approach will welcome the fact that some of the more quantitative material is placed in chapter supplements and that there are memo exercises, operations tours, and cases for assignment. Obviously, there are many possibilities between these two extremes.

ACKNOWLEDGMENTS

I would like to thank the reviewers of this edition who contributed significantly to the final product. They are: W. C. Benton, Ohio State University; Roy Boykin, California State University–Chico; Renato de Matta, University of Iowa; Ed Duplaga, Bowling Green State University; Frank Forst, Loyola University; Mark Gershon, Temple University; Nicholas A. Glaskowsky, University of Miami; Lisa Houts, California State University–Fresno; Tim Ireland, Oklahoma State University; Tom Johnson, University of South Florida; Xenophone Koufteras, University of Texas–El Paso; Arundhati Kumar, California State University–Northridge; Satish Mehra, University of Memphis; Mike Pesch, St. Cloud State University; Gregory J. Petrakis, Ottawa University and University of Missouri; Jeff Rummel, University of Connecticut; Buddhadev Roychoudhury, Mankato State University; Leonard Ross, California State Polytechnic University; Harold Schramm, University of Dubuque; and Victor Sower, Sam Houston State University.

I would also like to thank all of the reviewers of previous editions for their valuable contributions to the success of this book.

Many students and instructors offered valuable suggestions, and I want to thank them as well. And, Lee Tangedahl, Renato de Matta, and Frank Forst did a superb job of accuracy checking.

I would also like to thank the authors of the various CD-ROM supplements and stand-alone supplements that are designed to accompany the textbook. Lee Tangedahl developed the spreadsheet templates; Mehdi Kaighobadi put together the data files; Ceyhun Ozgur updated the Instructor's Manual; Ralph Butler developed Powerpoint presentations; Seung Lae Kim updated the Test Bank and the CompuTest; and Paul Van Ness coauthored the Study Guide.

Finally, I want to extend my thanks to all of the people at Irwin/McGraw-Hill for their efforts and support. It is always a pleasure to work with such a competent and professional group of people. Special thank you's go to Dick Hercher, Wanda Zeman, and Jean Lou Hess.

William J. Stevenson

NOTE TO THE STUDENT

The material in this text is part of the core knowledge in your education. Consequently, you will derive considerable benefit from your study of operations management, *regardless of your major.* Practically speaking, production and operations is a course in *management.*

This book describes principles and concepts of production and operations management. You should be aware that many of these principles and concepts are applicable to other aspects of your professional and personal life. Consequently, you should expect the benefits of your study of production and operations management to serve you in those other areas as well.

Some students approach this course with apprehension, and perhaps even some negative feelings. It may be that they have heard that the course contains a certain amount of quantitative material which they may feel uncomfortable with, or that the subject matter is dreary, or that the course is about "factory management." This is unfortunate, because the subject matter of this book is interesting and vital for all business students. While it is true that some of the material is quantitative, numerous examples, solved problems, and answers at the back of the book will help you with the quantitative material. As for "factory management," there is material on manufacturing as well as on services. Manufacturing is important, and something that you should know about for a number of reasons. Look around you. Most of the "things" you see were manufactured: cars, trucks, planes, clothing, shoes, computers, books, pens and pencils, stereos and cell phones. And these are just the tip of the iceberg. So it makes sense to know something about how these sorts of things are produced. Beyond all that is the fact that manufacturing is largely responsible for the high standard of living people have in industrialized countries.

After reading each chapter or supplement in the text, attending related classroom lectures, and completing assigned questions and problems, you should be able to do each of the following:

1. *Identify the key features* of that material.
2. *Define and use terminology.*
3. *Solve typical problems.*
4. *Recognize applications* of the concepts and techniques covered.
5. *Discuss the subject matter* in some depth, including its relevance, managerial considerations, and advantages and limitations.

You will encounter a number of chapter supplements. Check with your instructor to determine whether or not to study them.

This book places an emphasis on problem solving. There are many examples throughout the text illustrating solutions. In addition, at the end of most chapters and supplements you will find a group of solved problems. The examples within the chapter itself serve to illustrate concepts and techniques. Too much detail at those points would be counterproductive. Yet, later on, when you begin to solve the end-of-chapter problems, you will find the *solved problems* quite helpful. Moreover, those solved problems usually illustrate more and different details than the problems within the chapter.

I suggest the following approach for studying and problem solving:

1. Look over the chapter outline and learning objectives.
2. Read the chapter summary, and then skim the chapter.
3. Read the chapter and take notes using the study questions on the CD-ROM.
4. Look over and try to answer the discussion and review questions.
5. Solve the problems, referring to the solved problems and chapter examples as needed.

Note that the answers to many problems are given at the end of the book. Try to solve each problem before turning to the answer. Remember—tests don't come with answers.

A study guide is also available. If your bookstore does not stock it, you can ask them to order it for you.

Enjoy!

W.J.S.

BRIEF CONTENTS

CONTENTS

PART ONE INTRODUCTION

Introduction to production/operations management includes two chapters:

1. Production and operations management, Chapter 1
2. Productivity, competitiveness, and strategy, Chapter 2

Chapter 1 introduces you to the field of operations management. It describes the nature and scope of operations management, and how operations management relates to other parts of the organization. Among the important topics it covers are the different types of production systems, a comparison of manufacturing and service operations, a brief history of operations management, and a list of recent trends in operations. After you have read this chapter, you will have a good understanding of what the operations function of a business organization encompasses.

Chapter 2 discusses operations management in a broader context, and presents the issues of productivity, competition, and strategy. After you have read Chapter 2, you will understand the importance of the operations function relative to the goals of a business organization. This chapter also describes time-based strategies, which many organizations are now adopting as they seek to become more competitive and to better serve their customers. The supplement of Chapter 2 describes *decision theory*.

"If it ain't broke, keep looking."
BELL ATLANTIC WAY BEHAVIORS
BEST COST
TEAMWORK
COACH

LEARNING OBJECTIVES

After completing this chapter, you should be able to:

1. Define the term *production/operations management* (POM).
2. Identify the three major functional areas of organizations and describe how they interrelate.
3. Describe the operations function and the nature of the operations manager's job.
4. Differentiate between design and operation of production systems.
5. Provide a general description of the different types of operations.
6. Compare and contrast service and manufacturing operations.
7. Briefly describe the historical evolution of POM.
8. Describe the key aspects of operations management decision making.
9. Identify some of the current trends in operations management.
10. Describe the *Pareto phenomenon* and tell why it is important in problem solving.

CHAPTER ONE
PRODUCTION AND OPERATIONS MANAGEMENT

CHAPTER OUTLINE

This book is about production and operations management (POM), which involves the planning, coordinating, and executing of all activities that create goods or provide services. The subject matter is fascinating and timely: productivity, quality, foreign competition, and customer service are very much in the news. All are part of production and operations management. This first chapter presents an introduction and overview of operations management. Among the issues it addresses are: What is operations management? Why is it important? What does an operations manager do?

The chapter also provides a brief description of the historical evolution of operations management, and a discussion of the current trends that impact operations management.

INTRODUCTION

To many people, the term *production* conjures up images of factories, machines, and assembly lines. Interestingly enough, the field of production management in the past focused almost exclusively on manufacturing management, with a heavy emphasis on the methods and techniques used in operating a factory. In recent years, the scope of production management has broadened considerably. Production concepts and techniques are applied to a wide range of activities and situations *outside* manufacturing; that is, in *services* such as health care, food service, recreation, banking, hotel management, retail sales, education, transportation, and government. This broadened scope has given the field the name *production/operations management,* or more simply, **operations management,** a term that more closely reflects the diverse nature of activities to which its concepts and techniques are applied.

operations management The management of systems or processes that *create goods and/or provide services.*

We can use an airline company to illustrate a production/operations system. The system consists of the airplanes, airport facilities, and maintenance facilities, sometimes spread out over a wide territory. Most of the activities performed by management and employees fall into the realm of operations management:

Forecasting such things as weather and landing conditions, seat demand for flights, and the growth in air travel.

Capacity planning, essential for the airline to maintain the cash flow and make a reasonable profit. (Too few or too many planes, or even the right number of planes but in the wrong places, will hurt profits.)

Southwest Airlines workers at the maintenance hangar in Dallas performing a routine maintenance check on a Boeing 737 engine. Checks are periodically performed based on a maintenance schedule guided by FAA regulations.

www.iflyswa.com

Scheduling of planes for flights and for routine maintenance; scheduling of pilots and flight attendants; and scheduling of ground crews, counter staff, and baggage handlers.

Managing inventory of such items as foods and beverages, first-aid equipment, in-flight magazines, pillows and blankets, life preservers.

Assuring quality, essential in flying and maintenance operations, where the emphasis is on safety. Also important in dealing with customers at ticket counters, check-in, telephone reservations, and curb service, where the emphasis is on efficiency and courtesy.

Employee motivation and training in all phases of operations.

Location of facilities according to managers' decisions on which cities to provide service for, where to locate maintenance facilities, and where to locate major and minor hubs.

Now consider a bicycle factory. This might be primarily an *assembly* operation: buying components such as frames, tires, wheels, gears, and other items from suppliers, and then assembling bicycles. The factory might also do some of the *fabrication* work itself, forming frames, making the gears and chains, and buy mainly raw materials and a few parts and materials such as paint, nuts and bolts, and tires. Among the key management tasks in either case are scheduling production, deciding which components to make and which to buy, ordering parts and materials, deciding on the style of bicycle to produce and how many, purchasing new equipment to replace old or worn out equipment, maintaining equipment, motivating workers, and ensuring that quality standards are met.

Obviously, an airline company and a bicycle factory are completely different types of operations. One is primarily a service operation, the other a producer of goods. Nonetheless, these two operations have much in common. Both involve scheduling of activities, motivating employees, ordering and managing supplies, selecting and maintaining equipment, satisfying quality standards, and—above all—satisfying customers. In both systems, the success of the business depends on short- and long-term planning.

WHY STUDY OPERATIONS MANAGEMENT?

You may be wondering why you need to study operations management. Actually, there are a number of very good reasons. One is that operations management activities are at the core of *all* business organizations, regardless of what business they are in. Second, 35 percent or more of all jobs are in operations management-related areas—such areas as customer service, quality assurance, production planning and control, scheduling, job design, inventory management, and many more. Third, activities in all of the other areas of business organizations, such as finance, accounting, human resources, logistics, marketing, purchasing, as well as others are all interrelated with operations management activities, so it is essential for these people to have a basic understanding of operations management. But beyond all of this is the reality that production/operations management is about *management,* and *all managers* need to possess the knowledge and skills in the content areas you will learn about here. Among them are productivity, strategy, forecasting, quality, inventory control, and scheduling. Also, you will learn how to use a range of quantitative tools that enhance managerial decision making.

If you are thinking of a career in production/operations management, you can benefit by joining one or more of the professional societies.

American Production and Inventory Control Society (APICS)
500 West Annandale Road, Falls Church, Virginia 22046-4274

www.apics-stlouis.com/
www.apics-houston.org/
www.apicsaustin.com/

American Society for Quality (ASQ)
230 West Wells Street, Milwaukee, Wisconsin 53203

National Association of Purchasing Management (NAPM)
2055 E. Centennial Circle, Tempe, AZ 85284

www.napm.org

Association for Systems Management
P.O. Box 38370, Cleveland, Ohio 44130-0307

www.infoanalytic.com/asm/

APICS and NAPM both offer a practitioner certification examination that can enhance your qualifications. Information about job opportunities can be obtained from all of these societies as well as from other sources, such as the Decision Sciences Institute (University Plaza, Atlanta, Georgia, 30303), and the Institute of Industrial Engineers (25 Technology Park, Norcross, Georgia, 30092).

FUNCTIONS WITHIN BUSINESS ORGANIZATIONS

Organizations are formed to pursue goals that are achieved more efficiently by the concerted efforts of a group of people than by individuals working alone. Business organizations are devoted to producing goods and/or providing services. They may be for-profit or nonprofit organizations. Their goals, products, and services may be similar or quite different. Nonetheless, their functions and the way they operate are similar.

A typical business organization has three basic functions: finance, marketing, and production/operations (see Figure 1–1). These three functions, and other supporting functions, perform different but *related* activities necessary for the operation of the organization. The interdependency of the major functions is depicted by overlapping circles in Figure 1–2. The functions must interact to achieve the goals and objectives of the organization, and each makes an important contribution. Often the success of an organization depends not only on how well each area performs but also on how well the areas *interface* with each other. For instance, unless production and marketing work together, marketing may promote goods or services that production cannot profitably deliver, or production may turn out goods or services for which there is no demand. Similarly, unless finance and production people work closely, funds for expansion or new equipment may not be available when needed.

Let's take a closer look at these functions.

Operations

The operational function consists of all activities *directly* related to producing goods or providing services. The production function exists not only in manufacturing and as-

FIGURE 1–1

The three basic functions of business organizations

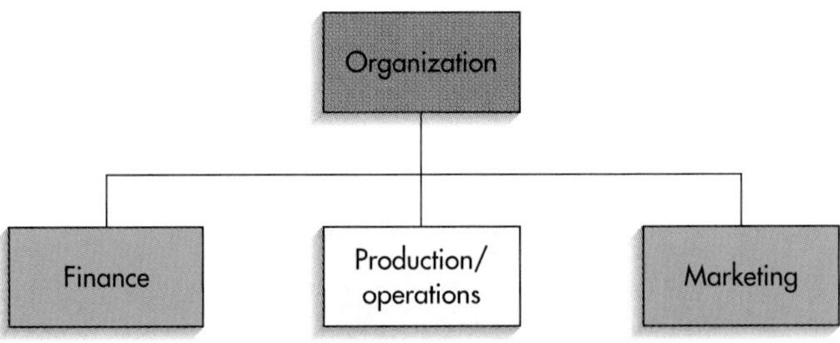

FIGURE 1–2

The three major functions of business organizations overlap

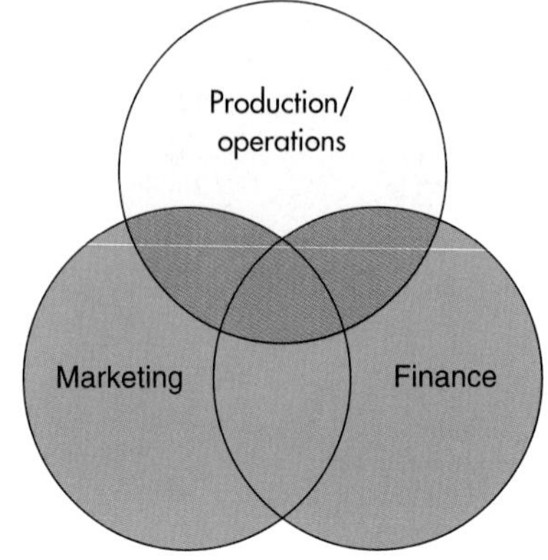

sembly operations, which are *goods-oriented,* but also in areas such as health care, transportation, food handling, and retailing, which are primarily *service-oriented.* Table 1–1 provides illustrations of the diversity of operations management settings.

The operations function is the core of most business organizations; it is responsible for the creation of an organization's goods or services. Inputs are used to obtain finished goods or services using one or more *transformation processes* (e.g., storing, transporting, cutting). To ensure that the desired outputs are obtained, measurements are taken at various points in the transformation process (*feedback*) and then compared with previously established standards to determine whether corrective action is needed (*control*). Figure 1–3 shows the conversion process.

Table 1–2 provides some examples of inputs, transformation processes, and outputs.

The essence of the operations function is to *add value* during the transformation process: **Value-added** is the term used to describe the difference between the cost of inputs and the value or price of outputs. In nonprofit organizations, the value of outputs (e.g., highway construction, police and fire protection) is their value to society; the greater the value added, the greater the effectiveness of these operations. In for-profit organizations, the value of outputs is measured by the prices that customers are willing to pay for those goods or services. Firms use the money generated by value-added for research and development, investment in new facilities and equipment, and *profits.* Consequently, the greater the value-added, the greater the amount of funds available for these purposes.

value-added The difference between the cost of inputs and the value or price of outputs.

One way that businesses attempt to become more productive is to examine critically whether the operations performed by their workers add value. Businesses consider those that do not add value wasteful. Eliminating or improving such operations decreases the cost of inputs or processing, thereby increasing the value-added. For instance, a firm may discover it is producing an item much earlier than the scheduled delivery date to a customer, thus requiring the storage of the item in a warehouse until delivery. In effect,

TABLE 1–1

Examples of types of operations

Type of Operations	Examples
Goods producing	Farming, mining, construction, manufacturing, power generation
Storage/transportation	Warehousing, trucking, mail service, moving, taxis, buses, hotels, airlines
Exchange	Retailing, wholesaling, banking, renting or leasing, library loans
Entertainment	Films, radio and television, plays, concerts, recording
Communication	Newspapers, radio and TV newscasts, telephone, satellites

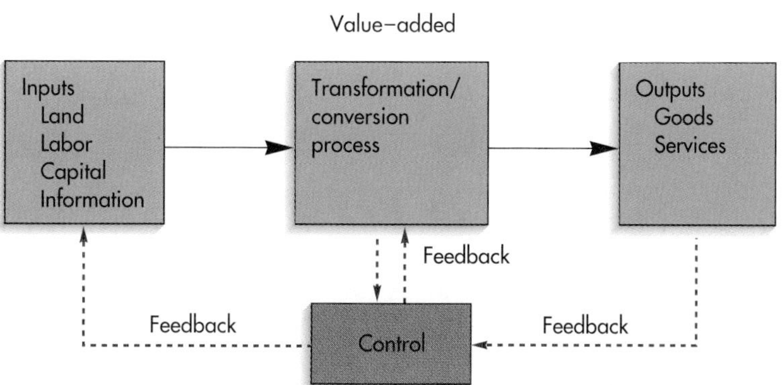

FIGURE 1–3

The operations function involves the conversion of inputs into outputs

TABLE 1–2

Examples of inputs, transformation, and outputs

Inputs	Transformation	Outputs
Land	Processes	Goods
Human	Cutting, drilling	Houses
Physical	Transporting	Automobiles
Intellectual	Teaching	Clothing
Raw materials	Farming	Computers
Energy	Mixing	Machines
Water	Packing	Televisions
Chemicals	Canning	Food products
Metals	Consulting	Textbooks
Wood	Copying, faxing	Magazines
Equipment		Shoes
Machines		CD players
Computers		Services
Trucks		Health care
Tools		Entertainment
Facilities		Car repair
Hospitals		Delivery
Factories		Gift wrapping
Offices		Legal
Retail stores		Banking
Other		Communication
Information		
Time		

TABLE 1–3

Illustrations of the transformation process

Food Processor	Inputs	Processing	Output
	Raw vegetables	Cleaning	Canned vegetables
	Metal sheets	Making cans	
	Water	Cutting	
	Energy	Cooking	
	Labor	Packing	
	Building	Labeling	
	Equipment		
Hospital	**Inputs**	**Processing**	**Output**
	Doctors, nurses	Examination	Healthy patients
	Hospital	Surgery	
	Medical supplies	Monitoring	
	Equipment	Medication	
	Laboratories	Therapy	

additional costs are incurred by storing the item without adding to the value of the item. Reducing storage time would reduce the transformation cost and, hence, increase the value-added.

Table 1–3 provides some specific illustrations of the transformation process.

Finance

The finance function comprises activities related to securing resources at favorable prices and allocating those resources throughout the organization. Finance and operations management personnel cooperate by exchanging information and expertise in such activities as:

1. *Budgeting.* Budgets must be periodically prepared to plan financial requirements. Budgets must sometimes be adjusted, and performance relative to a budget must be evaluated.

2. *Economic analysis of investment proposals.* Evaluation of alternative investments in plant and equipment requires inputs from both operations and finance people.

3. *Provision of funds.* The necessary funding of operations and the amount and timing of funding can be important and even critical when funds are tight. Careful planning can help avoid cash-flow problems. Most for-profit firms obtain the majority of their funds through the revenues generated by sales of goods and services.

Marketing

Marketing consists of selling and/or promoting the goods or services of an organization. Advertising and pricing decisions are made by marketing people. Marketing is also responsible for assessing customer wants and needs, and for communicating those to operations people (short term) and to design people (long term). That is, operations needs information about demand over the short-to-intermediate term so that it can plan accordingly (e.g., purchase materials or schedule work), while design people need information that relates to improving current products and services and designing new ones. Marketing, design, and production must work closely to successfully implement design changes and to develop and produce new products. Marketing can provide valuable insight on what competitors are doing. Marketing can also supply information on consumer preferences so that design will know the kinds of products and features needed; operations can supply information about capacities and judge the *manufacturability* of designs. Operations will also have advance warning if new equipment or skills will be needed for new products or services. Finance people should be included in these exchanges in order to provide information on what funds might be available (short term) and to learn what funds might be needed for new products or services (intermediate to long term). One important piece of information marketing needs from operations is the manufacturing or service **lead time** in order to give customers realistic estimates of how long it will take to fill their orders.

lead time The time necessary to deliver an order or perform a service.

Thus, marketing, operations, and finance must interface on product and process design, forecasting, setting realistic schedules, quality and quantity decisions, and keeping each other informed on the other's strengths and weaknesses.

Other Functions

There are a host of other supporting functions that interface with operations, finance, and marketing. Among them are accounting and purchasing. Also, depending on the nature of the organization, they may include personnel or human resources, product design and development, industrial engineering, and maintenance (see Figure 1–4).

Accounting has responsibility for preparing the financial statements, including the income statement and balance sheet. It also supplies information to management on costs of labor, materials, and overhead, and may provide reports on items such as scrap, downtime, and inventories. It must keep track of receivables, payables, and insurance costs, and prepare tax statements for the firm.

Purchasing has responsibility for procurement of materials, supplies, and equipment. Close contact with operations is necessary to ensure correct quantities and timing of purchases. The purchasing department is often called on to evaluate vendors for quality, reliability, service, price, and ability to adjust to changing demand. Purchasing is also involved in receiving and inspecting the purchased goods.

The *personnel* department is concerned with recruitment and training of personnel, labor relations, contract negotiations, wage and salary administration, assisting in manpower projections, and ensuring the health and safety of employees.

FIGURE 1-4

Operations interfaces with a number of supporting functions

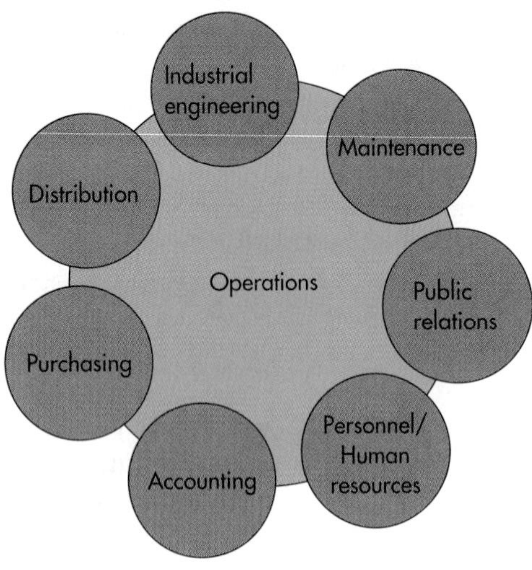

Public relations has responsibility for building and maintaining a positive public image of the organization. This involves news releases announcing promotions, and new products or services. It may also involve such things as sponsoring a Little League team, donating to cultural events, giving public tours of facilities, and sponsoring community affairs (e.g., marathons, bike races). Good public relations provides many potential benefits. An obvious one is in the marketplace. Other potential benefits include public awareness of the organization as a good place to work (labor supply), improved chances of approval of zoning change requests, community acceptance of expansion plans, and instilling a positive attitude among employees.

Industrial engineering is often concerned with scheduling, performance standards, work methods, quality control, and material handling. Manufacturing plants of medium and large firms typically include this function.

Distribution involves the shipping of goods to warehouses, retail outlets, or final customers.

Maintenance is responsible for general upkeep and repair of equipment, buildings and grounds, heating and air-conditioning; removing toxic wastes; parking; and perhaps security.

Many of these interfaces are elaborated on in later chapters.

The importance of production/operations management, both for organizations and society, should be fairly obvious: The consumption of goods and services is an integral part of our society. Production/operations management is responsible for creating those goods and services. Organizations exist primarily to provide services or create goods. Hence, production is the *core function* of an organization. Without this core, there would be no need for any of the other functions—the organization would have no purpose. Given the central nature of its function, it is not surprising that more than half of all employed people in this country have jobs in production and operations. Furthermore, the operations function is responsible for a major portion of the assets in most organizations.

DESIGNING AND OPERATING PRODUCTION SYSTEMS

We have already noted that the operations manager is responsible for the creation of goods and services. This encompasses acquisition of resources and the conversion of those inputs into outputs using one or more transformation processes. That involves planning, coordinating, and controlling the elements that make up the process, including workers, equipment, facilities, allocation of resources, and work methods. It also includes product and/or service design, a vital, ongoing process that most organizations must do. Operations performs this activity in conjunction with marketing. Marketing people can be a source of ideas concerning new products and services, and improvements to existing ones. Operations people can also be a source of new ideas for improvements in the processes that provide the goods or services. From a practical standpoint, product and service design and the processes that provide them are the lifeblood of a competitive organization.

A primary function of an operations manager is to guide the system by decision making. Certain decisions affect the *design* of the system, and others affect the *operation* of the system.

System design involves decisions that relate to system capacity, the geographic location of facilities, arrangement of departments and placement of equipment within physical structures, product and service planning, and acquisition of equipment. These decisions usually, but not always, require long-term commitments. **System operation** involves management of personnel, inventory planning and control, scheduling, project management, and quality assurance. In many instances, the operations manager is more involved in day-to-day operating decisions than with decisions relating to system design. However, the operations manager has a vital stake in system design because *system design essentially determines many of the parameters of system operation*. For example, costs, space, capacities, and quality are directly affected by design decisions. Even though the operations manager is not responsible for making all design decisions, he or she can provide to those decision makers a wide range of information that will have a bearing on their decisions.

system design Decisions concerning capacity, location, arrangement of departments, product and service planning, and acquisition and placement of equipment.

system operation Decisions concerning personnel, inventory, scheduling, project management, and quality assurance.

Take a moment to review the information in Table 1–4. It provides additional insights on the nature and scope of operations management.

DIFFERENTIATING FEATURES OF PRODUCTION SYSTEMS

There are a number of features that differentiate production systems. A brief discussion of some of these features will help you to develop a better understanding of the nature and scope of operations management. The three described are degree of standardization, type of operation, and production of goods versus service operations.

Degree of Standardization

The output of production systems can range from highly standardized to highly customized. *Standardized output* means that there is a high degree of uniformity in goods or services. Standardized goods include radios, televisions, computers, newspapers, canned foods, automobile tires, pens, and pencils. Standardized services include automatic car washes, televised newscasts, taped lectures, and commercial airline service. *Customized output* means that the product or service is designed for a specific case or individual. Customized goods include eyeglasses, custom-fitted clothing, window glass (cut to order), and customized draperies. Customized services include tailoring, taxi rides, and surgery.

Systems with standardized output can generally take advantage of standardized methods, materials, and mechanization, all of which contribute to higher volumes and lower unit costs. In custom systems, on the other hand, each job is sufficiently different so that workers must be more skilled, the work moves slower, and the work is less susceptible to mechanization.

Type of Operation

The degree of standardization and the volume of output of a product or service influence the way a firm organizes production. On one end of the scale is a single, large-scale product or service such as the launching of a space shuttle (service) or the construction

TABLE 1-4

Design and operating decisions

Decision Area	Basic Questions	Chapter
Forecasting	What will demand be?	3
Design		
Product and service design	What do customers want? How can products and services be improved?	4
Process selection	What processes should the organization use?	4, 5
Capacity (long range)	How much capacity will be needed? How can the organization best meet capacity requirements?	5
Layout	What is the best arrangement for departments, equipment, work flow, and storage in terms of cost, productivity?	6
Design of work systems	What is the best way to motivate employees? How can productivity be improved? How to measure work? How to improve work methods?	7
Location	What is a satisfactory location for a facility (factory, store, etc.)?	8
Operation		
Quality	How is quality defined?	9
Quality control	Are processes performing adequately? What standards should be used? Are standards being met?	10
Total quality management	How are quality goods and services achieved and improved?	11
Aggregate planning	How much capacity will be needed over the intermediate range? How can capacity needs best be met?	12
Inventory management	How much to order? When to reorder? Which items should get the most attention?	13
Materials requirements planning	What materials, parts, and subassemblies will be needed, and when?	14
Scheduling	How can jobs best be scheduled? Who will do which job? Which equipment to use?	17
Project management	Which activities are the most critical to the success of a project? What are the goals of a project? What resources will be needed, and when will they be needed?	18
Waiting lines	What capacity is appropriate?	19

of a skyscraper (product). On the other end is a continuous process, such as oil refining. In between these extremes are customized individual units of output, such as custom-made furniture, special-purpose machines, and auto repair; batches, such as paint and food products; and mass production, such as automobiles, personal computers, and appliances.

You will learn more about these different types of operations in later chapters.

Production of Goods versus Service Operations

Production of goods results in a *tangible output,* such as an automobile, a clock radio, a golf ball, a refrigerator—anything that we can see or touch. It may take place in a factory, but can occur elsewhere. Service, on the other hand, generally implies an *act.* A physician's examination, TV and auto repair, lawn care, and projecting a film in a theater are examples of services. The majority of service jobs fall into these categories:

Government (federal, state, local).

Wholesale/retail (clothing, food, appliances, stationery, toys, etc.).

Financial services (banking, stock brokerages, insurance, etc.).

Health care (doctors, dentists, hospitals, etc.).

Personal services (laundry, dry cleaning, hair/beauty, gardening, etc.).

Business services (data processing, delivery, employment agencies, etc.).

Education (schools, colleges, etc.).

Manufacturing and service are often similar in terms of *what* is done but different in terms of *how* it is done. For example, both involve design and operating decisions. Manufacturers must decide what size factory is needed. Service organizations (e.g., hospitals) must decide what size building is needed. Both must make decisions on location, schedule and control operations, and allocate scarce resources.

Manufacturing and service organizations differ chiefly because manufacturing is goods-oriented and service is act-oriented. The differences involve the following:

1. Customer contact
2. Uniformity of input
3. Labor content of jobs
4. Uniformity of output
5. Measurement of productivity
6. Quality assurance

Let us consider each of these differences.

1. By nature, service involves a much higher degree of customer contact than manufacturing. The performance of a service often occurs at the point of consumption. For example, repairing a leaky roof must take place where the roof is, and surgery requires the presence of the surgeon and the patient. On the other hand, manufacturing allows a separation between production and consumption, so that manufacturing may occur away from the consumer. This permits a fair degree of latitude in selecting work methods, assigning jobs, scheduling work, and exercising control over operations. Service operations, because of their contact with customers, can be much more limited in their range of options. Moreover, customers are sometimes a part of the system (e.g., self-service operations such as gas stations, shopping), so tight control is impossible. In addition, product-oriented operations can build up inventories of finished goods (e.g., cars, refrigerators), enabling them to absorb some of the shocks caused by varying demand. Service operations, however, cannot build up inventories of *time* and are much more sensitive to demand variability—banks and supermarkets alternate between lines of customers waiting for service and idle tellers or cashiers waiting for customers.

2. Service operations are subject to greater variability of inputs than typical manufacturing operations. Each patient, each lawn, and each auto repair presents a specific problem that often must be diagnosed before it can be remedied. Manufacturing operations often have the ability to carefully control the amount of variability of inputs and thus achieve low variability in outputs. Consequently, job requirements for manufacturing are generally more uniform than those for services.

3. Because of the on-site consumption of services and the high degree of variation of inputs, services require a higher labor content whereas manufacturing, with exceptions, can be more capital-intensive (i.e., mechanized).

4. Because high mechanization generates products with low variability, manufacturing tends to be smooth and efficient; service activities sometimes appear to be slow and awkward, and output is more variable.

5. Measurement of productivity is more straightforward in manufacturing due to the high degree of uniformity of most manufactured items. In service operations, variations in demand intensity and in requirements from job to job make productivity measurement considerably more difficult. For example, compare the productivity of two doctors. One may have a large number of routine cases while the other does not, so their productivity appears to differ unless a very careful analysis is made.

TABLE 1–5

Typical differences between goods and services

Characteristic	Goods	Services
Output	Tangible	Intangible
Customer contact	Low	High
Uniformity of input	High	Low
Labor content	Low	High
Uniformity of output	High	Low
Measurement of productivity	Easy	Difficult
Opportunity to correct quality problems before delivery to customer	High	Low

FIGURE 1–5

Percentage share of U.S. employment

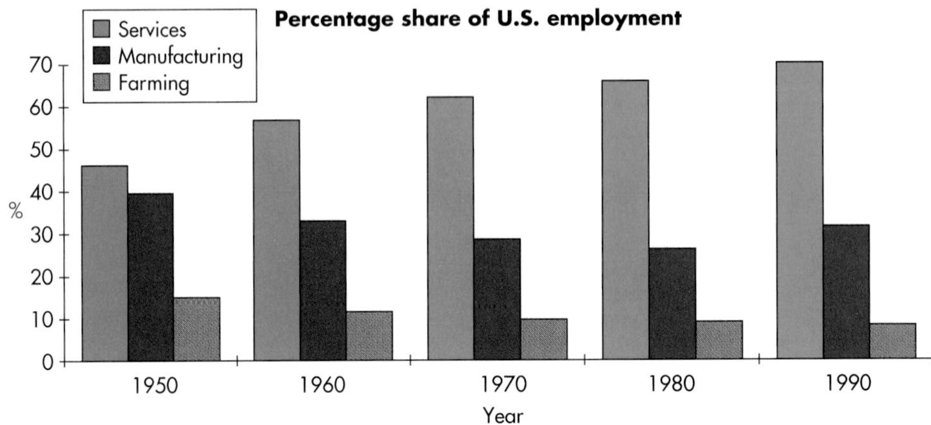

6. Quality assurance is more challenging in services when production and consumption occur at the same time. Moreover, the higher variability of input creates additional opportunity for the quality of output to suffer unless quality assurance is actively managed. Quality at the point of creation is typically more important for services than for manufacturing, where errors can be corrected before the customer receives the output.

Table 1–5 gives an overview of the differences between production of goods and service operations.

Although it is convenient to think in terms of systems devoted exclusively to goods or services, most real systems are a blend of both. For instance, maintenance and repair of equipment are services performed by virtually every manufacturing firm. Similarly, most service organizations typically sell goods that complement their services. Thus, a lawn care firm usually sells goods such as weed killers, fertilizers, and grass seed. Hospitals dispense medical and surgical supplies along with health services. Restaurants sell food. Movie theaters sell popcorn, candy, and beverages.

The service sector and the manufacturing sector are both important to the economy. The service sector now accounts for more than 70 percent of jobs in the United States. Moreover, the number of people working in services continues to increase, while the number of people working in manufacturing continues to decrease. (See Figure 1–5.) Unfortunately, a substantial portion of the expansion in the service sector is in lower-skill, lower-wage jobs that tend to have low productivity. Manufacturing is important in that it supplies a large proportion of exports that are beneficial for international trade. Moreover, many service jobs are *dependent* on manufacturing; that is, they support manufacturing. If the manufacturing sector continues to decline because of its inability to compete effectively with foreign companies, service jobs related to manufacturing (such as maintenance, security, and financial services) will also decline.

TABLE 1–6

Responsibilities of operations managers

Planning	Organizing
Capacity	Degree of centralization
Location	Subcontracting
Products and services	Staffing
Make or buy	Hiring/laying off
Layout	Use of overtime
Projects	Directing
Scheduling	Incentive plans
Controlling	Issuance of work orders
Inventory control	Job assignments
Quality control	

THE OPERATIONS MANAGER AND THE MANAGEMENT PROCESS

The operations manager is the key figure in the system: he or she has the ultimate responsibility for the creation of goods or provision of services.

The kinds of jobs that operations managers oversee vary tremendously from organization to organization largely because of the different products or services involved. Thus, managing a banking operation obviously requires a different kind of expertise than managing a steelmaking operation. However, in a very important respect, the jobs are the same: They are both essentially *managerial.* The same thing can be said for the job of any operations manager regardless of the kinds of goods or services being created. In every case, the operations manager must coordinate the use of resources through the management process of planning, organizing, staffing, directing, and controlling.

Examples of the responsibilities of operations managers according to these classifications are given in Table 1–6.

OPERATIONS MANAGERS AND DECISION MAKING

The chief role of an operations manager is that of planner and decision maker. In this capacity, the operations manager exerts considerable influence over the degree to which the goals and objectives of the organization are realized.

Throughout this book, you will encounter the broad range of decisions that operations managers must make, and you will be introduced to the tools necessary to handle those decisions. This section describes general approaches to decision making, including the use of quantitative methods, analysis of trade-offs, and the systems approach.

Quantitative Approaches

Quantitative approaches to problem solving often embody an attempt to obtain mathematically optimum solutions to managerial problems. Although quantitative techniques have traditionally been associated with production and operations management, it was not until World War II that major efforts were made to develop these techniques. In order to handle complex military logistics problems, interdisciplinary teams were assembled (e.g., psychologists, mathematicians, economists) to combine efforts in search of workable solutions. These efforts continued and expanded after the war, and many of the resulting techniques were applied to operations management. *Linear programming* and related mathematical techniques are widely used for optimum allocation of scarce resources. *Queuing techniques,* which originated around 1920 in the telephone industry but remained dormant until the 1950s and 1960s, are useful for analyzing situations in which waiting lines form. *Inventory models,* also popular after some early work, went through a long period of low interest but are now widely used to control inventories.

Project models such as PERT (program evaluation and review technique) and CPM (critical path method) are useful for planning, coordinating, and controlling large-scale projects. *Forecasting techniques* are widely used in planning and scheduling. *Statistical models* are currently used in many areas of decision making.

In large measure, quantitative approaches to decision making in operations management (and in other areas of decision making) have been accepted because of the introduction of calculators and the availability of high-speed computers capable of handling the required calculations. Computers have had an enormous influence on the practice of operations management, particularly in scheduling and inventory control. Because they are capable of rapid, error-free computations and keeping track of thousands of bits of information with instantaneous retrieval, computers have had a major impact on operations management. Moreover, the growing availability of software packages covering virtually every quantitative technique has greatly increased management's use of the computer. Many heretofore impractical techniques, such as multiple regression analysis and linear programming, can now be handled with ease.

Because of the emphasis on quantitative approaches in operations management decision making, it is important not to lose sight of the fact that managers typically use a combination of qualitative and quantitative approaches, and many important decisions are based on qualitative approaches.

Analysis of Trade-Offs

Operations managers encounter decisions that can be described as *trade-off* decisions. For example, in deciding on the amount of inventory to stock, the manager must take into account the trade-off between the increased level of customer service that the additional inventory would yield and the increased costs required to stock that inventory. In selecting a piece of equipment, a manager must evaluate the merits of extra features relative to the cost of those extra features. And in the scheduling of overtime to increase output, the manager must weigh the value of the increased output against the higher costs of overtime (e.g., higher labor costs, lower productivity, lower quality, and greater risk of accidents).

Throughout this book you will be presented with decision models that reflect these kinds of trade-offs. Managers sometimes deal with these decisions by listing the advantages and disadvantages—the pros and cons—of a course of action to better understand the consequences of the decisions they must make. In some instances, managers add weights to the items on their list that reflect the relative importance of various factors. This can help them "net out" the potential impacts of the trade-offs on their decision. An example of this is the factor-rating approach described in the chapter on facilities location.

A Systems Approach

system A set of interrelated parts that must work together.

A systems viewpoint is almost always beneficial in decision making. A **system** can be defined as a set of interrelated parts that must work together. In a business organization, the organization can be thought of as a system composed of subsystems (e.g., marketing subsystem, operations subsystem, finance subsystems), which in turn are composed of lower subsystems. The systems approach emphasizes interrelationships among subsystems, but its main theme is that *the whole is greater than the sum of its individual parts.* Hence, from a systems viewpoint, the output and objectives of the organization as a whole take precedence over those of any one subsystem. An alternative approach is to concentrate on efficiency within subsystems and thereby achieve overall efficiency. But that approach overlooks the fact that organizations must operate in an environment of scarce resources and that subsystems are often in direct competition for those scarce resources, so that an orderly approach to the allocation of resources is called for.

One undesirable result of the use of quantitative techniques is that many of the techniques tend to produce solutions that are optimal in a narrow sense but may not be optimal in a broader sense (e.g., in terms of a department, plant, division, or overall organization). Consequently, managers must evaluate "optimal" solutions produced by quantitative techniques in terms of the larger framework, and perhaps modify decisions accordingly.

A systems approach is essential whenever something is being designed, redesigned, implemented, improved, or otherwise changed. It is important to take into account the impact on all parts of the system. For example, if the upcoming model of an automobile will add antilock brakes, a designer must take into account how customers will view the change, instructions for using the brakes, chances for misuse, the cost of producing the new brakes, installation procedures, recycling worn-out brakes, and repair procedures. In addition, workers will need training to make and/or assemble the brakes, production scheduling may change, inventory procedures may have to change, quality standards will have to be established, advertising must be informed of the new features, and parts suppliers must be selected.

Establishing Priorities

In virtually every situation, managers discover that certain elements are more important than others. Recognizing this fact of life enables the managers to direct their efforts to where they will do the most good and to avoid wasting time and energy on insignificant elements.

Consider owning and operating an automobile. It has many parts and systems that can malfunction. Some of these are critical to the operation of the automobile: It would not function or would be dangerous to operate without them. Critical items include the engine and drive train, steering, brakes, tires, electrical system, and cooling system. In terms of maintaining and repairing the car, these items should receive the highest priority if the goal is to have reliable transportation.

There are other items that are of much less importance, such as scratches in the paint, minor dents, a missing piece of chrome, and worn seatcovers. In terms of transportation, these should receive attention only after other, more important items have been attended to.

Between these two extremes lies a range of items of intermediate priority. These should be given attention corresponding to their importance to the overall goal. The list might include soft tires, weak battery, wheel alignment, noisy muffler, body rust, inoperative radio, and headlights out of adjustment.

Obviously, certain parts of an automobile are more critical to its operation than others. The same concept applies to management. By recognizing this and setting priorities, a manager will be in a position to deal more effectively with problems as they arise and to prevent many others from arising at all.

Recognition of priorities means devoting more attention to what is most important. It is axiomatic that a relatively few factors are often most important, so that dealing with those factors will generally have a disproportionately large impact on the results achieved. This is referred to as the **Pareto phenomenon,** which means that all things are not equal; some things (a few) will be very important for achieving an objective or solving a problem, and other things (many) will not. The implication is that a manager should examine the situation, searching for the few factors that will have the greatest impact, and concentrate on those; little or nothing will be gained by focusing efforts on less important factors. This is one of the most important and pervasive concepts in operations management. In fact, this concept can be applied at all levels of management and to every aspect of decision making, both professional and personal.

Pareto phenomenon A few factors are very important; many factors are much less important.

Ethics

Operations managers, like all managers, have the responsibility to make ethical decisions. Ethical issues arise in many aspects of operations management, including:

- worker safety: providing adequate training, maintaining equipment in good working condition, maintaining a safe working environment;

- product safety: providing products that minimize the risk of injury to users or damage to property or the environment;
- quality: honoring warranties, avoiding hidden defects;
- the environment: obeying government regulations;
- the community: being a good neighbor;
- hiring and firing workers: don't hire under false pretenses (e.g., promising a long-term job when that is not what is intended);
- closing facilities: taking into account the impact on a community, and honoring commitments that have been made;
- workers' rights: respecting workers' rights, dealing with worker problems quickly and fairly.

In making decisions, managers must consider how their decisions will affect shareholders, management, employees, customers, the community at large, and the environment. Finding solutions that will be in the best interests of all of these stakeholders is not always easy, but it is a goal that all managers should strive to achieve. Furthermore, even managers with the best intentions will sometimes make mistakes. If mistakes do occur, managers should act responsibly to correct those mistakes as quickly as possible, and to address any negative consequences.

THE HISTORICAL EVOLUTION OF OPERATIONS MANAGEMENT

Systems for production have existed since ancient times. The Great Wall of China, the Egyptian pyramids, the ships of the Roman and Spanish empires, and the roads and aqueducts of the Romans provide examples of the human ability to organize for production. Even so, most of these examples could be classified as "public works" projects. The production of goods for sale, at least in the modern sense, and the modern factory system had their roots in the Industrial Revolution.

The Industrial Revolution

The Industrial Revolution began in the 1770s in England and spread to the rest of Europe and to the United States during the nineteenth century. Prior to that time, goods were produced in small shops by craftsmen and their apprentices. Under that system, it was common for one person to be responsible for making a product, such as a horse-drawn wagon or a piece of furniture, from start to finish. Only simple tools were available; the machines that we use today had not been invented.

Then, a number of inventions changed the face of production forever by substituting machine power for human power. Perhaps the most significant of these was the steam engine invented by James Watt in 1764, because it provided a source of power to operate machines in factories. James Hargreave's spinning jenny (1770) and Edmund Cartwright's power loom (1785) revolutionized the textile industry. Ample supplies of coal and iron ore provided materials for generating power and making machinery. The new machines, made of iron, were much stronger and more durable than the simple wooden machines they replaced.

craft production System in which highly skilled workers use simple, flexible tools to produce small quantities of customized goods.

In the earliest days of manufacturing, goods were produced using **craft production:** highly skilled workers using simple, flexible tools produced goods according to customer specifications, as described in the following excerpt:

> In 1894, the Honorable Evelyn Henry Ellis, a wealthy member of the English Parliament, set out to buy a car. . . . He didn't go to a car dealer—there weren't any. Nor did he contact an English automobile manufacturer—there weren't any of those either.
>
> Instead, he visited the noted Paris machine-tool company of Panhard et Levassor and commissioned an automobile. In 1894, it was the world's leading car company. . . .

When Ellis arrived at P&L, which was still primarily a manufacturer of metal-cutting saws rather than automobiles, he found in place the classic craft-production system. P&L's work force was overwhelmingly composed of skilled craftspeople who carefully hand-built cars in small numbers.

These workers thoroughly understood mechanical design principles and the materials with which they worked. What's more, many were their own bosses, often serving as independent contractors within the P&L plant or, more frequently, as independent machine-shop owners with whom the company contracted for specific parts or components.

The two company founders, Panhard and Levassor, and their immediate associates were responsible for talking to customers to determine the vehicle's exact specifications, ordering the necessary parts, and assembling the final product. Much of the work, though, including design and engineering, took place in individual craft shops scattered throughout Paris.

Different contractors, using slightly different gauges, made the parts. They then ran the parts through an oven to harden their surfaces enough to withstand heavy use. However, the parts frequently warped in the oven and needed further machining to regain their original shape.

When these parts eventually arrived at P&L's final assembly hall, their specifications could best be described as approximate. The job of the skilled fitters in the hall was to take the first two parts and file them down until they fit together perfectly.

Then they filed the third part until it fit the first two, and so on until the whole vehicle—with its hundreds of parts—was complete.

This sequential fitting produced what we today call *dimensional creep.* So, by the time the fitters reached the last part, the total vehicle could differ significantly in dimensions from the car on the next stand that was being built to the same blueprints.

Because P&L couldn't mass-produce identical cars, it didn't try. Instead, it concentrated on tailoring each product to the precise desires of individual buyers.

It also emphasized its cars' performance and their hand-fitted craftsmanship in which the gaps between individual parts were nearly invisible.

To the consumers Panhard was trying to woo, this pitch made perfect sense. These wealthy customers employed chauffeurs and mechanics on their personal staffs. Cost, driving ease, and simple maintenance weren't their primary concerns. Speed and customization were.

—*Source:* Reprinted with permission of Rawson Associates/Scribner, a division of Simon & Schuster, Inc., from *The Machine that Changed the World,* by James P. Womack, Daniel T. Jones, and Daniel Roos. Copyright 1990 by James P. Womack, Daniel T. Jones, Daniel Roos, and Donna Sammons Carpenter.

Craft production had major shortcomings. Because products were made by skilled craftsmen who custom fitted parts, production was slow and costly. And when parts failed, the replacements also had to be custom made, which was slow and costly. Another shortcoming was that production costs did not decrease as volume increased; there were no *economies of scale,* which would have provided a major incentive for companies to expand. Instead, many small companies emerged, each with its own set of standards.

A major change occurred that gave the industrial revolution a boost: the development of standard gauging systems. This greatly reduced the need for custom-made goods. Factories began to spring up and grow rapidly, providing jobs for countless people who were attracted in large numbers from rural areas.

Despite the major changes that were taking place, management theory and practice had not progressed much from early days. What was needed was an enlightened and more systematic approach to management.

Scientific Management

The scientific-management era brought widespread changes to the management of factories. The movement was spearheaded by the efficiency engineer and inventor, Frederick Winslow Taylor, who is often referred to as the father of scientific management. Taylor believed in a "science of management" based on observation, measurement, analysis and improvement of work methods, and economic incentives. He studied work methods in great detail to identify the best method for doing each job. Taylor also believed that management should be responsible for planning, carefully selecting and training workers, finding the best way to perform each job, achieving cooperation between management and workers, and separating management activities from work activities.

Taylor's methods emphasized maximizing output. They were not always popular with workers, who sometimes thought the methods were used to unfairly increase output without a corresponding increase in compensation. Certainly some companies did abuse workers in their quest for efficiency. Eventually, the public outcry reached the halls of Congress, and hearings were held on the matter. Taylor himself was called to testify in 1911, the same year in which his classic book, *The Principles of Scientific Management,* was published. The publicity from those hearings actually helped scientific management principles to achieve wide acceptance in industry.

A number of other pioneers also contributed heavily to this movement, including the following:

Frank Gilbreth was an industrial engineer who is often referred to as the father of motion study. He developed principles of motion economy that could be applied to incredibly small portions of a task.

Henry Gantt recognized the value of nonmonetary rewards to motivate workers, and developed a widely used system for scheduling, called a Gantt chart.

Harrington Emerson applied Taylor's ideas to organization structure, and encouraged the use of experts to improve organizational efficiency. He testified in a congressional hearing that railroads could save a million dollars a day by applying principles of scientific management.

Henry Ford, the great industrialist, employed scientific management techniques in his factories.

www.ford.com

During the early part of the 20th century, automobiles were just coming into vogue in the United States. Ford's Model T was such a success that the company had trouble keeping up with orders for the cars. In an effort to improve the efficiency of operations, Ford adopted the scientific management principles espoused by Frederick Winslow Taylor. He also introduced the *moving assembly line,* as described in the following excerpt:

> Charles Sorensen, who had become one of Ford's top production people, had a Model T chassis pulled slowly by a windlass across 250 feet of factory floor, timing the process all the while. Behind him walked six workers, picking up parts from carefully spaced piles on the floor and fitting them to the chassis. That was the birth of the assembly line, the very essence of what would become America's industrial revolution. Before, it had taken some 13 hours to make a car; now they had cut the time of assembly in half, to 5 hours and 50 minutes. Not satisfied, they pushed even harder, lengthening the line and bringing in more specialized workers for the final assembly. Within weeks they could complete a car in only two hours and 38 minutes. Now the breakthroughs came even more rapidly. In January 1914, Ford installed the first automatic conveyor belt. It was, he said, the first moving line ever used in an industrial plant, and it was fashioned after the overhead trolley that the Chicago meatpackers employed to move beef. Within two months of that innovation, Ford could assemble a car in an hour and a half. It was a stunning accomplishment, but it merely whetted their zeal. Everything now had to be timed, rationalized, broken down into smaller pieces, and speeded up. Just a few years before, in the days of stationary chassis assembly, the best record for putting a car together had been 728 hours of one man's work; with the new moving line it required only ninety-three minutes—*Source: Text of pages 80–81 from* THE RECKONING *by David Halberstam. Copyright © 1986 by David Halberstam. By permission of William Morrow & Company, Inc.*

Among Ford's many contributions was the introduction of **mass production** to the automotive industry, a system of production in which large volumes of standardized goods are produced by low-skilled or semiskilled workers using highly specialized, and often costly, equipment. Ford was able to do this by taking advantage of a number of important concepts. Perhaps the key concept that launched mass production was the concept of **interchangeable parts,** sometimes attributed to Ell Whitney, an American inventor who applied the concept to assembling muskets in the late 1700s. The basis for interchangeable parts is to standardize parts so that any part in a batch of parts would fit any automobile coming down the assembly line. This meant that parts did not have to be custom fitted, as they were in craft production. The standardized parts could also be used for replacement parts. The result was a tremendous decrease in assembly time and cost. Ford accomplished this by standardizing the gauges used to measure parts during production, and by using newly developed processes to produce uniform parts.

mass production System in which lower-skilled workers use specialized machinery to produce high volumes of standardized goods.

interchangeable parts Parts of a product made to such precision that they do not have to be custom fitted.

A second concept used by Ford was the **division of labor,** which Adam Smith wrote about in *The Wealth of Nations* (1776). Division of labor means that an operation, such as assembling an automobile, is divided up into a series of many small tasks, and workers are assigned to one of those tasks. Unlike craft production, where each worker was responsible for doing many tasks, and thus required skill, with division of labor the tasks were so narrow that virtually no skill was required.

division of labor Breaking up a production process into small tasks, so that each worker performs a small portion of the overall job.

Together, these concepts enabled Ford to tremendously increase the production rate at his factories using readily available inexpensive labor.

Ford not only perfected the interchangeable part, he perfected the interchangeable worker. By 1915, when the assembly lines at Highland Park were fully installed and output reached capacity, assembly workers numbered more than 7,000. Most were recent arrivals to Detroit, often coming directly from the farm. Many more were new to the United States.

A 1915 survey revealed that Highland Park workers spoke more than 50 languages and that many of them could barely speak English. . . . How could this army of strangers cooperate to produce a greater volume of a complex product (the Model T) than any company had previously imagined—and do it with consistent accuracy?

The answer lay in taking the idea of the division of labor to its ultimate extreme. The skilled fitter in Ford's craft-production plant of 1908 had gathered all the necessary parts, obtained tools from the tool room, repaired them if necessary, performed the complex fitting and assembly job for the entire vehicle, then checked over his work before sending the completed vehicle to the shipping department.

In stark contrast, the assembler on Ford's mass-production line had only one task—to put two nuts on two bolts or perhaps to attach one wheel to each car. He didn't even order parts, procure his tools, repair his equipment, inspect for quality, or even understand what the workers on either side of him were doing. Rather, he kept his head down and thought about other things. The fact that he might not even speak the same language as his fellow assemblers or the foreman was irrelevant to the success of Ford's system. (The use of "he," "him," and "his" is deliberate; until World War II, workers in auto factories in the United States and Europe were exclusively male.)

Someone, of course, did have to think about how all the parts came together and just what each assembler should do. This was the task for a newly created professional, the industrial engineer. Similarly, someone had to arrange for the delivery of parts to the line, usually a production engineer who designed conveyor belts or chutes to do the job. Housecleaning workers were sent around periodically to clean up work areas, and skilled repairmen circulated to refurbish the assemblers' tools. Yet another specialist checked quality. Work that was not done properly was not discovered until the end of the assembly line, where another group of workers was called into play—the rework men, who retained many of the fitters' skills.

With this separation of labor, the assembler required only a few minutes of training. Moreover, he was relentlessly disciplined by the pace of the line, which speeded up the slow and slowed down the speedy. The foreman—formerly the head of a whole area of the factory with wide-ranging duties and responsibilities, but now reduced to a semiskilled checker—could spot immediately any slacking off or failure to perform the assigned task. As a result, the workers on the line were as replaceable as the parts on the car.—*Source: Reprinted with permission of Rawson Associates/Scribner, a division of Simon & Schuster, Inc., from* The Machine That Changed the World, *by James P. Womack, Daniel T. Jones, and Daniel Roos. © 1990 by James P. Womack, Daniel T. Jones, Daniel Roos, and Donna Sammons Carpenter.*

The Human Relations Movement

Whereas the scientific-management movement heavily emphasized the technical aspects of work design, the human relations movement emphasized the importance of the human element in job design. Lillian Gilbreth, a psychologist and the wife of Frank Gilbreth, worked with her husband, focusing the human factor in work. (The Gilbreths were the subject of the classic 1950s film, *Cheaper by the Dozen.*) Many of her studies in the 1920s dealt with worker fatigue. In the following decades, there was much emphasis on motivation. During the 1930s, Elton Mayo conducted studies at the Hawthorne division of Western Electric. His studies revealed that in addition to the physical and technical aspects of work, worker motivation is critical for improving productivity. During the 1940s, Abraham Maslow developed motivational theories, which Frederick Hertzberg refined in the 1950s. Douglas McGregor added Theory X and Theory Y in the 1960s. These theories represented the two ends of the spectrum of how employees view work. Theory X, on the negative end, assumed that workers do not like to work, and have to be controlled—rewarded and punished—to get them to do good work. This attitude was quite common in the automobile industry and in some other industries, until the threat of global competition forced them to rethink that approach. Theory Y, on the other end of the spectrum, assumed that workers enjoy the physical and mental aspects of work and become committed to work. The Theory X approach resulted in an adversarial environment, whereas the Theory Y approach resulted in empowered workers and a more cooperative spirit. In the 1970s, William Ouchi added Theory Z, which combined the Japanese approach, with such features as lifetime employment, employee problem solving, and consensus building, and the traditional western approach that features short-term employment, specialists, and individual decision making and responsibility.

Decision Models and Management Science

The factory movement was accompanied by the development of several quantitative techniques. F. W. Harris developed one of the first models in 1915: a mathematical model for inventory management. In the 1930s, three coworkers at Bell Telephone Labs, H. F. Dodge, H. G. Romig, and W. Shewhart, developed statistical procedures for sampling and quality control. In 1935, L. H. C. Tippett conducted studies that provided the groundwork for statistical-sampling theory.

At first, these quantitative models were not widely used in industry. However, the onset of World War II changed that. The war generated tremendous pressures on manufacturing output, and specialists from many disciplines combined efforts to achieve advancements in the military and in manufacturing. After the war, efforts to develop and refine quantitative tools for decision making continued, resulting in decision models for forecasting, inventory management, project management, and other areas of operations management.

During the 1960s and 1970s, management science techniques were highly regarded; in the 1980s, they lost some favor. However, the widespread use of personal computers and user-friendly software in the workplace may cause a resurgence in the popularity of these techniques.

The Influence of Japanese Manufacturers

A number of Japanese manufacturers have developed or refined management practices that have increased the productivity of their operations and the quality of their products. This has made them very competitive, sparking interest in the new approaches by companies outside Japan. The new approaches emphasize quality and continual improvement, worker teams and empowerment, and achieving customer satisfaction. The Japanese can be credited with spawning the "quality revolution" that is occurring in industrialized countries, and with generating widespread interest in time-based management.

The influence of the Japanese on U.S. manufacturing and service companies has been enormous and promises to continue for the foreseeable future. Because of that influence, this book will provide considerable information about Japanese methods and successes.

Table 1–7 provides a chronological summary of some of the key developments in the evolution of operations management.

RECENT TRENDS

A number of trends currently command considerable attention in business, as they heavily influence planning and decision making. Many of the trends relate to competition, particularly foreign competition and the impact it has on manufacturing companies.

1. *Global marketplace.* Markets—and companies—are becoming increasingly global in nature. The North American Free Trade Agreement (NAFTA) has opened borders for trade between the United States, Canada, and Mexico. Even more far-reaching is the General Agreement on Tariffs and Trade (GATT). It involves 124 countries that have agreed to open their economies, reduce tariffs and subsidies, and expand protection of intellectual property. The restructuring of the Soviet Union, the reunification of Germany, and the European Union have created a European market that represents a

TABLE 1–7

Historical summary of operations management

Approximate Date	Contribution/Concept	Originator
1776	Division of labor	Adam Smith
1790	Interchangeable parts	Eli Whitney
1911	Principles of scientific management	Frederick W. Taylor
1911	Motion study; use of industrial psychology	Frank and Lillian Gilbreth
1912	Chart for scheduling activities	Henry Gantt
1913	Moving assembly line	Henry Ford
1915	Mathematical model for inventory management	F. W. Harris
1930	Hawthorne studies on worker motivation	Elton Mayo
1935	Statistical procedures for sampling and quality control	H. F. Dodge, H. G. Romig, W. Shewhart, L. H. C. Tippett
1940	Operations research applications in warfare	Operations research groups
1947	Linear programming	George Dantzig
1951	Commercial digital computers	Sperry Univac
1950s	Automation	Numerous
1960s	Extensive development of quantitative tools	Numerous
1975	Emphasis an manufacturing strategy	W. Skinner
1980s	Emphasis on quality, flexibility, time-based competition	Japanese manufacturers
1990s	Internet	Numerous

An electronic ordering system connected to over 800 suppliers helps global toy retailer Toys "R" Us to ensure that the right toy is on the right shelf at the right time.

major market for goods exceeding that of the United States. Numerous foreign manufacturing companies have or plan to have operations in Europe. Moreover, many foreign companies have established manufacturing plants in the United States. Asian markets, especially China, loom on the horizon. The result has been a marked increase in the level of competition throughout the world, a trend that shows no sign of ending in the foreseeable future.

NEWSCLIP

NAFTA OFFERS REAL OPPORTUNITIES

Serge Ratmiroff

The North American Free Trade Agreement (NAFTA) created enormous opportunities for U.S. manufacturers, but many companies have been slow to explore its potential benefits. This delay could be a mistake that eventually jeopardizes their long-term future in the global marketplace.

NAFTA gives both large and small companies a competitive advantage in pursuing opportunities in Mexico before companies outside North America gain greater access. Almost half of the tariffs on U.S. exports to Mexico were eliminated when NAFTA was signed. Most of the remaining tariffs will be phased out gradually over 5- or 10-year periods, but only for products that satisfy NAFTA criteria for being North American in origin.

The competitive advantage is real, but best serves those who act fast. Even companies that already export to Mexico should determine if NAFTA presents a benefit or a threat to their growth prospects. Forecasting an increase in sales from exports because of reduced tariffs overlooks the encroaching plans of aggressive companies. The agreement does not necessarily give a U.S. firm an advantage over its U.S. competitors who are interested in doing business south of the border.

Nor does it mean that Mexican companies are going to be caught napping while U.S. firms prepare to sell their products to the Mexican market.

True, many Mexican enterprises must lower their operating costs to keep their prices in line with the prices of U.S. goods that improve with declining tariffs. They also must invest heavily to bring their technology and product quality to world-class levels.

If they lack the capital to upgrade their operations and stay competitive, where will they get it? One likely option is foreign investors. In some cases, entire companies will be for sale. Such circumstances prompted by NAFTA represent an additional opportunity for U.S. firms, but Mexico can also locate investors and buyers from other countries.

Several Japanese and European companies are already investing in Mexican operations, or are building their own factories in anticipation of the projected growth and consumer demand in Mexico. They currently have tariff and non-tariff obstacles to selling their own goods in North America, but they are finding inroads that are preparing them for long-term opportunities.

Mexico is just the beginning. Companies that venture into Mexico will find themselves well-positioned to operate with a potential Latin American free-trade bloc. As early as next year, Chile is expected to be part of NAFTA. The economics, trading activities, and incomes throughout most of South America are among the fastest growing in the world.

U.S. companies should view Mexico as an ally in the war of global competition. They need to extend their perception of what constitutes the long term, and to anticipate their place in the world 10 to 20 years from now.

Any company that wants to enter Mexico must begin by determining if its products comply with NAFTA's rules of origin. Depending upon the industry and its products, only a limited percentage of components or raw materials may be sourced from outside the U.S., Mexico, or Canada.

Manufacturers that discover their products do not qualify and remain subject to full tariffs must make critical decisions. They can switch from Asian and European suppliers of raw materials and components to North American suppliers. Perhaps outsourced components could be manufactured in-house, or through a joint-venture partner in Mexico or Canada.

The decision to pursue the Mexican market or expand operations in that country requires studied evaluation and important long-term considerations. For example, many of the countries that supply low-cost components are also becoming significant markets for U.S. products, and maintaining strong business ties with them could be advantageous.

The point is, companies should remember that NAFTA was designed to strengthen the global positions of industries in North America. The opportunities should be thoroughly explored.

Mr. Ratmiroff is the senior manager of international services for Deloitte & Touche in Chicago.

Source: Reprinted with permission of *Industry Week*, October 3, 1994. Copyright Penton Publishing, Inc., Cleveland, Ohio.

2. *Operations Strategy.* During the 1970s and 1980s, many companies neglected to include operations strategy in their corporate strategy. Some of them paid dearly for that neglect. Now more and more companies are recognizing the importance of operations strategy on the overall success of their business, and the necessity for relating it to their overall business strategy.

3. *Total quality management (TQM).* Many firms are now adopting a total quality management approach to their businesses. Under this approach, the entire organization from the president down becomes committed to, and involved in, a never-ending quest to improve the quality of goods and services. Key features often include a team approach, finding and eliminating problems, emphasis on serving the customer, and continuously working to improve the system.

4. *Flexibility.* The ability to adapt quickly to changes in volume of demand, in the mix of products demanded, and in product design, has become a major competitive strategy. In manufacturing, the term *agile manufacturing* is sometimes used to connote flexibility. A short reading at the end of this chapter describes this.

flexibility The ability to adapt quickly to changes.

5. *Time reduction.* Many companies are focusing efforts on reducing the time needed to accomplish various tasks in order to gain a competitive edge. If two companies can provide the same product at the same price and quality, but one can deliver it four weeks earlier than the other, the quicker company will invariably get the sale. Time reductions are being achieved in processing, information retrieval, product design, and the response to customer complaints.

6. *Technology.* Technological advances have led to a vast array of new products and processes. Undoubtedly the computer has had—and will continue to have—the greatest impact on business organizations. It has truly revolutionized the way companies operate. Applications include product design, product features, processing technology, information processing, and communication. Technological advances in new materials, new methods, and new equipment have also made their mark on operations. Technological changes in products and processes can have major implications for production systems, affecting competitiveness and quality, but unless technology is carefully integrated into an existing system, it can do more harm than good by raising costs, reducing flexibility, and even reducing productivity.

READING
AGILE MANUFACTURING John Holusha

www.ford.com

Agile manufacturers emphasize flexibility of operations and speed to gain a competitive edge. An example of agile manufacturing can be found in Ford Motor Company's electronic components plant in Landsdale, Pennsylvania. The plant produces about 124,000 engine controllers, antilock brake sensors, and speed control units a day. And because each product has between 400 and 500 parts, this means that there are more than 5 million individual parts to keep track of daily. Nonetheless, according to the plant manager, when an order to change a product is received, the changed units can usually be shipped within 24 hours! The plant is able to respond to change orders quickly because it has highly flexible automated equipment, computer software that has been designed to accommodate changes, and lean inventories that don't have to be worked off before changed units can be shipped.

This agile philosophy of being able to switch quickly—and economically—from one design to another, with little disruption, is being implemented throughout the Ford Motor Company. It is also being implemented in many other companies, both large and small, as they shift away from more traditional mass production methods that have heavy emphasis on volume and cost reduction into production methods that emphasize speed and flexibility. This dramatic shift in the way things are made reflects a growing trend as American companies struggle to compete with foreign companies.

Perhaps central to the process of agile manufacturing is the rapid collection and processing of information. For instance, extensive use of bar codes provides information on the status of each product for Ford managers at the plant in Landsdale. Moreover, if defects are discovered, there is an electronic audit trail to trace the defect back to the exact spot in the process where it occurred, enabling management to focus efforts on correcting the cause of the problem.

Questions:

1. What is an agile manufacturer?
2. What is the value of being an agile manufacturer?
3. How could the managers at the plant keep track of the five million parts each day?
4. What are the keys to being an agile manufacturer?

Source: Adapted from "Agile Manufacturers Put Premium on Speed, Flexibility" by John Holusha, the New York Times Company, 1994.

7. *Worker involvement.* More and more companies are pushing the responsibility for decision making and problem solving to lower levels in the organization. The reasons for this trend include recognition of the knowledge workers possess about the production process and the contributions they can make to improving the production system. A key to worker involvement is the use of *teams* of workers who solve problems and make decisions on a consensus basis.

www.kodak.com
www.bell-atl.com

8. *Reengineering.* Some companies are taking drastic measures to improve their performance. They are conceptually starting from scratch in redesigning their processes. According to Michael Hammer, coauthor of *Reengineering the Corporation* (HarperBusiness), reengineering means starting over, asking why a company does things the way it does, and questioning basic rules and assumptions. Engineering focuses on significantly improving business processes, such as the steps required to fill a customer's request or the steps required to bring a new product to market. Kodak was able to cut in half the time needed to bring a new camera to market; Union Carbide was able to cut $400 million of fixed expenses, and Bell Atlantic was able to cut the time needed to hook up long-distance carriers from 15 days to less than 1 day, at a savings of $82 million. Of course, reengineering is not for everyone. The best candidates are firms that are in serious trouble and firms that see trouble on the horizon if they do not change. Reengineering is not a quick fix, and it doesn't always work. It takes team players, good communication, dedication, and paying attention to the "people aspects."

9. *Environmental issues.* Pollution control and waste disposal are key issues managers must contend with. There is increasing emphasis on reducing waste, using less toxic chemicals (e.g., lawncare services shifting to environmentally friendly approaches), recycling, making it easier for consumers to recycle products (e.g., including a shipping container for returning used laser printer cartridges), and designing products and parts that can be reused (remanufacturing products such as copying machines). The term *environmentally responsible manufacturing* is sometimes used to describe these policies. Regulations are increasing in number and complexity, and penalties for pollution and inadequate control of wastes can be severe. While this has placed an added burden on some industries, society should reap tremendous benefits in cleaner air and water, and less damage to the environment in general. Some of the consequences of *not* paying attention to environmental issues can be seen in the industrialized cities of the former Soviet Union and its East European satellites, where years of neglect have resulted in catastrophic damage to the environment, which will require many years and enormous amounts of money before recovery. Waste disposal regulation has led to the creation of opportunity for businesses that specialize in waste management and recycling.

10. *Corporate downsizing.* Squeezed by competition, lagging productivity, and stockholders calling for improved profits and share prices, many corporations have responded by reducing their labor forces. This has meant that operations managers often have to find ways to produce more with fewer workers.

11. *Supply-chain management.* Organizations are increasing their attention to managing the supply chain, from suppliers and buyers of raw materials all the way to final customers.

12. *Lean production.* This new approach to production emerged in the 1990s. It incorporates a number of the recent trends listed here, with an emphasis on quality, flexibility, time reduction, and teamwork. This has led to a *flattening* of the organizational structure, with fewer levels of management.

lean production System that uses minimal amounts of resources to produce a high volume of high-quality goods with some variety.

Lean production systems are so named because they use much less of certain resources than mass production systems use—less space, less inventory, and fewer workers—to produce a comparable amount of output. Lean production systems use a highly skilled work force and flexible equipment. In effect, they incorporate advantages of both mass production (high volume, low unit cost) and craft production (variety and flexibility). And quality is higher than in mass production.

The skilled workers in lean production systems are more involved in maintaining and improving the system than their mass production counterparts. They are taught to stop production if they discover a defect, and to work with other employees to find and correct the cause of the defect so that it won't recur. This results in an increasing level of quality over time, and eliminates the need to inspect and rework at the end of the line.

Because lean production systems operate with lower amounts of inventory, additional emphasis is placed on anticipating when problems might occur *before* they arise, and avoiding those problems through careful planning. Even so, problems still occur at times, and quick resolution is important. Workers participate in both the planning and correction stages. Technical experts are still used, but more as consultants rather than substitutes for workers. The focus is on designing a system (products and process) so that workers will be able to achieve high levels of quality and quantity.

Compared to workers in traditional systems, much more is expected of workers in lean production systems. They must be able to function in teams, playing active roles in operating and improving the system. Individual creativity is much less important than team success. Responsibilities also are much greater, which can lead to pressure and anxiety not present in traditional systems. Moreover, a flatter organizational structure means career paths are not as steep in lean production organizations. Workers tend to become generalists rather than specialists, another contrast to more traditional organizations.

	Craft Production	Mass Production	Lean Production
Description	High variety, customized output, with one or a few skilled workers responsible for an entire unit of output.	High volume of standardized output, emphasis on volume. Capitalizes on division of labor, specialized equipment, and interchangeable parts.	Moderate to high volume of output, with more variety than mass production. Fewer mass production buffers such as extra workers, inventory, or time. Emphasis on quality. Employee involvement and teamwork important.
Examples of Goods and Services	Home remodeling and landscaping, tailoring, portrait painting, diagnosis and treatment of injuries, surgery.	Automobiles, computers, calculators, sewing machines, compact discs, mail sorting, check clearing.	Similar to mass production.
Advantages	Wide range of choice, output tailored to customer needs.	Low cost per unit, requires mostly low-skilled workers.	Flexibility, variety, high quality of goods.
Disadvantages	Slow, requires skilled workers, few economies of scale, high cost, and low standardization.	Rigid system, difficult to accommodate changes in output volume, product design, or process design. Volume may be emphasized at the expense of quality.	No safety nets to offset any system breakdowns, fewer opportunities for employee advancement, more worker stress, requires higher-skilled workers than mass production.

TABLE 1–8

A comparison of craft, mass, and lean production

Unions often oppose conversion from a traditional system to a lean system because they view the added responsibility and multiple tasks as an expansion of job requirements without comparable increases in pay. In addition, workers sometimes complain that the company is the primary beneficiary of employee-generated improvements.

Table 1–8 provides a comparison of craft production, mass production, and lean production. Keep in mind that all three of these modes of production are in existence today.

Summary

Operations management is that part of a business organization responsible for planning and coordinating the use of the organization's resources to convert inputs into outputs. The operations function is one of three primary functions of business organizations; the other two are marketing and finance.

The operations function is present in both service-oriented and product-oriented organizations. Operations decisions involve design decisions and operating decisions. Design decisions relate to capacity planning, product design, process design, layout of facilities, and selecting locations for facilities. Operating decisions relate to quality assurance, scheduling, inventory management, and project management.

The chapter provides a brief discussion of the historical evolution of production/operations management and recent trends in the field. Among those trends are global competition; increasing emphasis on quality; integrating technology into production systems; increasing worker involvement in problem solving and decision making, particularly through the use of teams; increasing emphasis on flexibility and time reduction; increasing attention to environmental issues; supply chain management; and lean production.

Key Terms

craft production, 18
division of labor, 21
flexibility, 25
interchangeable parts, 21
lead time, 9
lean production, 27
mass production, 21
operations management, 4
Pareto phenomenon, 17
system, 16
system design, 11
system operation, 11
value-added, 7

Discussion and Review Questions

1. Briefly describe the term *production/operations management.*
2. Identify the three major functional areas of business organizations and briefly describe how they interrelate.
3. Describe the operations function and the nature of the operation manager's job.
4. List five important differences between goods production and service operations.
5. Briefly discuss each of these terms related to the historical evolution of production/operations management:
   a. Industrial Revolution
   b. Scientific management
   c. Interchangeable parts
   d. Division of labor
6. Why are services important? Why is manufacturing important?
7. Identify some of the current trends in operations management, and relate them to recent news items or to personal experience.
8. Describe the *Pareto phenomenon*.
9. List the trade-offs you would consider for each of these decisions:
   a. Driving your own car versus public transportation.
   b. Buying a computer now versus waiting for an improved model.
   c. Buying a new car versus buying a used car.
   d. Speaking up in class versus waiting to get called on by the instructor.
10. Describe each of these systems: craft production, mass production, and lean production.
11. Why might some workers prefer not to work in a lean production environment? Why might some managers resist a change from a more traditional mode of production to lean production?
12. Match each pioneer with the appropriate description.
   a. Henry Gantt — *i.* Mass production and the moving assembly line.
   b. F. W. Taylor — *ii.* Psychologist who emphasized the human factor in work design.
   c. Frank Gilbreth — *iii.* Father of scientific management.
   d. Lillian Gilbreth — *iv.* Motion study principles.
   e. Henry Ford — *v.* Charts used for scheduling.

13 Briefly explain these terms:
   a. environmentally responsible manufacturing   *b.* re-engineering

Memo Writing Exercises

Memo writing exercises are intended to help you to strengthen your written communication skills while applying what you have learned to issues of managerial concern. The following memo provides information on writing good memos. Use it as a guide.

MEMORANDUM

To: Students using this book

From: W. Stevenson

Date: April 4, 1998

Subject: Memo writing exercises

The purpose of this memo is to introduce you to an important mode of business communication, and to provide you with information on how to write a memo.

Memos are often used to convey information within an organization, but they are also used to communicate with people outside the organization, such as suppliers, distributors, and sometimes, customers.

Memos tend to be less formal than letters. Memos should communicate business information in an organized, concise format. Even so, their length and degree of formality depend in part on the purpose of the memo, the amount of information to convey, and the intended audience. Thus, a memo reporting financial information to the chief executive officer would be more formal and probably longer than a memo to employees about the company picnic.

A variety of memo styles can be used. One style is illustrated by this memo, which is informational.

1. Use the first paragraph to state the purpose of the memo. This might be to inform the recipient of a new policy or a decision that has been made, to offer suggestions or present a proposal, or to provide a response to a previous memo.
2. Use the second paragraph to provide major details of a proposal or major reasons for a decision or suggestion, or to indicate why you agree or disagree with a proposal or idea.
3. Use the third paragraph for minor details.
4. In a long memo, you may want to end by again stating your position on the subject.

1. Suppose that your boss, Tony Roman, who is the operations manager of your company, sends you a memo asking for your input on holding joint planning sessions with people from marketing and finance. A portion of that memo follows:

   Gail in Marketing and Jim in Finance have been bugging me lately about holding joint meetings. They're real gung-ho on this. But I'm a little concerned about agreeing to it. I don't want them to get the impression that they can come in here and tell us what to do. They don't really understand what it is we do or the kinds of problems we have. I'm sure they think we aren't doing as well as we could, but that's not any of their concern. I don't know why they just don't concentrate on what they do, and leave operations to us.

   Write a one-page memo that summarizes the potential benefits of holding joint planning meetings with Marketing and Finance.

2. Assume the role of a midlevel manager of a small company. The CEO, Rachael Barker, wants your views on a new computer network she is considering. The current network is adequate, but the new one will provide much greater capabilities for managing files, word processing, and accessing the database.

   Write a one-page memo in which you list (1) the key factors that you believe must be taken into account in making the decision, and (2) the trade-offs that must be considered.

3. Write a brief memo, responding to this one:

MEMORANDUM

To: (*Your name*), V.P. of Manufacturing

From: Deiter Smith, Production Supervisor, second shift

Date: Monday

Re: Accidental spill

We recently experienced an accidental discharge of mildly toxic chemicals into the creek that runs past one of our plants. Although there have been a few complaints of contamination from downstream property owners, I don't think that the spill can be traced back to our plant. Besides, other nearby companies have been getting away with more serious abuses for years, so let them take the heat.

I am confident that this is a one-time occurrence, so let's not blow the whole thing out of proportion. If we were to have to pay for a cleanup, this would have a big impact on quarterly profits, and our stockholders wouldn't like that one bit. Furthermore, in light of our recent efforts to promote an image of an environmental-friendly company, I say we should definitely keep quiet about this and let things die down.

What do you think?

CASE
HAZEL

Hazel had worked for the same Fortune 500 company for almost 15 years. Although the company had gone through some tough times, things were starting to turn around. Customer orders were up, and quality and productivity had improved dramatically from what they had been only a few years earlier due to a companywide quality improvement program. So it came as a real shock to Hazel and about 400 of her co-workers when they were suddenly terminated following the new CEO's decision to downsize the company.

After recovering from the initial shock, Hazel tried to find employment elsewhere. Despite her efforts, after eight months of searching she was no closer to finding a job than the day she started. Her funds were being depleted and she was getting more discouraged. There was one bright spot, though: She was able to bring in a little money by mowing lawns for her neighbors. She got involved quite by chance when she heard one neighbor remark that now that his children were on their own, nobody was around to cut the grass. Almost jokingly, Hazel asked him how much he'd be willing to pay. Soon Hazel was mowing the lawns of five neighbors. Other neighbors wanted her to work on their lawns, but she didn't feel that she could spare any more time from her job search.

However, as the rejection letters began to pile up, Hazel knew she had to make an important decision in her life. On a rainy Tuesday morning, she decided to go into business for herself—taking care of neighborhood lawns. She was relieved to give up the stress of job hunting, and she was excited about the prospects of being her own boss. But she was also fearful of being completely on her own. Nevertheless, Hazel was determined to make a go of it.

At first, business was a little slow, but once people realized Hazel was available, many asked her to take care of their lawns. Some people were simply glad to turn the work over to her; others switched from professional lawn care services. By the end of her first year in business, Hazel knew she could earn a living this way. She also performed other services such as fertilizing lawns, weeding gardens, and trimming shrubbery. Business became so good that Hazel hired two part-time workers to assist her and, even then, she believed she could expand further if she wanted to.

Questions

1. In what ways are Hazel's customers most likely to judge the quality of her lawn care services?
2. Hazel is the operations manager of her business. Among her responsibilities are forecasting, inventory management, scheduling, quality assurance, and maintenance.
   a. What kinds of things would likely require forecasts?
   b. What inventory items does Hazel probably have? Name one inventory decision she has to make periodically.
   c. What scheduling must she do? What things might occur to disrupt schedules and cause Hazel to reschedule?
   d. How important is quality assurance to Hazel's business? Explain.
   e. What kinds of maintenance must be performed?
3. What are some of the trade-offs that Hazel probably considered relative to:
   a. Working for a company instead of for herself?
   b. Expanding the business?
4. The town is considering an ordinance that would prohibit putting grass clippings at the curb for pickup because local landfills cannot handle the volume. What options might Hazel consider if the ordinance is passed? Name two advantages and two drawbacks of each option.

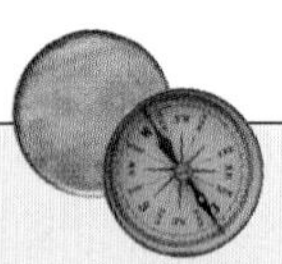

OPERATIONS TOUR
WEGMANS FOOD MARKETS

www.wegmans.com

Wegmans Food Markets, Inc., is one of the premier grocery chains in the United States. Headquartered in Rochester, NY, Wegmans operates over 70 stores, mainly in Rochester, Buffalo, and Syracuse. There are also a handful of stores elsewhere in New York State and Pennsylvania. The company employs over 23,000 people, and has annual sales of over $2.0 billion. In addition to supermarkets, the company operates Chase-Pitkin Home and Garden Centers and an egg farm.

Wegmans has a strong reputation for offering its customers high product quality and excellent service. Through a combination of market research, trial and error, and listening to its customers, Wegmans has evolved into a very successful organization. In fact, Wegmans is so good at what it does that grocery chains all over the country send representatives to Wegmans for a firsthand look at operations.

Superstores

Many of the company's stores are giant 100,000 square foot superstores, double or triple the size of average supermarkets. You can get an idea about the size of these stores from this: they usually have between 25 and 35 checkout lanes, and during busy periods, all of the checkouts are in operation. A superstore typically employs from 500 to 600 people.

Individual stores differ somewhat in terms of actual size and some special features. Aside from the features normally found in supermarkets, they generally have a full-service deli (typically a 40-foot display case), a 500 square foot fisherman's wharf that has perhaps 10 different fresh fish offerings most days, a large bakery section (each store bakes its own bread, rolls, cakes, pies, and pastries), and extra large produce sections. They also offer film processing, a complete pharmacy, a card shop, video rentals, and an Olde World Cheese™ section. In-store floral shops range in size up to 800 square feet of floor space, and offer a wide variety of fresh-cut flowers, flower arrangements, vases and plants. In-store card shops cover over 1,000 square feet of floor space. The bulk foods department provides customers with the opportunity to select what quantities they desire from a vast array of foodstuffs and some nonfood items such as birdseed and pet food.

Each store is a little different. Among the special features in some stores are a dry cleaning department, a wokery, and a salad bar. Some stores feature a Market Cafe™ that has different food stations, each devoted to preparing and serving a certain type of food. For example, one station will have pizza and other Italian specialties, and another oriental food, and still another chicken or fish. There will also be a sandwich bar, a salad bar, and a dessert station. Customers often wander among stations as they decide what to order. In some Market Cafes, diners can have wine with their meals and have brunch on Sundays. In several affluent locations, customers can stop in on their way home from work and choose from a selection of freshly prepared dinner entrees such as medallions of beef with herb butter, chicken Marsala, stuffed flank steak with mushrooms, grilled salmon, Cajun tuna, and crab cakes, and accompaniments such as roasted red potatoes, grilled vegetables, and Caesar salad. Many Wegmans stores offer ready-made sandwiches as well as made-to-order sandwiches during the lunch hour. Some stores have a coffee shop section with tables and chairs where shoppers can enjoy regular or specialty coffees and a variety of tempting pastries.

Each supermarket has an instore bakery.

Produce Department

The company prides itself on fresh produce. Produce is replenished as often as 12 times a day. The larger stores have produce sections that are four to five times the size of a produce section of an average supermarket. Wegmans offers locally grown produce in season. Wegmans uses a "farm to market" system whereby some local growers deliver their produce directly to individual stores, bypassing the main warehouse. That reduces the company's inventory holding costs and gets the produce into the stores as quickly as possible. Growers may use specially designed containers that go right onto the store floor instead of large bins. This avoids the bruising that often occurs when fruits and vegetables are transferred from bins to display shelves and the need to devote labor to transfer the produce to shelves.

Meat Department

In addition to large display cases of both fresh and frozen meat products, many stores have a full-service butcher shop that offers a variety of fresh meat products and where butchers are available to provide customized cuts of meat for customers.

Ordering

Each department handles its own ordering. Although sales records are available from records of items scanned at the checkouts, they are not used directly for replenishing stock. Other factors, such as pricing, special promotions, local circumstances (e.g., festivals, weather conditions) must all be taken into account. However, for seasonal periods, such as holidays, managers often check scanner records to learn what past demand was during a comparable period.

The superstores typically receive one truckload of goods per day from the main warehouse. During peak periods, a store may receive two truckloads from the main warehouse. The short lead time greatly reduces the length of time an item might be out of stock, unless the main warehouse is also out of stock.

The company exercises strict control over suppliers, insisting on product quality and on-time deliveries.

Market Café™

Wegmans' "farm to market" allows local growers to deliver produce directly to stores.

Employees

The company recognizes the value of good employees. It typically invests an average of $7,000 to train each new employee. In addition to learning about store operations, new employees learn the importance of good customer service and how to provide it. The employees are helpful, cheerfully answering customer questions or handling complaints. Employees are motivated through a combination of compensation, profit sharing, and benefits. In a *Fortune* survey of employees on the best companies to work for in the U.S., Wegmans ranked #16.[1]

[1]*Fortune*, January 12, 1998, p. 85.

Quality

Quality and customer satisfaction are utmost in the minds of Wegmans management and its employees. Private label food items as well as name brands are regularly evaluated in test kitchens, along with potential new products. Managers are responsible for checking and maintaining product and service quality in their department. Moreover, employees are encouraged to report problems to their managers.

If a customer is dissatisfied with an item, and returns it, or even a portion of the item, the customer is offered a choice of a replacement or a refund. If the item is a Wegmans brand food item, it is then sent to the test kitchen to determine the cause of the problem. If the cause can be determined, corrective action is taken.

Questions:

1. How do customers judge the quality of a supermarket?
2. Indicate how and why each of these factors is important to the successful operation of a supermarket:
   a. Customer satisfaction.
   b. Forecasting.
   c. Capacity planning.
   d. Location.
   e. Inventory management.
   f. Layout of the store.
   g. Scheduling.

Selected Bibliography

Bounds, Gregory M., Gregory H. Dobbins, and Oscar S. Fowler. *Management: A Total Quality Perspective.* Cincinnati: South-Western Publishing, 1995.

Cascio, Wayne F. *Managing Human Resources.* New York: McGraw-Hill, 1989, 2nd ed.

Cohen, Stephen S, and John Zysman. *Manufacturing Matters: The Myth of the Post-Industrial Economy.* New York: Basic Books, 1987.

Hammer, Michael, and James Champy. *Reengineering the Corporation.* New York: Harper Business, 1993.

Hopp, William J., and Mark Spearman. *Factory Physics: Foundations of Manufacturing Management.* Burr Ridge, Illinois: Irwin, 1996.

Womack, James P., Daniel Jones, and Daniel Roos. *The Machine that Changed the World.* New York: Harper Perrenial, 1991.

Wren, Daniel A. *The Evolution of Management Thought.* New York: Wiley, 3rd ed., 1987.

ProScan
MAGNAVOX
GENERAL ELECTRIC
SONY
RCA
SANYO
14

LEARNING OBJECTIVES

After completing this chapter, you should be able to:

1. Define the term *productivity* and explain why it is important to organizations and to countries.
2. List some of the reasons for poor productivity and some ways of improving it.
3. List and briefly discuss the primary ways that business organizations compete.
4. List five reasons for the poor competitiveness of some companies.
5. Define the term *strategy* and explain why strategy is important for competitiveness.
6. Contrast *strategy* and *tactics*.
7. Compare organization strategy and operations strategy, and explain why it is important to link them.
8. Describe and give examples of *time-based* strategies.

CHAPTER TWO

PRODUCTIVITY, COMPETITIVENESS, AND STRATEGY

CHAPTER OUTLINE

This chapter discusses productivity, competitiveness, and strategy: three separate but related topics that are vitally important to business organizations. *Productivity* relates to effective use of resources; *competitiveness* relates to how effective an organization is in the marketplace compared with other organizations that offer similar products or services; and *strategy* relates to the plans that determine the direction an organization takes in pursuing its goals.

Slumping productivity gains in the late 1980s and the impressive successes of foreign competition in the U.S. marketplace caused many U.S. companies not only to rethink their strategies, but to place increased emphasis on *operations strategy.*

INTRODUCTION

In this chapter you will learn what productivity is, why it is important, and some ways organizations can improve it. You will learn about the different ways companies compete, and why some firms do a very poor job of competing. And you will learn how effective strategies can lead to productive, competitive organizations.

PRODUCTIVITY

productivity A measure of the effective use of resources, usually expressed as the ratio of output to input.

One of the primary responsibilities of an operations manager is to achieve *productive use* of an organization's resources. The term *productivity* is used to describe this. **Productivity** is an index that measures output (goods and services) relative to the input (labor, materials, energy, and other resources) used to produce them. It is usually expressed as the ratio of output to input:

$$\text{Productivity} = \frac{\text{Output}}{\text{Input}} \qquad (2\text{–}1)$$

A productivity ratio can be computed for a single operation, a department, an organization, or even an entire country.

Productivity measures can be based on a single input (partial productivity), on more than one input (multifactor productivity), or on all inputs (total productivity). Table 2–1 lists some examples of productivity measures. The choice of productivity measure depends primarily on the purpose of the measurement. If the purpose is to track improvements in labor productivity, then labor becomes the obvious input measure.

Partial measures are often of greatest use in operations management. Table 2–2 provides some examples of partial productivity measures.

The units of output used in productivity measures depend on the type of job performed. The following are examples of labor productivity:

$$\frac{\text{Yards of carpet installed}}{\text{Labor hours}} = \text{Yards of carpet installed per labor hour}$$

$$\frac{\text{Number of offices cleaned}}{\text{Number of shifts}} = \text{Number of offices cleaned per shift}$$

$$\frac{\text{Board feet of lumber cut}}{\text{Number of weeks}} = \text{Board feet of lumber cut per week}$$

TABLE 2–1

Some examples of different types of productivity measures

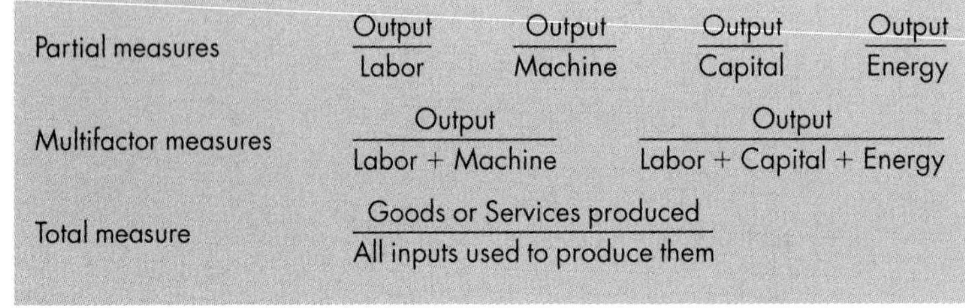

Partial measures	$\frac{\text{Output}}{\text{Labor}}$	$\frac{\text{Output}}{\text{Machine}}$	$\frac{\text{Output}}{\text{Capital}}$	$\frac{\text{Output}}{\text{Energy}}$
Multifactor measures	$\frac{\text{Output}}{\text{Labor + Machine}}$		$\frac{\text{Output}}{\text{Labor + Capital + Energy}}$	
Total measure	$\frac{\text{Goods or Services produced}}{\text{All inputs used to produce them}}$			

TABLE 2–2

Some examples of partial productivity measures

Labor Productivity	Units of output per labor hour Units of output per shift Value-added per labor hour Dollar value of output per labor hour
Machine Productivity	Units of output per machine hour Dollar value of output per machine hour
Capital Productivity	Units of output per dollar input Dollar value of output per dollar input
Energy Productivity	Units of output per kilowatt-hour Dollar value of output per kilowatt-hour

Similar examples can be listed for *machine productivity* (e.g., the number of pieces per hour turned out by a machine).

Example 1

Determine the productivity for these cases:

a. Four workers installed 720 square yards of carpeting in eight hours.

b. A machine produced 68 usable pieces in two hours.

Solution

$$a.\ \text{Productivity} = \frac{\text{Yards of carpet installed}}{\text{Labor hours worked}} = \frac{720 \text{ square yards}}{4 \text{ workers} \times 8 \text{ hours/worker}} = \frac{720 \text{ yards}}{32 \text{ hours}} = 22.5 \text{ yards/hour}$$

$$b.\ \text{Productivity} = \frac{\text{Usable pieces}}{\text{Production time}} = \frac{68 \text{ pieces}}{2 \text{ hours}} = 34 \text{ pieces/hour}$$

Calculations of multifactor productivity measure inputs and outputs using a common unit of measurement, such as cost or value. For instance, the measure might use cost of inputs and price of the output:

$$\frac{\text{Quantity of production at standard price}}{\text{Labor cost} + \text{Materials cost} + \text{Overhead}} \qquad (2\text{–}2)$$

Example 2

Determine the multifactor productivity for the combined input of labor and machine time using the following data:

Output: 1,760 units

Input

Labor: $1,000

Materials: $520

Overhead: $2,000

Solution

$$\text{Multifactor Productivity} = \frac{\text{Output}}{\text{Labor} + \text{Materials} + \text{Overhead}} = \frac{1{,}760 \text{ units}}{\$1{,}000 + \$520 + \$2{,}000} = .50 \text{ units/\$}$$

Productivity measures are useful on a number of levels. For an individual department or organization, productivity measures can be used to track performance *over time*. This allows managers to judge performance, and to decide where improvements are needed.

Productivity measures can also be used to judge the performance of an entire industry or the (national) productivity of a country as a whole. These productivity measures are *aggregate* measures determined by combining productivity measures of various companies or industries.

In essence, productivity measurements serve as scorecards of the effective use of resources. Business leaders are concerned with productivity as it relates to *competitiveness*: If two firms both have the same level of output, but one requires less input because of higher productivity, that one will be able to charge a lower price and consequently increase its share of the market. Or that firm might elect to charge the same price, thereby reaping a greater profit. Government leaders are concerned with national productivity because of the close relationship between productivity and a nation's *standard of living*. High levels of productivity are largely responsible for the relatively high standards of living enjoyed by people in industrial nations. Furthermore, wage and price increases not accompanied by productivity increases tend to create inflationary pressures on a nation's economy.

Productivity growth in the United States in the 1970s and 80s lagged behind that of other leading industrial countries, most notably Japan, Korea, the U.K., and West Germany. That caused concern among government officials and business leaders. Although U.S. productivity was still among the highest in the world, it was losing ground to other nations. Moreover, a significant portion of U.S. productivity could be attributed to high *agricultural* productivity; *manufacturing* productivity tended to be lower. That may be changing. U.S. productivity growth slowed during the 80s, and was actually negative in 1989. Since then, however, it has increased. (See Figure 2–1.)

The obvious question is: How can certain nations, industries, or companies achieve productivity gains while others cannot? Theorists and researchers cite negative impacts on productivity. Among the most prominent are:

1. A lower propensity to save and a higher propensity to consume, which slows capital formation and attracts foreign goods.
2. Increasing government regulations add to the administrative (and nonproductive) burden of many companies.
3. In the United States, there is an increasing demand for services, which are often less productive than manufacturing operations.

FIGURE 2–1

Percentage change in U.S. productivity (output per hour), 1989–1997

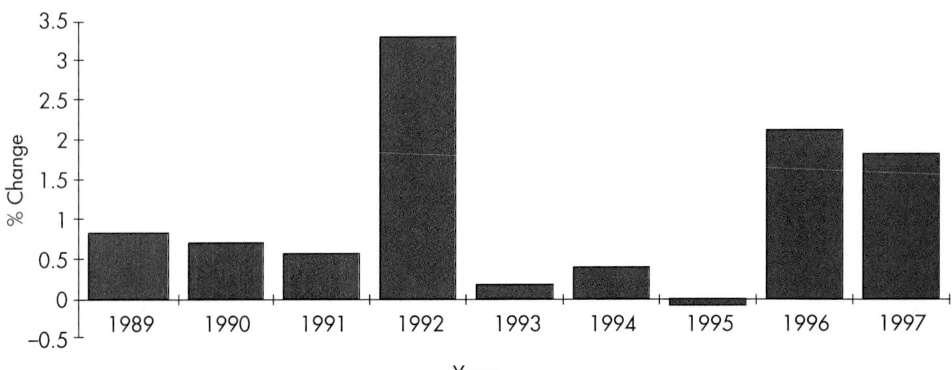

4. An emphasis on *short-run* performance (e.g., annual profits and sales) reduces the incentive to develop long-term solutions to problems. In addition, in periods of inflation and increased costs of borrowed money, managers are hesitant to commit funds for long periods of time because it reduces their flexibility to take advantage of other opportunities that might arise in the meantime.

Many authorities place much of the responsibility on management for the lagging U.S. productivity gains in the 70s and 80s. They point out that foreign competitors had to deal with similar (or worse) conditions than their U.S. counterparts and yet came out ahead on productivity improvements. Some authorities argue that most foreign competitors started out with productivity so low that the improvements merely reflect catching up with U.S. manufacturers. To a certain extent, that was true. Moreover, foreign competitors were able to learn much about quality and productivity improvements by studying Western methods. However, many of those companies achieved productivity levels that equal or surpass those of their U.S. counterparts. It seems that they were using the lessons learned to go beyond their teachers. Some cite the inability of U.S. managers to successfully integrate technological improvements into the production process; a common complaint was that U.S. managers seemed to be "throwing technology at the problem" rather than wisely analyzing their processes to see if and how technology could be used to gain competitive advantage.

Another managerial issue is the difficulty of accurately *measuring* productivity. For example, it can be difficult to measure white-collar productivity, particularly if jobs involve thinking or creative efforts. Also, variable job content in service industries can distort productivity measurements. Suppose a restaurant employs four waitresses, who serve 80 meals on Wednesday and 90 meals on Thursday during the same two-hour interval. It would appear the waitresses were more productive on Thursday, but perhaps the lower output on Wednesday reflected the fact that fewer people came into the restaurant. Or consider the case of two emergency room doctors, one of whom attended to six patients with minor injuries while the other spent an equal amount of time with one seriously injured patient. To say that the one who attended six patients had the higher productivity distorts the reality of the situation.

Quality differences may also distort productivity measurements. One way this can happen is when comparisons are made over time, such as comparing the productivity of a factory in the 1990s with one in the 1960s. Few would disagree that quality is now much higher than it was then, but there is no simple way to incorporate quality into productivity measurements.

All in all, productivity measurements must be viewed with an awareness of related factors, and of a certain amount of distortion in the figures. Consequently, it is best to treat productivities as approximate indicators rather than precise measurements.

Factors that Affect Productivity

Numerous factors affect productivity. Among them are methods, capital, quality, technology, and management.

Consider a student who plans to type a lengthy term paper. The student is an average typist and can turn out about three pages per hour. How could the student increase productivity (i.e., turn out more pages per hour)? One way would be to enroll in a short course offered by the college to improve typing skills (method). Another way might be to replace a typewriter with a more expensive computer and word-processing package (capital) to gain the speed of automatic features such as spell checking and error correction (quality). Still other productivity improvements might be achieved through improving organization and preparation for the actual typing (management). The incentive of receiving a good grade and the pride of doing a good job might also be important. The point is that all these factors are potential sources of productivity, not only for typing papers but for any kind of work, and it is generally up to the manager to see that they are fully exploited.

A commonly held misconception is that workers are the main determinant of productivity. According to that theory, the route to productivity gains involves getting employees to work harder. However, the fact is that many productivity gains in the past have come from *technological* improvements. Familiar examples include:

Paint rollers

Power lawn mowers

Electronic typewriters

Copying machines

Microwave ovens

Automatic washers, dryers, dishwashers, electric blenders

Long-distance direct dialing, pay-by-phone, cellular phones

Computerized billing and inventories

Automation

Calculators

Computers, personal computers

However, technology alone won't guarantee productivity gains; it must be used wisely and thoughtfully. Without careful planning, technology can actually *reduce* productivity, especially if it leads to inflexibility, high costs, or mismatched operations. Another current productivity pitfall results from employees' use of computers for nonwork-related activities (playing games, or checking stock prices or sports scores on the Internet). Beyond all of these is the dip in productivity that results while employees learn to use new equipment or procedures that will eventually lead to productivity gains after the learning phase ends.

Improving Productivity

A company or a department can take a number of key steps toward improving productivity:

1. Develop productivity measures for all operations; measurement is the first step in managing and controlling an operation.
2. Look at the system as a whole in deciding which operations are most critical; it is overall productivity that is important. This concept is illustrated in Figure 2–2, which shows several operations feeding their output into a *bottleneck* operation. The capacity of the bottleneck operation is less than the combined capacities of the operations that provide input, so units queue up waiting to be processed; hence the term bottleneck. Productivity improvements to any *non*bottleneck operation will not affect the productivity of the system. Improvements in the bottleneck operation *will* lead to increased productivity, up to the point where the output rate of the bottleneck equals the output rate of the operations feeding it.
3. Develop methods for achieving productivity improvements, such as soliciting ideas from workers (perhaps organizing teams of workers, engineers, and managers), study-

FIGURE 2–2

Bottleneck operation

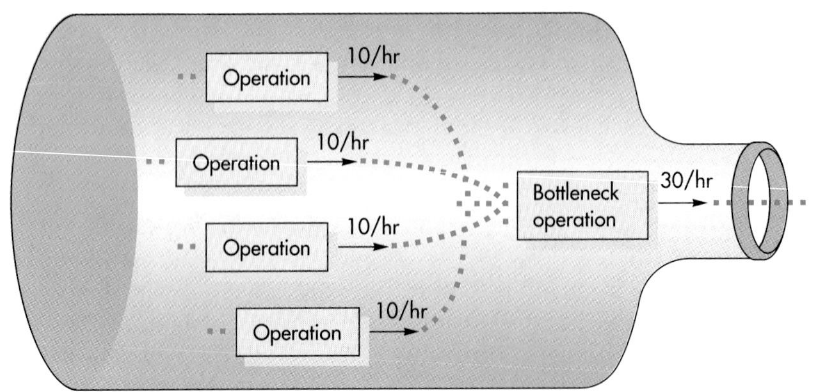

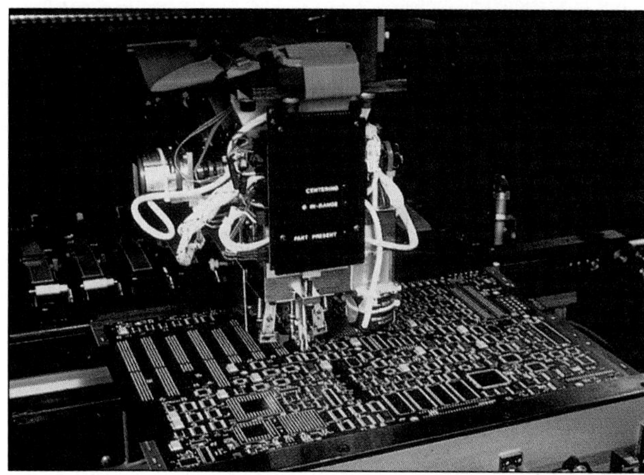

Robotic "pick and place" equipment is used at Texas Instruments to place components on the top and bottom of the printed circuit boards. This surface mount technology increases productivity.

ing how other firms have increased productivity, and reexamining the way work is done.

4. Establish reasonable goals for improvement.
5. Make it clear that management supports and encourages productivity improvement. Consider incentives to reward workers for contributions.
6. Measure improvements and publicize them.
7. Don't confuse productivity with *efficiency*. Efficiency is a narrower concept that pertains to getting the most out of a given set of resources; productivity is a broader concept that pertains to effective use of overall resources. For example, an efficiency perspective on mowing a lawn given a hand mower would focus on the best way to use the hand mower; a productivity perspective would include the possibility of using a power mower.

NEWSCLIP

PRODUCTIVITY GAINS CURB INFLATION

www.burgerking.com

Wage increases can lead to inflationary pressure—they can cause the prices consumers pay for products and services to rise. Unless, that is, they are offset by gains in productivity, which lead to an increase in profits. If that happens, a portion of the resulting profits can be used to cover the wage increases without having to raise prices.

Some Burger Kings were recently able to increase the starting pay of new workers by $1 by achieving productivity gains. The restaurants restructured the menu, combining items into meal packages such as a burger, fries, and soft drink. This enabled the counter staff to enter orders with a single key stroke instead of multiple key strokes on their point-of-sale machines, reducing the time needed to take an order. That, in turn, enabled them to take orders more quickly, increasing productivity and, consequently, reducing labor requirements, which produced higher profits.

Source: Based on "Despite Pay Increases, Gains in Productivity, Profits Curb Inflation," *The Wall Street Journal*, May 22, 1997, p. A1.

COMPETITIVENESS

Companies must be **competitive** to sell their goods and services in the marketplace. Competitiveness is an important factor in determining whether a company prospers, barely gets by, or fails. Business organizations compete with one another in a variety of ways. Key among them are price, quality, product or service differentiation, flexibility, and time to perform certain activities.

competitiveness How effectively an organization meets the needs of customers relative to others that offer similar goods or services.

1. **Price** is the amount a customer must pay for the product or service. If all other factors are equal, customers will choose the product or service that has the lower price. Organizations that compete on price may settle for lower profit margins, but most focus on lowering costs of goods or services.
2. **Quality** refers to materials and workmanship as well as design. Generally, it relates to the buyer's perceptions of how well the product or service will serve its purpose.
3. **Product differentiation** refers to any special features (e.g., design, cost, quality, ease of use, convenient location, warranty) that cause a product or service to be perceived by the buyer as more suitable than a competitor's product or service.
4. **Flexibility** is the ability to respond to changes. The better a company or department is at responding to changes, the greater its competitive advantage over another company that is not as responsive. The changes might relate to increases or decreases in volume demanded, or to changes in the design of goods or services.
5. **Time** refers to a number of different aspects of an organization's operations. One is how quickly a product or service is delivered to a customer. Another is how quickly new products or services are developed and brought to the market. Still another is the rate at which improvements in products or processes are made.

Organizations fail, or perform poorly, for a variety of reasons. Being aware of those reasons can help managers avoid making similar mistakes. Among the most common mistakes are the following:

1. Putting too much emphasis on short-term financial performance at the expense of research and development.
2. Failing to take advantage of strengths and opportunities, and/or failing to recognize competitive threats.
3. Neglecting operations strategy.
4. Placing too much emphasis on product and service design and not enough on process design.
5. Neglecting investments in capital and human resources.
6. Failing to establish good internal communications and cooperation among different functional areas.
7. Failing to consider customer wants and needs.

The key to successfully competing is to determine what customers want and then direct efforts toward meeting (or even exceeding) customer expectations. There are two basic questions that must be addressed. First: What do the customers want (i.e., *value*)? Second: How can we deliver that to them? One way to express customer value—and direct the organization's efforts in satisfying customers—is with the following relationship:[1]

$$\text{Value} = \frac{\text{Performance}}{\text{Cost}} = \frac{\text{Quality} + \text{Speed} + \text{Flexibility}}{\text{Cost}} \qquad (2\text{–}3)$$

In this model, product differentiation is included as part of quality. (In a later chapter, you will find that special features, i.e., product differentiation, is actually one of the dimensions of quality.) What the relationship says is that a customer will evaluate a product or service by its performance (as measured by the three factors) relative to its cost. As you might guess, depending on the nature of the product or service, and the customer, what is most important in the value relationship differs. Thus, in some instances, quality may be more (or less) important than speed and/or flexibility. A man-

[1]Described by Procter & Gamble CEO Edwin L. Artzt at Quality Forum VIII, October 1, 1992.

ager can incorporate these differences by *weighting* each performance factor according to its importance. This leads to the following formulation:

$$\text{Value} = \frac{w_1 \times \text{Quality} + w_2 \times \text{Speed} + w_3 \times \text{Flexibility}}{\text{Cost}} \qquad (2\text{–}4)$$

Understanding this value relationship can help managers develop successful strategies.

STRATEGY

An organization's *strategy* has a long-term impact on the nature and characteristics of the organization. In large measure, strategies affect the ability of an organization to compete or, in the case of a nonprofit organization, the ability to serve its intended purpose.

In this section, you will learn about both organization strategies and operations strategies, and how they influence an organization. The nature of an organization's strategies depends on its mission.

Mission

An organization's **mission** is the basis of the organization—the reason for its existence. Missions vary from organization to organization, depending on the nature of their business. A hospital's mission would be to provide health care; a construction firm's mission might be to construct new, single-family homes; and an insurance firm's mission might be to provide life insurance only or perhaps a full range of insurance (home, auto, life, accident). Part of the mission of nonprofit organizations is to provide services to clients, while part of the mission of for-profit organizations is to provide profits for owners (stockholders, partners).

mission The reason for existence of an organization.

It is important that an organization have a clear and simple **mission statement,** one which answers the question, "What business are we in?" The mission statement should serve to guide formulation of strategies for the organization as well as decision making at all levels. Not all organizations have mission statements; perhaps their leaders lack an awareness of how important such a statement is, or perhaps they are unclear about what the mission should be. Without a clear mission, an organization is unlikely to achieve its true potential, because there is little direction for formulating strategies. Table 2–3 provides some sample mission statements.

mission statement A clear statement of purpose that serves as a guide for strategy and decision making.

Strategies and Tactics. A mission statement provides a general direction for an organization and gives rise to organizational *goals,* which provide substance to the overall mission. For example, one goal of an organization may be to capture a certain percent of market share for a product; another goal may be to achieve a certain level of profitability. Taken together, the goals and the mission establish a destination for the organization.

Strategies are plans for achieving goals. If you think of goals as destinations, then strategies are the road maps for reaching the destinations. Strategies provide *focus* for decision making. Generally speaking, organizations have overall strategies called *organization strategies,* which relate to the entire organization, and they also have *functional strategies,* which relate to each of the functional areas of the organization. The functional strategies should support the overall strategies of the organization, just as the organizational strategies should support the goals and mission of the organization.

strategy A plan for achieving organizational goals.

Tactics are the methods and actions used to accomplish strategies. They are more specific in nature than strategies, and they provide guidance and direction for carrying out actual *operations,* which need the most specific and detailed plans and decision making in an organization. You might think of tactics as the "how to" part of the process

tactics The methods and actions taken to accomplish strategies.

TABLE 2–3

Selected company mission statements

www.ibm.com

www.unioncarbide.com

Globe Metallurgical	To provide products and services that lead the silicon metal and ferroalloy industries in the highest quality at the lowest manufacturing costs. To maintain our lead, we must continually improve our products and services. We will meet our customers' needs with products, services, and technologies that represent true value. This will ensure long-term profits for growth, job security for our employees, and give our owners and shareholders sustained, high investment returns. Reprinted by permission of Globe Metallurgical, Beverly, OH.
IBM	We create, develop, and manufacture the industry's most advanced information technologies, including computer systems, software, networking systems, storage devices, and microelectronics. We have two fundamental missions: • We strive to lead in the creation, development, and manufacture of the most advanced information technologies. • We translate advanced technologies into value for our customers as the world's largest information services company. Our professionals worldwide provide expertise within specific industries, consulting services, systems integration, and solution development and technical support. Reprinted by permission of IBM, Armonk, NY.
Skynet Worldwide Courier	The Skynet Worldwide Courier Network sets the standard for international delivery and distribution services by consistently exceeding customer expectations. Skynet delivers customer satisfaction by: • Integrating all aspects of the transportation process. • Personalizing service worldwide. • Investing in quality people and technology. • Innovating and adapting to meet customers' unique and changing requirements. Reprinted by permission of Skynet Worldwide Courier, Miami, FL.
Union Carbide	To grow the value of the Corporation by successfully pursuing strategies that capitalize on our business strengths in chemicals and polymers; To successfully execute wealth creation strategies that consistently deliver value to all stakeholders over the course of the business cycle. Reprinted by permission of Union Carbide, Danbury, CT.
Corporate Values	• Safety and environmental excellence • Customer focus • Technology leadership • People excellence • Simplicity and focus

(e.g., how to reach the destination, following the strategy road map) and operations as the actual "doing" part of the process.

It should be apparent that the overall relationship that exists from the mission down to actual operations is *hierarchical* in nature. This is illustrated in Figure 2–3.

A simple example may help to put this hierarchy into perspective.

Example 3

Rita is a high school student in Southern California. She would like to have a career in business, have a good job, and earn enough income to live comfortably.

A possible scenario for achieving her goals might look something like this:

Mission: Live a good life.

Goal: Successful career, good income.

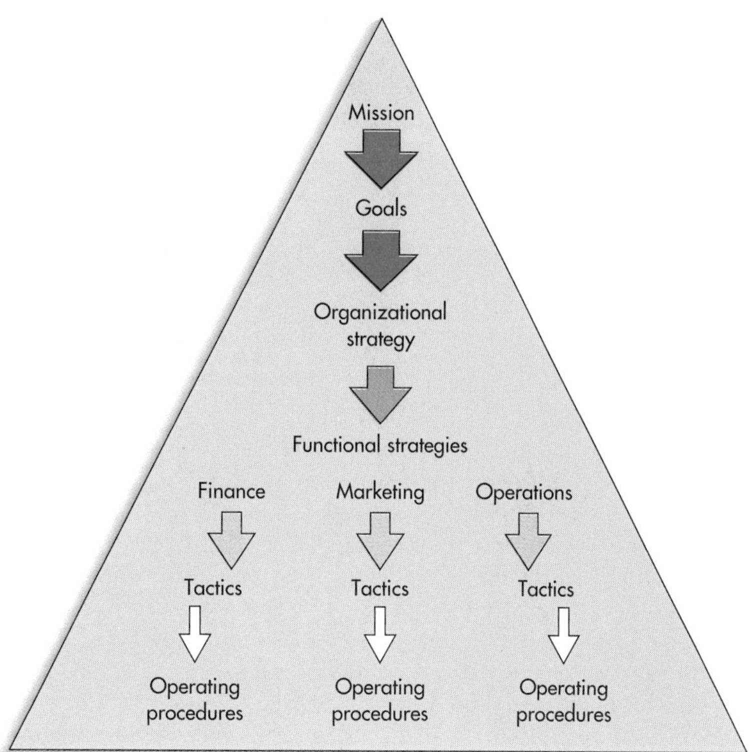

FIGURE 2–3

Planning and decision making is hierarchical in organizations

Strategy: Obtain a college education.

Tactics: Select a college and a major; decide how to finance college.

Operations: Register, buy books, take courses, study.

Operations Strategy

The organization strategy provides the overall direction for the organization. It is broad in scope, covering the entire organization. **Operations strategy** is narrower in scope, dealing primarily with the operations aspect of the organization. Operations strategy relates to products, processes, methods, operating resources, quality, costs, lead times, and scheduling. Table 2–4 provides a comparison of an organization's mission, its overall strategy, and its operations strategy, tactics, and operations.

operations strategy The approach, consistent with the organization strategy, that is used to guide the operations function.

In order for operations strategy to be truly effective, it is important to link it to organization strategy; that is, the two should not be formulated independently. Rather, formulation of organization strategy should take into account the realities of operations' strengths and weaknesses, capitalizing on strengths and dealing with weaknesses. Similarly, operations strategy must be consistent with the overall strategy of the organization, and formulated to support the goals of the organization. This requires that senior managers work with functional units to formulate strategies that will support, rather than conflict with, each other and the overall strategy of the organization. As obvious as this may seem, it doesn't always happen in practice. Instead, we may find power struggles between various functional units. These struggles are detrimental to the organization because they pit functional units against each other rather than focusing their energy on making the organization more competitive and better able to serve the customer. Some of the latest approaches in organizations, involving teams of managers and workers, may reflect a growing awareness of the synergistic effects of working together rather than competing internally.

		Management Level	Time Horizon	Scope	Level of Detail	Relates to
The Overall Organization	Mission	Top	Long	Broad	Low	Survival, profitability
	Strategy	Senior	Long	Broad	Low	Growth rate, market share
Production/ Operations	Strategic	Senior	Moderate to long	Broad	Low	Product design, choice of location, choice of technology, new facilities
	Tactical	Middle	Moderate	Moderate	Moderate	Employment levels, output levels, equipment selection, facility layout
	Operational	Low	Short	Narrow	High	Scheduling personnel, adjusting output rates, inventory management, purchasing

TABLE 2-4

Comparison of mission, organization strategy, and operations strategy

Operations strategy can have a major influence on the competitiveness of an organization. If it is well designed and well executed, there is a good chance that the organization will be successful; if it is not well designed or executed, the chances are much less that the organization will be successful.

In the 1970s and early 80s, operations strategy was often neglected in favor of marketing and financial strategies. That may have occurred because many chief executive officers did not come from operations backgrounds and perhaps did not fully appreciate the importance of the operations function. Mergers and acquisitions were common; leveraged buyouts were used, and conglomerates were formed that joined dissimilar operations. These did little to add value to the organization; they were purely financial in nature. Decisions were often made by individuals who were unfamiliar with the business, frequently to the detriment of that business. Meanwhile, foreign competitors began to fill the resulting vacuum with a careful focus on operations strategy.

In the late 1980s and early 90s, many companies began to realize this approach was not working. They recognized that they were less competitive than other companies. This caused them to focus attention on operations strategy. Toward that end, many firms are developing strategies that have *quality* or *time* as their central concern.

This correlates with a recent survey of manufacturing executives and managers who were asked to identify strategic and tactical issues that U.S. manufacturers must focus on to be competitive on a global basis in the 1990s.[2] The top two strategic issues were quality management and manufacturing strategy. The top two tactical issues were quality control, and manufacturing planning and control systems. A key element of both organization strategy and operations strategy is strategy formulation.

Strategy Formulation

order qualifiers Characteristics that customers perceive as minimum standards of acceptability to be considered as a potential for purchase.

order winners Characteristics of an organization's goods or services that cause it to be perceived as better than the competition.

To formulate an effective strategy, senior management must take into account the *distinctive competencies* of the organizations, and they must *scan the environment.* They must determine what competitors are doing, or planning to do, and take that into account. They must critically examine other factors that could have either positive or negative effects. This is sometimes referred to as the *SWOT* approach (strengths, weaknesses, opportunities, and threats).

In formulating a successful strategy, organizations must take into account both order qualifiers and order winners. Terry Hill describes **order qualifiers** as those characteristics that potential customers perceive as minimum standards of acceptability to be considered as a potential for purchase. However, that may not be sufficient to get a potential customer to purchase from the organization. **Order winners** are those characteristics of an organization's goods or services that cause them to be perceived as better than the competition.

[2]Manoj K. Malhotra, Daniel C. Steele, and Varun Grover, "Important Strategic and Tactical Manufacturing Issues in the 1990s." *Decision Sciences* 25, no. 2 (March/April 1994), pp. 189–214.

TABLE 2–5

Examples of distinctive competencies

	Competency	Examples of Companies or Services
Price	Low cost	U.S. first-class postage Motel-6, Red Roof Inns Mail-order computers
Quality	High-performance design and/or high quality	Sony TV Lexus, Cadillac Disneyland Five-star restaurants or hotels
	Consistent quality	Coca-Cola, Pepsi Cola Kodak, Xerox, Motorola Electrical power
Time	Rapid delivery	McDonald's restaurants Express Mail UPS Domino's Pizza
	On-time delivery	One-hour photo Federal Express Express Mail
Flexibility	Variety	Burger King ("Have it your way") Hospital emergency room
	Volume	McDonald's ("Buses welcome") Toyota Supermarkets (additional checkouts)
Service	Superior customer service	Disneyland Hewlett-Packard IBM Nordstroms
Location	Convenience	Supermarkets, dry cleaners Mall stores Service stations Banks, ATMs

Distinctive competencies are those special attributes or abilities possessed by an organization that give it a *competitive edge.* In effect, distinctive competencies relate to the ways that organizations compete. As noted previously, these can include price (based on some combination of low costs of resources such as labor and materials, low operating costs, and low production costs); quality (high performance or consistent quality); time (rapid delivery or on-time delivery); flexibility (variety or volume); customer service; and location. Table 2–5 lists the major distinctive competencies and examples of services and companies that exhibit those competencies.

distinctive competencies The special attributes or abilities that give an organization a competitive edge.

The most effective organizations seem to use an approach that develops distinctive competencies based on customer needs as well as on what the competition is doing. Marketing and operations work closely to match customer needs with operations capabilities. Competitor competencies are important for several reasons. For example, if a competitor is able to supply high-quality products, it may be necessary to meet that high quality as a baseline. However, merely *matching* a competitor is usually not sufficient to gain market share. It may be necessary to exceed the quality level of the competitor or gain an edge by excelling in one or more other dimensions, such as rapid delivery or service after the sale.

Some of the strategies various Japanese manufacturing companies have employed since World War II are:

- *Low labor cost strategy.* Immediately after the war, exploited the (then) inexpensive labor pool.

Toyota, Japan's largest automobile manufacturer, also markets the upscale, award winning Lexus line. Quality is highly valued, and workers have the authority to stop the production line if quality problems are encountered.

www.lexus.com

By offering dependable, on-time overnight delivery, Federal Express serves customers throughout the U.S. and internationally in the delivery of urgent packages.

www.fedex.com

Automatic teller machines (ATMs) are numerous, conveniently located, and available 24 hours. Networks continue to expand to serve international customers.

- *Scale-based strategy.* During the 1960s, used capital intensive methods to achieve higher labor productivity and lower unit costs.
- *Focused factories strategy.* During the 1970s, used smaller factories that focused on narrow product lines to take advantage of specialization and achieve higher quality.
- *Flexible factories strategy.* During the 1980s, reduced the time needed to incorporate new product and process designs. Used flexible equipment that allowed volume and design changes, as well as product variety. Continued to stress quality.[3]

In the 1990s, the leading Japanese manufacturers adopted an approach that incorporates introducing new product features and practicing continuous improvement of both products and processes.

environmental scanning The considering of events and trends that present threats or opportunities for a company.

Environmental scanning is the considering of events and trends that present either threats or opportunities for the organization. Generally these include competitors' activities; changing consumer needs; legal, economic, political, and environmental issues; the potential for new markets; and the like. Depending on the nature of an organization and the locations of its customers, these issues may be global, national, regional, or local. Thus, the disintegration of the former Soviet Union in 1991–92, the reunification of Germany, and the formation of the European Union would have major input into the strategic planning of such global companies as the Ford Motor Company, General Motors, Kodak, Coca-Cola, PepsiCo, and IBM, but would have little direct impact on local businesses such as supermarkets or landscaping firms. Companies that are *local suppliers* for global companies, however, may be affected by international events.

Another key factor to consider when developing strategies is technological change, which can present real opportunities and threats to an organization. Technological changes occur in products (high-definition TV, improved computer chips, improved cellular telephone systems, and improved designs for earthquakeproof structures); in services (faster order processing, faster delivery); and in processes (robotics, automation, computer-assisted processing, point-of-sale scanners, and flexible manufacturing systems). The obvious benefit is a competitive edge; the risk is that incorrect choices, poor execution, and higher-than-expected operating costs will create competitive *disadvantages*.

[3]George Stalk, Jr., and Thomas Hout, "Competing Against Time," *Research and Technology Management* 33, no. 2 (March–April 1990), pp. 19–24.

Barges on the Danube River carrying new vehicles from Ford's German plants to Austria help reduce road and rail traffic and fuel usage.

Important factors may be internal or external. The key external factors are:

1. *Economic conditions.* These include the general health and direction of the economy, inflation and deflation, interest rates, tax laws, and tariffs.
2. *Political conditions.* These include favorable or unfavorable attitudes toward business, political stability or instability, and wars.
3. *Legal environment.* This includes antitrust laws, government regulations, trade restrictions, minimum wage laws, product liability laws and recent court experience, labor laws, and patents.
4. *Technology.* This can include the rate at which product innovations are occurring, current and future process technology (equipment, materials handling), and design technology.
5. *Competition.* This includes the number and strength of competitors, the basis of competition (price, quality, special features), and the ease of market entry.
6. *Markets.* This includes size, location, brand loyalties, ease of entry, potential for growth, long-term stability, and demographics.

The organization must also take into account various *internal factors* that relate to possible strengths or weaknesses. Among the key internal factors are:

1. *Human resources.* These include the skills and abilities of managers and workers; special talents (creativity, designing, problem solving); loyalty to the organization; expertise; dedication; and experience.
2. *Facilities and equipment.* Capacities, location, age, and cost to maintain or replace can have significant impact on operations.
3. *Financial resources.* Cash flow, access to additional funding, existing debt burden, and cost of capital are important considerations.
4. *Customers.* Loyalty, existing relationships, and understanding of wants and needs are important.

5. *Products and services.* These include existing products and services, and the potential for new products and services.
6. *Technology.* This includes existing technology, the ability to integrate new technology, and the probable impact of technology on current and future operations.
7. *Suppliers.* Supplier relationships, dependability of suppliers, quality, flexibility, and service are typical considerations.
8. *Other.* Other factors include patents, labor relations, company or product image, distribution channels, relationships with distributors, maintenance of facilities and equipment, access to resources, and access to markets.

Quality and Time Strategies

Traditional strategies of business organizations have tended to emphasize cost minimization or product differentiation. While not abandoning those strategies, many organizations are now adopting strategies that are based on *quality* and/or *time.* These two approaches are rapidly gaining favor throughout the business world. They are exciting and challenging, for they promise to significantly change the way business organizations operate.

quality-based strategy Strategy that focuses on quality in all phases of an organization.

Quality-based strategies focus on satisfying the customer by integrating quality into all phases of the organization. This includes not only the final product or service provided to the customer but also the related *processes* such as design, production, and service after the sale.

time-based strategy Strategy that focuses on reduction of time needed to accomplish tasks.

Time-based strategies focus on reducing the time required to accomplish various activities (e.g., develop new products or services and market them, respond to a change in customer demand, or deliver a product or perform a service). By doing so, organizations seek to improve service to the customer and to gain a competitive advantage over rivals who take more time to accomplish the same tasks.

Time-based strategies focus on reducing the time needed to conduct the various activities in a process. The rationale is that by reducing time, costs are generally less, productivity is higher, quality tends to be higher, product innovations appear on the market sooner, and customer service is improved.

Organizations have achieved time reduction in some of the following:

Planning time: The time needed to react to a competitive threat, to develop strategies and select tactics, to approve proposed changes to facilities, to adopt new technologies, and so on.

Product/service design time: The time needed to develop and market new or redesigned products or services.

Processing time: The time needed to produce goods or provide services. This can involve scheduling, repairing equipment, wasted efforts, inventories, quality, training, and the like.

Changeover time: The time needed to change from producing one type of product or service to another. This may involve new equipment settings and attachments, different methods, equipment, schedules, or materials.

Delivery time: The time needed to fill orders.

Response time for complaints: These might be customer complaints about quality, timing of deliveries, and incorrect shipments. These might also be complaints from employees about working conditions (e.g., safety, lighting, heat or cold), equipment problems, or quality problems.

The following reading from the *Harvard Business Review* provides two examples of companies improving their competitiveness through time-based changes in the way they operate.

READING

TIME-BASED INNOVATION **George Stalk, Jr.**

A company that can bring out new products three times faster than its competitors enjoys a huge advantage. Today, in one industry after another, Japanese manufacturers are doing just that to their Western competition:

- In projection television, Japanese producers can develop a new television in one-third the time required by U.S. manufacturers.
- In custom plastic injection molds, Japanese companies can develop the molds in one-third the time of U.S. competitors and at one-third the cost.
- In autos, Japanese companies can develop new products in half the time—and with half as many people—as the U.S. and German competition.

To accomplish their fast-paced innovations, leading Japanese manufacturers have introduced a series of organizational techniques that precisely parallel their approach to flexible manufacturing:

- In manufacturing, the Japanese stress short production runs and small lot sizes. In innovation, they favor smaller increments of improvements in new products, but introduce them more often—versus the Western approach of more significant improvements made less often.
- In the organization of product development work, the Japanese use factory cells that are cross-functional teams. Most Western new product development activity is carried out by functional centers.
- In the scheduling of work, Japanese factories stress local responsibility, just as product development scheduling is decentralized. The Western approach to both requires plodding centralized scheduling, plotting, and tracking.

The effects of this time-based advantage are devastating; quite simply, American companies are losing leadership of technology and innovation—supposedly this country's source of long-term advantage. Unless U.S. companies reduce their new product development and introduction cycles from 36–48 months to 12–18 months, Japanese manufacturers will easily out-innovate and out-perform them. Taking the initiative in innovation will require even faster cycle times.

Residential air conditioners illustrate the Japanese ability to introduce more technological innovation in smaller increments—and how in just a few years these improvements add up to remarkably superior products. The Japanese introduce innovations in air conditioners four times faster than their American competitors; in technological sophistication the Japanese products are 7 to 10 years ahead of U.S. products.

Look at the changes in Mitsubishi Electric's three-horsepower heat pump between 1975 and 1985. From 1975 to 1979, the company did nothing to the product except change the sheet metal work, partly to improve efficiency but mostly to reduce materials cost. In 1979, the technological sophistication of the product was roughly equal to that of the U.S. competition. From this point on, the Japanese first established, and then widened the lead.

In 1980, Mitsubishi introduced its first major improvement: a new product that used integrated circuits to control the air-conditioning cycle. One year later, the company replaced the integrated circuits with microprocessors and added two important innovations to increase consumer demand. The first was "quick connect" freon lines. On the old product (and on the U.S. product), freon lines were made from copper tubing and cut to length, bent, soldered together, purged, and filled with freon—an operation requiring great skill to produce a reliable air conditioner. The Japanese substituted quick-connect freon lines—precharged hoses that simply clicked together. The second innovation was simplified wiring. On the old product (and still today on the U.S. product) the unit had six color-coded wires to connect. The advent of microprocessors made possible a two-wire connection with neutral polarity.

These two changes did not improve the energy-efficiency ratio of the product; nor were they intended to. Rather, the point was to fabricate a unit that would be simpler to install and more reliable, thereby broadening distribution and increasing demand. Because of these innovations, white-goods outlets could sell the new product, and local contractors could easily install it.

In 1982, Mitsubishi introduced a new version of the air conditioner featuring technological advances related to performance. A high-efficiency rotary compressor replaced the outdated reciprocating compressor. The condensing unit had louvered fins and inner fin tubes for better heat transfer. Because the balance of the system changed, all the electronics had to change. As a result, the energy-efficiency ratio improved markedly.

In 1983, Mitsubishi added sensors to the unit and more computing power, expanding the electronic control of the cycle and again improving the energy-efficiency ratio.

In 1984, Mitsubishi came out with another version of the product, this time with an inverter that made possible an even higher energy-efficiency ratio. The inverter, which requires additional electronics for the unit, allows unparalleled control over the speed of the electric motor, dramatically boosting the appliance's efficiency.

Using time-based innovation, Mitsubishi transformed its air conditioner. The changes came incrementally and steadily. Overall they gave Mitsubishi—and other Japanese companies on the same track—the position of technological leadership in the global residential air-conditioning industry.

In 1985, a U.S. air-conditioner manufacturer was just debating whether to use integrated circuits in its residential heat pump. In view of its four-to-five-year product development cycle, it could not have introduced the innovation until 1989 or 1990—putting the American company 10 years behind the Japanese. Faced with this situation, the U.S. air conditioner company followed the example of many U.S. manufacturers that have lost the lead in technology and innovation; it decided to source its air conditioners and components from its Japanese competition.

Consider the remarkable example of Atlas Door, a 10-year-old U.S. company. It has grown at an average annual rate of 15 percent in an industry with an overall annual growth rate of less than 5 percent. In recent years, its pretax earnings were 20 percent of sales, about five times the industry average. Atlas is debt free. In its 10th year the company achieved the number one competitive position in its industry.

The company's product: industrial doors. It is a product with almost infinite variety, involving limitless choices of width and height and material. Because of the importance of variety, inventory is almost useless in meeting customer orders; most doors can be manufactured only after the order has been placed.

Historically, the industry had needed almost four months to respond to an order for a door that was out of stock or customized. Atlas's strategic advantage was time; it could respond in weeks to any order. It had structured its order-entry, engineering, manufacturing, and logistics systems to move information and products quickly and reliably.

First, Atlas built just-in-time factories. These are fairly simple in concept. They require extra tooling and machinery to reduce changeover times and a fabrication process organized by product and scheduled to start and complete all of the parts at the same time. But even the performance of the factory—critical to the company's overall responsiveness—still only accounted for 2½ weeks of the completed product delivery cycle.

Second, Atlas compressed time at the front end of the system, where the order first entered and was processed. Traditionally, when customers, distributors, or salespeople called a door manufacturer with a request for price and delivery, they would have to wait more than one week for a response. If the desired door was not in stock, not in the schedule, or not engineered, the supplier's organization would waste even more time, pushing the search for an answer around the system.

Recognizing the opportunity to cut deeply into the time expenditure in this part of the system, Atlas first streamlined, then automated its entire order-entry, engineering, pricing, and scheduling processes. Today Atlas can price and schedule 95 percent of its incoming orders while the callers are still on the telephone. It can quickly engineer new special orders because it has preserved on computer the design and production data of all previous special orders—which drastically reduces the amount of reengineering necessary.

Third, Atlas tightly controlled logistics so that it always shipped only fully complete orders to construction sites. Orders require many components. Gathering all of them at the factory and making sure that they are with the correct order can be a time-consuming task. It is even more time-consuming, however, to get the correct parts to the job site *after* they have missed the initial shipment. Atlas developed a system to track the parts in production and the purchased parts for each order, ensuring arrival of all necessary parts at the shipping dock in time—a just-in-time logistics operation.

When Atlas started operations, distributors were uninterested in its product. The established distributors already carried the door line of a larger competitor; they saw no reason to switch suppliers except, perhaps, for a major price concession. But as a start-up, Atlas was too small to compete on price alone. Instead, it positioned itself as the door supplier of last resort, the company people came to if the established supplier could not deliver or missed a key date.

Of course, with industry lead times of almost four months, some calls inevitably came to Atlas. And when it did get a call, Atlas commanded a higher price because of its faster delivery. Atlas not only got a higher price but its time-based processes also yielded lower costs: it thus enjoyed the best of both worlds.

In 10 short years, the company replaced the leading door suppliers in 80 percent of the distributors in the country. With its strategic advantage the company could be selective, becoming the house supplier for only the strongest distributors.

In the wake of this indirect attack, the established competitors have not responded effectively. The conventional view is that Atlas is a "garage shop operator" that cannot sustain its growth: competitors expect the company's performance to degrade to the industry average as it grows larger. But this response—or nonresponse—only reflects a fundamental lack of understanding of time as the source of competitive advantage. The extra delay in responding only adds to the insurmountable lead the indirect time-based attack has created. While the traditional companies track costs and size, the new competitor derives advantage from time, staying on the cutting edge, leaving its rivals behind.

Source: Reprinted by permission of *Harvard Business Review*, "Time—The Next Source of Competitive Advantage," by George Stalk, Jr., July–August 1988. Copyright 1988 by the President and Fellows of Harvard College.

Summary

Productivity is one measure of an organization's effective use of its resources. Productivity comparisons help managers judge an organization's ability to compete and/or serve its customers. Organizations or countries that have relatively high productivity have a competitive edge.

Companies compete in a number of ways, including price, quality, special features or services, and time. Some companies suffer from poor competitiveness, often due to strategic weaknesses.

Strategies are the basic approaches used by an organization to achieve its goals. Strategies provide focus for planning and decision making. Organizations typically have overall strategies that pertain to the entire organization, and strategies for each of the functional areas. Functional strategies are narrower in scope and should be linked to the overall strategy. Many organizations have adopted either time-based strategies or quality-based strategies as they seek to become more competitive and/or they attempt to become more productive and better serve their customers.

Key Terms

competitiveness, 43
distinctive competencies, 49
environmental scanning, 50
mission, 45
mission statement, 45
operations strategy, 47
order qualifiers, 48
order winners, 48
productivity, 38
quality-based strategy, 52
strategy, 45
tactics, 45
time-based strategy, 52

Solved Problems

Problem 1

A company that processes fruits and vegetables is able to produce 400 cases of canned peaches in one-half hour with four workers. What is labor productivity?

Solution

$$\text{Labor productivity} = \frac{\text{Quantity produced}}{\text{Labor hours}} = \frac{400 \text{ cases}}{4 \text{ workers} \times 1/2 \text{ hour/worker}}$$

$$= 200 \text{ cases per hour per worker}$$

Problem 2

A wrapping paper company produced 2,000 rolls of paper one day. Labor cost was $160, material cost was $50, and overhead was $320. Determine the multifactor productivity.

Solution

$$\text{Multifactor productivity} = \frac{\text{Quantity produced @ standard price}}{\text{Labor cost} + \text{Material cost} + \text{Overhead}}$$

$$= \frac{2{,}000 \text{ rolls}}{\$160 + \$50 + \$320}$$

$$= 3.77 \text{ rolls per dollar}$$

A variation of the multifactor productivity calculation is to incorporate the standard price in the numerator by multiplying the units by the standard price.

Discussion and Review Questions

1. What is productivity and why is it important? Who is primarily responsible for productivity in an organization?
2. List some factors that can affect productivity and some ways that productivity can be improved.
3. It has been said that a typical Japanese automobile manufacturer produces more cars with fewer workers than its U.S. counterpart. What are some possible explanations for this, assuming that U.S. workers are as hardworking as Japanese workers?
4. From time to time, various groups clamor for import restrictions or tariffs on foreign-produced goods, particularly automobiles. How might these be helpful? Harmful?
5. List the key ways that organizations compete.
6. What are the key reasons for the poor competitiveness of some companies?
7. The United States has the highest agricultural productivity of any country in the world. What are the primary reasons for this high productivity?
8. Most experts seem to agree that workers are not primarily responsible for poor productivity. Who is?
9. What are distinctive competencies and how do they relate to strategy formulation?
10. Contrast the terms *strategies* and *tactics.*
11. Contrast *organization strategy* and *operations strategy.*
12. Explain the term *time-based strategies* and give three examples.

Memo Writing Exercises

1. Last year, your company initiated a major effort to improve productivity throughout the company. During the first few months, little improvement was made. Recently, however, major gains, exceeding 20 percent, have been made in both the packaging and equipment repair departments. You would like to commend them for their achievements, and announce to the other departments that gains are possible. You are also aware that some production employees believe that significant gains in productivity means the loss of some jobs because the company will be able to achieve the same output with fewer employees. Another local company laid off 5 percent of its workforce shortly after announcing productivity gains. Nonetheless, you believe that by achieving even modest productivity gains, your company will become more competitive and therefore be able to capture a larger share of the market. You believe that increased demand will more than offset productivity gains, and may even require hiring more employees. Naturally, you cannot guarantee this, although you are optimistic about the chances that this will occur.

   Write a one-page memo to employees that covers these points.
2. Assume the role of vice president of manufacturing of Eastern Products, Philadelphia. You are thinking about the forthcoming annual retreat of top-level managers of manufacturing, engineering, marketing, and product design to discuss strategic planning. Competitive pressures from several domestic and foreign companies give this year's meeting special significance, and you want to arrange for a truly productive session.

   In the past, some meetings haven't been as successful as you would have liked. You learned from experience that holding the meeting at corporate headquarters didn't work because the managers were too accessible to their subordinates. It's much better if the group can meet away from home territory, allowing them the opportunity to concentrate and develop camaraderie. You also found that traditional vacation spots are not conducive to achieving the best results. You want your administrative assistant to investigate agencies that specialize in business retreats.

   Twelve to 14 people will attend the conference. Each will require a separate hotel room. Exercise facilities should be on the premises. Audiovisual support will be needed for video presentations, and overhead projectors and flip charts will also be needed. Light lunches should be arranged. You need to get dietary preferences from those attending.

The location for the annual retreat should not require excessive travel costs or travel time. Arrangements must be completed within two weeks.

Write a one-page memo outlining these requirements to your assistant, Armin Joseph. It is his task to investigate possible retreat locations and make all the arrangements.

Problems

1. Suppose that a company produced 300 standard bookcases last week using eight workers and it produced 240 standard bookcases the week before using six workers. In which period was productivity higher? Explain.

Last week: 37.5 bookcases/worker.
Week before: 40 bookcases/worker.

2. The manager of a crew that installs carpeting has tracked the crew's output over the past several weeks, obtaining these figures:

Week	Crew Size	Yards Installed
1	4	960
2	3	702
3	4	968
4	2	500
5	3	696
6	2	500

Compute the labor productivity for each of the weeks. On the basis of your calculations, what can you conclude about crew size and productivity?

Week	Crew Size	Labor Productivity
1	4	240 yd.
2	3	234
3	4	242
4	2	250
5	3	232
6	2	250

Possibly even-sized crews are better than odd sizes, and a crew of 2 works best.

3. Compute the multifactor productivity measure for each of the weeks shown. What do the productivity figures suggest? Assume 40-hour weeks and an hourly wage of $12. Overhead is 1.5 times weekly labor cost. Material cost is $6 per pound. Standard price is $140 per unit.

Week	Output (units)	Workers	Material (lbs)
1	300	6	45
2	338	7	46
3	322	7	46
4	354	8	48
5	222	5	40
6	265	6	42
7	310	7	46

Week	Productivity
1	5.62
2	5.45
3	5.20
4	5.01
5	4.98
6	4.98
7	5.00

Multifactor productivity dropped steadily from an early high of 5.62, and then leveled off at about 5.00, starting at week 4.

4. A company that makes shopping carts for supermarkets and other stores recently purchased some new equipment that reduces the labor content of the jobs needed to produce the shopping carts. Prior to buying the new equipment, the company used five workers, who produced an average of 80 carts per hour. Labor cost was $10 per hour and machine cost was $40 per hour. With the new equipment, it was possible to transfer one of the workers to another department, and equipment cost increased by $10 per hour while output increased by four carts per hour.
   a. Compute labor productivity under each system. Use carts per worker per hour as the measure of labor productivity.
   b. Compute the multifactor productivity under each system. Use carts per dollar cost (labor plus equipment) as the measure.
   c. Comment on the changes in productivity according to the two measures, and on which one you believe is the more pertinent for this situation.

a. Before: 16 carts per worker per hour.
After: 21 carts per worker per hour.
b. Before: 89 carts/$.
After: 93 carts/$.
c. Labor productivity increased by 31%.
Multifactor productivity increased by only 4.5%.

READING
PRODUCTIVITY GAINS AT WHIRLPOOL

www.whirlpool.com

Workers and management at Whirlpool Appliance's Benton Harbor plant in Michigan have set an example of how to achieve productivity gains, which has benefited not only the company and its stockholders, but Whirlpool customers, and the workers themselves.

Things weren't always rosy at the plant. Productivity and quality weren't good. Neither were labor-management relations. Workers hid defective parts so management wouldn't find them, and when machines broke down, workers would simply sit down until sooner or later someone came to fix it. All that changed in the late 1980s. Faced with the possibility that the plant would be shut down, management and labor worked together to find a way to keep the plant open. The way was to increase productivity—producing more without using more resources. Interestingly, the improvement in productivity didn't come by spending money on fancy machines. Rather, it was accomplished by placing more emphasis on quality. That was a shift from the old way, which emphasized volume, often at the expense of quality. To motivate workers, the company agreed to gain sharing, a plan that rewarded workers by increasing their pay for productivity increases.

The company overhauled the manufacturing process, and taught its workers how to improve quality. As quality improved, productivity went up because more of the output was good, and costs went down because of fewer defective parts that had to be scrapped or reworked. Costs of inventory also decreased, because fewer spare parts were needed to replace defective output, both at the factory and for warranty repairs. And workers have been able to see the connection between their efforts to improve quality and productivity, and their pay.

Not only was Whirlpool able to use the productivity gains to increase workers' pay, it was also able to hold the lid on price increases and to funnel some of the savings into research, which added to cost saving and quality improvement.

Questions:

1. What were the two key things that Whirlpool management did to achieve productivity gains?
2. Who has benefited from the productivity gains?
3. How are productivity and quality related?
4. How can a company afford to pay its workers for productivity gains?

Source: Based on "A Whirlpool Factory Raises Productivity—And Pay of Workers," by Rick Wartzman, from *The Wall Street Journal*, 1992.

CASE
AN AMERICAN TRAGEDY: HOW A GOOD COMPANY DIED

Zachary Schiller

The Rust Belt is back. So say bullish observers as U.S. exports surge, long-moribund industries glow with newfound profits, and unemployment dips to lows not seen in a decade. But in the smokestack citadels, there's disquiet. Too many machine-tool and auto parts factories are silent; too many U.S. industries still can't hold their own.

What went wrong since the heyday of the 1960s? That's the issue Max Holland, a contributing editor of *The Nation*, takes up in his nutsy-boltsy but fascinating study, *When the Machine Stopped*.*

The focus of the story is Burgmaster Corp., a Los Angeles-area machine-tool maker founded in 1944 by Czechoslovakian immigrant Fred Burg. Holland's father worked there for 29 years, and the author interviewed 22 former employees. His shop-floor view of this small company is a refreshing change from academic treatises on why America can't compete.

The discussions of spindles and numerical control can be tough going. But Holland compensates by conveying the excitement and innovation of the company's early days and the disgust and cynicism accompanying its decline. Moreover, the fate of Burgmaster and its brethren is crucial to the U.S. industrial economy: Any manufactured item is either made by a machine tool or by a machine made by a machine tool.

*Max Holland, *When the Machine Stopped: A Contemporary Tale from Industrial America* (Boston, Mass.: Harvard Business School Press, 1988).

Producing innovative turret drills used in a wide variety of metalworking tasks, Burgmaster was a thriving enterprise by 1965, when annual sales amounted to about $8 million. The company needed backing to expand, however, so it sold out to Buffalo-based conglomerate Houdaille Industries Inc. Houdaille was in turn purchased in a 1979 leveraged buyout led by Kohlberg Kravis Roberts & Co. By 1982, when debt, competition, and a sickly machine-tool market had battered Burgmaster badly, Houdaille went to Washington with a petition to withhold the investment tax credit for certain Japanese-made machine tools.

Thanks to deft lobbying, the Senate passed a resolution supporting Houdaille's position, but President Reagan refused to go along. Houdaille's subsequent attempt to link Burgmaster up with a Japanese rival also failed, and Burgmaster was closed.

Holland uses Burgmaster's demise to explore some key issues of economic and trade policy. Houdaille's charge that a cartel led by the Japanese government had injured U.S. toolmakers, for example, became a rallying point for those who would blame a fearsome Japan Inc. for the problems of U.S. industry.

Holland describes the Washington wrangling over Houdaille in painful detail. But he does show that such government decisions are often made without much knowledge of what's going on in industry. He shows, too, that Japanese producers succeeded less because of government help than because they made better, cheaper machines.

For those who see LBOs as a symptom of what ails the U.S. economy, Holland offers plenty of ammunition. He argues persuasively that the LBO crippled Burgmaster by creating enormous pressure to generate cash. As Burgmaster pushed its products out as fast as possible, he writes, it routinely shipped defective machines. It promised customers features that engineers hadn't yet designed. And although KKR disputes the claim, Holland concludes that the LBO choked off Burgmaster's investment funds just when foreign competition made them most necessary. As for Houdaille, it was recapitalized and sold to Britain's Tube Investments Group.

But Burgmaster's problems had started even before the LBO. Holland's history of the company under Houdaille is a veritable catalog of modern management techniques that flopped. One of the most disastrous was a system for computerizing production scheduling that was too crude for complex machine-tool manufacturing. Holland gives a dramatic depiction of supply snafus that resulted in delays and cost increases.

As an independent company, "Burgmaster thrived because the Burgs knew their business," Holland writes. Their departure under Houdaille was followed by an "endless and ultimately futile search for a better formula." But, he concludes: "No formula was a substitute for management involvement on the shop floor."

In the end, however, Holland puts most of the blame for the industry's decline on government policy. He targets tax laws and macroeconomic policies that encourage LBOs and speculation instead of productive investment. He also criticizes Pentagon procurement policies for favoring exotic, custom machines over standard, low-cost models. This adds up to an industrial policy, Holland writes—a bad one.

The point is well taken, but Holland gives it excessive weight. Like their brethren in Detroit and Pittsburgh, domestic tool-makers in the 1970s were too complacent when imports seized the lower end of the product line. The conservatism that had for years served them in their cyclical industry left them ill-prepared for change. Even now some of the largest U.S. toolmakers are struggling to restructure. Blame the government, yes. But blame the industry, too.

Questions

1. Write a brief report that outlines the reasons (both internal and external) for Burgmaster's demise, and whether operations management played a significant role in the demise.

Source: Reprinted from April 17, 1989 issue of *Business Week* by special permission, copyright © 1989 by The McGraw-Hill companies.

CASE

HOME-STYLE COOKIES

The Company

The Lew-Mark Baking Company is located in a small town in western New York State. The bakery is run by two brothers, Lew and Mark, who formed the company after they purchased an Archway Cookie franchise. With exclusive rights in New York and New Jersey, it is the largest Archway franchise. The company employs fewer than 200 people, mainly blue-collar workers, and the atmosphere is informal.

The Product

The company's only product is soft cookies, of which it makes over 50 varieties. Larger companies, such as Nabisco, Sunshine, and Keebler, have traditionally produced biscuit cookies, in which most of the water has been baked out, resulting in crisp cookies. Archway cookies have no additives or preservatives. The high quality of the cookies has enabled the company to develop a strong market niche for its product.

The Customers

The cookies are sold in convenience stores and supermarkets throughout New York and New Jersey. Archway markets its cookies as "good food"—no additives or preservatives—and this appeals to a health-conscious segment of the market. Many customers are over 45 years of age, and prefer a cookie that is soft and not too sweet. Parents with young children also buy the cookies.

The Production Process

The company has two continuous band ovens that it uses to bake the cookies. The production process is called a batch processing system. It begins as soon as management gets orders from distributors. These orders are used to schedule production. At the start of each shift, a list of the cookies to be made that day is delivered to the person in charge of mixing. That person checks a master list, which indicates the ingredients needed for each type of cookie, and enters that information into the computer. The computer then determines the amount of each ingredient needed, according to the quantity of cookies ordered, and relays that information to storage silos located outside the plant where the main ingredients (flour, sugar, and cake flour) are stored. The ingredients are automatically sent to giant mixing machines where the ingredients are combined with proper amounts of eggs, water, and flavorings. After the ingredients have been mixed, the batter is poured into a cutting machine where it is cut into individual cookies. The cookies are then dropped onto a conveyor belt and transported through one of two ovens. Filled cookies, such as apple, date, and raspberry, require an additional step for filling and folding.

The nonfilled cookies are cut on a diagonal rather than round. The diagonal-cut cookies require less space than straight-cut cookies, and the result is a higher level of productivity. In addition, the company recently increased the length of each oven by 25 feet, which also increased the rate of production.

As the cookies emerge from the ovens, they are fed onto spiral cooling racks 20 feet high and 3 feet wide. As the cookies come off the cooling racks, workers place the cookies into boxes manually, removing any broken or deformed cookies in the process. The boxes are then wrapped, sealed, and labeled automatically.

Inventory

Most cookies are loaded immediately onto trucks and shipped to distributors. A small percentage are stored temporarily in the company's warehouse, but they must be shipped shortly because of their limited shelf life. Other inventory includes individual cookie boxes, shipping boxes, labels, and cellophane for wrapping. Labels are reordered frequently, in small batches, because FDA label requirements are subject to change, and the company does not want to get stuck with labels it can't use. The bulk silos are refilled two or three times a week, depending on how quickly supplies are used.

Cookies are baked in a sequence that minimizes downtime for cleaning. For instance, light-colored cookies (e.g., chocolate chip) are baked before dark-colored cookies (e.g., fudge), and oatmeal cookies are baked before oatmeal raisin cookies. This permits the company to avoid having to clean the processing equipment every time a different type of cookie is produced.

Quality

The bakery prides itself on the quality of its cookies. Cookies are sampled randomly by a quality control inspector as they come off the line to assure that their taste and consistency are satisfactory, and that they have been baked to the proper degree. Also, workers on the line are responsible for removing defective cookies when they spot them. The company has also installed an X-ray machine on the line that can detect small bits of metal filings that may have gotten into cookies during the production process. The use of automatic equipment for transporting raw materials and mixing batter has made it easier to maintain a sterile process.

Scrap

The bakery is run very efficiently and has minimal amounts of scrap. For example, if a batch is mixed improperly, it is sold for dog food. Broken cookies are used in the oatmeal cookies. These practices reduce the cost of ingredients and save on waste disposal costs. The company also uses heat reclamation: The heat that escapes from the two ovens is captured and used to boil the water that supplies the heat to the building. Also, the use of automation in the mixing process has resulted in a reduction in waste compared with the manual methods used previously.

New Products

Ideas for new products come from customers, employees, and observations of competitors' products. New ideas are first examined to determine whether the cookies can be made with existing equipment. If so, a sample run is made to determine the cost and time requirements. If the results are satisfactory, marketing tests are conducted to see if there is a demand for the product.

Potential Improvements

There are a number of areas of potential improvement at the bakery. One possibility would be to automate packing the cookies into boxes. Although labor costs are not high, automating the process might save some money and increase efficiency. So far, the owners have resisted making this change because they feel an obligation to the community to employ the 30 women who now do the boxing manually. Another possible improvement would be to use suppliers who are located closer to the plant. That would reduce delivery lead times and transportation costs, but the owners are not convinced that local suppliers could provide the same good quality. Other opportunities have been proposed in recent years, but the owners rejected them because they feared that the quality of the product might suffer.

Questions

1. Briefly describe the cookie production process.
2. What are two ways that the company has increased productivity? Why did increasing the length of the ovens result in a faster output rate?
3. Do you think that the company is making the right decision by not automating the packing of cookies? Explain your reasoning. What obligation does a company have to its employees in a situation such as this? What obligation does it have to the community? Is the size of the town a factor? Would it make a difference if the company was located in a large city? Is the size of the company a factor? What if it was a much larger company?
4. What factors cause Lew-Mark to carry minimal amounts of certain inventories? What benefits result from this policy?
5. As a consumer, what things do you consider in judging the quality of cookies you buy in a supermarket?
6. What advantages and what limitations stem from Lew-Mark's not using preservatives in cookies?
7. Briefly describe the company's strategy.

CASE

HAZEL REVISITED

(Refer to p. 31 for the Hazel Case.)

1. What competitive advantage does Hazel have over a professional lawn care service?
2. Hazel would like to increase her profits, but she doesn't believe that it would be wise to raise her prices considering the current state of the local economy. Instead, she has given some thought to increasing productivity.
   a. Explain how increased productivity could be an alternative to increased prices.
   b. What are some ways that Hazel could increase productivity?
3. Hazel is thinking about the purchase of new equipment. One would be power sidewalk edgers. She believes edgers will lead to an increase in productivity. Another would be a chain saw, which would be used for tree pruning. What trade-offs should she consider in her analysis?
4. Hazel has been fairly successful in her neighborhood, and now wants to expand to other neighborhoods, including some that are five miles away. What would be the advantages and disadvantages of doing this?
5. Hazel does not have a mission statement or a set of objectives. Take one of the following positions and defend it:
   a. Hazel doesn't need a formal mission statement and objectives. Many small businesses don't have them.
   b. She definitely needs a mission statement and a set of objectives. They would be extremely beneficial.
   c. There may be some benefit to Hazel's business, and she should consider developing one.

Selected Bibliography

Blackburn, Joseph D., ed. *Time-Based Competition.* Burr Ridge, IL: Business One Irwin, 1991.

Buffa, E. S. *Meeting the Competitive Challenge: Manufacturing Strategies for U.S. Companies.* Burr Ridge, IL: Richard D. Irwin, 1984.

Hall, Robert. *Attaining Manufacturing Excellence.* Burr Ridge, IL: Dow-Jones Irwin, 1987.

Hayes, Robert H., and Gary P. Pisano. "Beyond World Class: The New Manufacturing Strategy." *Harvard Business Review* 70, no. 1 (Jan.–Feb. 1994), pp. 77–86.

Hayes, Robert H., and Steve C. Wheelwright. *Restoring Our Competitive Edge: Competing through Manufacturing.* New York: John Wiley & Sons, 1984.

Hill, Terry. *Manufacturing Strategy.* 2nd ed. Burr Ridge, IL: Richard D. Irwin, 1994.

Peterson, Ronald S. "The Critical Issues for Manufacturing Management." *Operations Management Review* 2, no. 4 (1984), pp. 15–20.

Porter, Michael E. *Competitive Advantage: Creating and Sustaining Superior Performance.* New York: The Free Press, 1984.

Stalk, George, Jr. "Time—The Next Source of Competitive Advantage." *Harvard Business Review,* July–August 1988, pp. 41–51.

Wheelwright, Steven C. "Japan—Where Operations Really Are Strategic." *Harvard Business Review,* July–August 1981, pp. 67–74.

SUPPLEMENT TO CHAPTER TWO
DECISION MAKING

LEARNING OBJECTIVES

After completing this supplement, you should be able to:

1. State what the chief role of an operations manager is.
2. Outline the steps in the decision-making process.
3. List some causes of poor decisions.
4. Describe the different environments under which operations decisions are made.
5. Describe and use techniques that apply to decision making under uncertainty.
6. Describe and use the expected-value approach.
7. Construct a decision tree and use it to analyze a problem.
8. Compute the expected value of perfect information.
9. Conduct sensitivity analysis on a simple decision problem.

SUPPLEMENT OUTLINE

The chief role of the operations manager is that of decision maker. In this capacity, the operations manager has considerable influence over the degree to which the goals and objectives of the organization are achieved.

Throughout this book you will be exposed to a broad range of decisions that are encountered in operations management and to the tools necessary to handle those decisions. Decision theory is a general approach to decision making that has applications in many areas of operations management. This supplement examines the process of decision making, typical causes of poor decisions, and the classification of decision problems according to degree of uncertainty.

THE DECISION PROCESS

Decision making is a fundamental process of management. Unfortunately, decisions do not always turn out as planned. Most successful decision making follows a process that consists of these steps.

1. Specify objectives and the criteria for making the decision.
2. Develop alternatives.
3. Analyze and compare alternatives.
4. Select the best alternative.
5. Implement the chosen alternative.
6. Monitor the results to ensure that desired results are achieved.

Success or failure in decision making often depends on how well each of these steps is handled.

The decision maker must *identify the criteria* by which proposed solutions will be judged. Common criteria often relate to costs, profits, return on investment, increased productivity, risk, company image, impact on demand, or similar variables.

The ability to satisfactorily make decisions often depends on the degree of success one has in *developing suitable alternatives.* In the search for alternatives, there is always the danger that one or more potentially superior alternatives will be overlooked. Consequently, the optimal alternative may turn out to be less than optimum. Obviously, there is a limit to the number of possible alternatives that can be identified. Much depends on the experience and creativity of the decision maker as well as on the nature of the situation. However, as a general rule, efforts expended in carefully identifying alternatives can yield substantial dividends in terms of the overall decision. One alternative that is frequently overlooked, possibly because it seems too simple, is to do nothing. The beauty of doing nothing is that no time or effort is needed, no costs are incurred, and no implementations required.

Analyzing and comparing alternatives is often enhanced by the use of mathematical or statistical techniques. Many of these are described and demonstrated throughout this book.

Selection of the best alternative will depend on the objectives of the decision maker and the criteria that are being used to evaluate alternatives.

Implementing a solution simply means carrying out the actions indicated by the chosen alternative. Examples include buying the machine, refusing the loan application, beginning development of a new product, and authorizing the use of overtime. Of course, if the alternative selected is to do nothing, no action will be required to implement it. But some decision makers use this approach by default: By the time they get around to making a decision, it's too late!

Effective decision making requires that the results of the decision be *monitored* to make sure that the decision has achieved the desired consequences. If it has not, the decision maker may have to repeat the entire process. Or perhaps a review of the situation may reveal an error in implementation, an error in calculations, or a wrong assumption that will allow the situation to be remedied quickly.

The decision process is not always completed in a sequential manner. Instead, there is usually a certain amount of backtracking and feedback, especially in terms of developing and analyzing alternatives. If none of the alternatives can achieve the desired results, additional alternatives must be developed.

Causes of Poor Decisions

Despite the best efforts of a manager, a decision occasionally turns out poorly due to unforeseeable circumstances. Luckily, such occurrences are not common. Often, failures can be traced to some combination of mistakes in the decision process, bounded rationality, or suboptimization.

In many cases, managers fail to appreciate the importance of each step in the decision-making process. They may skip a step or not devote enough effort to completing it before jumping to the next step. Sometimes this happens owing to a manager's style of making quick decisions or a failure to recognize the consequences of a poor decision. The manager's ego can be a factor. This sometimes happens when the manager has experienced a series of successes—important decisions that turned out right. Some managers then get the impression that they can do no wrong. But they soon run into trouble, which is usually enough to bring them back down to earth. Other managers seem oblivious to negative results and continue the process they associate with their previous successes, not recognizing that some of that success may have been due more to luck than to any special abilities of their own. A part of the problem may be the manager's unwillingness to admit a mistake. Yet other managers demonstrate an inability to make a decision; they stall long past the time when the decision should have been rendered.

Of course, not all managers fall into these traps—it seems safe to say that the majority do not. Even so, this does not necessarily mean that every decision works out as expected. Another factor with which managers must contend is **bounded rationality,** or the limits imposed on decision making by costs, human abilities, time, technology, and the availability of information. Because of these limitations, managers cannot always expect to reach decisions that are optimal in the sense of providing the best possible outcome (e.g., highest profit, least cost). Instead, they must often resort to achieving a *satisfactory* solution.

bounded rationality The limitations on decision making caused by costs, human abilities, time, technology, and availability of information.

Still another cause of poor decisions is that organizations typically departmentalize decisions. Naturally, there is a great deal of justification for the use of departments in terms of overcoming span-of-control problems and human limitations. However, **suboptimization** can occur. This is a result of different departments attempting to reach a solution that is optimum for each. Unfortunately, what is optimal for one department may not be optimal for the organization as a whole.

suboptimization The result of different departments each attempting to reach a solution that is optimum for that department.

Decision Environments

Operations management decision environments are classified according to the degree of certainty present. There are three basic categories: certainty, risk, and uncertainty.

certainty Environment in which relevant parameters have known values.

Certainty means that relevant parameters such as costs, capacity, and demand have known values.

Risk means that certain parameters have probabilistic outcomes.

Uncertainty means that it is impossible to assess the likelihood of various possible future events.

risk Environment in which certain future events have probable outcomes.

uncertainty Environment in which it is impossible to assess the likelihood of various future events.

Consider these situations:

1. Profit per unit is $5. You have an order for 200 units. How much profit will you make? (This is an example of *certainty* since unit profits and total demand are known.)

2. Profit is $5 per unit. Based on previous experience, there is a 50 percent chance of an order for 100 units and a 50 percent chance of an order for 200 units. What is expected profit? (This is an example of *risk* since demand outcomes are probabilistic.)
3. Profit is $5 per unit. The probabilities of potential demands are unknown. (This is an example of *uncertainty*.)

The importance of these different decision environments is that they require different analysis techniques. Some techniques are better suited for one category than for others. You should make note of the environments for which each technique is appropriate.

DECISION THEORY

Decision theory represents a general approach to decision making. It is suitable for a wide range of operations management decisions. Among them are capacity planning, product and service design, equipment selection, and location planning. Decisions that lend themselves to a decision theory approach tend to be characterized by these elements:

1. A set of possible future conditions exists that will have a bearing on the results of the decision.
2. A list of alternatives for the manager to choose from.
3. A known payoff for each alternative under each possible future condition.

In order to use this approach, a decision maker would employ this process:

1. Identify the possible future conditions (e.g., demand will be low, medium, or high; the number of contracts awarded will be one, two, or three; the competitor will or will not introduce a new product). These are called *states of nature*.
2. Develop a list of possible *alternatives*, one of which may be to do nothing.
3. Determine or estimate the *payoff* associated with each alternative for every possible future condition.
4. If possible, estimate the *likelihood* of each possible future condition.
5. Evaluate alternatives according to some *decision criterion* (e.g., maximize expected profit), and select the best alternative.

payoff table Table showing the expected payoffs for each alternative in every possible state of nature.

The information for a decision is often summarized in a **payoff table,** which shows the expected payoffs for each alternative under the various possible states of nature. These tables are helpful in choosing among alternatives because they facilitate comparison of alternatives. Consider the following payoff table, which illustrates a capacity planning problem.

	POSSIBLE FUTURE DEMAND		
Alternatives	**Low**	**Moderate**	**High**
Small facility	$10*	$10	$10
Medium facility	7	12	12
Large facility	(4)	2	16

*Present value in $ millions.

The payoffs are shown in the body of the table. In this instance, the payoffs are in terms of present values, which represent equivalent current dollar values of expected future income less costs. This is a convenient measure because it places all alternatives on a comparable basis. If a small facility is built, the payoff will be the same for all three possible states of nature. For a medium facility, low demand will have a present value of $7 million, whereas both moderate and high demand will have present values of

$12 million. A large facility will have a loss of $4 million if demand is low, a present value of $2 million if demand is moderate, and a present value of $16 million if demand is high.

The problem for the decision maker is to select one of the alternatives, taking the present value into account.

Evaluation of the alternatives differs according to the degree of certainty associated with the possible future conditions. Again, there are three possibilities to consider: complete certainty, risk, and uncertainty.

Decision Making under Certainty

When it is known for certain which of the possible future conditions will actually happen, the decision is usually relatively straightforward: Simply choose the alternative that has the best payoff under that state of nature. Example 1 illustrates this.

Example 1

Determine the best alternative in the preceding payoff table for each of the cases: It is known with certainty that demand will be: (*a*) low, (*b*) moderate, (*c*) high.

Solution

Choose the alternative with the highest payoff. Thus, if we know demand will be low, we would elect to build the small facility and realize a payoff of $10 million. If we know demand will be moderate, a medium factory would yield the highest payoff ($12 million versus either $10 or $2 million). For high demand, a large facility would provide the highest payoff.

Although complete certainty is rare in such situations, this kind of exercise provides some perspective on the analysis. Moreover, in some instances, there may be an opportunity to consider allocation of funds to research efforts, which may reduce or remove some of the uncertainty surrounding the states of nature.

Decision Making under Uncertainty

At the opposite extreme is complete uncertainty: no information is available on how likely the various states of nature are. Under those conditions, four possible decision criteria are *maximin, maximax, Laplace,* and *minimax regret.* These approaches can be defined as follows:

maximin Choose the alternative with the best of the worst possible payoffs.

Maximin—Determine the worst possible *payoff* for each alternative, and choose the alternative that has the "best worst." The maximin approach is essentially a pessimistic one because it takes into account only the worst possible outcome for each alternative. The actual outcome may not be as bad as that, but this approach establishes a "guaranteed minimum."

maximax Choose the alternative with the best possible payoff.

Maximax—Determine the best possible payoff, and choose the alternative with that payoff. The maximax approach is an optimistic, "go for it" strategy; it does not take into account any payoff other than the best.

Laplace Choose the alternative with the best average payoff of any of the alternatives.

Laplace—Determine the average payoff for each alternative, and choose the alternative with the best average. The Laplace approach treats the states of nature as equally likely.

minimax regret Choose the alternative that has the least of the worst regrets.

Minimax regret—Determine the worst *regret* for each alternative, and choose the alternative with the "best worst." This approach seeks to minimize the difference between the payoff that is realized and the best payoff for each state of nature.

The next two examples illustrate these decision criteria.

Example 2

Referring to the preceding payoff table, determine which alternative would be chosen under each of these strategies:

a. Maximin

b. Maximax

c. Laplace

Solution

a. Using maximin, the worst payoffs for the alternatives are:

Small facility:	\$10 million
Medium facility:	7 million
Large facility:	−4 million

Hence, since \$10 million is the best, choose to build the small facility using the maximin strategy.

b. Using maximax, the best payoffs are:

Small facility:	\$10 million
Medium facility:	12 million
Large facility:	16 million

The best overall payoff is the \$16 million in the third row. Hence, the maximax criterion leads to building a large facility.

c. For the Laplace criterion, first find the row totals, and then divide each of those amounts by the number of states of nature (three in this case). Thus, we have:

	Row Total (in \$ millions)	Row Average (in \$ millions)
Small facility	\$30	\$10.00
Medium facility	31	10.33
Large facility	14	4.67

Because the medium facility has the highest average, it would be chosen under the Laplace criterion.

Example 3

Determine which alternative would be chosen using a minimax regret approach to the capacity planning program.

Solution

regret The difference between a given payoff and the best payoff for a state of nature.

The first step in this approach is to prepare a table of **opportunity losses,** or **regrets.** To do this, subtract every payoff *in each column* from the best payoff in that column. For instance, in the first column, the best payoff is 10, so each of the three numbers in that column must be subtracted from 10. Going down the column, the regrets will be $10 - 10 = 0$, $10 - 7 = 3$, and $10 - (-4) = 14$. In the second column, the best payoff is 12. Subtracting each payoff from 12 yields 2, 0, and 10. In the third column, 16 is the best payoff. The regrets are 6, 4, and 0. These results are summarized in a regret table:

	REGRETS (in \$ millions)			
Alternatives	**Low**	**Moderate**	**High**	**Worst**
Small facility	\$ 0	\$ 2	\$6	\$ 6
Medium facility	3	0	4	4
Large facility	14	10	0	14

The second step is to identify the worst regret for each alternative. For the first alternative, the worst is 6; for the second, the worst is 4; and for the third, the worst is 14.

The best of these worst regrets would be chosen using minimax regret. The lowest regret is 4, which is for a medium facility. Hence, that alternative would be chosen.

Solved Problem 6 at the end of the chapter illustrates decision making under uncertainty when the payoffs represent costs.

The main weakness of these approaches (except for Laplace) is that they do not take into account *all* of the payoffs. Instead, they focus on the worst or best, and so they lose some information. The weakness of Laplace is that it treats all states of nature as equally likely. Still, for a given set of circumstances, each has certain merits that can be helpful to a decision maker.

Decision Making under Risk

Between the two extremes of certainty and uncertainty lies the case of risk: The probability of occurrence for each state of nature is known. (Note that because the states are mutually exclusive and collectively exhaustive, these probabilities must add to 1.00.) A widely used approach under such circumstances is the *expected monetary value criterion*. The expected value is computed for each alternative, and the one with the highest expected value is selected. The expected value is the sum of the payoffs for an alternative where each payoff is *weighted* by the probability for the relevant state of nature. Thus, the approach is:

Expected monetary value criterion (EMV)—Determine the expected payoff of each alternative, and choose the alternative that has the best expected payoff.

expected monetary value (EMV) The best expected value among the alternatives.

Example 4

Using the expected monetary value criterion, identify the best alternative for the previous payoff table for these probabilities: low = .30, moderate = .50, and high = .20.

Solution

Find the expected value of each alternative by multiplying the probability of occurrence for each state of nature by the payoff for that state of nature and summing them:

$$EV_{small} = .30(\$10) + .50(\$10) + .20(\$10) = \$10$$

$$EV_{medium} = .30(\$7) + .50(\$12) + .20(\$12) = \$10.5$$

$$EV_{large} = .30(-4) + .50(\$2) + .20(\$16) = \$\ 3$$

Hence, choose the medium facility because it has the highest expected value.

The expected monetary value approach is most appropriate when a decision maker is neither risk averse nor risk seeking, but is risk neutral. Typically, well-established organizations with numerous decisions of this nature tend to use expected value because it provides an indication of the long-run, average payoff. That is, the expected-value amount (e.g., $10.5 million in the last example) is not an actual payoff but an expected or average amount that would be approximated if a large number of identical decisions were to be made. Hence, if a decision maker applies this criterion to a large number of similar decisions, the expected payoff for the total will approximate the sum of the individual expected payoffs.

Decision Trees

A **decision tree** is a schematic representation of the alternatives available to a decision maker and their possible consequences. The term gets its name from the treelike appearance of the diagram (see Figure 2S–1). Although tree diagrams can be used in place of a payoff table, they are particularly useful for analyzing situations that involve *sequential* decisions. For instance, a manager may initially decide to build a small facility only to discover that demand is much higher than anticipated. In this case, the manager may then be called upon to make a subsequent decision on whether to expand or build an additional facility.

decision tree A schematic representation of the available alternatives and their possible consequences.

A decision tree is composed of a number of *nodes* that have *branches* emanating from them (see Figure 2S–1). Square nodes denote decision points, and circular nodes denote

FIGURE 2S–1

Format of a decision tree

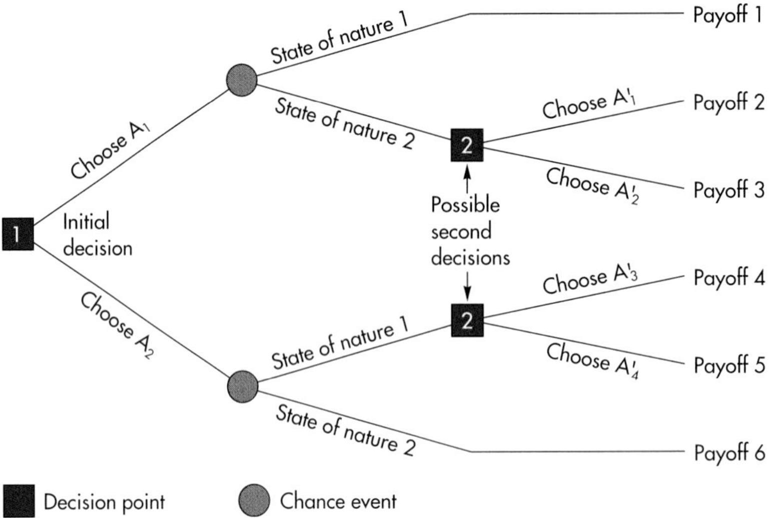

chance events. Read the tree from left to right. Branches leaving square nodes represent alternatives; branches leaving circular nodes represent chance events (i.e., the possible states of nature).

After the tree has been drawn, it is analyzed from *right to left*; that is, starting with the last decision that might be made. For each decision, choose the alternative that will yield the greatest return (or the lowest cost). If chance events follow a decision, choose the alternative that has the highest expected monetary value (or lowest expected cost).

Example 5

A manager must decide on the size of a video arcade to construct. The manager has narrowed the choices to two: large or small. Information has been collected on payoffs, and a decision tree has been constructed. Analyze the decision tree and determine which initial alternative (build small or build large) should be chosen in order to maximize expected monetary value.

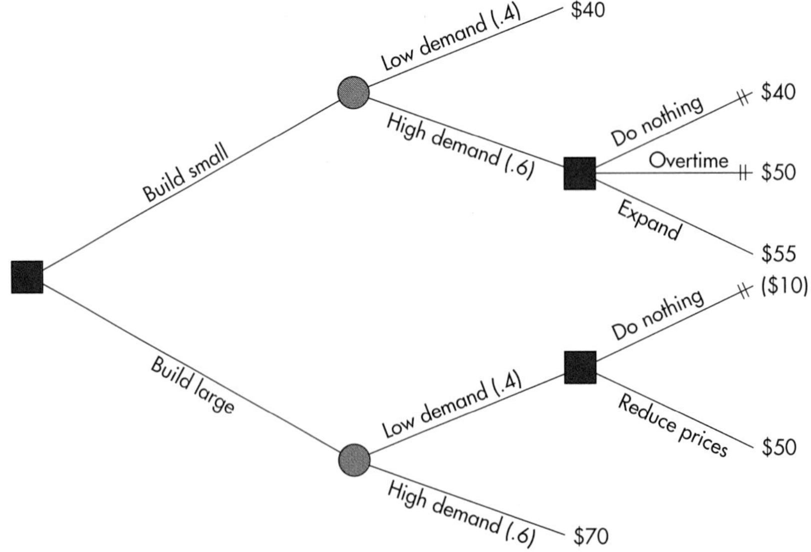

Solution

The dollar amounts at the branch ends indicate the estimated payoffs if the sequence of chance events and decisions that is traced back to the initial decision occurs. For example, if the initial decision is to build a small facility and it turns out that demand is low, the payoff will be $40 (thousand). Similarly, if a small facility is built, demand turns out high, and a later decision is made to expand, the payoff will be $55. The figures in parentheses on branches leaving the chance nodes indicate the probabilities of those states of nature. Hence, the probability of low demand is .4, and the probability of high demand is .6. Payoffs in parentheses indicate losses.

Analyze the decisions from right to left:

1. Determine which alternative would be selected for each possible second decision. For a small facility with high demand, there are three choices: *do nothing, work overtime,* and *expand.* Because *expand* has the highest payoff, you would choose it. Indicate this by placing a double slash through each of the other alternatives. Similarly, for a large facility with low demand, there are two choices: *do nothing* and *reduce prices.* You would choose *reduce prices* because it has the higher expected value, so a double slash is placed on the other branch.
2. Determine the product of the chance probabilities and their respective payoffs for the remaining branches:

Build small	
Low demand	.4($40) = $16
High demand	.6($55) = 33
Build large	
Low demand	.4($50) = 20
High demand	.6($70) = 42

3. Determine the expected value of each initial alternative:

Build small	$16 + $33 = $49
Build large	$20 + $42 = $62

Hence, the choice should be to build the large facility because it has a larger expected value than the small facility.

Expected Value of Perfect Information

In certain situations, it is possible to ascertain which state of nature will actually occur in the future. For instance, the choice of location for a restaurant may weigh heavily on whether a new highway will be constructed or whether a zoning permit will be issued. A decision maker may have probabilities for these states of nature; however, it may be possible to delay a decision until it is clear which state of nature will occur. This might involve taking an option to buy the land. If the state of nature is favorable, the option can be exercised; if it is unfavorable, the option can be allowed to expire. The question to consider is whether the cost of the option will be less than the expected gain due to delaying the decision (i.e., the expected payoff *above* the expected value). The expected gain is the *expected value of perfect information*, or EVPI.

Expected value of perfect information—the difference between the expected payoff with perfect information and the expected payoff under risk.

expected value of perfect information The difference between the expected payoff with perfect information and the expected payoff under risk.

Other possible ways of obtaining perfect information depend somewhat on the nature of the decision being made. Information about consumer preferences might come from market research, additional information about a product could come from product testing, or legal experts might be called on.

There are two ways to determine the EVPI. One is to compute the expected payoff under certainty and subtract the expected payoff under risk. That is,

$$\text{Expected value of perfect information} = \text{Expected payoff under certainty} - \text{Expected payoff under risk.} \qquad (2S\text{–}1)$$

Example 6

Using the information from Example 4, determine the expected value of perfect information using Formula 2S–1.

Solution

First, compute the expected payoff under certainty. To do this, identify the best payoff under each state of nature. Then combine these by weighting each payoff by the probability of that state of nature and adding the amounts. Thus, the best payoff under low demand is \$10, the best under moderate demand is \$12, and the best under high demand is \$16. The expected payoff under certainty is, then:

$$.30(\$10) + .50(\$12) + .20(\$16) = \$12.2$$

The expected payoff under risk, as computed in Example 4, is \$10.5. The EVPI is the difference between these:

$$\text{EVPI} = \$12.2 - \$10.5 = \$1.7$$

This figure indicates the upper limit on the amount the decision maker should be willing to spend to obtain perfect information in this case. Thus, if the cost equals or exceeds this amount, the decision maker would be better off not spending additional money and simply going with the alternative that has the highest expected payoff.

A second approach is to use the regret table to compute the EVPI. To do this, find the expected regret for each alternative. The minimum expected regret is equal to the EVPI.

Example 7

Determine the expected value of perfect information for the capacity-planning problem using the expected regret approach.

Solution

Using information from Examples 2, 3, and 4, we can compute the expected regret for each alternative. Thus:

Small facility	$.30(0) + .50(2) + .20(6) = 2.2$
Medium facility	$.30(3) + .50(0) + .20(4) = 1.7$ [minimum]
Large facility	$.30(14) + .50(10) + .20(0) = 9.2$

The lowest expected regret is 1.7, which is associated with the second alternative. Hence, the EVPI is \$1.7 million, which agrees with the previous example using the other approach.

Sensitivity Analysis

Generally speaking, both the payoffs and the probabilities in this kind of a decision problem are estimated values. Consequently, it can be useful for the decision maker to have some indication of how sensitive the choice of an alternative is to changes in one or more of these values. Unfortunately, it is impossible to consider all possible combinations of every variable in a typical problem. Nevertheless, there are certain things a decision maker can do to judge the sensitivity of probability estimates.

sensitivity analysis Determining the range of probability for which an alternative has the best expected payoff.

Sensitivity analysis provides a range of probability over which the choice of alternative would remain the same. The approach illustrated here is useful when there are two states of nature. It involves constructing a graph and then using algebra to determine a range of probabilities for which a given solution is best. In effect, the graph provides a visual indication of the range of probability over which the various alternatives are optimal, and the algebra provides exact values of the endpoints of the ranges. Example 8 illustrates the procedure.

Example 8

Given the following table, determine the range of probability for state of nature #2, that is, $P(2)$, for which each alternative is optimal under the expected-value approach.

		STATE OF NATURE	
		#1	#2
Alternative	A	4	12
	B	16	2
	C	12	8

Solution

First, plot each alternative relative to $P(2)$. To do this, plot the #1 value on the left side of the graph and the #2 value on the right side. For instance, for alternative A, plot 4 on the left side of the graph and 12 on the right side. Then connect these two points with a straight line. The three alternatives are plotted on the graph as shown below.

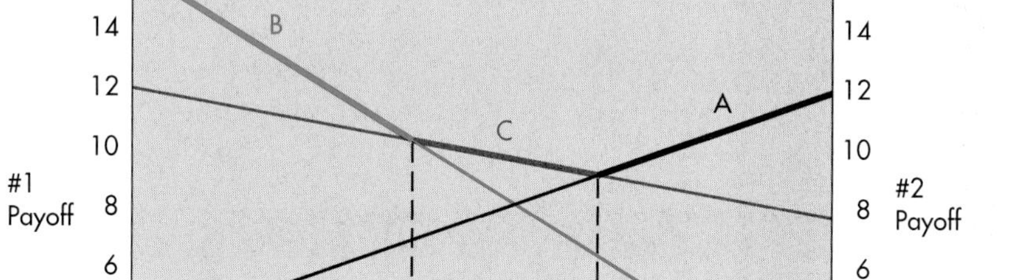
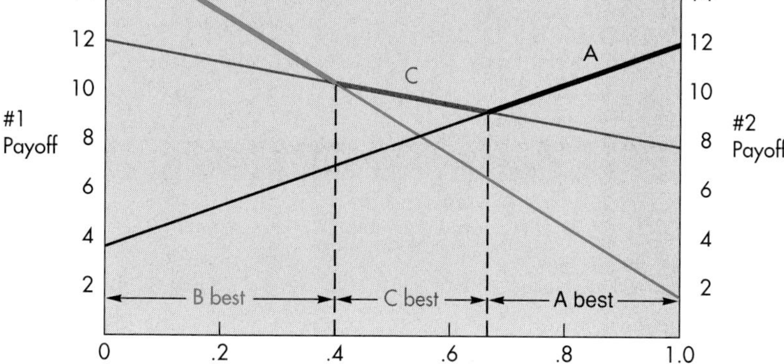

The graph shows the range of values of $P(2)$ over which each alternative is optimal. Thus, for low values of $P(2)$ [and thus high values of $P(1)$, since $P(1) + P(2) = 1.0$], alternative B will have the highest expected value; for intermediate values of $P(2)$, alternative C is best; and for higher values of $P(2)$, alternative A is best.

To find exact values of the ranges, determine where the upper parts of the lines intersect. Note that at the intersections, the two alternatives represented by the lines would be equivalent in terms of expected value. Hence, the decision maker would be indifferent between the two at that point. To determine the intersections, you must obtain the equation of each line. This is relatively simple to do. Because these are straight lines, they have the form $y = a + bx$, where a is the y-intercept value at the left axis, b is the slope of the line, and x is $P(2)$. Slope is defined as the change in y for a one-unit change in x. In this type of problem, the distance between the two vertical axes is 1.0. Consequently, the slope of each line is equal to the right-hand value minus the left-hand value. The slopes and equations are:

	#1	#2	Slope	Equation
A	4	12	$12 - 4 = +8$	$4 + 8P(2)$
B	16	2	$2 - 16 = -14$	$16 - 14P(2)$
C	12	8	$8 - 12 = -4$	$12 - 4P(2)$

From the graph, we can see that alternative B is best from $P(2) = 0$ to the point where that straight line intersects the straight line of alternative C, and that begins the region where C is better. To find that point, solve for the value of $P(2)$ at their intersection. This requires setting the two equations equal to each other and solving for $P(2)$. Thus,

$$16 - 14P(2) = 12 - 4P(2)$$

Rearranging terms yields

$$4 = 10P(2)$$

Solving yields $P(2) = .40$. Thus, alternative B is best from $P(2) = 0$ up to $P(2) = .40$. B and C are equivalent at $P(2) = .40$.

Alternative C is best from that point until its line intersects alternative A's line. To find that intersection, set those two equations equal and solve for $P(2)$. Thus,

$$4 + 8P(2) = 12 - 4P(2)$$

Rearranging terms results in

$$12P(2) = 8$$

Solving yields $P(2) = .67$. Thus, alternative C is best from $P(2) > .40$ up to $P(2) = .67$, where A and C are equivalent. For values of $P(2)$ greater than .67 up to $P(2) = 1.0$, A is best.

Note: If a problem calls for ranges with respect to $P(1)$, find the $P(2)$ ranges as above, and then subtract each $P(2)$ from 1.00 (e.g., .40 becomes .60, and .67 becomes .33).

Summary

Decision making is an integral part of operations management.

Decision theory is a general approach to decision making that is useful in many different aspects of operations management. Decision theory provides a framework for the analysis of decisions. It includes a number of techniques that can be classified according to the degree of uncertainty associated with a particular decision problem. Two visual tools useful for analyzing some decision problems are decision trees and graphical sensitivity analysis.

Key Terms

bounded rationality, 65
certainty, 65
decision tree, 69
expected monetary value (EMV) criterion, 69
expected value of perfect information (EVPI), 71
Laplace, 67
maximax, 67
maximin, 67
minimax regret, 67
opportunity losses, 68
payoff table, 66
regret, 68
risk, 65
sensitivity analysis, 72
suboptimization, 65
uncertainty, 65

Solved Problems

The following solved problems refer to this payoff table:

		New Bridge Built	No New Bridge
Alternative locations for new warehouse	A	1	14
	B	2	10
	C	4	6

Problem 1

Assume the payoffs represent profits. Determine the alternative that would be chosen under each of these decision criteria:

a. Maximin.

b. Maximax.

c. Laplace.

Solution

	New Bridge	No New Bridge	Maximin (worst)	Maximax (best)	Laplace (average)
A	1	14	1	14 [best]	15 ÷ 2 = 7.5 [best]
B	2	10	2	10	12 ÷ 2 = 6
C	4	6	4 [best]	6	10 ÷ 2 = 5

Thus, the alternatives chosen would be C under maximin, A under maximax, and A under Laplace.

Problem 2

Using graphical sensitivity analysis, determine the probability for no new bridge for which each alternative would be optimal.

Solution

Plot a straight line for each alternative. Do this by plotting the payoff for new bridge on the left axis and the payoff for no new bridge on the right axis and then connecting the two points. Each line represents the expected profit for an alternative for the entire range of probability of no new bridge. Because the lines represent expected profit, the line that is highest for a given value of P(no new bridge) is optimal. Thus, from the graph, you can see that for low values of this probability, alternative C is best, and for higher values, alternative A is best (B is never the highest line, so it is never optimal).

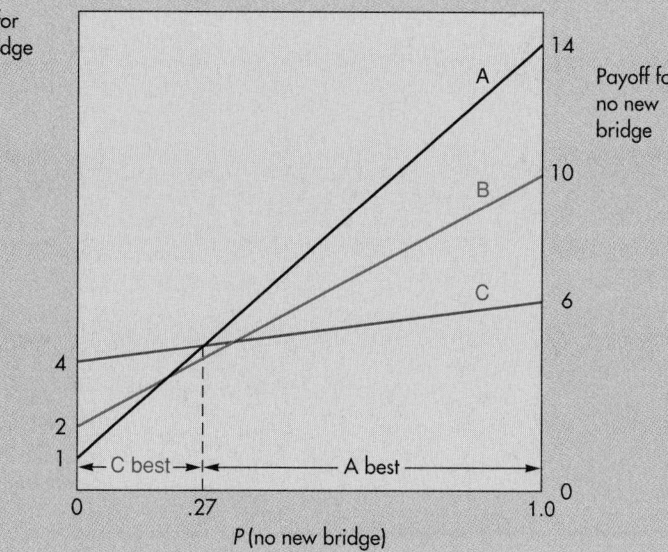

The dividing line between the ranges where C and A are optimal occurs where the two lines intersect. To find that probability, first formulate the equation for each line. To do this, let the intersection with the left axis be the y-intercept; the slope equals the right-side payoff minus the left-side payoff. Thus, for C you have $4 + (6 - 4)P$, which is $4 + 2P$. For A, $1 + (14 - 1)P$, which is $1 + 13P$. Setting these two equal to each other, you can solve for P:

$$4 + 2P = 1 + 13P$$

Solving, $P = .27$. Therefore, the ranges for P(no new bridge) are:

A: $.27 < P \leq 1.00$

B: never optimal

C: $0 \leq P < .27$

Problem 3

Using the information in the payoff table, develop a table of regrets, and then:

a. Determine the alternative that would be chosen under minimax regret.

b. Determine the expected value of perfect information using the regret table, assuming that the probability of a new bridge being built is .60.

Solution

To obtain the regrets, subtract all payoffs in each column from the best payoff in the column. The regrets are:

	New Bridge	No New Bridge
A	3	0
B	2	4
C	0	8

a. Minimax regret involves finding the worst regret for each alternative and then choosing the alternative that has the "best" worst. Thus, you would choose A:

	Worst
A	3 [best]
B	4
C	8

b. Once the regret table has been developed, you can compute the EVPI as the *smallest* expected regret. Since the probability of a new bridge is given as .60, we can deduce that the probability of no new bridge is 1.00 − .60 = .40. The expected regrets are:

A: .60(3) + .40(0) = 1.80

B: .60(2) + .40(4) = 2.80

C: .60(0) + .40(8) = 3.20

Hence, the EVPI is 1.80.

Problem 4

Using the probabilities of .60 for a new bridge and .40 for no new bridge, compute the expected value of each alternative in the payoff table, and identify the alternative that would be selected under the expected-value approach.

Solution

A: .60(1) + .40(14) = 6.20 [best]

B: .60(2) + .40(10) = 5.20

C: .60(4) + .40(6) = 4.80

Problem 5

Compute the EVPI using the information from the previous problem.

Solution

Using Formula (2–1), the EVPI is the expected payoff under certainty minus the maximum expected value. The expected payoff under certainty involves multiplying the best payoff in each column by the column probability and then summing those amounts. The best payoff in the first column is 4, and the best in the second is 14. Thus,

Expected payoff under certainty = .60(4) + .40(14) = 8.00

Then

EVPI = 8.00 − 6.20 = 1.80

(This agrees with the result obtained in Solved Problem 3*b*.)

Excel solution:

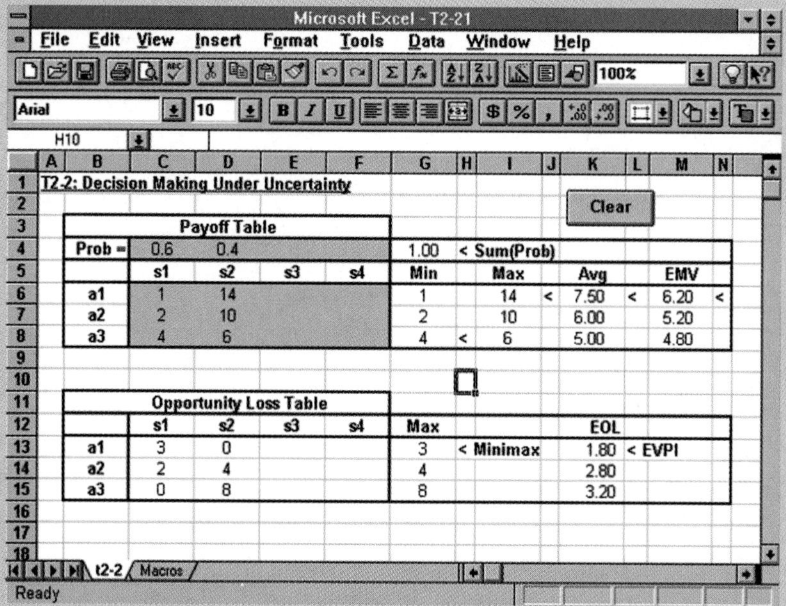

Placing the problem data in the cell positions shown, the expected monetary value (EMV) for each alternative is computed in column F by obtaining the product of the probability in cell C10 and the payoff in cell C6 and adding it to the product of the probability in cell D10 and the payoff in cell D6.

Then, the overall EMV is obtained in cell F12 as the maximum of the values in F6, F7, and F8.

The EVPI is obtained by summing the product of the maximum in column C and the probability in C10, and the product of the maximum in column D and the probability in D10.

Problem 6

Suppose that the values in the payoff table represent *costs* instead of profits.

a. Determine the choice that you would make under each of these strategies: maximin, *minimin*, and Laplace.

b. Develop the regret table, and identify the alternative chosen using minimax regret. Then find the EVPI if P(new bridge) = .60.

c. Using sensitivity analysis, determine the range of P(no new bridge) for which each alternative would be optimal.

d. If P(new bridge) = .60 and P(no new bridge) = .40, find the alternative chosen to minimize expected cost.

Solution

a. Note: *Minimin* is the reverse of maximax; for costs minimin identifies the lowest (best) cost.

	New Bridge	No New Bridge	Maximin (worst)	Minimin (best)	Laplace (average)
A	1	14	14	1 [best]	15 ÷ 2 = 7.5
B	2	10	10	2	12 ÷ 2 = 6
C	4	6	6 [best]	4	10 ÷ 2 = 5 [best]

b. Develop the regret table by subtracting the *lowest cost* in each column from each of the values in the column. (Note that none of the values is negative.)

	New Bridge	No New Bridge	Worst
A	0	8	8
B	1	4	4
C	3	0	3 [best]
	EVPI = .60(3) + .40(0) = 1,80		

c. The graph is identical to that shown in Solved Problem 2. However, the lines now represent expected *costs*, so the best alternative for a given value of P(no new bridge) is the *lowest* line. Hence, for very low values of P(no new bridge), A is best; for intermediate values, B is best; and for high values, C is best. You can set the equations of A and B, and B and C, equal to each other in order to determine the values of P (no new bridge) at their intersections. Thus,

A = B: $1 + 13P = 2 + 8P$; solving, $P = .20$

B = C: $2 + 8P = 4 + 2P$; solving, $P = .33$

Hence, the ranges are:

A best: $0 \leq P < .20$

B best: $.20 < P < .33$

C best: $.33 < P \leq 1.00$

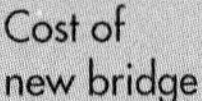

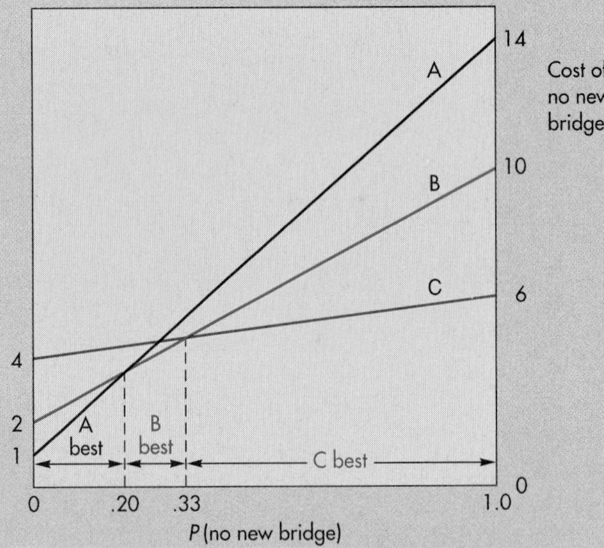

d. Expected-value computations are the same whether the values represent costs or profits. Hence, the expected payoffs for costs are the same as the expected payoffs for profits that were computed in Solved Problem 4. However, now you want the alternative that has the *lowest* expected payoff rather than the one with the highest payoff. Consequently, alternative C is the best because its expected payoff is the lowest of the three.

Discussion and Review Questions

1. What is the chief role of the operations manager?
2. List the steps in the decision-making process.
3. Explain the term *bounded rationality.*
4. Explain the term *suboptimization.*
5. What are some of the reasons for poor decisions?

6. What information is contained in a payoff table?
7. What is sensitivity analysis, and how can it be useful to a decision maker?
8. Contrast maximax and maximin decision strategies. Under what circumstances is each appropriate?
9. Under what circumstances is expected monetary value appropriate as a decision criterion? When isn't it appropriate?
10. Explain or define each of these terms:
    a. Laplace criterion.
    b. Minimax regret.
    c. Expected value.
    d. Expected value of perfect information.
11. What information does a decision maker need in order to perform an expected-value analysis of a problem? What options are available to the decision maker if the probabilities of the states of nature are unknown? Can you think of a way you might use sensitivity analysis in such a case?

Problems

1. A small building contractor has recently experienced two successive years in which work opportunities exceeded the firm's capacity. The contractor must now make a decision on capacity for next year. Estimated profits under each of the two possible states of nature are as shown in the table below. Which alternative should be selected if the decision criterion is:
    a. Maximax?
    b. Maximin?
    c. Laplace?
    d. Minimax regret?

	NEXT YEAR'S DEMAND	
Alternative	**Low**	**High**
Do nothing	\$50*	\$60
Expand	20	80
Subcontract	40	70

*Profit in \$ thousands.

a. Expand (\$80 is best payoff)
b. Do nothing 50
Expand 20
Subcontract 40
c. Do nothing 55
Expand 50
Subcontract 55
d.

	Low	High	Worst
Do nothing	0	20	20
Expand	30	0	30
Sub-contract	10	10	10

2. Refer to Problem 1. Suppose after a certain amount of discussion, the contractor is able to subjectively assess the probabilities of low and high demand: $P(\text{low}) = .3$ and $P(\text{high}) = .7$.
    a. Determine the expected profit of each alternative. Which alternative is best? Why?
    b. Analyze the problem using a decision tree. Show the expected profit of each alternative on the tree.
    c. Compute the expected value of perfect information. How could the contractor use this knowledge?

a. Expected profit
Do nothing 57
Expand 62
Subcontract 61
c. EPC: 71
Expected profit 62
EVPI: 9

3. Refer to Problems 1 and 2. Construct a graph that will enable you to perform sensitivity analysis on the problem. Over what range of $P(\text{high})$ would the alternative of doing nothing be best? Expand? Subcontract?

Do nothing 0 to < .50
Expand > .67 to 1.00
Subcontract > .50 to .67

4. A firm that plans to expand its product line must decide whether to build a small or a large facility to produce the new products. If it builds a small facility and demand is low, the net present value after deducting for building costs will be \$400,000. If demand is high, the firm can either maintain the small facility or expand it. Expansion would have a net present value of \$450,000, and maintaining the small facility would have a net present value of \$50,000.

a. Graph in IM
Large \$476,000
Small \$430,000

b. EPC $640,000
EPR 476,000
EVPI $164,000
c. Graph in IM
small: 0 to .46
large: .46 to 1

If a large facility is built and demand is high, the estimated net present value is $800,000. If demand turns out to be low, the net present value will be −$10,000.

The probability that demand will be high is estimated to be .60, and the probability of low demand is estimated to be .40.

a. Analyze using a tree diagram.

b. Compute the EVPI. How could this information be used?

c. Determine the range over which each alternative would be best in terms of the value of *P*(demand low).

a. See IM
b.

Staff	EP
0	3.6
2	4.4
4	4.8*

c.

Staff	EP
0	2.5*
2	2.3
4	1.5

5. Demand for movie rentals at a video store on Saturdays during summer months is related to the weather. If it is raining, or if the chance of rain is greater than 50%, demand tends to follow one distribution, whereas if it is not raining and the chance of rain does not exceed 50%, demand follows a different distribution. This is important to the video store because the manager must decide early Saturday how many employees to schedule for Saturday afternoon and early evening.

The two distributions are:

P(Rain) > 50%

Demand	Probability
Low	.10
Moderate	.20
High	.70

P(Rain) ≤ 50%

Demand	Probability
Low	.60
Moderate	.30
High	.10

The regular staff can handle Low demand. Moderate demand requires two additional employees, and High demand requires another two employees. The payoff table (profits in $000) is

		Demand		
		Low	Moderate	High
Extra staff	0	2	3	4
	2	1	4	5
	4	0	3	6

a. Construct a tree diagram showing the payoffs for this situation.

b. Determine the number of additional staff needed for a rainy Saturday.

c. Determine the number of staff needed when the chance of rain is 20% for a Saturday.

a. Renew $4,000,000
Relocate $5,000,000*
b. Renew $500,000*
Relocate $100,000
c. Renew $2,250,000
Relocate $2,550,000*
d. Renew $4,500,000
Relocate $3,900,000*

6. The lease of Theme Park, Inc., is about to expire. Management must decide whether to renew the lease for another 10 years or to relocate near the site of a proposed motel. The town planning board is currently debating the merits of granting approval to the motel. A consultant has estimated the net present value of Theme Park's two alternatives under each state of nature as shown below. What course of action would you recommend using:

a. Maximax?

b. Maximin?

c. Laplace?

d. Minimax regret?

Options	Motel Approved	Motel Rejected
Renew	$ 500,000	$4,000,000
Relocate	5,000,000	100,000

7. Refer to Problem 6. Suppose that the management of Theme Park, Inc., has decided that

there is a .35 probability that the motel's application will be approved.

a. If management uses maximum expected monetary value as the decision criterion, which alternative should it choose?

b. Represent this problem in the form of a decision tree.

c. If management has been offered the option of a temporary lease while the town planning board considers the motel's application, would you advise management to sign the lease? The lease will cost $24,000.

a. Renew $2,775,000
Relocate $1,815,000
Decision: Renew lease
b. Graph in IM
c. EVPI = $1,575,000
Sign lease

8. Construct a graph that can be used for sensitivity analysis for Problem 7.

a. How sensitive is the solution to the problem in terms of the probability estimate of .35?

b. Suppose that, after consulting with a member of the town planning board, management decides that an estimate of approval is approximately .45. How sensitive is the solution to this revised estimate? Explain.

c. Suppose the management is confident of all the estimated payoffs except for $4 million. If the probability of approval is .35, for what range of payoff for renew/rejected will the alternative selected using maximum expected value remain the same?

a, b: Break-even probability of 46%
c. Range is $2,523,077 or more

9. A firm must decide whether to construct a small, medium, or large stamping plant. A consultant's report indicates a .20 probability that demand will be low and an .80 probability that demand will be high.

If the firm builds a small facility and demand turns out to be low, the net present value will be $42 million. If demand turns out to be high, the firm can either subcontract and realize the net present value of $42 million or expand greatly for a net present value of $48 million.

The firm could build a medium-size facility as a hedge: if demand turns out to be low, its net present value is estimated at $22 million; if demand turns out to be high, the firm could do nothing and realize a net present value of $46 million, or it could expand and realize a net present value of $50 million.

If the firm builds a large facility and demand is low, the net present value will be −$20 million, whereas high demand will result in a net present value of $72 million.

a. Analyze this problem using a decision tree.

b. What is the maximin alternative?

c. Compute the EVPI and interpret it.

d. Perform sensitivity analysis on P(high).

a. Figure in IM
b. Build large: $53.6 million
Build small: $42 million
c. EPC = $66.0
EVPI = $12.4
d. $P(\text{high}) = x$
Small: $42 + 6x$
Medium: $22 + 28x$
Large: $-20 + 92x$
Small and large EV
Same for $x = 0.7209$

10. A manager must decide how many machines of a certain type to buy. The machines will be used to manufacture a new gear for which there is increased demand. The manager has narrowed the decision to two alternatives: buy one machine or buy two. If only one machine is purchased and demand is more than it can handle, a second machine can be purchased at a later time. However, the cost per machine would be lower if the two machines were purchased at the same time.

The estimated probability of low demand is .30, and the estimated probability of high demand is .70.

The net present value associated with the purchase of two machines initially is $75,000 if demand is low and $130,000 if demand is high.

The net present value for one machine and low demand is $90,000. If demand is high, there are three options. One option is to do nothing, which would have a net present value of $90,000. A second option is to subcontract; that would have a net present value of $110,000. The third option is to purchase a second machine. This option would have a net present value of $100,000.

How many machines should the manager purchase initially? Use a decision tree to analyze this problem.

Graph in IM
Buy 2 machines
$113,500

Alt. A: 22
Alt. B: 25
Choose Alt. B

11. Determine the course of action that has the highest EMV for the accompanying tree diagram.

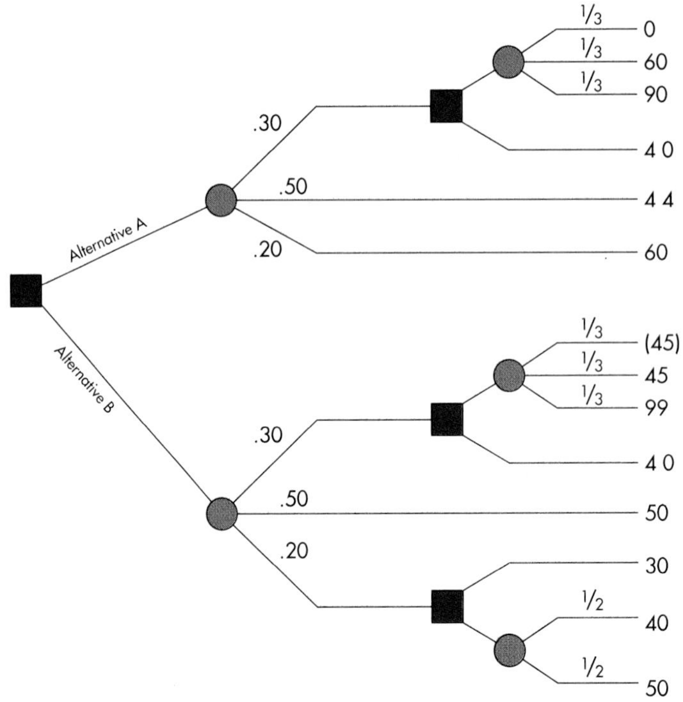

Design	EMV
001	12
002	12.5
003	20

EPC: 24
EVPI: 4

12. Space engineers have three alternative designs for the configuration of a component for an unmanned space shot. The space vehicle is likely to encounter one of four different conditions, which have probabilities of occurrence as listed in the following payoff table with the payoffs for each combination of design and state of nature. Additional data from previous flights are available but will require additional expenditures to analyze. However, the project director is confident that analysis of the data will clearly indicate which state of nature will be encountered. What amount would be justified for the data analysis?

		STATES OF NATURE			
		A	**B**	**C**	**D**
	Probability:	**.3**	**.4**	**.2**	**.1**
Design	001	20*	10	10	0
	002	15	10	0	40
	003	10	20	30	30

*In $ hundreds.

13. The director of social services of a county has learned that the state has mandated additional information requirements. This will place an additional burden on the agency. The director has identified three acceptable alternatives to handle the increased workload. One alternative is to reassign present staff members, the second is to hire and train two new workers, and the third is to redesign current practice so that workers can readily collect the information with little additional effort. An unknown factor is the caseload for the coming year when the new data will be collected on a trial basis. The estimated costs for various options and caseloads are shown in the following table:

	CASELOAD		
	Moderate	**High**	**Very High**
Reassign staff	$50*	60	85
New staff	60	60	60
Redesign collection	40	50	90

*Cost in $ thousands.

Assuming that past experience has shown the probabilities of various caseloads to be unreliable, what decision would be appropriate using each of the following criteria?

a. Maximin.

b. Maximax.

c. Minimax regret.

d. Laplace.

	Worst	**Best**	**Avg.**
Reassign	85	50	65
New staff	60	60	60
Redesign	90	40	60

a. New staff
b. Redesign
c. Table in IM
New staff
d. (tie) New staff or redesign

14. After contemplating the caseload question (see Problem 13), the director of social services has decided that reasonable caseload probabilities are .10 for moderate, .30 for high, and .60 for very high.

a. Which alternative will yield the minimum expected cost?

b. Construct a decision tree for this problem. Indicate the expected costs for the three decision branches.

c. Determine the expected value of perfect information using an opportunity loss table.

a. Reassign $74
New staff $60
Redesign $73
b. Figure in IM
c.

	EOL
Reassign	19
New staff	5
Redesign	18

15. Suppose the director of social services has the option of hiring an additional staff member if one staff member is hired initially and the caseload turns out to be high or very high. Under that plan, the first entry in row 2 of the cost table (see Problem 13) will be 40 instead of 60, the second entry will be 75, and the last entry will be 80. Assume the caseload probabilities are as noted in Problem 14. Construct a decision tree that shows the sequential nature of this decision, and determine which alternative will minimize expected cost.

Figure in IM

16. A manager has compiled estimated profits for various alternative courses of action but is reluctant to assign probabilities to the states of nature. The payoff table is:

		STATE OF NATURE	
		#1	**#2**
Alternative	A	$ 20*	140
	B	120	80
	C	100	40

*Cost in $ thousands.

a. Plot the expected-value lines on a graph.

b. Is there any alternative that would never be appropriate in terms of maximizing expected profit? Explain on the basis of your graph.

c. For what range of $P(2)$ would alternative A be the best choice if the goal is to maximize expected profit?

d. For what range of $P(1)$ would alternative A be the best choice if the goal is to maximize expected profit?

a. Figure in IM
b. C is always lower than B.
C is never appropriate.
c. $P(2) > .625$
d. $P(1) < .375$

17. Repeat all parts of Problem 16, assuming the values in the payoff table are estimated *costs* and the goal is to minimize expected costs.

18. The research staff of a marketing agency has assembled the following payoff table of estimated profits:

b. B is not appropriate
c. $P(2) < .444$
d. $P(1) > .556$
$EV_1 = 10 - 12P$
$EV_2 = 8 - 5P$
$EV_3 = 5 - 0P$
$EV_4 = 0 + 7P$
#1: $0 \leq P(2) < .286$
#2: $.286 < P(2) < .60$
#3: $.60 < P(2) < .714$
#4: $.714 < P(2) \leq 1.0$

		Receive Contract	Not Receive Contract
Proposal	#1	$10*	−2
	#2	8	3
	#3	5	5
	#4	0	7

*Cost in $ thousands.

Relative to the probability of not receiving the contract, determine the range of probability for which each of the proposals would maximize expected profit.

19. Given this payoff table:

a. A: $.70 < P(1) \leq 1.00$
B: never
C: $0 \leq P(1) < .20$
D: $.20 < P(1) < .70$
b. A: $0 \leq P(1) < .25$
B: $.25 < P(1) < .583$
C: $.583 < P(1) \leq 1.00$
D: never

		STATE OF NATURE	
		#1	#2
Alternative	A	$120*	20
	B	60	40
	C	10	110
	D	90	90

*Cost in $ thousands.

a. Determine the range of $P(1)$ for which each alternative would be best, treating the payoffs as profits.

b. Answer part *a* treating the payoffs as costs.

Selected Bibliography

Bierman, Harold; Charles P. Bonini; and Warren H. Hausman. *Quantitative Analysis for Business Decisions.* 8th ed. Burr Ridge, IL: Richard D. Irwin, 1991.

Eppen, G. D.; F. J. Gould; and C. P. Schmidt. *Introductory Management Science.* 3rd ed. Englewood Cliffs, NJ: Prentice Hall, 1991.

Levin, Richard I.; Charles A. Kirkpatrick; and David S. Rubin. *Quantitative Approaches to Management.* 6th ed. New York: McGraw-Hill, 1987.

Stevenson, William J. *Introduction to Management Science.* 2nd ed. Burr Ridge, IL: Richard D. Irwin, 1992.

Taylor, Bernard W. *Introduction to Management Science.* 3rd ed. Dubuque, Iowa: William C. Brown, 1990.

Ulvila, J. W., and R. V. Brown. "Decision Analysis Comes of Age." *Harvard Business Review,* September–October 1982, p. 130.

PART TWO
FORECASTING

This part is devoted solely to forecasting. It is presented early in the book because forecasts are the basis for a wide range of decisions that are described in the following chapters. In fact, forecasts are basic inputs for many kinds of decisions in business organizations. Consequently, it is important for *all* managers to be able to understand and use forecasts. Although forecasts are typically developed by the marketing function, the operations function is often called on to assist in forecast development. More important, though, is the reality that operations is a major user of forecasts.

Chapter 3 provides important insights on forecasting as well as information on how to develop and monitor forecasts.

LEARNING OBJECTIVES

After completing this chapter you should be able to:

1. List the elements of a good forecast.
2. Outline the steps in the forecasting process.
3. Describe at least four qualitative forecasting techniques and the advantages and disadvantages of each.
4. Compare and contrast qualitative and quantitative approaches to forecasting.
5. Briefly describe averaging techniques, trend and seasonal techniques, and regression analysis, and solve typical problems.
6. Describe two measures of forecast accuracy.
7. Describe two ways of evaluating and controlling forecasts.
8. Identify the major factors to consider when choosing a forecasting technique.

CHAPTER THREE FORECASTING

CHAPTER OUTLINE

Many new car buyers have a thing or two in common. Once they make the decision to buy a new car, they want it as soon as possible. They certainly don't want to order it and then have to wait six weeks or more for delivery. If the car dealer they visit doesn't have the car they want, they'll look elsewhere. Hence, it is important for a dealer to anticipate buyer wants and to have those models, with the necessary options, in stock. The dealer who can correctly forecast buyer wants, and have those cars available, is going to be much more successful than a competitor who guesses instead of forecasting—and guesses wrong—and gets stuck with cars customers don't want. So how does the dealer know how many cars of each type to stock? The answer is, the dealer *doesn't* know for sure, but based on analysis of previous buying patterns, and perhaps making allowances for current conditions, the dealer can come up with a reasonable *approximation* of what buyers will want.

Planning is an integral part of a manager's job. If uncertainties cloud the planning horizon, managers will find it difficult to plan effectively. Forecasts help managers by reducing some of the uncertainty, thereby enabling them to develop more meaningful plans. A **forecast** is a statement about the future.

forecast A statement about the future.

This chapter provides a survey of business forecasting. It describes the necessary steps in preparing a forecast, basic forecasting techniques, how to monitor a forecast, and elements of good forecasts.

INTRODUCTION

At one time or another, everyone has witnessed weather forecasts that were flops, even though most of the time forecasters are close in predicting the weather. Forecasting demand is a lot like forecasting the weather. In both cases, there is no such thing as a sure bet: Predictions usually turn out to be in the ballpark, but occasionally they miss the mark completely. Moreover, the forecasts in both cases serve as a *basis for planning.* Weather forecasts influence travel and recreation plans, choice of clothing, and whether to walk or ride to a friend's house. Farmers rely on weather forecasts to determine when to plant and harvest and when to take precautionary steps (e.g., protect against frost). In business, forecasts are the basis for budgeting and planning for capacity, sales, production and inventory, manpower, purchasing, and more. Forecasts play an important role in the planning process because they enable managers to anticipate the future so they can plan accordingly.

There are two uses for forecasts. One is to help managers *plan the system,* and the other is to help them *plan the use of the system.* Planning the system generally involves long-range plans about the types of products and services to offer, what facilities and equipment to have, where to locate, and so on. Planning the use of the system refers to short-range and intermediate-range planning, which involve tasks such as planning inventory and work force levels, planning purchasing and production, budgeting, and scheduling.

Business forecasting pertains to more than predicting demand. Forecasts are also used to predict profits, revenues, costs, productivity changes, prices and availability of energy and raw materials, interest rates, movements of key economic indicators (e.g., GNP, inflation, government borrowing), and prices of stocks and bonds. For the sake of simplicity, this chapter will focus on the forecasting of demand. Keep in mind, however, that the concepts and techniques apply equally well to the other variables.

In spite of its use of computers and sophisticated mathematical models, forecasting is not an exact science. Instead, successful forecasting often requires a skillful blending of art and science. Experience, judgment, and technical expertise all play a role in developing useful forecasts. Along with these, a certain amount of luck and a dash of humility can be helpful, because the worst forecasters occasionally produce a very good forecast, and even the best forecasters sometimes miss completely. Current forecasting techniques range from the mundane to the exotic. Some work better than others, but no single technique works all the time.

Generally speaking, the responsibility for preparing demand forecasts in business organizations lies with marketing or sales rather than operations. Nonetheless, operations people are often called on to make certain forecasts and to help others prepare forecasts. In addition, since forecasts are major inputs for many operations decisions, operations managers and staff must be knowledgeable about the kinds of forecasting techniques available, the assumptions that underlie their use, and their limitations. It is also important for managers to consider *how* forecasts affect operations. In short, forecasting is an integral part of operations management.

NEWSCLIP

HIGH FORECASTS CAN BE BAD NEWS

Overly optimistic forecasts by retail store buyers can easily lead retailers to over order, resulting in bloated inventories. When that happens, there is pressure on stores to cut prices in order to move the excess merchandise. Although customers delight in these markdowns, retailer profits generally suffer. Furthermore, retailers will naturally cut back on new orders while they work off their inventories, creating a ripple effect that hits the entire supply chain, from shippers, to producers, to suppliers of raw materials. The message is clear: Overly optimistic forecasts can be bad news.

Source: "Bloated Inventories at Retailers May Mean Trouble for Investors" by Susan Pulliam, *The Wall Street Journal*, May 22, 1997, p. C1.

FEATURES COMMON TO ALL FORECASTS

A wide variety of forecasting techniques are in use. In many respects, they are quite different from each other, as you shall soon discover. Nonetheless, certain features are common to all, and it is important to recognize them.

1. Forecasting techniques generally assume that the same underlying causal system that existed in the past will continue to exist in the future.

Comment. A manager cannot simply delegate forecasting to models or computers and then forget about it, because unplanned occurrences can wreak havoc with forecasts. For instance, weather-related events, tax increases or decreases, and changes in features or prices of competing products or services can have a major impact on demand. Consequently, a manager must be alert to such occurrences and be ready to override forecasts, which assume a stable causal system.

2. Forecasts are rarely perfect; actual results usually differ from predicted values. No one can predict *precisely* how an often large number of related factors will impinge upon the variable in question; this, and the presence of randomness, preclude a perfect forecast. Allowances should be made for inaccuracies.
3. Forecasts for groups of items tend to be more accurate than forecasts for individual items because forecasting errors among items in a group usually have a canceling effect. Opportunities for grouping may arise if parts or raw materials are used for multiple products or if a product or service is demanded by a number of independent sources.
4. Forecast accuracy decreases as the time period covered by the forecast—the *time horizon*—increases. Generally speaking, short-range forecasts must contend with fewer uncertainties than longer-range forecasts, so they tend to be more accurate.

An important consequence of the last point is that flexible business organizations—those that can respond quickly to changes in demand—require a shorter forecasting horizon and, hence, benefit from more accurate short-range forecasts than competitors who are less flexible and who must therefore use longer forecast horizons.

ELEMENTS OF A GOOD FORECAST

A properly prepared forecast should fulfill certain requirements:

1. The forecast should be *timely*. Usually, a certain amount of time is needed to respond to the information contained in a forecast. For example, capacity cannot be expanded overnight, nor can inventory levels be changed immediately. Hence, the forecasting horizon must cover the time necessary to implement possible changes.
2. The forecast should be *accurate* and the degree of accuracy should be stated. This will enable users to plan for possible errors and will provide a basis for comparing alternative forecasts.
3. The forecast should be *reliable*; it should work consistently. A technique that sometimes provides a good forecast and sometimes a poor one will leave users with the uneasy feeling that they may get burned every time a new forecast is issued.
4. The forecast should be expressed in *meaningful units*. Financial planners need to know how many *dollars* will be needed, production planners need to know how many *units* will be needed, and schedulers need to know what *machines* and *skills* will be required. The choice of units depends on user needs.
5. The forecast should be *in writing*. Although this will not guarantee that all concerned are using the same information, it will at least increase the likelihood of it. In addition, a written forecast will permit an objective basis for evaluating the forecast once actual results are in.
6. The forecasting technique should be *simple to understand and use*. Users often lack confidence in forecasts based on sophisticated techniques; they do not understand either the circumstances in which the techniques are appropriate or the limitations of the techniques. Misuse of techniques is an obvious consequence. Not surprisingly, fairly crude forecasting techniques enjoy widespread popularity because users are more comfortable working with them.

STEPS IN THE FORECASTING PROCESS

There are five basic steps in the forecasting process:

1. *Determine the purpose of the forecast.* What is its purpose and when will it be needed? This will provide an indication of the level of detail required in the forecast, the amount of resources (manpower, computer time, dollars) that can be justified, and the level of accuracy necessary.

2. *Establish a time horizon.* The forecast must indicate a time limit, keeping in mind that accuracy decreases as the time horizon increases.

3. *Select a forecasting technique.*

4. *Gather and analyze the appropriate data.* Before a forecast can be prepared, data must be gathered and analyzed. Identify any assumptions that are made in conjunction with preparing and using the forecast.

5. *Prepare the forecast.*

6. *Monitor the forecast.* A forecast has to be monitored to determine whether it is performing in a satisfactory manner. If it is not, reexamine the method, assumptions, validity of data, and so on; modify as needed; and prepare a revised forecast.

APPROACHES TO FORECASTING

There are two general approaches to forecasting: qualitative and quantitative. Qualitative methods consist mainly of subjective inputs, which often defy precise numerical description. Quantitative methods involve either the extension of historical data or the development of associative models that attempt to utilize *causal (explanatory) variables* to make a forecast.

Qualitative techniques permit inclusion of *soft* information (e.g., human factors, personal opinions, hunches) in the forecasting process. Those factors are often omitted or downplayed when quantitative techniques are used because they are difficult or impossible to quantify. Quantitative techniques consist mainly of analyzing objective, or *hard*, data. They usually avoid personal biases that sometimes contaminate qualitative methods. In practice, either or both approaches might be used to develop a forecast.

Forecasts Based on Judgment and Opinion

Judgmental forecasts rely on analysis of subjective inputs obtained from various sources, such as consumer surveys, the sales staff, managers and executives, and panels of experts. Quite frequently, these sources provide insights that are not otherwise available.

judgmental forecasts Forecasts that use subjective inputs such as opinions from consumer surveys, sales staff, managers, executives, and experts.

Forecasts Based on Time Series (Historical) Data

Some forecasting techniques depend on uncovering relationships between variables that can be used to predict future values of one of them; others simply attempt to project past experience into the future. The second approach exemplifies forecasts that use historical, or time series, data with the assumption that the future will be like the past. Some models merely attempt to smooth out random variations in historical data; others attempt to identify specific patterns in the data and project or extrapolate those patterns into the future, without trying to identify or measure those forces that cause the patterns.

Associative Forecasts

Associative models identify one or more *explanatory* variables that can be used to predict future demand. For example, demand for paint might be related to variables such as the price per gallon and the amount spent on advertising, as well as specific characteristics of the paint (e.g., drying time, ease of cleanup). The analysis in these cases yields a mathematical equation that enables the manager to predict volume of sales, for example, on the basis of given values of the "explaining" variable(s).

associative model Forecasting technique that uses explanatory variables to predict future demand.

FORECASTS BASED ON JUDGMENT AND OPINION

In some situations, forecasters rely solely on judgment and opinion to make forecasts. If management must have a forecast quickly, there is not always enough time to gather and analyze quantitative data. At other times, especially when political and economic conditions are changing, available data may be obsolete and more up-to-date information might not yet be available. Similarly, the introduction of new products and the redesign of existing products or packaging suffer from the absence of historical data that would be useful in forecasting. In such instances, forecasts are based on executive opinions, consumer surveys, opinions of the sales staff, and opinions of experts.

Executive Opinions

A small group of upper-level managers (e.g., in marketing, operations, and finance) may meet and collectively develop a forecast. This approach is often used as a part of long-range planning and new product development. It has the advantage of bringing together the considerable knowledge and talents of various managers. However, there is the risk that the view of one person will prevail, and the possibility that diffusing responsibility for the forecast over the entire group may result in less pressure to produce a good forecast.

Direct Customer Contact Composites

The sales staff or the customer service staff is often a good source of information because of their direct contact with consumers. They are often aware of any plans the customers may be considering for the future. There are, however, several drawbacks to this approach. One is that they may be unable to distinguish between what customers would *like* to do and what they actually *will* do. Another is that these people are sometimes overly influenced by recent experiences. Thus, after several periods of low sales, their estimates may tend to become pessimistic. After several periods of good sales, they may tend to be too optimistic. In addition, if forecasts are used to establish sales quotas, there will be a conflict of interest because it is in the salesperson's advantage to provide low sales estimates.

Consumer Surveys

Since it is the consumers who ultimately determine demand, it seems natural to solicit input from them. In some instances, every customer or potential customer can be contacted. However, there are usually too many customers or there is no way to identify all potential customers. Therefore, organizations seeking consumer input usually resort to consumer surveys, which enable them to *sample* consumer opinions. The obvious advantage of consumer surveys is that they can tap information that might not be available elsewhere. On the other hand, surveys require a considerable amount of knowledge and skill to handle correctly. A great deal of care is necessary to construct a survey, administer it, and correctly interpret the results for valid information. Surveys can be expensive and time-consuming. In addition, even under the best of conditions, surveys of the general public must contend with the possibility of irrational behavior patterns. For example, much of the consumer's thoughtful information gathering before purchasing a new car is often undermined by the glitter of a new car showroom or a high-pressure sales pitch. Along the same lines, low response rates to a mail survey should—but often don't—make the results suspect.

If these and similar pitfalls can be avoided, surveys can produce useful information.

Other Approaches

A manager may solicit opinions from a number of other managers and staff people. Occasionally, outside experts are needed to help with a forecast. These may include advice on political or economic conditions in the United States or a foreign country, or some other aspect of importance with which an organization lacks familiarity.

Delphi method Managers and staff complete a series of questionnaires, each developed from the previous one, to achieve a consensus forecast.

Another approach is the **Delphi method.** It involves circulating a series of questionnaires among individuals who possess the knowledge and ability to contribute meaningfully. Responses are kept anonymous, which tends to encourage honest responses. Each new questionnaire is developed using the information extracted from the previous one, thus enlarging the scope of information on which participants can base their judgments. The goal is to achieve a consensus forecast.

The Delphi method originated in the Rand Corporation in 1948. Since that time, it has been applied to a variety of situations, not all of which involve forecasting. The discussion here is limited to its use as a forecasting tool.

As a forecasting tool, the Delphi method is useful for *technological* forecasting; that is, the technique is a method for assessing changes in technology and their impact on an organization. Often the goal is to predict *when* a certain event will occur. For instance, the goal of a Delphi forecast might be to predict when video telephones might be installed in at least 50 percent of residential homes or when a vaccine for a disease might be developed and ready for mass distribution. For the most part, these are long-term, single-time forecasts, which usually have very little hard information to go by or data

NEWSCLIP

WHY TECHNOLOGY FORECASTS OFTEN FAIL

Professor Nathan Rosenberg presents an interesting explanation of why technology forecasting is so difficult. He identifies six general causes of uncertainty in long-term forecasting.

Potential Uses

All of the potential uses of new technologies are not known at the time of their development; thus, often the success of a new technology is dependent on the unknown, even though several other uses have been anticipated.

Complementary Innovations

The success of a new technology is often dependent on the emergence of a complementary technology. For example, the inventors of the laser at Bell Laboratories did not recognize its telecommunications importance because fiber optics had not yet been developed.

Systems Integration

Some new technologies do not yield spectacular breakthroughs quickly because the system use does not exist. Consider that cellular radio was developed in the 60s, but it did not become a market success until the microelectronics and telecommunications revolutions provided a cost-effective systems integration.

Problem-Solving Myopia

Inventions are often solutions to more problems than those that stimulated the discovery. Thus, developers often overlook other applications of new inventions. For example, the steam engine was invented in the 18th century to pump water out of mines, not to power locomotives or ocean liners.

Passing the "Needs Test"

A new technology must not only be technologically better but also economically so. If the new technology is not economical, then it will not become widespread. Unfortunately, changes in social customs are very difficult to predict; thus, new technologies can fail for many socioeconomic reasons.

Competing with the Past

The invention of a new way of doing something often stimulates improvements in the old technology, which may delay the adoption of the new technology.

Source: Delurgio, *Forecasting Principles and Applications,* Irwin/McGraw-Hill, © 1998.

Adapted from Nathan Rosenberg, "Why Technology Forecasts Often Fail," *The Futurist,* July–August 1995, pp. 15–21.

are costly to obtain, so the problem does not lend itself to analytical techniques. Rather, judgments of experts or others who possess sufficient knowledge to make predictions are used.

FORECASTS BASED ON TIME SERIES DATA

A **time series** is a time-ordered sequence of observations taken at regular intervals over a period of time (e.g., hourly, daily, weekly, monthly, quarterly, annually). The data may be measurements of demand, earnings, profits, shipments, accidents, output, precipitation, productivity, and the consumer price index. Forecasting techniques based on time series data are made on the assumption that future values of the series can be estimated from past values. Although no attempt is made to identify variables that influence the series, these methods are widely used, often with quite satisfactory results.

time series A time-ordered sequence of observations taken at regular intervals over time.

Analysis of time series data requires the analyst to identify the underlying behavior of the series. This can often be accomplished by merely *plotting* the data and visually examining the plot. One or more patterns might appear: trends, seasonal variations, cycles, and constant variations (variations around an average). In addition, there can be random or irregular variations. These behaviors can be described as follows:

1. **Trend** refers to a gradual, long-term upward or downward movement in the data. Population shifts, changing incomes, and cultural changes often account for such movements.

trend A long-term upward or downward movement in data.

FIGURE 3–1

Trend, seasonal, cyclical, random, and irregular variations

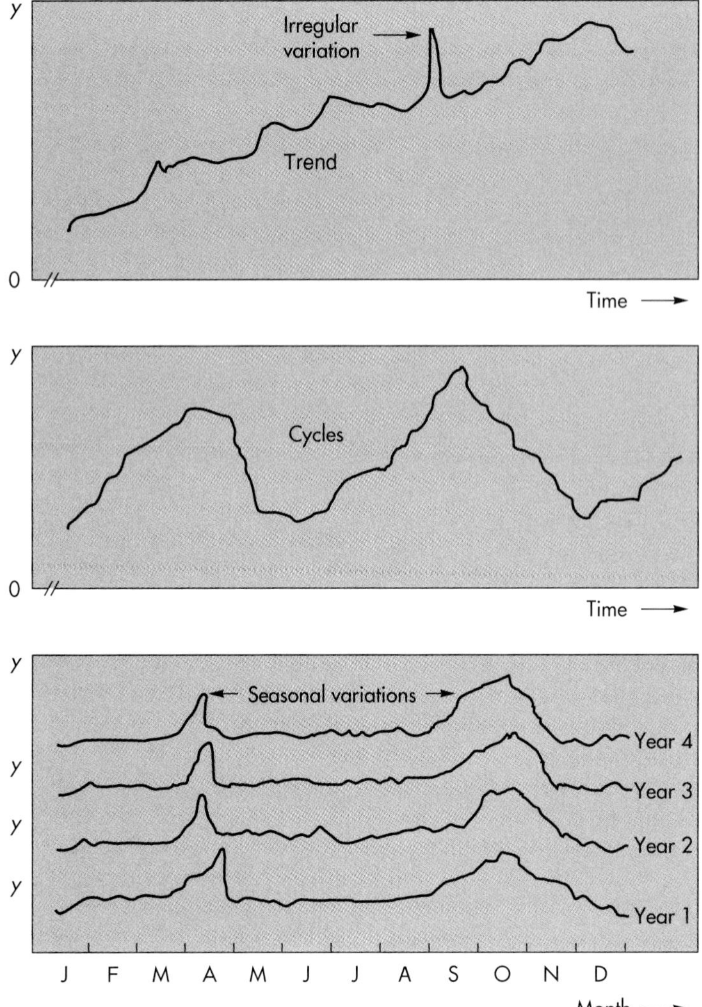

seasonality Short-term regular variations related to weather or other factors.

cycle Wavelike variation lasting more than one year.

irregular variation Caused by unusual circumstances, not reflective of typical behavior.

random variations Residual variations after all other behaviors are accounted for.

2. **Seasonality** refers to short-term, fairly regular variations generally related to factors such as weather, holidays, and vacations. Restaurants, supermarkets, and theaters experience weekly and even daily "seasonal" variations.
3. **Cycles** are wavelike variations of more than one year's duration. These are often related to a variety of economic, political, and even agricultural conditions.
4. **Irregular variations** are due to unusual circumstances such as severe weather conditions, strikes, or a major change in a product or service. They do not reflect typical behavior, and inclusion in the series can distort the overall picture. Whenever possible, these should be identified and removed from the data.
5. **Random variations** are residual variations that remain after all other behaviors have been accounted for.

These behaviors are illustrated in Figure 3–1. The small "bumps" in the plots represent random variability.

The remainder of this section has descriptions of the various approaches to the analysis of time series data. Before turning to those discussions, one point should be emphasized: A demand forecast should be based on a time series of past *demand* rather than sales. Sales would not truly reflect demand unless demand was less than the amount of a good or service available for sale. Similarly, shipments would not truly reflect demand if there are backlogs of orders; the timing of shipments would not correspond to the timing of demand.

FIGURE 3-2

Averaging applied to three possible patterns

Data
Forecast
Ideal
Step change
Gradual change

Techniques for Averaging

Historical data typically contain a certain amount of random variation, or *noise,* that tends to obscure systematic movements in the data. This randomness arises from the combined influence of many—perhaps a great many—relatively unimportant factors, and it cannot be reliably predicted. Ideally, it would be desirable to completely remove any randomness from the data and leave only "real" variations, such as changes in the demand. As a practical matter, however, it is usually impossible to distinguish between these two kinds of variations, so the best one can hope for is that the small variations are random and the large variations are "real."

Averaging techniques smooth fluctuations in a time series because the individual highs and lows in the data offset each other when they are combined into an average. A forecast based on an average thus tends to exhibit less variability than the original data (see Figure 3–2). This can be advantageous because many of these movements merely reflect random variability rather than a true change in level, or trend, in the series. Moreover, because responding to changes in expected demand often entails considerable cost (e.g., changes in production rate, changes in the size of a work force, inventory changes), it is desirable to avoid reacting to minor variations. Thus, minor variations are treated as random variations, whereas larger variations are viewed as more likely to reflect "real" changes, although these, too, are smoothed to a certain degree.

Averaging techniques generate forecasts that reflect recent values of a time series (e.g., the average value over the last several periods). These techniques work best when a series tends to vary around an average, although they can also handle step changes or gradual changes in the level of the series. Three techniques for averaging are described in this section:

1. Naive forecasts
2. Moving averages
3. Exponential smoothing

Naive Forecasts. The simplest forecasting technique is the naive method. A **naive forecast** for any period equals the previous period's actual value. For example, if demand last week was 50 units, the naive forecast for the upcoming week is 50 units. Similarly, if demand in the upcoming week turns out to be 54 units, the forecast for the following week would be 54 units.

naive forecast The forecast for any period equals the previous period's actual value.

Although at first glance the naive approach may appear *too* simplistic, it is nonetheless a legitimate forecasting tool. Consider the advantages: It has virtually no cost, it is quick and easy to prepare because data analysis is nonexistent, and it is easily understandable. The main objection to this method is its inability to provide highly accurate forecasts. However, if resulting accuracy is acceptable, this approach deserves serious consideration. Moreover, even if other forecasting techniques offer better accuracy, they will almost always involve a greater cost. The accuracy of a naive forecast can serve as a standard of comparison against which to judge the cost and accuracy of other techniques. Thus, managers must answer the question: Is the increased accuracy of another method worth the additional resources required to achieve that accuracy?

The naive forecast can also be applied to a series that exhibits seasonality or trend. For example, if monthly sales exhibit a seasonal pattern, demand for the current December can be based on demand for the preceding December, demand for January can be based on demand from the preceding January, and so on. Similarly, if trend is present, the increase (or decrease) from this period's actual demand to the next period's demand can be estimated as the same change observed between the last two periods. For instance, if demand in June is 90 units higher than in May, a naive forecast for July that allowed for trend would be June's actual demand plus an additional 90 units. Then, if demand in July was only 85 units greater than in June, the August forecast would be July's actual demand plus 85 units.

Moving Averages. One weakness of the naive method is that the forecast just *traces* the actual data, with a lag of one period; it does not smooth at all. But by expanding the amount of historical data a forecast is based on, this difficulty can be overcome. A **moving average** forecast uses a *number* of the most recent actual data values in generating a forecast. The moving average forecast can be computed using the following equation:

moving average Technique that averages a number of recent actual values, updated as new values become available.

$$MA_n = \frac{\sum_{i=1}^{n} A_i}{n} \qquad (3\text{–}1)$$

where

i = refers to the most recent period
n = Number of periods (data points) in the moving average
A_i = Actual value with age i
MA = Forecast

For example, MA_3 would refer to a three-period moving average forecast, and MA_5 would refer to a five-period moving average forecast.

Example 1

Compute a three-period moving average forecast given demand for shopping carts for the last five periods.

Period	Age	Demand
1	5	42
2	4	40
3	3	43
4	2	40 } the 3 most recent demands
5	1	41

Solution

$$MA_3 = \frac{43 + 40 + 41}{3} = 41.33$$

If actual demand in period 6 turns out to be 39, the moving average forecast for period 7 would be:

$$MA_3 = \frac{40 + 41 + 39}{3} = 40.00$$

Note that in a moving average, as each new actual value becomes available, the forecast is updated by adding the newest value and dropping the oldest and then recomputing the average. Consequently, the forecast "moves" by reflecting only the most recent values.

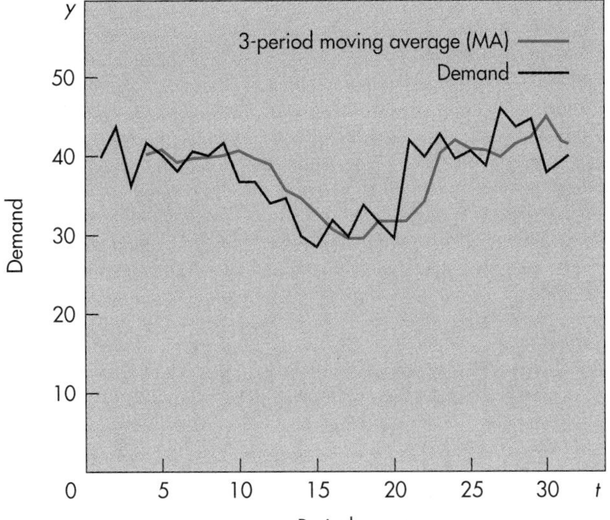

FIGURE 3–3

A moving average forecast tends to smooth and lag changes in the data

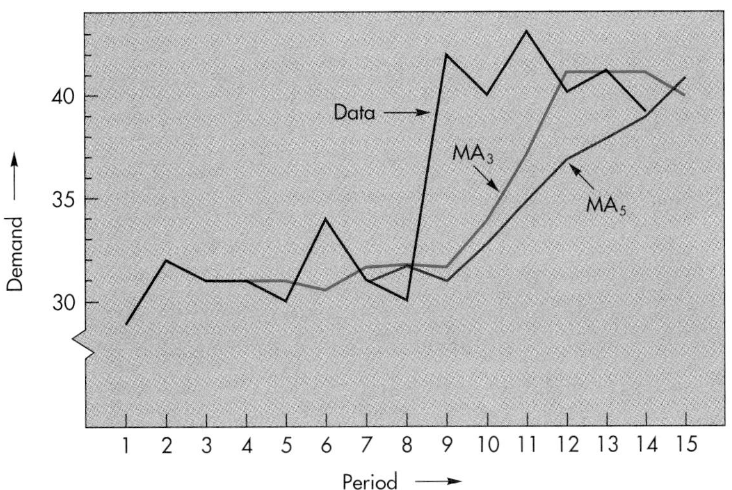

FIGURE 3–4

The more periods in a moving average, the greater the forecast will lag changes in the data

In computing a moving average, including a *moving total* column—which gives the sum of the *n* most current values from which the average will be computed—would aid computations. It is relatively simple to update the moving total: Subtract the oldest value from the newest value, and add that amount to the moving total for each update.

Figure 3–3 illustrates a three-period moving average forecast plotted against actual demand over 31 periods. Note how the moving average forecast *lags* the actual values and how smooth the forecasted values are compared with the actual values.

The moving average can incorporate as many data points as desired. In selecting the number of periods to include, the decision maker must take into account that the number of data points in the average determines its sensitivity to each new data point: The fewer the data points in an average, the more responsive the average tends to be. (See Figure 3–4.) If responsiveness is important, a moving average with relatively few data points should be used. This will permit quick adjustment to, say, a step change in the data, but it will also cause the forecast to be somewhat responsive even to random variations. Conversely, moving averages based on more data points will smooth more but be less responsive to "real" changes. Hence, the decision maker must weigh the cost of responding more slowly to changes in the data against the cost of responding to what might simply be random variations.

The advantages of a moving average forecast are that it is easy to compute and easy to understand. A possible disadvantage is that all values in the average are weighted equally. For instance, in a 10-period moving average, each value has a weight of 1/10. Hence, the oldest value has the *same weight* as the most recent value. If a change occurs in the series, a moving average forecast can be slow to react, especially if there are a large number of values in the average. Decreasing the number of values in the average increases the weight of more recent values, but it does so at the expense of losing potential information from less recent values.

A *weighted average* is similar to a moving average, except that it assigns more weight to the most recent values in a time series. For instance, the most recent value might be assigned a weight of .40, the next most recent value a weight of .30, the next after that a weight of .20, and the next after that a weight of .10. Note that the weights sum to 1.00, and that the heaviest weights are assigned to the most recent values.

Example 2

Given the following demand data,

a. Compute a weighted average forecast using a weight of .40 for the most recent period, .30 for the next most recent, .20 for the next, and .10 for the next.

b. If the actual demand for period 6 is 39, forecast demand for period 7 using the same weights as in part *a*.

Period	Demand
1	42
2	40
3	43
4	40
5	41

Solution

a. Forecast = .40(41) + .30(40) + .20(43) + .10(40) = 41.0

b. Forecast = .40(39) + .30(41) + .20(40) + .10(43) = 40.2

Note that if four weights are used, only the *four most recent* demands are used to prepare the forecast.

The advantage of a weighted average over a simple moving average is that the weighted average is more reflective of the most recent occurrences. However, the choice of weights is somewhat arbitrary and generally involves the use of trial and error to find a suitable weighting scheme.

exponential smoothing Weighted averaging method based on previous forecast plus a percentage of the forecast error.

Exponential Smoothing. **Exponential smoothing** is a sophisticated weight averaging method that is still relatively easy to use and understand. Each new forecast is based on the previous forecast plus a percentage of the difference between that forecast and the actual value of the series at that point. That is:

$$\text{Next forecast} = \text{Previous forecast} + \alpha(\text{Actual} - \text{Previous forecast})$$

where (Actual − Previous forecast) represents the forecast error and α is a percentage of the error. More concisely,

$$F_t = F_{t-1} + \alpha(A_{t-1} - F_{t-1}) \qquad (3\text{–}2a)$$

where

F_t = Forecast for period t
F_{t-1} = Forecast for period $t - 1$
α = Smoothing constant
A_{t-1} = Actual demand or sales for period $t - 1$

The smoothing constant α represents a percentage of the forecast error. Each new forecast is equal to the previous forecast plus a percentage of the previous error. For example, suppose the previous forecast was 42 units, actual demand was 40 units, and $\alpha = .10$. The new forecast would be computed as follows:

$$F_t = 42 + .10(40 - 42) = 41.8$$

Then, if the actual demand turns out to be 43, the next forecast would be:

$$F_t = 41.8 + .10(43 - 41.8) = 41.92$$

An alternate form of formula 3–2a reveals the weighting of the previous forecast and the latest actual demand:

$$F_t = (1 - \alpha)F_{t-1} + \alpha A_{t-1} \qquad (3\text{–}2b)$$

For example, if $\alpha = .10$, this would be

$$F_t = .90F_{t-1} + .10 A_{t-1}$$

The quickness of forecast adjustment to error is determined by the smoothing constant, α. The closer its value is to zero, the slower the forecast will be to adjust to forecast errors (i.e., the greater the smoothing). Conversely, the closer the value of α is to 1.00, the greater the responsiveness and the less the smoothing. This is illustrated in Example 3.

Example 3

Use exponential smoothing to develop a series of forecasts for the following data, and compute (Actual − Forecast) = Error, for each period.

a. Use a smoothing factor of .10.

b. Use a smoothing factor of .40.

c. Plot the actual data and both sets of forecasts on a single graph.

Period (t)	Actual Demand
1	42
2	40
3	43
4	40
5	41
6	39
7	46
8	44
9	45
10	38
11	40
12	

Solution

Period (t)	Actual Demand	α = .10 Forecast	α = .10 Error	α = .40 Forecast	α = .40 Error
1	42	—	—	—	—
2	40	42	−2	42	−2
3	43	41.8	1.2	41.2	1.8
4	40	41.92	−1.92	41.92	−1.92
5	41	41.73	−0.73	41.15	−0.15
6	39	41.66	−2.66	41.09	−2.09
7	46	41.39	4.61	40.25	5.75
8	44	41.85	2.15	42.55	1.45
9	45	42.07	2.93	43.13	1.87
10	38	42.35	−4.35	43.88	−5.88
11	40	41.92	−1.92	41.53	−1.53
12		41.73		40.92	

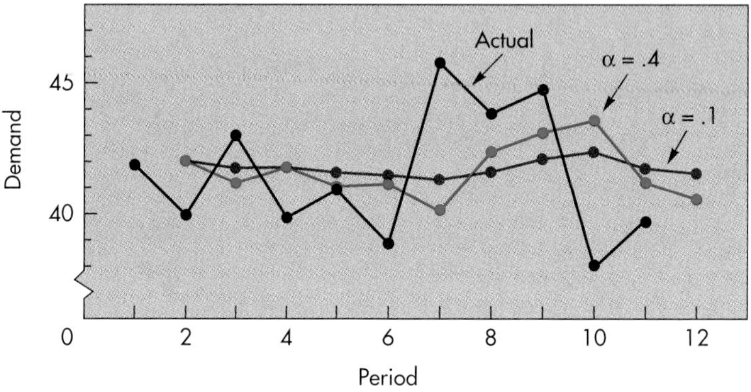

Selecting a smoothing constant is basically a matter of judgment or trial and error. The goal is to select a smoothing constant that balances the benefits of smoothing random variations with the benefits of responding to real changes if and when they occur. Commonly used values of α range from .05 to .50. Low values of α are used when the underlying average tends to be stable; higher values are used when the underlying average is susceptible to change.

Some computer packages include a feature that permits automatic modification of the smoothing constant if the forecast errors become unacceptably large.

Exponential smoothing is one of the most widely used techniques in forecasting, partly because of its ease of calculation, and partly because of the ease with which the weighting scheme can be altered—simply by changing the value of α.

A number of different approaches can be used to obtain a *starting forecast,* such as the average of the first several periods, a subjective estimate, or the first actual value as the forecast for period 2 (i.e., the naive approach). For simplicity, the naive approach is used in this book. In practice, using an average of, say, the first three values as a forecast for period 4 would provide a better starting forecast because that tends to be more representative.

Techniques for Trend

The trend component of a time series reflects the effects of any long-term factors on the series. Analysis of trend involves developing an equation that will suitably describe trend (assuming that trend is present in the data). The trend component may be linear, or it may not. Some commonly encountered nonlinear trend functions are illustrated in Figure 3–5. A simple plot of the data can often reveal the existence and nature of a

FIGURE 3-5

Graphs of some common nonlinear trends

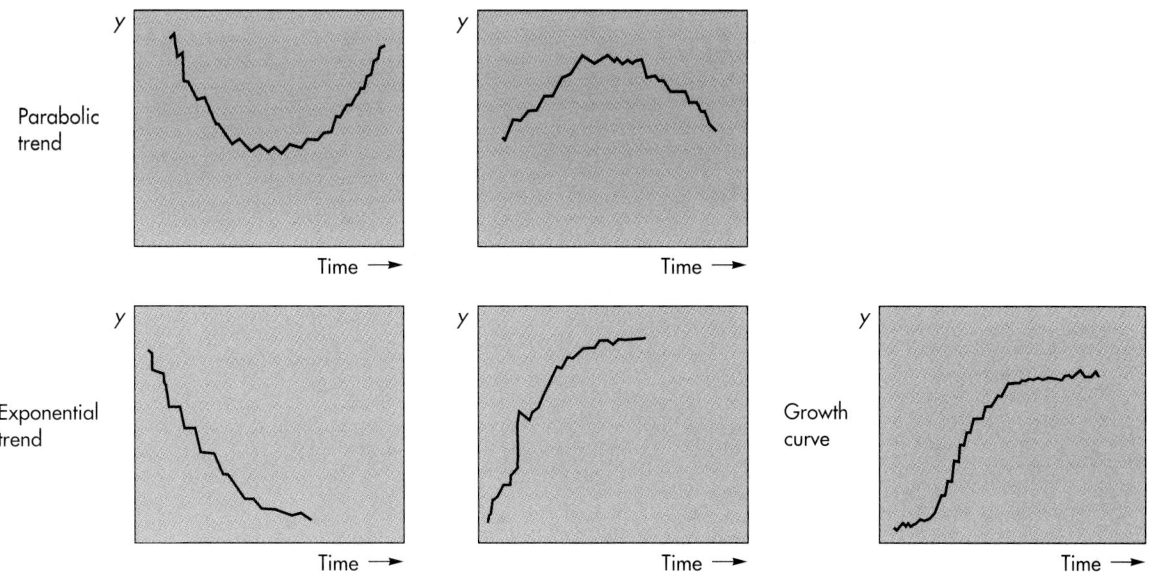

trend. The discussion here focuses exclusively on *linear* trends because these are fairly common and the easiest to work with.

There are two important techniques that can be used to develop forecasts when trend is present. One involves use of a trend equation; the other is an extension of exponential smoothing.

Trend Equation. A **linear trend equation** has the form

linear trend equation $y_t = a + bt$, used to develop forecasts when trend is present.

$$y_t = a + bt \tag{3–3}$$

where

t = Specified number of time periods from $t = 0$
y_t = Forecast for period t
a = Value of y_t at $t = 0$
b = Slope of the line

For example, consider the trend equation $y_t = 45 + 5t$. The value of y_t when $t = 0$ is 45, and the slope of the line is 5, which means that, on the average, the value of y_t will increase by five units for each time period. If $t = 10$, the forecast, y_t is $45 + 5(10) = 95$ units. The equation can be plotted by finding two points on the line. One can be found by substituting some value of t into the equation (e.g., $t = 10$) and then solving for y_t. The other point is a (i.e., y_t at $t = 0$). Plotting those two points and drawing a line through them yields a graph of the linear trend line.

The coefficients of the line, a and b, can be computed from historical data using these two equations:

$$b = \frac{n\Sigma ty - \Sigma t \Sigma y}{n\Sigma t^2 - (\Sigma t)^2} \tag{3–4}$$

$$a = \frac{\Sigma y - b\Sigma t}{n} \text{ or } \bar{y} - b\bar{t} \tag{3–5}$$

TABLE 3–1
Values of $\sum t$ and $\sum t^2$

n	$\sum t$	$\sum t^2$
1	1	1
2	3	5
3	6	14
4	10	30
5	15	55
6	21	91
7	28	140
8	36	204
9	45	285
10	55	385
11	66	506
12	78	650
13	91	819
14	105	1,015
15	120	1,240
16	136	1,496
17	153	1,785
18	171	2,109
19	190	2,470
20	210	2,870

where

n = Number of periods
y = Value of the time series

Note that these three equations are identical to those used for computing a linear regression line, except that t replaces x in the equations. Manual computation of the coefficients of a trend line can be simplified by use of Table 3–1, which lists values of $\sum t$ and $\sum t^2$ for up to 20 periods ($n = 20$).

Example 4

Calculator sales for a California-based firm over the last 10 weeks are shown in the following table. Plot the data, and visually check to see if a linear trend line would be appropriate. Then determine the equation of the trend line, and predict sales for weeks 11 and 12.

Week	Unit Sales
1	700
2	724
3	720
4	728
5	740
6	742
7	758
8	750
9	770
10	775

Solution

a. A plot suggests that a linear trend line would be appropriate:

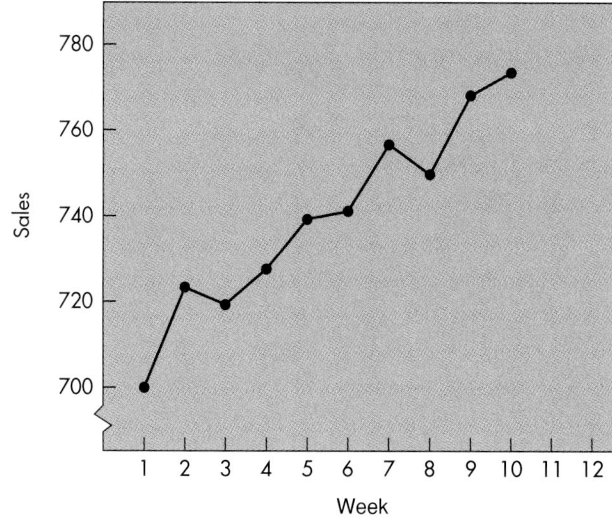

b.

Week (*t*)	*y*	*ty*
1	700	700
2	724	1,448
3	720	2,160
4	728	2,912
5	740	3,700
6	742	4,452
7	758	5,306
8	750	6,000
9	770	6,930
10	775	7,750
	7,407	41,358

From Table 3–1, for $n = 10$, $\Sigma t = 55$ and $\Sigma t^2 = 385$. Using Formulas 3–4 and 3–5, you can compute the coefficients of the trend line:

$$b = \frac{10(41{,}358) - 55(7{,}407)}{10(385) - 55(55)} = \frac{6{,}195}{825} = 7.51$$

$$a = \frac{7{,}407 - 7.51(55)}{10} = 699.40$$

Thus, the trend line is $y_t = 699.40 + 7.51t$, where $t = 0$ for period 0.

c. Substituting values of t into this equation, the forecasts for the next two periods (i.e., $t = 11$ and $t = 12$) are:

$$y_{11} = 699.40 + 7.51(11) = 782.01$$

$$y_{12} = 699.40 + 7.51(12) = 789.51$$

d. For purposes of illustration, the original data, the trend line, and the two projections (forecasts) are shown on the following graph:

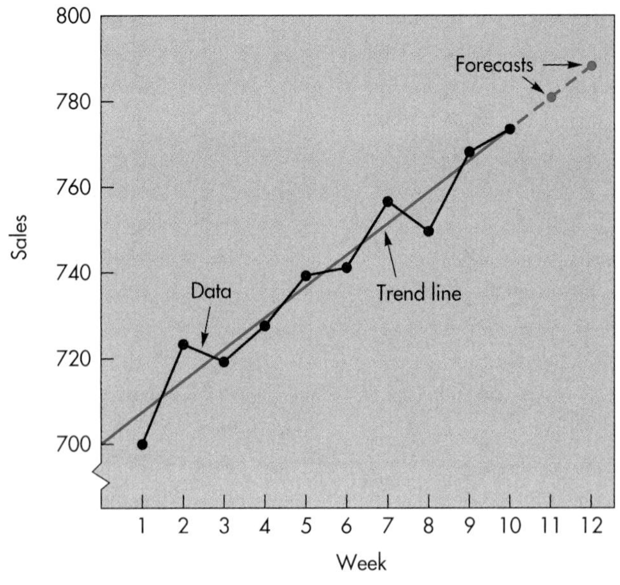

Trend-Adjusted Exponential Smoothing

trend-adjusted exponential smoothing Variation of exponential smoothing used when a time series exhibits trend.

A variation of simple exponential smoothing can be used when a time series exhibits trend. It is called **trend-adjusted exponential smoothing** or, sometimes, *double smoothing,* to differentiate it from simple exponential smoothing, which is appropriate only when data vary around an average or have step or gradual changes. If a series exhibits trend, and simple smoothing is used on it, the forecasts will all lag the trend: if the data are increasing, each forecast will be too low; if decreasing, each forecast will be too high. Again, plotting the data can indicate when trend-adjusted smoothing would be preferable to simple smoothing.

The trend-adjusted forecast (TAF) is composed of two elements: a smoothed error and a trend factor.

$$\text{TAF}_{t+1} = S_t + T_t \tag{3–6}$$

where

S_t = Smoothed forecast
T_t = Current trend estimate

and

$$S_t = \text{TAF}_t + \alpha(\text{A}_t - \text{TAF}_t)$$
$$T_t = T_{t-1} + \beta(\text{TAF}_t - \text{TAF}_{t-1} - T_{t-1}) \tag{3–7}$$

where α and β are smoothing constants. In order to use this method, one must select values of α and β (usually through trial and error) and make a starting forecast and an estimate of trend.

Example calculations are illustrated in Table 3–2. Notice that an initial estimate of trend is estimated from the first four values, and that the starting forecast (period 5) is developed using the previous (period 4) value of 728 plus the initial trend estimate:

Starting forecast = 728 + 9.30 = 737.30

Although manual computations are somewhat more involved for trend-adjusted smoothing than for a linear trend line, trend-adjusted smoothing has the ability to adjust to *changes* in trend. Of course, trend projections are much simpler with a trend line than

	t (*Period*)	A_t (*Actual*)		
	1	700	Trend estimate $= \dfrac{728 - 700}{3} = \dfrac{28}{3} = 9.33$	
Model development	2	724		
	3	720	Starting forecast = 728 + 9.3 = 737.3	
	4	728		
			$\text{TAF}_t + \alpha(A_t - \text{TAF}_t) = S_t$	$T_{t-1} + \beta(\text{TAF}_t - \text{TAF}_{t-1} - T_{t-1}) = T_t$
	5	740	737.3 + .4(740 − 737.3) = 738.38	9.3 + .3(0) = 9.30
	6	742	747.68 + .4(742 − 747.68) = 745.41	9.30 + .3(747.68 − 737.30 − 9.30) = 9.62
Model test	7	758	755.03 + .4(758 − 755.03) = 756.22	9.62 + .3(755.03 − 747.68 − 9.62) = 8.94
	8	750	765.16 + .4(750 − 765.16) = 759.10	8.94 + .3(765.16 − 755.03 − 8.94) = 9.30
	9	770	768.40 + .4(770 − 768.40) = 769.04	9.30 + .3(768.40 − 765.16 − 9.30) = 7.48
	10	775	776.52 + .4(775 − 776.52) = 775.91	7.48 + .3(776.52 − 768.40 − 7.48) = 7.67
Forecast	11		783.58 = 775.91 + 7.67	

TABLE 3-2

Examples of trend-adjusted forecast calculations

with trend-adjusted forecasts, so a manager must decide which benefits are most important when choosing between these two techniques for trend.

Example 5

Using the calculator data from the previous example (where it was concluded that the data exhibited a linear trend), use trend-adjusted exponential smoothing to prepare forecasts for periods 5 through 11, with $\alpha_1 = .4$ and $\alpha_2 = .3$.

Solution

The initial estimate of trend is based on the net change of 28 for the *three* changes from period 1 to period 4, for an average of 9.30. The data and calculations are shown in Table 3–2.

A plot of the actual data and predicted values is shown below.

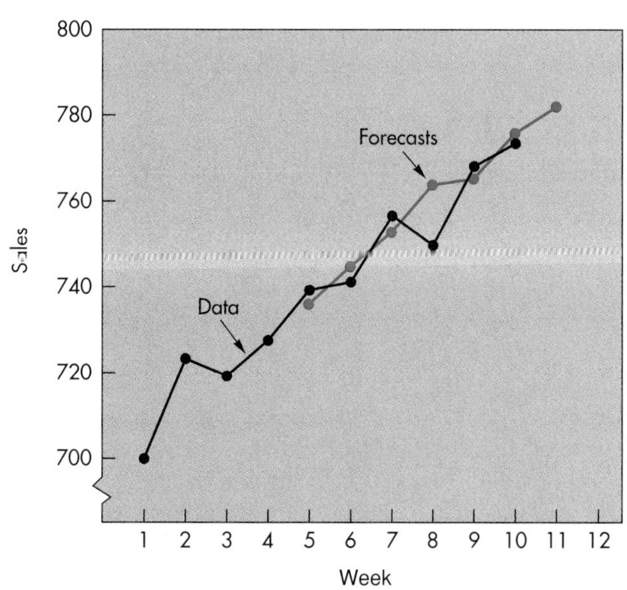

Demand for products such as lawnmowers and snowthrowers is subject to seasonal fluctuations. Toro matches these fluctuations by adjusting its manufacturing capacity for both products.

Techniques for Seasonality

seasonal variations Regularly repeating movements in series values that can be tied to recurring events.

Seasonal variations in time series data are regularly repeating upward or downward movements in series values that can be tied to recurring events. *Seasonality* may refer to regular annual variations. Familiar examples of seasonality are weather variations (e.g., sales of winter and summer sports equipment) and vacations or holidays (e.g., airline travel, greeting card sales, visitors at tourist and resort centers). The term *seasonal variation* is also applied to daily, weekly, monthly, and other regularly recurring patterns in data. For example, rush hour traffic occurs twice a day—incoming in the morning and outgoing in the late afternoon. Theaters and restaurants often experience weekly demand patterns, with demand higher later in the week. Banks may experience daily seasonal variations (heavier traffic during the noon hour and just before closing), weekly variations (heavier toward the end of the week), and monthly variations (heaviest around the end of the month because of social security, payroll, and welfare checks being cashed or deposited). Mail volume; sales of toys, beer, automobiles, and turkeys; highway usage; hotel registrations; and gardening also exhibit seasonal variations.

Seasonality in a time series is expressed in terms of the amount that actual values deviate from the *average* value of a series. If the series tends to vary around an average value, then seasonality is expressed in terms of that average (or a moving average); if trend is present, seasonality is expressed in terms of the trend value.

There are two different models of seasonality: additive and multiplicative. In the *additive* model, seasonality is expressed as a *quantity* (e.g., 20 units), which is added or subtracted from the series average in order to incorporate seasonality. In the *multiplicative* model, seasonality is expressed as a *percentage* of the average (or trend) amount (e.g., 1.10), which is then multiplied by the value of a series to incorporate seasonality. Figure 3–6 illustrates the two models for a linear trend line. In practice, businesses use the multiplicative model much more widely than the additive model, so we shall focus exclusively on the multiplicative model.

The seasonal percentages in the multiplicative model are referred to as *seasonal relatives* or *seasonal indexes.* Suppose that the seasonal relative for the quantity of toys sold in May at a store is 1.20. This indicates that toy sales for that month are 20 percent above the monthly average. A seasonal relative of .90 for July indicates that July sales are 90 percent of the monthly average.

Knowledge of seasonal variations is an important factor in retail planning and scheduling. Moreover, seasonality can be an important factor in capacity planning for systems that must be designed to handle peak loads (e.g., public transportation, electric power plants, highways, and bridges). Knowledge of the extent of seasonality in a time series can enable one to *remove* seasonality from the data (i.e., to seasonally adjust data) in order to discern other patterns or the lack of patterns in the series. Thus, one frequently

Demand
Additive model
Demand = Trend + Seasonality
Multiplicative model
Demand = Trend × Seasonality
Time

FIGURE 3–6

Seasonality: the additive and multiplicative models compared using a linear trend

FIGURE 3–7

Naive approaches with seasonality

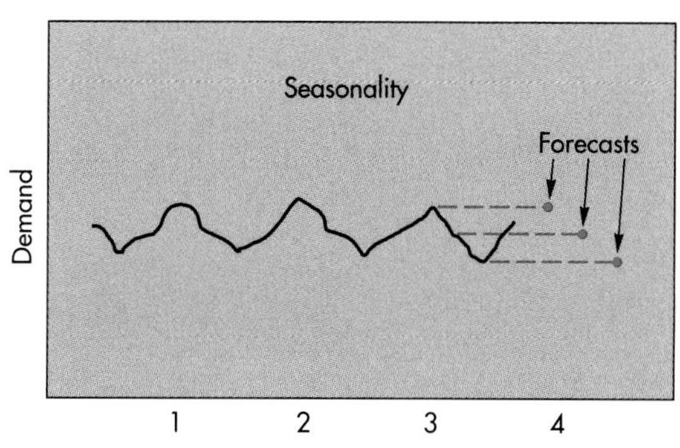

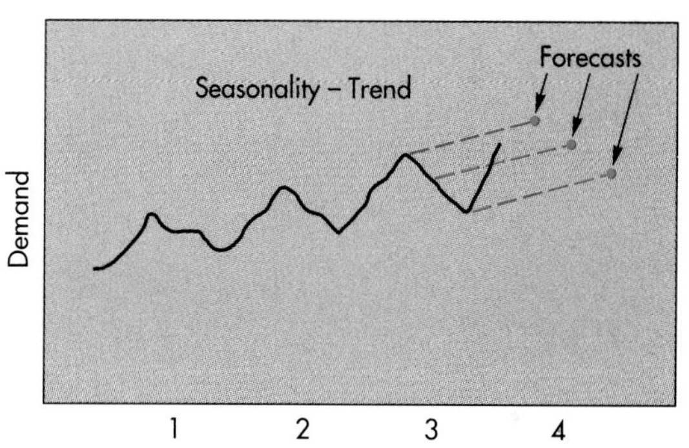

reads or hears about "seasonally adjusted unemployment" and "seasonally adjusted personal income."

The simplest seasonal model is a variation of the naive technique described for averages. Instead of using the actual demand of the last *period* as the forecast amount, the seasonal naive model uses the actual amount of the last *season* for the forecast. A forecast of theater attendance this Friday evening would equal last Friday's attendance using the seasonal naive model, and estimates of November toy sales in one year could be based on toy sales in the preceding November. If trend and seasonal variations are present, the naive model might be the demand of the last season plus 10 percent, or whatever increase or decrease is appropriate. (See Figure 3–7.)

Again, the naive approach can either be used alone or serve as a standard of comparison against which other, more refined techniques can be judged.

The next section briefly describes how seasonal relatives are used, and the following section describes how seasonal relatives are computed.

Using Seasonal Relatives. Seasonal relatives are used in two different ways in forecasting. One way is to *deseasonalize* data; the other way is to *incorporate seasonality* in a forecast.

To deseasonalize data is to remove the seasonal component from the data in order to get a clearer picture of the nonseasonal (e.g., trend) components. Deseasonalizing data is accomplished by *dividing* each data point by its corresponding seasonal relative (e.g., divide November demand by the November relative, divide December demand by the December relative, and so on).

Incorporating seasonality in a forecast is useful when demand has both trend (or average) and seasonal components. Incorporating seasonality can be accomplished in this way:

1. Obtain trend estimates for desired periods using a trend equation.
2. Add seasonality to the trend estimates by *multiplying* (assuming a multiplicative model is appropriate) these trend estimates by the corresponding seasonal relative (e.g., multiply the November trend estimate by the November seasonal relative, multiply the December trend estimate by the December seasonal relative, and so on).

Example 6 illustrates.

Example 6

A furniture manufacturer wants to predict quarterly demand for a certain loveseat for periods 15 and 16, which happen to be the second and third quarters of a particular year. The series consists of both trend and seasonality. The trend portion of demand is projected using the equation $y_t = 124 + 7.5t$. Quarter relatives are $Q_1 = 1.20$, $Q_2 = 1.10$, $Q_3 = 0.75$, and $Q_4 = 0.95$. Use this information to predict demand for periods 15 and 16.

Solution

The trend values at $t = 15$ and $t = 16$ are:

$$y_{15} = 124 + 7.5(15) = 236.5$$

$$y_{16} = 124 + 7.5(16) = 244.0$$

Multiplying the trend value by the appropriate quarter relative yields a forecast that includes both trend and seasonality. Given that $t = 15$ is a second quarter and $t = 16$ is a third quarter, the forecasts are:

Period 15: 236.5(1.10) = 260.15

Period 16: 244.0(0.75) = 183.00

centered moving average A moving average positioned at the center of the data that were used to compute it.

Computing Seasonal Relatives. A commonly used method for representing the trend portion of a time series involves a **centered moving average.** Computations and the resulting values are the same as those for a moving average forecast. However, the values are not projected as in a forecast instead, they are *positioned in the middle* of the periods used to compute the moving average. The implication is that the average is most representative of that point in the series. For example, assume the following time series data:

Period	Demand	Three-Period Centered Average	
1	40		
2	46	42.67	$\text{Average} = \frac{40 + 46 + 42}{3} = 42.67$
3	42		

The three-period average is 42.67. As a centered average, it is positioned at period 2; the average is most representative of the series at that point.

The ratio of demand at period 2 to this centered average at period 2 is an estimate of the seasonal relative at that point. Because the ratio is 46/42.67 = 1.08, the series is about 8 percent above average at that point. To achieve a reasonable estimate of seasonality for any season (e.g., Friday attendance at a theater), it is usually necessary to compute seasonal ratios for a number of seasons and then average these ratios. In the case of theater attendance, average the ratios of five or six Fridays for the Friday relative, average five or six Saturdays for the Saturday relative, and so on.

Example 7

The manager of a parking lot has computed daily relatives for the number of cars per day in the lot. The computations are repeated here (about three weeks are shown for illustration). A seven-period centered moving average is used because there are seven days (seasons) per week.

Day	Volume	Moving Total	Centered MA_7	Volume/MA
Tues	67			
Wed	75			
Thur	82			
Fri	98		71.86	98/71.86 = 1.36 (Friday)
Sat	90		70.86	90/70.86 = 1.27
Sun	36		70.57	36/70.57 = 0.51
Mon	55	503 ÷ 7 =	71.00	55/71.00 = 0.77
Tues	60	496 ÷ 7 =	71.14	60/71.14 = 0.84 (Tuesday)
Wed	73	494 etc.	70.57	73/70.57 = 1.03
Thur	85	497	71.14	85/71.14 = 1.19
Fri	99	498	70.71	99/70.71 = 1.40 (Friday)
Sat	86	494	71.29	86/71.29 = 1.21
Sun	40	498	71.71	40/71.71 = 0.56
Mon	52	495	72.00	52/72.00 = 0.72
Tues	64	499	71.57	64/71.57 = 0.89 (Tuesday)
Wed	76	502	71.86	76/71.86 = 1.06
Thur	87	504	72.43	87/72.43 = 1.20
Fri	96	501	72.14	96/72.14 = 1.33 (Friday)
Sat	88	503		
Sun	44	507		
Mon	50	505		

The estimated Friday relative is (1.36 + 1.40 + 133)/3 = 1.36. Relatives for other days can be computed in a similar manner. For example, the estimated Tuesday relative is: (0.84 + 0.89)/2 = 0.87.

The number of periods needed in a centered moving average is equal to the number of "seasons" involved. For example, with monthly data, a 12-period moving average is needed. When the number of periods is even, one additional step is needed because the middle of an even set falls between two periods. The additional step requires taking a centered two-period moving average of the even-numbered centered moving average, which results in averages that "line up" with data points and, hence, permit determination of seasonal ratios. (See Solved Problem 5 at the end of this chapter for an example.)

A centered moving average is used to obtain representative values because by virtue of its centered position—it "looks forward" and "looks backward"—it is able to closely follow data movements whether they involve trends, cycles, or random variability alone. Figure 3–8 illustrates how a three-period centered moving average closely tracks the data originally shown in Figure 3–3.

Techniques for Cycles

Cycles are up and down movements similar to seasonal variations but of longer duration—say, two to six years between peaks. When cycles occur in time series data, their frequent irregularity makes it difficult or impossible to project them from past data because turning points are difficult to identify. A short moving average or a naive approach may be of some value, although both will produce forecasts that lag cyclical movements by one or several periods.

The most commonly used approach is explanatory: Search for another variable that relates to, and *leads*, the variable of interest. For example, the number of housing starts

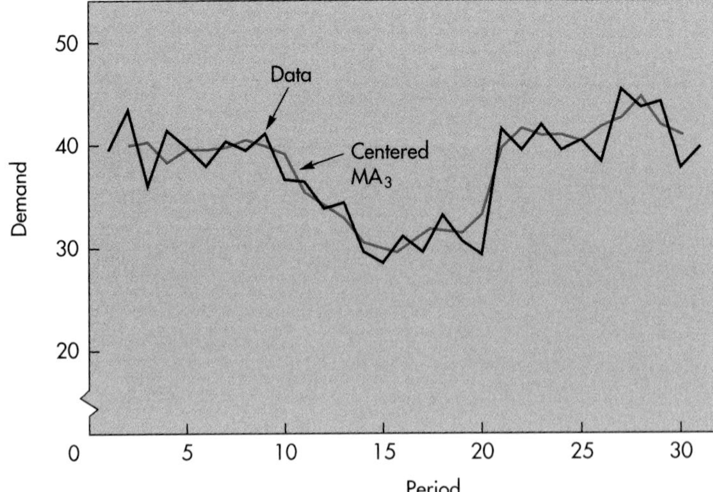

FIGURE 3-8

A centered moving average closely tracks the data

(i.e., permits to build houses) in a given month often is an indicator of demand a few months later for products and services directly tied to construction of new homes (landscaping; sales of washers and dryers, carpeting, and furniture; new demands for shopping, transportation, schools). Thus, if an organization is able to establish a high correlation with such a *leading variable* (i.e., changes in the variable precede changes in the variable of interest), it can develop an equation that describes the relationship, enabling forecasts to be made. It is important that a persistent relationship exists between the two variables. Moreover, the higher the correlation, the better the chances that the forecast will be on target.

ASSOCIATIVE FORECASTING TECHNIQUES

Associative techniques rely on identification of related variables that can be used to predict values of the variable of interest. For example, sales of beef may be related to the price per pound charged for beef and the prices of substitutes such as chicken, pork, and lamb; real estate prices are usually related to property location; and crop yields are related to soil conditions and the amounts and timing of water and fertilizer applications.

The essence of associative techniques is the development of an equation that summarizes the effects of **predictor variables.** The primary method of analysis is known as **regression.** A brief overview of regression should suffice to place this approach into perspective relative to the other forecasting approaches described in this chapter.

predictor variables Variables that can be used to predict values of the variable of interest.

regression Technique for fitting a line to a set of points.

Simple Linear Regression

The simplest and most widely used form of regression involves a linear relationship between two variables. A plot of the values might appear like that in Figure 3–9. The object in linear regression is to obtain an equation of a straight line that minimizes the sum of squared vertical deviations of data points from the line. This **least squares line** has the equation

least squares line Minimizes the sum of the squared deviations around the line.

$$y_c = a + bx \qquad (3\text{–}8)$$

where

y_c = Predicted (dependent) variable
x = Predictor (independent) variable
b = Slope of the line
a = Value of y_c when $x = 0$ (i.e., the height of the line at the y intercept)

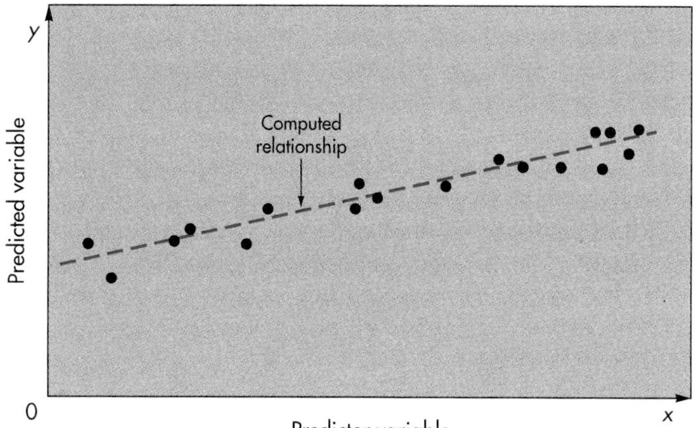

FIGURE 3-9

A straight line is fitted to a set of sample points

(Note: It is conventional to represent values of the predicted variable on the y axis and values of the predictor variable on the x axis.) Figure 3–10 shows a graph of a linear regression line.

The coefficients a and b of the line are computed using these two equations:

$$b = \frac{n(\Sigma xy) - (\Sigma x)(\Sigma y)}{n(\Sigma x^2) - (\Sigma x)^2} \tag{3–9}$$

$$a = \frac{\Sigma y - b\Sigma x}{n} \text{ or } \bar{y} - b\bar{x} \tag{3–10}$$

where

n = Number of paired observations

Example 8

Healthy Hamburgers has a chain of 12 stores in northern Illinois. Sales figures and profits for the stores are given in the following table. Obtain a regression line for the data, and predict profit for a store assuming sales of $10 million.

Sales, x	Profits, y
(in millions of dollars)	
$ 7	$0.15
2	0.10
6	0.13
4	0.15
14	0.25
15	0.27
16	0.24
12	0.20
14	0.27
20	0.44
15	0.34
7	0.17

Solution

First, plot the data and decide if a linear model is reasonable (i.e., do the points seem to scatter around a straight line? Figure 3–11 suggests they do). Next, compute the quantities Σx, Σy, Σxy, and Σx^2. Calculations are shown for these quantities in Table 3–3. One additional calculation, Σy^2, is included for later reference.

FIGURE 3-10

Equation of a straight line

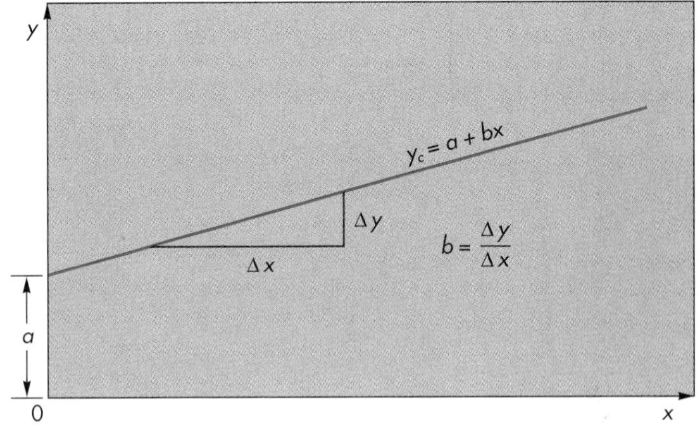

The line intersects the y axis where $y = a$. The slope of the line $= b$.

FIGURE 3-11

A linear model seems reasonable

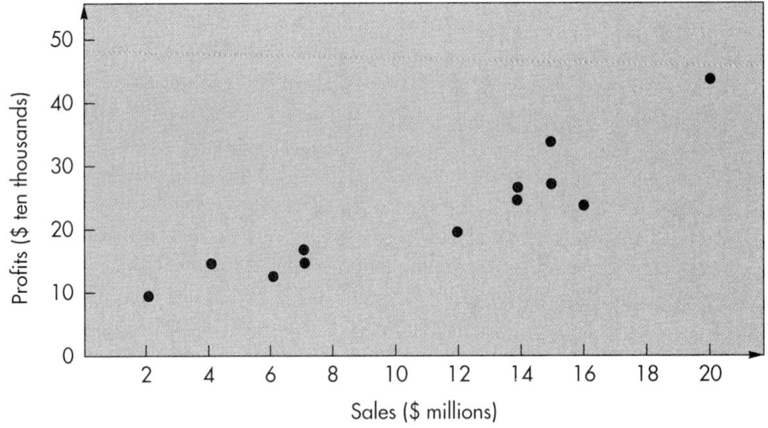

TABLE 3-3

Calculations for regression coefficients for Healthy Hamburgers

x	y	xy	x^2	y^2
7	0.15	1.05	49	0.0225
2	0.10	0.20	4	0.0100
6	0.13	0.78	36	0.0169
4	0.15	0.60	16	0.0225
14	0.25	3.50	196	0.0625
15	0.27	4.05	225	0.0729
16	0.24	3.84	256	0.0576
12	0.20	2.40	144	0.0400
14	0.27	3.78	196	0.0729
20	0.44	8.80	400	0.1936
15	0.34	5.10	225	0.1156
7	0.17	1.19	49	0.0289
132	2.71	35.29	1,796	0.7159

Substituting into the equation, you find:

$$b = \frac{n(\Sigma xy) - (\Sigma x)(\Sigma y)}{n(\Sigma x^2) - (\Sigma x)^2} = \frac{12(35.29) - 132(2.71)}{12(1{,}796) - 132(132)} = 0.01593$$

$$a = \frac{\Sigma y - b(\Sigma x)}{n} = \frac{2.71 - 0.01593(132)}{12} = 0.0506$$

Thus, the regression equation is: $y_c = 0.0506 + 0.01593x$. For sales of $x = 10$ (i.e., \$10 million), estimated profit is: $y_c = 0.0506 + 0.01593(10) = 0.2099$, or \$209,900. (It may appear strange that substituting $x = 0$ into the equation produces a predicted profit of \$50,600 because it seems to suggest that amount of profit will occur with no sales. However, the value of $x = 0$ is *outside of the range of observed values.* The regression line should only be used for the range of values from which it was developed; the relationship may be nonlinear outside that range. The purpose of the *a* value is simply to establish the height of the line where it crosses the *y* axis.)

One application of regression in forecasting relates to the use of indicators. These are uncontrollable variables that tend to lead or precede changes in a variable of interest. For example, changes in the Federal Reserve Board's discount rate may influence certain business activities. Similarly, an increase in housing starts during the spring and summer months may lead to increased demand for appliances, carpets, furniture, and similar items in the fall and winter months. Careful identification and analysis of indicators may yield insight into possible future demand in some situations. There are numerous published indexes from which to choose.[1]

Net change in inventories on hand and on order.

Interest rates for commercial loans.

Industrial output.

Consumer price index (CPI).

The wholesale price index.

Stock market prices.

Other potential indicators are population shifts, local political climates, and activities of other firms (e.g., the opening of a shopping center may result in increased sales for nearby businesses). Three conditions are required for an indicator to be valid:

1. The relationship between movements of an indicator and movements of the variable should have a logical explanation.
2. Movements of the indicator must precede movements of the dependent variable by enough time so that the forecast isn't outdated before it can be acted upon.
3. A fairly high correlation should exist between the two variables.

Correlation measures the strength and direction of relationship between two variables. Correlation can range from −1.00 to +1.00. A correlation of +1.00 indicates that changes in one variable are always matched by changes in the other; a correlation of −1.00 indicates that increases in one variable are matched by decreases in the other; and a correlation close to zero indicates little *linear* relationship between two variables. The correlation between two variables can be computed using the equation

correlation A measure of the strength and direction of relationship between two variables.

$$r = \frac{n(\Sigma xy) - (\Sigma x)(\Sigma y)}{\sqrt{n(\Sigma x^2) - (\Sigma x)^2} \cdot \sqrt{n(\Sigma y^2) - (\Sigma y)^2}} \qquad (3\text{–}11)$$

The square of the correlation coefficient, r^2, provides a measure of how well a regression line "fits" the data. The possible values of r^2 range from 0 to 1.00. The closer r^2 is to 1.00, the better the fit; r^2 indicates the percentage of variation in the dependent variable (the one we are trying to predict) that can be accounted for in our model by the independent variable. A high value of r^2, say .80 or more, would indicate that the independent variable is a good predictor of values of the dependent variable. A low value, say .25 or less, would indicate a poor predictor, and a value between .25 and .80 would indicate a moderate predictor.

[1]See, for example, *The National Bureau of Economic Research, The Survey of Current Business, The Monthly Labor Review,* and *Business Conditions Digest.*

Comments on the Use of Linear Regression Analysis

Use of simple regression analysis implies that certain assumptions have been satisfied. Basically, these are:

1. Variations around the line are random. If they are random, no patterns such as cycles or trends should be apparent when the line and data are plotted.
2. Deviations around the line should be normally distributed. A concentration of values close to the line with a small proportion of larger deviations supports the assumption of normality.
3. Predictions are being made only within the range of observed values.

If the assumptions are satisfied, regression analysis can be a powerful tool. Particularly useful are the confidence intervals for predicted values. To obtain the best results, observe the following:

1. Always plot the data to verify that a linear relationship is appropriate.
2. The data may be time-dependent. Check this by plotting the dependent variable versus time; if patterns appear, use analysis of time series instead of regression, or use time as an independent variable as part of a *multiple regression analysis.*
3. A small correlation may imply that other variables are important.

In addition, note these weaknesses of regression:

1. Simple linear regression applies only to linear relationships with *one* independent variable.
2. One needs a considerable amount of data to establish the relationship—in practice, 20 or more observations.
3. All observations are weighted equally.

Example 9

Sales of 19-inch color television sets and three-month lagged unemployment are shown in the following table. Determine if unemployment levels can be used to predict demand for 19-inch color TVs and, if so, derive a predictive equation.

Period	1	2	3	4	5	6	7	8	9	10	11
Units sold	20	41	17	35	25	31	38	50	15	19	14
Unemployment % (three-month lag)	7.2	4.0	7.3	5.5	6.8	6.0	5.4	3.6	8.4	7.0	9.0

Solution

1. Plot the data to see if a linear model seems reasonable. In this case, a linear model seems appropriate *for the range of the data.*

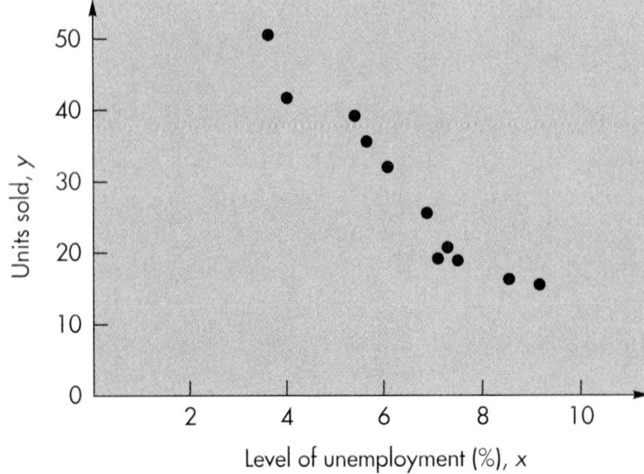

2. Compute the correlation coefficient to confirm that it is not close to zero.

x	y	xy	x^2	y^2
7.2	20	144.0	51.8	400
4.0	41	164.0	16.0	1,681
7.3	17	124.1	53.3	289
5.5	35	192.5	30.3	1,225
6.8	25	170.0	46.2	625
6.0	31	186.0	36.0	961
5.4	38	205.2	29.2	1,444
3.6	50	180.0	13.0	2,500
8.4	15	126.0	70.6	225
7.0	19	133.0	49.0	361
9.0	14	126.0	81.0	196
70.2	305	1,750.8	476.4	9,907

$$r = \frac{11(1{,}750.8) - 70.2(305)}{\sqrt{11(476.4) - (70.2)^2} \cdot \sqrt{11(9{,}907) - (305)^2}} = -.966$$

This is a fairly high negative correlation.

3. Compute the regression line:

$$b = \frac{11(1{,}750.8) - 70.2(305)}{11(476.4) - 70.2(70.2)} = -6.91$$

$$a = \frac{305 - (-6.9145)(70.2)}{11} = 71.85$$

$$y = 71.85 - 6.89x$$

Note that the equation pertains only to unemployment levels in the range 3.6 to 9.0, because sample observations covered only that range.

Curvilinear and Multiple Regression Analysis

Simple linear regression may prove inadequate to handle certain problems because a linear model is inappropriate or because more than one predictor variable is involved. When nonlinear relationships are present, you should employ curvilinear regression; models that involve more than one predictor require the use of multiple regression analysis. While these analyses are beyond the scope of this text, you should be aware that they are often used. The computations lend themselves more to computers than to hand calculation. Multiple regression forecasting substantially increases data requirements. In each case, it is necessary to weight the additional cost and effort against potential improvements in accuracy of predictions.

ACCURACY AND CONTROL OF FORECASTS

Accuracy and control of forecasts is a vital aspect of forecasting. The complex nature of most real-world variables makes it almost impossible to correctly predict future values of those variables on a regular basis. Consequently, it is important to include an indication of the extent to which the forecast might deviate from the value of the variable that actually occurs. This will provide the forecast user with a better perspective on how far off a forecast might be.

Moreover, because some techniques provide more accuracy than others in a given situation, the decision maker needs a measure of accuracy to use as a basis for comparison when choosing among different techniques.

Finally, some forecasting applications involve a series of forecasts (e.g., weekly revenues), whereas others involve a single forecast that will be used for a one-time decision (e.g., the size of a power plant). When making periodic forecasts, it is important to monitor forecast errors to determine if the errors are within reasonable bounds. If they are not, it will be necessary to take corrective action.

error Difference between the actual value and the value that was predicted for a given period.

Forecast **error** is the difference between the value that occurs and the value that was predicted for a given time period. Hence, Error = Actual − Forecast:

$$e_t = A_t - F_t \quad (3\text{–}12)$$

Positive errors result when the forecast is too low, negative errors when the forecast is too high. For example, if actual demand for a week is 100 units and forecast demand was 90 units, the forecast was too low; the error is 100 − 90 = +10.

Forecast errors influence decisions in two somewhat different ways. One is in making a choice between various forecasting alternatives, and the other is in evaluating the success or failure of a technique in use. We shall begin by examining ways to summarize forecast error over time, and see how that information can be applied to a choice among forecasting alternatives. Then we shall consider methods for controlling forecasts.

Summarizing Forecast Accuracy

Two aspects of forecast accuracy can have potential significance when deciding among forecasting alternatives. One is the historical error performance of a forecast, and the other is the ability of a forecast to respond to changes.

mean absolute deviation The average absolute error.

mean squared error The average of squared errors.

Two commonly used measures for summarizing historical errors are the **mean absolute deviation (MAD)** and the **mean squared error (MSE).** MAD is the average absolute error, and MSE is the average of squared errors. The formulas used to compute MAD[2] and MSE are:

$$\text{MAD} = \frac{\Sigma|\text{Actual} - \text{forecast}|}{n} \quad (3\text{–}13)$$

$$\text{MSE} = \frac{\Sigma(\text{Actual} - \text{forecast})^2}{n - 1} \quad (3\text{–}14)$$

Example 10 illustrates the computation of MAD and MSE.

Example 10

Compute MAD and MSE for the following data.

Period	Actual	Forecast	(A − F) Error	|Error|	Error²
1	217	215	2	2	4
2	213	216	−3	3	9
3	216	215	1	1	1
4	210	214	−4	4	16
5	213	211	2	2	4
6	219	214	5	5	25
7	216	217	−1	1	1
8	212	216	−4	4	16
			−2	22	76

[2]The absolute value, represented by the two vertical lines in Formula 3–13, ignores minus signs; all data are treated as positive values. For example, −2 becomes +2.

Solution

Using the figures shown in the table,

$$\text{MAD} = \frac{\Sigma|e|}{n} = \frac{22}{8} = 2.75$$

$$\text{MSE} = \frac{\Sigma e^2}{n-1} = \frac{76}{8-1} = 10.86$$

From a computational standpoint, the difference between these two measures is that MAD weights all errors evenly, and MSE weights errors according to their *squared* values.

One use for these measures is to compare alternative forecasting methods. For instance, using either MAD or MSE, a manager could compare the results of exponential smoothing with values of .1, .2, and .3, and select the one that yields the *lowest* MAD or MSE for a given set of data.

In some instances, historical error performance is secondary to the ability of a forecast to respond to changes in data patterns. Choice among alternative methods would then focus on the cost of not responding quickly to a change relative to the cost of responding to changes that are not really there (i.e., random fluctuations).

Overall, the operations manager must settle on the relative importance of historical performance versus responsiveness and whether to use MAD or MSE to measure historical performance.

Controlling the Forecast

It is necessary to monitor forecast errors to ensure that the forecast is performing adequately. This is accomplished by comparing forecast errors to predetermined values, or *action limits,* as illustrated in Figure 3–12. Errors that fall within the limits are judged acceptable, and errors outside of either limit signal that corrective action is needed.

There are a variety of possible sources of forecast errors, including the following:

1. The model may be inadequate due to (*a*) the omission of an important variable, (*b*) a change or shift in the variable that the model cannot deal with (e.g., sudden appearance of a trend or cycle), or (*c*) the appearance of a new variable (e.g., new competitor).
2. Irregular variations may occur due to severe weather or other natural phenomena, temporary shortages or breakdowns, catastrophes, or similar events.
3. The forecasting technique may be used incorrectly or the results misinterpreted.
4. There are always random variations in the data. Randomness is the inherent variation that remains in the data after all causes of variation have been accounted for.

A forecast is generally deemed to perform adequately when the errors exhibit only random variations. Hence, the key to judging when to reexamine the validity of a particular forecasting technique is whether forecast errors are random. If they are not random, investigate to determine which of the other sources is present and how to correct the problem.

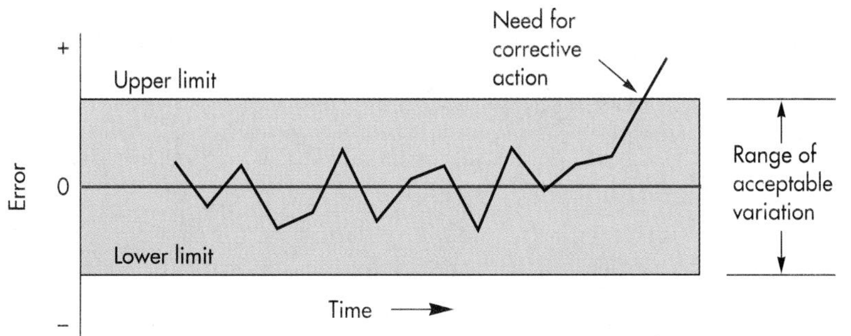

FIGURE 3–12

Monitoring forecast errors

FIGURE 3–13

Conceptual representation of a control chart

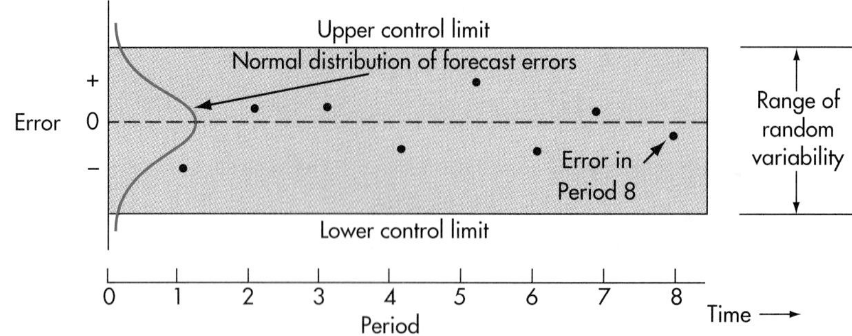

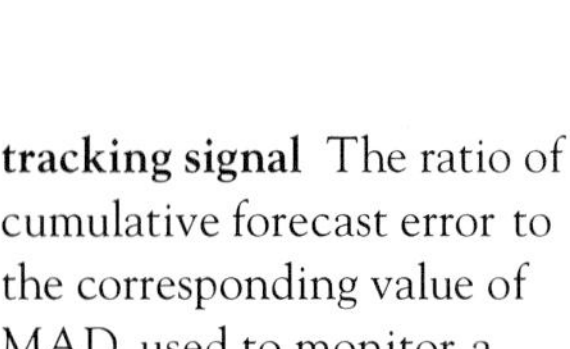

tracking signal The ratio of cumulative forecast error to the corresponding value of MAD, used to monitor a forecast.

Forecasts can be monitored using either *tracking signals* or *control charts*. A **tracking signal** focuses on the ratio of *cumulative* forecast error to the corresponding value of MAD:

$$\text{Tracking signal} = \frac{\Sigma(\text{Actual} - \text{Forecast})}{\text{MAD}} \quad (3\text{–}15)$$

bias Persistent tendency for forecasts to be greater or less than the actual values of a time series.

The cumulative forecast error reflects the **bias** in forecasts, which is the persistent tendency for forecasts to be greater or less than the actual values of a time series.

Tracking signal values are compared to predetermined limits based on judgment and experience. They often range from ± 3 to ± 8; for the most part, we shall use limits of ± 4, which are roughly comparable to three standard deviation limits. Values within the limits suggest—but do not guarantee—that the forecast is performing adequately. After an initial value of MAD has been computed, MAD can be updated using exponential smoothing:

$$\text{MAD}_t = \text{MAD}_{t-1} + \alpha(|\text{Actual} - \text{Forecast}|_t - \text{MAD}_{t-1}) \quad (3\text{–}16)$$

control chart Monitoring approach that sets limits for individual forecast errors; the limits are multiples of the square root of MSE.

The **control chart** approach involves setting upper and lower limits for *individual* forecast errors (instead of cumulative errors, as is the case with a tracking signal). The limits are multiples of the square root of MSE. This method assumes the following:

1. Forecast errors are randomly distributed around a mean of zero.
2. The distribution of errors is normal. See Figure 3–13.

The square root of MSE is used in practice as an estimate of the standard deviation of the distribution of errors.[3] That is,

$$s = \sqrt{\text{MSE}} \quad (3\text{–}17)$$

Recall that for a normal distribution, approximately 95 percent of the values (errors in this case) can be expected to fall within limits of $0 \pm 2s$, and approximately 99.7 percent of the values can be expected to fall within $\pm 3s$ of zero. Hence, if the forecast is "in control," 99.7 percent or 95 percent of the errors should fall within the limits, depending on whether $3s$ or $2s$ limits are used. Points that fall outside these limits should be regarded as evidence that corrective action is needed (i.e., the forecast is not performing adequately).

Example 11

Monthly sales of leather jackets at the Lucky Leather Shoppe for the past 24 months, and forecasts and errors for those months, are shown in the following table. Determine if the forecast is working using these approaches:

1. A tracking signal, beginning with month 10, updating MAD with exponential smoothing. Use limits of ± 4 and $\alpha = .2$.
2. A control chart with $2s$ limits. Use data from the first eight months to develop the control chart, then evaluate the remaining data with the control chart.

---

[3]The actual value could be computed as $s = \sqrt{\frac{\Sigma(e - \bar{e})^2}{n - 1}}$.

Month	A (sales)	F (forecast)	A − F (error)	|e|	Cumulative |e|
1	47	43	4	4	4
2	51	44	7	7	11
3	54	50	4	4	15
4	55	51	4	4	19
5	49	54	−5	5	24
6	46	48	−2	2	26
7	38	46	−8	8	34
8	32	44	−12	12	46
9	25	35	−10	10	56
10	24	26	−2	2	58
11	30	25	5	5	
12	35	32	3	3	
13	44	34	10	10	
14	57	50	7	7	
15	60	51	9	9	
16	55	54	1	1	
17	51	55	−4	4	
18	48	51	−3	3	
19	42	50	−8	8	
20	30	43	−13	13	
21	28	38	−10	10	
22	25	27	−2	2	
23	35	27	8	8	
24	38	32	6	6	
			−11		

Solution

a. The sum of absolute errors through the 10th month is 58. Hence, the initial MAD is 58/10 = 5.8. The subsequent MADs are updated using the formula $\text{MAD}_{new} = \text{MAD}_{old} + \alpha(|e| - \text{MAD}_{old})$. The results are shown in the following table.

The tracking signal for any month is:

$$\frac{\text{Cumulative error at that month}}{\text{Updated MAD at that month}}$$

t (month)	|e|	$\text{MAD}_t = \text{MAD}_{t-1} + .2(\|e\| - \text{MAD}_{t-1})$	Cumulative Error	Tracking Signal = Cumulative $\text{Error}_t \div \text{MAD}_t$
10			−20	−20/5.800 = −3.45
11	5	5.640 = 5.8 + .2(5 − 5.8)	−15	−15/5.640 = −2.66
12	3	5.112 = 5.640 + .2(3 − 5.64)	−12	−12/5.112 = −2.35
13	10	6.090 = 5.112 + .2(10 − 5.112)	−2	−2/6.090 = −0.33
14	7	6.272 = 6.090 + .2(7 − 6.090)	5	5/6.272 = 0.80
15	9	6.818 = 6.272 + .2(9 − 6.272)	14	14/6.818 = 2.05
16	1	5.654 = 6.818 + .2(1 − 6.818)	15	15/5.654 = 2.65
17	4	5.323 = 5.654 + .2(4 − 5.654)	11	11/5.323 = 2.07
18	3	4.858 = 5.323 + .2(3 − 5.323)	8	8/4.858 = 1.65
19	8	5.486 = 4.858 + .2(8 − 4.858)	0	0/5.486 = 0.00
20	13	6.989 = 5.486 + .2(13 − 5.486)	−13	−13/6.989 = −1.86
21	10	7.591 = 6.989 + .2(10 − 6.989)	−23	−23/7.591 = −3.03
22	2	6.473 = 7.591 + .2(2 − 7.591)	−25	−25/6.473 = −3.86
23	8	6.778 = 6.473 + .2(8 − 6.473)	−17	−17/6.778 = −2.51
24	6	6.622 = 6.778 + .2(6 − 6.778)	−11	−11/6.622 = −1.66

Because the tracking signal is within ±4 every month, there is no evidence of a problem.

b. (1) Make sure that the average error is approximately zero, because a large average would suggest a biased forecast.

$$\text{Average error} = \frac{\sum \text{errors}}{n} = \frac{-11}{24} = -0.46 \text{ [OK]}$$

(2) Compute the standard deviation:

$$s = \sqrt{\frac{\Sigma e^2}{n-1}}$$

$$= \sqrt{\frac{4^2 + 7^2 + 4^2 + 4^2 + (-5)^2 + (-2)^2 + (-8)^2 + (-12)^2}{8-1}} = 6.91$$

(3) Determine 2*s* control limits:

$$0 \pm 2s = 0 \pm 2(6.91) = -13.82 \text{ to } +13.82$$

(4) i. Check that all errors are within the limits. (They are.)

ii. Plot the data (see the following graph), and check for nonrandom patterns. Note the strings of positive and negative errors. This suggests nonrandomness (and that an improved forecast is possible). The tracking signal did not reveal this.

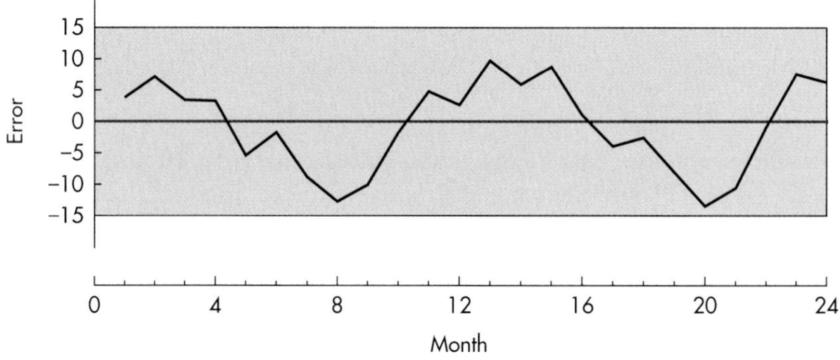

Plotting the errors with the help of a control chart can be very informative. A plot helps you to visualize the process and enables you to check for possible patterns *within the limits* that suggest an improved forecast is possible.[4]

Like the tracking signal, a control chart focuses attention on deviations that lie outside predetermined limits. With either approach, however, it is desirable to check for possible patterns in the errors, even if all errors are within the limits. Figure 3–14 illustrates some of the most common patterns. Checking is usually done by visual inspection although statistical tests are sometimes used. If there is a pattern, this means that errors are *predictable* and, thus, nonrandom. The implication is that the forecast can be improved. For example, trend in the errors means the errors are getting progressively worse. In a forecast based on time series data, adding or modifying a trend component may be needed. In an explanatory model, recomputing the slope or some other adjustment may be called for.

Incorporating the needed changes in the forecasting model will result in less variability in forecast errors, and thus, in narrower control limits. (Revised control limits

[4]The theory and application of control charts and the various methods for detecting patterns in the data are covered in more detail in Chapter 10, on quality control.

FIGURE 3-14

Examples of possible patterns

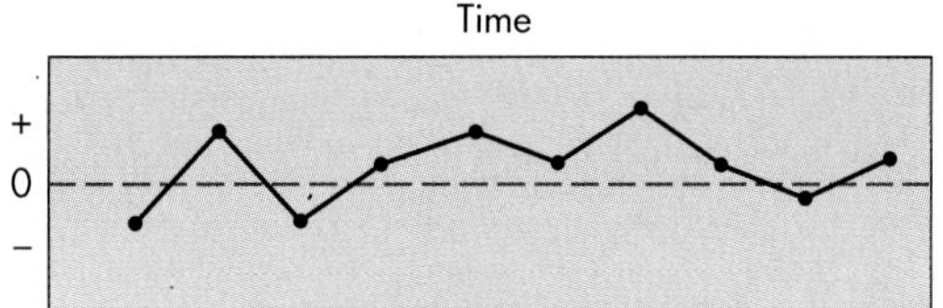

Bias (too many observations on one side of the zero line)

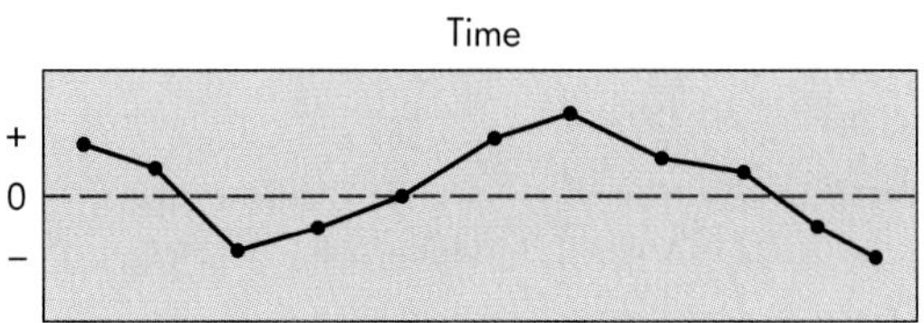

Cycling (periodic upward and downward movement)

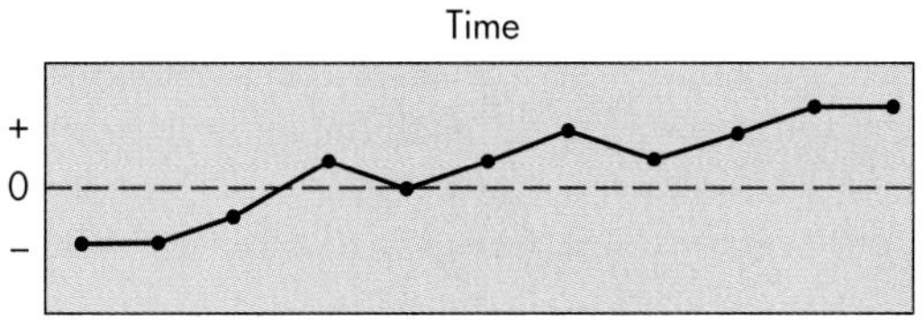

Trend (a persistent upward or downward movement)

NEWSCLIP

FORECASTING

Floods, hurricanes, earthquakes, and tornadoes cause many millions of dollars in property damage each year and take countless lives. Scientists around the world are working to develop new and improved ways of predicting these natural disasters, ways that will allow time for residents to be safely evacuated, if need be, and perhaps time to secure their property. Airports are paying special attention to wind shears, which are microbursts of wind that can be devastating for landings or takeoffs.

The tremendous toll in lost lives and property damage underscores the importance of forecasting, and some of the benefits that good forecasts can provide.

So far, efforts to predict flooding and hurricanes have been the most successful. In the case of flooding, forecasters obtain information from weather forecasts, topography, degree of ground saturation, and other factors and use computer models to predict the timing and degree of flooding that will occur. Weather forecasts use satellite data, and ground and aircraft radar to track the progress of hurricanes, and computer models use that data to predict the path and intensity of a hurricane. Good forecasts of earthquakes and tornadoes have been more elusive. Nonetheless, the potential benefits are so great that efforts will continue.

Major airports are installing detection equipment that will help them to detect the presence of microbursts and allow them to alert air traffic controllers and pilots, so that necessary precautions can be taken.

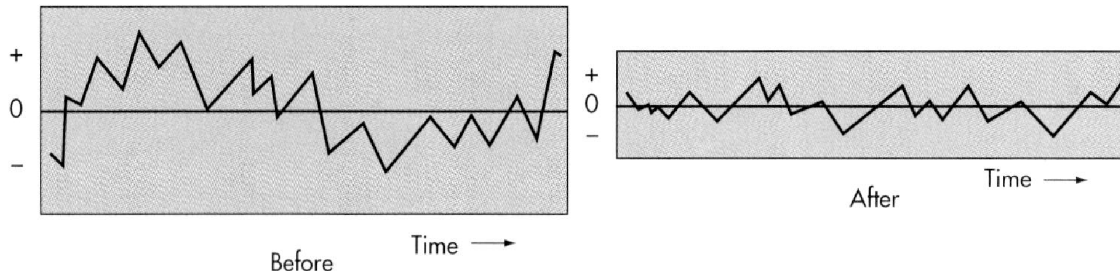

FIGURE 3–15

Removal of a pattern usually results in less variability, and hence, narrower control limits

must be computed using the resulting forecast errors.) Figure 3–15 illustrates the impact on control limits due to decreased error variability.

Comment. The control chart approach is generally superior to the tracking signal approach. A major weakness of the tracking signal approach is its use of cumulative errors: Individual errors can be obscured so that large positive and negative values cancel each other. Conversely, with control charts, every error is judged individually. Thus, it can be misleading to rely on a tracking-signal approach to monitor errors. In fact, the historical roots of the tracking-signal approach date from the first use of computers in business. At that time, it was much more difficult to compute standard deviations than to compute average deviations; for that reason, the concept of a tracking signal was developed. Now computers and calculators can easily provide standard deviations. Nonetheless, the use of tracking signals has persisted, probably because users are unaware of the superiority of the control chart approach.

CHOOSING A FORECASTING TECHNIQUE

Many different kinds of forecasting techniques are available, and no single technique works best in every situation. When selecting a technique for a given situation, the manager or analyst must take a number of factors into consideration.

The two most important factors are *cost* and *accuracy*. How much money is budgeted for generating the forecast? What are the possible costs of error, and what are the benefits that might accrue from an accurate forecast? Generally speaking, the higher the accuracy, the higher the cost, so it is important to weigh cost-accuracy trade-offs carefully. The best forecast is not necessarily the most accurate or the least costly; rather, it is some combination of accuracy and cost deemed best by management.

Other factors to consider in selecting a forecasting technique include the availability of historical data; the availability of computers; the ability of decision makers to utilize certain techniques; the time needed to gather and analyze data, and prepare the forecast; and any prior experience with a technique. The forecast horizon is important because some techniques are more suited to long-range forecasts while others work best for the short range. For example, moving averages and exponential smoothing are essentially short-range techniques, since they produce forecasts for the *next* period. Trend equations can be used to project over much longer time periods. When using time series data, *plotting the data* can be very helpful in choosing an appropriate method. Several of the qualitative techniques are well suited to long-range forecasts because they do not require historical data. The Delphi method and executive opinion methods are often used for long-range planning. New products and services lack historical data, so forecasts for them must be based on subjective estimates. In many cases, experience with similar items is relevant. Table 3–4 provides a guide for selecting a forecasting method. Table 3–5 provides additional perspectives on forecasts in terms of the time horizon.

Forecasting Method	Amount of Historical Data	Data Pattern	Forecast Horizon	Preparation Time	Personnel Background
Simple exponential smoothing	5 to 10 observations	Data should be stationary	Short	Short	Little sophistication
Trend-adjusted exponential smoothing	At least 4 or 5 observations per season	Trend	Short to medium	Short	Moderate sophistication
Trend models	10 to 20; for seasonality at least 5 per season	Trend	Short to medium	Short	Moderate sophistication
Seasonal	Enough to see 2 peaks and troughs	Handles cyclical and seasonal patterns	Short to medium	Short to moderate	Little sophistication
Causal regression models	10 observations per independent variable	Can handle complex patterns	Short, medium, or long	Long development time, short time for implementation	Considerable sophistication

Source: Adapted from J. Holton Wilson and Deborah Allison-Koerber, "Combining Subjective and Objective Forecasts Improves Results," *The Journal of Business Forecasting,* Fall 1992, p. 4.

TABLE 3–4

A guide to selecting an appropriate forecasting method

Factor	Short Range	Intermediate Range	Long Range
1. Frequency	Often	Occasional	Infrequent
2. Level of aggregation	Item	Product family	Total output Type of product/service
3. Type of model	Smoothing Projection Regression	Projection Seasonal Regression	Managerial Judgment
4. Degree of management involvement	Low	Moderate	High
5. Cost per forecast	Low	Moderate	High

TABLE 3–5

Forecast factors, by range of forecast

In some instances, a manager might use more than one forecasting technique to obtain independent forecasts. If the different techniques produced approximately the same predictions, that would give increased confidence in the results; disagreement among the forecasts would indicate that additional analysis may be needed.

USING FORECAST INFORMATION

A manager can take a *reactive* or a *proactive* approach to a forecast. A reactive approach views forecasts as probable descriptions of future demand, and a manager reacts to meet that demand (e.g., adjusts production rates, inventories, the workforce). Conversely, a proactive approach seeks to actively influence demand (e.g., by means of advertising, pricing, or product/service changes).

Generally speaking, a proactive approach requires either an explanatory model (e.g., regression) or a subjective assessment of the influence on demand. It is possible that a manager might use two forecasts: one to predict what will happen under the status quo and a second one based on a "what if" approach, if the results of the status quo forecast are unacceptable.

Operations Strategy

Forecasts are the basis for many decisions. Clearly, the more accurate an organization's forecasts, the better prepared it will be to take advantage of future opportunities and to reduce potential risks. Maintaining accurate, up-to-date information on prices, demand, and other variables can have a significant impact on forecast accuracy.

An organization also can do other things to improve forecasts. These do not involve searching for improved techniques, but relate to the inverse relation of accuracy to the forecast horizon; forecasts that cover shorter time frames tend to be more accurate than longer-term forecasts. Recognizing this, management might choose to devote efforts to *shortening the time horizon that forecasts must cover.* Essentially, this means shortening the *lead time* needed to respond to a forecast. This might involve building *flexibility* into operations to permit rapid response to changing demands for products and services, or to changing volumes in quantities demanded; shortening the lead time required to obtain supplies, equipment, and raw materials or the time needed to train or retrain employees; or shortening the time needed to *develop* new products and services.

Summary

Forecasts are vital inputs for the design and the operation of the productive systems because they help managers to anticipate the future.

TABLE 3-6
Forecasting approaches

	Approaches	Brief Description
Judgment/opinion:	Consumer surveys	Questioning consumers on future plans
	Direct-contact composites	Joint estimates obtained from salespeople or customer service people
	Executive opinion	Finance, marketing, and manufacturing managers join to prepare forecast
	Delphi technique	Series of questionnaires answered anonymously by knowledgeable people; successive questionnaires are based on information obtained from previous surveys
	Outside opinion	Consultants or other outside experts prepare the forecast
Statistical:	Time series:	
	Naive	Next value in a series will equal the previous value in a comparable period
	Moving averages	Forecast is based on an average of recent values
	Exponential smoothing	Sophisticated form of weighted moving average
	Associative models:	
	Simple regression	Values of one variable are used to predict values of another variable
	Multiple regression	Two or more variables are used to predict values of another variable

Forecasting techniques can be classified as qualitative or quantitative. Qualitative techniques rely on judgment, experience, and expertise to formulate forecasts; quantitative techniques rely on the use of historical data or associations among variables to develop forecasts. Some of the techniques are simple and others are complex. Some work better than others, but no technique works all the time. Moreover, all forecasts include a certain degree of inaccuracy, and allowance should be made for this. All techniques assume that the same underlying causal system that existed in the past will continue to exist in the future.

The qualitative techniques described in this chapter include consumer surveys, salesforce estimates, executive opinions, and manager and staff opinions. Two major quantitative approaches are described: analysis of time series data and associative techniques. The time series

TABLE 3-7
Summary of formulas

Technique	Formula	Definitions
Naive forecast	$F_t = A_{t-1}$	F = Forecast A = Actual data t = Current period
Moving average forecast	$F = \frac{\sum_{i=1}^{n} A_i}{n}$	n = Number of periods
Exponential smoothing forecast	$F_t = F_{t-1} + \alpha(A_{t-1} - F_{t-1})$	α = Smoothing factor
Linear trend forecast	$y_t = a + bt$ where $b = \frac{n\Sigma ty - \Sigma t\Sigma y}{n\Sigma t^2 - (\Sigma t)^2}$ $a = \frac{\Sigma y - b\Sigma t}{n}$ or $\bar{y} - bt$	a = y intercept b = Slope
Trend-adjusted forecast	$\text{TAF}_{t+1} = S_t + T_t$ where $S_t = \text{TAF}_t + \alpha(A_t - \text{TAF}_t)$ $T_t = T_{t-1} + \beta(\text{TAF}_t - \text{TAF}_{t-1} - T_{t-1})$	t = Current period TAF_{t+1} = Trend-adjusted forecast for next period S = Seasonal component T = Trend component
Linear regression forecast	$y_c = a + bx$ where $b = \frac{n(\Sigma xy) - (\Sigma x)(\Sigma y)}{n(\Sigma x^2) - (\Sigma x)^2}$ $a = \frac{\Sigma y - b\Sigma x}{n}$ or $\bar{y} - b\bar{x}$	y_c = Predicted (dependent) variable x = Predictor (independent) variable b = Slope of the line a = Value of y_c when $x = 0$
MAD	$\text{MAD} = \frac{\sum^{n} \lvert e \rvert}{n}$	MAD = Mean absolute deviation e = Error, $A - F$
MSE	$\text{MSE} = \frac{\sum^{n} e^2}{n-1}$	MSE = Mean squared error
Tracking signal	$\text{TS} = \frac{\sum^{n} e}{\text{MAD}}$	
Control limits	$\text{UCL} = 0 + zs$ $\text{LCL} = 0 - zs$	$s = \sqrt{\text{MSE}}$ z = Number of standard deviations; 2 and 3 are typical values.

techniques rely strictly on the examination of historical data; predictions are made by projecting past movements of a variable into the future without considering specific factors that might influence the variable. Associative techniques attempt to explicitly identify influencing factors and to incorporate that information into equations that can be used for predictive purposes.

All forecasts tend to be inaccurate; therefore, it is important to provide a measure of accuracy. It is possible to compute several measures of forecast accuracy that help managers to evaluate the performance of a given technique and to choose among alternative forecasting techniques. Control of forecasts involves deciding whether a forecast is performing adequately, using either a control chart or a tracking signal.

When selecting a forecasting technique, a manager must choose a technique that will serve the intended purpose at an acceptable level of cost and accuracy.

The various forecasting techniques are summarized in Table 3–6 (on page 124). Table 3–7 (on page 125) lists the formulas used in the forecasting techniques and in the methods of measuring their accuracy.

Key Terms

associative models, 91
bias, 118
centered moving average, 108
control chart, 118
correlation, 113
cycles, 94
Delphi method, 92
error, 116
exponential smoothing, 98
forecast, 88
irregular variation, 94
judgmental forecasts, 91
least squares line, 110
linear trend equation, 101
mean absolute deviation (MAD), 116
mean squared error (MSE), 116
moving average, 96
naive forecast, 95
predictor variable, 110
random variations, 94
regression, 110
seasonal variations, 106
seasonality, 94
time series, 93
tracking signal, 118
trend, 93
trend-adjusted exponential smoothing, 104

Solved Problems

Problem 1

Forecasts based on averages. Given the following data:

Period	Number of Complaints
1	60
2	65
3	55
4	58
5	64

Prepare a forecast using each of these approaches:

a. The naive approach.

b. A three-period moving average.

c. A weighted average using weights of .50 (most recent), .30, and .20.

d. Exponential smoothing with a smoothing constant of .40.

Solution

a. The most recent value of the series becomes the next forecast: 64.

b. $MA_3 = \frac{55 + 58 + 64}{3} = 59$

c. $F = .50(64) + .30(58) + .20(55) = 60.4$

d.

Period	Number of Complaints	Forecast	Calculations
1	60		
2	65	60	[Use previous value of series]
3	55	62	60 + .40(65 − 60) = 62
4	58	59.2	62 + .40(55 − 62) = 59.2
5	64	58.72	59.2 + .40(58 − 59.2) = 58.72
6		60.83	59.72 + .40(64 − 58.72) = 60.83

You can also obtain the forecasts and a plot using an Excel template, as shown:

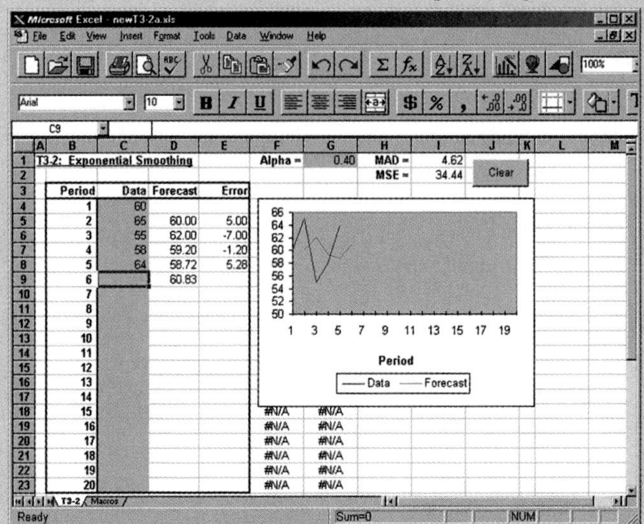

Problem 2

Time series analysis. Apple's Citrus Fruit Farm ships boxed fruit anywhere in the continental United States. Using the following information, forecast shipments for the first four months of next year.

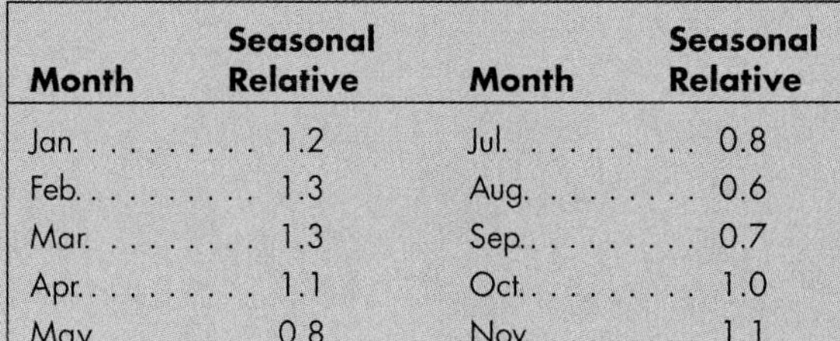

Month	Seasonal Relative	Month	Seasonal Relative
Jan.	1.2	Jul.	0.8
Feb.	1.3	Aug.	0.6
Mar.	1.3	Sep.	0.7
Apr.	1.1	Oct.	1.0
May	0.8	Nov.	1.1
Jun.	0.7	Dec.	1.4

The monthly forecast equation being used is:

$y_t = 402 + 3t$

where

t_0 = January of *last* year

y_t = Number of shipments

Solution

a. Determine trend amounts for the first four months of next year: January, $t = 24$; February, $t = 25$; etc. Thus,

$$Y_{Jan} = 402 + 3(24) = 474$$

$$Y_{Feb} = 402 + 3(25) = 477$$

$$Y_{Mar} = 402 + 3(26) = 480$$

$$Y_{Apr} = 402 + 3(27) = 483$$

b. Multiply each monthly trend by the corresponding seasonal relative for that month.

Month	Seasonal Relative	Forecast
Jan.	1.2	474(1.2) = 568.8
Feb.	1.3	477(1.3) = 620.1
Mar.	1.3	480(1.3) = 624.0
Apr.	1.1	483(1.1) = 531.3

Problem 3

Linear trend line. Plot the line and the data on a graph, and verify visually that a linear trend line is appropriate. Develop a line trend equation for the following data. Then use the equation to predict the next two values of the series.

Period	Demand
1	44
2	52
3	50
4	54
5	55
6	55
7	60
8	56
9	62

Solution

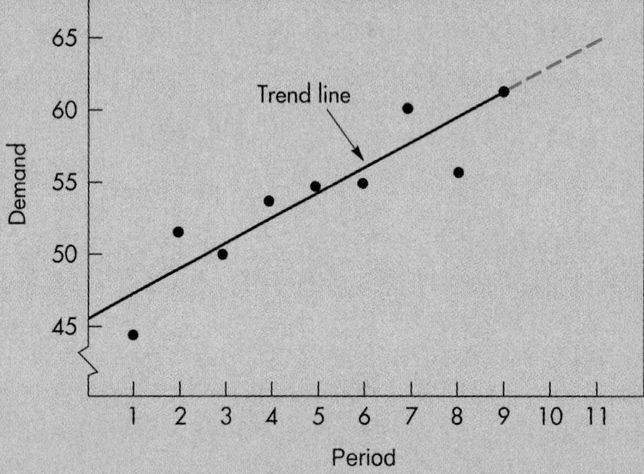

A plot of the data indicates that a linear trend line is appropriate:

Period, t	Demand, y	ty	
1	44	44	From Table 3–1, with $n = 9$,
2	52	104	
3	50	150	$\Sigma t = 45$ and $\Sigma t^2 = 285$
4	54	216	
5	55	275	
6	55	330	
7	60	420	
8	56	448	
9	62	558	
	488	2,545	

$$b = \frac{n\Sigma ty - \Sigma t\Sigma y}{n\Sigma t^2 - (\Sigma t)^2} = \frac{9(2{,}545) - 45(488)}{9(285) - 45(45)} = 1.75$$

$$a = \frac{\Sigma y - b\Sigma t}{n} = \frac{488 - 1.75(45)}{9} = 45.47$$

Thus, the trend equation is $y_t = 45.47 + 1.75t$. The next two forecasts are:

$$y_{10} = 45.47 + 1.75(10) = 62.97$$

$$y_{11} = 45.47 + 1.75(11) = 64.72$$

You can also use an Excel template to obtain the coefficients and a plot. Simply replace the existing data in the template with your data.

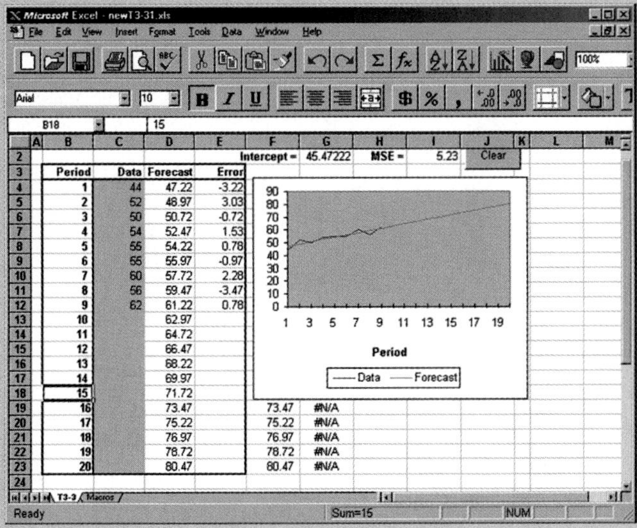

Period	Data	Forecast	Error
1	44	47.22	-3.22
2	52	48.97	3.03
3	50	50.72	-0.72
4	54	52.47	1.53
5	55	54.22	0.78
6	55	55.97	-0.97
7	60	57.72	2.28
8	56	59.47	-3.47
9	62	61.22	0.78
10		62.97	
11		64.72	
12		66.47	
13		68.22	
14		69.97	
15		71.72	
16		73.47	
17		75.22	
18		76.97	
19		78.72	
20		80.47	

Problem 4

Trend-adjusted smoothing. After plotting demand for four periods, a manager has concluded that a trend-adjusted exponential smoothing model is appropriate to predict future demand. Calculations are shown for the next five periods. The initial estimate of trend is based on the net change of 30 for the *three* periods from 1 to 4, for an average of +10 units. Use $\alpha = .5$ and $\beta = .4$.

Solution

	t Period	A_t Actual	TAF_t	TAF_t	$+\alpha(A_t - TAF_t)$	$= S_t$	T_{t-1}	$+\beta(TAF_{t-1} - TAF_{t-1} - T_{t-1}) = T_t$
Model Development	1	210						
	2	224						
	3	229						
	4	240						
Model Test	5	255	250*	250	+ .5(255 − 250)	= 252.5	10	+ .4(0) = 10
	6	265	262.5	262.5	+ .5(265 − 262.5)	= 263.75	10	+ .4(262.5 − 250 − 10) = 11.00
	7	272	274.75	274.75	+ .5(272 − 274.75)	= 273.37	11.00	+ .4(274.75 − 262.5 − 11.00) = 11.50
	8	285	284.87	284.87	+ .5(285 − 284.87)	= 284.94	11.50	+ .4(284.87 − 274.75 − 11.50) = 10.95
	9	294	295.89	295.89	+ .5(294 − 295.89)	= 294.95	10.95	+ .4(295.89 − 284.87 − 10.95) = 10.98
Next Forecast	→ 10	→	305.92					

*Estimated by the manager.

You can obtain the forecasts and a plot using the appropriate Excel template. Simply replace the data with your data.

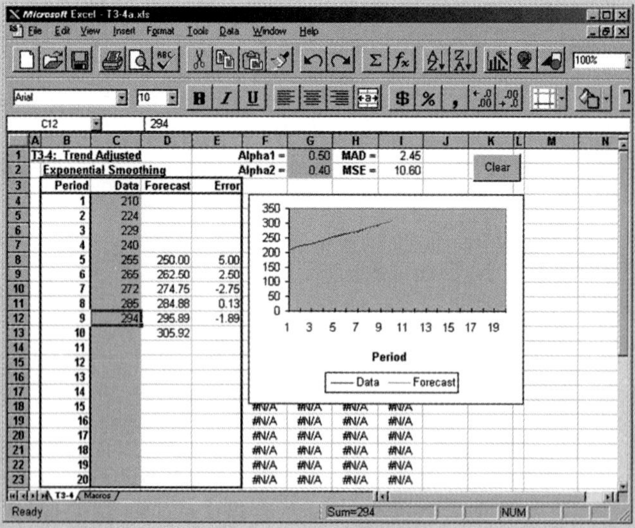

Problem 5

Seasonal relatives. Obtain estimates of quarter relatives for these data:

Year:	1				2				3				4
Quarter:	1	2	3	4	1	2	3	4	1	2	3	4	1
Demand:	14	18	35	46	28	36	60	71	45	54	84	88	58

Solution

Note that each season has an *even* number of data points. When an even-numbered moving average is used (in this case, a four-period moving average), the "centered value" will not correspond to an actual data point; the center of 4 is *between* the second and third data points. To correct for this, a *second* set of moving averages must be computed using the MA_4 values. The MA_2 values are centered between the MA_4 and "line up" with actual data points. For example, the first MA_4 value is 28.25. It is centered between 18 and 35 (i.e., between quarter 2 and quarter 3). When the average of the first two MA_4 values is taken (i.e., MA_2) and centered, it lines up with the 35 and, hence, with quarter 3.

So, whenever an even-numbered moving average is used as a centered moving average (e.g., MA_4, MA_{12}), a second moving average, a two-period moving average, is used to achieve correspondence with periods. This procedure is not needed when the number of periods in the centered moving average is odd. See Example 7 in this chapter for an example with an odd number of periods.

Year	Quarter	Demand	MA_4	MA_2	Demand/MA_2
1	1	14			
	2	18	28.25		
	3	35	31.75	30.00	1.17
	4	46	36.25	34.00	1.35
2	1	28	42.50	39.38	0.71
	2	36	48.75	45.63	0.79
	3	60	53.00	50.88	1.18
	4	71	57.50	55.25	1.29
3	1	45	63.50	60.50	0.74
	2	54	67.75	65.63	0.82
	3	84	71.00	69.38	1.21
	4	88			
4	1	58			

	QUARTER			
	1	**2**	**3**	**4**
	0.71	0.79	1.17	1.35
	0.74	0.82	1.18	1.29
	1.45	1.61	1.21	2.64
			3.56	
Average for the quarter:	0.725	0.805	1.187	1.320

The sum of these relatives is 4.037. Multiplying each by 4.00/4.037 will standardize the relatives, and make their total equal 4.00. The resulting relatives are: quarter 1, 0.718; quarter 2, 0.798; quarter 3, 1.176; quarter 4, 1.308.

Problem 6

Regression analysis. A large midwestern retailer has developed a graph that summarizes the effect of advertising expenditures on sales volume. Using the graph, determine an equation of the form $y = a + bx$ that describes this relationship.

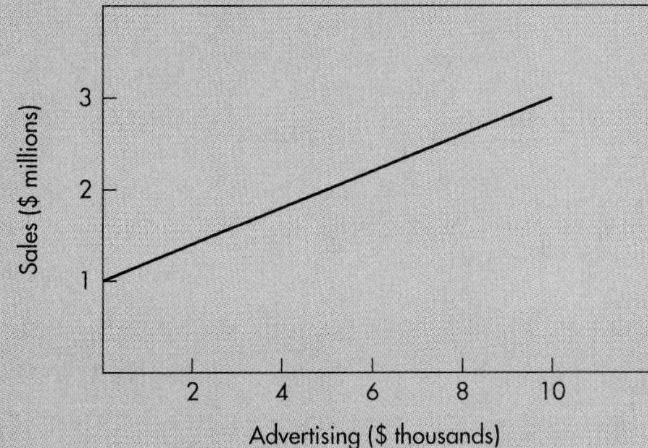

Solution

The linear equation has the form $y = a + bx$, where a is the value of y when $x = 0$ (i.e., where the line intersects the y axis) and b is the slope of the line (the amount by which y changes for a one-unit change in x).

Accordingly, $a = 1$ and $b = (3 - 1)/(10 - 0) = 0.2$, so $y = a + bx$ becomes $y = 1 + 0.2x$. [*Note:* $(3 - 1)$ is the change in y.]

Problem 7

Regression analysis. The owner of a small hardware store has noted a sales pattern for window locks that seems to parallel the number of break-ins reported each week in the newspaper. The data are:

Sales:	46	18	20	22	27	34	14	37	30
Break-ins:	9	3	3	5	4	7	2	6	4

a. Plot the data to determine which type of equation, linear or nonlinear, is appropriate.

b. Obtain a regression equation for the data.

c. Estimate sales when the number of break-ins is five.

Solution

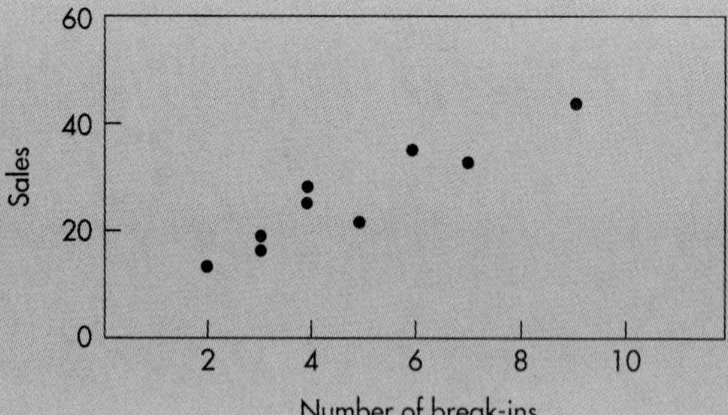

The graph supports a linear relationship.

b. The computations for a straight line are:

x	y	xy	x^2	y^2
9	 46	414	81	2,116
3	 18	54	9	324
3	 20	60	9	400
5	 22	110	25	484
4	 27	108	16	729
7	 34	238	49	1,156
2	 14	28	4	196
6	 37	222	36	1,369
4	 30	120	16	900
43	248	1,354	245	7,674

$$b = \frac{n(\Sigma xy) - (\Sigma x)(\Sigma y)}{n(\Sigma x^2) - (\Sigma x)^2} = \frac{9(1{,}354) - 43(248)}{9(245) - 43(43)} = 4.275$$

$$a = \frac{\Sigma y - b(\Sigma x)}{n} = \frac{248 - 4.275(43)}{9} = 7.129$$

Hence, the equation is: $y_c = 7.129 + 4.275x$.

You can obtain the regression coefficients using the appropriate Excel template. Simply replace the existing data for x and y with your data. Note: be careful to enter the values for the variable you want to predict as y values. In this problem, the objective is to predict sales, so the sales values are entered in the y column.

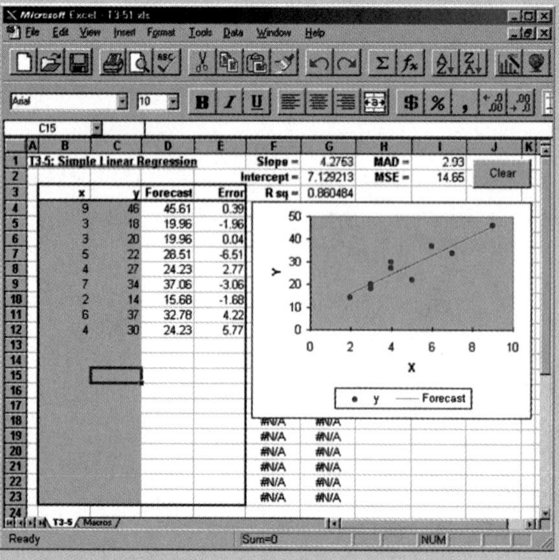

c. For $x = 5$, $y_c = 7.129 + 4.275(5) = 28.50$.

Problem 8

Accuracy and control of forecasts. The manager of a large manufacturer of industrial pumps must choose between two alternative forecasting techniques. Both techniques have been used to prepare forecasts for a six-month period. Using MAD as a criterion, which technique has the better performance record?

		FORECAST	
Month	**Demand**	**Technique 1**	**Technique 2**
1	492	488	495
2	470	484	482
3	485	480	478
4	493	490	488
5	498	497	492
6	492	493	493

Solution

Check that each forecast has an average error of approximately zero. (See computations that follow.)

Month	Demand	Technique 1	e	|e|	Technique 2	e	|e|
1	492	488	4	4	495	−3	3
2	470	484	−14	14	482	−12	12
3	485	480	5	5	478	7	7
4	493	490	3	3	488	5	5
5	498	497	1	1	492	6	6
6	492	493	−1	1	493	−1	1
			−2	28		+2	34

$$\text{MAD}_1 = \frac{\Sigma|e|}{n} = \frac{28}{6} = 4.67$$

$$\text{MAD}_2 = \frac{\Sigma|e|}{n} = \frac{34}{6} = 5.67$$

Technique 1 is superior in this comparison because its MAD is smaller, although six observations would generally be too few on which to base a realistic comparison.

Problem 9

Control chart. Given the demand data that follow, prepare a naive forecast for periods 2 through 10. Then determine each forecast error, and use those values to obtain 2s control limits. If the next two periods' demands turn out to be 125 and 130, can you conclude that the forecasts are in control?

Period:	1	2	3	4	5	6	7	8	9	10
Demand:	118	117	120	119	126	122	117	123	121	124

Solution

For a naive forecast, each period's demand becomes the forecast for the next period. Hence, the forecasts and errors are:

Period	Demand	Forecast	Error	Error^2
1	118	—	—	—
2	117	118	−1	1
3	120	117	3	9
4	119	120	−1	1
5	126	119	7	49
6	122	126	−4	16
7	117	122	−5	25
8	123	117	6	36
9	121	123	−2	4
10	124	121	3	9
			+6	150

$$s = \sqrt{\frac{\Sigma \text{error}^2}{n-1}} = \sqrt{\frac{150}{9-1}} = 4.33 \quad (n = \text{Number of errors})$$

The control limits are $2(4.33) = \pm 8.66$

The forecast for period 11 was 124. Demand turned out to be 125, for an error of $125 - 124 = +1$. This is within the limits of ± 8.66. If the next demand is 130 and the naive forecast is 125 (based on the period 11 demand of 125), the error is +5. Again, this is within the limits, so you cannot conclude the forecast is not working properly. With more values—at least five or six—you could plot the errors to see whether you could detect any patterns suggesting the presence of nonrandomness.

Discussion and Review Questions

1. What are the main advantages that quantitative techniques for forecasting have over qualitative techniques? What limitations do quantitative techniques have?
2. What are some of the consequences of poor forecasts? Explain.
3. List the specific weaknesses of each of these approaches to developing a forecast:
   a. Consumer surveys
   b. Salesforce composite
   c. Committee of managers or executives
4. Briefly describe the Delphi technique. What are its main benefits and weaknesses?
5. What is the purpose of establishing control limits for forecast errors?
6. What factors would you consider in deciding whether to use wide or narrow control limits for a forecast?
7. Contrast the use of MAD and MSE in evaluating forecasts.
8. What advantages as a forecasting tool does exponential smoothing have over moving averages?
9. How does the number of periods in a moving average affect the responsiveness of the forecast?
10. What factors enter into the choice of a value for the smoothing constant in exponential smoothing?
11. How accurate is your local five-day weather forecast? Support your answer with actual data.
12. Explain how using a centered moving average with a length equal to the length of a season eliminates seasonality from a time series.
13. Contrast the terms *sales* and *demand.*
14. Contrast the reactive and proactive approaches to forecasting. Give several examples of types of organizations or situations in which each type is used.

15. Explain how flexibility in production systems relates to the forecast horizon and forecast accuracy.
16. Explain how lean production impacts forecasting.

Memo Writing Exercises

1. You have received a call from the manager of a firm where you helped set up a forecasting system. The manager, Jill Rodgers, expressed concern that forecast errors, although within the control limits, were too large, and wondered if there was anything else that could be done, or whether they would "just have to live with it." What would you suggest? Write a memo to Jill.
2. Write a short memo to your boss, Jim Oliver, outlining the merits of using a control chart to monitor forecasts rather than a tracking signal.

Problems

1. The number of bushels of apples sold at a roadside fruit stand over a 12-day period were as follows:

Day	Number Sold	Day	Number Sold
1	25	7	35
2	31	8	32
3	29	9	38
4	33	10	40
5	34	11	37
6	37	12	32

See IM

a. If a two-period moving average had been used to forecast sales, what would the daily forecasts have been starting with the forecast for day 3?

b. If a four-period moving average had been used, what would the forecasts for each day have been starting with day 5?

c. Plot the original data and each set of forecasts on the same graph. Which forecast had the greater tendency to smooth? Which forecast had the better ability to respond quickly to changes?

d. What does use of the term *sales* instead of *demand* imply?

a–c: See IM.
d. Demand did not exceed supply. See IM.

2. National Mixer, Inc., sells can openers. Monthly sales for a seven-month period were as follows:

Month	Sales (000 units)
Feb.	19
Mar.	18
Apr.	15
May	20
Jun.	18
Jul.	22
Aug.	20

a. Plot the monthly data on a sheet of graph paper.

b. Forecast September sales volume using each of the following:
(1) A linear trend equation.
(2) A five-month moving average.
(3) Exponential smoothing with a smoothing constant equal to .20, assuming a March forecast of 19(000).

a. Graph in IM
b. (1) 20.86
(2) 19
(3) 19.26
(4) 20
(5) 20.4
c. Trend, because data vary around average of about 19.
d. Sales reflect demand.

(4) The naive approach.

(5) A weighted average using .60 for August, .30 for July, and .10 for June.

c. Which method seems least appropriate? Why?

d. What does use of the term *sales* rather than *demand* presume?

a. 88.16%
b. 88.54%

3. A dry cleaner uses exponential smoothing to forecast equipment usage at its main plant. August usage was forecast to be 88 percent of capacity; actual usage was 89.6 percent of capacity. A smoothing constant of .1 is used.

   a. Prepare a forecast for September.

   b. Assuming actual September usage of 92 percent, prepare a forecast for October usage.

a. 22
b. 20.75
c. 20.72

4. An electrical contractor's records during the last five weeks indicate the number of job requests:

Week:	1	2	3	4	5
Requests:	20	22	18	21	22

   Predict the number of requests for week 6 using each of these methods:

   a. Naive.

   b. A four-period moving average.

   c. Exponential smoothing with $\alpha = .30$.

a. Increasing by 15,000 bottles per year.
b. 275,000

5. A cosmetics manufacturer's marketing department has developed a linear trend equation that can be used to predict annual sales of its popular Hand & Foot Cream.

$$y_t = 80 + 15t$$

   where

$$y_t = \text{Annual sales (000 bottles)}$$
$$t_0 = 1988$$

   a. Are annual sales increasing or decreasing? By how much?

   b. Predict annual sales for the year 2001 using the equation.

a. $y_t = 500 - 20t$

6. From the following graph, determine the linear equation of the trend line using 1988 as the base year for Glib Sales, Inc.

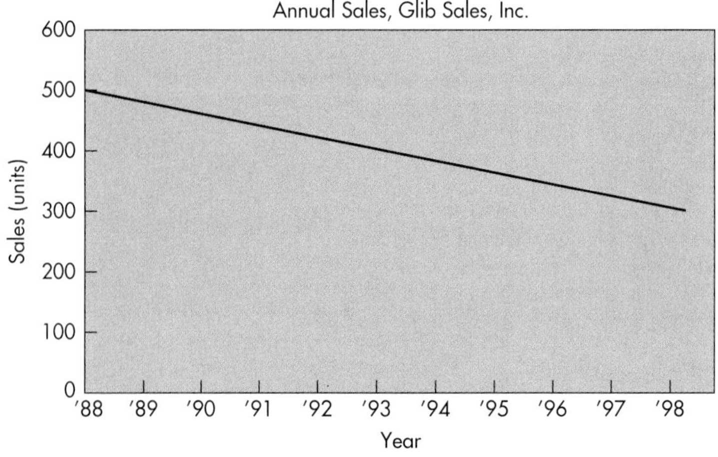

7. Freight car loadings over a 12-year period at a busy port are:

Year	Number (00)	Year	Number (000)
1	220	7	350
2	245	8	360
3	280	9	400
4	275	10	380
5	300	11	420
6	310	12	450

a. $y_t = 204.09 + 19.76t$
b. Graph in IM
c. 500.49

a. Compute a linear trend line for freight car loadings.

b. Plot the trend line and the original data on the same graph.

c. Use the trend equation to predict loadings for year 15.

8. Develop a linear trend equation for the following data on bread deliveries, and use it to predict deliveries for periods 11 through 14.

$y_t = 196.13 + 6.50t$
$y_{11} = 267.63$
$y_{12} = 274.13$
$y_{13} = 280.63$
$y_{14} = 287.13$

Period	Dozen Deliveries	Period	Dozen Deliveries
1	200	6	222
2	214	7	248
3	211	8	241
4	228	9	253
5	235	10	267

9. Use trend-adjusted smoothing with $\alpha = .3$ and $\beta = .2$ to smooth the bread delivery data in the preceding problem. 273.56

10. A manager of a store that sells and installs hot tubs wants to prepare a forecast for January, February, and March of next year. Her forecasts are a combination of trend and seasonality. She uses the following equation to estimate the trend component of monthly demand: $y_t = 70 + 5t$, where $t = 0$ in June of last year. Seasonal relatives are 1.10 for January, 1.02 for February, and .95 for March. What demands should she predict?

11. The following equation summarizes the trend portion of quarterly sales of automatic dishwashers over a long cycle. Sales also exhibit seasonal variations. Using the information given, prepare a forecast of sales for each quarter of 2001, and the first quarter of 2002.

Q_1: 157.85
Q_2: 175
Q_3: 126.3
Q_4: 325
Q_1: 322.85

$$y_t = 40 - 6.5t + 2t^2$$

where

y_t = Unit sales
t = 0 at the fourth quarter of 1998

Quarter	Relative
1	1.1
2	1.0
3	.6
4	1.3

12. A gift shop in a tourist center is open on weekends (Friday, Saturday, and Sunday). The owner-manager hopes to improve scheduling of part-time employees by determining seasonal relatives for each of these days. Data on recent activity at the store (sales transactions per day) have been tabulated and are shown in the table below.

Fri. = 0.79
Sat. = 1.34
Sun. = 0.87

	WEEK					
	1	**2**	**3**	**4**	**5**	**6**
Friday	149	154	152	150	159	163
Saturday	250	255	260	268	273	276
Sunday	166	162	171	173	176	183

Fri. = 163
Sat. = 276
Sun. = 183

a. Develop seasonal relatives for the shop.

b. Use a naive trend approach to predict sales transactions for the gift shop in the previous problem for the following week.

Wed = 0.60
Thurs = 0.80
Fri = 1.40
Sat = 1.20

13. The manager of a fashionable restaurant open Wednesday through Saturday says that the restaurant does about 35 percent of its business on Friday night, 30 percent on Saturday night, and 20 percent on Thursday night. What seasonal relatives would describe this situation?

a. Forecast will underestimate increasing data.
b. Trend line equation is
$y = 405.286 + 2.464t$
Forecasts are 425, 427.46, 429.93

14. Coal shipments from Mountain Coal Company's no. 4 mine for the past seven years are:

Year	Tons Shipped (in thousands)
1	405
2	410
3	420
4	415
5	412
6	420
7	424

a. Explain why an averaging technique would not be appropriate for forecasting.

b. Use an appropriate technique to develop a forecast for the next three years.

Day	Relative
1	0.902
2	0.836
3	0.919
4	1.034
5	1.416
6	1.487
7	0.427

15. Obtain estimates of daily relatives for the number of customers at a restaurant for the evening meal, given the following data. (*Hint:* Use a seven-day moving average.)

Day	Number Served	Day	Number Served
1	80	15	84
2	75	16	77
3	78	17	83
4	95	18	96
5	130	19	135
6	136	20	140
7	40	21	37
8	82	22	87
9	77	23	82
10	80	24	98
11	94	25	103
12	125	26	144
13	135	27	144
14	42	28	48

16. A pharmacist has been monitoring sales of a certain over-the-counter pain reliever. Daily sales during the last 15 days were:

Day:	1	2	3	4	5	6	7	8	9
Number sold:	36	38	42	44	48	49	50	49	52
Day:	10	11	12	13	14	15			
Number sold:	48	52	55	54	56	57			

a. Without doing any calculations, which method would you suggest using to predict future sales—a linear trend equation or trend-adjusted exponential smoothing? Why?

b. If you learn that on some days the store ran out of the specific pain reliever, would that knowledge cause you any concern? Explain.

c. Assume that the data refer to demand rather than sales. Using trend-adjusted smoothing with an initial forecast of 50 for week 8, an initial trend estimate of 2, and $\alpha = \beta = .3$, develop forecasts for days 9 through 16. What is the MSE for the eight forecasts for which there are actual data?

a. Trend-adjusted smoothing.
b. Yes. Wouldn't know demand.
c.

	TAF
9	51.7
10	53.7
11	53.93
12	54.77
13	56.09
14	56.74
15	57.60
16	58.43

MSE = 5.45

17. New car sales for a dealer in Cook County, Illinois, for the past year are shown in the following table, along with monthly (seasonal) relatives, which are supplied to the dealer by the regional distributor.

Month	Units Sold	Index	Month	Units Sold	Index
Jan.	640	0.80	Jul.	765	0.90
Feb.	648	0.80	Aug.	805	1.15
Mar.	630	0.70	Sept.	840	1.20
Apr.	761	0.94	Oct.	828	1.20
May.	735	0.89	Nov.	840	1.25
Jun.	850	1.00	Dec.	800	1.25

a. Plot the data. Does there seem to be a trend?

b. Deseasonalize car sales.

c. Plot the deseasonalized data on the same graph as the original data. Comment on the two graphs.

Jan	800	Jul	850
Feb	810	Aug	700
Mar	900	Sept	700
Apr	809.6	Oct	690
May	825.8	Nov	672
Jun	850	Dec	640

a. Graph in IM
b. Table in IM
c. Graph in IM

18. A quality control analyst has kept a record of the defective rate of a process she has been working to improve during a period of about four weeks. The following data (percentages) were recorded:

	WEEK			
	1	2	3	4
Monday	10.2	9.4	8.4	7.8
Tuesday	8.2	7.3	7.0	6.5
Wednesday	7.2	6.8	6.3	5.0
Thursday	6.8	6.0	5.4	4.8
Friday	9.4	9.0	8.2	7.1

a. Determine daily relatives for the defective rate. Then use the relatives to remove seasonality from the data.

b. Compute a linear trend equation for the deseasonalized data.

c. Using the trend equation and seasonal relatives, predict the defective rate for each day of week 5.

a. Mon = 1.17
Tue = 0.97
Wed = 0.86
Thu = 0.80
Fri = 1.19
b. $y_t = 8.88 - 0.142t$
c. Mon = 6.77
Tue = 5.48
Wed = 4.73
Thur = 4.28
Fri = 6.20

19. A farming cooperative manager wants to estimate quarterly relatives for grain shipments, based on the data shown (quantities are in metric tons):

$Q_1 = 0.83$
$Q_2 = 0.99$
$Q_3 = 0.83$
$Q_4 = 1.35$

Year	QUARTER 1	2	3	4
1	200	250	210	340
2	210	252	212	360
3	215	260	220	358
4	225	272	233	372
5	232	284	240	381

Determine quarter relatives. (*Hint:* Use a centered four-period moving average initially, and then use a centered two-period moving average of the four-period moving average.)

a. Graph in IM
b. $29,000

20. Long-Life Insurance has developed a linear model that it uses to determine the amount of straight life insurance a family of four should have, based on the current age of the head of the household. The equation is:

    $y = 32 - .1x$

    where

    y = Insurance needed ($000)
    x = Current age of head of household

    a. Plot the relationship on a graph.

    b. Use the equation to determine the amount of straight life insurance to recommend for a family of four if the head of the household is 30 years old.

a. w = weight
d = distance
y = delivery charge
$y = \$0.10w + \$0.15d_2 + \$10$
b. $17.90

21. Timely Transport provides local delivery service for a number of downtown and suburban businesses. Delivery charges are based on distance and weight involved for each delivery: 10 cents per pound and 15 cents per mile. Also, there is a $10 handling fee per parcel.

    a. Develop an expression that summarizes delivery charges.

    b. Determine the delivery charge for transporting a 40-pound parcel 26 miles.

22. The manager of a seafood restaurant was asked to establish a pricing policy on lobster dinners. Experimenting with prices produced the following data:

Average Number Sold per Day, y	Price, x
200	$6.00
190	6.50
188	6.75
180	7.00
170	7.25
162	7.50
160	8.00

a. Graph in IM
b. $r = -0.97$, high negative relationship

a. Plot the data and a regression line on the same graph.

b. Compute the correlation coefficient and interpret it.

c. Interpret r.

23. The following data were collected during a study of consumer buying patterns.

Observation	x	y	Observation	x	y
1	15	74	8	18	78
2	25	80	9	14	70
3	40	84	10	15	72
4	32	81	11	22	85
5	51	96	12	24	88
6	47	95	13	33	90
7	30	83			

a. Plot the data.

b. Fit a linear regression line to the data using the least squares method.

c. Is the line a good fit? Explain.

d. Use the equation determined in part *b* to predict the value of y for $x = 41$.

a. Graph in IM
b. $y = 66.33 + 0.584x$
c. $r^2 = .7553$
d. 90.27

24. Lovely Lawns, Inc., intends to use sales of lawn fertilizer to predict lawn mower sales. The store manager estimates a probable six-week lag between fertilizer sales and mower sales. The pertinent data are:

a. $r = +0.938$, yes
b. $y = 0.955 + 5.276x$
c. 11.51 mowers

Period	Fertilizer Sales (tons)	Number of Mowers Sold (six-week lag)
1	1.6	10
2	1.3	8
3	1.8	11
4	2.0	12
5	2.2	12
6	1.6	9
7	1.5	8

a. Compute the correlation between the two variables. Does it appear that a relationship between these variables will yield good predictions? Explain.

b. Use the least squares method to obtain a linear regression line for the data.

c. Predict lawn mower sales for the first week in August, given fertilizer sales six weeks earlier of 2 tons.

25. An analyst must decide between two different forecasting techniques for weekly sales of roller blades: a linear trend equation and the naive approach. The linear trend equation is $y_t = 124 + 2t$, and it was developed using data from periods 1 through 10. Based on data for periods 11 through 20 as shown in the table, which of these two methods has the greater accuracy?

25. Naive: MAD = 4.0
MSE = 25.25
Trend: MAD = 2.3
MSE = 10.11
Trend is better.

t	Units Sold	*t*	Units Sold
11	147	16	152
12	148	17	155
13	151	18	157
14	145	19	160
15	155	20	165

26. Two different forecasting techniques (F1 and F2) were used to forecast demand for cases of bottled water. Actual demand and the two sets of forecasts are as follows:

a. $MAD_1 = 4.0$
$MAD_2 = 3.0$
b. $MSE_1 = 25.14$
$MSE_2 = 15.14$
c. MAD easier to compute; use MAD with tracking signal. Use MAD with control chart.

		PREDICTED DEMAND	
Period	Demand	F1	F2
1	68	66	66
2	75	68	68
3	70	72	70
4	74	71	72
5	69	72	74
6	72	70	76
7	80	71	78
8	78	74	80

a. Compute MAD for each set of forecasts. Given your results, which forecast appears to be the most accurate? Explain.

b. Compute the MSE for each set of forecasts. Given your results, which forecast appears to be the most accurate?

c. In practice, *either* MAD *or* MSE would be employed to compute forecast errors. What factors might lead a manager to choose one rather than the other?

Period	(*a*)	(*b*)
5	5	1.40
6	5.9	−0.17
7	4.73	−0.63
8	3.911	−0.26
9	4.238	−1.42
10	3.267	−2.14
11	3.487	−0.86
12	3.941	0.51
13	3.659	−0.27
14	4.361	1.14

27. The manager of a travel agency has been using a seasonally adjusted forecast to predict demand for packaged tours. The actual and predicted values are:

Period	Demand	Predicted
1	129	124
2	194	200
3	156	150
4	91	94
5	85	80
6	132	140
7	126	128
8	126	124
9	95	100
10	149	150
11	98	94
12	85	80
13	137	140
14	134	128

a. Compute MAD for the fifth period, then update it period by period using exponential smoothing with $\alpha = .3$.

b. Compute a tracking signal for periods 5 through 14 using the initial and updated MADs. If limits of ±3 are used, what can you conclude?

a. Forecast 1
MSE = 10.44
MAD = 2.8
Forecast 2
MSE = 42.44
MAD = 3.6
b. #1: 4.29
#2: −4.44
c. #1: 0 ± 6.46
#2: 0 ± 13.03
d. MSE = 156
MAD = 10.67
$T.S._{10} = 1.87$
0 ± 25

28. Two independent methods of forecasting based on judgment and experience have been prepared each month for the past 10 months. The forecasts and actual sales are as follows.

Month	Sales	Forecast 1	Forecast 2
1	770	771	769
2	789	785	787
3	794	790	792
4	780	784	798
5	768	770	774
6	772	768	770
7	760	761	759
8	775	771	775
9	786	784	788
10	790	788	788

a. Compute the MSE and MAD for each forecast. Does either method seem superior? Explain.

b. Compute a tracking signal for the 10th month for each forecast. What does it show? (Use action limits of ±4.)

c. Compute 2*s* control limits for each forecast.

d. Prepare a naive forecast for periods 2 through 11 using the given sales data. Compute each of the following; (1) MSE, (2) MAD, (3) tracking signal at month 10, and (4) 2*s* control limits. How do the naive results compare with the other two forecasts?

29. The classified department of a monthly magazine has used a combination of quantitative and qualitative methods to forecast sales of advertising space. Results over a 20-month period are as follows:

Month	Error	Month	Error
1	−8	11	1
2	−2	12	6
3	4	13	8
4	7	14	4
5	9	15	1
6	5	16	−2
7	0	17	−4
8	−3	18	−8
9	−9	19	−5
10	−4	20	−1

a. Compute a tracking signal for months 11 through 20. Compute an initial value of MAD for month 11, and then update it for each month using exponential smoothing with $\alpha = .1$. What can you conclude? Assume limits of ±4.

b. Using the first half of the data, construct a control chart with 2*s* limits. What can you conclude?

c. Plot the last 10 errors on the control chart. Are the errors random? What is the implication of this?

a.

	MAD	T.S.
11	4.727	0
12	4.857	1.235
13	5.171	2.707
14	5.054	3.562
15	4.649	4.087
16	4.384	3.878
17	4.346	0.334
18	4.711	1.061
19	4.740	0
20	4.366	−0.229

Forecast is suspect beyond month 15.

b. 0 ± 12.38

All errors within limits

c. Graph in IM

Errors may be cyclical.

30. A textbook publishing company has compiled data on total annual sales of its business texts for the preceding nine years:

Year:	1	2	3	4	5	6	7	8	9
Sales (000):	40.2	44.5	48.0	52.3	55.8	57.1	62.4	69.0	73.7

a. Using an appropriate model, forecast textbook sales for each of the next five years.

b. Prepare a control chart for the forecast using the original data. Use 2*s* limits.

c. Suppose actual sales for the next five years turn out as follows:

Year:	10	11	12	13	14
Sales (000):	77.2	82.1	87.8	90.6	98.9

Is the forecast performing adequately? Explain.

a. $y = 35.8 + 4.02t$

$y_{10} = 75.98$, $y_{11} = 80.00$, $y_{12} = 84.02$, $y_{13} = 88.04$, $y_{14} = 92.05$

b. Control limits: 0 ± 2.74

c. No. Errors for years 12, 14 outside limits.

31. Sales of CD-ROMs over the past 12 weeks are shown in the following table. The manager has decided to use exponential smoothing to forecast sales, using a smoothing constant of either .10 or .40.

Week	1	2	3	4	5	6	7	8	9	10	11	12
Sales	40	44	46	43	45	44	40	43	44	42	46	45

a. Using the data from weeks 1 through 7, determine which smoothing constant would produce the smaller errors.

b. Construct a control chart based on the six errors for the better smoothing constant.

c. Using the better smoothing constant, continue on with the forecasts up to week 12. Are these forecast errors in control?

d. What is the forecast for week 13? If the actual sales turn out to be 48, is the forecast in control?

a.

	$\alpha = .10$	$\alpha = .40$
MSE	15.09	10.86

b. Control limits: $0 \pm 2\sqrt{\text{MSE}} = 0 \pm 6.59$

c. Yes

d. 44.4; yes

32. A manager uses this equation to predict demand: $y_t = 10 + 5t$. Over the past eight periods, demand has been as follows.

Period, *t*:	1	2	3	4	5	6	7	8
Demand:	15	21	23	30	32	38	42	47

Is the forecast performing adequately? Explain.

No. Demand is overestimated consistently.

33. A manager uses a trend equation plus quarterly relatives to predict demand. Quarter relatives are Q1 = .90, Q2 = .95, Q3 = 1.05, and Q4 = 1.10. The trend equation is: $y_t = 10 + 5t$. Over the past nine quarters, demand has been as follows.

Period, *t*:	1	2	3	4	5	6	7	8	9
Demand:	14	20	24	31	31	37	43	48	52

Is the forecast performing adequately? Explain.

CASE

M&L MANUFACTURING

M&L Manufacturing makes various components for printers and copiers. In addition to supplying these items to a major manufacturer of printers and copiers, the company distributes these and similar items to office supply stores and computer stores as replacement parts for printers and desktop copiers. In all, the company makes about 20 different items. The two markets (the major manufacturer and the replacement market) require somewhat different handling. For example, replacement products must be packaged individually whereas products are shipped in bulk to the major manufacturer.

The company does not use forecasts for production planning. Instead, the operations manager decides which items to produce, and the batch size, based partly on orders, and the amounts in inventory. The products that have the fewest amounts in inventory get the highest priority. Demand is uneven, and the company has experienced being overstocked on some items and out of others. Being understocked has occasionally created tensions with the managers of retail outlets. Another problem is that prices of raw materials have been creeping up, although the operations manager thinks that this might be a temporary condition.

Because of competitive pressures and falling profits, the manager has decided to undertake a number of changes. One change is to introduce more formal forecasting procedures in order to improve production planning and inventory management.

With that in mind, the manager wants to begin forecasting for two products. These products are important for several reasons. First, they account for a disproportionately large share of the company's profits. Second, the manager believes that one of these products will become increasingly important to future growth plans; and third, the other product has experienced periodic out-of-stock instances.

The manager has compiled data on product demand for the two products from order records for the previous 14 weeks. These are shown in the following table.

Week	Product 1	Product 2
1	50	40
2	54	38
3	57	41
4	60	46
5	64	42
6	67	41
7	70	41
8	76	47
9	79	42
10	82	43
11	85	42
12	87	49
13	92	43
14	96	44

Questions

1. What are some of the potential benefits of a more formalized approach to forecasting?
2. Prepare a weekly forecast for the next four weeks for each product. Briefly explain why you chose the methods you used. (*Hint:* For product 2, a simple approach, possibly some sort of naive/intuitive approach, would be preferable to a technical approach in view of the manager's disdain of more technical methods.)

Selected Bibliography

Bowerman, Bruce L., and Richard T. O'Connell. *Forecasting and Time Series: An Applied Approach.* 3rd ed. Belmont, CA: Duxbury Press, 1993.

Hanke, John E., and Arthur G. Reitsch. *Business Forecasting.* 4th ed. Boston: Allyn & Bacon, 1992.

Levenbach, Hans, and James P. Clary. *The Modern Forecaster: The Forecasting Process through Data Analysis.* Belmont, CA: Lifetime Learning Publications, 1984.

Makridakis, Spyros, and Steven Wheelright. *Forecasting Methods for Management.* 5th ed. New York: John Wiley and Sons, 1989.

Stevenson, William J. *Business Statistics: Concepts and Applications.* 2nd ed. New York: Harper & Row, 1985.

PART THREE
DESIGN OF PRODUCTION SYSTEMS

Production system design encompasses decisions involving:

1. Product and service design, Chapter 4
2. Process selection and capacity planning, Chapter 5
3. Facilities layout, Chapter 6
4. Design of work systems, Chapter 7
5. Location planning, Chapter 8

Satisfying the customer begins with product and service design. Moreover, decisions made in this area impact on operations and on the organization's overall success.

Similarly, process selection and capacity planning impact on the ability of the production system to perform and to satisfy customers. Flexibility, production time, and cost are key considerations in process design.

Process selection and layout are closely related. Layout decisions involve the arrangement of units and equipment within departments, which affects the flow of work through a system and impacts productivity, cost, and flexibility. Layout decisions are influenced by decisions made in product and service design.

Capacity and location decisions influence operating costs and the ability to respond to customer demand. Location decisions also impact on transportation costs, labor availability, material costs, and access to markets.

Work design focuses on the human element in production systems. Increasingly, managers are realizing that workers are a valuable asset and can contribute greatly to the organization's success. Strategic planning is beginning to incorporate employee participation to help improve production systems.

Design decisions have strategic significance for business organizations. Many of these decisions are not made by the operations manager. Nonetheless, because of the important links between operations and each strategic area, it is essential to the success of the organization to involve all of the functional areas of the organization in design decisions.

LEARNING OBJECTIVES

After completing this chapter, you should be able to:

1. Discuss the importance of product and service design.
2. Describe the design process.
3. Explain the concept of standardization and discuss its advantages and disadvantages.
4. Discuss the concept of modular design, including its advantages and disadvantages.
5. Describe the contributions of R&D to product and service design.
6. Define reliability and suggest possible ways of improving reliability.

CHAPTER FOUR
PRODUCT AND SERVICE DESIGN

CHAPTER OUTLINE

As more and more women join the workforce and more families rely on two incomes, the spending and eating habits of Americans are changing. Quick meals have replaced leisurely meals. There is an increased awareness of healthy foods. And spicy foods have replaced plain foods. Fast-food chains, food companies, and supermarkets, are scrambling to meet the challenge.

Spice giant McCormick is finding that sales of traditional spices are down. To compensate, the company is promoting seasoning mixes which are designed to save time. Salsa is becoming very popular, and Mexican restaurants are springing up all over. Supermarkets are offering a wide array of already prepared foods (see the Wegmans Tour in Chapter 1) as well as recipes for quick meals in their stores and on their web pages.

For these and other companies, from high tech to no tech, product and service design plays an important role in their profitability and their very survival.

The essence of any organization is the products or services it offers. There is an obvious link between the *design* of those products or services and the *success* of the organization. Organizations that have well-designed products or services are more likely to realize their goals than those with poorly designed products or services. Hence, organizations have a vital stake in achieving good product and service design.

This chapter presents the major aspects of product and service design: the objectives of product and service design, the design process, Taguchi methods for product design, research and development, and design of services.

INTRODUCTION

Product and service design plays a strategic role in the degree to which an organization is able to achieve its goals. It is a major factor in customer satisfaction, quality, and production costs. The customer connection is obvious: The main concern of the customer is the organization's products or services, which become the ultimate basis for judging the organization. The quality connection is twofold. Quality is clearly affected not only by design but also, during production, by the degree to which production conforms to the intent of design. A key factor in the production of goods is *manufacturability,* which refers to the ease with which design features can be achieved by production. Similarly, design affects cost—the cost of materials specified by design, and labor and equipment costs.

At the General Motors Tech Center in Warren, Michigan, designers are carving a clay model of a future car for prototype design styling.

www.gm.com

Product development involves nearly every functional area of an organization to some extent. However, three functions have major involvement here: marketing, design, and operations.

Reasons for Product or Service Design

Organizations become involved in product or service design for a variety of reasons. An obvious one is to be competitive by offering new products or services. Another one is to make the business grow and increase profits. Furthermore, the best organizations try to develop new products or services as an alternative to downsizing. When productivity gains result in the need for fewer workers, developing new products or services can mean adding jobs and retaining people instead of letting them go.

Sometimes product or service design is actually *redesign*. This, too, occurs for a number of reasons such as customer complaints, accidents or injuries, excessive warranty claims, or low demand. The desire to achieve cost reductions in labor or materials can also be a motivating factor.

NEWSCLIP

MORE CARS COME WITH A SHADE OF GREEN— RECYCLED MATERIALS

www.ford.com
www.gm.com
www.chryslercorp.com

Detroit's Big Three automakers, doing their best to build cars that don't fall apart, have a new goal: building cars that are easy to take apart.

The reason: Easy-to-remove parts are easy to recycle.

Car companies are putting the ability to recycle parts on the same level as safety, fuel economy, and costs when they design new vehicles.

For example, the 1995 Oldsmobile Aurora . . . uses scrap metal in its radiator mounting, and the bumper beams contain recycled copper and aluminum. Chrysler Corp. uses recycled tires for the splash guards on its midsize sedans.

Car parts have been recycled for years. But the auto industry only recently began to build cars with the idea of using recycled material. About 75 percent of new cars contain recycled material, mostly iron or steel used in the body.

Auto dismantlers usually buy a vehicle and remove all the parts that still work, such as seats, engines and headlights. The vehicle then goes to a shredder where it is reduced to small fragments and a huge magnet separates out the metal parts.

The challenge for auto companies is to find ways to separate the more than 20,000 different grades of plastic found in cars. About 24 percent of shredded material, known as "fluff," contains plastic, fluids, rubber, glass and other material. Most "fluff" can't be recycled.

Ford, GM and Chrysler have jointly formed the Vehicle Recycling Partnership in hopes of improving the technology to recover plastics and other material found in "fluff." Suppliers of material and the recycling industry are included in the partnership.

Manufacturers aren't suddenly becoming Friends of the Earth. "All of the recycling programs undertaken by Ford have been cost-effective," says Susan Day, vehicle recycling coordinator.

Source: *Rochester Democrat and Chronicle*, February 20, 1994, p. 11.

Trends in Product and Service Design

Over the last few years, the designing of products and services has increased emphasis on a number of aspects of design. Some of these are strictly related to product design, while others are related to both product and service design. Among them are the following:

1. Increased emphasis on customer satisfaction and increased pressure to be competitive. Total quality management programs, which have customer satisfaction as their primary focus, contribute to this.
2. Increased emphasis on reducing the time needed to *introduce* a new product or service.
3. Increased emphasis on reducing the time needed to *produce* a product or provide a service. Reduced production time usually results in lower cost and higher quality.

4. Greater attention to the capabilities of the organization to produce or deliver the item.
5. Greater attention to environmental concerns, including waste minimization, recycling parts, and disposal of worn-out products.
6. Increased emphasis on designing products and services that are user-friendly.
7. Increased effort to use less material for products (e.g., concentrated liquid detergents) and less packaging.

In a competitive environment, getting new or improved products or services to the market ahead of competitors gives an organization a competitive advantage that can lead to increased profits and an increased market share, and create an image of the organization as a leader. In the not-for-profit sector, getting new or improved products or services to the market as quickly as possible enables organizations to increase their level of customer service. For example, quick FDA approval of new drugs to combat cancer and AIDS results in a major benefit to society. Similarly, the quick approval and marketing of nonfat food products provides another health benefit to society.

Objectives of Product and Service Design

The objectives of product design and service design differ somewhat, but not as much as you might imagine. The overall objective for both is to satisfy the customer while making a reasonable profit.

It is important to note that although profit is generally the overall measure of design effectiveness, because the time interval between the design phase and profit realization is often considerable, more immediate measures come into play. These typically include development time and cost, the product or service cost, and the resulting product or service quality.

Quality, of course, is typically high on the list of priorities in product and service design. At one time, having high quality was enough for a product or service to stand out; now it is the norm, and those that fall below this norm are the ones that stand out.

Last, but certainly not least, it is *crucial* for designers to take into account the capabilities of the organization to produce or deliver a given product or service. This is sometimes referred to as **design for operations.**

design for operations Taking into account the capabilities of the organization in designing goods and services.

THE DESIGN PROCESS

The design process begins with the *motivation* for design. For a new business or a new product, the motivation may be obvious: to achieve the goals of the organization. For an existing business, in addition to that general motivation, there are more specific

Chrysler's 1.3 mile automated durability road simulates bad roads and is used for testing the integrity of their automobiles and trucks. Chrysler vehicles are "guided" by computer-controlled robots.

A milestone crash test performed at Ford Motor Corporation to plan for a new side-impact head and chest air bag.

factors to consider, such as government regulations, competitive pressures, customer needs, and the appearance of new technologies that have product or process applications.

Ultimately, the *customer* is the driving force for product and service design. Failure to satisfy customers can result in customer complaints, returns, warranty claims, and so on. Loss of market share becomes a potential problem if customer satisfaction is not achieved.

Vol. 4, Tape 1, Seg. 3
Detroit Diesel

For the design process to occur, a business must have *ideas* for new or improved designs. These come from a variety of sources, most obviously the customer. *Marketing* can tap this source of ideas in a number of ways, such as the use of focus groups, surveys, and analyses of buying patterns.

Some organizations have *research and development departments* (described later in the chapter) that also generate ideas for new or improved products and services.

Competitors are another important source of ideas. By studying a competitor's products or services, and how the competitor operates (e.g., pricing policies, return policies, warranties), an organization can learn a great deal about achieving design improvements. Beyond that, some companies buy a competitor's newly designed product the moment it appears on the market. Using a procedure called **reverse engineering,** they carefully dismantle and inspect the product. This may uncover product improvements that can be incorporated in their own product. The Ford Motor Company used this approach in developing its highly successful Taurus model; it examined competitors' automobiles, searching for the *best-in-class* products (e.g., the best hood release, the best dashboard, the best window handle). Sometimes reverse engineering can lead to a product that is *superior* to the one being examined; that is, designers conceive an improved design, which enables them to "leapfrog" the competition by quickly introducing an improved version of a competitor's product. This could enable the company to realize some of the rewards that normally accrue to the first to introduce a new product or feature.

reverse engineering Dismantling and inspecting a competitor's product to discover product improvements.

www.ford.com

Ideas for new or improved products and services cannot be entertained in a vacuum; production *capabilities* must be a basic consideration. Design needs to clearly understand the capabilities of production (e.g., equipment, skills, types of materials, schedules, technologies, special abilities). This will help in choosing designs that match capabilities. When opportunities and capabilities do not match, management must consider the potential for expanding or changing capabilities to take advantage of those opportunities.

Forecasts of future demand can be very useful, supplying information on the timing and volume of demand, and information on demands for new products and services.

The design of a product or service must take into account its cost, its target market, and its function. **Manufacturability** is a key concern for manufactured goods: Ease of fabrication and/or assembly is important for cost, productivity, and quality. With services, ease of providing the service, cost, productivity, and quality are of great concern.

manufacturability The ease of fabrication and/or assembly.

Figure 4–1 offers a humorous look at some of the ways various individuals and departments in the design process can interpret a "design." The point is that sufficient information must be available to determine the customer's wants, and communication and agreement among those responsible for designing, producing, and marketing the product or service.

In general, design, production or operations, and marketing must work closely together, keeping each other informed and taking into account the wants and needs of the customer. In addition, legal or regulatory consideration and *life-cycle* factors can influence the design function.

REGULATIONS AND LEGAL CONSIDERATIONS

Organizations have been faced with a large array of government agencies that regulate their activities. Among the more familiar ones are the Food and Drug Administration, the Environmental Protection Agency, the National Highway Safety Commission, and

FIGURE 4–1

Differing views of design created through lack of information

As proposed by the marketing department.

As specified in the product request.

As designed by the senior designer.

As produced by manufacturing.

As used by the customer.

What the customer wanted.

Source: *Educational Center Newsletter,* Minneapolis, Minnesota.

the Consumer Products Safety Commission. Bans on cyclamates, red food dye, phosphates, and asbestos have sent designers scurrying back to their drawing boards to find alternative designs that were acceptable to both government regulators and customers. Similarly, automobile pollution standards and safety features, such as seat belts, air bags, safety glass, and energy-absorbing bumpers and frames, have had a substantial impact on automotive design. Much attention also has been directed toward toy design to remove sharp edges, small pieces that can cause choking, and toxic materials. In construction, government regulations require the use of lead-free paint, safety glass in entranceways, access to public buildings for handicapped persons, and standards for insulation, electrical wiring, and plumbing.

product liability A manufacturer is liable for any injuries or damages caused by a faulty product.

Uniform Commercial Code Products carry an implication of merchantability and fitness.

Product liability can be a strong incentive for design improvements. **Product liability** means that a manufacturer is liable for any injuries or damages caused by a faulty product because of poor workmanship or design. Many business firms have faced lawsuits related to their products, including Firestone Tire & Rubber, Ford and General Motors, and toy manufacturers. Manufacturers also are faced with the implied warranties created by state laws under the **Uniform Commercial Code,** which says that products carry an implication of *merchantability* and *fitness*; that is, a product must be usable for its intended purposes.

The suits and potential suits have led to increased legal and insurance costs, expensive settlements with injured parties, and costly recalls. Moreover, increasing customer awareness of product safety can adversely affect product image and subsequent demand for a product.

Thus, it is extremely important to design products that are reasonably free of hazards. When hazards do exist, it is necessary to install safety guards or other devices for reducing accident potential, and to provide adequate warning notices of risks. Consumer groups, business firms, and various government agencies often work together to develop industrywide standards that help avoid some of the hazards.

RESEARCH AND DEVELOPMENT

research and development Organized efforts to increase scientific knowledge or product innovation.

Research and development (R&D) refers to organized efforts that are directed toward increasing scientific knowledge and product (or process) innovation. Most of the advances in semiconductors, medicine, communications, and space technology can be attributed to R&D efforts at colleges and universities, research foundations, government agencies, and private enterprises.

R&D efforts may involve *basic research, applied research*, or *development*.

Basic research has the objective of advancing the state of knowledge about a subject, without any near-term expectation of commercial applications.

Applied research has the objective of achieving commercial applications.

Development converts the results of applied research into useful commercial applications.

Basic research, because it does not lead to near-term commercial applications, is generally underwritten by the government and large corporations. Conversely, applied research and development, because of the potential for commercial applications, appeals to a wide spectrum of business organizations.

The benefits of successful R&D can be tremendous. Some research leads to patents, with the potential of licensing and royalties. However, many discoveries are not patentable, or companies don't wish to divulge details of their ideas following the patent route. Even so, the first organization to bring a new product or service to the market generally stands to profit from it before the others can catch up. Early products may be priced higher because a temporary monopoly exists until competitors bring their versions out.

www.kodak.com

The costs of R&D can be high. Kodak, for example, spends more than $1 million *a day* on R&D. Large companies in the automotive, computer, communications, and pharmaceutical industries spend even more. Even so, critics say that many U.S. companies spend too little on R&D, a factor often cited in the loss of competitive advantage.

It is interesting to note that some companies are now shifting from a focus primarily on *products* to a more balanced approach that explores both product and *process* R&D. One reason is that in too many instances, product innovations (e.g., for televisions, VCRs, and microwave ovens) made by U.S. companies have ended up being produced more competitively by foreign companies with better processes.

In certain instances, research may not be the best approach, as explained in the following reading.

READING

MANAGER'S JOURNAL: WHEN CUSTOMER RESEARCH IS A LOUSY IDEA Willard I. Zangwill

www.sony.com

Customer research is often touted as a necessary precursor to product introduction. The problem—especially for innovative products—is that it often proves wrong. For example, hair styling mousse is now a massive hit. Yet in its initial market tests in the U.S., it flopped. "Goopy and gunky" was what people said about it, and they did not like its feel when it "mooshed" through their hair.

Similarly, when the telephone answering machine was consumer tested, it faced an almost universally negative reaction. Back then, most individuals felt that using a mechanical device to answer a phone was rude and disrespectful. Today, of course, many people regard their answering machines as indispensable, and consider scheduling their daily activities without them as impossible. In the same vein, the computer mouse in its initial testing flunked, being evaluated by potential customers as awkward and unnecessary.

Because of these difficulties, some companies have gone so far as to eliminate customer research for their innovative products. According to Sony executive Kozo Ohsone, "When you introduce products that have never been invented before, what good is market research?" The Walkman was launched without the standard customer research, as is typical at Sony.

With customer research not only costly, but often in error, how can a manager determine the innovations customers want? The solution may be design-for-purpose, a new approach in which a firm uses speed and flexibility to gain customer information instead of, or in addition to, standard customer research.

To illustrate, Sony obtains information from the actual sales of various Walkman models and then quickly adjusts its product mix to conform to those sales patterns. Specifically, the process design of each Walkman model is based on a core platform containing the essential technology. But the platform is designed to be flexible, which allows a wide range of models to be easily built on it, such as a beach model, a child's model, one that attaches to the arm, and so on.

Depending upon which models sell, the models or features are changed, but the platform remains the same. If pink is a hot selling color, they make more pink models. If beach models sell well, they make more of the existing models and also expand the line. This technique is far more accurate than deciding what to make using traditional customer research.

Similarly, without customer research, every season Seiko "throws" into the market several hundred new models of its watches. Those that customers buy, it makes more of; the others, it drops. Capitalizing on the design-for-response strategy, Seiko has a highly flexible design and production process that lets it quickly and inexpensively introduce products. Do they worry if a high percentage of the watches they introduce fail, rejected by the customers? No (unless the failure rate is extremely high), because their fast, flexible product design process has slashed the cost of failure.

When creating a new magazine, Hearst Magazines also follows this approach. Hearst learned that it was almost impossible to customer test the magazine ideas, and that it was better to launch the magazine and see what happens. To do this, Hearst has created a special group of editors with the talent and flexibility to launch almost any new magazine. Based upon the initial sales of the new magazine, they will either revise the content and format or drop the publication. Any new magazine that proves successful is spun off to run independently.

Crucial to this approach, however, is reducing the cost of the failures by keeping expenses down. Hearst accomplishes this by initially hiring one overall editor on a short-time basis, using stringers as writers, and borrowing advertising people. Also, with experience it has discovered the tricks of launching new magazine products inexpensively. For example, it has learned how to test different cover designs efficiently, and how to test sales in different markets, such as newsstands or subscribers.

Many other firms also follow the strategy of using customer research data less and fast-flexible response more, with the food industry in the lead. One of the problems with customer research into foods is that a person's desire for food is powerfully influenced by the ambience, the dining companions and what foods were eaten recently, all of which confound and confuse the results of the customer research. Even more erratic are the results with children's food, say a new cereal or snack. The responses of kids are strongly swayed by how well they like the people doing the test and the playthings available. Worse, kids quickly change their minds, and in a taste test of several foods a child can judge one food the best but one hour later proclaim the same food as "icky."

Arthur D. Little & Co. discovered that of all new cereals introduced to the market, 92 percent had failed. Since using the full array of customer research techniques produces a success rate of only 8 percent, more and more companies are revising their thinking about doing customer research as usual. Innovative firms such as Keebler and the leading cereal makers are reducing their expenditure for customer research and instead are vigorously cutting the cost of launching new products, including making their manufacturing processes more flexible.

Design-for-response enables firms not only to employ customer research when beneficial but also to respond quickly to what the customers really want, keeping the firm on top of market shifts and surprises.

NOTE: Mr. Zangwill is a professor at the Graduate School of Business, University of Chicago, and author of "Lightning Strategies for Innovation" (Lexington, 1992).

(See related letter: "Letters to the Editor: Testing the Waters Before the Launch" WSJ—April 1, 1993)

Source: *The Wall Street Journal*, March 8, 1993, p. A12. Reprinted by permission of *The Wall Street Journal*,

STANDARDIZATION

standardization Extent to which there is absence of variety in a product, service, or process.

An important issue that often arises in both product/service design and process design is the degree of standardization. **Standardization** refers to the extent to which there is absence of variety in a product, service, or process. Standardized products are made in large quantities of identical items; calculators, computers, and 2 percent milk are examples. Standardized service implies that every customer or item processed receives essentially the same service. An automatic car wash is a good example; each car, regard-

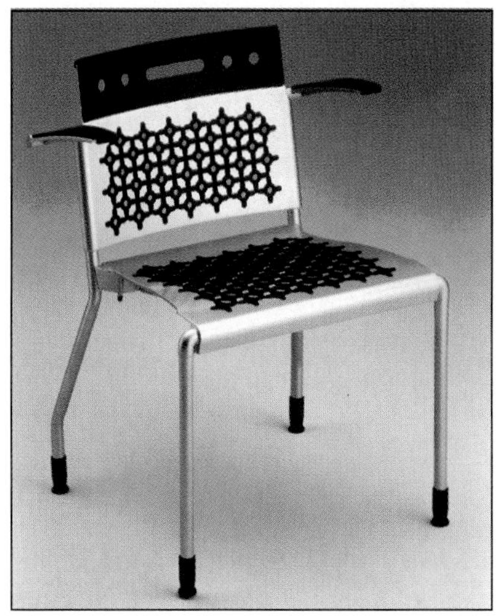

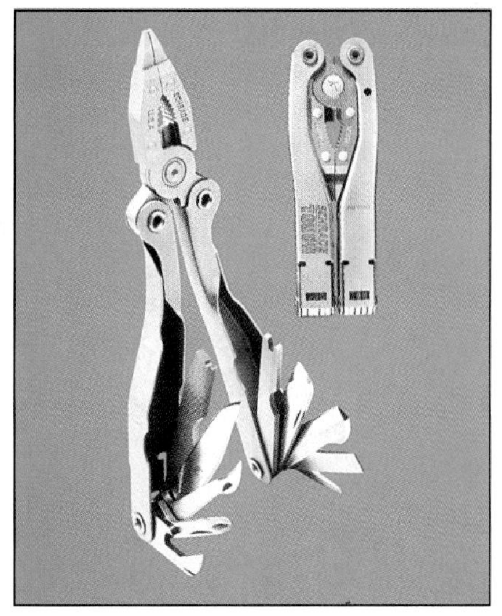

Winners of the 1997 Industrial Design Excellence Award sponsored by Business Week. *Itoki's indoor/outdoor chair; Shrade's Tough Tool has 21 pliers, screwdrivers, knives, and other implements; Toshiba's Infinia PC with its user friendly control panel.*

www.toshiba.com

less of how clean or dirty it is, receives the same service. Standardized processes deliver standardized service or produce standardized goods.

Standardization carries a number of important benefits as well as certain disadvantages. Standardized products mean *interchangeable parts,* which greatly lower the cost of production while increasing productivity and making replacement or repair relatively easy compared with that of customized parts. Design costs are generally lower. For example, General Motors recently has attempted to standardize key components of its automobiles across product lines; components such as brakes, electrical systems, and other "under-the-skin" parts would be the same for all GM car models. By reducing variety, GM saves time and money, and increases quality and reliability in its products.

www.gm.com

Another benefit of standardization is reduced time and cost to train employees, and reduced time to design jobs. Similarly, scheduling of work, inventory handling, and purchasing and accounting activities become much more routine.

Lack of standardization can at times lead to serious difficulties and competitive struggles, particularly when systems running under different conditions are incompatible. Consider a few examples: When VCRs were first introduced, there were two formats for tapes: VHS and Beta. Machines could play one or the other, but not both. This meant that producers needed to make two sets of tapes. High-definition television might have been introduced much earlier in the United States, but three competing—and incompatible—systems were proposed, which led to prolonged debate and study before one system could be agreed upon. The lack of standardization in computer software and operating systems (Macintosh versus IBM) has presented users with hard choices because of the difficulty in switching from one system to the other. And the use by U.S. manufacturers of the English system of measurement, while most of the rest of the world's manufacturers use the metric system, has led to problems in selling U.S. goods in foreign countries and in buying foreign machines for use in the United States. This may make it more difficult for U.S. firms to compete in the European Union. Similarly, U.S. auto manufacturers have complained for years about their inability to freely enter the Japanese market, but only recently have they begun to offer cars with steering wheels on the right side—the universal standard in Japan.

Standardization also has disadvantages. A major one relates to the reduction in variety. This can limit the range of customers to whom a product or service appeals. Customers may reluctantly accept a product only because nothing else suits their needs.

TABLE 4–1

Advantages and disadvantages of standardization

Advantages	1. Fewer parts to deal with in inventory and in manufacturing. 2. Reduced training costs and time. 3. More routine purchasing, handling, and inspection procedures. 4. Orders fillable from inventory. 5. Opportunities for long production runs and automation. 6. Need for fewer parts justifies increased expenditures on perfecting designs and improving quality control procedures.
Disadvantages	1. Designs may be frozen with too many imperfections remaining. 2. High cost of design changes increases resistance to improvements. 3. Decreased variety results in less consumer appeal.

But that creates a risk that a competitor will introduce a better product or greater variety (a feature of lean production), and realize a competitive advantage. Another disadvantage is that a manufacturer may freeze (standardize) a design prematurely and, once frozen, it may find compelling reasons to resist modification. A familiar example of this is the keyboard arrangement of typewriters and computer keyboards. Studies have demonstrated that another arrangement of keys would be more efficient, but the cost of replacing all of the equipment in existence and retraining millions of typists and word processors would not be worth the benefit. Similarly, U.S.-made color TV sets have a lower scanning rate (lines per inch) than foreign-made sets, resulting in poorer resolution. Early freezing of design was partly to blame for this.

Obviously, designers must consider important issues related to standardization when making choices. The major advantages and disadvantages of standardization are summarized in Table 4–1.

PRODUCT DESIGN

This section provides an overview of various approaches to product design, including product life cycles, manufacturing design, remanufacturing, robust design, concurrent engineering, computer-aided design, and modular design.

Product Life Cycles

product life cycle Incubation, growth, maturity, saturation, and decline.

Many new products go through a **product life cycle** in terms of demand. When an item is first introduced, it may be treated as a curiosity. Demand is generally low because potential buyers are not yet familiar with the item. Many potential buyers recognize that all of the bugs have probably not been worked out and that the price may drop after the introductory period. With the passage of time, production and design improvements usually create a more reliable and less costly product. Demand then grows for these reasons and because of increasing awareness of the product or service. At the next stage in the life cycle, the product reaches maturity: there are few, if any, design changes, and demand levels off. Eventually, the market becomes saturated, which leads to a decline in demand. These stages in the cycle are illustrated in Figure 4–2.

In the last stage of a life cycle, some firms adopt a defensive research posture whereby they attempt to prolong the useful life of a product or service by improving its reliability, reducing costs of producing it (and, hence, the price), redesigning it, or changing the packaging.

Consider the products in various stages of the life cycle in the music industry: Digital audio tapes are in the introductory stage, compact disks are in the growth stage, cassettes are moving from the maturity-saturation stage into the decline stage.

Some products do not exhibit life cycles: wooden pencils, paper clips, nails, knives, forks and spoons, drinking glasses, and similar items. However, most new products do.

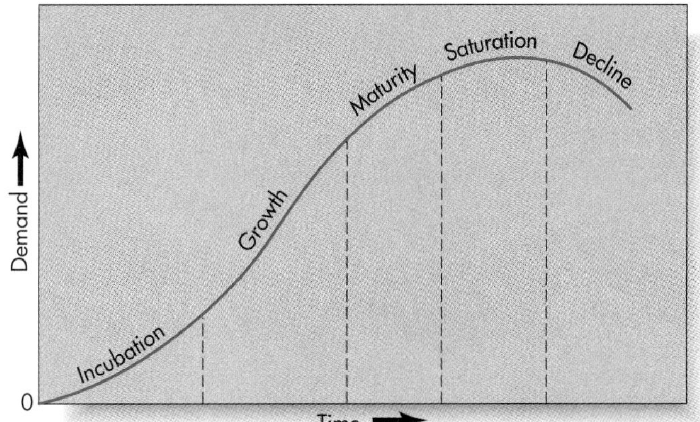

FIGURE 4-2

Products or services may exhibit life cycles over time

Wide variations exist in the amount of time a particular product takes to pass through a given phase of its life cycle: some products pass through various stages in a relatively short period; others take considerably longer. Often it is a matter of the basic *need* for the item and the *rate of technological change*. Some toys, novelty items, and style items have a life cycle of less than one year, whereas other, more useful items, such as clothes washers and dryers, may last for many years before yielding to technological change.

Manufacturing Design

The term **design for manufacturing (DFM)** is also used to indicate the designing of products that are compatible with an organization's capabilities. A related concept in manufacturing is **design for assembly (DFA).** A good design must take into account not only how a product will be fabricated, but also how it will be assembled. Design for assembly focuses on reducing the number of parts in an assembly, as well as the assembly methods and sequence that will be employed.

Environmental regulations and recycling have given rise to another concern for designers, **design for recycling (DFR).** Here the focus is on designing products to allow for dis-assembly of used products for the purpose of recovering components and materials for reuse.

design for manufacturing (DFM) Designers take into account the organization's capabilities when designing a product.

design for assembly (DFA) Design focuses on reducing the number of parts in a product and on assembly methods and sequence.

design for recycling (DFR) Design facilitates the recovery of materials and components in used products for reuse.

Remanufacturing

An emerging concept in manufacturing is to remanufacture products. **Remanufacturing** refers to removing some of the components of old products and reusing them in new products. This can be done by the original manufacturer, or another company. Among the products that have remanufactured components are automobiles, printers, copiers, cameras, computers, and telephones.

There are a number of important reasons for doing this. One is that a remanufactured product can be sold for about 50 percent of the cost of a new product. Another is that the process requires mostly unskilled and semi-skilled workers. And in the global market, European lawmakers are increasingly requiring manufacturers to take back used products, because this means fewer products end up in landfills and there is less depletion of natural resources such as raw materials and fuel.

remanufacturing Using some of the components of old products in the manufacture of new products.

Designing products so that they can be more easily taken apart has given rise to yet another design consideration: making it easier to take apart used products. **Design for dis-assembly (DFD)** includes using fewer parts and less material, and using snap-fits where possible instead of screws or nuts and bolts.

The reading "Making It (Almost) New Again" gives examples of what some companies are doing.

design for dis-assembly (DFD) Design so that used products can be easily taken apart.

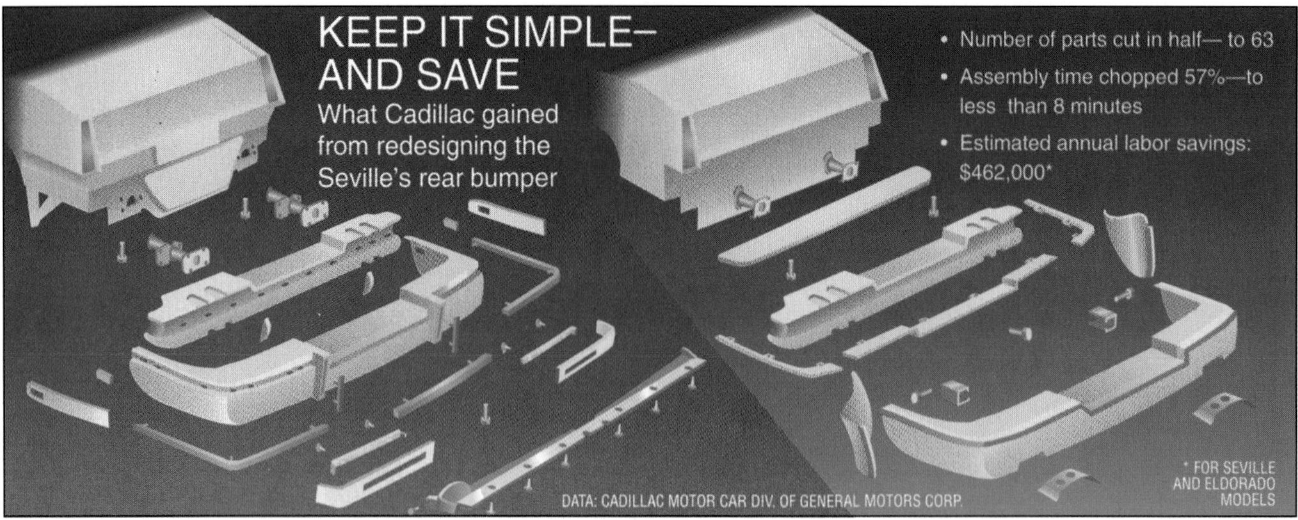

By redesigning the Seville's rear bumper, Cadillac cut the number of parts, thus reducing assembly time and labor costs. The new design also leads to high quality as there are few parts and steps that might be defective. Reprinted by special permission of Business Week *© 1991.*

READING

MAKING IT (ALMOST) NEW AGAIN **Phil Ebersole**

www.kodak.com

Tired: *Landfills.*
Wired: *Recycling.*
Inspired: *Remanufacturing.*

The symbol of 20th century industry was the assembly line. The symbol of 21st century industry may be the disassembly line.

Xerox Corp. and Eastman Kodak Co. design products to make them not only easy to put together, but easy to take apart.

That's because so many parts and components from their old products are refurbished and put into new ones.

Xerox and Kodak, along with Caterpillar Inc. are the leaders in a movement called remanufacturing, said Robert T. Lund, a professor of manufacturing engineering at Boston University and author of a 1996 study of the subject.

"The driving force behind remanufacturing is thrift," Lund said. "A remanufactured product can be sold for 45 to 60 percent of the cost of a new one. You have something enormously more valuable than if you ground it up as raw material."

But in a few years, remanufacturing may be more than just a good idea. European countries are developing rules to make manufacturers take back their products instead of allowing them to wind up in landfills. Europe's rules could set the standard for the world, just as California auto emissions laws set the standard for the U.S. auto industry.

The 15-nation European Union is considering a rule that would require 85 percent of a car by weight to be recycled or remanufactured by the year 2002. This would increase to 95 percent by 2015.

This goes beyond what's done now. Currently about 75 percent of the average U.S. car is recycled or remanufactured. About all the metal in a car is reused, but little plastic and other materials.

Fixing up used equipment for resale is nothing new, but Xerox and Kodak take remanufacturing to the point of breaking down the distinction between new and used.

Almost all their new copiers and single-use cameras contain remanufactured parts. Virgin and remanufactured components go through the same production lines and meet the same tests.

If you could find an all-virgin product, they say, you couldn't tell the difference between it and one that was 95 percent remanufactured.

It's a process that goes beyond recycling, because companies conserve not only raw materials, but the energy, labor and ingenuity that went into making the components.

Lund said there are 73,000 companies in 61 industries, ranging from computer chips to locomotives, who do remanufacturing. They have 480,000 employees and do $53 billion worth of business.

Rochester Institute of Technology is a leader in this movement. It operates a remanufacturing laboratory at its Center for Integrated Manufacturing

Studies and publishes a quarterly called *Remanufacturing Today.*

Remanufacturing isn't easy:

- Although companies ultimately may save money, the initial costs are higher. Remanufacturing is labor-intensive. Each remanufactured component is different, so the process can't be automated.
- Remanufacturers have to overcome a reputation for low quality. "People think remanufacturing is like repair, but it isn't," said Nabil Nasr, an RIT professor of manufacturing engineering.
- Designing products for remanufacturing makes it easier for other companies to refurbish your used products and sell them in competition with you.

For example, Kodak, along with Fuji Photo Film Co. and Konica Corp., battles "reloaders"—companies that sell poorly remade cameras under their own names with cheap Chinese film and used lenses and batteries, said David M. Snook, manager of worldwide recycling for Kodak single-use cameras.

The better Kodak designs its cameras for remanufacturing, the easier Kodak makes it for reloaders.

Remanufacturing is mainly carried on by small and mid-sized companies. Few large U.S. producers remanufacture their own products to the degree Xerox and Kodak do.

Some big companies still try to discourage remanufacturing, Lund said. They regard remanufactured products as competition for their virgin products. Others, like the Big 3 automakers, sanction or subcontract remanufacturing, but do little themselves.

Richard O. Carville, manager of design and manufacturing engineering for Xerox's print cartridge business unit, said he encountered skepticism in 1990 when he proposed remanufacturing print cartridges, the part of the copier that registers and prints the xerographic image.

After the first six months, the unit made a profit. It was able to cut prices as a result of the cost savings it had achieved.

One big challenge has been persuading customers to return the print cartridges, Carville said.

The first leaflets on cartridge return were ineffective. Now, when a customer opens the print engine package, the first thing he or she sees is a prepaid United Parcel Service or Canada Post mailer, shaped like an airplane.

"Environmental partnership" cartridges are sold at a discount if the customer promises to return them. "It's not a rebate," Carville said. "It's a prebate."

Xerox has a 60 percent return rate for cartridges. For comparison, Kodak has a 74 percent return rate for its Fun Saver cameras, which, Snook says, compares favorably to recycled aluminum cans.

The print cartridges are sent, at Xerox's expense, to centers in Nogales, Ariz., and Utica, NY, where the cartridges are dismantled, cleaned and inspected. Rejected parts are ground up as raw material. The rest are refurbished and shipped to Webster, NY, for remanufacturing.

All plastics in Xerox copiers are impregnated with a flame-retardant material. Carville's unit worked with Underwriters Laboratories in 1992 and 1993 to get an approval process for remanufactured plastic materials. Currently Xerox plastic is approved for up to five re-uses.

Carville said his unit doesn't want to franchise remanufacturers. Xerox wants to control the process so as to guarantee quality.

NOT A THROWAWAY

Kodak's Fun Saver shows remanufacturing in action.

Step 1
Buy Kodak Fun Saver Camera loaded with Kodak Gold Film. Take pictures and drop off the entire camera to a photofinisher.

Step 2
Photofinisher ships the used camera back to Kodak's sites at Elmgrove, Guadalajara, Mexico and Chalon, France. Kodak gives incentives and premiums to encourage returns.

Step 4
New Kodak Fun Saver cameras are made from virgin and refurbished parts, with new lenses and fresh film.

Step 3
Kodak and its subcontractors take the camera apart and inspect the parts. Worn out parts are ground up and used as raw material. Usable parts are refurbished for use in new cameras. 86 percent of the cameras by weight are recycled or refurbished.

But from the standpoint of the customer, you can get a remanufactured product quicker and cheaper from an outside company, said James D. Condon, president of Photikon Corp. of Fairport.

His company, originally a broker in copier or printer parts, started making photoreceptor belts in 1989 and now remanufactures entire printer cartridges.

Unlike Xerox, Photikon is a true remanufacturer. Its products are completely remade, not a blend of remade and virgin parts like Xerox's. For this reason, Photikon's products can be made cheaper than Xerox's, Condon said.

Photikon has worked on products remade up to 25 times, although not by Photikon every time.

The demand for remanufactured products is booming, Condon said. Photikon's 19 employees have been working overtime for the past three months, and he expects to hire 10 more in the next year.

Xerox remanufacturers about 1 million parts and 150,000 office machines each year. Kodak collects 50 million single-use cameras each year from 20 countries for remanufacturing, as well as reworking products ranging from microfilm machines to photographic film base.

Both companies use subcontractors extensively. Snook said that during the peak season, Outsource Enterprises of Rochester gets as many as 6 million single-use cameras in a week to be inspected, dis-assembled and sorted.

The most logical company to remanufacture a product is the original manufacturer, said Gordon H. McNeil, president of Magnetic Technologies Corp. of Pittsford, NY, which makes subsystems for Xerox and other companies. And looking at the used parts provides useful information in making new parts, he said.

He said about 35 percent of Magnetic Tech's output is remanufactured products, and this could grow to more than half in a few years.

Source: "Making It New Again and Again." Phil Ebersole *Rochester Democrat and Chronicle,* July 14, 1997, p. E1. Used with permission.

Robust Design

robust design Design that results in products or services that can function over a broad range of conditions.

Some products will perform as designed only within a narrow range of conditions, while other products will perform as designed over a much broader range of conditions. The latter have **robust design.** Consider a pair of fine leather boots—obviously not made for trekking through mud or snow. Now consider a pair of heavy rubber boots—just the thing for mud or snow. The rubber boots have a design that is more *robust* than the fine leather boots.

The more robust a product (or service), the less likely it will fail due to a change in the environment in which it is used or in which it is performed. Hence, the more designers can build robustness into the product or service, the better it should hold up, resulting in a higher level of customer satisfaction.

A similar argument can be made for robust design as it pertains to the production process. Environmental factors can have a negative effect on the quality of a product or service. The more resistant a design is to those influences, the less likely is the negative effect. For example, many products go through a heating process: food products, ceramics, steel, petroleum products, and pharmaceutical products. Furnaces often do not heat uniformly; heat may vary either by position in an oven or over an extended period of production. One approach to this problem might be to develop a superior oven; another might be to design a system that moves the product during heating to achieve uniformity. A robust-design approach would develop a product that is unaffected by minor variations in temperature during processing.

Taguchi's Approach. Japanese engineer Genichi Taguchi's approach is based on the robust design. His premise is that it is often easier to design a product that is insensitive to environmental factors, either in manufacturing or in use, than to control the environmental factors.

The central feature of Taguchi's approach—and the feature used most often by U.S. companies—is *parameter design.* This involves determining the specification settings for both the product and the process that will result in robust design in terms of manufacturing variations, product deterioration, and conditions during use.

The Taguchi approach modifies the conventional statistical methods of experimental design. Consider this example. Suppose a company will use 12 chemicals in a new product it intends to produce. There are two suppliers for these chemicals, but the chemical concentrations vary slightly between the two suppliers. Classical design of experiments would require $2^{12} = 4{,}096$ test runs to determine which combination of chemicals would be optimum. Taguchi's approach would involve only testing a portion of the possible combinations. Relying on experts to identify the variables that would be most likely to affect important performance, the number of combinations would be dramatically reduced, perhaps to, say, 32. Identifying the best combination in the smaller sample might be a near-optimal combination instead of the optimal combination. The value of this approach is its ability to achieve major advances in product or process design fairly quickly, using a relatively small number of experiments.

Critics charge that Taguchi's methods are inefficient and incorrect, and often lead to nonoptimal solutions. Nonetheless, his methods are widely used and have been credited with helping to achieve major improvements in U.S. products and manufacturing processes.

Concurrent Engineering

To achieve a smoother transition from product design to production, and to decrease product development time, many companies are using *simultaneous development*, or concurrent engineering. In its narrowest sense, **concurrent engineering** means bringing design and manufacturing engineering people together early in the design phase to simultaneously develop the product and the processes for creating the product. More recently, this concept has been enlarged to include manufacturing personnel (e.g., materials specialists) and marketing and purchasing personnel in loosely integrated, cross-functional teams. In addition, the views of suppliers and customers are frequently sought. The purpose, of course, is to achieve product designs that reflect customer wants as well as manufacturing capabilities.

concurrent engineering Bringing engineering design and manufacturing personnel together early in the design phase.

Traditionally, designers developed a new product without any input from manufacturing, and then turned over the design to manufacturing, which would then have to develop a process for making the new product. This "over-the-wall" approach created tremendous challenges for manufacturing, generating numerous conflicts and greatly increasing the time needed to successfully produce a new product. It also contributed to the "us versus them" mentality.

Vol. 3, Tape 1, Seg. 1
Caterpillar

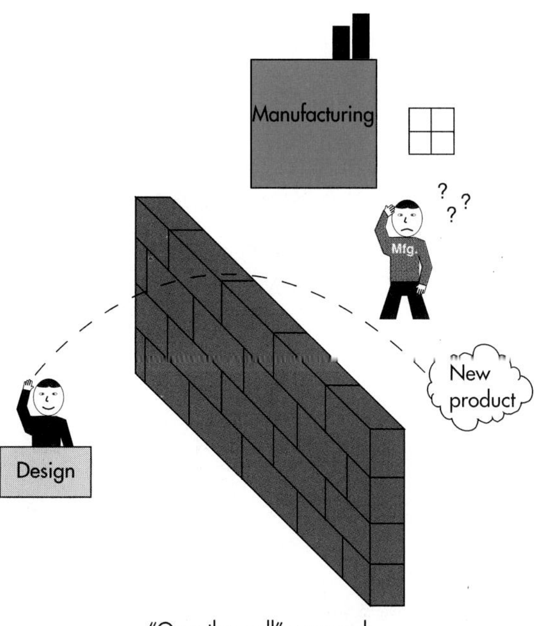

"Over-the-wall" approach

For these and similar reasons, the codevelopment approach has great appeal. Among the key advantages of this approach are the following:

1. Manufacturing personnel are able to identify production capabilities and capacities. Very often, they have some latitude in design in terms of selecting suitable materials and processes. Knowledge of production capabilities can help in the selection process. In addition, cost and quality considerations can be greatly influenced by design, and conflicts during production can be greatly reduced.
2. Early opportunities for design or procurement of critical tooling, some of which might have long lead times. This can result in a major shortening of the product development process, which could be a key competitive advantage.
3. Early consideration of the technical feasibility of a particular design or a portion of a design. Again, this can avoid serious problems during production.
4. The emphasis can be on *problem* resolution instead of *conflict* resolution.

However, a number of potential difficulties exist in this codevelopment approach. Two key ones are the following:

1. Long-standing existing boundaries between design and manufacturing can be difficult to overcome. Simply bringing a group of people together and thinking that they will be able to work together effectively is probably naive.
2. There must be extra communication and flexibility if the process is to work, and these can be difficult to achieve.

Hence, managers should plan to devote special attention if this approach is to work.

Computer-Aided Design (CAD)

computer-aided design (CAD) Product design using computer graphics.

Computers are increasingly used for product design. **Computer-aided design (CAD)** uses computer graphics for product design. The designer can modify an existing design or create a new one on a CRT by means of a light pen, a keyboard, a joystick, or a similar device. Once the design is entered into the computer, the designer can maneuver it on the screen: It can be rotated to provide the designer with different perspectives, it can be split apart to give the designer a view of the inside, and a portion of it can be enlarged for closer examination. The designer can obtain a printed version of the completed design and file it electronically, making it accessible to people in the firm who need this information (e.g., marketing).

A growing number of products are being designed in this way, including transformers, automobile parts, aircraft parts, integrated circuits, and electric motors.

A major benefit of CAD is the increased productivity of designers. No longer is it necessary to laboriously prepare mechanical drawings of products or parts and revise

At Ford's Engineering Computer Center in Dearborn, Michigan, this electronic version of a 31 mph impact helps designers program in structural and restraint systems design.

them repeatedly to correct errors or incorporate revisions. A rough estimate is that CAD increases the productivity of designers from 3 to 10 times. A second major benefit of CAD is the creation of a database for manufacturing that can supply needed information on product geometry and dimensions, tolerances, material specifications, and so on. It should be noted, however, that CAD needs this database to function and that this entails a considerable amount of effort.

Some CAD systems allow the designer to perform engineering and cost analyses on proposed designs. For instance, the computer can determine the weight and volume of a part and do stress analysis as well. When there are a number of alternative designs, the computer can quickly go through the possibilities and identify the best one, given the designer's criteria.

The reading about CAD–CAM (CAM stands for computer-aided manufacturing) gives some indication of the breadth of use of computer-aided design.

READING

COMPUTERS SPEED THE DESIGN OF MORE WORKADAY PRODUCTS **Bob Davis**

The same computer technology used to design cars and missiles now is being used to design sneakers and spoons. The reason: Computer-aided design and manufacturing systems have become cheaper and easier to use, encouraging even decidedly low-tech companies to trade their drafting tables for computer terminals.

"CAD–CAM companies have been beating down our doors for years," says Alfred Zeien, vice chairman of Gillette Co., whose Braun subsidiary in Germany uses computers to design hair dryers and electric shavers. "We were finally convinced they could deliver."

Other makers of workaday products also have been convinced. Hasbro Bradley Inc. uses computers to design parts for GI Joe aircraft carriers; Craft Patterns does the same for doghouses. ITT Corp.'s computers draw the fire hydrants; Oneida Ltd.'s make spoons and forks.

Computers take the drudgery out of design. Instead of laboriously drawing and redrawing blueprints, engineers create designs electronically and alter them at the push of a few buttons. Some computer programs also analyze the designs to see how well they handle stress or changes in temperature. Manufacturers say that encourages experimentation, yields better-quality goods, and helps them introduce products more quickly.

"It allows us to try blue-sky projects," says W. David Lee, who heads a product development group at the Arthur D. Little consulting firm. Using computerized design techniques, Mr. Lee's group produced a U.S. Army canteen that keeps coffee warm for eight hours in arctic temperatures and a gas stove that needs less insulation for Maytag Co.'s Hardwick Stove division.

Walter Stewart, a senior vice president at Oneida, says that adding CAD–CAM is "a matter of survival." Oneida estimates that it takes 70 weeks to design and produce a flatware set using conventional methods—nearly twice the time of Asian competitors who work longer hours and, Oneida contends, produce lower-quality goods. To compete better, Oneida began in 1983 to use computers to design flatware and make dies that punch out stainless steel forks and knives.

As a result, Oneida says it now can match the turnaround time of its Asian competitors on some items. It also can bid on rush jobs that it might otherwise lose. Last summer, for instance, it produced spoon designs using the hot-selling Care Bear characters for General Mills Inc., in only two days. Oneida also says that more precise tooling reduces the number of imperfections in the flatware which must be laboriously ground or polished out.

Manufacturers also use CAD–CAM to customize products. The Kennedy Valve unit of ITT designs fire hydrants by piecing together computer drawings of different size nozzles and pipes. "It's like working with building blocks," says Bruce Platusich, manager of product engineering.

In a more sophisticated application, privately owned Techmedica Inc., Camarillo, Calif., produces artificial bones by computer. Techmedica uses cross-sections of a patient's bone produced by a CT scan, or computer tomography, as a basis for the bone design. The computer then directs a machine that cuts the replacement out of a block of titanium alloy.

Anthony Hedley, director of surgical research at the Huntington Arthritis Center in Phoenix, says Techmedica can deliver an artificial bone in two weeks—a fraction of the time of suppliers that aren't computerized. As a result of the speeded up delivery and precision design, he says, some bone cancer patients are spared amputations.

Manufacturers also say CAD–CAM fattens their profit margins. By reducing product development time with computers, Gillette says it can quickly "abort" a project that doesn't meet cost goals. Hasbro Bradley says it can gear up faster for mass production by using computers to design molds.

"With some toys you can sell all you can make by Thanksgiving," says Hugh Maxwell, Hasbro Bradley's senior vice president for operations. If the company can produce a mold just one week faster, he estimates, it can sell another $300,000 worth of some toys.

But CAD–CAM has some shortcomings. Long-time employees sometimes resist learning the technology. And because the systems are so complex, they can be out of commission as much as 30 percent of the time, says Harley Shaiken, a Massachusetts Institute of Technology researcher. Among reasons for the downtime, he says, are computer breakdowns, inexperienced workers, and management failure to devise schedules that take advantage of the technology. (Confirms Henry Eichfield, an official at Computervision Corp., a large CAD–CAM maker: "Thirty percent downtime isn't atypical for the world of data processing.")

What's more, whatever cost savings result from the use of CAD–CAM don't necessarily translate into lower prices for consumers. Converse Inc., for instance, uses computers to design sneakers. But John O'Neil, president, says the company's large investment in CAD–CAM "means the cost of speeding up the process won't be reflected in the cost of the shoe."

Source: Bob Davis, "Computers Speed the Design of More Workaday Products," *The Wall Street Journal*, January 18, 1985, p. 25. Reprinted by permission of *The Wall Street Journal*,

www.oneida.com
www.hasbro.com

Modular Design

modular design A form of standardization in which component parts are subdivided into modules that are easily replaced or interchanged.

Modular design is another form of standardization. Modules represent groupings of component parts into subassemblies, usually to the point where the individual parts lose their separate identity. One familiar example of modular design is a television set with easily removed control panels. Computers, too, have modular parts that can be replaced if they become defective. By arranging modules in different configurations, different computer capabilities can be obtained. Modular design is also found in the construction industry. One firm in Rochester, New York, makes prefabricated motel rooms complete with wiring, plumbing, and even room decorations in its factory and then moves the complete rooms by rail to the construction site where they are integrated into the structure.

One advantage of modular design of equipment compared with nonmodular design is that failures are often easier to diagnose and remedy because there are fewer pieces to investigate. Similar advantages are found in ease of repair and replacement; the faulty module is conveniently removed and replaced with a good one. The manufacture and assembly of modules generally involves simplifications: fewer parts are involved, so purchasing and inventory control become more routine, fabrication and assembly operations become more standardized, and training costs often are less.

The main disadvantages of modular design stem from the decrease in variety: the number of possible configurations of modules is much less than the number of possible configurations based on individual components. Another disadvantage that is sometimes encountered is the inability to disassemble a module in order to replace a faulty part; the entire module must be scrapped—usually a more costly procedure.

SERVICE DESIGN

In certain cases, product design and service design go hand in hand. This stems from the fact that goods and services often exist in combination. For example, getting an oil change for your car involves a service (draining the old oil and putting in new oil), and a good (the new oil). Likewise, having new carpeting installed involves a service (the installation) and a good (the carpet). In some cases, what a customer receives is essentially a *pure* service, as in getting a haircut or your lawn mowed. However, the vast majority of cases involve some combination of goods and services. The proportion of service might be relatively low, as is the case in manufacturing, where the emphasis is on the production of goods. But even in manufacturing, there are services such as machine

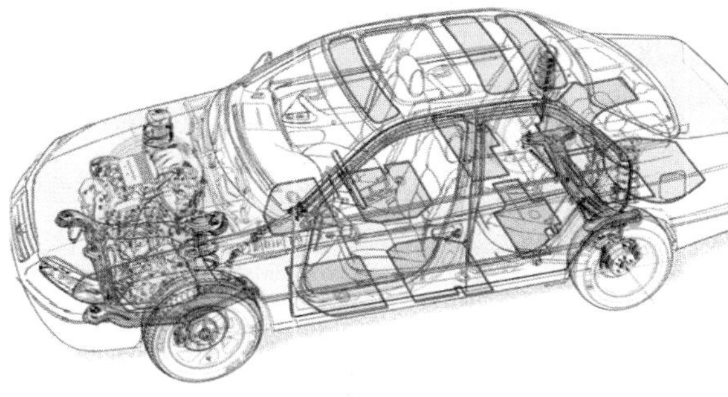

Toyota's Extra Care service contract protects the vehicle owner against costly mechanical and electrical repairs. This illustration uses design blueprint information to highlight covered parts and systems.

www.toyota.com

repair, employee training, safety inspections, and so on. Figure 4–3 provides some examples of where various operations fall on the goods–services spectrum. Because goods and services are so intertwined, managers must be knowledgeable about both in order to be able to manage effectively.

The discussion here relates primarily to items that have a high service content.

Differences between Service Design and Product Design

1. Products are generally tangible; services are generally intangible. Consequently, service design often focuses more on intangible factors (e.g., peace of mind, ambiance) than does product design.
2. In many instances services are created and delivered at the same time (e.g., a haircut, a car wash). In such instances there is less latitude in finding and correcting errors *before* the customer has a chance to discover them. Consequently, training, process design, and customer relations are particularly important.
3. Services cannot be inventoried. This poses restrictions on flexibility and makes capacity design very important.
4. Services are highly visible to consumers and must be designed with that in mind; this adds an extra dimension to process design, one that usually is not present in product design.
5. Some services have low barriers to entry and exit. This places additional pressures on service design to be innovative and cost-effective.
6. Location is often important to service design, with convenience as a major factor. Hence, design of services and choice of location are often closely linked.

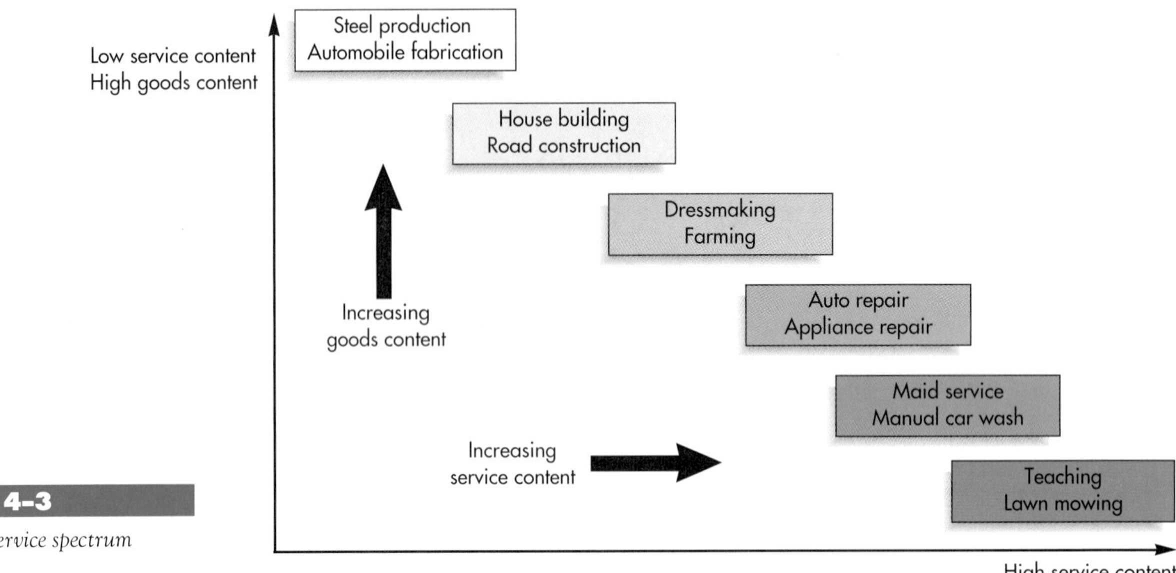

FIGURE 4-3

The goods–service spectrum

Let's consider some of these differences in more detail. One is the need to consider the degree of customer contact in service design. That can range from no contact to high contact. When there is little or no contact, service design can be very much like product design. However, the greater the degree of customer contact, the greater the difference between service and product design, and the more complex service design becomes. The element of customer contact means that service design must incorporate *process* design; when there is customer contact, the process *is* the service. Although it is desirable to consider the manufacturability of a product when designing products, the product and the process are nonetheless separate entities. The following example of service design illustrates the inseparable nature of the service/process connection when customers are a part of the system. If a refrigerator manufacturer changes the procedure it uses for assembling a refrigerator, that change will be transparent to the person who purchases the refrigerator. Conversely, if the bus company makes changes to the bus schedule, or the bus routes, those changes will not be transparent to the riders. Obviously, this service redesign could not be done realistically without considering the *process* for delivering the service.

Overview of Service Design

Service design begins with the choice of a service strategy, which determines the nature and focus of the service, and the target market. This requires an assessment by top management of the potential market and profitability (or need, in the case of a nonprofit organization) of a particular service, and an assessment of the organization's ability to provide the service. Once decisions on the focus of the service and the target market have been made, the customer requirements and expectations of the target market must be determined. The service designers then use that information to design the service delivery system (i.e., the facilities, processes, and personnel requirements needed to provide the service). Examples of possible service delivery systems include mail, telephone, electronic service (computer network, fax), and face-to-face contact.

Two key issues in service design are the degree of variation in service requirements, and the degree of customer contact and customer involvement in the delivery system. These have an impact on the degree to which service can be standardized or must be customized. The lower the degree of customer contact and service requirement variability, the more standardized the service can be. Service design with no contact and

FIGURE 4–4

Service variability and customer contact influence service design

Variability in Service Requirements				
High				Customized clothing purchase
Moderate			Department store purchase	
Low		Telephone purchase		
None	Internet purchase			
Degree of Contact with Customer	None	Low	Moderate	High

Highly customized (Customized clothing purchase); Highly standardized (Internet purchase)

little or no processing variability is very much like product design. Conversely, high variability and high customer contact generally mean the service must be highly customized. These concepts are illustrated in Figure 4–4.

A related consideration in service design is the opportunity for selling: The greater the degree of customer contact, the greater the opportunities for selling.

Design Guidelines

Vol. 2, Tape 1, Seg. 1
First National Bank
of Chicago

A number of simple but highly effective rules are often used to guide the development of service systems. The key rules are the following:

1. Have a single, unifying theme, such as convenience or speed. This will help personnel to work together rather than at cross-purposes.
2. Make sure the system has the capability to handle any expected variability in service requirements.
3. Include design features and checks to ensure that service will be reliable and will provide consistently high quality.
4. Design the system to be user-friendly. This is especially true for self-service systems.

Service Blueprinting

A useful tool in service design is a **service blueprint,** which is a method for describing and analyzing a service or a proposed service. A key element of a service blueprint is a flowchart of the service. The major steps in service blueprinting are the following:

service blueprint A method used in service design to describe and analyze a proposed service.

1. Establish boundaries for the process and decide on the level of detail that will be needed.
2. Identify the steps involved and describe them. If this is an existing process, get input from those who do it.
3. Prepare a flowchart of major process steps.
4. Identify potential failure points. Incorporate features that minimize the chances of failures.
5. Establish a timeframe for service execution, and an estimate of variability in processing time requirements. Time is a primary determinant of cost, so establishing a time standard for the service is important. Variability can also impact time, so an estimate of that is also important. Customers regard service time as a key concern—the shorter the service time, the better. However, there are exceptions, such as a

FIGURE 4–5

Flowchart of catalog call

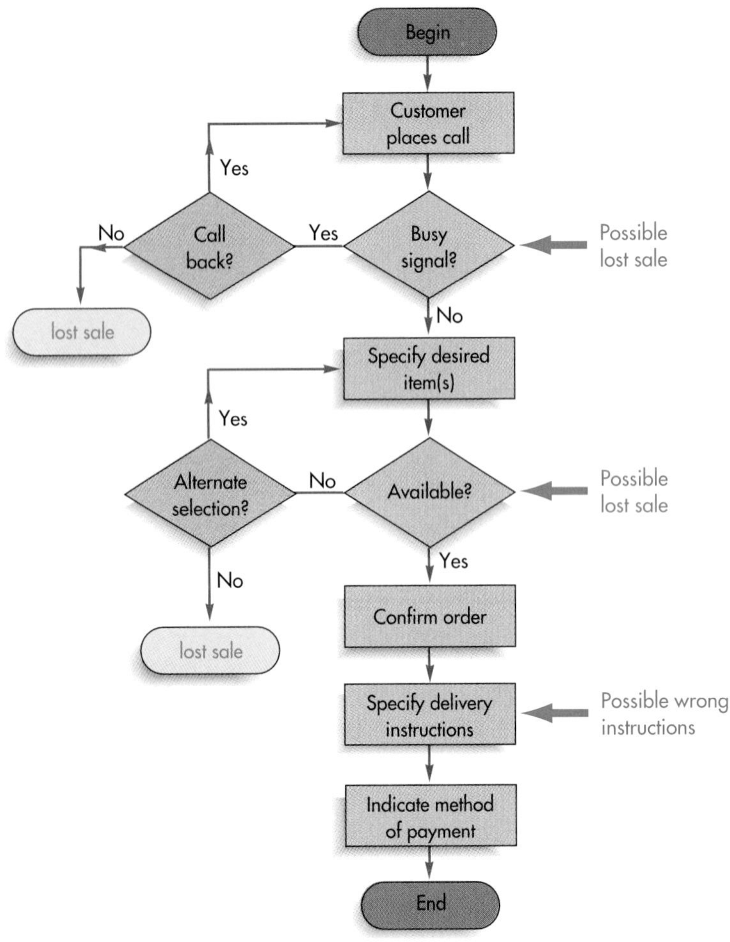

leisurely meal at a fine restaurant and a physician who takes the time to listen to a patient rather than rush to diagnosis or treatment.

6. Analyze profitability. Determine which factors can influence profitability, positively and negatively, and determine how sensitive profitability is to these factors. For example, customer waiting time is often a key factor. Concentrate design efforts on the key factors. Establish design features that protect against negative impacts, and maximize positive impacts.

Figure 4–5 illustrates a flowchart for receiving a telephone order from a catalog. The figure points out the main potential failure points.

QUALITY FUNCTION DEPLOYMENT

quality function deployment An approach that integrates the "voice of the customer" into the product development process.

Vol. 1, Tape 1, Seg. 2
Zytec, Motorola, Hewlett-Packard

Quality function deployment (QFD) is a structured approach for integrating the "voice of the customer" into the product or service development process. The purpose is to ensure that customer requirements are factored into every aspect of the process. Listening to and understanding the customer is the central feature of QFD. Requirements often take the form of a general statement such as, "It should be easy to adjust the cutting height of the lawn mower." Once the requirements are known, they must be translated into technical terms related to the product or service. For example, a statement about changing the height of the lawn mower may relate to the mechanism used to accomplish that, its position, instructions for use, tightness of the spring that controls the mechanism, or materials needed. For manufacturing purposes, these must be related to the materials, dimensions, and machinery used for processing.

The structure of QFD is based on a set of matrices. The main matrix relates customer requirements (what) and their corresponding technical requirements (how). This concept is illustrated in Figure 4–6.

Additional features are usually added to the basic matrix to broaden the scope of analysis. Typical additional features include importance weightings and competitive evaluations. A correlational matrix is usually constructed for technical requirements; this can reveal conflicting technical requirements. With these additional features, the set of matrices has the form illustrated in Figure 4–7. It is often referred to as the *house of quality* because of its houselike appearance.

An analysis using this format is shown in Figure 4–8. The data relate to a commercial printer (customer) and the company that supplies the paper. At first glance, the display appears complex. It contains a considerable amount of information for product and process planning. Therefore, let's break it up into separate parts and consider them one at a time. To start, a key part is the list of customer requirements on the left side of the figure. Next, note the technical requirements, listed vertically near the top. The important relationships, and their degree of importance, are shown in the center of the figure.

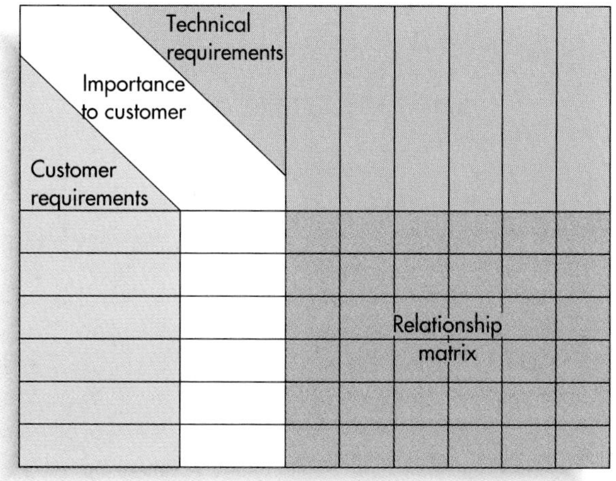

Technical requirements
Importance to customer
Customer requirements
Relationship matrix

FIGURE 4–6

The main QFD Matrix

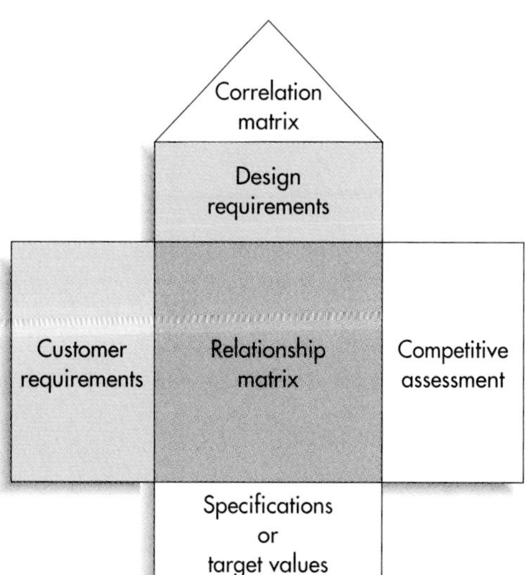

Source: V. Daniel Hunt, *Quality in America* (Homewood, Ill.: Business One Irwin, 1992), p. 270. Reprinted by permission.

FIGURE 4–7

The house of quality

FIGURE 4–8

An example of the house of quality

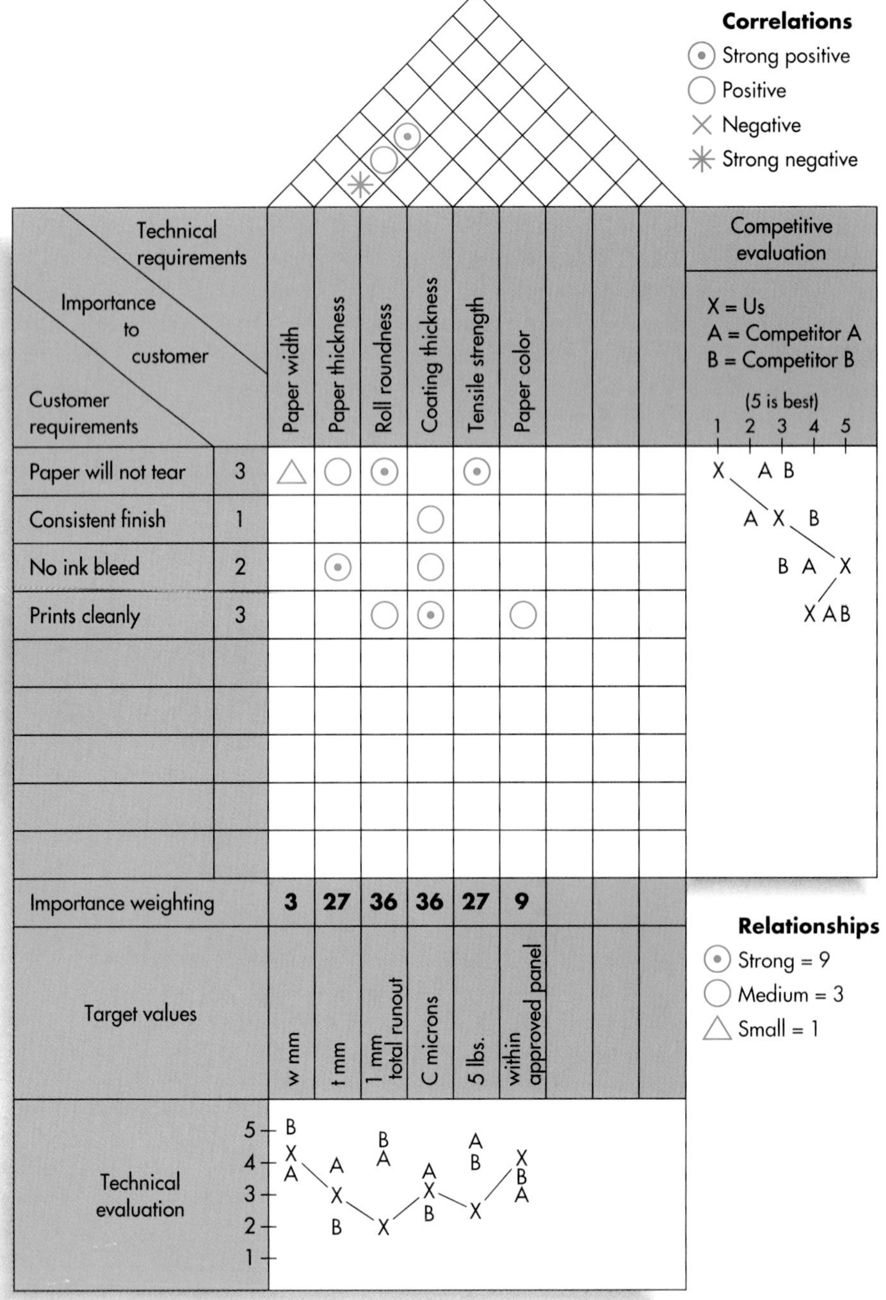

Source: Ernst and Young Consulting Group, *Total Quality* (Homewood, Ill.: Dow-Jones Irwin, 1991), p. 121. Reprinted by permission.

The circle with a dot inside indicates the strongest positive relationship; that is, it denotes the most important technical requirements for satisfying customer requirements. Now look at the "importance to customer" numbers that are shown next to each customer requirement (3 is the most important). Designers will take into account the importance values and the strength of correlation in determining where to focus the greatest effort.

Next, consider the correlation matrix at the top of the "house." Of special interest is the strong negative correlation between "paper thickness" and "roll roundness." Designers will have to find some way to overcome that or make a trade-off decision.

On the right side of the figure is a competitive evaluation comparing the supplier's performance on the customer requirements with each of the two key competitors (A and B). For example, the supplier (X) is worst on the first customer requirement and best on the third customer requirement. The line connects the X performances. Ideally, design will cause all of the Xs to be in the highest positions.

Across the bottom of Figure 4–8 are importance weightings, target values, and technical evaluations. The technical evaluations can be interpreted in a manner similar to that of the competitive evaluations (note the line connecting the Xs). The target values typically contain technical specifications, which we will not discuss. The importance weightings are the sums of values assigned to the relationships (see the lower right-hand key for relationship weights). The 3 in the first column is the product of the importance to the customer = 3 and the small (△) weight = 1. The importance weightings and target evaluations help designers focus on desired results. In this example, the first technical requirement has the lowest importance weighting, while the next four technical requirements all have relatively high importance weightings.

NEWSCLIP

A QFD SNAPSHOT

How a pencilmaker sharpened up its product by listening to "the voice of the customer" through quality function deployment.

Devised by Japan's Professor Yoji Akao, QFD has been winning adherents since it was transplanted to the U.S. in the late 1980s. In this example of how it works, Writesharp Inc. is imaginary, but the technique in the accompanying diagram is real.

First, Writesharp's customers were surveyed to determine what they value in a pencil and how they rate the leading brands. Each wish list item was correlated with a pencil's functional characteristics (see FUNCTIONAL CHARACTERISTICS matrix). "Reverse engineering"—tearing down a competitor's product to see what makes it tick—produced the competitive benchmark measurements for the various functions.

An analysis of the plots quickly revealed that the improvement with the biggest potential was a better-quality lead (see CUSTOMER SATISFACTION/CUSTOMER DEMANDS matrix). An interdepartmental team was assigned the task of evaluating new lead formulations that would last longer and generate less dust. Another team ran tests to determine whether substituting cedar for oak in the wood casing would improve shape quality, or hexagonality, and thus reduce the pencil's tendency to roll down slanted desktops.

The lead-formulation team organized its work with a similar matrix chart, segmented to show the functional contributions of the ingredients in pencil lead. This revealed that the binder, or glue, used in forming the lead was the key variable. Tests found a polymer that dramatically reduced dusting by retaining more moisture and also wore down more slowly. While this binder was more expensive, better production controls—going slightly beyond the performance of Competitor Y—promised to reduce waste enough

Japanese Professor Yoji devised the system and coined the phrase "quality function deployment." The QFD matrix combines benchmarking, customer demands, a product's characteristics, and customer satisfaction to measure and improve product quality.

to trim total per-pencil manufacturing costs by 1¢.

Changing the wood, meanwhile, yielded only marginal enhancements. So the company decided to upgrade the process controls used for cutting the wood and match the quality of Competitor X (see BENCHMARKS matrix).

Source: Reprinted from October 25, 1991, issue of *Business Week* by special permission, copyright © 1991 by The McGraw-Hill Companies, Inc.

Functional characteristics: △ Strong correlation ○ Some correlation □ Possible correlation

Customer satisfaction Scale: 1 to 5 (5 = best)

		Pencil length (inches)	Time between sharpenings (written lines)	Lead dust (particles per line)	Hexagonality	Importance rating (5 = highest)	Writesharp (now)	Competitor X (now)	Competitor Y (now)	Writesharp (target)
Customer demands	Easy to hold	○			○	3	4	3	3	4
	Does not smear		○	△		4	5	4	5	5
	Point lasts	□	△	○		5	4	5	3	5
	Does not roll	□			△	2	3	3	3	4
Benchmarks	Writesharp (now)	5	56	10	70%					
	Competitor X (now)	5	84	12	80%					
	Competitor Y (now)	4	41	10	60%					
	Writesharp (target)	5.5	100	6	80%					

	Writesharp (now)	Competitor X (now)	Competitor Y (now)	Writesharp (target)
Market price	15¢	18¢	14¢	16¢
Market share	16%	12%	32%	20%
Profit	2¢	3¢	2¢	4¢

RELIABILITY

reliability The ability of a product, part, or system to perform its intended function under a prescribed set of conditions.

failure Situation in which a product, part, or system does not perform as intended.

normal operating conditions The set of conditions under which an item's reliability is specified.

Reliability is a measure of the ability of a product, part, or system to perform its intended function under a prescribed set of conditions. The importance of reliability is underscored by its use by prospective buyers in comparing alternatives and by sellers as one determinant of price. Reliability also can have an impact on repeat sales, reflect on the product's image, and, if it is too low, create legal implications.

The term **failure** is used to describe a situation in which an item does not perform as intended. This includes not only instances in which the item does not function at all, but also instances in which the item's performance is substandard or it functions in a way not intended. For example, a smoke alarm might fail to respond to the presence of smoke (not operate at all), it might sound an alarm that is too faint to provide an adequate warning (substandard performance), or it might sound an alarm even though no smoke is present (unintended response).

Reliabilities are always specified with respect to certain conditions, called **normal operating conditions.** These can include load, temperature, and humidity ranges as well as operating procedures and maintenance schedules. Failure of users to heed these conditions often results in premature failure of parts or complete systems. For example, using a passenger car to tow heavy loads causes excess wear and tear on the drive train;

TABLE 4–2

Potential ways to improve reliability

1. Improve component design.
2. Improve production and/or assembly techniques.
3. Improve testing.
4. Use backup components.
5. Improve preventive maintenance procedures.
6. Improve user education.
7. Improve system design.

driving over potholes or curbs often results in untimely tire failure; and using a calculator to drive nails might have a marked impact on its usefulness for performing mathematical operations.

Improving Reliability

Reliability can be improved in a number of ways, some of which are listed in Table 4–2.

Because overall system reliability is a function of the reliability of individual components, improvements in their reliability can increase system reliability. Unfortunately, inadequate production or assembly procedures can negate even the best of designs, and this is often a source of failures. System reliability can be increased by the use of backup components. Failures in actual use can often be reduced by upgrading user education and refining maintenance recommendations or procedures. Finally, it may be possible to increase the overall reliability of the system by simplifying the system (thereby reducing the number of components that could cause the system to fail) or altering component relationships (e.g., increasing the reliability of interfaces).

A fundamental question concerning improving reliability is: How much reliability is needed? Obviously, the reliability that is needed for a light bulb isn't in the same category as the reliability that is needed for an airplane. So the answer to the question depends on the potential benefits of improvements and on the cost of those improvements. Generally speaking, reliability improvements become increasingly costly. Thus, although benefits initially may increase at a much faster rate than costs, the opposite eventually becomes true. The optimal level of reliability is the point where the incremental benefit received equals the incremental cost of obtaining it. In the short term, this trade-off is made in the context of relatively fixed parameters (e.g., costs). However, in the longer term, efforts to improve reliability and reduce costs will lead to higher optimal levels of reliability.

Operations Strategy

There are four major recommendations on operations strategy in the area of product and service design that can improve competitiveness:

1. Invest more in R&D.
2. Shift some emphasis away from short-term performance to long-term performance.
3. Work toward continual (albeit gradual) improvements instead of using a "big bang" approach.
4. Work to shorten the product development cycle.

Dollars invested in research and development can have a tremendous impact on a company's *future* competitiveness, affecting quality and reliability, technological innovation, and product improvement. The average Japanese company invests far more in R&D than the typical Western company. Western managers must be willing to initially sacrifice some short-term performance in favor of R&D that will eventually lead to *both*

long-term and (later) short-term performance. However, accomplishing this will require a different attitude than the one that still prevails in many companies.

A hallmark of Japanese success is the emphasis on *continual* improvement in products and process. In contrast, many Western managers appear bent on making a big splash—the tortoise and the hare? "Little" things such as product reliability improvements can have long-lasting effects on consumer attitudes and buying patterns.

Getting new products to the market before competitors usually results in substantial profits. Over the last decade Japanese producers of automobiles and major appliances have introduced new products and product innovations an average of one year earlier than their Western counterparts. The implication is clear: Western managers must shorten their product development cycles to be competitive.

Summary

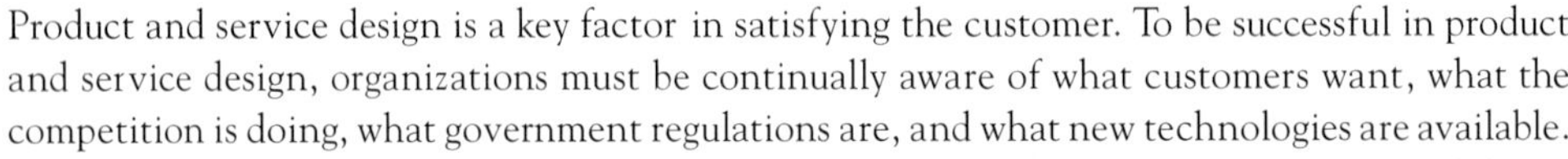

Product and service design is a key factor in satisfying the customer. To be successful in product and service design, organizations must be continually aware of what customers want, what the competition is doing, what government regulations are, and what new technologies are available.

The design process involves motivation, ideas for improvement, organizational capabilities, and forecasting. In addition to product life cycles, legal and regulatory considerations influence design choices. What degree of standardization designers should incorporate into designs is also an important consideration. Key objectives for designers are to achieve a product or service design that will meet or exceed customer expectations, that is within cost or budget, and that takes into account the capabilities of operations. Although product design and service design are similar in some respects, a number of key differences exist between products and services that influence the way they are designed.

Successful design often incorporates many of these basic principles: Determine what customers want as a starting point; minimize the number of parts needed to manufacture an item, or the number of steps to provide a service; simplify assembly or service, standardize as much as possible; and make the design robust. Trade-off decisions are common in design, and involve such things as development time and cost, product or service cost, special features/performance, and product or service complexity.

Research and development efforts can play a significant role in product and process innovations, although these are sometimes so costly that only large companies or governments can afford to underwrite them.

Reliability of a product or service is often a key dimension in the eyes of the customer. Measuring and improving reliability are important aspects of product and service design, although other areas of the organization also have an influence on reliability.

Key Terms

computer-aided design (CAD), 164
concurrent engineering, 163
design for assembly (DFA), 159
design for dis-assembly (DFD), 159
design for manufacturing (DFM), 159
design for operations, 152
design for recycling (DFR), 159
failure, 174
manufacturability, 153
modular design, 166
normal operating conditions, 174
product liability, 154
product life cycle, 158
quality function deployment (QFD), 170
research and development (R&D), 155
reliability, 174
remanufacturing, 159
reverse engineering, 153
robust design, 162
service blueprint, 169
standardization, 156
Uniform Commercial Code, 154

Discussion and Review Questions

1. What are some of the factors that cause organizations to redesign their products or services?
2. Contrast applied research and basic research.
3. What is CAD? Describe some of the ways a product designer can use it.
4. Name some of the main advantages and disadvantages of standardization.
5. What is modular design? What are its main advantages and disadvantages?
6. Explain the term *design for manufacturing* and briefly explain why it is important.
7. What are some of the competitive advantages of concurrent engineering?
8. Explain what *quality function deployment* is and how it can be useful.
9. What is meant by the term *product life cycle?* Why would this be a consideration in product design?
10. Why is R&D a key factor in productivity improvement? Name some ways R&D contributes to productivity improvements.
11. What is *service blueprinting?* Why is it of value in service design?
12. Name two factors that could make service design much different than product design.
13. Explain the term *robust design*.
14. What is *concurrent engineering,* and why is it important?
15. Explain the term *remanufacturing.*

Memo Writing Exercises

1. At a recent presentation, your company's CEO stated the company's intent to expand into the service sector. Currently, your company is devoted exclusively to manufacturing. Of particular interest to your supervisor, Tom Henry, were the following statements: "In all likelihood, we will use some of our own product designers for service design. They know our products and, besides, product design and service design are pretty much the same." Henry has asked you to look into this proposal. Write Tom a half-page memo indicating the circumstances under which this proposal might work and those under which it might not.
2. Suppose you have just received a memo questioning the merits of service blueprinting, a proposed new service to be offered by your company. The writer, Mary Barkley, a group leader in another department, is anxious to get started. She regards the blueprint as a "waste of time since everybody already knows the service." Write a half-page memo to her on the benefits of service blueprinting.

Problems

1. Prepare a table similar to Figure 4–4. Then place each of these banking transactions in the appropriate cell of the table:
   a. Make a cash withdrawal from an automatic teller machine (ATM).
   b. Make a savings deposit using a teller.
   c. Direct deposit by employer.
   d. Open a savings account.
   e. Apply for a home equity loan.
2. Prepare a table similar to Figure 4–4. Then place each of these post office transactions in the appropriate cell of the table:
   a. Buy stamps from a machine.
   b. Buy stamps from a postal clerk.
   c. Mail a package that involves checking first class and express rates.
   d. File a complaint.
3. Construct a flowchart for putting gasoline into your car for each case listed. Assume that paying cash is the only means of payment. For each flowchart, identify the potential trouble points and indicate a likely problem.

a. Self-service

b. Full service

4. Construct a flowchart for making a cash withdrawal from an automatic teller machine (ATM). Assume that the process begins at the ATM with your bank card in hand. Then identify the potential failure points (i.e., where problems might arise in the process). For each failure point, state one potential problem.

5. *a.* Refer to Figure 4–8. What two technical requirements have the highest impact on the customer requirement that the paper not tear?

   b. The following table presents technical requirements and customer requirements for the output of a laser printer. First, decide if any of the technical requirements relate to each customer requirement. Decide which technical requirement, if any, has the greatest impact on that customer requirement.

Customer Requirements	Technical Requirements		
	Type of Paper	Internal Paper Feed	Print Element
Paper doesn't wrinkle			
Prints cleanly			
Easy to use			

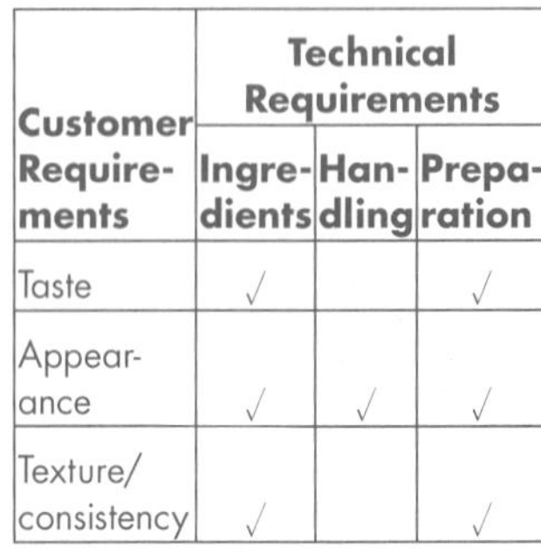

Customer Requirements	Technical Requirements		
	Ingredients	Handling	Preparation
Taste	✓		✓
Appearance	✓	✓	✓
Texture/consistency	✓		✓

6. Prepare a table similar to that shown in Problem 5b for cookies sold in a bakery. List what you believe are the three most important customer requirements (not including cost) and the three most relevant technical requirements (not including sanitary conditions). Next, indicate by a checkmark which customer requirements and which technical requirements are related.

CASE

TOYS, INC.

TOYS, Inc. is a 20-year-old company engaged in the manufacture and sale of toys and board games. The company has built a reputation on quality and innovation. Although the company is one of the leaders in its field, sales have leveled off in recent years. For the most recent six-month period, sales actually declined compared with the same period last year. The production manager, Ed Murphy, attributed the lack of sales growth to "the economy." He was prompted to undertake a number of belt-tightening moves that included cuts in production costs and layoffs in the design and product development departments. Although profits are still flat, he believes that within the next six months, the results of his decisions will be reflected in increased profits.

The vice president of sales, Joe Martin, has been concerned with customer complaints about the company's REALISTIC™ line of working model factories, farms, and service stations. The moving parts on certain models have become disengaged and fail to operate or they operate eratically. His assistant, Keith McNally, has proposed a trade-in program by which customers could replace malfunctioning models with new ones. McNally believes that this will demonstrate goodwill, and appease dissatisfied customers. He also proposes rebuilding the trade-ins and selling them at discounted prices in the company's retail outlet store. He doesn't think that this will take away from sales of new models. Under McNally's program, no new staff would be needed. Regular workers would perform needed repairs during periods of seasonal slowdowns, thus keeping production level.

When Steve Bukowski, a production assistant, heard Keith's proposal, he said that a better option would be to increase inspection of finished models before they were shipped. "With 100 percent inspection, we can weed out any defective models and avoid the problem entirely."

Take the role of a consultant who has been called in for advice by the company president, Marybeth Corbella. What do you recommend?

Selected Bibliography

Cohen, Morris A., and Uday M. Apte. *Manufacturing Automation*. McGraw-Hill, 1997.

Davidow, William H., and Bru Uttal. "Service Companies: Focus or Falter." *Harvard Business Review* (July–August 1989), pp. 77–85.

Dean, James W. Jr., and Gerald Susman. "Organizing for Manufacturing Design." *Harvard Business Review* (Jan.–Feb. 1989), pp. 28–36.

Groover, Mikell P. *Automation, Production Systems, and Computer-Aided Manufacturing.* Englewood Cliffs, NJ: Prentice Hall, 1990.

__________, and E. W. Zimmers, Jr. *CAD/CAM: Computer-Aided Design and Manufacturing.* Englewood Cliffs, NJ: Prentice Hall, 1984.

Heskett, James L., W. Earl Sasser, Jr., and Leonard A. Schlesinger. *The Service Profit Chain.* New York: The Free Press, 1997.

Lovelock, Christopher. *Managing Services: Marketing, Operations, and Human Resources.* Englewood Cliffs, NJ: Prentice Hall, 1988.

Prasad, Biren. *Concurrent Engineering Fundamentals: Integrated Product Development.* Englewood Cliffs, NJ: Prentice Hall PTR, 1997.

Rosenthal, Stephen R. *Effective Product Design and Development.* Burr Ridge, IL: Richard D. Irwin, 1992.

Shaw, John C. *The Service Focus.* Burr Ridge, IL: Richard D. Irwin, 1990.

Shostack, G. Lynn. "Designing Services that Deliver." *Harvard Business Review* (Jan.–Feb. 1984), pp. 133–39.

Ulrich, Karl T., and Steven D. Eppinger. *Product Design and Development.* New York: McGraw-Hill, 1995.

SUPPLEMENT TO CHAPTER FOUR

RELIABILITY

LEARNING OBJECTIVES

After completing this supplement, you should be able to

1 Define reliability.

2 Perform simple reliability computations.

3 Explain the purpose of redundancy in a system.

SUPPLEMENT OUTLINE

INTRODUCTION

Reliability is a measure of the ability of a product, part, or system to perform its intended function under a prescribed set of conditions. In effect, reliability is a *probability*.

reliability The ability of a product, part, or system to perform its intended function under a prescribed set of conditions.

Suppose that an item has a reliability of .90. This means that it has a 90 percent probability of functioning as intended. The probability it will fail is 1 − .90 = .10, or 10 percent. Hence, it is expected that, on the average, 1 of every 10 such items will fail or, equivalently, that the item will fail, on the average, once in every 10 trials. Similarly, a reliability of .985 implies 15 failures per 1,000 parts or trials.

QUANTIFYING RELIABILITY

Engineers and designers have a number of techniques at their disposal for assessing the reliability of parts, products, and systems. A discussion of those techniques is not within the scope of this text. Instead, let us turn to the issue of quantifying overall product or system reliability. Probability is used in two ways.

1. The probability that the product or system will function when activated.
2. The probability that the product or system will function for a given length of time.

The first of these focuses on *one point in time* and is often used when a system must operate for one time or a relatively few number of times. The second of these focuses on the *length of service*. The distinction will become more apparent as each of these approaches is examined in more detail.

The probability that a system or a product will operate as planned is an important concept in system and product design. Determining that probability when the product or system consists of a number of *independent* components requires the use of the rules of probability for independent events. **Independent events** have no relation to the occurrence or nonoccurrence of each other. What follows are three examples illustrating the use of probability rules to determine whether a given system will operate successfully.

independent events Events whose occurrence or nonoccurrence do not influence each other.

Rule 1. If two or more events are independent and "success" is defined as the probability that all of the events occur, then the probability of success is equal to the product of the probabilities of the events.

Example. Suppose a room has two lamps, but to have adequate light both lamps must work (success) when turned on. One lamp has a probability of working of .90, and the other has a probability of working of .80. The probability that both will work is .90 × .80 = .72. Note that the order of multiplication is unimportant: .80 × .90 = .72. Also note that if the room had three lamps, three probabilities would have been multiplied.

This system can be represented by the following diagram.

Even though the individual components of a system might have high reliabilities, the system as a whole can have considerably less reliability because all components that are in series (as are the ones in the preceding example) must function. As the number of components in a series increases, the system reliability decreases. For example, a system that has eight components in a series, each with a reliability of .99, has a reliability of only $.99^8 = .923$.

redundancy The use of backup components to increase reliability.

Obviously, many products and systems have a large number of component parts that must all operate, and some way to increase overall reliability is needed. One approach is to use **redundancy** in the design. This involves providing backup parts for some items.

Rule 2. If two events are independent and "success" is defined as the probability that *at least one* of the events will occur, the probability of success is equal to the probability of either one plus 1.00 minus that probability multiplied by the other probability.

Example. There are two lamps in a room. When turned on, one has a probability of working of .90 and the other has a probability of working of .80. Only a single lamp is needed to light for success. If one fails to light when turned on, the other lamp is turned on. Hence, one of the lamps is a backup in case the other one fails. Either lamp can be treated as the backup; the probability of success will be the same. The probability of success is $.90 + (1 - .90) \times .80 = .98$. If the .80 light is first, the computation would be $.80 + (1 - .80) \times .90 = .98$.

This system can be represented by the following diagram.

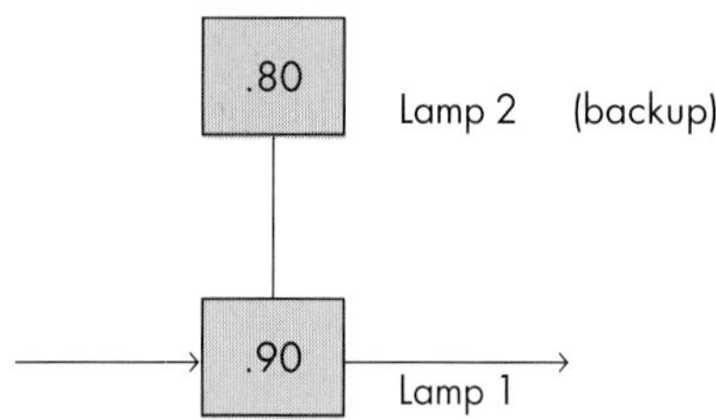

Rule 3. If three events are involved and success is defined as the probability that at least one of them occurs, the probability of success is equal to the probability of the first one (any of the events), plus the product of 1.00 minus that probability and the probability of the second event (any of the remaining events), plus the product of 1.00 minus each of the first two probabilities and the probability of the third event, and so on. This rule can be expanded to cover more than three events.

Example. Three lamps have probabilities of .90, .80, and .70 of lighting when turned on. Only one lighted lamp is needed for success; hence, two of the lamps are considered to be backups. The probability of success is:

[#1 operates]	[#1 fails and #2 operates]		[#1 fails and #2 fails and #3 operates]			
.90	+ (1 − .90) ×	.80	+ (1.00 − .90) ×	(1.00 − .80) ×	.70	= .994

This system can be represented by the following diagram:

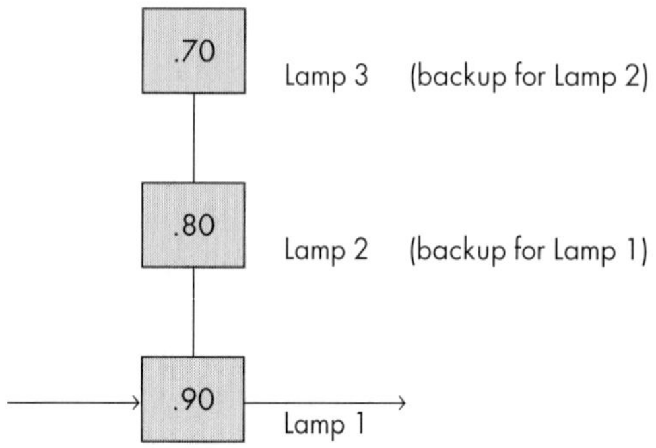

Example 1

Determine the reliability of the system shown below.

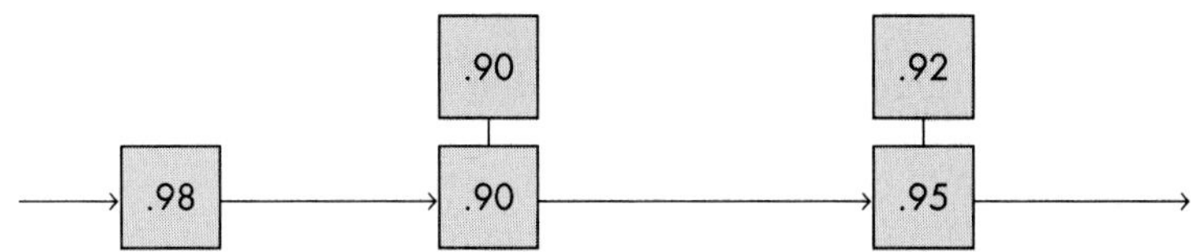

Solution

The system can be reduced to a series of three components:

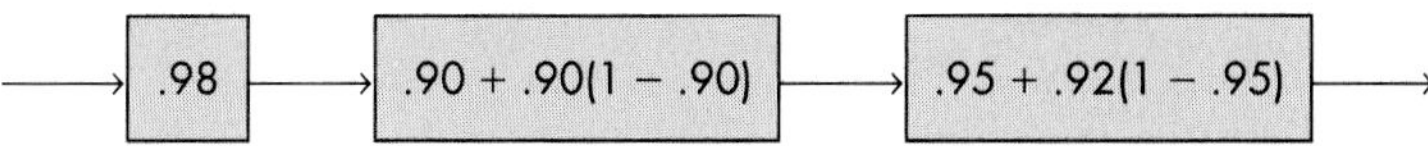

The system reliability is, then, the product of these:

$.98 \times .99 \times .996 = .966$

The second way of looking at reliability considers the incorporation of a time dimension: Probabilities are determined relative to a specified length of time. This approach is commonly used in product warranties, which pertain to a given period of time after purchase of a product.

A typical profile of product failure rate over time is illustrated in Figure 4S–1. Because of its shape, it is sometimes referred to as a bathtub curve. Frequently, a number of products fail shortly after they are put into service, not because they wear out, but because they are defective to begin with. The rate of failures decreases rapidly once the truly defective items are weeded out. During the second phase, there are fewer failures because most of the defective items have been eliminated, and it is too soon to encounter items that fail because they have worn out. In some cases, this phase covers a relatively long period of time. In the third phase, failures occur because the products are worn out, and the failure rate increases.

mean time between failures The average length of time between failures of a product or component.

Information on the distribution and length of each phase requires the collection of historical data and analysis of those data. It often turns out that the **mean time between failures (MTBF)** in the infant mortality phase can be modeled by a negative exponential distribution, such as that depicted in Figure 4S–2. Equipment failures as well as product failures may occur in this pattern. In such cases, the exponential distribution can be used to determine various probabilities of interest. The probability that equipment or a product put into service at time 0 will fail *before* some specified time, *T*, is equal to the area under the curve between 0 and *T*. Reliability is specified as the probability that a product will last *at least until* time *T*; reliability is equal to the area under the curve *beyond T*. (Note that the total area under the curve in each phase is treated as 100 percent for computational purposes.) Observe that as the specified length of service increases, the area under the curve to the right of that point (i.e., the reliability) decreases.

Determining values for the area under a curve to the right of a given point, *T*, becomes a relatively simple matter using a table of exponential values. An exponential distribution is completely described using a single parameter, the distribution mean, which reliability engineers often refer to as the mean time between failures. Using the symbol *T* to represent length of service, the probability that failure will not occur before time *T* (i.e., the area in the right tail) is easily determined:

$$P(\text{no failure before } T) = e^{-T/\text{MTBF}}$$

where

e = Natural logarithm, 2.7183 . . .
T = Length of service before failure
MTBF = Mean time between failures

FIGURE 4S-1

Failure rate is a function of time

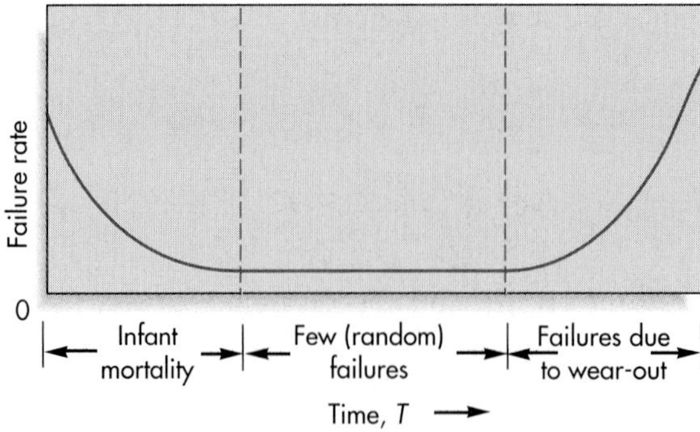

FIGURE 4S-2

An exponential distribution

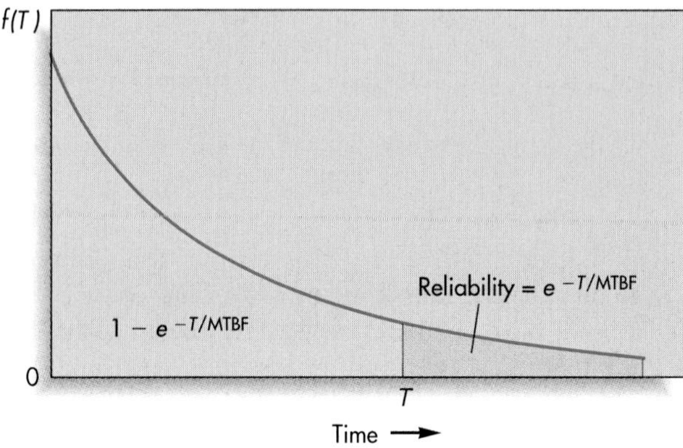

The probability that failure will occur before time T is 1.00 minus that amount:

$$P(\text{failure before } T) = 1 - e^{-T/\text{MTBF}}$$

Selected values of $e^{-T/\text{MTBF}}$ are listed in Table 4S–1.

Example 2

By means of extensive testing, a manufacturer has determined that its Super Sucker Vacuum Cleaner models have an expected life that is exponential with a mean of four years. Find the probability that one of these cleaners will have a life that ends:

a. After the initial four years of service.

b. Before four years of service are completed.

c. Not before six years of service.

Solution

MTBF = 4 years

a. $T = 4$ years:

$$T/\text{MTBF} = \frac{4 \text{ years}}{4 \text{ years}} = 1.0$$

From Table 4S–1, $e^{-1.0} = .3679$.

b. The probability of failure before $T = 4$ years is $1 - e^{-1}$, or $1 - .3679 = .6321$.

c. $T = 6$ years:

$$T/\text{MTBF} = \frac{6 \text{ years}}{4 \text{ years}} = 1.50$$

From Table 4S–1, $e^{-1.5} = .2231$.

TABLE 4S–1

Values of $e^{-T/MTBF}$

T/MTBF	$e^{-T/MTBF}$	T/MTBF	$e^{-T/MTBF}$	T/MTBF	$e^{-T/MTBF}$
0.10	.9048	2.60	.0743	5.10	.0061
0.20	.8187	2.70	.0672	5.20	.0055
0.30	.7408	2.80	.0608	5.30	.0050
0.40	.6703	2.90	.0550	5.40	.0045
0.50	.6065	3.00	.0498	5.50	.0041
0.60	.5488	3.10	.0450	5.60	.0037
0.70	.4966	3.20	.0408	5.70	.0033
0.80	.4493	3.30	.0369	5.80	.0030
0.90	.4066	3.40	.0334	5.90	.0027
1.00	.3679	3.50	.0302	6.00	.0025
1.10	.3329	3.60	.0273	6.10	.0022
1.20	.3012	3.70	.0247	6.20	.0020
1.30	.2725	3.80	.0224	6.30	.0018
1.40	.2466	3.90	.0202	6.40	.0017
1.50	.2231	4.00	.0183	6.50	.0015
1.60	.2019	4.10	.0166	6.60	.0014
1.70	.1827	4.20	.0150	6.70	.0012
1.80	.1653	4.30	.0136	6.80	.0011
1.90	.1496	4.40	.0123	6.90	.0010
2.00	.1353	4.50	.0111	7.00	.0009
2.10	.1255	4.60	.0101		
2.20	.1108	4.70	.0091		
2.30	.1003	4.80	.0082		
2.40	.0907	4.90	.0074		
2.50	.0821	5.00	.0067		

Product failure due to wear-out can sometimes be modeled by a normal distribution. Obtaining probabilities involves the use of a table (refer to Appendix Table B). The table provides areas under a normal curve from (essentially) the left end of the curve to a specified point z, where z is a *standardized* value computed using the formula

$$z = \frac{T - \text{Mean wear-out time}}{\text{Standard deviation of wear-out time}}$$

Thus, to work with the normal distribution, it is necessary to know the mean of the distribution and its standard deviation. A normal distribution is illustrated in Figure 4S–3. Appendix Table B contains normal probabilities (i.e., the area that lies to the left of z). To obtain a probability that service life will not exceed some value T, compute z and refer to the table. To find the reliability for time T, subtract this probability from 100 percent. To obtain the value of T that will provide a given probability, locate the nearest probability under the curve *to the left* in Table B. Then use the corresponding z in the preceding formula and solve for T.

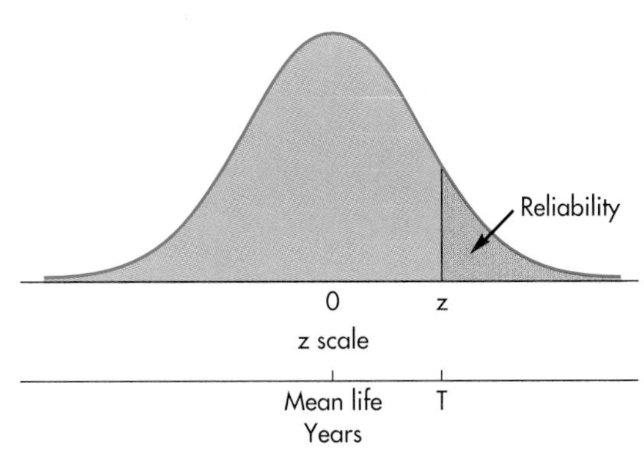

FIGURE 4S–3

A normal curve

Example 3

The mean life of a certain ball bearing can be modeled using a normal distribution with a mean of six years and a standard deviation of one year. Determine each of the following:

a. The probability that a ball bearing will wear out *before* seven years of service.

b. The probability that a ball bearing will wear out *after* seven years of service (i.e., find its reliability).

c. The service life that will provide a wear-out probability of 10 percent.

Solution

Wear-out life mean = 6 years
Wear-out life standard deviation = 1 year
Wear-out life is normally distributed

a. Compute z and use it to obtain the probability directly from Appendix Table B (see diagram).

$$z = \frac{7 - 6}{1} = +1.00$$

Thus, $P(T < 7) = .8413$.

b. Subtract the probability determined in part *a* from 100 percent (see diagram).

$$1.00 - .8413 = .1587$$

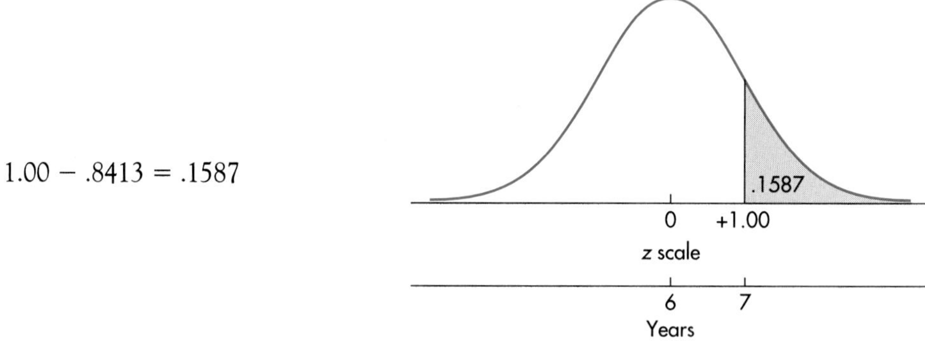

c. Use the normal table and find the value of z that corresponds to an area under the curve of 10% (see diagram).

$$z = -1.28 = \frac{T - 6}{1}$$

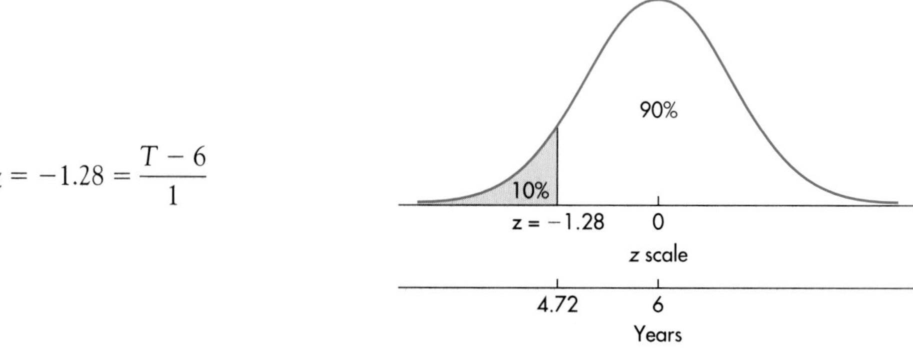

Solving for T, we find $T = 4.72$ years.

AVAILABILITY

A related measure of importance to customers, and hence to designers, is **availability.** It measures the fraction of time a piece of equipment is expected to be operational (as opposed to being down for repairs). Availability can range from zero (never available) to 1.00 (always available). Companies that can offer equipment with a high availability factor have a competitive advantage over companies that offer equipment with lower availability values. Availability is a function of both the mean time between failures and the mean time to repair. The availability factor can be computed using the following formula:

availability The fraction of time a piece of equipment is expected to be available for operation.

$$Availability = \frac{MTBF}{MTBF + MTR}$$

where

MTBF = Mean time between failures
MTR = Mean time to repair

Example 4

A copier is expected to be able to operate for 200 hours between repairs, and the mean repair time is expected to be two hours. Determine the availability of the copier.

Solution

MTBF = 200 hours, and MTR = 2 hours

$$Availability = \frac{MTBF}{MTBF + MTR} = \frac{200}{200 + 2} = .99$$

Two implications for design are revealed by the availability formula. One is that availability increases as the mean time between failures increases. The other is that availability also increases as the mean repair time decreases. It would seem obvious that designers would want to design products that have a long time between failures. However, some design options enhance repairability, which can be incorporated into the product. Laser printers, for example, are designed with print cartridges that can easily be replaced.

Key Terms

availability, 187
independent events, 181
mean time between failures (MTBF), 183
redundancy, 182
reliability, 181

Solved Problems

Problem 1

A product design engineer must decide if a redundant component is cost-justified in a certain system. The system in question has a critical component with a probability of .98 of operating. System failure would involve a cost of $20,000. For a cost of $100, a switch could be added that would automatically transfer the system to the backup component in the event of a failure. Should the backup be added if the backup probability is also .98?

Solution

Because no probability is given for the switch, we will assume its probability of operating when needed is 100 percent. The expected cost of failure (i.e., without the backup) is $20,000 (1 − .98) = $400.

With the backup, the probability of *not* failing would be:

$$.98 + .02(.98) = .9996$$

Hence, the probability of failure would be 1 − .9996 = .0004. The expected cost of failure with the backup would be the added cost of the backup plus the failure cost:

$\$100 + \$20{,}000(.0004) = \$108$

Because this is less than the cost without the backup, it appears that adding the backup is definitely cost justifiable.

Problem 2

Due to the extreme cost of interrupting production, a firm has two standby machines available in case a particular machine breaks down. The machine in use has a reliability of .94, and the backups have reliabilities of .90 and .80. In the event of a failure, either backup can be pressed into service. If one fails, the other backup can be used. Compute the system reliability.

Solution

$R_1 = .94$, $R_2 = .90$, and $R_3 = .80$

The system can be depicted in this way:

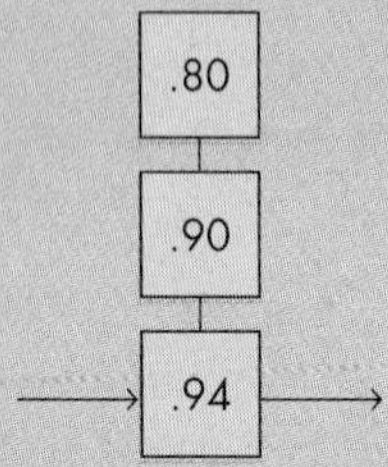

$$R_{system} = R_1 + R_2(1 - R_1) + R_3(1 - R_2)(1 - R_1)$$
$$= .94 + .90(1 - .94) + .80(1 - .90)(1 - .94) = .9988$$

Problem 3

A hospital has three *independent* fire alarm systems, with reliabilities of .95, .97, and .99. In the event of a fire, what is the probability that a warning would be given?

Solution

A warning would *not* be given if all three alarms failed. The probability that at least one alarm would operate is $1 - P(\text{none operate})$:

$$P(\text{none operate}) = (1 - .95)(1 - .97)(1 - .99) = .000015$$
$$P(\text{warning}) = 1 - .000015 = .999985$$

Problem 4

A weather satellite has an expected life of 10 years from the time it is placed into earth orbit. Determine its probability of no wear-out before each of the following lengths of service. (Assume the exponential distribution is appropriate.)

a. 5 years. *b.* 12 years. *c.* 20 years. *d.* 30 years.

Solution

MTBF = 10 years

Compute the ratio T/MTBF for $T = 5, 12, 20,$ and 30, and obtain the values of $e^{-T/\text{MTBF}}$ from Table 4S–1. The solutions are summarized in the following table.

T	MTBF	*T*/MTBF	$e^{-T/\text{MTBF}}$
a. 5	10	0.50	.6065
b. 12	10	1.20	.3012
c. 20	10	2.00	.1353
d. 30	10	3.00	.0498

Problem 5

What is the probability that the satellite described in Solved Problem 4 will fail between 5 and 12 years after being placed into earth orbit?

Solution

$$P(5 \text{ years} < \text{failure} < 12 \text{ years}) = P\ (\text{failure after 5 years}) - P(\text{failure after 12 years})$$

Using the probabilities shown in the previous solution, you obtain:

$$\begin{array}{r} P(\text{failure after 5 years}) = .6065 \\ -P(\text{failure after 12 years}) = \underline{.3012} \\ .3053 \end{array}$$

The corresponding area under the curve is illustrated as follows:

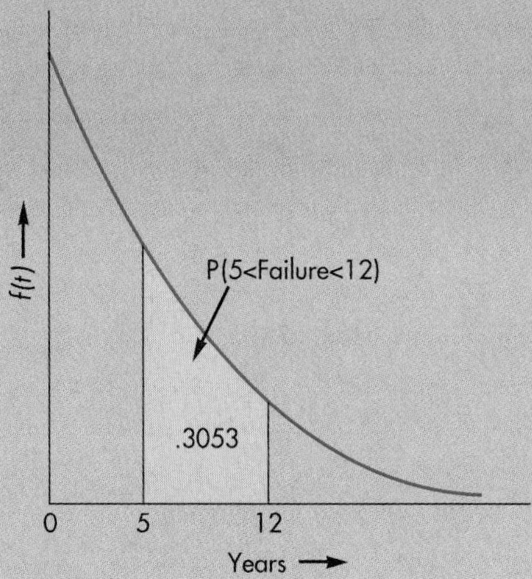

Problem 6

One line of radial tires produced by a large company has a wear-out life that can be modeled using a normal distribution with a mean of 25,000 miles and a standard deviation of 2,000 miles. Determine each of the following:

a. The percentage of tires that can be expected to wear out within ±2,000 miles of the average (i.e., between 23,000 miles and 27,000 miles).

b. The percentage of tires that can be expected to fail between 26,000 miles and 29,000 miles.

Solution

Notes: (1) Miles are analogous to time and are handled in exactly the same way; (2) the term *percentage* refers to a probability.

a. The phrase "within ±2,000 miles of the average" translates to within one standard deviation of the mean since the standard deviation equals 2,000 miles. Therefore the range of z is $z = -1.00$ to $z = +1.00$, and the area under the curve between those points is found as the difference between $P(z < +1.00)$ and $P(z < -1.00)$, using values obtained from Appendix Table B.

$$\begin{array}{r} P(z < +1.00) = .8413 \\ -P(z < -1.00) = .1587 \\ \hline P(-1.00 < z < +1.00) = .6826 \end{array}$$

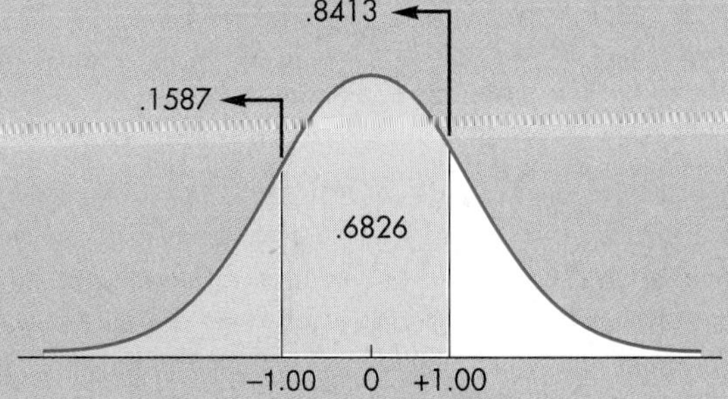

b. Wear-out mean = 25,000 miles
Wear-out standard deviation = 2,000 miles
$P(26{,}000 < \text{Wear-out} < 29{,}000) = P(z_{29,000}) - P(z < z_{26,000})$

$$z_{29,000} = \frac{29,000 - 25,000}{2,000} = +2.00 \rightarrow .9772$$

$$z_{26,000} = \frac{26,000 - 25,000}{2,000} = +0.50 \rightarrow .6915$$

The difference is .9772 − .6915 = .2857, which is the expected percent of tires that will wear out between 26,000 miles and 29,000 miles.

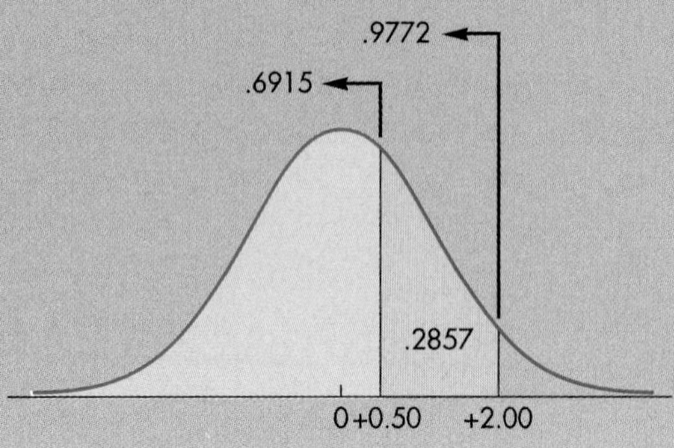

Discussion and Review Questions

1. Define the term *reliability*.
2. Explain why a product or system might have an overall reliability that is low even though it is comprised of components that have fairly high reliabilities.
3. What is redundancy and how can it improve product design?

Problems

1. Consider the following system:

Determine the probability that the system will operate under each of these conditions:

a. 0.81
b. 0.9801
c. 0.9783

a. The system as shown.

b. Each system component has a backup with a probability of .90 and a switch that is 100 percent reliable.

c. Backups with .90 probability and a switch that is 99 percent reliable.

0.9033

2. A product is composed of four parts. In order for the product to function properly in a given situation, each of the parts must function. Two of the parts each have a .96 probability of functioning, and two each have a probability of .99. What is the overall probability that the product will function properly?

0.9726

3. A system consists of three identical components. In order for the system to perform as intended, all of the components must perform. Each has the same probability of performance. If the system is to have a .92 probability of performing, what is the minimum probability of performing needed by each of the individual components?

0.93

4. A product engineer has developed the following equation for the cost of a system component: $C = (10P)^2$, where C is the cost in dollars and P is the probability that the component will operate as expected. The system is composed of two identical components, both of which must operate for the system to operate. The engineer can spend \$173 for the two components. To the nearest two decimal places, what is the largest component probability that can be achieved?

5. The guidance system of a ship is controlled by a computer that has three major modules. In order for the computer to function properly, all three modules must function. Two of the modules have reliabilities of .97, and the other has a reliability of .99.

a. What is the reliability of the computer?

b. A backup computer identical to the one being used will be installed to improve overall reliability. Assuming the new computer automatically functions if the main one fails, determine the resulting reliability.

c. If the backup computer must be activated by a switch in the event that the first computer fails, and the switch has a reliability of .98, what is the overall reliability of the system? (*Both* the switch and the backup computer must function in order for the backup to take over.)

a. 0.9315
b. 0.9953
c. 0.994

6. One of the industrial robots designed by a leading producer of servomechanisms has four major components. Components' reliabilities are .98, .95, .94, and .90. All of the components must function in order for the robot to operate effectively.

   a. Compute the reliability of the robot.

   b. Designers want to improve the reliability by adding a backup component. Due to space limitations, only one backup can be added. The backup for any component will have the same reliability as the unit for which it is the backup. Which component should get the backup in order to achieve the highest reliability?

   c. If one backup with a reliability of .92 can be added to any one of the main components, which component should get it to obtain the highest overall reliability?

a. 0.7876
b. 1st = 0.8034
2nd = 0.8270
3rd = 0.8349
4th = 0.8664
c. Add to component with 0.90 reliability.

7. A production line has three machines A, B, and C, with reliabilities of .99, .96, and .93, respectively. The machines are arranged so that if one breaks down, the others must shut down. Engineers are weighing two alternative designs for increasing the line's reliability. Plan 1 involves adding an identical backup *line,* and plan 2 involves providing a backup for each *machine.* In either case, three machines (A, B, and C) would be used with reliabilities equal to the original three.

   a. Which plan will provide the higher reliability?

   b. Explain why the two reliabilities are not the same.

   c. What other factors might enter into the decision of which plan to adopt?

a. Plan 1: 0.9865
Plan 2: 0.9934
b. See IM.
c. space, cost, etc.

8. A water treatment plant has five major components which must all function in order for the plant to operate as intended. Assuming that each component of the system has the same reliability, what is the minimum reliability each one must have in order for the overall system to have a reliability of .98?

0.996

9. Repeat Problem 8 under the condition that one of the components will have a backup with a reliability equal to that of any one of the other components.

0.995

10. Hoping to increase the chances of reaching a performance goal, the director of a research project has assigned three separate research teams the same task. The director estimates that the team probabilities are .9, .8, and .7 for successfully completing the task in the allotted time. Assuming that the teams work independently, what is the probability that the task will not be completed in time?

0.006

11. An electronic chess game has a useful life that is exponential with a mean of 30 months. Determine each of the following:

    a. The probability that any given unit will operate for at least: (1) 39 months, (2) 48 months, (3) 60 months.

    b. The probability that any given unit will fail sooner than: (1) 33 months, (2) 15 months, (3) 6 months.

    c. The length of service time after which the percentage of failed units will approximately equal: (1) 50 percent, (2) 85 percent, (3) 95 percent, (4) 99 percent.

11. *a.* (1): 0.2725
(2): 0.2019
(3): 0.1353
b. (1): 0.6671
(2): 0.3935
(3): 0.1813
c. (1): 21 months
(2): 57 months
(3): 90 months
(4): 138 months

12. *a.* 0.6321
b. 3 months

12. A manufacturer of programmable calculators is attempting to determine a reasonable free-service period for a model it will introduce shortly. The manager of product testing has indicated that the calculators have an expected life of 30 months. Assume product life can be described by an exponential distribution.
    a. If service contracts are offered for the expected life of the calculator, what percentage of those sold would be expected to fail during the service period?
    b. What service period would result in a failure rate of approximately 10 percent?

a. 0.3012
b. 0.1813
c. 0.5175

13. Lucky Lumen light bulbs have an expected life that is exponentially distributed with a mean of 5,000 hours. Determine the probability that one of these light bulbs will last
    a. At least 6,000 hours.
    b. No longer than 1,000 hours.
    c. Between 1,000 hours and 6,000 hours.

a. 0.2231
b. 0.8647
c. 0.0878
d. 0.0302

14. Planetary Communications, Inc., intends to launch a satellite that will enhance reception of television programs in Alaska. According to its designers, the satellite will have an expected life of six years. Assume the exponential distribution applies. Determine the probability that it will function for each of the following time periods:
    a. More than 9 years.
    b. Less than 12 years.
    c. More than 9 years but less than 12 years.
    d. At least 21 years.

a. 22.66%
b. 44%
c. 38.3%

15. An office manager has received a report from a consultant that includes a section on equipment replacement. The report indicates that scanners have a service life that is normally distributed with a mean of 41 months and a standard deviation of 4 months. On the basis of this information, determine the percentage of scanners that can be expected to fail in the following time periods:
    a. Before 38 months of service.
    b. Between 40 and 45 months of service.
    c. Within ±2 months of the mean life.

16. *a.* (1): $z = -2.00$
0.9772
(2): $z = 0.00$
0.5000
(3): $z = -3.00$
prob = 13/10,000
= .0013
b. $z = -4.00$
nearly zero

16. A major television manufacturer has determined that its 19-inch color TV picture tubes have a mean service life that can be modeled by a normal distribution with a mean of six years and a standard deviation of one-half year.
    a. What probability can you assign to service lives of at least: (1) Five years? (2) Six years? (3) Seven and one-half years?
    b. If the manufacturer offers service contracts of four years on these picture tubes, what percentage can be expected to fail from wear-out during the service period?

17. *a.* $z = -2.055$
4.97 years
b. $z = -1.645$
5.18 years

17. Refer to Problem 16. What service period would achieve an expected wear-out rate of:
    a. 2 percent?
    b. 5 percent?

18. *a.* .930
b. .980

18. Determine the availability for each of these cases:
    a. MTBF = 40 days, average repair time = 3 days.
    b. MTBF = 300 hours, average repair time = 6 hours.

19. .962

19. A machine can operate for an average of 10 weeks before it needs to be overhauled, a process which takes two days. The machine is operated five days a week. Compute the availability of this machine. (Hint: all times must be in the same units.)

20. A manager must decide between two machines. The manager will take into account each machine's operating costs and initial costs, and its breakdown and repair times. Machine A has a projected average operating time of 142 hours and a projected average repair time of seven hours. Projected times for machine B are an average operating time of 65 hours and a repair time of two hours. What are the projected availabilities of each machine?

20. Machine A = .953
Machine B = .970

21. A designer estimates that she can increase the average time between failures of a part by 5 percent at a cost of $450, or reduce the average repair time by 10 percent at a cost of $200. Which option would be more cost-effective? Currently, the average time between failures is 100 hours and the average repair time is four hours.

a. .9633
b. .9653

LEARNING OBJECTIVES

After completing this chapter, you should be able to:

1. Explain the importance of process selection and how process selection relates to product and service design and capacity planning.
2. Describe the different types of processing.
3. Describe automated approaches to processing.
4. Explain the need for management of technology.
5. Explain the importance of capacity planning.
6. Discuss ways of defining and measuring capacity.
7. Describe the factors that determine effective capacity alternatives.
8. Discuss the major considerations related to developing capacity alternatives.
9. Briefly describe approaches that are useful for evaluating process selection and capacity alternatives.

CHAPTER FIVE
PROCESS SELECTION AND CAPACITY PLANNING

CHAPTER OUTLINE

Product and service choices, process selection, capacity planning, and choices about location and layout are among the most basic decisions managers must make because those decisions have long-term consequences for the organization.

Processes transform inputs into outputs. Consequently, they are at the core of operations management. Not surprisingly, process planning plays a key role in the ability of an organization to achieve its mission.

In this chapter we examine process selection and capacity planning. Process selection is both exciting and challenging; technological advances in processing provide many new options for competitive advantage, but they also pose numerous risks for the unwary decision maker. The first part of the chapter is devoted to process selection, the remainder to capacity planning.

PROCESS SELECTION

Process selection refers to the way an organization chooses to produce its goods or provide its services. Essentially it involves choice of technology and related issues, and it has major implications for capacity planning, layout of facilities, equipment, and design of work systems. Process selection occurs as a matter of course when new products or services are being planned. However, it also occurs periodically due to technological changes in equipment. Figure 5–1 provides an overview of where process selection fits into system design.

How an organization approaches process selection is determined by the organization's *process strategy.* Key aspects include:

- Make or buy decisions: The extent to which the organization will produce goods or provide services in-house as opposed to relying on outside organizations to produce or provide them.
- Capital intensity: The mix of equipment and labor that will be used by the organization.
- Process flexibility: The degree to which the system can be adjusted to changes in processing requirements due to such factors as changes in product or service design, changes in volume processed, and changes in technology.

Make or Buy

The very first step in process planning is to consider whether to make or buy some or all of a product or to subcontract some or all of a service. A manufacturer might decide to purchase certain parts rather than make them; sometimes all parts are purchased, with the manufacturer simply performing assembly operations. Many firms contract out janitorial services, and some contract for repair services. If a decision is made to buy or contract, this lessens or eliminates the need for process selection.

In make or buy decisions, a number of factors are usually considered:

1. *Available capacity.* If an organization has available the equipment, necessary skills, and *time,* it often makes sense to produce an item or perform a service in-house. The additional costs would be relatively small compared with those required to buy items or subcontract services.
2. *Expertise.* If a firm lacks the expertise to do a job satisfactorily, buying might be a reasonable alternative.
3. *Quality considerations.* Firms that specialize can usually offer higher quality than an organization can attain itself. Conversely, special quality requirements or the ability to closely monitor quality may cause an organization to perform a job itself.
4. *The nature of demand.* When demand for an item is high and steady, the organization is often better off doing the work itself. However, wide fluctuations in demand or small orders are usually better handled by specialists who are able to combine orders

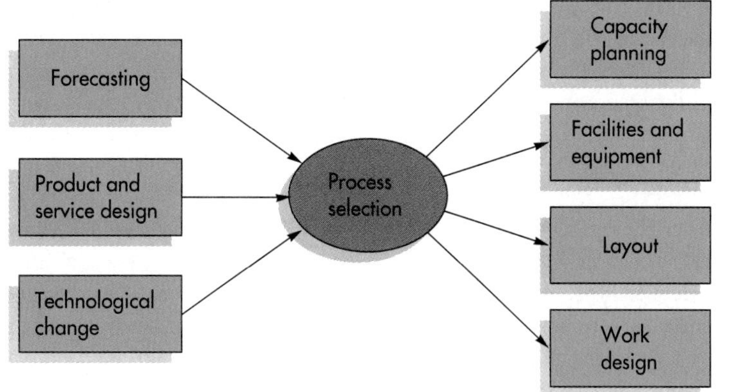

FIGURE 5-1

Process selection and system design

from multiple sources, which results in higher volume and tends to offset individual buyer fluctuations.

5. *Cost.* Any cost savings achieved from buying or making must be weighed against the preceding factors. Cost savings might come from the item itself or from transportation cost savings. If there are fixed costs associated with making an item that cannot be reallocated if the item is purchased, that has to be recognized in the analysis.

In some cases, a firm might choose to perform part of the work itself and let others handle the rest in order to maintain flexibility and to hedge against loss of a subcontractor. Moreover, this provides a bargaining tool in negotiations with contractors, or a head start if the firm decides later to take over the operation entirely.

If the organization decides to perform some or all of the processing, then the issue of process selection becomes important.

Type of Operation

The degree of standardization and the volume of output of a product or service influence the way production is organized. Output can range from high volume, highly standardized, to low volume, highly customized.

Continuous and intermittent processing systems have some key differences which affect how these systems are managed. The following sections highlight these key differences.

Continuous and Semicontinuous Processing. **Continuous processing** is employed when a highly uniform product or service is produced or rendered. Processing of chemicals, photographic film, newsprint, and oil products are all examples of this type of product. Machines often provide continuous services, such as air-conditioning, heating, continuous cleaning, and monitoring systems. Other examples include liquid and powder detergents, and water treatment. Generally, these products are measured on a continuous basis rather than counted as discrete units. Industries that use continuous processing are sometimes referred to as *process industries.* Operations are usually performed around the clock to avoid costly shutdowns and start-ups. The output of these systems is highly uniform (standardized). Semicontinuous processing systems, known as **repetitive processing,** produce output that allows for some variety; products are highly similar but not identical. Examples of goods include automobiles, televisions, computers, calculators, cameras, and video equipment. Typically, these products are produced in discrete units. Examples in the service sector include programs for mass inoculations, automatic car washes, mechanical harvesters, mail service, and fast-food operations. Note that applications in services are less plentiful because services tend to be more customized on a per-unit basis.

continuous processing A system that produces highly uniform products or continuous services, often performed by machines.

repetitive processing A production system that renders one or a few highly standardized products or services.

The highly standardized output of these systems lends itself to highly standardized methods and equipment. Skill requirements of workers are usually fairly low. Equipment tends to be highly specialized, which tends to make it expensive relative to more general-purpose equipment, but the high volumes of output result in a low cost per unit. As a general rule, goods produced by these systems are made for inventory rather than customer order.

intermittent processing A system that produces lower volumes of items or services with a greater variety of processing requirements.

batch processing A system used to produce moderate volumes of similar items.

Vol. 1, Tape 1, Seg. 3
Caterpillar, Ford,
Nucor Steel

job shop A system that renders unit or small lot production or service with varying specifications, according to customer needs.

Intermittent Processing. When systems handle a variety of processing requirements, **intermittent processing** is used. Volume is much lower than in continuous systems. Intermittent systems are characterized by general-purpose equipment that can satisfy a variety of processing requirements, semiskilled or skilled workers who operate the equipment, and a narrower span of supervision than for most continuous systems.

Batch processing is one form of intermittent processing. **Batch processing** is used when companies need to produce moderate volumes of similar products. Food processors (e.g., bakeries, canneries) typically produce in batches. For example, an ice cream producer will produce a batch of one type of ice cream, such as vanilla, and then a second type, such as strawberry, and so on. The processing requirements and equipment are the same, but some of the ingredients vary from one batch to the next. Similarly, a canning factory might process a variety of vegetables: one run may be sliced carrots, the next green beans, and the next corn or beets. All might follow a similar process of washing, sorting, slicing, cooking, and packing, but the equipment needs to be cleaned and adjusted between runs. Batch output may be standardized (e.g., paint, ice cream, canned vegetables) or customized. An example of customized output is printing (e.g., magazines, newspapers, textbooks, user manuals).

Job shops perform another type of intermittent processing. A job shop is designed to handle a greater variety of job requirements than batch processing. A **job shop** maintains the ability to perform certain types of operations but generally is not responsible for specific products. Instead, it performs to customer specifications; jobs tend to vary according to the needs of customers. Examples of job shops are organizations that do repair work (e.g., appliances, automobiles), health care, tool and die shops, and so on. Lot sizes vary from large to small, even a single unit. What distinguishes the job shop operation from batch processing is that the job requirements often vary considerably from job to job. This means that the sequence of processing steps and the job content of the steps also vary considerably. An auto repair shop is an example of a job shop. Each car is handled on an individual basis. Large repair shops may have specialists who deal in one kind of repair (e.g., brake jobs), but cars are still handled one at a time. For large jobs processing many of a single item or a few of many items, there is usually so much variety among successive jobs that the batch processing described for the canning factory would be too restrictive. Differences in job processing requirements add routing and scheduling complexities, as well as a frequent need to adjust equipment settings or make other alterations for successive jobs. Processing cost per unit is generally higher than it is under continuous or semicontinuous processing.

Further examples of intermittent processing are textbook publication, bakeries, health care systems, and educational systems. In some cases, the outputs are made for inventory (clothing, automobile tires); in others, they are designed to meet customer needs (health care) or specifications (special tools, parts, or equipment). Marketing efforts in these systems are often directed toward promoting system processing capabilities or customized services.

Table 5–1a provides a summary of the characteristics of the four types of operations. The processes can be organized into a product-process matrix as shown in Table 5–1b.

Notice that the examples all line up along the diagonal of the table. This is the most efficient alignment. If a producer chose some other combination (e.g., assembly line for a customized product or service), he or she would find that the highly customized requirements of the various products were in direct conflict with the more uniform require-

TABLE 5–1a

Type of Operation

	Job Shop	Batch	Repetitive/ Assembly	Continuous
Description	Customized goods or services	Semi-standardized goods or services	Standardized goods or services	Highly standardized goods or services
Examples of processes				
manufacturing	Machine shop	Bakery	Assembly line	Steel mill, paper mill, etc.
service	Beauty shop, barber shop	Classroom	Cafeteria line	Central heating system
Examples of				
goods	Specialty tools	Cookies	Automobiles	Steel, paper, flour, sugar
services	Hair styling	Education	Car wash	Heating, air conditioning
Volume	Low	Low-moderate	High	Very high
Output variety	Very high	Moderate	Low	Very low
Equipment flexibility	Very high	Moderate	Low	Very low
Advantages	Able to handle a wide variety of work	Flexibility	Low unit cost, high volume, efficient	Very efficient, very high volume
Disadvantages	Slow, high cost per unit, complex planning and scheduling	Moderate cost per unit, moderate scheduling complexity	Low flexibility, high cost of downtime	Very rigid, lack of variety, costly to change, very high cost of downtime

TABLE 5–1b

Matching the process with product variety, equipment flexibility, and volume requirements

Product variety	High	Moderate	Low	Very low
Equipment flexibility	High	Moderate	Low	Very low
Low volume	Job shop			
Moderate volume		Batch		
High volume			Repetitive assembly	
Very high volume				Continuous flow

Sources: Adapted from Robert H. Hayes and Stephen C. Wheelwright, "Link Manufacturing Product and Process Life Cycles," *Harvard Business Review*, Jan–Feb 1979, pp. 133–40; and Irwin Operations Management Video Series.

ments needed to effectively operate in the assembly-line mode. Similarly, a job shop arrangement (machines and personnel are capable of handling a wide variety of processing requirements) would be wasted on a highly standardized product; equipment and personnel need to be highly specialized.

A key concept in process selection is the need to match product requirements with process capabilities. The difference between success and failure in production can sometimes be traced to choice of process.

Table 5–1b can provide insights to managers selecting processes and managing existing operations. For new products, decision makers should make every attempt to achieve

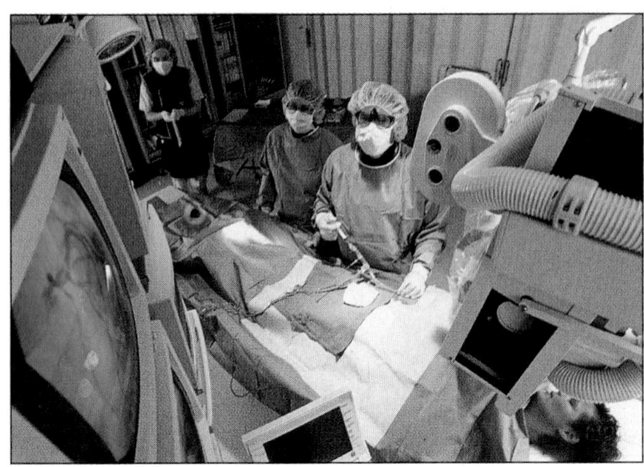

A job shop process: A Midwestern hospital medical team performs a diagnostic procedure which involves checking the opening of a cardiac catheterization.

A batch process: Ben and Jerry's produces ice cream one flavor at a time.

www.benjerry.com

A repetitive process: Robots put together Saturns on an automated assembly line. www.saturn.com

A continuous flow process: Pepsi-Cola is bottled at GEMEX in Mexico, Pepsi's largest non-U.S. bottler. www.pepsico.com

a matching of product and process requirements. For an ongoing operation, a manager should examine existing processes in light of the table to see how well processes and products are matched. Poor matches suggest the potential for improvement, perhaps with a substantial increase in efficiency and lowering of cost.

Another consideration is that products and services often go through *life cycles* that begin with low volume but which increase as products or services become better known. When that happens, a manager must know when to shift from one type of process (e.g., job shop) to the next (e.g., batch). Of course, some operations remain at a certain level (e.g., magazine publishing), while others increase (or decrease as markets become saturated) over time. Again, it is important for a manager to assess his or her products and services and make a judgment on whether to plan for changes in processing over time.

The four types of processes (job shop, batch, repetitive and continuous) are typically ongoing operations. However, there are situations that are not ongoing but instead are of limited durations. For example, upgrading a computer network, installing a new telephone system and developing a benefits package for employees are all limited duration. Indeed in similar situations, the work is organized as a **project.** A project is a set of activities directed toward a unique goal. Chapter 18 is devoted entirely to projects.

project A nonrepetitive set of activities directed toward a unique goal with a limited time frame.

Automation

Automation is machinery that has sensing and control devices that enable it to operate automatically. A key question in process planning is whether to automate. If a company decides to automate, the next question is how much. Automation can range from factories that are completely automated to a single automated operation.

automation Machinery that has sensing and control devices that enable it to operate automatically.

Automation offers a number of advantages over human labor. It has low variability, whereas it is difficult for a human to perform a task in exactly the same way, in the same amount of time, and on a repetitive basis. In a production setting, variability is detrimental to quality and to meeting schedules. Moreover, machines do not get bored or distracted, nor do they go out on strike, ask for higher wages, or file labor grievances. Still another advantage of automation is reduction of variable costs.

Automation is frequently touted as a strategy necessary for competitiveness. For example, the Nucor Steel plant in Crawfordsville, Indiana, can produce 800,000 tons of finished strip steel a day. The plant is automated, and produces the steel from molten metal to a hot band of finished steel in one uninterrupted process, giving the minimill a substantial productivity advantage over its foreign competitors. However, automation also has certain disadvantages and limitations compared with human labor. To begin with, it can be costly. Technology is expensive; usually it requires high volumes of output to offset high costs. In addition, automation is much less flexible than human labor. Once a process has been automated, there is substantial reason for not changing it. Moreover, workers sometimes fear automation because it might cause them to lose their jobs. That can have an adverse effect on morale and productivity.

Decision makers must carefully examine the issue of whether to automate or the degree in which to automate, so that they clearly understand all the ramifications. Also, much thought and careful planning are necessary to successfully *integrate* automation into a production system. Otherwise, it can lead to major problems. GM invested heavily in automation in the 1980s only to find its costs increasing while flexibility and productivity took a nosedive. Its market had shrunk while GM was increasing its capacity!

Generally speaking, there are three kinds of automation: fixed, programmable, and flexible.

Fixed automation is the most rigid of the three types. The concept was perfected by the Ford Motor Company in the early 1900s, and it has been the cornerstone of mass production in the auto industry. Sometimes referred to as Detroit-type automation, it uses high-cost, specialized equipment for a fixed sequence of operations. Low cost and high volume are its primary advantages; minimal variety and the high cost of making major changes in either product or process are its primary limitations.

Programmable automation is at the opposite end of the spectrum. It involves the use of high-cost, general-purpose equipment controlled by a computer program that provides both the sequence of operations and specific details about each operation. Changing the process is as easy (or difficult) as changing the computer program, and there is downtime while program changes are being made. This type of automation has the capability of economically producing a fairly wide variety of low-volume products in small batches. Numerically controlled (N/C) machines and some robots are applications of programmable automation.

computer-aided manufacturing The use of computers in process control.

Computer-aided manufacturing (CAM) refers to the use of computers in process control, ranging from robots to automated quality control. **Numerically controlled (N/C) machines** are programmed to follow a set of processing instructions based on mathematical relationships that tell the machine the details of the operations to be

numerically controlled machines Machines that perform operations by following mathematical processing instructions.

performed. The instructions are stored on a device such as a floppy disk, magnetic tape, or microprocessor. Although N/C machines have been used for many years, they are an important part of new approaches to manufacturing. Individual machines may have their own computer; this is referred to as *computerized numerical control (CNC)*. Or one computer may control a number of N/C machines, which is referred to as *direct numerical control (DNC)*.

N/C machines are best used in cases where parts are processed frequently and in small batches, part geometry is complex, close tolerances are required, mistakes are costly, and there is the possibility of frequent changes in design. The main limitations of N/C machines are the higher skill levels needed to program the machines and their inability to detect tool wear and material variation.

robot A machine consisting of a mechanical arm, a power supply, and a controller.

The use of robots in manufacturing is sometimes an option. A **robot** consists of three parts: a mechanical arm, a power supply, and a controller. Unlike movie versions of robots, which vaguely resemble human beings, industrial robots are much less glamorous and much less mobile; most robots are stationary except for their movable arms.

Robots can handle a wide variety of tasks, including welding, assembly, loading and unloading of machines, painting, and testing. They relieve humans from heavy or dirty work and often eliminate drudgery tasks.

Some uses of robots are fairly simple, others are much more complex. At the lowest level are robots that follow a fixed set of instructions. Next are programmable robots, which can repeat a set of movements after being led through the sequence. These robots "play back" a mechanical sequence much as a video recorder plays back a visual sequence. At the next level up are robots that follow instructions from a computer. At the top are robots that can recognize objects and make certain simple decisions.

Robots move in one of two ways. Point-to-point robots move to predetermined points and perform a specified operation; they then move on to the next point and perform another operation. Continuous-path robots follow a continuous (moving) path while performing an operation.

Robots can be powered pneumatically (air-driven), hydraulically (fluids under pressure), or electronically. Figures 5–2 and 5–3 illustrate robots.

FIGURE 5–2

Industrial robot

www.milacron.com

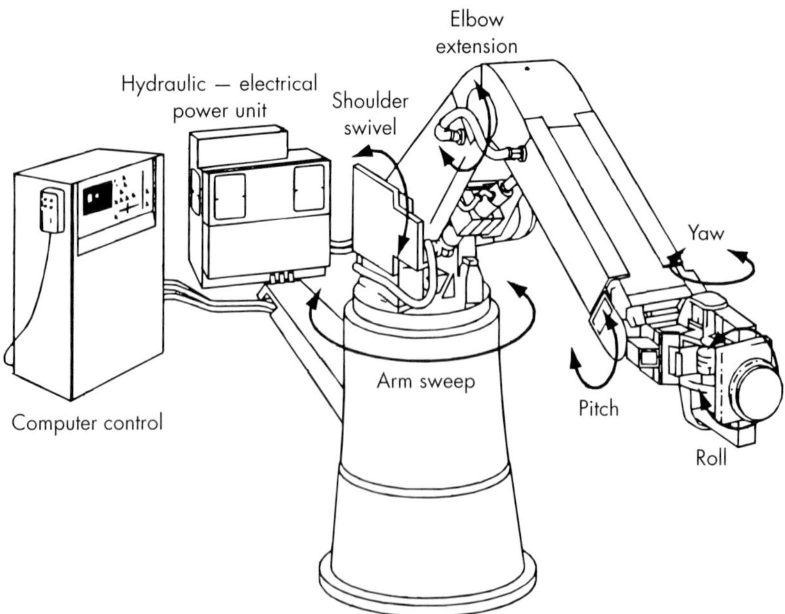

Cincinnati Milacron robot system

Source: Morris A. Cohen and Uday M. Apte, *Manufacturing Automation* (Burr Ridge, IL: Irwin/McGraw-Hill, 1997), p. 138.

FIGURE 5-3

Robots on an assembly line: Auto bodies on an assembly line at Ford being spot welded by an IRB 6000 robot

www.ford.com

Flexible automation evolved from programmable automation. It uses equipment that is more customized than that of programmable automation. A key difference between the two is that flexible automation requires significantly less changeover time. This permits almost continuous operation of equipment *and* product variety without the need to produce in batches.

READING

CUSTOM-MADE, DIRECT FROM THE PLANT **By Otis Port in New York**

The newest manufacturing combines state-of-the-art fabrication and product-delivery technologies with the old-world artisan's aim to please.

It is 2010, and you need a new suit—fast—for an overseas business trip. No problem. You head to the department store at the mall. You step into a kiosk-like contraption, and an optical scanner automatically measures your body. Seconds after you record your choice of style and fabric, the information is relayed to a plant where lasers cut the cloth precisely to your size. A few days later, the suit is ready.

You may not have to wait 15 years for this kind of high-tech couture. Manufacturing is fast entering a new age of industrial excellence: mass customization. Not only clothes, but a huge variety of goods, from autos to computers, will be manufactured to match each customer's taste, specifications, and budget. Mass customization will mark the culmination and synthesis of agile manufacturing, virtual companies, and total quality management. If ever the term "paradigm shift" was

THE EVOLUTION OF THE FACTORY

Factories will soon rival craftshops in their ability to customize

FACTORY FOCUS	**MASS PRODUCTION**	**FLEXIBLE PRODUCTION**	**MASS CUSTOMIZATION**
	1900 TO 1970	**1971 TO 2000**	**2001 TO 2020**
TYPICAL NUMBER OF:			
Machine tools	150	50 TO 30	25 TO 20
Products made	10 TO 15	100 TO 1,000	UNLIMITED
PRODUCTS REWORKED DUE TO POOR QUALITY	25% OR MORE	0.02% OR LESS	UNDER 0.0005%

appropriate, "this is it," says Roger N. Nagel, deputy director of Lehigh University's Iacocca Institute, the Bethlehem (Pa.) think tank where the notion of agile manufacturing was born three years ago.

Customized products will be made as fast and as cheaply as mass-produced products. "But factories won't sell things—they'll sell customer gratification," says Nagel. The price of products will be determined not by adding up the costs of all the parts in a finished product, but by the value of the knowhow and services that a company musters to assure utmost customer satisfaction. Once customers realize that agile suppliers can provide exactly what's needed on a tight schedule and eliminate the costs of maintaining inventories of parts, "price is no longer a factor," says David L. Ross, marketing director at Ross Operating Valve Co., a private 350-employee maker of pneumatic valves in Troy, Mich. "It is literally the last thing discussed."

This isn't just fantasy. More than 200 companies have joined the Agile Manufacturing Enterprise Forum, which Nagel co-founded. AMEF members include such heavyweights as Air Products & Chemicals, Chrysler, FMC, Honeywell, Milliken, Texas Instruments, and Westinghouse. This past summer, the National Science Foundation and the Pentagon's Advanced Research Projects Agency helped set up three agile manufacturing research institutes: at the University of Illinois, the University of Texas, and Rensselaer Polytechnic Institute. And the apparel industry's research consortium, the Textile/Clothing Technology Corp. in Cary, N.C., is leading the development of the optical-scanning "tailor" described above.

Agile manufacturing might do more for U.S. employment than the post-industrial work that was supposed to supplant manufacturing. Some of the production work that went offshore in the 1980s has been returning in the 1990s. One reason: Companies striving to become agile are forced to slash cycle time—the period between receiving an order and delivering the goods—and they often discover that moving products through an international pipeline eats up more time than the labor savings are worth. Items that have come back to agile wannabes range from some of Caterpillar Inc.'s huge earth-moving equipment, for a while made in Korea, to computer modems from U.S. Robotics Inc., which used to outsource the production of components in Mexico.

Valves by Phone

Nobody expects blue-collar jobs to hold the line, much less grow. Building on the pioneering work of Allen-Bradley Co. and others, tomorrow's factories will be smaller, with fewer but smarter machine tools that need only token human care—yet they'll turn out a far richer variety of goods. So, shop-floor employment is heading the way of farm labor. However, agile technology is also spawning new white-collar jobs in support industries, just as modern agriculture has created myriad jobs in package design, food engineering, marketing, and other services.

However, such growth is taxing agile companies. Take Ross Operating Valve. "I cannot train people quick enough to take care of all the potential new business," says Ross. At the Ross/Flex plant in Lavonia, Ga., customers phone to discuss what valves they need with company engineers called "integrators." The specs are entered into a CAD/CAM system to design a one-of-a-kind valve, and automated machine tools grind out the metal parts overnight. Finished valves are delivered in as little as 72 hours, at a typical cost of $3,000. That's about one-hundredth the time and one-tenth the cost of traditional methods.

Richard K. Dove, a consultant on agile manufacturing in Oakland, Calif., believes this new method may be the ticket to surpassing the Japanese. He sees a future where the U.S. exploits its lead in information technologies. "The Japanese miracle was based on the exquisite refinement of manufacturing," notes Steven L. Goldman, a Lehigh professor of humanities who is an expert in the history of technology. "But agility moves the center of value away from manufacturing itself. What counts now is the software, the knowledge—and that expertise is outside Japan."

Roving Robots

The goal is to link customers, suppliers, and manufacturers into a kind of superefficient confederation. What a factory produces today will be driven by yesterday's retail sales—or an order received moments ago from an on-line partner. By 2000, says Carl P. McCormick, president of electronics maker Group Technologies Corp., many factories will be tightly interlinked via nationwide networks such as the Enterprise Integration Network. EINet is being developed by Microelectronics & Computer Technology Corp. (MCC), a consortium of high-tech companies in Austin. Networked manufacturers will be able to dump their orders directly into their suppliers' factory computers, and these will in turn relay the suppliers' needs to their own suppliers.

Ultimately, the communications web may extend the connection down to individual machines on the shop floor. Already, researchers at Massachusetts Institute of Technology and Purdue University are jointly developing software for smart machine tools. Under the MIT-Purdue scheme, mobile robots would carry parts around the factory. When a robot rolled onto the floor, it would broadcast to the machine tools what work its part requires. The affected machine tools would then consult their production schedules to determine when they could be available, and the computers would collectively calculate which combination of machines would most efficiently polish off the necessary work.

Computers negotiating production schedules with other computers, taking care of routine business transactions without human intervention, and even programming each other: This agile world will take some getting used to. And it's coming faster than anyone expected. "Fundamental transformations of the terms of commerce have always been measured in decades," says Lehigh's Goldman. "In 1991, I figured it would take 15 years for agile manufacturing to become common practice. But now, I think it will be the norm within five years."

In practice, flexible automation is used in several different formats:

A **manufacturing cell** consists of one or a small number of N/C machines that produce a family of similar parts. The machines may be linked with automatic material handing devices.

manufacturing cell One or a few N/C machines that produce a wide variety of parts.

A **flexible manufacturing system (FMS)** is a group of machines that include supervisory computer control, automatic material handling, and robots or other automated processing equipment. Reprogrammable controllers enable these systems to produce a variety of *similar* products. Systems may range from three or four machines to more than a dozen. They are designed to handle intermittent processing requirements with some of the benefits of automation and some of the flexibility of individual, or stand-alone, machines (e.g., N/C machines). Flexible manufacturing systems offer reduced labor costs and more consistent quality compared with more traditional manufacturing methods, lower capital investment and higher flexibility than "hard" automation, and relatively quick changeover time. Flexible manufacturing systems appeal to managers, who hope to achieve both the flexibility of job shop processing and the productivity of repetitive processing systems.

flexible manufacturing system A group of machines designed to handle intermittent processing requirements and produce a variety of similar products.

Although these are important benefits, an FMS also has certain limitations. One is that this type of system can handle a relatively narrow range of part variety, so it must be used for a family of similar parts, which all require similar machining. Also, an FMS requires longer planning and development times than more conventional processing equipment because of its increased complexity and cost. Furthermore, companies sometimes prefer a gradual approach to automation, and FMS represents a sizable chunk of technology.

A computerized control room at the North Rhine Westphalia, Germany, Thyssen manufacturing plant, which produces electrolytically galvanized sheet metal.

computer integrated manufacturing A system for linking a broad range of manufacturing activities through an integrating computer system.

Computer-integrated manufacturing (CIM) is a system that uses an integrating computer system to link a broad range of manufacturing activities, including engineering design, flexible manufacturing systems, and production planning and control. Not all elements are absolutely necessary. For instance, CIM might be as simple as linking two or more FMSs by a host computer. More encompassing systems can link scheduling, purchasing, inventory control, shop control, and distribution. In effect, a CIM system integrates information from other areas of an organization with manufacturing.

The overall goal of using CIM is to link various parts of an organization to achieve rapid response to customer orders and/or product changes, to allow rapid production, and to reduce *indirect* labor costs.

A shining example of how process choices can lead to competitive advantages can be found at Allen-Bradley's computer-integrated manufacturing process in Milwaukee, Wisconsin. The company converted a portion of its factory to a fully automated "factory within a factory" to assemble contactors and relays for electrical motors. A handful of humans operate the factory, although once an order has been entered into the system, the machines do virtually all the work, including packaging and shipping, and quality control. Any defective items are removed from the line, and replacement parts are automatically ordered and scheduled to compensate for the defective items. The humans program the machines and monitor operations, and attend to any problems signaled by a system of warning lights.

As orders come into the plant, computers determine production requirements and schedules, and order the necessary parts. Bar-coded labels that contain processing instructions are automatically placed on individual parts. As the parts approach a machine, a sensing device reads the bar code and communicates the processing instructions to the machine. The factory can produce 600 units an hour.

The company has realized substantial competitive advantages from the system. Orders can be completed and shipped within 24 hours of entry into the system, direct labor costs and inventory costs have been greatly reduced, and quality is very high.

READING

ELECTRIC CAR DRIVES FACTORY INNOVATIONS

Rebecca Blumenstein, Staff Reporter of THE WALL STREET JOURNAL

www.gm.com

There are only about 150 of them on the road so far. Cold weather makes their batteries run down. They can go only 70 to 90 miles between rechargings, so drivers must choose their destinations carefully. Free towing is provided, however.

Even General Motors Corp. concedes that its EV1, the first electric vehicle now being mass-produced in the U.S., isn't the answer to the nation's current transportation needs. Still the teardrop-shaped vehicle may ultimately influence the way other cars are designed and built in the future. The superlight, aerodynamic car now in limited production is allowing GM to experiment with new technologies and new means of manufacturing. "We're learning from the car," says Maureen Midgley, plant manager of the Lansing Draft Centre in Lansing, Mich., where GM produces about 2,000 EV1s per year.

One big lesson has been working with new materials. Lightness is crucial to maximize the EV1's driving range, so the frame and many other parts are made of aluminum rather than steel used in most other cars. The entire frame weighs only 300 pounds, about 40 percent of a typical car frame.

That, in turn, dictates a different kind of assembly. Traditional welding doesn't work because heat disperses more rapidly through aluminum than steel and doesn't stay concentrated. Instead, GM glues the EV1's frame together with an adhesive ordinarily used by the aerospace industry and literally bakes the car in a giant oven at 375 degrees for 15 minutes.

GM says this use of light materials will likely show up in many of its future cars; removing weight to increase gas mileage is a chief goal for many auto engineers. "The EV1 allows you to put lightweight structures in all your vehicles," says Lawrence J. Oswald, a manager at GM's Alternative Propulsion Department.

The Lansing center itself bears little resemblance to a typical assembly plant, largely because it has taken outsourcing to the extreme. While traditional vehicle plants rely on hundreds of robots and thousands of workers along miles of assembly lines, Lansing has only 60 hourly workers who essentially build the car manually in a few stations. Suppliers deliver almost every part, premade, to the plant floor, from the 1,100-pound battery pack to the interior seats.

"Because of the technology, suppliers have had to have more involvement," says Arianna Kalian, manager of the body shop. The arrangement saves money by reducing labor and capital costs, a key hurdle the electric vehicle had to overcome to get the approval of GM's board.

Traditionally, auto workers' unions have resisted this level of outsourcing. Then again, jobs at the Lansing plant are very different from typical, high-volume plants in which workers must accomplish their assigned tasks in 60 seconds or less. The EV1 stays an average of 45 minutes at each of about a dozen stops along the line, and each worker performs dozens of tasks. The employees keep many of the functions in their head—120 to 150 times that of a regular assembly worker.

"It's like food for the brain," says Wayne Glaxner, one of the EV1 workers. Even Ms. Midgley, the plant manager and the daughter of a GM assembly-line worker, says she likes to work on the lines herself once a month to watch the car come together.

The EV1 is driving other innovations, as well. Its frame is so light that it can be wheeled into the general assembly area instead of lifted with the huge, expensive conveyors most plants use. The plant uses a lone robot to install a windshield containing an invisible electronic film that can defrost the glass without an energy-consuming blower. GM says the robot, being tested here for the first time, may some day be used in conventional assembly plants.

Once tires are mounted, the EV1 is driven to other stations. Assembly plants typically keep gas-powered cars on miles of conveyors because gasoline is too flammable to allow on a factory floor. Seats made of lightweight magnesium frames and a steering wheel are installed along with the rest of the car's futuristic-looking interior. Finally, the EV1's exterior panels—made of a lightweight composite material—are added, piece by piece, a change from traditional unit-body construction that integrates the frame with body panels made of sheets of steel.

The stations are so portable—unlike the huge monuments of automation that anchor most assembly plants—that they can be altered easily. Indeed, the entire plant can adapt quickly if new technology becomes available, and production can be expanded quickly from 2,000 cars per year to 10,000 or more if demand for electric cars warrants.

"Clearly, a manufacturer that learns either to be very flexible as a mass producer or learns to control the costs of low production is going to have a major leg up," says Mike Flynn, associate director of the University of Michigan's Office for the Study of Automotive Transportation. "This is a real significant opportunity."

To date, demand remains limited by the car's lack of range and practicality. GM is leasing the $34,000 cars for $480 to $680 per month only in Arizona and California and only through Saturn dealers, because of their reputation for friendly customer relations. Dealers need to be able to instruct drivers in the car's technical limitations, GM officials say, though they anticipate that a new nickel-metal hydride battery, available later this year, will increase the EV1's to as much as 270 miles.

Questions

1. Why is flexibility important in this instance; what advantages can it provide?
2. Name some of the major differences in the way these electric cars are produced compared to the traditional way cars are produced, and the reasons for these differences.
3. What is outsourcing, and what advantages does it provide to GM?

Operations Strategy

Process selection often requires engineering expertise. However, many managers charged with selecting a process may have little technical knowledge; their education and experience lie in marketing, finance, and the like. Consequently, they may tend to delegate technical decisions to engineers. Engineering white elephants are not uncommon, and neither are systems based on narrow viewpoints of problems and solutions. In the long run, the solution may be to hire and promote managers who have both managerial and technical skills and expertise. In the short run, managers must work with technical experts, asking questions and increasing their understanding of the benefits and limitations of sophisticated processing equipment and technology, and ultimately make decisions themselves. Thus, there is a real need for *management of technology*.

A key consequence of the increased use of automation in manufacturing is the impact on cost structures in organizations. The proportion of total costs represented by fixed costs is becoming larger, and the proportion of variable costs is becoming smaller. This means that volume of output has a decreasing effect on total cost, a situation that can be particularly burdensome during periods of slow demand. Moreover, automation creates additional requirements for the maintenance and repair of highly specialized equipment. Managers must proceed cautiously with decisions to automate, carefully weighing the benefits and risks before making what are usually long-term commitments.

Throughout this book, the importance of *flexibility* as a competitive strategy is stressed. However, flexibility does not always offer the best choice in processing decisions. Flexible systems or equipment are often more expensive and not as efficient as less flexible alternatives. In certain instances, flexibility is unnecessary because products are in mature stages, requiring few design changes, and there is a steady volume of output. Ordinarily, this type of situation calls for specialized processing equipment, with no need for flexibility. The implication is clear: Flexibility should be adopted with great care; its applications should be matched with situations in which a *need* for flexibility clearly exists.

In practice, decision makers choose flexible systems for either of two reasons: demands are varied, or uncertainty exists about demand. The second reason can be overcome through improved forecasting.

CAPACITY PLANNING

In this section, you will learn about the importance of capacity decisions, the measurement of capacity, how capacity requirements are determined, and the development and evaluation of capacity alternatives.

capacity The upper limit or ceiling on the load that an operating unit can handle.

Vol 5, Seg. 1
Tri-State

Capacity issues are important for all organizations, and at all levels of an organization. **Capacity** refers to an upper limit or ceiling on the load that an operating unit can handle. The operating unit might be a plant, department, machine, store, or worker.

The capacity of an operating unit is an important piece of information for planning purposes: It enables managers to quantify production capability in terms of inputs or outputs, and thereby make other decisions or plans related to those quantities. The basic questions in capacity planning of any sort are the following:

1. What kind of capacity is needed?
2. How much is needed?
3. When is it needed?

The question of what kind of capacity is needed depends on the products and services that management intends to produce or provide. Hence, in a very real sense, capacity planning is governed by those choices.

The most fundamental decisions in any organization concern the products and/or services it will offer. Virtually all other decisions pertaining to capacity, facilities, location, and the like are governed by product and service choices. Thus, a decision to produce high-quality steel will necessitate certain types of processing equipment and certain kinds of labor skills, and it will suggest certain types of arrangement of facilities. It will influence the size and type of building as well as the plant location. Notice how different each of these factors would be if the choice were to operate a family restaurant, and how still different they would be if the choice were to operate a hospital.

In some instances, capacity choices are made very infrequently; in others, they are made regularly, as part of an ongoing process. Generally, the factors that influence this frequency are the stability of demand, the rate of technological change in equipment and product design, and competitive factors. Other factors relate to the type of product or service and whether style changes are important (e.g., automobiles and clothing). In any case, management must review product and service choices periodically to ensure that the company makes changes when they are needed for cost, competitive effectiveness, or other reasons.

NEWSCLIP

LESS TRASH LEAVES LANDFILLS IN A BIND

www.gm.com

Not too long ago, dire predictions were made about the lack of landfill capacity to handle the growing amounts of trash companies and residences were generating. Now, some landfills around the country are not getting the trash (and the fees) they need to survive. What was once regarded as undercapacity has now turned into overcapacity.

The reasons for this turnaround can be found in strong efforts by the general public to recycle—stronger than most experts had predicted. Companies, too, are recycling more, a result of government regulations and cost-saving measures. They are also incorporating more recyclable and reusable parts and materials in their products, and they are reducing the amount of materials used to package their products.

But landfills, like many other kinds of operations, are designed to operate at a certain level. It is difficult (and in some states illegal) for them to operate above their design capacity, and it is inefficient to operate at levels much below design capacity. The shortfall that some landfills are experiencing underscores the risks involved in long-term capacity planning and the importance of good forecasts of future demand.

Based on: "Riga Landfill Strains to Survive," Michael Caputo, *Rochester Democrat and Chronicle*, July 28,1997, p. 1A.

Importance of Capacity Decisions

For a number of reasons, capacity decisions are among the most fundamental of all the design decisions that managers must make.

1. Capacity decisions have a real impact on the ability of the organization to meet future demands for products and services; capacity essentially limits the rate of output possible. Having capacity to satisfy demand can allow a company to take advantage of tremendous opportunities. When the Quigley (www.quigleyco.com) Corporation's zinc gluconate lozenges, sold under the name Cold-Eze™, attracted the public's interest during the height of the cold and flu season in 1997, drugstores and supermarkets quickly sold out. The product was so popular that the company couldn't keep up with demand. Because of this, the company was unable to take full advantage of the strong demand.
2. Capacity decisions affect operating costs. Ideally, capacity and demand requirements will be matched, which will tend to minimize operating costs. In practice, this is not always achieved because actual demand either differs from expected demand or tends to vary (e.g., cyclically). In such cases, a decision might be made to attempt to balance the costs of over- and undercapacity.
3. Capacity is usually a major determinant of initial cost. Typically, the greater the capacity of a productive unit, the greater its cost. This does not necessarily imply a one-for-one relationship; larger units tend to cost *proportionately* less than smaller units.

4. Capacity decisions often involve long-term commitment of resources and the fact that, once they are implemented, it may be difficult or impossible to modify those decisions without incurring major costs.
5. Capacity decisions can affect competitiveness. If a firm has excess capacity, or can quickly add capacity, that fact may serve as a barrier against entry by other firms.

Defining and Measuring Capacity

Capacity often refers to an upper limit on the *rate* of output. Even though this seems simple enough, there are subtle difficulties in actually measuring capacity in certain cases. These difficulties arise because of different interpretations of the term *capacity* and problems with identifying suitable measures for a specific situation.

In selecting a measure of capacity, it is important to choose one that does not require updating. For example, dollar amounts are often a poor measure of capacity (e.g., capacity of $30 million a year) because price changes necessitate continual updating of that measure.

Where only one product or service is involved, the capacity of the productive unit may be expressed in terms of that item. However, when multiple products or services are involved, as is often the case, using a simple measure of capacity based on units of output can be misleading. An appliance manufacturer may produce both refrigerators and freezers. If the output rates for these two products are different, it would not make sense to simply state capacity in units without reference to either refrigerators or freezers. The problem is compounded if the firm has other products. One possible solution is to state capacities in terms of each product. Thus, the firm may be able to produce 100 refrigerators per day *or* 80 freezers per day. Sometimes this approach is helpful, sometimes not. For instance, if an organization has many different products or services, it may not be practical to list all of the relevant capacities. This is especially true if there are frequent changes in the mix of output, because this would necessitate a continually changing composite index of capacity. The preferred alternative in such cases is to use a measure of capacity that refers to *availability of inputs.* Thus, a hospital has a certain number of beds, a factory has a certain number of machine hours available, and a bus has a certain number of seats and a certain amount of standing room.

No single measure of capacity will be appropriate in every situation. Rather, the measure of capacity must be tailored to the situation. Table 5–2 provides some examples of commonly used measures of capacity.

Up to this point, we have been using a working definition of capacity. Although it is functional, it can be refined into two useful definitions of capacity:

1. *Design capacity:* the maximum output that can possibly be attained.

TABLE 5-2

Measures of capacity

Business	Inputs	Outputs
Auto manufacturing	Labor hours, machine hours	Number of cars per shift
Steel mill	Furnace size	Tons of steel per day
Oil refinery	Refinery size	Gallons of fuel per day
Farming	Number of acres, number of cows	Bushels of grain per acre per year, gallons of milk per day
Restaurant	Number of tables, seating capacity	Number of meals served per day
Theater	Number of seats	Number of tickets sold per performance
Retail sales	Square feet of floor space	Revenue generated per day

2. *Effective capacity:* the maximum possible output given a product mix, scheduling difficulties, machine maintenance, quality factors, and so on.

Design capacity is the maximum rate of output achieved under ideal conditions. Effective capacity is usually less than design capacity (it cannot exceed design capacity) owing to realities of changing product mix, the need for periodic maintenance of equipment, lunch breaks, coffee breaks, problems in scheduling and balancing operations, and similar circumstances. Actual output cannot exceed effective capacity and is often less because of machine breakdowns, absenteeism, shortages of materials, and quality problems, as well as factors that are outside the control of the operations managers.

These different measures of capacity are useful in defining two measures of system effectiveness: efficiency and utilization. *Efficiency* is the ratio of actual output to effective capacity. *Utilization* is the ratio of actual output to design capacity.

$$\text{Efficiency} = \frac{\text{Actual output}}{\text{Effective capacity}} \tag{5–1}$$

$$\text{Utilization} = \frac{\text{Actual output}}{\text{Design capacity}} \tag{5–2}$$

It is common for managers to focus exclusively on efficiency, but in many instances, this emphasis can be misleading. This happens when effective capacity is low compared with design capacity. In those cases, high efficiency would seem to indicate effective use of resources when it does not. The following example illustrates this point.

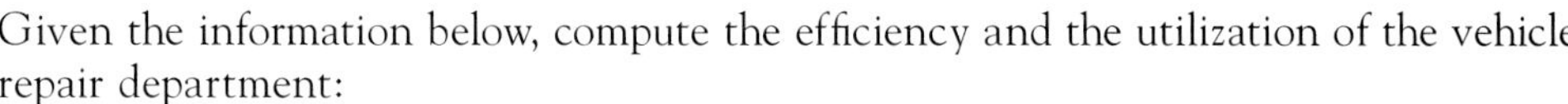

Example 1

Given the information below, compute the efficiency and the utilization of the vehicle repair department:

$$\begin{aligned} \text{Design capacity} &= 50 \text{ trucks per day} \\ \text{Effective capacity} &= 40 \text{ trucks per day} \\ \text{Actual output} &= 36 \text{ trucks per day} \end{aligned}$$

Solution

$$\text{Efficiency} = \frac{\text{Actual output}}{\text{Effective capacity}} = \frac{36 \text{ trucks per day}}{40 \text{ trucks per day}} = 90\%$$

$$\text{Utilization} = \frac{\text{Actual output}}{\text{Design capacity}} = \frac{36 \text{ trucks per day}}{50 \text{ trucks per day}} = 72\%$$

Thus, compared with the effective capacity of 40 units per day, 36 units per day looks pretty good. However, compared with the design capacity of 50 units per day, 36 units per day is much less impressive although probably more meaningful.

Because effective capacity acts as a lid on actual output, the real key to improving capacity utilization is to increase effective capacity by correcting quality problems, maintaining equipment in good operating condition, fully training employees, and fully utilizing bottleneck equipment.

Hence, increasing utilization depends on being able to increase effective capacity, and this requires a knowledge of what is constraining effective capacity.

The following section explores some of the main determinants of effective capacity. It is important to recognize that the benefits of high utilization are only realized in instances where there is demand for the output. When demand is not there, focusing exclusively on utilization can be counterproductive, because the excess output not only results in additional variable costs, it also generates the costs of having to carry the output as inventory. Another disadvantage of high utilization is that operating costs may increase because of increasing waiting time due to bottleneck conditions.

A mass-producer of large-frame computers, the Sujitsu plant in Numazu, Japan, conducts product testing. Building efficiency into product testing increases capacity.

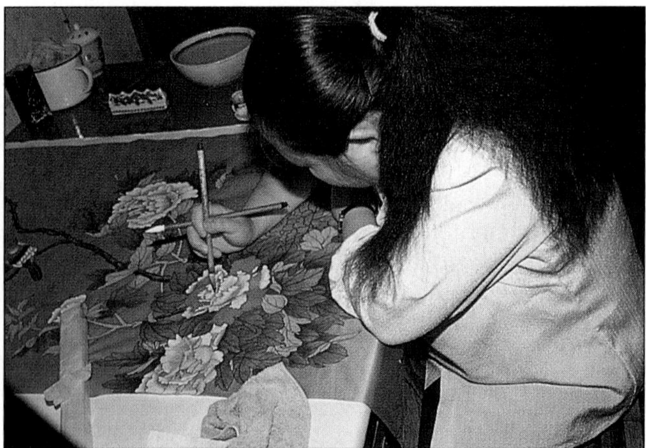

An independent company within the Republic of China produces customized scrolls. Capacity is very limited when items are specialized and made one at a time.

Determinants of Effective Capacity

Many decisions about system design have an impact on capacity. The same is true for many operating decisions. This section briefly describes some of these factors, which are then elaborated on elsewhere in the book. The main factors relate to the following:

1. Facilities
2. Products or services
3. Processes
4. Human considerations
5. Operations
6. External forces

Facilities Factors. The design of facilities, including size and provision for expansion, is key. Locational factors, such as transportation costs, distance to market, labor supply, energy sources, and room for expansion, are also important. Likewise, layout of the work area often determines how smoothly work can be performed, and environmental factors such as heating, lighting, and ventilation also play a significant role in determining whether personnel can perform effectively or whether they must struggle to overcome poor design characteristics.

Product/Service Factors. Product or service design can have a tremendous influence on capacity. For example, when items are similar, the ability of the system to produce those items is generally much greater than when successive items differ. Thus, a restaurant that offers a limited menu can usually prepare and serve meals at a faster rate than a restaurant with an extensive menu. Generally speaking, the more uniform the output, the more opportunities there are for standardization of methods and materials, which leads to greater capacity. The particular mix of products or services rendered must also be considered since different items will have different rates of output.

Process Factors. The quantity capability of a process is an obvious determinant of capacity. A more subtle determinant is the influence of output *quality*. For instance, if quality of output does not meet standards, the rate of output will be slowed by the need for inspection and rework activities.

TABLE 5–3

Factors that determine effective capacity

A. Facilities
   1. Design
   2. Location
   3. Layout
   4. Environment
B. Product/Service
   1. Design
   2. Product or service mix
C. Process
   1. Quantity capabilities
   2. Quality capabilities
D. Human factors
   1. Job content
   2. Job design
   3. Training and experience
   4. Motivation
   5. Compensation
   6. Learning rates
   7. Absenteeism and labor turnover
E. Operational
   1. Scheduling
   2. Materials management
   3. Quality assurance
   4. Maintenance policies
   5. Equipment breakdowns
F. External factors
   1. Product standards
   2. Safety regulations
   3. Unions
   4. Pollution control standards

Human Factors. The tasks that make up a job, the variety of activities involved, and the training, skill, and experience required to perform a job all have an impact on the potential and actual output. In addition, employee motivation has a very basic relationship to capacity, as do absenteeism and labor turnover.

Operational Factors. Scheduling problems may occur when an organization has differences in equipment capabilities among alternative pieces of equipment or differences in job requirements. Inventory stocking decisions, late deliveries, acceptability of purchased materials and parts, and quality inspection and control procedures also can have an impact on effective capacity.

For example, inventory problems had a negative impact on capacity when General Motors first introduced its front-wheel-drive cars. Unexpected high demand, created by shortages and rapid price increases of gasoline, exceeded the supply of cars. Company officials lamented that they could not take advantage of the opportunity to increase sales because of a shortage of component parts, which the company could not quickly overcome. Thus, insufficient capacity in one area affected overall capacity.

External Factors. Product standards, especially minimum quality and performance standards, can restrict management's options for increasing and using capacity. Thus, pollution standards on products and equipment often reduce effective capacity, as does paperwork required by government regulatory agencies by engaging employees in nonproductive activities. A similar effect occurs when a union contract limits the number of hours and type of work an employee may do.

Table 5–3 presents a summary of all these factors. In general, *inadequate planning* is a major limiting determinant of effective capacity.

Determining Capacity Requirements

Capacity planning decisions involve both long-term and short-term considerations. Long-term considerations relate to overall *level* of capacity, such as facility size; short-term considerations relate to probable *variations* in capacity requirements created by such things as seasonal, random, and irregular fluctuations in demand. Because the time intervals covered by each of these categories can vary significantly from industry to industry, it is misleading to put times on the intervals. Nevertheless, the distinction will serve as a framework within which to discuss capacity planning.

We determine *long-term* capacity needs by forecasting demand over a time horizon and then converting those forecasts into capacity requirements. Figure 5–4 illustrates some basic demand patterns that might be identified by a forecast. In addition to basic patterns there are more complex patterns, such as a combination of cycles and trends.

FIGURE 5–4

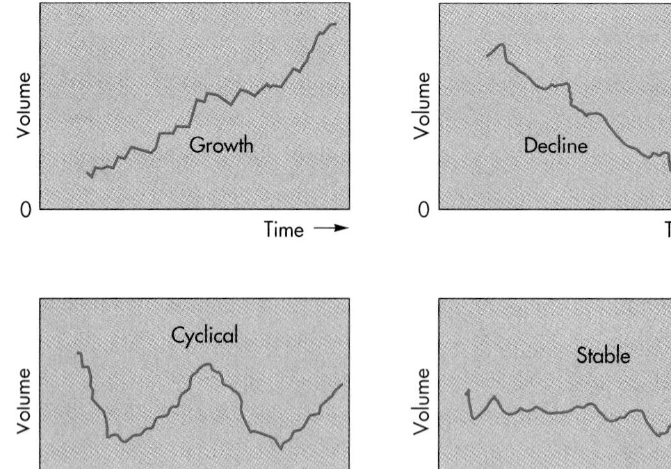

TABLE 5–4

Examples of seasonal demand patterns

Period	Items
Year.	Beer sales, toy sales, airline traffic, clothing, vacations, tourism, power usage, gasoline consumption, sports and recreation, education
Month	Welfare and social security checks, bank transactions
Week.	Retail sales, restaurant meals, automobile traffic, automotive rentals, hotel registrations
Day.	Telephone calls, power usage, automobile traffic, public transportation, classroom utilization, retail sales, restaurant meals

When trends are identified, the fundamental issues are (1) how long the trend might persist, because few things last forever, and (2) the slope of the trend. If cycles are identified, interest focuses on (1) the approximate length of the cycles, and (2) the amplitude of the cycles (i.e., deviation from average).

Short-term capacity needs are less concerned with cycles or trends than with seasonal variations and other variations from average. These deviations are particularly important because they can place a severe strain on a system's ability to satisfy demand at some times and yet result in idle capacity at other times.

An organization can identify seasonal patterns using standard forecasting techniques. Although commonly thought of as annual fluctuations, seasonal variations are also reflected in monthly, weekly, and even daily capacity requirements. Table 5–4 provides some examples of items that tend to exhibit seasonal demand patterns.

When time intervals are too short to have seasonal variations in demand, the analysis can often describe the variations by probability distributions such as a normal, uniform, or Poisson distribution. For example, we might describe the amount of coffee served during the midday meal at a luncheonette by a normal distribution with a certain mean and standard deviation. The number of customers who enter a bank branch on Monday mornings might be described by a Poisson distribution with a certain mean. It does not follow, however, that *every* instance of random variability will lend itself to description by a standard statistical distribution. Service systems in particular may experience a considerable amount of variability in capacity requirements unless requests for service can be scheduled. Manufacturing systems, because of their typical isolation from customers and the more uniform nature of production, are less likely to experience variations. Waiting-line models and simulation models can be useful when analyzing service systems. These models are described in Chapter 19.

Vol. 3, Seg. 3
Shouldice Hospital

Irregular variations are perhaps the most troublesome: They are virtually impossible to predict. They are created by such diverse forces as major equipment breakdowns, freak storms that disrupt normal routines, foreign political turmoil that causes oil shortages, discovery of health hazards (nuclear accidents, unsafe chemical dumping grounds, carcinogens in food and drink), and so on.

The link between marketing and operations is crucial to realistic determination of capacity requirements. Through customer contracts, demographic analyses, and forecasts, marketing can supply vital information to operations for ascertaining capacity needs for both the long term and the short term.

Developing Capacity Alternatives

Vol. 5, Seg. 2
TriState

Aside from the general considerations about the development of alternatives (i.e., conduct a reasonable search for possible alternatives, consider doing nothing, take care not to overlook nonquantitative factors), some specific considerations are relevant to developing capacity alternatives. The considerations to be discussed in this section include the following:

1. **Design flexibility into systems.** The long-term nature of many capacity decisions and the risks inherent in long-term forecasts suggest potential benefits from designing flexible systems. For example, provision for future expansion in the original design of a structure frequently can be obtained at a small price compared to what it would cost to remodel an existing structure that did not have such a provision. Hence, if future expansion of a restaurant seems likely, water lines, power hookups, and waste disposal lines can be put in place initially so that if expansion becomes a reality, modification to the existing structure can be minimized. Similarly, a new golf course may start as a nine-hole operation, but if provision is made for future expansion by obtaining options on adjacent land, it may progress to a larger (18-hole) course. Other considerations in flexible design involve layout of equipment, location, equipment selection, production planning, scheduling, and inventory policies, which will be discussed in later chapters.

2. **Take a "big picture" approach to capacity changes.** A consideration for managers contemplating capacity increases is whether the capacity is for a new product or service, or a mature one. Mature products or services tend to be more predictable in terms of capacity requirements, and they may have limited life spans. The predictable demand pattern means less risk of choosing an incorrect capacity, but the possible limited life span of the product or service may necessitate finding an alternate use for the additional capacity at the end of the life span. New products tend to carry higher risk because of the uncertainty often associated with predicting the quantity and duration of demand. That makes flexibility appealing to managers.

When developing capacity alternatives, it is important to consider how parts of the system interrelate. For example, when making a decision to increase the number of rooms in a motel, one should also take into account probable increased demands for parking, entertainment and food, and housekeeping. This is a "big picture" approach.

3. **Prepare to deal with capacity "chunks."** Capacity increases are often acquired in fairly large chunks rather than smooth increments, making it difficult to achieve a match between desired capacity and feasible capacity. For instance, the desired capacity of a certain operation may be 55 units per hour; but suppose that machines used for this operation are able to produce 40 units per hour each. One machine by itself would cause capacity to be 15 units per hour short of what is needed, but two machines would result in an excess capacity of 25 units per hour. The illustration becomes even more extreme if we shift the topic—to open-hearth furnaces or to the number of airplanes needed to provide a desired level of capacity.

4. **Attempt to smooth out capacity requirements.** Unevenness in capacity requirements also can create certain problems. For instance, during periods of inclement

weather, public transportation ridership tends to increase substantially relative to periods of pleasant weather. Consequently, the system tends to alternate between underutilization and overutilization. Increasing the number of buses or subway cars will reduce the burden during periods of heavy demand, but this will aggravate the problem of overcapacity at other times and certainly add to the cost of operating the system. Unfortunately, no simple solutions exist for these problems.

We can trace the unevenness in demand for products and services to a variety of sources. The bus ridership problem is weather related to a certain extent, but demand could be considered to be partly random (i.e., varying because of chance factors). Still another source of varying demand is seasonality. Seasonal variations are generally easier to cope with than random variations because they are *predictable*. Consequently, management can make allowances in planning and scheduling activities and inventories. However, seasonal variations can still pose problems because of their uneven demands on the system: At certain times the system will tend to be overloaded, while at other times it will tend to be underloaded. One possible approach to this problem is to identify products or services that have complementary demand patterns, that is, patterns that tend to offset each other. For instance, demand for snow skis and demand for water skis might complement each other: demand for water skis is greater in the spring and summer months, and demand for snow skis is greater in the fall and winter months. The same might apply to heating and air-conditioning equipment. The ideal case is one in which products or services with complementary demand patterns involve the use of the same resources but at different times, so that overall capacity requirements remain fairly stable. Figure 5–5 illustrates complementary demand patterns.

Variability in demand can pose a problem for managers. Simply adding capacity by increasing the size of the operation (e.g., increasing the size of the facility, the workforce, or the amount of processing equipment) is not always the best approach, because that reduces flexibility and adds to fixed costs. Consequently, managers often choose to respond to higher than normal demand in other ways. One way is through the use of overtime work. Another way is to subcontract some of the work. A third way is to draw down finished goods inventories during periods of high demand and replenish them during periods of slow demand. These options and others are discussed in detail in the chapter on aggregate planning.

5. **Identify the optimal operating level.** Production units typically have an ideal or optimal level of operation in terms of unit cost of output. At the ideal level, cost per unit is the lowest for that production unit; larger or smaller rates of output will result in a higher unit cost. Figure 5–6 illustrates this concept. Notice how unit costs rise as the rate of output varies from the optimal level.

The explanation for the shape of the cost curve is that at low levels of output, the costs of facilities and equipment must be absorbed (paid for) by very few units. Hence, the cost per unit is high. As output is increased, there are more units to absorb the "fixed" cost of facilities and equipment, so unit costs decrease. However, beyond a certain point, unit costs will start to rise. To be sure, the fixed costs are spread over even more units, so that does not account for the increase, but other factors now become important: worker fatigue; equipment breakdowns; the loss of flexibility, which leaves less of a margin for error; and, generally, greater difficulty in coordinating operations.

Both optimal operating rate and the amount of the minimum cost tend to be a function of the general capacity of the operating unit. For example, as the general capacity of a plant increases, the optimal output rate increases and the minimum cost for the optimal rate decreases. Thus, larger plants tend to have higher optimal output rates and lower minimum costs than smaller plants. Figure 5–7 illustrates these points.

In choosing the capacity of an operating unit, management must take these relationships into account along with the availability of financial and other resources and forecasts of expected demand. To do this, it is necessary to determine enough points for

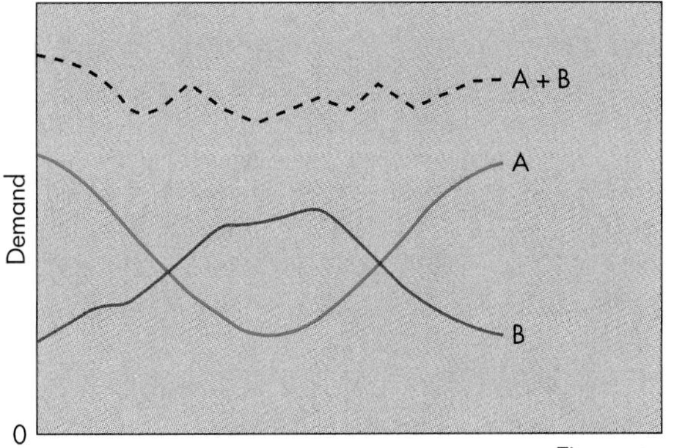

FIGURE 5-5

A and B have complementary demand patterns

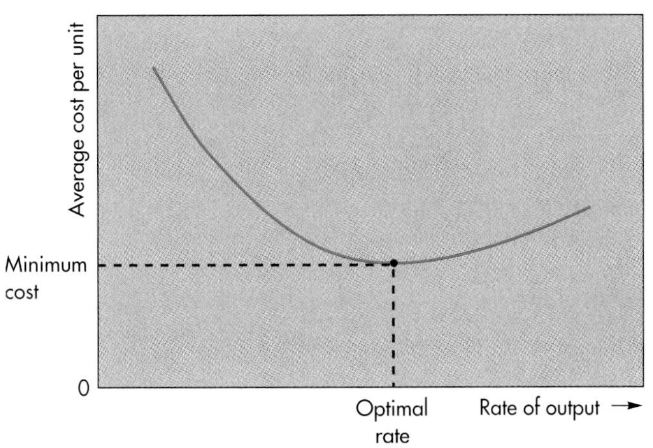

FIGURE 5-6

Production units have an optimal rate of output for minimum cost

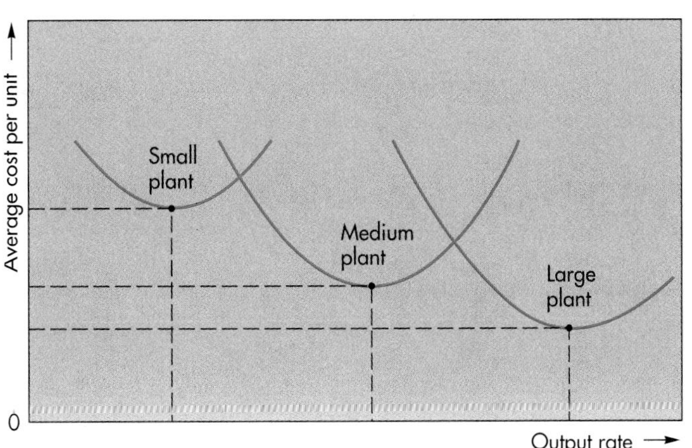

FIGURE 5-7

Minimum cost and optimal operating rate are functions of size of a production unit

each size facility to be able to make a comparison among different sizes. In some instances, facility sizes are givens, whereas in others, facility size is a continuous variable (i.e., any size can be selected). In the latter case, an ideal facility size can be selected. Usually, management must make a choice from given sizes, and none may have a minimum at the desired rate of output.

NEWSCLIP

"LEAN AND MEAN" COULD MEAN TROUBLE

The cost-cutting techniques many companies used in the last business downturn probably can't be replicated; there is little fat left to cut. This means that unless they can come up with some other way to cope with less demand, they'll be struck with excess capacity that will cut into their profits. Capital-intensive industries may be the most sensitive due to their high fixed costs and inability to quickly alter capacity. Part of the problem for all companies is that they tend to reinvest when they are flush with cash and don't give enough weight to timing reinvestments according to changes in demand.

Based on: "In Some Industries, Executives Foresee Tough Times Ahead," *The Wall Street Journal*, August 7, 1997, p. A1.

Evaluating Alternatives

An organization needs to examine alternatives for future capacity from a number of different perspectives. Most obvious are economic considerations: Will an alternative be economically feasible? How much will it cost? How soon can we have it? What will operating and maintenance costs be? What will its useful life be? Will it be compatible with present personnel and present operations?

Less obvious, but nonetheless important, is possible negative public opinion. For instance, the decision to build a new power plant is almost sure to stir up reaction, whether the plant is coal-fired, hydroelectric, or nuclear. Any option that could disrupt lives and property is bound to generate hostile reactions. Construction of new facilities may necessitate moving personnel to a new location. Embracing a new technology may mean retraining some people and terminating some jobs. Relocation can cause unfavorable reactions, particularly if a town is about to lose a major employer. Conversely, community pressure in a new location may arise if the presence of the company is viewed unfavorably (noise, traffic, pollution).

A number of techniques are useful for evaluating capacity alternatives from an economic standpoint. Some of the more common are cost–volume analysis, financial analysis, decision theory, and waiting-line analysis. Cost–volume analysis is described in this chapter. Financial analysis is mentioned briefly; decision analysis and waiting-line analysis are described in other chapters.

Calculating Processing Requirements. When evaluating capacity alternatives, a necessary piece of information is the capacity requirements of products that will be processed with a given alternative. To get this information, one must have reasonably accurate demand forecasts for each product, and know the standard processing time per unit for each product on each alternative machine, the number of work days per year, and the number of shifts that will be used.

Example 2

A department works one eight-hour shift, 250 days a year, and has these figures for usage of a machine that is currently being considered:

Product	Annual Demand	Standard Processing Time per Unit (Hr.)	Processing Time Needed (Hr.)
#1	400	5.0	2,000
#2	300	8.0	2,400
#3	700	2.0	1,400
			5,800

Working one eight-hour shift, 250 days a year provides an annual capacity of $8 \times 250 = 2{,}000$ hours per year. We can see that three of these machines would be needed to handle the required volume:

$$\frac{5{,}800 \text{ hours}}{2{,}000 \text{ hours/machine}} = 2.90 \text{ machines}$$

Cost–Volume Analysis. Cost–volume analysis focuses on relationships between cost, revenue, and volume of output. The purpose of cost–volume analysis is to estimate the income of an organization under different operating conditions. It is particularly useful as a tool for comparing capacity alternatives.

Use of the technique requires identification of all costs related to the production of a given product. These costs are then assigned to fixed costs or variable costs. *Fixed costs* tend to remain constant regardless of volume of output. Examples include rental costs, property taxes, equipment costs, heating and cooling expenses, and certain administrative costs. *Variable costs* vary directly with volume of output. The major components of variable costs are generally materials and labor costs. We will assume that variable cost per unit remains the same regardless of volume of output.

Table 5–5 summarizes the symbols used in the cost–volume formulas.

The total cost associated with a given volume of output is equal to the sum of the fixed cost and the variable cost per unit times volume:

$$TC = FC + VC \times Q \qquad (5\text{–}3)$$

Figure 5–8A shows the relationship between volume of output and fixed costs, total variable costs, and total (fixed plus variable) costs.

Revenue per unit, like variable cost per unit, is assumed to be the same regardless of quantity of output. Total revenue will have a linear relationship to output, as illustrated in Figure 5–8B. Assume that all output can be sold. The total revenue associated with a given quantity of output, Q, is:

$$TR = R \times Q \qquad (5\text{–}4)$$

Figure 5–8C describes the relationship between profit—which is the difference between total revenue and total (i.e., fixed plus variable) cost—and volume of output. The volume at which total cost and total revenue are equal is referred to as the **break-even point (BEP).** When volume is less than the break-even point, there is a loss rather than a profit; when volume is greater than the break-even point, there is a profit. The greater the deviation from this point, the greater the profit or loss. Total profit can be computed using the formula:

break-even point The volume of output at which total cost and total revenue are equal.

$$P = TR - TC = R \times Q - (FC + VC \times Q)$$

Rearranging terms, we have

$$P = Q(R - VC) - FC \qquad (5\text{–}5)$$

The required volume, Q, needed to generate a specified profit is:

$$Q = \frac{P + FC}{R - VC} \qquad (5\text{–}6)$$

A special case of this is the volume of output needed for total revenue to equal total cost. This is the break-even point, computed using the formula:

$$Q_{BEP} = \frac{FC}{R - VC} \qquad (5\text{–}7)$$

Example 3

The owner of Old-Fashioned Berry Pies, S. Simon, is contemplating adding a new line of pies, which will require leasing new equipment for a monthly payment of \$6,000. Variable costs would be \$2.00 per pie, and pies would retail for \$7.00 each.

TABLE 5–5

Cost–volume symbols

FC	= Fixed cost
VC	= Variable cost per unit
TC	= Total cost
TR	= Total revenue
R	= Revenue per unit
Q	= Quantity or volume of output
Q_{BEP}	= Break-even quantity
P	= Profit

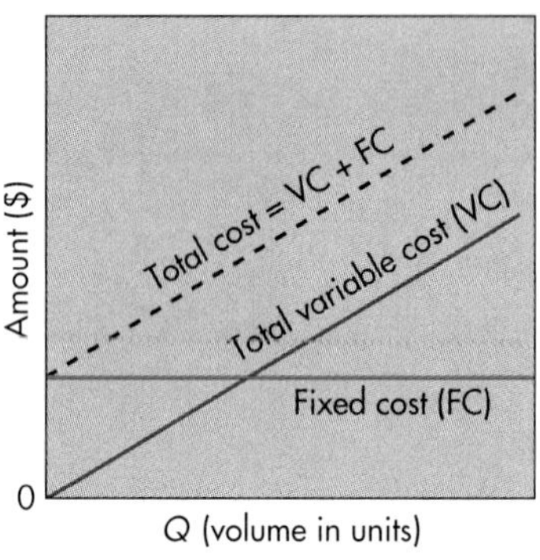

A. Fixed, variable, and total costs.

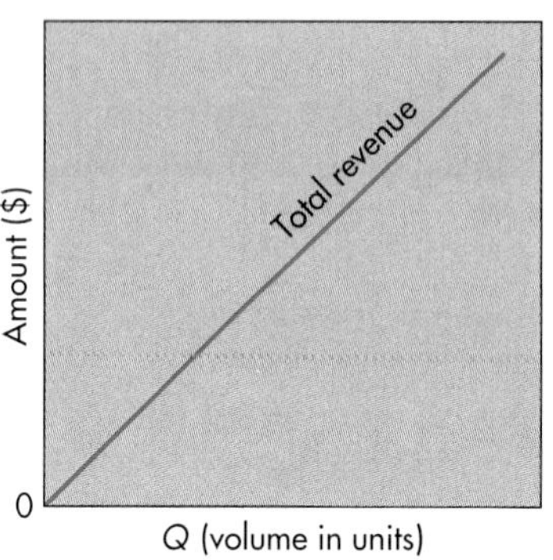

B. Total revenue increases linearly with output.

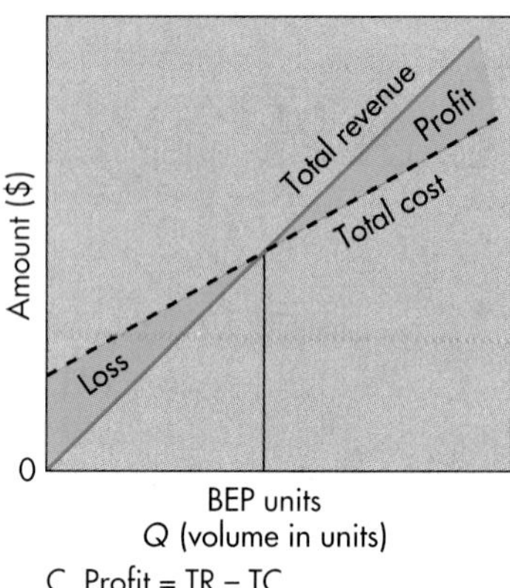

C. Profit = TR – TC.

FIGURE 5–8

Cost-volume relationships

a. How many pies must be sold in order to break even?

b. What would the profit (loss) be if 1,000 pies are made and sold in a month?

c. How many pies must be sold to realize a profit of $4,000?

Solution

$FC = \$6{,}000,\ VC = \$2 \text{ per pie},\ Rev = \$7 \text{ per pie}$

$$a.\ Q_{\text{BEP}} = \frac{FC}{Rev - VC} = \frac{\$6{,}000}{\$7 - \$2} = 1{,}200 \text{ pies/month}$$

b. For $Q = 1{,}000$, $P = Q(R - VC) - FC = 1{,}000(\$7 - \$2) - \$6{,}000 = -\$1{,}000$

c. $P = \$4{,}000$; solve for Q using equation 5–6:

$$Q = \frac{\$4{,}000 + \$6{,}000}{\$7 - \$2} = 2{,}000 \text{ pies}$$

Capacity alternatives may involve *step costs,* which are costs that increase stepwise as potential volume increases. For example, a firm may have the option of purchasing one, two, or three machines, with each additional machine increasing the fixed cost, although perhaps not linearly. (See Figure 5–9A.) Then fixed costs and potential volume would depend on the number of machines purchased. The implication is that *multiple break-even quantities* may occur, possibly one for each range. Note, however, that the total revenue line might not intersect the fixed-cost line in a particular range, meaning that there would be no break-even point in the first range. This possibility is illustrated in Figure 5–9B, where there is no break-even point in the first range. In order to decide how many machines to purchase, a manager must consider projected annual demand (volume) relative to the multiple break-even points and choose the most appropriate number of machines, as Example 4 shows.

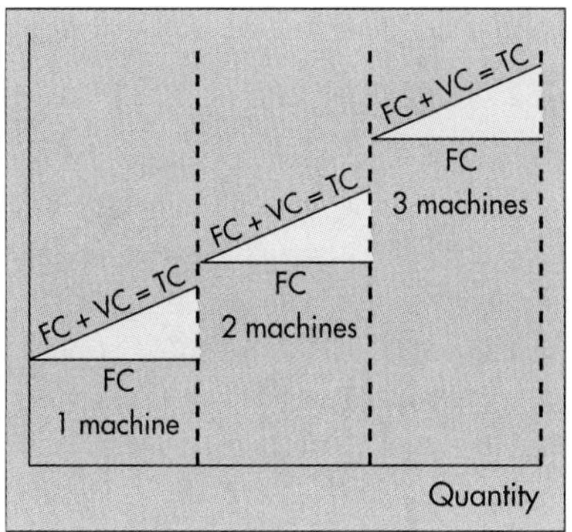

A. Step fixed costs and variable costs.

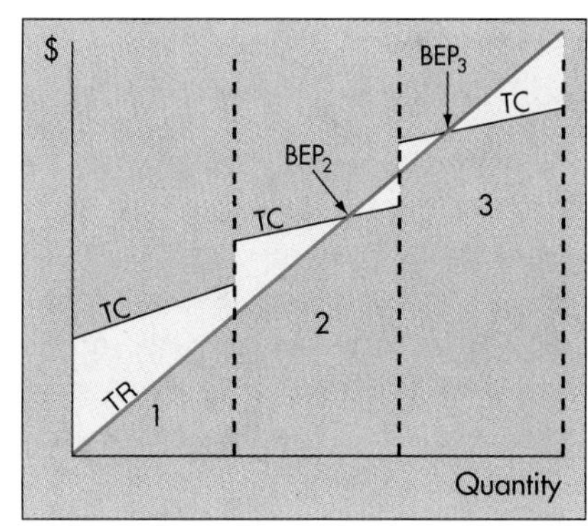

B. Multiple break-even points.

FIGURE 5-9

Break-even problem with step fixed costs

Example 4

A manager has the option of purchasing one, two, or three machines. Fixed costs and potential volumes are as follows:

Number of Machines	Total Annual Fixed Costs	Corresponding Range of Output
1	$ 9,600	0 to 300
2	15,000	301 to 600
3	20,000	601 to 900

Variable cost is $10 per unit, and revenue is $40 per unit.

a. Determine the break-even point for each range.

b. If projected annual demand is between 580 and 660 units, how many machines should the manager purchase?

Solution

a. Compute the break-even point for each range using the formula $Q_{BEP} = \frac{FC}{R - VC}$.

$$\text{For one machine } Q_{\text{BEP}} = \frac{\$9{,}600}{\$40/unit - \$10/unit} = 320 \text{ units [not in range].}$$

$$\text{For two machines } Q_{\text{BEP}} = \frac{\$15{,}000}{\$40/unit - \$10/unit} = 500 \text{ units.}$$

$$\text{For three machines } Q_{\text{BEP}} = \frac{\$20{,}000}{\$40/unit - \$10/unit} = 666.67 \text{ units.}$$

b. Comparing the projected range of demand to the two ranges for which a break-even point occurs, you can see that the break-even point is 500, which is in the range 301 to 600. This means that even if demand is at the low end of the range (i.e., 580), it would be above the break-even point and thus yield a profit. That is not true of range 601 to 900. At the top end of projected demand (i.e., 660), the volume would still be less than the break-even point for that range, so there would be no profit. Hence, the manager should choose two machines.

The assumptions of cost–volume analysis are:

1. One product is involved.
2. Everything produced can be sold.
3. The variable cost per unit is the same regardless of the volume.

4. Fixed costs do not change with volume changes, or they are step changes.
5. The revenue per unit is the same regardless of volume.

Cost–volume analysis can be a valuable tool for comparing capacity alternatives. As with any quantitative tool, it is important to verify that the assumptions on which the technique is based are reasonably satisfied for a particular situation. For example, revenue per unit or variable cost per unit are not always constant. In addition, fixed costs may not be constant over the range of possible output. If demand is subject to random variations, one must take that into account in the analysis. Also, cost–volume analysis requires that fixed and variable costs can be separated, and this is sometimes exceedingly difficult to accomplish.

Cost–volume analysis works best with one product or a few products that have the same cost characteristics. Nevertheless, a notable benefit of cost–volume considerations is the conceptual framework it provides for integrating cost, revenue, and profit estimates into capacity decisions. If a proposal looks attractive using cost–volume analysis, the next step would be to develop cash flow models to see how it fares with the addition of time and more flexible cost functions.

Financial Analysis. A problem that is universally encountered by managers is how to allocate scarce funds. A common approach is to use *financial analysis* to rank investment proposals.

Two important terms in financial analysis are *cash flow* and *present value*.

cash flow Difference between cash received from sales and other sources, and cash outflow for labor, material, overhead, and taxes.

present value The sum, in current value, of all future cash flows of an investment proposal.

Cash flow refers to the difference between the cash received from sales (of goods or services) and other sources (e.g., sale of old equipment) and the cash outflow for labor, materials, overhead, and taxes.

Present value expresses in current value the sum of all future cash flows of an investment proposal.

The three most commonly used methods of financial analysis are payback, present value, and internal rate of return.

Payback is a crude but widely used method that focuses on the length of time it will take for an investment to return its original cost. For example, an investment with an original cost of $6,000 and a monthly net cash flow of $1,000 has a payback period of six months. Payback ignores the *time value of money*. Its use is easier to rationalize for short-term than for long-term projects. However, some critics believe that use of this method has been one of the contributing factors in the failure of U.S. firms to match productivity increases achieved by firms in other nations.

The *present value (PV)* method summarizes the initial cost of an investment, its estimated annual cash flows, and any expected salvage value in a single value called the *equivalent current value*, taking into account the time value of money (i.e., interest rates).

The *internal rate of return (IRR)* summarizes the initial cost, expected annual cash flows, and estimated future salvage value of an investment proposal in an *equivalent interest rate*. In other words, this method identifies the rate of return that equates the estimated future returns and the initial cost.

These techniques are appropriate when there is a high degree of *certainty* associated with estimates of future cash flows. In many instances, however, operations managers and other managers must deal with situations better described as risky or uncertain. When conditions of risk or uncertainty are present, decision theory is often applied.

Decision Theory. Decision theory is a helpful tool for financial comparison of alternatives under conditions of risk or uncertainty. It is suited to capacity decisions and to a wide range of other decisions managers must make. Decision theory was described in the supplement to Chapter 2, but the same approaches can be applied to capacity decisions.

Waiting Line Analysis. Analysis of lines is often useful for designing service systems. Waiting lines have a tendency to form in a wide variety of service systems (e.g., airport ticket counters, telephone calls to a cable television company, hospital emergency rooms). The lines are symptoms of bottleneck operations. Analysis is useful in helping managers choose a capacity level that will be cost-effective through balancing the cost of having customers wait with the cost of providing additional capacity. It can aid in the determination of expected costs for various levels of service capacity.

This topic is described in Chapter 19.

Summary

Process selection is an important aspect in planning the design of a production system, and it can have major implications on productivity, cost, competitiveness, and flexibility. New products and services, changes in products or services, and technological changes are some of the reasons organizations become involved in process selection.

Basic issues in process selection include the type of processing to use (e.g., steady, high-volume; intermittent, lower-volume; or one-of-a-kind, large-scale projects), the level of automation or computer-assisted processing to use, the amount of flexibility to build in, and costs of processing.

Capacity refers to a system's potential for producing goods or delivering services over a specified time interval. Capacity decisions are important because capacity is a ceiling on output and a major determinant of operating costs.

A variety of factors can interfere with capacity utilization, so effective capacity is usually somewhat less than design capacity. These factors include facilities design and layout, human factors, product/service design, equipment failures, scheduling problems, and quality considerations.

Capacity planning involves long-term and short-term considerations. Long-term considerations relate to the overall level of capacity; short-term considerations relate to variations in capacity requirements due to seasonal, random, and irregular fluctuations in demand. Ideally, capacity will match demand. Thus, there is a close link between forecasting and capacity planning particularly in the long term. In the short term, emphasis shifts to describing and coping with variations in demand.

Development of capacity alternatives is enhanced by taking a systems approach to planning, by recognizing that capacity increments are often acquired in chunks, by designing flexible systems, and by considering product/service complements as a way of dealing with various patterns of demand.

In evaluating capacity alternatives. a manager must consider both quantitative and qualitative aspects. Quantitative analysis usually reflects economic factors, and qualitative considerations include intangibles such as public opinion and personal preferences of managers. Cost-volume analysis can be useful for analyzing alternatives.

Key Terms

automation, 201
batch processing, 198
break-even point (BEP), 219
capacity, 208
cash flow, 222
computer-aided manufacturing (CAM), 201
computer-integrated manufacturing (CIM), 206
continuous processing, 197
flexible manufacturing system (FMS), 205
intermittent processing, 198
job shop, 198
manufacturing cell, 205
numerically controlled machines (N/C), 201
present value, 222
project, 201
repetitive production, 197
robot, 202

Solved Problems

Problem 1

Make or buy. A firm's manager must decide whether to make or buy a certain item used in the production of vending machines. Cost and volume estimates are as follows:

	Make	Buy
Annual fixed cost	$150,000	None
Variable cost/unit	$60	$80
Annual volume (units)	12,000	12,000

a. Given these numbers, should the firm buy or make this item?

b. There is a possibility that volume could change in the future. At what volume would the manager be indifferent between making and buying?

Solution

a. Determine the annual cost of each alternative:

Total cost = Fixed cost + Volume × Variable cost

Make: $150,000 + 12,000($60) = $870,000
Buy: 0 + 12,000 ($80) = $960,000

Because the annual cost of making the item is less than the annual cost of buying it, the manager would reasonably choose to make the item.

b. To determine the volume at which the two choices would be equivalent, set the two total costs equal to each other, and solve for volume: $TC_{make} = TC_{buy}$. Thus, $150,000 + Q($60) = 0 + Q($80). Solving, Q = 7,500 units. Therefore, at a volume of 7,500 units a year, the manager would be indifferent between making and buying. For lower volumes, the choice would be to buy, and for higher volumes, the choice would be to make.

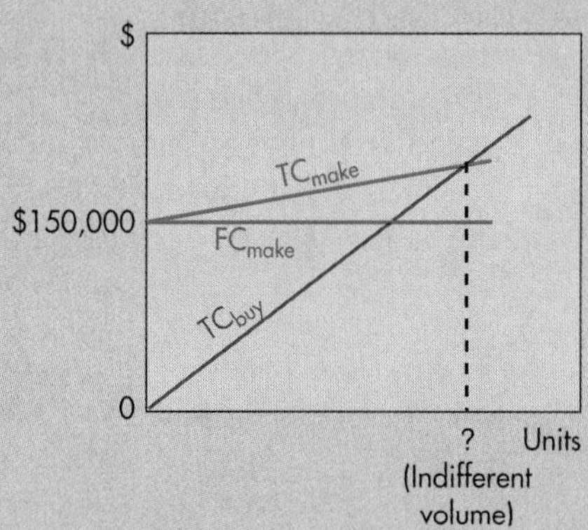

Problem 2

A small firm produces and sells novelty items in a five-state area. The firm expects to consolidate assembly of its electric turtle line at a single location. Currently, operations are in three widely scattered locations. The leading candidate for location will have a monthly fixed cost of $42,000 and variable costs of $3 per turtle. Turtles sell for $7 each. Prepare a table that shows total profits, fixed costs, variable costs, and revenues for monthly volumes of 10,000, 12,000, and 15,000 units. What is the break-even point?

Solution

Revenue = $7 per unit
Variable cost = $3 per unit
Fixed cost = $42,000 per month

Profit = $Q(R - VC) - FC$
Total cost = $FC + VC \times Q$

Volume	Total Revenue	Total VC	Fixed Cost	Total Cost	Total Profit
10,000	$ 70,000	$30,000	$42,000	$72,000	$ (2,000)
12,000	84,000	36,000	42,000	78,000	6,000
15,000	105,000	45,000	42,000	87,000	18,000

$$Q_{BEP} = \frac{FC}{R - VC} = \frac{\$42,000}{\$7 - \$3} = 10,500 \text{ units per month}$$

Problem 3

Refer to Problem 2. Develop an equation that can be used to compute profit for any volume. Use that equation to determine profit when volume equals 22,000 units.

Solution

$$\text{Profit} = Q(R - VC) - FC = Q(\$7 - \$3) - \$42,000 = \$4Q - \$42,000$$

For $Q = 22,000$, profit is

$$\$4(22,000) - \$42,000 = \$46,000$$

Discussion and Review Questions

1. Explain the importance of process selection in system design.
2. Briefly describe continuous processing, intermittent processing, and projects, and indicate the kinds of situations in which each would be used.
3. Briefly discuss the advantages and disadvantages of automation.
4. Briefly describe computer-assisted approaches to production.
5. What is a flexible manufacturing system, and under what set of circumstances is it most appropriate?
6. Why is management of technology important?
7. Why might the choice of equipment that provides flexibility sometimes be viewed as a management cop-out?
8. Contrast design capacity and effective capacity.
9. List and briefly explain three factors that may inhibit capacity utilization.
10. How do long-term and short-term capacity considerations differ?
11. Give an example of a good and a service that exhibit these seasonal demand patterns:
    a. Annual
    b. Monthly
    c. Weekly
    d. Daily
12. Give some examples of building flexibility into system design.
13. Why is it important to adopt a "big picture" approach to capacity planning?
14. What is meant by "capacity in chunks," and why is that a factor in capacity planning?
15. What kinds of capacity problems do many elementary and secondary schools periodically experience? What are some alternatives to deal with those problems?
16. How can a systems approach to capacity planning be useful?
17. How do capacity decisions influence productivity?
18. Identify some additional examples of products and services that use each of these processing systems: continuous, assembly, batch, job shop, and project.
19. Why is it important to match process capabilities with product requirements?
20. Briefly discuss how uncertainty affects process selection and capacity decisions.

21. How do volume and flexibility affect process selection? How does volume affect the degree of automation?

Memo Writing Exercises

1. Write a one-page memo to your supervisor, Sharon Williams, relating fixed automation and flexible automation to the product-process matrix (i.e., where do they fit?). Justify your choices.
2. Write a short memo to your boss, Al Thomas, outlining the general impact on break-even quantities of an increase in the proportion of automation in a process.
3. Write a one-page memo to Don Jones, a production supervisor. Don has questioned your practice of sometimes scheduling production that is below capacity, resulting in less than full utilization of personnel and equipment.

Problems

a. 46,000 units
b. (1) \$3,000
(2) \$8,200
c. 126,000 units
d. 25,556 units
e. Graph in IM

1. A producer of pottery is considering the addition of a new plant to absorb the backlog of demand that now exists. The primary location being considered will have fixed costs of \$9,200 per month and variable costs of 70 cents per unit produced. Each item is sold to retailers at a price that averages 90 cents.
   a. What volume per month is required in order to break even?
   b. What profit would be realized on a monthly volume of 61,000 units? 87,000 units?
   c. What volume is needed to obtain a profit of \$16,000 per month?
   d. What volume is needed to provide a revenue of \$23,000 per month?
   e. Plot the total cost and total revenue lines.

a. (A) 8,000 units
(B) 7,500 units
b. 10,000 units
c. P_A = \$20,000
P_B = \$18,000

2. A small firm intends to increase the capacity of a bottleneck operation by adding a new machine. Two alternatives, A and B, have been identified, and the associated costs and revenues have been estimated. Annual fixed costs would be \$40,000 for A and \$30,000 for B; variable costs per unit would be \$10 for A and \$12 for B; and revenue per unit would be \$15 for A and \$16 for B.
   a. Determine each alternative's break-even point in units.
   b. At what volume of output would the two alternatives yield the same profit?
   c. If expected annual demand is 12,000 units, which alternative would yield the higher profit?

a. 39,683 units
b. \$1.71

3. A producer of felt-tip pens has received a forecast of demand of 30,000 pens for the coming month from its marketing department. Fixed costs of \$25,000 per month are allocated to the felt-tip operation, and variable costs are 37 cents per pen.
   a. Find the break-even quantity if pens sell for \$1 each.
   b. At what price must pens be sold to obtain a monthly profit of \$15,000, assuming that estimated demand materializes?

a. (A) \$82
(B) \$92
(C) \$100
b. Graph in IM
c. A is optimal for 0 to <178 minutes. C is optimal for 178 or more.

4. A real estate agent is considering installing a cellular telephone in her car. There are three billing plans to choose from, all of which involve a weekly charge of \$20. Plan A has a cost of \$0.45 a minute for daytime calls and \$0.20 a minute for evening calls. Plan B has a charge of \$0.55 a minute for daytime calls and a charge of \$0.15 a minute for evening calls. Plan C has a flat rate of \$80 with 200 minutes of calls allowed per week and a cost of \$0.40 per minute beyond that, day or evening.
   a. Determine the total charge under each plan for this case: 120 minutes of day calls and 40 minutes of evening calls in a week.
   b. Prepare a graph that shows total weekly cost for each plan versus daytime call minutes.
   c. If the agent will use the service for daytime calls, over what range of call minutes will each plan be optimal?

33 percent daytime minutes

5. Refer to Problem 4. Suppose that the agent expects both daytime and evening calls. At what point (i.e., percent of call minutes for daytime calls) would she be indifferent between plans A and B?

6. A firm plans to begin production of a new small appliance. The manager must decide whether to purchase the motors for the appliance from a vendor at $7 each or to produce them in-house. Either of two processes could be used for in-house production; one would have an annual fixed cost of $160,000 and a variable cost of $5 per unit, and the other would have an annual fixed cost of $190,000 and a variable cost of $4 per unit. Determine the range of annual volume for which each of the alternatives would be best.

For Q less than 63,333, vendor is best. For larger quantities, best to produce in-house at $4 per unit.

7. A manager is trying to decide whether to purchase a certain part or to have it produced internally. Internal production could use either of two processes. One would entail a variable cost of $17 per unit and an annual fixed cost of $200,000; the other would entail a variable cost of $14 per unit and an annual fixed cost of $240,000. Three vendors are willing to provide the part. Vendor A has a price of $20 per unit for any volume up to 30,000 units. Vendor B has a price of $22 per unit for demand of 1,000 units or less, and $18 per unit for larger quantities. Vendor C offers a price of $21 per unit for the first 1,000 units, and $19 per unit for additional units.
   a. If the manager anticipates an annual volume of 10,000 units, which alternative would be best from a cost standpoint? For 20,000 units, which alternative would be best?
   b. Determine the range for which each alternative is best. Are there any alternatives that are never best? Which?

a. For 10,000 units:
Int. 1: $370,000
Int. 2: $380,000
Vend. A: $200,000
Vend. B: $180,000
Vend. C: $192,000
For 20,000 units:
Int. 1: $540,000
Int. 2: $520,000
Vend. A: $400,000
Vend. B: $360,000
Vend. C: S382,000

b.

Range	Best Alt.
1 to 999	A
1,000 to 59,999	B
60,000 or more	Int. 2

8. A company manufactures a product using two machine cells. Each cell has a design capacity of 250 units per day and an effective capacity of 230 units per day. At present, actual output averages 200 units per cell, but the manager estimates that productivity improvements soon will increase output to 225 units per day. Annual demand is currently 50,000 units. It is forecasted that within two years, annual demand will triple. The company could produce at the rate of 400 per day using available capacity. How many cells should the company plan to produce to satisfy predicted demand under these conditions? Assume 240 workdays per year.

3 cells

9. A manager must decide between these alternatives:

Machine	Cost
A	$40,000
B	$30,000
C	$80,000

Product forecasts and processing times on the machines are as follows:

Product	Annual Demand	PROCESSING TIME PER UNIT (MINUTES) A	B	C
1	16,000	3	4	2
2	12,000	4	4	3
3	6,000	5	6	4
4	30,000	2	2	1

Assume that only purchasing costs are being considered. Which machine would have the lowest total cost, and how many of that machine would be needed? Machines operate 10 hours a day, 250 days a year.

Total processing time per machine:
A: 186,000
B: 208,000
C: 122,000
Number of each machine needed and total cost:
A: 2, $80,000
B: 2, $60,000
C: 1, $80,000
Buy 2 Bs

10. Refer to Problem 9. Consider this additional information: The machines differ in terms of hourly operating costs: The A machines have an hourly operating cost of $10 each, B machines have an hourly operating cost of $11 each, and C machines have an hourly operating cost of $12 each. Which alternative would be selected, and how many machines, in order to minimize total cost while satisfying capacity processing requirements?

A. 111,000
B. 98,133
C. 104,400
Buy 2 machine B

11. A manager must decide how many machines of a certain type to purchase. Each machine can process 100 customers per hour. One machine will result in a fixed cost of $2,000 per

a. 80,152
b. 2

day, while two machines will result in a fixed cost of $3,800 per day. Variable costs will be $20 per customer, and revenue will be $45 per customer.

a. Determine the break-even point for each range.

b. If estimated demand is 90 to 120 customers per hour, how many machines should be purchased?

one line

12. The manager of a car wash must decide whether to have one or two wash lines. One line will mean a fixed cost of $6,000 a month, and two lines will mean a fixed cost of $10,500 a month. Each line would be able to process 15 cars an hour. Variable costs will be $3 per car, and revenue will be $5.95 per car. The manager projects an average demand of between 14 and 18 cars an hour. Would you recommend one or two lines? The car wash is open 300 hours a month.

Subcontract	$1.23
Expand	$1.57
Build	$1.35

13. Determine the course of action that has the highest expected payoff for the following decision tree.

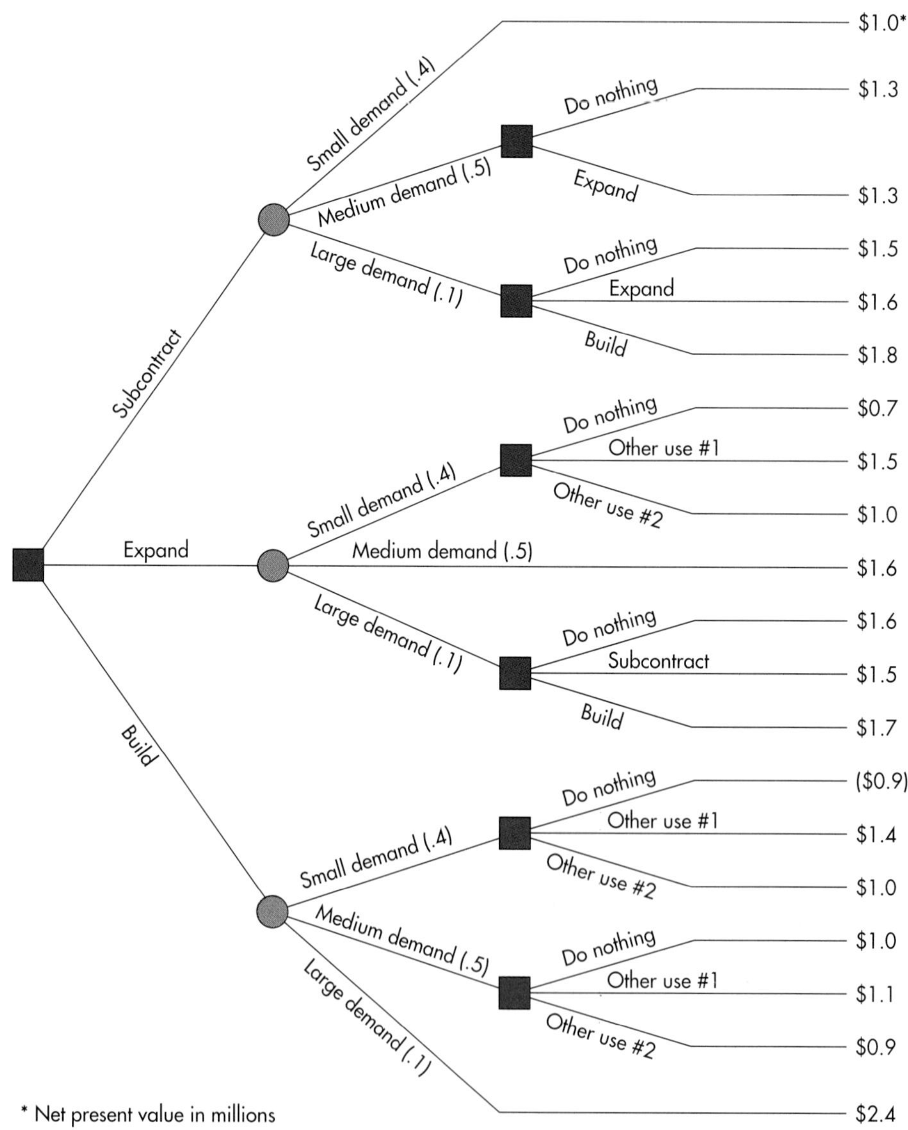

Graph in IM	
Large	$476,000
Small	$430,000

14. A firm that plans to expand its product line must decide whether to build a small or a large facility to produce the new products. If it builds a small facility and demand is low, the net present value after deducting for building costs will be $400,000. If demand is high, the firm

can either maintain the small facility or expand it. Expansion would have a net present value of $450,000, and maintaining the small facility would have a net present value of $50,000.

If a large facility is built and demand is high, the estimated net present value is $800,000. If demand turns out to be low, the net present value will be −$10,000.

The probability that demand will be high is estimated to be .60, and the probability of low demand is estimated to be .40.

Analyze using a tree diagram.

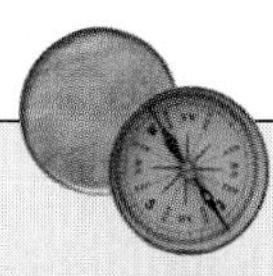

OPERATIONS TOUR

MORTON SALT

www.mortonintl.com

Introduction

Morton Salt is a subsidiary of Morton International, a manufacturer of specialty chemicals, air bags, and salt products. The Morton salt-processing facility in Silver Springs, New York, between Buffalo and Rochester, is one of six similar Morton salt-processing facilities in the United States. The Silver Springs plant employs about 200 people, ranging from unskilled to skilled. It produces salt products for water conditioning, grocery, industrial, and agricultural markets. The grocery business consists of 26 oz. round cans of iodized salt. Although the grocery business represents a relatively small portion of the total output (approximately 15 percent), it is the most profitable.

Salt Production

The basic raw material, salt, is obtained by injecting water into salt caverns that are located some 2,400 feet below the surface. There, the salt deposits dissolve in the water. The resulting brine is pumped to the surface where it is converted into salt crystals. The brine is boiled, and much of the liquid evaporates, leaving salt crystals and some residual moisture, which is removed in a drying process. The salt is then moved to storage silos. This process is run continuously for about six weeks at a time. Initially, salt is produced at the rate of 45 tons per hour. But the rate of output decreases due to scale build up, so that by the sixth week, output is only 75 percent of the initial rate. At that point, the process is halted to perform maintenance on the equipment and remove the scale, after which, salt production resumes.

The salt is stored in silos until it is needed for production, or it is shipped in bulk to industrial customers. Conveyors move the salt to each of the four dedicated production areas, one of which is round can production. (See diagram.) The discussion here focuses exclusively on round can production.

Round Can Production

Annual round can production averages roughly 3.8 million cans. Approximately 70 percent of the output is for the Morton label, and the rest is for private label. There are two parallel, high-speed production lines. The two lines share common processes at the beginning of the lines, and then branch out into two identical lines. Each line is capable of producing 9,600 cans per hour (160 cans per minute). The equipment is not flexible, so the production rate is fixed. The operations are completely standardized; the only variable is the brand label that is applied. One line requires 12 production workers, while both lines together can be operated by 18 workers because of the common processes. Workers on the line perform low-skilled, repetitive tasks.

The plant produces both the salt and the cans the salt is packaged in. The cans are essentially a cylinder with a top and a bottom; they are made of cardboard, except for an aluminum pour spout in the top. The cylinder portion is formed from two sheets of chip board that are glued together and then rolled into a continuous tube. The glue not only binds the material, it also provides a moisture barrier. The tube is cut in a two-step process: it is first cut into long sections, and those sections are then cut into can-size pieces. The top and bottom pieces for the cans are punched from a continuous strip of cardboard. The separate pieces move along conveyor belts to the lines where the components are assembled into cans and glued. The cans are then filled with salt and the pour spout is added. Finally, the cans are loaded onto pallets and placed into inventory, ready to be shipped to distributors.

Quality

Quality is checked at several points in the production process. Initially, the salt is checked for purity when it is obtained from the wells. Iodine and an anti-caking compound are added to the salt, and their levels are verified using chemical analysis. Crystal size is important. In order to achieve the desired size and to remove lumps, the salt is forced through a scraping screen, which can cause very fine pieces of metal to mix with the salt. However, these pieces are effectively removed by magnets that are placed at appropriate points in the process. If, for any reason, the salt is judged to be contaminated, it is diverted to a nonfood product.

Checking the quality of the cans is done primarily by visual inspection, including verifying the assembly operation is correct, checking filled cans for correct weight, inspecting cans to see that labels are properly aligned, and checking to see that metal pour spouts are correctly attached.

The equipment on the production line is sensitive to misshapen or damaged cans, and frequently jams, causing production delays. This greatly reduces the chance of a defective can getting through the process, but it reduces productivity, and the salt in the defective cans must be scrapped. The cost of quality is fairly high, owing to the amount of product that is scrapped, the large number of inspectors, and the extensive laboratory testing that is needed.

Production Planning and Inventory

The plant can sell all of the salt it produces. The job of the production scheduler is to distribute the salt that is stored in the silos to the various production areas, taking into account production capacities in each area and available inventory levels of those products. A key consideration is to make sure there is sufficient storage capacity in the silos to handle the incoming salt from brine production.

Equipment Maintenance and Repair

The equipment is 1950s vintage, and it requires a fair amount of maintenance to keep it in good working order. Even so, breakdowns occur as parts wear out. The plant has its own tool shop where skilled workers repair parts or make new parts because replacement parts are no longer available for the old equipment.

Questions

1. Briefly describe salt production, from brine production to finished round cans.
2. Briefly describe quality assurance efforts in round can production.
3. What are some of the possible reasons why the company continues to use the old processing equipment instead of buying new, more modern equipment?
4. Where would you place salt production in the product-process spectrum?
5. Determine the approximate number of tons of salt produced annually. Hints: one ton = 2,000 pounds, and one pound = 16 ounces.
6. What improvements can you suggest for the plant?

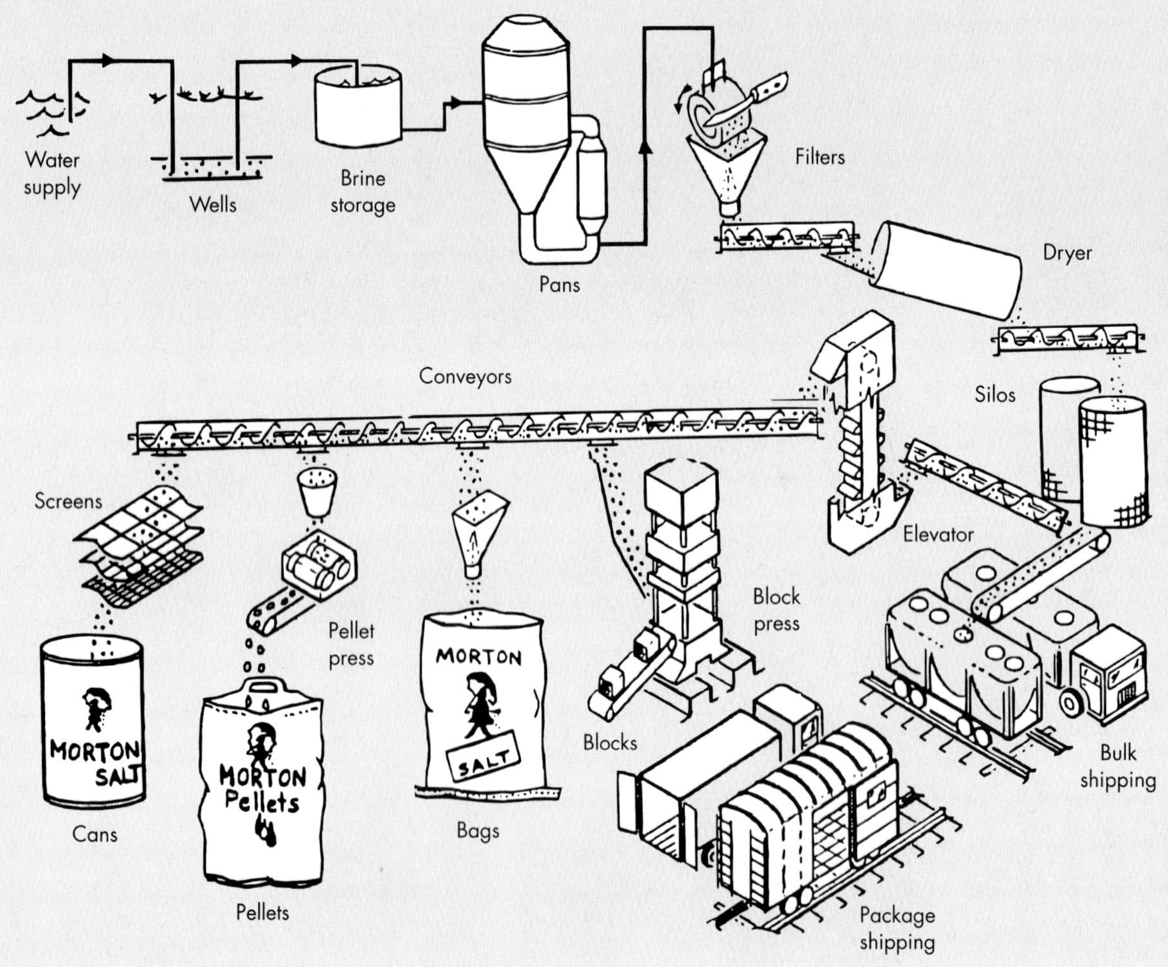

OPERATIONS TOUR

HIGH ACRES LANDFILL

The High Acres Landfill is located on a 70-acre site outside Fairport, New York. Opened in 1971, it is licensed to handle residential, commercial, and industrial nonhazardous waste. The landfill has 27 employees, and it receives approximately 900 tons of waste per day.

The public often has certain preconceived notions about a landfill, chief among them that landfills are dirty and unpleasant. However, a visit to the landfill dispelled some of those misconceptions. The entrance is nicely landscaped. Most of the site is planted with grass and a few trees. Although unpleasant odors can emanate from arriving trucks or at the dump site, the remainder of the landfill is relatively free of odors.

A major portion of the landfill consists of a large hill within which the waste is buried. Initially, the landfill began not as a hill but as a large hole in the ground. After a number of years of depositing waste, the hole eventually was filled. From that point on, as additional layers were added, the landfill began to take the shape of a flattop hill. Each layer is a little narrower than the preceding one, giving the hill a slope. The sides of the hill are planted with grass. Only the "working face" along the top remains unplanted. When the designated capacity is exhausted (this may take another 10 years), the landfill will be closed to further waste disposal. The site will be converted into a public park with hiking trails and picnic and recreation areas, and given to the town.

The construction and operation of landfills are subject to numerous state and federal regulations. For example, nonpermeable liners must be placed on the bottom and sides of the landfill to prevent leakage of liquids into the groundwater. (Independent firms monitor groundwater to determine if there is any leakage into wells placed around the perimeter of the hill.) Mindful of public opinion, every effort is made to minimize the amount of time that waste is left exposed. At the end of each day, the waste that has been deposited in the landfill is compacted and covered with six inches of soil.

The primary source of income for the landfill is the fees it charges users. The landfill also generates income from methane gas, a by-product of organic waste decomposition, that accumulates within the landfill. A collection system is in place to capture and extract the gas from the landfill, and it is then sold to the local power company. Also, the landfill has a composting operation in which leaves and other yard wastes are converted into mulch.

Part of the liner construction of a new landfill at the High Acres Landfill and Recycling Center in Fairport, New York. The hill in the background is a "closed" landfill, which has been through final cover.

Selected Bibliography

Bolwijn, P. T., and T. Kumpe. "Manufacturing in the 1990's—Productivity, Flexibility, and Innovation." *Long Range Planning* 23, no. 4 (1990), pp. 44–57.

Cohen, Morris, and Uday M. Apte. *Manufacturing Automation*. Burr Ridge, IL: Richard D. Irwin, 1997.

Ettlie, John, and Henry Stoll, *Managing the Design-Manufacturing Process*. New York: McGraw-Hill, 1990.

Groover, Mikell P. *Automation Production Systems, and Computer-Aided Manufacturing.* Englewood Cliffs, NJ: Prentice Hall, 1980.

Hill, Terry. *Manufacturing Strategy*. 2nd ed. Burr Ridge, IL: Richard D. Irwin, 1994.

Monroe, Joseph. "Strategic Use of Technology." *California Management Review,* Summer 1989, pp. 91–110.

Moore, Franklin, and Thomas E. Hendrick. *Production/Operations Management.* 9th ed. Burr Ridge, IL: Richard D. Irwin, 1985.

Shunk, Dan L. *Integrated Process Design and Development.* Burr Ridge, IL: Business One Irwin, 1992.

Stauffer, Robert. "Lessons Learned in Implementing New Technology." *Manufacturing Engineer,* June 1989.

Toward a New Era in U.S. Manufacturing: The Need for a National Vision. Washington, DC: National Academy Press, 1986.

Upton, David. "The Management of Manufacturing Flexibility." *California Management Review* 36, no. 2, 1994, pp. 72–89.

Upton, David. "What Really Makes Factories Flexible." *Harvard Business Review,* July–August 1995, pp. 74–84.

SUPPLEMENT TO CHAPTER FIVE

LINEAR PROGRAMMING

LEARNING OBJECTIVES

After completing this supplement, you should be able to:

1. Describe the type of problem that would lend itself to solution using linear programming.
2. Formulate a linear programming model from a description of a problem.
3. Solve simple linear programming problems using the graphical method.
4. Interpret computer solutions of linear programming problems.
5. Do sensitivity analysis on the solution of a linear programming problem.

SUPPLEMENT OUTLINE

Linear programming is a powerful quantitative tool used by operations managers and other managers to obtain optimal solutions to problems that involve restrictions or limitations, such as the available materials, budgets, and labor and machine time. These problems are referred to as constrained optimization problems. There are numerous examples of linear programming applications to such problems, including:

- Establishing locations for emergency equipment and personnel that will minimize response time
- Determining optimal schedules for airlines for planes, pilots, and ground personnel
- Developing financial plans
- Determining optimal blends of animal feed mixes
- Determining optimal diet plans
- Identifying the best set of worker-job assignments
- Developing optimal production schedules
- Developing shipping plans that will minimize shipping costs
- Identifying the optimal mix of products in a factory

INTRODUCTION

Linear programming (LP) techniques consist of a sequence of steps that will lead to an optimal solution to problems, in cases where an optimum exists. There are a number of different linear-programming techniques; some are special-purpose (i.e., used to find solutions for specific types of problems) and others are more general in scope. This supplement covers the two general-purpose solution techniques: graphical linear programming and computer solutions. Graphical linear programming provides a visual portrayal of many of the important concepts of linear programming. However, it is limited to problems with only two variables. In practice, computers are used to obtain solutions for problems, some of which involve a large number of variables.

LINEAR PROGRAMMING MODELS

Linear programming models are mathematical representations of constrained optimization problems. These models have certain characteristics in common. Knowledge of these characteristics enables us to recognize problems that can be solved using linear programming. In addition, it also can help us formulate LP models. The characteristics can be grouped into two categories: components and assumptions. First, let's consider the components.

Four components provide the structure of a linear programming model:

1. Objective.
2. Decision variables.
3. Constraints.
4. Parameters.

objective function Mathematical statement of total profit, cost, etc., for a given solution.

decision variables Amounts of either inputs or outputs.

Linear programming algorithms require that a single goal or *objective,* such as the maximization of profits, be specified. The two general types of objectives are maximization and minimization. A maximization objective might involve profits, revenues, efficiency, or rate of return. Conversely, a minimization objective might involve cost, time, distance traveled, or scrap. The **objective function** is a mathematical expression that can be used to determine the total profit (or cost, etc., depending on the objective) for a given solution.

Decision variables represent choices available to the decision maker in terms of amounts of either inputs or outputs. For example, some problems require choosing a combination of inputs to minimize total costs, while others require selecting a combination of outputs to maximize profits or revenues.

Constraints are limitations that restrict the alternatives available to decision makers. The three types of constraints are less than or equal to (≤), greater than or equal to (≥), and simply equal to (=). A ≤ constraint implies an upper limit on the amount of some scarce resource (e.g., machine hours, labor hours, materials) available for use. A ≥ constraint specifies a minimum that must be achieved in the final solution (e.g., must contain at least 10 percent real fruit juice, must get at least 30 MPG on the highway). The = constraint is more restrictive in the sense that it specifies *exactly* what a decision variable should equal (e.g., make 200 units of product A). A linear-programming model can consist of one or more constraints. The constraints of a given problem define the set of all feasible combinations of decision variables; this set is referred to as the **feasible solution space.** Linear-programming algorithms are designed to search the feasible solution space for the combination of decision variables that will yield an optimum in terms of the objective function.

constraints Limitations that restrict the available alternatives.

feasible solution space The set of all feasible combinations of decision variables as defined by the constraints.

An LP model consists of a mathematical statement of the objective and a mathematical statement of each constraint. These statements consist of symbols (e.g., x_1, x_2) that represent the decision variables and numerical values, called **parameters.** The parameters are fixed values; the model is solved *given* those values.

parameters Numerical constants.

In order for linear-programming models to be used effectively, certain *assumptions* must be satisfied. These are:

1. *Linearity:* the impact of decision variables is linear in constraints and the objective function.
2. *Divisibility:* noninteger values of decision variables are acceptable.
3. *Certainty:* values of parameters are known and constant.
4. *Nonnegativity:* negative values of decision variables are unacceptable.

Example 1 illustrates the components of an LP model.

Example 1

Decision variables
$$\begin{cases} x_1 = \text{Quantity of product 1 to produce} \\ x_2 = \text{Quantity of product 2 to produce} \\ x_3 = \text{Quantity of product 3 to produce} \end{cases}$$

Maximize $5x_1 + 8x_2 + 4x_3$ (profit) (Objective function)

Subject to

Labor $2x_1 + 4x_2 + 8x_3 \leq 250$ hours

Material $7x_1 + 6x_2 + 5x_3 \leq 100$ pounds (Constraints)

Product 1 $x_1 \geq 10$ units

$x_1, x_2, x_3 \geq 0$ (Nonnegativity constraints)

First, the model lists and defines the decision variables. These typically represent *quantities.* In this case, they are quantities of three different products that might be produced.

Next, the model states the objective function. It includes every decision variable in the model and the contribution (profit per unit) of each decision variable. Thus, product x_1 has a profit of \$5 per unit. The profit from product x_1 for a given solution will be 5 times the value of x_1 specified by the solution; the total profit from all products will be the sum of the individual product profits. Thus, if $x_1 = 10$, $x_2 = 0$, and $x_3 = 6$, the value of the objective function would be:

$$5(10) + 8(0) + 4(6) = 74$$

The objective function is followed by a list (in no particular order) of three constraints. Each constraint has a right-side numerical value (e.g., the labor constraint has a right-side value of 250) that indicates the amount of the constraint and a relation sign that indicates whether that amount is a maximum ($\leq$), a minimum ($\geq$), or an equality ($=$). The left side of each constraint consists of the variables subject to that particular constraint and a coefficient for each variable that indicates how much of the right-side quantity *one unit* of the decision variable represents. For instance, for the labor constraint, one unit of x_1 will require two hours of labor. The sum of the values on the left side of each constraint represents the amount of that constraint used by a solution. Thus, if $x_1 = 10$, $x_2 = 0$, and $x_3 = 6$, the amount of labor used would be:

$$2(10) + 4(0) + 8(6) = 68 \text{ hours}$$

Because this amount does not exceed the quantity on the right-hand side of the constraint, it is feasible.

Note that the third constraint refers to only a single variable; x_1 must be at least 10 units. Its coefficient is, in effect, 1, although that is not shown.

Finally, there are the nonnegativity constraints. These are listed on a single line; they reflect the condition that no decision variable is allowed to have a negative value.

Model Formulation

An understanding of the components of linear-programming models is necessary for model formulation. This helps provide organization to the process of assembling information about a problem into a model.

Naturally, it is important to obtain valid information on what constraints are appropriate, as well as on what values of the parameters are appropriate. If this is not done, the usefulness of the model will be questionable. Consequently, in some instances, considerable effort must be expended to obtain that information.

In formulating a model, use the format illustrated in Example 1. Begin by identifying the decision variables. Very often, decision variables are "the quantity of" something, such as $x_1 =$ the quantity of product 1. Generally, decision variables have profits, costs, times, or a similar measure of value associated with them. Knowing this can help you identify the decision variables in a problem.

Constraints are restrictions or requirements on one or more decision variables, and they refer to available amounts of resources such as labor, material, or machine time, or to minimal requirements, such as "make at least 10 units of product 1." It can be helpful to give a name to each constraint, such as "labor" or "material 1." Let's consider some of the different kinds of constraints you will encounter.

1. A constraint that refers to one or more decision variables. This is the most common kind of constraint. The constraints in Example 1 are of this type.

2. A constraint that specifies a ratio. For example, "the ratio of x_1 to x_2 must be at least 3 to 2." To formulate this, begin by setting up the ratio:

$$\frac{x_1}{x_2} \geq \frac{3}{2}$$

Then, cross multiply, obtaining

$$2x_1 \geq 3x_2$$

This is not yet in a suitable form because all variables in a constraint must be on the left side of the inequality (or equality) sign, leaving only a constant on the right side. To achieve this, we must subtract the variable amount that is on the right side from both sides. That yields:

$$2x_1 - 3x_2 \geq 0$$

[Note that the direction of the inequality remains the same.]

3. A constraint that specifies a percentage for one or more variables relative to one or more other variables. For example, "x_1 cannot be more than 20 percent of the mix." Suppose that the mix consists of variables x_1, x_2, and x_3. In mathematical terms, this would be:

$$x_1 \leq .20(x_1 + x_2 + x_3)$$

As always, all variables must appear on the left side of the relationship. To accomplish that, we can expand the right side, and then subtract the result from both sides. Thus,

$$x_1 \leq .20x_1 + .20x_2 + .20x_3$$

Subtracting yields

$$.80x_1 - .20x_2 - .20x_3 \leq 0$$

Once you have formulated a model, the next task is to solve it. The following sections describe two approaches to problem solution: graphical solutions and computer solutions.

GRAPHICAL LINEAR PROGRAMMING

Graphical linear programming is a method for finding optimal solutions to two-variable problems. This section describes that approach.

graphical linear programming Graphical method for finding optimal solutions to two-variable problems.

Outline of Graphical Procedure

The graphical method of linear programming plots the constraints on a graph and identifies an area that satisfies all of the constraints. The area is referred to as the *feasible solution space*. Next, the objective function is plotted and used to identify the optimal point in the feasible solution space. The coordinates of the point can sometimes be read directly from the graph, although generally an algebraic determination of the coordinates of the point is necessary.

The general procedure followed in the graphical approach is:

1. Set up the objective function and the constraints in mathematical format.
2. Plot the constraints.
3. Identify the feasible solution space.
4. Plot the objective function.
5. Determine the optimum solution.

The technique can best be illustrated through solution of a typical problem. Consider the problem described in Example 2.

Example 2

General description: A firm that assembles computers and computer equipment is about to start production of two new types of microcomputers. Each type will require assembly time, inspection time, and storage space. The amounts of each of these resources that can be devoted to the production of the microcomputers is limited. The manager of the firm would like to determine the quantity of each microcomputer to produce in order to maximize the profit generated by sales of these microcomputers.

Additional information: In order to develop a suitable model of the problem, the manager has met with design and manufacturing personnel. As a result of those meetings, the manager has obtained the following information:

	Type 1	Type 2
Profit per unit	\$60	\$50
Assembly time per unit	4 hours	10 hours
Inspection time per unit	2 hours	1 hour
Storage space per unit	3 cubic feet	3 cubic feet

The manager also has acquired information on the availability of company resources. These (daily) amounts are:

Resource	Amount Available
Assembly time	100 hours
Inspection time	22 hours
Storage space	39 cubic feet

The manager met with the firm's marketing manager and learned that demand for the microcomputers was such that whatever combination of these two types of microcomputers is produced, all of the output can be sold.

In terms of meeting the assumptions, it would appear that the relationships are *linear:* The contribution to profit per unit of each type of computer and the time and storage space per unit of each type of computer is the same regardless of the quantity produced. Therefore, the total impact of each type of computer on the profit and each constraint is a linear function of the quantity of that variable. There may be a question of *divisibility* because, presumably, only whole units of computers will be sold. However, because this is a recurring process (i.e., the computers will be produced daily, a noninteger solution such as 3.5 computers per day will result in 7 computers every other day), this does not seem to pose a problem. The question of *certainty* cannot be explored here; in practice, the manager could be questioned to determine if there are any other possible constraints and whether the values shown for assembly times, and so forth, are known with certainty. For the purposes of discussion, we will assume certainty. Lastly, the assumption of *nonnegativity* seems justified; negative values for production quantities would not make sense.

Because we have concluded that linear programming is appropriate, let us now turn our attention to constructing a model of the microcomputer problem. First, we must define the decision variables. Based on the statement, "The manager . . . would like to determine the quantity of each microcomputer to produce," the decision variables are the quantities of each type of computer. Thus,

x_1 = quantity of type 1 to produce
x_2 = quantity of type 2 to produce

Next, we can formulate the objective function. The profit per unit of type 1 is listed as \$60, and the profit per unit of type 2 is listed as \$50, so the appropriate objective function is

$$\text{Maximize} \quad Z = 60x_1 + 50x_2$$

where Z is the value of the objective function, given values of x_1 and x_2. Theoretically, a mathematical function requires such a variable for completeness. However, in practice, the objective function often is written without the Z, as sort of a shorthand version. That approach is underscored by the fact that computer input does not call for Z: It is understood. The output of a computerized model does include a Z, though.

Now for the constraints. There are three resources with limited availability: assembly time, inspection time, and storage space. The fact that availability is limited means that these constraints will all be $\leq$ constraints. Suppose we begin with the assembly constraint. The type 1 microcomputer requires 4 hours of assembly time per unit, whereas the type 2 microcomputer requires 10 hours of assembly time per unit. Therefore, with a limit of 100 hours available, the assembly constraint is

$$4x_1 + 10x_2 \leq 100 \text{ hours}$$

Similarly, each unit of type 1 requires 2 hours of inspection time, and each unit of type 2 requires 1 hour of inspection time. With 22 hours available, the inspection constraint is

$$2x_1 + 1x_2 \leq 22$$

(Note, the coefficient of 1 for x_2 need not be shown. Thus, an alternative form for this constraint is: $2x_1 + x_2 \leq 22$.) The storage constraint is determined in a similar manner:

$$3x_1 + 3x_2 \leq 39$$

There are no other system or individual constraints. The nonnegativity constraints are

$$x_1, x_2 \geq 0$$

In summary, the mathematical model of the microcomputer problem is

x_1 = quantity of type 1 to produce
x_2 = quantity of type 2 to produce

Maximize $60x_1 + 50x_2$

Subject to

Assembly	$4x_1 + 10x_2 \leq 100$ hours
Inspection	$2x_1 + 1x_2 \leq 22$ hours
Storage	$3x_1 + 3x_2 \leq 39$ cubic feet
	$x_1, x_2 \geq 0$

The next step is to plot the constraints.

Plotting Constraints

Begin by placing the nonnegativity constraints on a graph, as in Figure 5S–1. The procedure for plotting the other constraints is simple:

1. Replace the inequality sign with an equal sign. This transforms the constraint into an equation of a *straight line.*
2. Determine where the line intersects each axis.
   a. To find where it crosses the x_2 axis, set x_1 equal to zero and solve the equation for the value of x_2.
   b. To find where it crosses the x_1 axis, set x_2 equal to zero and solve the equation for the value of x_1.
3. Mark these intersections on the axes, and connect them with a straight line. (Note: If a constraint has only one variable, it will be a vertical line on a graph if the variable is x_1, or a horizontal line if the variable is x_2.)
4. Indicate by shading (or by arrows at the ends of the constraint line) whether the inequality is greater than or less than. (A general rule to determine which side of the line satisfies the inequality is to pick the point that is not on the line, such as 0,0, and see whether it is greater than or less than the constraint amount.)
5. Repeat steps 1–4 for each constraint.

Consider the assembly time constraint:

$$4x_1 + 10x_2 \leq 100$$

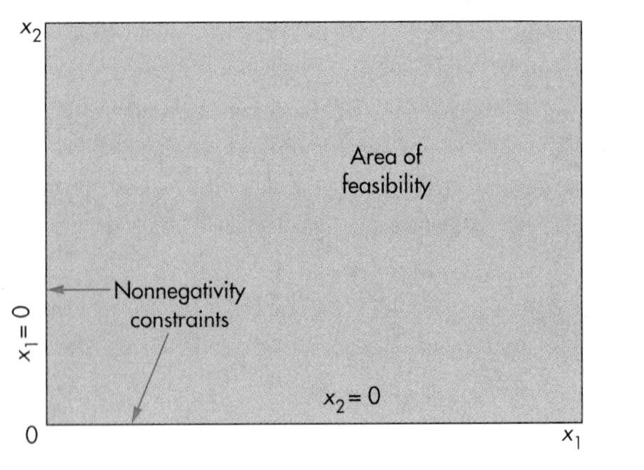

FIGURE 5S–1

Graph showing the nonnegativity constraints

Removing the inequality portion of the constraint produces this straight line:

$$4x_1 + 10x_2 = 100$$

Next, identify the points where the line intersects each axis, as step 2 describes. Thus with $x_2 = 0$, we find

$$4x_1 + 10(0) = 100$$

Solving, we find that $4x_1 = 100$, so $x_1 = 25$ when $x_2 = 0$. Similarly, we can solve the equation for x_2 when $x_1 = 0$:

$$4(0) + 10x_2 = 100$$

Solving for x_2, we find $x_2 = 10$ when $x_1 = 0$.

Thus, we have two points: $x_1 = 0, x_2 = 10$, and $x_1 = 25, x_2 = 0$. We can now add this line to our graph of the nonnegativity constraints by connecting these two points (see Figure 5S–2).

Next we must determine which side of the line represents points that are less than 100. To do this, we can select a test point that is not on the line, and we can substitute the x_1 and x_2 values of that point into the left side of the equation of the line. If the result is less than 100, this tells us that all points on that side of the line are less than the value of the line (e.g., 100). Conversely, if the result is greater than 100, this indicates that the other side of the line represents the set of points that will yield values that are less than 100. A relatively simple test point to use is the origin (i.e., $x_1 = 0, x_2 = 0$). Substituting these values into the equation yields

$$4(0) + 10(0) = 0$$

Obviously this is less than 100. Hence, the side of the line closest to the origin represents the "less than" area (i.e., the feasible region).

The feasible region for this constraint and the nonnegativity constraints then becomes the shaded portion shown in Figure 5S–3.

For the sake of illustration, suppose we try one other point, say $x_1 = 10, x_2 = 10$. Substituting these values into the assembly constraint yields

$$4(10) + 10(10) = 140$$

Clearly this is greater than 100. Therefore, all points on this side of the line are greater than 100 (see Figure 5S–4).

Continuing with the problem, we can add the two remaining constraints to the graph. For the inspection constraint:

1. Convert the constraint into the equation of a straight line by replacing the inequality sign with an equality sign:

$$2x_1 + 1x_2 \leq 22 \quad \text{becomes} \quad 2x_1 + 1x_2 = 22$$

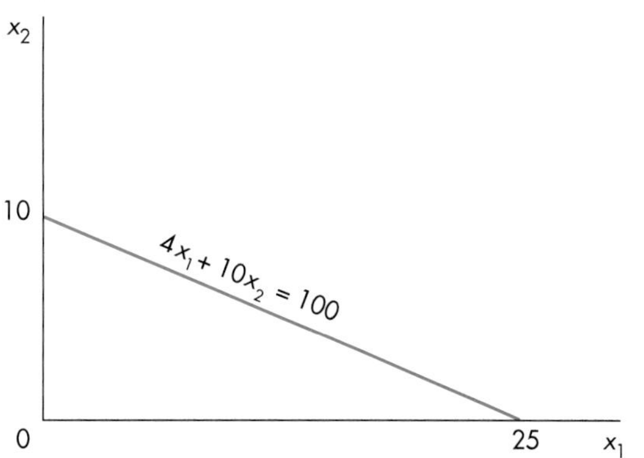

FIGURE 5S-2

Plot of the first constraint (assembly time)

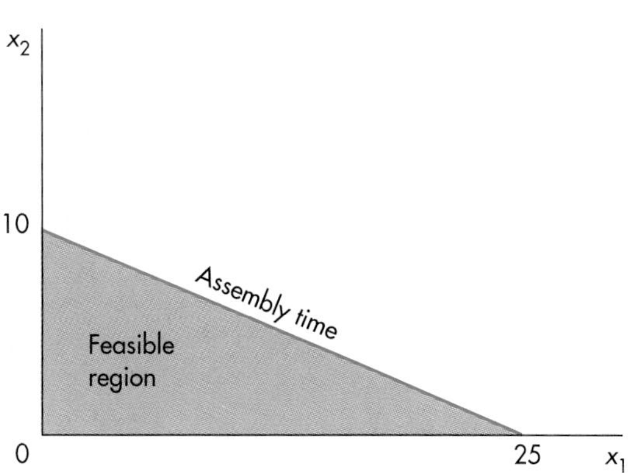

FIGURE 5S-3

The feasible region, given the first constraint and the nonnegativity constraints

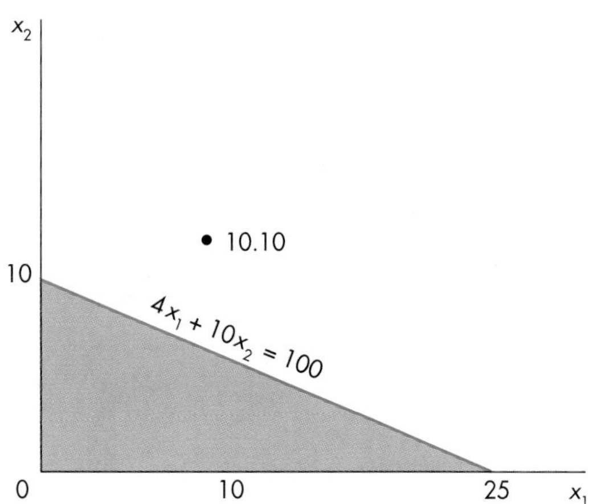

FIGURE 5S-4

The point 10.10 is above the constraint line

2. Set x_1 equal to zero and solve for x_2:

   $$2(0) + 1x_2 = 22$$

   Solving, we find $x_2 = 22$. Thus, the line will intersect the x_2 axis at 22.
3. Next, set x_2 equal to zero and solve for x_1:

   $$2x_1 + 1(0) = 22$$

   Solving, we find $x_1 = 11$. Thus, the other end of the line will intersect the x_1 axis at 11.
4. Add the line to the graph (see Figure 5S–5).

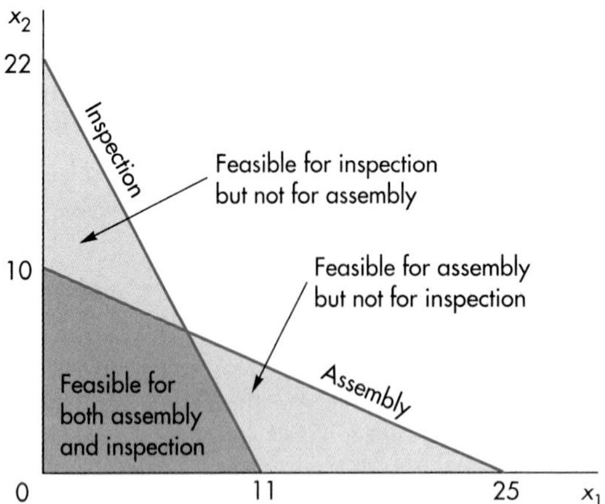

FIGURE 5S–5

Partially completed graph, showing the assembly, inspection, and nonnegativity constraints

Note that the area of feasibility for this constraint is below the line (Figure 5S–5). Again the area of feasibility at this point is shaded in for illustration, although when graphing problems, it is more practical to refrain from shading in the feasible region until all constraint lines have been drawn. However, because constraints are plotted one at a time, using a small arrow for each to indicate the direction of feasibility can be helpful.

The storage constraint is handled in the same manner:

1. Convert it into an equality:

   $$3x_1 + 3x_2 = 39$$

2. Set x_1 equal to zero and solve for x_2:

   $$3(0) + 3x_2 = 39$$

   Solving, $x_2 = 13$. Thus, $x_2 = 13$ when $x_1 = 0$.
3. Set x_2 equal to zero and solve for x_1:

   $$3x_1 + 3(0) = 39$$

   Solving, $x_1 = 13$. Thus, $x_1 = 13$ when $x_2 = 0$.
4. Add the line to the graph (see Figure 5S–6).

Identifying the Feasible Solution Space

The feasible solution space is the set of all points that satisfies *all* constraints. (Recall that the x_1 and x_2 axes form nonnegativity constraints.) The heavily shaded area shown in Figure 5S–6 is the feasible solution space for our problem.

The next step is to determine which point in the feasible solution space will produce the optimal value of the objective function. This determination is made using the objective function.

Plotting the Objective Function Line

Plotting an objective function line involves the same logic as plotting a constraint line: Determine where the line intersects each axis. Recall that the objective function for the microcomputer problem is

$$60x_1 + 50x_2$$

This is not an equation because it does not include an equal sign. We can get around this by simply setting it equal to some quantity. Any quantity will do, although one that is evenly divisible by both coefficients is desirable.

Suppose we decide to set the objective function equal to 300. That is,

$$60x_1 + 50x_2 = 300$$

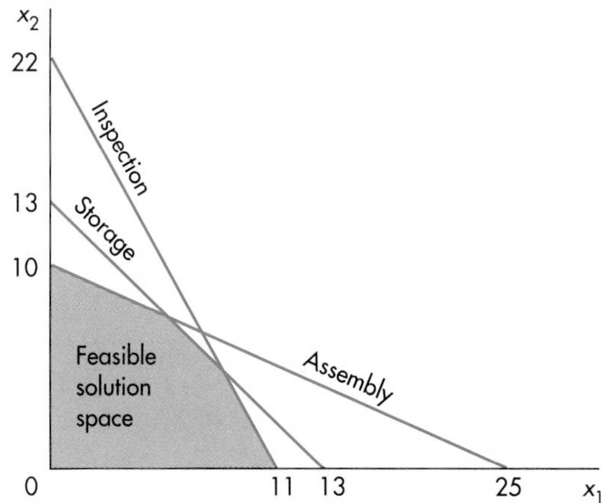

FIGURE 5S–6

Completed graph of the microcomputer problem showing all constraints and the feasible solution space

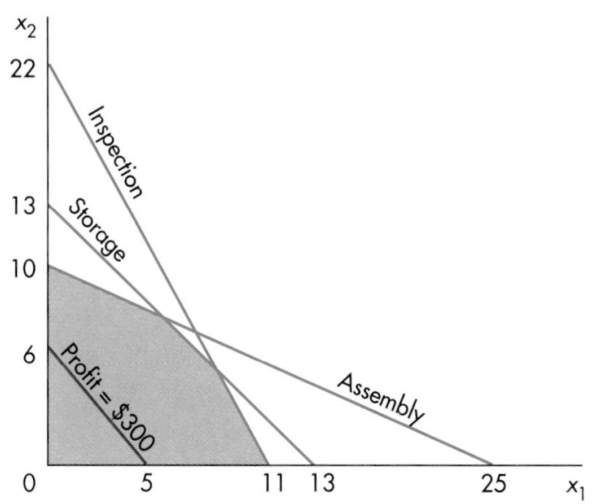

FIGURE 5S–7

Microcomputer problem with $300 profit line added

We can now plot the line of our graph. As before, we can determine the x_1 and x_2 intercepts of the line by setting one of the two variables equal to zero, solving for the other, and then reversing the process. Thus, with $x_1 = 0$, we have

$$60(0) + 50x_2 = 300$$

Solving, we find $x_2 = 6$. Similarly, with $x_2 = 0$, we have

$$60x_1 + 50(0) = 300$$

Solving, we find $x_1 = 5$. This line is plotted in Figure 5S–7.

The profit line can be interpreted in the following way. It is an *isoprofit* line; every point on the line (i.e., every combination of x_1 and x_2 that lies on the line) will provide a profit of $300. We can see from the graph many combinations that are both on the $300 profit line and within the feasible solution space. In fact, considering noninteger as well as integer solutions, the possibilities are infinite.

Suppose we now consider another line, say the $600 line. To do this, we set the objective function equal to this amount. Thus,

$$60x_1 + 50x_2 = 600$$

Solving for the x_1 and x_2 intercepts yields these two points:

x_1 intercept	x_2 intercept
$x_1 = 10$	$x_1 = 0$
$x_2 = 0$	$x_2 = 12$

FIGURE 5S–8

Microcomputer problem with profit lines of $300 and $600

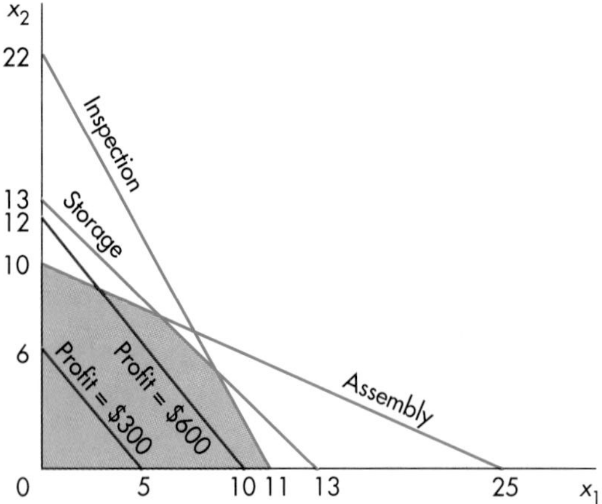

FIGURE 5S–9

Microcomputer problem with profit lines of $300, $600, and $900

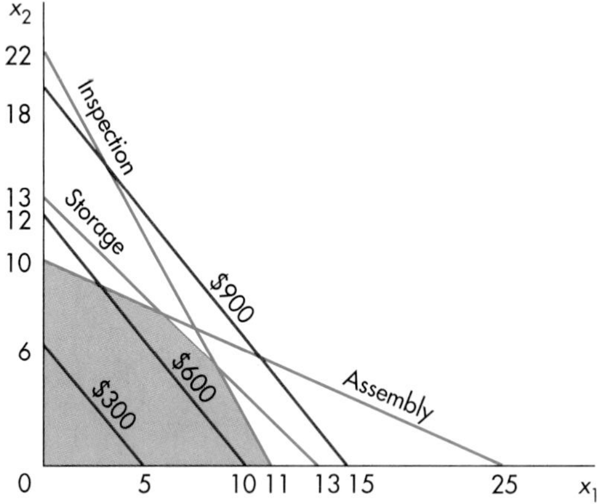

This line is plotted in Figure 5S–8, along with the previous $300 line for purposes of comparison.

Two things are evident in Figure 5S–8 regarding the profit lines. One is that the $600 line is *farther* from the origin than the $300 line; the other is that the two lines are *parallel.* The lines are parallel because they both have the same slope. The slope is not affected by the right side of the equation. Rather, it is determined solely by the coefficients 60 and 50. It would be correct to conclude that regardless of the quantity we select for the value of the objective function, the resulting line will be parallel to these two lines. Moreover, if the amount is greater than 600, the line will be even farther away from the origin than the $600 line. If the value is less than 300, the line will be closer to the origin than the $300 line. And if the value is between 300 and 600, the line will fall between the $300 and $600 lines.

Now consider a third line, one with the profit equal to $900. Figure 5S–9 shows that line along with the previous two profit lines. As expected, it is parallel to the other two, and even farther away from the origin. However, the line does not touch the feasible solution space at all. Consequently, there is no feasible combination of x_1 and x_2 that will yield that amount of profit. Evidently, the maximum possible profit is an amount between $600 and $900, which we can see by referring to Figure 5S–9. We could continue to select profit lines in this manner, and eventually, we could determine an amount that would yield the greatest profit. However, there is a much simpler alternative. We can plot just one line, say the $300 line. We know that all other lines will be parallel to it. Consequently, by moving this one line parallel to itself we can represent other profit lines. We also know that as we move away from the origin, the profits get larger. What

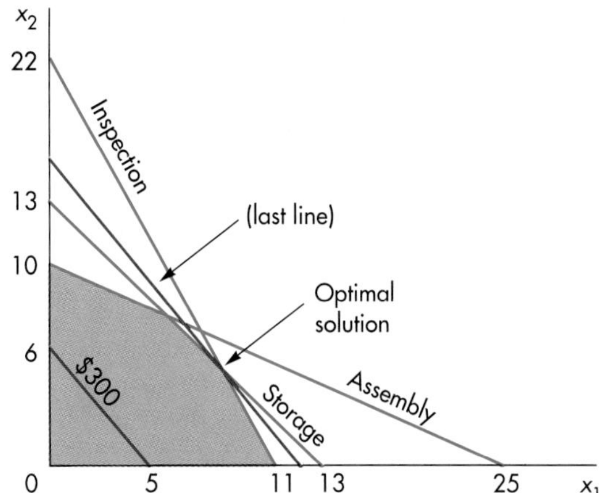

FIGURE 5S–10

Finding the optimal solution to the microcomputer problem

we want to know is how far the line can be moved out from the origin and still be touching the feasible solution space, and the values of the decision variables at that point of greatest profit (i.e., the optimal solution). Locate this point on the graph by placing a straight edge along the \$300 line (or any other convenient line) and sliding it away from the origin, being careful to keep it parallel to the line. This approach is illustrated in Figure 5S–10.

Once we have determined where the optimal solution is in the feasible solution space, we must determine the values of the decision variables at that point. Then, we can use that information to compute the profit for that combination.

Note that the optimal solution is at the intersection of the inspection boundary and the storage boundary (see Figure 5S–9). In other words, the optimal combination of x_1 and x_2 must satisfy both boundary (equality) conditions. We can determine those values by solving the two equations *simultaneously*. The equations are

Inspection $2x_1 + 1x_2 = 22$

Storage $3x_1 + 3x_2 = 39$

The idea behind solving two *simultaneous equations* is to algebraically eliminate one of the unknown variables (i.e., to obtain an equation with a single unknown). This can be accomplished by multiplying the constants of one of the equations by a fixed amount and then adding (or subtracting) the modified equation from the other. (Occasionally, it is easier to multiply each equation by a fixed quantity.) For example, we can eliminate x_2 by multiplying the inspection equation by 3 and then subtracting the storage equation from the modified inspection equation. Thus,

$$3(2x_1 + 1x_2 = 22) \quad \text{becomes} \quad 6x_1 + 3x_2 = 66$$

Subtracting the storage equation from this produces

$$\begin{array}{r} 6x_1 + 3x_2 = 66 \\ -\ (3x_1 + 3x_2 = 39) \\ \hline 3x_1 + 0x_2 = 27 \end{array}$$

Solving the resulting equation yields $x_1 = 9$. The value of x_2 can be found by substituting $x_1 = 9$ into either of the original equations or the modified inspection equation. Suppose we use the original inspection equation. We have

$$2(9) + 1x_2 = 22$$

Solving, we find $x_2 = 4$.

Hence, the optimal solution to the microcomputer problem is to produce nine type 1 computers and four type 2 computers per day. We can substitute these values into the objective function to find the optimal profit:

$$\$60(9) + \$50(4) = \$740$$

Hence, the last line—the one that would last touch the feasible solution space as we moved away from the origin parallel to the $300 profit line—would be the line where profit equaled $740.

In this problem, the optimal values for both decision variables are integers. This will not always be the case; one or both of the decision variables may turn out to be noninteger. In some situations noninteger values would be of little consequence. This would be true if the decision variables were measured on a continuous scale, such as the amount of water, sand, sugar, fuel oil, time, or distance needed for optimality, or if the contribution per unit (profit, cost, etc.) were small, as with the number of nails or ball bearings to make. In some cases, the answer would simply be rounded down (maximization problems) or up (minimization problems) with very little impact on the objective function. Here, we assume that noninteger answers are acceptable as such.

Let's review the procedure for finding the optimal solution using the objective function approach:

1. Graph the constraints.
2. Identify the feasible solution space.
3. Set the objective function equal to some amount that is divisible by each of the objective function coefficients. This will yield integer values for the x_1 and x_2 intercepts and simplify plotting the line. Often, the product of the two objective function coefficients provides a satisfactory line. Ideally, the line will cross the feasible solution space close to the optimal point, and it will not be necessary to slide a straight edge: The optimal solution is readily identified visually.
4. After identifying the optimal point, determine which two constraints intersect there. Solve their equations simultaneously to obtain the values of the decision variables at the optimum.
5. Substitute the values obtained in the previous step into the objective function to determine the value of the objective function at the optimum.

Redundant Constraints

redundant constraint A constraint that does not form a unique boundary of the feasible solution space.

In some cases, a constraint does not form a unique boundary of the feasible solution space. Such a constraint is called a **redundant constraint.** This constraint is illustrated in Figure 5S–11. Note that a constraint is redundant if it meets the following test: Its removal would not alter the feasible solution space.

When a problem has a redundant constraint, at least one of the other constraints in the problem is more restrictive than the redundant constraint.

Solutions and Corner Points

The feasible solution space in graphical linear programming is a polygon. Moreover, the solution to any problem will be at one of the corner points (intersections of constraints) of the polygon. It is possible to determine the coordinates of each corner point of the feasible solution space, and use those values to compute the value of the objective function at those points. Because the solution is always at a corner point, comparing the values of the objective function at the corner points and identifying the best one (e.g., the maximum value) is another way to identify the optimal corner point. Using the graphical approach, it is much easier to plot the objective function and use that to identify the optimal corner point. However, for problems that have more than two decision variables, and the graphical method isn't appropriate, this alternate approach is used to find the optimal solution.

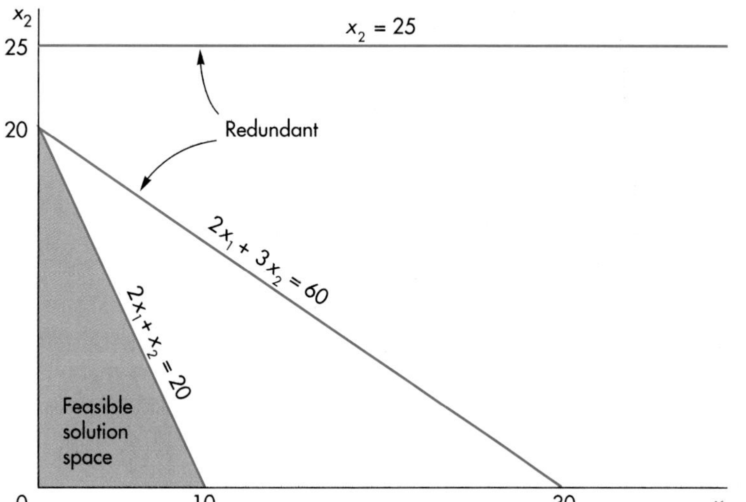

FIGURE 5S-11

Examples of redundant constraints

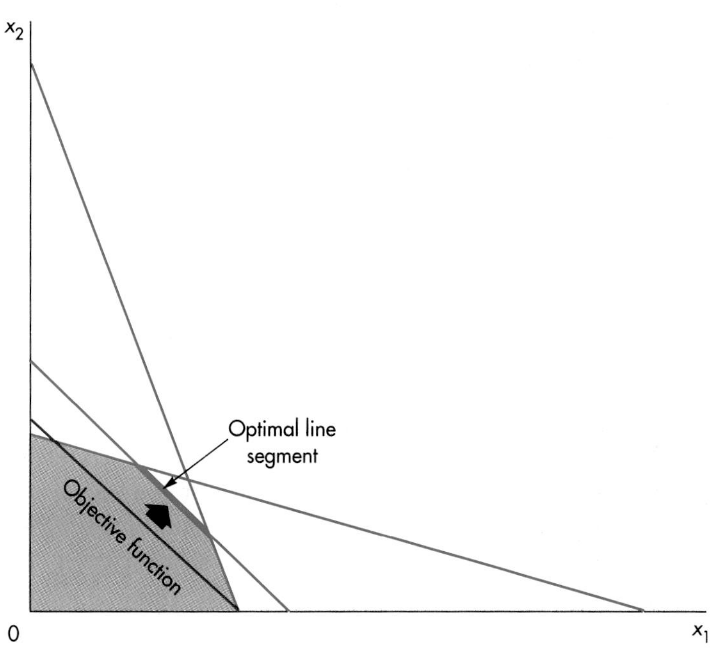

FIGURE 5S-12

Some LP problems have multiple optimal solutions

In some instances, the objective function will be parallel to one of the constraint lines *that forms a boundary of the feasible solution space.* When this happens, *every* combination of x_1 and x_2 on the segment of the constraint that touches the feasible solution space represents an optimal solution. Hence, there are multiple optimal solutions to the problem. Even in such a case, the solution will also be a corner point—in fact, the solution will be at *two* corner points: those at the ends of the segment that touches the feasible solution space. Figure 5S–12 illustrates an objective function line that is parallel to a constraint line.

Minimization

Graphical minimization problems are quite similar to maximization problems. There are, however, two important differences. One is that at least one of the constraints must be of the = or ≥ variety. This causes the feasible solution space to be away from the origin. The other difference is that the optimal point is the one closest to the origin. We

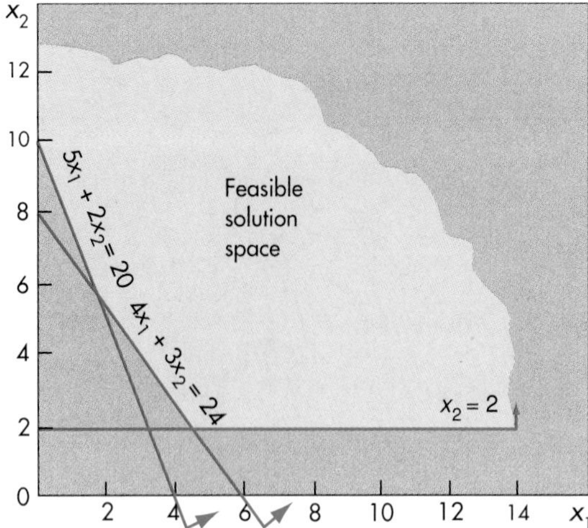

FIGURE 5S–13

The constraints define the feasible solution space

find the optimal corner point by sliding the objective function (which is an *isocost* line) *toward* the origin instead of away from it.

Example 3

Solve the following problem using graphical linear programming.

$$\begin{aligned} \text{Minimize} \quad & Z = 8x_1 + 12x_2 \\ \text{Subject to} \quad & 5x_1 + 2x_2 \geq 20 \\ & 4x_1 + 3x_2 \geq 24 \\ & x_2 \geq 2 \\ & x_1, x_2 \geq 0 \end{aligned}$$

Solution

1. Plot the constraints (shown in Figure 5S–13).
   a. Change constraints to equalities.
   b. For each constraint, set $x_1 = 0$ and solve for x_2, then set $x_2 = 0$ and solve for x_1.
   c. Graph each constraint. Note that $x_2 = 2$ is a horizontal line parallel to the x_1 axis and 2 units above it.
2. Shade the feasible solution space (see Figure 5S–13).
3. Plot the objective function.
   a. Select a value for the objective function that causes it to cross the feasible solution space. Try $8 \times 12 = 96$; $8x_1 + 12x_2 = 96$ (acceptable).
   b. Graph the line (see Figure 5S–14).
4. Slide the objective function toward the origin, being careful to keep it parallel to the original line.
5. The optimum (last feasible point) is shown in Figure 5S–14. The x_2 coordinate ($x_2 = 2$) can be determined by inspection of the graph. Note that the optimum point is at the intersection of the line $x_2 = 2$ and the line $4x_1 + 3x_2 = 24$. Substituting the value of x_2 into the latter equation will yield the value of x_1 at the intersection:

   $$4x_1 + 3(2) = 24 \qquad x_1 = 4.5$$

   Thus, the optimum is $x_1 = 4.5$ units and $x_2 = 2$.
6. Compute the minimum cost:

   $$8x_1 + 12x_2 = 8(4.5) + 12(2) = 60$$

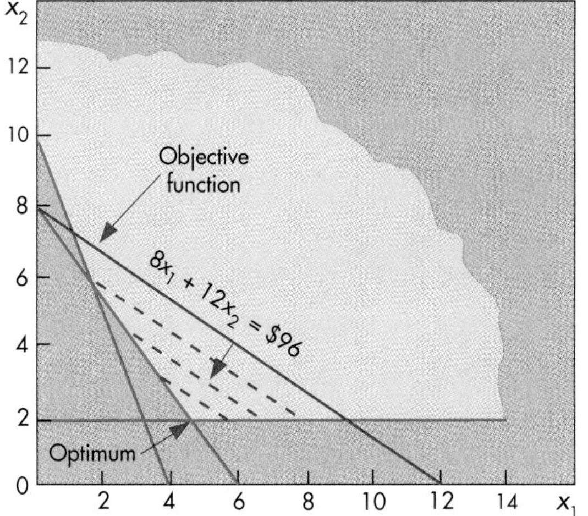

FIGURE 5S-14

The optimum is the last point the objective function touches as it is moved toward the origin

Slack and Surplus

If a constraint forms the optimal corner point of the feasible solution space, it is called a **binding constraint.** In effect, it limits the value of the objective function; if the constraint could be relaxed (less restrictive), an improved solution would be possible. For constraints that are not binding, making them less restrictive will have no impact on the solution.

binding constraint A constraint that forms the optimal corner point of the feasible solution space.

If the optimal values of the decision variables are substituted into the left side of a binding constraint, the resulting value will exactly equal the right-hand value of the constraint. However, there will be a difference with a nonbinding constraint. If the left side is greater than the right side, we say that there is **surplus**; if the left side is less than the right side, we say that there is **slack.** Slack can only occur in a $\leq$ constraint; it is the amount by which the left side is less than the right side when the optimal values of the decision variables are substituted into the left side. And surplus can only occur in a $\geq$ constraint; it is the amount by which the left side exceeds the right side of the constraint when the optimal values of the decision variables are substituted into the left side.

surplus When the values of decision variables are substituted into a $\geq$ constraint and the resulting value exceeds the right side value.

slack When the values of decision variables are substituted into a $\leq$ constraint, and the resulting value is less than the right side value.

For example, suppose the optimal values for a problem are $x_1 = 10$ and $x_2 = 20$. If one of the constraints is

$$3x_1 + 2x_2 \leq 100$$

substituting the optimal values into the left side yields

$$3(10) + 2(20) = 70$$

Because the constraint is $\leq$, the difference between the values of 100 and 70 (i.e., 30) is slack. Suppose the optimal values had been $x_1 = 20$ and $x_2 = 20$. Substituting these values into the left side of the constraint would yield $3(20) + 2(20) = 100$. Because the left side equals the right side, this is a binding constraint; slack is equal to zero.

Now consider this constraint:

$$4x_1 + x_2 \geq 50$$

Suppose the optimal values are $x_1 = 10$ and $x_2 = 15$; substituting into the left side yields

$$4(10) + 15 = 55$$

Because this is a $\geq$ constraint, the difference between the left and right side values is *surplus*. If the optimal values had been $x_1 = 12$ and $x_2 = 2$, substitution would result in the left side being equal to 50. Hence, the constraint would be a binding constraint, and there would be no surplus (i.e., surplus would be zero).

THE SIMPLEX METHOD

simplex A linear-programming algorithm that can solve problems having more than two decision variables.

The **simplex** method is a general-purpose linear-programming algorithm widely used to solve large-scale problems. Although it lacks the intuitive appeal of the graphical approach, its ability to handle problems with more than two decision variables makes it extremely valuable for solving problems often encountered in operations management.

Although manual solution of linear programming problems using simplex can yield a number of insights on how solutions are derived, in practice computers are used to solve these problems. For that reason, computer solutions will be illustrated here.

COMPUTER SOLUTIONS

The microcomputer problem will be used to illustrate computer solutions. We repeat it here for ease of reference.

Maximize $60x_1 + 50x_2$ where x_1 = the number of type 1 computers
x_2 = the number of type 2 computers

Subject to
Assembly $4x_1 + 10x_2 \leq 100$ hours
Inspection $2x_1 + 1x_2 \leq 22$ hours
Storage $3x_1 + 3x_2 \leq 39$ cubic feet
x_1 and $x_2 \geq 0$

Solving LP Models Using MS Excel

Solutions to linear programming models can be obtained from spreadsheet software such as Microsoft's Excel. Excel has a routine called *Solver* that performs the necessary calculations.

To use Solver:

1. First, enter the problem in a worksheet, as shown in Figure 5S–15. What is not obvious from the figure is the need to enter a formula for each spot where there is a zero (Solver automatically inserts the zero after you input the formula.) The formulas are for the value of the objective function and the constraints, in the appropriate cells. Before you enter the formulas, designate the cells where you want the optimal values of x_1 and x_2. Here, cells D4 and E4 are used. We want the optimal value of the objective function to appear in cell G4. For that cell, we enter the formula

```
=60*D4+50*E4
```

The constraint formulas, using cells C7, C8, and C9, are

for C7: `=4*D4+10*E4`
for C8: `=2*D4+1*E4`
for C9: `=3*D4+3*E4`

2. Now, click on Tools on the top of the worksheet, and in that menu, click on Solver. The Solver menu will appear as illustrated in Figure 5S–16. Begin by setting the Target Cell (i.e., indicating the cell where you want the optimal value of the objective function to appear). Note, if the activated cell is the cell designated for the value of Z when you click on the Tools menu, Solver will automatically set that cell as the target cell.

Highlight Max if it isn't already highlighted. The Changing Cells are the cells where you want the optimal values of the decision variables to appear. Here, they are cells D4 and E4. We indicate this by the range D4:E4 (Solver will add the $ signs).

Finally, add the constraints by clicking on Add . . . When that menu appears, for each constraint, enter the cell that contains the formula for the left side of the constraint, then select the appropriate inequality sign, and then enter either the right side

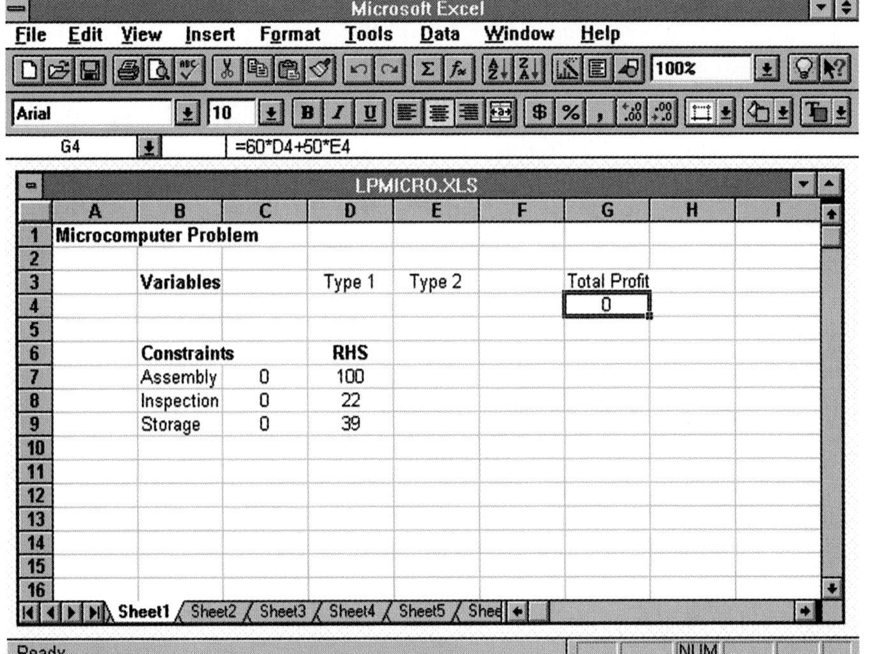

FIGURE 5S-15

MS Excel worksheet for microcomputer problem

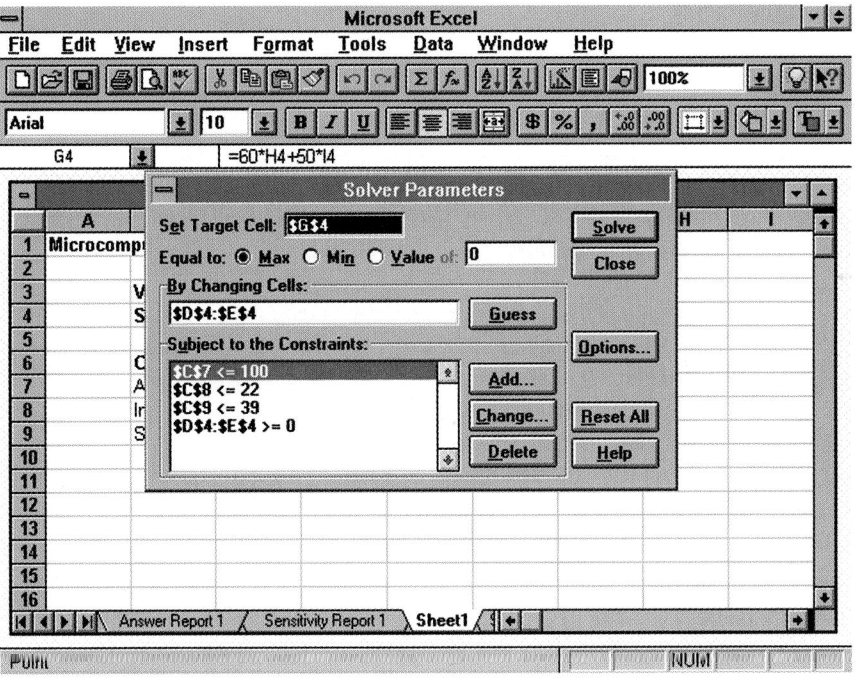

FIGURE 5S-16

MS Excel Solver parameters for microcomputer problem

amount, or the cell that has the right side amount. Here, the right side amounts are used. After you have entered each constraint, click on Add, and then enter the next constraint. (Note, constraints can be entered in any order.) For the nonnegativity constraints, enter the range of cells designated for the optimal values of the decision variables. After entered, click on OK rather than Add, and you will return to the Solver menu. Click on Options . . . , and in the Options menu, click on Assume Linear Model, and then click on OK. This will return you to the Solver Parameters menu. Click on Solve.

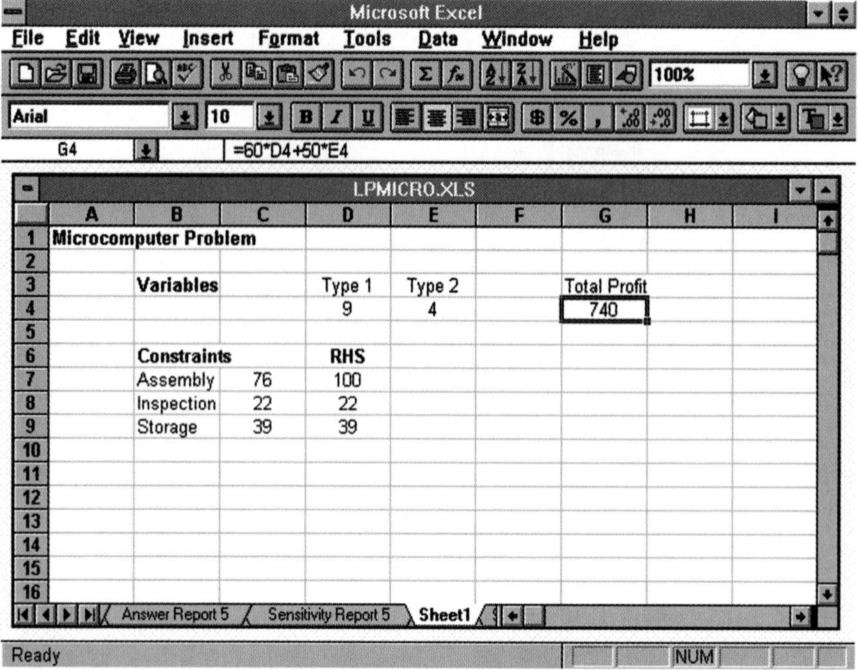

FIGURE 5S–17

MS Excel worksheet solution to microcomputer problem

3. The Solver Results menu will then appear, indicating that a solution has been found, or that an error has occurred. If there has been an error, go back to the Solver Parameters menu and check to see that your constraints refer to the correct changing cells, and that the inequality directions are correct. Make the corrections and click on Solve.

Assuming everything is correct, in the Solver Results menu, in the Reports box, highlight both Answer and Sensitivity, and then click on OK.

4. Solver will incorporate the optimal values of the decision variables and the objective function in your original layout on your worksheet (see Figure 5S–17). We can see that the optimal values are type 1 = 9 units and type 2 = 4 units, and the total profit is 740. The answer report will also show the optimal values of the decision variables (upper part of Figure 5S–18), and some information on the constraints (lower part of Figure 5S–18). Of particular interest here is the indication of which constraints have slack and how much slack. We can see that the constraint entered in cell C7 (assembly) has a slack of 24, and that the constraints entered in cells C8 (inspection) and C9 (storage) have slack equal to zero, indicating that they are binding constraints.

SENSITIVITY ANALYSIS

sensitivity analysis Assessing the impact of potential changes to the numerical values of an LP model.

Sensitivity analysis is a means of assessing the impact of potential changes to the parameters (the numerical values) of an LP model. Such changes may occur due to forces beyond a manager's control; or a manager may be contemplating making the changes, say, to increase profits or reduce costs.

There are three types of potential changes:

1. Objective function coefficients.
2. Right-hand values of constraints.
3. Constraint coefficients.

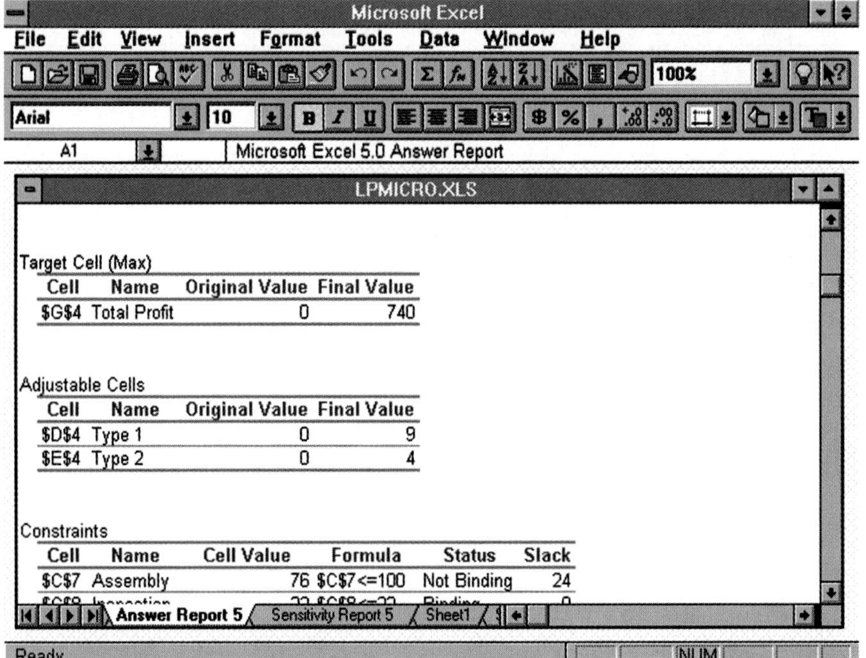

Microsoft Excel 5.0 Answer Report

Target Cell (Max)

Cell	Name	Original Value	Final Value
$G$4	Total Profit	0	740

Adjustable Cells

Cell	Name	Original Value	Final Value
$D$4	Type 1	0	9
$E$4	Type 2	0	4

Constraints

Cell	Name	Cell Value	Formula	Status	Slack
$C$7	Assembly	76	$C$7<=100	Not Binding	24

FIGURE 5S-18

MS Excel Answer Report for microcomputer problem

We will consider the first two of these here. We begin with changes to objective function coefficients.

Objective Function Coefficient Changes

A change in the value of an objective function coefficient can cause a change in the optimal solution of a problem. In a graphical solution, this would mean a change to another corner point of the feasible solution space. However, not every change in the value of an objective function coefficient will lead to a changed solution; generally there is a *range of values for which the optimal values of the decision variables will not change.* For example, in the microcomputer problem, if the profit on type 1 computers increased from $60 per unit to, say, $65 per unit, the optimal solution would still be to produce nine units of type 1 and four units of type 2 computers. Similarly, if the profit per unit on type 1 computers decreased from $60 to, say, $58, producing nine of type 1 and four of type 2 would still be optimal. These sorts of changes are not uncommon; they may be the result of such things as price changes in raw materials, price discounts, cost reductions in production, and so on. Obviously, when a change does occur in the value of an objective function coefficient, it can be helpful for a manager to know if that change will affect the optimal values of the decision variables. The manager can quickly determine this by referring to that coefficient's **range of optimality,** which is the range in possible values of that objective function coefficient over which the optimal values of the decision variables will not change. Before we see how to determine the range, consider the implication of the range. The range of optimality for the type 1 coefficient in the microcomputer problem is 50 to 100. That means that as long as the coefficient's value is in that range, the optimal values will be 9 units of type 1 and 4 units of type 2. Conversely, *if a change extends beyond the range of optimality, the solution will change.*

range of optimality Range of values over which the solution quantities of all the decision variables remains the same.

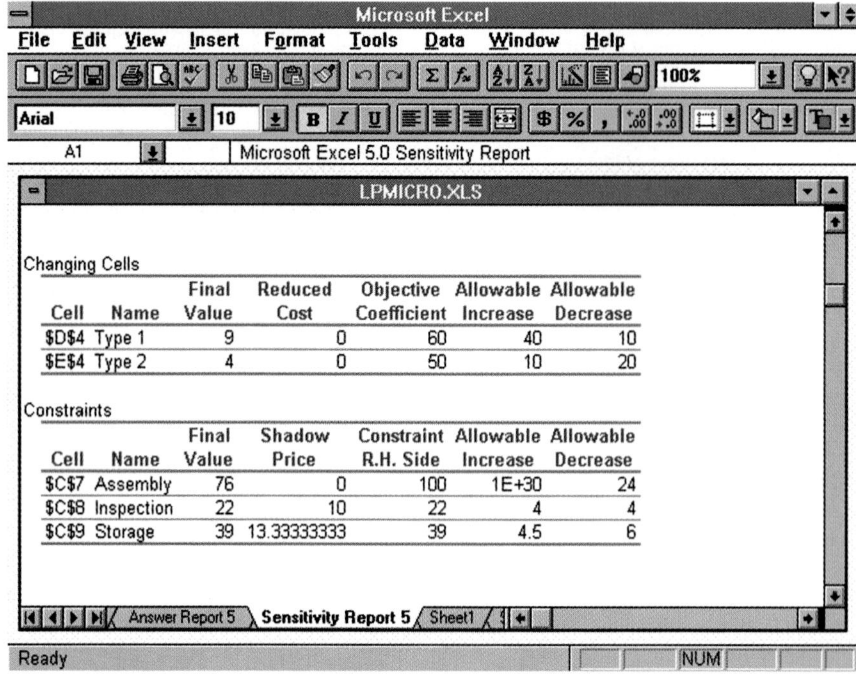

Changing Cells

Cell	Name	Final Value	Reduced Cost	Objective Coefficient	Allowable Increase	Allowable Decrease
$D$4	Type 1	9	0	60	40	10
$E$4	Type 2	4	0	50	10	20

Constraints

Cell	Name	Final Value	Shadow Price	Constraint R.H. Side	Allowable Increase	Allowable Decrease
$C$7	Assembly	76	0	100	1E+30	24
$C$8	Inspection	22	10	22	4	4
$C$9	Storage	39	13.33333333	39	4.5	6

FIGURE 5S–19

MS Excel sensitivity report for microcomputer problem

Similarly, suppose, instead, the coefficient of type 2 computers was to change. Its range of optimality is 30 to 60. As long as the value of the change doesn't take it outside of this range, nine and four will still be the optimal values. Note, however, even for changes that are *within* the range of optimality, the optimal value of the objective function *will* change. If the type 1 coefficient increased from $60 to $61, and nine units of type 1 is still optimum; profit would increase by $9: nine units times $1 per unit. Thus, for a change that is within the range of optimality, a revised value of the objective function must be determined.

Now let's see how we can determine the range of optimality using computer output.

Using MS Excel. There is a table for the Changing Cells (see Figure 5S–19). It shows the value of the objective function that was used in the problem for each type of computer (i.e., 60 and 50), and the allowable increase and allowable decrease for each coefficient. By subtracting the allowable decrease from the original value of the coefficient, and adding the allowable increase to the original value of the coefficient, we obtain the range of optimality for each coefficient. Thus, we find for type 1:

$$60 - 10 = 50 \quad \text{and} \quad 60 + 40 = 100$$

Hence, the range for the type 1 coefficient is 50 to 100. For type 2:

$$50 - 20 = 30 \quad \text{and} \quad 50 + 10 = 60$$

Hence the range for the type 2 coefficient is 30 to 60.

Remember that in this example, both of the decision variables are basic (i.e., nonzero). However, in other problems, one or more decision variables may be nonbasic (i.e., have an optimal value of zero). In such instances, unless the value of that variable's objective function coefficient increases by more than its reduced cost, it won't come into solution (i.e., become a basic variable). Hence, the range of optimality (sometimes referred to as the *range of insignificance*) for a nonbasic variable is from negative infinity to the sum of its current value and its reduced cost.

Now let's see how we can handle multiple changes to objective functions coefficients, that is, a change in more than one coefficient. To do this, divide each coefficient's change by the allowable change in the same direction. Thus, if the change is a decrease, divide that amount by the allowable decrease. Treat all resulting fractions as positive. Sum the fractions. If the sum does not exceed 1.00, then multiple changes are within the range of optimality and will not result in any change to the optimal values of the decision variables.

Changes in the Right-Hand Side Value of a Constraint

In considering right-hand side changes, it is important to know if a particular constraint is binding on a solution. A constraint is binding if substituting the values of the decision variables of that solution into the left side of the constraint results in a value that is equal to the RHS value. In other words, that constraint stops the objective function from achieving a better value (e.g., a greater profit or a lower cost). Each constraint has a corresponding **shadow price,** which is a marginal value that indicates the amount by which the value of the objective function would change if there were a one-unit change in the RHS value of that constraint. If a constraint is nonbinding, its shadow price is zero, meaning that increasing or decreasing its RHS value by one unit will have no impact on the value of the objective function. Nonbinding constraints have either slack (if the constraint is $\leq$) or surplus (if the constraint is $\geq$). Suppose a constraint has 10 units of slack in the optimal solution, which means 10 units that are unused. If we were to increase or decrease the constraint's RHS value by one unit, the only effect would be to increase or decrease its slack by one unit. But there is no profit associated with slack, so the value of the objective function wouldn't change. On the other hand, if the change is to the RHS value of a binding constraint, then the optimal value of the objective function would change. Any change in a binding constraint will cause the optimal values of the decision variables to change, and hence, cause the value of the objective function to change. For example, in the microcomputer problem, the inspection constraint is a binding constraint, it has a shadow price of 10. That means if there was one hour less of inspection time, total profit would decrease by $10, or if there was one more hour of inspection time available, total profit would increase by $10. In general, multiplying the amount of change in the RHS value of a constraint by the constraint's shadow price will indicate the change's impact on the optimal value of the objective function. However, this is only true over a limited range called the **range of feasibility.** In this range, the value of the shadow price remains constant. Hence, as long a change in the RHS value of a constraint is within its range of feasibility, the shadow price will remain the same, and one can readily determine the impact on the objective function.

shadow price Amount by which the value of the objective function would change with a one-unit change in value of a constraint.

range of feasibility Range of values for the RHS of a constraint over which the shadow price remains the same.

Let's see how to determine the range of feasibility from computer output.

Using MS Excel. In the sensitivity report there is a table labeled "Constraints" (see Figure 5S–19). The table shows the shadow price for each constraint, its RHS value, and the allowable increase and allowable decrease. Adding the allowable increase to the RHS value, and subtracting the allowable decrease will produce the range of feasibility for that constraint. For example, for the inspection constraint, the range would be

$$22 + 4 = 26; 22 - 4 = 18$$

Hence, the range of feasibility for Inspection is 18 to 26 hours. Similarly, for the storage constraint, the range is

$$39 - 6 = 33 \text{ to } 39 + 4.5 = 43.5$$

Table 5S–1

Summary of the impact of changes within their respective ranges

CHANGES TO OBJECTIVE FUNCTION COEFFICIENTS THAT ARE WITHIN THE RANGE OF OPTIMALITY

Component	Result
Values of decision variables	No change
Value of objective function	Will change

CHANGES TO RHS VALUES OF CONSTRAINTS THAT ARE WITHIN THE RANGE OF FEASIBILITY

Component	Result
Value of shadow price	No change
List of Basic variables	No change
Values of Basic variables	Will change
Value of objective function	Will change

The range for the assembly constraint is a little different; the assembly constraint is nonbinding (note the shadow price of 0) while the other two are binding (note their nonzero shadow prices). The assembly constraint has a slack of 24 (the difference between its RHS value of 100 and its final value of 76). With its slack of 24, its RHS value could be decreased by as much as 24 (to 76) before it would become binding. Conversely, increasing its right-hand side will only produce more slack. Thus, no amount of increase in the RHS value will make it binding, so there is no upper limit on the allowable increase. Excel indicates this by the large value (1E+30) shown for the allowable increase. So its range of feasibility has a lower limit of 76 and no upper limit.

If there are changes to more than one constraint's RHS value, analyze these in the same way as multiple changes to objective function coefficients. That is, if the change is an increase, divide that amount by that constraint's allowable increase; if the change is a decrease, divide the decrease by the allowable decrease. Treat all resulting fractions as positives. Sum the fractions. As long as the sum does not exceed 1.00, the changes are within the range of feasibility for multiple changes, and the shadow prices won't change.

Table 5S–1 summarizes the impacts of changes that fall within either the range of optimality or the range of feasibility.

Now let's consider what happens if a change goes beyond a particular range. In a situation involving the range of optimality, a change in an objective function that is beyond the range of optimality will result in a new solution. Hence, it will be necessary to recompute the solution. For a situation involving the range of feasibility, there are two cases to consider. The first case would be increasing the RHS value of a $\leq$ constraint to beyond the upper limit of its range of feasibility. This would produce slack equal to the amount by which the upper limit is exceeded. Hence, if the upper limit is 200, and the increase is 220, the result is that the constraint has a slack of 20. Similarly, for a $\geq$ constraint, going below its lower bound creates a surplus for that constraint. The second case for each of these would be exceeding the opposite limit (the lower bound for a $\leq$ constraint, or the upper bound for a $\geq$ constraint). In either instance, a new solution would have to be generated.

Key Terms

binding constraint, 249
constraints, 235
decision variables, 234
feasible solution space, 235
graphical linear programming, 237
objective function, 234
parameter, 235
range of feasibility, 255
range of optimality, 254
redundant constraint, 246
sensitivity analysis, 253
shadow price, 255
simplex, 250
slack, 249
surplus, 249

Solved Problems

Problem 1

A small construction firm specializes in building and selling single-family homes. The firm offers two basic types of houses, model A and model B. Model A houses require 4,000 labor hours, 2 tons of stone, and 2,000 board feet of lumber. Model B houses require 10,000 labor hours, 3 tons of stone, and 2,000 board feet of lumber. Due to long lead times for ordering supplies and the scarcity of skilled and semiskilled workers in the area, the firm will be forced to rely on its present resources for the upcoming building season. It has 400,000 hours of labor, 150 tons of stone, and 200,000 board feet of lumber. What mix of model A and B houses should the firm construct if model As yield a profit of \$1,000 per unit and model Bs yield \$2,000 per unit? Assume that the firm will be able to sell all the units it builds.

Solution

a. Formulate the objective function and constraints:[1]

Maximize $Z = 1{,}000A + 2{,}000B$

Subject to

Labor	$4{,}000A + 10{,}000B \leq 400{,}000$ labor hours
Stone	$2A + 3B \leq 150$ tons
Lumber	$2{,}000A + 2{,}000B \leq 200{,}000$ board feet
	$A, B \geq 0$

b. Graph the constraints and objective function, and identify the optimum corner point (see graph). Note that the lumber constraint is *redundant:* it does not form a boundary of the feasible solution space.

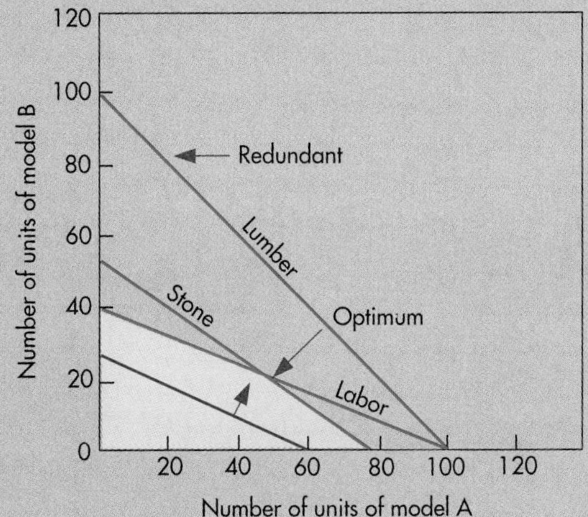

c. Determine the optimal quantities of models A and B, and compute the resulting profit. Because the optimum point is at the intersection of the stone and labor constraints, solve those two equations for their common point:

$$\begin{array}{rl} \text{Labor } 4{,}000A + 10{,}000B &= 400{,}000 \\ -2{,}000 \times (\text{Stone } \quad 2A + 3B &= 150) \\ \hline 4{,}000B &= 100{,}000 \\ B &= 25 \end{array}$$

Substitute $B = 25$ in one of the equations, and solve for A:

$$2A + 3(25) = 150 \quad A = 37.5$$
$$Z = 1{,}000(37.5) + 2{,}000(25) = 87{,}500$$

Problem 2

This LP model was solved by computer:

[1]For the sake of consistency, we will assign to the horizontal axis the first decision variable mentioned in the problem. In this case, variable A will be represented on the horizontal axis and variable B on the vertical axis.

Maximize $15x_1 + 20x_2 + 14x_3$ where x_1 = quantity of product 1
x_2 = quantity of product 2
x_3 = quantity of product 3

Subject to

Labor $5x_1 + 6x_2 + 4x_3 \leq 210$ hours

Material $10x_1 + 8x_2 + 5x_3 \leq 200$ pounds

Machine $4x_1 + 2x_2 + 5x_3 \leq 170$ minutes

x_1, x_2 and $x_3 \geq 0$

The following information was obtained from the output. The ranges were also computed based on the output, and they are shown as well.

```
Total profit = 548.00

Variable     Value    Reduced Cost    Range of Optimality

Product 1      0          10.6          0.00 to 25.60
Product 2      5           0            9.40 to 22.40
Product 3     32           0           12.50 to 50.00

Constraint    Slack   Shadow Price    Range of Feasibility

Labor          52          0.0        158.00 to unlimited
Material        0          2.4        170.00 to 270.91
Machine         0          0.4         50.00 to 200.00
```

a. Which decision variables are basic (i.e., in solution)?

b. By how much would the profit per unit of product 1 have to increase in order for it to have a nonzero value (i.e., for it to become a basic variable)?

c. If the profit per unit of product 2 increased by \$2 to \$22, would the optimal production quantities of products 2 and 3 change? Would the optimal value of the objective function change?

d. If the available amount of labor decreased by 12 hours, would that cause a change in the optimal values of the decision variables or the optimal value of the objective function? Would anything change?

e. If the available amount of material increased by 10 pounds to 210 pounds, how would that affect the optimal value of the objective function?

f. If profit per unit on product 2 increased by \$1 and profit per unit on product 3 decreased by \$.50, would that fall within the range of multiple changes? Would the values of the decision variables change? What would be the revised value of the objective function?

Solution

a. Products 2 and 3 are in solution (i.e., have nonzero values; the optimal value of product 2 is 5 units, and the optimal value of product 3 is 32 units.

b. The amount of increase would have to equal its reduced cost of \$10.60.

c. No, because the change would be within its range of optimality, which has an upper limit of \$22.40. The objective function value would increase by an amount equal to the quantity of product 2 and its increased unit profit. Hence, it would increase by 5(\$2) = \$10 to \$558.

d. Labor has a slack of 52 hours. Consequently, the only effect would be to decrease the slack to 40 hours.

e. Yes. The change is within the range of feasibility. The objective function value will increase by the amount of change multiplied by material's shadow price of \$2.40. Hence, the objective function value would increase by 10(\$2.40) = \$24.00. (Note: If the change had been a *decrease* of 10 pounds, which is also within the range of feasibility, the value of the objective function would have *decreased* by this amount.)

f. To determine if the changes are within the range for multiple changes, we first compute the ratio of the amount of each change to the end of the range *in the same direction*. For product 2, it is \$1/\$2.40 = .417; for Product 3, it is −\$.50/−\$1.50 = .333. Next, we compute the sum of these ratios: .417 +.333 = .750. Because this does not exceed 1.00, we conclude that these changes are within the range. This means that the optimal values of the decision variables will not change. We can compute the change to the value of the objective function by multiplying each product's optimal quantity by its changed profit per unit: 5(\$1) + 32(−\$.50) = −\$11. Hence, with these changes, the value of the objective function would decrease by \$11; its new value would be \$548 − \$11 = \$537.

Discussion and Review Questions

1. For which decision environment is linear programming most suited?
2. What is meant by the term *feasible solution space*? What determines this region?
3. Explain the term *redundant constraint*.
4. What is an isocost line? An isoprofit line?
5. What does sliding an objective function line toward the origin represent? Away from the origin?
6. Briefly explain these terms:
   a. Basic variable
   b. Shadow price
   c. Range of feasibility
   d. Range of optimality

Problems

1. Solve these problems using graphical linear programming and answer the questions that follow. Use simultaneous equations to determine the optimal values of the decision variables.

   a. Maximize $Z = 4x_1 + 3x_2$

   Subject to

Material	$6x_1 + 4x_2 \leq 48$ pounds
Labor	$4x_1 + 8x_2 \leq 80$ hours
	$x_1 + x_2 \geq 0$

   a. (1) $x_1 = 2$ $Z = 35$
   $x_2 = 9$
   (2) No
   (3) No
   (4) No

   b. Maximize $Z = 2x_1 + 10x_2$

   Subject to

R	$10x_1 + 4x_2 \geq 40$
S	$1x_1 + 6x_2 \geq 24$
T	$1x_1 + 2x_2 \leq 14$
	$x_1, x_2 \geq 0$

   b. (1) $x_1 = 1.5$ $Z = 65.5$
   $x_2 = 6.25$
   (2) No
   (3) Yes. S has surplus of 15.
   (4) No

   c. Maximize $Z = 6A + 3B$ (revenue)

   Subject to

Material	$20A + 6B \leq 600$ lb.
Machinery	$25A + 20B \leq 1{,}000$ hr.
Labor	$20A + 30B \leq 1{,}200$ hr.
	$A, B \geq 0$

   c. (1) $A = 24$
   $B = 20$
   $Z = \$204$
   (2) Yes. Labor is 420 hrs.
   (3) No
   (4) No

   (1) What are the optimal values of the decision variables and Z?
   (2) Do any constraints have (nonzero) slack? If yes, which one(s) and how much slack does each have?

(3) Do any constraints have (nonzero) surplus? If yes, which one(s) and how much surplus does each have?

(4) Are any constraints redundant? If yes, which one(s)? Explain briefly.

a. (1) $S = 8$ $T = 20$ $Z = \$58.40$
(2) No
(3) Yes. Protein = 92, $T = 10$
(4) Yes. Protein.
b. (1) $x_1 = 4.2$ $Z = 13.2$ $x_2 = 1.6$
(2) $s_2 = 4.6$
(3) No.
(4) No.

2. Solve these problems using graphical linear programming and then answer the questions that follow. Use simultaneous equations to determine the optimal values of the decision variables.

a. Minimize $Z = 1.80S + 2.20T$

Subject to

Potassium	$5S + 8T \geq 200$ gr.
Carbohydrate	$15S + 6T \geq 240$ gr.
Protein	$4S + 12T \geq 180$ gr.
T	$T \geq 10$ gr.
	$S, T \geq 0$

b. Minimize $Z = 2x_1 + 3x_2$

Subject to

D	$4x_1 + 2x_2 \geq 20$
E	$2x_1 + 6x_2 \geq 18$
F	$1x_1 + 2x_2 \leq 12$
	$x_1, x_2 \geq 0$

(1) What are the optimal values of the decision variables and Z?

(2) Do any constraints have (nonzero) slack? If yes, which one(s) and how much slack does each have?

(3) Do any constraints have (nonzero) surplus? If yes, which one(s) and how much surplus does each have?

(4) Are any constraints redundant? If yes, which one(s)? Explain briefly.

H = 132, W = 36
Z = \$6,360

3. An appliance manufacturer produces two models of microwave ovens: H and W. Both models require fabrication and assembly work; each H uses four hours of fabrication and two hours of assembly, and each W uses two hours of fabrication and six hours of assembly. There are 600 fabrication hours available this week and 480 hours of assembly. Each H contributes \$40 to profits, and each W contributes \$30 to profits. What quantities of H and W will maximize profits?

a. Deluxe = 90 bags
Standard = 60 bags
b. Profit = \$243

4. A small candy shop is preparing for the holiday season. The owner must decide how many bags of deluxe mix and how many bags of standard mix of Peanut/Raisin Delite to put up. The deluxe mix has ⅔ pound raisins and ⅓ pound peanuts, and the standard mix has ½ pound raisins and ½ pound peanuts per bag. The shop has 90 pounds of raisins and 60 pounds of peanuts to work with.

Peanuts cost \$.60 per pound and raisins cost \$1.50 per pound. The deluxe mix will sell for \$2.90 per pound, and the standard mix will sell for \$2.55 per pound. The owner estimates that no more than 110 bags of one type can be sold.

a. If the goal is to maximize profits, how many bags of each type should be prepared?

b. What is the expected profit?

a. 500 Apple
200 Grape
Revenue \$990
b. 50 cups sugar unused.
No flour or time unused.

5. A retired couple supplement their income by making fruit pies, which they sell to a local grocery store. During the month of September, they produce apple and grape pies. The apple pies are sold for \$1.50 to the grocer, and the grape pies are sold for \$1.20. The couple is able to sell all of the pies they produce owing to their high quality. They use fresh ingredients. Flour and sugar are purchased once each month. For the month of September, they have 1,200 cups of sugar and 2,100 cups of flour. Each apple pie requires 1½ cups of sugar and 3 cups of flour, and each grape pie requires 2 cups of sugar and 3 cups of flour.

a. Determine the number of grape and the number of apple pies that will maximize revenues if the couple working together can make an apple pie in six minutes and a grape pie in three minutes. They plan to work no more than 60 hours.

b. Determine the amounts of sugar, flour, and time that will be unused.

6. Solve each of these problems by computer and obtain the optimal values of the decision variables and the objective function.

   a. Maximize $4x_1 + 2x_2 + 5x_3$
   Subject to
   $1x_1 + 2x_2 + 1x_3 \le 25$
   $1x_1 + 4x_2 + 2x_3 \le 40$
   $3x_1 + 3x_2 + 1x_3 \le 30$
   $x_1, x_2, x_3 \ge 0$

   b. Maximize $10x_1 + 6x_2 + 3x_3$
   Subject to
   $1x_1 + 1x_2 + 2x_3 \le 25$
   $2x_1 + 1x_2 + 4x_3 \le 40$
   $1x_1 + 2x_2 + 3x_3 \le 40$
   $x_1, x_2, x_3 \ge 0$

   a. $x_1 = 4, x_2 = 0, x_3 = 18$
   $s_1 = 3, s_2 = 0, s_3 = 0$
   $Z = 106$
   b. $x_1 = 15, x_2 = 10, x_3 = 0$
   $s_1 = 0, s_2 = 0, s_3 = 5$
   $Z = 210$

7. For Problem 6*a*, determine the following:

   See IM

   a. The range of feasibility for each constraint.
   b. The range of optimality for the coefficients of the objective function.

8. For Problem 6*b*:

   See IM

   a. Find the range of feasibility for each constraint, and interpret your answers.
   b. Determine the range of optimality for each coefficient of the objective function. Interpret your results.

9. A small firm makes three similar products, which all follow the same three-step process, consisting of milling, inspection, and drilling. Product A requires 12 minutes of milling, 5 minutes for inspection, and 10 minutes of drilling per unit; product B requires 10 minutes of milling, 4 minutes for inspection, and 8 minutes of drilling per unit; and product C requires 8 minutes of milling, 4 minutes for inspection, and 16 minutes of drilling. The department has 20 hours available during the next period for milling, 15 hours for inspection, and 24 hours for drilling. Product A contributes $2.40 per unit to profit, B contributes $2.50 per unit, and C contributes $3.00 per unit. Determine the optimal mix of products in terms of maximizing contribution to profits for the period. Then, find the range of optimality for the profit coefficient of each variable.

   A = 0, B = 80, C = 50
   Z = 350
   C_a (insig.): $0 to $3.04
   C_b (opt.): $1.95 to $3.75
   C_c (opt.): $2 to $5

10. Formulate and then solve a linear programming model of this problem, to determine how many containers of each product to produce tomorrow to maximize profits. The company makes four juice products using orange, grapefruit, and pineapple juice.

    See IM

Product	Retail Price
Orange juice	$1.00
Grapefruit juice	.90
Pineapple juice	.80
All in One	1.10

The All-in-One juice has equal parts of orange, grapefruit, and pineapple juice. Each product is produced in a one-quart size (there are four quarts in a gallon). On hand are 400 gallons of orange juice, 300 gallons of grapefruit juice, and 200 gallons of pineapple juice. The cost per gallon is $2.00 for orange juice, $1.60 for grapefruit juice, and $1.40 for pineapple juice.

In addition, the manager wants grapefruit juice to be used for no more than 30 percent of the number of containers produced. She wants the ratio of the number of containers of orange juice to the number of containers of pineapple juice to be at least 7 to 5.

11. A wood products firm uses leftover time at the end of each week to make goods for stock. Currently, two products on the list of items are produced for stock: a chopping board and a

a. Board = 0, Holder = 50
b. Cutting = 16 minutes, gluing = 0, finishing = 210 minutes

knife holder. Both items require three operations: cutting, gluing, and finishing. The manager of the firm has collected the following data on these products:

		TIME PER UNIT (MINUTES)		
Item	**Profit/Unit**	**Cutting**	**Gluing**	**Finishing**
Chopping board	$2	1.4	5	12
Knife holder	$6	0.8	13	3

The manager has also determined that, during each week, 56 minutes are available for cutting, 650 minutes are available for gluing, and 360 minutes are available for finishing.

a. Determine the optimal quantities of the decision variables.

b. Which resources are not completely used by your solution? How much of each resource is unused?

a. Ham = 37.14, deli = 18, cost = $165.42
b. Ham = 20, deli = 84, profit = $376

12. The manager of the deli section of a grocery superstore has just learned that the department has 112 pounds of mayonnaise, of which 70 pounds is approaching its expiration date and must be used. To use up the mayonnaise, the manager has decided to prepare two items: a ham spread and a deli spread. Each pan of the ham spread will require 1.4 pounds of mayonnaise, and each pan of the deli spread will require 1.0 pound. The manager has received an order for 10 pans of ham spread and 8 pans of the deli spread. In addition, the manager has decided to have at least 10 pans of each spread available for sale. Both spreads will cost $3 per pan to make, but ham spread sells for $5 per pan and deli spread sells for $7 per pan.

    a. Determine the solution that will minimize cost.

    b. Determine the solution that will maximize profit.

See IM

13. A manager wants to know how many units of each product to produce on a daily basis in order to achieve the highest contribution to profit. Production requirements for the products are shown in the following table.

Product	**Material 1 (pounds)**	**Material 2 (pounds)**	**Labor (hours)**
A	2	3	3.2
B	1	5	1.5
C	6	–	2.0

Material 1 costs $5 a pound, material 2 costs $4 a pound, and labor costs $10 an hour. Product A sells for $80 a unit, product B sells for $90 a unit, and product C sells for $70 a unit. Available resources each day are 200 pounds of material 1, 300 pounds of material 2, and 150 hours of labor.

The manager must satisfy certain output requirements: The output of product A should not be more than one-third of the total number of units produced; the ratio of units of product A to units of product B should be 3 to 2; and there is a standing order for 5 units of product A each day. Formulate a linear-programming model for this problem, and then solve.

14. A chocolate maker has contracted to operate a small candy counter in a fashionable store. To start with, the selection of offerings will be intentionally limited. The counter will offer a regular mix of candy made up of equal parts of cashews, raisins, caramels, and chocolates, and a deluxe mix that is one-half cashews and one-half chocolates, which will be sold in one-pound boxes. In addition, the candy counter will offer individual one-pound boxes of cashews, raisins, caramels, and chocolates.

    A major attraction of the candy counter is that all candies are made fresh at the counter. However, storage space for supplies and ingredients is limited. Bins are available that can hold the amounts shown in the table:

Ingredient	Capacity (pounds per day)
Cashews	120
Raisins	200
Caramels	100
Chocolates	160

In order to present a good image and to encourage purchases, the counter will make at least 20 boxes of each type of product each day. Any leftover boxes at the end of the day will be removed and given to a nearby nursing home for goodwill.

The profit per box for the various items has been determined as follows:

Item	Profit per Box
Regular	$.80
Deluxe	.90
Cashews	.70
Raisins	.60
Caramels	.50
Chocolates	.75

a. Formulate the LP model.

b. Solve for the optimal values of the decision variables and the maximum profit.

a. See IM
b. Regular = 320
Deluxe = 40
Cashews = 20
Raisins = 120
Caramels = 20
Chocolates = 60
Profit = $433

15. Given this linear-programming model, solve the model and then answer the questions that follow.

Maximize $12x_1 + 18x_2 + 15x_3$ where x_1 = the quantity of product 1 to make etc.

Subject to

Machine	$5x_1 + 4x_2 + 3x_3 \le 160$ minutes
Labor	$4x_1 + 10x_2 + 4x_3 \le 288$ hours
Materials	$2x_1 + 2x_2 + 4x_3 \le 200$ pounds
Product 2	$x_2 \le 16$ units
	x_1, x_2 and $x_3 \ge 0$

a. Are any constraints binding? If so, which one(s)?

b. If the profit on product 3 was changed to $22 a unit, what would the values of the decision variables be? The objective function? Explain.

c. If the profit on product 1 was changed to $22 a unit, what would the values of the decision variables be? The objective function? Explain.

d. If 10 hours less of labor time were available, what would the values of the decision variables be? The objective function? Explain.

e. If the manager decided that as many as 20 units of product 2 could be produced (instead of 16), how much additional profit would be generated?

f. If profit per unit on each product increased by $1, would the optimal values of the decision variables change? Explain. What would the optimal value of the objective function be?

a. Machine and materials
b. Within range, so no change.
c. Still insignificant, so no change.
d. Only s_2 would change (decrease by 10).
e. None; s_4 is slack.
f. Yes; $844

16. A garden store prepares various grades of pine bark for mulch: nuggets (x_1), mini-nuggets (x_2), and chips (x_3). The process requires pine bark, machine time, labor time, and storage space. The following model has been developed.

$$\begin{array}{lll} \text{Maximize} & 9x_1 + 9x_2 + 6x_3 \text{ (profit)} \\ \text{Subject to} & \\ \quad\text{Bark} & 5x_1 + 6x_2 + 3x_3 \leq 600 \text{ pounds} \\ \quad\text{Machine} & 2x_1 + 4x_2 + 5x_3 \leq 660 \text{ minutes} \\ \quad\text{Labor} & 2x_1 + 4x_2 + 3x_3 \leq 480 \text{ hours} \\ \quad\text{Storage} & 1x_1 + 1x_2 + 1x_3 \leq 150 \text{ bags} \\ & x_1, x_2, x_3 \geq 0 \end{array}$$

a. $1.50; 510 to 750
b. $1.50 pound
c. $0; 375 to infinity
d. None. Machine time is slack.
e. 150 pounds of pine bark
f. Optimal quantities would not change; Z would increase by $75
g. Yes; $1,155
h. Yes; $1,110

a. What is the marginal value of a pound of pine bark? Over what range is this price value appropriate?

b. What is the maximum price the store would be justified in paying for additional pine bark?

c. What is the marginal value of labor? Over what range is this value in effect?

d. The manager obtained additional machine time through better scheduling. How much additional machine time can be effectively used for this operation? Why?

e. If the manager can obtain *either* additional pine bark or additional storage space, which one should she choose and how much (assuming additional quantities cost the same as usual)?

f. If a change in the chip operation increased the profit on chips from $6 per bag to $7 per bag, would the optimal quantities change? Would the value of the objective function change? If so, what would the new value(s) be?

g. If profits on chips increased to $7 per bag and profits on nuggets decreased by $.60, would the optimal quantities change? Would the value of the objective function change? If so, what would the new value(s) be?

h. If the amount of pine bark available decreased by 15 pounds, machine time decreased by 27 minutes, and storage capacity increased by five bags, would this fall in the range of feasibility for multiple changes? If so, what would the value of the objective function be?

CASE

SON, LTD.

Son, Ltd., manufactures a variety of chemical products used by photoprocessors. Son was recently bought out by a conglomerate, and managers of the two organizations have been working together to improve the efficiency of Son's operations.

Managers have been asked to adhere to weekly operating budgets and to develop operating plans using quantitative methods whenever possible. The manager of one department has been given a weekly operating budget of $11,980 for production of three chemical products, which for convenience shall be referred to as Q, R, and W. The budget is intended to pay for direct labor and materials. Processing requirements for the three products, on a per unit basis, are shown in the table.

The company has a contractual obligation for 85 units of product R per week.

Material A costs $4 per pound, as does material B. Labor costs $8 an hour.

Product Q sells for $122 a unit, product R sells for $115 a unit, and product W sells for $76 a unit.

The manager is considering a number of different proposals regarding the quantity of each product to produce. The manager is primarily interested in maximizing contribution. Moreover, the manager wants to know how much labor will be needed, as well as the amount of each material to purchase.

Questions

Prepare a report that addresses the following issues:

1. The optimal quantities of products and the necessary quantities of labor and materials.
2. One proposal is to make equal amounts of the products. What amount of each will maximize con-

tribution, and what quantities of labor and materials will be needed? How much less will total contribution be if this proposal is adopted?

3. How would you formulate the constraint for material A if it was determined that there is a 5 percent waste factor for material A and equal quantities of each product are required?

Product	Labor (hours)	Material A (pounds)	Material B (pounds)
Q	5	2	1
R	4	2	—
W	2	½	2

Selected Bibliography

Bierman, Harold; Charles P. Bonini; and Warren H. Hausman. *Quantitative Analysis for Business Decisions.* 7th ed. Burr Ridge, IL: Richard D. Irwin, 1986.

Cook, Thomas M. and Robert A. Russell. *Introduction to Management Science.* 3rd ed. Englewood Cliffs, NJ: Prentice Hall, 1985.

Eppen, G. D.; F. J. Gould; and C. P. Schmidt. *Introductory Management Science.* 2nd ed. Englewood Cliffs, NJ: Prentice Hall, 1987.

Levin, Richard; Charles A. Kirkpatrick; and David S. Rubin. *Quantitative Approaches to Management.* 6th ed. New York: McGraw-Hill, 1987.

Stevenson, W. J. *Introduction to Management Science.* 2nd ed. Burr Ridge, IL: Richard D. Irwin, 1992.

LEARNING OBJECTIVES

After completing this chapter, you should be able to:

1. List some of the different reasons for redesign of layouts.
2. Describe the basic types of layouts.
3. List the primary advantages and limitations of product layouts and process layouts.
4. Solve simple line balancing problems.
5. Develop simple process layouts.

CHAPTER SIX
FACILITIES LAYOUT

CHAPTER OUTLINE

www.mspaairport.com/

Layout refers to the configuration of departments, work centers, and equipment, with particular emphasis on movement of work (customers or materials) through the system. A change in the layout at the Minneapolis-St. Paul International Airport not long ago solved a problem that had plagued travelers. In the former layout, security checkpoints were located in the boarding area. That meant that arriving passengers who were simply changing planes had to pass through a security checkpoint before being able to board their connecting flight, along with other passengers whose journeys were originating at Minneapolis-St. Paul. This created excessive waiting times for both sets of passengers. The new layout relocated the security checkpoints, moving them from the boarding area to a position close to the ticket counters. That eliminated the need for passengers who were making connecting flights from having to pass through security, and in the process, considerably reduced the waiting time for passengers departing from Minneapolis-St. Paul.[1]

INTRODUCTION

This is the third in a series of chapters on design of production systems. The previous chapters dealt with product and service design, and process and capacity planning; this chapter deals with layout design. As in other areas of system design, layout decisions are important for three basic reasons: (1) they require substantial investments of money and effort, (2) they involve long-term commitments, which makes mistakes difficult to overcome, and (3) they have a significant impact on the cost and efficiency of short-term operations. This chapter describes the main types of layout designs and the models used to evaluate design alternatives.

The Need for Layout Decisions

The need for layout planning arises both in the process of designing new facilities and in redesigning existing facilities. The most common reasons for redesign of layouts include:

1. Inefficient operations (e.g., high cost, bottlenecks).
2. Accidents or safety hazards.
3. Changes in the design of products or services.
4. Introduction of new products or services.
5. Changes in the volume of output or mix of outputs.
6. Changes in methods or equipment.
7. Changes in environmental or other legal requirements.
8. Morale problems (e.g., lack of face-to-face contact).

BASIC LAYOUT TYPES

The three basic types of layouts are product, process, and fixed-position. *Product layouts* are most conducive to repetitive processing, *process layouts* are used for intermittent processing, and *fixed-position layouts* are used when projects require layouts. We will study the characteristics, advantages, and disadvantages of each layout type in this section. We'll also take a look at hybrid layouts, which are combinations of these pure types.

---

[1]Based on "Airport checkpoints moved to help speed travelers on their way," Minneapolis-St. Paul Star Tribune, January 13, 1995, p. 1B.

Product Layouts

Product layouts are used to achieve a smooth and rapid flow of large volumes of goods or customers through a system. This is made possible by highly standardized goods or services that require highly standardized (repetitive) processing operations. The work is divided into a series of standardized tasks, permitting specialization of both labor and equipment. The large volumes handled by these systems usually make it economical to invest substantial sums of money in equipment and job design. Because only one or a few very similar items are involved, it is feasible to arrange an entire layout to correspond to the technological processing requirements of the product or service. For instance, if a portion of a manufacturing operation required the sequence of cutting, sanding, and painting, the appropriate pieces of equipment would be arranged in that same sequence. And because each item follows the same sequence of operations, it is often possible to utilize fixed-path material-handling equipment such as conveyors to transport items between operations. The resulting arrangement forms a line like the one depicted in Figure 6–1. In manufacturing environments, the lines are referred to as **production lines** or **assembly lines,** depending on the type of activity involved. In service processes, the term *line* may or may not be used. It is common to refer to a cafeteria line as such but not a car wash, although from a conceptual standpoint the two are nearly identical. Figure 6–2 illustrates the layout of a typical cafeteria serving line. Examples of this type of layout are less plentiful in service environments because processing requirements usually exhibit too much variability to make standardization feasible. Without high standardization, many of the benefits of repetitive processing are lost. When lines are used, certain compromises may be made. For instance, an automatic car wash provides equal treatment to all cars—the same amount of soap, water, and scrubbing—even though cars may differ considerably in cleaning needs. As a result, very dirty cars may not come out completely clean, and relatively clean cars go through the same system with considerable waste of soap, water, and energy.

product layout Layout that uses standardized processing operations to achieve smooth, rapid, high-volume flow.

production line Standardized layout arranged according to a fixed sequence of production tasks.

assembly line Standardized layout arranged according to a fixed sequence of assembly tasks.

Product layouts achieve a high degree of labor and equipment utilization, which tends to offset their high equipment costs. Because items move quickly from operation to operation, the amount of work-in-process is often minimal. Consequently, operations are so closely tied to each other that the entire system is highly vulnerable to being shut down because of mechanical failure or high absenteeism. Maintenance procedures are

A car wash is a service assembly line with various cleaning tasks being performed in a specific order. Some facilities allow the customer options such as having the undercarriage washed or having protective coating applied.

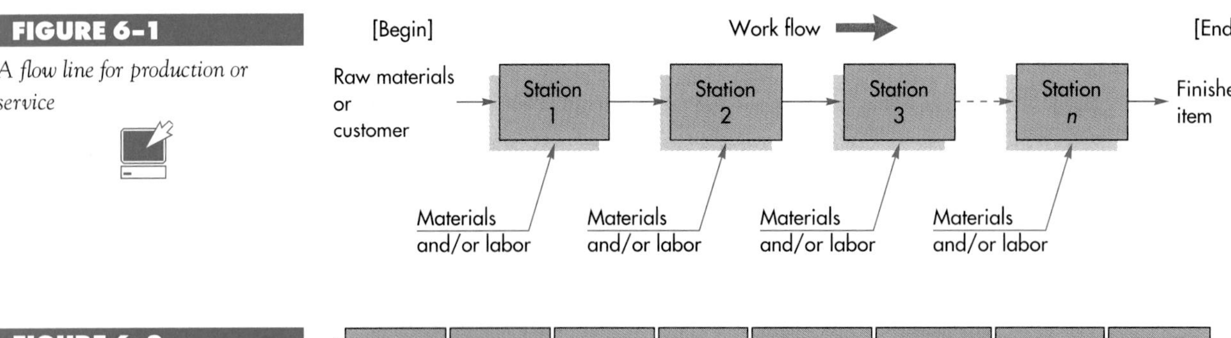

FIGURE 6-1
A flow line for production or service

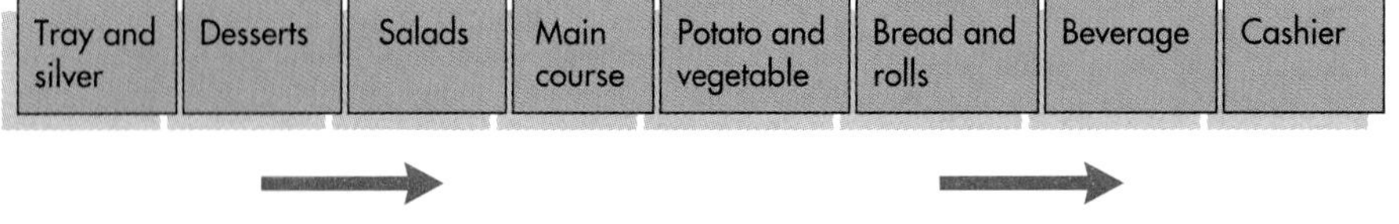

FIGURE 6-2
Cafeteria line

geared to this. Preventive maintenance—periodic inspection and replacement of worn parts or those with high failure rates—reduces the probability of breakdowns during the operations. Of course, no amount of preventive activity can completely eliminate failures, so management must take measures to provide quick repair. These include maintaining an inventory of spare parts and having repair personnel available to quickly restore equipment to normal operation. These procedures are fairly expensive; because of the specialized nature of equipment, problems become more difficult to diagnose and resolve, and spare-part inventories can be extensive.

The main advantages of product layouts are:

1. A high rate of output.
2. Low unit cost due to high volume; the high cost of specialized equipment is spread over many units.
3. Labor specialization reduces training costs and time, and results in a wide span of supervision.
4. Low material-handling cost per unit; material handling is simplified because units follow the same sequence of operations.
5. A high utilization of labor and equipment.
6. Routing and scheduling are established in the initial design of the system; they do not require much attention once the system is operating.
7. Accounting, purchasing, and inventory control are fairly routine.

The primary disadvantages of product layouts include:

1. The intensive division of labor usually creates dull, repetitive jobs that provide little opportunity for advancement and may lead to morale problems, and to repetitive stress injuries.
2. Poorly skilled workers may exhibit little interest in maintaining equipment or in the quality of output.
3. The system is fairly inflexible in response to changes in the volume of output or changes in product or process design.
4. The system is highly susceptible to shutdowns caused by equipment breakdowns or excessive absenteeism.
5. Preventive maintenance, the capacity for quick repairs, and spare-parts inventories are necessary expenses.
6. Incentive plans tied to individual output are impractical since they would cause variations among outputs of individual workers, which would adversely affect the smooth flow of work through the system.

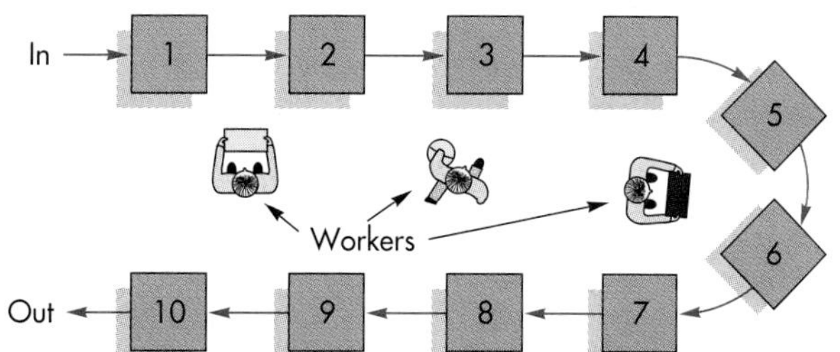

FIGURE 6–3
A U-shaped production line

U-Shaped Layouts. Although a straight production line may have intuitive appeal, a U-shaped line (see Figure 6–3) has a number of advantages that make it worthy of consideration. One disadvantage of a long, straight line is that it interferes with cross travel of workers and vehicles. A U-shaped line is more compact; it often requires approximately half the length of a straight production line. In addition, a U-shaped line permits increased communication among workers on the line because workers are clustered, thus facilitating teamwork. Flexibility in work assignments is increased because workers can handle not only adjacent stations but also stations on opposite sides of the line. Moreover, if materials enter the plant at the same point that finished products leave it, a U-shaped line minimizes material handling.

Of course, not all situations lend themselves to U-shaped layouts: on highly automated lines there is less need for teamwork and communication. And entry and exit points may be on opposite sides of the building. Also, operations may need to be separated because of noise or contamination factors.

Process Layouts

process layouts Layouts that can handle varied processing requirements.

Process layouts are designed to process items or provide services that involve a variety of processing requirements. The layouts feature departments or other functional groupings in which similar kinds of activities are performed. A manufacturing example of a process layout is the *machine shop,* which has separate departments for milling, grinding, drilling, and so on. Items that require those operations are frequently moved in lots to the departments in a sequence dictated by technical considerations. Different products may present quite different processing requirements and sequences of operations. Consequently, variable-path material-handling equipment (forklift trucks, jeeps, tote boxes) is needed to handle the variety of routes and items. The use of *general-purpose machines* provides the flexibility necessary to handle a wide range of processing requirements. Workers who operate the equipment are usually skilled or semiskilled. Figure 6–4 illustrates the departmental arrangement typical of a process layout.

Vol. 1, Tape 1, Seg. 3
Caterpillar, Ford, Nucor Steel

Process layouts are quite common in service environments. Examples include hospitals, colleges and universities, banks, auto repair shops, airlines, and public libraries. For instance, hospitals have departments or other units that specifically handle surgery, maternity, pediatrics, psychiatric, emergency, and geriatric care. And universities have separate schools or departments that concentrate on one area of study such as business, engineering, science, or math.

Because equipment in a process layout is arranged by type rather than by processing sequence, the system is much less vulnerable to shutdown caused by mechanical failure or absenteeism. In manufacturing systems especially, idle equipment is usually available to replace machines that are temporarily out of service. Moreover, because items are often processed in lots, there is considerably less interdependence between successive operations than with a product layout. Maintenance costs tend to be lower because the equipment is less specialized than that of product layouts, and the grouping of machinery permits repair personnel to become skilled in handling that type of equipment. Machine similarity reduces the necessary investment in spare parts. On the negative side, routing and scheduling must be done on a continual basis to accommodate the variety of processing demands typically imposed on these systems. Material handling is inefficient,

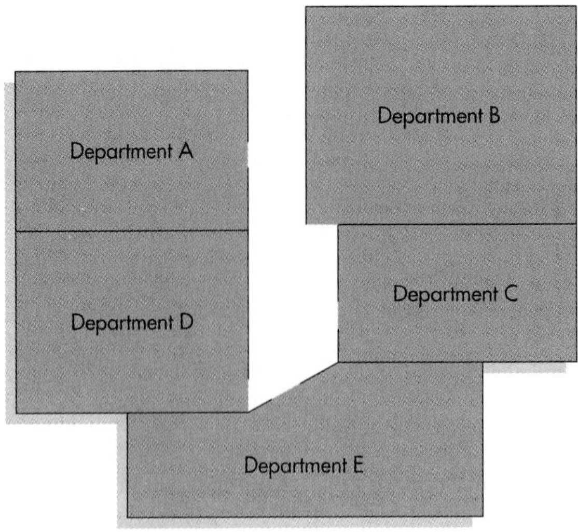

FIGURE 6–4

A typical process layout with departmental work centers

and unit handling costs are generally much higher than in product layouts. In-process inventories can be substantial due to batch processing. Furthermore, it is not uncommon for such systems to have equipment utilization rates under 50 percent because of routing and scheduling complexities related to the variety of processing demands being handled.

In sum, process layouts have both advantages and disadvantages. The advantages of process layouts include:

1. Systems can handle a variety of processing requirements.
2. Systems are not particularly vulnerable to equipment failures.
3. General-purpose equipment is often less costly than the specialized equipment used in product layouts and is easier and less costly to maintain.
4. It is possible to use individual incentive systems.

The disadvantages of process layouts include:

1. In-process inventory costs can be high if batch processing is used in manufacturing systems.
2. Routing and scheduling pose continual challenges.
3. Equipment utilization rates are low.
4. Material handling is slow and inefficient, and more costly per unit than in product layouts.
5. Job complexities often reduce the span of supervision and result in higher supervisory costs than with product layouts.
6. Special attention necessary for each product or customer (e.g., routing, scheduling, machine setups) and low volumes result in higher unit costs than with product layouts.
7. Accounting, inventory control, and purchasing are much more involved than with product layouts.

Fixed-Position Layouts

fixed-position layout Layout in which the product or project remains stationary, and workers, materials, and equipment are moved as needed.

In **fixed-position layouts,** the item being worked on remains stationary, and workers, materials, and equipment are moved about as needed. This is in marked contrast to product and process layouts. Almost always, the nature of the product dictates this kind of arrangement: weight, size, bulk, or some other factor makes it undesirable or extremely difficult to move the product. Fixed-position layouts are used in large construction projects (buildings, power plants, dams), shipbuilding, and production of large

Fixed position layout. When a large project must remain stationary, such as building a submarine at Electric Boat, a division of General Dynamics, workers and equipment come to the site. The Seawolf *weighs 9137 tons, is 353 feet long and 40 feet in diameter and takes more than five years to build.*

aircraft and space mission rockets. In those instances, attention is focused on timing of material and equipment deliveries so as not to clog up the work site and to avoid having to relocate materials and equipment around the work site. Lack of storage space can present significant problems, for example, at construction sites in crowded urban locations. Because of the many diverse activities carried out on large projects and because of the wide range of skills required, special efforts are needed to coordinate the activities, and the span of control can be quite narrow. For these reasons, the administrative burden is often much higher than it would be under either of the other layout types. Material handling may or may not be a factor: in many cases, there is no tangible product involved (e.g., designing a computerized inventory system). When goods and materials are involved, material handling often resembles process-type, variable-path, general-purpose equipment. Projects might require use of earth-moving equipment and trucks to haul materials to, from, and around the work site, for example.

Fixed-position layouts are widely used for farming, firefighting, road building, home building, remodeling and repair, and drilling for oil. In each case, compelling reasons bring workers, materials, and equipment to the "product's" location instead of the other way around.

Combination Layouts

The three basic layout types are ideal models, which may be altered to satisfy the needs of a particular situation. It is not hard to find layouts that represent some combination of these pure types. For instance, supermarket layouts are essentially process layouts, and yet we find most use fixed-path material-handling devices such as roller-type conveyors both in the stockroom, and belt-type conveyors at the cash registers. Hospitals also use the basic process arrangement, although frequently patient care involves more of a fixed-position approach, in which nurses, doctors, medicines, and special equipment are brought to the patient. By the same token, faulty parts made in a product layout may require off-line reworking, which involves customized processing. Moreover, conveyors are frequently observed in both farming and construction activities.

FIGURE 6–5

Comparison of functional and cellular manufacturing layouts

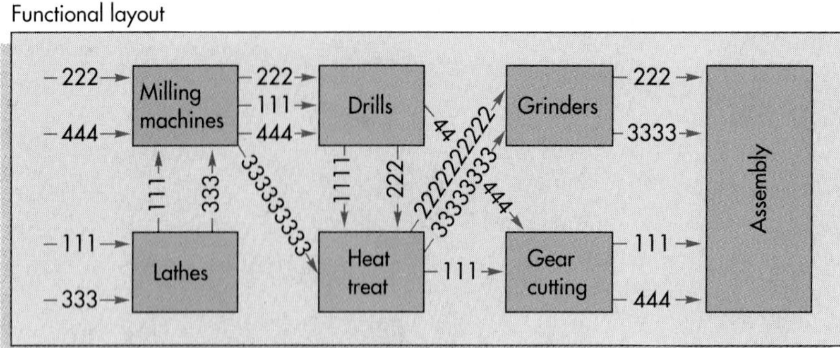

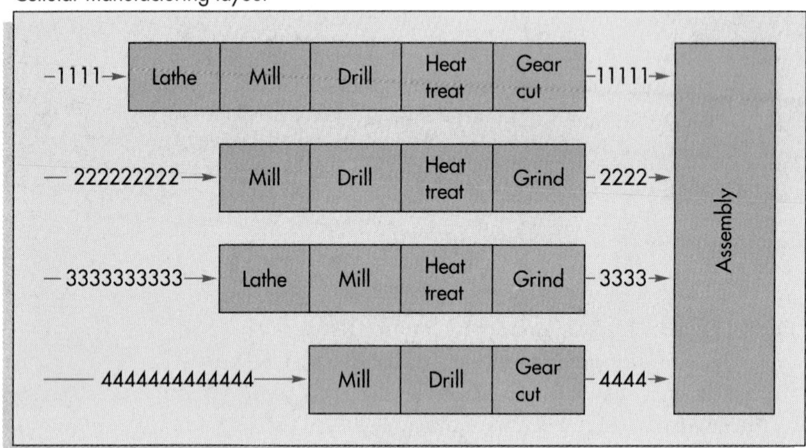

Source: Adapted from D. Fogarty and T. Hoffmann, *Production and Inventory Management* (Cincinnati: South-Western Publishing, 1983), p. 472.

Process layouts and product layouts represent two ends of a continuum from small batches to continuous production. Process layouts are conducive to the production of a wider range of products or services than product layouts, which is desirable from a customer standpoint where customized products are often in demand. However, process layouts tend to be less efficient and have higher unit production costs than product layouts. Some manufacturers are moving away from process layouts in an effort to capture some of the benefits of product layouts. Ideally, a system is flexible and yet efficient, with low unit production costs. Cellular manufacturing, group technology, and flexible manufacturing systems represent efforts to move toward this ideal.

Vol. 5, Seg. 2
TriState

CELLULAR LAYOUTS

Cellular Manufacturing

cellular manufacturing Layout in which machines are grouped into a cell that can process items that have similar processing requirements.

Cellular manufacturing is a type of layout in which machines are grouped into what is referred to as a *cell*. Groupings are determined by the operations needed to perform work for a set of similar items, or *part families*, that require similar processing. The cells become, in effect, miniature versions of product layouts. The cells may have no conveyorized movement of parts between machines, or may have a flow line connected by a conveyor (automatic transfer). Figure 6–5 compares a typical functional (process) layout

FIGURE 6–6

A group of parts with similar manufacturing process requirements but different design attributes

Source: Mikell P. Groover, *Automation, Production Systems, and Computer-Aided Manufacturing* © 1980, p. 540. Reprinted by permission of Prentice Hall, Inc., Englewood Cliffs, New Jersey.

and a cellular manufacturing layout. Observe that in the cellular layout, machines are arranged to handle all of the operations necessary for a group (family) of similar parts. Thus, all parts follow the same route although minor variations (e.g., skipping an operation) are possible. In contrast, the functional layout involves multiple paths for parts. Moreover, there is little effort or need to identify part families; the distinction in the figure is merely for purposes of comparison.

Vol. 2, Seg. 3
Bernard Welding

There are numerous benefits of cellular manufacturing. These relate to the grouping of equipment and include faster processing time, less material handling, less work-in-process inventory, and reduced setup time.

Group Technology

Effective cellular manufacturing must have groups of identified items with similar processing characteristics. The grouping process is known as **group technology** and involves identifying items with similarities in either *design characteristics* or *manufacturing characteristics,* and grouping them into part families. Design characteristics include size, shape, and function; manufacturing or processing characteristics involve the type and sequence of operations required. In many cases, design and processing characteristics are correlated, although this is not always the case. Thus, design families may be different from processing families. Figure 6–6 illustrates a group of parts with similar processing characteristics but different design characteristics.

group technology The grouping into part families of items with similar design or manufacturing characteristics.

Once similar items have been identified, items can be classified according to their families, then a system can be developed that facilitates retrieval from a database for purposes of design and manufacturing. For instance, a designer can use the system to determine if there is an existing part similar or identical to one that needs to be designed. It may happen that an existing part, with some modification, is satisfactory. This greatly enhances the productivity of design. Similarly, planning the manufacturing of a new part can include matching it with one of the part families in existence, thereby alleviating much of the burden of specific processing details.

The conversion to group technology and cellular manufacturing requires a systematic analysis of parts to identify the part families. This is often a major undertaking; it is a time-consuming job that involves the analysis of a considerable amount of data. Three primary methods for accomplishing this are visual inspection, examination of design and production data, and production flow analysis.

Visual inspection is the least accurate of the three but also the least costly and the simplest to perform. Examination of design and production data is more accurate but much more time-consuming; it is perhaps the most commonly used method of analysis. Production flow analysis has a manufacturing perspective and not a design perspective, because it examines operations sequences and machine routings to uncover similarities. Moreover, the operation sequences and routings are taken as givens; in reality the existing procedures may be far from optimal.

Conversion to cellular manufacturing can involve costly realignment of equipment. Consequently, a manager must weigh the benefits of a switch from a process layout to a cellular one against the cost of moving equipment as well as the cost and time needed for grouping parts.

Flexible Manufacturing Systems

As we discussed in Chapter 5, flexible manufacturing systems (FMSs) are more fully automated versions of cellular manufacturing: A computer controls the transfer of parts from machine to machine and the start of work at each machine. These systems are quite expensive, but they enable manufacturers to achieve some of the benefits of product layouts with small batch sizes and much greater flexibility because the system is capable of operating with little or no human intervention.

OTHER SERVICE LAYOUTS

In addition to the layouts already described, there are other layouts found in service environments, such as warehouse, retail, and office layouts.

Warehouse and Storage Layouts

The design of storage facilities presents a different set of factors than the design of factory layouts. Frequency of order is an important consideration; items that are ordered frequently should be placed near the entrance to the facility, and those ordered infrequently should be placed towards the rear of the facility. Any correlations between items are also significant (i.e., item A is usually ordered with item B), suggesting that placing those two items close together would reduce the cost and time of *picking* (retrieving) those items. Other considerations include the number and widths of aisles, the height of storage racks, rail and/or truck loading and unloading, and the need to periodically make a physical count of stored items.

Retail Layouts

The objectives that guide design of manufacturing layouts often pertain to cost minimization and product flow. However, with retail layouts such as department stores, supermarkets, and specialty stores, designers must take into account the presence of customers, and the opportunity to influence sales volume and customer attitudes through carefully designed layouts. Traffic patterns and traffic flow are important factors to consider. Some large retail chains use standard layouts for all or most of their stores. This has several advantages. Most obvious is the ability to save time and money by using one layout instead of custom designing one for each store. Another advantage is to avoid confusing consumers who visit more than one store. In the case of service retail outlets, especially small ones such as dry cleaners, shoe repair, and auto service centers, layout design is much simpler.

Office Layouts

Office layouts are undergoing transformations as the flow of paperwork is replaced with the increasing use of electronic communications. That means there is less need to place office workers in a layout that optimizes the physical transfer of information or paperwork. Another trend is to project an image of openness, and office walls are giving way to low-rise partitions and glass walls.

READING

DESIGNING SUPERMARKETS **David Schardt**

The produce is over here, the dairy's over there. The soft drink specials are at the end of the aisles, the candy's at the checkout. Always.

A visit to your local supermarket isn't as haphazard as it seems. It's been laid out so that you spend as much as possible on what the store wants you to buy. And that's often more than you came in for, as we learned when we spoke to supermarket industry insiders.

Here's how a typical supermarket is designed to maximize sales.

On the Edge

The more time you spend shopping along the sides and back of the supermarket, the more money the store makes. About half its profits come from perimeter items like fruits and veggies, milk and cheese, and meat, poultry, and fish. That's also where you'll find the bakery, the salad bar, and the deli. If a store wants to distinguish itself from its competitors, it's got to be here.

Space Eaters

Some foods are so profitable that they command their own aisles. Breakfast cereals bring in more dollars per foot of shelf space than any other product in the interior of the store. So most supermarkets give cereals plenty of space.

Soft drinks aren't as profitable . . . at least not on paper. But beverage manufacturers sweeten the pot with so much free merchandise and cash rebates that carbonated soft drinks end up being one of the biggest moneymakers in a typical store.

The Meating Place

Why are the meat, poultry, and seafood displays almost always along the back of the supermarket? So that you'll see them every time you emerge from an aisle. Not a bad place to put the most profitable sections of the store.

Going to the Dairy

Why are the dairy products usually as far away from the entrance as possible? Most everybody buys milk when they shop. To reach it, they've got to walk through a good chunk of the supermarket, often along the perimeter. That's right where the store wants shoppers.

Also, stores like to "anchor" a display by putting popular items at each end. That's why milk, for example, is often at one end of the dairy case and margarine and butter at the other. You've got to run the gauntlet of cheese, yogurts, dips, etc. to get what you came for.

Paying for Space

Every year, grocery chains are offered more than 15,000 new products, nearly all of which will fail. How do stores decide which ones to stock?

Moolah, in some cases. Large supermarkets often require manufacturers to pay for shelf space. "Slotting fees," as they're called, can range from $5,000 to $25,000 per supermarket chain for each new food. The small local tofu cheese plant seldom has that kind of money to throw around.

In "Prison"

Some supermarket insiders call the aisles of the store the "prison." Once you're in one, you're stuck until you come out the other end. The "prison" is where most of the less-profitable (for the store) national and regional name brands are, so the more time you spend there, the less time you'll spend along the perimeter . . . buying higher-profit items.

Productive Produce

Think it's a coincidence that you almost always have to walk through the produce department when you enter a supermarket? The look of those shiny, neatly stacked fruits and vegetables is *the* most important influence on where people decide to shop.

It also doesn't hurt that produce is the second most profitable section (meat is first). While it occupies a little over ten percent of the typical supermarket, it brings in close to 20 percent of the store's profits.

FIGURE 6–7

Auto assembly

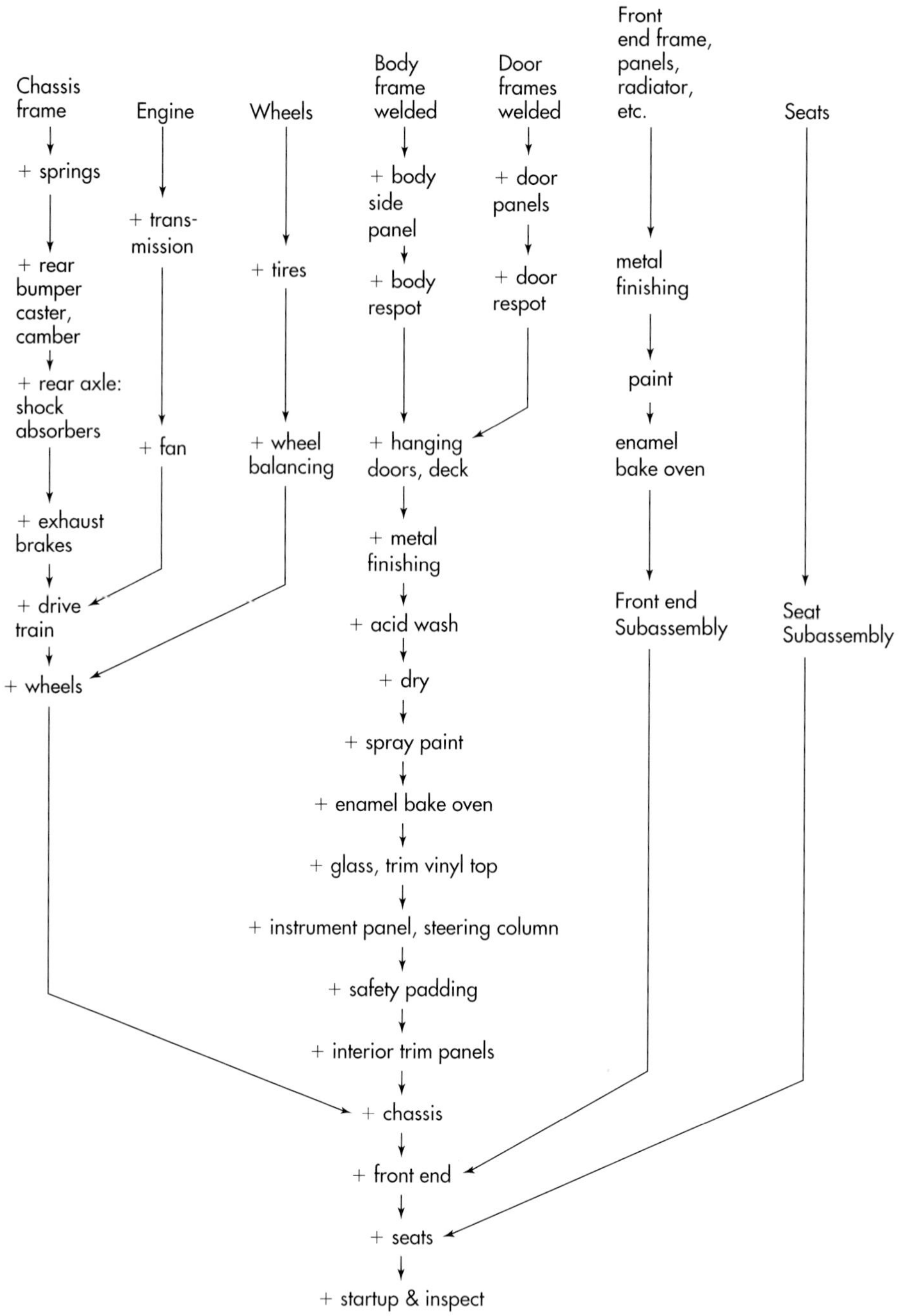

Chassis frame → + springs → + rear bumper caster, camber → + rear axle: shock absorbers → + exhaust brakes → + drive train → + wheels

Engine → + trans-mission → + fan → (+ drive train)

Wheels → + tires → + wheel balancing → (+ wheels)

Body frame welded → + body side panel → + body respot → + hanging doors, deck

Door frames welded → + door panels → + door respot → (+ hanging doors, deck)

+ hanging doors, deck → + metal finishing → + acid wash → + dry → + spray paint → + enamel bake oven → + glass, trim vinyl top → + instrument panel, steering column → + safety padding → + interior trim panels → + chassis → + front end → + seats → + startup & inspect

Front end frame, panels, radiator, etc. → metal finishing → paint → enamel bake oven → Front end Subassembly → (+ front end)

Seats → Seat Subassembly → (+ seats)

Source: "Computer Integrated Manufacturing," Vol. 1 in Revolution in Progress series, Chapman and Hall, London, 1990. Adapted from Ayres, R. U. Morris A. Cohen and Uday M. Apte. *Manufacturing Automation.* Burr Ridge, IL: McGraw-Hill, 1997, p. 175.

DESIGNING PRODUCT LAYOUTS: LINE BALANCING

Assembly lines range from fairly short, with just a few operations, to long lines that have a large number of operations. Automobile assembly lines are examples of long lines. At the assembly line for Ford Mustangs in Dearborn, Michigan, a Mustang travels about nine miles from start to finish!

www.ford.com

Figure 6–7 illustrates the major steps involved in assembling an automobile.

Many of the benefits of a product layout relate to the ability to divide required work into a series of elemental tasks (e.g., "assemble parts C and *D*") that can be performed quickly and routinely by low-skilled workers or specialized equipment. The durations of these elemental tasks typically range from a few seconds to 15 minutes or more. Most time requirements are so brief that it would be impractical to assign only one task to each worker. For one thing, most workers would quickly become bored by the limited job scope. For another, the number of workers required to complete even a simple product or service would be enormous. Instead, tasks are usually grouped into manageable bundles assigned to workstations manned by one or two operators.

The process of deciding how to assign tasks to workstations is referred to as **line balancing.** The goal of line balancing is to obtain task groupings that represent approximately equal time requirements. This minimizes the idle time along the line and results in a high utilization of labor and equipment. Idle time occurs if task times are not equal among workstations; some stations are capable of producing at higher rates than others. These "fast" stations will experience periodic waits for the output from slower stations or else be forced into idleness to avoid buildups of work between stations. Unbalanced lines are undesirable in terms of inefficient utilization of labor and equipment and because they may create morale problems at the slower stations for workers who must work continuously.

line balancing The process of assigning tasks to workstations in such a way that the workstations have approximately equal time requirements.

Lines that are perfectly balanced will have a smooth flow of work as activities along the line are synchronized to achieve maximum utilization of labor and equipment. The major obstacle to attaining a perfectly balanced line is the difficulty of forming task bundles that have the same duration. One cause of this is that it may not be feasible to combine certain activities into the same bundle, either because of differences in equipment requirements or because the activities are not compatible (e.g., risk of contamination of paint from sanding). Another cause of difficulty is that differences among elemental task lengths cannot always be overcome by grouping tasks. A third cause of an inability to perfectly balance a line is that a required technological sequence may prohibit otherwise desirable task combinations. Consider a series of three operations that have durations of two minutes, four minutes, and two minutes, as shown in the following diagram. Ideally, the first and third operations could be combined at one workstation and have a total time equal to that of the second operation. However, it may not be possible to combine the first and third operations. In the case of an automatic car wash, scrubbing and drying operations could not realistically be combined at the same workstation due to the need to rinse cars between the two operations.

Scrubbing 2 minutes → Rinsing 4 minutes → Drying 2 minutes

Line balancing involves assigning tasks to workstations. Usually, each workstation has one worker who handles all of the tasks at that station, although an option is to have several workers at a single workstation. For purposes of illustration, however, all of the examples and problems in this chapter have workstations with one worker. A manager could decide to use anywhere from one to five workstations to handle five tasks. With one workstation, all tasks would be done at that station; with five stations, one task would be assigned to each station. If two, three, or four workstations are used, some or all of the stations will have multiple tasks assigned to them. How does a manager decide how many stations to use?

The primary determinant is what the line's **cycle time** will be. The cycle time is the *maximum* time allowed at each workstation to perform assigned tasks before the work moves on. The cycle time also establishes the output rate of a line. For instance, if the cycle time is two minutes, units will come off the end of the line at the rate of one every two minutes.

cycle time The maximum time allowed at each workstation to complete its set of tasks on a unit.

We can gain some insight into task groupings and cycle time by considering a simple example.

Suppose that the work required to fabricate a certain product can be divided up into five elemental tasks, with the task times and precedence relationships as shown in the following diagram.

→ 0.1 min. → 0.7 min. → 1.0 min. → 0.5 min. → 0.2 min. →

The task times govern the range of possible cycle times. The *minimum* cycle time is equal to the *longest* task time (1.0 minute), and the *maximum* cycle time is equal to the sum of the task times (0.1 + 0.7 + 1.0 + 0.5 + 0.2 = 2.5 minutes). The minimum cycle time would apply if there were five workstations. The maximum cycle time would apply if all tasks were performed at a single workstation. The minimum and maximum cycle times are important because they establish the potential range of output for the line, which we can compute using the following formula:

$$\text{Output capacity} = \frac{\text{OT}}{\text{CT}} \tag{6–1}$$

where

OT = Operating time per day
CT = Cycle time

Assume that the line will operate for eight hours per day (480 minutes). With a cycle time of 1.0 minute, output would be

$$\frac{480 \text{ minutes per day}}{1.0 \text{ minute per unit}} = 480 \text{ units per day}$$

With a cycle time of 2.5 minutes, the output would be

$$\frac{480 \text{ minutes per day}}{2.5 \text{ minutes per unit}} = 192 \text{ units per day}$$

Assuming that no parallel activities are to be employed (e.g., two lines), the output selected for the line must fall in the range of 192 units per day to 480 units per day.

As a general rule, the cycle time is determined by the desired output; that is, a desired output level is selected, and the cycle time is computed. If the cycle time does not fall between the maximum and minimum bounds, the desired output rate must be revised. We can compute the cycle time using this formula:

$$\text{CT} = \frac{\text{OT}}{D} \tag{6–2}$$

where

D = Desired output rate

For example, suppose that the desired output rate is 480 units. Using Formula 6–2, the necessary cycle time is

$$\frac{480 \text{ minutes per day}}{480 \text{ units per day}} = 1.0 \text{ minute per unit}$$

The number of workstations that will be needed is a function of both the desired output rate and our ability to combine elemental tasks into workstations. We can determine the *theoretical minimum* number of stations necessary to provide a specified rate of output as follows:

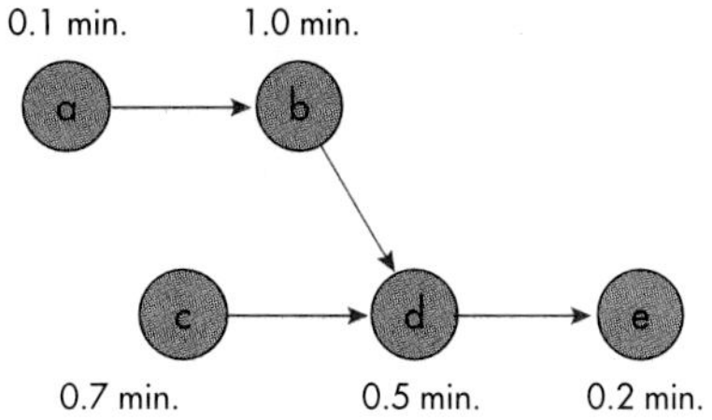

FIGURE 6–8

A simple precedence diagram

$$N_{\min} = \frac{\Sigma t}{\text{CT}} \tag{6–3}$$

where

$N_{\min}$ = Theoretical minimum number of stations
D = Desired output rate
Σt = Sum of task times

Suppose the desired rate of output is the maximum of 480 units per day.[2] (This will require a cycle time of 1.0 minute.) The minimum number of stations required to achieve this goal is:

$$N_{\min} = \frac{2.5 \text{ minutes per unit}}{1 \text{ minute per unit per station}} = 2.5 \text{ stations}$$

Because 2.5 stations is not feasible, it is necessary to *round up* (because 2.5 is the minimum) to three stations. Thus, the actual number of stations used will equal or exceed three, depending on how successfully the tasks can be grouped into work bundles.

A very useful tool in line balancing is a **precedence diagram.** Figure 6–8 illustrates a simple precedence diagram. It visually portrays the tasks that are to be performed along with the *sequential* requirements, that is, the *order* in which tasks must be performed. The diagram is read from left to right, so the initial task(s) are on the left and the final task is on the right. In terms of precedence requirements, we can see from the diagram, for example, that the only requirement to begin task b is that task a must be finished. However, in order to begin task d, tasks b and c must *both* be finished. Note that the elemental tasks are the same ones that we have been using.

precedence diagram A diagram that shows elemental tasks and their precedence requirements.

Now let's see how a line is balanced. This involves assigning tasks to workstations. Generally, no techniques are available that guarantee an optimal set of assignments. Instead, managers employ *heuristic (intuitive) rules,* which provide good and sometimes optimal sets of assignments. A number of line-balancing heuristics are in use, two of which are described here for purposes of illustration:

1. Assign tasks in order of most following tasks.
2. Assign tasks in order of greatest positional weight. Positional weight is the sum of each task's time and the times of all following tasks.

The general procedure used in line balancing is described in Table 6–1.

Example 1

Arrange the tasks shown in Figure 6–8 into three workstations. Use a cycle time of 1.0 minute. Assign tasks in order of the most number of followers.

[2]At first glance, it might seem that the desired output would logically be the maximum possible output. However, you will see why that is not always the best alternative.

TABLE 6–1

Line balancing procedure

1. Identify the cycle time and determine the minimum number of workstations.
2. Make assignments to workstations in order, beginning with Station 1. Tasks are assigned to workstations moving from left to right through the precedence diagram.
3. Before each assignment, use the following criteria to determine which tasks are *eligible* to be assigned to a workstation:
   a. All preceding tasks in the sequence have been assigned.
   b. The task time does not exceed the time remaining at the workstation.
   If no tasks are eligible, move on to the next workstation.
4. After each task assignment, determine the time remaining at the current workstation by subtracting the sum of times for tasks already assigned to it from the cycle time.
5. Break ties that occur using one of these rules:
   a. Assign the task with the longest task time.
   b. Assign the task with the greatest number of followers.
   If there is still a tie, choose one task arbitrarily.
6. Continue until all tasks have been assigned to workstations.
7. Compute appropriate measures (e.g., percent idle time, efficiency) for the set of assignments.

Solution

Workstation	Time Remaining	Eligible	Assign Task	Station Idle Time
1	1.0	a, c	a	
	0.9	c	c	
	0.2	none	-	0.2
2	1.0	b	b	0.0
3	1.0	d	d	
	0.5	e	e	
	0.3	-	-	0.3
				0.5

Comment: The initial "time remaining" for each workstation is equal to the cycle time. For a task to be eligible, tasks preceding it must have been assigned, and the task's time must not exceed the station's remaining time.

Example 1 was purposely simple; it was designed to illustrate the basic procedure. Later examples will illustrate tiebreaking, constructing precedence diagrams, and the positional weight method. Before considering those examples, let us first consider some measures of effectiveness that can be used for evaluating a given set of assignments.

Two widely used measures of effectiveness are:

balance delay Percentage of idle time of a line.

1. The *percentage of idle time* of the line. This is sometimes referred to as the **balance delay.** It can be computed as follows:

$$\text{Percentage of idle time} = \frac{\textit{Idle time per cycle}}{N_{\text{actual}} \times \textit{cycle time}} \times 100 \qquad (6\text{–}4)$$

where N_{actual} = Actual number of stations.
For the preceding example, the value is:

$$\text{Percentage of idle time} = \frac{0.5}{3 \times 1.0} \times 100 = 16.7\%$$

In effect, this is the average idle time divided by the cycle time, multiplied by 100. Note that cycle time refers to the actual cycle time that is achieved.

2. *Efficiency.* The efficiency of a line is computed as follows:

   Efficiency = 100 − percent idle time. (6–5)

   Efficiency = 100% − 16.7% = 83.3%

Now let's consider the question of whether the selected level of output should equal the maximum output possible. The minimum number of workstations needed is a function of the desired output rate; and therefore, the cycle time. Therefore, a lower rate of output (and hence, a longer cycle time) may result in a need for fewer stations. Hence, the manager must consider whether the potential savings realized by having fewer workstations would be greater than the decrease in profit resulting from producing fewer units.

The preceding examples serve to illustrate some of the fundamental concepts of line balancing. They are rather simple; in most real-life situations, the number of branches and tasks are often much greater. Consequently, the job of line balancing can be a good deal more complex. In many instances, the number of alternatives for grouping tasks is so great that it is virtually impossible to conduct an exhaustive review of all possibilities. For this reason, many real-life problems of any magnitude are solved using heuristic approaches. The purpose of a heuristic approach is to reduce the number of alternatives that must be considered, but it does not guarantee an optimal solution.

Some Guidelines for Line Balancing

In balancing an assembly line, risks are assigned *one at a time* to the line, starting at the first workstation. At each step, the unassigned tasks are checked to determine which are eligible for assignment. Next, the eligible tasks are checked to see which of them will fit in the workstation being loaded. Then a heuristic is used to select one of the tasks that will fit, and the task is assigned. This process is repeated until there are no eligible tasks that will fit. Then the next workstation can be loaded. This continues until all tasks are assigned. The objective is to minimize the idle time for the line subject to technological and output constraints.

Technological constraints tell us which elemental tasks are *eligible* to be assigned at a particular position on the line. Technological constraints can result from the precedence or ordering relationships among the tasks. The precedence relationships require that certain tasks must be performed before others (and so, must be assigned to workstations before others). Thus, in a car wash, the rinsing operation must be performed before the drying operation. The drying operation is not eligible for assignment until the rinsing operation is assigned. Technological constraints may also result from two tasks being "incompatible" (e.g., space restrictions or the nature of the operations may prevent their being placed in the same work center). For example, sanding and painting operations would not be assigned to the same work center because dust particles from the sanding operation could contaminate the paint.

Output constraints, on the other hand, determine the maximum amount of work that a manager can assign to each workstation, and this determines whether an eligible task *will fit* at a workstation. The desired output rate determines the cycle time, and the sum of the task times assigned to any workstation must not exceed the cycle time. If a task can be assigned to a workstation without exceeding the cycle time, then the task will fit.

Once it is known which tasks are *eligible* and *will fit,* the manager can select the task to be assigned (if there is more than one to choose from). This is where the heuristic rules come in to help us decide which task to assign from among those that are eligible and will fit.

To clarify the terminology, *following tasks* are all tasks that you would encounter by following all paths from the task in question through the precedence diagram. *Preceding tasks* are all tasks you would encounter by tracing all paths *backward* from the task in question. In the precedence diagram below, tasks *b, d, e,* and *f* are followers of task *a.* Tasks *a, b,* and *c* are preceding tasks for *e.*

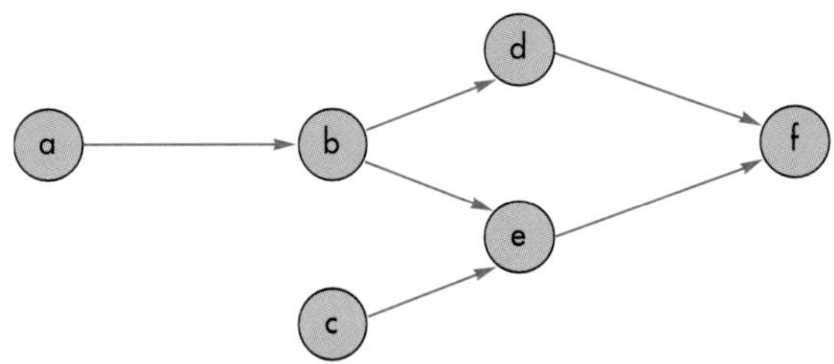

The *positional weight* for a task is the sum of the task times for itself and all its following tasks.

None of the heuristics *guarantees* the *best* solution, or even a good solution to the line-balancing problem, but they do provide guidelines for developing a solution. It may be useful to apply several different heuristics to the same problem and pick the best (least idle time) solution out of those developed.

Example 2

Using the information contained in the table shown, do each of the following:

1. Draw a precedence diagram.
2. Assuming an eight-hour workday, compute the cycle time needed to obtain an output of 400 units per day.
3. Determine the minimum number of workstations required.
4. Assign tasks to workstations using this rule: Assign tasks according to greatest number of following tasks. In case of a tie, use the tiebreaker of assigning the task with the longest processing time first.

Task	Immediate Follower	Task Time (in minutes)
a	b	0.2
b	e	0.2
c	d	0.8
d	f	0.6
e	f	0.3
f	g	1.0
g	h	0.4
h	end	0.3
		$\Sigma t = 3.8$

Solution

1. Drawing a precedence diagram is a relatively straightforward task. Begin with activities with no predecessors. We see from the list of Immediate Followers that tasks a and c do not appear. Hence, they have no immediate predecessors. We build from here.

Task b follows a, and d follows c.

Task e follows b.

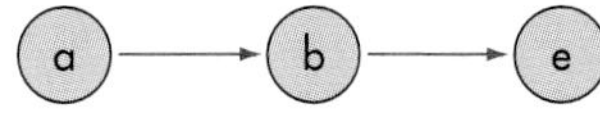

Task f follows e and d.

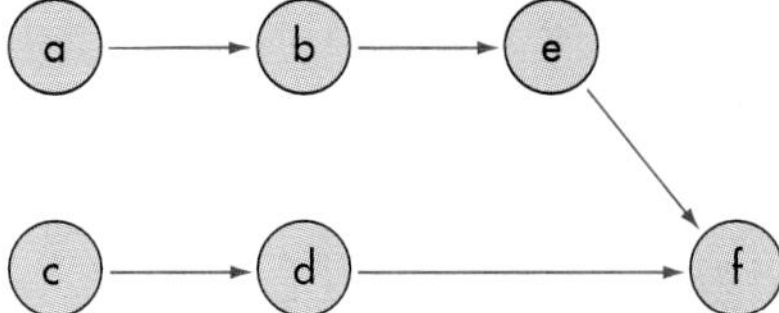

Task g follows f, and h follows g.

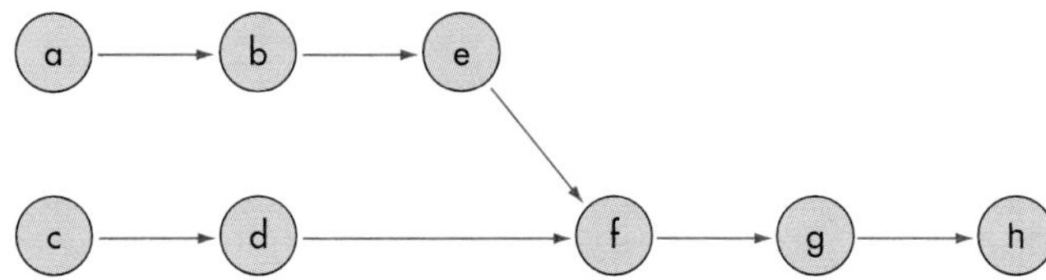

2. $CT = \frac{OT}{D} = \frac{480 \text{ minutes per day}}{400 \text{ units per day}} = 1.2 \text{ minutes per cycle}$

3. $N = \frac{\Sigma t}{CT} = \frac{3.8 \text{ minutes per unit}}{1.2 \text{ minutes per cycle per station}}$
$= 3.17 \text{ stations (round to 4)}$

4. Beginning with station 1, make assignments following this procedure: Determine from the precedence diagram which tasks are eligible for assignment. Then determine which of the eligible tasks will fit the time remaining for the station. Use the tie-breaker if necessary. Once a task has been assigned, remove it from consideration. When a station cannot take any more assignments, go on to the next station. Continue until all tasks have been assigned.

Station	Time Remaining	Eligible	Will Fit	Assign (task time)	Idle
1	1.2 min.	a, c*	a, c*	a (0.2)	
	1.0	c, b**	c, b**	c (0.8)	
	0.2	b, d	b	b (0.2)	
	0	e, d	None	—	0.0
2	1.2	e, d	e, d	d (0.6)	
	0.6	e	e	e (0.3)	
	0.3***	f	None	—	0.3
3	1.2	f	f	f (1.0)	
	0.2	g	None	—	0.2
4	1.2	g	g	g (0.4)	
	0.8	h	h	h (0.3)	
	0.5	—	—	—	0.5
					1.0 min.

*Neither a nor c have any predecessors, so both are eligible. Task a was assigned since it has more followers.

**Once a is assigned, b and c are now eligible. Both will fit in the time remaining of 1.0 minute. The tie cannot be broken by the "most followers" rule, so the longer task is assigned.

***Although f is eligible, this task will not fit, so station 2 is left with 0.3 minute of idle time per 1.2 minute cycle.

These assignments are shown in the following diagram. If you look carefully at this solution, you may discover that it can be improved upon. Thus, this solution is not necessarily optimal. One should not expect that heuristic approaches will always produce optimal solutions; they merely provide a practical way to deal with complex problems that may not lend themselves to optimizing techniques.

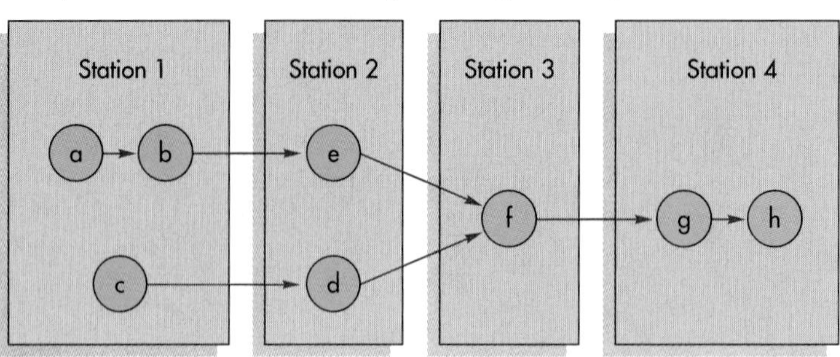

Other Factors

The preceding discussion on line balancing presents a relatively straightforward approach to approximating a balanced line. In practice, the ability to do this usually involves additional considerations, some of which are technical.

Technical considerations include skill requirements of different tasks. If skill requirements of tasks are quite different, it may not be feasible to place the tasks in the same workstation. Similarly, if the tasks themselves are incompatible (e.g., the use of fire and flammable liquids), it may not be feasible even to place them in stations that are near to each other.

Developing a workable plan for balancing a line may also require consideration of human factors, and equipment and space limitations.

Although it is convenient to treat assembly operations as if they occur at the same rate time after time, it is more realistic to assume that whenever humans are involved, task completion times will be variable. The reasons for the variations are numerous, including fatigue, boredom, and failure to concentrate on the task at hand. Absenteeism also can affect line balance.

For these reasons, lines that involve human tasks are more of an ideal than a reality. In practice, lines are rarely perfectly balanced. However, this is not entirely bad, because some unbalance means that slack exists at points along the line, which can reduce the impact of brief stoppages at some workstations. Also, workstations that have slack can be used for new workers who may not be "up to speed."

Other Approaches

There are a number of other approaches companies use to achieve a smooth flow of production. One approach is to use *parallel workstations.* These are beneficial for bottleneck operations which would otherwise disrupt the flow of product as it moves down the line. The bottlenecks may be the result of difficult or very long tasks. Parallel workstations increase the work flow and provide flexibility.

Consider this example.[3] A job has four tasks; task times are 1 minute, 1 minute, 2 minutes, and 1 minute. The cycle time for the line would be 2 minutes, and the output rate would be 30 units per hour:

$$\frac{\text{60 minutes per hour}}{\text{2 minutes per unit}} = \text{30 units per hour}$$

[3]Adapted from Mikell P. Groover, *Automation, Production Systems, and Computer-Aided Manufacturing,* Englewood Cliffs: Prentice-Hall, 1987, 2nd ed., Chapter 6.

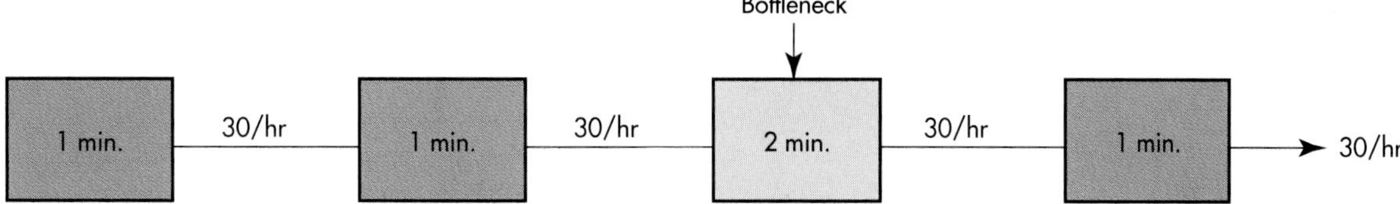

Using parallel stations for the third task would result in a cycle time of 1 minute because the output rate at the parallel stations would be equal to that of a single station, and allow an output rate for the line of 60 units per hour:

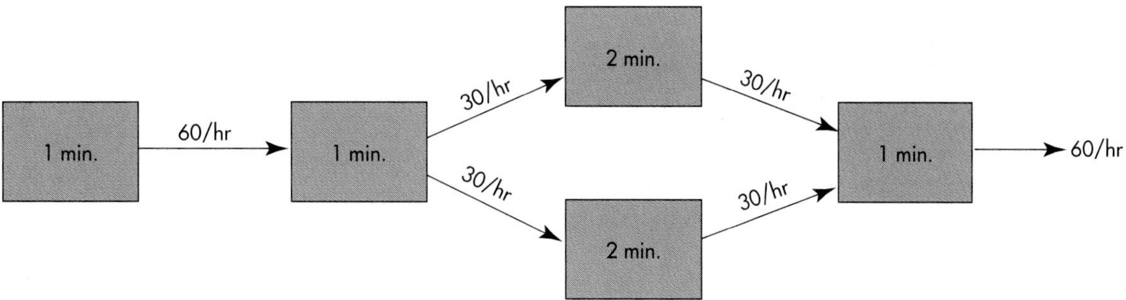

Another approach to achieving a balanced line is to *cross-train* workers so that they are able to perform more than one task. Then, when bottlenecks occur, the workers with temporarily increased idle time can assist other workers who are temporarily overburdened, thereby maintaining an even flow of work along the line. This is sometimes referred to as *dynamic line balancing,* and it is used most often in lean production systems.

Still another approach is to design a line to handle more than one product on the same line. This is referred to as a *mixed model line.* Naturally, the products have to be fairly similar, so that the tasks involved are pretty much the same for all products. This approach offers great flexibility in varying the amount of output of the products. The following newsclip describes one such line.

NEWSCLIP

TOYOTA MIXES AND MATCHES

Toyota, long famous for producing quality cars and light trucks, decided to produce minivans in the U.S. Wanting to get into production quickly, Toyota took an ambitious step at its Georgetown, Kentucky, manufacturing plant, deciding to produce Sienna minivans at the same time—and at the same workstations—that produce Camry automobiles.

"Although Camry and Sienna are built from the same basic chassis, and share 50 percent of their parts, there are key differences. Sienna is five inches longer, three inches wider and a foot taller than Camry. Each Sienna takes up more space on the assembly line and requires more and bigger parts.

"Another automaker might shut down a plant for months to make the changes. But Toyota needed to move quickly. Delay could jeopardize booming sales of Camry.

"Out of 300 stations on the assembly line, Sienna needs different parts at 26. But only seven new production steps are needed. . . . To save time, Toyota decided not to add workstations. Instead, it selected two teams of workers, one for each shift, that are responsible for attaching Sienna-only parts. Meanwhile, engineers, working with Toyota workers, designed equipment intended to make those duties easy to perform."

As soon as a Sienna approaches one of the seven spots on the assembly line, a member of the Sienna team is there to take over. Some team members climb inside, where they scoot around on wheeled carts that look like NASA's Sojourner Mars explorer. Others attach roof racks by standing on platforms that put the van's top waist high, eliminating the need to reach.

Toyota may have shortened the time needed to start production by three years by using this innovative approach to the assembly line.

Source: Based on "Camry Assembly Line Delivers New Minivan," p. B3. Copyright 1997, *USA TODAY.* Adapted with permission.

FIGURE 6–9

Work centers must be assigned to locations

Locations

A	B	C
D	E	F

Work centers to be assigned

1
2
3
4
5
6

DESIGNING PROCESS LAYOUTS

The main issue in design of process layouts concerns the relative positioning of the departments involved. As illustrated in Figure 6–9, departments must be assigned to locations. The problem is to develop a reasonably good layout; some combinations will be more desirable than others. For example, some departments may benefit from adjacent locations whereas others should be separated. A lab with delicate equipment would not be located near a department that had equipment with strong vibrations. Conversely, two departments that share some of the same equipment would benefit from being close together.

Layouts can also be influenced by external factors such as the location of entrances, loading docks, elevators, windows, and areas of reinforced flooring. Also important are noise levels, safety, and the size and locations of restrooms.

www.mcdonalds.com

In a few instances (e.g., the layouts of supermarkets, gas stations, and fast-food chains), a sufficient number of installations having similar characteristics justify the development of standardized layouts. For example, the use of the same basic patterns in McDonald's fast-food locations facilitates construction of new structures and employee training. Food preparation, order taking, and customer service follow the same pattern throughout the chain. Installation and service of equipment are also standardized. This same concept has been successfully employed in computer software products such as Microsoft Windows® and the Macintosh® Operating System. Different applications are designed with certain basic features in common, so that a user familiar with one application can readily use other applications without having to start from scratch with each new application.

The majority of layout problems involve single rather than multiple locations, and they present unique combinations of factors that do not lend themselves to a standardized approach. Consequently, these layouts require customized designs.

A major obstacle to finding the most efficient layout of departments is the large number of possible assignments. For example, there are more than 87 billion different ways that 14 departments can be assigned to 14 locations if the locations form a single line. Different location configurations (e.g., 14 departments in a two-by-seven grid) often reduce the number of possibilities, as do special requirements (e.g., the stamping department may have to be assigned to a location with reinforced flooring). Still, the remaining number of layout possibilities is quite large. Unfortunately, no algorithms exist to identify the best layout arrangement under all circumstances. Often planners must rely on heuristic rules to guide trial-and-error efforts for a satisfactory solution to each problem.

Measures of Effectiveness

One advantage of process layouts is their ability to satisfy a variety of processing requirements. Customers or materials in these systems require different operations and different sequences of operations, which causes them to follow different paths through the system. Material-oriented systems necessitate the use of variable-path material-handling equipment to move materials from work center to work center. In customer-oriented systems, people must travel or be transported from work center to work center. In both cases, transportation costs or time can be significant. Because of this factor, one of the

The profitable produce section of a grocery store is prominently located along the store's perimeter. This location assures that the customer has many opportunities to purchase additional items when coming out of the aisle locations.

The aisle locations contain the less profitable store and name brands, including canned goods and packaged items.

major objectives in process layout is to minimize transportation cost, distance, or time. This is usually accomplished by locating departments with relatively high interdepartmental work flow as close together as possible.

Other concerns in choosing among alternative layouts include initial costs in setting up the layout, expected operating costs, the amount of effective capacity created, and the ease of modifying the system.

In situations that call for improvement of an existing layout, costs of relocating any work center must be weighed against the potential benefits of the move.

Information Requirements

The design of process layouts requires the following information:

1. A list of departments or work centers to be arranged, their approximate dimensions, and the dimensions of the building or buildings that will house the departments.
2. A projection of future work flows between the various work centers.
3. The distance between locations and the cost per unit of distance to move loads between locations.
4. The amount of money to be invested in the layout.
5. A list of any special considerations (e.g., operations that must be close to each other or operations that must be separated).

The ideal situation is to first develop a layout and then design the physical structure around it, thus permitting maximum flexibility in design. This procedure is commonly followed when new facilities are constructed. Nonetheless, many layouts must be developed in existing structures where floor space, the dimensions of the building, location of entrances and elevators, and other similar factors must be carefully weighed in designing the layout. Note that multilevel structures pose special problems for layout planners.

Minimizing Transportation Costs or Distances

The most common goals in designing process layouts are minimization of transportation costs or distances traveled. In such cases, it can be very helpful to summarize the necessary data in *from-to charts* like those illustrated in Tables 6–2 and 6–3. Table 6–2 indicates the distance between each of the locations, and Table 6–3 indicates actual or projected work flow between each pair. For instance, the distance chart reveals that a

TABLE 6–2

Distance between locations (meters)

From \ To	A	B	C
	LOCATION		
A	—	20	40
B	20	—	30
C	40	30	—

TABLE 6–3

Interdepartmental work flow (loads per day)

From \ To	1	2	3
	DEPARTMENT		
1	—	10	80
2	20	—	30
3	90	70	—

trip from location A to location B will involve a distance of 20 meters. (Distances are often measured between department centers.) Oddly enough, the length of a trip between locations A and B may differ depending on the *direction* of the trip, due to one-way routes, elevators, or other factors. To simplify the discussion, assume a constant distance between any two locations regardless of direction. However, it is not realistic to assume that interdepartmental work flows are equal—there is no reason to suspect that department 1 will send as much work to department 2 as department 2 sends to 1. For example, several departments may send goods to packaging, but packaging may send only to the shipping department.

Transportation costs can also be summarized in from-to charts, but we shall avoid that complexity, assuming instead that costs are a direct, linear function of distance.

Example 3

Assign the three departments shown in Table 6–3 to locations A, B, and C, which are separated by the distances shown in Table 6–2, in such a way that transportation cost is minimized. Use this heuristic: Assign departments with the greatest interdepartmental work flow first to locations that are closest to each other.

Solution

Ranking departments according to highest work flow and locations according to highest interlocation distances helps in making assignments. If interlocation distances are independent of direction of flow, work flow between departments can be summed to achieve a clearer picture of the need for closeness. Thus:

Trip	Distance (meters)	Department Pair	Work Flow	
A–B	20	3–1	90	170
B–A	20	1–3	80	
B–C	30	3–2	70	100
C–B	30	2–3	30	
A–C	40	2–1	20	30
C–A	40	1–2	10	

From these listings, you can see that departments 1 and 3 have the highest interdepartmental work flow, and that locations A and B are the closest. Thus, it seems reasonable to consider assigning 1 and 3 to locations A and B, although it is not yet obvious which department should be assigned to which location. Further inspection of

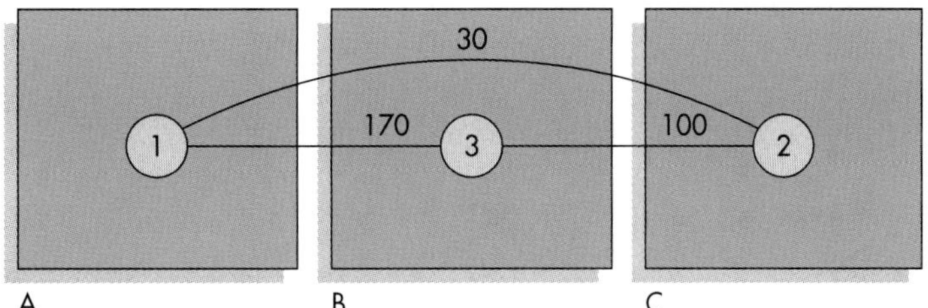

FIGURE 6–10

Interdepartmental work flows for assigned departments

the work flow list reveals that 2 and 3 have higher work flow than 1 and 2, so 2 and 3 should probably be located more closely than 1 and 2. Hence, it would seem reasonable to place 3 between 1 and 2, or at least centralize that department with respect to the other two. The resulting assignments might appear as illustrated in Figure 6–10.

Note that work flows between departments in the diagram are the sum of the flow each way (e.g., the 170 between 1 and 3 is the combined flow of 90 loads from 3 to 1 and 80 loads from 1 to 3).

If the cost per meter to move any load is $1, you can compute the total daily transportation cost for this assignment by multiplying each department's number of loads by the trip distance, and summing those quantities:

Department	Number of Loads to:	Location	Distance to:	Loads × Distance
1	2: 10	A	C: 40	10 × 40 = 400
	3: 80		B: 20	80 × 20 = 1,600
2	1: 20	C	A: 40	20 × 40 = 800
	3: 30		B: 30	30 × 30 = 900
3	1: 90	B	A: 20	90 × 20 = 1,800
	2: 70		C: 30	70 × 30 = 2,100
				7,600

At $1 per load meter, the cost for this plan is $7,600 per day. Even though it might appear that this arrangement yields the lowest transportation cost, you cannot be absolutely positive of that without actually computing the total cost for every alternative and comparing it to this one. Although this could be done in this example, where the number of alternatives is quite small (i.e., there are 3! = 6 possible arrangements), in problems with more departments, the number is likely to be too large to even consider examining every alternative. Instead, rely on the choice of reasonable heuristic rules such as those demonstrated above to arrive at a satisfactory, if not optimal, solution.

Closeness Ratings

Although the preceding approach is widely used, it suffers from the limitation of focusing on only one objective, and many situations involve multiple criteria. Richard Muther developed a more general approach to the problem, which allows for subjective input from analysis or managers to indicate the relative importance of each combination of department pairs.[4] That information is then summarized in a grid like that shown in Figure 6–11. Read the grid in the same way as you would read a mileage chart on a road map, except that letters rather than distances appear at the intersections. The letters represent the importance of closeness for each department pair, with A being the most important and X being an undesirable pairing. Thus, in the grid it is "absolutely necessary" to locate 1 and 2 close to each other because there is an A at the intersection of

[4]Richard Muther and John Wheeler, "Simplified Systematic Layout Planning," *Factory* 120, nos. 8, 9, and 10 (August, September, October, 1962) pp. 68–77, 111–19, 101–13.

FIGURE 6–11

A Muther grid

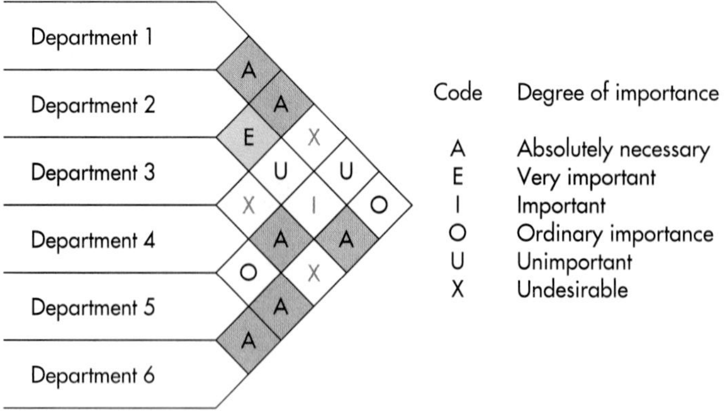

those departments on the grid. On the other hand, 1 and 4 should not be close together because their intersection has an X. In practice, the letters on the grid are often accompanied by numbers that indicate the reason for each assignment: they are omitted here to simplify the illustration. Muther suggests the following list:

1. Use same equipment or facilities.
2. Share the same personnel or records.
3. Sequence of work flow.
4. Ease of communication.
5. Unsafe or unpleasant conditions.
6. Similar work performed.

Example 4

Assign the six departments in Figure 6–10 to a 2 × 3 set of locations using the heuristic rule: Assign critical departments first.

Solution

Critical pairs of departments are those with A or X ratings. Prepare a list of those by referring to the grid:

As	Xs
1–2	1–4
1–3	3–6
2–6	3–4
3–5	
4–6	
5–6	

Next, form a cluster of A links, beginning with the department that appears most frequently in the A list (in this case, 6). For instance:

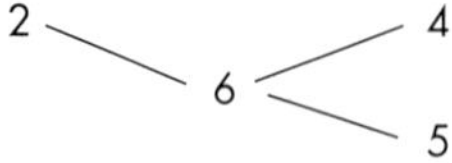

Take the remaining As in order, and add them to this main cluster where possible, rearranging the cluster as necessary. Form separate clusters for departments that do not link with the main cluster. In this case, all link with the main cluster.

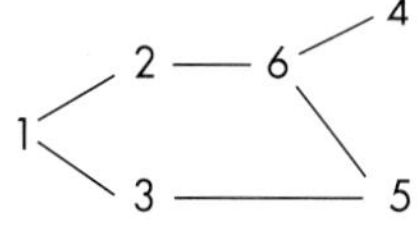

Next, graphically portray the Xs:

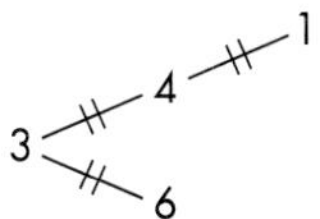

Observe that, as it stands, the cluster of As also satisfies the X separations. It is a fairly simple exercise to fit the cluster into a 2 × 3 arrangement:

1	2	6
3	5	4

Note that the lower-level ratings have also been satisfied with this arrangement, even though no attempt was made to explicitly consider the E and I ratings. Naturally, not every problem will yield the same results, so it may be necessary to do some additional adjusting to see if improvements can be made, keeping in mind that the A and X assignments deserve the greatest consideration.

Note that departments are considered close not only when they touch side to side, but also when they touch corner to corner.

The value of this rating approach is that it permits the use of multiple objectives and subjective inputs. Its limitations relate to the use of subjective inputs in general: They are imprecise and unreliable.

Computer Analysis

The size and complexity of process layout problems have led to the development of a number of computerized packages. The obvious advantage of computerized analyses of layout problems is the ability to handle large problems and to consider many different layout alternatives. Even so, in some instances, the results of computer analysis require additional manual adjustments before they can be used.

Summary

Layout decisions are an important aspect of the design of production systems, affecting operating costs and efficiency. Layout decisions are related to decisions involving location, product and service design, process selection, and capacity.

A major determinant of layout design is the type of processing required of a system. The two basic types are repetitive and intermittent. Repetitive processing is characterized by a high volume of one or a few similar products or services. Intermittent processing provides the processing capability needed to deal with a broader range of products or services than repetitive processing. Product layouts are typically used for repetitive processing, process layouts for intermittent processing, and fixed-position layouts for projects, although some projects do not require layouts.

Product layouts are geared to high-volume output of standardized items. Workers and equipment are arranged according to the technological sequence required by the product or service involved. Emphasis in design is on work flow through the system, and specialized processing and handling equipment is often used. Product layouts are highly vulnerable to breakdowns. Preventive maintenance reduces the occurrence of breakdowns.

Process layouts group similar activities into departments or other work centers. These systems can handle a wide range of processing requirements and are less susceptible to breakdowns. However, the variety of processing requirements necessitates continual routing and scheduling

TABLE 6–4

Product and process layouts compared

	Product Layout	Process Layout
Description	A sequential arrangement of personnel and/or equipment designed to provide standardized processing.	A functional arrangement of personnel and/or equipment designed to handle a variety of processing requirements.
Focus of layout	Balance operations to avoid bottlenecks, and attain a smooth flow of work.	Arrange equipment or departments to minimize transportation costs and/or congestion.
Processing	Repetitive or continuous.	Job shop or batch.
Examples of: goods	Automobiles, video tapes.	Furniture, baked goods.
services	Car wash, cafeteria line.	Auto repair, health care.
Product or service variety	Low.	Moderate to high.
Typical processing worker skill levels	Low, semiskilled.	Semiskilled to highly skilled.
Flexibility	Very low.	Moderate to high.
Volume	High.	Low to moderate.
Work-in-process inventory levels	Low.	High.
Material handling	Fixed path (e.g., conveyors).	Variable path (e.g., forklifts).
Primary approach to equipment maintenance	Preventive.	As needed.
Benefits	Low cost per unit, high productivity.	Can handle a variety of processing requirements.

and the use of variable-path material-handling equipment. The rate of output is generally much lower than that of product layouts.

Table 6–4 summarizes product and process layouts.

Fixed-position layouts are used when size, fragility, cost, or other factors make it undesirable or impractical to move a product through a system. Instead, workers, equipment, and materials are brought to the product.

The main design efforts in product layout development focus on dividing up the work required to produce a product or service into a series of tasks that are as nearly equal as possible. The goal is to achieve a high degree of utilization of labor and equipment. In process layout, design efforts often focus on the relative positioning of departments to minimize transportation costs or to meet other requirements concerning the proximity of certain department pairs.

The large number of possible alternatives to layout problems prevents an examination of each one. Instead, heuristic rules guide discovery of alternatives. The solutions thus obtained are usually satisfactory although not necessarily optimal. Computer packages are available to reduce the effort required to obtain solutions to layout problems, but these too rely largely on heuristic methods.

Key Terms

assembly line, 269
balance delay, 282
cellular manufacturing, 274
cycle time, 279
fixed-position layout, 272
group technology, 275
line balancing, 279
precedence diagram, 281
process layout, 271
product layout, 269
production line, 269

Solved Problems

Problem 1

The tasks shown in the following precedence diagram are to be assigned to workstations with the intent of minimizing idle time. Management has designed an output rate of 275 units per day. Assume 440 minutes are available per day.

a. Determine the appropriate cycle time.

b. What is the minimum number of stations possible?

c. Assign tasks using the "positional weight" rule: Assign tasks with highest following times (including a task's own time) first. Break ties using greatest number of following tasks.

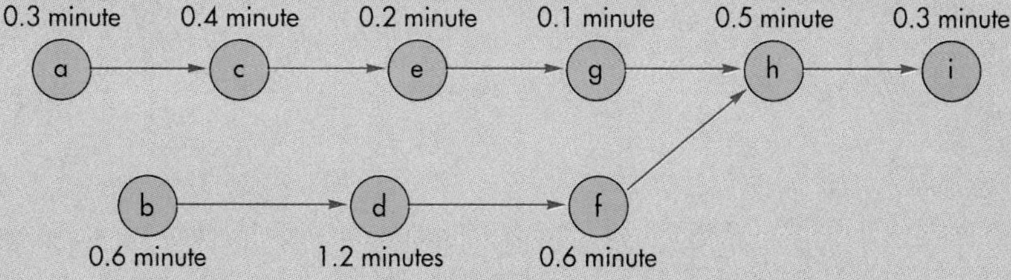

Solution

a. $\text{CT} = \dfrac{\text{Operating time}}{\text{Desired output}} = \dfrac{440 \text{ minutes per day}}{275 \text{ units per day}} = 1.6 \text{ minutes}$

b. $N = \dfrac{\Sigma t}{\text{Cycle time}} = \dfrac{4.2}{1.6 \text{ minutes}} = 2.625 \text{ (round to 3)}$

c. Add positional weights (following time including task time) to the diagram:

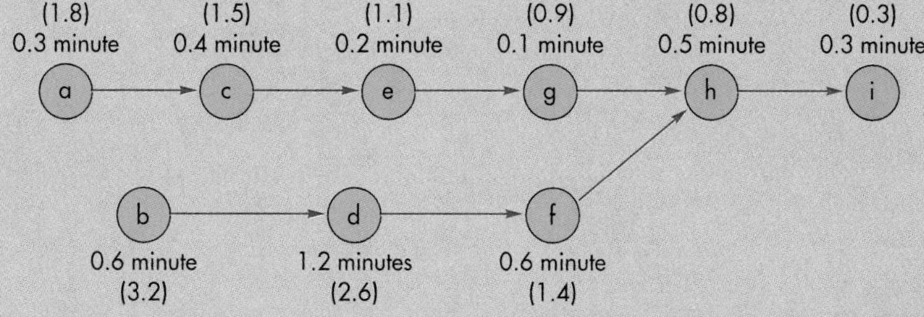

Station	Time Remaining*	Eligible	Will Fit	Assign Task/Time	Station Idle Time
1	1.6	a, b	a, b	b/0.6	
	1.0	a, d	a	a/0.3	
	0.7	c, d	c	c/0.4	
	0.3	e, d	e	e/0.2	
	0.1	g, d	g	g/0.1	
	0	—	—	—	0
2	1.6	d	d	d/1.2	
	0.4	f	none	none	0.4
3	1.6	f	f	f/0.6	
	1.0	h	h	h/0.5	
	0.5	i	i	i/0.3	
	0.2	—	—	—	0.2
					0.6

*The initial time for each station is the cycle time computed in part *a*.

The resulting assignments are shown below.

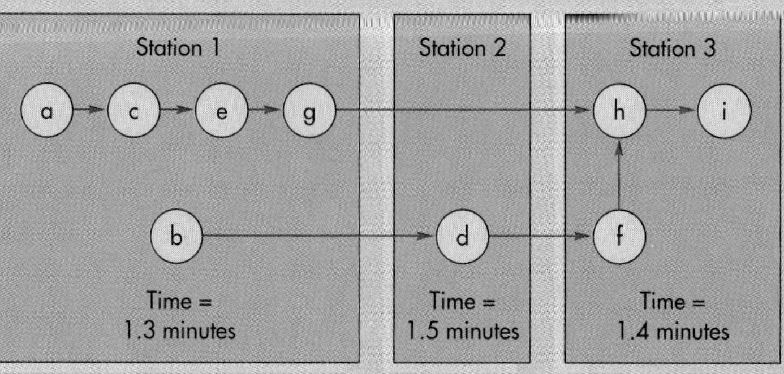

Problem 2

Assign nine automobile service departments to bays in a 3 × 3 grid so that the closeness ratings in the following matrix are satisfied. (The unimportant and ordinary-importance ratings have been omitted to simplify the example.) The location of department 4 must be in the upper right-hand corner of the grid to satisfy a town ordinance.

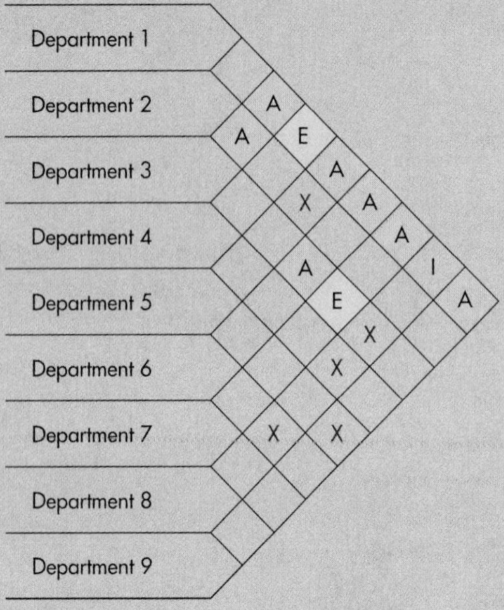

Solution

Note that department 1 has many A ratings, making it a strong candidate for the center position in the grid. We can form a cluster of departments that should be close together:

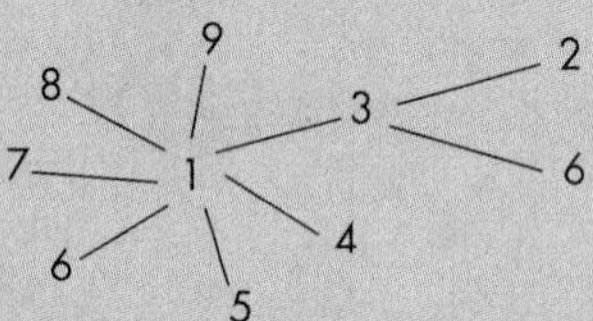

Next, we can identify departmental pairings that should be avoided:

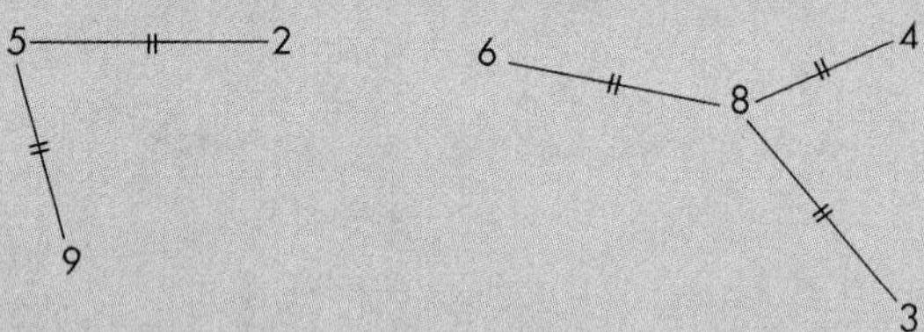

These departments should be spaced around the perimeter of the grid. After a bit of trial and error, the final grid shown below emerged. Check it against the rating matrix to see if it satisfies the ratings.

2	3	4
9	1	6
8	7	5

Problem 3

Five departments are to be assigned to locations B–F in the grid. (For technical reasons, department 6 must be assigned to location A.) Transportation cost is $2 per foot. The objective is to minimize total transportation cost. Information on interdepartmental work flows and distances between locations is shown in the following tables. Assign departments with the greatest interdepartmental work flow first.

DISTANCE BETWEEN LOCATIONS (FEET)

From \ To	A	B	C	D	E	F
A	—	50	100	50	80	130
B		—	50	90	40	70
C			—	140	60	50
D				—	50	120
E					—	50
F						—

NUMBER OF TRIPS PER DAY BETWEEN CENTERS

From \ To	1	2	3	4	5	6
1	—	90	25	23	11	18
2	35	—	8	5	10	16
3	37	2	—	1	0	7
4	41	12	1	—	4	0
5	14	16	0	9	—	3
6	32	38	13	2	2	—

A Dept. 6	B	C
D	E	F

Solution

a. First determine the interdepartmental work flows (e.g., for 1–2 the flow is 90 + 35 = 125). Then either rank them, or arrange them from high to low. Here they have been arranged from high to low.

Dept.	Workflow
1–2	125
1–4	64
1–3	62
2–6	54
1–6	50
2–5	26
1–5	25
3–6	20
2–4	17
4–5	13
2–3	10
5–6	5
3–4	2
4–6	2
3–5	0

From this, we can see that departments 1 and 2 have the greatest interdepartmental work flow, so they should be close, perhaps at B and E. Next, work flows for 1–3 and 1–4 are high. Note, though, that the work flow for 3–4 is low, suggesting that they need not be close. Instead, we would place them on either side of department 1. Note also that 3–4 is only 2, 5–6 is 0, while 3–6 is 20 and 4–5 is 13. Hence, place department 3 at location D, department 4 at location F, and department 5 at location C.

A Dept. 6	B Dept. 2	C Dept. 5
D Dept. 3	E Dept. 1	F Dept. 4

Total cost:

Trip	b Distance	c Frequency	(b × c × $2) Cost
1–2	(B–E) 40	125	$10,000
1–3	(D–E) 50	62	6,200
1–4	(F–E) 50	64	6,400
1–5	(E–C) 60	25	3,000
1–6	(A–E) 80	50	8,000
2–3	(B–D) 90	10	1,800
2–4	(B–F) 70	17	2,380
2–5	(B–C) 50	26	2,600
2–6	(A–B) 50	54	5,400
3–4	(F–D) 120	2	480
3–5	(D–C) 140	0	0
3–6	(A–D) 50	20	2,000
4–5	(C–F) 50	13	1,300
4–6	(A–F) 130	2	520
5–6	(A–C) 100	5	1,000
			$51,080

Discussion and Review Questions

1. List some common reasons for redesigning layouts.
2. Briefly describe the two main layout types.
3. What are the main advantages of a product layout? The main disadvantages?
4. What are the main advantages of a process layout? The main disadvantages?
5. What is the goal of line balancing? What happens if a line is unbalanced?
6. Why are routing and scheduling continual problems in process layouts?
7. Compare equipment maintenance strategies in product and process layouts.
8. Briefly outline the impact that job sequence has on each of the layout types.
9. The City Transportation Planning Committee must decide whether to begin a long-term project to build a subway system or to upgrade the present bus service. Suppose you are an expert in fixed-path and variable-path material-handling equipment, and the committee seeks your counsel on this matter. What are the advantages and limitations of the subway and bus systems?
10. Identify the fixed-path and variable-path material-handling equipment commonly found in supermarkets.
11. What are heuristic approaches, and why are they used in designing layouts?
12. Why are product layouts atypical in service environments?

13. According to a study by the Alliance of American Insurers, it costs more than three times the original purchase price in parts and labor to replace a totally wrecked Chevrolet. Explain the reasons for this large discrepancy in terms of the processes used to assemble the original car and those required to reconstruct the wrecked car.
14. Name some ways that a layout can help or hinder productivity.
15. What is cellular manufacturing? What are its main benefits and limitations?
16. What is group technology?

Memo Writing Exercises

1. The manager of financial services, Richard Winchester, remarked on a recent tour that a production line should be perfectly balanced, and that if done correctly, there should be little or no idle time along the line. Write a one-page memo to him, agreeing or disagreeing.
2. Write a memo to Karen Flint, plant manager, outlining the benefits of producing two products on the same assembly line, as Toyota did with its Sienna minivan and Camry automobile. Refer to the newsclip in the "Other Approaches" section of the chapter.

Problems

1. An assembly line with 17 tasks is to be balanced. The longest task is 2.4 minutes, and the total time for all tasks is 18 minutes. The line will operate for 450 minutes per day.
   a. What are the minimum and maximum cycle times?
   b. What range of output is theoretically possible for the line?
   c. What is the minimum number of workstations needed if the maximum output rate is to be sought?
   d. What cycle time will provide an output rate of 125 units per day?
   e. What output potential will result if the cycle time is (1) 9 minutes? (2) 15 minutes?

a. Min = 2.4 min.; max = 18 min.
b. 25 to 187.5 units
c. 8 stations
d. 3.6 min/cycle
e. (1) 50 units
(2) 30 units

2. A manager wants to assign tasks to workstations as efficiently as possible, and achieve an hourly output of 33⅓ units. Assume the shop works a 60-minute hour. Assign the tasks shown in the accompanying precedence diagram (times are in minutes) to workstations using the following rules:
   a. In order of most following tasks. Tie breaker: greatest positional weight.
   b. In order of greatest positional weight.
   c. What is the efficiency?

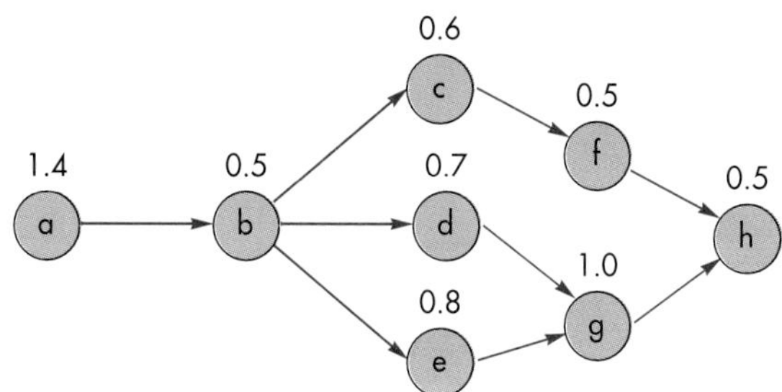

a.

Station	Tasks
1	a
2	b, e
3	d, c, f
4	g, h

b.

Station	Tasks
1	a
2	b, e
3	d, c, f
4	g, h

c. *a* and *b* are both 83.33%

3. A manager wants to assign tasks to workstations as efficiently as possible, and achieve an hourly output of 4 units. The department uses a working time of 56 minutes per hour. Assign the tasks shown in the accompanying precedence diagram (times are in minutes) to workstations using the following rules:
   a. In order of most following tasks. Tie breaker: greatest positional weight
   b. In order of greatest positional weight.
   c. What is the efficiency?

a.

Station	Tasks
1	f, a
2	d, g
3	b, c, e
4	h, i

b.

Station	Tasks
1	f, d
2	g, a, b
3	e, c
4	h, i

c. both are 80.36%

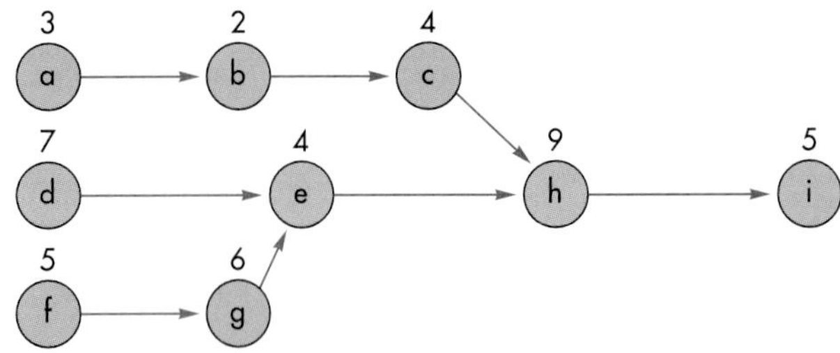

4. A large manufacturer of pencil sharpeners is planning to add a new line of sharpeners, and you have been asked to balance the process, given the following task times and precedence relationships. Assume that cycle time is to be the minimum possible.

Task	Length (minutes)	Immediate Follower
a	0.2	b
b	0.4	d
c	0.3	d
d	1.3	g
e	0.1	f
f	0.8	g
g	0.3	h
h	1.2	end

a. (1) See IM
(2) Furnish task assignments
(3) 11.54%
(4) 323 units per day
b. (1) 2.3 minutes, yes; assign a, b, c, d, e to Station 1; f, g, h to Station 2.
(2) 0%
(3) 182.6 units/day
(4) 91.3 units/day

a. Do each of the following:
   (1) Draw the precedence diagram.
   (2) Assign tasks to stations in order of greatest number of following tasks.
   (3) Determine the percentage of idle time.
   (4) Compute the rate of output that could be expected for this line assuming a 420-minute working day.

b. Answer these questions:
   (1) What is the shortest cycle time that will permit use of only two workstations? Is this cycle time feasible? Identify the tasks you would assign to each station.
   (2) Determine the percentage of idle time that would result if two stations were used.
   (3) What is the daily output under this arrangement?
   (4) Determine the output rate that would be associated with the maximum cycle time.

5. As part of a major plant renovation project, the industrial engineering department has been asked to balance a revised assembly operation to achieve an output of 240 units per eight-hour day. Task times and precedence relationships are is follows:

Task	Duration (minutes)	Precedes Task
a	0.2	b
b	0.4	c
c	0.2	f
d	0.4	e
e	1.2	g
f	1.2	g
g	1.0	end

Do each of the following:

a. See IM
b. 2.0 minutes
c. 3 stations

a. Draw the precedence diagram.

b. Determine the maximum cycle time.

c. Determine the minimum number of stations needed.

d. Assign tasks to workstations on the basis of greatest number of following tasks. Use longest processing time as a tiebreaker. If ties still exist, assume indifference in choice.

e. Compute the percentage of idle time for the assignment in part *d.*

d.

Station	Tasks
1	a, b, d, c
2	e
3	f
4	g

e. 4.17 percent

6. Twelve tasks, with times and precedence requirements as shown in the following table, are to be assigned to workstations using a cycle time of 1.5 minutes. Two heuristic rules will be tried: (1) greatest positional weight, and (2) greatest number of following tasks.

   In each case, the tiebreaker will be shortest task time.

Task	Length (minutes)	Follows Task
a	0.1	—
b	0.2	a
c	0.9	b
d	0.6	c
e	0.1	—
f	0.2	d, e
g	0.4	f
h	0.1	g
i	0.2	h
j	0.7	i
k	0.3	j
l	0.2	k

a. Draw the precedence diagram for this line.

b. Assign tasks to stations under each of the three rules.

c. Compute the percentage of idle time for each rule.

a. Diagram in IM

b. (1)

Station	Tasks
1	a, b, c, e
2	d, f, g, h, i
3	j, k, l

(2) Same as (1)

c. (1), (2) 11.11 percent

7. For the set of tasks given below, do the following:

   a. Develop the precedence diagram.

   b. Determine the maximum cycle time in seconds for a desired output of 500 units in a 7-hour day. Why might a manager use a cycle time of 50 seconds?

   c. Determine the minimum number of workstations for output of 500 units per day.

   d. Balance the line using the *largest positional weight* heuristic. Break ties with the *most following tasks* heuristic. Use a cycle time of 50 seconds.

   e. Calculate the percentage idle time for the line.

a. See IM.

b. $CT = \frac{OT}{D} = \frac{7(60)}{500}$

$= .84 \text{ min.}$

c. $N = \frac{(D)(\Sigma t)}{OT} = \frac{(500)(193)}{(420)(60)}$

$= 3.83$ or 4 stations

d. See IM.

e. Percentage of idle time $= \frac{\text{Total idle time}}{\text{N} \times \text{CT}} = \frac{57}{5(50)} =$ .228 or 22.8%

Task	Task Time (seconds)	Immediate Predecessors
A	45	—
B	11	A
C	9	B
D	50	—
E	26	D
F	11	E
G	12	C
H	10	C
I	9	F, G, H
J	10	I
	193	

8. A shop works a 400-minute day. The manager of the shop wants an output of 200 units per day for the assembly line that has the elemental tasks shown in the table. Do the following:

   a. Construct the precedence diagram.

a. See IM

b.

Station	Tasks
1	a, b
2	c, d
3	e, f, i
4	g, h, j
5	k, m

c.

Station	Tasks
1	a, d
2	b
3	c, e
4	f, i, g
5	j, h, k
6	m

d. b is 15%
c is 29.17%

b. Assign tasks according to the *most following tasks* rule.

c. Assign tasks according to the *greatest positional weight* rule.

d. Compute the balance delay for each rule. Which one yields the better set of assignments in this instance?

Task	Immediately Precedes Task(s)	Task Time
a	b, c, d	.5
b	e	1.4
c	e	1.2
d	f	.7
e	g, j	.5
f	i	1.0
g	h	.4
h	k	.3
i	j	.5
j	k	.8
k	m	.9
m	end	.3

1	4	3
2	5	6

9. Arrange six departments into a 2 × 3 grid so that these conditions are satisfied: 1 close to 2, 5 close to 2 and 6, 2 close to 5, and 3 not close to 1 or 2.

10. Using the information given in the preceding problem, develop a Muther-type grid using the letters A, O, and X. Assume that any pair of combinations not mentioned have an O rating.

Diagram in IM

11. Using the information in the following grid, determine if the department locations shown are appropriate. If not, modify the assignments so that the conditions are satisfied.

1	5	4
3	8	7
6	2	

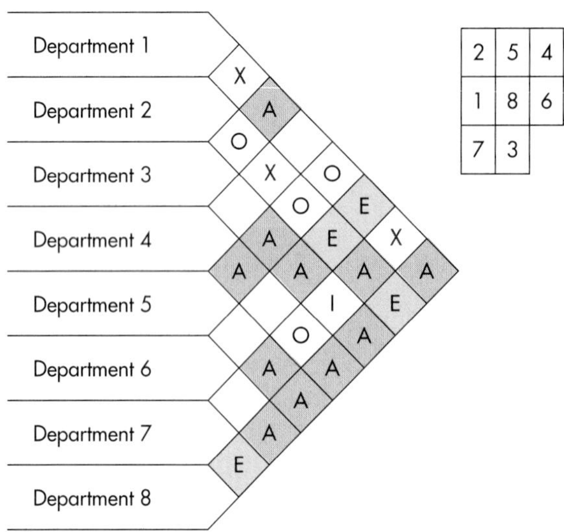

12. Arrange the eight departments shown in the accompanying Muther grid into a 2 × 4 format. *Note:* Department 1 must be in the location shown.

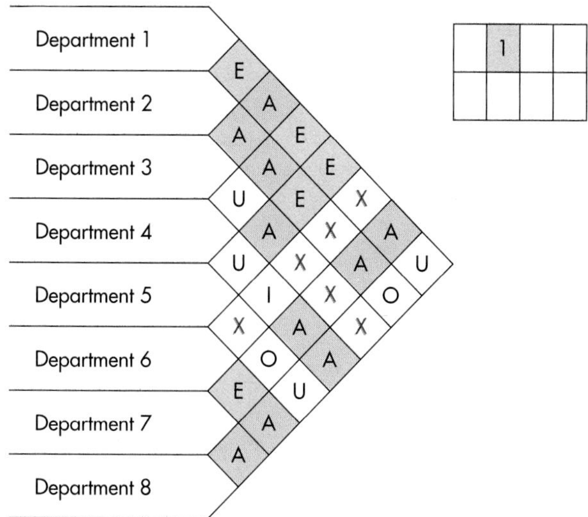

	1		

3	1	4	8
5	2	7	6

Or

5	1	4	8
3	2	7	6

13. Arrange the departments so they satisfy the conditions shown in the following rating grid into a 3 × 3 format. Place department 5 in the lower left corner of the 3 × 3 grid.

3	1	8
9	7	4
5	2	6

Department pair	Rating
1–2	X
1–3	A
1–4	E
1–5	U
1–6	X
1–7	A
1–8	A
1–9	E
2–3	X
2–4	I
2–5	A
2–6	A
2–7	A
2–8	U
2–9	E
3–4	X
3–5	U
3–6	X
3–7	E
3–8	X
3–9	A
4–5	X
4–6	A
4–7	A
4–8	E
4–9	X
5–6	X
5–7	A
5–8	X
5–9	A
6–7	E
6–8	U
6–9	X
7–8	A
7–9	A
8–9	X

14. Determine the placement of departments for a newly designed facility that will minimize total transportation costs using the data in the following tables. Assume that reverse distances are the same. The locations are shown in the grid. Use a cost of $1 per trip yard.

Location A	Location B	Location C
	Location D	

A #1	B #4	C #2
	D #3	

TC = $14,150

From \ To	DISTANCE BETWEEN LOCATIONS (YARDS)			
	A	B	C	D
A	—	40	80	70
B		—	40	50
C			—	60
D				—

From \ To	NUMBER OF TRIPS PER DAY BETWEEN DEPARTMENTS			
	1	2	3	4
1	—	10	20	30
2		—	40	40
3			—	25
4	50	50	30	—

15. Eight work centers must be arranged in an L-shaped building. The locations of centers 1 and 3 are assigned as shown in the accompanying diagram. Assuming transportation costs are $1 per load per meter, develop a suitable layout that minimizes transportation costs using the given information. Compute the total cost. (Assume the reverse distances are the same.)

A 1	B	
C	D	E 3
F	G	H

A #1	B #5	
C #7	D #4	E #3
F #6	G #2	H #8

Total cost = $143,650
Variations are reasonable as long as depts. 2, 4, 8 are close to 3. 4 is close to 5, and 5 is close to 1.

From \ To	DISTANCE (METERS)							
	A	B	C	D	E	F	G	H
A	—	40	40	60	120	80	100	110
B		—	60	40	60	140	120	130
C			—	45	85	40	70	90
D				—	40	50	40	45
E					—	90	50	40
F						—	40	60
G							—	40
H								—

From \ To	1	2	3	4	5	6	7	8
	LOADS PER DAY							
1	—	10	5	90	365	135	125	0
2	0	—	140	10	0	35	0	120
3	0	220	—	110	10	0	0	200
4	0	110	240	—	10	0	0	170
5	5	40	100	180	—	10	40	10
6	0	80	40	70	0	—	10	20
7	0	45	20	50	0	40	—	20
8	0	0	0	20	0	0	0	—

16. Develop a process layout that will minimize the total distance traveled by patients at a medical clinic, using the following information on projected departmental visits by patients and distance between locations. Assume a distance of 35 feet between the reception area and each potential location. Use the format shown.

A: 3; B: 5; C: 1; D: 4; E: 6; F: 2 Reverse order is an equivalent solution.

From \ To	A	B	C	D	E	F
	DISTANCE BETWEEN LOCATIONS (FEET)					
A	—	40	80	100	120	160
B		—	40	60	80	120
C			—	20	40	80
D				—	20	40
E					—	40
F						—

From \ To	1	2	3	4	5	6	Reception
	TRIPS BETWEEN DEPARTMENTS (PER DAY)						
Reception	10	10	200	20	0	100	—
1	—	0	0	80	20	40	10
2	0	—	0	0	0	20	40
3	40	0	—	10	190	10	10
4	30	50	0	—	10	70	0
5	60	40	60	30	—	20	10
6	10	100	0	20	0	—	30

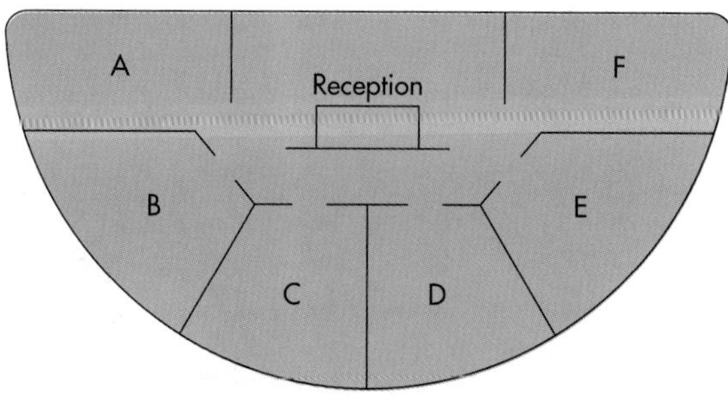

17. Ten labs will be assigned to the circular layout shown. Recalling a similar layout's congestion in the halls, the new lab manager has requested an assignment that will minimize traffic

A: 1; B: 2; C: 5; D: 4; E: 9; F: 8; G: 6; H: 10; I: 7; J: 3

between offices. In addition, movement in the halls is restricted to a counterclockwise route. Develop a suitable layout using the following information.

NUMBER OF TRIPS PER DAY BETWEEN DEPARTMENTS

From \ To	1	2	3	4	5	6	7	8	9	10
1	—	40	1	20	20	4	0	2	6	5
2	0	—	2	15	25	10	2	12	13	6
3	50	35	—	10	13	4	0	4	7	1
4	6	1	8	—	0	14	10	20	22	11
5	3	2	7	35	—	22	5	9	19	10
6	5	5	10	0	2	—	15	0	1	20
7	20	16	50	4	9	2	—	1	3	0
8	10	6	14	2	4	44	13	—	1	25
9	5	5	18	1	2	40	30	42	—	32
10	30	30	35	20	15	5	40	10	15	—

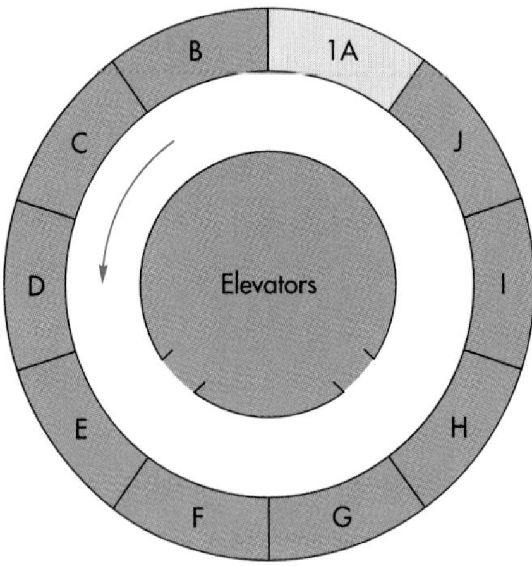

Station	Task
I	D
II	A
III	E, B, F
IV	C, G, H, I, J

Idle time = 3.5%

18. Rebalance the assembly line in Problem 7. This time, use the *longest operation time* heuristic. Break ties with the *most following tasks* heuristic. What is the percent idle time for your line?

Selected Bibliography

Cohen, Morris, and Uday M. Apte. *Manufacturing Automation*. Burr Ridge, IL: McGraw-Hill, 1997.

Francis, R. L., and J. A. White. *Facility Layout and Location: An Analytical Approach*. Englewood Cliffs, NJ: Prentice Hall, 1987.

Gerwin, D. "Do's and Don'ts of Computerized Manufacturing." *Harvard Business Review*, March–April 1982, pp. 107–16.

Groover, Mikell P. *Automation, Production Systems, and Computer-Aided Manufacturing*. Englewood Cliffs, NJ: Prentice-Hall, 1987, 2nd ed.

Hendrick, T. E., and F. G. Moore. *Production/Operations Management*. 9th ed. Burr Ridge, IL: Richard D. Irwin, 1985.

Kilbridge, M. D., and L. Wester. "A Heuristic Method of Assembly Line Balancing." *Journal of Industrial Engineering* 12 (July–August 1961).

Milas, Gene H. "Assembly Line Balancing . . . Let's Remove the Mystery." *Industrial Engineering*, May 1990, pp. 31–36.

Muther, R., and K. McPherson. "Four Approaches to Computerized Layout Planning." *Industrial Engineering* 2 (1970), pp. 39–42.

T177
FORK
FORK

LEARNING OBJECTIVES

After completing this chapter, you should be able to:

1 Explain the importance of work design.

2 Briefly describe the two basic approaches to job design.

3 Discuss the advantages and disadvantages of specialization

4 Explain the term *knowledge-based pay.*

5 Explain the purpose of methods analysis and describe how methods studies are performed.

6 Describe four commonly used techniques for motion study.

7 Discuss the impact of working conditions on job design.

8 Define a standard time.

9 Describe and compare time study methods and perform calculations.

10 Describe work sampling and perform calculations.

11 Compare stopwatch time study and work sampling.

12 Contrast time and output pay systems.

CHAPTER SEVEN DESIGN OF WORK SYSTEMS

CHAPTER OUTLINE

This chapter is devoted to design of work systems. Work design involves *job design, work measurement* and the establishment of time standards, and worker *compensation*.

As you read this chapter, note how decisions in other design areas have an impact on work systems, and how decisions on work design have an impact on the other areas. For example, product or service design decisions (e.g., operate a coal mine, offer computer dating service, sell sports equipment) in large measure determine the kinds of activities workers will be involved with. Similarly, layout decisions often influence work design. Process layouts tend to necessitate broader job content than product layouts. The implication of these interrelationships is that it is essential to adopt a systems approach to design; decisions in one area must be related to the overall system.

INTRODUCTION

The importance of work system design is underscored by the organization's dependence on human efforts (i.e., work) to accomplish its goals. Work design is one of the oldest aspects of operations management. In the past, it has often been de-emphasized in operations management courses in favor of other topics. Recent years, however, have seen renewed interest that has come from somewhat different directions: Some of the interest has resulted from studies that reveal a general dissatisfaction felt by many workers with their jobs. And some of the interest has been sparked by increasing concerns over productivity. It is perhaps ironic that one of the oldest fields of operations management is now an important key to productivity improvement and to continuous improvement.

Operations Strategy

It is important for management to make design of work systems a key element of its operations strategy. In spite of the major advances in computers and manufacturing technology, people are still the heart of a business; they can make or break it, regardless of the technology used. Technology is important, of course, but technology alone is not enough.

The topics described in this chapter—job design, methods analysis, motion study, work standards, and incentives—all have an impact on productivity. They lack the glamour of high tech; they are closer to the back-to-the-basics fundamentals of work improvement.

Workers can be a valuable source of insight and creativity because they actually perform the jobs and are closest to the problems that arise. All too often, managers have overlooked contributions and potential contributions of employees, sometimes from ignorance and sometimes from a false sense of pride. Union–management differences are also a factor. More and more, though, companies are attempting to develop a spirit of cooperation between employees and managers, based in part on the success of Japanese companies.

In the same vein, an increasing number of companies are focusing some of their attention on improving the quality of work life and instilling pride and respect among workers. Many organizations are reaping surprising gains through worker *empowerment*, giving workers more say over their jobs.

People work for a variety of reasons. Economic necessity is among the most important, but beyond that, people work for socialization, to give meaning and purpose to their lives, for status, for personal growth, and for other reasons. These motivations can play an important role in the lives of workers, and management should accord them serious consideration in the design of work systems.

JOB DESIGN

Job design involves specifying the content and methods of jobs. In general, the goal of job design is to create a work system that is productive and efficient. Job designers are concerned with *who* will do a job, *how* the job will be done, and *where* the job will be done.

job design The act of specifying the contents and methods of jobs.

To be successful, job design must be:

1. Carried out by experienced personnel who have the necessary training and background.
2. Consistent with the goals of the organization.
3. In written form.
4. Understood and agreed to by both management and employees.

The factors that affect job design and the implications of various alternatives are often so complex that a person without a good background in job design is likely to overlook important aspects of it. Workers and managers alike should be consulted in order to take advantage of their knowledge and to keep them informed. Because they are intimately involved with the work, employees are the source of valuable ideas for job improvements. Managerial support for job design depends on the commitment and involvement of managers. It is usually easier to sell a design to these two groups if they have been included in the process. Finally, establishing a written record of the job design can serve as a basis for referral if questions arise about it.

Current practice in job design contains elements of two basic schools of thought. One might be called the *efficiency* school because it emphasizes a systematic, logical approach to job design; the other is called the *behavioral* school because it emphasizes satisfaction of wants and needs.

The efficiency approach, a refinement of Frederick Winslow Taylor's scientific management concepts, received considerable emphasis in the past. The behavioral approach emerged during the 1950s and has continued to make inroads into many aspects of job design. A main contribution of the behavioral approach is that it has reminded managers of the complexity of human beings, and that the efficiency approach may not be appropriate in every instance.

The behavioral view received a shot in the arm in the 1970s when *Work in America* was released, a report detailing some of the problems that exist in our work systems.[1] The report revealed an apparently widespread dissatisfaction on the part of workers across the spectrum of jobs. Two points were of special interest to job designers. One was that many workers felt that their jobs were not interesting; the other was that workers wanted more control over their jobs. The central issue seemed to be the degree of specialization associated with jobs: high specialization appeared to generate the most dissatisfaction. It is noteworthy that specialization is a primary issue of disagreement between the efficiency and behavioral approaches.

Specialization

The term **specialization** describes jobs that have a very narrow scope. Examples range from assembly lines to medical specialties. College professors often specialize in teaching certain courses, some auto mechanics specialize in transmission repair, and some bakers specialize in wedding cakes. The main rationale for specialization is the ability to concentrate one's efforts and thereby become proficient at that type of work.

specialization Work that concentrates on some aspect of a product or service.

Sometimes the amount of knowledge or training required of a specialist and the complexity of the work suggest that individuals who choose such work are very happy with their jobs. This seems to be especially true in the "professions" (e.g., doctors, lawyers, professors). At the other end of the scale are assembly line workers, who are

[1]Upjohn Institute for Employment Research, *Work in America* (Cambridge, Mass.: MIT Press, 1973).

TABLE 7–1

Major advantages and disadvantages of specialization in business

Advantages	
For management:	For labor:
1. Simplifies training	1. Low education and skill requirements
2. High productivity	2. Minimum responsibilities
3. Low wage costs	3. Little mental effort needed
Disadvantages	
For management:	For labor:
1. Difficult to motivate quality	1. Monotonous work
2. Worker dissatisfaction, possibly resulting in absenteeism, high turnover, disruptive tactics, poor attention to quality	2. Limited opportunities for advancement
	3. Little control over work
	4. Little opportunity for self-fulfillment

also specialists, although much less glamorous. The advantage of these highly specialized jobs is that they yield high productivity and low unit costs, and are largely responsible for the high standard of living that exists today in industrial nations.

Unfortunately, many of these jobs can be described as monotonous or downright boring, and they are the source of much of the dissatisfaction among industrial workers today. Even so, it would be wrong to conclude that all workers oppose this type of work. Some workers undoubtedly prefer a job with limited requirements and responsibility for making decisions. Others are not capable of handling jobs with greater scopes. Nonetheless, many workers are frustrated and this manifests itself in a number of ways. Turnover and absenteeism are often high; in the automotive industry, for example, absenteeism runs as high as 20 percent, although not every absentee is a frustrated worker on an assembly line. Workers may also take out their frustrations through disruptive tactics, deliberate slowdowns, or poor attention to product quality.

The seriousness of these problems caused job designers and others to seek ways of alleviating them. Some of those approaches are discussed in the following sections. Before we turn to them, note that the advantages and disadvantages of specialization are summarized in Table 7–1.

Behavioral Approaches to Job Design

In an effort to make jobs more interesting and meaningful, job designers frequently consider job enlargement, job rotation, job enrichment, and increased use of mechanization.

job enlargement Giving a worker a larger portion of the total task, by horizontal loading.

Job enlargement means giving a worker a larger portion of the total task. This constitutes *horizontal loading*—the additional work is on the same level of skill and responsibility as the original job. The goal is to make the job more interesting by increasing the variety of skills required and by providing the worker with a more recognizable contribution to the overall output. For example, a production worker's job might be expanded so that he or she is responsible for a *sequence* of activities instead of only one activity.

job rotation Workers periodically exchange jobs.

Job rotation means having workers periodically exchange jobs. A firm can use this approach to avoid having one or a few employees stuck in monotonous jobs. It works best when workers can be transferred to more interesting jobs; there is little advantage in having workers exchange one boring job for another. Job rotation allows workers to broaden their learning experience and enables them to fill in for others in the event of sickness or absenteeism.

job enrichment Increasing responsibility for planning and coordination tasks, by vertical loading.

Job enrichment involves an increase in the level of responsibility for planning and coordination tasks. It is sometimes referred to as *vertical loading.* An example of this is to have stock clerks in supermarkets handle reordering of goods, thus increasing their responsibilities. The job enrichment approach focuses on the motivating potential of worker satisfaction.

At this pit stop at the Michigan International Speedway, the Pennzoil team takes no more than 15 seconds to do a thorough maintenance check of the race car's engine, fluid levels, and tires.

The importance of these approaches to job design is that they have the potential to increase the motivational power of jobs by increasing worker satisfaction through improvement in the *quality of work life*. Many firms are currently involved in or seriously considering programs related to quality of work life. In addition to the aforementioned approaches, organizations are also experimenting with choice of locations (e.g., medium-sized cities or campuslike settings), flexible work hours, and teams.

Teams

The efforts of business organizations to become more productive, competitive, and customer-oriented have caused them to rethink how work is accomplished. Significant changes in the structure of some work environments have been the increasing use of teams and the way workers are paid, particularly in lean production systems.

In the past, nonroutine job assignments, such as dealing with customer complaints or improving a process, were typically given to one individual or to several individuals who reported to the same manager. More recently, nonroutine assignments are being assigned to teams who develop and implement solutions to problems. Responsibility for the assignment is shared among team members, who often decide among themselves how the work is to be accomplished.

self-directed teams Empowered to make certain changes in their work processes.

Self-directed teams, sometimes referred to as *self-managed teams,* are designed to achieve a higher level of teamwork and employee involvement. Although such teams are not given absolute authority to make all decisions, they are typically empowered to make changes in the work processes under their control. The underlying concept is that the workers, who are close to the process and have the best knowledge of it, are better suited than management to make the most effective changes to improve the process. Moreover, because they have a vested interest and personal involvement in the changes, they tend to work harder to ensure that the desired results are achieved than they would if management had implemented the changes. For these teams to function properly, team members must be trained in quality, process improvement, and teamwork. Self-directed teams have a number of benefits. One is that fewer managers are necessary; very often one manager can handle several teams. Also, self-directed teams can provide improved responsiveness to problems, they have a personal stake in making the process work, and they require less time to implement improvements.

Generally, the benefits of teams include higher quality, higher productivity, and greater worker satisfaction. Moreover, higher levels of employee satisfaction can lead to less turnover and absenteeism, resulting in lower costs to train new workers and less need to fill in for absent employees. This does not mean that organizations will have no difficulties in applying the team concept. Managers, particularly middle managers, often feel threatened as teams assume more of the traditional functions of managers.

READING

TOUR DE FORCE By Gerald Scott

www.chryslercorp.com

A look at the goings-on inside Chrysler's plant that builds Viper and Prowler. It is the Motor City's "hottest ticket," but you can't just walk up to the front door, knock and go inside.

Rather, to tour Chrysler Corp.'s Conner Avenue Assembly Plant, you've got to be like Dorothy in "The Wizard of Oz"—persistent and resourceful.

After all, "it's not what you know but who you know" when you want to tour any auto plant. And as those things go, Chrysler's Conner Avenue plant, where the Dodge Viper and Plymouth Prowler are made on the city's east side, is easily the Motor City's equivalent of Disneyland's old "E-ticket" ride.

In Chrysler's Conner Avenue plant, the Dodge Viper and Plymouth Prowler are built by hand.

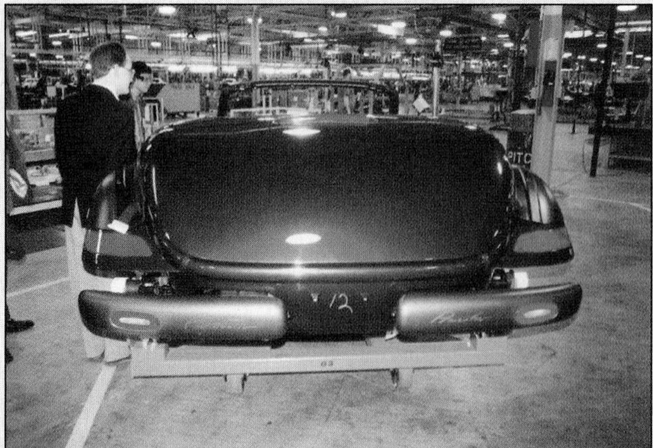

When each workstation completes its task, the entire line advances to the next station.

Public auto plant tours in Detroit are ambiguous and haphazard. The automakers say they frown on them because they distract the work force; and, with downsizing, they don't want to have to spend money on tour guides for each plant. But tours happen almost every day.

The flip side is that it's also a great way to draw attention to your product. The Chevrolet assembly plant making Corvettes in Bowling Green, Ky., for example, has hosted 40,000 people since reopening for tours in April after extensive renovations. In its one-year shutdown, attendance at the nearby National Corvette Museum dropped 25 percent.

In Detroit, the Viper-Prowler plant, as it is known, is the place to see. When comedian Jay Leno was in town to host the 1996 Auto Centennial celebrations, the Viper plant was his first stop. Ditto for Housing and Urban Development Secretary Andrew Cuomo on a whirlwind tour of the Motor City.

"We run plant tours every single day. We are the hot ticket for tours in Detroit," said John Hinckley, manager of the Conner Avenue plant.

"If you bring a group in for a tour, I'm not going to lead it. I won't have one of my management people lead it," Hinckley says. "I'll have one of the craftspersons from the floor lead it. The people who build the cars are the people who know what's going on around here."

Enhancing the Viper/Prowler plant is not your typical Detroit assembly operation. Instead of mass production techniques using robots, it's "craftsman-style" production for hand-building Vipers and Prowlers. With mass production, Chrysler can produce up to 75 cars per hour. With late shifts Chrysler's nearby Jefferson North Assembly Plant can crank out 1,114 Jeep Grand Cherokees in a 24-hour workday. By comparison, the Viper plant produces 13 Vipers a day and has a capacity for 20 Prowlers.

At 392,000 square feet, the Viper plant is a boutique compared to most massive auto plants, such as Saturn Corp.'s Spring Hill, Tenn., manufacturing complex, which is 5 million square feet. (Saturn has received so many "drop-in" visitors that it recently built a new Visitors Center.) The Jefferson North plant has 2.4 million square feet.

The Viper plant was built in 1966 as a Champion Spark Plug plant until 4-cylinder engines and 100,000-mile platinum-tipped spark plugs conspired to sharply reduce demand for replacement plugs. Chrysler acquired the vacant plant from Detroit in 1995 and moved Viper production there from a nearby plant.

"We're not the biggest plant in the area, but we've got the best work force and build the most exciting products—the Plymouth Prowler and Dodge Viper," said Hinckley, who has spent 33 years working in various Detroit auto plants for General Motors and Chrysler and is known for building and driving "kit car" racers. "We're expanding our plant, too. We also are part of the revitalization of Detroit.

"Our plant was about 380,000 square feet and we've just added another 10,000 square feet to improve our process flow and improve our quality. And it provides a little more space since we added the Prowler.

"It will allow us to do a better job of reaching the Prowler's ultimate capacity. It's about a $1 million expansion—for other large plants, that's nothing but for a facility this size it's a lot of money.

"For me it's a dream job. I've been a 'hot rodder' all my life and build race cars and racing engines, and how many hot rodders get to lead the team that runs the only hot-rod plant in the world?"

Conner Avenue is a throwback to early 20th Century, pre-mass assembly techniques. Vipers and Prowlers are built on parallel 720-foot assembly lines, each with a dozen or so workstations, where the cars are hand-assembled. In a rarity, there are no robots in this plant.

When each workstation completes its task, the entire line advances to the next station. So in those 45-minute stops, the employees are relatively free to grab a cup of coffee or talk to tour groups, something they could never do in a plant cranking out 73 units per hour.

Plus, Viper workers love the atten-

Viper seems to be built around its engine and transmission, which, according to employees, makes it a little harder to work on.

tion—having people clamoring to see the products you're assembling can be a boost to the production team's *esprit de corps*. And the automaker's flexible labor agreement with UAW Local 212 means everybody working in the plant is a "craftsman" and can solve any problem anywhere on the line—in most plants, job categories are sharply defined and protected.

Most large auto assembly plants still require 2,000 or more workers, while the Viper plant needs only 260.

"We do everything from tours to hi-lo [forklift] driving to mopping and sweeping, we do it all," says Andrew Stokes, a UAW craftsman who works in underbody and heavy repair.

"I'm one of the first 12 to work on the Prowler," he added. "The Prowler is a little easier to assemble but a little harder to repair. The Viper seems to be a lot more open than the Prowler is—the car seems to be built around the engine and trans and that makes it a little harder to work on. But they're two very good cars."

Thanks to such an interest from car buyers in this plant, Chrysler allows Viper customers to pick up their car as it comes out of final assembly, to meet the employees who built it and to drive it home from the plant instead of from the dealership.

Questions

1. What is different about this assembly plant compared to more typical auto assembly plants?
2. Why do you suppose there are no robots or other automation?

Source: Reprinted with permission from *Chicago Tribune*.

Methods Analysis

methods analysis Focuses on how a job is done.

Methods analysis focuses on how a job is done. Job design often begins with a methods analysis of an overall operation. It then moves from general to specific details of the job, concentrating on arrangement of the workplace and movements of materials and/or workers. Methods analysis can be a good source of productivity improvements.

The need for methods analysis can come from a number of different sources:

1. Changes in tools and equipment.
2. Changes in product design, or new products.
3. Changes in materials or procedures.
4. Government regulations or contractual agreements.
5. Other factors (e.g., accidents, quality problems).

Methods analysis is done both for existing jobs and new jobs. Although it might seem strange to analyze methods of a new job, it is needed to establish a method for a new job. For an existing job, the procedure usually is to have the analyst observe the job as it is currently being performed and then devise improvements. For a new job, the analyst must rely on a job description and an ability to visualize the operation.

The basic procedure in methods analysis is:

1. Identify the operation to be studied, and gather all pertinent facts about tools, equipment, materials, and so on.
2. For existing jobs, discuss the job with the operator and supervisor to get their input.
3. Study and document the present method of an existing job using process charts. For new jobs, develop charts based on information about the activities involved.
4. Analyze the job.
5. Propose new methods.
6. Install the new methods.
7. Follow up installation to assure that improvements have been achieved.

Selecting an Operation to Study. Sometimes a foreman or supervisor will request that a certain operation be studied. At other times, methods analysis will be part of an overall program to increase productivity and reduce costs. Some general guidelines for selecting a job to study are to consider jobs that:

1. Have a high labor content.
2. Are done frequently.
3. Are unsafe, tiring, unpleasant, and/or noisy.
4. Are designated as problems (e.g., quality problems, scheduling bottlenecks).

Documenting the Present Method. Use charts, graphs, and verbal descriptions of the way the job is now being performed. This will provide a good understanding of the job and serve as a basis of comparison against which revisions can be judged.

Analyzing the Job and Proposing New Methods. Job analysis requires careful thought about the what, why, when, where, and who of the job. Often, simply going through these questions will clarify the review process by encouraging the analyst to take a devil's advocate attitude toward both present and proposed methods.

Analyzing and improving methods is facilitated by the use of various charts such as *flow process charts* and *worker-machine charts.*

Flow process charts are used to review and critically examine the overall sequence of an operation by focusing on the movements of the operator or the flow of materials. These charts are helpful in identifying nonproductive parts of the process (e.g., delays, temporary storages, distances traveled). Figure 7–1 describes the symbols used in constructing a flow process chart, and Figure 7–2 illustrates a flow process chart.

flow process chart Chart used to examine the overall sequence of an operation by focusing on movements of the operator or flow of materials.

The uses for flow process charts include studying the flow of material through a department, studying the sequence that documents or forms take, analyzing movement and care of surgical patients, layout of department and grocery stores, and mail handling.

Experienced analysts usually develop a checklist of questions they ask themselves to generate ideas for improvements. Some representative questions are:

1. Why is there a delay or storage at this point?
2. How can travel distances be shortened or avoided?
3. Can materials handling be reduced?
4. Would a rearrangement of the workplace result in greater efficiency?
5. Can similar activities be grouped?
6. Would the use of additional or improved equipment be helpful?
7. Does the worker have any ideas for improvements?

A **worker-machine chart** is helpful in visualizing the portions of a work cycle during which an operator and equipment are busy or idle. The analyst can easily see when the operator and machine are working independently and when their work overlaps or is interdependent. One use of this type of chart is to determine how many machines or how much equipment the operator can manage. Figure 7–3 presents an example of a worker-machine chart. Among other things, the chart highlights worker and machine utilization.

worker-machine chart Chart used to determine portions of a work cycle during which an operator and equipment are busy or idle.

Installing the Improved Method. Successful implementation of proposed method changes requires convincing management of the desirability of the new method and obtaining the cooperation of the worker. If the worker has been consulted throughout the process and has made suggestions that are incorporated in the proposed changes, this part of the task will be considerably easier than if the analyst has assumed sole responsibility for the development of the proposal.

If the proposed method constitutes a major change from the way the job has been performed in the past, workers may have to undergo a certain amount of retraining, and full implementation may take some time to achieve.

FIGURE 7-1

Process chart symbols

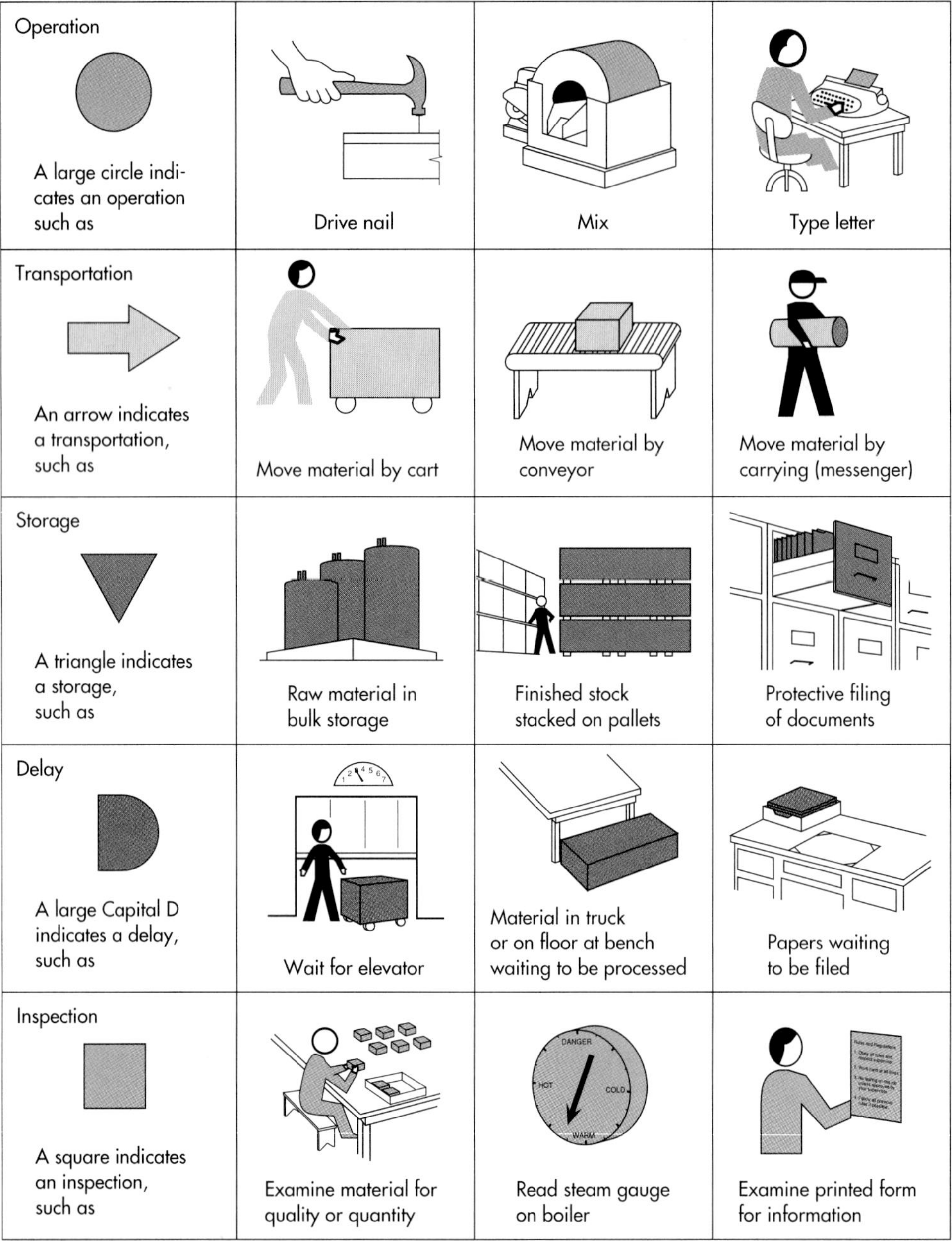

Source: Benjamin W. Niebel, *Motion and Time Study,* 8th ed. (Burr Ridge, IL: Richard D. Irwin, 1988), p. 33, © 1988 by Richard D. Irwin, Inc. Reprinted by permission.

The Follow-Up. In order to ensure that changes have been made and that the proposed method is functioning as expected, the analyst should review the operation after a reasonable period and consult again with the operator.

Motion Study

motion study Systematic study of the human motions used to perform an operation.

Motion study is the systematic study of the human motions used to perform an operation. The purpose is to eliminate unnecessary motions and to identify the best sequence of motions for maximum efficiency. Hence, motion study can be an important avenue for productivity improvements. Present practice evolved from the work of Frank Gilbreth, who originated the concepts in the bricklaying trade in the early 20th century. Through the use of motion study techniques, Gilbreth is generally credited with increasing the average number of bricks laid per hour by a factor of 3, even though he was not

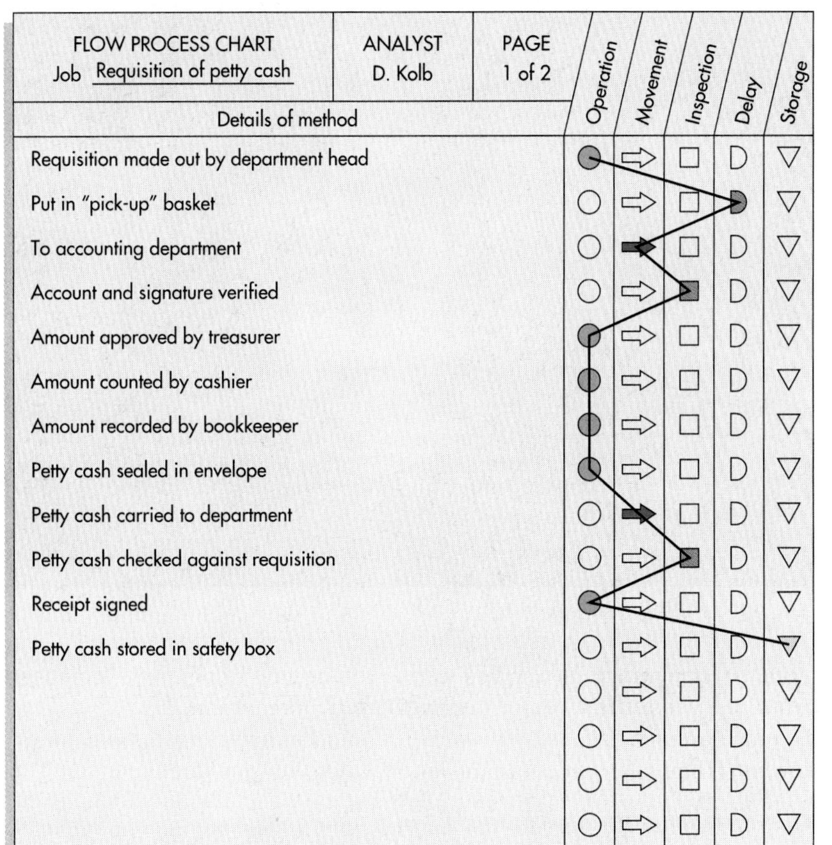

FLOW PROCESS CHART Job Requisition of petty cash	ANALYST D. Kolb	PAGE 1 of 2	Operation	Movement	Inspection	Delay	Storage
Details of method							
Requisition made out by department head							
Put in "pick-up" basket							
To accounting department							
Account and signature verified							
Amount approved by treasurer							
Amount counted by cashier							
Amount recorded by bookkeeper							
Petty cash sealed in envelope							
Petty cash carried to department							
Petty cash checked against requisition							
Receipt signed							
Petty cash stored in safety box							

Source: Elias M. Awad, *Systems Analysis and Design*, 4th ed. (Burr Ridge, IL: Richard D. Irwin, 1985), p. 113. © 1985 by Richard D. Irwin, Inc. Reprinted by permission.

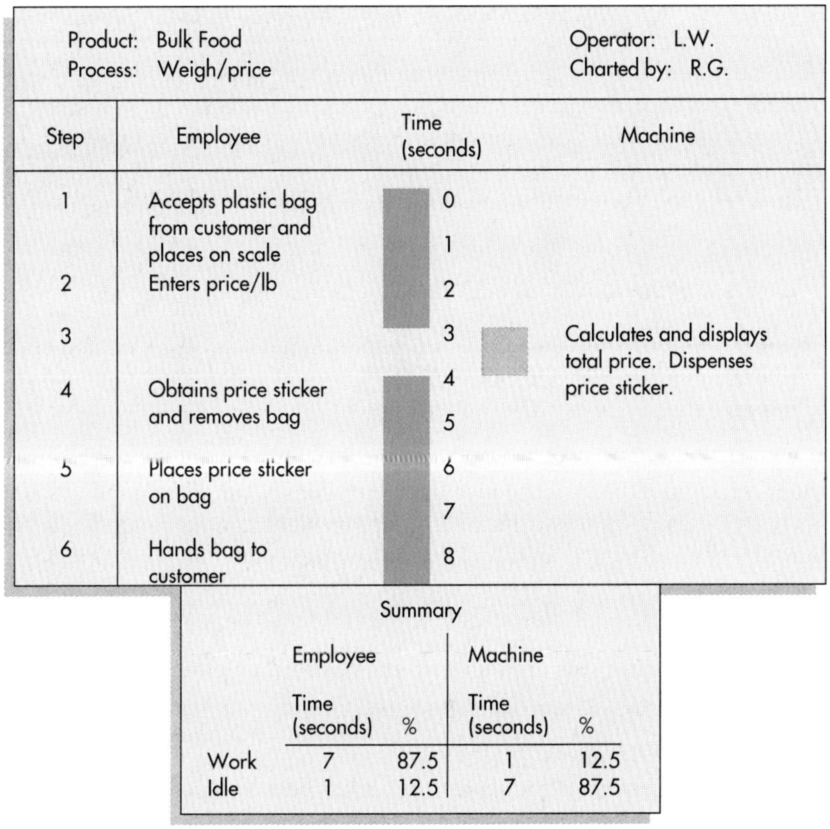

FIGURE 7-2

Format of a flow process chart

Product: Bulk Food Operator: L.W.
Process: Weigh/price Charted by: R.G.

Step	Employee	Time (seconds)	Machine
1	Accepts plastic bag from customer and places on scale	0 1	
2	Enters price/lb	2	
3		3	Calculates and displays total price. Dispenses price sticker.
4	Obtains price sticker and removes bag	4 5	
5	Places price sticker on bag	6 7	
6	Hands bag to customer	8	

Summary

	Employee Time (seconds)	%	Machine Time (seconds)	%
Work	7	87.5	1	12.5
Idle	1	12.5	7	87.5

FIGURE 7-3

Worker-machine chart

A manager checks a prospective employee on the time it takes to perform a technical operation at Circuit City.

www.circuitcity.com

a bricklayer by trade. When you stop to realize that bricklaying had been carried on for centuries, Gilbreth's accomplishment is even more remarkable.

There are a number of different techniques that motion study analysts can use to develop efficient procedures. The most-used techniques are:

1. Motion study principles
2. Analysis of therbligs
3. Micromotion study
4. Charts

motion study principles Guidelines for designing motion-efficient work procedures

Gilbreth's work laid the foundation for the development of **motion study principles,** which are guidelines for designing motion-efficient work procedures. The guidelines are divided into three categories: principles for use of the body, principles for arrangement of the workplace, and principles for the design of tools and equipment. Table 7–2 lists some examples of the principles.

In developing work methods that are motion efficient, the analyst tries to:

1. Eliminate unnecessary motions.
2. Combine activities.
3. Reduce fatigue.
4. Improve the arrangement of the workplace.
5. Improve the design of tools and equipment.

therbligs Basic elemental motions into which a job can be broken down.

Therbligs are basic elemental motions. The term *therblig* is Gilbreth spelled backwards (except for the *th*). The idea behind the development of therbligs is to break jobs down into minute elements and base improvements on an analysis of these basic elements by eliminating, combining, or rearranging them.

Although a complete description of therbligs is outside the scope of this text, a list of some common ones will illustrate the nature of these basic elemental motions:

Search implies hunting for an item with the hands and/or the eyes.

Select means to choose from a group of objects.

Grasp means to take hold of an object.

Hold refers to retention of an object after it has been grasped.

TABLE 7-2

Motion study principles

A. The use of the human body. Examples:
   1. Both hands should begin and end their basic divisions of accomplishment simultaneously and should not be idle at the same instant, except during rest periods.
   2. The motions made by the hands should be made symmetrically.
   3. Momentum should assist workers wherever possible and should be minimized if it must be overcome by muscular effort.
   4. Continuous curved motions are preferable to straight-line motions involving sudden and sharp changes in direction.
B. The arrangement and conditions of the workplace. Examples:
   1. Fixed locations for all tools and material should be provided to permit the best sequence and to eliminate or reduce the therbligs search and select.
   2. Gravity bins and drop delivery should reduce reach and move times; wherever possible, ejectors should remove finished parts automatically.
   3. All materials and tools should be located within the normal working area.
C. The design of tools and equipment. Examples:
   1. Multiple cuts should be taken whenever possible by combining two or more tools in one.
   2. All levers, handles, wheels, and other control devices should be readily accessible to the operator and designed to give the best possible mechanical advantage and to utilize the strongest available muscle group.
   3. Parts should be held in position by fixtures.

Source: Adapted from Benjamin W. Niebel, *Motion and Time Study* (Burr Ridge, IL: Richard D. Irwin, Inc., 1993), pp. 206–207.

Transport load means movement of an object after hold.

Release load means to deposit the object.

Some other therbligs are *inspect*, *position*, *plan*, *rest*, and *delay*.

Describing a job using therbligs often takes a substantial amount of work. However, for short, repetitive jobs, therbligs analysis may be justified.

Frank Gilbreth and his wife, Lillian, an industrial psychologist, were also responsible for introducing motion pictures for studying motions, called **micromotion study.** This approach is applied not only in industry but in many other areas of human endeavor, such as sports and health care. Use of the camera and slow-motion replay enables analysts to study motions that would otherwise be too rapid to see. In addition, the resulting films provide a permanent record that can be referred to, not only for training workers and analysts but also for settling job disputes involving work methods.

micromotion study Use of motion pictures and slow motion to study motions that otherwise would be too rapid to analyze.

This line at Mercedes produces multiple models. A computer supplies a distinct parts list and necessary explanation with each vehicle. You can see how this assembly line tilts and raises and lowers the cars for easy access.

www.daimler-benz.com

FIGURE 7–4

A simultaneous motion chart

SIMO CHART

OPERATOR: Ken Reisch
DATE: May 21,
OPERATION: Assembly
PART: Lace Finger
METHOD: Proposed
CHART BY: Joseph Riley

TIME SCALE (winks)	ELEMENT TIME	LEFT-HAND DESCRIPTION	SYMBOL	MOTION CLASS 1 2 3 4 5 5 4 3 2 1	SYMBOL	RIGHT-HAND DESCRIPTION	ELEMENT TIME	TIME SCALE (winks)
4548	12	Reach for finger	RE		RE	Reach for finger	12	4548
4560	19	Grasp finger	G		G	Grasp finger	19	4560
4579	31	Move finger	M		M	Move finger	31	4579
4610	75	Position and release finger	P RL		P RL	Position and release finger	75	4610
4685	15	Reach for clamp	RE		RE	Reach for clamp	15	4685
4700	15	Grasp clamp	G		G	Grasp clamp	15	4700
4715								4715
	12	Grasp assembly	G		G	Grasp assembly	12	
7541	18	Move and release assembly	M RL		M RL	Move and release assembly	18	7541
7559								7559

SUMMARY

%	TIME	LEFT-HAND SUMMARY	SYM.	RIGHT-HAND SUMMARY	TIME	%
8.56	249	Reach	RE	Reach	245	8.4
7.49	218	Grasp	G	Grasp	221	7.6
12.16	354	Move	M	Move	413	14.2
30.47	887	Position	P	Position	1124	38.6
39.33	1145	Use	U	Use	876	30.1
1.03	30	Idle	I	Idle	0	0.0
.96	28	Release	RL	Release	32	1.1
100.0	2911	TOTALS			2911	100.0

Source: Benjamin W. Niebel, *Motion and Time Study,* 8th ed. (Burr Ridge, IL: Richard D. Irwin, 1988), p. 229.

The cost of micromotion study limits its use to repetitive activities, where even minor improvements can yield substantial savings owing to the number of times an operation is repeated, or where other considerations justify its use (e.g., surgical procedures).

Motion study analysts often use charts as tools for analyzing and recording motion studies. Activity charts and process charts such as those described earlier can be quite helpful. In addition, analysts may use a *simo chart* (see Figure 7–4) to study simultaneous motions of the hands. (It is interesting that even though people tend to be either right-handed or left-handed, they have no difficulty using both hands to do tasks.) These

Mitsubishi manufacturing plant in Malaysia at the Proton factory. For this interior body welding plant, engineers and management must consider worker safety, ergonomics, time factors, and the use of technology.

www.mitsucars.com/

charts are invaluable in studying operations such as data entry, sewing, surgical and dental procedures, and certain assembly operations.

Working Conditions

Working conditions are an important aspect of job design. Physical factors such as temperature, humidity, ventilation, illumination, and noise can have a significant impact on worker performance in terms of productivity, quality of output, and accidents.

Temperature and Humidity. Although human beings can function under a fairly wide range of temperatures, work performance tends to be adversely affected if temperatures are outside a very narrow *comfort band*. That comfort band depends on how strenuous the work is; the more strenuous the work, the lower the comfort range.

Heating and cooling are less of a problem in offices than in factories and other work environments where high ceilings allow heat to rise and where there is often a constant flow of trucks and other moving and handling equipment through large doors. These conditions make it difficult to maintain a constant temperature. Solutions range from selection of suitable clothing to space heating or cooling devices.

Humidity is also an important variable in maintaining a comfortable working environment

Ventilation. Unpleasant and noxious odors can be distracting and dangerous to workers. Moreover, unless smoke and dust are periodically removed, the air can quickly become stale and unpleasant. Large fans and air-conditioning equipment are commonly used to exchange and recondition the air.

Illumination. The amount of illumination required depends largely on the type of work being performed; the more detailed the work, the higher the level of illumination needed for adequate performance. Other important considerations are the amount of glare and contrast. From a safety standpoint, good lighting in halls, stairways, and other dangerous points is important. However, because illumination is expensive, high illumination in all areas is not generally desirable.

Sometimes natural daylight can be used as a source of illumination. Not only is it free, it also provides some psychological benefits. Workers in windowless rooms may feel

FIGURE 7-5

Decibel values of typical sounds (db)

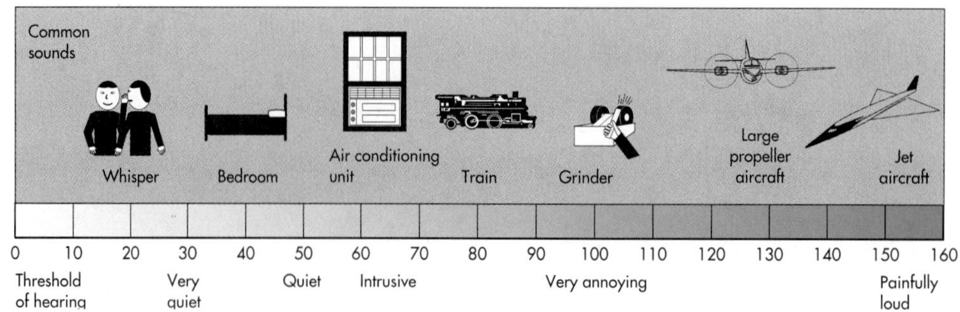

Source: Benjamin W. Niebel, *Motion and Time Study,* 8th ed. (Burr Ridge, IL: Richard D. Irwin, 1988), p. 248. © 1988 by Richard D. Irwin, Inc. Reprinted by permission.

cut off from the outside world and experience various psychological problems. On the down side, the inability to control natural light (e.g., cloudy days) can result in dramatic changes in light intensity.

Noise and Vibration. Noise is unwanted sound. It is caused by the vibrations of machines or equipment and by humans. Noise can be annoying or distracting, leading to errors and accidents. It can also damage or impair hearing if it is loud enough. Figure 7–5 illustrates loudness levels of some typical sounds.

Successful sound control begins with measurement of the offending sounds. In a new operation, selection and placement of equipment can eliminate or reduce many potential problems. In the case of existing equipment, it may be possible to redesign or substitute equipment. In some instances, the source of noise can be isolated from other work areas. If that isn't feasible, acoustical walls and ceilings or baffles that deflect sound waves may prove useful. Sometimes the only solution is to provide protective devices for those in the immediate vicinity (e.g., personnel who guide jet aircraft into landing gates wear protective devices over their ears).

Vibrations can be a factor in job design even without a noise component, so merely eliminating sound may not be sufficient in every case. Vibrations can come from tools, machines, vehicles, human activity, air-conditioning systems, pumps, and other sources. Corrective measures include padding, stabilizers, shock absorbers, cushioning, and rubber mountings.

Work Breaks. The frequency, length, and timing of work breaks can have a significant impact on both productivity and quality of output. One indication of the relationship between worker efficiency and work breaks is illustrated in Figure 7–6. It reveals that efficiency generally declines as the day wears on, but it also shows how breaks for lunch and rest can cause an upward shift in efficiency.

An important variable in the rate of decline of efficiency and potential effects of work breaks is the amount of physical and/or mental requirements of the job. Steelworkers, for instance, may need rest breaks of 15 minutes per hour due to the strenuous nature of their jobs. Physical effort is not the only condition that indicates the need for work breaks. People working at CRTs also need periodic breaks; and students need study breaks.

Safety. Worker safety is one of the most basic issues in job design. This area needs constant attention from management, employees, and designers. Workers cannot be effectively motivated if they feel they are in physical danger.

From an employer standpoint, accidents are undesirable because they are expensive (insurance and compensation); they usually involve damage to equipment and/or products; they require hiring, training, and makeup work; and they generally interrupt work. From a worker standpoint, accidents mean physical suffering, mental anguish, potential loss of earnings, and disruption of the work routine.

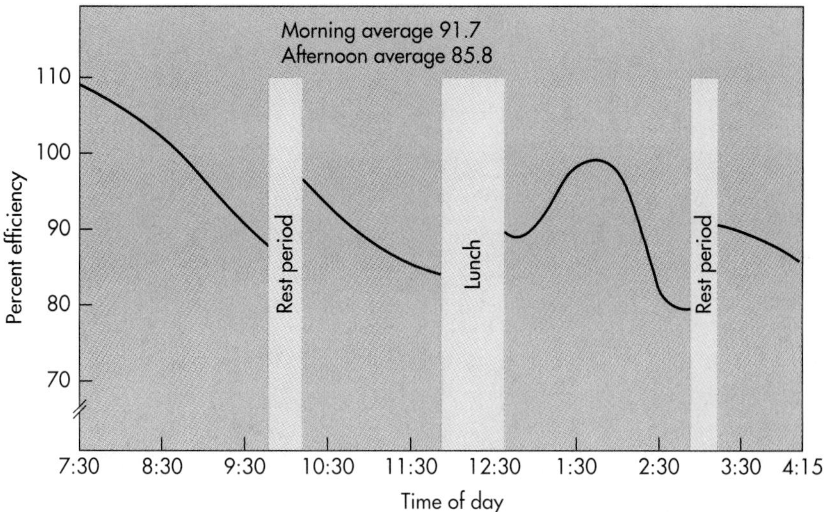

Source: Benjamin W. Niebel, *Motion and Time Study,* 8th ed. (Burr Ridge, IL: Richard D. Irwin, 1988), p. 270. © 1988 by Richard D. Irwin, Inc. Reprinted by permission.

FIGURE 7-6

A typical relationship between worker efficiency and time of day

Causes of Accidents. The two basic causes of accidents are worker *carelessness* and accident *hazards*. Under the heading of carelessness come unsafe acts. Examples include driving at high speeds, drinking and driving, failure to use protective equipment, overriding safety controls (e.g., taping control buttons down), disregarding safety procedures (e.g., running, throwing objects, cutting through, failing to observe one-way signs), improper use of tools and equipment, and failure to use reasonable caution in danger zones. Unsafe conditions include unprotected pulleys, chains, material-handling equipment, machinery, and so on. Also, poorly lit walkways, stairs, and loading docks constitute hazards. Toxic wastes, gases and vapors, and radiation hazards must be contained. In many instances, these cannot be detected without special equipment, so they would not be obvious to workers or emergency personnel. Protection against hazards involves use of proper lighting, clearly marked danger zones, use of protective equipment (hardhats, goggles, earmuffs, gloves, heavy shoes and clothing), safety devices (machine guards, dual control switches that require an operator to use both hands), emergency equipment (emergency showers, fire extinguishers, fire escapes), and thorough instruction in safety procedures and use of regular and emergency equipment. Housekeeping (clean floors, open aisles, waste removal) is another important safety factor.

An effective program of safety and accident control requires the cooperation of both workers and management. Workers must be trained in proper procedures and attitudes, and they can contribute to a reduction in hazards by pointing out hazards to management. Management must enforce safety procedures and use of safety equipment. If supervisors allow workers to ignore safety procedures or look the other way when they see violations, workers will be less likely to take proper precautions. Some firms use contests that compare departmental safety records. However, accidents cannot be completely eliminated, and a freak accident may seriously affect worker morale and might even contribute to additional accidents. Posters can be effective, particularly if they communicate in specific terms how to avoid accidents. For example, the admonition to "be careful" is not nearly as effective as "wear hardhats," "walk, don't run," or "hold on to rail."

The enactment of the Occupational Safety and Health Act (**OSHA**) in 1970, and the creation of the Occupational Safety and Health Administration, emphasized the importance of safety considerations in systems design. The law was intended to ensure that workers in all organizations have healthy and safe working conditions. It provides specific safety regulations with inspectors to see that they are adhered to. Inspections are carried out both at random and to investigate complaints of unsafe conditions.

OSHA Occupational Safety and Health Act of 1970; Occupational Safety and Health Administration.

Personal protection equipment, such as hard hats and earmuffs, is worn at the Allegheny Ludlum stainless steel rolls manufacturing plant in Pittsburgh. Note that protective gloves are not worn where finger dexterity is required or where there is a danger of getting caught in the equipment.

OSHA officials are empowered to issue warnings, to impose fines, and even to invoke court-ordered shutdowns for unsafe conditions.

OSHA must be regarded as a major influence on operations management decisions in all areas relating to worker safety. OSHA has promoted the welfare and safety of workers in its role as a catalyst, spurring companies to make changes that they knew were needed but "hadn't gotten around to making."

WORK MEASUREMENT

work measurement Determining how long it should take to do a job.

Job design determines the *content* of a job, and methods analysis determines *how* a job is to be performed. **Work measurement** is concerned with determining the *length of time* it should take to complete the job. Job times are vital inputs for manpower planning, estimating labor costs, scheduling, budgeting, and designing incentive systems. Moreover, from the workers' standpoint, time standards provide an indication of expected output. Time standards reflect the amount of time it should take an average worker to do a given job working under typical conditions. The standards include expected activity time plus allowances for probable delays.

standard time The amount of time it should take a qualified worker to complete a specified task, working at a sustainable rate, using given methods; tools and equipment, raw materials, and workplace arrangements.

A **standard time** is the amount of time it should take a qualified worker to complete a specified task, working at a sustainable rate, using given methods, tools and equipment, raw material inputs, and workplace arrangement. Whenever a time standard is developed for a job, it is essential to provide a complete description of the parameters of the job because the actual time to do the job is sensitive to all of these factors; changes in any one of the factors can materially affect time requirements. For instance, changes in product design or changes in job performance brought about by a methods study should trigger a new time study to update the standard time. As a practical matter, though, minor changes are occasionally made that do not justify the expense of restudying the job. Consequently, the standards for many jobs may be slightly inaccurate. Periodic time studies may be used to update the standards.

Organizations develop time standards in a number of different ways. Although some small manufacturers and service organizations rely on subjective estimates of job times, the most commonly used methods of work measurement are:

1. Stopwatch time study
2. Historical times
3. Predetermined data
4. Work sampling

The following pages describe each of these techniques in some detail.

Stopwatch Time Study

Stopwatch time study was formally introduced by Frederick Winslow Taylor in the late 19th century. Today it is the most widely used method of work measurement. It is especially appropriate for short, repetitive tasks.

Stopwatch time study is used to develop a time standard based on observations of one worker taken over a number of cycles. That is then applied to the work of all others in the organization who perform the same task. The basic steps in a time study are:

stopwatch time study Development of a time standard based on observations of one worker taken over a number of cycles.

1. Define the task to be studied, and inform the worker who will be studied.
2. Determine the number of cycles to observe.
3. Time the job, and rate the worker's performance.
4. Compute the standard time.

The analyst who studies the job should be thoroughly familiar with it since it is not unusual for workers to attempt to include extra motions during the study in hope of gaining a standard that allows more time per piece (i.e., the worker will be able to work at a slower pace and still meet the standard). Furthermore, the analyst will need to check that the job is being performed efficiently before setting the time standard.

In most instances, an analyst will break all but very short jobs down into basic elemental motions (e.g., reach, grasp) and obtain times for each element. There are several reasons for this: One is that some elements are not performed in every cycle, and the breakdown enables the analyst to get a better perspective on them. Another is that the worker's proficiency may not be the same for all elements of the job. A third reason is to build a file of elemental times that can be used to set times for other jobs. This use will be described later.

It is important to inform the observed worker about the study in order to avoid suspicion or misunderstandings. Workers sometimes feel uneasy about being studied and fear changes that might result. The analyst should make an attempt to discuss these things with the worker prior to studying the operations to allay such fears and to enlist the cooperation of the worker.

Vol. 5, Tape 2
TriState

The number of cycles that must be timed is a function of three things: (1) the variability of observed times, (2) the desired accuracy, and (3) the desired level of confidence for the estimated job time. Very often the desired accuracy is expressed as a percentage of the mean of the observed times. For example, the goal of a time study may be to achieve an estimate that is within 10 percent of the actual mean. The sample size needed to achieve that goal can be determined using this formula:

$$n = \left(\frac{zs}{a\bar{x}}\right)^2 \qquad (7\text{–}1)$$

where

z = Number of normal standard deviations needed for desired confidence
s = Sample standard deviation
a = Desired accuracy percentage
$\bar{x}$ = Sample mean

Typical values of z used in this computation are:[2]

Desired Confidence (%)	z Value
90	1.65
95	1.96
95.5	2.00
98	2.33
99	2.58

Of course, the value of z for any desired confidence can be obtained from the normal table in Appendix B (Table A or Table B).

An alternate formula used when the desired accuracy is stated as an *amount* (e.g., within one minute of the true mean) instead of a percentage is:

$$n = \left(\frac{zs}{e}\right)^2 \tag{7–2}$$

where

e = Accuracy or maximum acceptable error

To make a preliminary estimate of sample size, it is typical to take a small number of observations (i.e., 10 to 20) and compute values of $\bar{x}$ and s to use in the formula for n. Toward the end of the study, the analyst may want to recompute n using revised estimates of $\bar{x}$ and s based on the increased data available.

Note: These formulas may or may not be used in practice, depending on the person doing the time study. Often, an experienced analyst will rely on his or her judgment in deciding on the number of cycles to time.

Example 1

A time study analyst wants to estimate the time required to perform a certain job. A preliminary study yielded a mean of 6.4 minutes and a standard deviation of 2.1 minutes. The desired confidence is 95 percent. How many observations will he need (including those already taken) if the desired maximum error is:

a. ±10 percent of the sample mean?

b. One-half minute?

Solution

a. s = 2.1 minutes $\quad z = 1.96$

$\bar{x}$ = 6.4 minutes $\quad a = 10\%$

$$n = \left(\frac{zs}{a\bar{x}}\right)^2 = \left(\frac{1.96(2.1)}{.10(6.4)}\right)^2 = 41.36 \text{ (round up to 42)}$$

b. $e = .5$

$$n = \left(\frac{zs}{e}\right)^2 = \left(\frac{1.96(2.1)}{.5}\right)^2 = 67.77 \text{ (round up to 68)}$$

Development of a time standard involves computation of three times: the *observed time* (OT), the *normal time* (NT), and the *standard time* (ST).

[2]Theoretically, a t value rather than a z should be used because the population standard deviation is unknown. However the use of z is simpler and provides reasonable results when the number of observations is 30 or more, as it generally is. In practice. z is used almost exclusively.

Observed Time. The observed time is simply the average of the recorded times. Thus,

$$OT = \frac{\Sigma x_i}{n} \tag{7–3}$$

where

OT = Observed time
Σx_i = Sum of recorded times
n = Number of observations

Note: If a job element does not occur each cycle, its average time should be determined separately and that amount should be included in the observed time, OT.

Normal Time. The normal time is the observed time adjusted for worker performance. It is computed by multiplying the observed time by a *performance rating.* That is,

$$NT = OT \times PR \tag{7–4}$$

where

NT = Normal time
PR = Performance rating

This assumes that a single performance rating has been made for the entire job. If ratings are made on an element-by-element basis, the normal time is obtained by multiplying each element's average time by its performance rating and summing those values:

$$NT = \Sigma(\bar{x}_j \times PR_j) \tag{7–5}$$

where

$\bar{x}_j$ = Average time for element j
PR_j = Performance rating for element j

The reason for including this adjustment factor is that the worker being observed may be working at a rate different from a "normal" rate, either to deliberately slow the pace or because his or her natural abilities differ from the norm. For this reason, the observer assigns a performance rating, to adjust the observed times to an "average" pace. With a normal rating of 1.00, a performance rating of 0.9 indicates a pace that is 90 percent of normal, whereas a rating of 1.05 indicates a pace that is slightly faster than normal. For long jobs, each element may be rated; for short jobs, a single rating may be made for an entire cycle.

When assessing performance, the analyst must compare the observed performance to his or her concept of normal. Obviously, there is room for debate about what constitutes normal performance, and performance ratings are sometimes the source of considerable conflict between labor and management. Although no one has been able to suggest a way around these subjective evaluations, sufficient training and periodic recalibration by analysts using training films can provide a high degree of consistency in the ratings of different analysts.

Standard Time. Normal time is the length of time a worker should take to perform a job if there are no delays or interruptions. It does not take into account such factors as personal delays (getting a drink of water or going to the restroom), unavoidable delays (machine adjustments and repairs, talking to a supervisor, waiting for materials), or rest breaks. The standard time for a job is the normal time plus an *allowance* for these delays. The standard time is

$$ST = NT \times AF \tag{7–6}$$

where

ST = Standard time
AF = Allowance factor

Allowances can be based on either job time or time worked (e.g., a workday). If allowances are based on the *jot time,* the allowance factor must be computed using the formula

$$AF_{job} = 1 + A \tag{7–7}$$

where

A = Allowance percentage based on job time

This is used when different jobs have different allowances. If allowances are based on a percentage of the time worked (i.e., the *workday*), the appropriate formula is

$$AF_{day} = \frac{1}{1 - A} \tag{7–8}$$

where

A = Allowance percentage based on workday

This is used when jobs are the same or similar, and have the same allowance factors.

Example 2

Compute the allowance factor for these two cases:

a. The allowance is 20 percent of *job* time.

b. The allowance is 20 percent of *work* time.

Solution

$A = .20$

a. $AF = 1 + A = 1.20$, or 120%

b. $AF = \frac{1}{1 - A} = \frac{1}{1 - .20} = 1.25$, or 125%

Table 7–3 illustrates some typical allowances. In practice, allowances may be based on the judgment of the time study analyst, work sampling (described later in the chapter), or negotiations between labor and management.

Example 3 illustrates the time study process from observed times to the standard time.

Example 3

A time study of an assembly operation yielded the following observed times for one element of the job, for which the analyst gave a performance rating of 1.13 Using an allowance of 20 percent of *job* time, determine the appropriate standard time for this operation.

i Observation	Time, x (minutes)	Observation	Time (minutes)
1	1.12	6	1.18
2	1.15	7	1.14
3	1.16	8	1.14
4	1.12	9	1.19
5	1.15	Total	10.35

Solution

$n = 9 \qquad PR = 1.13 \qquad A = .20$

TABLE 7–3

Typical allowance percentages for working conditions

	Percent
A. Constant allowances:	
1. Personal allowance	5
2. Basic fatigue allowances	4
B. Variable allowances:	
1. Standing allowance	2
2. Abnormal position allowance:	
a. Slightly awkward	0
b. Awkward (bending)	2
c. Very awkward (lying, stretching)	7
3. Use of force or muscular energy (lifting, pulling, or pushing):	
Weight lifted (in pounds):	
5	0
10	1
15	2
20	3
25	4
30	5
35	7
40	9
45	11
50	13
60	17
70	22
4. Bad light:	
a. Slightly below recommended	0
b. Well below	2
c. Very inadequate	5
5. Atmospheric conditions (heat and humidity)—variable	0–10
6. Close attention:	
a. Fairly fine work	0
b. Fine or exacting	2
c. Very fine or very exacting	5
7. Noise level:	
a. Continuous	0
b. Intermittent—loud	2
c. Intermittent—very loud	5
d. High-pitched—loud	5
8. Mental strain:	
a. Fairly complex process	1
b. Complex or wide span of attention	4
c. Very complex	8
9. Monotony:	
a. Low	0
b. Medium	1
c. High	4
10. Tediousness:	
a. Rather tedious	0
b. Tedious	2
c. Very tedious	5

Source: Benjamin W. Niebel, *Motion and Time Study,* 8th ed. (Burr Ridge, IL: Richard D. Irwin, 1988), p. 416.

a. $\text{OT} = \frac{\Sigma x_i}{n} = \frac{10.35}{9} = 1.15$ minutes.

b. $\text{NT} = \text{OT} \times \text{PR} = 1.15(1.13) = 1.30$ minutes.

c. $\text{ST} = \text{NT} \times (1 + A)\ 1.30(1.20) = 1.56$ minutes.

Note: If an abnormally short time has been recorded, it typically would be assumed to be the result of observational error and thus discarded. If one of the observations in Example 3 had been 0.10, it would have been discarded. However, if an abnormally *long* time has been recorded, the analyst would want to investigate that observation to determine whether some irregularly occurring aspect of the task (e.g., retrieving a dropped tool or part) exists, which should legitimately be factored into the job time.

Despite the obvious benefits that can be derived from work measurement using time study, some limitations also must be mentioned. One limitation is the fact that only those jobs that can be observed can be studied. This eliminates most managerial and creative jobs, because these involve mental as well as physical aspects. Also, the cost of the study rules out its use for irregular-operations and infrequently occurring jobs. Finally, it disrupts the normal work routine, and workers resent it in many cases.

Standard Elemental Times

standard elemental times Time standards derived from a firm's historical time data.

Standard elemental times are derived from a firm's own historical time study data. Over the years, a time study department can accumulate a file of elemental times that are common to many jobs. After a certain point, many elemental times can be simply retrieved from the file, eliminating the need for analysts to go through a complete time study to obtain them.

The procedure for using standard elemental times consists of the following steps:

1. Analyze the job to identify the standard elements.
2. Check the file for elements that have historical times, and record them. Use time study to obtain others, if necessary.
3. Modify the file times if necessary (explained below).
4. Sum the elemental times to obtain the normal time, and factor in allowances to obtain the standard time.

In some cases, the file times may not pertain exactly to a specific task. For instance, standard elemental times might be on file for "move the tool 3 centimeters" and "move the tool 9 centimeters," when the task in question involves a move of 6 centimeters. However, it is often possible to interpolate between values on file to obtain the desired time estimate.

One obvious advantage of this approach is the potential savings in cost and effort created by not having to conduct a complete time study for each job. A second advantage is that there is less disruption of work, again because the analyst does not have to time the worker. A third advantage is that performance ratings do not have to be done; they are generally *averaged* in the file times. The main disadvantage of this approach is that times may not exist for enough standard elements to make it worthwhile, and the file times may be biased or inaccurate.

The method described in the following section is a variation of this approach, which helps avoid some of these problems.

Predetermined Time Standards

predetermined time standards Published data based on extensive research to determine standard elemental times.

Predetermined time standards involve the use of published data on standard elemental times. A commonly used system is *methods-time measurement* (MTM), which was developed in the late 1940s by the Methods Engineering Council. The MTM tables are based on extensive research of basic elemental motions and times. To use this approach, the analyst must divide the job into its basic elements (reach, move, turn, disengage), measure the distances involved (if applicable), rate the difficulty of the element, and then refer to the appropriate table of data to obtain the time for that element. The standard time for the job is obtained by adding the times for all of the basic elements. Times of the basic elements are measured in time measurement units (TMUs); one TMU equals 0.0006 minutes. One minute of work may cover quite a few basic elements; a typical job may involve several hundred or more of these basic elements. The analyst requires a considerable amount of skill to adequately describe the operation and develop realistic time estimates. Table 7–4 presents a portion of the MTM tables, to give you an idea of the kind of information they provide.

A high level of skill is required to generate a predetermined time standard. Analysts generally take training or certification courses to develop the necessary skills to do this kind of work.

Among the advantages of predetermined time standards are the following:

1. They are based on large numbers of workers under controlled conditions.
2. The analyst is not required to rate performance in developing the standard.
3. There is no disruption of the operation.
4. Standards can be established even before a job is done.

Although proponents of predetermined standards claim that the approach provides much better accuracy than stopwatch studies, not everyone agrees with that claim. Some argue that many activity times are too specific to a given operation to be generalized from published data. Others argue that different analysts perceive elemental activity breakdowns in different ways, and that this adversely affects the development of times and produces varying time estimates among analysts. Still others claim that analysts differ on the degree of difficulty they assign a given task and thereby obtain different time standards.

Work Sampling

Work sampling is a technique for estimating the proportion of time that a worker or machine spends on various activities.

work sampling Technique for estimating the proportion of time that a worker or machine spends on various activities.

Unlike time study, work sampling does not require timing an activity, nor does it even involve continuous observation of the activity. Instead, an observer makes brief observations of a worker or machine at random intervals and simply notes the nature of the activity. For example, a machine may be busy or idle; a secretary may be typing, filing, talking on the telephone, and so on; and a carpenter may be carrying supplies, taking measurements, cutting wood, and so on. The resulting data are *counts* of the number of times each category of activity or nonactivity was observed.

Although work sampling is occasionally used to set time standards, its two primary uses are in (1) ratio-delay studies, which concern the percentage of a worker's time that involves unavoidable delays or the proportion of time a machine is idle, and (2) analysis of nonrepetitive jobs. In a ratio-delay study, a hospital administrator, for example, might want to estimate the percentage of time that a certain piece of X-ray equipment is not in use. In a nonrepetitive job, such as secretarial work or maintenance, it can be important to establish the percentage of time an employee spends doing various tasks.

Nonrepetitive jobs typically involve a broader range of skills than repetitive jobs, and workers in these jobs are often paid on the basis of the highest skill involved. Therefore, it is important to determine the proportion of time spent on the high-skill level. For example, a secretary may take dictation, do word processing, file, answer the telephone, schedule appointments, and do other routine office work. If the secretary spends a high percentage of time filing instead of doing word processing or taking shorthand, the compensation will be lower than for a secretary who spends a high percentage of time doing word processing and taking shorthand. Work sampling can be used to verify those percentages and can therefore be an important tool in developing the job description. In addition, work sampling can be part of a program for validation of job content that is needed for "bona fide occupational qualifications"—that is, advertised jobs requiring the skills that are specified.

Work sampling estimates include some degree of error. Hence, it is important to treat work sampling estimates as *approximations* of the actual proportion of time devoted to a

TABLE 7-4

A portion of the MTM tables

Table II—Move—M

Distance Moved (inches)	TIME (TMU) A	B	C	Hand in Motion B	WEIGHT ALLOWANCE Weight (pounds) up to:	Dynamic Factor	Static Constant TMU	Case and Description
3/4 or less	2.0	2.0	2.0	1.7				
1	2.5	2.9	3.4	2.3	2.5	1.00	0	
2	3.6	4.6	5.2	2.9				
3	4.9	5.7	6.7	3.6	7.5	1.06	2.2	**A. Move object to other hand or against stop.**
4	6.1	6.9	8.0	4.3				
5	7.3	8.0	9.2	5.0	12.5	1.11	3.9	
6	8.1	8.9	10.3	5.7				
7	8.9	9.7	11.1	6.5	17.5	1.17	5.6	
8	9.7	10.6	11.8	7.2				
9	10.5	11.5	12.7	7.9	22.5	1.22	7.4	
10	11.3	12.2	13.5	8.6				**B. Move object to approximate or indefinite location.**
12	12.9	13.4	15.2	10.0	27.5	1.28	9.1	
14	14.4	14.6	16.9	11.4				
16	16.0	15.8	18.7	12.8	32.5	1.33	10.8	
18	17.6	17.0	20.4	14.2				
20	19.2	18.2	22.1	15.6	37.5	1.39	12.5	
22	20.8	19.4	23.8	17.0				
24	22.4	20.6	25.5	18.4	42.5	1.44	14.3	**C. Move object to exact location.**
26	24.0	21.8	27.3	19.8				
28	25.5	23.1	29.0	21.2	47.5	1.50	16.0	
30	27.1	24.3	30.7	22.7				
Additional	0.8	0.6	0.85		TMU per inch over 30 inches			

given activity. The goal of work sampling is to obtain an estimate that provides a specified confidence not differing from the true value by more than a specified error. For example, a hospital administrator might request an estimate of X-ray idle time that will provide a 95 percent confidence of being within 4 percent of the actual percentage.

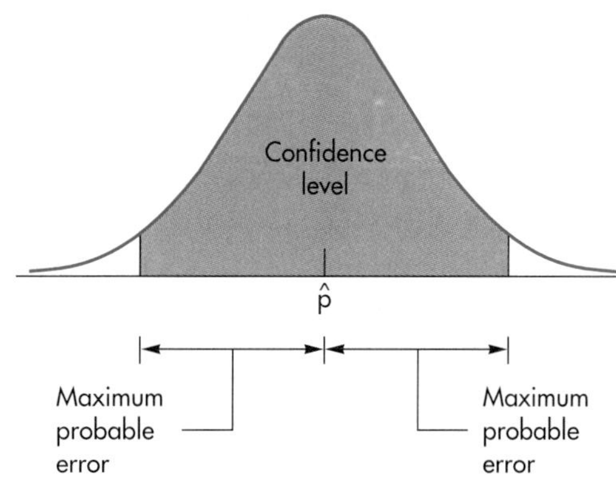

FIGURE 7–7

A confidence interval for the estimate of the true proportion is based on a normal distribution

Hence, work sampling is designed to produce a value, $\hat{p}$, which estimates the true proportion, p, within some allowable error, e: $\hat{p} \pm e$. The variability associated with sample estimates of p tends to be approximately normal for large sample sizes. Consequently, the normal distribution can be used to construct a confidence interval and to determine required sample sizes. One (nonclassical) way to view the sampling distribution is to imagine all of the *possible* true proportions that could produce a given sample value of $\hat{p}$. The result is a distribution of possible true proportions that is (1) normal and (2) centered around the sample proportion. The most likely values of the actual proportion are close to the sample proportion; a true value close to the center of the distribution is much more likely than a value in one of the tails. This concept is illustrated in Figure 7–7. The degree to which this distribution is narrow or spread out is a function of the sample size and p. The amount of maximum probable error is a function of both the sample size and the desired level of confidence.

For large samples, the maximum error e can be computed using the formula

$$e = z\sqrt{\frac{\hat{p}(1-\hat{p})}{n}} \qquad (7\text{–}9)$$

where

z = Number of standard deviations needed to achieve desired confidence
$\hat{p}$ = Sample proportion (the number of occurrences divided by the sample size)
n = Sample size

In most instances, management will specify the desired confidence level and amount of allowable error, and the analyst will be required to determine a sample size sufficient to obtain these results. The appropriate value for n can be determined by solving Formula 7–9 for n, which yields

$$n = \left(\frac{z}{e}\right)^2 \hat{p}(1-\hat{p}) \qquad (7\text{–}10)$$

Example 4

The manager of a small supermarket chain wants to estimate the proportion of time stock clerks spend making price changes on previously marked merchandise. The manager wants a 98 percent confidence that the resulting estimate will be within 5 percent of the true value. What sample size should she use?

Solution

$e = .05$ $\quad z = 2.33$ (see p. 328) $\quad \hat{p}$ is unknown

When no sample estimate of p is available, a preliminary estimate of sample size can be obtained using $\hat{p} = .50$. After 20 or so observations, a new estimate of $\hat{p}$ can be obtained from those observations and a revised value of n computed using the new $\hat{p}$. It would be prudent to recompute the value of n at two or three points during the study to obtain a better indication of the necessary sample size. Thus, the initial estimate of n is

$$n = \left(\frac{2.33}{.05}\right)^2 .50(1 - .50) = 542.89, \text{ or } 543 \text{ observations}$$

Suppose that, in the first 20 observations, stock clerks were found to be changing prices twice, making $p = 2/20 = .10$. The revised estimate of n at that point would be

$$n = \left(\frac{2.33}{.05}\right)^2 .10(1 - .10) = 195.44, \text{ or } 196$$

Suppose a second check is made after a total of 100 observations, and $\hat{p} = .11$ at this point (including the initial 20 observations). Recomputing n yields

$$n = \left(\frac{2.33}{.05}\right)^2 .11(.89) = 212.60, \text{ or } 213$$

Perhaps the manager might make one more check to settle on a final value for n. If the computed value of n is less than the number of observations already taken, sampling would be terminated at that point.

Note: If the resulting value of n is noninteger, round *up*.

Determining the sample size is only one part of work sampling. The overall procedure consists of the following steps:

1. Clearly identify the worker(s) or machine(s) to be studied.
2. Notify the workers and supervisors of the purpose of the study to avoid arousing suspicions.
3. Compute an initial estimate of sample size using a preliminary estimate of p, if available (e.g., from analyst experience or past data). Otherwise, use $\hat{p} = .50$.
4. Develop a random observation schedule.
5. Begin taking observations. Recompute the required sample size several times during the study.
6. Determine the estimated proportion of time spent on the specified activity.

Careful problem definition can prevent mistakes such as studying the wrong worker or wrong activity. Similarly, it is important to inform related parties of the purpose and scope of the study to reduce unnecessary fears that might be generated by unannounced data collection. It is also important to obtain random observations to achieve valid results.

Observations must be spread out over a period of time so that a true indication of variability is obtained. If observations are bunched too closely in time, the behaviors observed during that time may not genuinely reflect typical performance. The degree to which observations should be spread out will depend in part on the nature of the activity studied; a decision on this is usually best left to the analyst.

Table 7–5 presents a comparison of work sampling and time study. It suggests that a work sampling approach to determining job times is less formal and less detailed, and best suited to nonrepetitive jobs.

COMPENSATION

Compensation is a significant issue for the design of work systems. It is important for organizations to develop suitable compensation plans for their employees. If wages are too low, organizations may find it difficult to attract and hold competent workers and managers. If wages are too high, the increased costs may result in lower profits, or may

TABLE 7-5

Work sampling compared with stopwatch time study

Advantages
1. Observations are spread out over a period of time, making results less susceptible to short-term fluctuations.
2. There is little or no disruption of work.
3. Workers are less resentful.
4. Studies are less costly and less time-consuming, and the skill requirements of the analyst are much less.
5. Studies can be interrupted without affecting the results.
6. Many different studies can be conducted simultaneously.
7. No timing device is required.
8. It is well suited for nonrepetitive tasks.
Disadvantages
1. There is much less detail on the elements of a job.
2. Workers may alter their work patterns when they spot the observer, thereby invalidating the results.
3. In many cases, there is no record of the method used by the worker.
4. Observers may fail to adhere to a random schedule of observations.
5. It is not well suited for short, repetitive tasks.
6. Much time may be required to move from one workplace to another and back to satisfy the randomness requirement.

force the organization to increase its prices, which might adversely affect demand for the organization's products or services.

Organizations use two basic systems for compensating employees: *time-based systems* and *output-based systems.* **Time-based systems,** also known as *hourly* and *measured day-work* systems, compensate employees for the time the employee has worked during a pay period. Salaried workers also represent a form of time-based compensation. **Output-based (incentive) systems** compensate employees according to the amount of output they produce during a pay period, thereby tying pay directly to performance.

time-based system Compensation based on time an employee has worked during a pay period.

output-based system Compensation based on amount of output an employee produced during a pay period.

Time-based systems are more widely used than incentive systems, particularly for office, administrative, and managerial employees, but also for blue-collar workers. One reason for this is that computation of wages is straightforward and managers can readily estimate labor costs for a given employee level. Employees often prefer time-based systems because the pay is steady and they know how much compensation they will receive for each pay period. In addition, employees may resent the pressures of an output-based system.

Another reason for using time-based systems is that many jobs do not lend themselves to the use of incentives. In some cases, it may be difficult or impossible to measure output. For example, jobs that require creative or mental work cannot be easily measured on an output basis. Other jobs may include irregular activities or have so many different forms of output that measuring output and determining pay are fairly complex. In the case of assembly lines, the use of *individual* incentives could disrupt the even flow of work; however, *group* incentives are sometimes used successfully in such cases. Finally, *quality* considerations may be as important as *quantity* considerations. In health care, for example, emphasis is generally placed on both the quality of patient care and the number of patients processed.

On the other hand, situations exist where incentives are desirable. Incentives reward workers for their output, presumably causing some workers to produce more than they might under a time-based system. The advantage is that certain (fixed) costs do not vary with increases in output, so the overall cost per unit decreases if output increases. Workers may prefer incentive systems because they see a relationship between their efforts and their pay: An incentive system presents an opportunity for them to earn more money.

On the negative side, incentive systems involve a considerable amount of paperwork, computation of wages is more difficult than under time-based systems, output has to be

TABLE 7–6

Comparison of time-based and output-based pay systems

	Management	Worker
TIME-BASED		
Advantages	1. Stable labor costs 2. Easy to administer 3. Simple to compute pay 4. Stable output	1. Stable pay 2. Less pressure to produce than under output system
Disadvantages	1. No incentive for workers to increase output	1. Extra efforts not rewarded
OUTPUT-BASED		
Advantages	1. Lower cost per unit 2. Greater output	1. Pay related to efforts 2. Opportunity to earn more
Disadvantages	1. Wage computation more difficult 2. Need to measure output 3. Quality may suffer 4. Difficult to incorporate wage increases 5. Increased problems with scheduling	1. Pay fluctuates 2. Workers may be penalized because of factors beyond their control (e.g., machine breakdown)

measured and standards set, cost-of-living increases are difficult to incorporate into incentive plans, and contingency arrangements for unavoidable delays have to be developed.

Table 7–6 lists the main advantages and disadvantages of time-based and output-based plans.

In order to obtain maximum benefit from an incentive plan, the plan should be:

1. Accurate
2. Easy to apply
3. Consistent
4. Easy to understand
5. Fair

In addition, there should be an obvious relationship between effort and reward, and no limit on earnings.

Incentive systems may focus on the output of each individual or on that of a group.

Individual Incentive Plans

Individual incentive plans take a variety of forms. The simplest plan is *straight piecework.* Under this plan, a worker's pay is a direct linear function of his or her output. In the past, piecework plans were fairly popular. Now minimum wage legislation makes them somewhat impractical. Even so, many of the plans currently in use represent variations of the straight piecework plan. They typically incorporate a base rate that serves as a floor: Workers are guaranteed that amount as a minimum, regardless of output. The base rate is tied to an output standard; a worker who produces less than the standard will be paid at the base rate. This protects workers from pay loss due to delays, breakdowns, and similar problems. In most cases, incentives are paid for output above standard, and the pay is referred to as a *bonus.*

Group Incentive Plans

A variety of group incentive plans, which stress sharing of productivity gains with employees, are in use. Some focus exclusively on output, while others reward employees for output and for reductions in material and other costs. The following four plans reflect the main features of most of the plans currently in operation.

Scanlon Plan. The main feature of the plan is to encourage reductions in labor costs by allowing workers to share in the gains from any reductions achieved. The plan includes formation of worker committees to actively seek out areas for improvement.

Kaiser Plan. Like the Scanlon plan, it uses committees to suggest ways of reducing costs, with savings shared by employees. However, in addition to reductions in labor costs, it also provides for workers to share in the gains from reductions in material and supply costs.

Lincoln Plan. The Lincoln Electric Company in Cleveland, Ohio, developed this plan. It includes profit sharing, job enlargement, and participative management. Like the other plans, it uses evaluation committees to generate suggestions. The three main components of the plan are a piecework system, an annual bonus, and a stock purchase provision.

Kodak Plan. This plan uses a combination of premium wage levels and an annual bonus related to company profits instead of more traditional incentives. Workers are encouraged to help set goals and to decide on reasonable performance levels. The idea is that their involvement will make workers more apt to produce at a premium rate.

One form of group incentive is the *team approach,* which many companies are now using for problem solving and continuous improvement. The emphasis is on *team,* not *individual,* performance.

Knowledge-Based Pay Systems

As companies shift toward lean production, a number of changes have had a direct impact on the work environment. One is that many of the buffers that previously existed are gone. Another is that fewer managers are present. Still another is increased emphasis on quality, productivity, and flexibility. Consequently, workers who can perform a variety of tasks are particularly valuable. Organizations are increasingly recognizing this, and they are setting up pay systems to reward workers who undergo training that increases their skill levels. This is sometimes referred to as **knowledge-based pay.** It is a portion of a worker's pay that is based on the knowledge and skill that the worker possesses. Knowledge-based pay has three dimensions: *Horizontal skills* reflect the variety of tasks the worker is capable of performing; *vertical skills* reflect management tasks the worker is capable of; and *depth skills* reflect quality and productivity results.

knowledge-based pay A pay system used by organizations to reward workers who undergo training that increases their skills.

Management Compensation

Many organizations that traditionally rewarded managers and senior executives on the basis of *output* are now seriously reconsidering that approach. With the new emphasis on customer service and quality, reward systems are being restructured to reflect new dimensions of performance. In addition, executive pay in many companies is being more closely tied to the success of the company or division that executive is responsible for. This is quite different than the 1980s policy of increasing the compensation of top executives even as workers were being laid off and the company was losing large amounts of money!

Summary

The design of work systems involves job design, work measurement, and compensation.

Job design is concerned with the content of jobs and work methods. In the past, job design tended to focus on efficiency, but now there seems to be an increasing awareness and consideration of the behavioral aspects of work and worker satisfaction. Current concern about productivity has thrust job design into the limelight. However, the jobs usually associated with high productivity are often the same jobs that are the greatest source of worker dissatisfaction, creating somewhat of a paradox for job designers.

Analysts often use methods analysis and motion study techniques to develop the "efficiency" aspects of jobs, but these do not directly address behavioral aspects. Nonetheless, they are an important part of job design. Working conditions are also a notable aspect of job design, not only because of the behavioral and efficiency factors but also because of concern for the health and safety of workers.

Work measurement is concerned with specifying the length of time needed to complete a job. Such information is vital for personnel planning, cost estimating, budgeting, scheduling, and worker compensation. Commonly used approaches include stopwatch time study and predetermined times. A related technique is work sampling, which can also be used to obtain data on activity times. More commonly, work sampling is used to estimate the proportion of time a worker spends on a certain aspect of the job. Table 7–7 provides a summary of the formulas used in time studies.

Organizations can choose from a variety of compensation plans. It is important to do so carefully, for compensation is key to both the worker and the organization, and, once adopted, it is usually difficult to substantially change a compensation plan.

Key Terms

flow process chart, 317
job design, 311
job enlargement, 312
job enrichment, 312
job rotation, 312
knowledge-based pay, 339
methods analysis, 316
micromotion study, 321
motion study, 318
motion study principles, 320
OSHA, 325
output-based (incentive) system, 337
predetermined time standards, 332
self-directed teams, 313
specialization, 311
standard elemental times, 332
standard time, 326
stopwatch time study, 327
therblig, 320
time-based system, 337
work measurement, 326
work sampling, 333
worker-machine chart, 317

Solved Problems

Problem 1

A time study analyst timed an assembly operation for 30 cycles, and then computed the average time per cycle, which was 18.75 minutes. The analyst assigned a performance rating of .96, and decided that an appropriate allowance was 15 percent. Assume the allowance factor is based on the *workday*. Determine the following: the observed time (OT), the normal time (NT), and the standard time (ST).

Solution

$$\text{OT} = \text{Average time} = 18.75 \text{ minutes}$$
$$\text{NT} = \text{OT} \times \text{Performance rating} = 18.75 \text{ minutes} \times .96 = 18 \text{ minutes}$$
$$\text{AF} = \frac{1}{1 - \text{A}} = \frac{1}{1 - .15} = 1.176$$
$$\text{ST} = \text{NT} \times \text{AF} = 18 \times 1.176 = 21.17 \text{ minutes}$$

Problem 2

A time study analyst wants to estimate the number of observations that will be needed to achieve a specified maximum error, with a confidence of 95.5%. A preliminary study yielded

TABLE 7-7

Summary of formulas

Time Study

A. Sample size

$$n = \left(\frac{zs}{a\bar{x}}\right)^2 \quad (7\text{–}1)$$

$$n = \left(\frac{zs}{e}\right)^2 \quad (7\text{–}2)$$

B. Observed time

$$\text{OT} = \frac{\Sigma x_i}{n} \quad (7\text{–}3)$$

C. Normal time

$$\text{NT} = \text{OT} \times \text{PR} \quad (7\text{–}4)$$

$$\text{NT} = \Sigma(\bar{x}_j \times \text{PR}_j) \quad (7\text{–}5)$$

D. Standard time

$$\text{ST} = \text{NT} \times \text{AF} \quad (7\text{–}6)$$

E. Allowance factor

$$\text{AF}_{\text{job}} = 1 + \text{A} \quad (7\text{–}7)$$

$$\text{AF}_{\text{day}} = \frac{1}{1 - \text{A}} \quad (7\text{–}8)$$

Work Sampling

A. Maximum error

$$e = z\sqrt{\frac{\hat{p}(1-\hat{p})}{n}} \quad (7\text{–}9)$$

B. Sample size

$$n = \left(\frac{z}{e}\right)^2 \hat{p}(1-\hat{p}) \quad (7\text{–}10)$$

Symbols:

- a = Allowable error as percentage of average time
- A = Allowance percentage
- e = Maximum acceptable error
- n = Number of observations needed
- NT = Normal time
- OT = Observed, or average, time
- s = Standard deviation of observed times
- ST = Standard time
- x_i = Time for ith observation ($i = 1, 2, 3, \ldots, n$)

a mean of 5.2 minutes and a standard deviation of 1.1 minutes. Determine the total number of observations needed for these two cases:

a. A maximum error of ±6% of the sample mean.

b. A maximum error of .40 minute.

Solution

a. $x = 5.2$ minutes $z = 2.00$ for 95.5%
$x = 1.1$ minutes $a = .06$

$$n = \left(\frac{zs}{ax}\right)^2 = \left(\frac{2.00(1.1)}{.06(5.2)}\right)^2 = 49.72 \text{ (round to 50 observations)}$$

b. $e = .40$

$$n = \left(\frac{zs}{e}\right)^2 = \left(\frac{2.00(1.1)}{.40}\right)^2 = 30.25 \text{ (round to 31 observations)}$$

Problem 3

Work sampling. An analyst has been asked to prepare an estimate of the proportion of time that a turret lathe operator spends adjusting the machine, with a 90 percent confidence level.

Based on previous experience, the analyst believes the proportion will be approximately 30 percent.

a. If the analyst uses a sample size of 400 observations, what is the maximum possible error that will be associated with the estimate?

b. What sample size would the analyst need in order to have the maximum error be no more than ± 5 percent?

$\hat{p} = .30 \qquad z = 1.65$ (for 90 percent confidence)

Solution

$$a.\ e = z\sqrt{\frac{\hat{p}(1-\hat{p})}{n}} = 1.65\sqrt{\frac{.3(.7)}{400}} = .038$$

$$b.\ n = \left(\frac{z}{e}\right)^2 \hat{p}(1-\hat{p}) = \left(\frac{1.65}{.05}\right)^2 (.3)(.7) = 228.69, \text{ or } 229$$

Problem 4

Work sampling. The owner of a large gift shop wants to estimate the proportion of time shoppers spend in a certain part of the store. The shop is open nine hours a day, six days a week. Twelve observations will be taken over a 30-day period. Determine the random observation times using Table 7–5 and these instructions:

a. For days, read the first two digits of column 2, going down, and then the first two digits of column 3, going down.

b. For hours, read one-digit numbers in row 4 from left to right.

c. For minutes, read two-digit numbers using the first two digits in column 1, going down, and then the second two digits in column 1, going down.

Solution

a. **Days: Random Number**	*b.* **Hours: Random Number**	*c.* **Minutes: Random Number**
72*	1	69†
11	6	34
24	3	47
15	2	16
72*	1	85†
13	5	39
77*	4	78†
13	6	27
82*	1	29
99*	9	85†
16	5	05
01	0‡	29
28	1	12
05		91‡
27		15
19		32
09		10

*Since there will be 30 days in the study, random numbers larger than 30 are not relevant.

†There are 60 minutes per hour. Larger numbers do not correspond, so they are discarded.

‡The store is open nine hours per day. The numbers 1–9 correspond to the hours 1–9; 0 is not applicable.

Combining these results, in the order they appeared, produces the following results:

Day	Hour	Minute
11	1	34
24	6	47
15	3	16
13	2	39
13	1	27
16	5	29
01	4	05
28	6	29
05	1	12
27	9	15
19	5	32
09	1	10

Arranging the results chronologically by day, hour, and minute (but maintaining the same *combinations*) produces this list of observation times:

Observation	Day	Hour	Minute
1	01	4	05
2	05	1	12
3	09	1	10
4	11	1	34
5	13	1	27
6	13	2	39
7	15	3	16
8	16	5	29
9	19	5	32
10	24	6	47
11	27	9	15
12	28	6	29

Discussion and Review Questions

1. What is job design, and why is it important?
2. What are some of the main advantages and disadvantages of specialization from a management perspective? From a worker's perspective?
3. Contrast the meanings of the terms *job enlargement* and *job enrichment*.
4. What is the purpose of job enlargement and job enrichment?
5. Explain the term *knowledge-based pay system*.
6. What are self-directed work teams? What are some potential benefits of using these teams?
7. Some Japanese firms have a policy of rotating their managers among the different managerial jobs. In contrast, American managers are more likely to specialize in a certain area (e.g., finance or operations). Discuss the advantages and disadvantages of each of these approaches. Which do you prefer? Why?
8. What are motion study principles? How are they classified?
9. Name some reasons why methods analyses are needed. How is methods analysis linked to productivity improvements?
10. How are devices such as flow process charts and worker-machine charts useful?
11. What is a time standard? What factors must be taken into account when developing standards?
12. What are the main uses of time study information?
13. Could performance rating be avoided by studying a group of workers and averaging their times? Explain briefly.

14. If an average worker could be identified, what advantage would there be in using that person for a time study? What are some reasons why an average worker might not be studied?
15. What are the main limitations of time study?
16. Comment on the following. "At any given instant, the standard times for many jobs will not be strictly correct."
    a. Why is this so?
    b. Does this mean that those standards are useless? Explain.
17. Why do workers sometimes resent time studies?
18. What are the key advantages and disadvantages of:
    a. Time-based pay plans?
    b. Incentive plans?
19. What is work sampling? How does it differ from time study?

Memo Writing Exercises

1. Write a one-page memo to your subordinate, Joe Brown, about self-directed work teams. Explain what self-directed teams are, and their potential benefits and risks.
2. Your supervisor, the production manager of a company that recently shifted operations from a traditional manufacturing mode to a lean production mode, has asked you to write a memo to H. Penny, the chief financial officer of the company, explaining knowledge-based pay systems. Penny is particularly interested in the rationale for adopting this pay system. In your memo, be sure to include the three categories of knowledge or skills that are covered in this system.

Problems

1. An analyst has timed a metal-cutting operation for 50 cycles. The average time per cycle was 10.40 minutes, and the standard deviation was 1.20 minutes for a worker with a performance rating of 125 percent. Assume an allowance of 16 percent of job time. Find the standard time for this operation.

15.08 minutes

2. A job was timed for 60 cycles and had an average of 1.2 minutes per piece. The performance rating was 95 percent, and workday allowances are 10 percent. Determine each of the following:
    a. Observed time
    b. Normal time
    c. Standard time

a. 1.2 minutes
b. 1.14 minutes
c. 1.27 minutes

3. A. time study was conducted on a job that contains four elements. The observed times and performance ratings for six cycles are shown in the following table.

Element	Performance Rating	OBSERVATIONS (MINUTES PER CYCLE) 1	2	3	4	5	6
1	90%	0.44	0.50	0.43	0.45	0.48	0.46
2	85	1.50	1.54	1.47	1.51	1.49	1.52
3	110	0.84	0.89	0.77	0.83	0.85	0.80
4	100	1.10	1.14	1.08	1.20	1.16	1.26

a. Determine the average cycle time for each element.
b. Find the normal time for each element.
c. Assuming an allowance factor of 15 percent of job time, compute the standard time for this job.

Element	*a*	*b*	*c*
1	0.46	0.414	0.476
2	1.505	1.280	1.472
3	0.83	0.913	1.050
4	1.16	1.160	1.334

4. Given these observed times (in minutes) for four elements of a job, determine the observed time (OT) for each element. Note: The second element only occurs every other cycle.

See IM.

	CYCLE					
	1	2	3	4	5	6
Element 1	4.1	4.0	4.2	4.1	4.1	4.1
Element 2	—	1.5	—	1.6	—	1.4
Element 3	3.2	3.2	3.3	3.2	3.3	3.3
Element 4	2.7	2.8	2.7	2.8	2.8	2.8

5. Given these observed times (in minutes) for five elements of a job, determine the observed time (OT) for each element. Note. Some of the elements occur only periodically.

See IM.

	CYCLE					
	1	2	3	4	5	6
Element 1	2.1	2.0	2.2	2.1	2.1	—
Element 2	—	1.1	—	1.0	—	1.2
Element 3	3.4	3.5	3.3	3.5	3.4	3.3
Element 4	4.0	—	—	4.2	—	—
Element 5	1.4	1.4	1.5	1.5	1.5	1.4

6. Using Table 7–3, develop an allowance percentage for a job element that requires the worker (1) to lift a weight of 10 pounds while standing in a slightly awkward position, (2) in light that is slightly below recommended standards, and (3) with intermittent loud noises occurring. The monotony for this element is high. Include a personal allowance of 5 percent and a basic fatigue allowance of 4 percent of job time.

Factor	Percent
Personal	5%
Fatigue	4
Standing	2
Lifting	1
Bad light	0
Noise level	2
Monotony	4
	18%

7. A worker-machine operation was found to involve 3.3 minutes of machine time per cycle in the course of 40 cycles of stopwatch study. The worker's time averaged 1.9 minutes per cycle, and the worker was given a rating of 120 percent (machine rating is 100 percent). Midway through the study, the worker took a 10-minute rest break. Assuming an allowance factor of 12 percent, determine the standard time for this job.

5.85 minutes

8. A recently negotiated union contract allows workers in a shipping department 24 minutes for rest, 10 minutes for personal time, and 14 minutes for delays for each four hours worked. A time study analyst observed a job that is performed continuously and found an average time of 6.0 minutes per cycle for a worker she rated at 95 percent. What standard time is applicable for that operation?

7.125 minutes

9. The data in the table below represent time study observations for a woodworking operation.
   a. Based on the observations, determine the standard time for the operation, assuming an allowance of 15 percent.
   b. How many observations would be needed to estimate the mean time for element 2 within 1 percent of its true value with a 95.5 percent confidence?
   c. How many observations would be needed to estimate the mean time for element 2 within .01 minute of its true value with a 95.5 percent confidence?

a.

Element	ST
1	1.505
2	1.098
3	0.676

b. 67
c. 47

Element	Performance Rating	OBSERVATIONS (MINUTES PER CYCLE) 1	2	3	4	5	6
1	110%	1.20	1.17	1.16	1.22	1.24	1.15
2	115	0.83	0.87	0.78	0.82	0.85	1.32*
3	105	0.58	0.53	0.52	0.59	0.60	0.54

*Unusual delay, disregard time.

10. How many observations should a time study analyst plan for in an operation that has a standard deviation of 1.5 minutes per piece if the goal is to estimate the mean time per piece to within 0.4 minutes with a confidence of 95.5 percent?

57

11. How many work cycles should be timed to estimate the average cycle time to within 2 percent of the sample mean with a confidence of 99 percent if a pilot study yielded these times (minutes): 5.2, 5.5, 5.8, 5.3, 5.5, and 5.1?

37

a. 12%
b. 163

12. In an initial survey designed to estimate the percentage of time air-express cargo loaders are idle, an analyst found that loaders were idle in 6 of the 50 observations.
    a. What is the estimated percent of idle time?
    b. Based on the initial results, approximately how many observations would you require to estimate the actual percentage of idle time to within 5 percent with a confidence of 95 percent?

377

13. A job in an insurance office involves telephone conversations with policyholders. The office manager estimates that about half of the employee's time is spent on the telephone. How many observations are needed in a work sampling study to estimate that time percentage to within 6 percent and have a confidence of 98 percent?

READING

MAKING HOTPLATES **Edgar F. Huse**

A group of 10 workers were responsible for assembling hotplates (instruments for heating solutions to a given temperature) for hospital and medical laboratory use. A number of different models of hotplates were being manufactured. Some had a vibrating device so that the solution could be mixed while being heated. Others heated only test tubes. Still others could heat solutions in a variety of different containers.

With the appropriate small tools, each worker assembled part of a hotplate. The partially completed hotplate was placed on a moving belt, to be carried from one assembly station to the next. When the hotplate was completed, an inspector would check it over to ensure that it was working properly. Then the last worker would place it in a specially prepared cardboard box for shipping.

The assembly line had been carefully balanced by industrial engineers, who had used a time and motion study to break the job down into subassembly tasks, each requiring about three minutes to accomplish. The amount of time calculated for each subassembly had also been "balanced" so that the task performed by each worker was supposed to take almost exactly the same amount of time. The workers were paid a straight hourly rate.

However, there were some problems. Morale seemed to be low, and the inspector was finding a relatively high percentage of badly assembled hotplates. Controllable rejects—those "caused" by the operator rather than by faulty materials—were running about 23 percent.

After discussing the situation, management decided to try something new. The workers were called together and asked if they would like to build the hotplates individually. The workers decided they would like to try this approach, provided they could go back to the old program if the new one did not work well. After several days of training, each worker began to assemble the entire hotplate.

The change was made at about the middle of the year. Productivity climbed quickly. By the end of the year, it had leveled off at about 84 percent higher than during the first half of the year, although no other changes had been made in the department or its personnel. Controllable rejects had dropped from 23 percent to 1 percent during the same period. Absenteeism had dropped from 8 percent to less than 1 percent. The workers had responded positively to the change, and their morale was higher. As one person put it, "Now, it is *my* hotplate." Eventually, the reject rate dropped so low that all routine final inspection was done by the assembly workers themselves. The full-time inspector was transferred to another job in the organization.

Questions

1. What changes in the work situation might account for the increase in productivity and the decrease in controllable rejects?
2. What might account for the drop in absenteeism and the increase in morale?
3. What were the major changes in the situation? Which changes were under the control of the manager? Which were controlled by workers?
4. What might happen if the workers went back to the old assembly line method?

Selected Bibliography

Barnes, Ralph M. *Motion and Time Study: Design and Measurement of Work,* 8th ed. New York: John Wiley & Sons, 1980.

Buffa, Elwood Spencer. *Operations Management: The Management of Productive Systems.* 2d ed. New York: John Wiley & Sons, 1980.

Carlisle, Brian. "Job Design Implications for Operations Managers." *International Journal of Operations and Production Management* 3, no. 3 (1983), pp. 40–48.

Cunningham, J. Barton, and Ted Eberle. "A Guide to Job Enrichment and Redesign." *Personnel,* February 1990, pp. 56–61.

Dumaine. Brian. "Who Needs a Boss?" *Fortune,* May 7, 1990, pp. 52–60.

Huchinson, R. D. *New Horizons for Human Factors in Job Design.* New York: McGraw-Hill, 1981.

Konz, S. *Work Design: Industrial Ergonomics.* 2nd ed. New York: John Wiley & Sons, 1981.

Larson, Carl E., and Frank M. LoFasto. *Teamwork: What Must Go Right/What Can Go Wrong.* Newbury Park, CA: Sage, 1989.

Mundel, Marvin E. *Motion and Time Study.* 5th ed. Englewood Cliffs, NJ: Prentice Hall, 1978.

Nadler, G. *Work Design.* Rev. ed. Burr Ridge, IL: Richard D. Irwin, 1970.

Niebel, Benjamin W. *Motion and Time Study.* 9th ed. Burr Ridge, IL: Richard D. Irwin, 1993.

Osborn, Jack D.; Linda Moran; Ed Musslewhite; and John H. Zenger. *Self-Directed Work Teams.* Burr Ridge, IL: Richard D. Irwin, 1990.

Sakamoto, Shigeyasu. "Key to Productivity: Work Measurement, An International Survey Report." *MTM Journal of Methods—Time Measurement* 13, (1987), pp. 68–75.

SUPPLEMENT TO CHAPTER SEVEN

LEARNING CURVES

LEARNING OBJECTIVES

After completing this supplement, you should be able to:

1. Explain the concept of a learning curve.
2. Make time estimates based on learning curves.
3. List and briefly describe some of the main applications of learning curves.
4. Outline some of the cautions and criticisms of learning curves.

SUPPLEMENT OUTLINE

Learning is usually occurring when humans are involved; this is a basic consideration in the design of work systems. It is important to be able to predict how learning will affect task times and costs. This supplement addresses those issues.

THE CONCEPT OF LEARNING CURVES

Human performance of activities typically shows improvement when the activities are done on a repetitive basis: The time required to perform a task decreases with increasing repetitions. *Learning curves* summarize this phenomenon. The degree of improvement and the number of tasks needed to realize the major portion of the improvement is a function of the task being done. If the task is short and somewhat routine, only a modest amount of improvement is likely to occur, and it generally occurs during the first few repetitions. If the task is fairly complex and has a longer duration, improvements will occur over a longer interval (i.e., a larger number of repetitions). Therefore, learning factors have little relevance for planning or scheduling routine activities, but they do have relevance for complex repetitive activities.

Figure 7S–1 illustrates the basic relationship between increasing repetitions and a decreasing time per repetition. It should be noted that the curve will never touch the horizontal axis; that is, the time per unit will never be zero.

T. P. Wright drew attention to learning curves in 1936, when he described how the direct labor cost of producing a certain airframe decreased with experience.[1] Since that time, other studies have confirmed the general relationship that Wright described, which is now alternately referred to as an experience curve, a progress function, or an improvement function. Experts now widely agree that the learning effect is the result of other factors in addition to actual worker learning. Some of the improvement can be traced to preproduction factors, such as selection of tooling and equipment, product design, methods analysis, and, in general, the amount of effort expended prior to the start of the work. Other contributing factors may involve changes after production has begun, such as changes in methods, tooling, and design. In addition, management input can be an important factor through improvements in planning, scheduling, motivation, and control.

Changes that are made once production is under way can cause a temporary *increase* in time per unit until workers adjust to the change, even though they eventually lead to an increased output rate. If a number of changes are made during production, the learning curve would be more realistically described by a series of scallops instead of a smooth curve, as illustrated in Figure 7S–2. Nonetheless, it is convenient to work with a smooth curve, which can be interpreted as the *average* effect.

From an organizational standpoint, what makes the learning effect more than an interesting curiosity is its *predictability*, which becomes readily apparent if the relationship is plotted on a log-log scale (see Figure 7S–3). The straight line that results reflects a constant learning percentage, which is the basis of learning curve estimates: every *doubling* of repetitions results in a *constant percentage* decrease in the time per repetition. This applies both to the *average* and to the *unit* time. Typical decreases range from 10 percent to 20 percent. By convention, learning curves are referred to in terms of the *complements* of their improvement rates. For example, an 80 percent learning curve denotes a 20 percent decrease in unit (or average) time with each doubling of repetitions, and a 90 percent curve denotes a 10 percent improvement rate. Note that a 100 percent curve would imply no improvement at all.

Example S-1

An activity is known to have an 80 percent learning curve. It has taken a worker 10 hours to produce the first unit. Determine expected completion times for these units: the 2nd, 4th, 8th, and 16th (note successive doubling of units).

[1]T. P. Wright, "Factors Affecting the Cost of Airplanes," *Journal of the Aeronautical Sciences* 3 (February 1936).

FIGURE 7S–1

The learning effect: time per repetition decreases as the number of repetitions increases

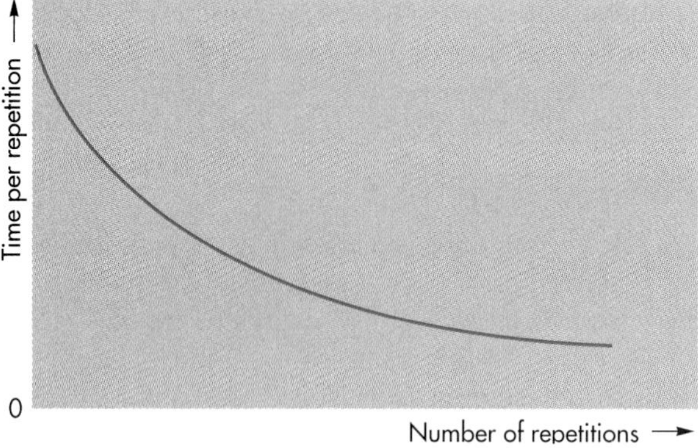

FIGURE 7S–2

Improvements may create a scallop effect in the curve

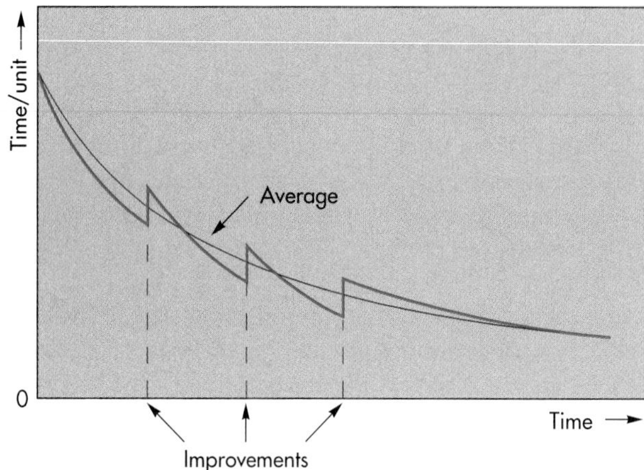

FIGURE 7S–3

On a log-log graph, learning curves are straight lines

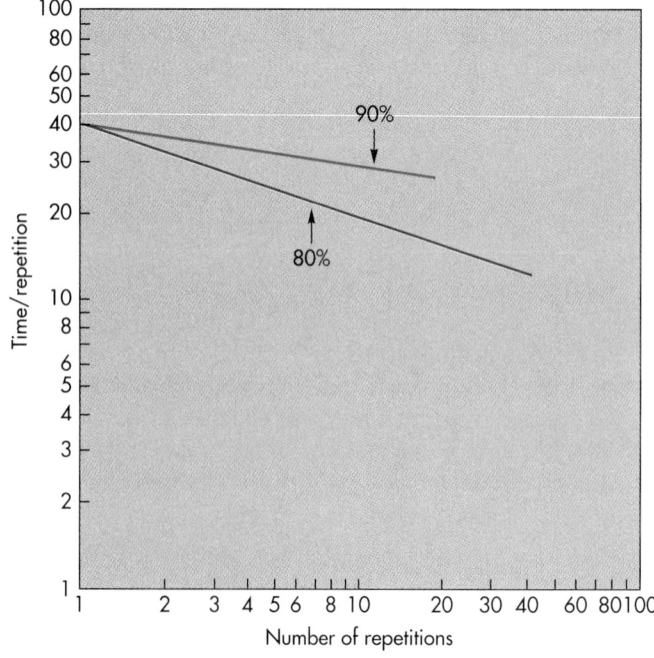

Solution

Each time the cumulative output doubles, the time per unit for that amount should be approximately equal to the previous time multiplied by the learning percentage (80 percent in this case). Thus:

Unit	Unit Time (hours)
1.	= 10
2.	0.8(10) = 8
4.	0.8(8) = 6.4
8.	0.8(6.4) = 5.12
16	0.8(5.12) = 4.096

Example S-1 illustrates an important point and also raises an interesting question. The point is that the time reduction *per unit* becomes less and less as the number of repetitions increase. For example, the second unit required two hours less time than the first, and the improvement from the 8th to the 16th unit was only slightly more than one hour. The question raised is: How are times computed for values such as three, five, six, seven, and other units that don't fall into this pattern?

There are two ways to obtain the times. One is to use a formula; the other is to use a table of values.

First, consider the formula approach. The formula is based on the existence of a linear relationship between the time per unit and the number of units when these two variables are expressed in logarithms.

The unit time (i.e., the number of direct labor hours required) for the *n*th unit can be computed using the formula

$$T_n = T_1 \times n^b \qquad (7S\text{–}1)$$

where

T_n = Time for *n*th unit

T_1 = Time for first unit

b = ln learning percent/ln 2; ln stands for the natural logarithm

To use the formula, you need to know the time for the first unit and the learning percentage. For example, for an 80 percent curve with $T_1 = 10$ hours, the time for the third unit would be computed as

$$T_3 = 10(3^{\ln .8/\ln 2}) = 7.02$$

The second approach is to use a "learning factor" obtained from a table such as Table 7S–1.

The table shows two things for some selected learning percentages. One is a unit value for each of the outputs listed. This enables us to easily determine how long any unit will take to produce. The other is a cumulative value, which enables us to compute the total number of hours needed to complete any given number of repetitions. The computation for both is a relatively simple operation: Multiply the table value by the time required for the first unit.

Example S-2

Production Airplanes is negotiating a contract for the production of 20 small jet aircraft. The initial jet required the equivalent of 400 days of direct labor. Estimate the expected number of days of direct labor for:

a. The 20th jet.

b. All 20 jets.

c. The average time for 20 jets.

TABLE 7S-1

Learning curve coefficients

	70%		75%		80%		85%		90%	
Unit Number	Unit Time	Total Time	Unit Time	Total Time	Unit Time	Total Time	Unit Time	Total Time	Unit Time	Total Time
1	1.000	1.000	1.000	1.000	1.000	1.000	1.000	1.000	1.000	1.000
2	.700	1.700	.750	1.750	.800	1.800	.850	1.850	.900	1.900
3	.568	2.268	.634	2.384	.702	2.502	.773	2.623	.846	2.746
4	.490	2.758	.562	2.946	.640	3.142	.723	3.345	.810	3.556
5	.437	3.195	.513	3.459	.596	3.738	.686	4.031	.783	4.339
6	.398	3.593	.475	3.934	.562	4.299	.657	4.688	.762	5.101
7	.367	3.960	.446	4.380	.534	4.834	.634	5.322	.744	5.845
8	.343	4.303	.422	4.802	.512	5.346	.614	5.936	.729	6.574
9	.323	4.626	.402	5.204	.493	5.839	.597	6.533	.716	7.290
10	.306	4.932	.385	5.589	.477	6.315	.583	7.116	.705	7.994
11	.291	5.223	.370	5.958	.462	6.777	.570	7.686	.695	8.689
12	.278	5.501	.357	6.315	.449	7.227	.558	8.244	.685	9.374
13	.267	5.769	.345	6.660	.438	7.665	.548	8.792	.677	10.052
14	.257	6.026	.334	6.994	.428	8.092	.539	9.331	.670	10.721
15	.248	6.274	.325	7.319	.418	8.511	.530	9.861	.663	11.384
16	.240	6.514	.316	7.635	.410	8.920	.522	10.383	.656	12.040
17	.233	6.747	.309	7.944	.402	9.322	.515	10.898	.650	12.690
18	.226	6.973	.301	8.245	.394	9.716	.508	11.405	.644	13.334
19	.220	7.192	.295	8.540	.388	10.104	.501	11.907	.639	13.974
20	.214	7.407	.288	8.828	.381	10.485	.495	12.402	.634	14.608
21	.209	7.615	.283	9.111	.375	10.860	.490	12.892	.630	15.237
22	.204	7.819	.277	9.388	.370	11.230	.484	13.376	.625	15.862
23	.199	8.018	.272	9.660	.364	11.594	.479	13.856	.621	16.483
24	.195	8.213	.267	9.928	.359	11.954	.475	14.331	.617	17.100
25	.191	8.404	.263	10.191	.355	12.309	.470	14.801	.613	17.713
26	.187	8.591	.259	10.449	.350	12.659	.466	15.267	.609	18.323
27	.183	8.774	.255	10.704	.346	13.005	.462	15.728	.606	18.929
28	.180	8.954	.251	10.955	.342	13.347	.458	16.186	.603	19.531
29	.177	9.131	.247	11.202	.338	13.685	.454	16.640	.599	20.131
30	.174	9.305	.244	11.446	.335	14.020	.450	17.091	.596	20.727

Solution

Using Table 7S–1 with $n = 20$ and an 80 percent learning percentage, you find these factors; Unit time = .381. Total time = 10.485.

a. Expected time for 20th jet: 400(.381) = 152.4 labor days.

b. Expected total time for all 20: 400(10.485) = 4,194 labor days.

c. Average time for 20: 4,194 ÷ 20 = 209.7 labor days.

Use of Table 7S–1 requires a time for the first unit. If for some reason the completion time of the first unit is not available, or if the manager believes the completion time for some later unit is more reliable, the table can be used to obtain an estimate of the initial time.

Example S-3

The manager in Example S-2 believes that some unusual problems were encountered in producing the first jet and would like to revise that estimate based on a completion time of 276 days for the third jet.

Solution

The unit value for $n = 3$ and an 80 percent curve is .702 (Table 7S–1). Divide the actual time for unit 3 by the table value to obtain the revised estimate for unit 1's time: 276 days ÷ .702 = 393.2 labor days.

APPLICATIONS OF LEARNING CURVES

Learning curve theory has found useful applications in a number of areas, including:

1. Manpower planning and scheduling.
2. Negotiated purchasing.
3. Pricing new products.
4. Budgeting, purchasing, and inventory planning.
5. Capacity planning.

Knowledge of output projections in learning situations can help managers make better decisions about how many workers they will need than they could determine from decisions based on initial output rates. Of course, managers obviously recognize that improvement will occur; what the learning curve contributes is a method for quantifying expected future improvements.

Negotiated purchasing often involves contracting for specialized items that may have a high degree of complexity. Examples include aircraft, computers, and special-purpose equipment. The direct labor cost per unit of such items can be expected to decrease as the size of the order increases. Hence, negotiators first settle on the number of units and then negotiate price on that basis. The government requires learning curve data on contracts that involve large, complex items. For contracts that are terminated before delivery of all units, suppliers can use learning curve data to argue for an increase in the unit price for the smaller number of units. Conversely, the government can use that information to negotiate a lower price per unit on follow-on orders on the basis of projected additional learning gains.

Managers must establish prices for their new products and services, often on the basis of production of a few units. Generalizing from the cost of the first few units would result in a much higher price than can be expected after a greater number of units have been produced. Actually, the manager needs to use the learning curve to avoid underpricing as well as overpricing. The manager may project initial costs by using the learning progression known to represent an organization's past experience, or else do a regression analysis of the initial results.

The learning curve projections help managers to plan costs, and labor, purchasing, and inventory needs. For example, initial cost per unit will be high and output will be fairly low, so purchasing and inventory decisions can reflect this. As productivity increases, purchasing and/or inventory actions must allow for increased usage of raw materials and purchased parts to keep pace with output. Because of learning effects, the usage rate will increase over time. Hence, failure to refer to a learning curve would lead to substantial *overestimates* of labor needs and *underestimates* of the rate of material usage.

The learning principles can sometimes be used to evaluate new workers during training periods. This is accomplished by measuring each worker's performance, graphing the results, and comparing them to an expected rate of learning. The comparison reveals which workers are underqualified, average, and overqualified for a given type of work (see Figure 7S-4). Moreover, measuring a worker's progress can help predict whether the worker will make a quota within a required period of time.

Operations Strategy

Learning curves often have strategic implications for market entry, when an organization hopes to rapidly gain market share. The use of time-based strategies can contribute to this. An increase in market share creates additional volume, enabling operations to quickly move down the learning curve, thereby decreasing costs and, in the process, gaining a competitive advantage. In some instances, the volumes are sufficiently large

FIGURE 7S–4

Worker learning curves can help guide personnel job placement

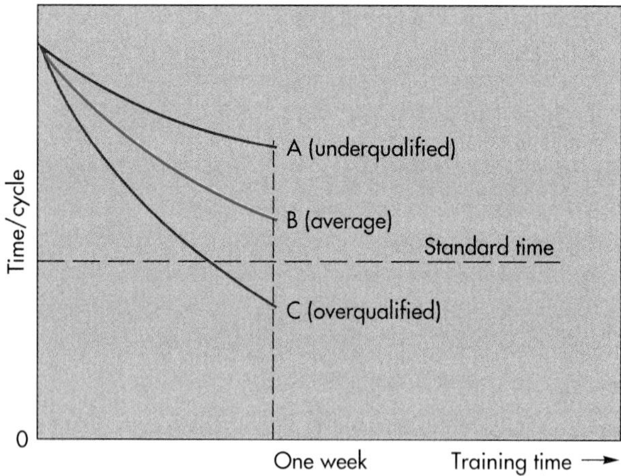

that operations will shift from batch mode to repetitive operation, which can lead to further cost reductions.

Learning curve projections can be useful for capacity planning. Having realistic time estimates based on learning curve theory, managers can translate that information into actual capacity needs, and plan on that basis.

CAUTIONS AND CRITICISMS

Managers using learning curves should be aware of their limitations and pitfalls. This section briefly outlines some of the major cautions and criticisms of learning curves.

1. Learning rates may differ from organization to organization and by type of work. Therefore, it is best to base learning rates on empirical studies rather than assumed rates where possible.
2. Projections based on learning curves should be regarded as *approximations* of actual times and treated accordingly.
3. Because time estimates are based on the time for the first unit, considerable care should be taken to ensure that the time is valid. It may be desirable to revise the base time as later times become available. Since it is often necessary to estimate the time for the first unit prior to production, this caution is very important.
4. It is possible that at some point the curve might level off or even tip upward, especially near the end of a job. The potential for savings at that point is so slight that most jobs do not command the attention or interest to sustain improvements. Then, too, some of the better workers or other resources may be shifted into new jobs that are starting up.
5. Some of the improvements may be more apparent than real: Improvements in times may be caused in part by *increases* in *indirect* labor costs.
6. In mass production situations, learning curves may be of initial use in predicting how long it will take before the process stabilizes. For the most part, however, the concept does *not* apply to mass production because the decrease in time per unit is imperceptible for all practical purposes (see Figure 7S–5).
7. Users of learning curves sometimes fail to include carryover effects; previous experience with similar activities can reduce activity times, although it should be noted that the *learning rate* remains the same.
8. Shorter product life cycles, flexible manufacturing, and cross-functional workers can affect the ways in which learning curves may be applied.

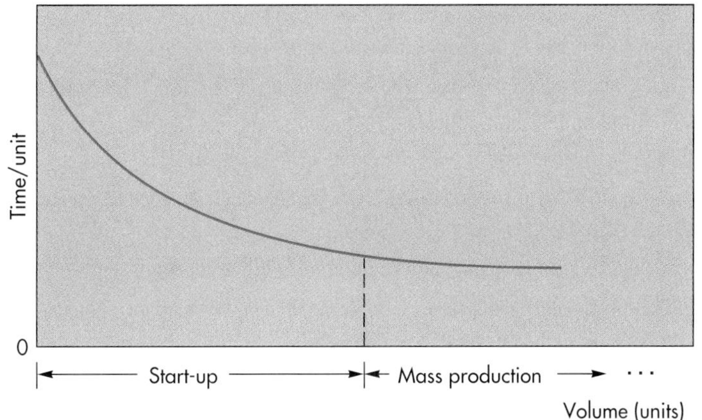

FIGURE 7S-5

Learning curves are useful for production start-up, but not usually for mass production

Solved Problems

Problem 1

An assembly operation has a 90 percent learning curve. The line has just begun work on a new item; the initial unit required 28 hours. Estimate the time that will be needed to complete:

a. The first five units.

b. Units 20 through 25.

Solution

Use the total time factor in the 90 percent column of Table 7S–1.

a. Table value: 4.339.
Estimated time for five units: 28(4.339) = 121.49 hours.

b. The total time for units 20 through 25 can be determined by subtraction:

	Hours
Total time for 25 units:	28(17.713) = 495.96
−Total time for 19 units:	28(13.974) = 391.27
Total time for 20 through 25	104.69

Problem 2

A manager wants to determine an appropriate learning rate for a new type of work his firm will undertake. He has obtained completion times for the initial six repetitions of a job of this type. What learning rate is appropriate?

Unit	Completion Time (hours)
1	15.9
2	12.0
3	10.1
4	9.1
5	8.4
6	7.5

Solution

According to theory, the time per unit decreases at a constant rate each time the output *doubles* (e.g., unit 1 to 2, 2 to 4, and 3 to 6). The ratios of these observed times will give us an approximate rate. Thus,

$$\frac{\text{Unit 2}}{\text{Unit 1}} = \frac{12.0}{15.9} = .755 \quad \frac{\text{Unit 4}}{\text{Unit 2}} = \frac{9.1}{12.0} = .758 \quad \frac{\text{Unit 6}}{\text{Unit 3}} = \frac{7.5}{10.1} = .743$$

Not surprisingly, there is some variability; the rate is usually a smoothed approximation. Even so, the ratios are fairly close—a rate of 75 percent seems reasonable in this case.

Discussion and Review Questions

1. If the learning phenomenon applies to all human activity, why isn't the effect noticeable in mass production or high-volume assembly work?
2. Under what circumstances might a manager prefer a learning rate of approximately 100 percent (i.e., no "learning")?
3. What would a learning percentage of 120 percent imply?
4. Explain how an increase in indirect labor cost can contribute to a decrease in direct labor cost per unit.
5. List the kinds of factors that create the learning effect.
6. Explain how changes in a process, once it is under way, can cause scallops in a learning curve.
7. Name some areas in which learning curves are useful.
8. What factors might cause a learning curve to tip up toward the end of a job?
9. Users of learning curves sometimes fail to include carryover effects; previous experience with similar activities can reduce initial activity times, although it should be noted that the *learning rate* remains the same. What is the implication of this item from the list of cautions and criticisms?

Problems

1. An aircraft company has an order to refurbish the interiors of 18 jet aircraft. The work has a learning curve percentage of 80. On the basis of experience with similar jobs, the industrial engineering department estimates that the first plane will require 300 hours to refurbish. Estimate the amount of time needed to complete:

   a. The fifth plane.

   b. The first five planes.

   c. All 18 planes.

   (*a.* 178.8 hr. *b.* 1,121.4 hr. *c.* 2,914.8 hr.)

2. Estimate the time it will take to complete the 4th unit of a 12-unit job involving a large assembly if the initial unit required approximately 80 hours for each of these learning percentages:

   a. 72 percent

   b. 87 percent

   c. 95 percent

   (*a.* 41.47 hr. *b.* 60.55 hr. *c.* 72.20 hr.)

3. A small contractor intends to bid on a job installing 30 in-ground swimming pools. Because this will be a new line of work for the contractor, he believes there will be a learning effect for the job. After reviewing time records from a similar type of activity, the contractor is convinced that an 85 percent curve is appropriate. He estimates that the first pool will take his crew eight days to install. How many days should the contractor budget for:

   a. The first 10 pools?

   b. The second 10 pools?

   c. The final 10 pools?

   (*a.* 56.928 days *b.* 42.288 days *c.* 37.512 days)

4. A job is known to have a learning percentage equal to .82. If the first unit had a completion time of 20 hours, estimate the times that will be needed to complete the third and fourth units.

   (T_3 = 14.60 hours, T_4 = 13.448 hours)

5. A manager wants to determine an appropriate learning percentage for a certain activity. Toward that end, times have been recorded for completion of each of the first six repetitions. They are:

Repetition	Time (minutes)
1	46
2	39
3	35
4	33
5	32
6	30

a. Determine the approximate learning percentage. *a.* 85%

b. Using your answer from part *a*, estimate the average completion time per repetition assuming a total of 30 repetitions are planned. *b.* 26.21 minutes

6. Students in an operations management class have been assigned four similar homework problems. One student noted that it took her 50 minutes to complete the first problem. Assume that the four problems are similar and that a 70 percent learning curve is appropriate. How much time can this student plan to spend solving the remaining problems? 87.9 minutes

7. A subcontractor is responsible for outfitting six satellites that will be used for solar research. Four of the six have been completed in a total of 600 hours. If the crew has a 75 percent learning curve, how long should it take them to finish the last two units? 201.26 hr.

8. The 5th unit of a 25-unit job took 14.5 hours to complete. If a 90 percent learning curve is appropriate: *a.* 11.35 hr. *b.* 13.05 hr. *c.* 13.12 hr.

   a. How long should it take to complete the last unit?

   b. How long should it take to complete the 10th unit?

   c. Estimate the average time per unit for the 25 units.

9. The labor cost to produce a certain item is $8.50 per hour. Job setup costs $50 and material costs are $20 per unit. The item can be purchased for $88.50 per unit. The learning rate is 90 percent. Overhead is charged at a rate of 50 percent of labor, materials, and setup costs. *a.* $80.31 *b.* 10 units

   a. Determine the unit cost for 20 units.

   b. What is the minimum production quantity necessary to make production cost less than purchase cost?

10. A firm has a training program for a certain operation. The progress of trainees is carefully monitored. An established standard requires a trainee to be able to complete the sixth repetition of the operation in six hours or less. Those who are unable to do this are assigned to other jobs.

    Currently, three trainees have each completed two repetitions. Trainee A had times of 9 hours for the first and 8 hours for the second repetition; trainee B had times of 10 hours and 8 hours for the first and second repetitions; and trainee C had times of 12 and 9 hours.

    Which trainee(s) do you think will make the standard? Explain your reasoning.

    B and C will meet the standard. At this rate of improvement, A will take 6.86 hrs., which does not meet standard.

11. The first unit of a job took 40 hours to complete. The work has a learning percentage of .88. The manager wants time estimates for units 2, 3, 4, and 5. Develop those time estimates. $T_2 = 35.20$ $T_3 = 32.66$ $T_4 = 30.98$ $T_5 = 29.73$

12. A manager wants to estimate the remaining time that will be needed to complete a five-unit job. The initial unit of the job required 12 hours, and the work has a learning percentage of .77. Estimate the total time remaining to complete the job. 30.82 hours

13. A job is supposed to have a learning percentage of .82. Times for the first four units were 30.5, 28.4, 27.2, and 27.0 minutes. Does a learning percentage of .82 seem reasonable? Justify your answer using appropriate calculations. No.

14. The 5th unit of a 10-unit job took five hours to complete. The sixth unit has been worked on for two hours, but is not yet finished. Estimate the *additional* amount of time needed to finish the 10-unit job if the work has a 75 percent learning rate. 18.76 hours

CASE

PRODUCT RECALL

An automobile manufacturer is conducting a product recall after it was discovered that a possible defect in the steering mechanism could cause loss of control in certain cars. The recall covers a span of three model years. The company sent out letters to car owners promising to repair the defect at no cost at any dealership.

The company's policy is to pay the dealer a fixed amount for each repair. The repair is somewhat complicated, and the company expected learning to be a factor. In order to set a reasonable rate for repairs, company engineers conducted a number of repairs themselves. It was then decided that a rate of $88 per repair would be appropriate, based on a flat hourly rate of $22 per hour and a 90 percent learning rate.

Shortly after dealers began making repairs, the company received word that several dealers were encountering resistance from workers who felt the flat rate was much too low and who were threatening to refuse to work on those jobs. One of the dealers collected data on job times and sent that information to the company: Three mechanics each completed two repairs. Average time for the first unit was 9.6 hours, and average time for the second unit was 7.2 hours. The dealer has suggested a rate of $110 per repair.

You have been asked to investigate the situation and to prepare a report.

Questions

1. Prepare a list of questions that you will need to have answered in order to analyze this situation.
2. Prepare a list of observations regarding the information provided in the case.
3. What preliminary thoughts do you have on solutions/partial solutions to the points you have raised?

Selected Bibliography

Abernathy, W. J. "The Limits of the Learning Curve." *Harvard Business Review,* September–October 1974, pp. 109–19.

Andress, Frank J. "The Learning Curve as a Production Tool." *Harvard Business Review,* January–February 1954, pp. 87–95.

Argote, Linda, and Dennis Epple. "Learning Curves in Manufacturing." *Science* 247 (February, 1990), pp. 920–24.

Fabrycky, W. J.; P. M. Ghare; and P. E. Torgersen. *Industrial Operations Research.* Englewood Cliffs, NJ: Prentice Hall, 1972.

Holdham, J. H. "Learning Curves—Their Applications in Industry." *Production and Inventory Management* (fourth quarter 1970), pp. 40–55.

LEARNING OBJECTIVES

After completing this chapter, you should be able to:

1 List some of the main reasons organizations need to make location decisions.

2 Explain why location decisions are important.

3 Discuss the options that are available for location decisions.

4 Describe some of the major factors that affect location decisions.

5 Outline the decision process for making these kinds of decisions.

6 Use the techniques presented to solve typical problems.

CHAPTER EIGHT
LOCATION PLANNING AND ANALYSIS

CHAPTER OUTLINE

When a well-known real estate broker was asked what the three most important determinants of the value of a property are, he said, "That's easy. Location, location, and location."

In the residential real estate market, location is an important factor. Although the style of house, number of bedrooms and bathrooms, level of maintenance, and how modern the kitchen is undoubtedly enter into the picture, some locations are just more desirable than others.

In many respects, the choice of location for a business organization is every bit as important as it is for a house, although for different reasons.

Location decisions represent an integral part of the strategic planning process of virtually every organization. Although it might appear that location decisions are one-time problems pertaining to new organizations, existing organizations often have a bigger stake in these kinds of decisions than new organizations.

This chapter examines location analysis. It begins with a brief overview of the reasons firms must make location decisions, the nature of these decisions, and a general procedure for developing and evaluating location alternatives.

THE NEED FOR LOCATION DECISIONS

Existing organizations may need to make location decisions for a variety of reasons. Firms such as banks, fast-food chains, supermarkets, and retail stores view locations as part of marketing strategy, and they look for locations that will help them to expand their markets. Basically, the location decisions in those cases reflect the *addition* of new locations to an existing system.

A similar situation occurs when an organization experiences a growth in demand for its products or services that cannot be satisfied by expansion at an existing location. The addition of a new location to complement an existing system is often a realistic alternative.

Some firms face location decisions through depletion of basic inputs. For example, fishing and logging operations are often forced to relocate due to the temporary exhaustion of fish or forests at a given location. Mining and petroleum operations face the same sort of situation, although usually with a longer time horizon.

For other firms, a shift in markets causes them to consider relocation, or the costs of doing business at a particular location reach a point where other locations begin to look more attractive.

THE NATURE OF LOCATION DECISIONS

Location decisions for many types of businesses are made infrequently, but they tend to have a significant impact on the organization. In this section we look at the importance of location decisions, the usual objectives managers have when making location choices, and some of the options that are available to them.

Importance of Location Decisions

There are two primary reasons that location decisions are a highly important part of production systems design. One is that they entail a long-term commitment, which makes mistakes difficult to overcome. The other is that location decisions often have an impact on investment requirements, operating costs and revenues, and operations. A poor choice of location might result in excessive transportation costs, a shortage of qualified labor, loss of competitive advantage, inadequate supplies of raw materials, or some similar condition that is detrimental to operations. For services, a poor location could result in lack of customers and/or high operating costs. For both manufacturing and services, location decisions can have a significant impact on competitive advantage.

Objectives of Location Decisions

As a general rule, profit-oriented organizations base their decisions on profit potential, whereas nonprofit organizations strive to achieve a balance between cost and the level of customer service they provide. It would seem to follow that all organizations attempt to identify the "best" location available. However, this is not necessarily the case.

In many instances, no single location may be significantly better than the others. There may be numerous acceptable locations from which to choose, as shown by the wide variety of locations where successful organizations can be found. Furthermore, the number of possible locations that would have to be examined to find the best location may be too large to make an exhaustive search practical. Consequently, most organizations do not set out with the intention of identifying the *one best* location; rather, they hope to find a number of *acceptable* locations from which to choose.

Location Options

Managers generally consider four options in location planning. One is to expand an existing facility. This option can be attractive if there is adequate room for expansion, especially if the location has desirable features that are not readily available elsewhere. Expansion costs are often less than those of other alternatives.

Another option is to add new locations while retaining existing ones, as is done in many retail operations. In such cases, it is essential to take into account what the impact will be on the total system. Opening a new store in a shopping mall may simply draw customers who already patronize an existing store in the same chain, rather than expand the market. On the other hand, adding locations can be a defensive strategy designed to maintain a market share or to prevent competitors from entering a market.

A third option is to shut down at one location and move to another. An organization must weigh the costs of a move and the resulting benefits against the costs and benefits of remaining in an existing location. A shift in markets, exhaustion of raw materials, and the cost of operations often cause firms to consider this option seriously.

Finally, organizations have the option of doing nothing. If a detailed analysis of potential locations fails to uncover benefits that make one of the previous three alternatives attractive, a firm may decide to maintain the status quo, at least for the time being.

GENERAL PROCEDURE FOR MAKING LOCATION DECISIONS

The way an organization approaches location decisions often depends on its size and the nature or scope of its operations. New or small organizations tend to adopt a rather informal approach to location decisions. New firms typically locate in a certain area simply because the owner lives there. Similarly, managers of small firms often want to keep operations in their backyard, so they tend to focus almost exclusively on local alternatives. Large established companies, particularly those that already operate in more than one location, tend to take a more formal approach. Moreover, they usually consider a wider range of geographic locations. The discussion here pertains mainly to a formal approach to location decisions.

The general procedure for making location decisions usually consists of the following steps:

1. Decide on the criteria to use for evaluating location alternatives, such as increased revenues or community service.
2. Identify important factors, such as location of markets or raw materials.
3. Develop location alternatives:
   a. Identify the general region for a location.

b. Identify a small number of community alternatives.
c. Identify site alternatives among the community alternatives.
4. Evaluate the alternatives and make a selection.

Step (1) is simply a matter of managerial preference. Steps (2) through (4) may need some elaboration.

FACTORS THAT AFFECT LOCATION DECISIONS

Many factors influence location decisions. However, it often happens that one or a few factors are so important that they dominate the decision. For example, in manufacturing, the potentially dominating factors usually include availability of an abundant energy and water supply and proximity to raw materials. Thus, nuclear reactors require large amounts of water for cooling, heavy industries such as steel and aluminum production need large amounts of electricity, and so on. Transportation costs can be a major factor. In service organizations, possible dominating factors are market related and include traffic patterns, convenience, and competitors' locations, as well as proximity to the market. For example, car rental agencies locate near airports and midcity where their customers are.

Once an organization has determined the most important factors, it will try to narrow the search for suitable alternatives to one geographic region. Then a small number of community-site alternatives are identified and subjected to detailed analysis. Community and site factors are often so intertwined that it makes sense to consider them jointly.

NEWSCLIP

INNOVATIVE MCI UNIT FINDS CULTURE SHOCK IN COLORADO SPRINGS

By Alex Markels

www.mci.com

Convinced this town's spectacular setting would inspire his workers, Richard Liebhaber figured "build it, and they will come."

In 1991, the chief technology officer of MCI Communications Corp. decided to relocate MCI's brain trust—the 4,000-employee Systems Engineering division that created numerous breakthrough products—from MCI's Washington, D.C., headquarters to Colorado Springs. An avid skier, he believed the mountains, low crime rate, healthy climate and rock-bottom real-estate prices would be "a magnet for the best and brightest" computer software engineers.

He rejected warnings from at least half a dozen senior executives that Colorado Springs' isolated and politically conservative setting would actually repel the eclectic, ethnically diverse engineers MCI hoped to attract. Mr. Liebhaber argued that new hires would jump at the chance to live in ski country, while veterans would stay longer, reducing MCI's more than 15% annual turnover rate in Washington. The move, he contended, would also save money by cutting MCI's facilities, labor and recruiting costs. Besides, four other high-tech companies—including Digital Equipment Corp. and Apple Computer Inc.—had recently moved there. "One of the things that gave me more comfort was the fact that these other guys had selected Colorado Springs," Mr. Liebhaber says.

He was mistaken.

While many rank-and-file MCI employees, buoyed by generous relocation packages, made the move, numerous key executives and engineers, and hundreds of the division's 51% minority population, said no, or fled Colorado Springs soon after relocating.

Living in 'Wonder Bread'

"It was like living in a loaf of Wonder Bread," says James Finucane, who is of Japanese descent and whose wife is from Argentina. A veteran senior engineer, Mr. Finucane was considered MCI's top engineer until he took a job with a competitor back east in 1994. "There's no culture, no diversity, no research university, no vitality or resiliency to the job market."

The move isolated MCI's engineers from top management and from marketing colleagues at headquarters, undermining the spontaneous collaborations that had generated some of company's most innovative products. Meanwhile, the professionals Mr. Liebhaber hoped to recruit from outside proved difficult and expensive to woo, pushing the move's total cost to about $200 million—far more than MCI officials anticipated. "Most of the savings we had hoped for never materialized," says LeRoy Pingho, a senior executive who oversaw the relocation.

As numerous companies consider relocating to smaller cities and towns, MCI's move shows the perils of transplanting urban professionals to the nation's heartland and segregating key operations. Dozens of former and current employees say Systems Engineering has lost its innovative and productive edge at a time when competition in telecommunications is fiercer than ever.

Moving Expenses

When the move was announced in March 1991, many rank-and-file workers were enthusiastic. MCI's relocation policy paid for every expense imaginable. Costing an average of $100,000 per employee, it included up to six months of temporary housing and living expenses, private-school tuition for workers' children and a full month's pay for miscellaneous expenses. And there were exceptional housing bargains. "In Alexandria, [Va.,] we had a tiny place on a 50-by-112-foot lot," says Jerome Sabolik, a senior software engineer. "For the same money, we got a 3,000-square-foot house on 2½ acres." Thousands of workers—far more than Mr. Liebhaber expected—took advantage of the offer, undercutting his plans to recruit lower-cost employees in Colorado.

But there was far less enthusiasm among senior managers. James Zucco, Mr. Ditchfield's successor and the head of Systems Engineering, stayed behind and eventually left to join AT&T Corp. Also staying put was Gary Wiesenborn, the division's No. 2 executive, who later moved to Bell Atlantic Corp. Mr. Pingho, who oversaw the division's financial planning and budgeting, declined to move and quit in 1993.

There was also significant fallout among the division's minority population. Although MCI declines to provide specific numbers, it confirms there was a reduction. According to former employees who had access to Equal Employment Opportunity Commission data, there were roughly 1,300 African-Americans on Systems Engineering's staff and a combined 700 Asians and Hispanics before the relocation. Since the relocation, minority representation has been cut almost in half, to about 600 blacks and a combined 500 Asians and Hispanics. "It was a disaster for diversity," Mr. Ditchfield says.

But MCI officials say that despite the reduction, its Colorado division is still more ethnically diverse than other local companies. "We think that we have numbers that are significantly better than the available work force there," says William D. Wooten, a senior vice president of human resources.

Among those who opted out: Tony Martin, a vice president of operations who is Asian-American, and Rod Avery, who designed the complex billing system for MCI's successful "Friends & Family" long-distance program. One of the company's highest ranking African-Americans, Mr. Avery moved to AT&T.

Regional Factors

The primary regional factors involve raw materials, markets, and labor considerations.

Location of Raw Materials. Firms locate near or at the source of raw materials for three primary reasons: necessity, perishability, and transportation costs. Mining operations, farming, forestry, and fishing fall under *necessity*. Obviously, such operations must locate close to the raw materials. Firms involved in canning or freezing of fresh fruit and vegetables, processing of dairy products, baking, and so on must consider *perishability* when considering location. *Transportation costs* are important in industries where processing eliminates much of the bulk connected with a raw material, making it much less expensive to transport the product or material after processing. Examples include aluminum reduction, cheese making, and paper production. Where inputs come from different locations, some firms choose to locate near the geographic center of the sources. For instance, steel producers use large quantities of both coal and iron ore, and many are located somewhere between the Appalachian coal fields and iron ore mines. Transportation costs are often the reason that vendors locate near their major customers. Moreover, regional warehouses are used by supermarkets and other retail operations to supply multiple outlets. Often the choice of new locations and additional warehouses reflects the locations of existing warehouses or retail outlets.

Tropicana, an orange juice producer, has processing plants located in Florida and California. Being located near orange groves has the benefit of time and cost savings in the transportation of perishable fruit.

Vol. 5, Seg. 3
American President Line

Location of Markets. Profit-oriented firms frequently locate near the markets they intend to serve as part of their competitive strategy, whereas nonprofit organizations choose locations relative to the needs of the users of their services. Other factors include distribution costs or the perishability of a finished product.

Retail sales and services are usually found near the center of the markets they serve. Examples include fast-food restaurants, service stations, dry cleaners, and supermarkets. Quite often their products and those of their competitors are so similar that they rely on convenience to attract customers. Hence, these businesses seek locations with high population densities or high traffic. The competition/convenience factor is also important in locating banks, hotels and motels, auto repair shops, drugstores, newspaper kiosks, and shopping centers. Similarly, doctors, dentists, lawyers, barbers, and beauticians typically serve clients who reside within a limited area.

Competitive pressures for retail operations can be extremely vital factors. In some cases, a market served by a particular location may be too small to justify two or more competitors (e.g., one hamburger franchise per block), so that a search for potential locations tends to concentrate on locations without competitors. The opposite also might be true; it could be desirable to locate near competitors. Large department stores often locate near each other, and small stores like to locate in shopping centers that have large department stores as anchors. The large stores attract large numbers of shoppers who become potential customers in the smaller stores or in the other large stores.

Some firms must locate close to their markets because of the perishability of their products. Examples include bakeries, flower shops, and fresh seafood stores. For other types of firms, distribution costs are the main factor in closeness to market. For example, sand and gravel dealers usually serve a limited area because of the high distribution costs associated with their products. Still other firms require close customer contact, so they too tend to locate within the area they expect to serve. Typical examples are tailor shops, home remodelers, home repair services, cabinetmakers, rug cleaners, and lawn and garden services.

Locations of many government services are near the markets they are designed to serve. Hence, post offices are typically scattered throughout large metropolitan areas. Police and emergency health care locations are frequently selected on the basis of client needs. For instance, police patrols often concentrate on high crime areas, and emergency health care facilities are usually found in central locations to provide ready access from all directions.

Many foreign manufacturing companies have located manufacturing operations in the United States, because it is a major market for their products. Chief among them are automobile manufacturers, most notably Japanese, but other nations are also represented. Another possible reason that Japanese producers decided to locate in the United States was to offset possible negative consumer sentiment related to job losses of U.S. workers. Thousands of U.S. auto workers are now employed in U.S. manufacturing plants of Japanese and other foreign companies.

Software can be helpful in location analysis. One example is Anc View's Geographic Information System which helps to develop location information based on demographic data.

Labor Factors. Primary labor considerations are the cost and availability of labor, wage rates in an area, labor productivity and attitudes toward work, and whether unions are a serious potential problem.

Labor costs are very important for labor-intensive organizations. The shift of the textile industry from the New England states to southern states was due partly to labor costs.

Skills of potential employees may be a factor, although some companies prefer to train new employees rather than rely solely on previous experience. Increasing specialization in many industries makes this possibility even more likely than in the past. Although most companies concentrate on the supply of blue-collar workers, some firms are more interested in scientific and technical people as potential employees, and they look for areas with high concentrations of those types of workers.

Worker attitudes toward turnover, absenteeism, and similar factors may differ among potential locations—workers in large urban centers may exhibit different attitudes than workers in small towns or rural areas. Furthermore, worker attitudes in different parts of the country or in different countries may be markedly different.

Some companies offer their current employees jobs if they move to a new location. However, in many instances, employees are reluctant to move, especially when it means leaving families and friends. Furthermore, in families with two wage earners, relocation would require that one wage earner give up a job and then attempt to find another job in the new location.

Other Factors. Climate and taxes sometimes play a role in location decisions. For example, a string of unusually severe winters in northern states may cause some firms to seriously consider moving to a milder climate, especially if delayed deliveries and work disruptions caused by inability of employees to get to work have been frequent. Similarly, the business and personal income taxes in some states reduce their attractiveness to companies seeking new locations. Many companies have been attracted to some Sun Belt states by ample supplies of low-cost energy or labor, the climate, and tax considerations.

The growth in multinational operations over the past several decades is evidence of the importance of foreign locations. Some firms are attracted to foreign locations because of nearby deposits of aluminum, copper, timber, oil, or other natural resources. Other firms view foreign locations as a way to expand their markets, and still others are attracted by ample supplies of labor. Some countries may offer financial incentives to companies to create jobs for their people. Developing countries may establish tariffs to protect their young industries from outside competition, which may also reduce the amount of "foreign" competition a firm must face if it locates in such a country. Until the North American Free Trade Agreement eliminated restrictions, the Fisher-Price Toy Company factory in Matamoros, Mexico, was not allowed to sell in Mexico the Muppet toys it makes. U.S. companies with factories in Mexico could import raw materials duty-free, but they were required to export all of their output.

Many developing countries offer an abundant supply of cheap labor. For example, a considerable amount of the clothes sold in the United States bear labels indicating they were made in Korea, China, or Taiwan. In some instances, it is less expensive to ship raw materials or semifinished goods to foreign countries for fabrication and assembly, and then ship them to their final destinations, than it is to produce them in the United States. However, the final cost per unit is the most important factor. In some cases, the low cost of labor in a foreign country can be negated by low productivity and shipping costs.

High production costs in Germany have contributed to a number of German companies locating some of their production facilities in lower-cost countries. Among them are industrial products giant Siemens, A.G. (a semi-conductor plant in Britain), drug makers Bayer, A.G. (a plant in Texas) and Hoechst, A.G. (a plant in China), and auto makers Mercedes (plants in Spain, France, and Alabama), and BMW (a plant in Spartansburg, South Carolina).

A firm contemplating a foreign location must carefully weigh the potential benefits against the potential problems. A major factor is the stability of a country's government and its attitude toward American firms. Import restrictions can pose problems with bringing in equipment and spare parts.

Some of the problems of a foreign location can be caused by language and cultural differences between the United States and the host country. U.S. firms often find it necessary to use U.S. technical personnel but find it difficult to convince workers to move to a foreign country; workers may have to leave their families behind or else subject them to substandard housing or educational systems. Companies are now exerting additional efforts to reduce these obstacles. Some provide housing allowances and have schools for U.S. children. They are also improving their efforts to see that the employees they send abroad are familiar with local customs and have a reasonable facility with the language of the host country.

Community Considerations

Many communities actively try to attract new businesses because they are viewed as potential sources of future tax revenues and new job opportunities. However, communities do not, as a rule, want firms that will create pollution problems or otherwise lessen the quality of life in the community. Local groups may actively seek to exclude certain companies on such grounds, and a company may have to go to great lengths to convince local officials that it will be a "responsible citizen." Furthermore, some organizations discover that even though overall community attitude is favorable, there may still be considerable opposition to specific sites from nearby residents who object to possible increased levels of noise, traffic, or pollution. Examples of this include community resistance to airport expansion, changes in zoning, construction of nuclear facilities, and highway construction.

From a company standpoint, a number of factors determine the desirability of a community as a place for its workers and managers to live. They include facilities for education, shopping, recreation, transportation, religious worship, and entertainment; the quality of police, fire, and medical services; local attitudes toward the company; and the size of the community. Community size can be particularly important if a firm will be a major employer in the community; a future decision to terminate or reduce operations in that location could have a serious impact on the economy of a small community.

Other community-related factors are the cost and availability of utilities, environmental regulations, taxes (state and local, direct and indirect), and often a laundry list of enticements offered by state or local governments that can include bond issues, tax abatements, low-cost loans, grants, and worker training.

Site-Related Factors

The primary considerations related to sites are land, transportation, and zoning or other restrictions.

Evaluation of potential sites may require consulting with engineers or architects, especially in the case of heavy manufacturing or the erection of large buildings or

TABLE 8–1

Factors affecting location decisions

Level	Factors	Considerations
Regional	Location of raw materials or supplies	Proximity, modes and costs of transportation, quantity available
	Location of markets	Proximity, distribution costs, target market, trade practices/restrictions
	Labor	Availability (general and for specific skills), age distribution of workforce, work attitudes, union or nonunion, productivity, wage scales, unemployment compensation laws
Community	Quality of life	Schools, churches, shopping, housing, transportation, entertainment, recreation, cost of living
	Services	Medical, fire, and police
	Attitudes	Pro/con
	Taxes	State/local, direct and indirect
	Environmental regulations	State/local
	Utilities	Cost and availability
	Development support	Bond issues, tax abatement, low-cost loans, grants
Site	Land	Cost, degree of development required, soil characteristics and drainage, room for expansion, parking
	Transportation	Type (access roads, rail spurs, air freight)
	Environmental/legal	Zoning restrictions

facilities with special requirements. Soil conditions, load factors, and drainage rates can be critical and often necessitate certain kinds of expertise in evaluation.

Because of the long-term commitment usually required, land costs may be secondary to other site-related factors, such as room for future expansion, current utility and sewer capacities—and any limitations on these that could hinder future growth—and sufficient parking space for employees and customers. In addition, for many firms access roads for trucks or rail spurs are important.

Industrial parks may be worthy alternatives for firms involved in light manufacturing or assembly, warehouse operations, and customer service facilities. Typically, the land is already developed—power, water, and sewer hookups have been attended to, and zoning restrictions do not require special attention. On the negative side, industrial parks may place restrictions on the kinds of activities that a company can conduct, which can limit options for future development of a firm's products and services as well as the processes it may consider. Sometimes stringent regulations governing the size, shape, and architectural features of buildings limit managerial choice in these matters. Also, there may not be an adequate allowance for possible future expansion.

For firms with executives who travel frequently, the size and proximity of the airport or train station as well as travel connections can be important, although schedules and connections are subject to change.

Table 8–1 provides a summary of the factors that affect location decisions.

Multiple Plant Manufacturing Strategies

When companies have multiple manufacturing facilities, they can organize operations in several ways. One is to assign different product lines to different plants. Another is to assign different market areas to different plants. And a third is to assign different processes to different plants. Each strategy carries certain cost and managerial implications, as well as competitive advantages.

Product Plant Strategy. With this strategy, entire products or product lines are produced in separate plants, and each plant usually supplies the entire domestic market. This is essentially a decentralized approach, with each plant focusing on a narrow set of requirements that entails specialization of labor, materials, and equipment along product lines. Specialization often results in economies of scale and, compared with multipurpose plants, lower operating costs. Plant locations may be widely scattered or clustered relatively close to one another.

Market Area Plant Strategy. With this strategy, plants are designed to serve a particular geographic segment of a market (e.g., the West Coast, the Northeast). Individual plants produce most if not all of a company's products and supply a limited geographical area. Although operating costs tend to be higher than those of product plants, significant savings on shipping costs for comparable products can be made. This arrangement is particularly desirable when shipping costs are high due to volume, weight, or other factors. Such arrangements have the added benefit of rapid delivery and response to local needs. This approach requires centralized coordination of decisions to add or delete plants, or to expand or downsize current plants due to changing market conditions.

Process Plant Strategy. With this strategy, different plants concentrate on different aspects of a process. Automobile manufacturers often use this approach, with different plants for engines, transmissions, body stamping, and even radiators. This approach is best suited to products that have numerous components; separating the production of components results in less confusion than if all production was carried out at the same location.

When an organization uses process plants, coordination of production throughout the system becomes a major issue and requires a highly informed, centralized administration to achieve effective operation. A key benefit is that individual plants are highly specialized and generate volumes that yield economies of scale.

Multiple plants have an additional benefit: the increase in learning opportunities that occurs when similar operations are being done in different plants. Similar problems tend to arise, and solutions to those problems as well as improvements in general in products and processes made at one plant can be shared with other plants.

Service and Retail Locations

Service and retail are typically governed by somewhat different considerations than manufacturing organizations in making location decisions. For one thing, nearness to raw materials is usually not a factor, nor is concern about processing requirements. But customer access is usually a prime consideration.

Retail and service organizations typically place traffic volume and convenience high on the list of important factors. Specific types of retail or service businesses may pay more attention to certain factors due to the nature of their business or their customers. If a business is unique, and has its own drawing power, nearness to competitors may not be a factor. However, generally retail businesses prefer locations that are near other retailers (although not necessarily competitors) because of the higher traffic volumes and convenience to customers. Thus, large department stores tend to locate near each other, often as anchor stores in mall settings, and smaller retailers occupy spaces between the larger stores. Restaurants and specialty stores often locate in and around malls, benefiting from the high traffic.

Medical services are often located near hospitals for convenience of patients. Doctors' offices may be located near hospitals, or grouped in other, centralized areas with other doctors' offices. Available public transportation is often a consideration.

Good transportation and/or parking facilities can be vital to retail establishments. Downtown areas have a competitive disadvantage in attracting shoppers compared to malls because malls offer ample free parking and nearness to residential areas.

Mall of America in Minnesota is a four-level shopping mall, as well as a tourist destination with features unique to a mall, including amusement park rides, comedy clubs, an interactive Chrysler Showcase, and an 18-hole golf course. Anchor stores, such as Bloomingdales, and a large variety of specialty shops are attracted to this high-traffic location.

Customer safety and security can be key factors, particularly in urban settings, for all types of services that involve customers coming to the service location (as opposed, say, to in-home services such as home repair and rug cleaning).

www.mallofamerica.com

Competitors' locations can be of significance. In some cases, firms will want to locate near competitors to benefit from the concentration of potential customers. Mall stores and auto dealers are good examples. In other cases, locating near similar businesses is not important because there are no competitors. And in some cases, it is important *not* to be near a competitor (e.g., another franchise operation of the same fast-food chain).

NEWSCLIP

VYING FOR PATIENTS. HOSPITALS THINK LOCATION, LOCATION **Doreen Carvajal**

The mood is as vital as the medicine at the elegant Upper East Side clinic squeezed among the fashionable addresses of Calvin Klein, Ralph Lauren and Barneys New York.

Silk magnolias bloom in the marble lobby, and abstract oil paintings dominate the pale rose corridors of Columbia-Presbyterian Medical Center's new satellite at East 60th Street and Madison Avenue.

The clinic is one of many ways for Columbia-Presbyterian to draw new patients for routine services as well as to feed its main hospital in Washington Heights for more complicated ones. A free shuttle regularly ferries the clinic's patients northwest along the Hudson River, an indirect route skirting Harlem on the way to the hospital, where the neighbors are not elegant boutiques, but the less fabled Self-Serve Laundromat and El Presidente cafeteria.

Location is becoming fiercely important to hospitals in New York's competitive medical marketplace as managed-care systems, so popular elsewhere, begin to take hold here. In the new world of fixed reimbursements, shorter hospital stays and tighter rationing of highly specialized services, hospitals must make their money from volume rather than from service fees.

The resulting battle for market share—and for new patients—is being waged not with scalpel and stethoscope, but with hammer and nails, in the neighborhoods where potential customers live.

"New York has been relatively slow to follow the rest of the country toward managed health care because it was so regulated," said Kenneth E. Raske, president of the Greater New York Hospital Association. "But they're here now. And they're going to have a profound influence on redrafting the map of New York health care."

Emerging as hospitals' most promising source of new patients are clinics like Columbia-Presbyterian Eastside, which, in health care jargon, are called "centers," lest potential patients confuse these gleaming outposts with conventional clinics that cater to the poor and uninsured. To attract middle-class patients wary of leaving their protected blocks, the city's huge hospitals are branching out to ethnic enclaves, upscale New York neighborhoods, affluent suburban communities, and even distant American expatriate communities in Eastern Europe.

Four months ago, Columbia-Presbyterian created a satellite in Moscow and more centers are planned for Warsaw, Prague, St. Petersburg, Budapest and possibly Beijing.

"Our feeling is that there will be no hospitals in the future," said Dr. William T. Speck, president of Columbia-Presbyterian. "And probably, in the next 10 or 20 years most of the activity will take place in a center or maybe even in homes."

Hospital executives contend that satellites, which typically have no beds and no provision for round-the-clock care, are cheaper to operate than highly specialized hospitals, which provide all the incidental services of a hotel. Babies, for instance, can be delivered at alternative sites for half a hospital's $9,000 fee, Dr. Speck said.

Not surprisingly, Columbia-Presbyterian is planning a birthing center just a block away from its Madison Avenue location.

GLOBAL LOCATIONS

Recent trends in locating facilities, particularly manufacturing facilities, reflect a combination of competitive and technological factors. One trend has been that of foreign producers, especially automotive firms, to locate plants in the United States. The United States represents a tremendous market for cars, trucks, and recreational vehicles. By locating in the United States, these firms can shorten delivery time and reduce delivery costs. Furthermore, they can avoid any future tariffs or quotas that might be applied to imports.

A development that affects location decisions was the passage of GATT in 1994 (described in Chapter 1). One of its provisions was the reduction and elimination of various tariffs. Consequently, location within the borders of a country to escape tariffs is now much less of an issue.

NEWSCLIP

JAPANESE AUTOMAKER PRODUCTION IN U.S. UP: SURPASSES IMPORTS FOR FIRST TIME EVER **The Associated Press**

WASHINGTON—For the first time ever, Japanese automakers last year produced more vehicles in the United States than they shipped to this country, a Japanese auto association said.

More than 1.8 million vehicles were produced in the United States, for the fiscal year ended in January 1994, while Japan exported 1.5 million vehicles to the United States, according to the Japanese Automobile Manufacturers Association.

"The Japanese automotive companies have invested more than $11 billion in the U.S., creating more than 300,000 American jobs and becoming an important, positive economic force in communities across America," said the association's general director, William C. Duncan.

Japan has been under U.S. pressure to produce more cars in the United States, with more U.S. labor and parts in their vehicles, because Japanese exports account for $60 billion—more than 40 percent—of the U.S. trade deficit. More than half of that comes from sales of Japanese autos and auto parts in the United States.

The trend has been to produce more Japanese cars in the United States, using American labor. The number of Americans employed by Japanese automakers in the United States more than tripled from 11,236 in 1987 to 37,435 in fiscal 1993, according to the association.

However, Steve Beckman, an international economist with the United Auto Workers, said the Japanese association's figures fail to address the low U.S. part content in Japanese automobiles.

Only 50 percent of the parts in Japanese cars manufactured in the United States are American, on average, Beckman said, and parts on autos shipped from Japan have near-zero U.S. content.

"We'd like to see 70 percent of the parts be U.S. parts on Japanese cars made in America," he said.

He predicted that the U.S. trade imbalance with Japan, and its auto imbalance with Japan this year would be "larger probably than ever before." He said the U.S. trade deficit with Japan in autos probably would be $20 billion, while auto parts would add more than $12 billion to the deficit.

The Japanese association said that Japanese automakers' purchases of U.S. auto parts increased more than sixfold between 1986 and 1993—from $2.49 billion to $15.54 billion. The number of U.S. parts suppliers approximately quadrupled during the same period, from 298 in 1986 to 1,245 in March 1994, the association said.

Japanese automakers own and operate seven auto manufacturing plants in the United States, which sometimes manufacture parts for American cars.

Source: Copyright 1994 by the Associated Press.

Another trend is just-in-time manufacturing techniques (see Chapter 15), which encourage suppliers and customers to locate near each other to reduce supplier lead times. For this reason, some U.S. firms are reconsidering decisions to locate offshore. Moreover, in light manufacturing (e.g., electronics), low-cost labor is becoming less important than nearness to markets; users of electronics components want suppliers that are close to their manufacturing facilities. One offshoot of this is the possibility that the future will see a trend toward smaller factories located close to markets. In some industries, small, automated **microfactories** with narrow product focuses will be located near major markets to reduce response time.

microfactory Small factory with a narrow product focus, located near major markets.

It is likely that advances in information technology will enhance the ability of manufacturing firms to gather, track, and distribute information that links purchasing, marketing, and distribution with design, engineering, and manufacturing. This will reduce the need for these functions to be located close together, thereby permitting a strategy of locating production facilities near major markets.

READING

NEW U.S. FACTORY JOBS AREN'T IN THE FACTORY

Stephen Baker and James B. Treece

www.sony.com/

An hour's drive east from Pittsburgh in the hills of Westmoreland County, Pa., stands a symbol of America's manufacturing renaissance. In 1988, the sprawling former auto assembly plant was abandoned, a Rust Belt relic, its workers scattered to other jobs or the unemployment rolls. Then two years ago, Sony Corp. arrived. Today, Westmoreland is one of the world's most advanced electronics plants, churning out thousands of big-screen television sets. And this is no border-style *maquiladora,* with Americans merely snapping together Japanese-made parts. Fully 80 percent of the components are U.S. made.

Why does Sony make televisions in Pennsylvania? The U.S. is the world's largest TV market by far, especially for big-screen versions, and the closer Sony gets to its customers, the less it has to spend on shipping.

Moreover, the U.S. has a strong industrial structure—wide roads, clean water, educated workers, and plenty of suppliers—and a superb technical base. Drawing on resources at nearby Carnegie-Mellon University helped Sony engineers build a robot that automatically fine-tunes the televisions before shipping.

Sony's plant is part of a fundamental shift in manufacturing—one that will leave the U.S. well positioned for global competition in the coming decades. Taking advantage of great economies of scale, manufacturers long built huge mass-production plants, then shipped their products to distant markets. But today, the trend is toward smaller, more flexible, less labor-intensive operations near the customers.

That's as true for steel and autos as it is for TV sets. Built close to their industrial customers the low-cost U.S. minimills run by Nucor Corp., Birmingham Steel Corp., and others are among the world's most efficient steel plants. In autos, U.S. production this year will surpass Japan's for the first time since 1979, thanks in part to more than a million Hondas, Nissans, and Toyotas pouring out of factories in Ohio, Tennessee, and Kentucky.

And it's not just currency swings and local-content concerns that lure manufacturers to the U.S. In no other market in the world do American, Asian, and European manufacturers compete so directly and freely. "If you are successful in the U.S., you are proving that you are a global player," says Helmut Werner, chairman of Mercedes-Benz. The company is setting up an assembly plant in Alabama.

Manuals in German

Don't expect the manufacturing renaissance in the U. S. to spark a boom in factory jobs, however. Just as the mechanization of agriculture a century ago all but emptied the farms, the new factory model means fewer workers are needed to put the final product together. The number of U.S. workers employed in manufacturing has fallen by 3.2 million since the peak of 21 million in 1979. Yet manufacturing output has remained stable as a percentage of gross national product. "We did not see any trends toward lower manufacturing activity," says economist Gordan Richards of the National Association of Manufacturers.

Yet as these factory jobs vanish, a new manufacturing-driven economic sector will emerge—industries such as computer software, robot making, and countless services that will add jobs to supply the leaner manufacturers. At Carnegie Group Inc. in Pittsburgh, technicians are putting together software to guide Ford Motor Co.'s engineers through circuit board design. And they devise programs that translate Caterpillar Inc.'s mountains of technical manuals into German, Russian, and other languages.

Carnegie Group is expanding from 125 to 200 workers this year alone. Says Chief Executive Dennis Yablonsky: "It's a wide universe of technology companies that service manufacturing. The jobs add up pretty quickly."

These support industries, with their high component of knowledge skills, constitute nothing less than a second tier of the manufacturing industries. "A smaller percentage of our workforce will be in production," comments Carnegie Mellon University professor Richard L. Florida. "But a much larger percentage will be supporting that."

These U.S. systems analysts, robot makers, and software engineers will increasingly be selling their wares around the globe. Indeed, 20 years after the much-ballyhooed deindustrialization of America, the U.S. keeps discovering new ways to make things.

Source: Baker, Stephen, and Treece, James B., "New U.S. Factory Jobs Aren't in the Factory," *Business Week,* Special 1994 Bonus Issue, p. 160. Reprinted with Permission.

Questions

1. What are three important reasons that foreign companies want to locate in the United States?
2. Does the addition of foreign manufacturers setting up manufacturing plants in the United States mean there will be a boom in factory jobs? Explain.
3. What kind of jobs and industries will increase due to the increase in manufacturing? What implications does this have for workers in the near future in terms of education?

Table 8–2 provides a checklist of considerations pertaining to decisions on global operations.

Two other factors are causing companies to locate manufacturing facilities in countries that contain their markets (e.g., Japanese auto companies establishing factories in the United States). One is to counter negative sentiments such as "not made in this country." Thus, Japanese factories in the United States produce cars made by U.S. workers. The second factor relates to currency fluctuations and devaluations. These changes can have a significant impact on demand and, hence, on profits. Changes in currency value alter the price of foreign goods, but not the price of goods produced within a country. For instance, if the value of a country's currency falls relative to that of other countries, prices within the country don't change, but foreign-produced goods become more expensive. If demand is elastic, then demand for those foreign goods will fall. Furthermore, currency changes may result in increased costs of parts supplied by foreign producers. By locating and selling within a country, and buying from suppliers in that country, manufacturers can avoid the impact of currency changes.

NEWSCLIP

1.2 BILLION NEW CUSTOMERS

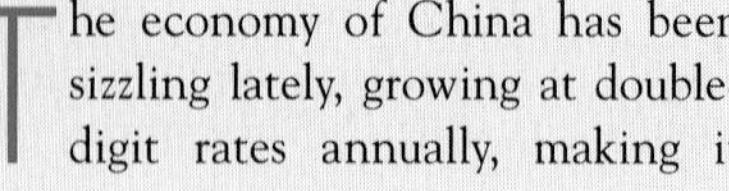

The economy of China has been sizzling lately, growing at double-digit rates annually, making it perhaps the fastest-growing major economy in the world. Not surprisingly, investment by foreign companies in China has burgeoned as companies seek the rapid growth that developed countries cannot provide.

U.S. companies such as Kodak, Bausch & Lomb, Xerox, and many others signed more than 6,000 contracts worth almost $7 billion in direct investment in 1993. Although the amount invested by U.S. companies is significant, it is only a small portion of the investment dollars that are pouring in from foreign companies. Altogether, foreign companies signed more than 80,000 contracts worth over $100 billion. This exceeds the total amount invested by foreign companies in China for the entire decade of the 1980s.

These companies recognize the possibility of erratic near-term results if Chinese officials decide to tighten control and credit. However, they envision tremendous long-term growth potential in the world's most populous nation.

Source: Based on "Chasing 1.2 Billion New Customers," *Rochester Democrat and Chronicle/Times-Union*, July 25, 1994.

TABLE 8-2

Factors relating to foreign locations

Foreign Government	*a.* Policies on foreign ownership of production facilities Local content requirements Import restrictions Currency restrictions Environment regulations Local product standards *b.* Stability issues
Cultural Differences	Living circumstances for foreign workers and their dependents Religious holidays/traditions
Customer Preferences	Possible buy locally sentiment
Labor	Level of training and education of workers Work practices Possible regulations limiting number of foreign employees Language differences
Resources	Availability and quality of raw materials, energy, transportation

EVALUATING LOCATION ALTERNATIVES

There are a number of techniques that are helpful in evaluating location alternatives: locational cost-profit-volume analysis, factor rating, and the center of gravity method.

Locational Cost-Profit-Volume Analysis

The economic comparison of location alternatives is facilitated by the use of cost-volume-profit analysis. The analysis can be done numerically or graphically. The graphical approach will be demonstrated here because it enhances understanding of the concept and indicates the ranges over which one of the alternatives is superior to the others.

The procedure for **locational cost-profit-volume analysis** involves these steps:

locational cost-profit-volume analysis Technique for evaluating location choices in economic terms.

1. Determine the fixed and variable costs associated with each location alternative.
2. Plot the total-cost lines for all location alternatives on the same graph.
3. Determine which location will have the lowest total cost for the expected level of output. Alternatively, determine which location will have the highest profit.

This method assumes the following:

1. Fixed costs are constant for the range of probable output.
2. Variable costs are linear for the range of probable output.
3. The required level of output can be closely estimated.
4. Only one product is involved.

For a cost analysis, compute the total cost for each location:

$$\text{Total cost} = FC + VC \times Q \qquad (8\text{–}1)$$

where

FC = Fixed cost
VC = Variable cost per unit
Q = Quantity or volume of output

Example 1

Fixed and variable costs for four potential plant locations are shown below:

Location	Fixed Cost per Year	Variable Cost per Unit
A	$250,000	$11
B	100,000	30
C	150,000	20
D	200,000	35

a. Plot the total-cost lines for these locations on a single graph.

b. Identify the range of output for which each alternative is superior (i.e., has the lowest total cost).

c. If expected output at the selected location is to be 8,000 units per year, which location would provide the lowest total cost?

Solution

a. To plot the total-cost lines, select an output that is approximately equal to the expected output level (e.g., 10,000 units per year). Compute the total cost for each location at that level:

	Fixed Cost	+	Variable Cost	=	Total Cost
A	$250,000	+	$11(10,000)	=	$360,000
B	100,000	+	30(10,000)	=	400,000
C	150,000	+	20(10,000)	=	350,000
D	200,000	+	35(10,000)	=	550,000

Plot each location's fixed cost (at Output = 0) and the total cost at 10,000 units; and connect the two points with a straight line. (See the accompanying graph.)

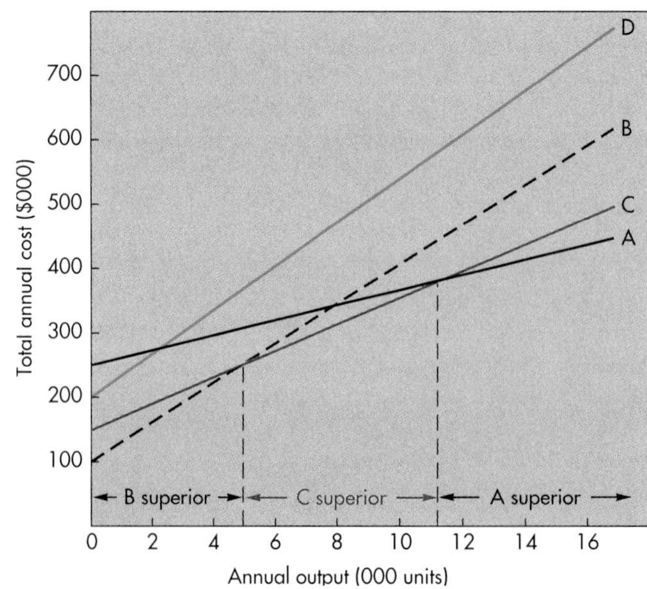

b. The *approximate* ranges for which the various alternatives will yield the lowest costs are shown on the graph. Note that location D is never superior. The *exact* ranges can be determined by finding the output level at which lines B and C and lines C and A cross. To do this, set their total cost equations equal and solve for Q, the break-even output level. Thus, for B and C:

$$\underset{(B)}{\$100{,}000 + \$30Q} = \underset{(C)}{\$150{,}000 + \$20Q}$$

Solving, you find $Q = 5{,}000$ units per year.

For C and A:

$$\underset{(C)}{\$150{,}000 + \$20Q} = \underset{(A)}{\$250{,}000 + \$11Q}$$

Solving, $Q = 11{,}111$ units per year.

c. From the graph, you can see that for 8,000 units per year, location C provides the lowest total cost.

For a profit analysis, compute the total profit for each location:

$$\text{Total profit} = Q(R - VC) - FC \tag{8–2}$$

where

R = Revenue per unit

Solved Problem 2 at the end of the chapter illustrates profit analysis.

Where the expected level of output is close to the middle of the range over which one alternative is superior, the choice is readily apparent. If the expected level of output is very close to the edge of a range, it means that the two alternatives will yield comparable annual costs, so management would be indifferent in choosing between the two *in terms of total cost*. However, it is important to recognize that, in most situations, other factors besides costs must also be considered. Later in this section, a general scheme for including a broad range of factors is described. First, let's consider another kind of cost often considered in location decisions: transportation costs.

The Transportation Model

Transportation costs sometimes play an important role in location decisions. These can stem from the movement of either raw materials or finished goods. If a facility will be the sole source or destination of shipments, the company can include the transportation costs in a locational cost-volume analysis by incorporating the transportation cost per unit being shipped into the variable cost per unit. (If raw materials are involved, the transportation cost must be converted into cost per unit of *output* in order to correspond to other variable costs.)

When a problem involves shipment of goods from multiple sending points to multiple receiving points, and a new location (sending or receiving point) is to be added to the system, the company should undertake a separate analysis of transportation. In such instances the *transportation model* of linear programming is very helpful. It is a special-purpose algorithm used to determine the minimum transportation cost that would result if a potential new location were to be added to an existing system. It can also be used if a *number* of new facilities are to be added or if an entire new system is being developed. The model is used to analyze each of the configurations considered, and it reveals the minimum costs each entails. This information can then be included in the evaluation of location alternatives. Solved Problem 1 illustrates how results of a transportation analysis can be combined with the results of a locational cost-volume analysis. See also the chapter supplement.

Factor Rating

A typical location decision involves both qualitative and quantitative inputs, which tend to vary from situation to situation depending on the needs of each organization. **Factor rating** is a general approach that is useful for evaluating a given alternative and comparing alternatives. The value of factor rating is that it provides a rational basis for evaluation and facilitates comparison among alternatives by establishing a *composite* value for each alternative that summarizes all related factors. Factor rating enables decision makers to incorporate their personal opinions and quantitative information into the decision process.

factor rating General approach to evaluating locations that includes quantitative and qualitative inputs.

The following procedure is used to develop a factor rating:

1. Determine which factors are relevant (e.g., location of market, water supply, parking facilities, revenue potential).
2. Assign a weight to each factor that indicates its relative importance compared with all other factors. Typically, weights sum to 1.00.
3. Decide on a common scale for all factors (e.g., 0 to 100).
4. Score each location alternative.
5. Multiply the factor weight by the score for each factor, and sum the results for each location alternative.
6. Choose the alternative that has the highest composite score.

This procedure is illustrated in the next example.

Example 2

A photo-processing company intends to open a new branch store. The table below contains information on two potential locations.

Solution

		SCORES (OUT OF 100)		WEIGHTED SCORES	
Factor	**Weight**	**Alt. 1**	**Alt. 2**	**Alternative 1**	**Alternative 2**
Proximity to existing store	.10	100	60	.10(100) = 10.0	.10(60) = 6.0
Traffic volume	.05	80	80	.05(80) = 4.0	.05(80) = 4.0
Rental costs	.40	70	90	.40(70) = 28.0	.40(90) = 36.0
Size	.10	86	92	.10(86) = 8.6	.10(92) = 9.2
Layout	.20	40	70	.20(40) = 8.0	.20(70) = 14.0
Operating costs	.15	80	90	.15(80) = 12.0	.15(90) = 13.5
	1.00			70.6	82.7

Alternative 2 is better because it has the higher composite score.

In some cases, managers may prefer to establish minimum *thresholds* for composite scores. If an alternative fails to meet that minimum, they can reject it without further consideration. If none of the alternatives meets the minimum, this means that either additional alternatives must be identified and evaluated or the minimum threshold must be reevaluated.

The Center of Gravity Method

center of gravity method Method for locating a distribution center that minimizes distribution cost.

The **center of gravity method** is a method to determine the location of a distribution center that will minimize distribution costs. It treats distribution cost as a linear function of the distance and the quantity shipped. The quantity to be shipped to each destination is assumed to be fixed (i.e., will not change over time). An acceptable variation is that quantities are allowed to change, as long as their relative amounts remain the same (e.g., seasonal variations).

FIGURE 8–1

Center of gravity method

a. Map showing destination.

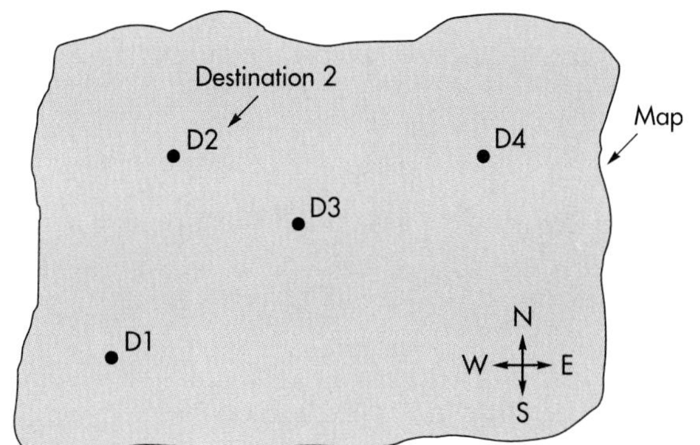

b. Add a coordinate system.

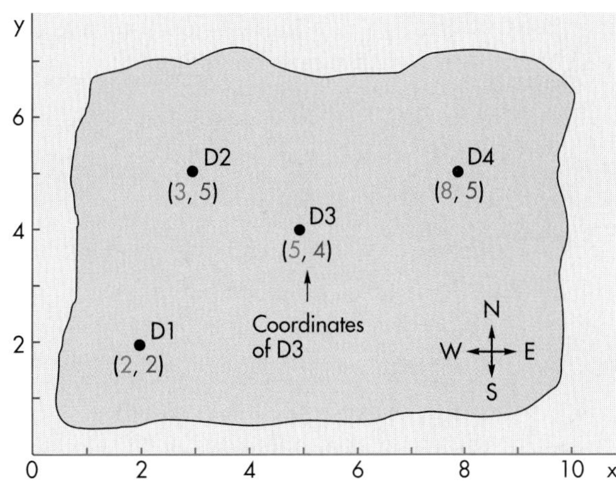

c. Center of gravity.

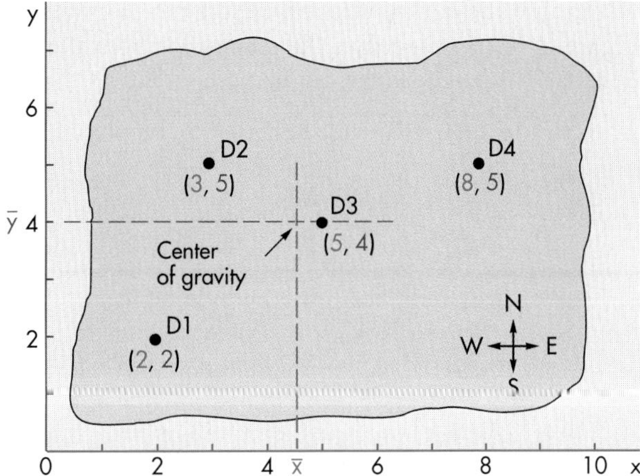

The method includes the use of a map that shows the locations of destinations. The map must be accurate and drawn to scale. A coordinate system is overlaid on the map to determine relative locations. The location of the 0,0 point of the coordinate system, and its scale, is unimportant. Once the coordinate system is in place, you can determine the coordinates of each destination. (See Figure 8–1, parts a and b.)

If the quantities to be shipped to every location are *equal,* you can obtain the coordinates of the center of gravity (i.e., the location of the distribution center) by finding the average of the x coordinates and the average of the y coordinates (see Figure 8–1). These averages can be easily determined using the following formulas:

$$\bar{x} = \frac{\Sigma x_i}{n} \qquad \bar{y} = \frac{\Sigma y_i}{n} \tag{8–3}$$

where

x_i = x coordinate of destination i
y_i = y coordinate of destination i
n = number of destinations

When the number of units to be shipped are not the same for all destinations (usually the case), *weighted average* must be used to determine the center of gravity, with the weights being the *quantities* to be shipped.

The appropriate formulas are:

$$\bar{x} = \frac{\Sigma x_i Q_i}{\Sigma Q_i} \qquad \bar{y} = \frac{\Sigma y_i Q_i}{\Sigma Q_i} \tag{8–4}$$

where

Q_i = Quantity to be shipped to destination
x_i = x coordinate of destination i
y_i = y coordinate of destination i

Example 3

Determine the coordinates of the center of gravity for the problem that is depicted in Figure 8–1c. Assume that the shipments from the center of gravity to each of the four destinations will be equal quantities.

Solution

The coordinates of the destinations can be obtained from Figure 8–1b:

$$\bar{x} = \frac{\Sigma x_i}{n} = \frac{18}{4} = 4.5 \qquad \bar{y} = \frac{\Sigma y_i}{n} = \frac{16}{4} = 4$$

Destination	x, y
D1	2, 2
D2	3, 5
D3	5, 4
D4	8, 5
	18 16

Hence, the center of gravity is at (4.5, 4), which places it just west of destination D3 (see Figure 8–1).

Example 4

Suppose that the shipments for the problem depicted in Figure 8–1a are not all equal, but instead are the following:

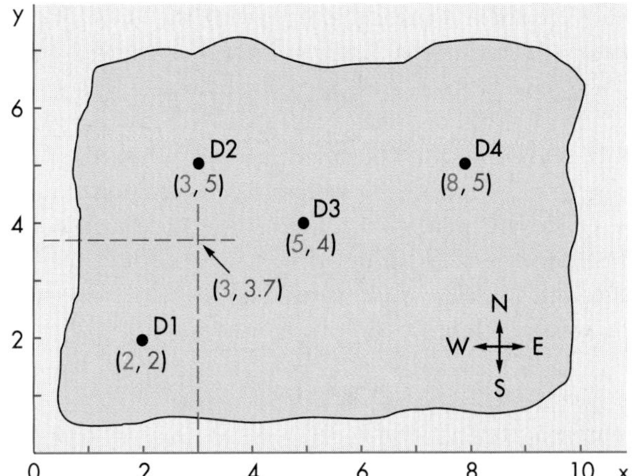

FIGURE 8–2

Center of gravity for Example 4

Destination	x, y	Weekly Quantity
D1	2, 2	800
D2	3, 5	900
D3	5, 4	200
D4	8, 5	100
		2,000

Determine the center of gravity.

Solution

Because the quantities to be shipped differ among destinations, you must use the weighted average formulas.

$$\bar{x} = \frac{\Sigma x_i Q_i}{\Sigma Q_i} = \frac{2(800) + 3(900) + 5(200) + 8(100)}{2{,}000} = \frac{6{,}100}{2{,}000} = 3.05 \text{ [round to 3]}$$

$$\bar{y} = \frac{\Sigma y_i Q_i}{\Sigma Q_i} = \frac{2(800) + 5(900) + 4(200) + 5(100)}{2{,}000} = \frac{7{,}400}{2{,}000} = 3.7$$

Hence, the coordinates of the center of gravity are approximately (3, 3.7). This would place it south of destination D2, which has coordinates of 3, 5). (See Figure 8–2.)

Summary

Location decisions confront both new and existing organizations. Growth, market shifts, depletion of raw materials, and the introduction of new products and services are among the reasons organizations are concerned with location decisions. The importance of these decisions is underscored by the long-term commitment they typically involve and by their potential impact on the operating system.

The primary location options available to existing organizations are to expand an existing location, move to a new location, maintain existing facilities while adding another facility in a new location, or do nothing.

In practice, the major influences on location decisions are location of raw materials, labor supply, market considerations, community-related factors, site-related factors, and climate. Foreign locations may be attractive in terms of labor costs, abundance of raw materials, or as potential markets for a firm's products or services. Problems organizations sometimes encounter in foreign countries include language differences, cultural differences, bias, and political instability.

A common approach to narrowing the range of location alternatives is to first identify a country or region that seems to satisfy overall needs and then identify a number of community-site alternatives for more in-depth analysis. A variety of methods are used to evaluate location alternatives. Those described in the chapter include locational cost-volume analysis, factor rating, and the center of gravity method. We briefly noted the transportation model; the chapter supplement contains a more complete description of that subject.

There are numerous commercial software packages available for location analysis.[1] In addition to the models just described, many packages employ linear programming or mixed integer programming algorithms. In addition, some software packages use heuristic approaches to obtain reasonable solutions to location problems.

Key Terms

center of gravity method, 378
factor rating, 378
locational cost-profit-volume analysis, 375
microfactory, 373

Solved Problems

Problem 1

Cost analysis. A farm implements dealer is seeking a fourth warehouse location to complement three existing warehouses. There are three potential locations: Charlotte, N.C.; Atlanta, Ga.; and Columbia, S.C. Charlotte would involve a fixed cost of $4,000 per month and a variable cost of $4 per unit; Atlanta would involve a fixed cost of $3,500 per month and a variable cost of $5 per unit; and Columbia would involve a fixed cost of $5,000 per month and a variable cost of $6 per unit. Use of the Charlotte location would increase system transportation costs by $19,000 per month, Atlanta by $22,000 per month, and Columbia by $18,000 per month. Which location would result in the lowest total cost to handle 800 units per month?

Solution

Given: Volume = 800 units per month

	FC per Month	VC per Unit	Transportation Cost per Month
Charlotte	$4,000	$4	$19,000
Atlanta	3,500	5	22,000
Columbia	5,000	6	18,000

Monthly total cost = FC + VC + Transportation cost

Charlotte: $4,000 + $4 per unit × 800 units + $19,000 = $26,200
Atlanta: 3,500 + 5 per unit × 800 units + 22,000 = 29,500
Columbia: 5,000 + 6 per unit × 800 units + 18,000 = 27,800

Hence, Charlotte would have the lowest total cost for this monthly volume.

Problem 2

Profit analysis. A manufacturer of staplers is about to lose its lease, so it must move to another location. Two sites are currently under consideration. Fixed costs would be $8,000 per month at site A and $9,400 per month at site B. Variable costs are expected to be $5 per unit at site A and $4 per unit at site B. Monthly demand has been steady at 8,800 units for the last several years and is not expected to deviate from that amount in the foreseeable future. Assume staplers sell for $6 per unit. Determine which location would yield the higher profit under these conditions.

Solution

$\text{Profit} = Q(R - VC) - FC$

[1]See Ronald H. Ballou and James M. Masters, "Commercial Software for Locating Warehouses and Other Facilities," *Journal of Business Logistics* 14, no. 2 (1993), 71–92.

Site	Revenue	FC	VC	Monthly Profit
A	$52,800	$8,000	$44,000	$ 800
B	$52,800	$9,400	$35,200	$8,200

Hence, site B is expected to yield the higher monthly profit.

Problem 3

Factor rating. Determine which location has the highest factor rating given the following information:

		LOCATION	
Factor	**Weight**	**A**	**B**
Labor cost	.50	20	40
Material cost	.30	10	30
Supplier base	.20	50	10
	1.00		

Solution

Combining the weights with the location scores, we can see that location B has the higher score:

		LOCATION		WEIGHTED SCORES	
Factor	**Weight**	**A**	**B**	**A**	**B**
Labor cost	.50	20	40	.50(20) = 10	.50(40) = 20
Material cost	.30	10	30	.30(10) = 3	.30(30) = 9
Supplier base	.20	50	10	.20(50) = 10	.20(10) = 2
	1.00			23	31

Discussion and Review Questions

1. In what ways can the location decision have an impact on the production system?
2. Respond to this statement: "The importance of the location decision is often vastly overrated; the fact that virtually every type of business is located in every section of the country means there should be no problem in finding a suitable location."
3. What community factors influence location decisions?
4. How are manufacturing and nonmanufacturing location decisions similar? Different?
5. What are the potential benefits of locating in foreign countries? Potential drawbacks?
6. What is factor rating, and how does it work?
7. Outline the general approach for developing location alternatives.
8. What are the basic assumptions in locational cost-volume analysis?
9. Discuss recent trends in location and possible future strategies.

Memo Writing Exercises

1. You must decide which of several sites will be best for the location of a third restaurant in the company's chain in a suburb of a medium-sized city. Write a one-page memo to your employee, Jim Watson, outlining the information you will need from him before you can analyze the location alternatives.
2. You have been asked to provide a location analysis on whether your company should expand an existing manufacturing plant or close the plant and move to a larger plant recently

vacated by a bankrupt firm. Write a one-page memo to your supervisor, Karen Saunders, outlining the factors you will consider in your analysis.

Problems

Omaha: −$8,000
K.C.: $256,000
Choose K.C.

1. A newly formed firm must decide on a plant location. There are two alternatives under consideration: locate near the major raw materials or locate near the major customers. Locating near the raw materials will result in lower fixed and variable costs than locating near the market, but the owners believe there would be a loss in sales volume because customers tend to favor local suppliers. Revenue per unit will be $185 in either case. Using the following information, determine which location would produce the greater profit.

	Omaha	Kansas City
Annual fixed costs ($ millions)	$1.2	$1.4
Variable cost per unit	$36	$47
Expected annual demand (units)	8,000	12,000

	A	B	C
a.	16,854	17,416	17,753
b.	$13,690	$14,080	$14,670

2. The owner of Genuine Subs, Inc., hopes to expand the present operation by adding one new outlet. She has studied three locations. Each would have the same labor and materials costs (food, serving containers, napkins, etc.) of $1.76 cents per sandwich. Sandwiches sell for $2.65 each in all locations. Rent and equipment costs would be $5,000 per month for location A, $5,500 per month for location B, and $5,800 per month for location C.
   a. Determine the volume necessary at each location to realize a monthly profit of $10,000.
   b. If expected sales at A, B, and C are 21,000 per month, 22,000 per month, and 23,000 per month, respectively, which location would yield the greatest profits?

a. 120 units
b. A: 0 to 119
B: 121 or more

3. A small producer of machine tools wants to move to a larger building, and has identified two alternatives. Location A has annual fixed costs of $800,000 and variable costs of $14,000 per unit; location B has annual fixed costs of $920,000 and variable costs of $13,000 per unit. The finished items sell for $17,000 each.
   a. At what volume of output would the two locations have the same total cost?
   b. For what range of output would location A be superior? For what range would B be superior?

a. B: 0 to 33
C: 34 to 400
A: 400 or more
b. C
c. Subcontracting prices more precise; subcontracting provides another source of supply; expansion allows more control and flexibility.

4. A company that produces pleasure boats has decided to expand one of its lines. Current facilities are insufficient to handle the increased workload, so the company is considering three alternatives, A (new location), B (subcontract), and C (expand existing facilities).
   Alternative A would involve substantial fixed costs but relatively low variable costs: fixed costs would be $250,000 per year, and variable costs would be $500 per boat. Subcontracting would involve a cost per boat of $2,500, and expansion would require an annual fixed cost of $50,000 and a variable cost of $1,000 per boat.
   a. Find the range of output for each alternative that would yield the lowest total cost.
   b. Which alternative would yield the lowest total cost for an expected annual volume of 150 boats?
   c. What other factors might be considered in choosing between expansion and subcontracting?

A: $329,000
B: $400,000
C: $270,000
C is best

5. Rework Problem 4*b* using this additional information: Expansion would result in an increase of $70,000 per year in transportation costs, subcontracting would result in an increase of $25,000 per year, and adding a new location would result in an increase of $4,000 per year.

	Total cost
Memphis	$170,000
Biloxi	$160,000
Birmingham	$175,000

6. A firm that has recently experienced an enormous growth rate is seeking to lease a small plant in either Memphis, Tenn.; Biloxi, Miss.; or Birmingham, Ala. Prepare an economic analysis of the three locations given the following information: Annual costs for building, equipment, and administration would be $40,000 for Memphis, $60,000 for Biloxi, and $100,000 for Birmingham. Labor and materials are expected to be $8 per unit in Memphis,

$4 per unit in Biloxi, and $5 per unit in Birmingham. The Memphis location would increase system transportation costs by $50,000 per year, the Biloxi location by $60,000 per year, and the Birmingham location by $25,000 per year. Expected annual volume is 10,000 units.

7. A retired auto mechanic hopes to open a rustproofing shop. Customers would be local new-car dealers. Two locations are being considered, one in the center of the city and one on the outskirts. The central city location would involve fixed monthly costs of $7,000 and labor, materials, and transportation costs of $30 per car. The outside location would have fixed monthly costs of $4,700 and labor, materials, and transportation costs of $40 per car. Dealer price at either location will be $90 per car.

   a. Which location will yield the greatest profit if monthly demand is (1) 200 cars? (2) 300 cars?

   b. At what volume of output will the two sites yield the same monthly profit?

a.

	City	Outside
P_{200}:	$5,000	$5,300
P_{300}:	$11,000	$10,300

b. 230 cars

8. For each of the four types of organizations shown, rate the importance of each factor in terms of making location decisions using L = low importance, M = moderate importance, and H = high importance.

Factor	Local Bank	Steel Mill	Food Ware-house	Public School
Convenience for customers	______	______	______	______
Attractiveness of building	______	______	______	______
Nearness to raw materials	______	______	______	______
Large amounts of power	______	______	______	______
Pollution controls	______	______	______	______
Labor cost and availability	______	______	______	______
Transportation costs	______	______	______	______
Construction costs	______	______	______	______

H	L	M-H	M-H
H	L	M	M-H
L	H	L	M
L	H	L	L
L	H	L	L
L	M	L	L
L	M-H	M-H	M
M	H	M	M-H

9. Using the following factor ratings, determine which location alternative should be chosen on the basis of maximum composite score, A, B, or C.

A: 87.02
B: 82.62
C: 80.90
A is best

Factor (100 points each)	Weight	LOCATION A	B	C
Convenient	.15	80	70	60
Parking facilities	.20	72	76	92
Display area	.18	88	90	90
Shopper traffic	.27	94	86	80
Operating costs	.10	98	90	82
Neighborhood	.10	96	85	75
	1.00			

10. A manager has received an analysis of several cities being considered for a new office complex. The data (10 points maximum) are

a. A: 39/7
B: 44/7
C: 44/7
B or C best
b. A: 53/9
B: 55/9
C: 54/9
B is best, A is worst

	LOCATION		
Factor	**A**	**B**	**C**
Business services	9	5	5
Community services	7	6	7
Real estate cost	3	8	7
Construction costs	5	6	5
Cost of living	4	7	8
Taxes	5	5	4
Transportation	6	7	8

a. If the manager weights the factors equally, how would the locations stack up?

b. If business services and construction costs are given weights that are double the weights of the other factors, how would the locations stack up?

$\bar{x} = 5, \bar{y} = 4$
Location optimal at (5, 4)

11. A toy manufacturer produces toys in five locations throughout the country. Raw materials (primarily barrels of powdered plastic) will be shipped from a new, centralized warehouse whose location is to be determined. The monthly quantities to be shipped to each location are the same. A coordinate system has been established, and the coordinates of each location have been determined as shown. Determine the coordinates of the centralized warehouse.

Location	(x, y)
A	3, 7
B	8, 2
C	4, 6
D	4, 1
E	6, 4

$\bar{x} = 6, \bar{y} = 7$
Location optimal at (6, 7)

12. A clothing manufacturer produces women's clothes at four locations in Mexico. Relative locations have been determined, as shown in the table below. The location of a central shipping point for bolts of cloth must now be determined. Weekly quantities to be shipped to each location are shown below. Determine the coordinates of the location that will minimize distribution costs.

Location	(x, y)	Weekly Quantity
A	5, 7	15
B	6, 9	20
C	3, 9	25
D	9, 4	30

$\bar{x} = 5.97, \bar{y} = 5.95$
Optimal location is (6, 6)

13. A company that handles hazardous waste wants to minimize the shipping cost for shipments to a disposal center from five receiving stations it operates. Given the locations of the receiving stations and the volumes to be shipped daily, determine the location of the disposal center.

Location of Processing Station, (x, y)	Volume, Tons per Day
10, 5	26
4, 1	9
4, 7	25
2, 6	30
8, 7	40

14. Determine the center of gravity for the destinations shown on the following map. Monthly shipments will be the quantities listed in the table.

$\bar{x} = 3.03, \bar{y} = 2.21$
Center of gravity is (3, 2.2)

Destination	Quantity
D1	900
D2	300
D3	700
D4	600
D5	800

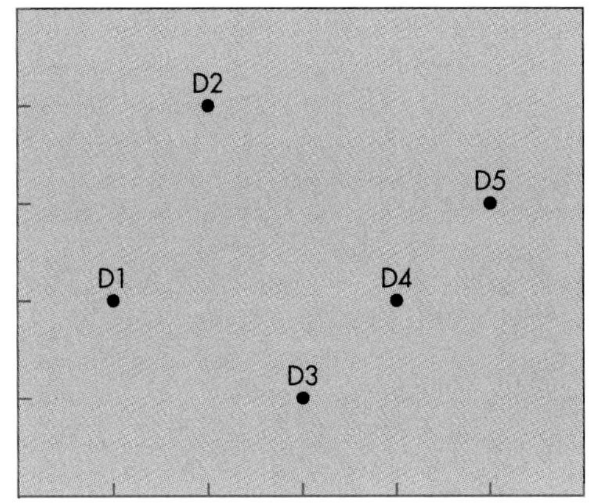

READING

www.gm.com

GLOBAL STRATEGY: GM IS BUILDING PLANTS IN DEVELOPING NATIONS TO WOO NEW MARKETS

Rebecca Blumenstein

ROSARIO, Argentina—On the outskirts of this city more than 6,000 miles from Detroit, bulldozers have been running around the clock moving the fertile topsoil to make way for a General Motors Corp. auto-assembly plant of tomorrow.

The state-of-the-art plant, on a remote road in what had been a remote part of the automotive world, is at the center of GM's aggressive shift away from its roots in North America to the developing world. GM is quietly building so many plants in so many corners of the world that the No. 1 automaker has decided to save money by building essentially the same plant in Argentina, Poland, China and Thailand—simultaneously.

This "four-plant strategy" is GM's biggest international expansion, costing at least $2.2 billion. The company has designed the plants to look so much alike that engineers may mistake which country they are in. And the assembly lines are being set up so that a glitch in a robot in Thailand, rather than turning into an expensive engineering problem that requires an expert for each machine at each plant, may well be solved by a quick call to Rosario or to Shanghai, China.

Expandable and Efficient

But the significance of GM's new plants extends much further. They are being built to be easily expandable as demand in developing markets grows. And they are being laid out in a giant U-shape so that suppliers can cart in an increasing array of already assembled parts to cut GM's costs, something GM can't do in the U.S. because of union resistance.

The new factories illustrate, more than anything else, how the nature of multinational corporations is changing as today's marketplace turns global. Just a few years ago, GM's South American plants were churning out Chevy Chevettes that hadn't been produced in the U.S. for years. The auto industry viewed the world largely as a dumping ground for obsolete technology and outdated models.

Now, the industry is seeking future growth by concentrating its investments in the developing world and turning it into a showcase for the latest in technology and lean manufacturing. And GM is pursuing that global strategy more vigorously than any of its rivals, though every other major auto maker sees the need to tap into emerging markets, and some have a substantial foreign presence. Nobody, however, is building plants on the scale of GM, which recently announced plans for a fifth and even more advanced new plant, in southern Brazil.

The entire Rosario plant is designed to be lean. Its total cost, at $350 million, is one of the lowest for a new GM plant; $1 billion had been the rule-of-thumb price tag for a new assembly plant of any size. Moreover, the investment here includes a high-tech stamping machine that will press metal parts more quickly than any other GM press and an engine plant that will deliver to the assembly line engines with air conditioning, transmission and belts already installed.

Workers will do many tasks, a key principle of lean manufacturing, and will be assigned to teams working autonomously. Every worker will be responsible for an entire process in the assembly operation, including even the cleaning and basic maintenance of machines. In Japan, that is a common approach, but it is inconceivable in a GM plant in the U.S. because of union work rules, which allow only specified categories of skilled workers to repair and maintain machines.

Although the four plants are designed to be as similar as possible, they do have differences, ranging from efforts to protect machines from rusting in Thailand's humid climate to China's poor transportation system. There, Mr. Stevens says, "We are going to have things delivered to our Shanghai plant by bicycle."

Questions

1. Briefly describe GM's four-plant strategy.
2. What are some signs of GM's shift in global thinking?
3. List the lean manufacturing elements of the Rosario plant.
4. What are some nonlabor local conditions that had to be taken into consideration in designing individual plants?

Selected Bibliography

Ballou, Ronald H. *Business Logistics Management: Planning and Control.* Englewood Cliffs, NJ: Prentice Hall, 1985, 2nd ed.

Buffa, E. S. *Meeting the Competitive Challenge.* Burr Ridge, Ill.: Richard D. Irwin, 1984, pp. 65–82.

Francis, R. L.; J. A. White; and L. McGinniss. *Facilities Layout and Location: An Analytical Approach.* Englewood Cliffs, NJ: Prentice Hall, 2nd ed. 1991.

Hayes, Robert H., and Stephen Wheelright. *Restoring Our Competitive Edge: Competing through Manufacturing.* New York: John Wiley & Sons, 1984.

Lyne, Jack. "Quality-of-Life Factors Dominate Many Facility Location Decisions." *Site Selection Handbook.* August, 1988, pp. 868–70.

Schmenner, R. W. *Making Business Location Decisions.* Englewood Cliffs, NJ: Prentice Hall, 1982.

Toward a New Era in U.S. Manufacturing: The Need for a National Vision. Washington: National Academy Press, 1986.

SUPPLEMENT TO CHAPTER EIGHT

THE TRANSPORTATION MODEL

LEARNING OBJECTIVES

After completing this supplement, you should be able to:

1. Describe the nature of a transportation problem.
2. Solve transportation problems manually and interpret the results.
3. Set up transportation problems in the general linear-programming format.
4. Interpret computer solutions.

SUPPLEMENT OUTLINE

The transportation problem involves finding the lowest-cost plan for distributing stocks of goods or *supplies* from multiple origins to multiple destinations that *demand* the goods. For instance, a firm might have three factories, all of which are capable of producing identical units of the same product, and four warehouses that stock or demand those products, as depicted in Figure 8S–1. The *transportation model* can be used to determine how to allocate the supplies available from the various factories to the warehouses that stock or demand those goods, in such a way that total shipping cost is minimized. Usually, analysis of the problem will produce a shipping plan that pertains to a certain period of time (day, week), although once the plan is established, it will generally not change unless one or more of the parameters of the problem (supply, demand, unit shipping cost) changes.

INTRODUCTION

Although Figure 8S–1 illustrates the nature of the transportation problem, in real life managers must often deal with allocation problems that are considerably larger in scope. A beer maker may have four or five breweries and hundreds or even thousands of distributors, and an automobile manufacturer may have eight assembly plants scattered throughout the United States and Canada and thousands of dealers that must be supplied with those cars. In such cases, the ability to identify the optimal distribution plan makes the transportation model very important.

The shipping (supply) points can be factories, warehouses, departments, or any other place from which goods are sent. Destinations can be factories, warehouses, departments, or other points that receive goods. The information needed to use the model consists of the following:

1. A list of the origins and each one's capacity or supply quantity per period.
2. A list of the destinations and each one's demand per period.
3. The unit cost of shipping items from each origin to each destination.

This information is arranged into a *transportation table* (see Table 8S–1).

The transportation model is one of a class of linear-programming models, so named because of the linear relationships among variables. In the transportation model, transportation costs are treated as a direct linear function of the number of units shipped.

Use of the transportation model implies that certain assumptions are satisfied. The major ones are:

1. The items to be shipped are homogeneous (i.e., they are the same regardless of their source or destination).
2. Shipping cost per unit is the same regardless of the number of units shipped.
3. There is only one route or mode of transportation being used between each origin and each destination.

The transportation model starts with the development of a feasible solution, which is then sequentially tested and improved until an optimal solution is obtained. The description of the technique on the following pages focuses on each of the major steps in the process in this order:

1. Obtaining an initial solution.
2. Testing the solution for optimality.
3. Improving suboptimal solutions.

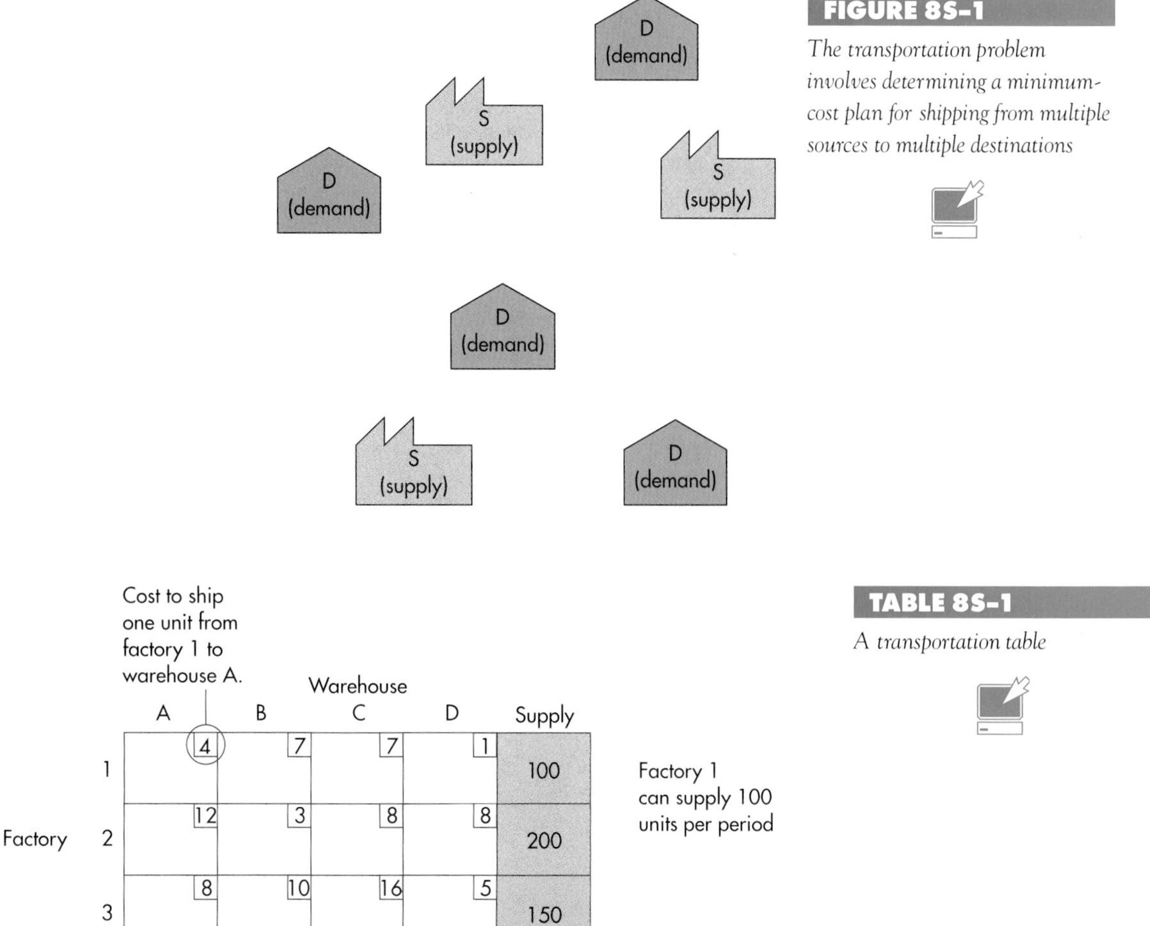

FIGURE 8S-1

The transportation problem involves determining a minimum-cost plan for shipping from multiple sources to multiple destinations

TABLE 8S-1

A transportation table

Factory	A	B	C	D	Supply
1	4	7	7	1	100
2	12	3	8	8	200
3	8	10	16	5	150
Demand	80	90	120	160	450 / 450

Warehouse (columns A–D)

Cost to ship one unit from factory 1 to warehouse A. (points to 4)

Factory 1 can supply 100 units per period

Total supply capacity per period (points to 450)

Warehouse B can use 90 units per period (points to 90)

Total demand per period (points to 450)

OBTAINING AN INITIAL SOLUTION

To begin the process, it is necessary to develop a feasible distribution plan. A number of different methods are available for obtaining such a plan. The discussion here will focus on the **intuitive approach,** a heuristic approach that yields an initial solution that is often optimal or near optimal. The method will be demonstrated using Table 8S–1.

intuitive approach Heuristic approach to obtaining an initial solution.

The method requires that total supply equal total demand. If you refer to Table 8S–1, you will see that they are equal. Later you will learn what to do if they are not equal.

The Intuitive Lowest-Cost Approach

With the intuitive approach, cell allocations are made according to cell cost, beginning with the lowest cost. The procedure involves these steps:

1. Identify the cell with the lowest cost.
2. Allocate as many units as possible to that cell, and cross out the row or column (or both) that is exhausted by this.

3. Find the cells with the next lowest cost from among the *feasible* cells.
4. Repeat steps (2) and (3) until all units have been allocated.

Cell 1-D has the lowest cost ($1) (see Table 8S–2). The factory 1 supply is 100, and the warehouse D demand is 160. Therefore, the most we can allocate to this cell is 100 units. Since the supply of 100 is exhausted, we cross out the costs in the first row along with the supply of 100. In addition, we must adjust the column total to reflect the allocation, which leaves 60 units of unallocated demand in column D.

The next lowest cost is $3 in cell 2-B. Allocating 90 units to this cell exhausts the column total and leaves 110 units for the supply of factory 2 (see Table 8S–3). Also, we cross out the costs for column B.

The next lowest cost (that is not crossed out) is the $5 in cell 3-D. Allocating 60 units to this cell exhausts the column total and leaves 90 units for row 3. We now cross out the costs in column D (see Table 8S–4).

At this point, there is a *tie* for the next lowest cell cost: cell 3-A and cell 2-C each has a cost of $8. Break such a tie arbitrarily. Suppose we choose cell 3-A. The demand is 80 units, and the remaining supply is 90 units in the row. The smaller of these is 80, so that amount is placed in cell 3-A (see Table 8S–5). This exhausts the column, so we cross out the cell costs in column A. Also, the remaining supply in factory 3 is now 10.

The other cell in the tie, cell 2-C, still remains. The remaining supply in the row is 110 and demand in the column is 120. The smaller of these two amounts, 110, is assigned to the cell, and the row costs are crossed out. The row total is changed to zero, and the column total is changed to 10 units. The only remaining cell with a cost that has not been crossed out is cell 3-C. Both supply and demand are 10 units, so 10 units are assigned to the cell. This completes the initial allocation. (See Table 8S–6.)

Now let's determine if this initial solution obtained using the intuitive method is optimum.

TABLE 8S–2

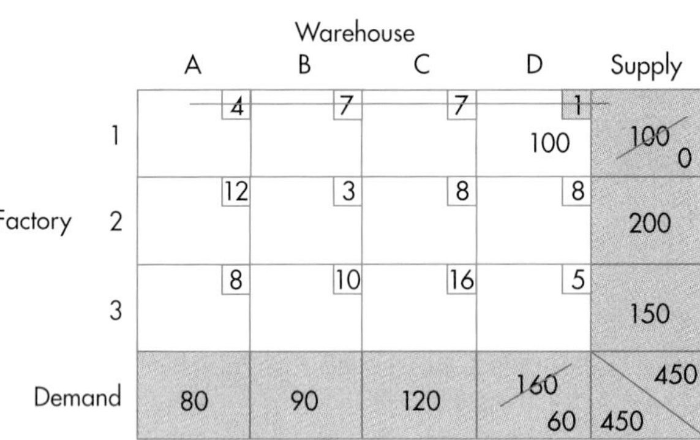

	Warehouse A	B	C	D	Supply
Factory 1	[4]	[7]	[7]	[1] 100	~~100~~ 0
Factory 2	[12]	[3]	[8]	[8]	200
Factory 3	[8]	[10]	[16]	[5]	150
Demand	80	90	120	~~160~~ 60	450 / 450

TABLE 8S–3

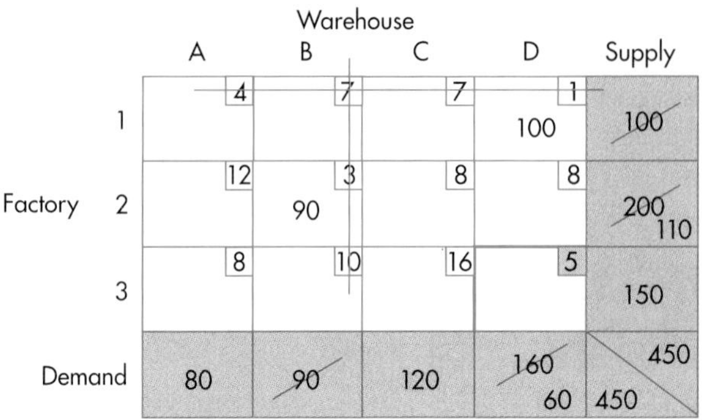

	Warehouse A	B	C	D	Supply
Factory 1	[4]	[7]	[7]	[1] 100	~~100~~
Factory 2	[12]	[3] 90	[8]	[8]	~~200~~ 110
Factory 3	[8]	[10]	[16]	[5]	150
Demand	80	~~90~~	120	~~160~~ 60	450 / 450

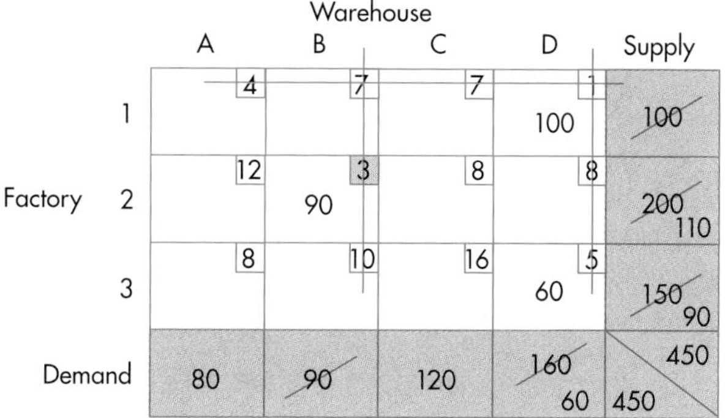

Factory	A	B	C	D	Supply
1	[4]	[7]	[7]	[1] 100	100
2	[12]	[3] 90	[8]	[8]	200 110
3	[8]	[10]	[16]	[5] 60	150 90
Demand	80	90	120	160 60	450 / 450

TABLE 8S-4

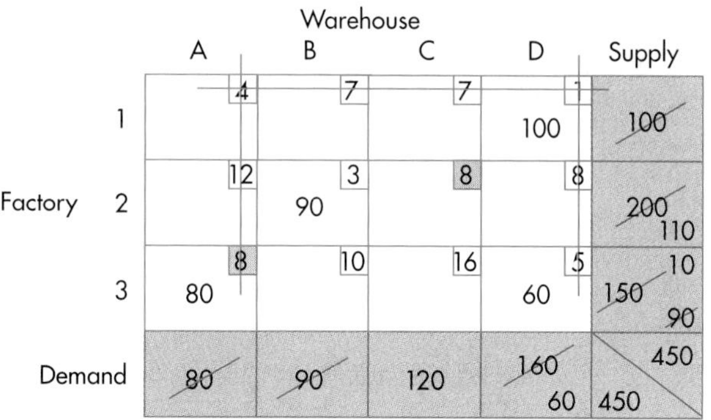

Factory	A	B	C	D	Supply
1	[4]	[7]	[7]	[1] 100	100
2	[12]	[3] 90	[8]	[8]	200 110
3	[8] 80	[10]	[16]	[5] 60	150 90 10
Demand	80	90	120	160 60	450 / 450

TABLE 8S-5

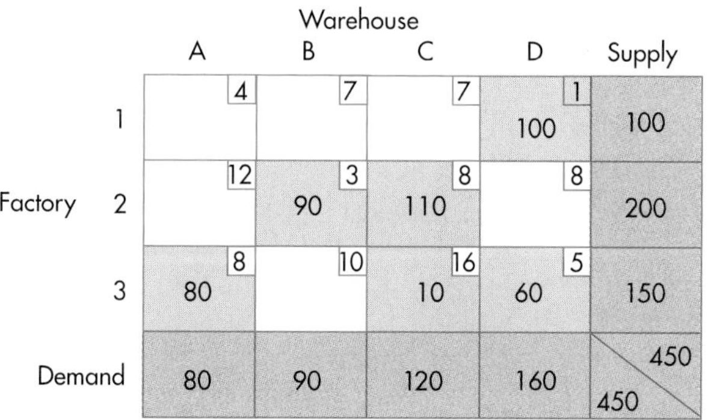

Factory	A	B	C	D	Supply
1	[4]	[7]	[7]	[1] 100	100
2	[12]	[3] 90	[8] 110	[8]	200
3	[8] 80	[10]	[16] 10	[5] 60	150
Demand	80	90	120	160	450 / 450

TABLE 8S-6

Initial allocation using intuitive approach

TESTING FOR OPTIMALITY

Testing for optimality and revising suboptimal solutions involves analysis of each unused (empty) cell to determine the potential for reducing the total cost of the solution. This is accomplished by transferring one unit into an empty cell and noting its impact on costs. If costs are increased, that implies that using the cell would increase total costs. If costs remain the same, that implies the existence of an alternative option with the same total cost as the current plan. However, if analysis reveals a cost decrease, the implication is that an improved solution is possible. The test for optimality requires that *every* unused cell be evaluated for potential improvement. Either of two methods can be used: stepping-stone or MODI.

Evaluating Empty Cells: The Stepping-Stone Method

In the **stepping-stone method,** cell evaluation proceeds by borrowing one unit from a full cell and using it to assess the impact of shifting units into the empty cell. For instance, if a shift of one unit causes an increase of \$5, total costs would be increased by \$5 times the number of units shifted into the cell. Obviously, such a move would be unwise since the objective is to decrease costs.

stepping-stone method Cell evaluation by borrowing one unit from a full cell to assess the impact of shifting units into the empty cell.

TABLE 8S–7

Evaluation path for cell 1-A

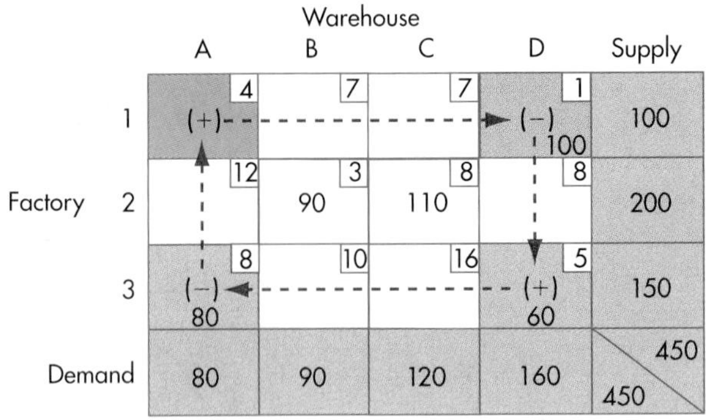

The name *stepping-stone* derives from early descriptions of the method that likened the procedures to crossing a shallow pond by stepping from stone to stone. Here, the occupied cells are the "stones"; shifting units into empty cells requires borrowing units from occupied cells. To maintain the balance of supply and demand for every row and column, a shift of one unit into an empty cell requires a series of shifts from other occupied cells.

The best way to understand the evaluation process is to consider a simple example. Suppose we want to evaluate cell 1-A (see Table 8S–7). We must shift one unit into that cell, which implies that one unit will be shipped from factory 1 to warehouse A. However, factory 1 has only 100 units it can ship, and all are allocated to warehouse D. Therefore, to ship one unit to warehouse A, we borrow a unit from cell 1-D. This creates two problems: Column A now has one extra unit, and Column D is short one unit (i.e., A has 1 + 80 = 81 units but needs only 80 units, and D has 99 + 60 = 159 but needs 160 units). We remedy these problems by subtracting one unit from cell 3-A and adding it to cell 3-D; this gives column A a total of 1 + 79 = 80 and column D a total of 99 + 61 = 160. (Note: Instead of making each addition or subtraction of the single unit, a + or − sign is simply inserted in the appropriate cell to signify the operation.)

Let's see what effect such a shift would have on costs. For each cell to which a unit was added, the cost will increase by the transportation cost for that cell; for cells that gave up a unit, costs will decrease by the cell's transportation cost. We can summarize this with a *T account* as follows.

Cell 1-A

+		−	
(1-A)	+4	(1-D)	−1
(3-D)	+5	(3-A)	−8
	+9		−9

0

Thus, shifting units around this path would have no impact whatsoever on transportation costs: use of cell 1-A would be an equivalent-cost alternative. Management may prefer to use it for other reasons, or management may prefer to stay with the original solution. In any case, knowledge of such alternatives adds a certain degree of flexibility to decision making.

So far, we have not discovered any improvement potential, but five unused cells still must be analyzed. Before we evaluate the remaining empty cells, it is important to mention that constructing paths using the stepping-stone method requires a minimum number of occupied cells and unless that number exists, it will be impossible to evaluate all the empty cells in this manner. The number of occupied cells must equal the sum of the number of rows and columns minus 1, or $R + C - 1$, where R = Number of rows and C = Number of columns. In this example, $R = 3$ and $C = 4$, so we must have

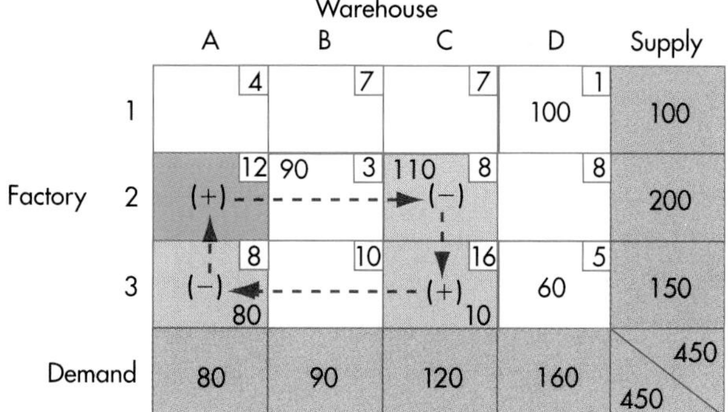

Factory \ Warehouse	A	B	C	D	Supply
1	[4]	[7]	[7]	[1] 100	100
2	[12] (+)	[3] 90	[8] 110 (−)	[8]	200
3	[8] 80 (−)	[10]	[16] 10 (+)	[5] 60	150
Demand	80	90	120	160	450 / 450

TABLE 8S–8

Evaluation path for cell 2-A

3 + 4 − 1 = 6 used or completed cells (which we do have). If there are too few occupied cells, the matrix is said to be *degenerate.* A method for overcoming this problem is explained later in this supplement.

For now, let's continue evaluating the unused cells. Suppose we now consider cell 2-A. Begin by adding a unit to the empty cell. Moving to the right in row 2, we have what seems to be a choice: borrow a unit from 90 or from 110. However, borrowing from 90 will leave column B one unit short, and since adding and borrowing must involve occupied cells and there are no others in the B column, the 90 cannot be used. Instead, we must borrow from the 110 and add one unit to the 10 in cell 3-C. We can complete our path back to the original cell by subtracting a unit from the 80 units in cell 3-A. The +/− path is shown in Table 8S–8. The impact on total cost for the path associated with cell 2-A would be

Cell 2-A

+		−	
(2-A)	+12	(2-C)	−8
(3-C)	+16	(3-A)	−8
	+28		−16

+12

This means that for every unit shifted into cell 2-A, the total cost would *increase* by $12. Thus we should avoid shipping units from factory 2 to warehouse A.

At this point, let's consider some helpful rules for obtaining evaluation paths, since you may still be unsure of how to do it.

1. Start by placing a + sign in the cell you wish to evaluate.
2. Move horizontally (or vertically) to a completed cell (a cell that has units assigned to it). It is OK to pass *through* an empty cell or a completed cell without stopping. Choose a cell that will permit your next move to another completed cell. Assign a minus sign (−) to the cell.
3. Change direction and move to another completed cell. Again, choose one that will permit your next move. Assign a plus sign (+) to the cell.
4. Continue this process of moving to completed cells and alternating + and − signs until you can complete a closed path back to the original cell. Make only horizontal and vertical moves.
5. You may find it helpful to keep track of cells that have been evaluated by placing the cell evaluation value in the appropriate cell with a circle around it.

Let's try another one, say, 3-B. Start by placing a + sign in cell 3-B. Move to the 90 in cell 2-B, and place a − sign there. Move to the 110 in cell 2-C, and place a + sign

TABLE 8S-9

Evaluation path for cell 3-B

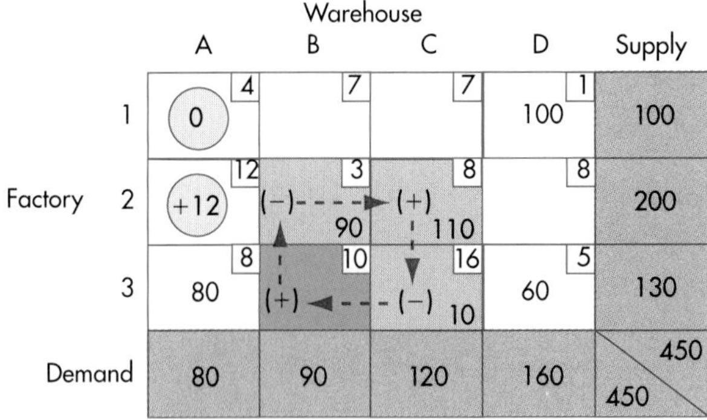

TABLE 8S-10

Evaluation path for cell 2-D

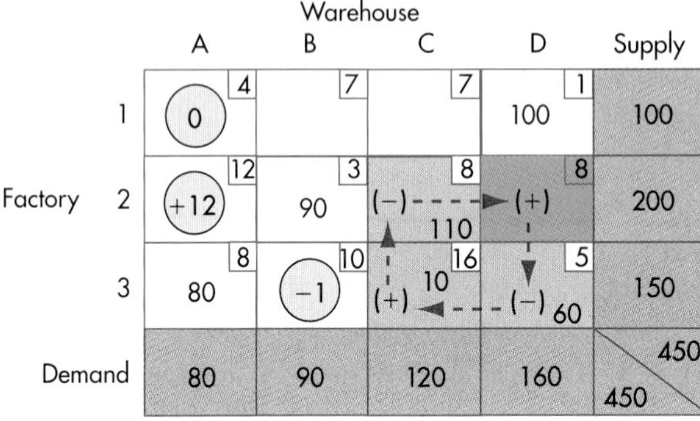

there. Move to cell 3-C, and give it a − sign. The path is shown in Table 8S–9. This one was fairly easy. Let's see what the impact on cost would be:

Cell 3-B

+		−	
(3-B)	+10	(2-B)	−3
(2-C)	+8	(3-C)	−16
	+18		−19

(−1)

The −1 indicates that for each unit we can shift into cell 3-B, our cost will decrease by \$1. If we could shift 100 units, we could save \$100 over the present solution. However, at this point we are not ready to make any changes since some other empty cell might yield even greater savings, and we can only make one shift per table.

Evaluation of cell 2-D is similar to the evaluation of cell 3-B. The evaluation path is shown in Table 8S–10. The cost impact would be:

Cell 2-D

+		−	
(2-D)	+8	(3-D)	−5
(3-C)	+16	(2-C)	−8
	+24		−13

+11

Thus, no improvement is possible.

There are two cells remaining to be evaluated: 1-B and 1-C. The evaluation of cell 1-B is a little more involved than the previous paths. Begin by placing a + in cell 1-B. Move to cell 1-D (100 units) and give it a −. Move vertically to 3-D (60 units) and give

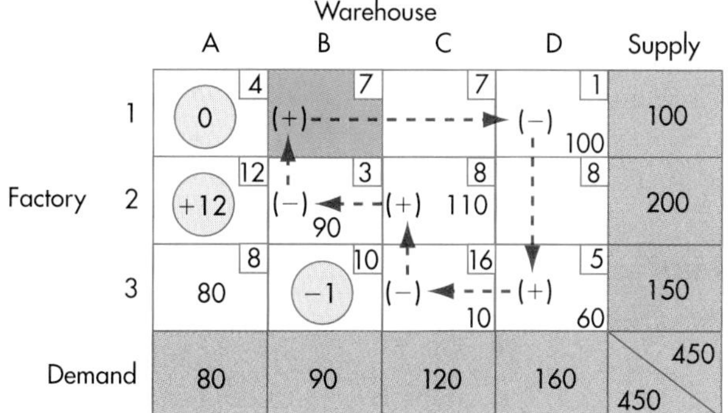

	A	B	C	D	Supply
Factory 1	[4] (0)	[7] (+)	[7]	[1] (−) 100	100
Factory 2	[12] (+12)	[3] (−) 90	[8] (+) 110	[8]	200
Factory 3	[8] 80	[10] (−1)	[16] (−) 10	[5] (+) 60	150
Demand	80	90	120	160	450 / 450

TABLE 8S–11

Evaluation path for cell 1-B

	A	B	C	D	Supply
Factory 1	[4] (0)	[7] (0)	[7] (+)	[1] (−) 100	100
Factory 2	[12] (+12)	[3] 90	[8] 110	[8] (+11)	200
Factory 3	[8] 80	[10] (−1)	[16] (−) 10	[5] (+) 60	150
Demand	80	90	120	160	450 / 450

TABLE 8S–12

Evaluation path for cell 1-C

it a +. Move to the left and put a − in cell 3-C (10 units). Move up to 110 and put a + in that cell. Move to the left to 90 units (cell 2-B) and give it a − sign, completing the path. The path is shown in Table 8S–11. The cost impact of the path would be zero.

Cell 1-B

+		−	
(1-B)	+7	(1-D)	−1
(3-D)	+5	(3-C)	−16
(2-C)	+8	(2-B)	−3
	+20		−20

0

Thus, no improvement is possible using this cell path.

The last empty cell is 1-C. Its evaluation path is shown in Table 8S–12. Begin with 1-C as +, 1-D as −, 3-D as +, and 3-C as −. The cost impact would be:

Cell 1-C

+		−	
(1-C)	+7	(3-D)	−1
(3-D)	+5	(3-C)	−16
	+12		−17

(−5)

Hence, for each unit we can transfer into this empty cell, we improve the cost by $5.

At this point, each of the unused cells has been evaluated. The resulting costs are summarized below.

Cell	1-A	2-A	3-B	2-D	1-B	1-C
Cost.	0	+12	(−1)	+11	0	(−5)

Cells that have a positive or zero evaluation present no opportunities for improvement. The fact that 1-A and 1-B are both zero indicates that at least one other equivalent alternative exists. However, they are of no interest because two cells have negative cell evaluations, indicating that the present solution is not an optimum. In a later section, you will learn how to develop an improved solution.

Evaluating Empty Cells: The MODI Method

modified distribution method (MODI) Evaluating empty cells by computing row and column index numbers.

Another method for evaluating empty cells is the **modified distribution method (MODI).** It involves computing row and column index numbers that can be used for cell evaluation. In many respects, it is simpler than the stepping-stone method because it avoids having to trace cell evaluation paths. Nevertheless, the cell evaluations it produces are identical to those obtained using the stepping-stone method. Note, however, that if a solution is not optimal, *one stepping-stone path* must be traced to obtain an improved solution. This is explained in detail in the next section.

The MODI procedure consists of these steps:

1. Make an initial allocation using the intuitive method.
2. Obtain an index number for each row and column. Do this using only *completed* cells. Note that there will always be at least one completed cell in each row and in each column.
   a. Begin by assigning a zero to the first row.
   b. Determine the column index for any completed cells in row 1 using the relationship: Column index = Cell cost − Row index. For example, if a cell cost is $8 per unit, the column index will be 8 − 0 = 8.
   c. Each new column value will permit the calculation of *at least* one new row value, and vice versa. Continue until all rows and columns have index numbers, using only completed cells.
3. Obtain cell evaluations for *empty* cells using the relationship:

   Cell evaluation = Cell cost − (Row index + Column index)

Comments:

1. Row or column values may be positive, negative, or zero.
2. A reallocation requires that new row and column cost values be calculated.

Let's see how the MODI method can be used for cell evaluation.

We begin by assigning a zero to row 1 (see Table 8S–13). Since we can only work with occupied cells and since cell 1-D is the only such cell in row 1, we focus on it. Since the sum of the row and column index numbers must add to the cell cost (which is 1), we can see that the index number for column D must be 1. There are no other occupied cells in row 1, so we shift to Column D, cell 3-D. Again, the sum of the row and column index numbers must add to the cell cost (which is 5). Since the column index for D is 1, the index number for row 3 must be 4, as 5 − 1 = 4.

We can use this row 3 index number and the other two occupied cells in row 3 to obtain index numbers for columns A and C. For column A, the A index number plus 4 must equal the 3-A cell cost of 8. Hence, the index for A is 4: 8 − 4 = 4. For C, the index plus 4 must equal 16; hence, the index number is 12.

Next, for row 2, using cell 2-C and a C index of 12, the row index is −4. Then, using cell 2-B, the column index is 7: 3 − (−4) = 7. This completes the row and column index numbers.

The cell evaluation for each empty cell is determined by subtracting from the cost of the empty cell the sum of the row and column index numbers:

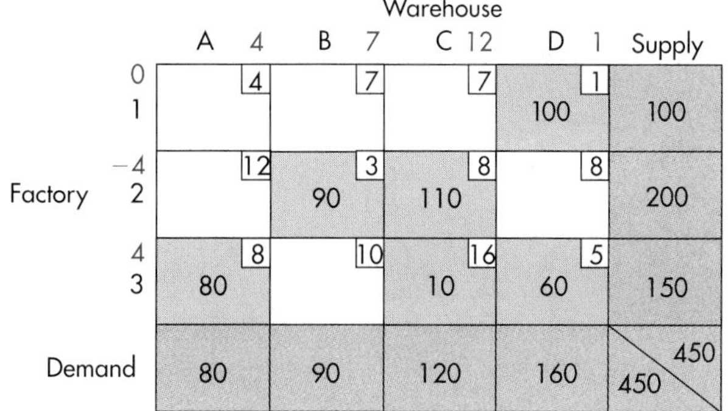

		A (4)	B (7)	C (12)	D (1)	Supply
Factory	1 (0)	[4]	[7]	[7]	[1] 100	100
	2 (−4)	[12]	[3] 90	[8] 110	[8]	200
	3 (4)	[8] 80	[10]	[16] 10	[5] 60	150
Demand		80	90	120	160	450 / 450

TABLE 8S–13

MODI index numbers

Cell	Evaluation
1-A	4 − (0 + 4) = 0
1-B	7 − (0 + 7) = 0
1-C	7 − (0 + 12) = −5
2-A	12 − (−4 + 4) = 12
2-D	8 − (−4 + 1) = 11
3-B	10 − (4 + 7) = −1

TABLE 8S–14

MODI cell evaluation

Cell evaluation = Cell cost − (Row index + Column index)

Table 8S–14 shows all of the evaluations. Note that these agree with the evaluations obtained using the stepping-stone method. Because some evaluations are negative, the solution is not optimal.

OBTAINING AN IMPROVED SOLUTION

The presence of negative cell evaluations is evidence that an improved solution is possible. By the same token, if *no* negatives appear, an optimum has been achieved.

In this case, there are two cells with negative evaluations: 3-B, with a value of −1, and 1-C, with a value of −5. Select the value that implies more improvement (i.e., −5), and ignore the other.

The implication of the −5 is that a savings of $5 per unit will be realized for every unit that can be shifted into cell 1-C. Recall the path used to evaluate the cell; it is reproduced in Table 8S–15A. In the course of the evaluation, a single unit was added or subtracted to certain cells, as denoted by the + and − signs. We want to repeat that procedure, but instead of shifting *one* unit, we want to shift *as many as possible.* The *quantities* in the cells that have − signs (i.e., 10 in cell 3-C and 100 in cell 1-D) are potential candidates for shifting, since the − sign implies units are subtracted from those cells. It is the *smaller* of these quantities that is the limiting factor. That is, we can subtract 10 from both numbers (10 and 100), but if we try to subtract 100 from both, we end up with 0 and −90. The negative value would imply shipping from the warehouse back to the factory! Hence, 10 units can be shifted. Always shift the minimum quantity that occupies a negative position in the ± path.

TABLE 8S-15

Evaluation path for cell 1-C (see A) and reallocation of 10 units around the path (see B)

A.

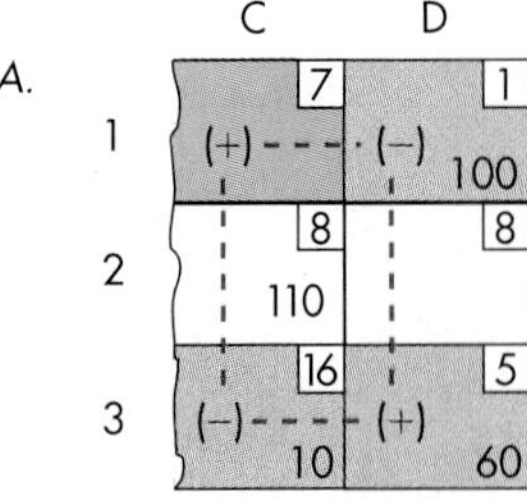

B.

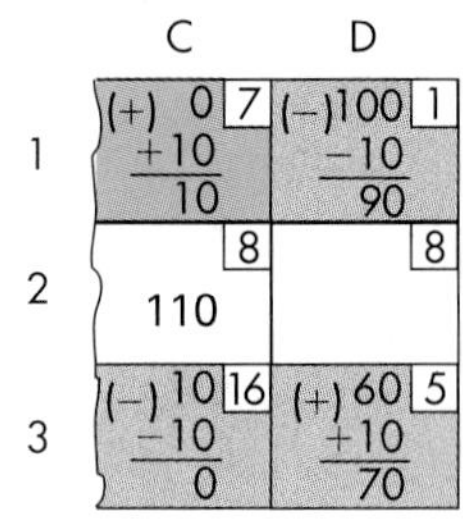

TABLE 8S-16

Revised solution

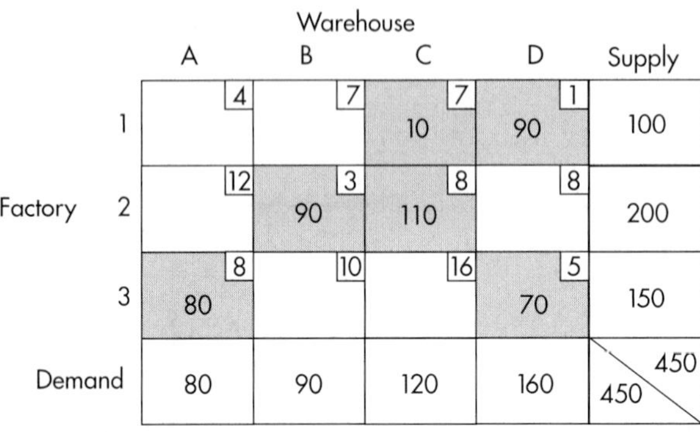

Factory \ Warehouse	A	B	C	D	Supply
1	[4]	[7]	[7] 10	[1] 90	100
2	[12]	[3] 90	[8] 110	[8]	200
3	[8] 80	[10]	[16]	[5] 70	150
Demand	80	90	120	160	450 / 450

To accomplish the shift, add 10 units to each cell that has a + sign and subtract 10 units from each cell with a − sign. (Because the signs alternate, the row and column totals will still be satisfied.) This is shown in Table 8S–15.

The improved solution is shown in Table 8S–16. The total costs are:

			Total Cost
1-C	10($7)	=	$ 70
1-D	90($1)	=	90
2-B	90($3)	=	270
2-C	110($8)	=	880
3-A	80($8)	=	640
3-D	70($5)	=	350
			$2,300

Note that the revised solution has a total cost that is $50 less than the previous solution ($2,300 versus $2,350), which is exactly equal to the cell evaluation value times the number of units shifted: −5 × 10 units = −$50.

Is this the optimum? Once again, we must evaluate each empty cell to determine if further improvements are possible. Recall that we must check to see if there are enough completed cells to allow all empty cells to have evaluation paths. Thus,

$$R + C - 1 = 3 + 4 - 1 = 6$$

Since there are six completed cells, the second round of cell evaluations can proceed. The cells paths and their values are shown in Tables 8S–17 to 8S–22. Since all are either zero or positive, the solution is the optimum. (Note: MODI could have been used instead of stepping-stone.)

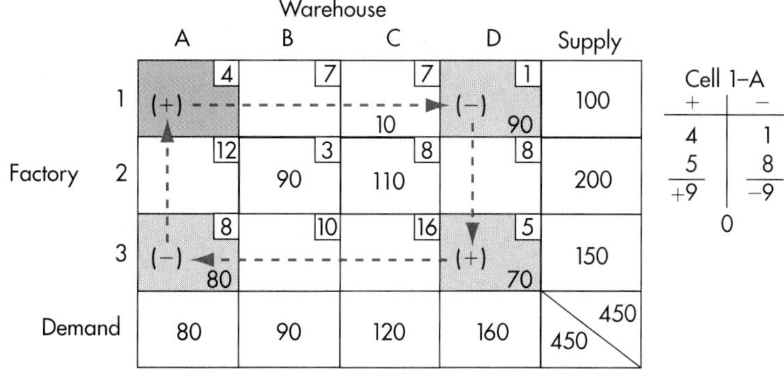

Factory \ Warehouse	A	B	C	D	Supply
1	4 (+)	7	7 10	1 (−) 90	100
2	12	3 90	8 110	8	200
3	8 (−) 80	10	16	5 (+) 70	150
Demand	80	90	120	160	450 / 450

Cell 1–A

+	−
4	1
5	8
+9	−9

0

TABLE 8S-17

Evaluation of cell 1-A

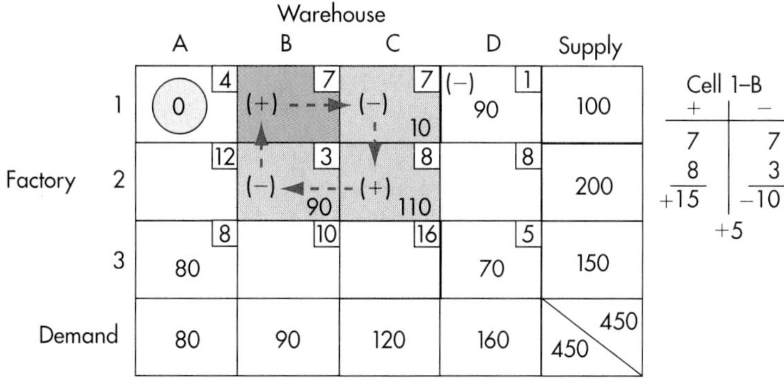

Factory \ Warehouse	A	B	C	D	Supply
1	4 (0)	7 (+)	7 (−) 10	1 (−) 90	100
2	12	3 (−) 90	8 (+) 110	8	200
3	8 80	10	16	5 70	150
Demand	80	90	120	160	450 / 450

Cell 1–B

+	−
7	7
8	3
+15	−10

+5

TABLE 8S-18

Evaluation of cell 1-B

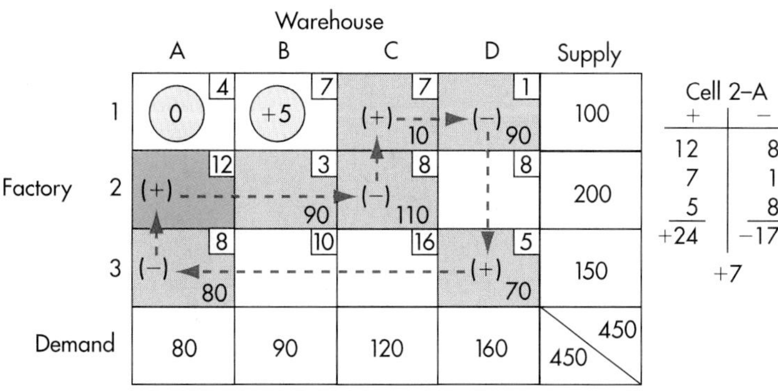

Factory \ Warehouse	A	B	C	D	Supply
1	4 (0)	7 (+5)	7 (+) 10	1 (−) 90	100
2	12 (+)	3 90	8 (−) 110	8	200
3	8 (−) 80	10	16	5 (+) 70	150
Demand	80	90	120	160	450 / 450

Cell 2–A

+	−
12	8
7	1
5	8
+24	−17

+7

TABLE 8S-19

Evaluation of cell 2-A

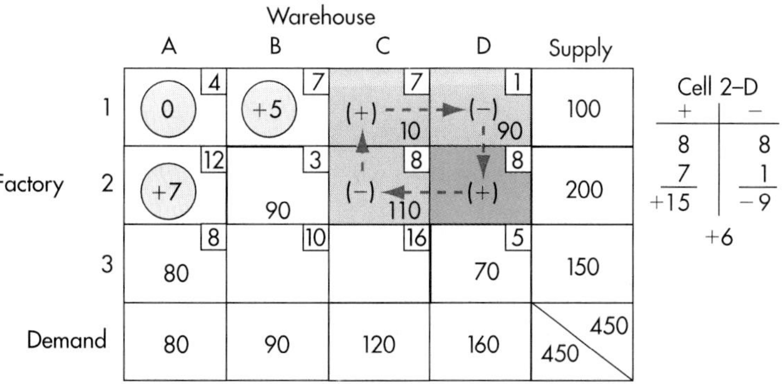

Factory \ Warehouse	A	B	C	D	Supply
1	4 (0)	7 (+5)	7 (+) 10	1 (−) 90	100
2	12 (+7)	3 90	8 (−) 110	8 (+)	200
3	8 80	10	16	5 70	150
Demand	80	90	120	160	450 / 450

Cell 2–D

+	−
8	8
7	1
+15	−9

+6

TABLE 8S-20

Evaluation of cell 2-D

TABLE 8S–21

Evaluation of cell 3-B

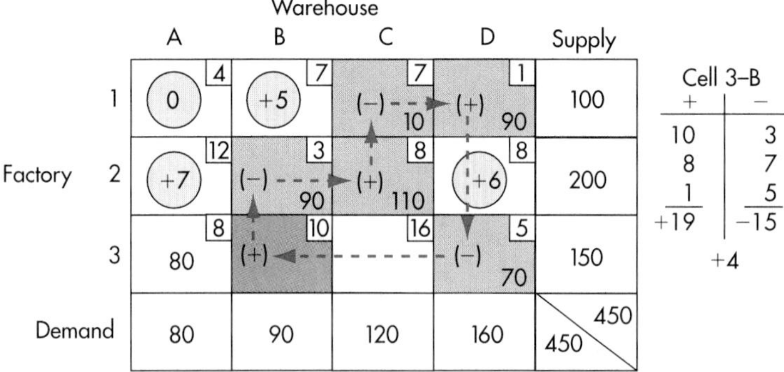

TABLE 8S–22

Evaluation of cell 3-C

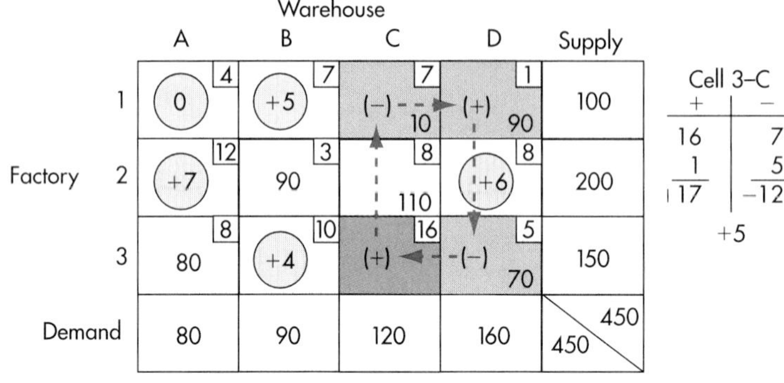

SPECIAL PROBLEMS

Not all transportation problems are as straightforward as the one just presented. Various irregularities may occur, which require certain adjustments before a solution can be obtained. Two of the most common irregularities are unequal supply and demand, and degeneracy, a term previously noted as referring to an insufficient number of completed cells to allow evaluation of every empty cell.

Unequal Supply and Demand

A case in which supply and demand are not equal is illustrated in Table 8S–23. Total supply (capacity) of the two sources is 200 units, but total demand of the destinations is only 170. Thus, the deficiency in demand is 30 units. This condition can be remedied by adding a **dummy** destination with a demand equal to the difference between supply and demand. Since there is no such destination, no units will actually be shipped, and unit shipping costs for each dummy cell are $0. Table 8S–24 illustrates this.

dummy Imaginary number added equal to the difference between supply and demand when these are unequal.

Once the dummy has been added, the transportation method can then be applied in the usual manner. The final solution will have a total number of units assigned to the dummy that equals the original difference between supply and demand (see Table 8S–25). These numbers indicate which source(s) (e.g., factories) will hold the extra units or will have excess capacity. In the example shown, factory 1 will ship only 70 units even though it has the capacity to ship 100 units.

The total transportation costs for this distribution plan is:

			Total Cost
1-A	70($5)	=	$350
1-Dummy	30($0)	=	0
2-A	10($4)	=	40
2-B.	90($2)	=	180
			$570

From: \ To:	A	B	Supply
1	[5]	[9]	100
2	[4]	[2]	100
Demand	80	90	200 / 170

TABLE 8S-23

Supply exceeds demand by 30 units

From: \ To:	A	B	Dummy	Supply
1	[5]	[9]	[0]	100
2	[4]	[2]	[0]	100
Demand	80	90	30	200 / 200

TABLE 8S-24

A dummy column with a demand of 30 units is added to the matrix; dummy cell costs are zero

From: \ To:	A	B	Dummy	Supply
1	[5] 70	[9]	[0] 30	100
2	[4] 10	[2] 90	[0]	100
Demand	80	90	30	200 / 200

TABLE 8S-25

Units in the dummy column are not shipped; factory 1 will ship only 70 units

A similar situation exists when demand exceeds supply. In those cases a dummy *row* (e.g., factory) with a supply equal to the difference between supply and demand must be added. Again, units will not actually be shipped, so transportation costs are $0 for each cell in the dummy row. Units in the dummy row in the final solution indicate the destinations that will not receive all the units they desire and the amount of the shortage for each.

Note: When using the intuitive approach, if a dummy row or column exists, make assignments to dummy cell(s) *last*.

Degeneracy

The transportation method involves evaluation of empty cells using completed cells as stepping stones. **Degeneracy** exists when there are too few completed cells to allow all necessary paths to be constructed. The condition occurs if an allocation (other than the final one) exhausts *both* the row (supply) and the column (demand) quantities. It can occur in an initial solution or in subsequent solutions, so it is necessary to test for degeneracy after each iteration using $R + C - 1$.

degeneracy The condition of too few completed cells to allow all necessary paths to be constructed.

An example of a matrix with too few completed cells is shown in Table 8S–26. Note that there are only four completed cells, although the necessary number is five: $3 + 3 - 1 = 5$. Because of this, it is impossible to develop an evaluation path for cells 1-B, 1-C, 2-A, or 3-A. The situation can be remedied by placing a very small quantity, represented by the symbol ε, into one of the empty cells and then treating it as a completed cell. The quantity is so small that it is negligible; it will be ignored in the final solution.

TABLE 8S–26

There are too few completed cells; some empty cells cannot be evaluated

	A	B	C	Supply
1	[3] 40	[2]	[5]	40
2	[8]	[1] 50	[4] 10	60
3	[7]	[7]	[6] 20	20
Demand	40	50	30	120 / 120

TABLE 8S–27

Cell 3-A becomes a completed cell with the addition of a very small quantity, represented by the symbol ε

	A	B	C	Supply
1	[3] 40	[2]	[5]	40
2	[8]	[1] 50	[4] 10	60
3	[7] ε	[7]	[6] 20	20
Demand	40	50	30	120 / 120

The idea is to select a position for ε that will permit evaluation of all remaining empty cells. For example, putting it in cell 3-A will permit evaluation of all empty cells (see Table 8S–27). Some experimentation may be needed to find the best spot for ε, because not every cell will enable construction of evaluation paths for the remaining cells. Moreover, avoid placing ε in a minus position of a cell path that turns out to be negative because reallocation requires shifting the smallest quantity in a minus position. Since the smallest quantity is ε, which is essentially zero, no reallocation will be possible.

The cell evaluations (not shown) are all positive, so this solution is optimal. (Note that cell 3-A is considered to be a completed cell and is therefore *not* evaluated.)

SUMMARY OF PROCEDURE

1. Make certain that supply and demand are equal. If they are not, determine the difference between the two amounts. Create a dummy source or destination with a supply (or demand) equal to the difference so that demand and supply are equal.
2. Develop an initial solution using the intuitive, lowest-cost approach.
3. Check that the number of completed cells is equal to $R + C - 1$. If it isn't, the solution is degenerate and will require insertion of a minute quantity, ε, in one of the empty cells to make it a completed cell.
4. Evaluate each of the empty cells. If any evaluations turn out to be negative, an improved solution is possible. Identify the cell that has the largest negative evaluation. Reallocate units around its evaluation path.
5. Repeat steps (3) and (4) until all cells have zero or positive values. When that occurs, you have achieved the optimal solution.

LOCATION DECISIONS

The transportation model can be used to compare location alternatives in terms of their impact on the total distribution costs for a system. The procedure involves working through a separate problem for each location being considered and then comparing the resulting total costs. Solved Problem 2 provides an example of this.

TABLE 8S–28

		Warehouse A	Warehouse B	Warehouse C	Warehouse D	Supply
Factory	1	4	7	7	1	100
	2	12	3	8	8	200
	3	8	10	16	5	150
Demand		80	90	120	160	450 / 450

If other costs, such as production costs, differ among locations, these can easily be included in the analysis, provided they can be determined on a per-unit basis. In this regard, note that merely adding or subtracting a constant to all cost values in any row or column will not affect the optimum solution; any additional costs should only be included if they have a varying effect within a row or column.

OTHER APPLICATIONS

We have seen how the transportation model can be used to minimize the costs associated with distributing goods, and we have seen how the model can be used for comparing location alternatives. The model is also used in a number of other ways. For example, in a slight variation of the model, profits can be used in place of costs. In such cases, each of the cell profits can be subtracted from the largest profit, and the remaining values (opportunity costs) can be treated in the same manner as shipping costs.

Some of the other uses of the model include production planning and scheduling (see Chapter 12), problems involving assignment of personnel or jobs to certain departments or machines (see Chapter 15), capacity planning, and transshipment problems.[1]

The use of the transportation model for capacity planning parallels its use for location decisions. An organization can subject proposed capacity alternatives to transportation analysis to determine which one would generate the lowest total shipping cost. For example, it is perhaps intuitively obvious that a factory or warehouse that is close to its market—or has low transportation costs for some other reason—should probably have a larger capacity than other locations. Of course, many problems are not so simple, and they require actual use of the model.

COMPUTER SOLUTIONS

Although manual solution of transportation problems is fairly straightforward, computer solutions are generally preferred, particularly for moderate or large problems. Many software packages call for data input in the same tabular form used throughout this chapter. A more general approach is to format the problem as a standard linear-programming model (i.e., specify the objective function and a set of constraints). That approach enables one to use the more general version of an LP package to solve a transportation problem. Let's consider this general approach.

Table 8S–28 repeats the transportation model presented earlier in this supplement. The decision variables for a transportation model are the quantities to be shipped. Because each cell represents a potential transportation route, each must have a decision

[1]Transshipment relates to problems with major distribution centers that in turn redistribute to smaller market destinations. See, for example, W. J. Stevenson, *Introduction to Management Science*, 2nd ed. (Burr Ridge, IL: Richard D. Irwin, 1992).

FIGURE 8S–2

Excel template for Table 8S–28

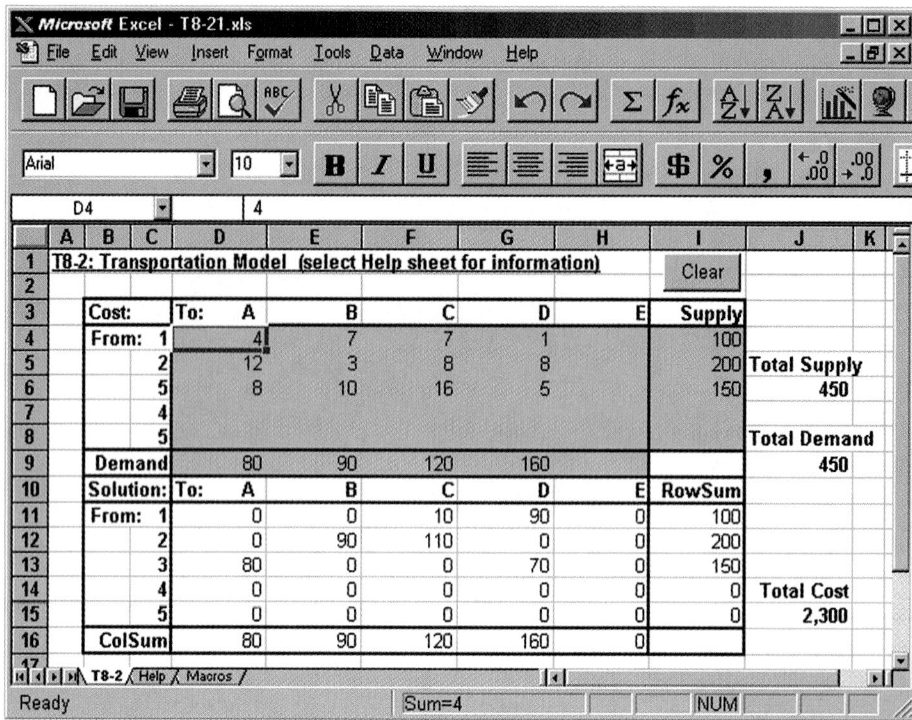

variable. We can use the symbol x_{1A} to represent the decision variable for cell 1-A, x_{1B} for cell 1-B, and so on. The objective function consists of the cell costs and these cell symbols:

$$\text{Minimize } 4x_{1A} + 7x_{1B} + 7x_{1C} + 1x_{1D} + 12x_{2A} + 3x_{2B} + 8x_{2C} + 8x_{2D} + 8x_{3A} + 10x_{3B} + 16x_{3C} + 5x_{3D}$$

Because the amounts allocated in any row or column must add to the row or column total, each row and column must have a constraint. Thus, we have

$$\begin{aligned} \text{Supply (rows)} \quad & x_{1A} + x_{1B} + x_{1C} + x_{1D} = 100 \\ & x_{2A} + x_{2B} + x_{2C} + x_{2D} = 200 \\ & x_{3A} + x_{3B} + x_{3C} + x_{3D} = 150 \\ \text{Demand (columns)} \quad & x_{1A} + x_{2A} + x_{3A} = 80 \\ & x_{1B} + x_{2B} + x_{3B} = 90 \\ & x_{1C} + x_{2C} + x_{3C} = 120 \\ & x_{1D} + x_{2D} + x_{3D} = 160 \end{aligned}$$

We do not need a constraint for the total; the row and column constraints take care of this.

If supply and demand are not equal, add the appropriate dummy row or column to the table before writing the constraints.

Another approach to transportation problems is to use spreadsheet software. The Excel Templates can also be used to solve transportation problems. Figure 8S–2 illustrates the Excel worksheet for the preceding problem.

Key Terms

degeneracy, 403
dummy, 402
intuitive lowest-cost approach, 391
modified distribution method (MODI), 398
stepping-stone method, 393

Solved Problems

Problem 1

Use the intuitive lowest-cost approach to develop an initial solution to the following problem, and then determine if the solution is optimal. If it is not, develop the optimal solution.

From \ To:	A	B	Supply
1	[9]	[6]	75
2	[5]	[3]	75
Demand	80	90	

Solution

a. First check to see if supply and demand are equal. Total demand = 170 and total supply = 150; therefore, it will be necessary to create a dummy origin (supply) with zero transportation costs in each of its cells (see following table).

From \ To:	A	B	Supply
1	[9]	[6]	75
2	[5]	[3]	75
Dummy	[0]	[0]	20
Demand	80	90	170 / 170

b. Find the cell that has the lowest unit transportation cost. When a dummy is present and you are using the intuitive method, leave the dummy allocation until last. Thus, aside from the \$0 costs in the dummy row, the lowest cost is \$3 for cell 2-B. Assign as many units as possible to this cell. The column total is 90, and the row total is 75. Because 75 is the smaller of the two, assign 75 to cell 2-B. This exhausts the row total, so cross out the 75, and cross out the cell costs for the second row. Revise the column total to 15.

From \ To:	A	B	Supply
1	[9]	[6]	75
2	~~[5]~~	~~[3]~~ 75	~~75~~
Dummy	[0]	[0]	20
Demand	80	~~90~~ 15	170 / 170

The next smallest cell cost is \$6 in cell 1-B. We want to assign as many units as possible to this route (cell). The remaining demand for column B is 15, and the row total is 75. We assign the smaller of these, 15, to cell 1-B, thereby exhausting the column and reducing the row supply to 60. Next we draw a line through the cell costs of column B.

From \ To:	A	B	Supply
1	[9]	[6] 15	~~75~~ 60
2	[5]	[3] 75	~~75~~
Dummy	[0]	[0]	20
Demand	80	~~90~~ ~~15~~	170 \ 170

Still ignoring the dummy row, the next lowest cell cost is $9 in 1-A, where a supply of 60 and a demand of 80 exist. We assign the smaller of these, 60 units, to cell 1-A, so row 1 costs must be crossed out. Also, the revised column A total becomes 20.

From \ To:	A	B	Supply
1	[9] 60	[6] 15	~~75~~ ~~60~~
2	[5]	[3] 75	~~75~~
Dummy	[0]	[0]	20
Demand	~~80~~ 20	~~90~~ ~~15~~	170 \ 170

At this point, only the dummy row remains. Cell dummy-A has a supply and a demand of 20; assigning this amount to the cell exhausts both the supply and the demand, finishing the initial allocation.

From \ To:	A	B	Supply
1	[9] 60	[6] 15	~~75~~ ~~60~~
2	[5]	[3] 75	~~75~~
Dummy	[0] 20	[0]	~~20~~
Demand	~~80~~ ~~20~~	~~90~~ ~~15~~	170 \ 170

c. The next step is to evaluate this solution to determine if it is the optimal solution. First we must check for degeneracy: $R + C - 1 = 3$ (rows) + 2 (columns) $- 1 = 4$, the required number of filled or completed cells. Because there are four completed cells, the solution is not degenerate.

The evaluation can use either the stepping-stone method or MODI. Suppose we use MODI. We always begin by placing a zero at the left edge of row 1. This is the row 1 index number. We can use it to find the column index number for any cell in row 1 that is a completed cell. Because both cell 1-A and cell 1-B are completed, we can obtain both of their column index numbers. For completed cells, the sum of the row and the column index numbers must equal the cell cost. For cell 1-A, 0 + Column index = 9, so the column A index must be 9. Similarly, for cell 1-B, 0 + Column index = 6, so the column index must be 6. Next, these index numbers are used to find values of row index numbers for completed cells in their respective columns. Thus, for cell 2-B, 6 + Row index = 3, so the row index must be −3. For dummy-A, 9 + Row index = 0, so the row index must be −9.

From: \ To:	A 9	B 6	Supply
0 1	[9] 60	[6] 15	75
−3 2	[5]	[3] 75	75
−9 Dummy	[0] 20	[0]	20
Demand	80	90	170 / 170

Next, we use these index numbers to evaluate the empty cells. We determine a cell evaluation by subtracting from the cell cost the sum of its row and column index numbers. Thus, for cell 2-A, we have $5 - (-3 + 9) = -1$, and for cell dummy-B, we have $0 - (-9 + 6) = 3$.

From: \ To:	A 9	B 6	Supply
0 1	[9] 60	[6] 15	75
−3 2	[5] (−1)	[3] 75	75
−9 Dummy	[0] 20	[0] (+3)	20
Demand	80	90	170 / 170

The appearance of a negative cell evaluation tells us this solution can be improved on. When negatives appear, we select the cell with the largest negative as the focal point of our effort to improve the solution. Because there is only one such cell, it is the one we select.

d. The −1 in cell 2-A tells us that for every unit we can move into this cell, we can reduce the total cost of the solution by $1. Obviously, we would like to put as many units as possible into this cell, but we are limited by the row and column totals (row 2 and column A). We are also constrained by the need for all row and column totals to be maintained. To put units into cell 2-A, we must take them from some other cell, which will be in another row or column. Fortunately, this matter is simple to resolve: Determine the stepping-stone path for cell 2-A, and use it to guide the reallocation. That path is illustrated in the following table.

From: \ To:	A	B	Supply
1	[9] 60 (−)	[6] 15 (+)	75
2	[5] (+)	[3] 75 (−)	75
Dummy	[0] 20	[0]	20
Demand	80	90	170 / 170

The amount that can be shifted into cell 2-A is the *smallest* quantity that appears in a *negative position* of the path. There are two such quantities, 60 and 75. Because 60 is the smaller of these, it is shifted around the entire path (i.e., added where there is a + sign and subtracted where there is a − sign). The result is our revised solution:

From: \ To:	A	B	Supply
1	[9]	[6] 75	75
2	[5] 60	[3] 15	75
Dummy	[0] 20	[0]	20
Demand	80	90	170 \ 170

e. This improved solution may be optimal, or it may need further revision. To find out, we must evaluate it.

First, we check for degeneracy: $R + C - 1 = 3 + 2 - 1 = 4$. Because there are four completed cells, the solution is not degenerate.

Next, we develop MODI index numbers using the completed cells. We always begin by assigning an index of 0 to the first row. Then, for any completed cell in the first row, the column index number is the same as the cell cost. Because cell 1-B has a cell cost of 6, the column B index number is 6. Using this value and the cell cost for completed cell 2-B enables us to find the index number for row 2: Cell cost − Column index = Row index. Thus, $3 - 6 = -3$. This row 2 index number can be used with the cell cost of completed cell 2-A to find the column A index number, $5 - (-3) = 8$, and the index of 8 can be used with the cell cost of dummy-A to find the dummy index number: $0 - 8 = -8$. These index numbers are shown in the following table.

From: \ To:	A (8)	B (6)	Supply
1 (0)	[9]	[6] 75	75
2 (−3)	[5] 60	[3] 15	75
Dummy (−8)	[0] 20	[0]	20
Demand	80	90	170 \ 170

Evaluations for the empty cells can now be determined.

Cell 1-A: $9 - (0 + 8) = 1$

Cell dummy-B: $0 - (-8 + 6) = 2$

Because there are no negative evaluations, the current solution is optimal: Ship 75 units from 1 to B, 60 units from 2 to A, and 15 units from 2 to B. The 20 units in dummy-A means that destination A will be 20 units short of satisfying its demand of 80 units.

The total cost of the optimal solution is $795:

$$
\begin{aligned}
75(6) &= 450 \\
60(5) &= 300 \\
15(3) &= 45 \\
20(0) &= \underline{0} \\
 &\ 795
\end{aligned}
$$

Problem 2

A firm that specializes in nonferrous casting must decide between locating a new foundry in Chicago or Detroit. Transportation costs from each proposed location to existing warehouses and transportation costs for existing locations for a certain type of casting are shown in the following table. Each will be able to supply 2,500 units per month. Explain how to determine which location should be chosen on a transportation cost basis.

From Detroit to	Cost per Unit	From Chicago to	Cost per Unit
A	$10	A	$12
B	8	B	13
C	15	C	5

From: Foundry \ To: Warehouse	A	B	C	Supply (hundreds of units/month)
1	17	10	6	30
2	7	12	14	20
Demand (hundreds of units/month)	25	10	40	

Solution

This decision requires solving two transportation problems, one using Detroit costs and the other using Chicago costs.

a. Set up the two transportation matrices.

	Warehouse A	B	C	Supply
1	17	10	6	30
2	7	12	14	20
Detroit	10	8	15	25
Demand	25	10	40	

	Warehouse A	B	C	Supply
1	17	10	6	30
2	7	12	14	20
Chicago	12	13	5	25
Demand	25	10	40	

b. Solve each problem.

c. The problem that yields the lowest total cost will indicate the preferred location.

Discussion and Review Questions

1. What information do you need to use the transportation model?
2. What check must you make before proceeding to develop an initial solution? What corrective action might need to be taken?
3. Would it ever make sense to have a situation that required both a dummy row and a dummy column? Explain briefly.
4. What is linear about the objective function in a transportation problem?
5. Why do the + and − signs alternate in cell evaluation paths?
6. How do you know if a solution is optimal?
7. What does a zero value for an evaluation path indicate?
8. If a solution is not optimal:
   a. How do you decide which empty cell to shift units into?
   b. How do you know which cells to include in the shifting operation?
   c. How do you decide how many units to shift?
9. What is meant by the term *degeneracy*? How is it overcome?
10. How is the transportation model useful in location decisions? How does use of the transportation model in location decisions differ from its use as a tool to develop an optimal distribution plan for a given set of sources and destinations?
11. How are total costs determined for a given distribution plan?
12. What interpretation is given to quantities allocated to dummy destinations? Dummy sources?
13. What is the MODI method? How does it differ from the stepping-stone method?

Problems

1. Obtain the optimal distribution plan for the following transportation problem. Develop the initial solution using the intuitive lowest-cost approach.
   Use the stepping-stone method for cell evaluations.

From \ To:	A	B	C	Supply
1	3	4	2	40
2	5	1	7	60
3	8	7	4	50
Demand	30	45	75	150 / 150

a, b.

	A	B	C
1	15		25
2	15	45	
3			50

TC = $415

2. Use the following transportation table to do the following:
   a. Find the optimal solution. Use the intuitive method for the initial solution and MODI for evaluation.
   b. What is the total cost?
   c. Is there an alternate optimal solution? If so, what is it?

a.

40		
	50	
15	5	45

b. TC = $485
c. No alternate solution

From: \ To:	1	2	3	Supply
1	[3]	[6]	[2]	40
2	[3]	[1]	[3]	50
3	[7]	[6]	[4]	65
Demand	55	55	45	

3. Given the following transportation problem, do each of the following:
   a. Develop an initial feasible solution using the intuitive approach.
   b. Use the stepping-stone method to evaluate the solution.
   c. Repeat part *b* using MODI.
   d. Find the optimal solution.

a.

	A	B	C	D
1		40		
2	80			
3	10	40	30	50

TC = $6,330

b, c, d.

	A	B	C	D
1	10	30		
2	80			
3		50	30	50

TC = $6,310

From: \ To:	A	B	C	D	Supply
1	[18]	[12]	[14]	[16]	40
2	[23]	[24]	[27]	[33]	80
3	[42]	[34]	[31]	[26]	130
Demand	90	80	30	50	

4. Determine the optimal distribution plan for the following problem, and compute the total cost for that plan. Use the MODI method for cell evaluation.

	A	B	C	D	Dummy
1	13		35		
2	28	2		20	6
3		32			

TC = $2,156

From: \ To:	A	B	C	D	Supply
1	[14]	[24]	[18]	[28]	48
2	[17]	[18]	[25]	[16]	56
3	[30]	[16]	[22]	[30]	32
Demand	41	34	35	20	

	A	B	C	D
1		40		
2	80			
3		30	30	50
Dummy	10	10		

TC = $5,570

5. Refer to problem 3. Due to a temporary condition, origin 3 has experienced a capacity reduction of 20 units per period. Using the intuitive method to develop an initial solution, determine a temporary distribution plan that will minimize transportation costs.

6. Refer to problem 4. The market supplied by destination D is experiencing a period of rapid growth. Projected demand is for 60 units per period. A new factory with a capacity of 50 units is planned for one of two locations, Baltimore or Philadelphia. Transportation costs are as follows. Which location would result in the lower total cost? Use the MODI method for cell evaluation.

	A	B	C	D	Dummy
1	41		7		
2				56	
3		32			
Balt.		2	28	4	16

TC = $2,864

	A	B	C	D	Dummy
1	41		7		
2		2		54	
3		32			
Ph.			28	6	16

TC = $2,764

From Baltimore to	Cost per Unit	From Philadelphia to	Cost per Unit
A	$18	A	$31
B	16	B	25
C	22	C	19
D	27	D	20

a.

70		30
	90	10
		80

b. TC = $1,310
c. No alternate solution

7. Use the following transportation table to do the following:
   a. Find the optimal solution. Use MODI for evaluation.
   b. What is the total cost?
   c. Is there an alternate optimal solution? If so, what is it?

From: \ To:	Rochester	Detroit	Toronto	Supply
Cleveland	6	4	8	100
Chicago	7	2	7	100
Buffalo	4	4	5	80
Demand	70	90	120	

	A	B	C	Dummy
1		500		
2	400			
N1		100	350	50

TC = $6,750

	A	B	C	Dummy
1		500		
2	400			
N2		100	350	50

TC = $5,500

8. A toy manufacturer wants to open a third warehouse that will supply three retail outlets. The new warehouse will supply 500 units of backyard playsets per week. Two locations are being studied, N1 and N2. Transportation costs for location N1 to stores A, B, and C are $6, $8, and $7, respectively; for location N2, the costs are $10, $6, and $4, respectively. The existing system is shown in the following table. Which location would result in the lower transportation costs for the system?

Warehouse From: \ Store To:	A	B	C	Capacity (Units/week)
1	8	3	7	500
2	5	10	9	400
Demand (Units/week)	400	600	350	

	A	B	C
1			210
2	140		
3	80	60	10
Tol.		160	

TC = $6,720

9. A large firm is contemplating construction of a new manufacturing facility. The two leading locations are Toledo and Cincinnati. The new factory would have a supply capacity of 160 units per week. Transportation costs from each potential location and existing locations are shown in the following table. Determine which location would provide the lower transportation costs.

From Toledo to	Cost per Unit	From Cincinnati to	Cost per Unit
A.	$18	A	$ 7
B.	8	B	17
C.	13	C.	13

	A	B	C	Supply (Units/week)
1	10	14	10	210
2	12	17	20	140
3	11	11	12	150
Demand (Units/week)	220	220	220	

	A	B	C
1			210
2	60	80	
3		140	10
Cin.	160		

TC = $6,960

10. A large retailer is planning to open a new store. Three locations in California are currently under consideration: South Coast Plaza (SCP), Fashion Island (FI), and Laguna Hills (LH). Transportation costs for the locations, and costs, demands, and supplies for existing locations and warehouses (origins), are shown below. Each of the locations has a demand potential of 300 units per week. Which location would yield the lowest transportation costs for the system?

South Coast:

	A	B	C
1		500	160
2	340		
3	60		140

TC = $10,080

From Warehouse	TO SCP	FI	LH
1	$ 4	$7	$5
2	11	6	5
3	5	5	6

Fashion Island:

	A	B	C
1	60	500	100
2	340		
3			200

TC = $10,500

Laguna Hill:

Same as South Coast, but TC = $10,380

	A	B	Supply (Units/week)
1	15	9	660
2	10	7	340
3	14	18	200
Demand (Units/week)	400	500	

11. Set up problem 1 in a general linear programming format using an objective function and a set of constraints. See IM

12. Set up problem 3 in a general linear programming format using an objective function and a set of constraints. See IM

13. A soft drink manufacturer has recently begun negotiations with brokers in areas where it intends to distribute new products. Before making final agreements, however, the firm wants to determine shipping routes and costs. The firm has three plants with capacities as follows:

Plant	Capacity (cases per week)
Metro	40,000
Ridge	30,000
Colby	25,000

a.

	RS1	RS2	RS3	RS4	RS5
Metro		1	23	16	
Ridge		20			10
Colby	24	1			

TC = 67.8

b.

	RS1	RS2	RS3	RS4	RS5
Metro		17	23		
Ridge		4		16	10
Colby	24	1			

TC = 67

c. .8 ($800)

Estimated demands in each of the warehouse localities are:

Warehouse	Demand (cases per week)
SR1	24,000
SR2	22,000
SR3	23,000
SR4	16,000
SR5	10,000

The estimated shipping costs per case for the various routes are:

	TO				
From	**RS1**	**RS2**	**RS3**	**RS4**	**RS5**
Metro	.80	.75	.60	.70	.90
Ridge	.75	.80	.85	.70	.85
Colby	.70	.75	.70	.80	.80

Determine the optimal shipping plan that will minimize total shipping cost under these conditions:

a. Route Ridge-RS4 is unacceptable.

b. All routes are acceptable.

c. What is the additional cost of the Ridge-RS4 route not being acceptable?

	1	2	3
1		15	75
2	105		
3	45	60	

TC = $1,050

14. Solve this LP problem using the transportation method. Find the optimal transportation plan and the minimum cost. Also, decide if there is an alternate solution. If there is one, identify it.

$$\text{Minimize } 8x_{11} + 2x_{12} + 5x_{13} + 2x_{21} + x_{22} + 3x_{23} + 7x_{31} + 2x_{32} + 6x_{33}$$

$$\begin{aligned} \text{Subject to } x_{11} + x_{12} + x_{13} &= 90 \\ x_{21} + x_{22} + x_{23} &= 105 \\ x_{31} + x_{32} + x_{33} &= 105 \\ x_{11} + x_{21} + x_{31} &= 150 \\ x_{12} + x_{22} + x_{32} &= 75 \\ x_{13} + x_{23} + x_{33} &= 75 \\ \text{All variables} &\geq 0 \end{aligned}$$

15. Write an objective function and a set of constraints for problem 2.

Selected Bibliography

Bierman, Harold; Charles P. Bonini; and Warren H. Hausman. *Quantitative Analysis for Business Decisions.* 8th ed. Burr Ridge, IL: Richard D. Irwin, 1991.

Eppen, G. D.; F. J. Gould; and C. P. Schmidt. *Introductory Management Science.* 2nd ed. Englewood Cliffs, NJ: Prentice Hall, 1987.

Levin, Richard I.; Charles A. Kirkpatrick; and David Rubin. *Quantitative Approaches to Management.* 6th ed. New York: McGraw-Hill, 1987.

Stevenson, William J. *Introduction to Management Science.* 2nd ed. Burr Ridge, IL: Richard D. Irwin, 1992.

PART FOUR
QUALITY

There are three chapters in this part:

1. Introduction to Quality, Chapter 9
2. Quality Control, Chapter 10
3. Total Quality Management (TQM), Chapter 11

The topics in this part relate to quality. Chapter 9 introduces quality concepts and the thinking of quality "gurus." Chapter 10 explains quality control procedures. Chapter 11 describes total quality management, problem solving, and process improvement.

the things you've

You know that feeling, when you can rely on your family, or put your trust in an old friend. That confident,

come to rely on the most

assured feeling of knowing they will be there for you. This unique feeling comes standard with every Camry,

are the things you

along with a comfortable interior, advanced safety design features and quality workmanship. It is a feeling

think about the least.

that comes from reliability and, once you own a Camry, it is one you will never forget.

CAMRY

TOYOTA | everyday

1800-GO-TOYOTA ✦ www.toyota.com

LEARNING OBJECTIVES

After completing this chapter, you should be able to:

1. Define the term *quality.*
2. Explain why quality is important, and consequences of poor quality.
3. Discuss the determinants of quality.
4. List and describe the various costs associated with quality.
5. Discuss ISO 9000 and ISO 14000.
6. Discuss the philosophies of quality gurus.

CHAPTER NINE
INTRODUCTION TO QUALITY

CHAPTER OUTLINE

This chapter is the first of three chapters on quality. In this chapter you will learn about the evolution of quality management, definitions of quality, the costs of quality and the consequences of poor quality, and you will learn about some quality awards and standards.

INTRODUCTION

quality The ability of a product or service to consistently meet or exceed customer expectations.

Broadly defined, **quality** refers to the ability of a product or service to consistently meet or exceed customer expectations. Quality means getting what you pay for. Prior to the increased level of Japanese competition in the U.S. marketplace in the 1970s and 1980s, customers often got less than they paid for with U.S.-made products; quality was not uppermost in the minds of U.S. business organizations. They tended to focus on cost and productivity rather than on quality. It wasn't that quality was *un*important, it just wasn't *very* important.

www.honda.co.jp
www.nissan.co.jp
www.toyota.co.jp

Partly because of this thinking, foreign companies, many of them Japanese, captured a significant share of the U.S. market. In the automotive sector, leading Japanese manufacturers, Honda, Nissan, and Toyota, became major players in the auto sales market in the United States. Both Honda and Toyota have built a reputation for quality and reliability in their cars.

Many U.S. companies changed their views about quality after that, and changed them drastically. Stung by the success of foreign competitors, they embraced quality in a big way. They hired consultants, sent their people (including top executives) to seminars, and initiated a vast array of quality improvement programs. These companies clearly have recognized the importance of quality and that quality isn't something that is tacked on as a special feature, but an *integral part* of a product or service. In the 1990s, U.S. automakers began to close the quality gap.

THE EVOLUTION OF QUALITY MANAGEMENT

Prior to the Industrial Revolution, skilled craftsmen performed all stages of production. Pride of workmanship and reputation often provided the motivation to see that a job was done right. Lengthy guild apprenticeships caused this attitude to carry over to new workers. Moreover, one person or a small group of people were responsible for an entire product.

A division of labor accompanied the Industrial Revolution; each worker was then responsible for only a small portion of each product. Pride of workmanship became less meaningful because workers could no longer identify readily with the final product. The responsibility for quality control shifted to the foremen. Inspection was either nonexistent or haphazard, although in some instances 100 percent inspection was used.

Frederick Winslow Taylor, the "Father of Scientific Management," gave new emphasis to quality by including product inspection and gauging in his list of fundamental areas of manufacturing management. G. S. Radford improved Taylor's methods. Two of his most significant contributions were the notions of involving quality considerations early in the product design stage and making connections between high quality, increased productivity, and lower costs.

In 1924, W. Shewhart of Bell Telephone Laboratories introduced statistical control charts that could be used to monitor production. Around 1930, H. F. Dodge and H. G. Romig, also of Bell Labs, introduced tables for acceptance sampling. Nevertheless, statistical quality control procedures were not widely used until World War II, when the U.S. government began to require vendors to use them.

World War II caused a dramatic increase in emphasis on quality control. The U.S. Army refined sampling techniques for dealing with large shipments of arms from many suppliers. By the end of the 1940s, the U.S. Army, Bell Labs, and major universities were training engineers in other industries in the use of statistical sampling techniques. About the same time, professional quality organizations were emerging throughout the country. One of these organizations was the American Society for Quality Control (ASQC, now known as ASQ). Over the years, the society has promoted quality with its publications, seminars and conferences, and training programs.

During the 1950s, the quality movement evolved into quality assurance. Quality guru W. Edwards Deming introduced statistical quality control methods to Japanese manufacturers, promising that this would help them to rebuild their manufacturing base and to compete in world markets. Prior to that time, much output of Japanese factories was of inferior quality.

At about the same time, another quality guru, Joseph Juran, began his "cost of quality" approach, emphasizing accurate and complete identification and measurement of the costs of quality. He stressed the desirability of lowering costs associated with prevention, and advocated using quality control techniques to accomplish this.

In the mid-1950s, Armand Feigenbaum proposed *total quality control,* which enlarged the realm of quality efforts from its primary focus on manufacturing to also include product design and incoming raw materials. One important feature of his work was greater involvement of upper management in quality.

During the 1960s, the concept of "zero defects" gained favor. Championed by quality guru Philip Crosby, this approach focused on employee motivation and awareness, and the expectation of perfection from each employee. It evolved from the success of the Martin Company in producing a "perfect" missile for the U.S. Army.

In the 1970s, quality assurance methods gained increasing emphasis in services including government operations, health care, banking, and the travel industry.

The evolution of quality took a dramatic shift from quality assurance to a strategic approach to quality in the late 1970s. Up until that time, the main emphasis had been on finding and correcting defectives before they reached the market. It was still a reactive approach. The strategic approach, advocated by Harvard professor David Garvin and others, is proactive, focusing on preventing mistakes from occurring altogether. Quality and profits are more closely linked. This approach also places greater emphasis on customer satisfaction, and involves all levels of management as well as workers in a continuing effort to increase quality.

QUALITY: THE BASICS

Any serious effort to deal with quality must begin with a clear understanding of the meaning of the word *quality.*

The Dimensions of Quality

The term *quality* is used in a variety of ways. Sometimes it refers to the grade of a product, such as "USDA Choice," or "grade A" eggs. At other times, it refers to materials, workmanship, or special features, such as "waterproof," or "subtle aroma." And sometimes it is related to price, as in "cheap" or "expensive."

The implication in these various connotations of quality is that customers value certain aspects of a product or service, and therefore associate those aspects with the quality that they perceive a product or service has. In fact, it may be obvious that from a customer perspective, quality does not pertain to a single aspect of a product or service, but to a number of different *dimensions* of the product or service.

Although they may vary somewhat from product to product, or between a product and a service, generally speaking, the **dimensions of quality** include:*

dimensions of quality Performance, aesthetics, special features, conformance, safety, reliability, durability, perceived quality, and service after sale.

Performance—main characteristics of the product or service.

Aesthetics—appearance, feel, smell, taste.

Special features—extra characteristics.

Conformance—how well a product or service corresponds to the customer's expectations.

Safety—risk of injury or harm.

Reliability—consistency of performance.

Durability—the useful life of the product or service.

*Adapted from David Garvin, "What does quality really mean?" *Sloan Management Review* 26, no. 1 (1984), pp. 25–43.

TABLE 9–1

Examples of some of the quality dimensions for a product and a service

Dimension	(Product) Automobile	(Service) Automobile Repair
1. Performance	Everything works, fit and finish	All work done, at agreed price
	Ride, handling, grade of materials used	Friendliness, courtesy
		Competency, quickness
2. Aesthetics	Interior design, soft touch	Clean work/waiting areas
3. Special features		
Convenience	Placement of gauges and controls	Location, call when ready
High tech	Cellular phone, CD player	Computer diagnostics
4. Safety	Antilock brakes, airbags	Separate waiting area
5. Reliability	Infrequency of breakdowns	Work done correctly, ready when promised
6. Durability	Useful life in miles, resistance to rust and corrosion	Work holds up over time
7. Perceived quality	Top-rated car	Award-winning service department
8. Service after sale	Handling of complaints and/or requests for information	Handling of complaints

Perceived quality—indirect evaluation of quality (e.g., reputation).

Service after sale—handling of complaints or checking on customer satisfaction.

These dimensions are further described by the examples presented in Table 9–1. When referring to a product, a customer sometimes judges the first four dimensions by its *fitness for use.*

The Determinants of Quality

The degree to which a product or a service successfully satisfies its intended purpose has four primary determinants. They are:

1. Design.
2. How well it conforms to the design.
3. Ease of use.
4. Service after delivery.

quality of design Intention of designers to include or exclude features in a product or service.

The design phase is the starting point for the level of quality eventually achieved. Design involves decisions about the specific characteristics of a product or service such as size, shape, and location. **Quality of design** refers to the intention of designers to include or exclude certain features in a product or service. For example, many different models of automobiles are on the market today. They differ in size, appearance, roominess, fuel economy, comfort, and materials used. These differences reflect choices made by designers that determine the quality of design. Design decisions must take into account customer wants, production or service capabilities, safety and liability (both during production and after delivery), costs, and other similar considerations.

focus group In market research, a group of consumers who express their opinions of a product or service.

Designers may determine customer wants from information provided by marketing, perhaps through the use of consumer surveys or other market research. Marketing may organize **focus groups** of consumers to express their views on a product or service (what they like and don't like, and what they would like to have).

Designers must work closely with representatives of operations to ascertain that designs are manufacturable; that is, that production or service has the equipment, capacity, and skills necessary to produce or provide a particular design.

A poor design can result in difficulties in production or service. For example, materials might be difficult to obtain, specifications difficult to meet, or procedures difficult

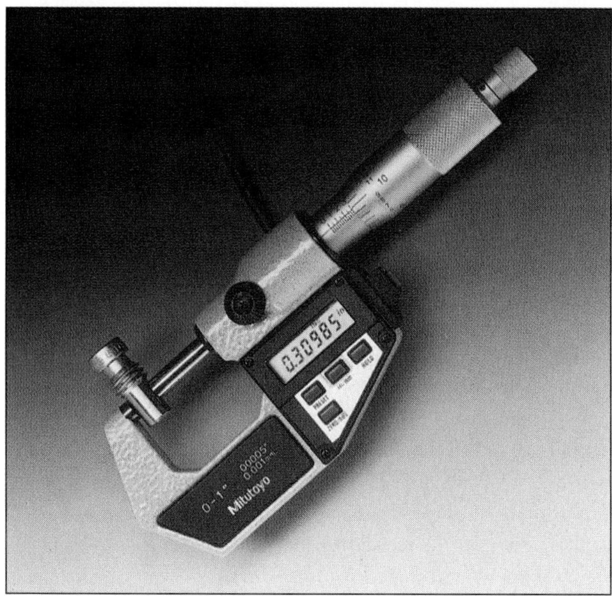

Digital micrometers collect precise measurements and send the information directly into a computer database.

www.mitutoyo.com.sg/profile.htm

to follow. Moreover, if a design is inadequate or inappropriate for the circumstances, the best workmanship in the world may not be enough to achieve the desired quality. Also, we cannot expect a worker to achieve good results if the given tools or procedures are inadequate. Similarly, a superior design usually cannot offset poor workmanship.

Quality of conformance refers to the degree to which goods and services conform to (i.e., *achieve*) the intent of the designers. This is affected by factors such as the capability of equipment used; the skills, training, and motivation of workers; the extent to which the design lends itself to production; the monitoring process to assess conformance; and the taking of corrective action (e.g., through problem solving) when necessary.

quality of conformance The degree to which goods or services conform to the intent of the designers.

The determination of quality does not stop once the product or service has been sold or delivered. *Ease of use* and user instructions are important. They increase the chances, but do not guarantee, that a product will be used for its intended purposes and in such a way that it will continue to function properly and safely. (Companies involved in liability litigation often argue that injuries and damages occurred because the user misused the product.) Much of the same reasoning can be applied to services. Customers, patients, clients, or other users must be clearly informed on what they should or should not do; otherwise, there is the danger that they will take some action that will adversely affect quality. Some examples include the doctor who fails to specify that a medication should be taken *before* meals and *not* with orange juice and the attorney who neglects to inform a client of a deadline for filing a claim.

Much consumer education takes the form of printed instructions and labeling. Thus, manufacturers must ensure that directions for unpacking, assembling, using, maintaining, and adjusting the product—and what to do if something goes wrong (e.g., flush eyes with water, call a physician, induce vomiting, do not induce vomiting, disconnect set immediately)—are *clearly visible* and *easily understood.*

For a variety of reasons, products do not always perform as expected, and services do not always yield the desired results. Whatever the reason, it is important from a quality standpoint to remedy the situation—through recall and repair of the product, adjustment, replacement or buyback, or reevaluation of a service—and do whatever is necessary to bring the product or service up to standard.

The Consequences of Poor Quality

It is important for management to recognize the different ways that the quality of a firm's products or services can affect the organization and to take these into account in developing and maintaining a quality assurance program. Some of the major ways that quality affects an organization are:

1. Loss of business
2. Liability
3. Productivity
4. Costs

Poor designs or defective products or services can result in *loss of business.* Failure to devote adequate attention to quality can damage a profit-oriented organization's image and lead to a decreased share of the market, or it can lead to increased criticism and/or controls for a government agency or nonprofit organization.

> A potentially . . . devastating consequence to the bottom line is the reaction of the consumer who receives a defective or otherwise unsatisfactory product or service. A recent study showed that, while a satisfied customer will tell a few people about his or her experience, a dissatisfied person will tell an average of 19 others.
>
> Unfortunately, the company is usually the last to know of dissatisfaction. People rarely complain directly to the provider of poor quality goods and services. In fact, studies suggest that people usually complain, if at all, to their most immediate contact (e.g., a salesperson or service manager) and that these complaints are rarely transmitted further. A more common response is simply to switch to a competing product or service. Typically, formal complaints are received from less than 5 percent of dissatisfied customers.—*Source: The Ernst & Young Quality Improvement Consulting Group,* Total Quality: An Executive's Guide for the 1990s *(Burr Ridge, IL: Irwin Professional Publishing: 1990), pp. 6–7.*

Organizations must pay special attention to their potential *liability* due to damages or injuries resulting from either faulty design or poor workmanship. This applies to both products and services. Thus, a poorly designed steering arm on a car might cause the driver to lose control of the car, but so could improper assembly of the steering arm. However, the net result is the same. Similarly, a tree surgeon might be called to cable a tree limb. If the limb later falls and causes damage to a neighbor's car, the accident might be traced to a poorly designed procedure for cabling or to improper workmanship. Liability for poor quality has been well established in the courts. An organization's liability costs can often be substantial, especially if large numbers of items are involved, as in the automobile industry, or if potentially widespread injury or damage is involved (e.g., an accident at a nuclear power plant). Express written warranties as well as implied warranties generally guarantee the product as safe when used as intended. The courts have tended to extend this to *foreseeable* uses, even if these uses were not intended by the producer. In the health care field, medical malpractice claims and insurance costs are contributing to skyrocketing costs and have become a major issue nationwide.

Productivity and quality are often closely related. Poor quality can adversely affect productivity during the manufacturing process if parts are defective and have to be reworked or if an assembler has to try a number of parts before finding one that fits properly. Also, poor quality in tools and equipment can lead to injuries and defective output, which must be reworked or scrapped, thereby reducing the amount of usable output for a given amount of input. Similarly, poor service can mean having to re-do the service and reduce service productivity. Conversely, improving and maintaining good quality can have a positive effect on productivity.

Poor quality increases certain *costs* incurred by the organization. The following section provides further detail on costs associated with quality.

The Costs of Quality

Any serious attempt to deal with quality issues must take into account the costs associated with quality. Those costs can be classified into four categories:

Internal failure costs
External failure costs
Appraisal costs
Prevention costs

TABLE 9–2

Summary of quality costs

Category	Description	Examples
Internal failure costs	Costs related to defective products or services before they are delivered to customers	Rework costs, problem solving, material and product losses, scrap, and downtime
External failure costs	Costs related to delivering substandard products or services to customers	Returned goods, reworking costs, warranty costs, loss of goodwill, liability claims, and penalties
Appraisal costs	Costs related to measuring, evaluating, and auditing materials, parts, products, and services to assess conformance with quality standards	Inspection equipment, testing, labs, inspectors, and the interruption of production to take samples
Prevention costs	Costs related to reducing the potential for quality problems	Quality improvement programs, training, monitoring, data collection and analysis, and design costs

Failure costs are incurred by defective parts or products, or faulty services. **Internal failures** are those discovered during the production process; **external failures** are those discovered after delivery to the customer. Internal failures occur for a variety of reasons, including defective material from vendors, incorrect machine settings, faulty equipment, incorrect methods, incorrect processing, carelessness, and faulty or improper material handling procedures. The costs of internal failures include lost production time, scrap and rework, investigation costs, possible equipment damage, and possible employee injury. Rework costs involve the salaries of workers and the additional resources needed to perform the rework (e.g., equipment, energy, raw materials). Beyond those costs are items such as inspection of reworked parts, disruption of schedules, the added costs of parts and materials in inventory waiting for reworked parts, and the paperwork needed to keep track of the items until they can be reintegrated into the process. External failures are defectives or poor service that go undetected by the producer. Resulting costs include warranty work, handling of complaints, replacements, liability/litigation, payments to customers or discounts used to offset the inferior quality, loss of customer goodwill, and opportunity costs related to lost sales. Table 9–2 summarizes failure costs.

failure costs Costs caused by defective parts or products, or faulty services.

internal failures Failures discovered during production.

external failures Failures discovered after delivery to the customer.

Appraisal costs relate to inspection, testing, and other activities intended to uncover defective products or services, or to assure that there are no defectives. They include the cost of inspectors, testing, test equipment, labs, quality audits, and field testing.

appraisal costs Costs of activities designed to ensure quality or uncover defects.

Prevention costs relate to attempts to prevent defects from occurring. They include costs such as planning and administration systems, working with vendors, training, quality control procedures, and extra attention in both the design and production phases to decrease the probability of defective workmanship.

prevention costs Costs of preventing defects from occurring.

The traditional view of quality cost for a long time was that there is an optimal level of quality costs—a level that will balance the cost of allowing some poor quality to exist and the cost associated with achieving that degree of quality. The idea was that the cost of achieving a higher level of quality would exceed the benefit of that higher level. The current view is that the cost of failure is much higher than the cost of prevention, so that the optimal quality level is zero defects. In this view, no matter how high the level of quality, unless it is perfect, the cost of achieving an even higher level will be less than the gain realized by doing so.

QUALITY GURUS

A core of quality experts has shaped modern quality practices. Among the most famous of this core of "gurus" are Deming, Juran, Feigenbaum, Ishikawa, Taguchi, and Crosby. Together, these great thinkers have had a tremendous impact on the management and control of quality, and the way companies operate.

TABLE 9-3
Deming's 14 points

1. Create constancy of purpose toward improvement of product and service with a plan to become competitive and to stay in business. Decide to whom top management is responsible.
2. Adopt the new philosophy. We are in a new economic age. We can no longer live with commonly accepted levels of delays, mistakes, defective materials, and defective workmanship.
3. Cease dependence on mass inspection. Require, instead, statistical evidence that quality is built in. (*Prevent* defects rather than *detect* defects.)
4. End the practice of awarding business on the basis of price tag. Instead, depend on meaningful measures of quality along with price. Eliminate suppliers that cannot qualify with statistical evidence of quality.
5. Find problems. It is management's job to work continually on the system (design, incoming materials, composition of material, maintenance, improvement of machine, training, supervision, retraining).
6. Institute modern methods of training on the job.
7. The responsibility of foremen must be changed from sheer numbers to quality . . . [which] will automatically improve productivity. Management must prepare to take immediate action on reports from foremen concerning barriers such as inherent defects, machines not maintained, poor tools, and fuzzy operational definitions.
8. Drive out fear, so that everyone may work effectively for the company.
9. Break down barriers between departments. People in research, design, sales, and production must work as a team to foresee problems of production that may be encountered with various materials and specifications.
10. Eliminate numerical goals, posters, and slogans for the work force, asking for new levels of productivity without providing methods.
11. Eliminate work standards that prescribe numerical quotas.
12. Remove barriers that stand between the hourly worker and his right to pride of workmanship.
13. Institute a vigorous program of education and retraining.
14. Create a structure in top management that will push every day on the above 13 points.

Source: W. Edwards Deming, *Quality, Productivity, and Competitive Position* (Cambridge, MA: MIT, Center for Advanced Engineering Study, 1982), pp. 16–17.

W. Edwards Deming

Deming was the senior guru. A statistics professor at New York University in the 1940s, he went to Japan after World War II to assist the Japanese in improving quality and productivity. The Japanese were so impressed that in 1951, after a series of lectures presented by Deming, they established the **Deming Prize,** which is awarded annually to firms that distinguish themselves with quality management programs.

Deming Prize Prize established by the Japanese and awarded annually to firms that distinguish themselves with quality management programs.

Although the Japanese revered Deming, he was largely unknown to business leaders in the United States. In fact, he worked with the Japanese for almost 30 years before he gained recognition in his own country. Before his death in 1993, U.S. companies turned their attention to Deming, embraced his philosophy, and requested his assistance in setting up quality improvement programs.

Deming compiled a famous list of 14 points he believed were the prescription needed to achieve quality in an organization (see Table 9–3). His message is that the cause of inefficiency and poor quality is the *system*, not the employees. *Management's responsibility* is to correct the system to achieve the desired results. In addition to the 14 points, Deming stressed the need to reduce variation in output (deviation from a standard), which can be accomplished by distinguishing between *special causes* of variation (i.e., correctable) and *common causes* of variation (i.e., random).

The key elements of Deming's 14 points are constancy of purpose, continual improvement, and profound knowledge. *Profound knowledge* involves (1) an appreciation for a system, (2) a theory of variation, (3) a theory of knowledge, and (4) psychology. Deming's concept of profound knowledge incorporates the beliefs and values about learning that guided Japan's rise to a world economic power.

Appreciation for the system is the starting point, and it refers to everyone in the organization working to achieve optimization. Towards that end, management must eliminate internal competition. Reducing *variation* is an important key to quality

W. Edwards Deming

Joseph Juran

Philip B. Crosby

improvement, but it is necessary to differentiate between random variation and correctable variation, and to focus on the latter. Deming believed that *knowledge* comes from theory, and that learning cannot occur within an organization without a theory of knowledge. Deming felt that *psychology* was the most powerful element of profound knowledge. He believed that workers want to create and learn, but that management unintentionally often does things such as establishing rating systems that rob them of their internal motivation. He believed that management's greatest challenge in achieving quality was in motivating workers to contribute their collective efforts to achieve a common goal. Lastly, Deming felt that in order to benefit from the concept of profound knowledge, it was necessary to embrace the concept in its entirety.

Joseph M. Juran

Juran, like Deming, taught Japanese manufacturers how to improve the quality of their goods, and he, too, can be regarded as a major force in Japan's success in quality. He made his first trip to Japan a few years after the publication in 1951 of his *Quality Control Handbook.* Juran's approach to quality may be the closest to Deming's of all the gurus, although his approach differs on the importance of statistical methods and what an organization must do to achieve quality. Whereas Deming's work envisioned a "transformation," Juran believes that an organization can manage for quality. He doesn't think that managing quality is as difficult as Deming thought, although he does admit that most quality programs that fail do so because the companies do not realize how difficult developing new processes can be. Juran also places less emphasis on statistical methods than Deming. It is his view that quality begins by knowing what customers want.

Juran views quality as fitness-for-use. He also believes that roughly 80 percent of quality defects are management controllable; thus, management has the responsibility to correct this deficiency. He describes quality management in terms of a *trilogy* consisting of quality planning, quality control, and quality improvement.[1] According to Juran, quality planning is necessary to establish processes that are *capable* of meeting quality standards; that quality control is necessary in order to know when corrective action is needed; and that quality improvement will help to find better ways of doing things. A key element of Juran's philosophy is the commitment of management to continual improvement.

Juran is credited as one of the first to measure the cost of quality, and he demonstrated the potential for increased profits that would result if the costs of poor quality could be reduced. Also, he made a series of videotapes entitled "Juran on Quality," which are available from the Juran Institute in Wilton, Connecticut.

Juran proposed 10 steps for quality improvement. These are shown in Table 9–4.

[1] J.M. Juran, "The Quality Trilogy," *Quality Progress* 10, no. 8 (August 1986, pp. 19–24).

TABLE 9–4

Juran's 10 steps for quality improvement

1. Build awareness for the need and opportunity for improvement.
2. Set goals for improvement.
3. Organize people to reach the goals.
4. Provide training throughout the organization.
5. Carry out projects to solve problems.
6. Report progress.
7. Give recognition.
8. Communicate results.
9. Keep score.
10. Maintain momentum by making annual improvement part of the regular systems and processes of the company.

Source: Joseph M. Juran, ed. *Quality Control Handbook* (New York: McGraw-Hill, 1951).

Armand Feigenbaum

Feigenbaum was instrumental in advancing the "cost of nonconformance" approach as a reason for management to commit to quality. At the age of 24, he was General Electric's top expert on quality. He recognized that quality was not simply a collection of tools and techniques, but a "total field." He saw that when improvements were made in a process, other areas of the company also achieved improvements. Feigenbaum's understanding of systems theory led him to create an environment in which people could learn from each other's successes, and his leadership and open work environment led to cross-functional teamwork.

In 1961, his book *Total Quality Control* was published, in which he layed out quality principles into 40 steps. Table 9–5 lists some of the key ideas that differentiate him from the other gurus.

According to Feigenbaum, it is the customer who defines quality. Deming would disagree; in his philosophy, companies should get to know their customers so well that they can anticipate their future needs.

Philip Crosby

Crosby worked at Martin Marietta in the 1960s. While he was there, he developed the concept of *zero defects*, and popularized the phrase, "Do it right the first time." He stressed prevention, and argued against the idea that "there will always be some level of defectives." He was the corporate vice president for quality for ITT in the 1970s and was instrumental in making quality a concern of top company executives. In 1979, his book *Quality Is Free* was published. The title is based on a quote from the CEO of ITT. The book explains quality concepts in simple terms.

In accordance with the concept of zero defects, Crosby believes that any level of defects is too high, and that management must install programs that help the organization move toward that goal. Among some of his key points are the following.[2]

1. Top management must demonstrate its commitment to quality and its willingness to give support to achieve good quality.
2. Management must be persistent in efforts to achieve good quality.
3. Management must spell out clearly what it wants in terms of quality and what workers must do to achieve that.
4. Make it (or do it) right the first time.

Unlike the other gurus, Crosby maintains that achieving quality can be relatively easy. His book *Quality without Tears: The Art of Hassle-Free Management* was published in

[2]Philip Crosby, *Quality without Tears: The Art of Hassle-Free Management* (New York: McGraw-Hill, 1984).

TABLE 9–5

Key elements of Feigenbaum's philosophy of quality control

1. Total quality control is a system for integrating quality development, maintenance, and improvement efforts in an organization that will enable engineering, marketing, production, and service to function at optimal economic levels while achieving customer satisfaction.
2. The "control" aspect of quality control should involve setting quality standards, appraising performance relative to these standards, taking corrective action when the standards are not met, and planning for improvement in the standards.
3. Factors that affect quality can be divided into two major categories: technological and human. The human factor is the more important one.
4. Operating quality costs can be divided into four categories: prevention costs, appraisal costs, internal failure costs, and external failure costs.
5. It is important to control quality at the source.

Source: Adapted from Peter Capezio and Debra Morehouse, *Taking the Mystery Out of TQM*, 2nd ed., © 1995, published by Career Press, Franklin Lakes, NJ, pp. 100–101.

1984. The quality-is-free concept is that the costs of poor quality are much greater than traditionally defined. According to Crosby, these costs are so great that rather than viewing quality efforts as costs, organizations should view them as a way to reduce costs, because the improvements generated by quality efforts will more than pay for themselves.

Kaoru Ishikawa

The late Japanese expert on quality was strongly influenced by both Deming and Juran, although he made significant contributions of his own to quality management. Among his key contributions were the development of the cause-and-effect diagram (also known as a fishbone diagram) for problem solving, and for implementing quality circles that involve workers in quality improvement. He was the first quality expert to call attention to the *internal customer*—the next person in the process. He was a strong proponent of the need for companies to have a shared vision in order to unite everyone in the organization in a common goal, and he is widely recognized for his efforts to make quality control "user friendly" for workers.

Genichi Taguchi

Taguchi is best known for the Taguchi loss function, which involves a formula for determining the cost of poor quality. The idea is that the deviation of a part from a standard causes a loss, and the combined effect of deviations of all parts from their standards can be large, even though each individual deviation is small. In contrast to Taguchi, Deming believed that it is impossible to determine the actual cost of the lack of quality, and Crosby believes that it would be difficult to apply the concept in most U.S. firms. Nonetheless, Taguchi's method is credited with helping the Ford Motor Company to reduce its warranty losses by achieving less variation in the output of transmissions.

Table 9–6 provides a summary of the important contributions of the gurus to modern quality management.

QUALITY AWARDS

Quality awards have been established to generate awareness and interest in quality. The Malcolm Baldrige Award and the Deming Prize are two well-known awards given annually to firms that have integrated quality management in their operations.

The Baldrige Award

In 1987, Congress passed the Malcolm Baldrige National Quality Improvement Act. This legislation was designed to inspire increased efforts on the part of businesses located in the United States to improve the quality of their products and services. Named

Vol. 1, Tape 1, Seg. 2

TABLE 9-6

A summary of key contributors to quality management

Contributor	Key Contributions
Deming	14 points; special versus common causes of variation
Juran	Quality is fitness-for-use; quality trilogy
Feigenbaum	Quality is a total field; the customer defines quality
Crosby	Quality is free; zero defects
Ishikawa	Cause-and-effect diagrams; quality circles
Taguchi	Taguchi loss function

after the late Malcolm Baldrige, an industrialist and former secretary of commerce, the annual **Baldrige Award** is administered by the National Institute of Standards and Technology.

Baldrige Award Annual award given by the U.S. government to recognize quality achievements of U.S. companies.

The purpose of the award competition is to stimulate efforts to improve quality, to recognize quality achievements of U.S. companies, and to publicize successful programs. A maximum of two awards are given annually in each of three categories: large manufacturer, large service organization, and small business (500 or fewer employees).

www.amesrubber.com

READING

PERCEPTION IS THE REALITY **Brian S. Moskal**

Ames Rubber Corp., a 1993 Malcolm Baldrige winner in the small-business category, wasn't always in a league of its own.

For almost 40 years since its founding in 1949 as a maker of typewriter rollers, it relied on the same "proven" approach to customer satisfaction and gauging its customers' needs. Informal surveys of customers by sales representatives and knowledge gained from being close to customers were the hallmarks of the program.

"Back in the 1980s," recalls Robert D. Dondero, vice president of sales and marketing, "we largely used reports from customers. Our customers would track such things as on-time delivery and quality in terms of percent good or defects per million. I'm not sure we did a good job of tracking their concerns and complaints."

Five years ago Ames realized this approach was outmoded. If the firm's then four-year-old, companywide Total Quality Management effort was to succeed, it would require a formal gauge of customer satisfaction.

"We had only a superficial view of our customers, and even that wasn't being used properly," says Mr. Dondero. "We didn't know how little we knew until we got educated. At the time we didn't know we weren't doing a good job in customer satisfaction."

Education came in 1990, when Ames hired Walker: Customer Satisfaction Measurements, Indianapolis. Although Ames understood that customer-satisfaction measurement was a key ingredient in its future business success, it needed a more comprehensive approach to measuring and interpreting customer-satisfaction requirements.

"It was clear to us," says Chuck Roberts, vice president of total quality, "that customer satisfaction was an imperative for us. We needed to know if we were meeting our customers' requirements. We also needed to know if our customers were afraid to tell us the truth about our performance and upset us."

Since Hamburg, N.J.-based Ames hired Walker it has been surprised to learn *perception is reality.*

"Your customers' perception of how you are doing is how they rate you, whether they truly understand or not," says Mr. Roberts.

After Ames made formal total-quality presentations and addressed issues of customer satisfaction with its customer base, follow-up surveys by Walker found that Ames' customer-satisfaction performance went up in the eyes of its customers.

"We learned about perceptions," says Mr. Dondero. "After the formal presentations, our actual delivery performance improved from 97 percent on-time to 98 percent on-time, but our customers said our overall delivery performance went from just *good* in a 1991 Walker customer satisfaction survey to *excellent* in a similar survey in 1992. Perception is reality."

Customer-satisfaction grades for Ames, a manufacturer of precision, high-performance products for the reproduction graphics industry, including rollers used in photocopying machines and rubber seals found in the axles of front-wheel-drive vehicles, improved in all categories in 1992 over 1991. Part of that improvement Mr. Roberts attributes to the survey process itself. "A well-thought-out, valid, and objective survey is good in itself to help improve customer satisfaction. Just the act of surveying says to the customer that you need them and their business is important to you."

Ames has also learned that too many surveys can annoy customers. "Too much surveying has a built-in inconvenience factor for the customer," stresses Mr. Dondero. Now routine customer contact surveys by sales representatives are done on a monthly basis, Ames' own customer survey is done semi-annually versus quarterly, and Walker's independent survey is carried out every 18 months.

Ames' customer surveys ask questions about quality, cost, delivery, and service, and take an in-depth approach to its line of 2,000 custom-made products.

On the technical-service side, the survey asks customers to rate Ames' ability to deliver prototypes from blueprints, translate those into preproduction models, and turn those into products.

This information has led Ames to expand its definition of customer service. "We underestimated the elements that go into customer satisfaction," admits Mr. Dondero. "We largely thought we did well in customer satisfaction if we delivered product on time with no defects."

Adds Mr. Roberts, "Before we revamped our survey techniques, we were largely surveying ourselves to learn how well we were satisfying our customers' needs. Now we survey the customer on how well we are doing as their supplier."

The collaboration between Ames and Walker changed Ames' understanding of customer wants.

"Walker helped us to define the value components of our customer satisfaction," says Mr. Dondero. "We learned how to define value, measure it, and improve the value of our products in our customers' eyes. Companies sell and market their products from a whole host of different directions. We sell on the basis of value. We may not be the lowest price. But we do bring to the table good technology, supported by good manufacturing capability. That's our value message."

The $50 million, 450-employee company has a customer-satisfaction team that takes results from the surveys and melds them into "actionable activities" to improve performance. Ames takes all of the statistical data and plots out performance ratios that tell how important each issue is to its customers and then uses that information to allocate resources for continuous improvement.

For example, "We thought we were a leader in applying advanced technology, but our customers didn't think so in the surveys," recalls Mr. Roberts. "Our customers perceived that we weren't as good as we were until we took steps to inform them about our use of technology in meeting their requirements."

In addition, although the quantitative data suggested that price was the third most important driver of Ames' customer loyalty—behind total quality and innovation—the qualitative analysis offered more insight. Ames learned that product price emerges at the forefront because customers expect a high level of quality and innovation before even considering precision-component suppliers. The research also showed that in addition to their immediate requirements, customers looked to Ames for innovative solutions to such long-term needs as parts refabrication and the development of environmentally friendly, water-based adhesives.

This information led Ames to divide its technical-services group into two distinct units. One group of 15 to 18 people works to satisfy evolutionary, day-to-day customer needs. A second group is now charged with product planning and development and with making revolutionary, next-generation steps to meet customer needs.

"We separated them to be proactive to our customers," says Mr. Roberts. "Now, product planning is housed in a separate building and does not get involved in everyday firefighting. In the past technical services could crash on a one-day problem and R&D would be put on a back burner."

Ames' product-planning-and-development group has partnered with several customers to create a host of new products and processes, leading to several key patents in the reprographics market area. "We solved some key technology problems—ways of applying material more effectively and reducing cost, cycle time, and scrap as a result," observes Mr. Dondero.

So what's the connection between the Baldrige Award and the customer surveys?

"There is a direct connection between customer satisfaction and the Baldrige criteria," says Mr. Roberts. "We couldn't have won the award unless we scored high in the customer-satisfaction area."

Businesses that compete for the award are required to submit an application of no more than 75 pages documenting their quality systems. Those who pass an initial screening must undergo a more intense evaluation by examiners from government and industry, and consultants. The examination includes an on-site visit. Applicants are evaluated in seven main areas: leadership, information and analysis, strategic planning, human resource management, quality assurance of products and services, quality results, and customer satisfaction. Table 9–7 contains a list of the items included in each area and the maximum number of points that evaluators can assign. Note that customer satisfaction has a large number of points.

Examiners check the extent to which top management incorporates quality values in daily management; whether products or services are at least as good as those of competitors; whether employees receive training in quality techniques; if the business works with suppliers to improve quality; and if customers are satisfied. All applicants receive a written summary of the strengths and weaknesses of their quality management and suggestions for improvement.

The Baldrige Award has been both praised and criticized.[3] Among the praises are: It has raised the quality-consciousness of U.S. businesses, those who compete find the process motivating, and some contenders have made giant strides in quality and competitiveness. Among the criticisms: The award process involves tremendous amounts of time and effort for both employees and top management at substantial cost (sometimes millions of dollars); applicants nominate themselves (as opposed to being nominated by satisfied customers); winning does not mean that a business has top-quality products or has solved all of its quality problems; and winning may be followed by relaxation under the belief that "we have it made." Some critics say that the number of awards (a maximum of six per year) is too few, considering the large number of organizations that are potential candidates.

The Deming Prize

The Deming Prize, named in honor of the late American statistician, W. Edwards Deming, is Japan's highly coveted award recognizing successful quality efforts. It is given annually to any company that meets the award's standards. Although typically given to Japanese firms, in 1989, Florida Power and Light became the first U.S. company to win the award.

The major focus of the judging is on statistical quality control, making it much narrower in scope than the Baldrige Award, which focuses more on customer satisfaction. Companies that win the Deming Prize tend to have quality programs that are detailed and well-communicated throughout the company. Their quality improvement programs also reflect the involvement of senior management and employees, customer satisfaction, and training.

QUALITY CERTIFICATION

Many firms that do business internationally recognize the importance of quality certification.

ISO 9000

The purpose of the International Organization for Standardization (ISO) is to promote worldwide standards that will improve operating efficiency, improve productivity, and reduce costs. The ISO is composed of the national standards bodies of 91 countries. The U.S. representative body is the American National Standards Institute (ANSI). The

[3]See, for example, "Is the Baldrige Overblown?" *Fortune,* July 1, 1991, pp. 61–65; and "Does the Baldrige Award Really Work?" *Harvard Business Review,* January–February 1992, pp. 126–47.

Solectron Corporation, a worldwide provider of customized electronics manufacturing services, won the Malcolm Baldrige National Quality Award for Manufacturing twice, most recently in December 1997. President Bill Clinton presented the award to Ko Nishimura, Solectron's chairman, president, and CEO.

www.solectron.com

TABLE 9-7

Baldrige Award criteria

Categories	Points	
1 Leadership		**110**
1.1 Leadership System	80	
1.2 Company Responsibility and Citizenship	30	
2 Strategic Planning		**80**
2.1 Strategy Development Process	40	
2.2 Company Strategy	40	
3 Customer and Market Focus		**80**
3.1 Customer and Market Knowledge	40	
3.2 Customer Satisfaction and Relationship Enhancement	40	
4 Information and Analysis		**80**
4.1 Selection and Use of Information and Data	25	
4.2 Selection and Use of Comparative Information and Data	15	
4.3 Analysis and Review of Company Performance	40	
5 Human Resource Development and Management		**100**
5.1 Work Systems	40	
5.2 Employee Education, Training, and Development	30	
5.3 Employee Well-Being and Satisfaction	30	
6 Process Management		**100**
6.1 Management of Product and Service Processes	60	
6.2 Management of Support Processes	20	
6.3 Management of Supplier and Partnering Processes	20	
7 Business Results		**450**
7.1 Customer Satisfaction Results	130	
7.2 Financial and Market Results	130	
7.3 Human Resource Results	35	
7.4 Supplier and Partner Results	25	
7.5 Community-Specific Results	130	
TOTAL POINTS		**1000**

work of the ISO is conducted by some 180 technical committees. ISO 9000 is the work of the Quality Management and Quality Assurance Committee.

ISO 9000 A set of international standards on quality management and quality assurance, critical to international business.

The **ISO 9000** series is a set of international standards on quality management and quality assurance. These standards are critical to doing business internationally, particularly in Europe. They must go through a process that involves documenting quality procedures and on-site assessment. The process often takes 12 to 18 months. With certification comes *registration* in an ISO directory that companies seeking suppliers can refer to for a list of certified companies. They are generally given preference over unregistered companies. More than 40,000 companies are registered worldwide, three-fourths of which are located in Europe.

A key requirement for registration is that a company review, refine, and map functions such as process control, inspection, purchasing, training, packaging, and delivery. Similar to the Baldrige Award, the review process involves considerable self-appraisal, resulting in problem identification and improvement. Unlike the Baldrige Award, registered companies face an ongoing series of audits, and they must be reregistered every three years.

There are essentially five standards associated with the ISO 9000 series. Table 9–8 outlines them.

In addition to the obvious benefits of certification for companies who want to deal with the European Union, the ISO 9000 certification and registration process is particularly helpful for companies that do not currently have a quality management system; it provides guidelines for establishing the system and making it effective.

READING

AESOP ON QUALITY SYSTEMS

Denise E. Robitaille

What can Aesop's fable about the goat and the goatherd teach us about quality auditing?

Imagine that the master of a medium-sized goat farm is preparing for a third-party audit to be conducted on behalf of a local dairy that wishes to purchase some of his goats. He takes great care with his goats, knowing well-tended animals produce the best milk. He is also proud that his beautiful goats have won numerous prizes at town fairs.

Two days before the audit, the master's goatherd, whose evening task is to gather all the goats and return them to their pen for the night, has trouble with one recalcitrant goat, which refuses to budge. The goatherd, after much fruitless cajoling, throws several stones at the goat to frighten her into the pen. In the process, he breaks one of her horns. The goatherd begs the goat not to tell the master. The goat replies, "I may be silent. But the broken horn will speak for itself."

When the auditor arrives, the master regales him with numerous examples of how well his goats are treated. The auditor proceeds to assess the herd and finds the goat with the broken horn. Despite all he has been told, the auditor is required to report that although the master has insisted the animals are well treated, the objective evidence is a goat with a broken horn. The data clearly indicate inconsistencies between this vendor's procedures and the actual practice.

The client, an avid proponent of ISO 9000, concludes, through his vendor assessment process, that the goatherd will have to implement corrective action and provide verification before he can be qualified to sell the dairy any goats.

Aesop, a fabulist who lived in the sixth century B.C., traveled through Greece and the Mediterranean area as an instructor, philosopher, and diplomat. His advice was sought on matters of state. Through his fables, he created an amusing vehicle for illustrating the foibles of human beings. Filled with object lessons that survive into our time, his tales provide insight into the breakdowns and pitfalls 20th century businesses contend with every day.

The Ant and the Grasshopper

A Grasshopper happened upon an Ant who was toiling in preparation for the long winter ahead. She asked of the Ant, "Why do you work so hard on such a pleasant day? Why not dance and sing as I do?" The Ant responded, "I need to plan and prepare myself for the cold weather to come." The Grasshopper laughed and danced away. When the first snows fell the Grasshopper perished. The Ant, snug and secure in her well-provisioned nest, survived and, indeed, prospered oblivious to the tempest outside her home.

More than two centuries ago, Aesop identified the value of good planning and the dire consequences of failing to do so. Today, that value is articulated in the second clause of the ISO 9000 standard. The first clause of the ISO 9000 standard specifically addresses the need to make available necessary resources to ensure the fulfillment of the quality objectives. And we have found that planning is the first step of W. Edwards Deming's plan-do-check-act cycle. Careful planning is an indisputable key element of any quality system.

The Wild Boar and the Fox

A Fox observed a Wild Boar sharpening his tusks. There appearing to be no imminent danger, the Fox asked, "Why do you sharpen your teeth when your enemies are not around?" The Wild Boar responded, "I would have no time to sharpen them if my enemies were upon me."

Aesop gives us a powerful case for the ninth clause of the ISO 9000 standard, which clearly mandates equipment be maintained so it is reliable when needed. How many of us have missed an important ship date because a vital piece of equipment failed at a critical period due to lack of maintenance?

The Farmer and the Stork

A Farmer cast a net upon his fields in an attempt to capture the cranes that had been eating his seed. In with the cranes, he snared a Stork. The Stork pleaded for his life, saying, "Honorable Farmer, I am not like these others who came to steal your seed. I am a Stork." The Farmer replied, "You may very well be a stork, but from the evidence here, I can only assume that you, too, have come to eat my seeds." And with that he slew the Stork along with the cranes.

The ISO 10011 standard addresses the guidelines for auditing a quality system. An auditor's role is to report findings objectively. Documentation and records should be accurate and reflect actual practices. Evidence that contradicts the auditee will result in an auditor finding inconsistencies between procedures and practices.

Belling the Cat

Once there were a colony of mice inhabiting a large house. The mice were constantly attacked by a cunning and overzealous Cat. The mice held a council in which they discussed how to solve their problem. A clever young mouse rose and offered the following solution, "If we could hear the Cat coming, we would be able to flee in time. Therefore, I propose that we hang a bell around the Cat's neck." The mice cheered wildly over the brilliant solution until an aged mouse stepped to the podium. The elder mouse spoke. "I have only one question. Who shall bell the Cat?"

Returning to Deming's cycle and to any plans for corrective/preventive action, it is wisest to consider the stakeholders and the organization's capability to implement a plan. Many plans fail due to this failure to assess human resources and adequately address constraints. It is always easiest to come up with a plan when someone else is accountable for its implementation.

The Ass and the Mule

An Ass and a Mule were both heavily laden with wares for the monthly fair. The Ass, being smaller and less agile on steep slopes, asked the Mule to relieve him of his burden. The Mule ignored his pleas. Shortly thereafter, the Ass collapsed under his burden and died. The master skinned the dead animal and then placed both his load and the Ass's hide on the Mule's back. The Mule was obliged to carry not only his load, but that of his dead companion.

Imagine the ass is an overburdened department. The mule represents management. Failing to recognize that the department is over-taxed and in need of assistance, management actually contributes to the collapse of an essential process and is left with the need to assume the consequences of that breakdown. Had management fulfilled the mandate to provide adequate resources, the breakdown would have been avoided.

The Pig, the Sheep, and the Goat

The Pig, the Sheep, and the Goat were all penned in for the night. The Pig whined of his condition. The Sheep and the Goat, wearied by his constant complaining, finally said, "We are in the same pen with you. Yet you do not hear us whining." The Pig answered, "The master takes from you only your wool and your milk. It is a different matter with me."

This last fable reiterates the need for management's involvement in the implementation of a quality system. If managers attempt to detach themselves from accountability for the achievement of a goal, they may very well find that their staff assigns to itself the role of the pig and identifies executive managers as the sheep and goats. The fable also serves as an object lesson for Deming's tenet: "Drive out fear."

The Dying Father and His Sons

A Father on his death bed told his Sons that there was a great treasure to be found in his vineyard. After he had died the Sons raced out into the fields and overturned all the soil in search of the prize. No treasure was found, but, due to the revitalization of the soil, the vineyard yielded a bountiful harvest.

Many companies pursue registration to ISO 9000 or another quality standard under pressure from their customers or as a marketing strategy. Along the way to implementation, they discover they have so improved their procedures and processes that the benefits exceed all their expectations, and their company prospers.

The Boy Bathing and the Traveler

A Traveler happened upon a drowning Boy who had foolishly chosen to bathe in a river with a swift current. The Traveler chastised the Boy for his imprudence and then left him to drown. As he sank beneath the water the Boy thought, "How worthless it is to offer criticism without aid."

None of us would leave a drowning victim to perish. We might, however, leave a process owner to tread water by failing to provide problem-solving skills or necessary resources. Consider the unproductive nature of a quality system that does not seamlessly integrate the audit function, corrective action, and the mandate for management to provide necessary resources. In such an instance, management only looks at a list of nonconformances and does not provide a vehicle for root-cause analysis or direction for corrective action. The auditee is left without the adequate resources to properly address his problem.

A further lesson that may be derived from all of Aesop's fables is a reminder that quality systems should reflect common sense. ISO 9000 or any quality standard should only exist as a framework that facilitates the fulfillment of the quality objectives. But, the fundamentals are not new. Good planning, accountability, objective evidence, and maintenance are concepts with validity that have long survived the test of time. Indeed, as has been illustrated, they have been around for over two millennia.

Bibliography

Aesop's Fables, translated by George F. Townsend (Garden City, NY: Doubleday & Co., 1968).

Source: Reprinted with permission from *Quality Progress,* December 1997.

TABLE 9-8

The ISO 9000 series

ISO 9000	Helps companies determine which standard of ISO 9001, 9002, and 9003 applies.
ISO 9001	Outlines guidelines for companies that engage in design, development, production, installation, and servicing of products or service.
ISO 9002	Similar to ISO 9001, but excludes companies engaged in design and development.
ISO 9003	Covers companies engaged in final inspection and testing.
ISO 9004	The guidelines for applying the elements of the Quality Management System.

ISO 14000

ISO 14000 A set of international standards for assessing a company's environmental performance.

The International Organization for Standardization introduced a new set of standards in 1996: **ISO 14000** is intended to assess a company's performance in terms of environmental responsibility. Initially, ISO 14000 began as a program of voluntary guidelines and certification. The standards for certification bear upon three major areas:

Management systems—Systems development and integration of environmental responsibilities into business planning.

Operations—Consumption of natural resources and energy.

Environmental systems—Measuring, assessing, and managing emissions, effluents, and other waste streams.

Proponents hope that the new standards will receive the same high interest that the ISO 9000 standards have in the international business community, and will result in businesses giving more attention to environmental responsibilities.

Summary

The success of foreign companies in North American markets has shown that they are formidable competitors. They have built reputations for quality products. Moreover, they have achieved their success using management techniques quite different from those traditionally practiced in the United States. This has caused American managers to reexamine their own approaches, and to adopt some of those different approaches in order to improve quality and become more competitive.

Quality is defined according to various dimensions that pertain to customer satisfaction. The consequences of poor quality include loss of market share, liability claims, a decrease in productivity, and an increase in costs. Determinants of quality are design, conformance to design, ease of use, and service after delivery.

Modern quality management is directed at preventing mistakes rather than finding them after they occur. Currently, the business community shows widespread interest in improving quality and competitiveness.

The chapter includes a description of the key contributors to quality management, and it outlines the ISO 9000 international quality standards.

Two awards of distinction, the Baldrige Award and the Deming Prize, are given annually to organizations that have shown great achievement in quality management.

Key Terms

appraisal costs, 425
Baldrige Award, 430
Deming Prize, 426
dimensions of quality, 421
external failure costs, 425
failure costs, 425
focus groups, 422
internal failure costs, 425
ISO 9000, 434
ISO 14000, 436
prevention costs, 425
quality, 420
quality of conformance, 423
quality of design, 422

Discussion and Review Questions

1. List and briefly explain the dimensions of quality.
2. Explain the terms *quality of design* and *quality of conformance.*
3. What are some possible consequences of poor quality?
4. Use the dimensions of quality to describe typical characteristics of these products and services:
   a. A television set
   b. A restaurant meal (product)
   c. A restaurant meal (service)
   d. Painting a house
5. List the four determinants of quality.
6. How does modern quality management differ from the "inspection" approach?
7. Select one of the quality gurus and briefly describe his major contributions to quality management.
8. What is ISO 9000, and why is it important for global businesses to have ISO 9000 certification?
9. Briefly explain how a company can achieve lower production costs by improving the quality of its products or services.

Memo Writing Exercises

1. Write a one-page memo to your supervisor, Penny Lane, summarizing the benefits of doing a task right the first time.
2. Suppose that your company is considering entering the competition for a Baldrige Award. Prepare a one-page memo outlining both the pros and cons of doing so to the head of operations, Tom Cartwright.

Selected Bibliography

Capezio, Peter, and Debra Morehouse. *Taking the Mystery Out of TQM: A Practical Guide to Total Quality Management.* Franklin Lakes, NJ: National Press Publications, 2nd ed., 1995.

Cartin, Thomas J. *Principles and Practices of TQM.* Milwaukee, WI: ASQC Press, 1993.

Crosby, Philip B. *Quality without Tears: The Art of Hassle Free Management.* New York: McGraw-Hill, 1984.

Deming, W. E. *Out of Crises.* Cambridge, Mass.: MIT Center for Advanced Engineering Study, 1982.

Garvin, David A. *Managing Quality.* New York: Free Press, 1988.

Gitlow, Howard S., and Shelly J. Gitlow. *The Deming Guide to Quality and Competitive Position.* Englewood Cliffs, NJ: Prentice Hall, 1987.

Hall, Robert. *Attaining Manufacturing Excellence.* Burr Ridge, IL: Dow-Jones Irwin, 1987.

Hunt, V. Daniel. *Quality in America.* Burr Ridge, IL: Business One Irwin, 1992.

Johnson, Richard S. *TQM: Leadership for the Quality Transformation.* Vols 1-4, Milwaukee: ASQC Quality Press, 1993.

Juran, J. M. "The Quality Trilogy." *Quality Progress* 19, no. 8, August 1986, pp. 19–24.

Shores, Richard A. *Reengineering the Factory: A Primer for World-Class Manufacturing.* Milwaukee: ASQC Press, 1994.

Weimershirch, Arnold, and Stephen George. *Total Quality Management: Strategies and Techniques Proven at Today's Most Successful Companies.* New York: John Wiley & Sons, 1994.

LEARNING OBJECTIVES

After completing this chapter, you should be able to:

1. List and briefly explain the elements of the control process.
2. Explain how control charts are used to monitor a process, and the concepts that underlie their use.
3. Use and interpret control charts.
4. Use run tests to check for nonrandomness in process output.

CHAPTER TEN
QUALITY CONTROL

CHAPTER OUTLINE

FIGURE 10–1

Approaches to quality assurance

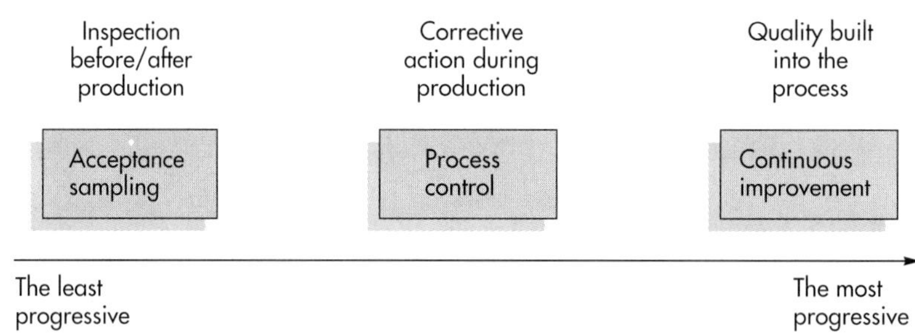

This chapter covers quality control. The purpose of quality control is to assure that processes are performing in an acceptable manner. Companies accomplish this by monitoring process output using statistical techniques. If the results are acceptable, no further action is required; unacceptable results call for corrective action.

INTRODUCTION

The best companies emphasize designing quality *into the process,* thereby greatly reducing the need for inspection or control efforts. As you might expect, different business organizations are in different stages of this evolutionary process: The least progressive rely heavily on inspection. Many occupy a middle ground that involves some inspection and a great deal of process control. The most progressive have achieved an inherent level of quality that is sufficiently high that they can avoid wholesale inspection activities and process control activities. That is the ultimate goal. Figure 10–1 illustrates these phases of quality assurance.

Quality assurance that relies primarily on inspection after production is referred to as *acceptance sampling.* It is described in the chapter supplement. Quality control efforts that occur during production are referred to as *statistical process control,* and these we examine in the following sections.

INSPECTION

Monitoring in the production process can occur at three points: before production, during production, and after production. The logic of checking conformance before production is to make sure that inputs are acceptable. The logic of checking conformance during production is to make sure that the conversion of inputs into outputs is proceeding in an acceptable manner. The logic of checking conformance of output is to make a final verification of conformance before passing goods on to customers.

Monitoring before and after production involves *acceptance sampling* procedures; monitoring during the production process is referred to as *process control.* Figure 10–2 gives an overview of where these two procedures are applied in the production process.

To determine whether a process is functioning as intended or to verify that a shipment of raw materials or final products does not contain more than a specified percentage of defectives, it is necessary to physically examine at least some of the items in question. The purpose of inspection is to provide information on the degree to which items conform to a standard. The basic issues are:

1. How much to inspect and how often.
2. At what points in the process inspection should occur.
3. Whether to inspect in a centralized or on-site location.
4. Whether to inspect attributes or variables.

Consider, for example, inspection at an intermediate step in the manufacture of personal computers. Because inspection costs are often significant, questions naturally arise on whether one needs to inspect every computer or whether a small sample of

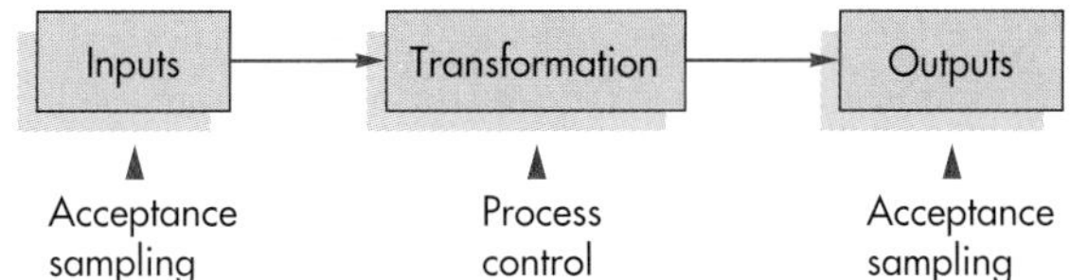

FIGURE 10–2

Acceptance sampling and process control

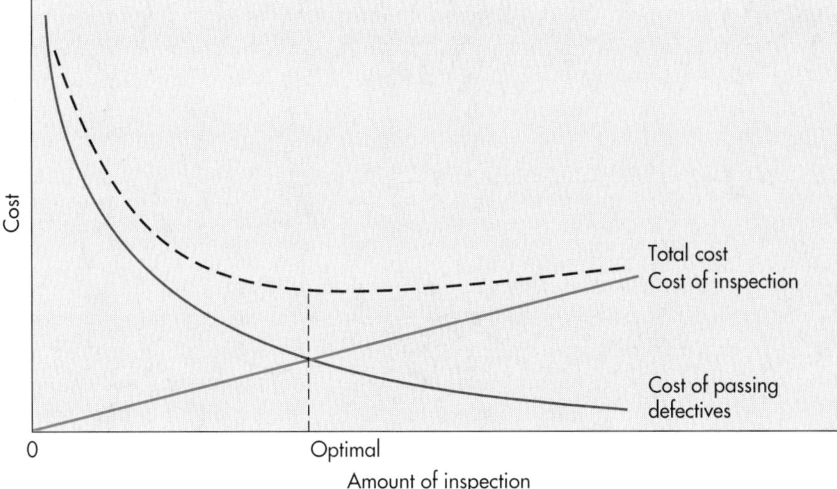

FIGURE 10–3

Traditional view: The amount of inspection is optimal when the sum of the costs of inspection and passing defectives is minimized

computers will suffice. Moreover, although inspections could be made at numerous points in the production process, it is not generally cost-effective to make inspections at every point. Hence, the question comes up of which points one should make inspections. Once these points have been identified, a manager must decide whether to remove the computers from the line and take them to a lab, where specialized equipment might be available to run certain tests, or to test them where they are being made. We will examine these points in the following sections.

How Much to Inspect and How Often

The amount of inspection can range from no inspection whatsoever to inspecting each item numerous times. Low-cost, high-volume items such as paper clips, roofing nails, and wooden pencils often require little inspection because (1) the cost associated with passing defectives is quite low and (2) the processes that produce these items are usually highly reliable, so that defectives are rare. Conversely, high-cost, low-volume items that have large costs associated with passing defectives often require more intensive inspections. Thus, critical components of a manned-flight space vehicle are closely scrutinized because of the risk to human safety and the high cost of mission failure. In high-volume systems, *automated* inspection is one option that may be employed.

The majority of quality control applications lies somewhere between the two extremes. Most require some inspection, but it is neither possible nor economically feasible to critically examine every part, or portion of a service for control purposes. The cost of inspection, resulting interruptions of a process or delays caused by inspection, and the manner of testing typically outweigh the benefits of 100 percent inspection. However, the cost of letting undetected defectives slip through is sufficiently high that inspection cannot be completely ignored. The amount of inspection needed is governed by the costs of inspection and the expected costs of passing defective items. As illustrated in Figure 10–3, if inspection activities increase, inspection costs increase, but the costs of undetected defectives decrease. The traditional goal was to minimize the sum of these two costs. In other words, it may not pay to attempt to catch every defective, particularly if the cost of inspection exceeds the penalties associated with letting some defectives get through. Current thinking is that every reduction in defective output reduces costs.

As a rule, operations with a high proportion of human involvement necessitate more inspection effort than mechanical operations, which tend to be more reliable.

The frequency of inspection depends largely on the rate at which a process may go out of control or the number of lots being inspected. A stable process will require only infrequent checks, whereas an unstable one or one that has recently given trouble will require more frequent checks. Likewise, many small lots will require more samples than a few large lots because it is important to obtain sample data from each lot.

NEWSCLIP

OOPS!

The Federal Aviation Administration (FAA) has an interesting device it uses to test the strength of airplane windshields in collisions with birds. The device propels dead chickens at the windshield at typical velocities to determine if the windshield can withstand the impact without breaking.

Officials in another country asked to borrow the device to test windshields of a high-speed train that country was developing.

The results of the initial test were astounding. Not only did the windshield break, the dead chicken went through the back of the engineer's chair and lodged in the wall behind it! The inspectors reduced the velocity of the next chicken, but the results were similar, much to the dismay of both the inspectors and the designers. After additional testing, the inspectors wondered if perhaps the device was defective, or perhaps they were using it incorrectly.

When FAA officials were informed of the results, they were dumbfounded. They had never experienced similar results through numerous trials with the device. Further investigation led to a solution to the problem—and a slight addition to the instructions for future users of the machine: *Thaw* chicken before using.

Source: Based on "Inside the Beltway" by John McCaslin, *The Washington Times,* May 15, 1996.

Where to Inspect in the Process

Many operations have numerous possible inspection points. Because each inspection adds to the cost of the product or service, it is important to restrict inspection efforts to the points where they can do the most good. In manufacturing, some of the typical inspection points are:

1. *Raw materials and purchased parts.* There is little sense in paying for goods that do not meet quality standards and in expending time and effort on material that is bad to begin with.
2. *Finished products.* Customer satisfaction and the firm's image are at stake here, and repairing or replacing products in the field is usually much more costly than doing it at the factory. Likewise, the seller is usually responsible for shipping costs on returns, and payments may be held up pending delivery of satisfactory goods.
3. *Before a costly operation.* The point is to not waste costly labor or machine time on items that are already defective.
4. *Before an irreversible process.* In many cases, items can be reworked up to a certain point; beyond that point they cannot. For example, pottery can be reworked prior to firing. After that, defectives must be discarded or sold as seconds at a lower price.
5. *Before a covering process.* Painting, plating, and assemblies often mask defects.

In the service sector, inspection points are incoming purchased materials and supplies, personnel, service interfaces (e.g., service counter), and outgoing completed work (e.g., repaired appliances, automobiles). Table 10–1 illustrates a number of examples.

TABLE 10-1

Examples of inspection points in service organizations

Type of Business	Inspection Points	Characteristics
Fast food	Cashier	Accuracy
	Counter area	Appearance, productivity
	Eating area	Cleanliness, no loitering
	Building and grounds	Appearance, safety hazards
	Kitchen	Cleanliness, purity of food, food storage, health regulations
	Parking lot	Safety, good lighting
Hotel/motel	Accounting/billing	Accuracy, timeliness
	Building and grounds	Appearance and safety
	Main desk	Appearance, waiting times, accuracy of bills
	Maid service	Completeness, productivity
	Personnel	Appearance, manners, productivity
	Reservations/occupancy	Over/underbooking, percent occupancy
	Restaurants	Kitchen, menus, meals, bills
	Room service	Waiting time, quality of food
	Supplies	Ordering, receiving, inventories
Supermarket	Cashiers	Accuracy, courtesy, productivity
	Deliveries	Quality, quantity
	Produce	Freshness, ample stock
	Aisles and stockrooms	Uncluttered layout
	Inventory control	Stock-outs
	Shelf stock	Ample supply, rotation of perishables
	Shelf displays	Appearance
	Checkouts	Waiting time
	Shopping carts	Good working condition, ample supply, theft/vandalism
	Parking lot	Safety, good lighting
	Personnel	Appearance, productivity

Centralized versus On-Site Inspection

Some situations require that inspections be performed *on site*. For example, inspecting the hull of a ship for cracks requires inspectors to visit the ship. At other times, specialized tests can best be performed in a lab (e.g., medical tests, analyzing food samples, testing metals for hardness, running viscosity tests on lubricants).

The central issue in the decision concerning on-site or lab inspections is whether the advantages of specialized lab tests are worth the time and interruption needed to obtain the results. Reasons favoring on-site inspection include quicker decisions and avoidance of introduction of extraneous factors (e.g., damage or other alteration of samples during transportation to the lab). On the other hand, specialized equipment and a more favorable test environment (less noise and confusion, lack of vibrations, absence of dust, and workers "helping" with inspections) offer strong arguments for using a lab.

Some companies rely on self-inspections by operators if errors can be traced back to specific operators. This places responsibility for errors at their source.

STATISTICAL PROCESS CONTROL

Quality control is concerned with the **quality of conformance** of a process: Does the output of a process conform to the intent of design? Toward that end, managers use **statistical process control** to evaluate the output of a process to determine its acceptability. To do this, they take periodic samples from the process and compare them with a predetermined standard. If the sample results are not acceptable, they stop the process and take corrective action. If the sample results are acceptable, they allow the process to continue.

quality of conformance A product or service conforms to specifications.

statistical process control Statistical evaluation of the output of a process during production.

Two statistical tools are used for quality control: *control charts* and *run tests*. Often, they are used together.

In its product testing, Sherwin-Williams, North America's largest manufacturer of paints and varnishes, tests its paints for longevity and durability. The company operates more than 2,100 paint stores in the U.S., Canada, and Puerto Rico and is a major supplier for Sears and Wal-Mart.

www.sherwin-williams.com

The Control Process

Sampling and corrective action are only a part of the control process. Effective control requires the following steps:

1. Define.
2. Measure.
3. Compare to a standard.
4. Evaluate.
5. Take corrective action if necessary.
6. Evaluate corrective action.

The first step means to define in sufficient detail what is to be controlled. It is not enough, for example, to simply refer to a painted surface. The paint can have a number of important characteristics such as its thickness, hardness, and resistance to fading or chipping. Different characteristics may require different approaches for control purposes.

Only those characteristics that can be counted or measured are candidates for control. Thus, it is important to consider how measurement will be accomplished.

There must be a standard of comparison that can be used to evaluate the measurements. This will relate to the level of quality being sought.

Management must establish a definition of *out of control*. Even a process that is functioning as it should will not yield output that conforms exactly to a standard, simply because of the natural (i.e., random) variations inherent in all processes, manual or mechanical—a certain amount of variation is inevitable. The main task of quality control is to distinguish random from *nonrandom* variability, because nonrandom variability means that a process is out of control.

When a process is judged out of control, corrective action must be taken. This involves uncovering the cause of nonrandom variability (e.g., worn equipment, incorrect methods, failure to follow specified procedures) and correcting it.

To ensure that corrective action is effective, the output of a process must be monitored for a sufficient period of time to verify that the problem his been eliminated.

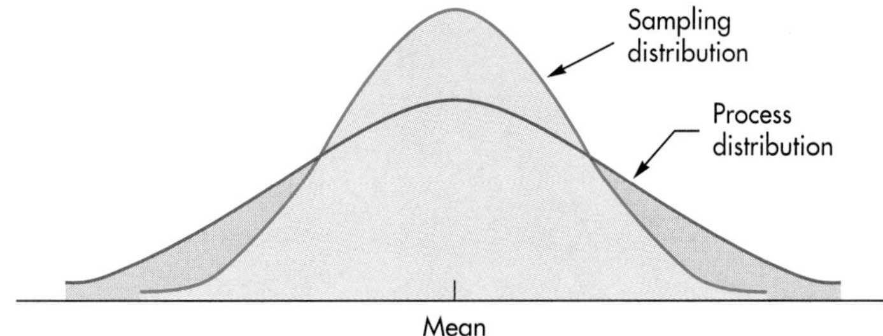

FIGURE 10–4

The sampling distribution of means is normal, and it has less variability than the process

In a nutshell, control is achieved by inspecting a portion of the goods or services, comparing the results to a predetermined standard, evaluating departures from the standard, taking corrective action when necessary, and following up to ensure that problems have been corrected.

Variations and Control

All processes that provide a good or a service exhibit a certain amount of "natural" variation in their output. The variations are created by the combined influences of countless minor factors, each one so unimportant that even if it could be identified and eliminated, the decrease in process variability would be negligible. In effect, this variability is inherent in the process. It is often referred to as *chance* or **random variation,** although it sometimes is referred to as *common variability.* The amount of inherent variability differs from process to process. For instance, older machines generally exhibit a higher degree of natural variability than newer machines, partly because of worn parts and partly because new machines may incorporate design improvements that lessen the variability in their output.

random variation Natural variations in the output of process, created by countless minor factors.

A second kind of variability in process output is called **assignable variation,** or *special variation.* Unlike natural variation, the main sources of assignable variation can usually be identified (assigned to a specific cause) and eliminated. Tool wear, equipment that needs adjustment, defective materials, human factors (carelessness, fatigue, noise and other distractions, failure to follow correct procedures, and so on) and problems with measuring devices are typical sources of assignable variation.

assignable variation In process output, a variation whose cause can be identified.

When samples of process output are taken, and sample *statistics* such as the sample mean or range are computed, they exhibit the same kind of variability; that is, there is variation in the values of sample means, and variation in the values of sample ranges. The variability of a sample statistic can be described by its *sampling distribution,* which is a theoretical distribution that describes the *random* variability of sample statistics.

Vol. 1, Tape 1, Seg. 2
Zytec, Motorola, Hewlett-Packard

The goal of sampling is to determine whether nonrandom—and thus, correctable—sources of variation are present in the output of a process. The sampling distribution provides the theoretical basis for accomplishing this. Let's see how this is done.

Suppose there is a process for filling bottles with soft drink. If the amount of soft drink in a large number of bottles is measured accurately, we would discover slight differences among the bottles. If these amounts were arranged on a graph, the frequency distribution would reflect the *process variability.* The values would be clustered close to the process average (e.g., 16 ounces), but some of the values would vary somewhat from the mean.

If we return to the process and take *samples* of 10 bottles each and compute the *mean* amount of soft drink in each sample, we would discover that these values also vary, just as the *individual* values varied. They, too, would have a distribution of values.

If the process contained *only* random variability, the distribution of process values would represent the inherent process variability, and the distribution of sample means would represent the random variability of all possible sample means.

The two distributions are illustrated in Figure 10–4. The sampling distribution exhibits much less variability (i.e., it is less spread out) than the process distribution. This

FIGURE 10-5

Percentage of values within given ranges in a normal distribution

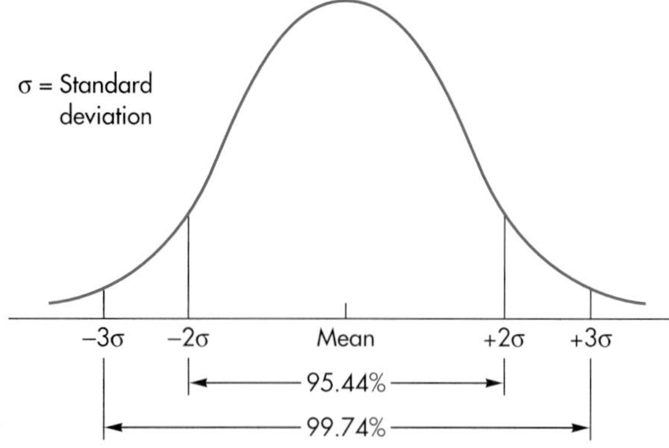

reflects the *averaging* that occurs in computing the sample means: High and low values in samples tend to offset each other, resulting in less variability among sample means than among individual values. Note that both distributions have the same mean; the mean of the sampling distribution is exactly equal to the mean of the process. Finally, note that the sampling distribution is a *normal* distribution, even if the process distribution isn't normal. The *central limit theorem* provides the basis for the assumption that the sampling distribution will be normal or at least approximately normal, even if the population (i.e., the process) is not.

The normal distribution can be used to help judge whether a process is performing adequately. If the output reflects only random variability, one would conclude that the process is *stable* (i.e., in control). But if there is evidence of nonrandom variability, one would conclude that the process is unstable (i.e., out of control). To understand how the normal distribution is used, consider the following: approximately 95.5 percent of the area under a normal curve (and, hence, 95.5 percent of the sample means) will have values that are within ±2 standard deviations of the distribution mean, and approximately 99.7 percent of the sample means will have values that are within ±3 standard deviations of the distribution mean. (See Figure 10–5.) These values are typically used for control limits.

Control Charts

control chart A time-ordered plot of sample statistics, used to distinguish between random and nonrandom variability.

A **control chart** is a *time-ordered* plot of sample statistics. It is used to distinguish between random variability and nonrandom variability. The basis for the control chart is the sampling distribution, which essentially describes random variability. There is, however, one minor difficulty relating to the use of a normal sampling distribution. The theoretical distribution extends in either direction to *infinity*. Therefore, *any* value is theoretically possible, even one that is a considerable distance from the mean of the distribution. However, as a practical matter, we know that, say, 99.7 percent of the values will be within ±3 standard deviations of the mean of the distribution. Therefore, we could decide to draw the line, so to speak, at values that represent ±3 standard deviations from the mean, and conclude that any value that was farther away than these limits was a nonrandom variation. In effect, these limits are **control limits:** the dividing lines between random deviations from the mean of the distribution and nonrandom deviations from the mean of the distribution. Figure 10–6 illustrates how control limits are based on the sampling distribution.

control limits The dividing lines between random and nonrandom deviations from the mean of the distribution.

Control charts have two limits that separate random variation and nonrandom variation. The larger value is the *upper control limit* (UCL), and the smaller value is the *lower control limit* (LCL). A sample statistic that falls between these two limits suggests (but does not prove) randomness, while a value outside or on either limit suggests (but does not prove) nonrandomness.

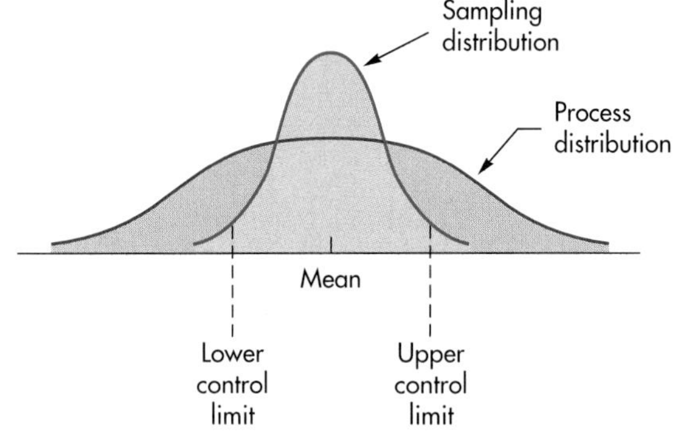

FIGURE 10–6

Control limits are based on the sampling distribution

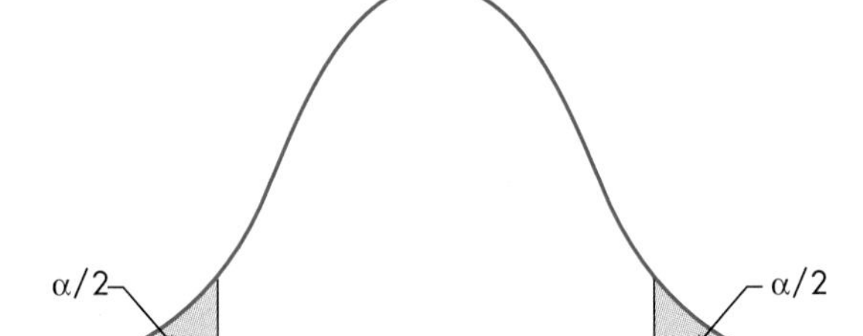

FIGURE 10–7

The probability of a Type I error

It is important to recognize that because any limits will leave some area in the *tails* of the distribution, there is a small probability that a value will fall outside the limits *even though only random variations are present.* For example, if ±2 sigma (standard deviation) limits are used, they would include 95.5 percent of the values. Consequently, the complement of that number (100 percent − 95.5 percent = 4.5 percent) would not be included. That percentage (or *probability*) is sometimes referred to as the probability of a **Type I error,** where the "error" is concluding that nonrandomness is present when only randomness is present. It is also referred to as an *alpha* risk, where alpha (α) is the sum of the probabilities in the two tails. Figure 10–7 illustrates this concept.

Type I error Concluding a process is not in control when it actually is.

Using wider limits (e.g., ±3 sigma limits) reduces the probability of a Type I error because it decreases the area in the tails. However, wider limits make it more difficult to detect nonrandom variations if they *are* present. For example, the mean of the process might shift (an assignable cause of variation) enough to be detected by two-sigma limits, but not enough to be readily apparent using three-sigma limits. That could lead to a second kind of error, known as a **Type II error,** which is concluding that a process is in control when it is really out of control (i.e., nonrandom variations are not present, when they are). In theory, the costs of making each error should be balanced by their probabilities. However, in practice, two-sigma limits and three-sigma limits are commonly used without specifically referring to the probability of a Type II error.

Type II error Concluding a process is in control when it is not.

Figure 10–8 illustrates the components of a control chart. Each sample is represented by a single value (e.g., the sample mean) on the chart. Moreover, each value is compared to the extremes of the sampling distribution (the control limits) to judge if it is within the acceptable (random) range. Figure 10–9 illustrates this concept.

There are four commonly used control charts. Two are used for **variables,** and two are used for **attributes.** Attribute data are *counted* (e.g., the number of defective parts in a sample, the number of calls per day); variables data are *measured,* usually on a continuous scale (e.g., amount of time needed to complete a task, length or width of a part).

variables Generate data that are *measured.*

attributes Generate data that are *counted.*

The two control charts for variables data are described in the next section, and the two control charts for attribute data are described in the section following that.

FIGURE 10-8

Example of a control chart

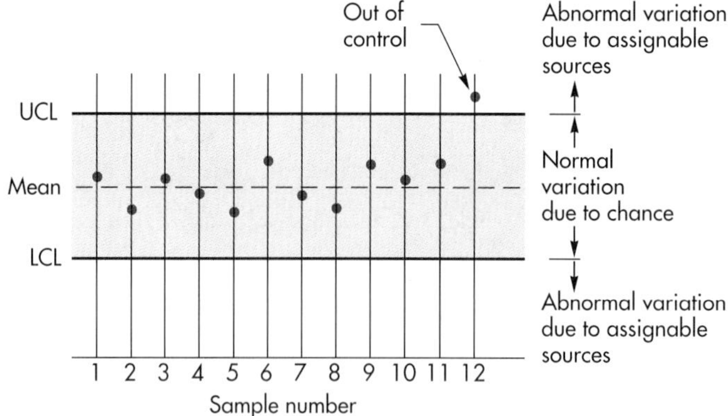

FIGURE 10-9

Each observation is compared to the selected limits of the sampling distribution

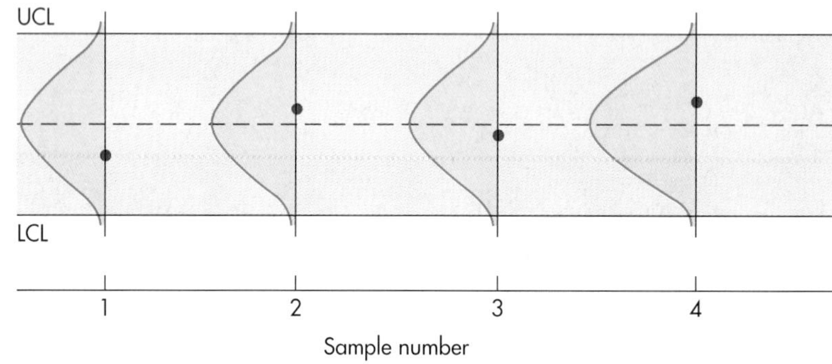

Control Charts for Variables

Mean and range charts are used to monitor variables. Control charts for means monitor the *central tendency* of a process, and range charts monitor the *dispersion* of a process.

mean control chart Control chart used to monitor the central tendency of a process.

Mean Charts. A **mean control chart,** sometimes referred to as an $\bar{x}$ ("x-bar") chart, can be constructed in one of two ways. The choice depends on what information is available. Although the value of the standard deviation of a process, σ, is often unknown, if a reasonable estimate is available, one can compute control limits using these formulas:

$$\text{Upper control limit (UCL):} = \bar{\bar{x}} + z\sigma_{\bar{x}} \qquad (10\text{–}1)$$
$$\text{Lower control limit (LCL):} = \bar{\bar{x}} - z\sigma_{\bar{x}}$$

where

$\sigma_{\bar{x}} = \sigma/\sqrt{n}$
$\sigma_{\bar{x}}$ = Standard deviation of distribution of sample means
σ = Process standard deviation
n = Sample size
z = Standard normal deviate
$\bar{\bar{x}}$ = Average of sample means

The following example illustrates the use of these formulas.

Example 1

A quality inspector took five samples, each with four observations, of the length of time to process a loan application at a credit union. The analyst computed the mean of each sample and then computed the grand mean. All values are in minutes. Use this information to obtain three-sigma (i.e., $z=3$)control limits for means of future times. It is known from previous experience that the standard deviation of the process is .02 minutes.

		Sample 1	2	3	4	5
Observation	**1**	12.11	12.15	12.09	12.12	12.09
	2	12.10	12.12	12.09	12.10	12.14
	3	12.11	12.10	12.11	12.08	12.13
	4	12.08	12.11	12.15	12.10	12.12
Mean		12.10	12.12	12.11	12.10	12.12

Solution

$$\bar{\bar{x}} = \frac{12.10 + 12.12 + 12.11 + 12.10 + 12.12}{5} = 12.11$$

Using Formula 10–1, with $z = 3$, $n = 4$ observations per sample, and $\sigma = .02$, we find

$$\text{UCL: } 12.11 + 3\left(\frac{.02}{\sqrt{4}}\right) = 12.14$$

$$\text{LCL: } 12.11 - 3\left(\frac{.02}{\sqrt{4}}\right) = 12.08$$

Note: If one applied these control limits to this data, one would judge the process to be *in control* because all of the sample means have values that fall within the control limits. The fact that some of the *individual* measurements fall outside of the control limits (e.g., the first observation in Sample 2 and the last observation in Sample 3) is irrelevant. You can see why by referring to Figure 10–7: individual values are represented by the process distribution, a large portion of which lies outside of the control limits.

A second approach is to use the sample *range* as a measure of process variability. The appropriate formulas for control limits are

$$\text{UCL} = \bar{\bar{x}} + A_2\bar{R} \qquad (10\text{–}2)$$
$$\text{LCL} = \bar{\bar{x}} - A_2\bar{R}$$

where

$\bar{R}$ = Average of sample ranges

Example 2

Twenty samples of $n = 8$ have been taken from a cleaning operation. The average sample range for the 20 samples was 0.016 minutes, and the average mean was 3 minutes. Determine three-sigma control limits for this process.

Solution

$\bar{\bar{x}} = 3$ cm. $\bar{R} = 0.016$, $A_2 = 0.37$ for $n = 8$ (from Table 10–2)

$$\text{UCL} = \bar{\bar{x}} + A_2\bar{R} = 3 + 0.37(0.016) = 3.006 \text{ minutes}$$
$$\text{LCL} = \bar{\bar{x}} - A_2\bar{R} = 3 + 0.37(0.016) = 2.994 \text{ minutes}$$

Note that this approach assumes that the range is in control.

range control chart Control chart used to monitor process dispersion.

Range Charts. **Range control charts** (R-charts) are used to monitor process dispersion; they are sensitive to changes in process dispersion. Although the underlying sampling distribution is not normal, the concepts for use of range charts are much the same as those for use of mean charts. Control limits for range charts are found using the average sample range in conjunction with these formulas:

TABLE 10–2

Factors for three-sigma control limits for $\bar{x}$ and R charts

Number of Observations in Subgroup, n	Factor for $\bar{x}$ Chart, A_2	FACTORS FOR *R* CHARTS: Lower Control Limit, D_3	Upper Control Limit, D_4
2	1.88	0	3.27
3	1.02	0	2.57
4	0.73	0	2.28
5	0.58	0	2.11
6	0.48	0	2.00
7	0.42	0.08	1.92
8	0.37	0.14	1.86
9	0.34	0.18	1.82
10	0.31	0.22	1.78
11	0.29	0.26	1.74
12	0.27	0.28	1.72
13	0.25	0.31	1.69
14	0.24	0.33	1.67
15	0.22	0.35	1.65
16	0.21	0.36	1.64
17	0.20	0.38	1.62
18	0.19	0.39.	1.61
19	0.19	0.40	1.60
20	0.18	0.41	1.59

Source: Adapted from Eugene Grant and Richard Leavenworth, *Statistical Quality Control*, 5th ed. (New York: McGraw-Hill, 1980).

$$\text{UCL}_R = D_4\bar{R} \qquad (10\text{–}3)$$
$$\text{LCL}_R = D_3\bar{R}$$

where values of D_3 and D_4 are obtained from Table 10–2.

Example 3

Twenty-five samples of $n = 10$ observations have been taken from a milling process. The average sample range was 0.01 centimeter. Determine upper and lower control limits for sample ranges.

Solution

$\bar{R} = 0.01$ cm, $n = 10$

From Table 10–2, for $n = 10$, $D_4 = 1.78$ and $D_3 = 0.22$.

$$\text{UCL}_R = 1.78(0.01) = 0.0178 \text{ or } 0.018$$
$$\text{LCL}_R = 0.22(0.01) = 0.0022 \text{ or } 0.002$$

In Example 3, a sample range of 0.018 centimeters or more would suggest that the process variability had increased. A sample range of 0.002 or less would imply that the process variability had decreased. In the former case, this means that the process was producing too much variation; we would want to investigate this in order to remove the cause of variation. In the latter case, even though decreased variability is desirable, we would want to determine what was causing it: Perhaps an improved method has been used, in which case we would want to identify it. Possibly the improved quality has come at the expense of productivity, or this was only a random occurrence. Hence, it can be beneficial to investigate points beyond the lower limit as well as points beyond the upper limit in a range chart.

Using Mean and Range Charts. Mean control charts and range control charts provide different perspectives on a process. As we have seen, mean charts are sensitive to

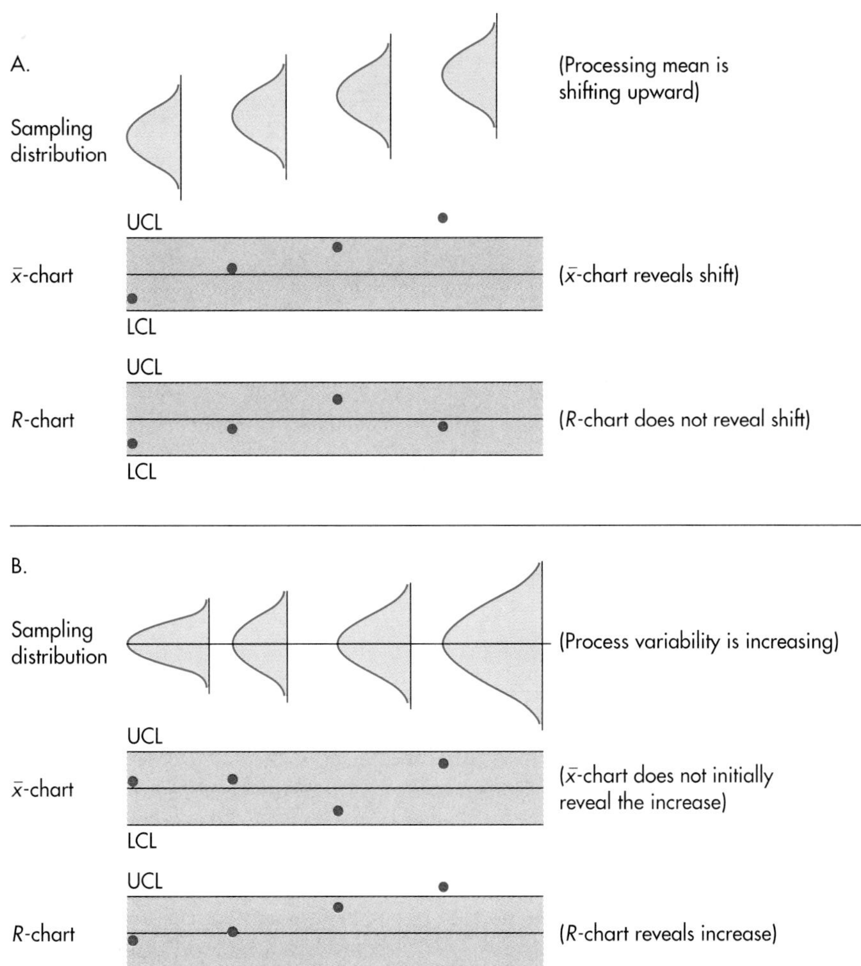

FIGURE 10–10

Mean and range charts used together complement each other

shifts in the process mean, whereas range charts are sensitive to changes in process dispersion. Because of this difference in perspective, both types of charts might be used to monitor the same process. The logic of using both is readily apparent in Figure 10–10. In 10–10A, the mean chart picks up the shift in the process mean, but because the dispersion is not changing, the range chart fails to indicate a problem. Conversely, in 10–10B, a change in process dispersion is less apt to be detected by the mean chart than by the range chart. Thus, use of both charts provides more complete information than either chart alone. Even so, a single chart may suffice in some cases. For example, a process may be more susceptible to changes in the process mean than to changes in dispersion, so it might be unnecessary to monitor dispersion. Because of the time and cost of constructing control charts, gathering the necessary data, and evaluating the results, only those aspects of a process that tend to cause problems should be monitored.

Once control charts have been set up, they can serve as a basis for deciding when to interrupt a process and search for assignable causes of variation. To determine initial control limits, one can use the following procedure.

1. Obtain 20 to 25 samples. Compute the appropriate sample statistic(s) for each sample (e.g., mean).
2. Establish preliminary control limits using the formulas, and graph them.
3. Plot the sample statistics on the control chart(s), and note whether any points fall outside the control limits.

At Hertz, there is no paperwork and no long line once the customer is enrolled in car rental services. Car keys to a pre-selected vehicle are provided, along with a copy of the rental agreement, as soon as a driver's license is shown.

www.hertz.com

4. If you find no out-of-control signals, assume that the process is in control. If not, investigate and correct assignable causes of variation. Then resume the process and collect another set of observations upon which control limits can be based.

Control Charts for Attributes

Control charts for attributes are used when the process characteristic is *counted* rather than measured. For example, the number of defective items in a sample is counted, whereas the length of each item is measured. There are two types of attribute control charts, one for the fraction of defective items in a sample (a *p*-chart) and one for the number of defects per unit (a *c*-chart). A *p-chart* is appropriate when the data consist of two categories of items. For instance, if glass bottles are inspected for chipping and cracking, both the good bottles and the defective ones can be counted. However, one can count the number of accidents that occur during a given period of time but *not* the number of accidents that did not occur. Similarly, one can count the number of scratches on a polished surface, the number of bacteria present in a water sample, and the number of crimes committed during the month of August, but one cannot count the number of nonoccurrences. In such cases, a *c-chart* is appropriate. See Table 10-3.

***p*-chart** Control chart for attributes, used to monitor the proportion of defectives in a process.

***p*-Chart.** A ***p*-chart** is used to monitor the proportion of defectives generated by a process. The theoretical basis for a *p*-chart is the binomial distribution, although for large sample sizes, the normal distribution provides a good approximation to it. Conceptually, a *p*-chart is constructed and used in much the same way as a mean chart.

The center line on a *p*-chart is the average fraction defective in the population, *p*. The standard deviation of the sampling distribution when *p* is known is

$$\sigma_p = \sqrt{\frac{p(1-p)}{n}}$$

Control limits are computed using the formulas

$$UCL_p = p + z\sigma_p \qquad (10\text{–}4)$$
$$LCL_p = p - z\sigma_p$$

If *p* is unknown, it can be estimated from samples. That estimate, $\bar{p}$, replaces *p* in the preceding formulas, as illustrated in Example 4.

Note: Because the formula is an approximation, it sometimes happens that the computed LCL is negative. In those instances, zero is used as the lower limit.

Example 4

An inspector counted the number of defective computer chips in each of 20 samples. Using the following information, construct a control chart that will describe 99.74

TABLE 10-3

p-chart or c-chart?

The following tips should help you select the type of control chart, a *p*-chart or a *c*-chart, that is appropriate for a particular application:

Use a *p*-chart:

1. When observations can be placed into one of *two categories*. Examples include items (observations) that can be classified as:
   a. Good or bad.
   b. Pass or fail.
   c. Operate or don't operate.
2. When the data consist of multiple samples of *n* observations each (e.g., 15 samples of $n = 20$ observations each).

Use a *c*-chart:

When only the number of occurrences per unit of measure can be counted; nonoccurrences cannot be counted. Examples of occurrences and units of measure include:

a. Scratches, chips, dents, or errors per item.
b. Cracks or faults per unit of distance (e.g., meters, miles).
c. Breaks or tears, per unit of area (e.g., square yard, square meter).
d. Bacteria or pollutants per unit of volume (e.g., gallon, cubic foot, cubic yard).
e. Calls, complaints, failures, equipment breakdowns, or crimes per unit of time (e.g., hour, day, month, year).

percent of the chance variation in the process when the process is in control. Each sample contained 100 chips.

Sample	Number of Defectives	Sample	Number of Defectives
1	14	11	8
2	10	12	12
3	12	13	9
4	13	14	10
5	9	15	11
6	11	16	10
7	10	17	8
8	12	18	12
9	13	19	10
10	10	20	16
			220

Solution

z for 99.74 percent is 3.00 (from Appendix Table A).

$$\bar{p} = \frac{\text{Total number of defectives}}{\text{Total number of observations}} = \frac{220}{20(100)} = .11$$

$$\hat{\sigma}_p = \sqrt{\frac{\bar{p}(1-\bar{p})}{n}} = \sqrt{\frac{.11(1-.11)}{100}} = .03$$

Control limits are:

$$\text{UCL}_p = \bar{p} + z(\hat{\sigma}_p) = .11 + 3.00(.03) = .20$$

$$\text{LCL}_p = \bar{p} - z(\hat{\sigma}_p) = .11 - 3.00(.03) = .02$$

Plotting the control limits and the sample fraction defective, you can see that the process is initially in control.

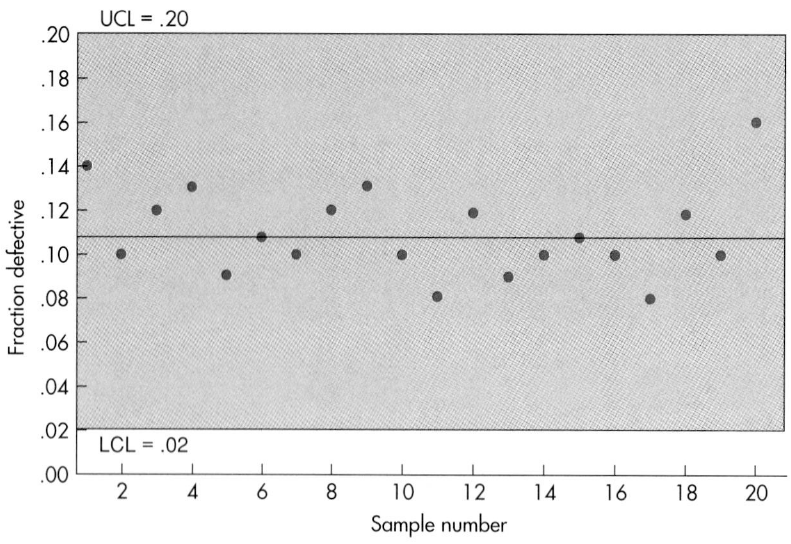

c-chart Control chart for attributes, used to monitor the number of defects per unit.

c-Chart. When the goal is to control the number of *defects per unit,* a **c-chart** is used. Units might be automobiles, hotel rooms, typed pages, or rolls of carpet. The underlying sampling distribution is the Poisson distribution. Use of the Poisson distribution assumes that defects occur over some *continuous* region and that the probability of more than one defect at any particular spot is negligible. The mean number of defects per unit is c and the standard deviation is $\sqrt{c}$. For practical reasons, the normal approximation to the Poisson is used. The control limits are

$$UCL_c = c + z\sqrt{c}$$
$$LCL_c = c - z\sqrt{c} \qquad (10\text{–}5)$$

Example 5

Rolls of coiled wire are monitored using a *c*-chart. Eighteen rolls have been examined, and the number of defects per roll has been recorded in the following table. Is the process in control? Plot the values on a control chart using three standard deviation control limits.

Sample	Number of Defects	Sample	Number of Defects
1	3	10	1
2	2	11	3
3	4	12	4
4	5	13	2
5	1	14	4
6	2	15	2
7	4	16	1
8	1	17	3
9	2	18	1
			45

Solution

$$\bar{c} = 45/18 = 2.5$$
$$UCL_c = \bar{c} + 3\sqrt{\bar{c}} = 2.5 + 3\sqrt{2.5} = 7.24$$
$$LCL_c = c - 3\sqrt{\bar{c}} = 2.5 - 3\sqrt{2.5} = -0.66 \rightarrow 0$$

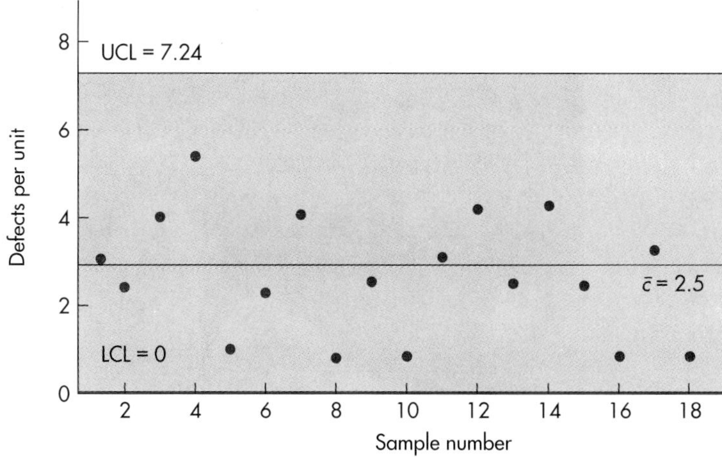

If the process average is unknown, it can be estimated from sample data, using $\bar{c}$ = Number of defects ÷ Number of samples.

When the computed lower control limit is negative, the effective lower limit is zero. The calculation sometimes produces a negative lower limit due to the use of the normal distribution to approximate the Poisson distribution: the normal is symmetrical whereas the Poisson is not symmetrical when c is close to zero.

Managerial Considerations concerning Control Charts

Using control charts adds to the cost and time needed to obtain output. Ideally a process is so good that the desired level of quality could be achieved without the use of any control charts. The best organizations strive to reach this level, but many are not yet at that level, so they employ control charts at various points in their processes. In those organizations, managers must make a number of important decisions about the use of control charts:

1. At what points in the process to use control charts.
2. What size samples to take.
3. What type of control chart to use (i.e., variables or attribute).

The decision about where to use control charts should focus on those aspects of the process that (1) have a tendency to go out of control and (2) are critical to the successful operation of the product or service.

Sample size is important for two reasons. One is that cost and time are functions of sample size; the greater the sample size, the greater the cost to inspect those items (and the greater the lost product if destructive testing is involved) and the longer the process must be held up while waiting for the results of sampling. The second reason is that smaller samples are more likely to reveal a change in the process than larger samples because a change is more likely to take place *within* the large sample, but *between* small samples. Consequently, a sample statistic such as the sample mean in the large sample could combine both "before-change" and "after-change" observations, whereas in two smaller samples, the first could contain "before" observations and the second "after" observations, making detection of the change more likely.

In some instances, a manager can choose between using a control chart for variables (a mean chart) and a control chart for attributes (a p-chart). If the manager is monitoring the diameter of a drive shaft, either the diameter could be measured and a mean chart used for control, or the shafts could be inspected using a *go, no-go gauge*—which simply indicates whether a particular shaft is within specification without giving its

Standardization is the key component of McDonald's strategy and all foods are prepared according to a strict reference manual. Farmers even have special guidelines to follow in raising potatoes for french fries. Here, quality assurance staff checks all aspects of french fries, from the sweetness of the ketchup to the length of the fries.

www.mcdonalds.com

exact dimensions—and a p-chart could be used. Measuring is more costly and time-consuming per unit than the yes-no inspection using a go, no-go gauge, but because measuring supplies more information than merely counting items as good or bad, one needs a much smaller sample size for a mean chart than a p-chart. Hence, a manager must weigh the time and cost of sampling against the information provided.

Run Tests

run test A test for randomness.

Analysts often supplement control charts with a **run test,** which is another kind of test for randomness. This enables an analyst to do a better job of detecting abnormalities in a process and provides insights into correcting a process that is out of control. A variety of run tests are available; this section describes two that are widely used.

When a process is stable or in statistical control, the output it generates will exhibit random variability over a period of time. The presence of patterns, such as trends, cycles, or bias in the output, indicates that assignable, or nonrandom, causes of variation exist. Hence, a process that produces output with such patterns is not in a state of statistical control. This is true even though all points on a control chart may be within the control limits. For this reason, it is usually prudent to subject control chart data to run tests to determine whether patterns can be detected.

run Sequence of observations with a certain characteristic.

A **run** is defined as a sequence of observations with a certain characteristic, followed by one or more observations with a different characteristic. The characteristic can be anything that is observable. For example, in the series A A A B, there are two runs: a run of three As followed by a run of one B. Underlining each run helps in counting them. In the series A A B B B A, the underlining indicates three runs.

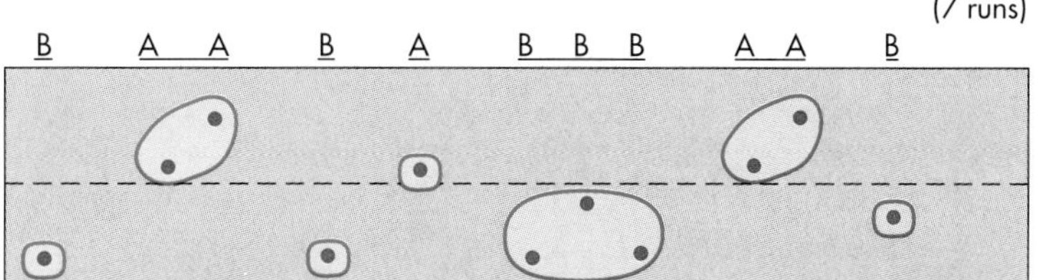

FIGURE 10–11

Counting above/below median runs

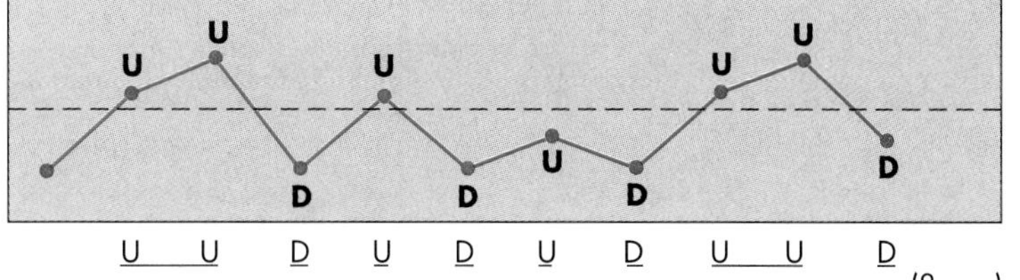

FIGURE 10–12

Counting up/down runs

Two useful run tests involve examination of the number of runs *up and down* and runs above and below the *median*.[1] In order to count these runs, the data are transformed into a series of Us and Ds (for *up* and *down*) and into a series of As and Bs (for *above* and *below* the median). Consider the following sequence, which has a median of 36.5. The first two values are below the median, the next two are above it, the next to last is below, and the last is above. Thus, there are four runs:

25	29	42	40	35	38
B	B	A	A	B	A

In terms of up and down, there are three runs in the same data. The second value is up from the first value, the third is up from the second, the fourth is down from the third, and so on:

25	29	42	40	35	38
—	U	U	D	D	U

(The first value does not receive either a U or a D because nothing precedes it.)

If a plot is available, the runs can be easily counted directly from the plot, as illustrated in Figures 10–11 and 10–12.

To determine whether any patterns are present in control chart data, one must transform the data into both As and Bs and Us and Ds, and then count the number of runs in each case. These numbers must then be compared with the number of runs that would be expected in a completely random series. For both the median and the up/down run tests, the expected number of runs is a function of the number of observations in the series. The formulas are

$$E(r)_{\text{med}} = \frac{N}{2} + 1 \tag{10–6a}$$

$$E(r)_{u/d} = \frac{2N - 1}{3} \tag{10–7a}$$

where N is the number of observations.

[1]The median and mean are approximately equal for control charts. The use of the median depends on its ease of determination; use the mean instead of the median if it is given.

FIGURE 10–13

A sampling distribution for runs is used to distinguish chance variation from patterns

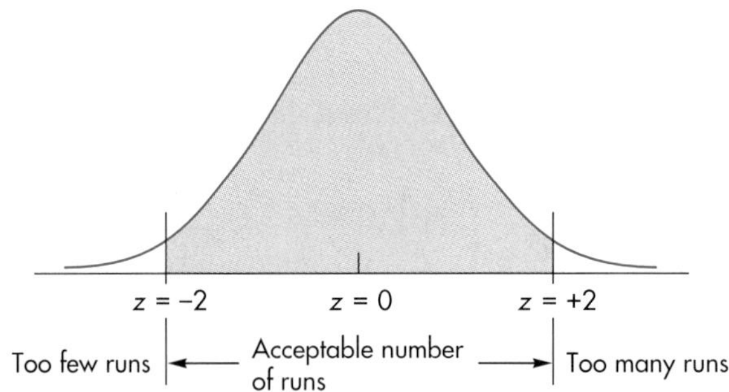

The actual number of runs in any given set of observations will vary from the expected number, due to chance and any patterns that might be present. Chance variability is measured by the standard deviation of runs. The formulas are

$$\sigma_{\text{med}} = \sqrt{\frac{N-1}{4}} \tag{10–6b}$$

$$\sigma_{\text{u/d}} = \sqrt{\frac{16N-29}{90}} \tag{10–7b}$$

Distinguishing chance variability from patterns requires use of the sampling distributions for median runs and up/down runs. Both distributions are approximately normal. Thus, for example, 95.5 percent of the time a random process will produce an observed number of runs within two standard deviations of the expected number. If the observed number of runs falls in that range, there are probably no nonrandom patterns; for observed numbers of runs beyond such limits, we begin to suspect that patterns are present. Too few or too many runs can be an indication of nonrandomness.

In practice, it is often easiest to compute the number of standard deviations, z, by which an observed number of runs differs from the expected number. This z value would then be compared to the value ± 2 (z for 95.5 percent) or some other desired value (e.g., ± 1.96 for 95 percent, ± 2.33 for 98 percent). A test z that exceeds the desired limits indicates patterns are present. (See Figure 10–13.) The computation of z takes the form

$$z_{\text{test}} = \frac{\text{Observed number of runs} - \text{Expected number of runs}}{\text{Standard deviation of number of runs}}$$

For the median and up/down tests, one can find z using these formulas:

$$\text{Median:} \quad z = \frac{r - [(N/2) + 1]}{\sqrt{(N-1)/4}} \tag{10–8}$$

$$\text{Up and down:} \quad z = \frac{r - [(2N-1)/3]}{\sqrt{(16N-29)/90}} \tag{10–9}$$

where

N = Total number of observations
r = Observed number of runs of either As and Bs or Us and Ds, depending on which test is involved.

It is desirable to apply both run tests to any given set of observations because each test is different in terms of the types of patterns it can detect. Sometimes both tests will pick up a certain pattern, but sometimes only one will detect nonrandomness. If either does, the implication is that some sort of nonrandomness is present in the data.

Example 6

Twenty sample means have been taken from a process. The means are shown in the following table. Use median and up/down run tests with $z = 2$ to determine if assignable causes of variation are present. Assume the median is 11.0.

Solution

The means are marked according to above/below the median and up/down. The solid lines represent the runs.

Sample	A/B	Mean	U/D	Sample	A/B	Mean	U/D
1.	B	10.0	—	11.	B	10.7	D
2.	B	10.4	U	12.	A	11.3	U
3.	B	10.2	D	13.	B	10.8	D
4.	A	11.5	U	14.	A	11.8	U
5.	B	10.8	D	15.	A	11.2	D
6.	A	11.6	U	16.	A	11.6	U
7.	A	11.1	D	17.	A	11.2	D
8.	A	11.2	U	18.	B	10.6	D
9.	B	10.6	D	19.	B	10.7	U
10.	B	10.9	U	20.	A	11.9	U

A/B: 10 runs U/D: 17 runs

The expected number of runs for each test is:

$$E(r)_{\text{med}} = \frac{N}{2} + 1 = \frac{20}{2} + 1 = 11$$

$$E(r)_{\text{u/d}} = \frac{2N - 1}{3} = \frac{2(20) - 1}{3} = 13$$

The standard deviations are

$$\sigma_{\text{med}} = \sqrt{\frac{N - 1}{4}} = \sqrt{\frac{20 - 1}{4}} = 2.18$$

$$\sigma_{\text{u/d}} = \sqrt{\frac{16N - 29}{90}} = \sqrt{\frac{16(20) - 29}{90}} = 1.80$$

The z_{test} values are

$$z_{\text{med}} = \frac{10 - 11}{2.18} = -0.46$$

$$z_{\text{u/d}} = \frac{17 - 13}{1.80} = +2.22$$

Although the median test does not reveal any pattern, since its z_{test} value is less than ± 2, the up/down test does; its value exceeds $+2$. Consequently, nonrandom variations are probably present in the data and, hence, the process is not in control.

If ties occur in either test (e.g., a value equals the median or two values in a row are the same), assign A/B or U/D in such a manner that that z_{test} is as large as possible. If z_{test} still does not exceed ± 2 (± 1.96, etc.), you can be reasonably confident that a conclusion of randomness is justified.

PROCESS CAPABILITY

Vol. 2, seg 2
Bernard Welding

The variability of a process can significantly impact quality. Three commonly used terms refer to the variability of process output. Each term relates to a slightly different aspect of that variability, so it is important to differentiate these terms.

tolerances Specifications for range of acceptable values established by engineering design or customer requirements.

Tolerances or *specifications* are established by engineering design or customer requirements. They indicate a range of values in which individual units of output must fall in order to be acceptable.

Control limits are statistical limits that reflect the extent to which *sample statistics* such as means and ranges can vary due to randomness alone.

process variability Natural or inherent variability in a process.

Process variability reflects the natural or inherent (i.e., random) variability in a process. It is measured in terms of the process standard deviation.

Control limits and process variability are directly related: control limits are based on sampling variability, and sampling variability is a function of process variability. On the other hand, there is no direct link between tolerances and either control limits or process variability. Tolerances are specified in terms of a product or service, not in terms of the *process* by which the product or service is generated. Hence, in a given instance, the output of a process may or may not conform to specifications, even though the process may be statistically in control. That is why it is also necessary to take into account the *capability* of a process. The term **process capability** refers to the inherent variability of process output *relative to* the variation allowed by the design specifications. The following section describes capability analysis.

process capability The inherent variability of process output relative to the variation allowed by the design specification.

Capability Analysis

Capability analysis means determining whether the inherent variability of the process output falls within the acceptable range of variability allowed by the design specifications for the process output. If it is within the specifications, the process is said to be "capable." If it is not, the manager must decide how to correct the situation.

Consider the three cases illustrated in Figure 10–14. In the first case, process capability and output specifications are well matched, so that nearly all of the process output can be expected to meet the specifications. In the second case, the process variability is much less than what is called for, so that virtually 100 percent of the output should be within tolerance. In the third case, however, the specifications are tighter than what the process is capable of, so that even when the process is functioning as it should, a sizable percentage of the output will fail to meet the specifications. In other words, the process could be in control and still generate unacceptable output. Thus, we cannot automatically assume that a process that is in control will provide desired output. Instead, we must specifically check whether a process is *capable* of meeting specifications and not simply set up a control chart to monitor it. A process should be in control and within specifications *before* production begins—in essence, "Set the toaster correctly at the start. Don't burn the toast and then scrape it!"

In instances such as case C in Figure 10–14, a manager might consider a range of possible solutions: (1) redesign the process so that it can achieve the desired output, (2) use an alternate process that can achieve the desired output, (3) retain the current process but attempt to eliminate unacceptable output using 100 percent inspection, and (4) examine the specifications to see whether they are necessary or could be relaxed without adversely affecting customer satisfaction.

Obviously, process variability is the key factor in process capability. It is measured in terms of the process standard deviation; process capability is typically deemed to be ± 3 standard deviations from the process mean. To determine whether the process is capable, compare this ± 3 standard deviations value to the specifications that are expressed as an allowance deviation from an ideal value. For example, suppose the ideal

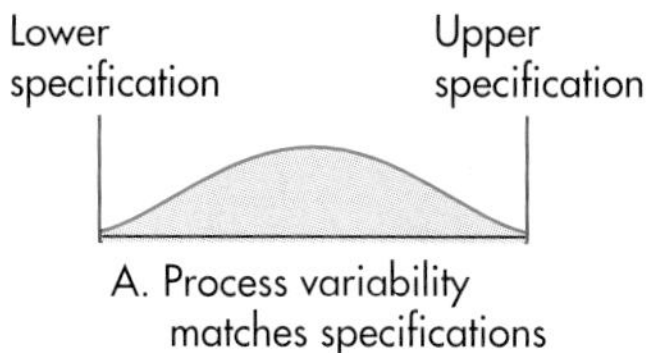

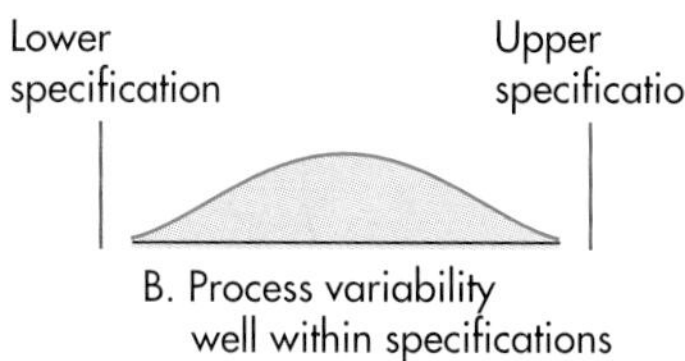

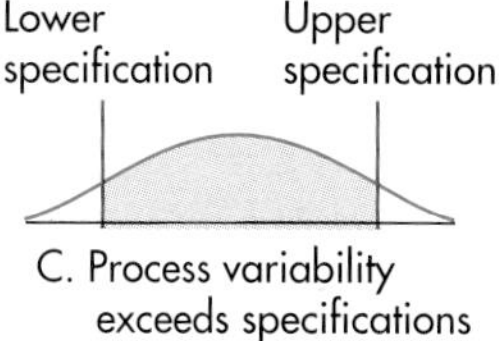

FIGURE 10–14

Process capability and specifications may or may not match

length of time to perform a service is 10 minutes, and an acceptable range of variation around this time is ±1 minute. If the process has a standard deviation of 0.5 minutes, it would not be capable because ±3 standard deviations would be ±1.5 minutes, exceeding the specification of ±1 minute.

Example 7

A manager has the option of using any one of three machines for a job. The machines and their standard deviations are listed below. Determine which machines are capable if the specification is 1.00 mm to 1.60 mm.

Machine	Standard Deviation (mm)
A	0.10
B	0.08
C	0.13

Solution

Determine the capability of each machine (i.e., 6 standard deviations) and compare that value to the specification of ±0.60 cm.

Machine	Standard Deviation (mm)	Machine Capability
A	0.10	0.60
B	0.08	0.48
C	0.13	0.78

We can see that both machine A and machine B are capable of producing output that is within the specs, but that machine C is not, because it exceeds the specification width of 0.60. See Figure 10–14 for a visual portrayal of these results.

To express the capability of a machine or process, some companies use the ratio of the specification width to the process capability. It can be computed using the following formula:

$$\text{Process capability ratio, } C_p = \frac{\text{specification width}}{\text{process width}} = \frac{\text{Upper specification} - \text{lower specification}}{6\sigma} \quad (10\text{–}10)$$

Example 8

Compute the process capability index for each machine in Example 7.

Solution

The specification width in Example 7 was 0.60 mm. Hence, to determine the capability index for each machine, you divide 0.60 by the process width (i.e., 6 standard deviations) of each machine. The results are shown in the following table:

FIGURE 10–15

Three-sigma quality and six-sigma quality

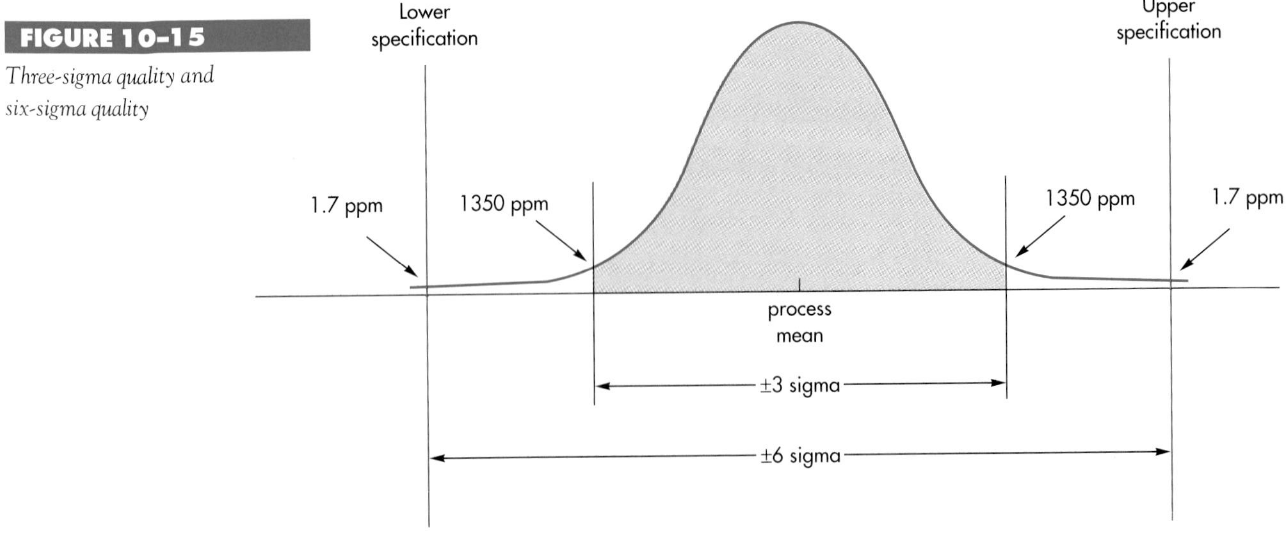

Machine	Standard Deviation (mm)	Machine Capability	C_p
A	0.10	0.60	0.60/0.60 = 1.00
B	0.08	0.48	0.60/0.48 = 1.25
C	0.13	0.78	0.60/0.78 = 0.77

Using the capability ratio, you can see that for a process to be capable, it must have a capability ratio of at least 1.00. Moreover, the greater the capability ratio, the greater the probability that the output of a machine or process will fall within design specifications.

The Motorola Corporation is well known for its use of the term *six-sigma,* which refers to its goal of achieving a process variability so small that the design specifications represent six standard deviations of the process. That means a process capability ratio equal to 2.00, resulting in an extremely small probability of getting any output not within the design specifications. This is illustrated in Figure 10–15.

Operations Strategy

It is neither necessary nor desirable to use control charts on every production process. Some processes are highly stable; other processes that previously produced unacceptable output may have been subjected to improvements that resulted in quality being *designed into* the process or product, making continual monitoring unnecessary. To use a "shotgun" approach that puts control charts on many processes would be wasteful.

Managers should use control charts on processes that tend to go out of control and, even then, only on a temporary basis until the causative problems have been permanently corrected. Similarly, when a process is changed significantly or a new process is introduced, it makes sense to monitor the output until it becomes apparent that the process is stable. After that, unless there is some compelling reason to believe that the process is apt to become unstable, there would be no reason to continue with a control chart.

READING

SIX-SIGMA QUALITY

www.motorola.com

Motorola refers to its quality program as *six-sigma quality,* where *sigma* is the standard deviation of a normal distribution, and *six-sigma* quality is an error rate of about 3 defectives per million units. This means that the percentage of the distribution that is more than six standard deviations from the mean of the distribution is 0.0003 percent! In some organizations, an error-free rate of 99 percent is regarded as very good; only one defective in every hundred units. However, consider that at that rate, a high-volume operation such as the U.S. Postal Service would lose or mishandle five million pieces of mail every day. Or suppose that the 99 percent error-free referred to health care. That would mean that one out of every 100 patients was misdiagnosed, or offered the wrong treatment or the wrong prescription. In Motorola's case, the six sigma is a goal the company is striving for; actual defective rates are currently higher. But it serves as a reminder that no matter how good quality is, there is always room for improvement.

Motorola's "six-sigma" quality program is one of the best in the world. Motorola works closely with and trains its suppliers to improve the quality of parts as shown here.

Summary

This chapter describes inspection and statistical process control. Inspection means examining the output of a process to determine whether it is acceptable. Key issues in inspection include where to inspect in the process, how often to inspect, and whether to inspect on-site or in a laboratory.

Statistical process control focuses on detecting departures from randomness in a process. Two basic tools of process control are control charts and run tests. The general theory of control charts is discussed, and four types of control charts—two for variables and two for attributes—and two types of run tests are described in the chapter. The chapter ends with a discussion of process capability. Table 10–4 provides a summary of formulas.

Key Terms

assignable variation, 445
attributes, 447
c-chart, 454
control chart, 446
control limits, 446
mean control chart, 448
p-chart, 452
process capability, 460
process variability, 460
quality of conformance, 443
random variation, 445
range control charts, 449
run, 456
run test, 456
statistical process control, 443
tolerances, 460
Type I error, 447
Type II error, 447
variables, 447

TABLE 10-4

Summary of formulas

CONTROL CHARTS		
Name	**Symbol**	**Control Limits**
Mean	$\bar{x}$	$\bar{\bar{x}} \pm z\frac{\sigma}{\sqrt{n}}$ or $\bar{\bar{x}} \pm A_2\bar{R}$
Range	R	UCL $= D_4\bar{R}$, LCL $= D_3\bar{R}$
Fraction defective	p	$\bar{p} \pm z\sqrt{\frac{\bar{p}(1-\bar{p})}{n}}$
Number of defects	c	$\bar{c} \pm z\sqrt{\bar{c}}$

RUN TESTS				
	NUMBER OF RUNS			
Name	**Observed**	**Expected**	**Standard Deviation**	**z**
Median	r	$\frac{N}{2}+1$	$\sqrt{\frac{N-1}{4}}$	$\frac{r-[(N/2)+1]}{\sqrt{(N-1)/4}}$
Up/down	r	$\frac{2N-1}{3}$	$\sqrt{\frac{16N-29}{90}}$	$\frac{r-[(2N-1)/3]}{\sqrt{(16N-29)/90}}$

PROCESS CAPABILITY		
Name	**Symbol**	**Formula**
Process Capability Index	C_p	$\frac{\text{specification width}}{3\sigma \text{ of process}}$

Solved Problems

Problem 1

Process distribution and sampling distribution. An industrial process that makes 3-foot sections of plastic pipe produces pipe with an average inside diameter of 1 inch and a standard deviation of 0.05 inches.

a. If you randomly select one piece of pipe, what is the probability that its inside diameter will exceed 1.02 inches, assuming the population is normal?

b. If you select a random sample of 25 pieces of pipe, what is the probability that the sample *mean* will exceed 1.02 inches?

Solution

$\mu = 1.00, \sigma = 0.05$

a. $z = \frac{x-\mu}{\sigma} = \frac{1.02 - 1.00}{0.05} = 0.4$

Using Appendix Table A, $P(z > 0.4) = .5000 - .1554 = .3446$.

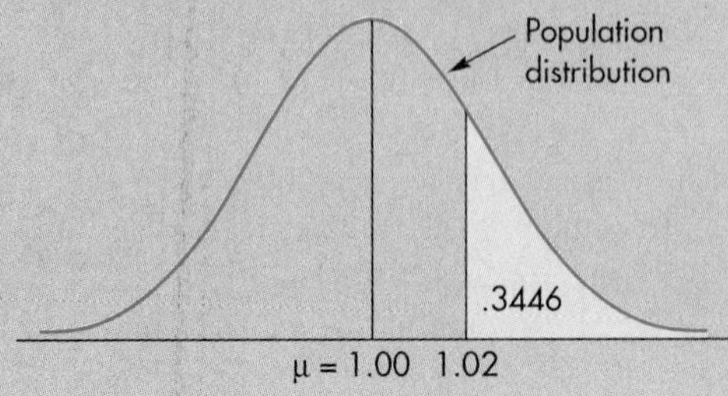

b. $z = \frac{\bar{x} - \mu}{\sigma/\sqrt{n}} = \frac{1.02 - 1.00}{0.05/\sqrt{25}} = 2.00$

Using Appendix Table A, $P(z>2.00) = .5000 - .4772 = .0228$

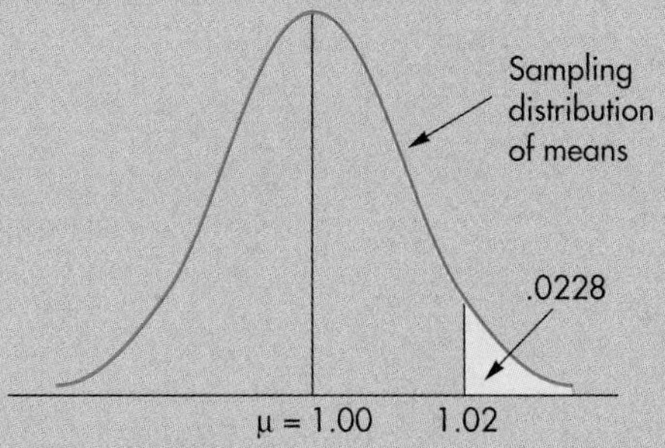

Problem 2

Control charts for means and ranges. Drop-forged steel handles have a designed weight of 10 ounces. Five samples of four observations each have been taken. Use the sample data in conjunction with Table 10–2 to construct upper and lower control limits for both a mean chart and a range chart. Do the results suggest that the process is in control?

	Sample 1	Sample 2	Sample 3	Sample 4	Sample 5
	10.2	10.3	9.7	9.9	9.8
	9.9	9.8	9.9	10.3	10.2
	9.8	9.9	9.9	10.1	10.3
	10.1	10.4	10.1	10.5	9.7
Totals	40.0	40.4	39.6	40.8	40.0

Solution

a. Determine the mean and range of each sample.

$\bar{x} = \frac{\sum x}{n}$, Range = Largest − Smallest

Sample	Mean	Range
1	40.0/4 = 10.0	10.2 − 9.8 = 0.4
2	40.4/4 = 10.1	10.4 − 9.8 = 0.6
3	39.6/4 = 9.9	10.1 − 9.7 = 0.4
4	40.8/4 = 10.2	10.5 − 9.9 = 0.6
5	40.0/4 = 10.0	10.3 − 9.7 = 0.6

b. Compute the average mean and average range:

$$\bar{\bar{x}} = \frac{10.0 + 10.1 + 9.9 + 10.2 + 10.0}{5} = \frac{50.2}{5} = 10.04$$

$$\bar{R} = \frac{0.4 + 0.6 + 0.4 + 0.6 + 0.6}{5} = \frac{2.6}{5} = 0.52$$

c. Obtain factors A_2, D_4, and D_3 from Table 10–2 for $n = 4$: $A_2 = 0.73$, $D_4 = 2.28$, $D_3 = 0$

d. Compute upper and lower limits:

$$UCL_{\bar{x}} = \bar{\bar{x}} + A_2\bar{R} = 10.04 + 0.73(0.52) = 10.42$$

$$LCL_{\bar{x}} = \bar{\bar{x}} - A_2\bar{R} = 10.04 + 0.73(0.52) = 9.66$$

$$UCL_R = D_4\bar{R} = 2.28(0.52) = 1.19$$

$$LCL_R = D_3\bar{R} = 0(0.52) = 0$$

e. Plot sample means and ranges on their respective control charts, or otherwise verify that points are within limits.

The smallest sample mean is 9.9, and the largest is 10.2. Both are well within the control limits. Similarly, the largest sample range is 0.6, which is also within the control limits. Hence, the results suggest that the process is in control. Note, however, that for illustrative purposes, the number of samples is deliberately small; 20 or more samples would give a clearer indication of control limits and whether the process is in control.

Problem 3

Type I error (alpha risk). After several investigations of points outside control limits revealed nothing, a manager began to wonder about the probability of a Type I error for the control limits used ($z = 1.90$).

a. Determine the alpha risk for this value of z.

b. What z would provide an alpha risk of about 2 percent?

Solution

a. Using Appendix Table A, find that the area under the curve between $z = 0$ and $z = +1.90$ is .4713. Therefore, the area (probability) of values *within* -1.90 to $+1.90$ is $2(.4713) = .9426$, and the area *beyond* these values is $1 - .9426 = .0574$.

b. The alpha risk (Type I error probability is always specified as an *area* in the tail(s) of a distribution. With control charts, you are involved with two-sided control limits. Consequently, half of the risk lies in each tail. Hence, the area in the right tail is 1 percent, or .0100. This means that .4900 should be the area under the curve between $z = 0$ and the value of z you are looking for. The closest value is .4901 for $z = \pm 2.33$. Thus, control limits based on $z = \pm 2.33$ provide an alpha risk of about 2 percent.

Problem 4

p*-chart and* c*-chart.* Using the appropriate control chart, determine two-sigma control limits for each case:

a. An inspector found an average of 3.9 scratches in the exterior paint of each of the automobiles being prepared for shipment to dealers.

b. Before shipping lawnmowers to dealers, an inspector attempts to start each mower and notes any that do not start on the first try. The lot size is 100 mowers, and an average of 4 did not start (4 percent).

Solution

The choice between these two types of control charts relates to whether *two* types of results can be counted (*p*-chart) or whether *only occurrences* can be counted (*c*-chart).

a. The inspector can only count the scratches that occurred, not the ones that did not occur. Consequently, a *c*-chart is appropriate. The sample average is 3.9 scratches per car. Two-sigma control limits are found using the formula

$$\text{UCL} = \bar{c} + z\sqrt{\bar{c}}$$
$$\text{LCL} = \bar{c} - z\sqrt{\bar{c}}$$

where $\bar{c} = 3.9$ and $z = 2$. Thus,

$$\text{UCL} = 3.9 + 2\sqrt{3.9} = 7.85 \text{ scratches}$$
$$\text{LCL} = 3.9 - 2\sqrt{3.9} = -0.05 \text{ (0 scratches)}$$

(Note: Round to zero only if the computed lower limit is negative.)

b. The inspector can count both the lawnmowers that started and those that did not start. Consequently, a *p*-chart is appropriate. Two-sigma control limits can be computed using the following:

$$UCL = \bar{p} + z\sqrt{\frac{\bar{p}(1-\bar{p})}{n}}$$

$$LCL = \bar{p} - z\sqrt{\frac{\bar{p}(1-\bar{p})}{n}}$$

where

$\bar{p} = .04$
$n = 100$
$z = 2$

Thus,

$$UCL = .04 + 2\sqrt{\frac{.04(.96)}{100}} = 0.079$$

$$LCL = .04 - 2\sqrt{\frac{.04(.96)}{100}} = 0.001$$

Problem 5

Run tests. The number of defectives per sample for 11 samples is shown below. Determine if nonrandom patterns are present in the sequence.

	SAMPLE										
	1	**2**	**3**	**4**	**5**	**6**	**7**	**8**	**9**	**10**	**11**
Number of defectives	22	17	19	25	18	20	21	17	23	23	24

Solution

Since the median isn't given, it must be estimated from the sample data. To do this, array the data from low to high; the median is the middle value. (In this case, there are an odd number of values. For an even number of values, average the middle two to obtain the median.) Thus,

17 17 18 19 20 21 22 23 23 24 25
(5 below) ↑ (5 above)
median

The median is 21.

Next, code the observations using A/B and U/D:

Sample	**A/B**	**Number of Defectives**	**U/D**
1	A	22	—
2	B	17	D
3	B	19	U
4	A	25	U
5	B	18	D
6	B	20	U
7	tie	21	U
8	B	17	D
9	A	23	U
10	A	23	tie
11	A	24	U

Note that each test has tied values. How these are resolved can affect the number of observed runs. Suppose that you adhere to this rule: Assign a letter (A or B, U or D) so that the resulting difference between the observed and expected number of runs is as large as possible. To accomplish this, it is necessary to initially ignore ties and count the runs to see whether there are too many or too few. Then return to the ties and make the assignments. The rationale for this rule is that it is a conservative method for retaining data; if you conclude that the data are random using this approach, you can be reasonably confident that the method has not "created" randomness. With this in mind, assign a B to sample 7 since the expected number of runs is

$$E(r)_{med} = N/2 + 1 = 11/2 + 1 = 6.5$$

and the difference between the resulting number of runs, 5, and 6.5 is greater than between 6.5 and 7 (which occurs if A is used instead of B). Similarly, in the up/down test, a U for sample 10 produces 6 runs, whereas a D produces 8 runs. Since the expected number of runs is

$$E(r)_{u/d} = (2N - 1) \div 3 = (22 - 1) \div 3 = 7$$

it makes no difference which one is used: both yield a difference of 1. For the sake of illustration, a D is assigned.

The computations for the two tests are summarized below. Each test has a z value that is greater than 2.00. Because neither test reveals nonrandomness, you may conclude that the data are random.

	Runs Observed	$E(r)$	σ_r	z	Conclude
Median	5	6.5	1.58	−0.95	Random
Up/down	8	7.0	1.28	0.78	Random

Discussion and Review Questions

1. List the steps in the control process.
2. What are the key concepts that underlie the construction and interpretation of control charts?
3. What is the purpose of a control chart?
4. Why is order of observation important in process control?
5. Briefly explain the purpose of each of these control charts:
   a. x-bar
   b. Range
   c. *p*-chart
   d. *c*-chart
6. What is a run? How are run charts useful in process control?
7. If all observations are within control limits, does that guarantee that the process is random? Explain.
8. Why is it usually desirable to use both a median run test and an up/down run test on the same data?
9. If both run tests are used, and neither reveals nonrandomness, does that prove that the process is random? Explain.
10. Define and contrast control limits, tolerances, and process variability.
11. A customer has recently tightened the specs for a part your company supplies. The specs are now much tighter than the machine being used for the job is capable of. Briefly identify alternatives you might consider to resolve this problem. (See Figure 10–14C.)
12. A new order has come into your department. The capability of the machine used for this type of work will enable virtually all of the output to be well within the specs. (See Figure 10–14B.)

a. What benefits might be derived from this situation?

b. What alternatives might be considered by the manager?

13. Answer these questions about inspection:

a. What level of inspection is optimal?

b. What factors guide the decision of how much to inspect?

c. What are the main considerations in choosing between centralized inspection and on-site inspection?

d. What points are potential candidates for inspection?

Memo Writing Exercises

1. Suppose you recently received a memo from your manager, Sid Twickenham, in which he asks for your input on the possibility of switching from the use of a *p*-chart to an *x*-bar chart to monitor a process. He wants to know the circumstances under which such a change would be acceptable, and the advantages and costs of making the switch.

   Write a one-page memo in response.

2. You receive a complaint about defective explosion-proof fittings found coming from a process. The fittings are produced during the first and second shifts, but defectives occur only during the second shift. You must decide whether to proceed with continuous monitoring of the process or to use a control chart to periodically monitor the process in order to pinpoint the cause of the defectives.

   Write a one-page memo to your subordinate, Arnie Green, asking him to research the complaint. Indicate to Green what information he should obtain in order for you to make your decision.

Problems

1. Manufacturing specifications call for motor housings that weigh between 24 and 25 pounds. The housings are cast using a process that produces castings with a mean of 24.5 pounds and a standard deviation of 0.2 pounds. The distribution of output is normal.

   a. What percentage of housings will not meet the weight specs?

   b. Within what values will 99.74 percent of sample means of this process fall, if samples of $n = 16$ are taken and the process is in control (random)?

   a. $z = 2.5$; .0124
   b. 24.5±.15 lbs.

2. An automatic filling machine is used to fill 1-liter bottles of cola. The machine's output is approximately normal with a mean of 1.0 liter and a standard deviation of 0.01 liter. Output is monitored using means of samples of 25 observations.

   a. Determine upper and lower control limits that will include roughly 95.5 percent of the sample means when the process is in control.

   b. Graph the control chart and plot these sample means: 1.005, 1.001, 0.998, 1.002, 0.995, and 0.999. Is the process in control?

   a. $z = 2.0$;
   LCL = .996 liters
   UCL = 1.004 liters
   b. Out of control.

3. A grinding and deburring operation is monitored using a mean and a range chart. Six samples of $n = 20$ observations have been obtained and the sample means and ranges computed:

Sample	Mean	Range
1	3.06	0.42
2	3.15	0.50
3	3.11	0.41
4	3.13	0.46
5	3.06	0.46
6	3.09	0.45

$\bar{\bar{x}} = 3.10$
$\bar{R} = 0.45$

a. mean:
UCL = 3.181
LCL = 3.019
range:
UCL = 0.7155
LCL = 0.1845
b. In control.

a. Using the factors in Table 10–2, determine upper and lower limits for mean and range charts.

b. Is the process in control?

Mean chart:
UCL = 81.04
LCL = 78.88
Range chart:
UCL = 4.009
LCL = 0
Process is in control.

4. A continuous process cuts plastic tubing into nominal lengths of 80 centimeters. Samples of five observations each have been taken, and the results are as listed. Using factors from Table 10–2, determine upper and lower control limits for mean and range charts, and decide if the process is in control.

SAMPLE					
1	**2**	**3**	**4**	**5**	**6**
79.2	80.5	79.8	78.9	80.5	79.7
78.8	78.7	79.4	79.4	79.6	80.6
80.0	81.0	80.4	79.7	80.4	80.5
78.4	80.4	80.3	79.4	80.8	80.0
81.0	80.1	80.8	80.6	78.8	81.1

5. Using samples of 200 observations each, a quality inspector found the following:

Sample	1	2	3	4
Number of defectives	4	2	5	9

a. 1 .020
2 .010
3 .025
4 .045
b. 2.5%
c. Mean = .025
Std. dev. = .011
d. $z = 2.17$; .0011 to .0489
e. $z = 2$; alpha = .0456
f. Yes.
g. Mean = .02
Std. dev. = .0099
h. 0 to .04
Last sample beyond UCL.

a. Determine the fraction defective in each sample.

b. If the true fraction defective for this process is unknown, what is your estimate of it?

c. What is your estimate of the mean and standard deviation of the sampling distribution of fractions defective for samples of this size?

d. What control limits would give an alpha risk of .03 for this process?

e. What alpha risk would control limits of 0.047 and 0.003 provide?

f. Using control limits of 0.047 and 0.003, is the process in control?

g. Suppose that the long-term fraction defective of the process is known to be 2 percent. What are the values of the mean and standard deviation of the sampling distribution?

h. Construct a control chart for the process, assuming a fraction defective of 2 percent, using two-sigma control limits. Is the process in control?

UCL = .0234
LCL = 0
Sample 10 is too large.
New UCL = .0197
New LCL = 0

6. An automatic screw machine produces 2-inch hex nuts. Using the following sample data and $n = 200$, construct a control chart for the fraction defective using two-sigma limits. Is the process in control? If not, eliminate any values that are outside the limits and compute revised limits.

	SAMPLE												
	1	**2**	**3**	**4**	**5**	**6**	**7**	**8**	**9**	**10**	**11**	**12**	**13**
Number of defectives	1	2	2	0	2	1	2	0	2	7	3	2	1

UCL = 16.266
LCL = 0
In control.

7. The postmaster of a small western city receives a certain number of complaints each day about mail delivery. Assume that the distribution of daily complaints is Poisson. Construct a control chart with three sigma limits using the following data. Is the process in control?

	DAY													
	1	**2**	**3**	**4**	**5**	**6**	**7**	**8**	**9**	**10**	**11**	**12**	**13**	**14**
Number of complaints	4	10	14	8	9	6	5	12	13	7	6	4	2	10

8. Construct a control chart with three-sigma limits for the number of defects per spool of cable, given the following data. Is the process in control?

UCL = 5.17
LCL = 0
In control.

Observation	Number of Defects	Observation	Number of Defects
1	2	8	0
2	3	9	2
3	1	10	1
4	0	11	3
5	1	12	1
6	3	13	2
7	2	14	0

9. During long runs of canned vegetables through a labeling machine, a few labels invariably fail to adhere properly and eventually fall off. Using the following sample data, which are based on samples of 100 observations each, construct a control chart for the fraction defective using 95 percent control limits, and determine if the process is in control.

UCL = .098
LCL = .010
In control.

Sample	Number of Defectives	Sample	Number of Defectives
1	5	9	5
2	3	10	8
3	6	11	3
4	7	12	4
5	4	13	5
6	6	14	6
7	8	15	6
8	4	16	7

10. Tolerances for a metal shaft are much wider than the machine used to make the shafts is capable of. Consequently, the decision has been made to allow the cutting tool to wear a certain amount before replacement. The tool wears at the rate of 0.004 centimeters per piece. The process has a natural variation, σ, of 0.01 centimeters and is normally distributed. Tolerances are 15.0 to 15.2 centimeters, and $n = 1$. For three-sigma limits, how many shafts can the process turn out before tool replacement becomes necessary? (See diagram.)

$$\frac{15.2 - \text{Initial UCL}}{\text{wear rate}} = \frac{.14}{.004} = 35 \text{ pieces}$$

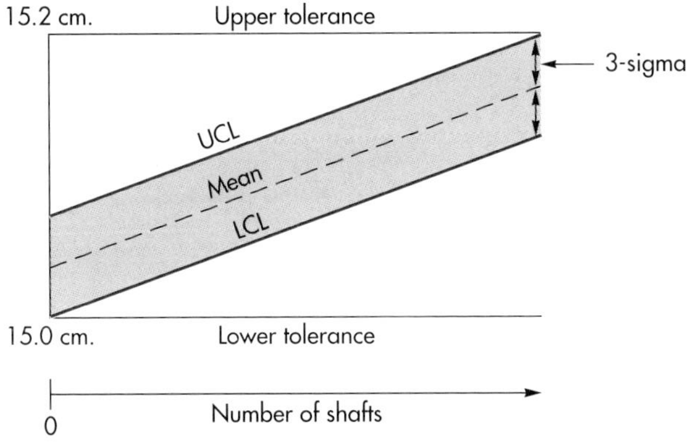

11. Tolerances for the plastic tubing mentioned in Problem 4 are 78 centimeters and 81 centimeters. Based on the data given in the problem, are the tolerances being met? Estimate the percentage of process output that can be expected to fall within the tolerances.

Only 1 of 30 (3.3%) exceeds tolerances.
97% will be acceptable.

12. A process that produces bars of soap is to be investigated. Historically, the process has had a standard deviation equal to 0.146. The means of 39 samples of $n = 14$ are:

a. UCL = 3.97
LCL = 3.73
Out of control.

b.

	Median	U/D
obs.	18	29
exp.	20.5	25.7
σ	3.08	2.57
z	−.81	1.28
Concl.	random	random

Sample	Mean	Sample	Mean	Sample	Mean
1	3.86	14	3.81	27	3.81
2	3.90	15	3.83	28	3.86
3	3.83	16	3.86	29	3.98
4	3.81	17	3.82	30	3.96
5	3.84	18	3.86	31	3.88
6	3.83	19	3.84	32	3.76
7	3.87	20	3.87	33	3.83
8	3.88	21	3.84	34	3.77
9	3.84	22	3.82	35	3.86
10	3.80	23	3.89	36	3.80
11	3.88	24	3.86	37	3.84
12	3.86	25	3.88	38	3.79
13	3.88	26	3.90	39	3.85

a. Construct an $\bar{x}$-chart for this process with three-sigma limits. Is the process in control?

b. Analyze the data using a median run test and an up/down run test. What can you conclude?

Median and up/down: random for A. Median and up/down: nonrandom for B.

13. For each of the accompanying control charts, analyze the data using both median and up/down run tests with $z = \pm 1.96$ limits. Are nonrandom variations present? Assume the center line is the long-term median.

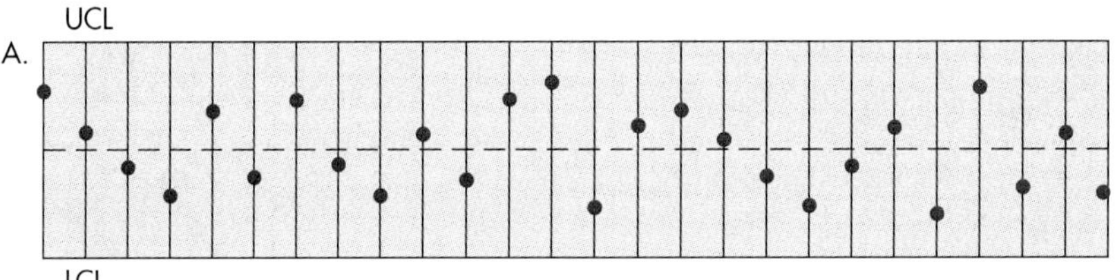

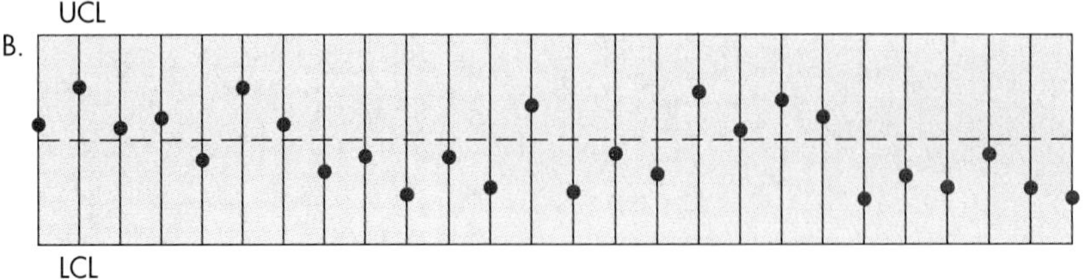

	Median	Up/down
a.	z = −0.52	z = −1.47
b.	z = −1.11	z = −1.36

14. Analyze the data in the problems listed below using median and up/down run tests with $z = \pm 2$.

a. Problem 9.

b. Problem 7.

15. Use both types of run tests to analyze the daily expense voucher listed. Assume a median of $31.

Day	Amount	Day	Amount	Day	Amount
1.....	$27.69	21	$28.60	41	$26.76
2.....	28.13	22	20.02	42	30.51
3.....	33.02	23	26.67	43	29.35
4.....	30.31	24	36.40	44	24.09
5.....	31.59	25	32.07	45	22.45
6.....	33.64	26	44.10	46	25.16
7.....	34.73	27	41.44	47	26.11
8.....	35.09	28	29.62	48	29.84
9.....	33.39	29	30.12	49	31.75
10	32.51	30	26.39	50	29.14
11	27.98	31	40.54	51	37.78
12	31.25	32	36.31	52	34.16
13	33.98	33	27.14	53	38.28
14	25.56	34	30.38	54	29.49
15	24.46	35	31.96	55	30.81
16	29.65	36	32.03	56	30.60
17	31.08	37	34.40	57	34.46
18	33.03	38	25.67	58	35.10
19	29.10	39	35.80	59	31.76
20	25.19	40	32.23	60	34.90

	Median	Up/down
obs.	22	35
exp.	31	39.67
σ	3.84	3.22
z	−2.34	−1.45
Concl.	non-random	random

Since one test suggests nonrandom, conclusion is process not in control.

16. A company has just negotiated a contract to produce a part for another firm. In the process of manufacturing the part, the inside diameter of successive parts becomes smaller and smaller as the cutting tool wears. However, the specs are so wide relative to machine capabilities that it is possible to set the diameter initially at a large value and let the process run for a while before replacing the cutting tool.

    The inside diameter decreases at an average rate of 0.001 cm per part, and the process has a standard deviation of 0.01 cm. The variability is approximately normal. Assuming three-sigma control limits, how frequently must the tool be replaced if the process specs are 3 cm and 3.5 cm, and the initial setting of the UCL is at the upper spec? Use $n = 1$.

    440 pieces

17. (Refer to Solved Problem 2.) Suppose the process specs are 9.65 and 10.35 ounces. Based on the data given, does it appear that the specs are being met? If not, what should one look for?

    10% of obs. above limit; lower process mean, or reduce variability.

18. A production process consists of a three-step operation. The scrap rate is 10 percent for the first step and 6 percent for the other two steps.

    a. If the desired daily output is 450 units, how many units must be started to allow for loss due to scrap?

    b. If the scrap rate for each step could be cut in half, how many units would this save in terms of the scrap allowance?

    c. If the scrap represents a cost of $10 per unit, how much is it costing the company per day for the original scrap rate?

    a. 556 units
    b. 62 units
    c. 116 units; $1,160.00

19. (Refer to the data in Example 5.) Two additional observations have been taken. The first resulted in three defects, and the second had four defects. Using the set of 20 observations, perform run tests on the data. What can you conclude about the data?

    Median $z = .9177$
    Up/down $z = .5561$
    Both suggest randomness

20. A machine operator has obtained the following measurements:

a. 1 = 4.3
2 = 4.5
3 = 4.5
4 = 4.7
b. Mean = 4.5
Std. dev. = .192
c. Mean = 4.5
Std. dev. = .086
d. UCL = 4.758
LCL = 4.242
Risk = .0026
e. z = 4.19; Risk = 0
f. None
g. Means: 4.3115 to 4.6885
First mean below LCL Last mean above UCL; ranges: 0 to .6875
All within limits.
h. Two different measures of dispersion used.
i. UCL = 4.64
LCL = 4.16
Last value above UCL.

SAMPLE			
1	**2**	**3**	**4**
4.5	4.6	4.5	4.7
4.2	4.5	4.6	4.6
4.2	4.4	4.4	4.8
4.3	4.7	4.4	4.5
4.3	4.3	4.6	4.9

a. Determine the mean of each sample.

b. If the process parameters are unknown, estimate its mean and standard deviation.

c. Estimate the mean and standard deviation of the sampling distribution.

d. What would three-sigma control limits for the process be? What alpha risk would they provide?

e. What alpha risk would control limits of 4.14 and 4.86 provide?

f. Using limits of 4.14 and 4.86, are any sample means beyond the control limits? If so, which one(s)?

g. Construct control charts for means and ranges using Table 10–2. Are any samples beyond the control limits? If so, which one(s)?

h. Explain why the control limits are different for means in parts *d* and *g*.

i. If the process has a known mean of 4.4 and a known standard deviation of 0.18, what would three-sigma control limits be for a mean chart? Are any sample means beyond the control limits? If so, which one(s)?

a. $C_p = \frac{\text{specification width}}{\text{process width}} = \frac{.01}{.009} = 1.11$

b. In order to be capable, the process capability index must be at least 1.00. In this instance, the index is 1.11, so the process is capable.

21. A process that produces computer chips has a mean of .03 defective chips and a standard deviation of .003 chips. The allowable variation is from .02 to .04 defectives.

a. Compute the capability ratio for the process.

b. Is the process capable?

See IM.

22. Given the following list of machines, the standard deviation for each, and specifications for a job that may be processed on that machine, determine which machines are capable of performing the given jobs.

Machine	Standard Deviation (gr)	Job Specification (±gr)
001	0.02	0.05
002	0.04	0.07
003	0.10	0.18
004	0.05	0.15
005	0.01	0.04

See IM.

23. Suppose your manager presents you with the following information about machines that could be used for a job, and wants your recommendation on which one to choose. The specification width is 0.24 mm. In this instance, you can narrow the set of choices, but you probably wouldn't make a recommendation without an additional piece of information. Explain the logic of the last statement.

Machine	Cost per Unit ($)	Standard Deviation (mm)
A	20	0.079
B	12	0.080
C	11	0.084
D	10	0.081

CASE

TIGER TOOLS

Tiger Tools, a division of Drillmore Industries, was about to launch a new product. Production Manager Michelle York asked her assistant, Jim Peterson, to check the capability of the oven used in the process. Jim obtained 18 random samples of 20 pieces each. The results of those samples are shown in the following table. After he analyzed the data, he concluded that the process was not capable based on a specification width of 1.20 cm.

Michelle was quite disappointed when she heard this. She had hoped that with the introduction of the new product her operation could run close to full capacity and regain some of its lost luster. The company has a freeze on capital expenditures of more than $10,000, and a replacement oven would cost many times that amount. Jim Peterson worked with the oven crew to see if perhaps different settings could produce the desired results, but they were unable to achieve any meaningful improvements.

Still not ready to concede, Michelle contacted one of her former professors and explained the problem. The professor suggested obtaining another set of samples, this time using a smaller sample size and taking more samples. Michelle then conferred with Jim and they agreed that he would take 27 samples of five observations each. The results are shown in the following table. Consider the following questions, and then write a brief report to Michelle summarizing your findings.

Sample	Mean	Range
1	45.01	.85
2	44.99	.89
3	45.02	.86
4	45.00	.91
5	45.04	.87
6	44.98	.90
7	44.91	.86
8	45.04	.89
9	45.00	.85
10	44.97	.91
11	45.11	.84
12	44.96	.87
13	45.00	.86
14	44.92	.89
15	45.06	.87
16	44.94	.86
17	45.00	.85
18	45.03	.88

Sample	Mean	Range
1	44.96	.42
2	44.98	.39
3	44.96	.41
4	44.97	.37
5	45.02	.39
6	45.03	.40
7	45.04	.39
8	45.02	.42
9	45.08	.38
10	45.12	.40
11	45.07	.41
12	45.02	.38
13	45.01	.41
14	44.98	.40
15	45.00	.39
16	44.95	.41
17	44.94	.43
18	44.94	.40
19	44.87	.38
20	44.95	.41
21	44.93	.39
22	44.96	.41
23	44.99	.40
24	45.00	.44
25	45.03	.42
26	45.04	.38
27	45.03	.40

1. How did Jim conclude that the process was not capable based on his first set of samples? (Hint: Estimate the process standard deviation, σ, using $A_2R \approx 3\frac{\sigma}{\sqrt{n}}$.)
2. Does the second set of samples show anything that the first set didn't? Explain what and why.
3. Assuming the problem can be found and corrected, what impact do you think this would have on the capability of the process? Compute the potential process capability.
4. If small samples can reveal something that large samples might not, why not just take small samples in every situation?

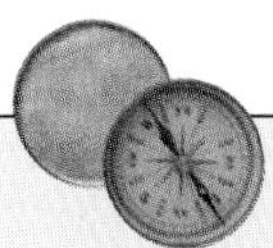

OPERATIONS TOUR

IN THE CHIPS AT JAYS **Neil Steinberg**

A potato chip is a delicate thing. Fragile. A pound of pressure will crush it. So when you're moving 250 tons of chips through your plant, as they do every day at Jays Foods, you need to have a system.

"You don't buy potato crumbs, you buy potato chips," said Tom Howe, CEO and co-owner of the Chicago company, at 99th and Cottage Grove. Jays makes 125 different types and brands of chips and several hundred varieties of popcorn, puffs, twists, pretzels and assorted bagged munchies.

Jays combats the tendency of potato chips to crush into flinders with a variety of conveyor belts, radial filling chutes and gently vibrating slides, where masses of chips, a yard deep, are gradually massaged forward, the outer layer of chips shearing away like the face of a glacier.

The raw material is far easier to handle. An entire semi-trailer of sturdy North Dakota "chipping" potatoes can be emptied in a matter of minutes, by backing the trailer onto a hydraulic lift, tilting it 45 degrees and letting the potatoes—grown for their thin skins and low moisture—tumble out.

About a dozen semi-trailers' worth of potatoes arrive every day. The potatoes are immediately separated into big and small sizes for a purpose both reasonable and extraordinary: Big potatoes make big chips that go into large bags; small potatoes make small chips for lunch-size bags.

"Nobody wants to open a small bag and find three big potato chips in it," Howe said.

Computers keep track of everything, shunting potatoes to 15,000-pound holding bins. Each bin feeds into a pipe containing a turning screw—a version of the ancient Archimedes screw used to pump water—that moves the potatoes from the bin to conveyor belts, to where they are washed and skinned—the skin scrubbed off by metal bristle brushes.

No machine can detect if a potato is rotten inside. So a pair of human inspectors reach into the passing brown parade and give the potatoes a quick squeeze. Occasionally, they snatch one and slice it open, usually revealing black areas of rot, a skill they attribute to experience.

"I *know,*" said Alicia Jimenez, asked to explain what about a potato tips her off to slice it open and find rot.

The naked potatoes are sent into high-speed chippers—spinning brass rings, each with eight blades inside, straight blades for straight chips, ripple blades for ripple chips.

The blades cut the potatoes, but the potatoes take their revenge. Every three hours the blades are dulled and the line must be stopped so the old rings can be replaced by new rings with sharpened blades.

The sheer quantity of slicing spews big foamy banks of starch from either side of the chipper, which calls to mind a washing machine gone berserk.

Potato chips account for about 55 percent of Jays' business.

The raw chips spend three minutes cooking in hot corn oil, which is constantly circulated and filtered. Then they are salted, and flavorings—barbecue, for instance, or sour cream and onion—are added.

After the chips are fried, there is another quality check, in which women pluck burned and deformed chips out of the masses passing by. The chips are conveyed on a link grid, wide enough to let broken chips fall through.

The chips also are laser-inspected, rushing, in a single layer, over a complex device called an Opti-Sort Scanner. Chips with dark spots or holes are detected by a laser, which instructs one of 82 small tubes to fire a puff of air that knocks the substandard chip off the line, into a discard bin.

The discards—about 3 percent of production—are gathered up and used: Starch is drawn out and sold to cornstarch makers; the rest goes to hog feed. Just as the stockyards were said to use every part of the pig but the squeal, at Jays every part of the potato is used but the rich, earthy smell.

Jays even tried to sell burnt chips to the public once, about 20 years ago. "Consumers kept telling us they liked the brown chips," said Len Japp Jr., recalling the "Brownies" variety. "It went over like a lead balloon." Japp and his father, now 93 and honorary chairman of the board, sold the company to Borden in 1986. "They almost ruined it," Howe said, citing a slump in product quality and neglect of the Jays distribution system. "They lost the connection with the consumer."

By 1994, Jays was on the rocks and the Japps, allied with Howe, bought the company back. "Not too many people have a second chance in life," said Japp, whose children are in the company.

Getting the chips in the bags is another challenge: You can't just fill up bags and seal them; the chips would be smashed. Rather, a conveyor pours chips—gently—on the central hub of a large, wheel-like device, where the chips scatter into 15 buckets that are, basically, scales. A computer monitors the weight of each bucket and opens up the exact combination that, in this case, will fill a 14-ounce bag. The bags are packed into boxes that read: "HANDLE LIKE EGGS."

While not exactly perishable, potato chips do have a shelf life of about eight weeks, only one day of which is spent at the plant.

"Potatoes that are in this morning will be in our branches tomorrow morning, ready to hit the streets," Howe said. Jays is still a regional brand, sold in Illinois, Indiana, Michigan, Wisconsin and Missouri. But business has grown 50 percent in the past two years.

"We connect to people's lifestyle," Howe said. "People treat themselves with Jays. We're in the fun food business."

Source: Reprinted with permission of *Chicago Sun-Times*, December 26, 1997.

Selected Bibliography

Baker, Thomas R. "Quality Engineering by Design." *Quality Progress*, December 1986, pp. 32–42.

Duncan, A. J. *Quality Control and Industrial Statistics*. 5th ed. Burr Ridge, IL: Richard D. Irwin, 1986.

Enrick, Norbert L. *Quality, Reliability, and Process Improvement*. 8th ed. New York: Industrial Press, 1985.

Ernst & Young Consulting Group. *Total Quality*. Burr Ridge, IL: Dow Jones-Irwin, 1990.

Garvin, David A. "Quality on the Line." *Harvard Business Review*, September–October 1983, p. 64.

Gitlow, Howard S.; Shelly Gitlow; Alan Oppenheim; and Rosa Oppenheim. *Tools and Methods for the Improvement of Quality*. Burr Ridge, IL: Richard D. Irwin, 1989.

Gitlow, Howard S., and Paul T. Hertz. "Product Defects and Productivity." *Harvard Business Review*, September–October 1983, p. 131.

Grant, Eugene L., and Richard S. Leavenworth. *Statistical Quality Control*. 5th ed. New York: McGraw-Hill, 1980.

Hall, Robert W. *Attaining Manufacturing Excellence*. Burr Ridge, IL: Dow Jones-Irwin, 1987.

Hunt, V. Daniel. *Quality in America*. Burr Ridge, IL: Business One Irwin, 1992.

Juran, J. M. "The Quality Trilogy." *Quality Progress* 19, no. 8 (August 1986), pp. 19–24.

Juran, J. M., and F. M. Gryma, Jr. *Quality Planning and Analysis*. 2nd ed. New York: McGraw-Hill, 1980.

Kume, Hitoshi. *Statistical Methods for Quality Improvement*. Tokyo: AOTS Chosakai, 1985.

Leonard, Frank S., and Earl Sasser. "The Incline of Quality." *Harvard Business Review*, September–October 1982, pp. 163–71.

Reddy, Jack, and Abe Berger. "Three Essentials of Product Quality." *Harvard Business Review*, July–August 1983, pp. 153–59.

Schein, Lawrence, and Melissa A. Berman, eds. *Total Quality Performance*. Research Report No. 909. New York Conference Board, 1988.

SUPPLEMENT TO CHAPTER TEN

ACCEPTANCE SAMPLING

LEARNING OBJECTIVES

After completing this supplement, you should be able to:

1. Explain the purpose of acceptance sampling.
2. Contrast acceptance sampling and process control.
3. Compare and contrast single and multiple sampling plans.
4. Determine the average outgoing quality of inspected lots.

SUPPLEMENT OUTLINE

INTRODUCTION

Acceptance sampling is a form of inspection that is applied to lots or batches of items either before or after a process instead of during the process. In the majority of cases, the lots represent incoming purchased items or final products awaiting shipment to warehouses or customers. The purpose of acceptance sampling is to decide whether a lot satisfies predetermined standards. Lots that satisfy these standards are passed or *accepted*; those that do not are *rejected*. Rejected lots may be subjected to 100 percent inspection, or purchased goods may be returned to the supplier for credit or replacement (especially if destructive testing is involved).

acceptance sampling Form of inspection applied to lots or batches of items before or after a process, to judge conformance with predetermined standards.

Acceptance sampling procedures are most useful when one or more of the following conditions exist:

1. A large number of items must be processed in a short time.
2. The cost consequences of passing defectives are low.
3. Destructive testing is required.
4. Fatigue or boredom caused by inspecting large numbers of items leads to inspection errors.

Acceptance sampling procedures can be applied to both attribute (counts) and variable (measurements) inspection. However, inspection of attributes is perhaps more widely used, so for purposes of illustration, the discussion here focuses exclusively on attribute sampling plans.

SAMPLING PLANS

A key element of acceptance sampling is the sampling plan. **Sampling plans** specify the lot size, N; the sample size, n; the number of samples to be taken; and the acceptance/rejection criteria. A variety of sampling plans can be used. Some plans call for selection of a single sample, and others call for two or more samples, depending on the nature of the plan. The following paragraphs briefly describe some of the different kinds of plans.

sampling plans Plans that specify lot size, sample size, number of samples, and acceptance/rejection criteria.

Single-Sampling Plan

In this plan, one random sample is drawn from each lot, and every item in the sample is examined and classified as either "good" or "defective." If any sample contains more than a specified number of defectives, c, that lot is rejected.

Double-Sampling Plan

A double-sampling plan allows for the opportunity to take a second sample if the results of the initial sample are inconclusive. For example, if the quality of the initial sample is high, the lot can be accepted without need for a second sample. If the quality in the initial sample is poor, sampling can also be terminated and the lot rejected. For results between those two cases, a second sample is then taken and the items inspected, after which the lot is either accepted or rejected on the basis of the evidence obtained from both samples. A double-sampling plan specifies the lot size, the size of the initial sample, accept/reject criteria for the initial sample, the size of the second sample, and a single acceptance number.

With a double-sampling plan, two values are specified for the number of defective items, a lower level, c_1, and an upper level, c_2. For instance, the lower level might be two defectives and the upper level might be five defectives. Using those values as decision rules, the first sample is taken. If the number of defective items in the first sample is less than or equal to the lower value (i.e., c_1), the lot is judged to be good and sampling is terminated. Conversely, if the number of defectives exceeds the upper value (i.e., c_2), the lot is rejected. If the number of defectives falls somewhere in between, a second sample is taken and the number of defectives in both samples is compared to a third value, c_3. For example, c_3 might be six. If the combined number of defectives does not exceed that value, the lot is accepted; otherwise, the lot is rejected.

Multiple-Sampling Plan

A multiple-sampling plan is similar to a double-sampling plan except that more than two samples may be required. A sampling plan will specify each sample size and two limits for each sample. The values increase with the number of samples. If, for any sample, the cumulative number of defectives found (i.e., those in the present sample plus those found in all previous samples) exceeds the upper limit specified for that sample, sampling is terminated and the lot is rejected. If the cumulative number of defectives is less than or equal to the lower limit, sampling is terminated and the lot is passed. If the number is between the two limits, another sample is taken. The process continues until the lot is either accepted or rejected.

Choosing a Plan

The cost and time required for inspection often dictate the kind of sampling plan used. The two primary considerations are the number of samples needed and the total number of observations required. Single-sampling plans involve only a single sample, but the sample size is large relative to the total number of observations taken under double- or multiple-sampling plans. Where the cost to obtain a sample is relatively high compared with the cost to analyze the observations, a single-sampling plan is more desirable. For instance, if a sample of moon soil is needed, clearly the cost of returning for a second or third sample far outweighs the cost of analyzing a single large sample. Conversely, where item inspection costs are relatively high, such as destructive testing, it may be better to use double or multiple sampling because the average number of items inspected per lot will be lower. This stems from the fact that a very good or very poor lot quality will often show up initially, and sampling can be terminated.

OPERATING CHARACTERISTIC CURVE

operating characteristic (OC) curve Probability curve that shows the probabilities of accepting lots with various fractions defective.

An important feature of a sampling plan is how it discriminates between lots of high and low quality. The ability of a sampling plan to discriminate is described by its **operating characteristic (OC) curve.** A typical curve for a single-sampling plan is shown in Figure 10S–1. The curve shows the probability that use of the sampling plan will result in lots with various *fractions defective* being accepted. For example, the graph shows that a lot with 3 percent of defectives (a fraction defective of .03) would have a probability of about .90 of being accepted (and a probability of 1.00 − .90 = .10 of being rejected). Note the downward relationship: As lot quality decreases, the probability of acceptance decreases, although the relationship is not linear.

A sampling plan does not provide perfect discrimination between good and bad lots; some low-quality lots will invariably be accepted, and some lots with very good quality will invariably be rejected. Even lots containing more than 20 percent defectives still have some probability of acceptance, whereas lots with as few as 3 percent defectives have some chance of rejection.

The degree to which a sampling plan discriminates between good and bad lots is a function of the steepness of the graph's OC curve: the steeper the curve, the more discriminating the sampling plan. (See Figure 10S–2.) Note the curve for an ideal plan (i.e., one that can discriminate perfectly between good and bad lots). To achieve that, you need to inspect 100 percent of each lot. Obviously, if you are going to do that, theoretically, *all* of the defectives can be eliminated (although errors and boredom might result in a few defectives remaining). However, the point is that 100 percent inspection provides a perspective from which to view the OC curves of other sampling plans.

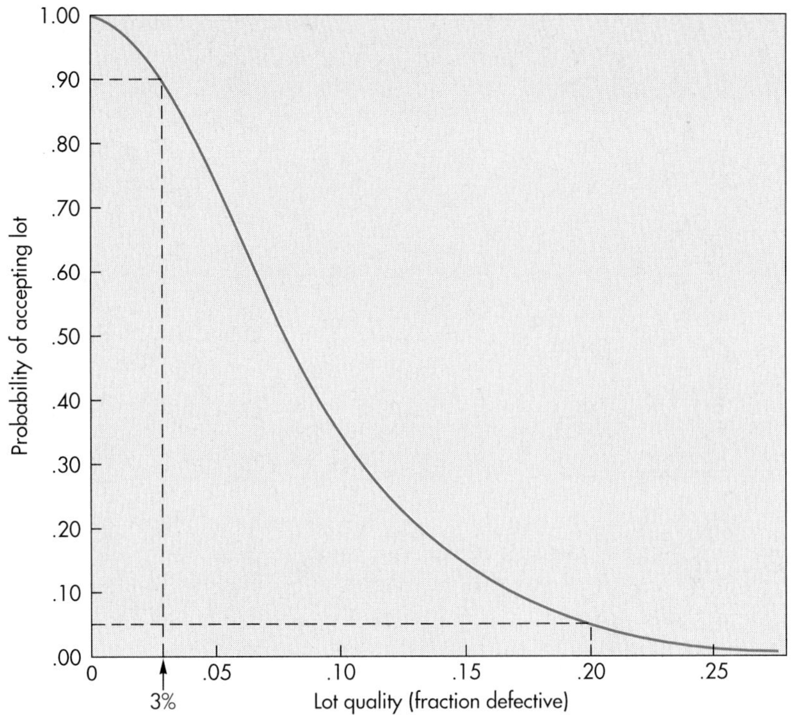

FIGURE 10S-1

A typical OC curve for proportions

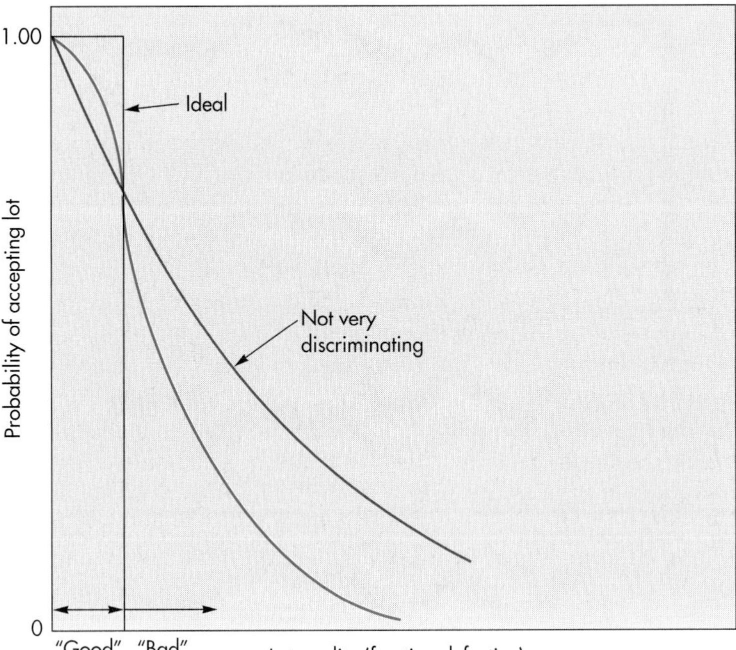

FIGURE 10S-2

The steeper the OC curve, the more discriminating the sampling plan

Be aware that the cost and time needed to conduct 100 percent inspection often rule out 100 percent inspection, as does destructive testing, leaving acceptance sampling as the only viable alternative.

For these reasons, buyers ("consumers") are generally willing to accept lots that contain small percentages of defective items as "good," especially if the cost related to a few defectives is low. Often this percentage is in the neighborhood of 1 percent to 2 percent defective. This figure is known as the **acceptable quality level (AQL).**

acceptable quality level The percentage level of defects at which consumers are willing to accept lots as "good."

FIGURE 10S-3

The AQL indicates good lots, and the LTPD indicates bad lots

Probability of accepting lot
1.00
.90
.80
.70
.60
.50
.40
.30
.20
.10
.00
α = .10
β = .10
0 .05 .10 .15 .20 .25
AQL
LTPD
Fraction defective in the lot
Indifferent
"Good"
"Bad"

lot tolerance percent defective The upper limit on the percentage of defects that a consumer is willing to accept.

consumer's risk The probability that a lot containing defectives exceeding the LTPD will be accepted.

producer's risk The probability that a lot containing the acceptable quality level will be rejected.

Because of the inability of random sampling to clearly identify lots that contain more than this specified percentage of defectives, consumers recognize that some lots that actually contain more will be accepted. However, there is usually an upper limit on the percentage of defectives that a consumer is willing to tolerate in accepted lots. This is known as the **lot tolerance percent defective (LTPD).** Thus, consumers want quality equal to or better than the AQL, and are willing to live with some lots with quality as poor as the LTPD, but they prefer not to accept any lots with a defective percentage that exceeds the LTPD. The probability that a lot containing defectives exceeding the LTPD will be accepted is known as the **consumer's risk,** or beta (β), or the probability of making a *Type II error.* The probability that a lot containing the acceptable quality level will be rejected is known as the **producer's risk,** alpha (α), or the probability of making a *Type I error.* Many sampling plans are designed to have a producer's risk of 5 percent and a consumer's risk of 10 percent, although other combinations are also used. It is possible by trial and error to design a plan that will provide selected values for alpha and beta given the AQL and the LTPD. However, standard references such as the government MIL-STD tables are widely used to obtain sample sizes and acceptance criteria for sampling plans. Figure 10S–3 illustrates an OC curve with the AQL, LTPD, producer's risk, and consumer's risk.

A certain amount of insight is gained by actually constructing an OC curve. Suppose you want the curve for a situation in which a sample of $n = 10$ items is drawn from lots containing $N = 2{,}000$ items, and a lot is accepted if no more than $c = 1$ defective is found. Because the sample size is small relative to the lot size, it is reasonable to use the binomial distribution to obtain the probabilities that a lot will be accepted for various lot qualities.[1] A portion of the cumulative binomial table found in Appendix Table D is reproduced here to facilitate the discussion.

[1]Since sampling is generally performed "without replacement," if the ratio n/N is 5 percent or more, the hypergeometric distribution is more appropriate since the probability of finding a defective would vary from observation to observation. We shall consider only the more general case of the binomial distribution (i.e., $n/N < 5$ percent).

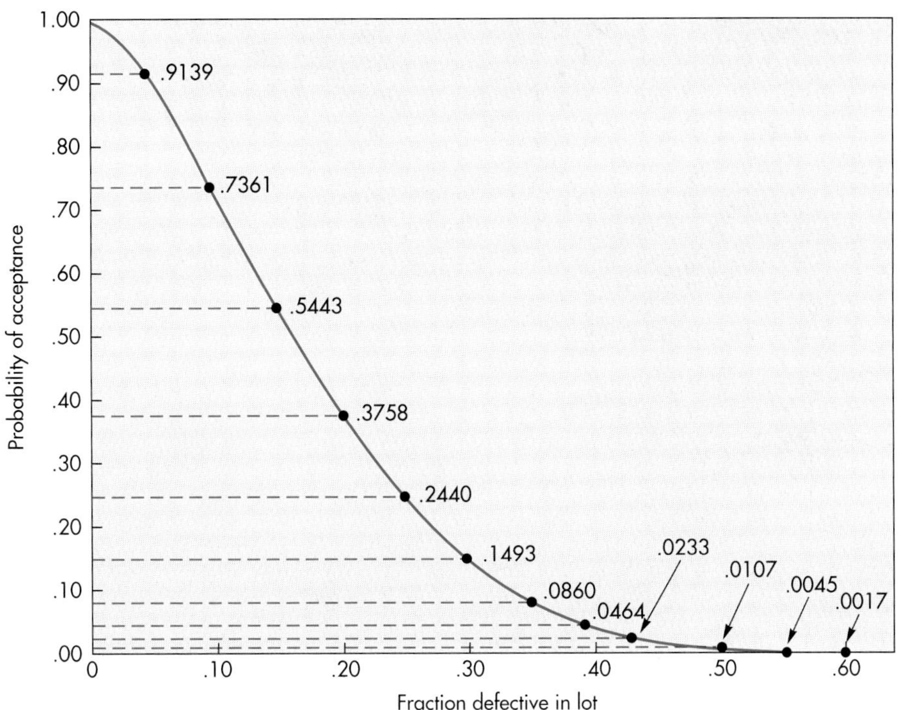

FIGURE 10S-4

OC curve for n = 10, c = 1

n	x	.05	.10	.15	.20	.25	.30	.35	.40	.45	.50	.55	.60
		FRACTION DEFECTIVE, p											
10	0	.5987	.3487	.1969	.1074	.0563	.0282	.0135	.0060	.0025	.0010	.0003	.0001
c = 1 →	1	.9139	.7361	.5443	.3758	.2440	.1493	.0860	.0464	.0233	.0107	.0045	.0017
	2	.9885	.9298	.8202	.6778	.5256	.3828	.2616	.1673	.0996	.0547	.0274	.0123
	3	.9990	.9872	.9500	.8791	.7759	.6496	.5138	.3823	.2660	.1719	.1020	.0548

To use the table, select various lot qualities (values of p listed across the top of the table), beginning with .05, and find the probability that a lot with that percentage of defectives would be accepted (i.e., the probability of finding zero or one defective in this case). For $p = .05$, the probability of one or no defectives is .9139. For a lot with 10 percent defectives (i.e., a fraction defective of .10), the probability of one or fewer defectives drops to .7361, and for 15 percent defectives, the probability of acceptance is .5443. In effect, you simply read the probabilities across the row for $c = 1$. By plotting these points (e.g., .05 and .9139, .10 and .7361) on a graph and connecting them, you obtain the OC curve illustrated in Figure 10S–4.

When $n > 20$ and $p < .05$, the Poisson distribution is useful in constructing operating characteristic curves for proportions. In effect, the Poisson distribution is used to approximate the binomial distribution. The Poisson approximation involves treating the mean of the binomial distribution (i.e., np) as the mean of the Poisson (i.e., μ):

$$\mu = np \tag{10S–1}$$

As with the binomial distribution, you select various values of lot quality, p, and then determine the probability of accepting a lot (i.e., finding two or fewer defectives) by referring to the cumulative Poisson table. Values of p in increments of.01 are often used in this regard. Example 1 illustrates this use of the Poisson table.

Example 1

Use the cumulative Poisson table to construct an OC curve for this sampling plan:

$N = 5{,}000, \quad n = 80, \quad c = 2$

Solution

Selected Values of p	$\mu = np$	$P_{ac}[P(x \le 2)$ from Appendix Table C]
.01	80(.01) = 0.8	.953
.02	80(.02) = 1.6	.783
.03	80(.03) = 2.4	.570
.04	80(.04) = 3.2	.380
.05	80(.05) = 4.0	.238
.06	80(.06) = 4.8	.143
.07	80(.07) = 5.6	.082
.08	80(.08) = 6.4	.046

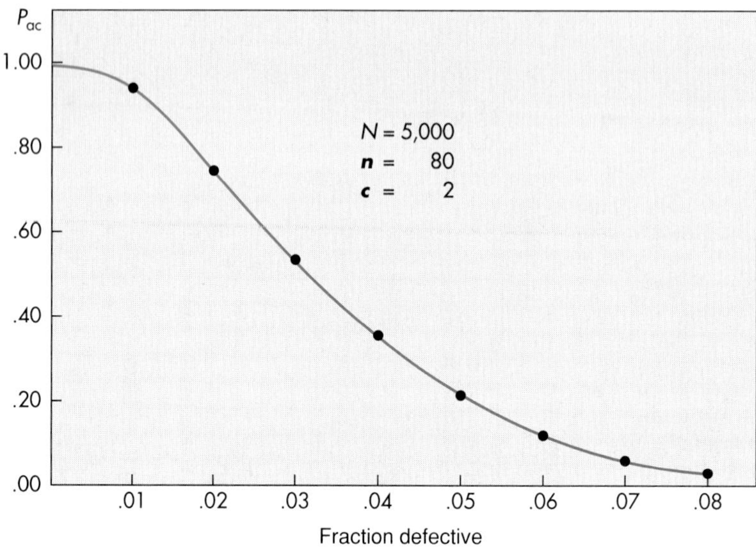

Operating characteristic curves can be constructed for variables sampling plans as well as for attributes sampling plans. To go into detail is beyond the scope of this presentation. The purpose here is merely to illustrate the concept of an OC curve and to show how its construction is based on an underlying *sampling distribution*.

AVERAGE QUALITY OF INSPECTED LOTS

An interesting feature of acceptance sampling is that the level of inspection automatically adjusts to the quality of lots being inspected, assuming rejected lots are subjected to 100 percent inspection. The OC curve reveals that the greater the percentage of defectives in a lot, the less likely the lot is to be accepted. Generally speaking, good lots have a high probability and bad lots have a low probability of being accepted. If the lots inspected are mostly good, few will end up going through 100 percent inspection. The poorer the quality of the lots, the greater the number of lots that will come under close scrutiny. This tends to improve overall quality of lots by weeding out defectives. In this way, the level of inspection is affected by lot quality.

average outgoing quality Average of rejected lots (100 percent inspection) and accepted lots (a sample of items inspected).

If all lots have some given fraction defective, p, the **average outgoing quality (AOQ)** of the lots can be computed using the following formula, assuming defectives are replaced with good items.

$$AOQ = P_{ac} \times p\left(\frac{N - n}{N}\right) \tag{10S–2}$$

where

P_{ac} = Probability of accepting the lot
p = Fraction defective
N = Lot size
n = Sample size

In practice, the last term is often omitted since it is usually close to 1.0 and therefore has little effect on the resulting values. The formula then becomes

$$AOQ = P_{ac} \times p \qquad (10S\text{–}3)$$

Example 2

Construct the AOQ curve for this situation:

$$N = 500, \quad n = 10, \quad c = 1$$

Solution

Let values of p vary from .05 to .40 in steps of .05. You can read the probabilities of acceptance, P_{ac} from Appendix Table D.

$$AOQ = P_{ac} \times p$$

p	**P_{ac}**	**AOQ**
.05	.9139	.046
.10	.7361	.074
.15	.5443	.082
.20	.3758	.075
.25	.2440	.061
.30	.1493	.045
.35	.0860	.030
.40	.0464	.019

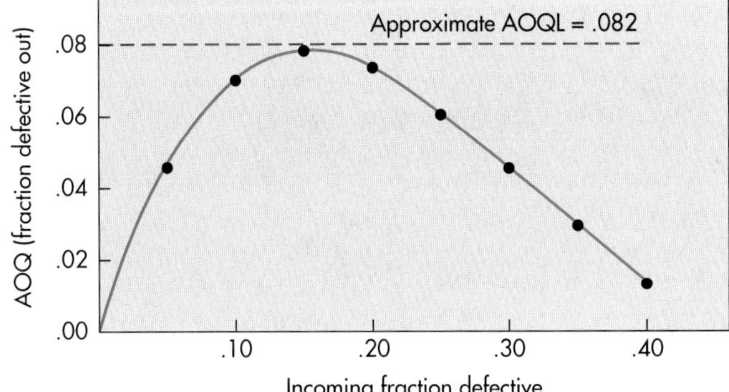

The average outgoing quality limit (AOQL) appears to be just above 8 percent.

Use this formula instead of 10S–2 for computing AOQ values.

By allowing the percentage, p, to vary, a curve such as the one in Example 2 can be constructed in the same way that an OC curve is constructed. The curve illustrates the point that if lots are very good or very bad, the average outgoing quality will be high. The maximum point on the curve becomes apparent in the process of calculating values for the curve.

There are several managerial implications of the graph in Example 2. First, a manager can determine the worst possible outgoing quality. Second, the manager can determine the amount of inspection that will be needed by obtaining an estimate of the incoming quality. Moreover, the manager can use this information to establish the relationship between inspection cost and the incoming fraction defective, thereby underscoring the benefit of implementing process improvements to reduce the incoming fraction defective rather than trying to weed out bad items through inspection.

Key Terms

acceptable quality level (AQL), 481
average outgoing quality (AOQ), 484
acceptance sampling, 479
consumer's risk, 482
lot tolerance percent defective (LTPD), 482
operating characteristic (OC) curve, 480
producer's risk, 482
sampling plans, 479

Solved Problems

Problem 1

Inspection. A process for manufacturing shock absorbers for light trucks produces 5 percent defectives. Inspection cost per shock is $0.40, and 100 percent inspection generally catches all defectives, due to the nature of the inspection and the small volume produced. Any defectives installed on trucks must eventually be replaced at a cost of $12 per shock. Is 100 percent inspection justified?

Solution

Five percent of the output is defective. The expected cost per shock for replacement is thus 0.05($12) = 60 cents. Since this is greater than the inspection cost per shock of 40 cents, 100 percent inspection is justified.

Problem 2

Acceptance sampling. Shipments of 300 boxes of glassware are received at a warehouse of a large department store. Random samples of five boxes are checked, and the lot is rejected if more than one box reveals breakage. Construct the OC curve for this plan.

Solution

When the sample size is less than 5 percent of the lot size, the binomial distribution can be used to obtain P_{ac} for various lot percentages defective. Here, $n/N = 5/300 = .017$, so the binomial can be used. A portion of the cumulative binomial table is shown below. Note that $c = 1$.

CUMULATIVE BINOMIAL PROBABILITIES

***p* = FRACTION DEFECTIVE**

	n	*x*	.05	.10	.15	.20	.25	.30
	5	0	.7738	.5905	.4437	.3277	.2373	.1681
c = 1 →		1	.9974	.9185	.8352	.7373	.6328	.5282
		2	.9988	.9914	.9734	.9421	.8965	.8369
		3	1.0000	.9995	.9978	.9933	.9844	.9692
		4	1.0000	1.0000	.9999	.9997	.9990	.9976
		5	1.0000	1.0000	1.0000	1.0000	1.0000	1.0000

.35	.40	.45	.50	.55	.60	.65	.70	.75	.80
.1160	.0778	.0503	.0313	.0185	.0102	.0053	.0024	.0010	.0003
.4284	.3370	.2562	.1875	.1312	.0870	.0540	.0308	.0156	.0067
.7648	.6826	.5931	.5000	.4069	.3174	.2352	.1631	.1035	.0579
.9460	.9130	.8688	.8125	.7438	.6630	.5716	.4718	.3672	.2627
.9947	.9898	.9815	.9688	.9497	.9222	.8840	.8319	.7627	.6723
1.0000	1.0000	1.0000	1.0000	1.0000	1.0000	1.0000	1.0000	1.0000	1.0000

The table indicates that $P_{ac} = .9974$ when lot quality is 5 percent defective, .9185 for 10 percent defective, .8352 for 15 percent, and so on. The resulting curve is:

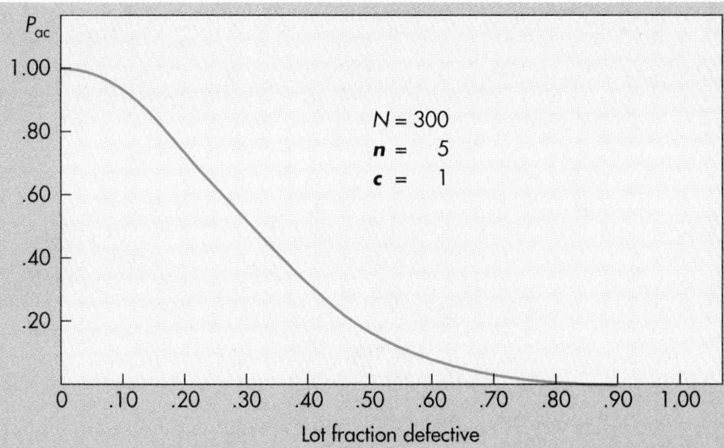

Develop the AOQ curve for the previous problem.

$$AOQ = P_{ac} \times p$$

(Values of lot quality, p, are taken from the portion of the binomial table shown.)

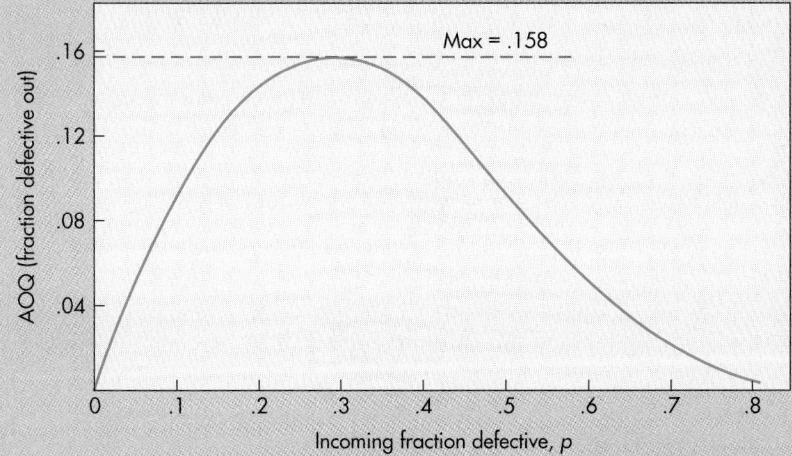

p	P_{ac}	AOQ	p	P_{ac}	AOQ
.05. . . .	.9974	.050	.45. . . .	.2562	.115
.10. . . .	.9185	.092	.50. . . .	.1875	.094
.15. . . .	.8352	.125	.55. . . .	.1312	.072
.20. . . .	.7373	.147	.60. . . .	.0870	.052
.25. . . .	.6328	.158	.65. . . .	.0540	.035
.30. . . .	.5258	.158	.70. . . .	.0380	.027
.35. . . .	.4284	.150	.75. . . .	.0156	.012
.40. . . .	.3370	.135	.80. . . .	.0067	.005

Discussion and Review Questions

1. What is the purpose of acceptance sampling?
2. How does acceptance sampling differ from process control?
3. What is an operating characteristic curve, and how is it useful in acceptance sampling?
4. What general factors govern the choice between single-sampling plans and multiple-sampling plans?
5. Briefly explain or define each of these terms.

a. AOQ
b. AOQL
c. LTPD
d. Alpha risk
e. Beta risk

Problems

1. An assembly operation for trigger mechanisms of a semiautomatic spray gun produces a small percentage of defective mechanisms. Management must decide whether to continue the current practice of 100 percent inspection or to replace defective mechanisms after final assembly when all guns are inspected. Replacement at final assembly costs $30 each; inspection during trigger assembly costs $12 per hour for labor and overhead. The inspection rate is one trigger per minute.
   a. Would 100 percent inspection during trigger assembly be justified if there are (1) 4 percent defective? (2) 1 percent defective?
   b. At what point would management be indifferent between 100 percent inspection of triggers and only final inspection?

   a. Expected replacement costs: (1) $1.20/*p* (2) $.30/*p* Both justified
   b. .0067

2. Random samples of $n = 20$ circuit breakers are tested for damage caused by shipment in each lot of 4,000 received. Lots with more than one defective are pulled and subjected to 100 percent inspection.
   a. Construct the OC curve for this sampling plan.
   b. Construct the AOQ curve for this plan, assuming defectives found during 100 percent inspection are replaced with good parts. What is the approximate AOQL?

   a. See graph in IM.
   b. See graph in IM. AOQL ≈ .039

3. Auditors use a technique called *discovery sampling* in which a random sample of items is inspected. If any defectives are found, the entire lot of items sampled is subjected to 100 percent inspection.
   a. Draw an OC curve for the case where a sample of 15 credit accounts will be inspected out of a total of 8,000 accounts.
   b. Draw an OC curve for the case where 150 accounts out of 8,000 accounts will be examined. (Hint: Use $p = .001, .002, .003, \ldots$)
   c. Draw the AOQ curve for the preceding case, and estimate the AOQL.

   a, b, c. See graphs in IM.
   c. AOQL ≈ .0024

4. Random samples of lots of textbooks are inspected for defectives just prior to shipment to the warehouse. Each lot contains 3,000 books.
   a. On a single graph, construct OC curves for $n = 100$ and (1) $c = 0$, (2) $c = 1$, and (3) $c = 2$. (Hint: Use $p = .001, .002, .003, \ldots$)
   b. On a single graph, construct OC curves for $c = 2$ and (1) $n = 5$, (2) $n = 20$, and (3) $n = 120$.

   a, b. See graphs in IM.

5. A manufacturer receives shipments of several thousand parts from a supplier every week. The manufacturer has the option of conducting a 100 percent inspection before accepting the parts. The decision is based on a random sample of 15 parts. If parts are not inspected, defectives become apparent during a later assembly operation, at which time replacement cost is $6.25 per unit. Inspection cost for 100 percent inspection is $1 per unit.
   a. At what fraction defective would the manufacturer be indifferent between 100 percent inspection and leaving discovery of defectives until the later assembly operation?
   b. For the sample size used, what is the maximum number of sample defectives that would cause the lot to be passed without 100 percent inspection, based on your answer to part *a*?

   a. .16
   b. 2
   c. (1) Accept (2) .0362 (3) .9638 (4) Type I = .0362, Type II = 0
   d. (1) .3980 (2) .6020 (3) Reject (4) Type I = 0, Type II = .3980

c. If the shipment actually contains 5 percent defectives:
   (1) What is the correct decision?
   (2) What is the probability it would be rejected in favor of 100 percent inspection?
   (3) What is the probability that it would be accepted without 100 percent inspection?
   (4) What is the probability of a Type I error? A Type II error?

d. Answer the questions in part *c* for a shipment that contains 20 percent defectives.

6. (Refer to Problem 5c.) Suppose there are two defectives in the sample.

   a. If the acceptance number is $c = 1$, what decision should be made? What type of error is possible?

   b. If the acceptance number is $c = 3$, what decision should be made? What type of error is possible?

   c. Determine the average outgoing quality for each of these percent defectives if $c = 1$.
   (1) 5 percent.
   (2) 10 percent.
   (3) 15 percent.
   (4) 20 percent.

a. Reject; Type I possible
b. Accept; Type II possible
c. (1) .0415
(2) .0549
(3) .0478
(4) .0334

Selected Bibliography

Duncan, A. J. *Quality Control and Industrial Statistics*, 5th ed. Burr Ridge, Ill.: Richard D. Irwin, 1986.

Enrick, Norbert L. *Quality Reliability and Process Improvement*. 8th ed. New York: Industrial Press, 1985.

Juran, J. M., and F. M. Gryma, Jr. *Quality Planning and Analysis*. 2nd ed. New York: McGraw-Hill, 1980.

www.trekbikes.com
01) New stronger OCLV™ frame
The new self-reinforced mainframe gives 60% more lateral stiffness and 100% more vertical stiffness.
1998 Y Frame
1997 Y Frame
02) Lightweight OCLV carbon fiber
Only with OCLV carbon fiber could we build a full suspension bike that weighs 24.3 lbs.
OCLV
03) New geometry
Rider weight shifted back and down, for more stability downhill, better traction uphill.
04) New Responsive Steering™
New geometry incorporates a longer top tube and shorter stem that provides a tight turning radius around the steer tube. Wider handlebars give you the upper hand.
05) Fully Active Uni-Link™
Unified rear triangle isolates suspension from drivetrain. Chain tension won't activate the shock. Fully active all the time—standing up or sitting down.
06) Front and rear shocks in balance
Trek-exclusive Fox Racing Shox get custom valving to optimize the Y's unique compression trajectories, while harmonizing the front and rear shock.
07) Tunable Shock Compression™
Adjustable shock mount plates fine-tune your rear shock's compression rate, customizing performance to suit all conditions.
08) New OCLV shock coupling
Sleeve-style coupling insures the straight-on alignment of shock mount. Precision-machined so shock mount fits inside within of an inch!
09) New shock mount
New shock mount loads residual shock force onto frame as shear rather than as compression. Less stress means a longer frame life.
10) Stiffer rear stays
10% increase in diameter adds 30% more stiffness. Truer tracking, more efficient transfer of pedaling energy.
old new
Call for a free catalog 1.800.369.TREK
TREK
Y33
10 new reasons Y.

LEARNING OBJECTIVES

After completing this chapter, you should be able to:

1. Describe TQM.
2. Give an overview of problem solving.
3. Give an overview of process improvement.
4. Describe and use various quality tools.

CHAPTER ELEVEN
TOTAL QUALITY MANAGEMENT (TQM)

CHAPTER OUTLINE

A primary role of management is to lead an organization in its daily operation and to maintain it as a viable entity into the future. Quality has become an important factor to success in this latter, strategic responsibility.

Although ostensibly always an objective of business, customer satisfaction, *in customer terms,* became a specific goal in the late 1980s. Providing high quality was recognized as a key element for success. Most large corporations taking that path have documented their success. First, they survived the strong overseas competition that had set the high quality levels and now have regained some of their former markets. Smaller companies are also adopting similar goals. The Xerox Corporation filled 85 percent of the copier market up to about 1979, when its share dropped to 15 percent. Its Japanese competitors could *sell* an equivalent but higher quality machine for the Xerox *manufacturing* costs. Xerox was close to losing it all. Through its subsidiary in Japan it determined that the Japanese application of TQM made the biggest difference. Xerox also implemented TQM and after years of hard work regained 60 percent of the copier market. Management, with a new approach, has played the critical role. The new approach is reflected first in the expressed changes in policy. The Ford Motor Company *Operating Philosophy* is a good example:

"The operating philosophy of Ford Motor Company is to meet customer needs and expectations by establishing and maintaining an environment which encourages all employees to pursue never-ending improvement in the quality and productivity of products and services throughout the corporation, its supply base, and its dealer organization."[1]

As the 20th century ends, business organizations are involved in what has become a "quality revolution." It began in Japan and has now spread to North America and other parts of the world. It involves an entirely new way of thinking about, and dealing with, quality that encompasses the entire organization. This new approach has been given a variety of names, but the one we shall use is *total quality management,* or TQM.

INTRODUCTION

total quality management A philosophy that involves everyone in an organization in a continual effort to improve quality and achieve customer satisfaction.

Vol. 4, tape 1, seg. 1
Trek

www.trekbikes.com

The term **total quality management (TQM)** refers to a quest for quality that involves everyone in an organization. There are two key philosophies in this approach. One is a never-ending push to improve, which is referred to as *continuous improvement*; the other is a goal of *customer satisfaction,* which involves meeting or exceeding customer expectations.

We can describe the TQM approach as follows:

1. Find out what customers want. This might involve the use of surveys, focus groups, interviews, or some other technique that integrates the customer's voice in the decision-making process. Be sure to include the *internal customer* (the next person in the process) as well as the *external customer* (the final customer).
2. Design a product or service that will meet (or exceed) what customers want. Make it easy to use and easy to produce.
3. Design a production process that facilitates doing the job right the first time. Determine where mistakes are likely to occur and try to prevent them. When mistakes do occur, find out why so that they are less likely to occur again. Strive to make the process "mistake-proof."
4. Keep track of results, and use those to guide improvement in the system. Never stop trying to improve.
5. Extend these concepts to suppliers and to distribution.

Successful TQM programs are built through the dedication and combined efforts of everyone in the organization. As noted, top management must be committed and involved. If it isn't, TQM will become just another fad that quickly dies and fades away.

---

[1]Source: Reprinted from Thomas J. Cartin, *Principles and Practices of TQM,* ASQC Press, 1993, p. 29.

The preceding description provides a good idea of what TQM is all about, but it doesn't tell the whole story. A number of other elements of TQM are important, including:

1. *Continual improvement.* The *philosophy* that seeks to improve all factors related to the process of converting inputs into outputs on an ongoing basis is called **continuous improvement.** It covers equipment, methods, materials, and people. Under continuous improvement, the old adage, "If it ain't broke, don't fix it" gets transformed into "Just because it isn't broke doesn't mean it can't be improved."

continuous improvement Philosophy that seeks to make never-ending improvements to the process of converting inputs into outputs.

kaizen Japanese term for continuous improvement.

The concept of continuous improvement is not new, but it did not receive much interest in the United States until recently, even though it originated here. However, many Japanese companies have used it for years, and it has become a cornerstone of the Japanese approach to production. The Japanese use the term **kaizen** to refer to continuous improvement. The successes of Japanese companies have caused other companies to reexamine many of their approaches. This has resulted in a strong interest in the continuous improvement approach.

2. *Competitive benchmarking.* This involves identifying companies or other organizations that are the best at something and studying how they do it to learn how to improve your operation. The company need not be in the same line of business as yours. For example, Xerox used the mail-order company, L. L. Bean, to benchmark order filling.

3. *Employee empowerment.* Giving workers the responsibility for improvements and the authority to make changes to accomplish them provides strong motivation for employees. This puts decision making into the hands of those who are closest to the job and have considerable insight into problems and solutions.

4. *Team approach.* The use of teams for problem solving and to achieve consensus takes advantage of group synergy, gets people involved, and promotes a spirit of cooperation and shared values among employees.

5. *Decisions based on facts rather than opinions.* Management gathers and analyzes data as a basis for decision making.

6. *Knowledge of tools.* Employees and managers are trained in the use of quality tools.

7. *Supplier quality.* Suppliers must be included in quality assurance and quality improvement efforts so that their processes are capable of delivering quality parts and materials in a timely manner.

quality at the source The philosophy of making each worker responsible for the quality of his or her work.

The term **quality at the source** refers to the philosophy of making each worker responsible for the quality of his or her work. This incorporates the notions of "do it right" and "if it isn't right, fix it." Workers are expected to provide goods or services that meet specifications and to find and correct mistakes that occur. In effect, each worker becomes a quality inspector for his or her work. When the work is passed on to the next operation in the process (the internal customer) or, if that step is the last in the process, to the ultimate customer, the worker is "certifying" that it meets quality standards.

This accomplishes a number of things: (1) it places direct responsibility for quality on the person(s) who directly affect it; (2) it removes the adversarial relationship that often exists between quality control inspectors and production workers; and (3) it motivates workers by giving them control over their work as well as pride in it.

Sign on the wall of a company cafeteria

> Sometimes they can be cranky, and it may sometimes seem like they expect too much, but they do provide our paychecks and our benefits, such as sick leave, maternity leave, health insurance, and three weeks of paid vacation time each year. And what about all the new equipment we've been getting lately? They pay for that, too. And a lot more. So the next time you see them, give them a great big smile to show how much you appreciate them—our *customers*!

Suppliers are partners in the process, and long-term relationships are encouraged. This gives suppliers a vital stake in providing quality goods and services. Suppliers, too, are expected to provide quality at the source, thereby reducing or eliminating the need to inspect deliveries from suppliers.

TABLE 11–1

Comparing the cultures of TQM and traditional organizations

Aspect	Traditional	TQM
Overall mission	Maximize return on investment	Meet or exceed customer satisfaction
Objectives	Emphasis on short term	Balance of long term and short term
Management	Not always open; sometimes inconsistent objectives	Open; encourages employee input; consistent objectives
Role of manager	Issue orders; enforce	Coach, remove barriers, build trust
Customer requirements	Not highest priority; may be unclear	Highest priority; important to identify and understand
Problems	Assign blame; punish	Identify and resolve
Problem solving	Not systematic; individuals	Systematic; teams
Improvement	Erratic	Continual
Suppliers	Adversarial	Partners
Jobs	Narrow, specialized; much individual effort	Broad, more general; much team effort
Focus	Product oriented	Process oriented

It would be incorrect to think of TQM as merely a collection of techniques. Rather, TQM reflects a whole new attitude toward quality. It is about the *culture* of an organization. To truly reap the benefits of TQM, the culture of an organization must change.

Table 11–1 illustrates the differences between cultures of a TQM organization and a more traditional organization.

TQM programs are touted as a way for U.S. companies to regain their competitiveness, which is a very worthwhile objective. Nonetheless, TQM programs are not without criticism. Some of the major ones are:

1. Blind pursuit of TQM programs: Overzealous advocates may focus attention on quality even though other priorities may be more important (e.g., responding quickly to a competitor's advances).

2. Programs may not be linked to the strategies of the organization in a meaningful way.

3. Quality-related decisions may not be tied to market performance. For instance, customer satisfaction may be carried to the extent that its cost far exceeds any direct or indirect benefit of doing so.

4. Failure to carefully plan a program before embarking on it can lead to false starts, employee confusion, and meaningless results.

Note that there is nothing inherently wrong with TQM, it is how some individuals or organizations misuse it. Let's turn our attention to problem solving and process improvement.

www.calcomp.com

READING

CALCOMP: DISASTER BECOMES SUCCESS **Michelle Vranizan**

Flash back to the early 1980s. The only thing world-class about CalComp was the mess it was in.

The company that practically invented the computer plotter—a device engineers and architects use to print intricate, oversized schematics—had become arrogant, inattentive and lazy.

Every last plotter that rolled off the assembly line didn't work well enough to ship without some rejiggering. A legion of field technicians was needed to make house calls on installed machines that malfunctioned every few weeks. Competitors such as Hewlett-Packard jumped into the breach, stealing dissatisfied customers.

Flash forward to the CalComp of the 1990s, recognized as a leader in world-class manufacturing. No more mass assembly lines. No more bugs. No more field technicians.

Hewlett-Packard still sells more pen plotters. But CalComp is a strong No. 2 and has a tight grip on other segments of the plotter and digitizer business. Now bursting with confidence, the $525 million subsidiary of Lockheed Corp. is plowing into other areas of the multibillion dollar computer graphics industry.

The difference between the early 1980s and the 1990s?

Quality.

Specifically, a quality program spearheaded by then CalComp President William Conlin that permeates every square inch of the company's grassy Anaheim headquarters and the attitudes of its 2,800 employees.

At CalComp, quality boils down to pleasing customers with gracefully built, innovative products that work from the start, rarely break down, are competitively priced and are upgraded faster than the other guy's.

"We built a manufacturing system that improves every day," said Bernard Masson, senior vice president of the plotter division, which accounts for about 60 percent of CalComp's revenue. "The product we build tomorrow will be better than the ones we build today."

In the early 1970s and 1980s, CalComp's manufacturing process was fractured. Product design and manufacturing weren't coordinated. The company stockpiled parts and the plotter division alone had more than 650 suppliers. A plotter was only tested for imperfections after it was made.

The system worked as long as CalComp only sold 1,000 or so plotters a year to customers who used mainframes or minicomputers, also very temperamental machines prone to breakdowns.

Then along came personal computers, which worked straight out of the box. When personal computers became the preferred tools for engineering and graphics, CalComp managers realized the old way of doing things had to go.

Conlin signed on in late 1983 and CalComp took its first steps toward world-class manufacturing soon after.

CalComp embraced the teachings of quality gurus such as W. Edwards Deming and Richard J. Schonberger, focusing first on the factory.

In the beginning, changes were simple, such as writing down the steps it took to assemble a plotter—something CalComp had never done before, Masson said.

Next, assembly workers were taught quality principles, given a say in how things were done and encouraged to catch mistakes on the line.

Other steps CalComp took:

1. When a new product is conceived, a team of more than a dozen people representing virtually every department shepherds it from development to delivery. With this kind of teamwork engineers don't end up designing parts that factory workers can't put together, said Linda Gronski, an operating unit manager in the plotter division.

2. Instead of stockpiling parts, CalComp cut inventories to the bare minimum and now takes delivery of just enough parts for the next week's work. Some suppliers even deliver parts on a daily basis.

3. CalComp put strict quality controls on suppliers and reduced its vendors in the plotter division to about 180. CalComp started a three-tier preferred-supplier program, dubbed "a mini-Malcolm Baldrige award" contest by one supplier. Parts from suppliers on the higher tier are expected to be so perfect CalComp doesn't even inspect them before building them into products.

By keeping inventories down, CalComp cut its overhead costs. It also can avoid writing off obsolete parts should demand shift suddenly and can direct energies to new products more quickly, company executives said.

Readjusting attitudes toward work was the most important part of the jump to quality, CalComp managers said.

Today, front-line workers are treated with as much respect as executives. Assembly workers are never reprimanded for stopping the line if they can't solve a problem. Gronski and other production managers wear beepers and are expected to attend to problems immediately.

Suppliers said they benefited from classes at CalComp on quality processes: They use the information to transform their businesses and attract other customers.

"When Xerox found out where we were with CalComp, they decided to send a team of people out to visit us," said Harry Esayian, owner of Zac Precision in Anaheim, which makes machine parts for dozens of CalComp products.

Draftsmen and engineers who use CalComp plotters and digitizers might be unfamiliar with the company's quality programs. But they appreciate the results.

"I'm always talking up the plotter because it's a good deal," said customer David Terry, who has a Santa Ana engineering firm that designs waste water-treatment plants. "They're low-cost and no-maintenance. We're real happy."

Questions

1. In general terms, what was wrong in the early 1980s at CalComp?
2. Specifically, what was wrong with:
   a. design and manufacturing?
   b. inventory?
   c. suppliers?
3. In general terms, what change helped to turn things around at CalComp?
4. What steps were taken to improve the situation?
5. What benefits have resulted from the changes?

Source: *The Orange County Register,* March 1, 1992.

L.L. Bean has been in business for over 80 years and prides itself on its attention to quality at a reasonable price. The company is committed to regularly looking for materials that last longer and making improvements in manufacturing that ensure durability.

www.llbean.com/

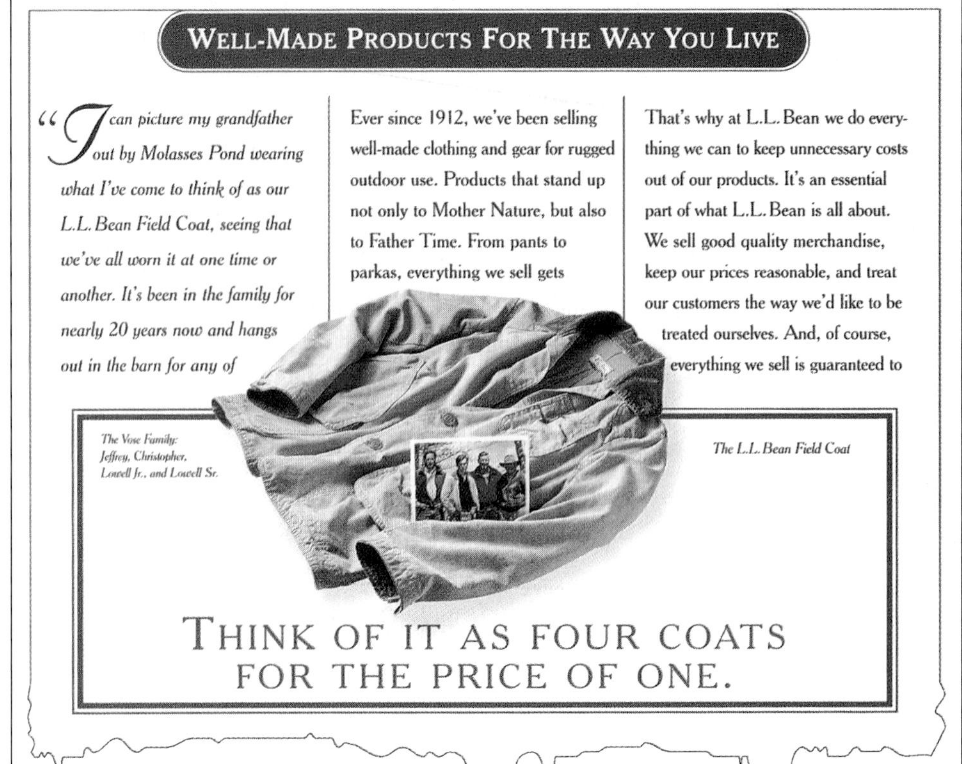

PROBLEM SOLVING

Problem solving is one of the basic procedures of TQM. In order to be successful, problem-solving efforts should follow a standard approach. Table 11–2 describes the basic steps in the TQM problem-solving process.

An important aspect of problem solving in the TQM approach is *eliminating* the cause so that the problem does not reoccur. This is why users of the TQM approach often like to think of problems as "opportunities for improvement."

TABLE 11-2

Basic steps in problem solving

Step 1	**Define the problem and establish an improvement goal.** Give problem definition careful consideration; don't rush through this step because this will serve as the focal point of problem-solving efforts.
Step 2	**Collect data.** The solution must be based on *facts*. Possible tools include check sheet, scatter diagram, histogram, run chart, and control chart.
Step 3	**Analyze the problem.** Possible tools include Pareto chart, cause-and-effect diagram.
Step 4	**Generate potential solutions.** Methods include brainstorming, interviewing, and surveying.
Step 5	**Choose a solution.** Be sure what the criteria are for choosing a solution. (Refer to the goal established in Step 1.) Apply criteria to potential solutions and select the best one.
Step 6	**Implement the solution.** Keep everyone informed.
Step 7	**Monitor the solution to see if it accomplishes the goal.** If not, modify the solution, or return to Step 1. Possible tools include control chart and run chart.

PROCESS IMPROVEMENT

Process improvement is a *systematic* approach to improving a process. It involves documentation, measurement, and analysis for the purpose of improving the functioning of a process. Typical goals of process improvement include increasing customer satisfaction, achieving higher quality, reducing waste, reducing cost, increasing productivity, and speeding up the process.

process improvement A systematic approach to improving a process.

Table 11–3 provides an overview of process improvement.

TABLE 11–3

Overview of process improvement

A. Process mapping

1. Collect information about the process; identify each step in the process. For each step, determine:
   The inputs and outputs.
   The people involved.
   All decisions that are made.
   Document such measures as time, cost, space used, waste, employee morale and any employee turnover, accidents and/or safety hazards, working conditions, revenues and/or profits, quality, and customer satisfaction, as appropriate.
2. Prepare a flowchart that *accurately* depicts the process; note that too little detail will not allow for meaningful analysis, and too much detail will overwhelm analysts and be counterproductive. Make sure that key activities and decisions are represented.

B. Analyze the process

1. Ask these questions about the process:
   Is the flow logical?
   Are any steps or activities missing?
   Are there any duplications?
2. Ask these questions about each step:
   Is the step necessary? Could it be eliminated?
   Does the step add value?
   Does any waste occur at this step?
   Could the time be shortened?
   Could the cost to perform the step be reduced?
   Could two (or more) steps be combined?

C. Redesign the process

Using the results of the analysis, redesign the process. Document the improvements; potential measures include reductions in time, cost, space, waste, employee turnover, accidents, safety hazards, and increases/improvements in employee morale, working conditions, revenues/profits, quality, and customer satisfaction.

The Plan-Do-Study-Act Cycle

The **plan-do-study-act (PDSA) cycle,** also referred to as either the Shewhart cycle or the Deming wheel, is the conceptual basis for continuous improvement activities. The cycle is illustrated in Figure 11–1. Representing the process with a circle underscores its continuing nature.

plan-do-study-act cycle A framework for improvement activities.

There are four basic steps in the cycle:

Plan. Begin by studying the current process. Document that process. Then collect data to identify problems. Next, survey data and develop a plan for improvement. Specify measures for evaluating the plan.

Do. Implement the plan, on a small scale if possible. Document any changes made during this phase. Collect data systematically for evaluation.

Study. Evaluate the data collection during the *do* phase. Check how closely the results match the original goals of the *plan* phase.

Act. If the results are successful, *standardize* the new method and communicate the new method to all people associated with the process. Implement training for the new method. If the results are unsuccessful, revise the plan and repeat the process or cease this project.

FIGURE 11-1

The PDSA cycle

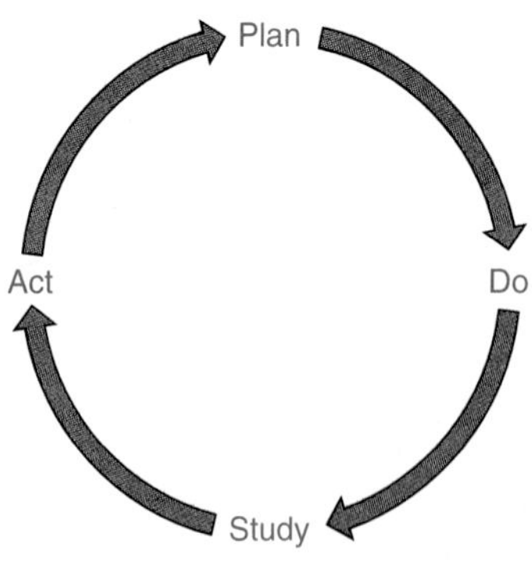

In replicating successful results elsewhere in the organization, the cycle is repeated. Similarly, if the plan was unsuccessful and you wish to make further modifications, repeat this cycle.

Employing this sequence of steps provides a systematic approach to continuous improvement.

TOOLS

There are a number of tools that an organization can use for problem solving and process improvement. This section describes eight of these tools. The tools aid in data collection and interpretation, and provide the basis for decision making.

The first seven tools are often referred to as the *seven basic quality tools*. Figure 11–2 provides a quick overview of the seven tools.

check sheet A tool for recording and organizing data to identify a problem.

Check Sheets. A **check sheet** is a simple tool frequently used for problem identification. Check sheets provide a format that enables users to record and organize data in a way that facilitates collection and analysis. This format might be one of simple checkmarks. Check sheets are designed on the basis of what the users are attempting to learn by collecting data.

Many different formats can be used for a check sheet and there are many different types of sheets. One frequently used form of check sheet deals with type of defect, another with location of defects. These are illustrated in Figures 11–3 and 11–4 (on page 500).

Figure 11–3 shows tallies that denote the type of defect and the time of day each occurred. Problems with missing labels tend to occur early in the day and smeared print tends to occur late in the day, whereas offcenter labels are found throughout the day. Identifying types of defects and when they occur can help in pinpointing causes of the defects.

Figure 11–4 makes it easy to see where defects on the product are occurring. In this case, defects seem to be occurring on the tips of the thumb and first finger, in the finger valleys (especially between the thumb and first finger), and in the center of the gloves. Again, this may help determine why the defects occur and lead to a solution.

flowchart A diagram of the steps in process.

Flowcharts. A **flowchart** is a visual representation of a process. As a problem-solving tool, a flowchart can help investigators in identifying possible points in a process where problems occur. Figure 11–5 (on page 501) illustrates a flowchart.

Check sheet

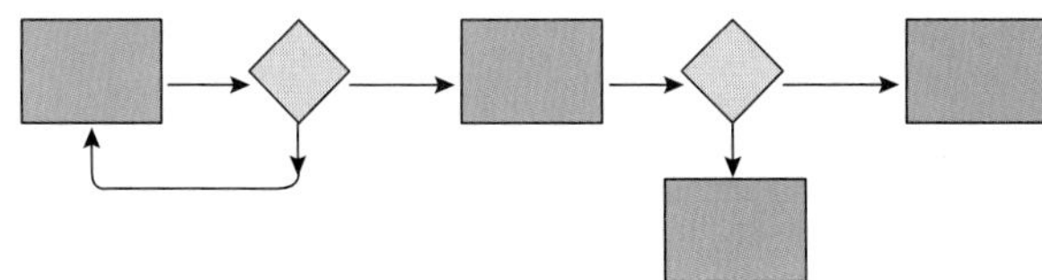

	Day			
Defect	1	2	3	4
A	///		////	/
B	//	/	//	///
C	/	////	//	////

A tool for organizing and collecting data; a tally of problems or other events by category

Flowchart

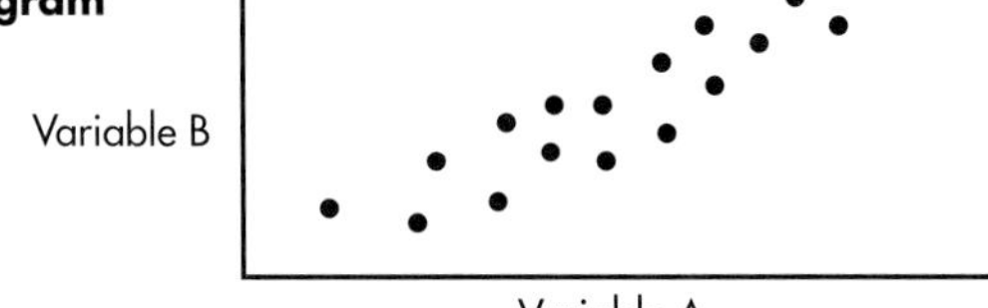

A diagram of the steps in a process

Scatter diagram

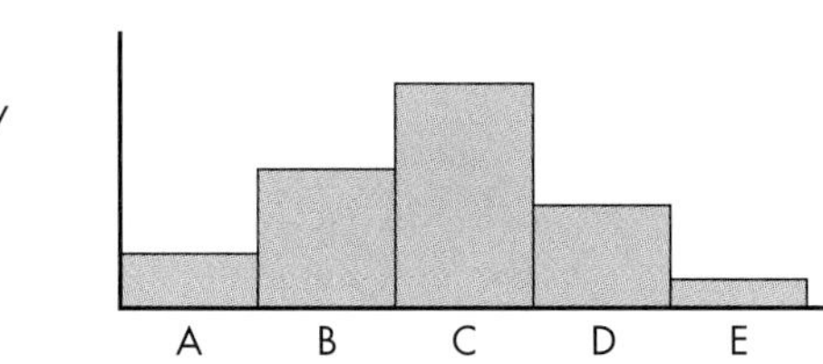

A graph that shows the degree and direction of relationship between two variables

Histogram

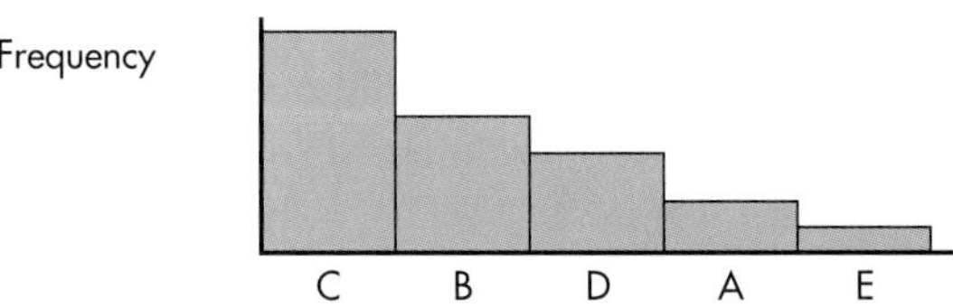

A chart that shows an empirical frequency distribution

Pareto chart

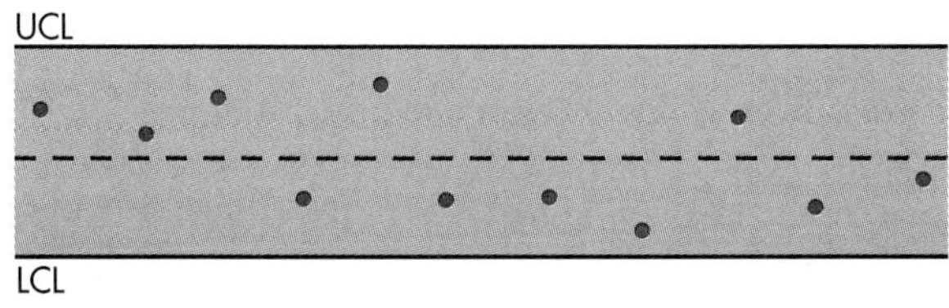

A diagram that arranges categories from highest to lowest frequency of occurrence

Control chart

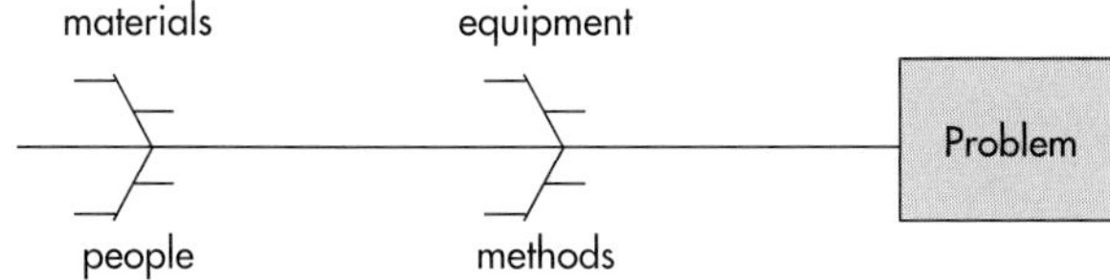

A statistical chart of time-ordered values of a sample statistic (e.g., sample means)

Cause-and-effect diagram

A diagram used to organize a search for the cause(s) of a problem; also known as a *fishbone* diagram

FIGURE 11-2

The seven basic quality tools

FIGURE 11–3

A typical check sheet

Day	Time	Type of Defect: Missing label	Off-center	Smeared print	Loose or folded	Other	Total
M	8-9	IIII	II				6
	9-10		III				3
	10-11	I	III	I			5
	11-12		I		I	I (Torn)	3
	1-2		I				1
	2-3		II	III	I		6
	3-4		II	~~IIII~~ I			8
Total		5	14	10	2	1	32

FIGURE 11–4

A special purpose check sheet

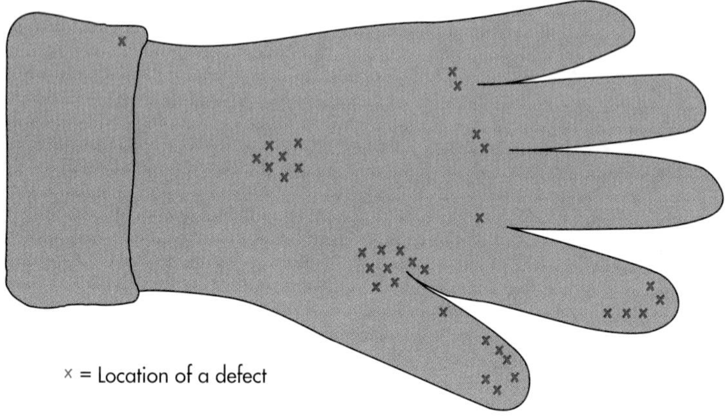

The diamond shapes in the flowchart represent decision points in the process, and the rectangular shapes represent procedures. The arrows show the direction of "flow" of the steps in the process.

To construct a simple flowchart, begin by listing the steps in a process. Then classify each step as either a procedure or a decision (or check) point. Try to not make the flowchart too detailed, or it may be overwhelming, but be careful not to omit any key steps.

scatter diagram A graph that shows the degree and direction of relationship between two variables.

Scatter Diagrams. A **scatter diagram** can be useful in deciding if there is a correlation between the values of two variables. A correlation may point to a cause of a problem. Figure 11–6 shows an example of a scatter diagram. In this particular diagram, there is a *positive* (upward sloping) relationship between the humidity and the number of errors per hour. High values of humidity correspond to high numbers of errors, and vice versa. On the other hand, a *negative* (downward sloping) relationship would mean that when values of one variable are low, values of the other variable are high, and vice versa.

The higher the correlation between the two variables, the less scatter in the points; the points will tend to line up. Conversely, if there were little or no relationship between two variables, the points would be completely scattered. Here, the correlation between humidity and errors seems strong, because the points appear to scatter along an imaginary line.

histogram A chart of an empirical frequency distribution.

Histograms. A **histogram** can be useful in getting a sense of the distribution of observed values. Among other things, one can see if the distribution is symmetrical, what the range of values is, and if there are any unusual values. Figure 11–7 (on page 502)

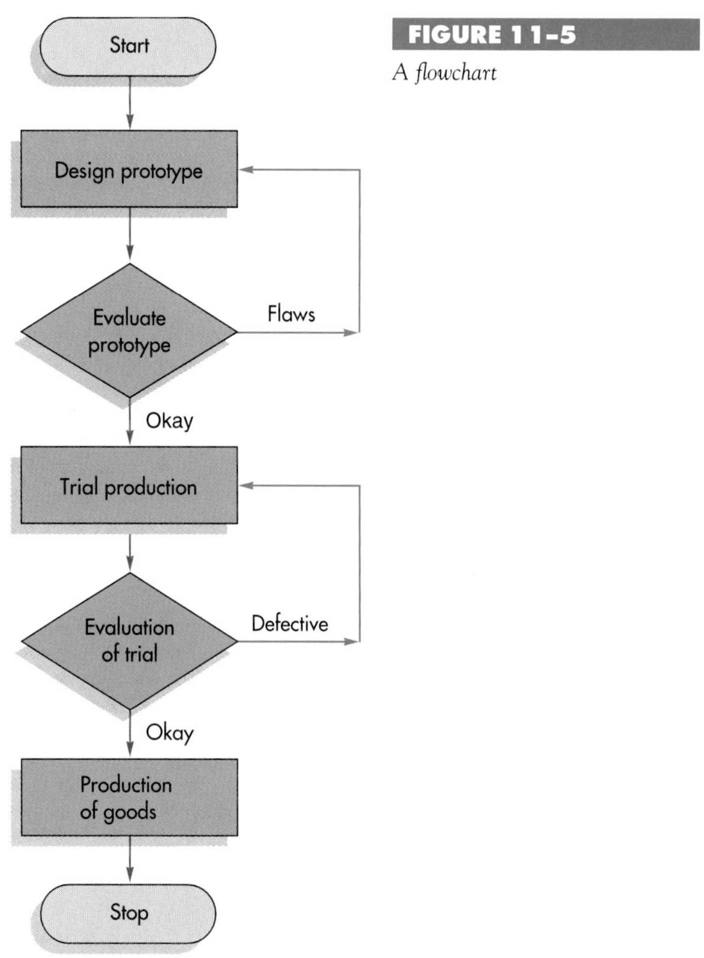

FIGURE 11-5

A flowchart

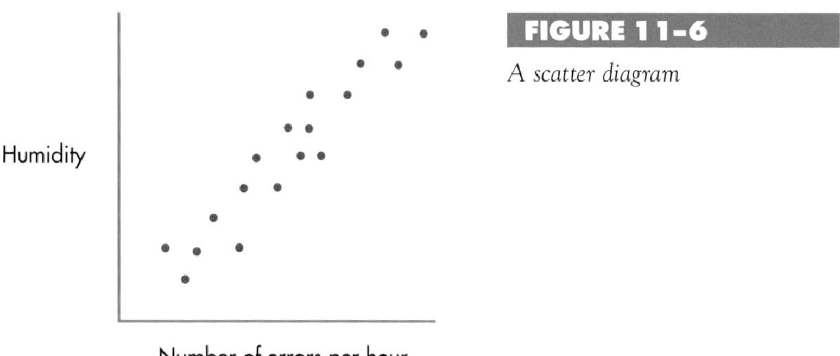

FIGURE 11-6

A scatter diagram

FIGURE 11-7

A histogram

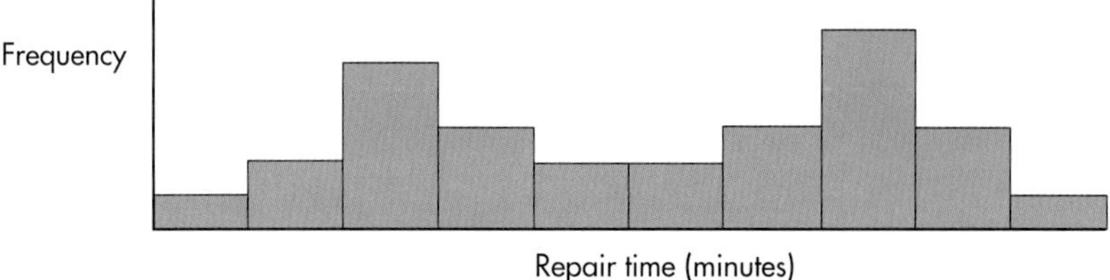

FIGURE 11-8

A Pareto diagram based on data in Figure 11–3

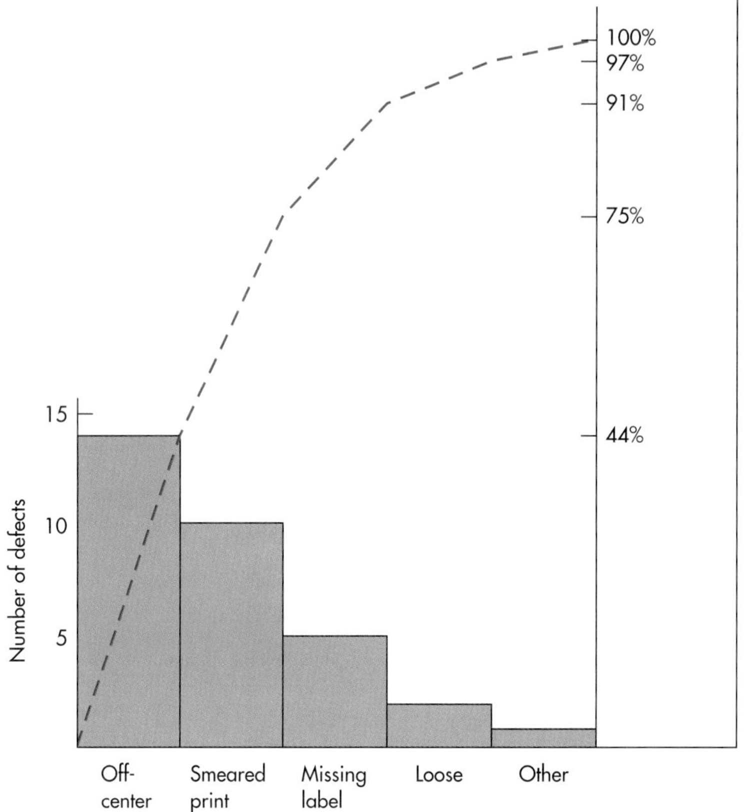

illustrates a histogram. Note the two peaks. This suggests the possibility of *two* distributions with different centers. Possible causes might be two different workers or two different types of work.

Pareto analysis Technique for classifying problem areas according to degree of importance, and focusing on the most important.

Pareto Analysis. **Pareto analysis** is a technique for focusing attention on the most important problem areas. The Pareto concept, named after the 19th-century Italian economist Vilfredo Pareto, is that a relatively few factors generally account for a large percentage of the total cases (e.g., complaints, defects, problems). The idea is to classify the cases according to degree of importance, and focus on resolving the most important, leaving the less important. Often referred to as the 80–20 rule, the Pareto concept states that approximately 80 percent of the problems come from 20 percent of the items. For instance, 80 percent of machine breakdowns come from 20 percent of the machines, and 80 percent of the product defects come from 20 percent of the causes of defects.

Often, it is useful to prepare a chart that shows the number of occurrences by category, arranged in order of frequency. Figure 11–8 illustrates such a chart corresponding to the check sheet shown in Figure 11–3. The dominance of the problem with offcenter labels becomes apparent. Presumably, the manager and employees would focus on trying to resolve this problem. Once they accomplished that, they could address the remaining

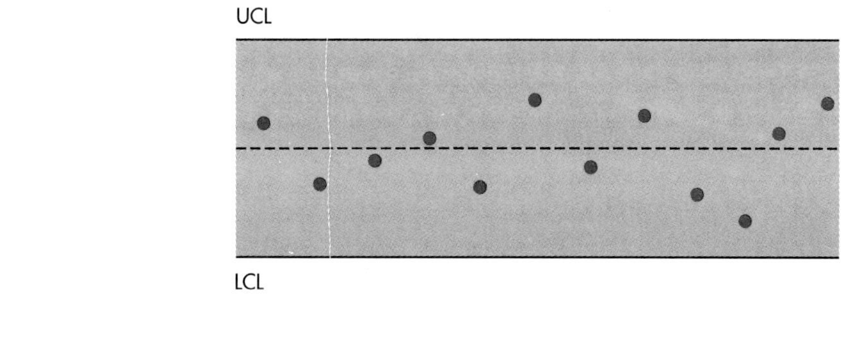

A control chart

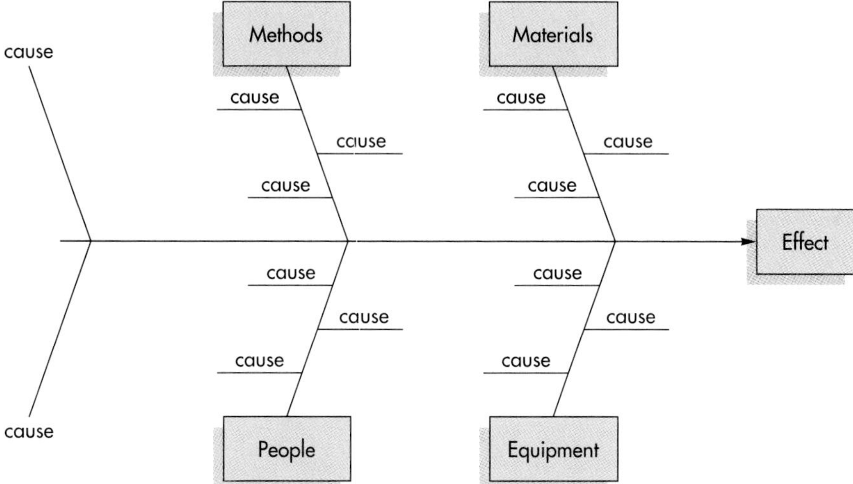

FIGURE 11-10

One format of a cause-and-effect diagram

defects in similar fashion; "smeared print" would be the next major category to be resolved, and so on. Additional check sheets would be used to collect data to verify that the defects in these categories have been eliminated or greatly reduced. Hence, in later Pareto diagrams, categories such as "offcenter" may still appear but would be much less prominent.

Control Charts. A **control chart** can be used to monitor a process to see if the process output is random. It can help detect the presence of *correctable* causes of variation. Figure 11–9 illustrates a control chart. Control charts can also indicate when a problem occurred and give insight into what caused the problem. Control charts are described in detail in the preceding chapter.

control chart A statistical chart of time-ordered values of a sample statistic.

Cause-and-Effect Diagrams. A **cause-and-effect diagram** offers a structured approach to the search for the possible cause(s) of a problem. It is also known as a **fishbone diagram** because of its shape, or an *Ishikawa diagram,* after the Japanese professor who developed the approach to aid workers overwhelmed in problem solving by the number of possible sources of problems. This tool helps to organize problem-solving efforts by identifying *categories* of factors that might be causing problems. Often this tool is used after brainstorming sessions, to organize the ideas generated. Figure 11–10 illustrates one form of a cause-and-effect diagram.

cause-and-effect diagram Used to search for the cause(s) of a problem; also called *fishbone diagram.*

An example of an application of such a cause-and-effect diagram is shown in Figure 11–11. Each of the factors listed in the diagram is a potential source of ticket errors. Some are more likely causes than others, depending on the nature of the errors. If the cause is still not obvious at this point, additional investigation into the *root cause* may be necessary, involving a more in-depth analysis. Often, more detailed information can be obtained by asking *who, what, where, when, why,* and *how* questions about factors that appear to be the most likely sources of problems.

FIGURE 11–11

Cause-and-effect diagram for airline ticket errors

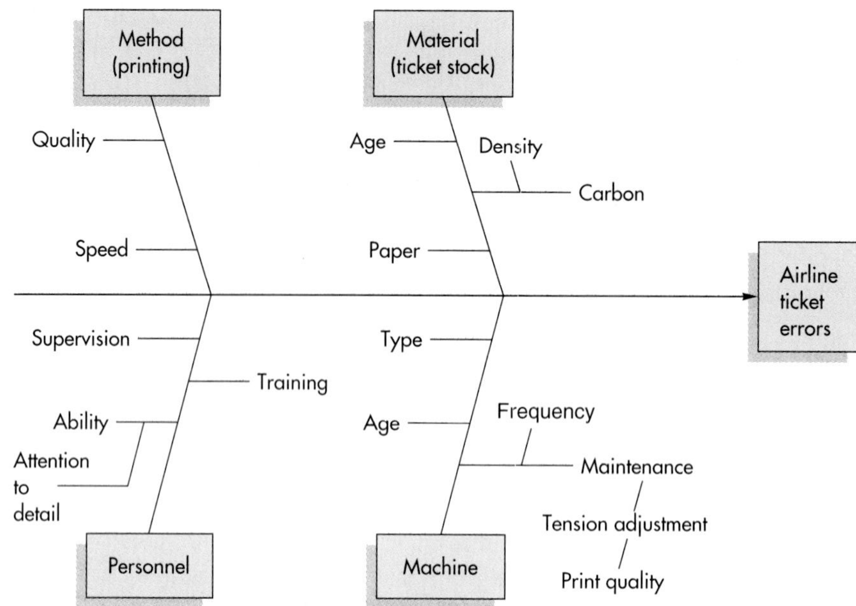

Source: Reprinted from Howard Gitlow, Shelly Gitlow, Alan Oppenheim, and Rosa Oppenheim, *Tools and Methods for the Improvement of Quality* (Burr Ridge, Ill.: Richard D. Irwin, 1989), p. 384, by permission.

FIGURE 11–12

A run chart shows performance over time

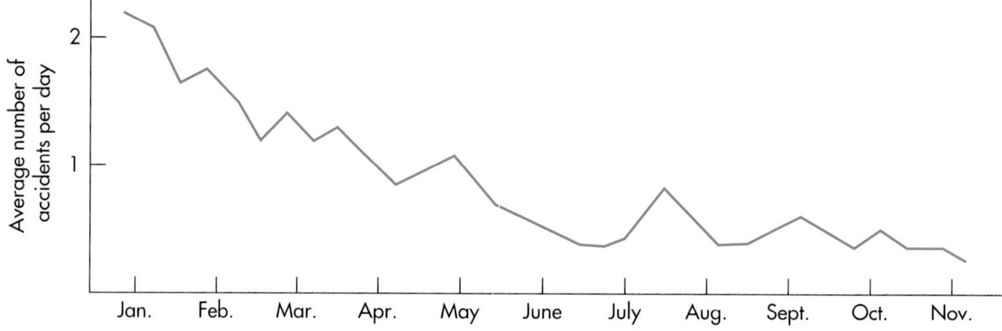

run chart Tool for tracking results over a period of time.

Run Charts. A **run chart** can be used to track the values of a variable over time. This can aid in identifying trends or other patterns that may be occurring. Figure 11–12 provides an example of a run chart showing a decreasing trend in accident frequency over time. Important advantages of run charts are ease of construction and ease of interpretation.

Some Illustrations of the Use of Graphical Tools. This section presents some illustrations of the use of graphical tools in process or product improvement. Figure 11–13 begins with a check sheet that can be used to develop a Pareto chart of the types of errors found. That leads to a more focused Pareto diagram of the most frequently occurring type of error, followed (moving right) by a cause-and-effect diagram of the second most frequently occurring error. Additional cause-and-effect diagrams, such as errors by location, might also be used.

Figure 11–14 shows how Pareto charts measure the amount of improvement achieved in a before-and-after scenario of errors.

Figure 11–15 illustrates how control charts track two phases of improvement in a process that was initially out of control.

FIGURE 11-13

Employing graphical tools in problem solving

Check sheet: Type of error by location

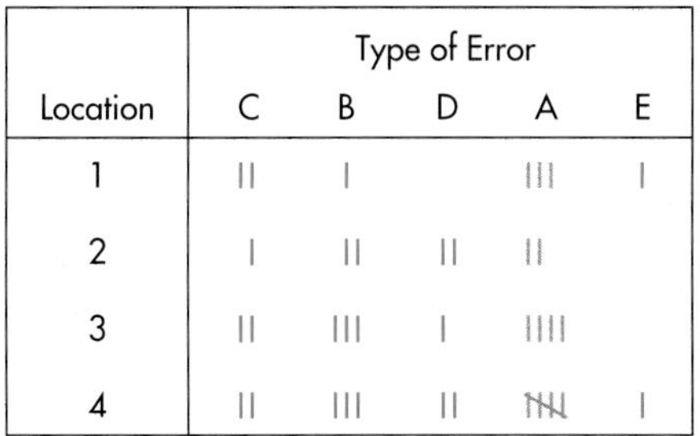

Location	C	B	D	A	E
1	II	I		III	I
2	I	II	II	II	
3	II	III	I	IIII	
4	II	III	II	~~IIII~~	I

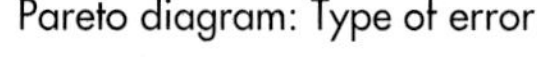

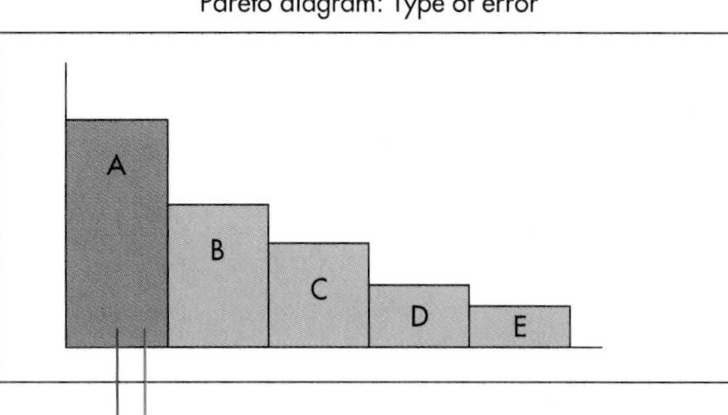

Pareto diagram: A errors by location

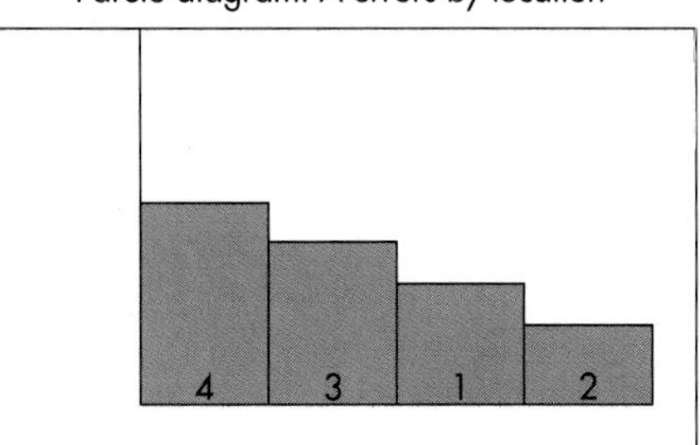

Cause-and-Effect diagram: A errors

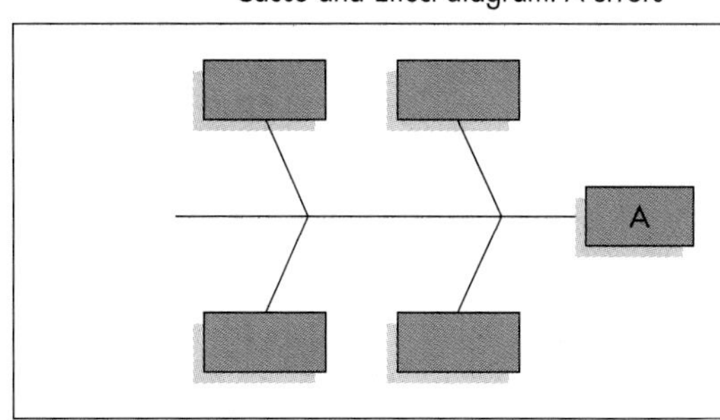

Before

100%

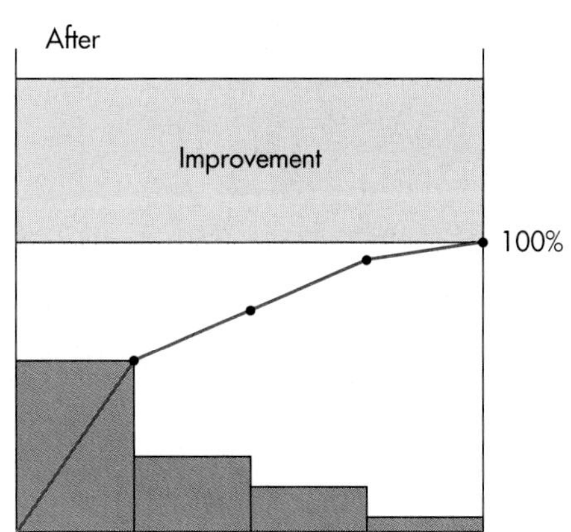

FIGURE 11-14

Comparison of before and after using Pareto charts

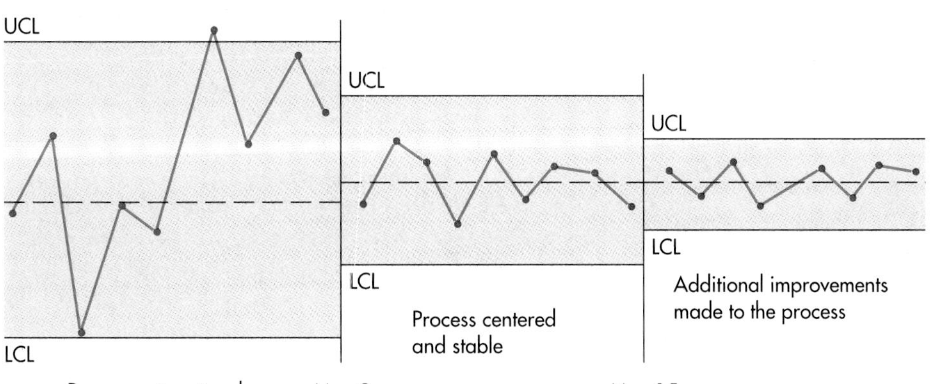

Process not centered and not stable

May 3

May 15

UCL = Upper Control Limit
LCL = Lower Control Limit

FIGURE 11-15

Using a control chart to track improvements

READING

CONTINUOUS IMPROVEMENT ON THE FREE-THROW LINE

Timothy Clark and Andrew Clark

In 1924, Walter Shewhart developed a problem-solving method to continually improve quality by reducing variation (the difference between the ideal outcome and the actual situation). To help guide improvement efforts, Shewhart outlined a process referred to as the plan-do-study-act (PDSA) cycle. The PDSA cycle combined with the traditional concepts of decision making and problem solving are what my son and I used to continuously improve his basketball free-throw shooting.

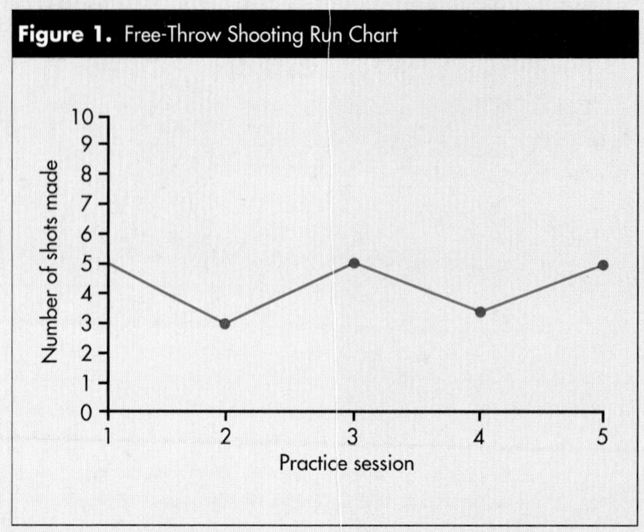

Figure 1. Free-Throw Shooting Run Chart

Recognizing the Problem

Identify the facts. I had observed over a three-year period from 1991 to 1993 that in basketball games, my son Andrew's free-throw shooting percentage averaged between 45 percent and 50 percent.

Identify and define the process. Andrew's process for shooting free throws was simple: Go to the free-throw line, bounce the ball four times, aim, and shoot.

The desired outcome was a higher free-throw shooting percentage. An ideal outcome, or perfection, would be one in which 100 percent of the shots fall through the middle of the rim, land at the same spot on the floor every time, and roll straight back in the shooter's direction after landing.

Plot the points. To confirm my observations on the results of the current process, we went to the YMCA and Andrew shot five sets of 10 free throws for a total of 50 shots. His average was 42 percent. Results were recorded on a run chart (see Figure 1). Based on this information as well as on past observations, I estimated the process was stable.

Decision Making

Identify the causes. Causes of variation in any process can be identified through the general categories of people, equipment, materials, methods, environment, and measurement. A cause-and-effect diagram is used to graphically illustrate the relationship between the effect—a low free-throw shooting percentage—and the principal causes (see Figure 2).

In analyzing my son's process, I noticed that he did not stand at the same place on the free-throw line every time. I believed his inconsistent shooting position affected the direction of the shot. If the shot goes left or right, there is a smaller probability that the ball will have a lucky bounce and go in. I also

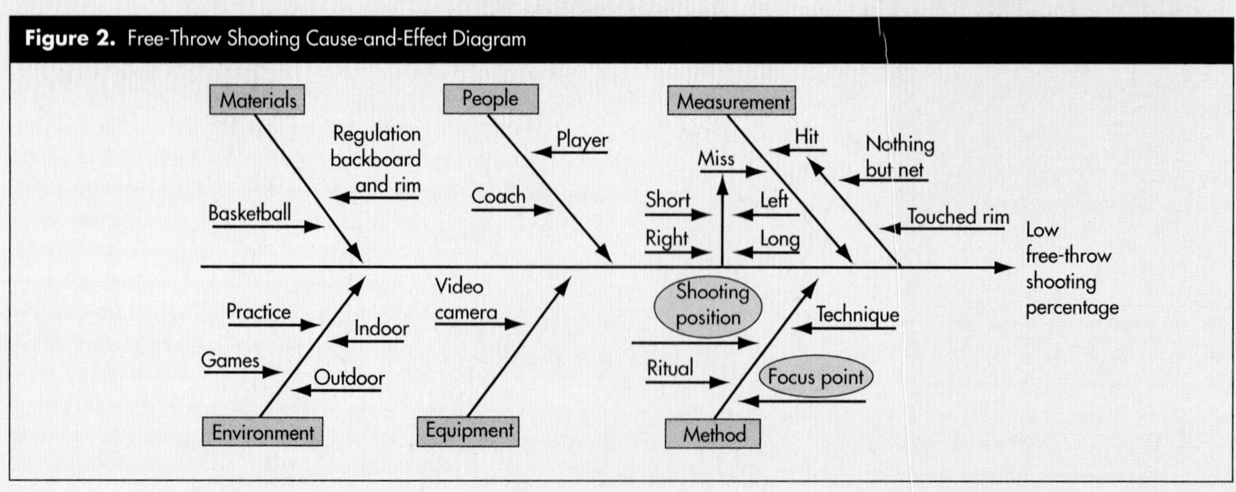

Figure 2. Free-Throw Shooting Cause-and-Effect Diagram

noticed that he didn't seem to have a consistent focal point.

Develop, analyze, and select alternatives. The alternatives selected for Andrew, a right-handed shooter, were for him to line up his right foot on the middle of the free-throw line, focus on the middle of the front part of the rim, and visualize the perfect shot before he released the ball. The modified process is:

1. Stand at the center of the free-throw line.
2. Bounce the ball four times.
3. Focus on the middle of the front part of the rim, and visualize a perfect shot.
4. Shoot.

Develop an action plan. The course of action at this point was for Andrew to shoot five more sets of 10 free throws to test the effectiveness of the changes.

Problem Solving

Implement the selected alternative and compare actual with expected results. The new process resulted in a 36 percent improvement in Andrew's average free-throw percentage at basketball practice, which raised his average to 57 percent (see Figure 3). The new process was first implemented in games toward the end of the 1994 season, and in the last three games, Andrew hit nine of his 13 free throws for a free-throw shooting average of 69 percent.

During the 1995 season, Andrew made 37 of his 52 free throws in games for an average of 71 percent. In one extremely close game where the other team was forced to foul Andrew's team in an effort to get the ball back, Andrew hit seven of his seven shots, which helped his team win the game. In team practices, the coaches had the players shoot two free throws and then rotate. For the entire season, Andrew hit 101 of 169 of his team practice free throws for an average of 60 percent.

As we monitored Andrew's process from March 17, 1994, to Jan. 18, 1996, we plotted the total number of practice shots made out of 50, using Shewhart's number-of-affected-units control chart (see Figure 4). A control chart is a trend chart with upper and lower control limits, the variation in the process is due to normal or common causes of variation, and the conclusion can be predictable. A special cause might represent a temporary or fleeting event and might require little or no action to resolve. Variation due to a common cause requires a permanent change in the process. In this case, the process is stable, which will make it easier to validate future improvement efforts.

Figure 3. Free-Throw Shots Made Before and After Implementing the PDSA Cycle (March 17, 1994, to Nov. 23, 1994)

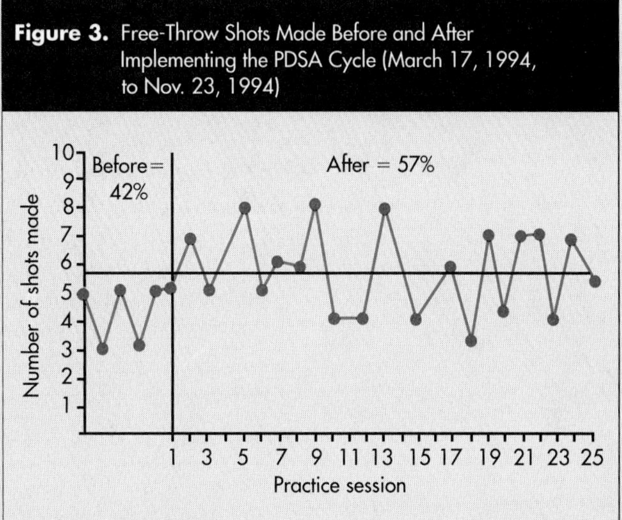

Figure 4. Determining Whether the Free-Throw Process Is Stable (March 17, 1994, to Jan. 18, 1996)

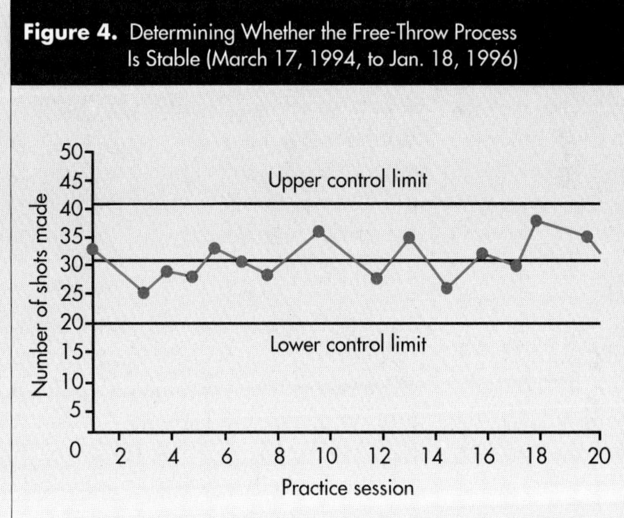

made that the process is stable or predictable. In other words, if you always do what you always did, on average, you will usually get what you always got.

If any points fall outside the limits, the variation is due to a special cause that makes the process unstable or un-

In the late summer of 1995, Andrew went to a basketball camp where he was advised to change his shooting technique. This change to his process reduced his shooting percentage during the 1996 season to 50 percent. This caused him to lose confidence in his

shooting ability, and consequently, he took fewer shots. We then reinstalled his old process, and his shooting percentage returned to its former level. In one series of 50 practice free throws, he hit 35 of 50 shots for an average of 70 percent and in another set, he hit 32 of 50 for an average of 64 percent. During the remaining team practices, Andrew hit 14 of 20 of his practice free throws for an average of 70 percent. During the final three games, he hit two of three free throws for an average of 67 percent.

During the 1996 and 1997 seasons, Andrew was a point guard and was responsible for controlling and distributing the ball. In this position, he had fewer opportunities to shoot free throws. Therefore, during the 1997 season, he had the opportunity to shoot only 12 free throws, but he made nine of them for an average of 75 percent.

Overall benefits. In addition to the tangible results, such as improved free-throw shooting, the intangible benefits were also significant. For example, Andrew's confidence improved, and he learned how to determine when changes to his shooting technique resulted in improvement. W. Edwards Deming referred to this type of knowledge as profound.

Continuous Improvement

Take appropriate action based on study results. In preparation for the 1998 season, Andrew's priorities for improvement are to continue to monitor his free-throw shooting to ensure it remains stable and to work on improving the shooting percentage of his two- and three-point shots.

Knowledge Changes How People Look at the World

Shewhart's methodology represents a problem-solving or decision-making process that requires a fundamental change in thinking. Traditionally, people are trained to make decisions based on gut feelings or on relatively few facts. The PDSA process requires people to first determine if the process outcome is due to a common cause or a special cause. This knowledge becomes the foundation for making decisions, which can only be developed by plotting points. Developing a knowledge and understanding of variation will change the way you look at the world forever and can lead to unprecedented levels of quality.

Methods For Generating Ideas

Some additional tools that are useful for problem solving and/or for process improvement are run charts, brainstorming, quality circles, interviewing, and benchmarking. We begin with run charts.

brainstorming Technique for generating a free flow of ideas in a group of people.

Brainstorming. **Brainstorming** is a technique in which a group of people share thoughts and ideas on problems in a relaxed atmosphere that encourages unrestrained collective thinking. The goal is to generate a free flow of ideas on identifying problems, and finding causes, solutions, and ways to implement solutions. In successful brainstorming, criticism is absent, no single member is allowed to dominate sessions, and all ideas are welcomed.

quality circles Groups of workers who meet to discuss ways of improving products or processes.

Quality Circles. One way companies have tapped employees for ideas concerning quality improvement is through **quality circles.** The circles comprise a number of workers who get together periodically to discuss ways of improving products and processes. Not only are quality circles a valuable source of worker input, they also can motivate workers, if handled properly, by demonstrating management interest in worker ideas. Quality circles are usually less structured and more informal than teams involved in continuous improvement, but in some organizations quality circles have evolved into continuous improvement teams. Perhaps a major distinction between quality circles and teams is the amount of authority given to the teams. Typically, quality circles have had very little authority to implement any but minor changes; continuous improvement teams are sometimes given a great deal of authority. Consequently, continuous improvement teams have the added motivation generated by *empowerment.*

The team approach works best when it reaches decisions based on consensus. This may involve one or more of the following methods:

1. *List reduction* is applied to a list of possible problems or solutions. Its purpose is to clarify items, and in the *process,* reduce the list of items by posing questions about affordability, feasibility, and likelihood of solving the problem for each item.

Keeping a watchful eye on copy quality, Xerox technicians check the performance of a newly manufactured Xerox 5090 duplicator. The duplicator, the flagship of the Xerox 50 Series line, is manufactured in Webster, outside Rochester, N.Y. More than two million copiers and duplicators have been built in the plant since Xerox broke ground there in the early 1960s.

www.xerox.com

TABLE 11–4

The benchmarking approach

1. What organization does it the best?
2. How do they do it?
3. How do we do it now?
4. How can we change to match or exceed the best?

2. A *balance sheet* approach lists the pros and cons of each item, and focuses discussion on important issues.

3. *Paired comparisons* is a process by which each item on a list is compared with every other item, two at a time. For each pair, team members select the preferred item. This approach forces a choice between items. It works best when the list of items is small: say, five or less.

Interviewing. Another technique a firm can use to identify problems or collect information about a problem is **interviewing.** Internal problems may require interviewing employees; external problems may require interviewing external customers.

interviewing Technique for identifying problems and collecting information.

Ideas for improvement can come from a number of sources: research and development, customers, competitors, and employees. Customer satisfaction is the ultimate goal of improvement activities, and customers can offer many valuable suggestions about products and the service process. They are less apt to have suggestions for manufacturing processes.

Benchmarking. **Benchmarking** is an approach that can inject new energy into improvement efforts. Summarized in Table 11–4, benchmarking is the process of measuring an organization's performance on a key customer requirement against the best in the industry, or against the best in any industry. Its purpose is to establish a standard against which performance is judged, and to identify a model for learning how to improve. A benchmark demonstrates the degree to which customers of other organizations are satisfied. Once a benchmark has been identified, the goal is to meet or exceed that standard through improvements in appropriate processes.

benchmarking Process of measuring performance against the best in the same or another industry.

TABLE 11–5
The 5W2H approach

Category	5W2H	Typical Questions	Goal
Subject	What?	What is being done?	Identify the focus of analysis.
Purpose	Why?	Why is this necessary?	Eliminate unnecessary tasks.
Location	Where?	Where is it being done? Why is it done there? Would it be better to do it someplace else?	Improve the location.
Sequence	When?	When is it done? Would it be better to do it at another time?	Improve the sequence.
People	Who?	Who is doing it? Could someone else do it better?	Improve the sequence or output.
Method	How?	How is it being done? Is there a better way?	Simplify tasks, improve output.
Cost	How much?	How much does it cost now? What would the new cost be?	Select an improved method.

Source: Adapted from Alan Robinson, *Continuous Improvement in Operations, A Systematic Approach to Waste Reduction* (Cambridge, Mass.: Productivity Press, 1991), p. 246.

The benchmarking process usually involves these steps:

1. Identify a critical process that needs improvement (e.g., order entry, distribution, service after sale).
2. Identify an organization that excels in the process, preferably the best.
3. Contact the benchmark organization, visit it, and study the benchmark activity.
4. Analyze the data.
5. Improve the critical process at your own organization.

Selecting an industry leader provides insight into what competitors are doing; but competitors may be reluctant to share this information. Several organizations are responding to this difficulty by conducting benchmarking studies and providing that information to other organizations without revealing the sources of the data.

Vol. 1, Tape 1, seg. 1
Caterpillar, Cummins Engine, Navistar

Selecting organizations that are world leaders in different industries is another alternative. For example, the Xerox Corporation uses many benchmarks: For employee involvement, Procter & Gamble; for quality process, Florida Power and Light, Toyota, and Fuji Xerox; for high-volume production, Kodak and Canon; for billing collection, American Express; for research and development, AT&T and Hewlett-Packard; for distribution, L.L. Bean and Hershey Foods, and for daily scheduling, Cummins Engine.

5W2H approach A method of asking questions about a process that includes what, why, where, when, who, how, and how much.

The 5W2H Approach. Asking questions about the current process can lead to important insights about why the current process isn't working as well as it could, as well as potential ways to improve it. One method is called the **5W2H** (5 "w" words and 2 "h" words) approach (see Table 11–5).

Operations Strategy

In order for total quality management to be successful, it is essential that a majority of those in an organization "buy in" to the idea. Otherwise, there is a risk that a significant portion of the benefits of the approach will not be realized. Therefore, it is important to give this sufficient attention, and to confirm that concordance exists before plunging ahead. Also important is education of managers and workers in the concepts, tools, and procedures of quality. Again, if education is incomplete, there is the risk that TQM will not produce the desired benefits.

Summary

Total quality management is a never-ending pursuit of quality that involves everyone in an organization. The driving force is customer satisfaction; a key philosophy is continuous improvement. Training of managers and workers in quality concepts, tools, and procedures is an important aspect of the approach. Teams are an integral part of TQM.

Two major aspects of the TQM approach are problem solving and process improvement.

Key Terms

benchmarking, 509
brainstorming, 508
cause-and-effect diagram, 503
check sheet, 498
continuous improvement, 493
control chart, 503
5W2H approach, 510
fishbone diagram, 503
flowchart, 498
histogram, 500
interviewing, 509
kaizen, 493
Pareto analysis, 502
plan-do-study-act cycle, 497
process improvement, 497
quality at the source, 493
quality circles, 508
run chart, 504
scatter diagram, 500
total quality management, 492

Discussion and Review Questions

1. What are the key elements of the TQM approach? What is the driving force behind TQM?
2. Briefly describe each of the seven quality tools.
3. Briefly define or explain each of these tools:
   a. Brainstorming
   b. Benchmarking
   c. Run charts
4. Explain each of these methods:
   a. The plan-do-study-act cycle
   b. The 5W2H approach
5. List the steps of problem solving.
6. Select four tools and describe how they could be used in problem solving.
7. List the steps of process improvement.
8. Select four tools and describe how they could be used for process improvement.

Memo Writing Exercises

1. The vice president of manufacturing of your company, June Seymour, has asked you to write her a memo that lists the key elements of a total quality management approach and outlines the benefits and risks of adopting a TQM approach for the company.
   Write a one- to two-page memo to her.
2. You are an assistant to the production manager of a large company. The company wants to give worker teams in several departments additional authority and responsibility to see whether quality and productivity in those areas are increased. However, the proposal has met with resistance from both managers and worker teams.
   Write a one-page memo to Jeff Rogers, the production manager, discussing the probable causes of the resistance and potential solutions for overcoming it.
3. Select a task that you do on a regular basis, such as taking notes in class, doing a homework assignment, grocery shopping, or another task. Write a one-page, quality-at-the-source memo on the difficulties you might encounter by not doing it correctly the first time.

Problems

Work	Frequency
Lube & oil	12
Brakes	7
Tires	6
Battery	4
Transmission	1
	30

1. Make a check sheet and then a Pareto diagram for this car repair shop data.

Ticket No.	Work	Ticket No.	Work
1	Tires	16	Tires
2	Lube & oil	17	Lube & oil
3	Tires	18	Brakes
4	Battery	19	Tires
5	Lube & oil	20	Brakes
6	Lube & oil	21	Lube & oil
7	Lube & oil	22	Brakes
8	Brakes	23	Transmission
9	Lube & oil	24	Brakes
10	Tires	25	Lube & oil
11	Brakes	26	Battery
12	Lube & oil	27	Lube & oil
13	Battery	28	Battery
14	Lube & oil	29	Brakes
15	Lube & oil	30	Tires

2. An air-conditioning repair department manager has compiled data on the primary reason for 41 service calls for the previous week, as shown in the table. Using the data, make a check sheet for the problem types for each customer type, and then construct a Pareto diagram for each type of customer.

	Res.	Com.	
Noisy	10	3	13
Failed	7	2	9
Odor	5	7	12
Warm	3	4	7
	25	16	41

Job Number	Problem/ Customer Type	Job Number	Problem/ Customer Type	Job Number	Problem/ Customer Type
301	F/R	315	F/C	329	O/C
302	O/R	316	O/C	330	N/R
303	N/C	317	W/C	331	N/R
304	N/R	318	N/R	332	W/R
305	W/C	319	O/C	333	O/R
306	N/R	320	F/R	334	O/C
307	F/R	321	F/R	335	N/R
308	N/C	322	O/R	336	W/R
309	W/R	323	F/R	337	O/C
310	N/R	324	N/C	338	O/R
311	N/R	325	F/R	339	F/R
312	F/C	326	O/R	340	N/R
313	N/R	327	W/C	341	O/C
314	W/C	328	O/C		

Key: Problem type:
N = Noisy
F = Equipment failure
W = Runs warm
O = Odor

Customer type:
C = Commercial customer
R = Residential customer

3. Prepare a run chart for the occurrences of defective computer monitors based on the following data, which an analyst obtained from the process for making the monitors. Workers are given a 15-minute break at 10:15 A.M. and 3:15 P.M., and a lunch break at noon. What can you conclude?

Interval Start Time	Number of Defectives
8:00	1
8:15	0
8:30	0
8:45	1
9:00	0
9:15	1
9:30	1
9:45	2
10:00	3
10:30	1
10:45	0
11:00	0
11:15	0
11:30	1
11:45	3

Interval Start Time	Number of Defectives
1:00	1
1:15	0
1:30	0
1:45	1
2:00	1
2:15	0
2:30	2
2:45	2
3:00	3
3:30	0
3:45	1
4:00	0
4:15	0
4:30	1
4:45	3

Occurrence of defectives is highest just before breaks and lunch.

4. Prepare a run diagram for this 911 call data. Use five-minute intervals (i.e., count the calls received in each five-minute interval. Use intervals of 0-4, 5-9, etc.). Note: Two or more calls may occur in the same minute; there were three operators on duty this night. What can you conclude from the run chart?

Call	Time
1	1:03
2	1:06
3	1:09
4	1:11
5	1:12
6	1:17
7	1:21
8	1:27
9	1:28
10	1:29
11	1:31
12	1:36
13	1:39
14	1:42
15	1:43
16	1:44
17	1:47
18	1:48
19	1:50
20	1:52
21	1:53

Call	Time
22	1:56
23	1:56
24	2:00
25	2:00
26	2:01
27	2:02
28	2:03
29	2:03
30	2:04
31	2:06
32	2:07
33	2:08
34	2:08
35	2:11
36	2:12
37	2:12
38	2:13
39	2:14
40	2:14
41	2:16
42	2:19

The call rate was very high from 2 to 2:15 a.m.

5. Suppose that a table lamp fails to light when turned on. Prepare a simple cause-and-effect diagram to analyze possible causes.

6. Prepare a cause-and-effect diagram to analyze the possible causes of late delivery of parts ordered from a supplier.

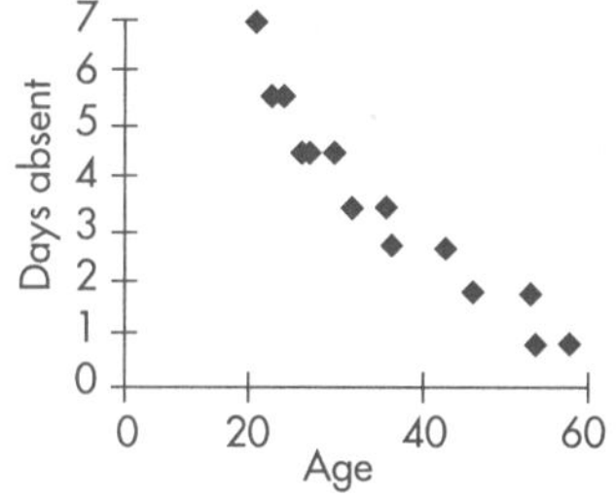

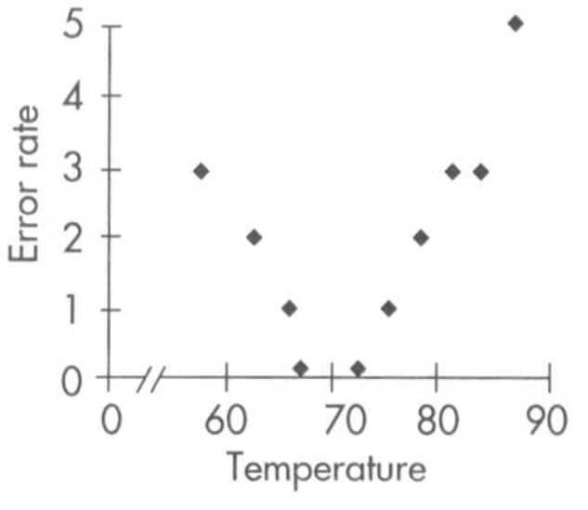

7. Prepare a cause-and-effect diagram to analyze why a machine has produced a large run of defective parts.

8. Prepare a scatter diagram for each of these data sets and then express in words the apparent relationship between the two variables. Put the first variable on the horizontal axis and the second variable on the vertical axis.

a.

Age	24	30	22	25	33	27	36	58	37	47	54	28	42	55
Absenteeism rate	6	5	7	6	4	5	4	1	3	2	2	5	3	1

b.

Temperature (F)	65	63	72	66	82	58	75	86	77	65	79
Error rate	1	2	0	0	3	3	1	5	2	1	3

9. Prepare a flowchart that describes going to the library to study for an exam. Your flowchart should include these items: finding a place at the library to study, checking to see if you have your book, paper, highlighter, etc., traveling to the library, and the possibility of moving to another location if the place you chose to study starts to get crowded.

CASE
CHICK-N-GRAVY DINNER LINE

The operations manager of a firm that produces frozen dinners had received numerous complaints from supermarkets about the firm's Chick-n-Gravy dinners. The manager then asked her assistant, Ann, to investigate the matter and to report her recommendations.

Ann's first task was to determine what problems were generating the complaints. The majority of complaints centered on five defects: underfilled packages, a missing item, spills/mixed items, unacceptable taste, and improperly sealed packages.

Next, she took samples of dinners from the two production lines and examined each sample, making note of any defects that she found. A summary of those results is shown in the table.

The data resulted from inspecting approximately 800 frozen dinners. What should Ann recommend to the manager?

			DEFECT OBSERVED				
Date	**Time**	**Line**	**Underfilled**	**Missing Item**	**Spill/ Mixed**	**Unacceptable Taste**	**Improperly Sealed**
5/12	0900	1		√√	√	√√√	
5/12	1330	2			√√		√√
5/13	1000	2				√	√√√
5/13	1345	1	√√		√√		
5/13	1530	2		√√	√√√		√
5/14	0830	1		√√√		√√√	
5/14	1100	2	√		√	√√	
5/14	1400	1			√		√
5/15	1030	1		√√√		√√√√√	
5/15	1145	2			√	√√	
5/15	1500	1	√		√		
5/16	0845	2				√√	√√
5/16	1030	1		√√√	√	√√√	
5/16	1400	1					
5/16	1545	2	√	√√√√√	√	√	√√

CASE
TIP TOP MARKETS

Tip Top Markets is a regional chain of supermarkets located in the Southeastern United States. Karen Martin, manager of one of the stores, was disturbed by the large number of complaints from customers at her store, particularly on Tuesdays, so she obtained complaint records from the store's customer service desk for the last eight Tuesdays. These are shown below.

Assume you have been asked to help analyze the data and to make recommendations for improvement. Analyze the data using a check sheet, a Pareto diagram, and run charts. Then construct a cause-and-effect diagram for the leading category on your Pareto diagram. Based on your analysis, prepare a list of recommendations that will address customer complaints.

June 1

out of orange yogurt	cashier not friendly
bread stale	produce not fresh
checkout lines too long	lemon yogurt past sell date
overcharged	couldn't find rice
double charged	milk past sell date
meat smelled strange	stock clerk rude
charged for item not purchased	out of maple walnut ice cream
couldn't find the sponges	something green in meat
meat tasted strange	didn't like music
store too cold	checkout lines too slow
light out in parking lot	

June 8

fish smelled funny	undercharged
out of diet bread	out of roses
dented can	meat spoiled
out of hamburger rolls	overcharged on two items
fish not fresh	store too warm
cashier helpful	out of ice
meat tasted bad	telephone out of order
ATM ate card	overcharged
slippery floor	rolls stale
music too loud	bread past sale date

June 15

wanted smaller size	overcharged on special
too cold in store	couldn't find aspirin
out of Wheaties	undercharged
out of Minute Rice	checkout lines too long
cashier rude	out of diet cola
fish tasted fishy	meat smelled bad
ice cream thawed	overcharged on eggs
double charged on hard rolls	bread not fresh
long wait at checkout	didn't like music
wrong price on item	lost wallet
overcharged	overcharged on bread
fish didn't smell right	

June 22

milk past sales date	couldn't find oatmeal
store too warm	out of Bounty paper towels
foreign object in meat	overcharged on orange juice
store too cold	lines too long at checkout
eggs cracked	couldn't find shoe laces
couldn't find lard	out of Smucker's strawberry jam
out of 42 oz. Tide	out of Frosty Flakes cereal
fish really bad	out of Thomas English Muffins
windows dirty	

June 29

checkout line too long	restroom not clean
out of Dove soap	couldn't find sponges
out of Bisquick	checkout lines slow
eggs cracked	out of 18 oz. Tide
store not clean	out of Campbell's turkey soup
store too cold	out of pepperoni sticks
cashier too slow	checkout lines too long
out of skimmed milk	meat not fresh

July 6

out of straws	out of Dial soap
out of bird food	store too warm
overcharged on butter	price not as advertised
out of masking tape	need to open more checkouts
stockboy was helpful	shopping carts hard to steer
lost child	debris in aisles
meat looked bad	out of Drano
overcharged on butter	out of Chinese cabbage
out of Swiss chard	store too warm
too many people in store	floors dirty and sticky
out of bubble bath	out of Diamond chopped walnuts

July 13

wrong price on spaghetti	undercharged
water on floor	out of brown rice
store looked messy	out of mushrooms
store too warm	overcharged
checkout lines too long	checkout wait too long

cashier not friendly	shopping cart broken	store too warm	lights not on in parking lot
out of Cheese Doodles	couldn't find aspirin	found money in aisle	checkout lines too long
triple charged	out of brown lunch bags	wrong price on meat	telephone out of order
out of Saran Wrap	out of straws	overcharged on bread	leak in roof
out of Dove bars		nails on floor	spill in soft drink aisle
		out of 18 oz. Tide	out of jelly doughnuts
July 20		out of decorator napkins	out of wild bird seed
couldn't find frozen foods	out of Minute Rice	out of Heinz catsup	out of dog biscuits
checkout lines too slow	restrooms not clean	out of peach jam	out of bug spray

READING

MAKING QUALITY PAY: RETURN ON QUALITY

Start with an effective quality program. Companies that don't have the basics, such as process and inventory controls and other building blocks, will find a healthy return on quality elusive.

Calculate the cost of current quality initiatives. Cost of warranties, problem prevention, and monitoring activities all count. Measure these against the returns for delivering a product or service to the customer.

Determine what key factors retain customers—and what drives them away. Conduct detailed surveys. Forecast market changes, especially quality and new-product initiatives of competitors.

Focus on quality efforts most likely to improve customer satisfaction at a reasonable cost. Figure the link between each dollar spent on quality and its effect on customer retention and market share.

Roll out successful programs after pilot-testing the most promising efforts and cutting the ones that don't have a big impact. Closely monitor results. Build word of mouth by publicizing success stories.

Improve programs continually. Measure results against anticipated gains. Beware of the competition's initiative and don't hesitate to revamp programs accordingly. Quality never rests.

When the "total quality" mantra swept U.S. boardrooms in the 1980s, few companies responded with the fervor and dedication of Varian Associates Inc. The scientific-equipment maker put 1,000 of its managers through a four-day course on quality. The company's Silicon Valley headquarters buzzed with quality-speak. Talk of work teams and cycle times replaced discussion of electrons and X-rays. There was even a mascot, Koala T—a manager who wore a koala costume and roamed Varian's cafeteria handing out homilies about quality.

And it wasn't just buzzwords and bear suits. Varian went about virtually reinventing the way it did business—with what seemed to be stunning results. A unit that makes vacuum systems for computer clean rooms boosted on-time delivery from 42 percent to 92 percent. The radiation-equipment-service department ranked No. 1 in its industry for prompt customer visits. The semiconductor unit cut the time it took to put out new designs by 14 days. W. Edwards Deming and J. M. Juran, the famous management consultants and leading prophets of quality, would have been proud.

But while Varian thought it was playing quality by the book, the final chapter didn't feature the happy ending the company expected. Obsessed with meeting production schedules, the staff in that vacuum-equipment unit didn't return customers' phone calls and the operation ended up losing market share. Radiation-repair people were so rushed to meet deadlines that they left before explaining their work to customers. Sure, Varian could boast about quality. But in 1990, its sales grew by a paltry 3 percent, to $1.3 billion. And Varian posted a $4.1 million loss after a $32 million profit in 1989. "All of the quality-based charts went up and to the right, but everything else went down," says Richard M. Levy, executive vice-president for quality.

Levy isn't the only one who's dismayed. Countless other managers have heeded the siren song of total quality management, or TQM, only to discover that quality doesn't necessarily pay. At Johnson & Johnson, quality teams for several product lines crisscrossed the country, benchmarking against other companies, but costs skyrocketed. In 1990, Wallace Co. won the Malcolm Baldrige National Quality Award. Two years later, the oil equipment company filed for Chapter 11 as the cost of its quality programs soared and oil prices collapsed.

Rallying Cry. Of course, the quest for quality doesn't always have unhappy results. Detroit, for instance, finally caught the quality wave in the 1980s, and it's hard not to shudder at the thought of how the Big Three would be faring today if they were still turning out Chevy Citations instead of Saturns. And much of the rest of U.S. industry would be locked out of the game in today's global economy without the quality strides of the past few years.

But at too many companies, it turns out, the push for quality can be as badly misguided as it is well-intended. It can be popular with managers and their consultants, but as at Varian, it can devolve into a mechanistic exercise that proves meaningless to customers. And quality that means little to customers doesn't produce a payoff in improved sales, profits, or market share. It's wasted effort and expense.

That's why a growing number of companies and management thinkers are starting to refine the notion. Today's rallying cry: return on quality. Concepts such as better product designs and swifter manufacturing aren't being rejected, but advocates of the new theory are abandoning the narrow statistical benchmarks worshipped by some TQM acolytes. Instead, managers are trying to make sure that the quality they offer is the quality their customers want. And they're starting to use sophisticated financial tools to ensure that quality programs have a payoff. Roland Rust, a Vanderbilt University professor of management and one of ROQ's chief apostles, says executives have to worry about only one thing: "If we're not going to make money off of it, we're not going to do it."

The ROQ revisionism is attracting a growing number of corporate devotees across a wide spectrum of industries. Banking giant NationsBank Corp., for example, now measures every improvement in service, from adding tellers to offering new mortgage products, in terms of added revenue. Telecommunications powerhouse GTE Corp. is looking for quality at reasonable costs. Even companies that were in the vanguard of the 1980s quality push are considering the benefits of ROQ. "We're trying to isolate quality improvements that just don't add any value to the service that is delivered to the customer," says Michael E. Reed, managing director of operations at Federal Express Corp.

For FedEx, a 1990 Baldrige recipient, that has meant rethinking its original quality goals. In its sorting operation, for example, FedEx stressed speed over accuracy. Workers met schedules, but the number of misdirected packages soared as they scrambled to meet deadlines. FedEx eventually fixed most errors, but redirecting each wayward package cost it some $50. Now, the Memphis-based shipper has eased the sorting crunch by investing $100 million in new equipment that routes packages to various destinations.

ROQ is more than just a new twist on an old theme. Many companies believe that applying a bottom-line discipline to quality is crucial at a time when the economy is rebounding and competition is growing. AT&T CEO Robert E. Allen, for example, receives a quarterly report from each of the company's 53 business units that spells out quality improvements and their subsequent financial impact.

Return Threshold. Everything from the installation of new technology to methods of improving billing accuracy is held up against an array of financial yardsticks, such as potential sales gains and return on capital. Based on its experience, AT&T has found that when customers perceive improved quality, it shows up in better financial results three months later. "This is the most important thing that AT&T has ever done," Allen told a meeting of top managers the day before his June board presentation.

To win approval from AT&T's top management these days, proponents of any new quality initiative must first demonstrate that the effort will yield at least a 30 percent drop in defects and a 10 percent return on investment. Ma Bell used those criteria last year to maintain its supremacy in the toll-free 800-number market. To reduce service outages—its customers' biggest complaint—AT&T mulled a vast modernization program. But it seemed unlikely that the staggering $1 billion-plus project would net enough new customers to clear the 10 percent investment-return threshold. Instead, Ma Bell invested $300 million in backup power equipment to guard against failures in its 800-number system. "It isn't the old 'Give me money and I'll fix it' stuff," says Phillip M. Scanlan, a corporate quality officer. "We're taking the cost out of making our system better."

Chasing Prizes. Of course, quality was always supposed to make bottom-line sense. In the Deming and Juran doctrines, empowered employees would make quicker and more market-based decisions. Faster and better manufacturing processes would lead to improved products and broader market share. That message was popularized by Deming in the 1950s, and it soon became the cornerstone of Japanese management theory. The quality theory emigrated to the U.S. in the 1980s as American companies tried to duplicate the Japanese miracle.

For some of them, including Motorola, Intel, Hewlett-Packard, and General Electric, excellence became the norm. But others among the legions who followed Deming came to confuse process with purpose. Quality devotees grew obsessed with methodology—cost-cutting, defect reduction, quicker cycle times, continual improvement. Before too long, customer concerns seemed to fall by the wayside.

Quality became its own reward. Standards were more important than sales. And companies appeared more interested in chasing prizes than profits. Pleasing the International Standards Organization, which sets European quality standards, became a paramount concern for some companies. Meanwhile, Baldrige wannabes often tripped and fell as they tried to complete an obstacle course of requirements that emphasizes process over proceeds. "There's been an insufficient focus on the aspect of quality improvements that will make the largest contribution to overall financial performance," admits Curt W. Reimann, director of the Baldrige Quality Award.

The new focus on the relationship between quality and financial returns does have its detractors. Critics say it's just a smokescreen behind which companies are cutting back on their quality efforts. A healthier economy and rising sales may be prompting them to slack off on the costly discipline of TQM. And some companies—Hewlett-Packard among them—argue it's a mistake to take a bean-counter's view of something as fundamental as quality. Yes, HP makes its decisions about quality based on sound business considerations. But that doesn't mean it takes out a calculator every time it launches a quality program. "Saying that this is a quality move and this is what it's worth is like saying, 'What's my left lung worth?'" says Richard LeVitt, director of corporate quality. "Quality is intrinsic to our whole business."

Ironing It Out. To its advocates, ROQ is about getting companies back to something that's equally intrinsic to everyone's business: customer focus. Instead of talking about attracting new customers with dazzling statistical displays of quality, ROQ emphasizes customer retention. After all, selling more to existing customers is a cheaper way to build market share than luring business away from competitors. "Customers are an economic asset. They're not on the balance sheet, but they should be," says Claess Fornell, a University of Michigan professor who is a leading ROQ advocate. Extensive surveying, perhaps even inviting customers into design and production processes, helps companies identify the key factors that affect customers' buying decisions.

Questions

1. According to the ROQ approach, how should a company decide which quality activities to fund?
2. What criticisms does the author level at those who don't use ROQ to guide quality efforts?
3. What risks are there in rigidly applying the ROQ approach?

Source: Reprinted from the August 8, 1994 issue of *Business Week* by special permission.

READING
QUALITY PROGRAMS DON'T GUARANTEE RESULTS

U.S. companies have poured millions of dollars into quality programs in the last few years. Unfortunately, the programs do not always achieve the results companies expect. The McKinsey Consulting Group has developed several useful guidelines for executives concerning quality programs:

Don't promote continuous improvement if dramatic results are needed. Sagging sales and profits often imply the need for something more than incremental improvement. Continuous improvement programs should be reserved for those instances where an organization has already achieved substantial quality results but still wants to improve its operations.

Link quality programs to strategic planning. Then, set goals for the program, and evaluate senior management, based on how well those goals are met. However, let lower-level employees set their own goals in order to get them involved, and to motivate their best performance.

Focus programs on market "break points." Customers may not be able to perceive a difference between an on-time delivery performance of 90 percent and one of say, 95 percent, although they would perceive a difference in 90 percent versus 99 percent on-time delivery. So determine what the break points are, and don't waste resources on improving performance that does not achieve a higher break point.

Choose a single theme. It is important for everybody to be rowing in the same direction. Note, however, that a single focus does carry a risk: It may become an end in itself.

Emphasize results as well as the process. Focusing exclusively on the process carries the risk of diverting attention from results, and may also lead to excessive buildup in staff associated with the program. Instead, set specific goals in terms of *measurable* results.

Questions

1. List some of the ways a company can judge whether its quality program is working.
2. Explain the importance of measurements for quality programs.
3. For each guideline, explain the rationale.

Source: Based on "When Quality Control Gets in the Way of Quality," by Graham Sharman, *The Wall Street Journal*, ©1992.

READING

SWIMMING UPSTREAM **Theodore B. Kinni**

An early convert to the philosophy of *kaizen,* Richard Chang has been preaching the gospel of continuous improvement for two decades. Founder of Irvine, Calif.-based Richard Chang Associates, Inc., a diversified organizational-consulting firm and publishing house, Mr. Chang currently serves as a senior examiner for the Malcolm Baldrige National Quality Award and judge for California's Governor's Golden State Quality Award.

He is also a prolific writer, having penned 14 books himself. He is the publisher of The Practical Guide Book Series, a collection of four series including the eight-volume Quality Improvement Series—which covers process-improvement methodology, techniques, and tools. In the following interview he explains for IW readers how the spirit of kaizen is successfully infused throughout a small company.

IW: *What are the environmental prerequisites of kaizen?*

CHANG: Typically, you need to run a lean or even understaffed operation. . . . You have to make an investment in training and must also hire people who are open to continuous improvement—they are hard to find. . . . Finally, especially in a small company, the philosophy must come from the top. The CEO can't be a reactive firefighter, but must behave in a manner consistent with continuous improvement. Also, instead of taking a punishment approach, you have to look at problems and mistakes as learning opportunities.

IW: *If we can create such an environment, what's next?*

CHANG: Oftentimes, you would then create a sense of vision and core values that includes the principles of kaizen. You must also build reinforcement into your reward systems and align corporate goals with continuous improvement. Remember, kaizen has to be the way you run your business; it can't just be an extra program you tack on to the operation.

If you think of that as Phase One, Phase Two is then about building capacity. This is when you concentrate on the skills that make you capable of continuously improving the business. People tend to skip this—they want to get right to the results.

IW: *What would a typical continuous-improvement-initiative rollout look like?*

CHANG: We use a technique called discovery meetings. We simply gather groups of employees and ask them to think about areas that need improvement. It's just some downtime to sit and think. From that we can begin to look at opportunities for improvement.

IW: *Do discovery meetings turn into giant gripe sessions?*

CHANG: In the beginning, yes, you tend to get a lot of ideas connected to dissatisfaction. The first meeting or two may be mostly discovery baggage, but you have to let the emoting happen. We suggest making some quick fixes to help build confidence that we really plan to act on the results of these meetings. Soon, with some coaching and leading, the groups get to work processes and develop worthy opportunities.

IW: *How much of the workforce is involved in these meetings?*

CHANG: At the early stages, no more than 15 percent to 25 percent. I'm not an advocate of the blanket approach.

IW: *When improvement opportunities are established, what happens?*

CHANG: Put the teams to work on the problems they identified. Now is the perfect time to start training. This gives people the opportunity to learn and apply the learning at the same time.

When the projects are complete, and before the groups disband, have the teams collect feedback and data, and then transfer ownership of the process to them. Now they take responsibility for continuously improving the process and other members of the workforce can start their training.

IW: *Don't opportunities for improvement eventually dry up?*

CHANG: There is a certain point when the process capability might max out and the benefit you get from incremental improvements is not worth the cost. When your processes top out, that's when reengineering comes in. A complete overhaul, in turn, leads to more rounds of continuous improvement. That's how continuous improvement and reengineering complement each other.

IW: *Can you tip us off to the pitfalls?*

CHANG: There are five that I like to describe in medical terms. "Widespread implementation rash" is when companies try to do too much, too fast. This is a philosophy; you don't get in and get out. Then there is "key process selection deficiency"—don't improve the petty cash process, make the work important. "Elevated doses of training" is when people train the whole workforce before they have established the environmental conditions for success. "High count of quality teams" is when we get team mania. The goal is not to implement teams; it's to improve processes. Last is "persistent process measurement cost," which occurs when organizations start improving processes before they are measured. The result is that they have no way of knowing whether they got anything worthwhile accomplished.

IW: *If you only had a moment to sum up continuous improvement, what would you tell us?*

CHANG: I usually end my speeches with some version of this thought: Implementing continuous-improvement initiatives is like swimming upstream against the downward flow of habit.

Selected Bibliography

Brassard, Michael and Diane Ritter. *The Memory Jogger™ II.* Methuen, MA: GOAL/QPC, 1994.

Capezio, Peter, and Debra Morehouse. *Taking the Mystery Out of TQM.* 2nd ed. Franklin Lakes, NJ: Career Press, 1995.

Cartin, Thomas J. *Principles and Practices of TQM.* Milwaukee: ASQC Quality Press, 1993.

Costin, H. *Readings in Total Quality Management.* New York: Dryden Press, 1994.

Evans, James R., and William M. Lindsay. *The Management and Control of Quality.* 3rd ed. St. Paul, MN: West Publishing, 1996.

Galloway, Dianne. *Mapping Work Processes.* Milwaukee: ASQC Press, 1994.

Gitlow, Howard; Shelly Gitlow; Alan Oppenheim; and Rosa Oppenheim. *Tools and Methods for Improvement of Quality.* Burr Ridge, IL: Richard D. Irwin, 1989.

Juran, Joseph M., and Frank M. Gryna. *Quality Planning and Analysis.* 3rd ed. New York: McGraw-Hill, 1993.

Shores, Richard A. *Reengineering the Factory: A Primer for World-Class Manufacturing.* Milwaukee: ASQC Press, 1994.

Walton, Mary. *The Deming Management Method.* New York: Dodd, Mead, 1986.

PART FIVE

OPERATING AND CONTROLLING THE SYSTEM

The chapters in this section cover the following topics:

The topics in this part relate to operating and controlling the production of goods and services. Aggregate planning involves intermediate planning to balance expected demand and capacity. Inventory management involves determining order quantities and timing of orders of finished goods inventories. Material requirements planning involves ordering and scheduling of parts and materials for assembly operations. Just-in-time is a lean production approach that strives for smooth flow of work through the system. Scheduling is concerned with determining short-term schedules for jobs. Supply chain management is concerned with integrating the flows of materials, information, and finance from suppliers to final customers. Project management deals with planning and coordinating projects. The last chapter describes the analysis of waiting lines, which are often present in service systems that experience random requests for service.

BOEING 777
777
BOEING 747-400

LEARNING OBJECTIVES

After completing this chapter, you should be able to:

1. Explain what aggregate planning is and how it is useful.
2. Identify the variables decision makers have to work with in aggregate planning and some of the possible strategies they can use.
3. Describe some of the graphical and quantitative techniques planners use.
4. Prepare aggregate plans and compute their costs.

CHAPTER TWELVE
AGGREGATE PLANNING

CHAPTER OUTLINE

aggregate planning Intermediate-range capacity planning, usually covering 2 to 12 months.

In the spectrum of production planning, **aggregate planning** is intermediate-range capacity planning that typically covers a time horizon of 2 to 12 months, although in some companies it may extend to as much as 18 months. It is particularly useful for organizations that experience seasonal or other fluctuations in demand or capacity. The goal of aggregate planning is to achieve a production plan that will effectively utilize the organization's resources to satisfy expected demand. Planners must make decisions on output rates, employment levels and changes, inventory levels and changes, back orders, and subcontracting.

The purpose of this chapter is to introduce the concept of aggregate planning, discuss the pertinent costs and possible strategies, and illustrate some of the different approaches currently in use.

INTRODUCTION

Intermediate Planning in Perspective

Organizations make capacity decisions on three levels: long term, intermediate term, and short term. Long-term decisions relate to product and service selection (i.e., determining which products or services to offer), facility size and location, equipment decisions, and layout of facilities. These long-term decisions essentially define the capacity constraints within which intermediate planning must function. Intermediate decisions, as noted above, relate to general levels of employment, output, and inventories, which in turn define the boundaries within which short-range capacity decisions must be made. Thus, short-term decisions essentially consist of deciding the best way to achieve desired results within the constraints resulting from long-term and intermediate-term decisions. Short-term decisions involve scheduling jobs, workers and equipment, and the like. The three levels of capacity decisions are depicted in Table 12–1. Long-term capacity decisions were covered in Chapter 5, and scheduling and related matters are covered in Chapter 17. This chapter covers intermediate capacity decisions.

Many business organizations develop a *business plan* that encompasses both long-term and intermediate-term planning. The business plan establishes guidelines for the organization, taking into account the organization's strategies and policies; forecasts of

TABLE 12–1

Overview of planning levels

Short-range plans	Intermediate plans	Long-range plans
Detailed plans:	General levels of:	Long-term capacity
Machine loading	Employment	Location
Job assignments	Output	Layout
Job sequencing	Finished-goods inventories	Product design
Production lot size	Subcontracting	Work system design
Order quantities	Backorders	
Work schedules		

Long range

Intermediate

Short range

Now 2 months 1 year Planning horizon

demand for the organization's products or services; and economic, competitive, and political conditions. A key objective in business planning is to coordinate the intermediate plans of various organization functions, such as marketing, operations, and finance. In manufacturing companies, coordination also includes engineering and materials management. The plans must fit into the framework established by the organization's long-term goals and strategies, and the limitations established by long-term facility and capital budget decisions.

The business plan guides the planning process of each functional area. In the operations function, a *production plan*—service organizations may refer to this as an *operations plan*—is developed to guide the more detailed planning that eventually leads to a *master schedule.* Figure 12–1 illustrates the planning sequence.

The Concept of Aggregation

Aggregate planning is essentially a "big picture" approach to planning. Planners usually try to avoid focusing on individual products or services—unless of course the organization has only one major product or service. Instead, they focus on a group of similar products or sometimes an entire product line. For example, for purposes of aggregate planning, planners in a company producing television sets would not concern themselves with 21-inch sets versus 25-inch or 27-inch sets. Instead, planners would lump all models together and deal with them as though they were a single product; hence, the term *aggregate* planning. For purposes of aggregate planning, a bicycle company might lump all the different sizes and styles of bikes it produced into a single category of "bikes." Similarly, when fast-food companies such as McDonalds, Burger King or Wendy's plan employment and output levels, they don't try to determine how demand will be broken down into the various options they offer; they focus on overall demand and the overall capacity they want to provide.

Finally, consider how aggregate planning might work in a large department store. Space allocation is often an aggregate decision. That is, a manager might decide to allocate 20 percent of the available space in the clothing department to women's sportswear, 30 percent to juniors, and so on, without regard for what brand names will be

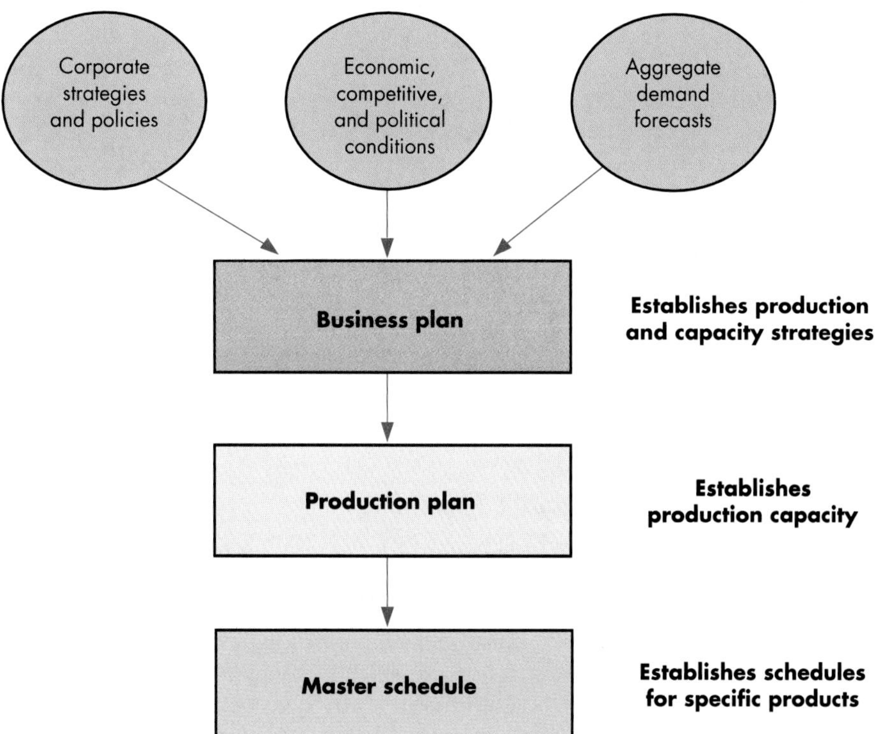

FIGURE 12-1
Planning sequence

Rubbermaid Incorporated, a multinational manufacturer and marketer of commercial and household food containers and other items, produces more than 4,000 products sold in 100 countries at retail establishments and direct to customers. At its corporate headquarters in Wooster, Ohio, planners discuss product performance.

www.rubbermaid.com

offered or how much of juniors will be slacks. The aggregate measure might be square feet of space or racks of clothing.

In each of these examples, an aggregate approach permits planners to make general decisions about intermediate-range capacity without having to deal with highly specific details. They can instead concern themselves with overall decisions on levels of output, employment, and inventories. They do this by lumping demand for all products into one or a few categories, and planning on that basis.

For purposes of aggregate planning, it is often convenient to think of capacity in terms of labor hours or machine hours per period, or output rates (barrels per period, units per period), without worrying about how much of a particular item will actually be involved. This approach frees planners to make general decisions about the use of resources without having to get into the complexities of individual product or service requirements. Product groupings make the problem of obtaining an acceptable unit of aggregation easier because product groupings may lend themselves to the same aggregate measures.

Why do organizations need to do aggregate planning? The answer is twofold. One part is related to *planning:* It takes time to implement plans. For instance, if plans call for increasing the size of facilities and/or hiring (and training) new workers, that will take time. The second part is strategic: *aggregation* is important because it is not possible to predict with any degree of accuracy the timing and volume of demand for individual items. So if an organization were to "lock in" on individual items, it would lose the flexibility to respond to the market.

An Overview of Aggregate Planning

Aggregate planning begins with a forecast of aggregate demand for the intermediate range. This is followed by a general plan to meet demand requirements by setting output, employment, and finished-goods inventory levels. Managers might consider a number of plans, each of which must be examined in light of feasibility and cost. If a plan is reasonably good but has minor difficulties, it may be reworked. Conversely, a poor plan should be discarded and alternative plans considered until an acceptable one is uncovered. The production plan is essentially the output of aggregate planning.

TABLE 12-2

Aggregate planning inputs

Resources	Costs
Workforce/production rates	Inventory carrying cost
Facilities and equipment	Back orders
Demand forecast	Hiring/firing
Policy statements on workforce changes	Overtime
Subcontracting	Inventory changes
Overtime	Subcontracting
Inventory levels/changes	
Back orders	

THE PURPOSE AND SCOPE OF AGGREGATE PLANNING

In this section, we examine the basic problem addressed by aggregate planning—the balancing of supply and demand—along with the purpose of aggregate planning, the primary decision variables available to planners, and associated costs.

If supply and demand aren't in balance, that will cost the organization. There will be added costs of adjusting the system as well as opportunity costs.

Demand and Capacity

Aggregate planners are concerned with the *quantity* and the *timing* of expected demand. If total expected demand for the planning period is much different from available capacity over that same period, the major approach of planners will be to try to achieve a balance by altering capacity, demand, or both. On the other hand, even if capacity and demand are approximately equal for the planning horizon as a whole, planners may still be faced with the problem of dealing with *uneven* demand within the planning interval. In some periods, expected demand may exceed projected capacity, in others expected demand may be less than projected capacity, and in some periods the two may be equal. The task of aggregate planners is to achieve rough equality of demand and capacity over the entire planning horizon. Moreover, planners are usually concerned with minimizing the cost of the production plan, although cost is not the only consideration.

Inputs to Aggregate Planning

There are a number of important informational needs for effective aggregate planning. First, the available resources over the planning period must be known. Then, a forecast of expected demand must be available. Finally, planners must take into account any policies regarding changes in employment levels (e.g., some organizations view layoffs as extremely undesirable, so they would use that only as a last resort).

Table 12–2 lists the major inputs to aggregate planning.

Demand and Capacity Options

Management has a wide range of decision options at its disposal for purposes of aggregate planning. These include changing prices, promotion, backlogging orders, using overtime, using part-time workers, subcontracting, adding or deleting extra shifts, and stockpiling inventories. Some of these, such as pricing and promotion, represent options that are intended to alter the pattern of demand and are usually handled by marketing. Using part-time workers, overtime, and subcontracting, represent options that are intended to alter capacity or supply. We can examine these options in more detail.

Demand Options. The basic demand options are the following:

1. *Pricing.* Pricing differentials are commonly used to shift demand from peak periods to off-peak periods. Some hotels, for example, offer lower rates for weekend stays, and

some airlines offer lower fares for night travel. Movie theaters may offer reduced rates for matinees, and some restaurants offer "early bird specials" in an attempt to shift some of the heavier dinner demand to an earlier time that traditionally has less traffic. Some restaurants also offer smaller portions at reduced rates, and most have smaller portions and prices for children. The smaller portions act to decrease demand. To the extent that pricing is effective, demand will be shifted so that it corresponds more closely to capacity, albeit for an *opportunity cost* that represents the lost profit stemming from capacity insufficient to meet demand during certain periods.

An important factor to consider is the *degree* of price elasticity for the product or service; the more the elasticity, the more effective pricing will be in influencing demand patterns.

2. *Promotion.* Advertising and other forms of promotion, such as displays and direct marketing, can sometimes be very effective in shifting demand so that it conforms more closely to capacity. Obviously, timing of these efforts and knowledge of response rates and response patterns will be needed to achieve the desired results. Unlike pricing policy, there is much less control over the timing of demand; there is always the risk that promotion can worsen the condition it was intended to improve.

3. *Back orders.* An organization can shift demand to other periods by allowing back orders. Hence, orders are taken in one period and deliveries promised for a later period. The success of this approach depends on how willing customers are to wait for delivery. Moreover, the costs associated with back orders can be difficult to pin down since it would include lost sales, annoyed or disappointed customers, and perhaps additional paperwork.

4. *New demand.* Many organizations are faced with the problem of having to provide products or services for peak demand in situations where demand is very uneven. For instance, demand for bus transportation tends to be more intense during the morning and late afternoon rush hours but much lighter at other times. Creating new demand for buses at other times (e.g., trips by schools, clubs, and senior citizen groups) would make use of the excess capacity during those slack times. Similarly, many fast-food restaurants are open for breakfast to use their capacities more fully, and some landscaping firms in northern climates use their equipment during the winter months for snow removal. Manufacturing firms that experience seasonal demands for certain products (e.g., snowblowers) are sometimes able to develop a demand for a complementary product (e.g., lawn mowers, garden equipment) that makes use of the same production processes. They thereby achieve a more consistent use of labor, equipment, and facilities.

Capacity Options. The basic capacity options are the following:

Vol. 5, Seg. 1 and 2
TriState

1. *Hire and lay off workers.* The extent to which operations are labor intensive determines the impact that changes in the workforce level will have on capacity. The resource requirements of each worker can also be a factor. For instance, if a supermarket usually has 10 of 14 checkout lines operating, an additional four checkout workers could be added. Hence, the ability to add workers is constrained at some point by other resources needed to support the workers. Conversely, there may be a lower limit on the number of workers needed to maintain a viable operation (e.g., a skeleton crew).

Union contracts may restrict the amount of hiring and laying off a company can do. Moreover, because laying off can present serious problems for workers, some firms have policies that either prohibit or limit downward adjustments to a workforce. On the other hand, hiring presumes an available supply of workers. This may change from time to time and, at times of low supply, has an impact on the ability of an organization to pursue this approach.

Use of hiring and laying off entails certain costs. Hiring costs include recruitment, screening, and training to bring new workers "up to speed." And quality may suffer. Some savings may occur if workers who have recently been laid off are rehired. Layoff costs include severance pay, the cost of realigning the remaining workforce, potential bad feelings toward the firm on the part of workers who have been laid off, and some

At this naval shipyard in Charleston, South Carolina, workers leave the afternoon shift between 3 and 4 p.m.

loss of morale for workers who are retained (i.e., in spite of company assurances, some workers will believe that in time they too may be laid off).

An increasing number of organizations view workers as assets rather than as variable costs, and would not consider this approach. Instead, they might use slack time for other purposes.

2. *Overtime/slack time.* Use of overtime or slack time is a less severe method for changing capacity than hiring and laying off workers, and it can be used across the board or selectively as needed. It can also be implemented more quickly than hiring and laying off and allows the firm to maintain a steady base of employees. The use of overtime can be especially attractive in dealing with seasonal demand peaks by reducing the need to hire and train people who will have to be laid off during the off-season. Overtime also permits the company to maintain a skilled workforce and employees to increase earnings. Moreover, in situations with crews, it is often necessary to use a full crew rather than to hire one or two additional people. Thus, having the entire crew work overtime would be preferable to hiring extra people.

It should be noted that some union contracts allow workers to refuse overtime. In those cases, it may be difficult to muster a full crew to work overtime or to get an entire production line into operation after regular hours. Although workers often like the additional income overtime can generate, they may not appreciate having to work on short notice or the fluctuations in income that result. Still other considerations relate to the fact that overtime often results in lower productivity, poorer quality, more accidents, and increased payroll costs, whereas idle time results in less efficient use of machines and other fixed assets.

The use of slack when demand is less than capacity can be an important consideration. Some organizations use this time for traning. It can also give workers time for problem solving and process improvement, while retraining skilled workers.

3. *Part-time workers.* In certain instances, the use of part-time workers is a viable option—much depends on the nature of the work, training and skills needed, and union agreements. Seasonal work requiring low-to-moderate job skills lends itself to part-time workers, who generally cost less than regular workers in hourly wages and fringe benefits. However, unions may regard such workers unfavorably because they typically do not pay union dues and may lessen the power of unions. Department stores, restaurants, and

supermarkets make use of part-time workers. So do parks and recreation departments, resorts, travel agencies, hotels and other service organizations with seasonal demands. In order to be successful, these organizations must be able to hire part-time employees when they are needed.

4. *Inventories.* The use of finished-goods inventories allows firms to produce goods in one period and sell or ship them in another period, although this involves holding or carrying those goods as inventory until they are needed. The cost includes not only storage costs and the cost of money tied up that could be invested elsewhere, but also the cost of insurance, obsolescence, deterioration, spoilage, breakage, and so on. In essence, inventories can be built up during periods when production capacity exceeds demand and drawn down in periods when demand exceeds production capacity.

This method is more amenable to manufacturing than to service industries since manufactured goods can be stored whereas services generally cannot. However, an analogous approach used by services is to make efforts to streamline services (e.g., standard forms) or otherwise do a portion of the service during slack periods (e.g., organize the workplace). In spite of these possibilities, services tend not to make much use of inventories to alter capacity requirements.

5. *Subcontracting.* Subcontracting enables planners to acquire temporary capacity, although it affords less control over the output and may lead to higher costs and quality problems. The question of whether to make or buy (i.e., in manufacturing) or to perform a service or hire someone else to do the work generally depends on factors such as available capacity, relative expertise, quality considerations, cost, and the amount and stability of demand.

In some cases, a firm might choose to perform part of the work itself and let others handle the rest in order to maintain flexibility and as a hedge against loss of a subcontractor. Moreover, this gives the organization a bargaining tool in negotiations with contractors and a head start if it decides at a later date to take over the operation entirely.

BASIC STRATEGIES FOR MEETING UNEVEN DEMAND

As you see, managers have a wide range of decision options they can consider for achieving a balance of demand and capacity in aggregate planning. Since the options that are most suited to influencing demand fall more in the realm of marketing than in operations (with the exception of backlogging), we shall concentrate on the capacity options, which are in the realm of operations but include the use of back orders.

Aggregate planners might adopt a number of strategies. Some of the more prominent ones are:

1. Maintain a level workforce.
2. Maintain a steady output rate.
3. Match demand period by period.
4. Use a combination of decision variables.

level capacity strategy Maintaining a steady rate of regular-time output while meeting variations in demand by a combination of options.

chase demand strategy Matching capacity to demand; the planned output for a period is set at the expected demand for that period.

While other strategies might be considered, these will suffice to give you a sense of how aggregate planning operates in a vast number of organizations. The first three strategies are "pure" strategies because each has a single focal point; the last strategy is "mixed" because it lacks the single focus. Under a **level capacity strategy,** variations in demand are met by using some combination of inventories, overtime, part-time workers, subcontracting, and back orders. Matching capacity to demand implies a **chase demand strategy**; the planned output for any period would be the expected demand for that period.

Many organizations regard a level workforce as very appealing. Since workforce changes through hiring and laying off can have a major impact on the lives and morale of employees and can be disruptive for managers, organizations often prefer to handle uneven demand in other ways. Moreover, changes in workforce size can be very costly, and there is always the risk that there will not be a sufficient pool of workers with the appropriate skills when needed. Aside from these considerations, such changes can involve a significant amount of paperwork. Unions tend to favor a level workforce because the freedom to hire and lay off workers diminishes union strengths.

To maintain a constant level of output and still meet demand requirements, an organization must resort to some combination of subcontracting, backlogging, and use of inventories to absorb fluctuations. Subcontracting requires an investment in evaluating sources of supply as well as possible increased costs, less control over output, and perhaps quality considerations. Backlogs can lead to lost sales, increased record keeping, and lower levels of customer service. Allowing inventories to absorb fluctuations can entail substantial costs by having money tied up in inventories, having to maintain relatively large storage facilities, and incurring other costs related to inventories. Furthermore, inventories are not usually an alternative for service-oriented organizations. However, there are certain advantages, such as minimum costs of recruitment and training, minimum overtime and idle-time costs, fewer morale problems, and stable use of equipment and facilities.

A chase demand strategy presupposes a great deal of ability and willingness on the part of managers to be flexible in adjusting to demand. A major advantage of this approach is that inventories can be kept relatively low, which can yield substantial savings for an organization. A major disadvantage is the lack of stability in operations—the atmosphere is one of dancing to demand's tune. Also, when forecast and reality differ, morale can suffer, since it quickly becomes obvious to workers and managers that efforts have been wasted. Figure 12–2 provides a comparison of the two strategies.

Organizations may opt for a strategy that involves some combination of the pure strategies. This allows managers greater flexibility in dealing with uneven demand and perhaps in experimenting with a wide variety of approaches. However, the absence of a clear focus may lead to an erratic approach and confusion on the part of employees.

Choosing a Strategy

Whatever strategy an organization is considering, two important factors are *company policy* and *costs*. Company policy may set constraints on the available options or the extent to which they can be used. For instance, company policy may discourage layoffs except under extreme conditions. Subcontracting may not be a viable alternative due to the desire to maintain secrecy about some aspect of the manufacturing of the product (e.g., a secret formula or blending process). Union agreements often impose restrictions. For example, a union contract may specify both minimum and maximum numbers of hours part-time workers can be used.

As a rule, aggregate planners seek to match supply and demand within the constraints imposed on them by policies or agreements and at minimum cost. They usually evaluate alternatives in terms of their overall costs. In the next section, a number of techniques for aggregate planning are described and presented with some examples of cost evaluation of alternative plans.

TECHNIQUES FOR AGGREGATE PLANNING

Numerous techniques help decision makers with the task of aggregate planning. Generally, they fall into one of two categories: informal trial-and-error techniques and mathematical techniques. In practice, informal techniques are more commonly used. However, a considerable amount of research has been devoted to mathematical techniques, and even though they are not as widely used, they often serve as a basis for comparing the effectiveness of alternative techniques for aggregate planning. Thus, it will be instructive to briefly examine them as well as the informal techniques.

A general procedure for aggregate planning consists of the following steps:

1. Determine demand for each period.
2. Determine capacities (regular time, overtime, subcontracting) for each period.

A possible uneven demand pattern

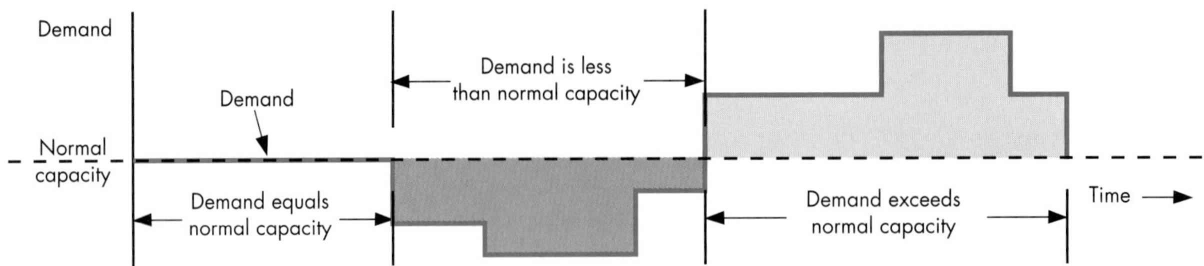

Chase demand strategy

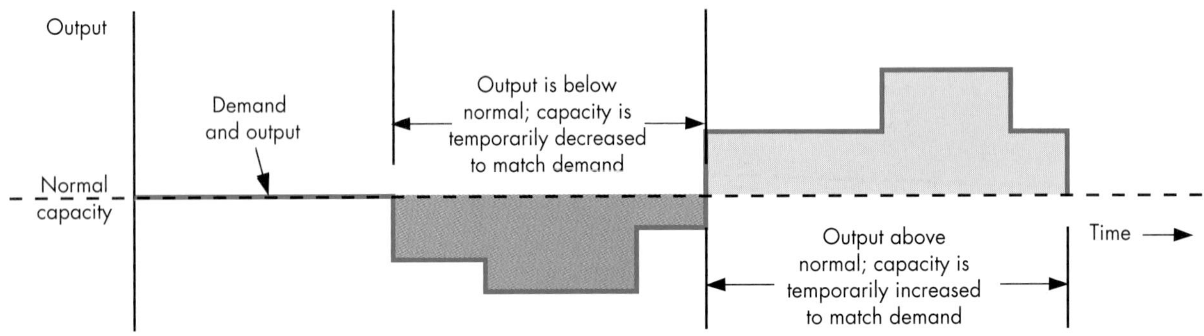

Level output strategy

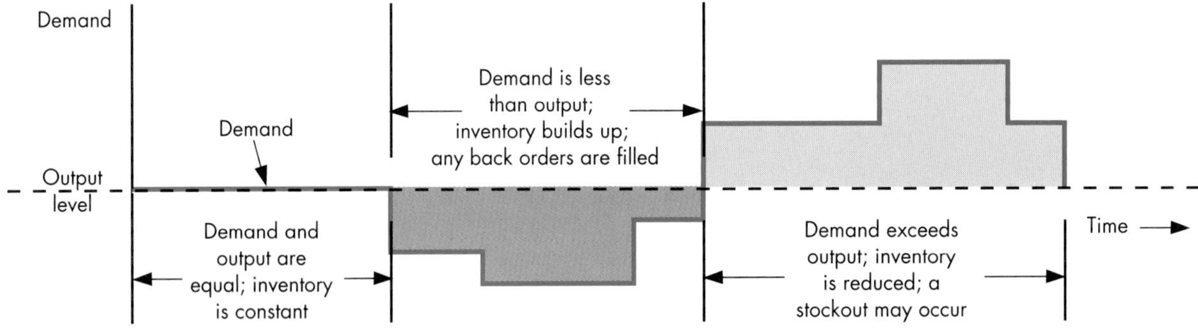

FIGURE 12–2

Comparison of chase and level strategies

3. Identify company or departmental policies that are pertinent (e.g., maintain a safety stock of 5 percent of demand, maintain a reasonably stable workforce).
4. Determine unit costs for regular time, overtime, subcontracting, holding inventories, back orders, layoffs, and other relevant costs.
5. Develop alternative plans and compute the cost for each.
6. If satisfactory plans emerge, select the one that best satisfies objectives. Otherwise, return to step 5.

It can be helpful to use a worksheet that summarizes demand, capacity, and cost for each plan, such as the one illustrated in Table 12–3. In addition, graphs can be used to guide development of alternatives.

Informal Techniques

Informal approaches consist of developing simple tables or graphs that enable planners to visually compare projected demand requirements with existing capacity. Alternatives

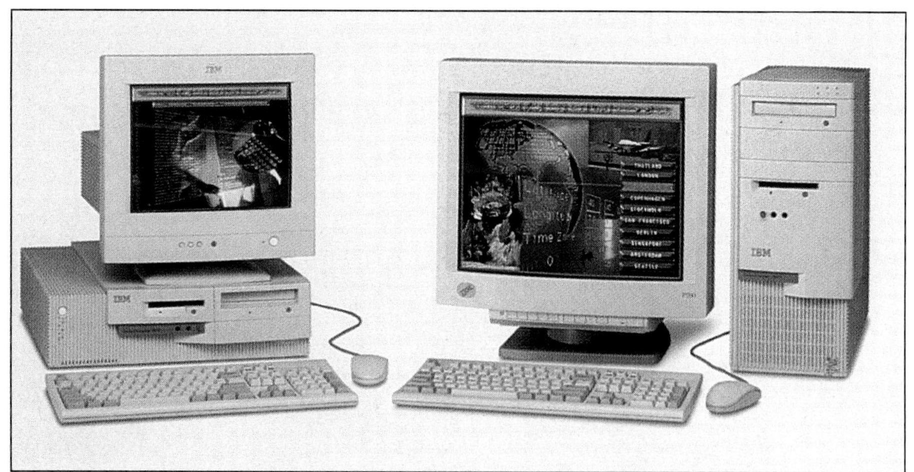

IBM planners need to balance large product lines. These computers use some of the same parts, labor, and assembly facilities so planning is necessary to avoid parts shortages and match demand. Shown here is the PC Model 300PL with G72 and P200 monitors.

www.ibm.com/

TABLE 12–3

Worksheet

Period	1	2	3	4	5		Total
Forecast							
Output							
Regular time							
Overtime							
Subcontract							
Output – Forecast							
Inventory							
Beginning							
Ending							
Average							
Backlog							
Costs							
Output							
Regular							
Overtime							
Subcontract							
Hire/Lay off							
Inventory							
Back orders							
Total							

are usually evaluated in terms of their overall costs. The chief disadvantage of such techniques is that they do not necessarily result in the optimal aggregate plan.

Very often, graphs can be used to guide the development of alternatives. Some planners prefer cumulative graphs while others prefer to see a period-by-period breakdown of a plan. For instance, Figure 12–3 shows a cumulative graph for a plan with steady

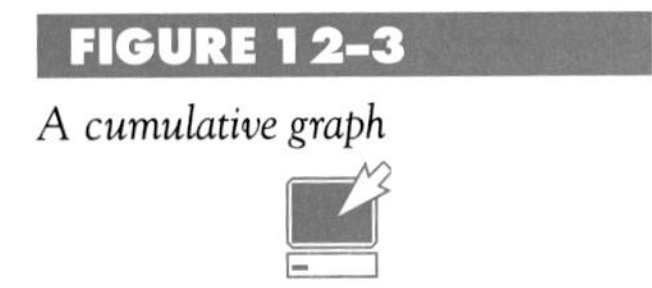

FIGURE 12–3

A cumulative graph

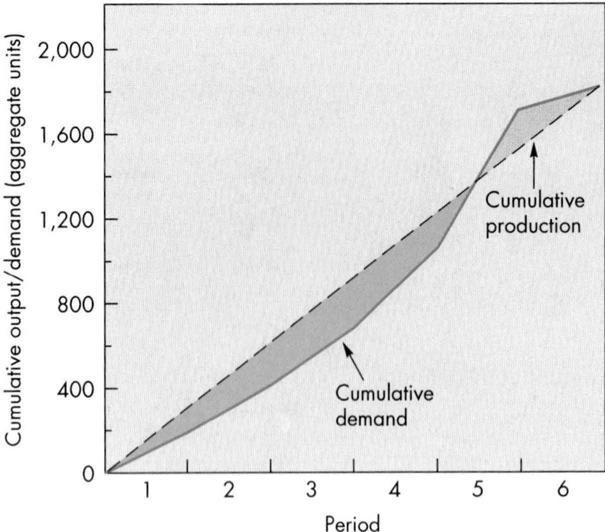

output (the slope of the dashed line represents the production rate) and inventory absorption of demand variations. Figure 12–3 is an example of a period-by-period graph. The obvious advantage of a graph is that it provides a visual portrayal of a plan. The preference of the planner determines which of these two types of graphs is chosen.

Two examples illustrate the development and comparison of aggregate plans. In the first example, regular output is held steady, with inventory absorbing demand variations. In the second example, a lower rate of regular output is used, supplemented by use of overtime. In both examples, some backlogs are allowed to build up.

These examples and other examples and problems in this chapter are based on the following assumptions:

1. The regular output capacity is the same in all periods. No allowance is made for holidays, different numbers of workdays in different months, and so on. This assumption simplifies computation.
2. Cost (back order, inventory, subcontracting, etc.) is a linear function composed of unit cost and number of units. This often has a reasonable approximation to reality, although there may be only narrow ranges over which this is true. Cost is sometimes more of a step function.
3. Plans are feasible; that is, sufficient inventory capacity exists to accommodate a plan, subcontractors with appropriate quality and capacity are standing by, and changes in output can be made as needed.
4. All costs associated with a decision option can be represented by a lump sum or by unit costs that are independent of the quantity involved. Again, a step function may be more realistic; but for purposes of illustration and simplicity, this assumption is appropriate.
5. Cost figures can be reasonably estimated and are constant for the planning horizon.
6. Inventories are built up and drawn down at a uniform rate and output occurs at a uniform rate throughout each period. However, backlogs are treated as if they exist for an entire period, even though in periods where they initially appear, they would tend to build up toward the end of the period. Hence, this assumption is a bit unrealistic for some periods, but it simplifies computations.

In the examples and problems in this chapter, we use the following relationships to determine the number of workers, the amount of inventory, and the cost of a particular plan.

The number of workers available in any period is

Number of workers in a period	=	Number of workers at end of the previous period	+	Number of new workers at start of the period	−	Number of laid-off workers at start of the period

Note: An organization would not hire and lay off simultaneously, so at least one of the last two terms will equal zero.

The amount of inventory at the end of a given period is

Inventory at the end of a period	=	Inventory at end of the previous period	+	Production in the current period	−	Amount used to satisfy demand in the current period

The average inventory for a period is equal to

$$\frac{\text{Beginning inventory} + \text{Ending inventory}}{2}$$

The cost of a particular plan for a given period can be determined by summing the appropriate costs:

$$\text{Cost for a period} = \text{Output cost (Reg + OT + Subcontract)} + \text{Hire/lay off cost} + \text{Inventory cost} + \text{Back-order cost}$$

The appropriate costs are calculated as follows:

Type of Cost	How to Calculate
Output	
Regular	Regular cost per unit × Quantity of regular output
Overtime	Overtime cost per unit × Overtime quantity
Subcontract	Subcontract cost per unit × Subcontract quantity
Hire/lay off	
Hire	Cost per hire × Number hired
Lay off	Cost per layoff × Number laid off
Inventory	Carrying cost per unit × Average inventory
Back order	Back order cost per unit × Number of back-order units

The following examples are only three of many possible options that could be tried. Perhaps some of the others would result in a lower cost. With trial and error, you can never be completely sure you have identified the lowest-cost alternative unless every possible alternative is evaluated. Of course, the purpose of these examples is to illustrate the process of developing and evaluating an aggregate plan rather than to find the lowest-cost plan. Problems at the end of the chapter cover still other alternatives.

In practice, successful achievement of a good plan depends on the resourcefulness and persistence of the planner. A good computer package can eliminate the computational burden of trial-and-error techniques.

Example 1

Planners for a company that makes several models of tractors are about to prepare the aggregate plan that will cover six periods. They have assembled the following information:

Period	1	2	3	4	5	6	Total
Forecast	200	200	300	400	500	200	1,800

Costs
Output
Regular time = $2 per tractor
Overtime = $3 per tractor
Subcontract = $6 per tractor
Inventory = $1 per tractor per period on average inventory
Back orders = $5 per tractor per period

They now want to evaluate a plan that calls for a steady rate of regular-time output, mainly using inventory to absorb the uneven demand but allowing some backlog. They intend to start with zero inventory on hand in the first period. Prepare an aggregate plan and determine its cost using the preceding information. Assume a level output rate of 300 units (tractors) per period with regular time (i.e., 1,800/6 = 300). Note that the planned ending inventory is zero. There are 15 workers.

Solution

Period	1	2	3	4	5	6	Total
Forecast	200	200	300	400	500	200	1,800
Output							
Regular	300	300	300	300	300	300	1,800
Overtime	—	—	—	—	—	—	
Subcontract	—	—	—	—	—	—	
Output − Forecast	100	100	0	(100)	(200)	100	0
Inventory							
Beginning	0	100	200	200	100	0	
Ending	100	200	200	100	0	0	
Average	50	150	200	150	50	0	600
Backlog	0	0	0	0	100	0	100
Costs							
Output							
Regular	$600	600	600	600	600	600	$3,600
Overtime	—	—	—	—	—	—	
Subcontract	—	—	—	—	—	—	
Hire/Lay off	—	—	—	—	—	—	
Inventory	$ 50	150	200	150	50	0	$ 600
Back orders	$ 0	0	0	0	500	0	$ 500
Total	$650	750	800	750	1,150	600	$4,700

Note that the total regular-time output of 1,800 units equals the total expected demand. Ending inventory equals beginning inventory plus or minus the quantity Output − Forecast. If Output − Forecast is negative, inventory is decreased in that period by that amount. If insufficient inventory exists, a backlog equal to the shortage amount appears, as in period 5. This is taken care of using the excess output in period 6.

The costs were computed as follows. Regular cost in each period equals 300 units × $2 per unit or $600. Inventory cost equals average inventory × $1 per unit. Back-order cost is $5 per unit. The total cost for this plan is $4,700.

Note that the first two quantities in each column are givens. The remaining quantities in the upper portion of the table were determined working down each column, beginning with the first column The costs were then computed based on the quantities in the upper part of the table.

Example 2

After reviewing the plan developed in the preceding example, planners have decided to develop an alternative plan They have learned that one person is about to retire from the company. Rather than replace that person, they would like to stay with the smaller workforce and use overtime to make up for the lost output. The reduced regular-time output is 280 units per period. The maximum amount of overtime output per period is 40 units. Develop a plan and compare it to the previous one.

Solution

Period	1	2	3	4	5	6	Total
Forecast	200	200	300	400	500	200	1,800
Output							
Regular	280	280	280	280	280	280	1,680
Overtime	0	0	40	40	40	0	120
Subcontract	—	—	—	—	—	—	
Output − Forecast	80	80	20	(80)	(180)	80	0
Inventory							
Beginning	0	80	160	180	100	0	
Ending	80	160	180	100	0	0	
Average	40	120	170	140	50	0	520
Backlog	0	0	0	0	80	0	80
Costs							
Output							
Regular	$560	560	560	560	560	560	$3,360
Overtime	0	0	120	120	120	0	$ 360
Subcontract	—	—	—	—	—	—	
Hire/Lay off	—	—	—	—	—	—	
Inventory	$ 40	120	170	140	50	0	$ 520
Back orders	$ 0	0	0	0	400	0	$ 400
Total	$600	680	850	820	1,130	560	$4,640

The amount of overtime that must be scheduled has to make up for lost output of 20 units per period for six periods, which is 120. This is scheduled toward the center of the planning horizon since that is where the bulk of demand occurs. Scheduling it earlier would increase inventory carrying costs; scheduling it later would increase the backlog cost.

Overall, the total cost for this plan is $4,640, which is $60 less than the previous plan. Regular-time production cost and inventory cost are down, but there is overtime cost. However, this plan achieves savings in back-order cost, making it somewhat less costly overall than the plan in Example 1.

Example 3

A third option is to use temporary workers to fill in during months of high demand. Suppose that it costs an additional $100 to hire and train a temporary worker, and that a temporary worker can produce at the rate of 15 units per period (compared with 20 units per period for regular workers).

Dividing the number of units needed (120) by the output rate of 15 per temporary worker, you find that eight worker-periods are needed (e.g., two workers for four months each, four workers for two months each).

Noting that periods 4 and 5 have the heaviest demand, using four temporary workers for two months each (i.e., periods 4 and 5) seems to be reasonable. The results are summarized in the following plan.

Solution

Period	1	2	3	4	5	6	Total
Forecast	200	200	300	400	500	200	1,800
Output							
Regular	280	280	280	340	340	280	1,800
Overtime	—	—	—	—	—	—	
Subcontract	—	—	—	—	—	—	
Output − Forecast	80	80	(20)	(60)	(160)	80	0
Inventory							
Beginning	0	80	160	140	80	0	
Ending	80	160	140	80	0	0	
Average	40	120	150	110	40	0	460
Backlog	0	0	0	0	80	0	80
Costs							
Output							
Regular	$560	560	560	680	680	560	$3,600
Overtime	—	—	—	—	—	—	
Subcontract	—	—	—	—	—	—	
Hire/Lay off	$ 0	0	0	400	400	0	$ 800
Inventory	$ 40	120	150	110	40	0	$ 460
Back orders	$ 0	0	0	0	400	0	$ 400
Total	$600	680	710	1,190	1,520	560	$5,260

Overall, the total cost for this plan is $5,260, making it the most expensive of the three alternatives examined.

Mathematical Techniques

A number of mathematical techniques have been developed to handle aggregate planning. They range from mathematical programming models to heuristic and computer search models. This section briefly describes some of the better-known techniques.

Linear Programming. Linear-programming models are methods for obtaining optimal solutions to problems involving the allocation of scarce resources in terms of cost minimization or profit maximization. With aggregate planning, the goal is usually to minimize the sum of costs related to regular labor time, overtime, subcontracting, inventory holding costs, and costs associated with changing the size of the workforce. Constraints involve the capacities of the workforce, inventories, and subcontracting.

E. H. Bowman proposed formulating the problem in terms of a transportation-type programming model (described in detail in the supplement to Chapter 8) as a way to obtain aggregate plans that would match capacities with demand requirements and minimize costs.[1] In order to use this approach, planners must identify capacity (supply) of regular time, overtime, subcontracting, and inventory on a period-by-period basis, as well as related costs of each variable.

Table 12–4 shows the notation and setup of a transportation table. Note the systematic way that costs change as you move across a row from left to right. Regular cost, overtime cost, and subcontracting cost are at their lowest when the output is consumed (i.e., delivered, etc.) in the same period it is produced (at the intersection of period 1 row and column for regular cost, at the intersection of period 2 row and column for regular cost, and so on). If goods are made available in one period but carried over to

[1]E. H. Bowman, "Production Planning by the Transportation Method of Linear Programming," *Journal of Operations Research Society* 4 (February 1956), pp. 100–103.

r = Regular production cost per unit
t = Overtime cost per unit
s = Subcontracting cost per unit
h = Holding cost per unit per period
b = Backorder cost per unit per period
n = Number of periods in planning horizon

TABLE 12-4

Transportation notation for aggregate planning

Period		Period 1	Period 2	Period 3	...	Ending inventory period n	Unused capacity	Capacity
	Beginning inventory	0	h	$2h$	...	nh	0	I_0
1	Regular time	r	$r+h$	$r+2h$	...	$r+nh$	0	R_1
	Overtime	t	$t+h$	$t+2h$	...	$t+nh$	0	O_1
	Subcontract	s	$s+h$	$s+2h$	...	$s+nh$	0	S_1
2	Regular time	$r+b$	r	$r+h$	...	$r+(n-1)h$	0	R_2
	Overtime	$t+b$	t	$t+h$	...	$t+(n-1)h$	0	O_2
	Subcontract	$s+b$	s	$s+h$	...	$s+(n-1)h$	0	S_2
3	Regular time	$r+2b$	$r+b$	r	...	$r+(n-2)h$	0	R_3
	Overtime	$t+2b$	$t+b$	t	...	$t+(n-2)h$	0	O_3
	Subcontract	$s+2b$	$s+b$	s	...	$s+(n-2)h$	0	S_3
	Demand				...			Total

later periods (i.e., moving across a row), holding costs are incurred at the rate of h per period. Thus, holding goods for two periods results in a unit cost of $2h$, whether or not the goods came from regular production, overtime, or subcontracting. Conversely, with back orders, the unit cost increases as you move across a row from right to left, beginning at the intersection of a row and column for the same period (e.g., period 3). For instance, if some goods are produced in period 3 to satisfy back orders from period 2, a unit back-order cost of b is incurred. And if goods in period 3 are used to satisfy back orders two periods earlier (e.g., from period 1), a unit cost of $2b$ is incurred. Unused capacity is generally given a unit cost of 0, although it is certainly possible to insert an actual cost if that is relevant. Finally, beginning inventory, is given a unit cost of 0 if it is used to satisfy demand in period 1. However, if it is held over for use in later periods, a holding cost of h per unit is added for each period. If the inventory is to be held for the entire planning horizon, a total unit cost of h times the number of periods, n, will be incurred.

Example 4 illustrates the setup and final solution of a transportation model of an aggregate planning problem.

Example 4

Given the following information set up the problem in a transportation table and solve for the minimum-cost plan:

TABLE 12-5

Transportation solution

Period	Supply from	Demand for: Period 1	Period 2	Period 3	Unused capacity (dummy)	Total capacity available (supply)
	Beginning inventory	0 100	1	2	0	100
1	Regular time	60 450	61 50	62	0	500
	Overtime	80	81 50	82	0	50
	Subcontract	90	91 30	92	0 90	120
2	Regular time	63	60 500	61	0	500
	Overtime	83	80 50	81	0	50
	Subcontract	93	90 20	91 100	0	120
3	Regular time	66	63	60 500	0	500
	Overtime	86	83	80 50	0	50
	Subcontract	96	93	90 100	0	100
	Demand	550	700	750	90	2,090

	PERIOD 1	2	3
Demand	550	700	750
Capacity			
Regular	500	500	500
Overtime	50	50	50
Subcontract	120	120	100
Beginning inventory	100		
Costs			
Regular time	$60 per unit		
Overtime	80 per unit		
Subcontract	90 per unit		
Inventory carrying cost	$1 per unit per month		
Backorder cost	$3 per unit per month		

Solution

The transportation table and solution are shown in Table 12–5. Some of the entries require additional explanation:

a. In this example, inventory carrying costs are $1 per unit per period (costs are shown in the upper right-hand corner of each cell in the table). Hence, units produced in one period and carried over to a later period will incur a holding cost that is a linear function of the length of time held.

b. Linear programming models of this type require that supply (capacity) and demand be equal. A dummy column has been added (unused capacity) to satisfy that requirement. Since it does not "cost" anything extra to not use capacity in this case, cell costs of $0 have been assigned.

c. No backlogs were needed in this example.

d. The quantities (e.g., 100 and 450 in column 1) are the amounts of output or inventory that will be used to meet demand requirements. Thus, the demand of 550 units in period 1 will be met using 100 units from inventory and 450 obtained from regular-time output.

Where backlogs are not permitted, the cell costs for the backlog positions can be made prohibitively high so that no backlogs will appear in the solution.

The main limitations of LP models are the assumptions of linear relationships among variables, the inability to continuously adjust output rates, and the need to specify a single objective (e.g., minimize costs) instead of using multiple objectives (e.g., minimize cost while stabilizing the workforce.)

Linear Decision Rule. Another optimizing technique, the **linear decision rule,** was developed by Charles Holt, Franco Modigliani, John Muth, and Herbert Simon.[2] It seeks to minimize the combined costs of regular payroll, hiring and layoffs, overtime, and inventory using a set of cost-approximating functions, three of which are *quadratic* (contain squared terms), to obtain a single quadratic equation. Using calculus, two linear equations (hence, the name *linear decision rule*) can be derived from the quadratic equation. One of these equations can be used to plan the output for each period in the planning horizon, and the other can be used to plan the workforce for each period.

linear decision rule Optimizing technique that seeks to minimize combined costs, using a set of cost-approximating functions to obtain a single quadratic equation.

Although the model has found some applications, its chief function seems to be as a benchmark against which proposed techniques can be evaluated. In practice, the model suffers from three limitations: (1) a specific type of cost function is assumed, (2) considerable effort must usually be expended to obtain relevant cost data and develop cost functions for each organization, and (3) the method can produce solutions that are unfeasible or impractical.

Simulation Models. A number of **simulation models** have been developed for aggregate planning. (Simulation is described in detail in the supplement to Chapter 19.) The essence of simulation is the development of computerized models that can be tested under a variety of conditions in an attempt to identify reasonably acceptable (although not always optimal) solutions to problems.

simulation models Computerized models that can be tested under different scenarios to identify acceptable solutions to problems.

Table 12–6 summarizes the mathematical techniques.

Aggregate planning techniques other than trial and error do not appear to be widely used. Instead, in the majority of organizations, aggregate planning seems to be accomplished more on the basis of experience along with trial-and-error methods. It is difficult to say exactly why some of the mathematical techniques mentioned are not used to any great extent. Perhaps the level of mathematical sophistication discourages greater use; or the assumptions required in certain models appear unrealistic; or the models may be too narrow in scope. Whatever the reasons, none of the techniques to date have captured the attention of aggregate planners on a broad scale. Simulation is one technique that seems to be gaining in favor. Research on improved approaches to aggregate planning is continuing.

[2]Charles C. Holt, Franco Modigliani, John F. Muth, and Herbert A. Simon, *Planning Production, Inventories and Work Force* (Englewood Cliffs, NJ: Prentice Hall, 1960).

TABLE 12-6
Summary of planning techniques

Technique	Solution Approach	Characteristics
Graphical/charting	Trial and error	Intuitively appealing, easy to understand; solution not necessarily optimal
Linear programming	Optimizing	Computerized; linear assumptions not always valid
Linear decision rule	Optimizing	Complex, requires considerable effort to obtain pertinent cost information and to construct model; cost assumptions not always valid
Simulation	Trial and error	Computerized models can be examined under a variety of conditions.

AGGREGATE PLANNING IN SERVICES

Aggregate planning for manufacturing and aggregate planning for services share similarities in some respects, but there are some important differences—related in general to the differences between manufacturing and services:

1. *Services occur when they are rendered.* Unlike manufacturing output, most services can't be inventoried. Services such as financial planning, tax counseling, and oil changes can't be stockpiled. This removes the option of building up inventories during a slow period in anticipation of future demand. On the other hand, service capacity that goes unused is essentially wasted. Consequently, it becomes important to be able to match capacity and demand.

2. *Demand for service can be difficult to predict.* The volume of demand for services is often quite variable. In some situations, customers may need prompt service (e.g., police, fire, medical emergency), while in others, they simply *want* prompt service and may be willing to go elsewhere if their wants are not met. These factors place a greater burden on service providers to anticipate demand. Consequently, service providers must pay careful attention to planned capacity levels.

3. *Capacity availability can be difficult to predict.* Processing requirements for services can sometimes be quite variable, similar to the variability of work in a job shop setting. Moreover, the variety of tasks required of servers can be great, again similar to the variety of tasks in a job shop. However, in services, the types of variety are more pervasive than they are in manufacturing. This makes it more difficult to establish simple measures of capacity. For example, what would be the capacity of a person who paints interiors of houses? The number of rooms per day or the number of square feet per hour are possible measures, but rooms come in many different sizes, and because the level of detail (and, thus, the painting implements that can be used) vary tremendously, a suitable measure for planning purposes can be quite difficult to arrive at. Similarly, bank tellers are called upon to handle a wide variety of transactions and requests for information, again making it difficult to establish a suitable measure of their capacity.

4. *Labor flexibility can be an advantage in services.* Labor often comprises a significant portion of service compared to manufacturing. That, coupled with the fact that service providers are often able to handle a fairly wide variety of service requirements means that to some extent, planning is easier than it is in manufacturing. Of course, manufacturers recognize this advantage, and many are cross-training their employees to achieve the same flexibility. Moreover, in both manufacturing and service systems, the use of part-time workers can be an important option.

In self-service systems, the (customer) labor automatically adjusts to changes in demand!

A dry cleaner can plan for heavier demand in the fall and winter months when customers have a greater need to clean garments, such as coats or suits, that cannot be laundered.

DISAGGREGATING THE AGGREGATE PLAN

For the production plan to be translated into meaningful terms for production, it is necessary to *disaggregate* the aggregate plan. This involves breaking down the aggregate plan into specific product requirements in order to determine labor requirements (skills, size of work force), materials, and inventory requirements. This process is described in Chapter 14, Material Requirements Planning. At this stage, however, it will be helpful for you to have some understanding of the need for disaggregation and what the term implies.

Working with aggregate units facilitates intermediate planning. However, to put the production plan into operation, one must convert, or decompose, those aggregate units into units of actual products or services that are to be produced or offered. For example, a lawn mower manufacturer may have an aggregate plan that calls for 200 lawn mowers in January, 300 in February, and 400 in March. That company may produce push mowers, self-propelled mowers, and riding mowers. Although all the mowers probably contain some of the same parts and involve some similar or identical operations for fabrication and assembly, there would be some differences in the materials, parts, and operations that each type requires. Hence, the 200, 300, and 400 aggregate lawn mowers that are to be produced during those three months must be translated into specific numbers of mowers of each type prior to actually purchasing the appropriate materials and parts, scheduling operations, and planning inventory requirements.

The result of disaggregating the aggregate plan is a **master schedule** showing the quantity and timing of *specific* end items for a scheduled horizon, which often covers about six to eight weeks ahead. A master schedule shows the planned output for individual products rather than an entire product group, along with the timing of production. The master schedule contains important information for marketing as well as for production. It reveals when orders are scheduled for production and when completed orders are to be shipped.

master schedule The result of disaggregating an aggregate plan; shows quantity and timing of specific end items for a scheduled horizon.

Once a *tentative* master schedule has been developed a planner can do the **rough-cut capacity planning** to test the feasibility of a proposed master schedule relative to available capacities, to assure that no obvious capacity constraints exist. This means checking capacities of production and warehouse facilities, labor, and vendors to ensure that no gross deficiencies exist that will render the master schedule unworkable. The master schedule then serves as the basis for *short-range* planning. It should be noted that whereas the aggregate plan covers an interval of, say, 12 months, the master schedule covers only a portion of this. In other words, the aggregate plan is disaggregated in stages, or phases, that may cover a few weeks to two or three months. Moreover, the master schedule may be updated monthly, even though it covers two or three months. For instance, the lawn mower master schedule would probably be updated at the end of January to include any

rough-cut capacity planning Approximate balancing of capacity and demand to test the feasibility of a master schedule.

FIGURE 12–4

Disaggregating the aggregate plan

Aggregate plan	**Month**	**Jan.**	**Feb.**	**Mar.**
	Planned output*	200	300	400

*Aggregate units

Master schedule	**Month**	**Jan.**	**Feb.**	**Mar.**
	Planned output*			
	Push	100	100	100
	Self-propelled	75	150	200
	Riding	25	50	100
	Total	200	300	400

*Actual units

revisions in planned output for February and March as well as new information on planned output for April.

Figure 12–4 illustrates the concept of disaggregating the aggregate plan. The illustration makes a simple assumption in order to clearly show the concept of disaggregation: The totals of the aggregate and the disaggregated units are equal. In reality, that is not always true. As a consequence, it can require considerable effort to disaggregate the aggregate plan.

MASTER SCHEDULING

A master schedule indicates the quantity and timing (i.e., delivery times) for a product, or a group of products, but it does not show planned *production*. For instance, a master schedule may call for delivery of 50 cases of cranberry-apple juice to be delivered on May 1. But this may not require any production; there may be 200 cases in inventory. Or it may require *some* production: If there were 40 cases in inventory, an additional 10 cases would be needed to achieve the specified delivery amount. Or it may involve production of 50 or more cases: In some instances, it is more economical to produce large amounts rather than small amounts, with the excess temporarily placed in inventory until needed. Thus, the *production lot size* might be 70 cases, so if additional cases were needed (e.g., 50 cases), a run of 70 cases would be made.

master production schedule This schedule indicates the quantity and timing of planned production.

The **master production schedule (MPS)** indicates the quantity and timing of planned production, taking into account desired delivery quantity and timing as well as on-hand inventory. The master production schedule is one of the primary outputs of the master scheduling process, as illustrated in Figure 12–5.

Inputs

Vol. 3, seg. 2
Washburn Guitar

The master schedule has three inputs: the beginning inventory, which is the actual quantity on hand from the preceding period; forecasts for each period of the schedule; and customer orders, which are quantities already *committed* to customers.

Outputs

available-to-promise inventory Uncommitted inventory.

The master scheduling process uses this information on a period-by-period basis to determine the protected inventory, production requirements, and the resulting uncommitted inventory, which is referred to as **available-to-promise (ATP) inventory.** Knowledge of the uncommitted inventory can enable marketing to make realistic promises to customers about deliveries of new orders.

The master scheduling process begins with a preliminary calculation of projected on-hand inventory. This reveals when additional inventory (i.e., production) will be needed. Consider this example. A company that makes industrial pumps wants to pre-

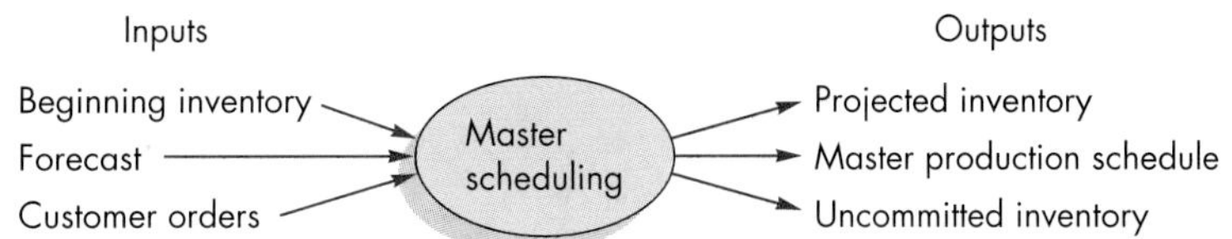

FIGURE 12-5

The master scheduling process

	June				July			
	1	2	3	4	5	6	7	8
Forecast	30	30	30	30	40	40	40	40

FIGURE 12-6

Weekly forecast requirements for industrial pumps

Beginning inventory	June				July			
64	1	2	3	4	5	6	7	8
Forecast	30	30	30	30	40	40	40	40
Customer orders (committed)	33	20	10	4	2			

FIGURE 12-7

Eight-week schedule showing forecasts, customer orders, and beginning inventory

pare a master production schedule for June and July. Marketing has forecasted demand of 120 pumps for June and 160 pumps for July. These have been evenly distributed over the four weeks in each month: 30 per week in June and 40 per week in July, as illustrated in Figure 12–6.

Now, suppose that there are currently 64 pumps in inventory (i.e., beginning inventory is 64 pumps), and that there are customer orders that have been committed (booked) and must be filled (see Figure 12–7).

Figure 12–7 contains the three primary inputs to the master scheduling process: the beginning inventory, the forecast, and the customer orders that have been booked or committed. This information is necessary to determine three quantities: the projected on-hand inventory, the master production schedule, and the uncommitted (ATP) inventory. The first step is to calculate the projected on-hand inventory, one week at a time, until it falls below a specified limit. In this example, the specified limit will be zero. Hence, we will continue until the projected on-hand inventory becomes negative.

The projected on-hand inventory is calculated as follows:

$$\text{Projected on-hand inventory} = \text{Inventory from previous week} - \text{Current week's requirements} \quad (12\text{–}1)$$

where the current week's requirements is the *larger* of forecast and customer orders (committed).

For the first week, projected on-hand inventory equals beginning inventory minus the larger of forecast and customer orders. Because customer orders (33) is larger than the forecast (30), the customer orders amount is used. Thus, for the first week, we obtain:

$$\text{Projected on-hand inventory} = 64 - 33 = 31$$

Projected on-hand inventories are shown in Figure 12–8 for the first three weeks (i.e., until the projected on-hand amount becomes negative).

FIGURE 12-8

Projected on-hand inventory is computed week by week until it becomes negative

Beginning inventory 64	June 1	2	3	4	July 5	6	7	8
Forecast	30	30	30	30	40	40	40	40
Customer orders (committed)	33	20	10	4	2			
Projected on-hand inventory	31	1	−29					

Customer orders are larger than forecast in week 1; projected on-hand inventory is 64 − 33 = 31

Forecast is larger than customer orders in week 2; projected on-hand inventory is 31 − 30 = 1

Forecast is larger than customer orders in week 3; projected on-hand inventory is 1 − 30 = −29

FIGURE 12-9

Determining the MPS and projected on-hand inventory

Week	Inventory from Previous Week	Require-ments*	Net Inventory before MPS		(70) MPS		Projected Inventory
1	64	33	31				31
2	31	30	1				1
3	1	30	−29	+	70	=	41
4	41	30	11				11
5	11	40	−29	+	70	=	41
6	41	40	1				1
7	1	40	−39	+	70	=	31
8	31	40	−9	+	70	=	61

*Requirements equals the larger of forecast and customer orders in each week.

When the projected on-hand inventory becomes negative, this is a signal that production will be needed to replenish inventory. Hence, a negative projected on-hand inventory will require planned production. Suppose that a production lot size of 70 pumps is used, so that whenever production is called for, 70 pumps will be produced. (We discuss determination of lot size in Chapter 13.) Hence, the negative projected on-hand inventory in the third week will require production of 70 pumps, which will meet the projected shortfall of 29 pumps, and leave 41 (i.e., 70 − 29 = 41) pumps for future demand.

These calculations continue for the entire schedule. Every time projected inventory becomes negative, another production lot of 70 pumps is added to the schedule. Figure 12–9 illustrates the calculations. The result is the master schedule and projected on-hand inventory for each week of the schedule. These can now be added to the master schedule (see Figure 12–10).

It is now possible to determine the amount of inventory that is uncommitted, and hence, available to promise. Several methods are used in practice. The one we shall employ involves a "look-ahead" procedure: Sum booked customer orders week by week until (but not including) a week in which there is an MPS amount. For example, in the first week, this procedure results in summing customer orders of 33 (week 1) and 20

	June				July			
64	**1**	**2**	**3**	**4**	**5**	**6**	**7**	**8**
Forecast	30	30	30	30	40	40	40	40
Customer orders (committed)	33	20	10	4	2			
Projected on-hand inventory	31	1	41	11	41	1	31	61
MPS			70		70		70	70

FIGURE 12–10

Projected on-hand inventory and MPS are added to the master schedule

	June				July			
64	**1**	**2**	**3**	**4**	**5**	**6**	**7**	**8**
Forecast	30	30	30	30	40	40	40	40
Customer orders (committed)	33	20	10	4	2			
Projected on-hand inventory	31	1	41	11	41	1	31	61
MPS			70		70		70	70
Available-to-promise inventory (uncommitted)	11		56		68		70	70

FIGURE 12–11

The available-to-promise inventory quantities have been added to the master schedule

(week 2) to obtain 53. In the first week, this amount is subtracted from the beginning inventory of 64 pumps plus the MPS (zero in this example) to obtain the amount that is available to promise. Thus,

$$64 + 0 - (33 + 20) = 11$$

This inventory is uncommitted, and can be delivered in either week 1 or 2, or part can be delivered in week 1 and part in week 2. (Note that the ATP quantity is only calculated for the first week, and for other weeks in which there is an MPS quantity. Hence, it is calculated for weeks 1, 3, 5, 7, 8.) See Figure 12–11.

For weeks other than the first week, the beginning inventory drops out of the computation, and ATP is the look-ahead quantity subtracted from the MPS quantity.

Thus, for week 3, the promised amounts are $10 + 4 = 14$, and the ATP is $70 - 14 = 56$.

For week 5, customer orders are 2 (future orders have not yet been booked). The ATP is $70 - 2 = 68$.

For weeks 7 and 8, there are no customer orders, so for the present, all of the MPS amount is available to promise.

As additional orders are booked, these would be entered in the schedule, and the ATP amounts would be updated to reflect those orders. Marketing can use the ATP amounts to provide realistic delivery dates to customers.

FIGURE 12-12

Time fences in the MPS

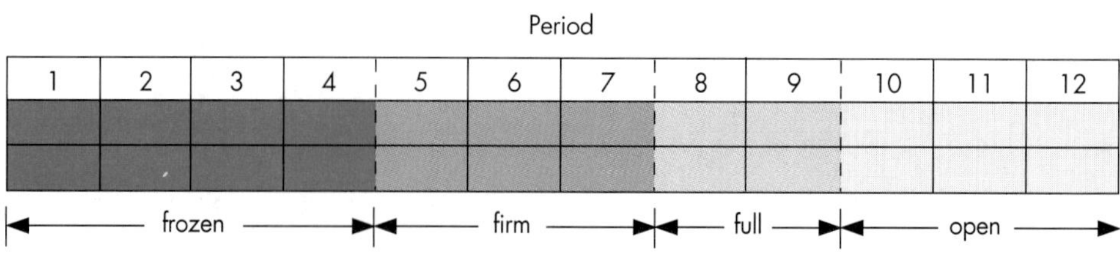

Stabilizing the Master Schedule

Changes to a master schedule can be disruptive, particularly changes to the early, or near, portions of the schedule. Typically, the further out in the future a change is, the less the tendency to cause problems.

Master production schedules are often divided into four stages, or phases. The dividing lines between phases are sometimes referred to as *time fences*. In the first phase, usually the first few periods of the schedule, changes can be quite disruptive. Consequently, once established, that portion of the schedule is generally *frozen,* which implies that all but the most critical changes cannot be made without permission from the highest levels in an organization. The purpose, of course, is to achieve a high degree of *stability* in the production system. In the next stage, perhaps the next two or three periods, changes are still disruptive, but not to the extent that they are in the first phase of the schedule. Management views the schedule as *firm,* and only exceptional changes are made. In the third stage, management views the schedule as *full,* meaning that all available capacity has been allocated. Although changes do impact the schedule, their effect is less dramatic, and they are usually made if there is a good reason for doing so. In the final phase, management views the schedule as *open,* meaning that not all capacity has been allocated. This is where new orders are usually entered in the schedule.

Figure 12–12 illustrates time fences.

Summary

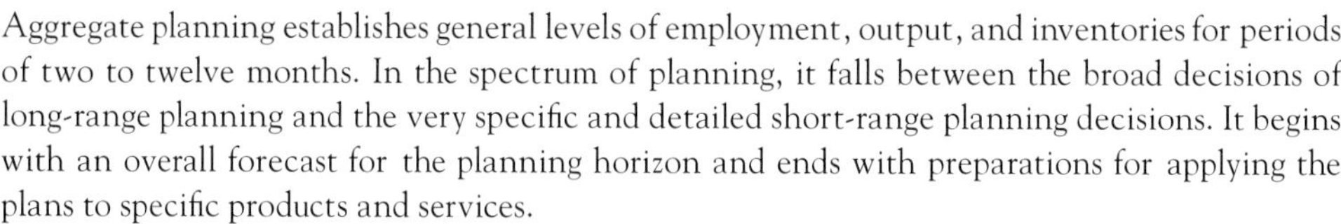

Aggregate planning establishes general levels of employment, output, and inventories for periods of two to twelve months. In the spectrum of planning, it falls between the broad decisions of long-range planning and the very specific and detailed short-range planning decisions. It begins with an overall forecast for the planning horizon and ends with preparations for applying the plans to specific products and services.

The essence of aggregate planning is the aggregation of products or services into one "product" or "service." This permits planners to consider overall levels of employment and inventories without having to become involved with specific details that are better left to short-range planning. Planners often use informal graphic and charting techniques to develop plans, although various mathematical techniques have been suggested. It appears that the complexity and the restrictive assumptions of these techniques limit their widespread use in practice.

After the aggregate plan has been developed, it is disaggregated or broken down into specific product requirements. This leads to a master schedule, which indicates the planned quantities and timing of specific outputs. Inputs to the master schedule are on-hand inventory amounts, forecasts of demand, and customer orders. The outputs are projected production and inventory requirements, and the projected uncommitted inventory, which is referred to as available-to-promise (ATP) inventory.

Key Terms

aggregate planning, 524
available-to-promise (ATP) inventory, 544
chase demand strategy, 530
level capacity strategy, 530
linear decision rule, 541
master production schedule, (MPS), 544
master schedule, 543
rough-cut capacity planning, 543
simulation models, 541

Solved Problems

Problem 1

A manager is attempting to put together an aggregate plan for the coming nine months. She has obtained a forecast of expected demand for the planning horizon. The plan must deal with highly seasonal demand; demand is relatively high in periods 3 and 4 and again in period 8, as can be seen from the following forecasts:

Period	1	2	3	4	5	6	7	8	9	Total
Forecast	190	230	260	280	210	170	160	260	180	1,940

The department now has 20 full-time people, each of whom can produce 10 units of output per period at a cost of $6 per unit. Inventory carrying cost is $5 per unit per period, and backlog cost is $10 per unit per period. The manager is considering a plan that would involve hiring two people to start working in period 1, one on a temporary basis who would work only through period 5. This would cost $500 in addition to unit production costs.

a. What is the rationale for this plan?

b. Determine the total cost of the plan, including production, inventory, and back order costs.

Solution

a. With the current workforce of 20 people each producing 10 units per period, regular capacity is 1,800 units. That is 140 units less than expected demand. Adding one worker would increase regular capacity to 1,800 + 90 = 1,890 units. That would still be 50 units short, or just the amount one temporary worker could produce in five periods. Since one of the two seasonal peaks is quite early, it would make sense to start the temporary worker right away to avoid some of the back-order cost.

b. The production plan for this strategy is as follows:

Period	1	2	3	4	5	6	7	8	9	Total
Forecast	190	230	260	280	210	170	160	260	180	1,940
Output										
Regular	220	220	220	220	220	210	210	210	210	1,940
Overtime	—	—	—	—	—	—	—	—	—	—
Subcontract	—	—	—	—	—	—	—	—	—	—
Output − Forecast	30	(10)	(40)	(60)	10	40	50	(50)	30	0
Inventory										
Beginning	0	30	20	0	0	0	0	20	0	
Ending	30	20	0	0	0	0	20	0	0	
Average	15	25	10	0	0	0	10	10	0	70
Backlog	0	0	20	80	70	30	0	30	0	230
Costs										
Output										
Regular @ $6	$1,320	1,320	1,320	1,320	1,320	1,260	1,260	1,260	1,260	$11,640
Overtime										
Subcontract										
Inventory @ $5	$ 75	125	50	0	0	0	50	50	0	$ 350
Backorder @ $10	0	0	200	800	700	300	0	300	0	$ 2,300
Total	$1,395	1,445	1,570	2,120	2,020	1,560	1,310	1,610	1,260	$14,290

The total cost for this plan is $14,290, plus the $500 cost for hiring and for the layoff, giving a total of $14,790. This plan may or may not be good. The manager would need information on other costs and options before settling on one plan.

Although the calculations are relatively straightforward, the backlogs can sometimes seem difficult to obtain. Consider these rules for computing the backlog:

1. Start with the Output − Forecast value. If this is positive and there was a backlog in the preceding period, reduce the backlog by this amount. If the amount exceeds the backlog, the difference becomes the ending inventory for the period. If they are exactly equal, the backlog and the ending inventory will both be equal to zero.
2. If Output − Forecast is negative, subtract it from the beginning inventory. If this produces a negative value, that value becomes the backlog for that period.

You can also use the appropriate Excel Template to obtain the solution:

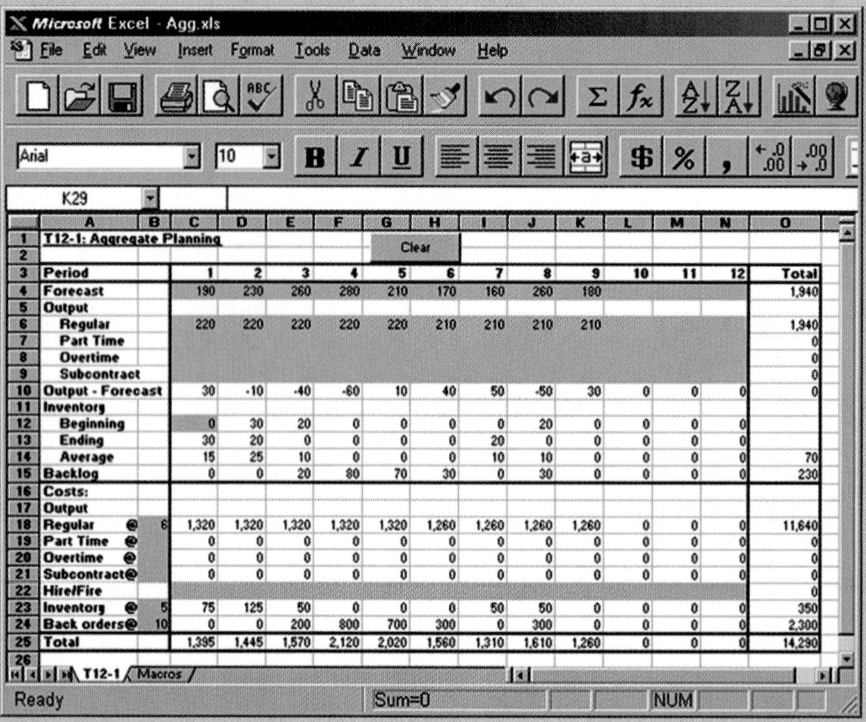

Problem 2

Prepare a schedule like that shown in Figure 12–10 for the following situation. The forecast for each period is 70 units. The starting inventory is zero. The MPS rule is to schedule production if the projected inventory on hand is negative. The production lot size is 100 units. The following table shows committed orders.

Period	Customer Orders
1	80
2	50
3	30
4	10

Solution

Period	(A) Inventory from Previous Period	(B) Requirements*	(C = B − A) Net Inventory before MPS	MPS	(MPS + C) Projected Inventory
1	0	80	(80)	100	20
2	20	70	(50)	100	50
3	50	70	(20)	100	80
4	80	70	10	0	10

*Requirements equals the larger of forecast and customer orders in each period.

Starting Inv. = 0	1	2	3	4
Forecast	70	70	70	70
Customer orders	80	50	30	10
Projected on-hand inventory*	20	50	80	10
MPS	100	100	100	0

ATP	20	50	60	0

Discussion and Review Questions

1. What three levels of planning involve operations managers? What kinds of decisions are made at the various levels?
2. What are the three phases of intermediate planning?
3. What is aggregate planning? What is its purpose?
4. Why is there a need for aggregate planning?
5. What are the most common decision variables for aggregate planning in a manufacturing setting? In a service setting?
6. What aggregate planning difficulty that might confront an organization offering a variety of products and/or services would not confront an organization offering one or a few similar products or services?
7. Briefly discuss the advantages and disadvantages of each of these planning strategies:
   a. Maintain a level rate of output and let inventories absorb fluctuations in demand.
   b. Vary the size of the workforce to correspond to predicted changes in demand requirements.
   c. Maintain a constant workforce size, but vary hours worked to correspond to predicted demand requirements.
8. What are the primary advantages and limitations of informal graphic and charting techniques for aggregate planning?
9. Briefly describe the planning techniques listed below, and give an advantage and disadvantage for each:
   a. Linear programming
   b. Linear decision rule
   c. Simulation
10. What are the inputs to master scheduling? What are the outputs?

Memo Writing Exercises

1. J. B. Ronks is a marketing manager at your company. He is teaching an introductory marketing course at a local college one evening a week and needs a concise explanation of the "what and why of aggregate planning." Because of your knowledge of operations management, he has sent you a memo asking for your help. Write a brief memo providing the explanation Ronks needs.
2. Your immediate supervisor, Rhonda Waters, has raised the possibility of switching from a chase demand strategy to a level capacity strategy for aggregate planning, and wants your opinion on the merits of one versus the other. Write her a one-page memo on this matter.

Problems

1. Refer to Example 1. The president of the firm has decided to shut down the plant for vacation and installation of new equipment in period 4. After installation, the cost per unit will remain the same, but the output rate for regular time will be 450. Regular output is the same as in Example 1 for periods 1, 2, and 3; 0 for period 4; and 450 for each of the remaining periods. Note, though, that the forecast of 400 units in period 4 must be dealt with. Prepare the aggregate plan, and compute its total cost.

   Total cost: $6,350

2. Refer to Example 1. Suppose that the regular output rate will drop to 290 units per period due to an expected change in production requirements. Costs will not change. Prepare an aggregate plan and compute its total cost for each of these alternatives:

   a. Total cost: $4,670
   b. Total cost: $4,800

   a. Use overtime at a fixed rate of 20 units per period as needed. Plan for an ending inventory of zero for period 6. Backlogs cannot exceed 90 units per period.
   b. Use subcontracting at a *maximum* rate of 50 units per period; the usage need not be the same in every period. Have an ending inventory of zero in the last period. Again backlogs cannot exceed 90 units in any period. Compare these two plans.
3. Refer to Example 2. Suppose you can use a combination of overtime and subcontracting, but you cannot use subcontracting in more than two periods. Up to 50 units of subcontracting and either 0 or 40 units of overtime are allowed per period. Subcontracting is $6 per

   Minimum total cost: $4,640

unit, and overtime is $3 per unit. (*Hint:* Use subcontracting only when overtime units are not sufficient to decrease backlogs to 80 units or less.) Plan for an ending inventory balance of 0 for period 6. Prepare a plan that will minimize total cost.

Total cost: $4,970

4. Refer to Example 2. Determine whether a plan to use subcontracting at a maximum rate of 50 units per period as needed with no overtime would achieve a lower total cost than the plan shown in Example 2. Again, plan for a zero inventory balance at the end of period 6.

a. TC = $63,300
b. TC = $62,945

5. Manager T. C. Downs of Plum Engines, a producer of lawn mowers and leaf blowers, must develop an aggregate plan given the forecast for engine demand shown in the table. The department has a normal capacity of 130 engines per month. Normal output has a cost of $60 per engine. The beginning inventory is zero engines. Overtime has a cost of $90 per engine.

   a. Develop a chase plan that matches the forecast and compute the total cost of your plan.

   b. Compare the costs to a level plan that uses inventory to absorb fluctuations. Inventory carrying cost is $2 per engine per month. Backlog cost is $90 per engine per month.

Month	**1**	**2**	**3**	**4**	**5**	**6**	**7**	**8**	**Total**
Forecast	120	135	140	120	125	125	140	135	1,040

a. TC = $78,500
b. TC = $77,350

6. Manager Chris Channing of Fabric Mills, Inc., has developed the forecast shown in the table for bolts of cloth. The figures are in hundreds of bolts. The department has a normal capacity of 275(00) bolts per month, except for the seventh month, when capacity will be 250(00) bolts. Normal output has a cost of $40 per hundred bolts. Workers can be assigned to other jobs if production is less than normal. The beginning inventory is zero bolts.

   a. Develop a chase plan that matches the forecast and compute the total cost of your plan. Overtime is $60 per hundred bolts.

   b. Would the total cost be less with regular production with no overtime, but using a subcontractor to handle the excess above normal capacity at a cost of $50 per hundred bolts? Backlogs are not allowed. The inventory carrying cost is $2 per hundred bolts.

Month	**1**	**2**	**3**	**4**	**5**	**6**	**7**	**Total**
Forecast	250	300	250	300	280	275	270	1,925

a. TC = $31,250
b. TC = $31,520

7. SummerFun, Inc., produces a variety of recreation and leisure products. The production manager has developed an aggregate forecast:

Month	**Mar**	**Apr**	**May**	**Jun**	**Jul**	**Aug**	**Sep**	**Total**
Forecast	50	44	55	60	50	40	51	350

Use the following information to develop aggregate plans.

Regular production cost	$80 per unit.
Overtime production cost	$120 per unit
Regular capacity	40 units per month
Overtime capacity	8 units per month
Subcontracting cost	$140 per unit
Subcontracting capacity	12 units per month
Holding cost	$10
Back-order cost	$20 per unit
Beginning inventory	0 units
	5 workers

Develop an aggregate plan using each of the following guidelines and compute the total cost for each plan. Which plan has the lowest total cost?

   a. Use regular production. Supplement using inventory, overtime, and subcontracting as needed. No backlogs allowed.

b. Use a level strategy. Use a combination of backlogs, subtracting, and inventory to handle variations in demand.

8. Waterlovers produces and distributes bottled water in a variety of flavors. A planner has developed an aggregate forecast for demand for the next six months (quantities are tankloads of water):

a. TC = $ 4,600
b. TC = $ 4,680
c. TC = $ 4,660

Month	May	Jun	Jul	Aug	Sep	Oct	Total
Forecast	50	60	70	90	80	70	420

Use the following information to develop aggregate plans. One unit represents 10,000 bottles, which is one tankload.

Regular production cost	$10 per unit
Regular production capacity	60 units
Overtime production cost	$16 per unit
Subcontracting cost	$18 per unit
Holding cost	$2
Back-ordering cost	$50 per month per unit
Hiring cost	$200
Layoff cost	$500
Beginning inventory	0 units

Develop an aggregate plan using each of the following guidelines and compute the total cost for each plan. Which plan has the lowest total cost?

a. Use level production. Supplement using overtime as needed.

b. Use a combination of overtime (10 units per period maximum), inventory, and subcontracting (10 units per period maximum) to handle variations in demand.

c. Use overtime up to 15 units per period and inventory to handle variations in demand.

9. Wormwood, Ltd., produces a variety of furniture products. The planning committee wants to prepare an aggregate plan for the next six months using the following information:

See IM

	MONTH					
	1	2	3	4	5	6
Demand	160	150	160	180	170	140
Capacity						
Regular	150	150	150	150	160	160
Overtime	10	10	0	10	10	10

COST PER UNIT	
Regular time	$50
Overtime	$75
Subcontract	$80
Inventory, per period	$4

Subcontracting can handle a maximum of 10 units per month. Beginning inventory is zero. Develop a plan that minimizes total cost. No back orders allowed.

10. Refer to Solved Problem 1. Prepare two additional aggregate plans. Call the one in the solved problem plan A. For plan B, hire one more worker at a cost of $200 in that period. Make up any shortfall using subcontracting at $8 per unit, with a maximum of 20 units per period (i.e., use subcontracting to reduce back orders when the forecast exceeds regular output). Note that the ending inventory in period 9 should be zero. Therefore, Total forecast = Total output + Quantity subcontracted. An additional constraint is that back orders cannot exceed 80 units in any period. For plan C, assume no workers are hired (so regular output is 200 units per period instead of 210 as in plan B). Use subcontracting as needed, but no more than 20 units per period. Compute the total cost of each plan. Which plan has the lowest cost?

Plan B: $14,340
Plan C: $14,370
Plan B has the lowest cost.

11. Refer to Solved Problem 1. Suppose another option is to use part-time workers to assist during seasonal peaks. The cost per unit, including hiring and training, is $11. The output

$13,475

rate is 10 units per worker per period for all workers. A maximum of 10 part-time workers can be used, and the same number of part-time workers must be used in all periods that have part-time workers. The ending inventory in period 9 should be 10 units. The limit on backlogs is 20 units per period. Try to make up backlogs as soon as possible. Compute the total cost for this plan, and compare it to the cost of the plan used in the solved problem.

$13,885

12. Refer to Solved Problem 1. Prepare an aggregate plan that uses overtime ($9 per unit, maximum output 25 units per period) and inventory variation. Try to minimize backlogs. The ending inventory in period 9 should be zero, and the limit on backlogs is 60 units per period. Note that Total forecast = Total regular output + Quantity overtime. Compute the total cost of your plan, and compare it to the total cost of the plan used in the solved problem.

$12,930

13. Refer to Solved Problem 1. Prepare an aggregate plan that uses some combination of laying off ($100 per worker), subcontracting ($8 per unit, maximum of 20 units per period, must use for three consecutive periods), and overtime ($9 per unit, maximum of 20 per period, maximum of 50 for the planning horizon). Compute the total cost, and compare it with any of the other plans you have developed. Which plan has the lowest total cost? Assume you start with 21 workers.

See IM

14. Verify the transportation solution shown in Example 4.

Total cost: $124,960

15. Refer to Example 4. Suppose that an increase in warehousing costs and other costs brings inventory carrying costs to $2 per unit per month. All other costs and quantities remain the same. Determine a revised solution to this transportation problem.

Total cost: $126,650
Additional cost: $230

16. Refer to Example 4. Suppose that regular-time capacity will be reduced to 440 units in period 3 to accommodate a companywide safety inspection of equipment. What will the additional cost of the optimal plan be as compared to the one shown in Example 4? Assume all costs and quantities are the same as given in Example 4 except for the regular-time output in period 3.

Total cost: $127,000
Additional cost: $580

17. Solve Problem 16 using an inventory carrying cost of $2 per unit per period.

18. Dundas Bike Components Inc. of Wheelville, Illinois, manufactures bicycle wheels in two different sizes for the Big Bike Co. assembly plant located across town. David Dundas, the firm's owner-manager, has just received Big Bike's order for the next six months.

	20-inch Wheels	24-inch Wheels
Nov.	1,000 units	500 units
Dec.	900	500
Jan.	600	300
Feb.	700	500
Mar.	1,100	400
Apr.	1,100	600

a. TC = $63,300
b. TC = $63,945
c. TC = $78,500
d. TC = $77,550

a. Under what circumstances will it be possible for David to develop just one aggregate plan rather than two (one for each size wheel)? Explain in two to three sentences without calculations.

b. Currently Dundas employs 28 full-time, highly skilled employees, each of whom can produce 50 wheels per month. Because skilled labor is in short supply in the Wheelville area, David would like to develop a pure level output plan. There is no inventory of finished wheels on hand at present, but David would like to have 300 on hand at the end of April. Big Bike will tolerate back orders of up to 200 units per month. Show your level plan in tabular form.

c. Calculate the total annual cost of your plan using these costs:

Regular	$5.00	Hiring	$300
Overtime	$7.50	Layoff	$400
Part-time	NA	Inventory	$1.00
Subcontract	NA	Backorder	$6.00

19. Prepare a master production schedule for industrial pumps in the manner of Figure 12–10 in the chapter. Use the same inputs as the example, but change the MPS rule from "schedule production when the projected on-hand inventory would be negative without production" to "schedule production when the projected on-hand inventory would be less than 10 without production." See IM
20. Update the master schedule shown in Figure 12–10 given these updated inputs: It is now the end of week 1; customer orders are 25 for week 2, 16 for week 3, 11 for week 4, 8 for week 5, and 3 for week 6. Use the MPS rule of ordering production when projected on-hand inventory would be negative without production. See IM
21. Prepare a master schedule like that shown in Figure 12–10 given this information: The forecast for each week of an eight-week schedule is 50 units. The MPS rule is to schedule production if the projected on-hand inventory would be negative without it. Customer orders (committed) are: See IM

Week	Customer Orders
1	52
2	35
3	20
4	12

Use a production lot size of 75 units and no beginning inventory.

22. Determine the available-to-promise (ATP) quantities for each period for Solved Problem 2. See IM
23. Prepare a schedule like that shown in Figure 12–11 for the following situation: The forecast is 80 units for each of the first two periods and 60 units for each of the next three periods. The starting inventory is 20 units. The company uses a chase strategy for determining the production lot size, except there is an upper limit on the lot size of 70 units. Also, the desired safety stock is 10 units. Note: the ATP quantities are based on maximum allowable production, and do not include safety stock. Committed orders are: See IM

Period	Customer Orders
1	82
2	80
3	60
4	40
5	20

Selected Bibliography

Fogarty, Donald W.; John H. Blackstone, Jr.; and Thomas R. Hoffmann. *Production and Inventory Management.* Cincinnati: South-Western Publishing, 1991.

Krajewski, L., and L. Ritzman. "Disaggregation in Manufacturing and Service Organizations." *Decision Sciences* 8, no. 1 (1977), pp. 1–18.

Leon, Robert A., and John R. Meyer. "Capacity Strategies for the 1980s." *Harvard Business Review,* November–December 1980, p. 133.

Posner, M. E., and W. Szwarc. "A Transportation Type Aggregate Production Model with Backordering." *Management Science* 29, no. 2 (February 1983), pp. 188–99.

Sipper, Daniel, and Robert Bulfin, Jr. *Production: Planning, Control, and Integration.* New York: McGraw-Hill, 1997.

Vollmann, Thomas E.; William L. Berry; and D. Clay Whybark. *Manufacturing Planning and Control Systems.* 4th ed. Burr Ridge, IL: Richard D. Irwin, 1997.

Ware, Norman, and Donald Fogarty. "Master Schedule/Master Production Schedule: The Same or Different?" *Production and Inventory Management Journal,* First Quarter, 1990, pp. 34–37.

LEARNING OBJECTIVES

After completing this chapter, you should be able to:

1. Define the term *inventory* and list the major reasons for holding inventories.
2. Contrast independent and dependent demand.
3. List the main requirements for effective inventory management.
4. Discuss periodic and perpetual review systems.
5. Describe the A-B-C approach and explain how it is useful.
6. Discuss the objectives of inventory management.
7. Describe the basic EOQ model and its assumptions and solve typical problems.
8. Describe the economic run size model and solve typical problems.
9. Describe the quantity discount model and solve typical problems.
10. Describe reorder point models and solve typical problems.
11. Describe situations in which the single-period model would be appropriate, and solve typical problems.

CHAPTER THIRTEEN
INVENTORY MANAGEMENT

CHAPTER OUTLINE

Good inventory management is essential to the successful operation of most organizations for a number of reasons. One is the amount of money inventory represents; another is the impact that inventories have on the daily operations of an organization. Some organizations have excellent inventory management and many have satisfactory inventory management. Too many, however, have unsatisfactory inventory management, which sometimes is a sign that management does not recognize the importance of inventories. More often than not, though, the recognition is there. What is lacking is an understanding of what needs to be done and how to do it. This chapter presents the concepts that underlie good inventory management.

This chapter focuses on management of finished goods, raw materials, purchased parts, and retail items. The different functions of inventories, requirements for effective inventory management, objectives of inventory control, and techniques for determining *how much* to order and *when* to order are explained.

INTRODUCTION

inventory A stock or store of goods.

An **inventory** is a stock or store of goods. Firms typically stock hundreds or even thousands of items in inventory, ranging from small things such as pencils, paper clips, screws, nuts, and bolts to large items such as machines, trucks, construction equipment, and airplanes. Naturally, many of the items a firm carries in inventory relate to the kind of business it engages in. Thus, manufacturing firms carry supplies of raw materials, purchased parts, partially finished items, and finished goods, as well as spare parts for machines, tools, and other supplies. Department stores carry clothing, furniture, carpeting, stationery, appliances, gifts, cards, and toys. Some also stock sporting goods, paints, and tools. Hospitals stock drugs, surgical supplies, life-monitoring equipment, sheets and pillow cases, and more. Supermarkets stock fresh and canned foods, packaged and frozen foods, household supplies, magazines, baked goods, dairy products, produce, and other items.

NEWSCLIP

$$$

We proceed as follows. First look for a five-by-five-by-three-foot bin of gears or parts that looks like it has been there awhile. Pick up a gear and ask, casually, "How much is this worth?" You then ask, "How many of these are in the bin?" followed by, "How long has this bin been here?" and, "What's your cost of money for this company?" I recall one case in a nameless South American country where the unit cost times the number of parts times the time it had been there times the interest rate resulted in a cost-per-day figure that would insure comfortable retirement for the plant manager on the bank of the Rio de la Plata at one of the better resorts to be found there. The plant manager suddenly realized that what he was holding was not just a chunk of high-test steel, but was *real money*. He then pointed out that *he* now understood the value of the inventory but could I suggest a way to drive the point home to upper management? I suggested that he go to the accounting department and borrow enough money to be equal to the bin's value for as long as it had been sitting there, and pile it on the top of the bin. I further suggested that he do that for every bin on the production line. We rapidly figured out that by the time we had the money piled up on the bin, you would not even be able to *see* the bin. My opinion was that if the upper managers were given a tour of the line with the money piled up, they would *never* forget it.

Source: Gene Woolsey, "On Doing Good Things and Dumb Things in Production and Inventory Control," *Interfaces* 5, no. 3 (May 1975). Copyright 1975, The Institute of Management Science. Reprinted by permission.

Independent and Dependent Demand

A major distinction in the way inventory planning and control are managed is whether demand for items in inventory is *independent* or *dependent*. **Dependent demand** items are typically subassemblies or component parts that will be used in the production of a final or finished product. Demand (i.e., usage) of subassemblies and component parts is *derived* from the number of finished units to be produced. A classic example of this is demand for wheels for new cars. If each car is to have five wheels, then the total number of wheels required for a production run is simply a function of the number of cars that are to be produced in that run. For example, 200 cars would require $200 \times 5 = 1{,}000$ wheels. **Independent demand** items are the finished goods or other *end items* that are sold to someone. There is usually no way to determine precisely how many of these items will be demanded during any given time period because demand typically includes an element of randomness. Therefore, forecasting plays an important role in stocking decisions, whereas stock requirements for dependent demand items are determined by reference to the production plan. This chapter focuses on management of independent demand items. Chapter 14 focuses on dependent demand.

dependent demand Demand for items in inventory that are subassemblies or component parts to be used in the production of finished goods.

independent demand Demand for items that are finished items.

Vol. 1, Tape 2, seg. 2
Navistar, Caterpillar

THE NATURE AND IMPORTANCE OF INVENTORIES

Inventories are a vital part of business. Not only are they necessary for operations, but also they contribute to customer satisfaction. To get a sense of the significance of inventories, consider the following: Although the amounts and dollar values of inventories carried by different types of firms varies widely, a typical firm probably has about 30 percent of its current assets and perhaps as much as 90 percent of its working capital invested in inventory.[1]

A typical manufacturing firm carries different types of inventories, including the following:

Raw materials and purchased parts.

Partially completed goods, called *work-in-process* (WIP).

Finished-goods inventories (manufacturing firms) or merchandise (retail stores).

Replacement parts, tools, and supplies.

Goods-in-transit to warehouses or customers.

Service firms do not carry these types of inventories, although they do carry inventories of supplies and equipment. To understand why firms have inventories at all, you need to be aware of the various functions of inventory.

Functions of Inventory

Inventories serve a number of functions. Among the most important are the following:

1. To meet anticipated demand.
2. To smooth production requirements.
3. To decouple components of the production-distribution system.
4. To protect against stockouts.
5. To take advantage of order cycles.

[1]One widely used measure of managerial performance relates to *return on investment* (ROI), which is profit after taxes divided by total assets. Because inventories may represent a significant portion of total assets, a reduction of inventories can result in a significant increase in ROI.

6. To hedge against price increases or to take advantage of quantity discounts.
7. To permit operations.

Let's take a look at each of these.

1. *To meet anticipated customer demand.* A customer can be a person who walks in off the street to buy a new stereo system, a mechanic who requests a tool at a tool crib, or a manufacturing operation. These inventories are referred to as *anticipation stocks* because they are held to satisfy expected *average* demand.

2. *To smooth production requirements.* Firms that experience seasonal patterns in demand often build up inventories during off-season periods to meet overly high requirements during certain seasonal periods. These inventories are aptly named *seasonal inventories.* Companies that process fresh fruits and vegetables deal with seasonal inventories. So do stores that sell greeting cards, skis, snowmobiles, or Christmas trees.

3. *To decouple operations.* Historically, manufacturing firms have used inventories as buffers between successive operations to maintain continuity of production that would otherwise be disrupted by events such as breakdowns of equipment and accidents that cause a portion of the operation to shut down temporarily. The buffers permit other operations to continue temporarily while the problem is resolved. Similarly, firms have used buffers of raw materials to insulate production from disruptions in deliveries from suppliers, and finished goods inventory to buffer sales operations from manufacturing disruptions. More recently, companies have taken a closer look at buffer inventories, recognizing the cost and space they require, and realizing that finding and eliminating sources of disruptions can greatly decrease the need for decoupling operations.

4. *To protect against stockouts.* Delayed deliveries and unexpected increases in demand increase the risk of shortages. Delays can occur because of weather conditions, supplier stockouts, deliveries of wrong materials, quality problems, and so on. The risk of shortages can be reduced by holding *safety stocks,* which are stocks in excess of average demand to compensate for *variabilities* in demand and lead time.

5. *To take advantage of order cycles.* To minimize purchasing and inventory costs, a firm often buys in quantities that exceed immediate requirements. This necessitates storing some or all of the purchased amount for later use. Similarly, it is usually economical to produce in large rather than small quantities. Again, the excess output must be stored for later use. Thus, inventory storage enables a firm to buy and produce in *economic lot sizes* without having to try to match purchases or production with demand requirements in the short run. This results in *periodic* orders, or order *cycles.* The resulting stock is known as *cycle stock.* Order cycles are not always due to economic lot sizes. In some instances, it is practical or economical to group orders and/or to order at fixed intervals.

6. *To hedge against price increases.* Occasionally a firm will suspect that a substantial price increase is about to be made and purchase larger-than-normal amounts to avoid the increase. The ability to store extra goods also allows a firm to take advantage of price discounts for larger orders.

7. *To permit operations.* The fact that production operations take a certain amount of time (i.e., they are not instantaneous) means that there will generally be some work-in-process inventory. In addition, intermediate stocking of goods—including raw materials, semifinished items, and finished goods at production sites, as well as goods stored in warehouses—leads to *pipeline* inventories throughout a production-distribution system.

Objectives of Inventory Control

Inadequate control of inventories can result in both under- and overstocking of items. Understocking results in missed deliveries, lost sales, dissatisfied customers, and production bottlenecks; overstocking unnecessarily ties up funds that might be more productive elsewhere. Although overstocking may appear to be the lesser of the two evils, the price tag for excessive overstocking can be staggering when inventory holding costs are high—as illustrated by the little story about the bin of gears at the beginning of the chapter—and matters can easily get out of hand. It is not unusual for managers to discover that their firm has a 10-year supply of some item. (No doubt the firm got a good buy on it!)

Inventory management has two main concerns. One is the *level of customer service,* that is, to have the right goods, in sufficient quantities, in the right place, at the right time. The other is the costs of ordering and carrying inventories.

The overall objective of inventory management is to achieve satisfactory levels of customer service while keeping inventory costs within reasonable bounds. Toward this end, the decision maker tries to achieve a balance in stocking. He or she must make two fundamental decisions: the *timing* and *size* of orders (i.e., when to order and how much to order). The greater part of this chapter is devoted to models that can be applied to assist in making those decisions.

Managers have a number of measures of performance they can use to judge the effectiveness of inventory management. The most obvious, of course, is customer satisfaction, which they might measure by the number and quantity of backorders and/or customer complaints. A widely used measure is inventory turnover, which is the ratio of annual cost of goods sold to average inventory investment. The turnover ratio indicates how many times a year the inventory is sold. Generally, the higher the ratio, the better, because that implies more efficient use of inventories. A benefit of this measure is that it can be used to compare companies of different size in the *same* industry. Another useful measure is days of inventory, a number that indicates the expected number of days of sales that can be supplied from existing inventory. Here, a balance is desirable; a high number of days might imply excess inventory, while a low number might imply a risk of running out of stock.

REQUIREMENTS FOR EFFECTIVE INVENTORY MANAGEMENT

Management has two basic functions concerning inventory. One is to establish a system of keeping track of items in inventory, and the other is to make decisions about how much and when to order. To be effective, management must have the following:

1. A system to *keep track of the inventory* on hand and on order.
2. A reliable *forecast of demand* that includes an indication of possible *forecast error.*
3. Knowledge of *lead times* and *lead time variability.*
4. Reasonable estimates of inventory *holding costs, ordering costs,* and *shortage costs.*
5. A *classification system* for inventory items.

Let's take a closer look at each of these requirements.

Inventory Counting Systems

Inventory counting systems can be periodic or perpetual. Under a **periodic system,** a physical count of items in inventory is made at periodic intervals (e.g., weekly, monthly) in order to decide how much to order of each item. Many small grocers use this approach: a manager periodically checks the shelves and stockroom to determine the quantity on hand. Then the manager estimates how much will be demanded prior to the next delivery period and bases the order quantity on that information. An advantage of this type of system is that orders for many items occur at the same time, which can result in economies in processing and shipping orders. There are also several disadvantages of periodic reviews. One is a lack of control between reviews. Another is the need to protect against shortages between review periods by carrying extra stock. A third disadvantage is the need to make a decision on order quantities at each review.

periodic system Physical count of items in inventory made at periodic intervals (weekly, monthly).

A General Electric Supply employee uses a bar code scanner to quickly and accurately record inventory items. GE Supply, one of the many companies that are part of General Electric Company, is a full service electrical distributor selling GE and non-GE products to electrical contractors, industrial users, utilities, and commercial establishments.

www.ge.com

perpetual inventory system System that keeps track of removals from inventory continuously, thus monitoring current levels of each item.

A **perpetual inventory system** (also known as a *continual* system) keeps track of removals from inventory on a continuous basis, so the system can provide information on the current level of inventory for each item. When the amount on hand reaches a predetermined minimum, a fixed quantity, Q, is ordered. An obvious advantage of this system is the control provided by the continuous monitoring of inventory withdrawals. Another advantage is the fixed-order quantity; management can justify an *economic order size* (discussed in detail later in the chapter). One disadvantage of this approach is the added cost of record keeping. Moreover, a physical count of inventories must still be performed periodically to verify records because of errors, pilferage, spoilage, and other factors that can reduce the effective amount of inventory. Bank transactions such as customer deposits and withdrawals are examples of continuous recording of inventory changes.

two-bin system Two containers of inventory; reorder when the first is empty.

Perpetual systems range from very simple to very sophisticated. A **two-bin system,** a very elementary system, uses two containers for inventory. Items are withdrawn from the first bin until its contents are exhausted. It is then time to reorder. Sometimes an order card is placed at the bottom of the first bin. The second bin contains enough stock to satisfy expected demand until the order is filled, plus an extra cushion of stock that will reduce the chance of a stockout if the order is late or if usage is greater than expected. The advantage of this system is that there is no need to record each withdrawal from inventory; the disadvantage is that the reorder card may not be turned in for a variety of reasons (e.g., misplaced, the person responsible forgets to turn it in).

Perpetual systems can be either *batch* or *on-line.* In batch systems, inventory records are collected periodically and entered into the system. In on-line systems, the transactions are recorded instantaneously. The advantage of on-line systems is that they are always up-to-date. In batch systems, a sudden surge in demand could result in reducing the amount of inventory below the reorder point between the periodic read-ins. Frequent batch collections can minimize that problem.

universal product code Bar code printed on a label that has information about the item to which it is attached.

Supermarkets, discount stores, and department stores have always been major users of periodic accounting systems. Today, most have switched to computerized checkout systems using a laser scanning device that reads a **universal product code** (UPC), or bar code, printed on an item tag or on packaging. A typical grocery product code is illustrated on the next page.

The zero on the left of the bar code identifies this as a grocery item, the first five numbers (14800) indicate the manufacturer (Mott's), and the last five numbers (23208) indicate the specific item (natural-style applesauce). Items in small packages, such as candy and gum, use a six-digit number.

UPC scanners represent a major change in the inventory systems of stores that use them. In addition to their increase in speed and accuracy, these systems give managers continuous information on inventories, reduce the need for periodic inventories and order-size determinations, and improve the level of customer service by indicating the price and quantity of each item on the customer's receipt, as illustrated below.

```
           BRACO CAPELLINI          .79
           BUB YUM DBL LIME         .30 T
           2/LO FAT MILK H G       1.03
           EUROP ROLLS              .91
           HUNTS TOMATO             .55
           NEWSPAPER                .35
           KR CAS BRICK CHEES      1.59
           GRAPES-GREEN
  .91 LB @ .89 PER LB               .81
           TAX DUE                  .02
           TOTAL                   6.35

           CASH                   20.00*
           CHANGE                 13.65

           8/07/98 18:01 21  16  23100  2570
```

Bar coding represents an important development for other sectors of business besides retailing. Manufacturing and service industries also benefit from the simplified production and inventory control it provides. In manufacturing, bar codes attached to parts, subassemblies, and finished goods greatly facilitate counting and monitoring activities. Automatic routing, scheduling, sorting, and packaging can also be done using bar codes.

Demand Forecasts and Lead Time Information

Inventories are used to satisfy demand requirements, so it is essential to have reliable estimates of the amount and timing of demand. Similarly, it is essential to know how long it will take for orders to be delivered. In addition, managers need to know the extent to which demand and **lead time** (the time between submitting an order and receiving it) might vary; the greater the potential variability, the greater the need for additional stock to reduce the risk of a shortage between deliveries. Thus, there is a crucial link between forecasting and inventory management.

lead time Time interval between ordering and receiving the order.

Cost Information

Three basic costs are associated with inventories: holding, transaction (ordering), and shortage costs.

Holding or **carrying costs** relate to physically having items in storage. Costs include interest, insurance, taxes (in some states), depreciation, obsolescence, deterioration, spoilage, pilferage, breakage, and warehousing costs (heat, light, rent, security). They also include opportunity costs associated with having funds which could be used elsewhere tied up in inventory. Note that it is the *variable* portion of these costs that is pertinent.

holding (carrying) cost Cost to carry an item in inventory for a length of time, usually a year.

NEWSCLIP

FORD TRIPLES ITS BILLION-DOLLAR COST-CUTTING GOAL

The Ford Motor Company exceeded its original goal of cutting out $1 billion in costs in 1997. One major chunk of savings was the result of worker suggestions; in one case teams from two assembly plants visited each others' plants and offered suggestions on ways to improve production. Another major chunk came from the use of standard parts—using the same parts in different models. Not only did this reduce design and assembly costs, it reduced the number of different parts carried in inventory, and simplified record keeping. At the same time, it increased production flexibility. Beyond that, standard parts help lessen the chance of experiencing out-of-stock incidents, which can result in costly shutdowns at assembly plants; and standard parts mean dealers don't have to carry as many different replacement parts.

Source: Based in part on "Ford Triples Its Billion-Dollar Cost-Cutting Goal," Micheline Maynard, *USA TODAY,* December 15, 1997, p. 1B.

www.ford.com

The significance of the various components of holding cost depends on the type of item involved, although taxes, interest, and insurance are generally based on the dollar value of an inventory. Items that are easily concealed (e.g., pocket cameras, transistor radios, calculators) or fairly expensive (cars, TVs) are prone to theft. Fresh seafood, meats and poultry, produce, and baked goods are subject to rapid deterioration and spoilage. Dairy products, salad dressings, medicines, batteries, and film also have limited shelf lives.

Holding costs are stated in either of two ways: as a percentage of unit price or as a dollar amount per unit. In any case, typical annual holding costs range from 20 percent to 40 percent of the value of an item. In other words, to hold a $1 item for one year could cost from 20 cents to 40 cents.

ordering costs Costs of ordering and receiving inventory.

Ordering costs are the costs of ordering and receiving inventory. They are the costs that vary with the actual placement of an order. These include determining how much is needed, preparing invoices, shipping costs, inspecting goods upon arrival for quality and quantity, and moving the goods to temporary storage. Ordering costs are generally expressed as a fixed dollar amount per order, regardless of order size.

When a firm produces its own inventory instead of ordering it from a supplier, the costs of machine setup (e.g., preparing equipment for the job by adjusting the machine, changing cutting tools) are analogous to ordering costs; that is, they are expressed as a fixed charge per run, regardless of the size of the run.

shortage costs Costs resulting when demand exceeds the supply of inventory on hand; often unrealized profit per unit.

Shortage costs result when demand exceeds the supply of inventory on hand. These costs can include the opportunity cost of not making a sale, loss of customer goodwill, late charges, and similar costs. Furthermore, if the shortage occurs in an item carried for internal use (e.g., to supply an assembly line), the cost of lost production or downtime is considered a shortage cost. Such costs can easily run into hundreds of dollars a minute or more. Shortage costs are sometimes difficult to measure, and they may be subjectively estimated.

Classification System

An important aspect of inventory management is that items held in inventory are not of equal importance in terms of dollars invested, profit potential, sales or usage volume, or stockout penalties. For instance, a producer of electrical equipment might have electric generators, coils of wire, and miscellaneous nuts and bolts among the items carried in inventory. It would be unrealistic to devote equal attention to each of these items. Instead, a more reasonable approach would be to allocate control efforts according to the *relative importance* of various items in inventory.

Weyerhaeuser employees use a computer to gauge timber-cutting costs at the cut site. Entering and processing data on-site allows for maintaining more accurate, up-to-date records of costs as well as cut-tree inventory. Weyerhaeuser is an international forest products company whose principal businesses are the growing and harvesting of trees; the manufacture, distribution, and sale of forest products; and real estate construction and development.

www.weyerhaeuser.com

The **A-B-C approach** classifies inventory items according to some measure of importance, usually annual dollar usage (i.e., dollar value per unit multiplied by annual usage rate), and then allocates control efforts accordingly. Typically, three classes of items are used: A (very important), B (moderately important), and C (least important). However, the actual number of categories may vary from organization to organization, depending on the extent to which a firm wants to differentiate control efforts. With three classes of items, A items generally account for about 15 to 20 percent of the *number* of items in inventory but about 60 to 70 percent of the *dollar usage*. At the other end of the scale, C items might account for about 60 percent of the number of items but only about 10 percent of the dollar usage of an inventory. These percentages vary from firm to firm, but in most instances a relatively small number of items will account for a large share of the value or cost associated with an inventory, and these items should receive a relatively greater share of control efforts. For instance, A items should receive close attention through frequent reviews of amounts on hand and control over withdrawals, where possible, to make sure that customer service levels are attained. The C items should receive only loose control (two-bin system, bulk orders), and the B items should have controls that lie between the two extremes.

A-B-C approach Classifying inventory according to some measure of importance, and allocating control efforts accordingly.

Note that C items are not necessarily *un*important; incurring a stockout of C items such as the nuts and bolts used to assemble manufactured goods can result in a costly shutdown of an assembly line. However, due to the low annual dollar volume of C items, there may not be much additional cost incurred by ordering larger quantities of some items, or ordering them a bit earlier.

Example 1

Classify the inventory items as A, B, or C based on annual dollar value, given the following information:

Item	Annual Demand	×	Unit Cost	=	Annual Dollar Value
1	1,000		$4,300		$4,300,000
2	5,000		720		3,600,000
3	1,900		500		950,000
4	1,000		710		710,000
5	2,500		250		625,000
6	2,500		192		480,000
7	400		200		80,000
8	500		100		50,000
9	200		210		42,000
10	1,000		35		35,000
11	3,000		10		30,000
12	9,000		3		27,000

Solution

The first two items have a relatively high annual dollar value, so it seems reasonable to classify them as A items. The next four items appear to have moderate annual dollar values and should be classified as B items. The remainder are C items, based on their relatively low dollar value.

Although annual dollar volume may be the primary factor in classifying inventory items, a manager may take other factors into account in making exceptions for certain items (e.g., changing the classification of a C item to an A item). Factors may include the risk of obsolescence, the risk of a stockout, the distance of a supplier, and so on.

Figure 13–1 illustrates the A-B-C concept.

Managers use the A-B-C concept in many different settings to improve operations. One key use occurs in customer service, where a manager can focus attention on the most important aspects of customer service by categorizing different aspects as very important, important, or of only minor importance. The point is to not overemphasize minor aspects of customer service at the expense of major aspects.

cycle counting A physical count of items in inventory.

Another application of the A-B-C concept is as a guide to **cycle counting,** which is a physical count of items in inventory. The purpose of cycle counting is to reduce discrepancies between the amounts indicated by inventory records and the actual quantities of inventory on hand. Accuracy is important because inaccurate records can lead to disruptions in production, poor customer service, and unnecessarily high inventory carrying costs.

FIGURE 13–1

A typical A-B-C breakdown in relative annual dollar value of items and number of items by category

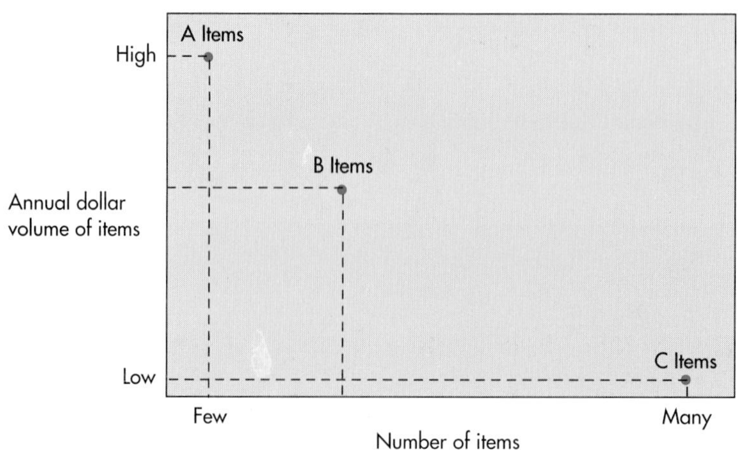

The key questions concerning cycle counting for management are:

1. How much accuracy is needed?
2. When should cycle counting be performed?
3. Who should do it?

The American Production and Inventory Control Society (APICS) recommends the following guidelines for inventory record accuracy: ±0.2 percent for A items, ±1 percent for B items, and ±5 percent for C items.

www.apics~stlouis.com

Some companies use certain events to trigger cycle counting, whereas others do it on a periodic (scheduled) basis. Events that can trigger a physical count of inventory include an out-of-stock report written on an item indicated by inventory records to be in stock, an inventory report that indicates a low or zero balance of an item, and a specified level of activity (e.g., every 2,000 units sold).

Some companies use regular stockroom personnel to do cycle counting during periods of slow activity while others contract with outside firms to do it on a periodic basis. Use of an outside firm provides an independent check on inventory and may reduce the risk of problems created by dishonest employees. Still other firms maintain full-time personnel to do cycle counting.

HOW MUCH TO ORDER: ECONOMIC ORDER QUANTITY MODELS

The question of how much to order is frequently determined by using an **economic order quantity (EOQ)** model. EOQ models identify the optimal order quantity by minimizing the sum of certain annual costs that vary with order size. Three order size models are described here:

economic order quantity The order size that minimizes total cost.

1. The economic order quantity model.
2. The economic order quantity model with noninstantaneous delivery.
3. The quantity discount model.

Basic Economic Order Quantity (EOQ) Model

The basic EOQ model is the simplest of the three models. It is used to identify the order size that will minimize the sum of the annual costs of holding inventory and ordering inventory. The unit purchase price of items in inventory is not generally included in the total cost because the unit cost is unaffected by the order size unless quantity discounts are a factor. If holding costs are specified as a percentage of unit cost, then unit cost is indirectly included in the total cost as a part of holding costs.

The basic model involves a number of assumptions. They are listed in Table 13–1.

Inventory ordering and usage occur in cycles. Figure 13–2 illustrates several inventory cycles. A cycle begins with receipt of an order of Q units, which are withdrawn at a constant rate over time. When the quantity on hand is just sufficient to satisfy demand during lead time, an order for Q units is submitted to the supplier. Because it is assumed that both the usage rate and the lead time do not vary, the order will be received at the

TABLE 13–1

Assumptions of the basic EOQ model

1. Only one product is involved.
2. Annual demand requirements are known.
3. Demand is spread evenly throughout the year so that the demand rate is reasonably constant.
4. Lead time does not vary.
5. Each order is received in a single delivery.
6. There are no quantity discounts.

FIGURE 13-2

The inventory cycle: profile of inventory level over time

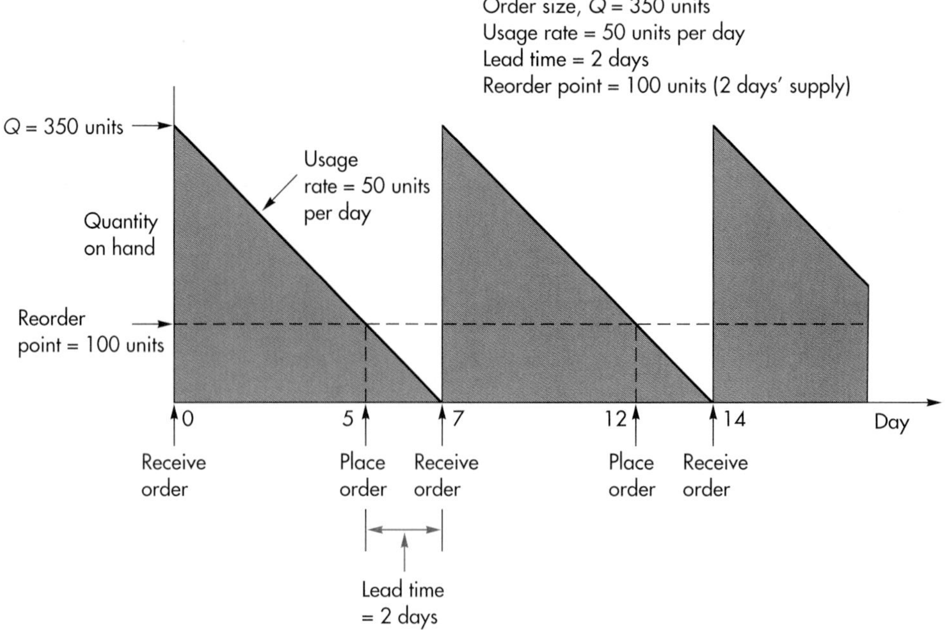

precise instant that the inventory on hand falls to zero. Thus, orders are timed to avoid both excess stock and stockouts (i.e., running out of stock).

The optimal order quantity reflects a trade-off between carrying costs and ordering costs: As order size varies, one type of cost will increase while the other decreases. For example, if the order size is relatively small, the average inventory will be low, resulting in low carrying costs. However, a small order size will necessitate frequent orders, which will drive up annual ordering costs. Conversely, ordering large quantities at infrequent intervals can hold down annual ordering costs, but that would result in higher average inventory levels and therefore increased carrying costs. Figure 13–3 illustrates these two extremes.

Thus, the ideal solution is an order size that causes neither a few very large orders nor many small orders, but one that lies somewhere between. The exact amount to order will depend on the relative magnitudes of carrying and ordering costs.

Annual carrying cost is computed by multiplying the average amount of inventory on hand by the cost to carry one unit for one year, even though any given unit would not be held for a year. The average inventory is simply half of the order quantity: The amount on hand decreases steadily from Q units to 0, for an average of $(Q + 0)/2$, or $Q/2$. Using the symbol H to represent the average annual carrying cost per unit, the *total annual carrying cost* is

$$\text{Annual carrying cost} = \frac{Q}{2}H$$

where

Q = Order quantity in units
H = holding (carrying) cost per unit

Carrying cost is thus a linear function of Q: Carrying costs increase or decrease in direct proportion to changes in the order quantity Q, as Figure 13–4A illustrates.

On the other hand, annual ordering cost will decrease as order size increases because for a given annual demand, the larger the order size, the fewer the number of orders needed. For instance, if annual demand is 12,000 units and the order size is 1,000 units per order, there must be 12 orders over the year. But if Q = 2,000 units, only six orders will be needed; if Q = 3,000 units, only four orders will be needed. In general, the

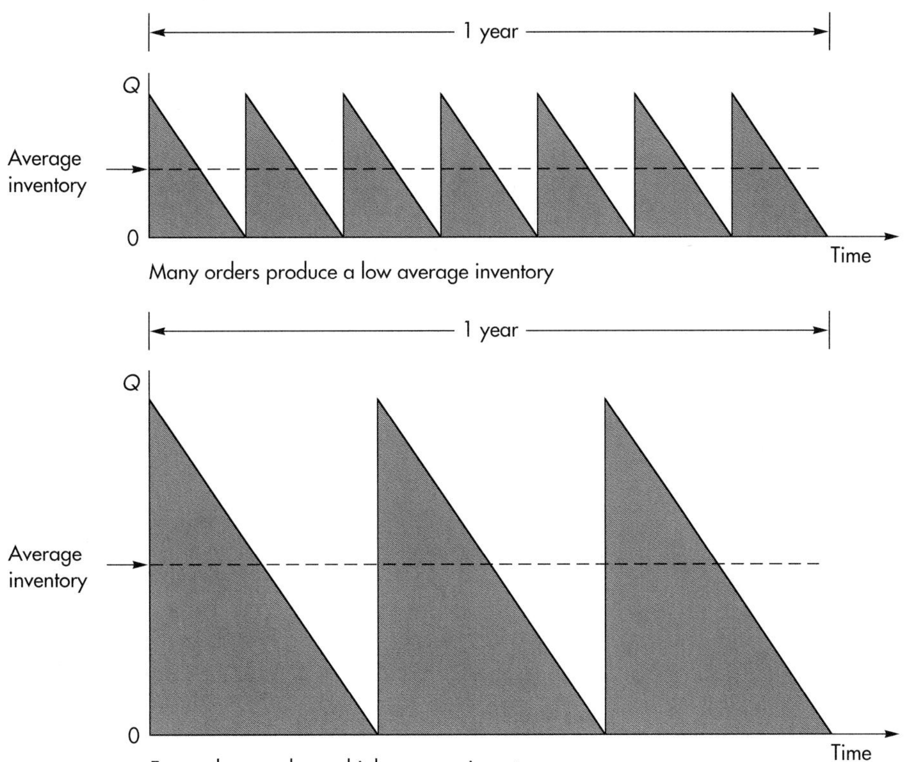

FIGURE 13-3

Average inventory level and number of orders per year are inversely related: as one increases, the other decreases

A. Carrying costs are linearly related to order size.

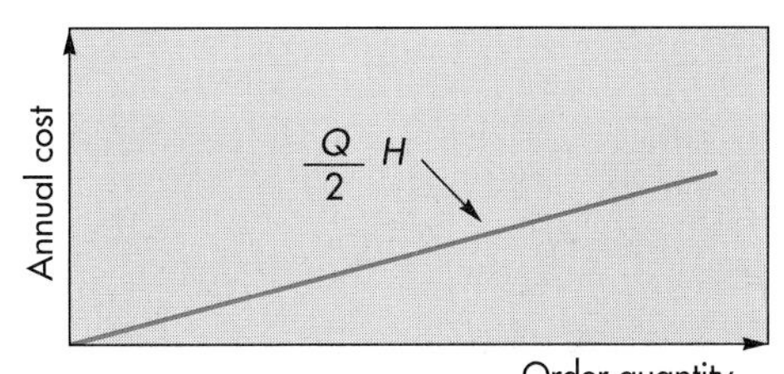

FIGURE 13-4

Carrying cost, ordering cost, and total cost curve

B. Ordering costs are inversely and nonlinearly related to order size.

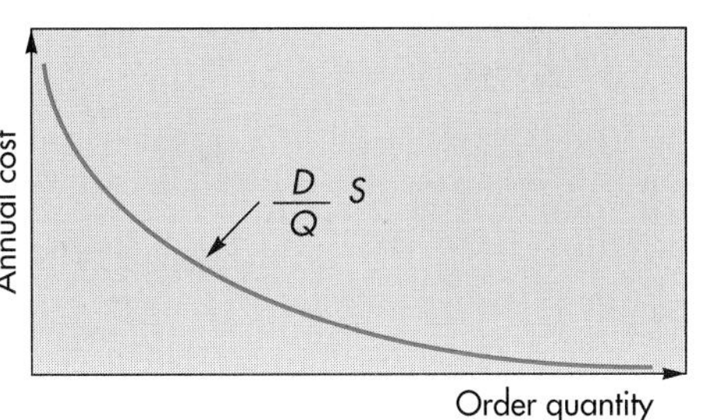

C. The total-cost curve is U-shaped.

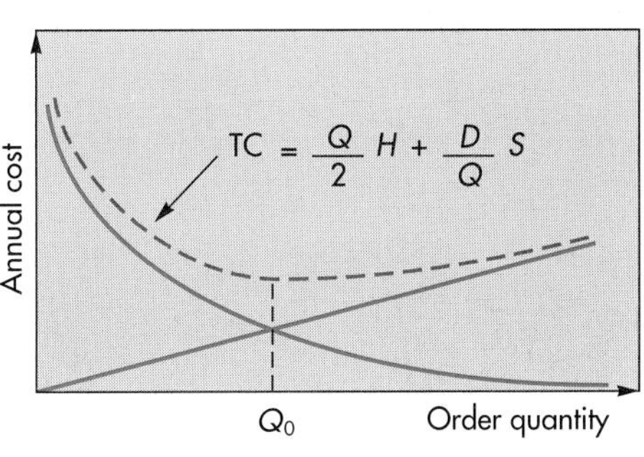

number of orders per year will be D/Q, where D = Annual demand and Q = Order size. Unlike carrying costs, ordering costs are relatively insensitive to order size; regardless of the amount of an order, certain activities must be done, such as determine how much is needed, periodically evaluate sources of supply, and prepare the invoice. Even inspection of the shipment to verify quality and quantity characteristics is not strongly influenced by order size since large shipments are sampled rather than completely inspected. Hence, there is a fixed ordering cost. *Annual ordering cost* is a function of the number of orders per year and the ordering cost per order.

$$\text{Annual ordering cost} = \frac{D}{Q}S$$

where

D = Demand, usually in units per year
S = Ordering cost

Because the number of orders per year, D/Q, decreases as Q increases, annual ordering cost is inversely related to order size, as Figure 13–4B illustrates.

The total annual cost associated with carrying and ordering inventory when Q units are ordered each time is

$$TC = \underset{\text{cost}}{\overset{\text{Annual}}{\text{carrying}}} + \underset{\text{cost}}{\overset{\text{Annual}}{\text{ordering}}} = \frac{Q}{2}H + \frac{D}{Q}S \qquad (13\text{–}1)$$

(Note that D and H must be in the same units, e.g. months, years). Figure 13–4C reveals that the total-cost curve is U-shaped (i.e., convex, with one minimum) and that *it reaches its minimum at the quantity where carrying and ordering costs are equal.* An expression for the optimal order quantity, Q_0, can be obtained using calculus.[2] The result is the formula

$$Q_0 = \sqrt{\frac{2DS}{H}} \qquad (13\text{–}2)$$

Thus, given annual demand, the ordering cost per order, and the annual carrying cost per unit, one can compute the optimal (economic) order quantity. The minimum total cost is then found by substituting Q_0 for Q in Formula 13–1.

The length of an order cycle (i.e., the time between orders) is

$$\text{Length of order cycle} = \frac{Q_0}{D} \qquad (13\text{–}3)$$

Example 2

A local distributor for a national tire company expects to sell approximately 9,600 steel-belted radial tires of a certain size and tread design next year. Annual carrying cost is \$16 per tire, and ordering cost is \$75. The distributor operates 288 days a year.

a. What is the EOQ?

b. How many times per year does the store reorder?

---

[2]We can find the minimum point of the total-cost curve by differentiating TC with respect to Q, setting the result equal to zero, and solving for Q. Thus,

1. $\frac{dTC}{dQ} = \frac{dQ}{2}H + d(D/Q)S = H/2 - DS/Q^2$

2. $0 = H/2 - DS/Q^2$, so $Q^2 = \frac{2DS}{H}$ and $Q = \sqrt{\frac{2DS}{H}}$

Note that the second derivative is positive, which indicates a minimum has been obtained.

c. What is the length of an order cycle?

d. What is the total annual cost if the EOQ quantity is ordered?

Solution

D = 9,600 tires per year
H = \$16 per unit per year
S = \$75

a. $Q_0 = \sqrt{\frac{2DS}{H}} = \sqrt{\frac{2(9{,}600)75}{16}} = 300$ tires.

b. Number of orders per year: $D/Q_0 = \frac{9{,}600 \text{ tires}}{300 \text{ tires}} = 32$.

c. Length of order cycle: $Q_0/D = \frac{300 \text{ tires}}{9{,}600 \text{ tires/yr.}} = 1/32$ of a year, which is $1/32 \times 288$, or nine workdays.

d. TC = Carrying cost + Ordering cost

$$
\begin{aligned}
TC &= (Q_0/2)H + (D/Q_0)S \\
&= (300/2)16 + (9{,}600/300)75 \\
&= \$2{,}400 + \$2{,}400 \\
&= \$4{,}800.
\end{aligned}
$$

Note that the ordering and carrying costs are equal at the EOQ, as illustrated in Figure 13–4C.

Carrying cost is sometimes stated as a percentage of the purchase price of an item rather than as a dollar amount per unit. However, as long as the percentage is converted into a dollar amount, the EOQ formula is still appropriate.

Example 3

Piddling Manufacturing assembles security monitors. It purchases 3,600 black-and-white cathode ray tubes a year at \$65 each. Ordering costs are \$31, and annual carrying costs are 20 percent of the purchase price. Compute the optimal quantity and the total annual cost of ordering and carrying the inventory.

Solution

D = 3,600 cathode ray tubes per year
S = \$31
H = .20(\$65) = \$13

$$Q_0 = \sqrt{\frac{2DS}{H}} = \sqrt{\frac{2(3{,}600)(31)}{13}} = 131 \text{ cathode ray tubes}$$

TC = Carrying costs + Ordering costs

$$
\begin{aligned}
&= (Q_0/2)H + (D/Q_0)S \\
&= (131/2)13 + (3{,}600/131)31 \\
&= \$852 + \$852 = \$1{,}704
\end{aligned}
$$

Comment. Holding and ordering costs, and annual demand, are typically estimated values rather than values that can be precisely determined, say, from accounting records. Holding costs are sometimes *designated* rather than computed by managers. Consequently, the EOQ should be regarded as an *approximate* quantity rather than an exact quantity. Thus, rounding the calculated value is perfectly acceptable; stating a value to several decimal places would tend to give an unrealistic impression of the precision involved. An obvious question is: How good is this "approximate" EOQ in terms of minimizing cost? The answer is that the EOQ is fairly robust; the total cost curve is relatively flat near the EOQ, especially to the right of the EOQ. In other words, even if

FIGURE 13–5

The total-cost curve is relatively flat near the EOQ

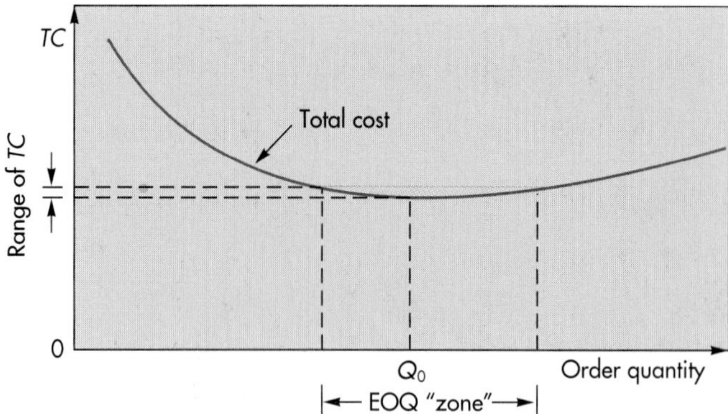

the resulting EOQ differs from the actual EOQ, total costs will not increase much at all. This is particularly true for quantities larger than the real EOQ, because the total cost curve rises very slowly to the right of the EOQ. (See Figure 13–5.)

EOQ with Incremental Replenishment

The basic EOQ model assumes that each order is delivered at a single point in time (instantaneous replenishment). In some instances, however, such as when a firm is both a producer and user, or when deliveries are spread over time, inventories are replenished over time instead of instantaneously.

If usage and production (or delivery) rates are equal, there will be no inventory buildup, because all output will be used immediately and the issue of lot size doesn't come up. In the more typical case, the production or delivery rate *exceeds* the usage rate, creating the situation depicted in Figure 13–6. In the production case, production occurs over only a portion of each cycle because the production rate is greater than the usage rate, and usage occurs over the entire cycle. During the production phase of the cycle, inventory builds up at a rate equal to the difference between production and usage rates. For example, if the daily production rate is 20 units and the daily usage rate is 5 units, inventory will build up at the rate of $20 - 5 = 15$ units per day. As long as production occurs, the inventory level will continue to build; when production ceases, the inventory level will begin to decrease. Hence, the inventory level will be maximum at the point where production ceases. When the amount of inventory on hand is exhausted, production is resumed, and the cycle repeats itself.

If a company makes the product itself, there are no ordering costs as such. Nonetheless, with every run there are setup costs—the costs required to prepare the equipment for the job, such as cleaning, adjusting, and changing tools and fixtures. Setup costs are analogous to ordering costs because they are independent of the lot (run) size. They are treated in the formula in exactly the same way. The larger the run size, the fewer the number of runs needed and, hence, the lower the annual setup cost. The number of runs is D/Q, and the annual setup cost is equal to the number of runs per year times the setup cost per run: $(D/Q)S$.

Total cost is

$$TC_{min} = \text{Carrying cost} + \text{Setup cost} \qquad (13\text{–}4)$$

$$= \left(\frac{I_{max}}{2}\right)H + (D/Q_0)S$$

where

I_{max} = Maximum inventory

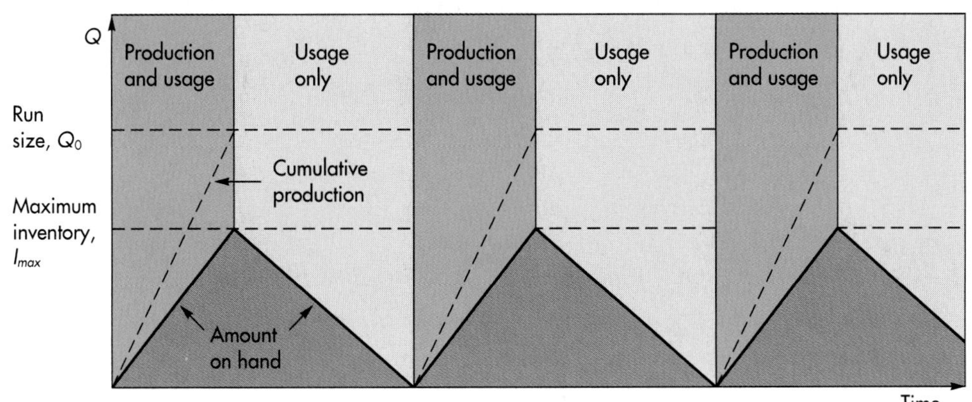

FIGURE 13-6

EOQ with incremental replenishment

The economic run quantity is

$$Q_0 = \sqrt{\frac{2DS}{H}}\sqrt{\frac{p}{p-u}} \tag{13–5}$$

where

p = Production or delivery rate
u = Usage rate

The cycle time (the time between orders or between the beginnings of runs) for the economic run size model is a function of the run size and usage (demand) rate:

$$\text{Cycle time} = \frac{Q_0}{u} \tag{13–6}$$

Similarly, the run time (the production phase of the cycle) is a function of the run size and the production rate:

$$\text{Run time} = \frac{Q_0}{p} \tag{13–7}$$

The maximum and average inventory levels are

$$I_{\text{max}} = \frac{Q_0}{p}(p-u) \quad \text{and} \quad I_{\text{average}} = \frac{I_{\text{max}}}{2} \tag{13–8}$$

Example 4

A toy manufacturer uses 48,000 rubber wheels per year for its popular dump truck series. The firm makes its own wheels, which it can produce at a rate of 800 per day. The toy trucks are assembled uniformly over the entire year. Carrying cost is $1 per wheel a year. Setup cost for a production run of wheels is $45. The firm operates 240 days per year. Determine the:

a. Optimal run size
b. Minimum total annual cost for carrying and setup
c. Cycle time for the optimal run size
d. Run time

Solution

D = 48,000 wheels per year
S = $45
H = $1 per wheel per year
p = 800 wheels per day
u = 48,000 wheels per 240 days, or 200 wheels per day

a. $Q_0 = \sqrt{\frac{2DS}{H}}\sqrt{\frac{p}{p-u}} = \sqrt{\frac{2(48{,}000)45}{1}}\sqrt{\frac{800}{800-200}}$

$= 2{,}400$ wheels

b. TC_{min} = Carrying cost + Setup cost = $\left(\frac{I_{max}}{2}\right)H + (D/Q_0)S$

Thus, you must first compute I_{max}:

$$I_{max} = \frac{Q_0}{p}(p-u) = \frac{2{,}400}{800}(800-200) = 1{,}800 \text{ wheels}$$

$$TC = \frac{1{,}800}{2} \times \$1 - \frac{48{,}000}{2{,}400} \times \$45 = \$900 + \$900 = \$1{,}800$$

Note again the equality of cost (in this example, setup and carrying costs) at the EOQ.

c. Cycle time $= \frac{Q_0}{u} = \frac{2{,}400 \text{ wheels}}{200 \text{ wheels per day}} = 12$ days

Thus, a run of wheels will be made every 12 days.

d. Run time $= \frac{Q_0}{p} = \frac{2{,}400 \text{ wheels}}{800 \text{ wheels per day}} = 3$ days.

Thus, each run will require three days to complete.

Quantity Discounts

quantity discounts Price reductions for large orders.

Quantity discounts are price reductions for large orders offered to customers to induce them to buy in large quantities. For example, a Chicago surgical supply company publishes the price list shown in Table 13–2 for boxes of gauze strips. Note that the price per box decreases as order quantity increases.

If quantity discounts are offered, the customer must weigh the potential benefits of reduced purchase price and fewer orders that will result from buying in large quantities against the increase in carrying costs caused by higher average inventories. The buyer's goal with quantity discounts is to select the order quantity that will minimize total cost, where total cost is the sum of carrying cost, ordering cost, *and* purchasing cost:

$$\text{TC} = \text{Carrying cost} + \text{Ordering cost} + \text{Purchasing cost}$$

$$= \left(\frac{Q}{2}\right)H + \left(\frac{D}{Q}\right)S + PD \qquad (13\text{–}9)$$

where

P = Unit price

Recall that in the basic EOQ model, determination of order size does not involve the purchasing cost. The rationale for not including unit price is that under the assumption of no quantity discounts, price per unit is the same for all order sizes. Inclusion of unit price in the total-cost computation in that case would merely increase the total cost by the amount P times D. A graph of total annual purchase cost versus quantity would be a horizontal line. Hence, including purchasing costs would merely raise the total-cost curve by the same amount (PD) at every point. That would not change the EOQ. (See Figure 13–7.)

When quantity discounts are offered, there is a separate U-shaped total-cost curve for each unit price. Again, including unit prices merely raises each curve by a constant amount. However, because the unit prices are all different, each curve is raised by a different amount: Smaller unit prices will raise a total-cost curve less than larger unit

Order Quantity	Price per Box
1 to 44	$2.00
45 to 69	1.70
70 or more	1.40

TABLE 13–2

Price list for extra-wide gauze strips

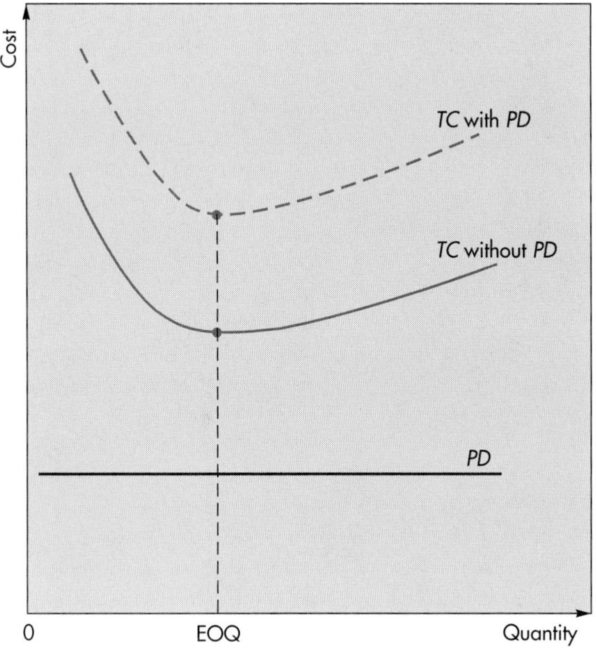

FIGURE 13–7

Adding PD doesn't change the EOQ

prices. Note that no one curve applies to the entire range of quantities; each curve applies to only a *portion* of the range. (See Figure 13–8.) Hence, the applicable or *feasible* total cost is initially on the curve with the highest unit price and then drops down, curve by curve, at the *price breaks*, which are the minimum quantities needed to obtain the discounts. Thus, in Table 13–2, the price breaks for gauze strips are at 45 and 70 boxes. The result is a total-cost curve with *steps* at the price breaks.

Even though each curve has a minimum, those points are not necessarily feasible. For example, the minimum point for the $1.40 curve in Figure 13–8 appears to be about 65 units. However, the price list shown in Table 13–2 indicates that an order size of 65 boxes will involve a unit price of $1.70. The actual total-cost curve is denoted by the solid lines; only those price-quantity combinations are feasible. The objective of the quantity discount model is to identify an order quantity that will represent the lowest total cost for the entire set of curves.

There are two general cases of the model. In one, carrying costs are constant (e.g., $2 per unit); in the other, carrying costs are stated as a percentage of purchase price (e.g., 20 percent of unit price). When carrying costs are constant, there will be a single EOQ, which is the same for all of the cost curves. Consequently, the total-cost curves line up vertically, differing only in that the lower unit prices are reflected by lower total-cost curves as shown in Figure 13–9A on page 577. For purposes of illustration, the horizontal purchasing cost lines have been omitted.

When carrying costs are specified as a percentage of unit price, each curve will have a different EOQ. Because carrying costs are a percentage of price, lower prices will mean lower carrying costs and larger EOQs. Thus, as price decreases, each curve's EOQ will be to the right of the next higher curve's EOQ. (See Figure 13–9B.)

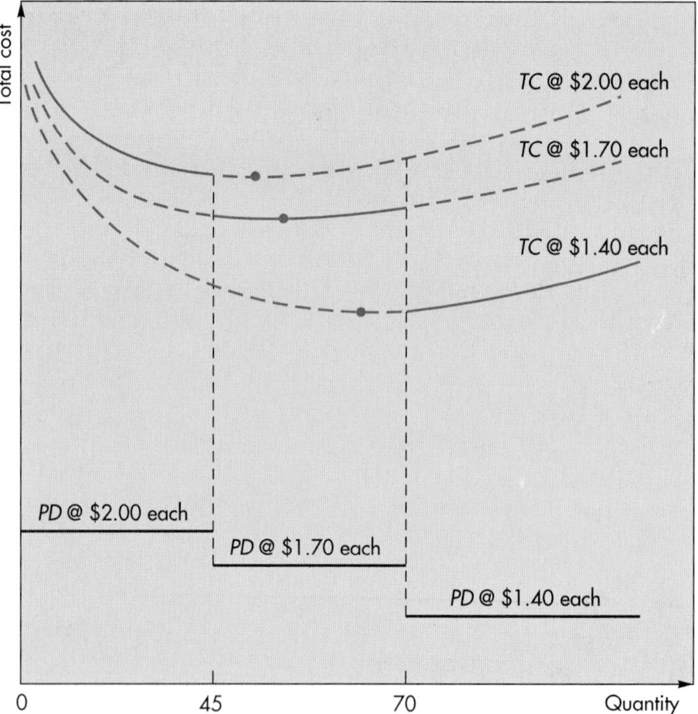

FIGURE 13-8

The total-cost curve with quantity discounts is composed of a portion of the total-cost curve for each price

The procedure for determining the overall EOQ differs slightly, depending on which of these two cases is relevant. For carrying costs that are constant, the procedure is as follows:

1. Compute the common EOQ.
2. Only one of the unit prices will have the EOQ in its feasible range since the ranges do not overlap. Identify that range.
   a. If the feasible EOQ is on the lowest price range, that is the optimal order quantity.
   b. If the feasible EOQ is in any other range, compute the total cost for the EOQ and for the price breaks of all *lower* unit costs. Compare the total costs; the quantity (EOQ or price break) that yields the lowest total cost is the optimal order quantity.

Example 5

The maintenance department of a large hospital uses about 816 cases of liquid cleanser annually. Ordering costs are $12, carrying costs are $4 per case a year, and the new price schedule indicates that orders of less than 50 cases will cost $20 per case, 50 to 79 cases will cost $18 per case, 80 to 99 cases will cost $17 per case, and larger orders will cost $16 per case. Determine the optimal order quantity and the total cost.

Solution

See Figure 13–10.

$D = 816$ cases per year $S = \$12$ $H = \$4$ per case per year

Range	Price
1 to 49	$20
50 to 79	18
80 to 99	17
100 or more	16

1. Compute the common EOQ: $= \sqrt{\dfrac{2DS}{H}} = \sqrt{\dfrac{2(816)12}{4}} = 70$ cases

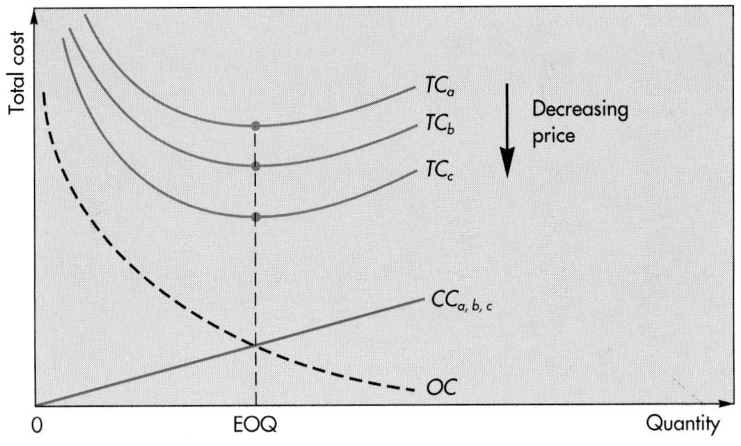

FIGURE 13-9

Comparison of TC curves for constant carrying costs and carrying costs that are a percentage of unit costs

A. When carrying costs are constant, all curves have the same EOQ.

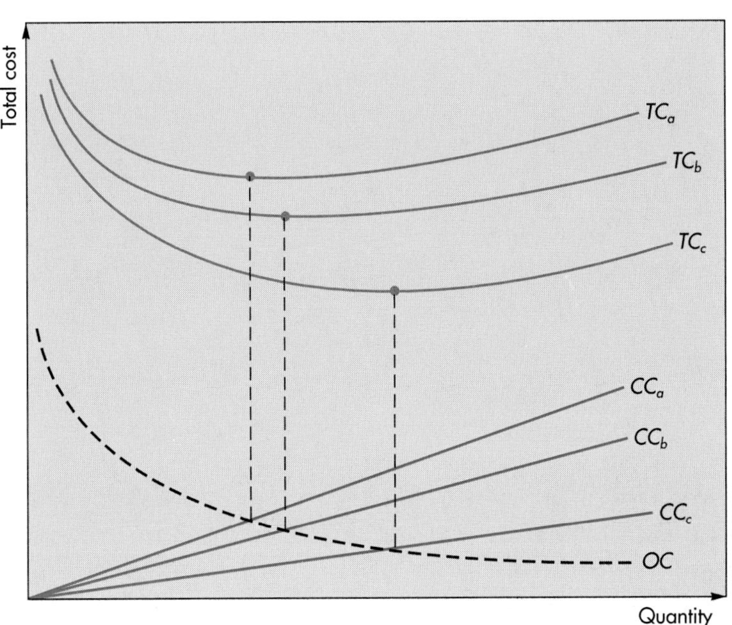

B. When carrying costs are given as a percentage of unit price, price decreases reduce carrying costs and that causes an increase in the EOQ.

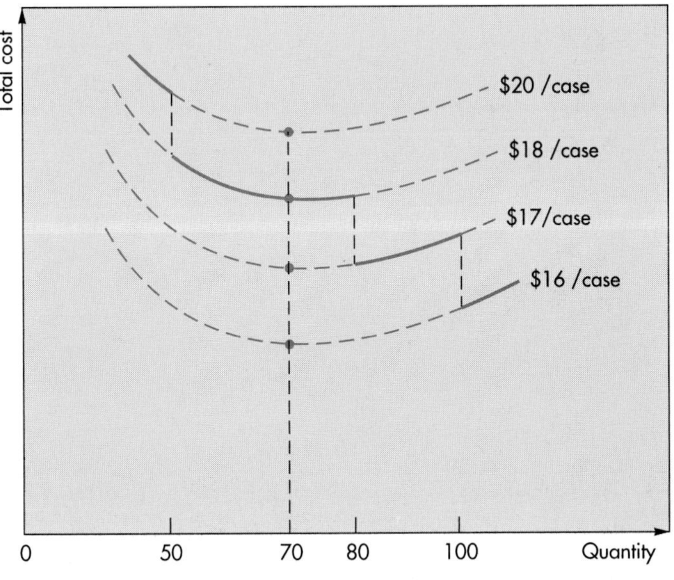

FIGURE 13-10

Total-cost curves for Example 5

2. The 70 cases can be bought at \$18 per case because 70 falls in the range of 50 to 79 cases. The total cost to purchase 816 cases a year, at the rate of 70 cases per order, will be

$$
\begin{aligned}
TC_{70} &= \text{Carrying cost} + \text{Order cost} + \text{Purchase cost} \\
&= (Q/2)H + (D/Q_0)S + PD \\
&= (70/2)4 + (816/70)12 + 18(816) = \$14{,}968
\end{aligned}
$$

Because lower cost ranges exist, each must be checked against the minimum cost generated by 70 cases at \$18 each. In order to buy at \$17 per case, at least 80 cases must be purchased. (Because the *TC* curve is rising, 80 cases will have the lowest *TC* for that curve's feasible region.) The total cost at 70 cases will be

$$TC_{80} = (80/2)4 + (816/80)12 + 17(816) = \$14{,}154$$

To obtain a cost of \$16 per case, at least 100 cases per order are required, and the total cost will be

$$TC_{100} = (100/2)4 + (816/100)12 + 16(816) = \$13{,}354$$

Therefore, because 100 cases per order yields the lowest total cost, 100 cases is the overall optimal order quantity.

When carrying costs are expressed as a percentage of price, determine the best purchase quantity with the following procedure:

1. Beginning with the lowest unit price, compute the EOQs for each price range until you find a feasible EOQ (i.e., until an EOQ falls in the quantity range for its price).
2. If the EOQ for the lowest unit price is feasible, it is the optimal order quantity. If the EOQ is not feasible in the lowest price range, compare the total cost at the price break for all *lower* prices with the total cost of the largest feasible EOQ. The quantity that yields the lowest total cost is the optimum.

Example 6

Surge Electric uses 4,000 toggle switches a year. Switches are priced as follows: 1 to 499, 90 cents each; 500 to 999, 85 cents each; and 1,000 or more, 80 cents each. It costs approximately \$30 to prepare an order and receive it, and carrying costs are 40 percent of purchase price per unit on an annual basis. Determine the optimal order quantity and the total annual cost.

Solution

See Figure 13–11.

$D = 4{,}000$ switches per year $\quad S = \$30 \quad H = 0.40P$

Range	Unit Price	H
1 to 499	\$0.90	.40(0.90) = .36
500 to 999	\$0.85	.40(0.85) = .34
1,000 or more	\$0.80	.40(0.80) = .32

Find the EOQ for each price, starting with the lowest price, until you locate a feasible EOQ.

$$\text{EOQ}_{0.80} = \sqrt{\frac{2DS}{H}} = \sqrt{\frac{2(4{,}000)30}{.32}} = 866 \text{ switches}$$

Because an order size of 988 switches will cost \$0.85 each rather than \$0.80 each, 866 is not a feasible EOQ for \$.80 per switch. Next, try \$0.85 per unit.

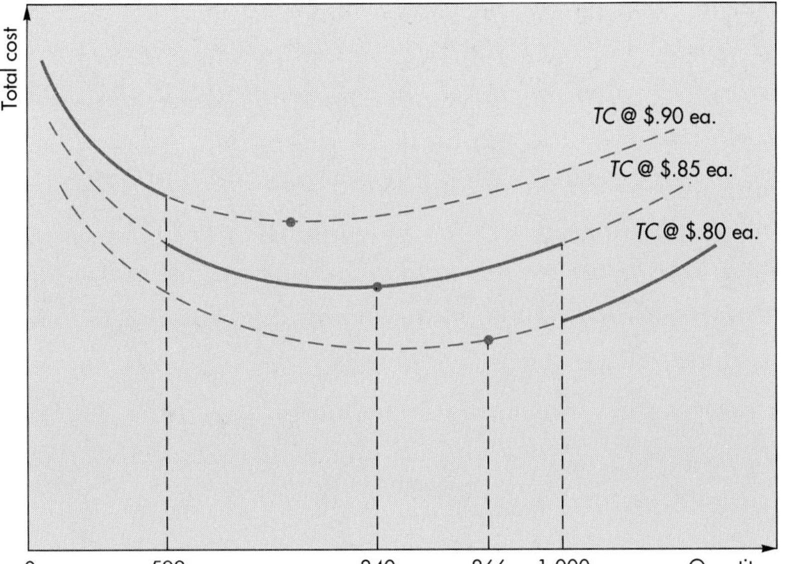

FIGURE 13-11

Total-cost curves for Example 6

$$EOQ_{0.85} = \sqrt{\frac{2(4{,}000)30}{.34}} = 840 \text{ switches}$$

This is feasible; it falls in the $0.85 per switch range of 500 to 999.

Now compute the total cost for 840, and compare it to the total cost of the minimum quantity necessary to obtain a price of $0.80 per switch.

$$TC = \text{Carrying costs} + \text{Ordering costs} + \text{Purchasing costs}$$

$$= \left(\frac{Q}{2}\right)H + \left(\frac{D}{Q}\right)S + PD$$

$$TC_{840} = \frac{840}{2}(.34) + \frac{4{,}000}{840}(30) + 0.85(4{,}000) = \$3{,}686$$

$$TC_{1{,}000} = \frac{1{,}000}{2}(.32) + \frac{4{,}000}{1{,}000}(30) + .80(4{,}000) = \$3{,}480$$

Thus, the minimum-cost order size is 1,000 switches.

WHEN TO REORDER WITH EOQ ORDERING

EOQ models answer the question of how much to order, but not the question of when to order. The latter is the function of models that identify the **reorder point (ROP)** in terms of a *quantity:* the reorder point occurs when the quantity on hand drops to a predetermined amount. That amount generally includes expected demand during lead time and perhaps an extra cushion of stock, which serves to reduce the probability of experiencing a stockout during lead time. Note that in order to know when the reorder point has been reached, a *perpetual* inventory is required.

reorder point When the quantity on hand of an item drops to this amount, the item is reordered.

The basic concern of the manager is to place an order when the amount of inventory on hand is sufficient to satisfy demand during the time it takes to receive that order (i.e., lead time). There are four determinants of the reorder point quantity:

1. The rate of demand (usually based on a forecast).
2. The lead time.

3. The extent of demand and/or lead time variability.
4. The degree of stockout risk acceptable to management.

If demand and lead time are both constant, the reorder point is simply

$$\text{ROP} = d \times \text{LT} \qquad (13\text{–}10)$$

where

d = Demand rate (units per day or week)
LT = Lead time in days or weeks

Note: Demand and lead time have the same time units.

Example 7

Tingly takes Two-a-Day vitamins, which are delivered to his home by a routeman seven days after an order is called in. At what point should Tingly reorder?

Solution

Usage = 2 vitamins a day
Lead time = 7 days
ROP = Usage × Lead time
= 2 vitamins per day × 7 days = 14 vitamins

Thus, Tingly should reorder when 14 vitamin tablets are left.

safety stock Stock that is held in excess of expected demand due to variable demand rate and/or lead time.

When variability is present in demand or lead time, it creates the possibility that actual demand will exceed expected demand. Consequently, it becomes necessary to carry additional inventory, called **safety stock,** to reduce the risk of running out of inventory (a stockout) during lead time. The reorder point then increases by the amount of the safety stock:

$$\text{ROP} = \begin{matrix}\text{Expected demand} \\ \text{during lead time}\end{matrix} + \text{Safety stock} \qquad (13\text{–}11)$$

For example, if expected demand during lead time is 100 units, and the desired amount of safety stock is 10 units, the ROP would be 110 units.

Figure 13–12 illustrates how safety stock can reduce the risk of a stockout during lead time (LT). Note that stockout protection is needed only during lead time. If there is a sudden surge at any point during the cycle, this will trigger another order. Once that order is received, the danger of an imminent stockout is negligible.

service level Probability that demand will not exceed supply during lead time.

Service Level. Because it costs money to hold safety stock, a manager must carefully weigh the cost of carrying safety stock against the reduction in stockout risk it provides. The customer *service level* increases as the risk of stockout decreases. Order cycle **service level** can be defined as the probability that demand will not exceed supply during lead time (i.e., that the amount of stock on hand will be sufficient to meet demand). Hence, a service level of 95 percent implies a probability of 95 percent that demand will not exceed supply during lead time. An equivalent statement that demand will be satisfied in 95 percent of such instances does *not* mean that 95 percent of demand will be satisfied. The risk of a stockout is the complement of service level; a customer service level of 95 percent implies a stockout risk of 5 percent. That is,

Service level = 100 percent − Stockout risk

Later you will see how the order cycle service level relates to the *annual* service level.

The amount of safety stock that is appropriate for a given situation depends on the following factors:

1. The average demand rate and average lead time.
2. Demand and lead time variability.
3. The desired service level.

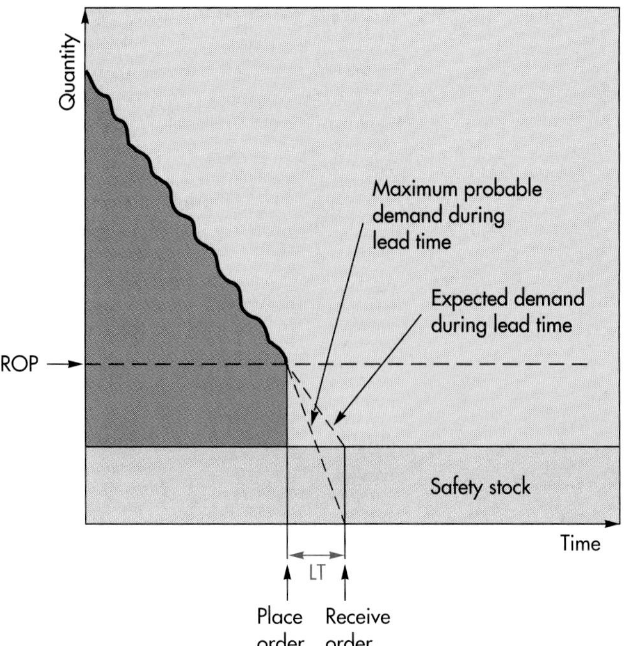

FIGURE 13-12

Safety stock reduces risk of stockout during lead time

For a given order cycle service level, the greater the variability in either demand rate or lead time, the greater the amount of safety stock that will be needed to achieve that service level. Similarly, for a given amount of variation in demand rate or lead time, achieving an increase in the service level will require increasing the amount of safety stock. Selection of a service level may reflect stockout costs (e.g., lost sales, customer dissatisfaction) or it might simply be a policy variable (e.g., the manager wants to achieve a specified service level for a certain item).

Let us look at several models that can be used in cases when variability is present. The first model can be used if an estimate of expected demand during lead time and its standard deviation are available. The formula is

$$\text{ROP} = \frac{\text{Expected demand}}{\text{during lead time}} + z\sigma_{\text{dLT}} \qquad (13\text{–}12)$$

where

z = Number of standard deviations
σ_{dLT} = The standard deviation of lead time demand

The models generally assume that any variability in demand rate or lead time can be adequately described by a normal distribution. However, this is not a strict requirement; the models provide approximate reorder points even where actual distributions depart from normal.

The value of z (see Figure 13–13) used in a particular instance depends on the stockout risk that the manager is willing to accept. Generally, the smaller the risk the manager is willing to accept, the greater the value of z. Use Appendix Table B to obtain the value of z, given a desired service level for lead time.

Example 8

Suppose that the manager of a construction supply house determined from historical records that demand for sand during lead time averages 50 tons. In addition, suppose the manager determined that demand during lead time could be described by a normal distribution that has a mean of 50 tons and a standard deviation of 5 tons. Answer these questions, assuming that the manager is willing to accept a stockout risk of no more than 3 percent:

FIGURE 13–13

The ROP based on a normal distribution of lead time demand

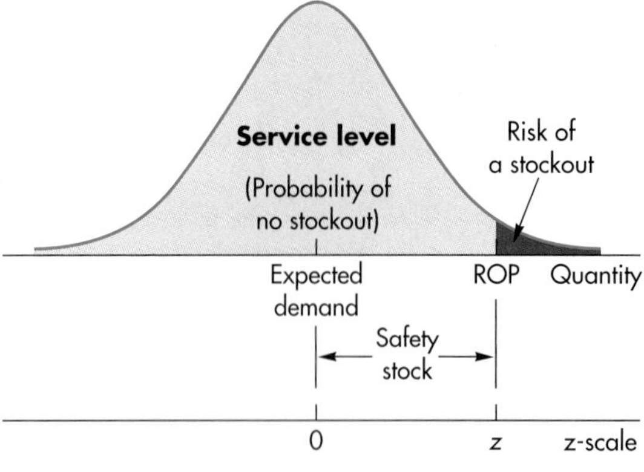

a. What value of z is appropriate?

b. How much safety stock should be held?

c. What reorder point should be used?

Solution

Expected lead time demand = 50 tons

$\sigma_{dLT} = 5$ tons

Risk = 3 percent

a. From Appendix Table B, using a service level of 1 − .03 = .9700, you obtain a value of $z = +1.88$.

b. Safety stock $= z\sigma_{dLT} = 1.88(5) = 9.40$ tons

c. $\text{ROP} = \frac{\text{expected lead}}{\text{time demand}} + \frac{\text{safety}}{\text{stock}} = 50 + 9.40 = 59.40$ tons

When data on lead time demand are not readily available, Formula 13–12 cannot be used. Nevertheless, data are generally available on daily or weekly demand, and on the length of lead time. Using that data, a manager can determine whether demand and/or lead time is variable, and if variability exists in one or both, the related standard deviation(s). For those situations, one of the following formulas can be used:

If only demand is variable, then $\sigma_{dLT} = \sqrt{LT}\sigma_d$, and the reorder point is

$$\text{ROP} = \bar{d} \times \text{LT} + z\sqrt{\text{LT}}\sigma_d \qquad (13\text{–}13)$$

where

$\bar{d}$ = *Average* daily or weekly demand
σ_d = Standard deviation of demand per day or week
LT = Lead time in days or weeks

If only lead time is variable, then $\sigma_{dLT} = d\sigma_{LT}$, and the reorder point is

$$\text{ROP} = d \times \overline{\text{LT}} + zd\sigma_{LT} \qquad (13\text{–}14)$$

where

d = Daily or weekly demand
$\overline{\text{LT}}$ = *Average* lead time in days or weeks
σ_{LT} = Standard deviation of lead time in days or weeks

If both demand and lead time are variable, then

$$\sigma_{dLT} = \sqrt{\overline{\text{LT}}\sigma_d^2 + \bar{d}^2\sigma_{LT}^2},$$

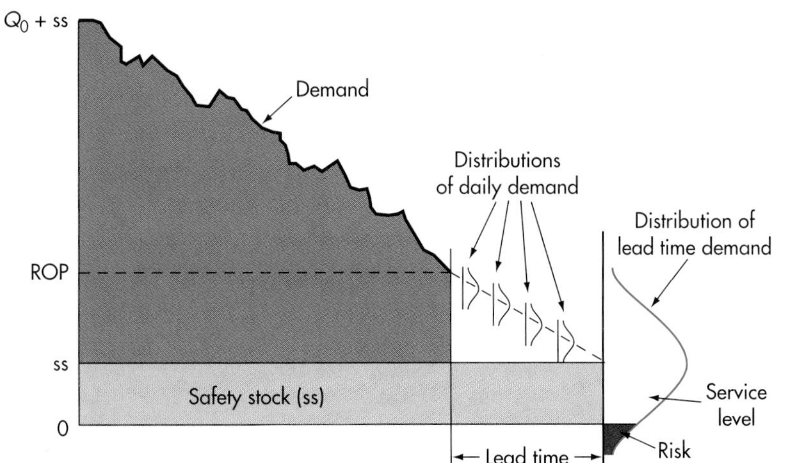

FIGURE 13–14

Lead time demand

and the reorder point is

$$\text{ROP} = \bar{d} \times \overline{\text{LT}} + z\sqrt{\overline{\text{LT}}\sigma_d^2 + \bar{d}^2\sigma_{\text{LT}}^2} \tag{13–15}$$

Note: Each of these models assumes that demand and lead time are *independent*.

Example 9

A restaurant uses an average of 50 jars of a special sauce each week. Weekly usage of sauce has a standard deviation of 3 jars. The manager is willing to accept no more than a 10 percent risk of stockout during lead time, which is two weeks. Assume the distribution of usage is normal.

a. Which of the above formulas is appropriate for this situation? Why?

b. Determine the value of z.

c. Determine the ROP.

Solution

$\bar{d}$ = 50 jars per week LT = 2 weeks

σ_d = 3 jars per week Acceptable risk = 10 percent, so service level is .90.

a. Because only demand is variable (i.e., has a standard deviation), Formula 13–13 is appropriate.

b. From Appendix Table B, using a service level of .9000, you obtain $z = +1.28$.

c. $\text{ROP} = \bar{d} \times \text{LT} + z\sqrt{\text{LT}}\sigma_d = 50 \times 2 + 1.28\sqrt{2}(3) = 100 + 5.43 = 105.43$

Comment. The logic of the three formulas for the reorder point may not be immediately obvious. The first part of each formula is the expected demand, which is the product of daily (or weekly) demand and the number of days (or weeks) of lead time. The second part of the formula is z times the standard deviation of lead time demand. For the formula in which only demand is variable, daily (or weekly) demand is assumed to be normally distributed and has the same mean and standard deviation (see Figure 13–14). The standard deviation of demand for the entire lead time is found by summing the *variances* of daily (or weekly) demands, and then finding the square root of that number because, unlike variances, standard deviations are not additive. Hence, if the daily standard deviation is σ_d, the *variance* is σ_d^2, and if lead time is four days, the variance of lead time demand will equal the sum of the 4 variances, which is $4\sigma_d^2$. The standard deviation of lead time demand will be the square root of this, which is equal to $2\sigma_d$. In general, this becomes $\sqrt{\text{LT}}\,\sigma_d$ and, hence, the last part of Formula 13–13.

When only lead time is variable, the explanation is much simpler. The standard deviation of lead time demand is equal to the constant daily demand multiplied by the standard deviation of lead time.

z	Lead Time Service Level	*E*(*z*)	*z*	Lead Time Service Level	*E*(*z*)	*z*	Lead Time Service Level	*E*(*z*)	*z*	Lead Time Service Level	*E*(*z*)
−2.40	.0082	2.403	−0.80	.2119	0.920	0.80	.7881	0.120	2.40	.9918	0.003
−2.36	.0091	2.363	−0.76	.2236	0.889	0.84	.7995	0.112	2.44	.9927	0.002
−2.32	.0102	2.323	−0.72	.2358	0.858	0.88	.8106	0.104	2.48	.9934	0.002
−2.28	.0113	2.284	−0.68	.2483	0.828	0.92	.8212	0.097	2.52	.9941	0.002
−2.24	.0125	2.244	−0.64	.2611	0.798	0.96	.8315	0.089	2.56	.9948	0.002
−2.20	.0139	2.205	−0.60	.2743	0.769	1.00	.8413	0.083	2.60	.9953	0.001
−2.16	.0154	2.165	−0.56	.2877	0.740	1.04	.8508	0.077	2.64	.9959	0.001
−2.12	.0170	2.126	−0.52	.3015	0.712	1.08	.8599	0.071	2.68	.9963	0.001
−2.08	.0188	2.087	−0.48	.3156	0.684	1.12	.8686	0.066	2.72	.9967	0.001
−2.04	.0207	2.048	−0.44	.3300	0.657	1.16	.8770	0.061	2.76	.9971	0.001
−2.00	.0228	2.008	−0.40	.3446	0.630	1.20	.8849	0.056	2.80	.9974	0.0008
−1.96	.0250	1.969	−0.36	.3594	0.597	1.24	.8925	0.052	2.84	.9977	0.0007
−1.92	.0274	1.930	−0.32	.3745	0.576	1.28	.8997	0.048	2.88	.9980	0.0006
−1.88	.0301	1.892	−0.28	.3897	0.555	1.32	.9066	0.044	2.92	.9982	0.0005
−1.84	.0329	1.853	−0.24	.4052	0.530	1.36	.9131	0.040	2.96	.9985	0.0004
−1.80	.0359	1.814	−0.20	.4207	0.507	1.40	.9192	0.037	3.00	.9987	0.0004
−1.76	.0392	1.776	−0.16	.4364	0.484	1.44	.9251	0.034	3.04	.9988	0.0003
−1.72	.0427	1.737	−0.12	.4522	0.462	1.48	.9306	0.031	3.08	.9990	0.0003
−1.68	.0465	1.699	−0.08	.4681	0.440	1.52	.9357	0.028	3.12	.9991	0.0002
−1.64	.0505	1.661	−0.04	.4840	0.419	1.56	.9406	0.026	3.16	.9992	0.0002
−1.60	.0548	1.623	0.00	.5000	0.399	1.60	.9452	0.023	3.20	.9993	0.0002
−1.56	.0594	1.586	0.04	.5160	0.379	1.64	.9495	0.021	3.24	.9994	0.0001
−1.52	.0643	1.548	0.08	.5319	0.360	1.68	.9535	0.019	3.28	.9995	0.0001
−1.48	.0694	1.511	0.12	.5478	0.342	1.72	.9573	0.017	3.32	.9995	0.0001
−1.44	.0749	1.474	0.16	.5636	0.324	1.76	.9608	0.016	3.36	.9996	0.0001
−1.40	0.808	1.437	0.20	.5793	0.307	1.80	.9641	0.014	3.40	.9997	0.0001
−1.36	.0869	1.400	0.24	.5948	0.290	1.84	.9671	0.013			
−1.32	.0934	1.364	0.28	.6103	0.275	1.88	.9699	0.012			
−1.28	.1003	1.328	0.32	.6255	0.256	1.92	.9726	0.010			
−1.24	.1075	1.292	0.36	.6406	0.237	1.96	.9750	0.009			
−1.20	.1151	1.256	0.40	.6554	0.230	2.00	.9772	0.008			
−1.16	.1230	1.221	0.44	.6700	0.217	2.04	.9793	0.008			
−1.12	.1314	1.186	0.48	.6844	0.204	2.08	.9812	0.007			
−1.08	.1401	1.151	0.52	.6985	0.192	2.12	.9830	0.006			
−1.04	.1492	1.117	0.56	.7123	0.180	2.16	.9846	0.005			
−1.00	.1587	1.083	0.60	.7257	0.169	2.20	.9861	0.005			
−0.96	.1685	1.049	0.64	.7389	0.158	2.24	.9875	0.004			
−0.92	.1788	1.017	0.68	.7517	0.148	2.28	.9887	0.004			
−0.88	.1894	0.984	0.72	.7642	0.138	2.32	.9898	0.003			
−0.84	.2005	0.952	0.76	.7764	0.129	2.36	.9909	0.003			

TABLE 13-3

Normal distribution service levels and unit normal loss function

When both demand and lead time are variable, the formula appears truly impressive. However, it is merely the result of squaring the standard deviations of the two previous formulas to obtain their variances, summing them, and then taking the square root.

Shortages and Service Levels

The ROP computation does not reveal the expected *amount* of shortage for a given lead time service level. The expected number of units short can, however, be very useful to a manager. This quantity can easily be determined from the same information used to compute the ROP, with one additional piece of information (see Table 13–3). Use of the table assumes that the distribution of lead time demand can be adequately represented by a normal distribution. If it can, the expected number of units short in each order cycle is given by this formula:

$$E(n) = E(z)\sigma_{dLT} \qquad (13\text{–}16)$$

where

$E(n)$ = Expected number of units short per order cycle
$E(z)$ = Standardized number of units short obtained from Table 13–3
σ_{dLT} = Standard deviation of lead time demand

Example 10

Suppose the standard deviation of lead time demand is known to be 20 units. Lead time demand is approximately normal.

a. For a lead time service level of 90 percent, determine the expected number of units short for any order cycle.

b. What lead time service level would an expected shortage of 2 units imply?

Solution

σ_{dLT} = 20 units

a. Lead time (cycle) service level = .90. From Table 13–3, $E(z) = 0.048$. Using Formula 13–16, $E(n) = 0.048(20 \text{ units}) = 0.96$, or about 1 unit.

b. For the case where $E(n) = 2$, you must solve for $E(z)$ and then use Table 13–3 to determine the lead time service that implies. Thus, $E(n) = E(z)\sigma_{dLT}$, so $E(z) = E(n)/\sigma_{dLT} = 2/20 = 0.100$. From Table 13–3, this implies a service level of approximately 81.5 percent (interpolating).

The expected number of units short is just that—an expected or *average* amount; the exact number of units short in any given cycle will be an amount close to that. Moreover, if discrete items are involved, the actual number of units short in any cycle will be an integer.

Having determined the expected number of units short for an order cycle, you can determine the expected number of units short per year. It is simply the expected number of units short per cycle multiplied by the number of cycles (orders) per year. Thus,

$$E(N) = E(n)\frac{D}{Q} \qquad (13\text{–}17)$$

where

$E(N)$ = Expected number of units short per year

Example 11

Given the following information, determine the expected number of units short per year.

$D = 1{,}000 \quad Q = 250 \quad E(n) = 2.5$

Solution

Using the formula $E(N) = E(n)\frac{D}{Q}$

$$E(N) = 2.5\left(\frac{1{,}000}{250}\right) = 10.0 \text{ units per year}$$

It is sometimes convenient to think of service level in annual terms. One definition of annual service level is the percentage of demand filled directly from inventory. This is also known as the *fill rate*. Thus, if D = 1,000, and 990 units were filled directly from inventory (shortages totaling 10 units over the year were recorded), the annual service level (fill rate) would be 990/1,000 = 99 percent. The annual service level and the lead time service level can be related using the following formula:

$$SL_{annual} = 1 - \frac{E(N)}{D} \tag{13–18}$$

Using Formulas 13–17 and 13–16,

$$E(N) = E(n)D/Q = E(z)\sigma_{dLT}D/Q$$

Thus,

$$SL_{annual} = 1 - \frac{E(z)\sigma_{dLT}}{Q} \tag{13–19}$$

Example 12

Given a lead time service level of 90, $D = 1{,}000$, $Q = 250$, and $\sigma_{dLT} = 16$, determine the annual service level, and the amount of cycle safety stock that would provide an annual service level of .98. From Table 13–3, $E(z) = 0.048$ for a 90 percent lead time service level.

Solution

a. Using Formula 13–19, you find:

$$SL_{annual} = 1 - \frac{0.048(16)}{250} = .997$$

b. Using Formula 13–19, and an annual service level of .98, solve for $E(z)$:

$$.98 = 1 - \frac{E(z)(16)}{250}$$

Solving, $E(z) = 0.312$. From Table 13–3, with $E(z) = 0.312$, you can see that this value of $E(z)$ is a little more than the value of 0.307. So it appears that an acceptable value of z might be 0.19. The necessary safety stock to achieve the specified annual service level is equal to $z\sigma_{dLT}$. Hence, the safety stock is $0.19(16) = 3.04$, or approximately 3 units.

Note that in the preceding example, a lead time service level of 90 percent provided an annual service level of 99.7 percent. Naturally, different values of D, Q, and σ_{dLT} will tend to produce different results for a cycle service level of 90 percent. Nonetheless, the annual service level will usually be greater than the cycle service level. In addition, since the annual service level as defined relates to the percentage of units short per year, it makes sense to base cycle service levels on a specified annual service level. This means setting the annual level, using Formula 13–19 to solve for $E(z)$, and then using that value to obtain the service level for the order cycles.

HOW MUCH TO ORDER: FIXED-ORDER-INTERVAL MODEL

fixed-order-interval model Orders are placed at fixed time intervals.

The fixed-order-interval (FOI) model is used when orders must be placed at fixed time intervals (weekly, twice a month, etc.). The question to be answered at each order point is: How much should be ordered for the next (fixed) interval? If demand is variable, the order size will tend to vary from cycle to cycle. This is quite different from an EOQ/ROP approach in which the order size generally remains fixed from cycle to cycle, while the length of the cycle varies (shorter if demand is higher than average, and longer if demand is less than average).

Finished goods inventory.

Parts inventory.

Reasons to Use the Fixed-Order-Interval Model

In some cases, a supplier's policy might encourage orders at fixed intervals. Even when that is not the case, grouping orders for items from the same supplier can produce savings in shipping costs. Furthermore, some situations do not readily lend themselves to continuous monitoring of inventory levels. Many retail operations (e.g., drugstores, small grocery stores) fall into this category. The alternative for them is to use fixed-interval ordering, which requires only periodic checks of inventory levels.

Determining the Amount to Order

If both the demand rate and lead time are constant, the fixed-interval model and the fixed-quantity model function identically. The differences in the two models become apparent only when examined under conditions of variability. Like the ROP model, the fixed-interval model can have variations in demand only, in lead time only, or in both demand and lead time. However, for the sake of simplicity and because it is perhaps the most frequently encountered situation, the discussion here will focus only on *variable demand* and *constant lead time*.

Figure 13–15 provides a comparison of the fixed-quantity and fixed-interval systems. In the fixed-quantity arrangement, orders are triggered by a *quantity* (ROP), while in the fixed-interval arrangement orders are triggered by a *time*. Therefore, the fixed-interval system must have stockout protection for lead time plus the next order cycle, but the fixed-quantity system needs protection only during lead time because additional orders can be placed at any time and will be received shortly (lead time) thereafter. Consequently, there is a greater need for safety stock in the fixed-interval model than in the fixed-quantity model. Note, for example, the large dip into safety stock during the second order cycle with the fixed-interval model.

Both models are sensitive to demand experience just prior to reordering, but in somewhat different ways. In the fixed-quantity model, a higher-than-normal demand causes a *shorter time* between orders, whereas in the fixed-interval model, the result is a *larger order size*. Another difference is that the fixed-quantity model requires close monitoring of inventory levels in order to know *when* the amount on hand has reached the

FIGURE 13-15

Comparison of fixed-quantity and fixed-interval ordering

Fixed quantity

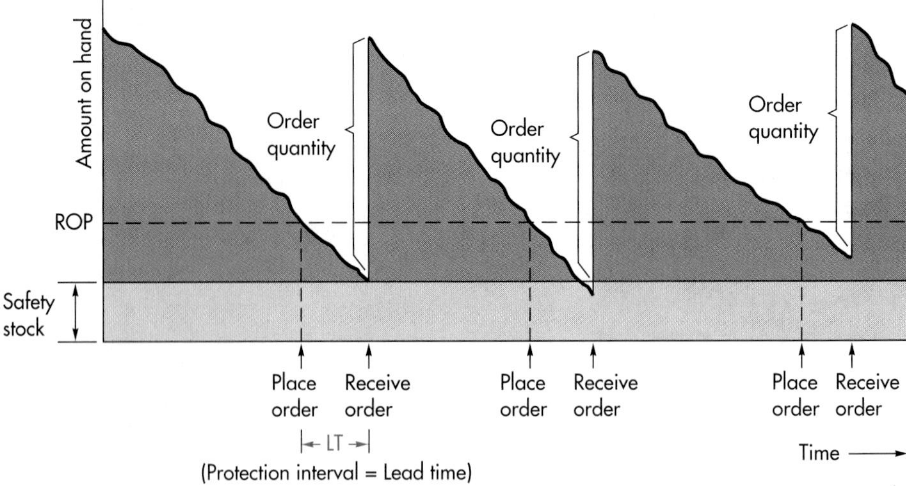

Fixed interval

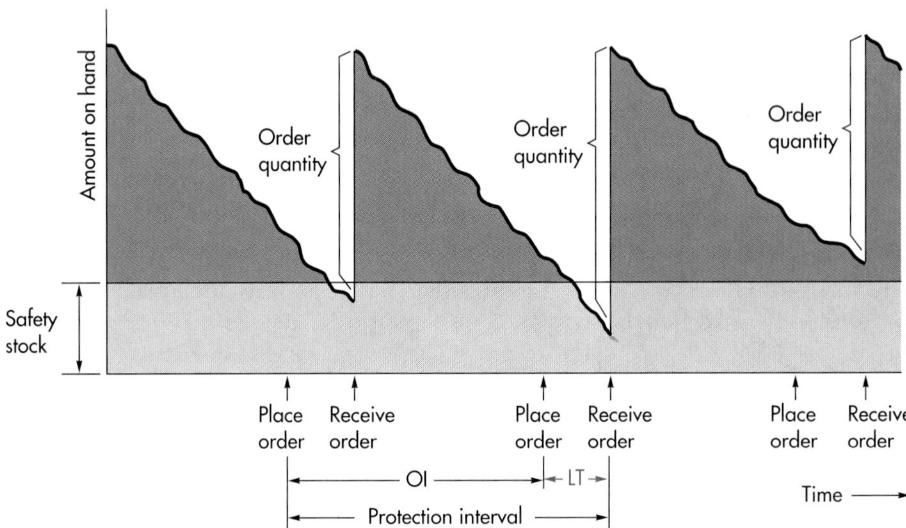

reorder point. The fixed-interval model requires only a periodic review (i.e., physical inspection) of inventory levels just prior to placing an order to determine how much is needed.

Order size in the fixed-interval model is determined by the following computation:

$$\begin{aligned}\text{Amount to order} &= \text{Expected demand during protection interval} + \text{Safety stock} - \text{Amount on hand at reorder time} \\ &= \bar{d}(\text{OI} + \text{LT}) + z\sigma_d\sqrt{\text{OI} + \text{LT}} - \text{A}\end{aligned} \qquad (13\text{–}20)$$

where

OI = Order interval (length of time between orders)
A = Amount on hand at reorder time

As in previous models, we assume that demand during the protection interval is normally distributed.

Example 13

Given the following information, determine the amount to order.

$\bar{d}$ = 30 units per day   Desired service level = 99 percent

σ_d = 3 units per day   Amount on hand at reorder time = 71 units

LT = 2 days   OI = 7 days

Solution

$z = 2.33$ for 99 percent service level

$$\text{Amount to order} = \bar{d}(\text{OI} + \text{LT}) + z\sigma_d\sqrt{\text{OI} + \text{LT}} - A$$

$$= 30(7 + 2) + 2.33(3)\sqrt{7 + 2} - 71 = 220 \text{ units}$$

Benefits and Disadvantages

The fixed-interval system results in the tight control needed for A items in an A-B-C classification due to the periodic reviews it requires. In addition, when two or more items come from the same supplier, grouping orders can yield savings in ordering, packing, and shipping costs. Moreover, it may be the only practical approach if inventory withdrawals cannot be closely monitored.

On the negative side, the fixed-interval system necessitates a larger amount of safety stock for a given risk of stockout because of the need to protect against shortages during an entire order interval plus lead time (instead of lead time only), and this increases the carrying cost. Also, there are the costs of the periodic reviews.

THE SINGLE-PERIOD MODEL

The **single-period model** is used to handle ordering of perishables (fresh fruits, vegetables, seafood, cut flowers) and items that have a limited useful life (newspapers, magazines, spare parts for specialized equipment. The *period* for spare parts is the life of the equipment, assuming that the parts cannot be used for other equipment.) What sets unsold or unused goods apart is that they are not typically carried over from one period to the next, at least not without penalty. Day-old baked goods, for instance, are often sold at reduced prices, leftover seafood may be discarded, and out-of-date magazines may be offered to used book stores at bargain rates. There may even be some cost associated with disposal of leftover goods.

single-period model Model for ordering of perishables and other items with limited useful lives.

Analysis of single-period situations generally focuses on two costs: shortage and excess. Shortage cost may include a charge for loss of customer goodwill as well as the opportunity cost of lost sales. Generally, **shortage cost** is simply unrealized profit per unit. That is,

shortage cost Generally, the unrealized profit per unit.

$$C_{\text{shortage}} = C_s = \text{Revenue per unit} - \text{Cost per unit}$$

If a shortage or stockout relates to an item used in production or to a spare part for a machine, then shortage cost refers to the actual cost of lost production.

Excess cost pertains to items left over at the end of the period. In effect, excess cost is the difference between purchase cost and salvage value. That is,

excess cost Difference between purchase cost and salvage value of items left over at the end of a period.

$$C_{\text{excess}} = C_e = \text{Original cost per unit} - \text{Salvage value per unit}$$

If there is cost associated with disposing of excess items, the salvage will be negative and will therefore *increase* the excess cost per unit.

The goal of the single-period model is to identify the order quantity, or stocking level, that will minimize the long-run excess and shortage costs.

There are two general categories of problems that we will consider: those for which demand can be approximated using a continuous distribution (perhaps a theoretical one

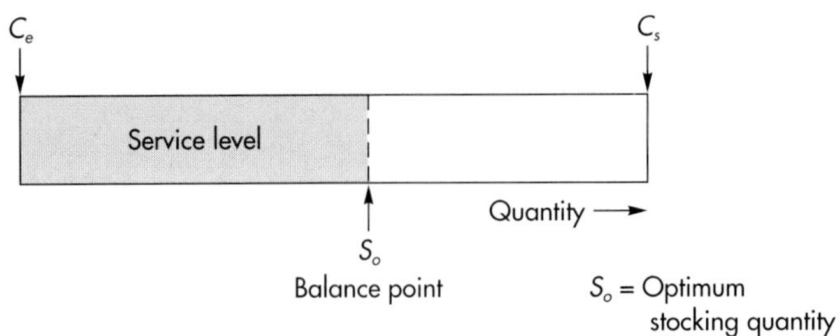

FIGURE 13–16

The optimal stocking level balances unit shortage and excess costs

such as a uniform or normal distribution) and those for which demand can be approximated using a discrete distribution (say, historical frequencies or a theoretical distribution such as the Poisson). The kind of inventory can indicate which type of model might be appropriate. For example, demand for petroleum, liquids, and gases tends to vary over some *continuous scale,* thus lending itself to description by a continuous distribution. Demand for tractors, cars, and computers is expressed in terms of the *number of units* demanded and lends itself to description by a discrete distribution.

Continuous Stocking Levels

The concept of identifying an optimal stocking level is perhaps easiest to visualize when demand is *uniform.* Choosing the stocking level is similar to balancing a seesaw, but instead of a person on each end of the seasaw, we have excess cost per unit (C_e) on one end of the distribution and shortage cost per unit (C_s) on the other. The optimal stocking level is analogous to the fulcrum of the seesaw; the stocking level equalizes the cost weights, as illustrated in Figure 13–16.

The *service level* is the *probability* that demand will not exceed the stocking level, and computation of the service level is the key to determining the optimal stocking level, S_o.

$$\text{Service level} = \frac{C_s}{C_s + C_e} \qquad (13\text{–}21)$$

where

C_s = Shortage cost per unit
C_e = Excess cost per unit

If actual demand exceeds S_o, there is a shortage; hence, C_s is on the right end of the distribution. Similarly, if demand is less than S_o, there is an excess, so C_e is on the left end of the distribution. When $C_e = C_s$, the optimal stocking level is halfway between the endpoints of the distribution. If one cost is greater than the other, S_o will be closer to the larger cost.

A similar approach applies when demand is normally distributed.

Example 14

Sweet cider is delivered weekly to Cindy's Cider Bar. Demand varies uniformly between 300 liters and 500 liters per week. Cindy pays 20 cents per liter for the cider and charges 80 cents per liter for it. Unsold cider has no salvage value and cannot be carried over into the next week due to spoilage. Find the optimal stocking level and its stockout risk for that quantity.

Solution

C_e = Cost per unit − Salvage value per unit
= \$0.20 − \$0
= \$0.20 per unit

C_s = Revenue per unit − Cost per unit
= \$0.80 − \$0.20

$$= \$0.60 \text{ per unit}$$

$$SL = \frac{C_s}{C_s + C_e} = \frac{\$0.60}{\$0.60 + \$0.20} = .75$$

Thus, the optimal stocking level must satisfy demand 75 percent of the time. For the uniform distribution, this will be at a point equal to the minimum demand plus 75 percent of the difference between maximum and minimum demands:

$$S_o = 300 + 0.75(500 - 300) = 450 \text{ liters}$$

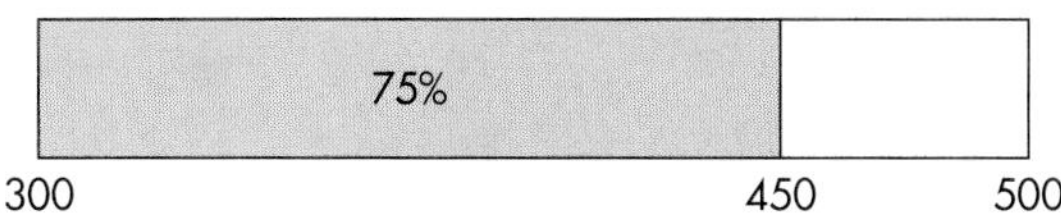

The stockout risk is $1.00 - 0.75 = 0.25$.

Example 15

Cindy's Cider Bar also sells a blend of cherry juice and apple cider. Demand for the blend is approximately normal, with a mean of 200 liters per week and a standard deviation of 10 liters per week. C_s = 60 cents per liter, and C_e = 20 cents per liter. Find the optimal stocking level for the apple-cherry blend.

Solution

$$SL = \frac{C_s}{C_s + C_e} = \frac{\$0.60}{\$0.60 + \$0.20} = .75.$$

This indicates that 75 percent of the area under the normal curve must be to the left of the stocking level. Appendix Table B shows that a value of z between +0.67 and +0.68, say, +0.675, will satisfy this. Thus,

$$S_o = 200 \text{ liters} + 0.675(10 \text{ liters}) = 206.75 \text{ liters}$$

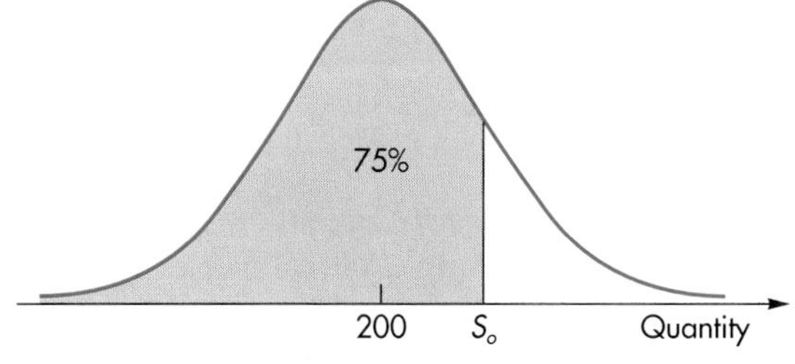

Discrete Stocking Levels

When stocking levels are discrete rather than continuous, the service level computed using the ratio $C_s/(C_s + C_e)$ usually does not coincide with a feasible stocking level (e.g., the optimal amount may be *between* five and six units). The solution is to stock at the *next higher level* (e.g., six units). In other words, choose the stocking level so that the desired service level is equaled or *exceeded*. Figure 13–17 illustrates this concept.

The next example illustrates the use of an empirical distribution, followed by an example that illustrates the use of a Poisson distribution.

Example 16

Historical records on the use of spare parts for several large hydraulic presses are to serve as an estimate of usage for spares of a newly installed press. Stockout costs involve downtime expenses and special ordering costs. These average $4,200 per unit short. Spares cost $800 each, and unused parts have zero salvage. Determine the optimal stocking level.

FIGURE 13-17

The service level achievement must equal or exceed the ratio $\frac{C_s}{C_s + C_e}$

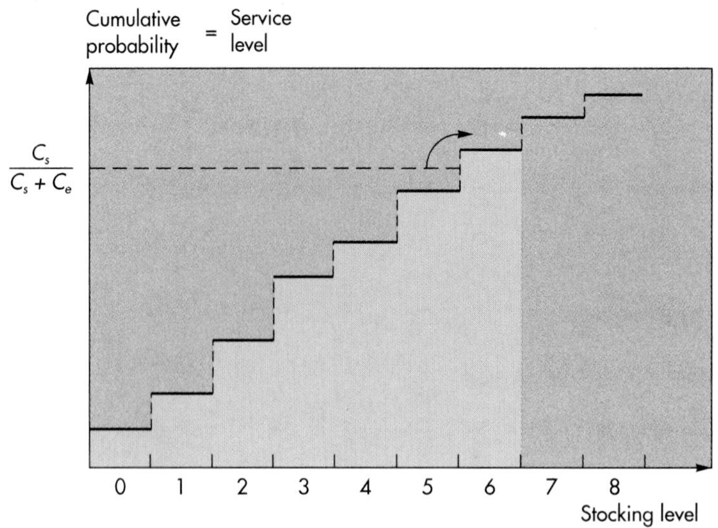

Number of Spares Used	Relative Frequency	Cumulative Frequency
0	.20	.20
1	.40	.60
2	.30	.90
3	.10	1.00
4 or more	.00	
	1.00	

$$C_s = \$4{,}200 \quad C_e = \$800$$

Solution

$$SL = \frac{C_s}{C_s + C_e} = \frac{\$4{,}200}{\$4{,}200 + \$800} = .84$$

The cumulative-frequency column indicates the percentage of time that demand did not exceed (was equal to or less than) some amount. For example, demand does not exceed one spare 60 percent of the time, or two spares 90 percent of the time. Thus, in order to achieve a service level of *at least* 84 percent, it will be necessary to stock two spares (i.e., to go to the next higher stocking level).

Example 17

Demand for long-stemmed red roses at a small flower shop can be approximated using a Poisson distribution that has a mean of four dozen per day. Profit on the roses is \$3 per dozen. Leftover flowers are marked down and sold the next day at a loss of \$2 per dozen. Assume that all marked-down flowers are sold. What is the optimal stocking level?

Solution

$$C_s = \$3 \quad C_e = \$2 \quad SL = \frac{C_s}{C_s + C_e} = \frac{\$3}{\$3 + \$2} = .60$$

Obtain the cumulative frequencies from the Poisson table (Appendix Table C) for a mean of 4.0:

Demand (dozen per day)	Cumulative Frequency
0	.018
1	.092
2	.238
3	.434
4	.629
5	.785
⋮	⋮

Compare the service level to the cumulative frequencies. In order to attain a service level of at least .60, it is necessary to stock four dozen.

One final point about discrete stocking levels: If the computed service level is *exactly* equal to the cumulative probability associated with one of the stocking levels, there are *two* equivalent stocking levels in terms of minimizing long-run cost—the one with equal probability and the next higher one. In the preceding example, if the ratio had been equal to .629, we would be indifferent between stocking four dozen and stocking five dozen roses each day.

Operations Strategy

Inventories are a necessary part of doing business, but having too much inventory is not good. One reason is that inventories tend to hide problems; they make it easier to live with problems rather than eliminate them. Another reason is that inventories are costly to maintain. Consequently, a wise operations strategy is to work toward cutting back inventories by (1) reducing lot sizes and (2) reducing safety stocks.

Japanese manufacturers use smaller lot sizes than their Western counterparts because they have a different perspective on inventory carrying costs. In addition to the usual components (e.g., storage, handling, obsolescence), the Japanese recognize the opportunity costs of disrupting the work flow, inability to place machines and workers closer together (which encourages cooperation, socialization, and communication), and hiding problems related to product quality and equipment breakdown. When these are factored in, carrying costs become higher—perhaps much higher—than before. The impact of an EOQ with revised (higher) costs is illustrated in Figure 13–18. Recall that carrying costs and ordering costs are equal at the EOQ. A higher carrying cost results in a steeper carrying-cost line, and the resulting intersection with the ordering-cost line occurs at a smaller quantity; hence, a smaller EOQ.

Although it can be argued that these costs are not readily quantifiable, Western manufacturers probably should reassess holding costs and revise them upward.

The second factor in the EOQ model that can contribute to smaller lot sizes is the setup or ordering processing cost. Numerous cases can be cited where these costs have been reduced through research efforts. However, while reduction related to carrying costs stems from a reassessment of those costs, a reduction related to ordering or setup costs must come from actively pursuing improvement. Together, these cost reductions can lead to even smaller lot sizes (see Figure 13–19).

Western companies can achieve additional reductions in inventory by reducing the amount of safety stock carried. Important factors in safety stock are lead time and lead time variability, reductions of which will result in lower safety stocks. Firms can often

FIGURE 13-18

Revised (increased) carrying costs result in a smaller EOQ

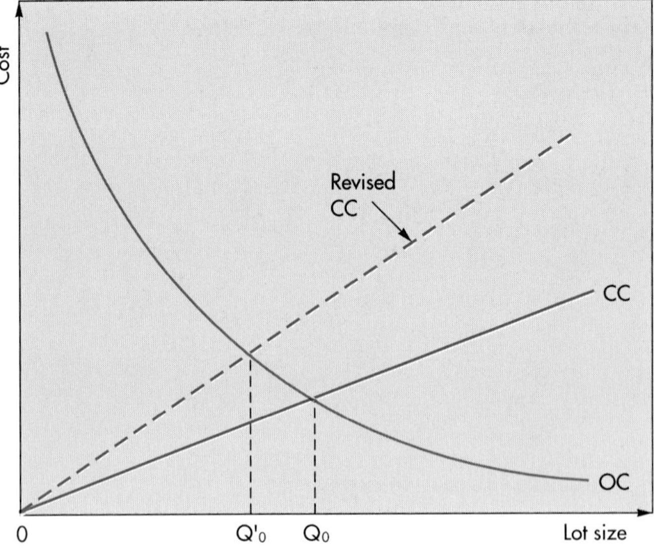

FIGURE 13-19

Reductions in both ordering/setup cost and carrying cost result in much smaller lot sizes

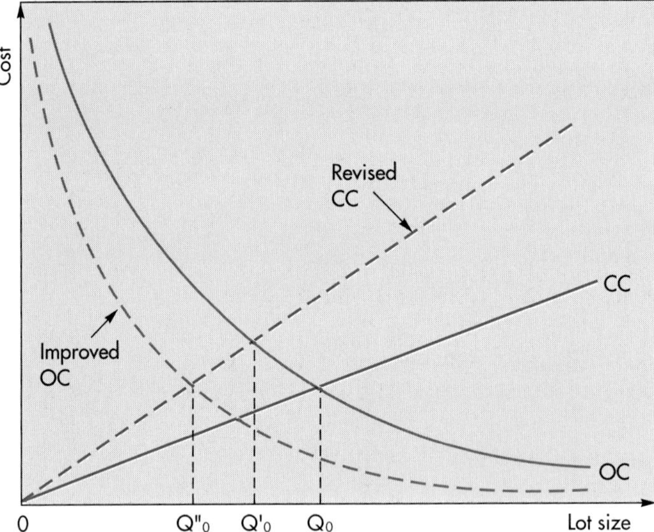

realize these reductions by working with suppliers, choosing suppliers located close to the buyer, and shifting to smaller lot sizes.

To achieve these reductions, an A-B-C approach is very beneficial. This means that all phases of operation should be examined, and those showing the greatest potential for improvement should be attacked first. Early results will demonstrate the benefits of this strategy to both management and workers, making both parties more willing to lend their support to additional efforts.

Last, it is important to make sure that inventory records be kept *accurate* and *up-to-date*. Estimates of holding costs, ordering costs, setup costs, and lead times should be reviewed periodically, and updated as necessary.

Summary

Good inventory management is often the mark of a well-run organization. Inventory levels must be planned carefully in order to balance the cost of holding inventory and the cost of providing reasonable levels of customer service. Successful inventory management requires a system to keep track of inventory transactions, accurate information about demand and lead times, realistic estimates of certain inventory-related costs, and a priority system for classifying the items in inventory and allocating control efforts.

The models described in this chapter are relevant for instances where demand for inventory items is independent. Four classes of models are described: EOQ, ROP, fixed-interval, and the single-period model. The first three are appropriate if unused items can be carried over into subsequent periods. The single-period model is appropriate when items cannot be carried over. EOQ models address the question of how much to order. The ROP models address the question of when to order and are particularly helpful in dealing with situations that include variations in either demand rate or lead time. ROP models involve service level and safety stock considerations. When the time between orders is fixed, the FOI model is useful. The single-period model is used for items which have a "shelf life" of one period. The models presented in this chapter are summarized in Table 13–4 on page 596.

Key Terms

A-B-C approach, 565
cycle counting, 566
dependent demand, 559
economic order quantity (EOQ), 567
excess cost, 589
fixed-order-interval (FOI) model, 586
holding carrying costs, 563
independent demand, 559
inventory, 558
lead time, 563
ordering costs, 564
periodic system, 561
perpetual inventory system, 562
quantity discounts, 574
reorder point (ROP), 579
safety stock, 580
service level, 580
shortage costs, 564, 589
single-period model, 589
two-bin system, 562
universal product code, 562

TABLE 13–4

Summary of inventory formulas

Model	Formula		Symbols
1. Basic EOQ	$Q_0 = \sqrt{\frac{2DS}{H}}$	(13–2)	Q_0 = Economic order quantity
	$TC = \frac{Q}{2}H + \frac{D}{Q}S$	(13–1)	D = Annual demand
	Length of order cycle $= \frac{Q_0}{D}$	(13–3)	S = Order cost; H = Annual carrying cost per unit
2. Economic run size	$Q_0 = \sqrt{\frac{2DS}{H}}\sqrt{\frac{p}{p-u}}$	(13–5)	Q_0 = Optimal run or order size
	$TC = \frac{I_{max}}{2}H + \frac{D}{Q}S$	(13–4)	p = Production or delivery rate
	Cycle time $= \frac{Q}{u}$	(13–6)	u = Usage rate
	Run time $= \frac{Q}{p}$	(13–7)	I_{max} = Maximum inventory level
	$I_{max} = \frac{Q_0}{p}(p-u)$	(13–8)	
3. Quantity discounts	$TC = \frac{Q}{2}H + \frac{D}{Q}S + PD$	(13–9)	P = Unit price
4. Reorder point under:			ROP = Quantity on hand at reorder point
a. Constant demand and lead time	ROP = $d(\text{LT})$		d = Demand rate; LT = Lead time
b. Variable demand rate	ROP = $\bar{d}\text{LT} + z\sqrt{\text{LT}}\,(\sigma_d)$	(13–11)	$\bar{d}$ = Average demand rate; σ_d = Standard deviation of demand rate
c. Variable lead time	ROP = $d\overline{\text{LT}} + zd(\sigma_{\text{LT}})$	(13–12)	z = Standard normal deviation
d. Variable lead time and demand	ROP = $\bar{d}\overline{\text{LT}} + z\sqrt{\overline{\text{LT}}\sigma_d^2 + \bar{d}^2\sigma_{\text{LT}}^2}$	(13–15)	$\overline{\text{LT}}$ = Average lead time; σ_{LT} = Standard deviation of lead time
5. ROP shortages			$E(n)$ = Expected number short per cycle
a. Units short per cycle	$E(n) = E(z)\sigma_{d\text{LT}}$	(13–16)	$E(z)$ = Standardized number short
b. Units short per year	$E(N) = E(n)\frac{D}{Q}$	(13–17)	$\sigma_{d\text{LT}}$ = Standard deviation of lead time demand
c. Annual service level	$\text{SL}_{annual} = 1 - \frac{E(z)\sigma_{d\text{LT}}}{Q}$	(13–19)	$E(N)$ = Expected number short per year; SL_{annual} = Annual service level
6. Fixed interval	$Q = \bar{d}(\text{OI} + \text{LT}) + z\sigma_d\sqrt{\text{OI} + \text{LT}} - A$	(13–20)	OI = Time between orders; A = Amount on hand at order time
7. Single period	$\text{SL} = \frac{C_s}{C_s + C_e}$		SL = Service level; C_s = Shortage cost per unit; C_e = Excess cost per unit

Solved Problems

Problem 1

Basic EOQ. A toy manufacturer uses approximately 32,000 silicon chips annually. The chips are used at a steady rate during the 240 days a year that the plant operates. Annual holding cost is 60 cents per chip, and ordering cost is \$24. Determine:

a. The optimal order size.

b. The number of workdays in an order cycle.

Solution

$D = 32{,}000$ chips per year $\quad S = \$24$

$H = \$.60$ per unit per year

a. $Q_0 = \sqrt{\frac{2DS}{H}} = \sqrt{\frac{2(32{,}000)\$24}{\$.60}} = 1{,}600$ chips.

b. $\frac{Q_0}{D} = \frac{1{,}600 \text{ chips}}{32{,}000 \text{ chips/yr.}} = \frac{1}{20}$ year (i.e., $\frac{1}{20} \times 240$ days), or 12 days.

Problem 2

Incremental replenishment. The Dine Corporation is both a producer and a user of brass couplings. The firms operates 220 days a year and uses the couplings at a steady rate of 50 per day. Couplings can be produced at a rate of 200 per day. Annual storage cost is \$1 per coupling, and machine setup cost is \$35 per run.

a. Determine the economic run size.

b. Approximately how many runs per year will there be?

c. Compute the maximum inventory level.

d. Determine the length of the *pure consumption* portion of the cycle.

Solution

$D = 50$ units per day $\times$ 220 days per year $= 11{,}000$ units per year

$S = \$35$

$H = \$1$ per unit per year

$p = 200$ units per day

$u = 50$ units per day

a. $Q_0 = \sqrt{\frac{2DS}{H}}\sqrt{\frac{p}{p-u}} = \sqrt{\frac{2(11{,}000)35}{1}}\sqrt{\frac{200}{200-50}} = 1{,}013$ units.

b. Number of runs per year: $D/Q_0 = 11{,}000/1{,}013 = 10.86$, or 11.

c. $I_{max} = \frac{Q_0}{p}(p-u) = \frac{1{,}013}{200}(200-50) = 759.75$ or 760 units.

d. Length of cycle $= \frac{Q_0}{u} = \frac{1{,}013 \text{ units}}{50 \text{ units per day}} = 20.26$ days

Length of run $= \frac{Q_0}{p} = \frac{1{,}013 \text{ units}}{200 \text{ units per day}} = 5.065$ days

Length of pure consumption portion = Length of cycle − Length of run

$= 20.26 - 5.065 = 15.20$ days.

Problem 3

Quantity discounts. A small manufacturing firm uses roughly 3,400 pounds of chemical dye a year. Currently the firm purchases 300 pounds per order and pays \$3 per pound. The supplier has just announced that orders of 1,000 pounds or more will be filled at a price of \$2 per pound. The manufacturing firm incurs a cost of \$100 each time it submits an order and assigns an annual holding cost of 17 percent of the purchase price per pound.

a. Determine the order size that will minimize the total cost.

b. If the supplier offered the discount at 1,500 pounds instead of 1,000 pounds, what order size would minimize total cost?

Solution

$D = 3{,}400$ pounds per year $\quad S = \$100 \quad H = 0.17P$

a. Compute the EOQ for $2 per pound:
The quantity ranges are

Range	Unit Price
1 to 999	$3
1,000+	$2

$$Q_{\$2/\text{pound}} = \sqrt{\frac{2DS}{H}} = \sqrt{\frac{2(3{,}400)100}{0.17(2)}} = 1{,}414 \text{ pounds.}$$

Because this quantity is feasible at $2 per pound, it is the optimum.

b. When the discount is offered at 1,500 pounds, the EOQ for the $2 per pound range is no longer feasible. Consequently, it becomes necessary to compute the EOQ for $3 per pound and compare the total cost for that order size with the total cost using the price break quantity (i.e., 1,500).

$$Q_{\$3/\text{pound}} = \sqrt{\frac{2DS}{H}} = \sqrt{\frac{2(3{,}400)100}{0.17(3)}} = 1{,}155 \text{ pounds}$$

$$TC = \left(\frac{Q}{2}\right)H + \left(\frac{D}{Q}\right)S + PD$$

$$TC_{1,155} = \left(\frac{1{,}155}{2}\right)0.17(3) + \left(\frac{3{,}400}{1{,}155}\right)100 + 3(3{,}400)$$

$$= \$294.53 + \$294.37 + \$10{,}200 = \$10{,}789$$

$$TC_{1,500} = \left(\frac{1{,}500}{2}\right)0.17(2) + \left(\frac{3{,}400}{1{,}500}\right)100 + 2(3{,}400)$$

$$= \$255 + \$226.67 + \$6{,}800 = \$7{,}282$$

Hence, because it would result in a lower total cost, 1,500 is the optimal order size.

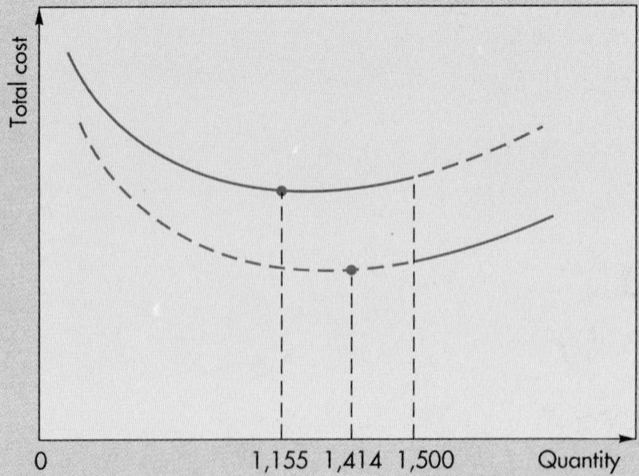

Problem 4

ROP for variable demand and constant lead time. The housekeeping department of a motel uses approximately 400 washcloths per day. The actual amount tends to vary with the number of guests on any given night. Usage can be approximated by a normal distribution that has a mean of 400 and a standard deviation of 9 washcloths per day. A linen supply company delivers towels and washcloths with a lead time of three days. If the motel policy is to maintain a stockout risk of 2 percent, what is the minimum number of washcloths that must be on hand at reorder time, and how much of that amount can be considered safety stock?

Solution

$\bar{d}$ = 400 washcloths per day   LT = 3 days

σ_d = 9 washcloths per day   Risk = 2 percent, so service level = 98 percent

From Appendix Table B, the z value that corresponds to an area under the normal curve to the left of z for 98 percent is about +2.055.

$$\text{ROP} = \bar{d}\text{LT} + z\sqrt{\text{LT}}\sigma_d = 400(3) + 2.055\sqrt{3}(9)$$

$$= 1{,}200 + 32.03, \text{ or approximately } 1{,}232 \text{ washcloths}$$

Safety stock is approximately 32 washcloths.

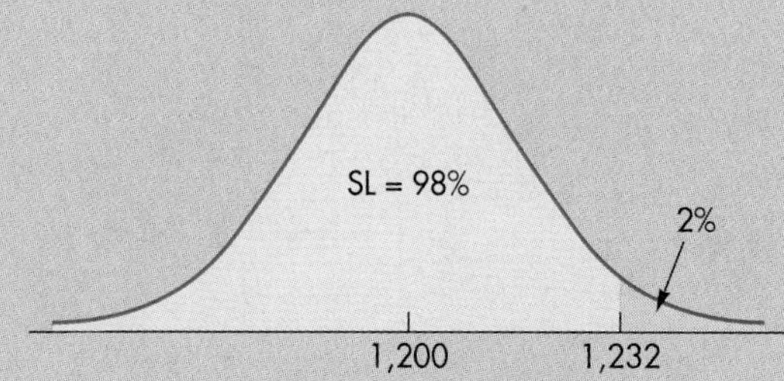

Problem 5

ROP for constant demand and variable lead time. The motel in the preceding example uses approximately 600 bars of soap each day, and this tends not to vary by more than a few bars either way. Lead time for soap delivery is normally distributed with a mean of six days and a standard deviation of two days. A service level of 90 percent is desired. Find the ROP.

Solution

d = 600 bars per day

SL = 90 percent, so z = +1.28 (from Appendix Table B)

$\overline{\text{LT}}$ = 6 days

σ_{LT} = 2 days

$$\text{ROP} = d\overline{\text{LT}} + zd(\sigma_{\text{LT}}) = 600(6) + 1.28(600)2$$

$$= 5{,}136 \text{ bars of soap}$$

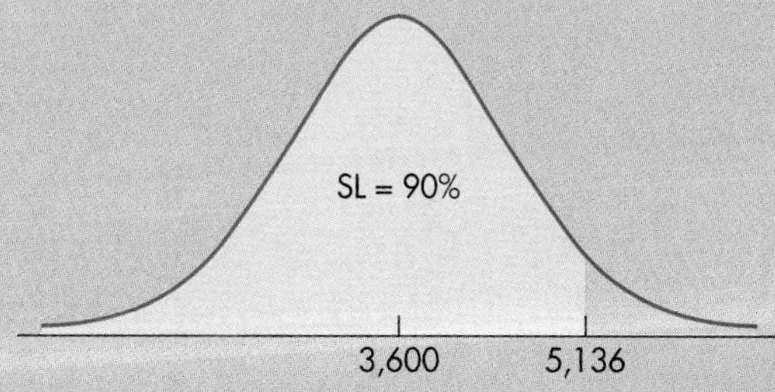

Problem 6

ROP for variable demand rate and variable lead time. The motel replaces broken glasses at a rate of 25 per day. In the past, this quantity has tended to vary normally and have a standard deviation of 3 glasses per day. Glasses are ordered from a Cleveland supplier. Lead time is normally distributed with an average of 10 days and a standard deviation of 2 days. What ROP should be used to achieve a service level of 95 percent?

Solution

$\bar{d} = 25$ glasses per day $\quad \overline{\text{LT}} = 10$ days

$\sigma_d = 3$ glasses per day $\quad \sigma_{\text{LT}} = 2$ days

SL = 95 percent, so $z = +1.65$ (Appendix Table B)

$$\text{ROP} = \bar{d}\,\overline{\text{LT}} + z\sqrt{\overline{\text{LT}}\sigma_d^2 + \bar{d}^2\sigma_{\text{LT}}^2}$$

$$= 25(10) + 1.65\sqrt{10(3)^2 + (25)^2(2)^2} = 334 \text{ glasses}$$

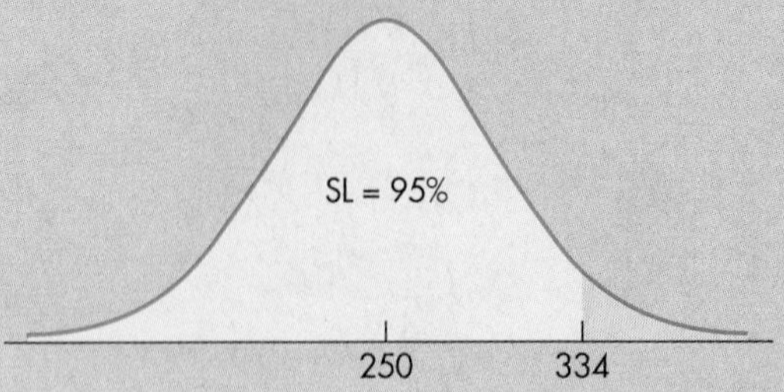

Problem 7

Shortages and service levels. The manager of a store that sells office supplies has decided to set an annual service level of 96 percent for a certain model of telephone answering equipment. The store sells approximately 300 of this model a year. Holding cost is \$5 per unit annually, ordering cost is \$25, and $\sigma_{d\text{LT}} = 7$.

a. What average number of units short per year will be consistent with the specified annual service level?

b. What average number of units short per cycle will provide the desired annual service level?

c. What lead time service level is necessary for the 96 percent annual service level?

Solution

$\text{SL}_{\text{annual}} = 96$ percent $\quad D = 300$ units $\quad H = \$5 \quad S = \$25 \quad \sigma_{d\text{LT}} = 7$

a. $E(N) = (1 - \text{SL}_{\text{annual}})\,D = (1 - .96)(300) = 12$ units.

b. $E(N) = E(n)\dfrac{D}{Q}$. Solving for $E(n)$, you have

$$E(n) = E(N) \div \left(\frac{D}{Q}\right) = 12 \div \left(\frac{300}{Q}\right)$$

$$Q = \sqrt{\frac{2DS}{H}} = \sqrt{\frac{2(300)(35)}{5}} = 54.77 \text{ (round to 55)}$$

$$\text{Then } E(n) = 12 \div \left(\frac{300}{55}\right) = 2.2.$$

c. In order to find the lead time service level, you need the value of $E(z)$. Because the value of $E(n)$ is 2.2 and $E(n) = E(z)\sigma_{d\text{LT}}$, you have $2.2 = E(z)(7)$. Solving gives $E(z) = 2.2 \div 7 = 0.314$. Interpolating in Table 13–3 gives the approximate lead time service level. Thus,

$$\frac{0.307 - 0.314}{0.307 - 0.324} = \frac{.5793 - x}{.5793 - .5636}$$

Solving,

$$x = .5728.$$

[To interpolate, find the two values between which the computed number falls in the $E(z)$ column. Then find the difference between the computed value and one end of the range, and divide by the difference between the two ends of the range. Perform the corresponding calculation on the two service levels using x for the unknown value, and solve for x. Often, simply "eyeballing" the unknown value will suffice.]

Problem 8

Fixed-order-interval model. A lab orders a number of chemicals from the same supplier every 30 days. Lead time is five days. The assistant manager of the lab must determine how much of one of these chemicals to order. A check of stock revealed that eleven 25-ml jars are on hand. Daily usage of the chemical is approximately normal with a mean of 15.2 ml per day and a standard deviation of 1.6 ml per day. The desired service level for this chemical is 95 percent.

a. How many jars of the chemical should be ordered?

b. What is the average amount of safety stock of the chemical?

Solution

$\bar{d} = 15.2$ ml per day, OI = 30 days, SL = 95% requires $z = 1.65$

$\sigma_d = 1.6$ ml per day, LT = 5 days, A = 11 jars × 25 ml per jar = 275 ml

a. $$\text{Amount to order} = \bar{d}(\text{OI} + \text{LT}) + z\sigma_d\sqrt{\text{OI} + \text{LT}} - \text{A}$$

$$= 15.2(30 + 5 + 1.65(1.6)\sqrt{30 + 5} - 275 = 272.62 \text{ ml}$$

Convert this to number of jars:

$$\frac{272.62 \text{ ml}}{25 \text{ ml per jar}} = 10.90 \text{ or } 11 \text{ jars.}$$

b. Safety stock $= z\sigma_d\sqrt{\text{OI} + \text{LT}} = 1.65(1.6)\sqrt{30 + 5} = 15.62$ ml.

Problem 9

A firm that installs cable TV systems uses a certain piece of equipment for which it carries two spare parts. The parts cost $500 each and have no salvage value. Part failures can be modeled by a Poisson distribution with a mean of two failures during the useful life of the equipment. Holding and disposal costs are negligible. Estimate the apparent range of shortage cost.

Solution

C_s is unknown $C_e = \$500$

The Poisson table (Appendix Table C) provides these values for a mean of 2.0:

Number of Failures	Cumulative Probability
0	.135
1	.406
2	.677
3	.857
4	.947
5	.983
⋮	⋮

For the optimal stocking level, the service level must usually be rounded up to a feasible stocking level. Hence, you know that the service level must have been between .406 and .677 in order to make two units the optimal level. By setting the service level equal first to .406 and then to .677, you can establish bounds on the possible range of shortage costs.

$$\frac{C_s}{C_s + \$500} = .406, \text{ so } C_s = .406(\$500 + C_s)$$

Solving, you find $C_s = \$341.75$.

Similarly,

$$\frac{C_s}{C_s + \$500} = .677, \text{ so } C_s = .677(\$500 + C_s)$$

Solving, you find $C_s = \$1{,}047.99$. Hence, the apparent range of shortage cost is \$341.75 to \$1,047.99.

Discussion and Review Questions

1. What are the primary reasons for holding inventory?
2. What are the requirements for effective inventory management?
3. Briefly describe each of the costs associated with inventory.
4. Contrast independent and dependent demand with respect to inventories.
5. Why might it be inappropriate to use inventory turnover ratios to compare inventory performance of companies that are in different industries?
6. List the major assumptions of the EOQ model.
7. How would you respond to the criticism that EOQ models tend to provide misleading results because values of *D*, *S*, and *H* are, at best, educated guesses?
8. What are quantity discounts? What three costs enter into the order quantity decision when discounts are available?
9. What is safety stock, and what is its purpose?
10. Under what circumstances would the amount of safety stock held be

    a. Large? *b.* Small? *c.* Zero?
11. What is meant by the term *service level?* Generally speaking, how is service level related to the amount of safety stock held?
12. Describe briefly the A-B-C approach to inventory control.
13. The purchasing agent for a company that assembles and sells air-conditioning equipment in a Latin American country has noted that the cost of compressors has increased significantly each time they have reordered. The company uses an EOQ model to determine order size. What are the implications of this price escalation with respect to order size? What factors other than price must be taken into consideration?
14. Explain how a decrease in setup time can lead to a decrease in the average amount of inventory a firm holds, and why that would be beneficial.
15. What is the single-period model, and under what circumstances is it appropriate?
16. Can the optimal stocking level in the single-period model ever be less than expected demand? Explain briefly.
17. What are some ways that a company can reduce the need for inventories?

Memo Writing Exercises

1. A fellow manager has just returned from a conference with a revolutionary idea: Eliminate all inventory. Write a memo to this manager in which you outline the reasons why this is not a desirable idea.
2. In an effort to become more efficient and build better supplier relations, your manager, Jane Furman, has examined the company's current practice of using eight different suppliers for eight different items and considered switching to a single supplier who would supply all eight items. However, she discovered that the new supplier will handle orders only on a fixed-interval basis. Write a one-page memo to Furman outlining the cost factors that might increase, and those that might decrease, along with a brief explanation for each factor, under the proposed system.

Problems

1. The manager of an automobile repair shop hopes to achieve a better allocation of inventory control efforts by adopting an A-B-C approach to inventory control. Given the monthly usages in the following table, classify the items in A, B, and C categories according to dollar usage.

Item	Usage	Unit Cost
4021	50	$1,400
9402	300	12
4066	40	700
6500	150	20
9280	10	1,020
4050	80	140
6850	2,000	15
3010	400	20
4400	7,000	5

Item	Category
4021	A
9402	C
4066	B
6500	C
9280	C
4050	C
6850	B
3010	C
4400	B

2. The following table contains figures on the monthly volume and unit costs for a random sample of 16 items from a list of 2,000 inventory items at a health care facility.

Item	Unit Cost	Usage
K34	10	200
K35	25	600
K36	36	150
M10	16	25
M20	20	80
Z45	80	200
F14	20	300
F95	30	800
F99	20	60
D45	10	550
D48	12	90
D52	15	110
D57	40	120
N08	30	40
P05	16	500
P09	10	30

a. Develop an A-B-C classification for these items.

b. How could the manager use this information?

c. After reviewing your classification scheme, suppose that the manager decides to place item P05 into the A category. What are some possible explanations for this decision?

a.

Item	Category
K34	C
K35	A
K36	B
M10	C
M20	C
Z45	A
F14	B
F95	A
F99	C
D45	B
D48	C
D52	C
D57	B
N08	C
P05	B
P09	C

b. Degree of control.
c. Important for a reason other than annual dollar volume.

3. A large bakery buys flour in 25-pound bags. The bakery uses an average of 4,860 bags a year. Preparing an order and receiving a shipment of flour involves a cost of $4 per order. Annual carrying costs are $30 per bag.

a. Determine the economic order quantity.

b. What is the average number of bags on hand?

c. How many orders per year will there be?

d. Compute the total cost of ordering and carrying flour.

e. If annual ordering cost were to increase by $1 per order, how much would that affect the minimum total annual cost?

a. 36 bags
b. 18 bags
c. 135
d. $1,080
e. Increase by $127.48

4. A large law firm uses an average of 40 packages of copier paper a day. The firm operates 260 days a year. Storage and handling costs for the paper are $3 a year per pack, and it costs approximately $6 to order and receive a shipment of paper.

a. 204 packages
b. $611.74
c. Yes
d. No. TC = $612; only save $.26

a. What order size would minimize total annual ordering and carrying costs?

b. Compute the total annual cost using your order size from part *a*.

c. Except for rounding, are annual ordering and carrying costs always equal at the EOQ?

d. The office manager is currently using an order size of 200 packages. The partners of the firm expect the office to be managed "in a cost-efficient manner." Would you recommend that the office manager use the optimal order size instead of 200 packages? Justify your answer.

a. 1,039 pots; $519.62
b. Increase $44.55

5. Garden Variety Flower Shop uses 750 clay pots a month. The pots are purchased at $2 each. Annual carrying costs are estimated to be 25 percent of cost, and ordering costs are $30 per order.

   a. Determine the economic order quantity and the total annual carrying and ordering costs.

   b. Suppose an analysis shows the actual carrying costs are roughly double the current estimate. If the order size wasn't changed, how much extra cost would the firm incur?

Approximately $364.29

6. A produce distributor uses 800 packing crates a month, which it purchases at a cost of $10 each. The manager has assigned an annual carrying cost of 35 percent of the purchase price per crate. Ordering costs are $28. Currently the manager orders once a month. How much could the firm save annually in ordering and carrying costs by using the EOQ?

a. 1–6: 75 units
7–12: 91 units
b. EOQ model requires
c. 1–6: 50 units
7–12: 100 units

7. A manager receives a forecast for next year. Demand is projected to be 600 units for the first half of the year and 900 units for the second half. The monthly holding cost is $2 per unit, and it costs an estimated $55 to process an order.

   a. Assuming that monthly demand will be level during each of the six-month periods covered by the forecast (e.g., 100 per month for each of the first six months), determine an order size that will minimize the sum of ordering and carrying costs that can be used for each of the six-month periods.

   b. Why is it important to be able to assume that demand will be level during each six-month period?

   c. If the vendor is willing to offer the manager a discount of $10 *per order* for ordering in multiples of 50 units (e.g., 50, 100, 150), would you advise the manager to take advantage of the offer in either period? If so, what order size would you recommend?

a. $1.32
b. $24.30

8. A food processor uses approximately 27,000 glass jars a month for its fruit juice product. Because of storage limitations, a lot size of 4,000 jars has been used. Monthly holding cost is 18 cents per jar, and reordering cost is $60 per order. The company operates an average of 20 days a month.

   a. What penalty is the company incurring by its present order size?

   b. The manager would prefer ordering 10 times each month but would have to justify any change in order size. One possibility is to simplify order processing to reduce the ordering cost. What ordering cost would enable the manager to justify ordering every other day?

   c. Suppose that after investigating ordering cost, the manager is able to reduce it to $50. How else could the manager justify using an order size that would be consistent with ordering every other day?

a. 4,812
b. 15.59 (approximately 16)
c. .96

9. The Friendly Sausage Factory (FSF) can produce hot dogs at a rate of 5,000 per day. FSF supplies hot dogs to local stores and restaurants at a steady rate of 250 per day. The cost to prepare the equipment for producing hot dogs is $22. Annual holding costs are 15 cents per hot dog. The factory operates 300 days a year. Find

   a. The optimal run size.

   b. The number of runs per year.

   c. The length (in days) of a run.

10. A chemical firm produces sodium bisulfate in 100-pound bags. Demand for this product is 20 tons per day. The capacity for producing the product is 50 tons per day. Setup costs $100,

and storage and handling costs are $5 per ton a year. The firm operates 200 days a year. (Note: 1 ton = 2,000 pounds.)

a. How many bags per run are optimal?

b. What would the average inventory be for this lot size?

c. Determine the approximate length of a production run, in days.

d. About how many runs per year would there be?

e. How much could the company save annually if the setup cost could be reduced to $25 per run?

a. 10,328 bags
b. 3,098 bags
c. 10.33 days
d. 7.75 (approximately 8)
e. $774.50

11. A company is about to begin production of a new product. The manager of the department that will produce one of the components for the product wants to know how often the machine used to produce the item will be available for other work. The machine will produce the item at a rate of 200 units a day. Eighty units will be used daily in assembling the final product. Assembly will take place five days a week, 50 weeks a year. The manager estimates that it will take almost a full day to get the machine ready for a production run, at a cost of $60. Inventory holding costs will be $2 a year.

    a. What run quantity should be used to minimize total annual costs?

    b. What is the length of a production run in days?

    c. During production, at what rate will inventory build up?

    d. If the manager wants to run another job between runs of this item, and needs a minimum of 10 days per cycle for the other work, will there be enough time?

a. Q_0 = 1,414 units
b. 7.07 days
c. 120 units per day
d. No. Run + setup leaves 9.6 days

12. A company produces plastic powder in lots of 2,000 pounds, at the rate of 250 pounds per hour. The company uses the powder in an injection molding process at the steady rate of 50 pounds per hour for an eight-hour day, five days a week. The manager has indicated that the setup cost is $100 for this product, but "We really haven't determined what the holding cost is."

    a. What weekly holding cost per pound does the lot size imply, assuming the lot size is optimal?

    b. Suppose the figure you compute for holding cost has been shown to the manager, and the manager says that it is not that high. Would that mean the lot size is too large or too small? Explain.

a. $.125
b. too small

13. A mail-order house uses 18,000 boxes a year. Carrying costs are 20 cents per box a year, and ordering costs are $32. The following price schedule applies. Determine:

    a. The optimal order quantity.

    b. The number of orders per year.

a. 10,000 boxes
b. 1.8

Number of Boxes	Price per Box
1,000 to 1,999	$1.25
2,000 to 4,999	1.20
5,000 to 9,999	1.18
10,000 or more	1.15

14. A jewelry firm buys semiprecious stones to make bracelets and rings. The supplier quotes a price of $8 per stone for quantities of 600 stones or more, $9 per stone for orders of 400 to 599 stones, and $10 per stone for lesser quantities. The jewelry firm operates 200 days per year. Usage rate is 25 stones per day, and ordering costs are $48.

    a. If carrying costs are $2 per year for each stone, find the order quantity that will minimize total annual cost.

    b. If annual carrying costs are 30 percent of unit cost, what is the optimal order size?

    c. If lead time is six working days, at what point should the company reorder?

a. 600 stones
b. 600 stones
c. 150 stones

tie:
495 pulleys,
or 1,000 pulleys

15. A manufacturer of exercise equipment purchases the pulley section of the equipment from a supplier who lists these prices: less than 1,000, $5 each; 1,000 to 3,999, $4.95 each; 4,000 to 5,999, $4.90 each; and 6,000 or more, $4.85 each. Ordering costs are $50, annual carrying costs are 40 percent of purchase cost, and annual usage is 4,900 pulleys. Determine an order quantity that will minimize total cost.

a. A
b. 400 units

16. A company will begin stocking remote control devices. Expected monthly demand is 800 units. The controllers can be purchased from either supplier A or supplier B. Their price lists are as follows:

SUPPLIER A		SUPPLIER B	
Quantity	**Unit Price**	**Quantity**	**Unit Price**
1–199	$4.00	1–149	$4.10
200–399	3.80	150–349	3.90
400+	3.60	350+	3.70

Ordering cost is $40 and annual holding cost is $6.00 per unit.

a. Which supplier should be used?

b. What order quantity is optimal if the intent is to minimize total annual costs?

Buy 300 from Vendor X.
TC = $6,685

17. The owner of a diner has summarized the price lists from four potential vendors (see the following table) who want to supply her with cooking oil. Monthly usage is 300 gallons, ordering cost is $10 per order, and the monthly carrying cost is $.50 a gallon.

Which vendor should she use, and what order quantity is best if she wants to minimize total monthly costs?

VENDOR W		VENDOR X		VENDOR Y		VENDOR Z	
Quantity	**Price**	**Quantity**	**Price**	**Quantity**	**Price**	**Quantity**	**Price**
1–99	$25	1–79	$25	1–25	$27	1–59	$26
100–399	24	80–139	24	26–89	25	60–139	25
400+	22	140–299	23	90–199	24	140–249	24
		300+	22	200+	23	250+	23

6,600 ft.

18. A newspaper publisher uses roughly 800 feet of baling wire each day to secure bundles of newspapers while they are being distributed to carriers. The paper is published Monday through Saturday. Lead time is six workdays. What is the appropriate reorder point quantity, given that the company desires a service level of 95 percent, if that stockout risk for various levels of safety stock are as follows: 1,500 feet, 0.10; 1,800 feet, 0.05; 2,100 feet, 0.02; and 2,400 feet, 0.01?

a. $z = 2.33$
ROP = 370 units
b. 70 units
c. Less safety stock;
ROP would be smaller

19. Given this information:

Expected demand during lead time = 300 units
Standard deviation of lead time demand = 30 units

Determine each of the following, assuming that lead time demand is distributed normally:

a. The ROP that will provide a risk of stockout of 1 percent during lead time.

b. The safety stock needed to attain a 1 percent risk of stockout during lead time.

c. Would a stockout risk of 2 percent require more or less safety stock than a 1 percent risk? Explain. Would the ROP be larger, smaller, or unaffected if the acceptable risk was 2 percent instead of 1 percent? Explain.

a. Safety stock = 91
b. ROP = 691

20. Given this information:

Lead-time demand = 600 pounds
Standard deviation of lead-time demand = 52 pounds
Acceptable stockout risk during lead time = 4 percent

a. What amount of safety stock is appropriate?

b. When should this item be reordered?

21. Demand for walnut fudge ice cream at the Sweet Cream Dairy can be approximated by a normal distribution with a mean of 21 gallons per week and a standard deviation of 3.5 gallons per week. The new manager desires a service level of 90 percent. Lead time is two days, and the dairy is open seven days a week. (Hint: Work in terms of weeks.)

    a. If an ROP model is used, what ROP would be consistent with the desired service level?

    b. If a fixed-interval model is used instead of an ROP model, what order size would be needed for the 90 percent service level with an order interval of 10 days and a supply of 2 gallons on hand at the order time?

    c. Suppose the manager is using the ROP model described in part *a.* One day after placing an order with the supplier, the manager receives a call from the supplier that the order will be delayed because of problems at the supplier's plant. The supplier promises to have the order there in two days. After hanging up, the manager checks the supply of walnut fudge ice cream and finds that 2 gallons have been sold since the order was placed. Assuming the supplier's promise is valid, what is the probability that the dairy will run out of this flavor before the shipment arrives?

    a. 8.39 gallons
    b. 40 gallons
    c. $z = .208$, Risk = .4168

22. The injection molding department of a company uses an average of 30 gallons of special lubricant a day. The supply of the lubricant is replenished when the amount on hand is 170 gallons. It takes four days for an order to be delivered. Safety stock is 50 gallons, which provides a stockout risk of 9 percent. What amount of safety stock would be needed if the acceptable risk of a stockout is 3 percent?

    70.14 gal.

23. A company uses 85 circuit boards a day in a manufacturing process. The person who orders the boards follows this rule: Order when the amount on hand drops to 625 boards. Orders are delivered approximately six days after being placed. The delivery time is normal with a mean of six days and a standard deviation of 1.10 days. What is the probability that the supply of circuit boards will be exhausted before the order is received if boards are reordered when the amount on hand drops to 625 boards?

    Risk = .1093

24. One item a shoe store sells is supplied by a vendor who handles only that item. Demand for that item recently changed, and the store manager must determine when to replenish it. The manager wants a probability of at least 96 percent of not having a stockout during lead time. The manager expects demand to average a dozen units a day and have a standard deviation of 2 units a day. Lead time is variable, averaging four days with a standard deviation of one day. Assume normality and that seasonality is not a factor.

    a. When should the manager reorder to achieve the desired probability?

    b. Why might the model not be appropriate if seasonality was present?

    a. ROP = 70.14
    b. Average demand would not be the same in different periods.

25. The manager of a car wash receives a revised price list from the vendor who supplies soap, and a promise of a shorter lead time for deliveries. Formerly the lead time was four days, but now the vendor promises a reduction of 25 percent in that time. Annual usage of soap is 4,500 gallons. The car wash is open 360 days a year. Assume that daily usage is normal, and that it has a standard deviation of 2 gallons per day. The ordering cost is $30 and annual carrying cost is $3 a gallon. The revised price list (cost per gallon) is shown in the table.

Quantity	Unit Price
1–399	$2.00
400–799	1.70
800+	1.62

    a. What order quantity is optimal?

    b. What ROP is appropriate if the acceptable risk of a stockout is 1.5 percent?

    a. 400 gal.
    b. 45.02 gal.

26. A small copy center uses five 500-sheet boxes of copy paper a week. Experience suggests that usage can be well approximated by a normal distribution with a mean of five boxes per

week and a standard deviation of one-half box per week. Two weeks are required to fill an order for letterhead stationery. Ordering cost is $2, and annual holding cost is 20 cents per box.

a. 72 boxes
b. .0023
c. .0228

a. Determine the economic order quantity for minimizing ordering and carrying costs, assuming a 52-week year.

b. If the copy center reorders when the supply on hand is 12 boxes, compute the risk of a stockout.

c. If a fixed interval of seven weeks instead of an ROP is used for reordering, what risk of stockout does the copy center incur if it orders 36 boxes when the amount on hand is 12 boxes?

a. 749 units
b. 4.073 units

27. Ned's Natural Foods sells unshelled peanuts by the pound. Historically, Ned has observed that daily demand is normally distributed with a mean of 80 pounds and a standard deviation of 10 pounds. Lead time also appears normally distributed with a mean of eight days and a standard deviation of one day.

    a. What ROP would provide stockout risk of 10 percent during lead time?

    b. What is the expected number of units (pounds) short per cycle?

a. 134
b. 36.1
c. .0554/cycle, 1.5 rolls/year
d. .9996

28. Regional Supermarket is open 360 days per year. Daily use of cash register tape averages 10 rolls. Usage appears normally distributed with a standard deviation of 2 rolls per day. The cost of ordering tape is $1, and carrying costs are 40 cents per roll a year. Lead time is three days.

    a. What is the EOQ?

    b. What ROP will provide a lead time service level of 96 percent?

    c. What is the expected number of units short per cycle with 96 percent? Per year?

    d. What is the annual service level?

a. $Q_0 = 179$
$E(z) = .0358$
Safety stock = .40 cases
b. Risk = .4681

29. A service station uses 1,200 cases of oil a year. Ordering cost is $40, and annual carrying cost is $3 per case. The station owner has specified an *annual* service level of 99 percent.

    a. What level of safety stock is appropriate if lead time demand is normally distributed with a mean of 80 cases and a standard deviation of 5 cases?

    b. What is the risk of a stockout during lead time?

a. −0.40 gal. Order less than expected demand.
b. −1.98 gal.

30. Weekly demand for diesel fuel at a department of parks depot is 250 gallons. The depot operates 52 weeks a year. Weekly usage is normal, and has a standard deviation of 14 gallons. Holding cost for the fuel is $1 a month, and it costs $20 in administrative time to submit an order for more fuel. It takes one-half week to receive a delivery of diesel fuel. Determine the amount of safety stock that would be needed if the manager wants

    a. An annual service level of 98 percent. What is the implication of negative safety stock?

    b. The expected number of units short per order cycle to be no more than 5 gallons.

Item	Order Size
K033	581
K144	458
L700	0

31. A drugstore uses fixed-order cycles for many of the items it stocks. The manager wants a service level of .98. Determine the order size that will be consistent with this service level for the items in the table for an order interval of 14 days and a lead time of 2 days:

Item	Average Daily Demand	Standard Deviation	Quantity on Hand
K033	60	5	420
K144	50	4	375
L700	8	2	160

32. A manager must set up inventory ordering systems for two new production items, P34 and P35. P34 can be ordered at any time, but P35 can only be ordered once every four weeks.

The company operates 50 weeks a year, and the weekly usage rates for both items are normally distributed. The manager has gathered the following information about the items:

	Item P34	Item P35
Average weekly demand	60 units	70 units
Standard deviation	4 units per week	5 units per week
Unit cost	$15	$20
Annual holding cost	30%	30%
Ordering cost	$70	$30
Lead time	2 weeks	2 weeks
Acceptable stockout risk	2.5%	2.5%

a. When should the manager reorder each item?

b. Compute the order quantity for P34.

c. Compute the order quantity for P35 if 110 units are on hand at the time the order is placed.

a. P34: ROP = 131.1 units
P35: every 4 weeks
b. Q = 306 units
c. Q = 334 units

33. Given the following list of items,

a. Classify the items as A, B, or C.

b. Determine the economic order quantity for each item (round to the nearest whole unit).

Item	Estimated Annual Demand	Ordering Cost	Holding Cost (%)	Unit Price
H4-010	20,000	50	20	2.50
H5-201	60,200	60	20	4.00
P6-400	9,800	80	30	28.50
P6-401	16,300	50	30	12.00
P7-100	6,250	50	30	9.00
P9-103	4,500	50	40	22.00
TS-300	21,000	40	25	45.00
TS-400	45,000	40	25	40.00
TS-041	800	40	25	20.00
V1-001	26,100	25	35	4.00

a. Item	Classification	*b.* EOQ
H4-010	C	2,000
H5-201	B	3,005
P6-400	B	428
P6-401	B	635
P7-100	C	481
P9-103	C	292
TS-300	A	386
TS-400	A	600
TS-041	C	113
V1-001	C	1,087

34. Demand for jelly doughnuts on Saturdays at Don's Doughnut Shoppe is shown in the following table. Determine the optimal number of doughnuts, in dozens, to stock if labor, materials, and overhead are estimated to be 80 cents per dozen, doughnuts are sold for $1.20 per dozen, and leftover doughnuts at the end of each day are sold the next day at half price. What is the *resulting* service level?

25 dozen; service level is .73

Demand (dozens)	Relative Frequency
19	.01
20	.05
21	.12
22	.18
23	.13
24	.14
25	.10
26	.11
27	.10
28	.04
29	.02

35. A public utility intends to buy a turbine as part of an expansion plan and must now decide on the number of spare parts to order. One part, no. X135, can be purchased for $100 each.

Carrying and disposal costs are estimated to be 145 percent of the purchase price over the life of the turbine. A stockout would cost roughly $88,000 due to downtime, ordering, and "special purchase" factors. Historical records based on the performance of similar equipment operating under similar conditions suggest that demand for spare parts will tend to approximate a Poisson distribution with a mean of 3.2 parts for the useful life of the turbine.

a. 9 spares
b. $C_s \leq \$78.03$

a. What is the optimal number of spares to order?

b. Carrying no spare parts would be the best strategy for what range of shortage cost?

78.9 lb.

36. Skinner's Fish Market buys fresh Boston bluefish daily for $1.40 per pound and sells it for $1.90 per pound. At the end of each business day, any remaining bluefish is sold to a producer of cat food for 80 cents per pound. Daily demand can be approximated by a normal distribution with a mean of 80 pounds and a standard deviation of 10 pounds. What is the optimal stocking level?

$4.89/quart; customers may buy other items with the berries that they wouldn't buy without them.

37. A small grocery store sells fresh produce, which it obtains from a local farmer. During the strawberry season, demand for fresh strawberries can be reasonably approximated using a normal distribution with a mean of 40 quarts per day and a standard deviation of 6 quarts per day. Excess costs run 35 cents per quart. The grocer orders 49 quarts per day.

a. What is the implied cost of shortage per quart?

b. Why might this be a reasonable figure?

5 cakes

38. Demand for devil's food whipped-cream layer cake at a local pastry shop can be approximated using a Poisson distribution with a mean of six per day. The manager estimates it costs $3 to prepare each cake. Fresh cakes sell for $4. Day-old cakes sell for $3 each. What stocking level is appropriate if one-half of the day-old cakes are sold and the rest thrown out?

421.5 lb.

39. Burger Prince buys top-grade ground beef for $1.00 per pound. A large sign over the entrance guarantees that the meat is fresh daily. Any leftover meat is sold to the local high school cafeteria for 80 cents per pound. Four hamburgers can be prepared from each pound of meat. Burgers sell for 60 cents each. Labor, overhead, meat, buns, and condiments cost 50 cents per burger. Demand is normally distributed with a mean of 400 pounds per day and a standard deviation of 50 pounds per day. What daily order quantity is optimal? (Hint: Shortage cost must be in dollars per pound.)

a. $.53 to $1.76
b. Decrease
c. $56.67 to $190.00

40. Demand for rug-cleaning machines at Clyde's U-Rent-It is shown in the following table. Machines are rented by the day only. Profit on the rug cleaners is $10 per day. Clyde has four rug-cleaning machines.

Demand	Frequency
0	.30
1	.20
2	.20
3	.15
4	.10
5	.05
	1.00

a. Assuming that Clyde's stocking decision is optimal, what is the implied range of excess cost per machine?

b. Your answer from part *a* has been presented to Clyde, who protests that the amount is too low. Does this suggest an increase or a decrease in the number of rug machines he stocks? Explain.

c. Suppose now that the $10 mentioned as profit is instead the excess cost per day for each machine and that the shortage cost is unknown. Assuming that the optimal number of machines is four, what is the implied range of shortage cost per machine?

CASE

HARVEY INDUSTRIES

Background

Harvey Industries, a Wisconsin company, specializes in the assembly of high-pressure washer systems and in the sale of repair parts for these systems. The products range from small portable high-pressure washers to large industrial installations for snow removal from vehicles stored outdoors during the winter months. Typical uses for high-pressure water cleaning include:

Automobiles
Airplanes
Building maintenance
Barns
Engines
Ice cream plants
Lift trucks
Machinery
Swimming pools

Industrial customers include General Motors, Ford, Chrysler, Delta Airlines, United Parcel Service, and Shell Oil Company.

Although the industrial applications are a significant part of its sales, Harvey Industries is primarily an assembler of equipment for coin operated self-service car wash systems. The typical car wash is of concrete block construction with an equipment room in the center, flanked on either size by a number of bays. The cars are driven into the bays where the owner can wash and wax the car, utilizing high-pressure hot water and liquid wax. A dollar bill changer is available to provide change for the use of the equipment and the purchase of various products from dispensers. The products include towels, tire cleaner, and upholstery cleaner.

In recent years Harvey Industries has been in financial difficulty. The company has lost money for three of the last four years, with the last year's loss being $17,174 on sales of $1,238,674. Inventory levels have been steadily increasing to their present levels of $124,324.

The company employs 23 people with the management team consisting of the following key employees: president, sales manager, manufacturing manager, controller, and purchasing manager. The abbreviated organization chart reflects the reporting relationship of the key employees and the three individuals who report directly to the manufacturing manager.

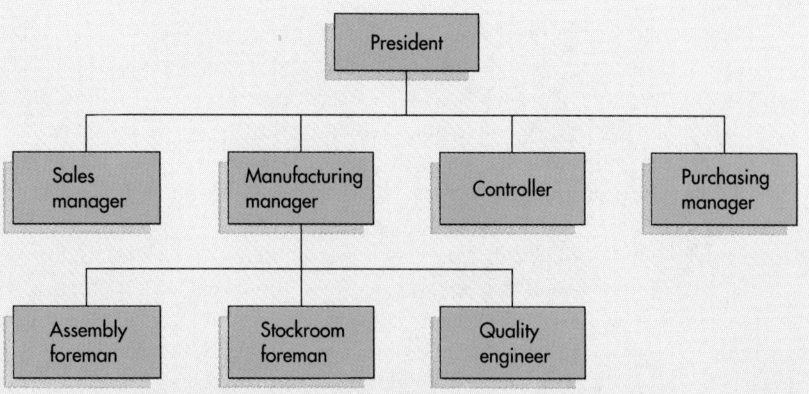

Current Inventory Control System

The current inventory control "system" consists of orders for stock replenishment being made by the stockroom foreman, the purchasing manager, or the manufacturing manager whenever one of them notices that the inventory is low. An order for replenishment of inventory is also placed whenever someone (either a customer or an employee in the assembly area) wants an item and it is not in stock.

Some inventory is needed for the assembly of the high-pressure equipment for the car wash and industrial applications. There are current and accurate bills of material for these assemblies. The material needs to support the assembly schedule are generally known well in advance of the build schedule.

The majority of inventory transactions are for repair parts and for supplies used by the car washes, such as paper towels, detergent, and wax concentrate. Because of the constant and rugged use of the car wash equipment, there is a steady demand for the various repair parts.

The stockroom is well organized, with parts stored in locations according to each vendor. The number of vendors is relatively limited, with each vendor generally supplying many different parts. For example, the repair parts from Allen Bradley, a manufacturer of electrical motors, are stocked in the same location. These repair parts will be used to provide service for the many electrical motors that are part of the high-pressure pump and motor assembly used by all of the car washes.

Because of the heavy sales volume of repair parts, there are generally two employees working in the stockroom—a stockroom foreman who reports to the manufacturing manager and an assistant to the foreman. One of these two employees will handle customer orders. Many customers stop by and order the parts and supplies they need. Telephone orders are also received and are shipped by United Parcel Service the same day.

The assembly area has some inventory stored on the shop floor. This inventory consists of low-value items that are used every day, such as nuts, bolts, screws, and washers. These purchased items do not amount to very much dollar volume throughout the year. Unfortunately, oftentimes the assembly area is out of one of these basic items and this causes a significant amount of downtime for the assembly lines.

Paperwork is kept to a minimum. A sales slip listing the part numbers and quantities sold to a customer is generally made out for each sale. If the assembly department needs items that are not stocked on the assembly floor, someone from that department will enter the stockroom and withdraw the necessary material. There is no paperwork made out for the items needed on the assembly floor.

There were 973 different part numbers purchased for stock last year and those purchases amounted to $314,673. Although the company does not utilize a computer, it does have some accurate records on how much money was spent on each part number last year. An analysis of the data shows that $220,684 was spent on just 179 of the part numbers.

Fortunately for Harvey Industries, most of the items they purchase are stocked by either the manufacturer or by a wholesaler. When it is discovered that the company is out of stock on an item, it generally takes only two or three days to replenish the stock.

Due to the company's recent losses, its auditing firm became concerned about the company's ability to continue in business. Recently the company sold off excess vacant land adjoining its manufacturing facility to generate cash to meet its financial obligations.

New President

Because of the recent death of the owner, the trust department of a Milwaukee Bank (as trustee for the state) has taken over the company's affairs and has appointed a new company president. The new president has identified many problem areas—one of which is improper inventory control. He has retained you as a consultant to make specific recommendations concerning a revised inventory control system. What are your recommendations and their rationale?

Source: This case was prepared by Donald F. Condit of the Lawrence Technological University, Southfield, Michigan, as a basis for class discussion rather than to illustrate either effective or ineffective organizational practices. It was presented at the Midwest Case Writers Association Workshop, 1984, and accepted by referees of the Midwest Case Writers Association for international distribution. Reprinted by permission.

CASE

THE DEWEY STAPLER COMPANY

From: Martin Crane,
Sales Manager

To: Allen Grace, President

Dear Allen:

Well, it has been a very disappointing year. We've missed our quota by 10 or 15 percent in virtually every district and this is the year I had such high hopes for. When we decided to open up four branch warehouses rather than shipping from our central location only, I was convinced that this would give us better customer service. The last of the warehouses was opened up last May, just before our peak summer season, so perhaps some of the problem is just not having enough experience with branch warehouses. But I think it goes deeper than that.

Our warehouse people are authorized to keep a one-month supply of inventory on hand. While I know you feel strongly that the substantial increase in inventory we had during the year was due to the warehouse program, I can't see why it requires any more inventory to keep a month's supply on hand in four branches and a main location than it did to keep a month's supply on hand back at the main location. A month's supply is a month's supply no matter how you look at it.

To my way of thinking, the real problem is customer service. Our salespeople are demoralized. They simply can't get the stock shipped out of the warehouses because the warehouses don't have it on hand. Forty percent of our customer orders, of course, are still being shipped out of the main location. Our warehouse people tell me that these customer orders get preference and their stock replenishment orders are pushed aside.

Allen, we've got to solve this problem. There's no sense in having a sales force if we can't have the stock to back them up. I propose the plant location be required to ship warehouse stocking orders just the same as they ship customer orders. They should treat the warehouses like a customer. In fact, the warehouses are their *biggest* customer and should be serviced accordingly. I propose also that the one-month inventory guideline be removed. Let the warehouse people stock whatever they think they need to support the sales force. I would volunteer to have my district managers sit with the branch warehouse people to give them some idea of what they should really be ordering.

Allen, this branch warehouse program has been as big a disappointment to me as it has to you. I know you're concerned about the fact that inventories have been going up, but I frankly attribute that to poor management back at the plant. And, quite honestly, Allen, I don't think that people at the plant realize our problems out here in the field and are giving us the kind of support we need. Without it, we have no chance of making the sales quota. Instead of selling, I'm spending most of my time playing chaplain to a bunch of demoralized salespeople.

Sincerely,

Martin

From: Robert Ellers,
Inventory Manager

To: Allen Grace, President

Dear Mr. Grace:

You asked me what my plans were to respond to Martin Crane's letter of January 5. I don't know where to start. This warehouse program has really torn us apart.

We thought that when branch warehouses were added, we would simply have to split some of the stock we had among the warehouses. Instead, we've had to build up the inventory very substantially. We don't get any plans from the warehouses at all. All we see is orders. We have no idea what their inventory position is when we get the orders and we only get them two to three weeks before we have to ship them. Then comes the moment of truth. We have a shortage on a particular item. Here's a customer order and also a warehouse stock replenishment order. Does the warehouse really need it? We know the customer does. In practice, I must admit we wind up waiting until the warehouse screams although we know we may very well be hurting customer service at the branch warehouses.

Mr. Grace, I'm more worried about *this* year than I was about last. Some warehouses showed a disturbing tendency to keep their inventories low during the off-season so that they could boast about their inventory turnover. Then during the peak season, they expect me to turn the faucet on back at the plant. We don't have enough storage space at the plant to build up the inventory required during the off-season in order to keep people working at a steady rate. We need this inventory buildup in order to give good service during the peak season. I've been told repeatedly by plant management that we must keep people working at a steady rate.

All this squabbling about inventory levels prompts me to suggest an approach. We normally would manufacture in a lot size that would be equivalent to about a three-months' supply. When we do that, why don't I just ship a three-months' supply out to each branch warehouse, and then we won't have to bother worrying about them until the next lot is manufactured. Then they can't complain that they're not getting their fair share.

One of the disturbing elements that you may not have heard about is that Frank, our traffic manager, has now suggested that we ship to the West Coast warehouses by sea. This would mean going through the Panama Canal and would substantially increase our lead time and reduce our flexibility. He says, "Flexibility is like motherhood. I'm talking a $50,000 savings in transportation costs, and if you guys have to work a little harder to make that happen, so be it."

Mr. Grace, I really am almost at my wit's end. Perhaps one of the things we ought to consider would be a computer system for tying all of the warehouses together so we could cover a shortage at one warehouse by shipping from another warehouse. Last September, I checked on items that were out of stock in the Atlanta warehouse and I found that virtually every one of them was in adequate supply throughout the system; i.e., we either had them in Dallas, Los Angeles, Chicago, or back at the main plant. This type of computer system would be expensive, but perhaps this is the answer to our service problem.

Sincerely,

Robert Ellers

Question

The Dewey Stapler Company has some very serious problems. A number of misconceptions about inventory management in the company need to be corrected. Take the position of a consultant called in by Allen Grace, the president. You have enough information in these letters to give him some very helpful recommendations. Write a memo outlining your thoughts. Include a discussion of what will happen to service levels if total inventory investment remains constant.

Source: Reprinted by permission from Oliver W. Wright, *Production and Inventory Management in the Computer Age* (Boston: Cahners, 1974).

OPERATIONS TOUR

BRUEGGER'S BAGEL BAKERY

Bruegger's Bagel Bakery makes and sells a variety of bagels, including plain, onion, poppyseed, and cinnamon raisin, as well as assorted flavors of cream cheese. Bagels are the major source of revenue for the company.

The bagel business is a $3 billion dollar industry. Bagels are very popular with consumers. Not only are they relatively low in fat, they are filling, and they taste good! Investors like the bagel industry because it can be highly profitable: it only costs about $.10 to make a bagel, and they can be sold for $.50 each or more. Although some bagel companies have done poorly in recent years, due mainly to poor management, Bruegger's business is booming; it is number one nationally, with over 450 shops that sell bagels, coffee, and bagel sandwiches for takeout or on-premise consumption. Many stores in the Bruegger's chain generate an average of $800,000 in sales annually.

Production of bagels is done in batches, according to flavor, with each flavor being produced on a daily basis. Production of bagels at Bruegger's begins at a processing plant, where the basic ingredients of flour, water, yeast, and flavorings are combined in a special mixing machine. After the dough has been thoroughly mixed, it is transferred to another machine that shapes the dough into individual bagels. Once the bagels have been formed, they are loaded onto refrigerated trucks for shipping to individual stores. When the bagels reach a store, they are unloaded from the trucks and temporarily stored while they rise. The final two steps of processing involve boiling the bagels in a kettle of water and malt for one minute, and then baking the bagels in an oven for approximately 15 minutes.

The process is depicted in Figure 1.

Quality is an important feature of a successful business. Customers judge the quality of bagels by their appearance (size, shape, and shine), taste, and consistency. Customers are also sensitive to the service they receive when they make their purchases. Bruegger's devotes careful attention to quality at every stage of operation, from choosing suppliers of ingredients, careful monitoring of ingredients, and keeping equipment in good operating condition to monitoring output at each step in the process. At the stores, employees are instructed to watch for deformed bagels and to remove them when they find them. (Deformed bagels are returned to the main plant where they are sliced into bagel chips, packaged, and then taken back to the stores for sale, thereby reducing the scrap rate.) Employees who work in the stores are carefully chosen and then trained so that they are competent to operate the necessary equipment in the stores and to provide the desired level of service to customers.

The company operates with minimal inventories of raw materials and inventories of partially completed bagels at the plant and very little inventory of bagels at the stores. One reason for this is to maintain a high degree of freshness in the final product by continually supplying fresh product to the stores. A second reason is to keep costs down; minimal inventories mean less space is needed for storage.

Questions

1. Bruegger's maintains relatively little inventory at either its plants or its retail stores. List the benefits and risks of this policy.
2. Quality is very important to Bruegger's.
   a. What features of bagels do customers look at to judge their quality of bagels?
   b. At what points in the production process do workers check bagel quality?
   c. List the steps in the production process, beginning with purchasing ingredients, and ending with the sale, and state how quality can be positively affected at each step.
3. Which inventory models could be used for ordering the ingredients for bagels? Which model do you think would be most appropriate for deciding how many bagels to make in a given batch?
4. Bruegger's has bagel-making machines at its plants. Another possibility would be to have a bagel-making machine at each store. What advantages does each alternative have?

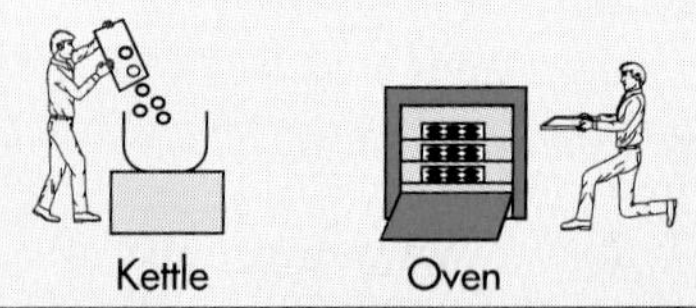

Selected Bibliography

Brooks, Roger B., and Larry W. Wilson. *Inventory Record Accuracy: Unleashing the Power of Cycle Counting.* Essex Junction, Vermont: Oliver Wright, 1993.

Fogarty, Donald W.; John H. Blackstone; and Thomas R. Hoffmann. *Production and Inventory Management.* 2nd ed. Cincinnati, Ohio: South-Western Publishing Co., 1991.

Peterson, R., and E. A. Silver. *Decision Systems for Inventory Management and Production Planning.* 2nd ed. New York: John Wiley & Sons, 1984.

Tersine, Richard J. *Principles of Inventory and Materials Management.* 3rd ed. New York: Elsevier North-Holland, 1987.

Vollman, Thomas E.; William L. Berry; and D. Clay Whybark. *Manufacturing Planning and Control Systems.* 5th ed. Burr Ridge, IL: Richard D. Irwin, 1997.

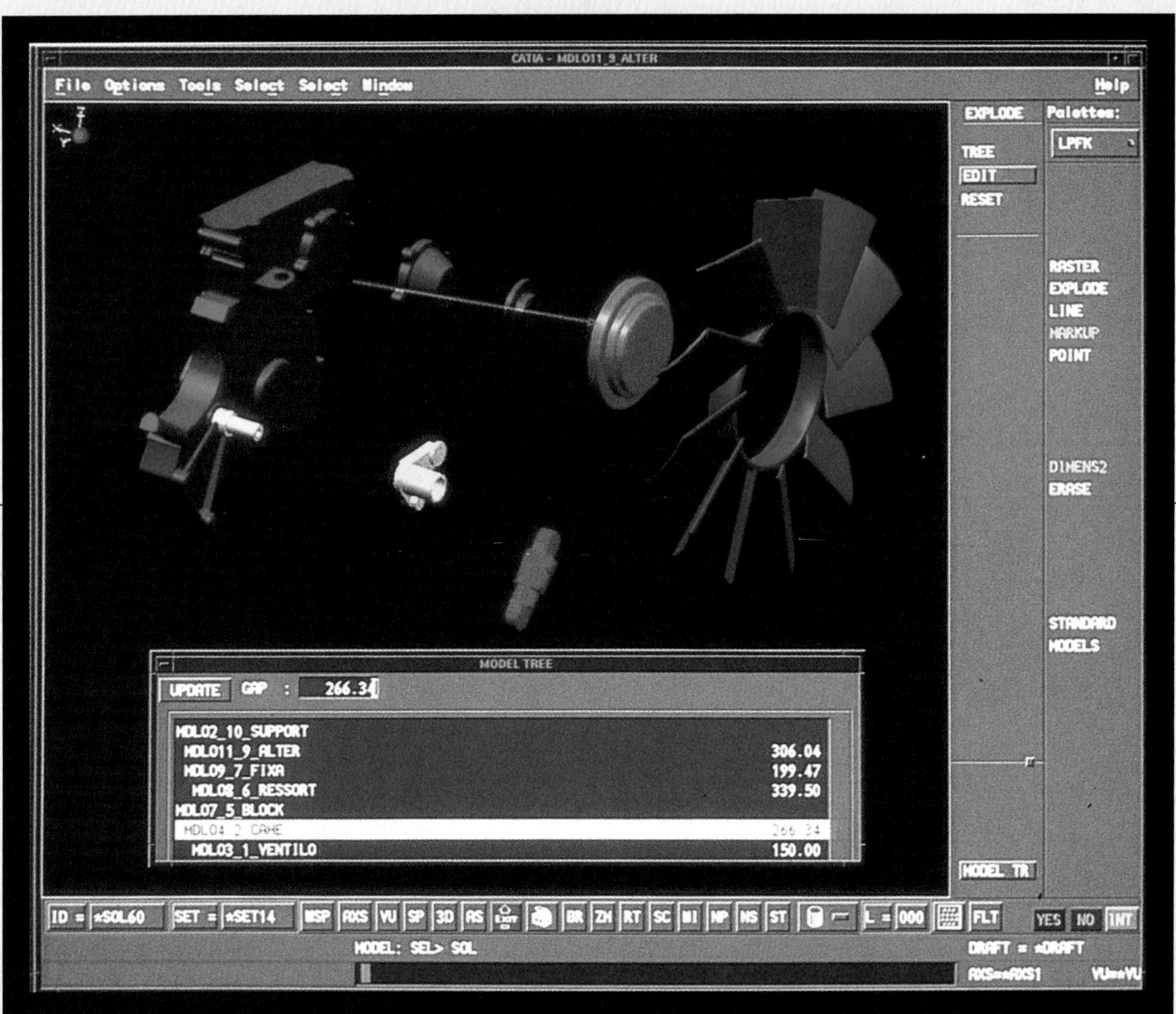
CATIA - MDL011_9_ALTER
File
Options
Tools
Select
Select
Window
Help
EXPLODE
Palettes:
LPFK
TREE
EDIT
RESET
RASTER
EXPLODE
LINE
MARKUP
POINT
DIMENS2
ERASE
STANDARD
MODELS
MODEL TREE
UPDATE
GAP : 266.34
MDL02_10_SUPPORT
MDL011_9_ALTER 306.04
MDL09_7_FIXA 199.47
MDL08_6_RESSORT 339.50
MDL07_5_BLOCK
MDL04_2_CAME 266.34
MDL03_1_VENTILO 150.00
MODEL TR
ID = *SOL60
SET = *SET14
YES
NO
INT
MODEL: SEL> SOL
DRAFT = *DRAFT
AXS=*AXS1

LEARNING OBJECTIVES

After completing this chapter, you should be able to:

1 Describe the conditions under which MRP is most appropriate.

2 Describe the inputs, outputs, and nature of MRP processing.

3 Explain how requirements in a master production schedule are translated into material requirements for lower-level items.

4 Discuss the benefits and requirements of MRP.

5 Explain how an MRP system is useful in capacity requirements planning.

6 Outline the potential benefits and some of the difficulties users have encountered with MRP.

7 Describe MRP II and how it relates to MRP.

CHAPTER FOURTEEN
MATERIAL REQUIREMENTS PLANNING

CHAPTER OUTLINE

Two managers commented on each of their company's experiences with material requirements planning:

I don't know how we managed to get along without MRP for as long as we did. Our scheduling has gone from a state of turmoil to relative calm. We have achieved a substantial reduction in inventory size and cost, and yet our customer-service levels have actually improved quite a bit. We are more than satisfied with MRP; it allows us to *plan* instead of having to spend all of our time trying to *react* to one crisis after another. I would strongly recommend that all manufacturing firms adopt MRP or some other similar approach.

We had heard so many good things about MRP that we couldn't wait to get started, but so far, it seems like we've just been spinning our wheels. We're not even close to going online with it, and yet we've already exceeded our original time estimates and our costs are out of sight. Everyone is discouraged at this point, especially the brass. There is even some talk now of just scrapping the whole darn idea.

These two views reflect the range of reactions that typifies efforts of manufacturing firms to implement and use material requirements planning. In this chapter, we will examine the major issues of this topic, including the need for this approach, the basic concepts of planning requirements, the benefits and limitations, and some reasons for the difficulties some firms have encountered. These issues will be followed by a description of MRP II, an expanded approach to resource planning that incorporates MRP.

Chapter 13 focused on inventory management under conditions of *independent* demand. In this chapter, the focus is on managing inventories that have *dependent* or *derived* demand.

DEPENDENT VERSUS INDEPENDENT DEMAND

dependent demand Demand for items that are subassemblies or component parts to be used in the production of finished goods.

A major distinction in the way inventories are managed results from the nature of demand for those items. When demand for items is derived from plans to make certain products, as it is with raw materials, parts, and assemblies used in producing a finished product, those items are said to have **dependent demand.** The parts and materials that go into the production of an automobile are examples of dependent demand because the total amount of parts and raw materials needed during any time period is a function of the number of cars that will be produced. Conversely, demand for the *finished* cars is independent—a car is not a component of another item.

Independent demand is fairly stable once allowances are made for seasonal variations, but dependent demand tends to be sporadic or "lumpy"; large quantities are used at specific points in time with little or no usage at other times. For example, a firm that produces lawn and garden equipment might make a variety of items, such as trimmers, lawn mowers, and small tractors. Suppose that the various products are produced periodically—in one month, push mowers; in the next month, mulching mowers; and in the third month, tractors. Some components may be used in most of the items (e.g., nuts and bolts, screws). It makes sense to have a continual inventory of these parts because they are always needed. On the other hand, some parts might be used for only one item. Consequently, demand for those parts occurs only when that item is being produced, which might be once every eight or nine weeks; the rest of the time, demand is zero. Thus, demand is "lumpy." Because of these tendencies, independent-demand items must be carried on a continual basis, but dependent-demand items need only be stocked just prior to the time they will be needed in the production process. Moreover, the predictability of usage of dependent-demand items means that there is little or no need for safety stock. Figure 14–1 illustrates key differences in independent- and dependent-demand inventories.

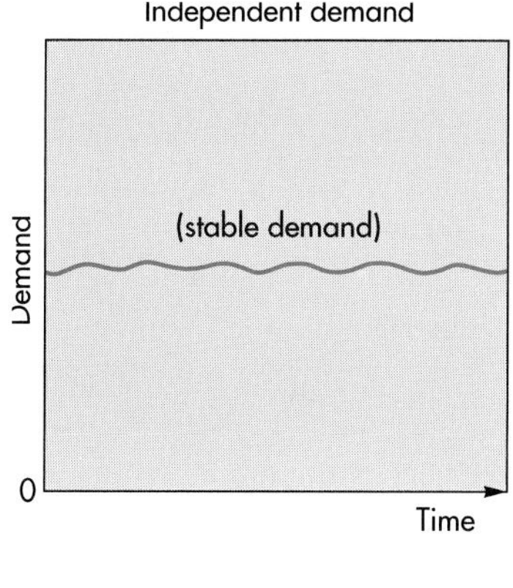

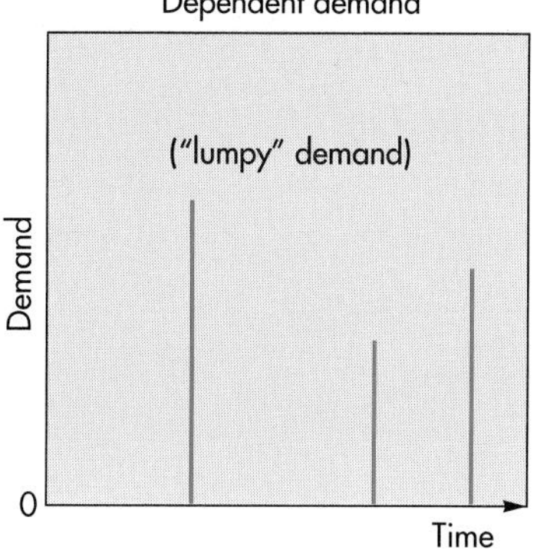

FIGURE 14-1

Comparison of independent and dependent demand

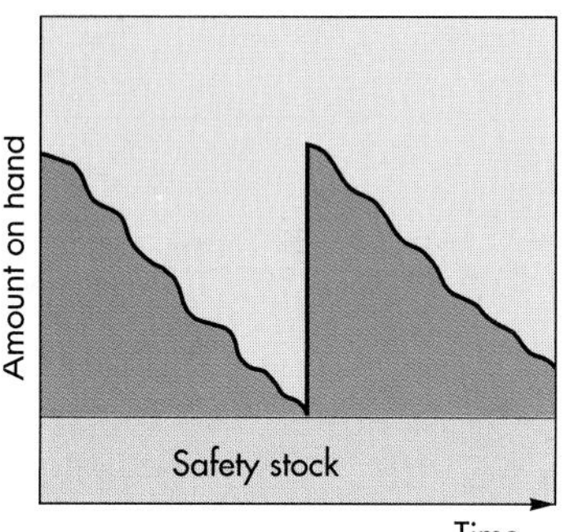

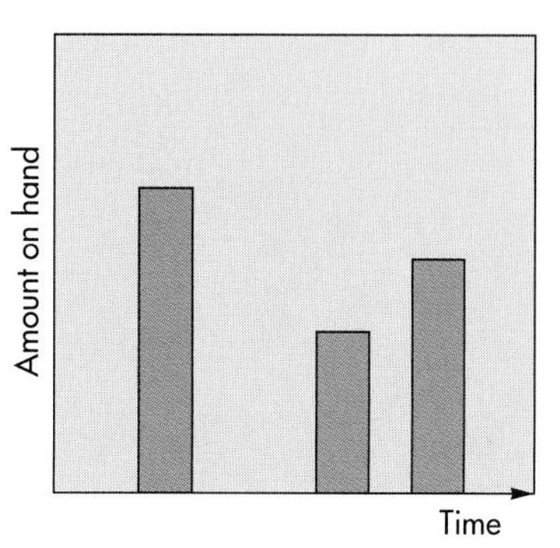

AN OVERVIEW OF MRP

material requirements planning Computer-based information system for ordering and scheduling of dependent-demand inventories.

Material requirements planning (MRP) is a computer-based information system designed to handle ordering and scheduling of dependent-demand inventories (e.g., raw materials, component parts, and subassemblies). A production plan for a specified number of finished products is translated into requirements for component parts and raw materials working backward from the due date, using lead times and other information to determine when and how much to order. Hence, requirements for end items generate requirements for lower-level components, which are broken down by planning periods (e.g., weeks) so that ordering, fabrication, and assembly can be scheduled for timely completion of end items while inventory levels are kept reasonably low.

Material requirements planning is as much a philosophy as it is a technique, and as much an approach to scheduling as it is to inventory control.

Historically, ordering and scheduling of assembled products suffered from two difficulties. One was the enormous task of setting up schedules, keeping track of large numbers of parts and components, and coping with schedule and order changes. The other was a lack of differentiation between independent demand and dependent demand. All too often, techniques designed for independent-demand items were used to handle assembled items, which resulted in excessive inventories. Consequently, inventory planning and scheduling presented major problems for manufacturers.

In the 1970s, manufacturers began to recognize the importance of the distinction between independent- and dependent-demand items and to approach these two categories in different ways. Much of the burden of record keeping and determining material requirements in many firms has now been transferred to computers, using techniques such as MRP. A great deal of the credit for publicizing MRP and educating potential users about MRP goes to Joseph Orlicky,[1] George Plossl, Oliver Wight, and the

[1] Orlicky is the author of *Material Requirements Planning* (New York: McGraw-Hill, 1975).

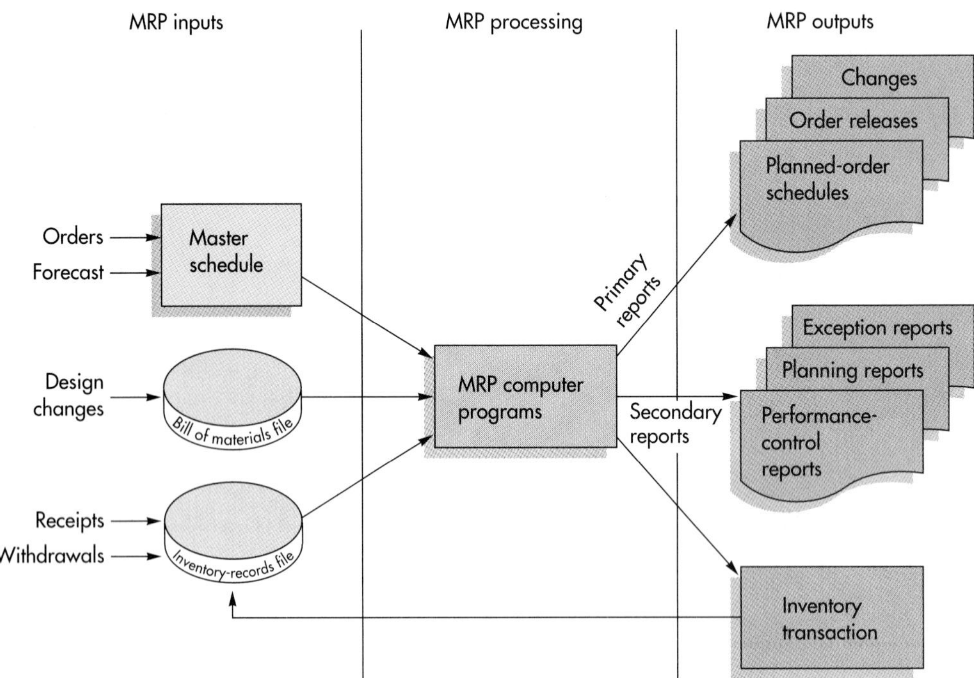

FIGURE 14–2
Overview of MRP

American Production and Inventory Control Society (APICS). APICS offers certification information and exams.

MRP begins with a schedule for finished goods that is converted into a schedule of requirements for the subassemblies, component parts, and raw materials needed to produce the finished items in the specified time frame. Thus, MRP is designed to answer three questions: *what* is needed; *how much* is needed? and *when* is it needed?

The primary inputs of MRP are a bill of materials, which tells the composition of a finished product; a master schedule, which tells how much finished product is desired and when; and an inventory records file, which tells how much inventory is on hand or on order. The planner processes this information to determine the *net* requirements for each period of the planning horizon.

Outputs from the process include planned-order schedules, order releases, changes, performance-control reports, planning reports, and exception reports. These topics are discussed in more detail in subsequent sections. Figure 14–2 provides an overview of an MRP system.

MRP INPUTS

An MRP system has three major sources of information: a master schedule, a bill-of-materials file, and an inventory records file (see Figure 14–2). Let's consider each of these inputs.

The Master Schedule

master schedule One of three primary inputs in MRP; states which end items are to be produced, when these are needed, and in what quantities.

The **master schedule,** also referred to as the *master production schedule,* states which end items are to be produced, when they are needed, and in what quantities. Figure 14–3 illustrates a portion of a master schedule that shows planned output for end item X for the planning horizon. The schedule indicates that 100 units of X will be needed (e.g., for shipments to customers) at the *start* of week 4 and that another 150 units will be needed at the *start* of week 8.

The quantities in a master schedule come from a number of different sources, including customer orders, forecasts, orders from warehouses to build up seasonal inventories, and external demand.

FIGURE 14-3

A master schedule for end item X

Week number

Item: X	1	2	3	4	5	6	7	8
Quantity				100				150

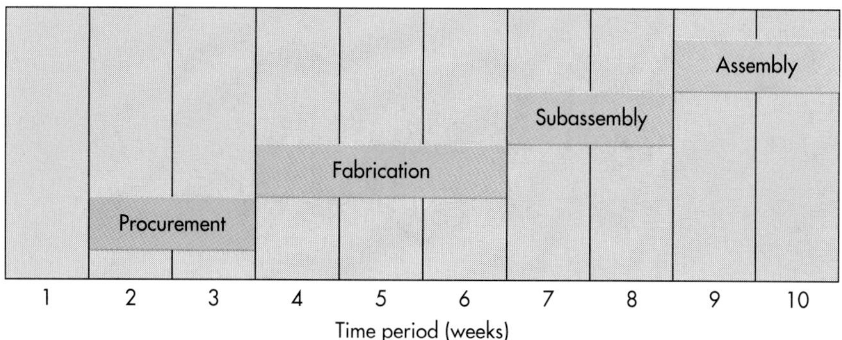

FIGURE 14-4

The planning horizon must cover the cumulative lead time

The master schedule separates the planning horizon into a series of time periods or time *buckets,* which are often expressed in weeks. However, the time buckets need not be of equal length. In fact, the near-term portion of a master schedule may be in weeks, but later portions may be in months or quarters. Usually, plans for those more distant time periods are more tentative than near-term requirements.

Although a master production schedule has no set time period that it must cover, most managers like to plan far enough into the future so they have some general idea of probable upcoming demands for the near term. It is important, though, that the master schedule cover the *stacked* or **cumulative lead time** necessary to produce the end items.This amounts to the sum of the lead times that sequential phases of the production or assembly process require, as illustrated in Figure 14–4, where a total of nine weeks of lead time is needed from ordering parts and raw materials until final assembly is completed.

cumulative lead time The sum of the lead times that sequential phases of a process require, from ordering of parts or raw materials to completion of final assembly.

The Bill-of-Materials File

A **bill of materials (BOM)** contains a listing of all of the assemblies, subassemblies, parts, and raw materials that are needed to produce one unit of a finished product. Thus, each finished product has its own bill of materials.

bill of materials One of the three primary inputs of MRP; a listing of all of the raw materials, parts, subassemblies, and assemblies needed to produce one unit of a product.

The listing in the bill-of-materials file is hierarchical; it shows the quantity of each item needed to complete one unit of the following level of assembly. The nature of this aspect of a bill of materials is clear when you consider a **product structure tree,** which provides a visual depiction of the subassemblies and components needed to assemble a product. Figure 14–5 shows an assembly diagram for a chair and a product structure tree for the chair. This chart is a simple product structure tree for a chair. The end item (in this case, the chair, the finished product) is shown at the top of the tree. Just beneath it are the subassemblies, or major components, that must be put together to make up the end item. Beneath each major component are the necessary lesser components. At each stage moving down the tree are shown the components (parts, materials) needed to make one unit of the next higher item in the tree.

product structure tree Visual depiction of the requirements in a bill of materials, where all components are listed by levels.

A product structure tree is useful in illustrating how the bill of materials is used to determine the quantities of each of the ingredients (requirements) needed to obtain a desired number of end items.

Let's consider the product structure tree shown in Figure 14–6. End item X is composed of two Bs and one C. Moreover, each B requires three Ds and one E, and each D

FIGURE 14–5

Assembly diagram and product structure tree for chair assembly

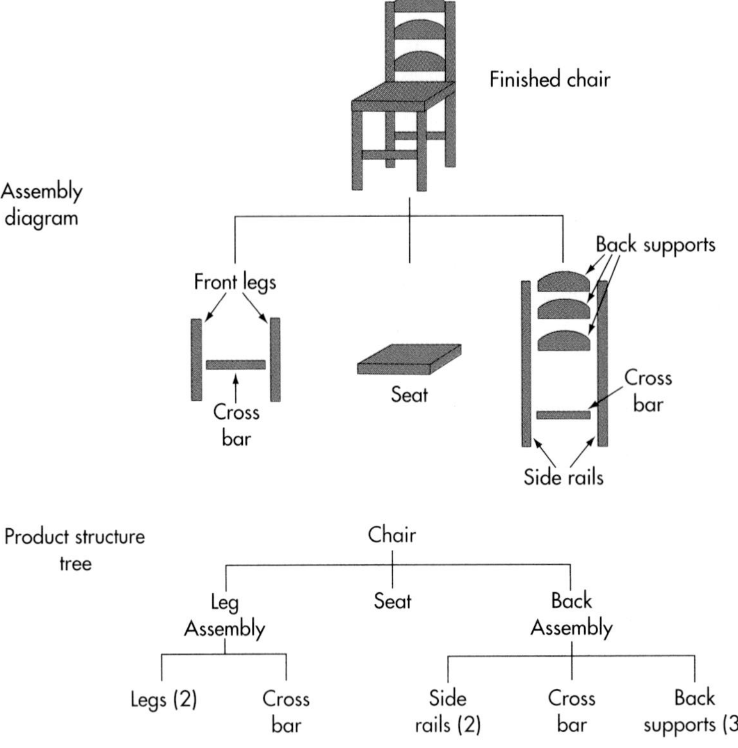

FIGURE 14–6

A product structure tree for end item X

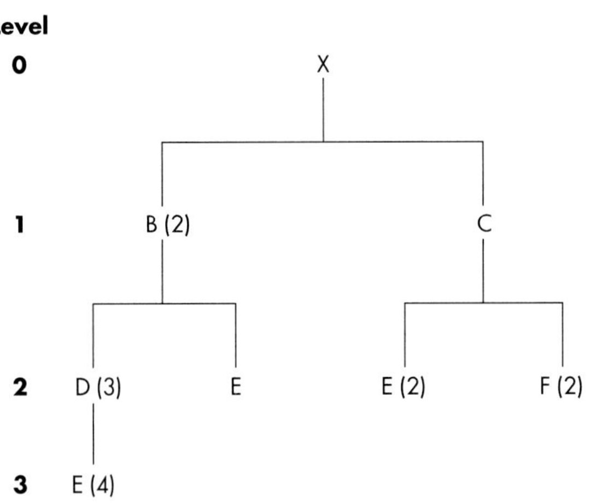

requires four Es. Similarly, each C is made up of two Es and two Fs. These *requirements* are listed by *level*, beginning with level 0 for the end item, then level 1 for the next level, and so on. The items at each level are *components* of the next level up and, as in a family tree, are *parents* of their respective components. Note that the quantities of each item in the product structure tree refer only to the amounts needed to complete the assembly in the next higher level.

Example 1

Use the information presented in Figure 14–6 to do the following:

a. Determine the quantities of B, C, D, E, and F needed to assemble one X.

b. Determine the quantities of these components that will be required to assemble 200 Xs.

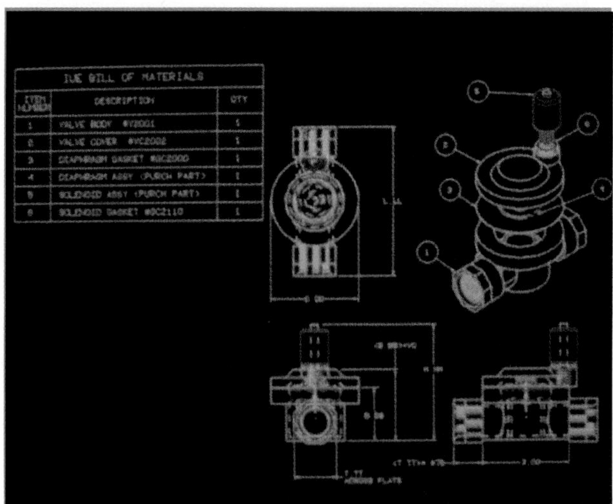

IBM CADAM software can be used to help prepare a bill of materials. This diagram shows a bill of materials for a pump, along with several different views of the complete product with the subassemblies.

Solution

a.

Component		Quantity
B	2 Bs per X	= 2
D	3 Ds per B × 2 Bs per X	= 6
E	4 Es per D × 3 Ds per B × 2 Bs per X	= 24
E	1 E per B × 2 Bs per X	= 2
C	1 C per X	= 1
E	2 Es per C × 1 C per X	= 2
F	2 Fs per C × 1 C per X	= 2

Note that E appears in three separate places. We can determine its total requirements by summing the separate amounts, which yields 28.

b. In order to assemble 200 units of X, the quantities of each component must be multiplied by 200. Hence, there must be 200(2) = 400 Bs, 200(6) = 1,200 Ds, 200(28) = 5,600 Es, 200(1) = 200 Cs, and 200(2) = 400 Fs.

Determining total requirements is usually more complicated than Example 1 might suggest. For one thing, many products have considerably more components. For another, the issue of *timing* is essential (i.e., when must the components be ordered or made) and must be included in the analysis. Finally, for a variety of reasons, some of the components/subassemblies may be on hand (i.e., currently in inventory). Consequently, in determining total requirements, the amounts on hand must be *netted out* (i.e., subtracted from the apparent requirements) to determine the true requirements.

When requirements are calculated in an MRP system, the computer scans the product structure level by level, starting at the top. When a component (such as E in Figure 14–6) appears on more than one level, its total requirements cannot be determined until all levels have been scanned. From a computational standpoint, this is somewhat inefficient. A simplification sometimes used to increase efficiency is **low-level coding,** which involves restructuring the BOM so that all occurrences of an item are made to coincide with the lowest level in which the item appears. Figure 14–7 illustrates how to rearrange component E, which appears in two different levels of the tree, so that it appears at only one level. When the same component is used in multiple products, low-level coding may involve additional adjustment to get the component on the same level across all products. Many commercial software packages have a BOM processor that automatically assigns codes.

low-level coding A simplified restructuring of the BOM so that all occurrences of an item coincide with the lowest level in which it appears.

Comment. You may be wondering what happens if a component that appears at different levels is "buried" at an upper level and can't be lowered? The answer has two parts.

FIGURE 14-7

Low-level coding for Component E

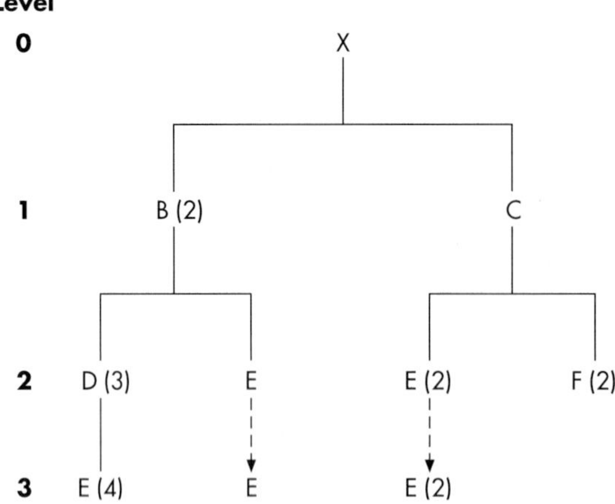

FIGURE 14-8

One component G seems to be buried, apparently preventing low-level coding

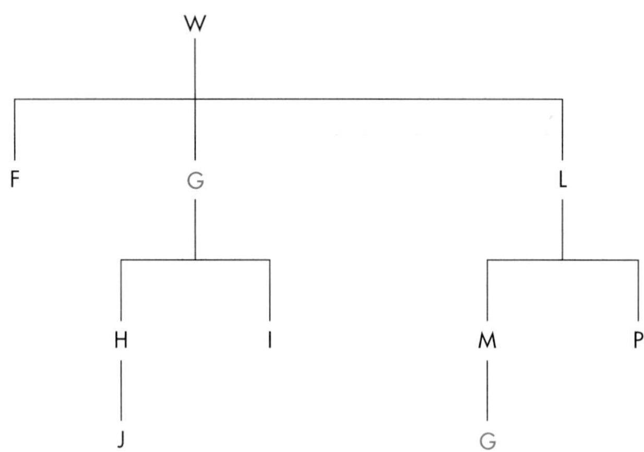

One is that low-level coding is not limited to items at the end of "branches"; see Figure 14–8. Parent G and its components (and subcomponents, if necessary) can all be lowered, if necessary, to achieve low-level coincidence. The second part of the answer lies in the fact that parent G should have the *same* components anywhere it appears in the product structure. In other words, components H and I would always be required to assemble one G. If, for some reason there is an exception to this, one of the Gs should be given a different name. Each name should carry a unique set of component requirements no matter where it appears in the product structure. Similarly, *there should be only one MRP record for each part number.*

It is extremely important that the bill of materials accurately reflect the composition of a product, particularly since errors at one level become magnified by the multiplication process used to determine quantity requirements. As obvious as this might seem, many companies find themselves with incorrect bills-of-material records. These make it impossible to effectively determine material requirements; moreover, the task of correcting, these records can be complex and time-consuming. Accurate records are a prerequisite for effective MRP.

The Inventory-Records File

inventory records file One of the three primary inputs in MRP; includes information on the status of each item by time period.

The **inventory records file** is used to store information on the status of each item by time period. This includes gross requirements, scheduled receipts, and expected amount on hand. It also includes other details for each item, such as supplier, lead time, and lot size. Changes due to stock receipts and withdrawals, canceled orders, and similar events also are recorded in this file.

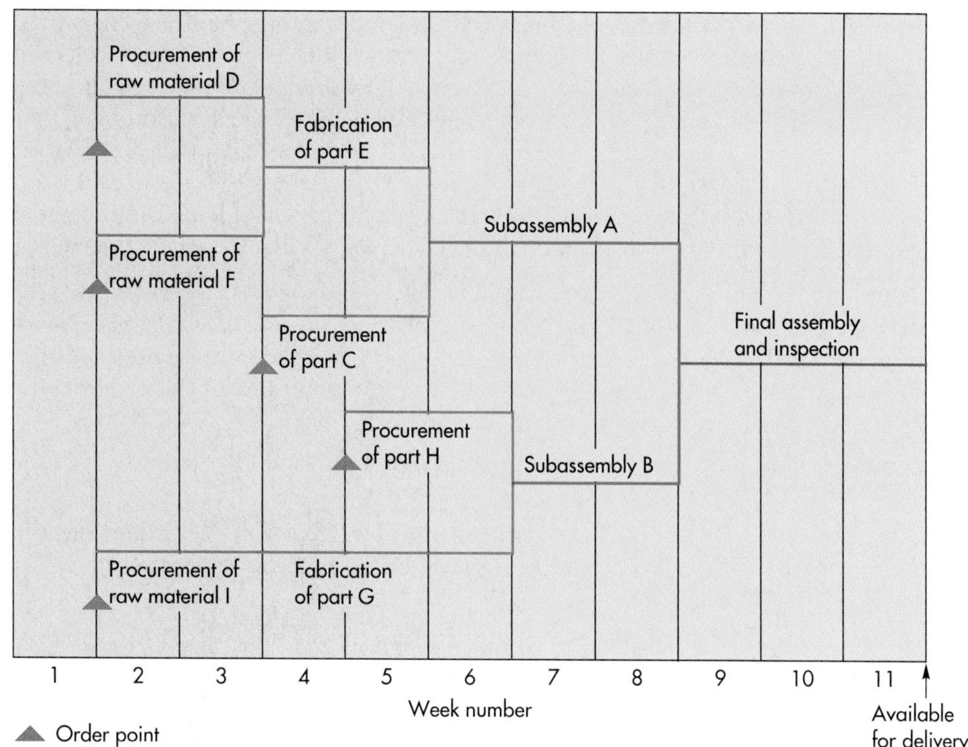

FIGURE 14–9

Assembly time chart showing material order points needed to meet scheduled availability of the end item

Like the bill of materials, the inventory records file must be accurate. Erroneous information on requirements or lead times can have a detrimental impact on MRP and create turmoil when incorrect quantities are on hand or expected delivery times are not met.

MRP PROCESSING

MRP processing takes the end item requirements specified by the master schedule and "explodes" them into *time-phased* requirements for assemblies, parts, and raw materials using the bill of materials offset by lead times. You can see the time-phasing of requirements in the assembly time chart in Figure 14–9. For example, raw materials D, F, and I must be ordered at the start of week 2, part C at the start of week 4, and part H at the start of week 5 in order to be available for delivery as planned.

The quantities that are generated by exploding the bill of materials are *gross requirements*; they do not take into account any inventory that is currently on hand or due to be received. The materials that a firm must actually acquire to meet the demand generated by the master schedule are the *net material requirements.*

The determination of the net requirements (*netting*) is the core of MRP processing. One accomplishes it by subtracting from gross requirements the sum of inventory on hand and any scheduled receipts, and then adding in safety stock requirements, if applicable:

$$\begin{matrix}\text{Net}\\ \text{requirements}\\ \text{in period } t\end{matrix} = \begin{matrix}\text{Gross}\\ \text{requirements}\\ \text{in period } t\end{matrix} - \begin{matrix}\text{Projected}\\ \text{inventory}\\ \text{in period } t\end{matrix} + \begin{matrix}\text{Safety}\\ \text{stock}\end{matrix} \qquad (14\text{–}1)$$

For simplicity, we will omit safety stock from computations in examples and most problems. Net requirements are sometimes adjusted to include an allowance for waste; but for simplicity, this, too, we will not include in examples or most problems.

The timing and sizes of orders (i.e., materials ordered from suppliers or work started within the firm) are determined by *planned-order releases*. The timing of the receipts of these quantities is indicated by *planned-order receipts*. Depending on ordering policy, the planned-order releases may be multiples of a specified quantity (e.g., 50 units), or they may be equal to the quantity needed at that time. Although there are other possibilities, these two seem to be the most widely used. Example 2 illustrates the difference between these two ordering policies as well as the general concepts of time-phasing material requirements in MRP. As you work through the example, you may find the following list of terms helpful.

gross requirements Total expected demand for an item or raw material in a time period.

scheduled receipts Open orders scheduled to arrive from vendors or elsewhere in the pipeline.

projected on hand Expected amount of inventory that will be on hand at the beginning of each time period.

net requirements The actual amount needed in each time period.

planned-order receipts Quantity expected to be received by the beginning of the period in which it is shown.

planned-order releases Planned amount to order in each time period; planned-order receipts offset by lead time.

Gross requirements: The total expected demand for an item or raw material *during* each time period without regard to the amount on hand. For end items, these quantities are shown in the master schedule; for components, these quantities are derived from the planned-order releases of their immediate "parents."

Scheduled receipts: Open orders scheduled to arrive from vendors or elsewhere in the pipeline by the *beginning* of a period.

Projected on hand: The expected amount of inventory that will be on hand at the *beginning* of each time period: scheduled receipts plus available inventory from last period.

Net requirements: The actual amount needed in each time period.

Planned-order receipts: The quantity expected to be received by the *beginning* of the period in which it is shown. Under lot-for-lot ordering, this quantity will equal net requirements. Under lot-size ordering, this quantity may exceed net requirements. Any excess is added to available inventory in the *next* time period for simplicity, although in reality, it would be available in that period.

Planned-order releases: Indicates a *planned* amount to order in each time period; equals planned-order receipts offset by lead time. This amount generates gross requirements at the next level in the assembly or production chain. When an order is executed, it is removed from "planned-order releases" and entered under "scheduled receipts."

These quantities are used in a time-phased plan in this format:

Week number:	1	2	3	4	5	6	7	8
Item:								
Gross requirements								
Scheduled receipts								
Projected on hand								
Net requirements								
Planned-order receipts								
Planned-order releases								

Example 2

A firm that produces wood shutters and bookcases has received two orders for shutters: one for 100 shutters and one for 150 shutters. The 100-unit order is due for delivery at the start of week 4 of the current schedule, and the 150-unit order is due for delivery at the start of week 8. Each shutter consists of four slatted wood sections and two frames. The wood sections are made by the firm, and fabrication takes one week. The frames are ordered, and lead time is two weeks. Assembly of the shutters requires one week. There is a scheduled receipt of 70 wood sections in (i.e., at the beginning of) week 1. Determine the size and timing of planned-order releases necessary to meet delivery requirements under each of these conditions:

Computer-based MRP procedures at Kirby Refrigeration integrate the manufacturing, financial, and sales systems—enabling optimum business control. Kirby, a manufacturer and wholesaler of air conditioning and refrigeration components, is a wholly owned Australian company with facilities in Australia and New Zealand.

www.kirby.co.nz/

1. Lot-for-lot ordering (i.e., order size equal to net requirements).
2. Lot-size ordering with a lot size of 320 units for frames and 70 units for wood sections.

Solution

a. Develop a master schedule:

Week number:	1	2	3	4	5	6	6	8
Quantity:				100				150

b. Develop a product structure tree:

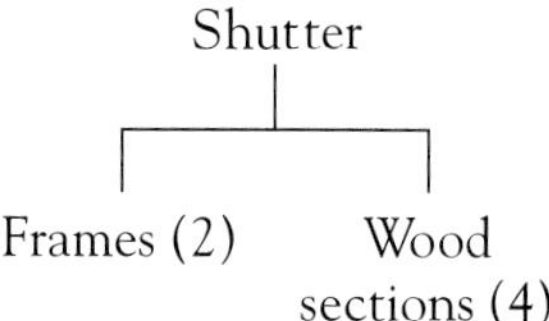

c. Using the master schedule, determine gross requirements for shutters. Next, compute net requirements. Assuming *lot-for-lot ordering,* determine planned-order receipt quantities and the planned-order release timing to satisfy the master schedule (see Figure 14–10).

The master schedule calls for 100 shutters to be ready for delivery, and no shutters are projected to be on hand at the start of week 4, so the net requirements are also 100 shutters. Therefore, planned receipts for week 4 equal 100 shutters. Because shutter assembly requires one week, this means a planned-order release at the start of week 3. Using the same logic, 150 shutters must be assembled during week 7 in order to be available for delivery at the start of week 8.

The planned-order release of 100 shutters at the start of week 3 means that 200 frames (gross requirements) must be available at that time. Because none are expected to be on hand, this generates net requirements of 200 frames and necessitates planned receipts of 200 frames by the start of week 3. With a two-week lead time, this means that the firm must order 200 frames at the start of week 1. Similarly, the planned-order release of 150 shutters at week 7 generates gross and net requirements of 300 frames for week 7 as well as planned receipts for that time. The two-week lead time means the firm must order frames at the start of week 5.

FIGURE 14-10

MRP schedule with lot-for-lot ordering

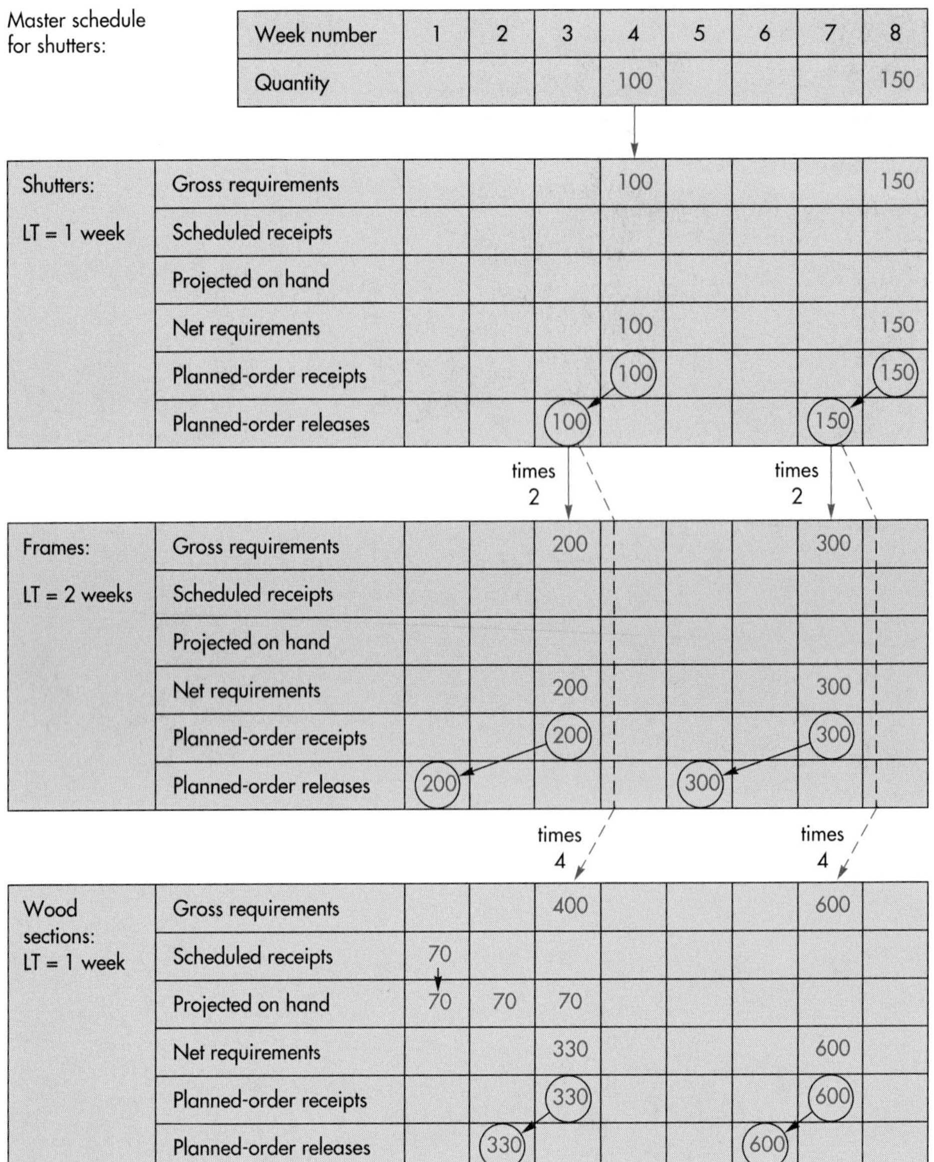

Master schedule for shutters:

Week number	1	2	3	4	5	6	7	8
Quantity				100				150

Shutters: LT = 1 week	1	2	3	4	5	6	7	8
Gross requirements				100				150
Scheduled receipts								
Projected on hand								
Net requirements				100				150
Planned-order receipts				100				150
Planned-order releases			100				150	

Frames: LT = 2 weeks	1	2	3	4	5	6	7	8
Gross requirements			200				300	
Scheduled receipts								
Projected on hand								
Net requirements			200				300	
Planned-order receipts			200				300	
Planned-order releases	200				300			

Wood sections: LT = 1 week	1	2	3	4	5	6	7	8
Gross requirements			400				600	
Scheduled receipts	70							
Projected on hand	70	70	70					
Net requirements			330				600	
Planned-order receipts			330				600	
Planned-order releases		330				600		

The planned-order release of 100 shutters at the start of week 3 also generates gross requirements of 400 wood sections at that time. However, because 70 wood sections are expected to be on hand, net requirements are 400 − 70 = 330. This means a planned receipt of 330 by the start of week 3. Since fabrication time is one week, the fabrication must start (planned-order release) at the beginning of week 2.

Similarly, the planned-order release of 150 shutters in week 7 generates gross requirements of 600 wood sections at that point. Because no on-hand inventory of wood sections is projected, net requirements are also 600, and planned-order receipt is 600 units. Again, the one-week lead time means 600 sections are scheduled for fabrication at the start of week 6.

d. Under lot-size ordering, the only difference is the possibility that planned receipts will exceed net requirements. The excess is recorded as projected inventory in the following period. For example, in Figure 14–11, the order size for frames is 320 units, Net requirements for week 3 are 200; thus, there is an excess of 320 − 200 = 120 units, which become projected inventory in the next week. Similarly, net frame

FIGURE 14-11

MRP schedule with lot sizes for components

Master schedule for shutters:

Week number	1	2	3	4	5	6	7	8
Quantity				100				150

Shutters: LT = 1 week; Lot size = lot-for-lot		1	2	3	4	5	6	7	8
	Gross requirements				100				150
	Scheduled receipts								
	Projected on hand								
	Net requirements				100				150
	Planned-order receipts				100				150
	Planned-order releases			100				150	

times 2   times 2

Frames: LT = 2 weeks; Lot size = multiples of 320		1	2	3	4	5	6	7	8
	Gross requirements			200				300	
	Scheduled receipts								
	Projected on hand				120	120	120	120	140
	Net requirements			200				180	
	Planned-order receipts			320				320	
	Planned-order releases	320				320			

times 4   times 4

Wood sections: LT = 1 week; Lot size = multiples of 70		1	2	3	4	5	6	7	8
	Gross requirements			400				600	
	Scheduled receipts	70							
	Projected on hand	70	70	70	20	20	20	20	50
	Net requirements			330				580	
	Planned-order receipts			350				630	
	Planned-order releases		350				630		

requirements of 180 units are 140 less than the 320 order size; again, the excess becomes projected inventory in week 8. The same thing happens with wood sections; an excess of planned receipts in weeks 3 and 7 is added to projected inventory in weeks 4 and 8. Note that the order size must be in *multiples* of the lot size; for week 3 it is 5 times 70, and for week 7 it is 9 times 70.

MRP provides plans for the end item and each of its subassemblies and components. Conceptually, this amounts to what is depicted in Figure 14–12. Practically speaking, however, the number of components in even a relatively simple product would make the width of the resulting spreadsheet far too wide to handle. Consequently, the plans for the individual components are stacked, as illustrated in the preceding example. Because of this, it is important to refer to the product tree in order to track relationships between components.

Example 2 is useful for describing some of the main features of MRP processing, but it understates the enormity of the task of keeping track of material requirements, especially in situations where the same subassemblies, parts, or raw materials are used in a

Master schedule for shutters:

Week number	1	2	3	4	5	6	7	8
Quantity								

Item: Shutters								
Gross requirements								
Scheduled receipts								
Projected on hand								
Net requirements								
Planned-order receipts								
Planned-order releases								

Item: Frames								
Gross requirements								
Scheduled receipts								
Projected on hand								
Net requirements								
Planned-order receipts								
Planned-order releases								

Item: Wood sections								
Gross requirements								
Scheduled receipts								
Projected on hand								
Net requirements								
Planned-order receipts								
Planned-order releases								

FIGURE 14–12

Net requirements at each level determine gross requirements at the next

number of different products. Differences in timing of demands and quantities needed, revisions caused by late deliveries, high scrap rates, and canceled orders all have an impact on processing.

Consider, for example, bracket G, which is used for end items W, X, and Y, as shown by the product trees in Figure 14–13. For simplicity, assume that all lead times are one week. Figure 14–14 shows the master production schedule for these products. Note that these are the *gross requirements* for these end items: *net requirements* will be less by the amount of any on-hand inventories. Again, for simplicity, suppose that there are no on-hand inventories (or scheduled receipts). This results in a schedule of net requirements that is the same as the gross requirements, but one week earlier (lead time is one week), as depicted at the top of Figure 14–15A on page 632. Referring to the tree diagrams in Figure 14–13, we can see that each W requires one N, each X requires one M, and each Y requires one H. Consequently, the gross requirements for N, M, and H are identical to the net requirements of W, X, and Y, and their net requirements are also identical, but one week earlier (again, lead times are one week), as depicted in the middle of Figure 14–15A.

Again referring to the tree diagrams in Figure 14–13, we can see the number of Gs needed for each N (1), M (3), and H (2). Applying the multipliers to the net requirements for N, M, and H, we obtain the gross requirements for G as depicted at the bottom of Figure 14–15A. Figure 14–15B computes the net requirements for G, where we see that there is a projected on-hand inventory of 100 brackets that must be taken into account.

Finally, note that G also has a lead time of one week, which means that unless the 610 brackets were made in the previous week, the initial portion of this schedule will be late!

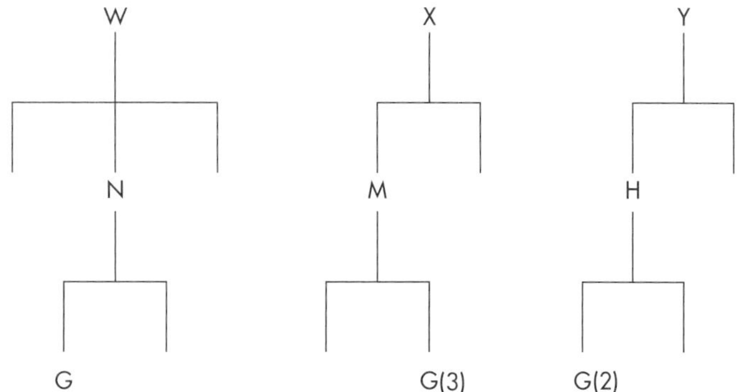

FIGURE 14-13

Bracket G is used in three different products

FIGURE 14-14

Master production schedule for end items that use bracket G

Week number

End item	1	2	3	4	5	6	7
W				50		70	
X				200			100
Y				30		80	90

The term **pegging** denotes working this process in reverse; that is, identifying the parent items that have generated a given set of material requirements for some item such as G. Although the process may appear simple enough given the product trees and schedules shown in this chapter, when multiple products are involved, the process is more complex. Pegging enables managers to determine which product(s) will be affected if orders are late due to late deliveries, quality problems, or other problems.

pegging The process of identifying the parent items that have generated a given set of material requirements for an item.

The importance of the computer becomes evident when you consider that a typical firm would have not one but many end items for which it needs to develop material requirements plans, each with its own set of components. Inventories on hand and on order, schedules, order releases, and so on must all be updated as changes and rescheduling occur. Without the aid of a computer, the task would be almost hopeless; with the computer, planners can accomplish all of these things with much less difficulty.

Updating the System

A material requirements plan is not a static document. As time passes, some orders will have been completed, other orders will be nearing completion, and new orders will have been entered. In addition, there may have been changes to orders, such as changes in quantity, delays, missed deliveries of parts or raw materials, and so on. Hence, a material requirements plan is a "living" document, one that changes over time. And what we refer to as "Period 1" (i.e., the current period) is continually moving ahead. So what is now Period 2 will soon be Period 1. In a sense, schedules such as these have a *rolling horizon,* which means that plans are updated and revised so that they reflect the next set number of periods.

The two basic systems used to update MRP records are *regenerative* and *net change.* A **regenerative system** is updated periodically; a **net-change system** is continuously updated.

regenerative system Approach that updates MRP records periodically.

A regenerative system is essentially a batch-type system, which compiles all changes (e.g., new orders, receipts) that occur within the time interval (e.g., week) and periodically updates the system. Using that information, a revised production plan is developed in the same way that the original plan was developed (e.g., exploding the bill of materials, level by level).

net-change system Approach that updates MRP records continuously.

In a net-change system, the production plan is modified to reflect changes as they occur. If some defective purchased parts had to be returned to a vendor, the manager

FIGURE 14-15

Gross and net requirements for bracket G

A. Gross requirements.

Week number

Net requirements:	1	2	3	4	5	6	7
W LT=1			50		70		
X LT=1			200			100	
Y LT=1			30		80	90	

*No on-hand quantities for W, X, or Y

Week number

Net requirements*	1	2	3	4	5	6	7
N LT=1		50		70			
M LT=1		200			100		
H LT=1		30		80	90		

*No on-hand quantities for N, M, or H

Week number

Gross requirements for G* due to:	1	2	3	4	5	6	7
N	50		70				
M(3)	600			300			
H(2)	60		160	180			
Total	710		230	480			

*LT=1 week

B. Net requirements

Week number

	1	2	3	4	5	6	7
Gross requirements	710		230	480			
Scheduled receipts	100						
Net requirements	610		230	480			

can enter this information into the system as soon as it becomes known. Only the *changes* are exploded through the system, level by level; the entire plan would not be regenerated.

The regenerative system is best suited to fairly stable systems, whereas the net-change system is best suited to systems that have frequent changes. The obvious disadvantage of a regenerative system is the potential amount of lag between the time information becomes available and the time it can be incorporated into the material requirements plan. On the other hand, processing costs are typically less using regenerative systems; changes that occur in a given time period could ultimately cancel each other, thereby avoiding the need to modify and then remodify the plan. The disadvantages of the net-change system relate to the computer processing costs involved in continuously updating the system and the constant state of flux in a system caused by many small changes. One way around this is to enter minor changes periodically and major changes immediately. The primary advantage of the net-change system is that management can have up-to-date information for planning and control purposes.

MRP OUTPUTS

MRP systems have the ability to provide management with a fairly broad range of outputs. These are often classified as *primary reports,* which are the main reports, and *secondary reports,* which are optional outputs.

Primary Reports. Production and inventory planning and control are part of primary reports. These reports normally include the following:

1. **Planned orders,** a schedule indicating the amount and timing of future orders.
2. **Order releases,** authorizing the execution of planned orders.
3. **Changes** to planned orders, including revisions of due dates or order quantities and cancellations of orders.

Secondary Reports. Performance control, planning, and exceptions belong to secondary reports.

1. **Performance-control reports** evaluate system operation. They aid managers by measuring deviations from plans, including missed deliveries and stockouts, and by providing information that can be used to assess cost performance.
2. **Planning reports** are useful in forecasting future inventory requirements. They include purchase commitments and other data that can be used to assess future material requirements.
3. **Exception reports** call attention to major discrepancies such as late and overdue orders, excessive scrap rates, reporting errors, and requirements for nonexistent parts.

The wide range of outputs generally permits users to tailor MRP to their particular needs.

planned orders Schedule indicating the amount and timing of future orders.

order releases Authorization for the execution of planned orders.

changes Revisions of due dates or order quantities, or cancellations of orders.

performance-control reports Evaluation of system operation, including deviations from plans and cost information.

planning reports Data useful for assessing future material requirements.

exception reports Data on any major discrepancies encountered.

OTHER CONSIDERATIONS

Aside from the main details of inputs, outputs, and processing, managers must be knowledgeable about a number of other aspects of MRP. These include the holding of safety stock, lot-sizing choices, and the possible use of MRP for unfinished products.

Safety Stock

Theoretically, inventory systems with dependent demand should not require safety stock below the end item level. This is one of the main advantages of an MRP approach. Supposedly, safety stock is not needed because the manager can project usage quantities once the master schedule has been established. Practically, however, there may be exceptions. For example, a bottleneck process or one with varying scrap rates can cause shortages in downstream operations. Furthermore, shortages may occur if orders are late or fabrication or assembly times are longer than expected. On the surface, these conditions lend themselves to the use of safety stock to maintain smooth operations; but the problem becomes more complicated when dealing with multiechelon items (i.e., multiple-level arenas such as assembled products) because a shortage of *any* component will prevent manufacture of the final assembly. However, a major advantage of MRP is lost by holding safety stock for all lower-level items.

MRP systems deal with these problems in several ways. The manager's first step is to identify activities or operations that are subject to variability and to determine the extent of that variability. When lead times are variable, the concept of safety *time* instead of safety *stock* is often used. This results in scheduling orders for arrival or completion sufficiently ahead of the time they are needed in order to eliminate or substantially reduce the element of chance in waiting for those items. When quantities tend to vary, some safety stock may be called for, but the manager must carefully weigh the need and cost of carrying extra stock. Frequently, managers elect to carry safety stock for end items, which are subject to random demand, and for selected lower-level operations when safety time is not feasible.

It is important in general to make sure that lead times are accurate, particularly when the objective is to have incoming shipments of parts and materials arrive shortly before they are needed. Early arrivals increase on-hand inventory and carrying costs, but late

FIGURE 14-16

Demand for part K

	Period 1	2	3	4	5
Demand	70	50	1	80	4
Cumulative demand	70	120	121	201	205

arrivals can raise havoc, possibly delaying all following operations. Knowing this, managers may inflate lead times (i.e., use safety time) and cause early arrivals, defeating the objective of matching the arrival of orders with production schedules.

If safety stock is needed, planned order release amounts can be increased by the safety stock quantities for the designated components.

Lot Sizing

lot sizing Choosing a lot size for ordering or production.

Determining a lot size to order or to produce is an important issue in inventory management for both independent- and dependent-demand items. This is called **lot sizing.** For independent-demand items, managers often use economic order sizes and economic run sizes. For dependent-demand systems, however, a much wider variety of plans is used to determine lot sizes, mainly because no single plan has a clear advantage over the others. Some of the most popular plans for lot sizing are described in this section.

A primary goal of inventory management for both independent- and dependent-demand systems is to minimize the sum of ordering cost (or setup cost) and holding cost. With independent demand, that demand is frequently distributed uniformly throughout the planning horizon (e.g., six months, year). Demand tends to be much more lumpy for dependent demand, and the planning horizon shorter (e.g., three months), so that economic lot sizes are usually much more difficult to identify. Consider the situation depicted in Figure 14–16. Period demands vary from 1 to 80 units, and no demand size repeats over the horizon shown.

Managers can realize economies by grouping order or run sizes. This would be the case if the additional cost incurred by holding the extra units until they were used led to a savings in setup or ordering cost. This determination can be very complex at times, for several reasons. First, combining period demands into a single order, particularly for middle-level or end items, has a cascading effect down through the product tree; that is, in order to achieve this grouping, you must also group items at lower levels in the tree and incorporate their setup and holding costs into the decision. Second, the uneven period demand and the relatively short planning horizon require a continual recalculation and updating of lot sizes. Not surprisingly, the methods used to handle lot sizing range from the complex, which attempt to include all relevant costs, to the very simple, which are easy to use and understand. In certain cases, the simple models seem to approach cost minimization although generalizations are difficult. Let's consider some of these models.

Lot-for-Lot Ordering. Perhaps the simplest of all the methods is lot-for-lot ordering. The order or run size for each period is set equal to demand for that period. Example 2 demonstrated this method. Not only is the order size obvious, it also virtually eliminates holding costs for parts carried over to other periods. Hence, lot-for-lot ordering minimizes investment in inventory. Its two chief drawbacks are that it usually involves many different order sizes and thus cannot take advantage of the economies of fixed order size (e.g., standard containers and other standardized procedures), and it requires a new setup for each run. If setup costs can be significantly reduced, this method may approximate a minimum-cost lot size.

Economic Order Quantity Model. Sometimes economic order quantity models (EOQ) are used. They can lead to minimum costs if usage is fairly uniform. This is sometimes the case for lower-level items that are common to different parents and for raw materials. However, the more lumpy demand is, the less appropriate such an approach is. Since demand tends to be most lumpy at the end item level, EOQ models tend to be less useful for end items than for items and materials at the lowest levels.

Fixed-Period Ordering. This type of ordering provides coverage for some predetermined number of periods (e.g., two or three). In some instances, the span is simply arbitrary; in other cases, a review of historical demand patterns may lead to a more rational designation of a fixed period length. A simple rule is: Order to cover a two-period interval. The rule can be modified when common sense suggests a better way. For example, take a look at the demands shown in Figure 14–16. Using a two-period rule, an order size of 120 units would cover the first two periods. The next two periods would be covered by an order size of 81 units. However, the demands in periods 3 and 5 are so small, it would make sense to combine them both with the 80 units and order 85 units.

Part-Period Model. The part-period model represents another attempt to balance setup and holding costs. The term *part period* refers to holding a part or parts over a number of periods. For instance, if 10 parts were held for two periods, this would be $10 \times 2 = 20$ part periods. The economic part period (EPP) can be computed as the ratio of setup costs to the cost to hold a unit for one period. Thus, the formula for computing the EPP is:

$$\text{EPP} = \frac{\text{Setup cost}}{\text{Unit holding cost per period}} \tag{14–2}$$

In order to determine an order size that is consistent with the EPP, various order sizes are examined for a planning horizon, and each one's number of part periods is determined. The one that comes closest to the EPP is selected as the best lot size. The order sizes that are examined are based on cumulative demand. Example 3 illustrates this approach.

The choice of a lot-sizing technique must take into account the nature of demand (degree of uniformity), the relative importance of carrying costs versus ordering costs, and any other considerations that affect ordering. It appears that no single method is suited to all conditions.

Regardless of the lot-sizing method in use, there is always the possibility of adjustments in order sizes due to allowance for shrinkage or scrap, minimum and maximum order quantities established by management (e.g., do not order more than five months' supply), operating or shipping constraints (e.g., 200 pieces per run or 12 dozen per shipment) that require rounding of order sizes (usually up) to these amounts, and similar factors.

Example 3

Use the part-period method to determine order sizes for this demand schedule:

	Period							
	1	**2**	**3**	**4**	**5**	**6**	**7**	**8**
Demand	60	40	20	2	30	—	70	50
Cumulative demand	60	100	120	122	152	152	222	272

Setup cost is $80 per run for this item, and unit holding cost is $.95 per period.

Solution

1. First compute the EPP: EPP = $80/$.95 = 84.21, which rounds to 84 part periods. This is the target quantity.
2. Next, try the cumulative lot sizes, beginning with 60, until the part periods approximate the EPP. Continue this process for the planning horizon. This leads to the following:

Period When Order Is Placed	Lot Size	Extra Inventory Carried	×	Periods Carried	=	Part Periods	Cumulative Part Periods
1	60	0		0		0	0
	100	40		1		40	40
	120	20		2		40	80
	122	2		3		6	86*
5	30	0		0		0	0
	100	70		2		140	140*
8	50	0		0		0	0

*Closest to 84

The computations of part periods indicate that 122 units should be ordered to be available at period 1, and 100 units should be ordered to be available at period 5. The next lot will be ordered for period 8, but there is insufficient information now to determine its size.

The lot sizes considered for period 1 correspond to cumulative demand. Once the best lot size has been identified, the cumulative demand is set equal to zero and then summed beginning with the next period. In this case, the lot size of 122 covers the first four periods, so cumulative demand is started next for period 5. The next lot size covers through period 7, and the count begins again at period 8. Hence, that lot size will be at least 50 units.

The process works well for the first lot size because the cumulative number of part periods is close to the EPP, but the effect of lumpy demand is apparent for the second lot size of 100 (140 part periods is not very close to 84 part periods).

CAPACITY REQUIREMENTS PLANNING

One of the most important features of MRP is its ability to aid managers in capacity planning.

capacity requirements planning The process of determining short-range capacity requirements.

Capacity requirements planning is the process of determining short-range capacity requirements. The necessary inputs include planned-order releases for MRP, the current shop load, routing information, and job times. Outputs include load reports for each work center. When variances (underloads or overloads) are projected, managers might consider remedies such as alternative routings, changing or eliminating lot sizing or safety stock requirements, and lot splitting. Moving production forward or backward can be extremely challenging because of precedence requirements and the availability of components.

A firm usually generates a master schedule initially in terms of what is needed and not what is possible. The initial schedule may or may not be feasible given the limits of the production system and availability of materials when end items are translated into requirements for procurement, fabrication, and assembly. Unfortunately, the MRP system cannot distinguish between a feasible master schedule and a nonfeasible one. Consequently, it is often necessary to run a proposed master schedule through MRP processing in order to obtain a clearer picture of actual requirements, which can then be compared to available capacity and materials. If it turns out that the current master schedule is not feasible, management may make a decision to increase capacity (e.g., through overtime or subcontracting) or to revise the master schedule. In the latter case, this may entail several revisions, each of which is run through the system until a feasible plan is obtained. At that point, the master schedule is *frozen,* at least for the near term, thus establishing a firm schedule from which to plan requirements.

Stability in short-term production plans is very important; without it, changes in order quantity and/or timing can render material requirements plans almost useless. The term *system nervousness* describes the way a system might react to changes. The reaction can sometimes be greater than the original change. For example, a small change near the top of a product tree can reverberate throughout much of the lower parts of the tree, causing changes to order quantities and production schedules of many components. That, in turn, might cause queues to form at various portions of the system, leading to late orders, increased work in process, and added carrying costs.

To minimize such problems, many firms establish a series of time intervals, called **time fences,** during which changes can be made to orders. For example, a firm might specify time fences of 4, 8, and 12 weeks, with the nearest fence being the most restrictive and the farthest fence being the least restrictive. Beyond 12 weeks, changes are expected; from 8 to 12 weeks, substitutions of one end item for another may be permitted as long as the components are available and the production plan is not compromised; from 4 to 8 weeks, the plan is fixed, but small changes may be allowed; and the plan is frozen out to the four-week fence.

time fences Series of time intervals during which order changes are allowed or restricted; the nearest fence is most restrictive to change, the farthest is least restrictive.

Some companies use two fences: one is a near-term *demand* fence, and the other is a long-term *planning* fence. For example, the demand fence might be four weeks from the present time while the planning fence might be 10 weeks away. In the near term, customer orders receive precedence over the forecast. The time beyond the planning fence is available for inserting new orders into the master schedule. Between the demand fence and the planning fence, management must make trade-offs when changes are introduced unless excess capacity is expected to be available.

In establishing time fences, a manager must weigh the benefits of stability in the production plan against the possible negative impact on the competitive advantage of being able to quickly respond to new orders.

Figure 14–17 presents an overview of the capacity planning process. The process begins with a proposed or tentative master production schedule that must be tested for feasibility and possibly adjusted before it becomes permanent. The proposed schedule is processed using MRP to ascertain the material requirements the schedule would generate. These are then translated into resource (i.e., capacity) requirements, often in the form of a series of **load reports** for each department or work center, which compares known and expected future capacity requirements with projected capacity availability. Figure 14–18 illustrates the nature of a load report. It shows expected resource requirements (i.e., usage) for jobs currently being worked on, planned orders, and expected orders for the planning horizon. Given this sort of information, the manager can more easily determine whether capacity is sufficient to satisfy these requirements. If there is enough capacity, he or she can freeze the portion of the master production schedule that generates these requirements. In the load report illustrated in Figure 14–18, planned-order releases in time period 4 will cause an overload. However, it appears possible to accommodate demand by slightly shifting some orders to adjacent periods. Similarly, an overload appears likely in period 11, but that too can be handled by shifting some jobs to adjacent time periods. In cases where capacity is insufficient, a manager may be able to increase capacity (by scheduling overtime, transferring personnel from other areas, or subcontracting some of the work) if this is possible and economical, or else revise the master production schedule and repeat the process until an acceptable production schedule is obtained.

load reports Department or work center reports that compare known and expected future capacity requirements with projected capacity availability.

If the master production schedule must be revised, this generally means that the manager must assign priorities to orders, since some orders will be finished later than originally planned.

FIGURE 14-17

Using MRP to assist in planning capacity requirements

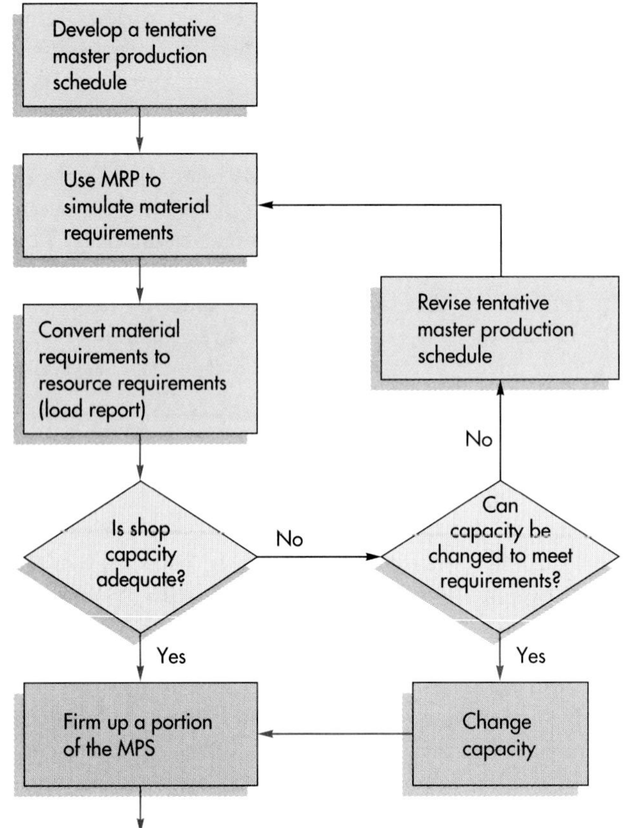

Source: Stephen Love, *Inventory Control* (New York: McGraw-Hill, 1979), p. 164. Reprinted by permission.

FIGURE 14-18

A hypothetical department load report

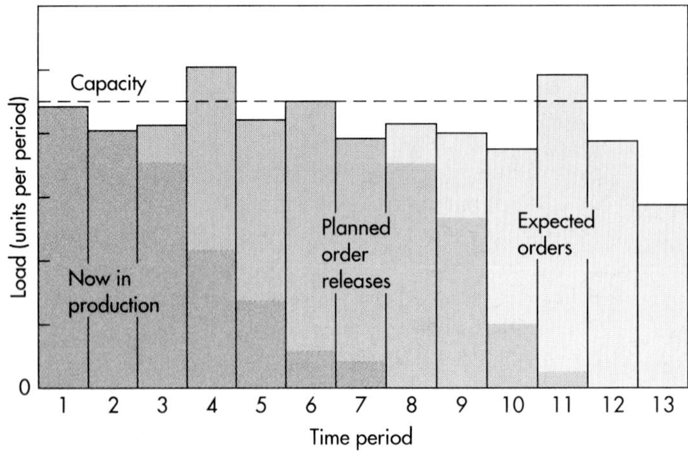

One note of caution is in order concerning capacity load reports. Often, the load reports are only approximations, and may not give a true picture because the loading does not take into account scheduling and queuing delays. Consequently, it is possible to experience system backups even though a load report implies sufficient capacity to handle projected loads.

An important aspect of capacity requirements planning is the conversion of quantity requirements into labor and machine requirements. One accomplishes this by multiplying each period's quantity requirements by standard labor and/or machine requirements per unit. For instance, if 100 units of product A are scheduled in the fabrication department, and each unit has a labor standard time of 2 hours and a machine standard time of 1.5 hours, then 100 units of A convert into these capacity requirements.

$$\text{Labor:} \quad 100 \text{ units} \times 2 \text{ hours/unit} = 200 \text{ labor hours}$$
$$\text{Machine:} \quad 100 \text{ units} \times 1.5 \text{ hours/unit} = 150 \text{ machine hours}$$

One can then compare these capacity requirements with available department capacity to determine the extent to which this product utilizes capacity. For example, if the department has 200 labor hours and 200 machine hours available, labor utilization will be 100 percent because all of the labor capacity will be required by this product. However, machine capacity will be underutilized.

$$\frac{\text{Required}}{\text{Available}} \times 100 = \frac{150 \text{ hours}}{200 \text{ hours}} \times 100 = 75 \text{ percent}$$

Underutilization may mean that unused capacity can be used for other jobs; overutilization indicates that available capacity is insufficient to handle requirements. To compensate, production may have to be rescheduled or overtime may be needed.

BENEFITS AND REQUIREMENTS OF MRP

Benefits. MRP offers a number of benefits for the typical manufacturing or assembly type of operation, including:

1. Low levels of in-process inventories.
2. The ability to keep track of material requirements.
3. The ability to evaluate capacity requirements generated by a given master schedule.
4. A means of allocating production time.

A range of people in a typical manufacturing company are important users of the information provided by an MRP system. Production planners are obvious users of MRP. Production managers, who must balance work loads across departments and make decisions about scheduling work, and plant foremen, who are responsible for issuing work orders and maintaining production schedules, also rely heavily on MRP output. Other users include customer service representatives, who must be able to supply customers with projected delivery dates, purchasing managers, and inventory managers. The benefits of MRP depend in large measure on the use of a computer to maintain up-to-date information on material requirements.

Requirements. In order to implement and operate an effective MRP system, it is necessary to have:

1. A computer and the necessary software programs to handle computations and maintain records. (Table 14–1 shows a sample of software systems.)
2. Accurate and up-to-date
   a. Master schedules
   b. Bills of materials
   c. Inventory records
3. Integrity of file data.

Vol. 4, Tape 2, Seg. 1
Federal Signal

Accuracy is absolutely essential for a successful MRP system. Inaccuracies in inventory record files or bills-of-material files can lead to unpleasant surprises, ranging from missing parts, ordering too many of some items and too few of others, and failure to stay on schedule, all of which contribute to inefficient use of resources, missed delivery dates, and poor customer service. Moreover, implementing MRP can be arduous and costly.

TABLE 14-1

1997 manufacturing resource planning software survey— the top ten (ordered by number of installed users)

Product	Vendor	Installed User Base	Average Price
Caliach MRP	Manufacturing and Computer Systems	71,000	$20,000
pc/MRP for Windows	Software Arts	8,700	$395
BPCS Client/Server	System Software Associates	8,000+	—
NRS	NRS Consulting	6,000	$40,000
Impact Encore/Award	Syspro Group	5,000	$2,000/mod
R/3	SAP America	>5,000	—
BAAN IV	Baan Company	4,100	$3,500–$5,000 per user
ADD+ON Software	ADD+ON Software, Inc.	4,000+	Reseller
JBA System 21	JBA International	4,000+	—
MOVEX	Intentia	4,000	—

APICS THE PERFORMANCE ADVANTAGE, September 1997, Volume 7, Number 9.

Consequently, it is important for companies considering an MRP system to be aware of this and to carefully weigh these factors against the benefits of MRP.

Unfortunately, some firms that have attempted to install an MRP system have seriously underestimated the importance of these items. In many cases, bills of materials are outdated because design changes were not incorporated into the records, leading to parts lists that did not correspond to actual requirements for assembly of the finished product. It is not unusual for a firm to discover that the same part is carried under different part numbers, making it difficult to develop meaningful records. Moreover, some firms have encountered resistance from foremen and others, who argue, "We've managed for 30 years without the stuff, so why bother with it now?"

These obstacles can cause the implementation of an MRP system to take a year or more, taking into account employee education, training, and *convincing,* and correction of record-keeping deficiencies.

On the whole, the introduction of MRP has led to major improvements in scheduling and inventory management, but it has not proved to be the cure-all that many hoped it would be. Consequently, manufacturers are beginning to take a much broader approach to resource planning. One such approach is referred to as MRP II.

NEWSCLIP

SAP R/3 LEADS PACK OF ENTERPRISE RESOURCE PLANNING SOFTWARE PACKAGES

SAP (Systeme, Anwendungen, Produkt), a German company, distributes software designed to fully link all functional areas within a company, from financial accounting to sales and human resource planning including many of the manufacturing and planning and control functions.

According to SAP information, the Production Planning and Control module can draw expected demand data directly from the SOP (sales and operations planning) module, which can then be used to create master schedules. Most any kind of production is "covered" within the production planning and control program, from flow/process type systems down to repetitive manufacturing and a KANBAN specific support system which actually links to a barcode reader. The MRP system within R/# will calculate quantities and procurement dates down to the raw material level.

The ERP idea is also reflected in the Product Data Management part of R/3, which is basically a database system containing a "material master record" for parts and/or products. This record would typically include part number or component data, design information—possibly including CAD specs. This data object is "central to the R/3 system."

SAP is widely used globally and the company has announced the release of R/3 Release 4. SAP Japan began shipping release 4 in April 1998 after successful pilots in late 1997 at Yodobashi Camera and Mitsubishi. The latest version provides extended global trade functions which are based on the "requirements of Japanese sogo shosha, powerful international conglomerates that trade a wide variety of products including chemicals, oil, machinery, textiles and food."

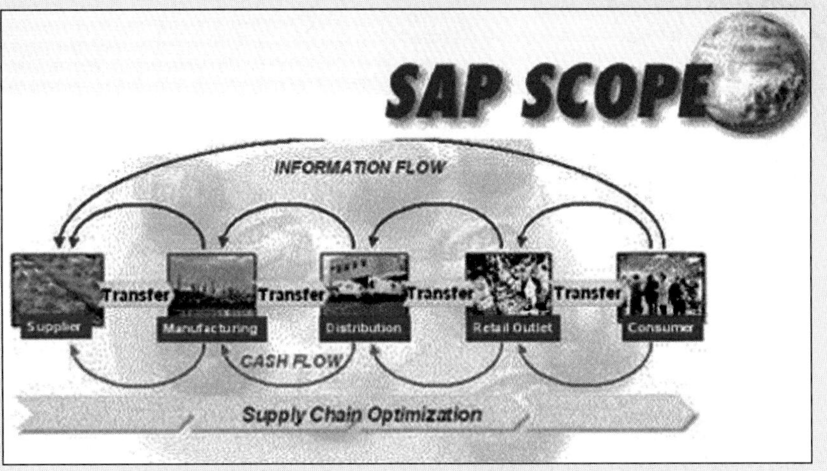

Source: SAP Company press releases.

www.sap.ag.de/

MRP II

In the early 1980s, material requirements planning was expanded into a much broader approach for planning and scheduling the resources of manufacturing firms. This expanded approach has been dubbed MRP II, which refers to **manufacturing resources planning.** It has not replaced MRP, nor is it an improved version of it. Rather, it represents an effort to expand the scope of production resource planning and to involve other functional areas of the firm in the planning process. Marketing and finance are the two most notable areas that are affected by and have an impact on the manufacturing plan.

manufacturing resources planning Expanded approach to production resource planning, involving other areas of a firm in the planning process, such as marketing and finance.

In too many instances, production, marketing, and finance operate without complete knowledge or apparent regard for what other areas of the firm are doing. To be most effective, all areas of the firm need to focus on a common set of goals. A major purpose of MRP II is to integrate primary functions and other functions such as personnel, engineering, and purchasing in the planning process.

Material requirements planning is at the heart of the process (see Figure 14–19). The process begins with an aggregation of demand from all sources (e.g., firm orders, forecasts, safety stock requirements). Production, marketing, and finance personnel work toward developing a master production schedule. Although manufacturing people will have a major input in determining that schedule and a major responsibility for making it work, marketing and finance will also have important inputs and responsibilities. The rationale for having these functional areas work together is the increased likelihood of developing a plan that works and with which everyone can live. Moreover, because each of these functional areas has been involved in formulating the plan, they will have reasonably good knowledge of the plan and more reason to work toward achieving it.

In addition to the obvious manufacturing resources needed to support the plan, financing resources will be needed and must be planned for, both in amount and timing. Similarly, marketing resources will also be needed in varying degrees throughout the process. In order for the plan to work, the firm must have all of the necessary resources available as needed. Often, an initial plan must be revised based on an assessment of the availability of various resources. Once these have been decided, the master production schedule can be firmed up.

At this point, material requirements planning comes into play, generating material and schedule requirements. Next, management must make more detailed capacity requirements planning to determine whether these more specific capacity requirements

FIGURE 14-19
An overview of MRP II

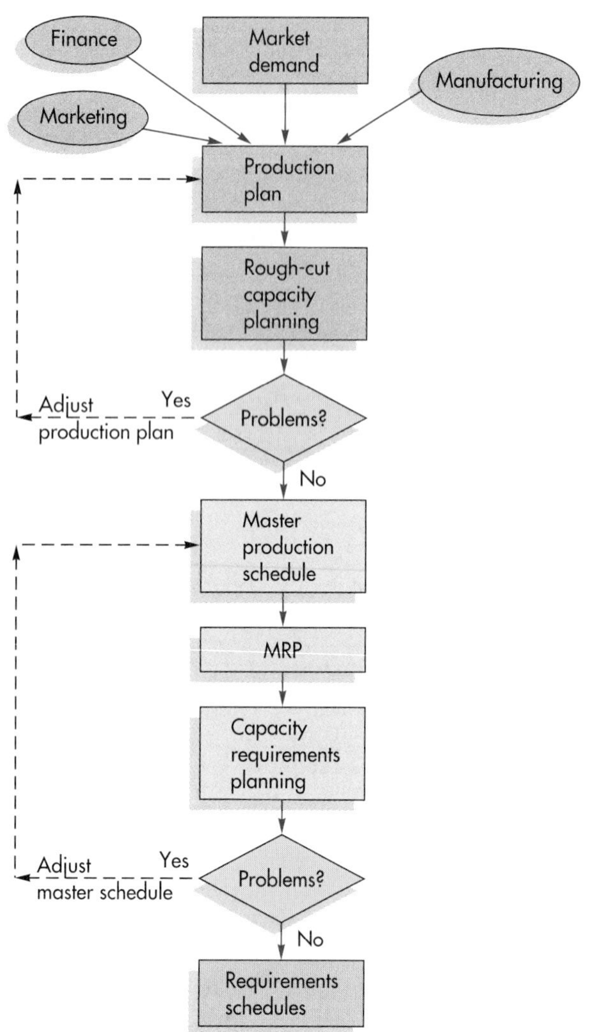

can be met. Again, some adjustments in the master production schedule may be required.

As the schedule unfolds and actual work begins, a variety of reports help managers to monitor the process and to make any necessary adjustments to keep operations on track.

In effect, this is a continuing process, where the master production schedule is updated and revised as necessary to achieve corporate goals. The business plan that governs the entire process usually undergoes changes too, although these tend to be less frequent than the changes made at lower levels (i.e., the master production schedule).

Most MRP II systems have the capability of performing simulation, enabling managers to answer a variety of "what if" questions so they can gain a better appreciation of available options and their consequences.

An extension of MRP II is *enterprise resource planning (ERP),* a term coined by the Gartner Group of Stamford, Connecticut. It is intended to improve resource planning by extending the scope of planning to include more of the supply chain than MRP II.

Software is an essential component of MRP II systems. There is a wide range of software available, with differing features and capabilities. One of the leading products is Caliach MRP, offered by Manufacturing and Computer Systems. Caliach MRP has an installed user base of over 70,000 users, with an average price of $20,000, according to a survey of manufacturing resource planning software done by APICS. The survey is available at *http://lionhrtub.com/apics/surveys/MRPII97/*ordered-by-install.html.

Summary

Material requirements planning (MRP) is an information system used to handle ordering of dependent-demand items (i.e., components of assembled products). The planning process begins with customer orders, which are used along with any back orders to develop a master schedule that indicates the timing and quantity of finished items. The end items are exploded using the bill of materials, and material requirements plans are developed that show quantity and timing for ordering or producing components.

The main features of MRP are the time-phasing of requirements, calculating component requirements, and planned-order releases. To be successful, MRP requires a computer program and accurate master production schedules, bills of materials, and inventory data. Firms without reasonably accurate records or schedules have experienced major difficulties in trying to implement MRP.

MRP II is a second-generation approach to planning that adds a broader scope to manufacturing resource planning because it links business planning, production planning, and the master production schedule. ERP systems are being implemented which build even further on these linkages.

Key Terms

Solved Problems

Problem 1

The following product structure tree indicates the components needed to assemble one unit of product W. Redraw the tree so that it conforms to low-level coding. Then determine the quantities of each component needed to assemble 100 units of W.

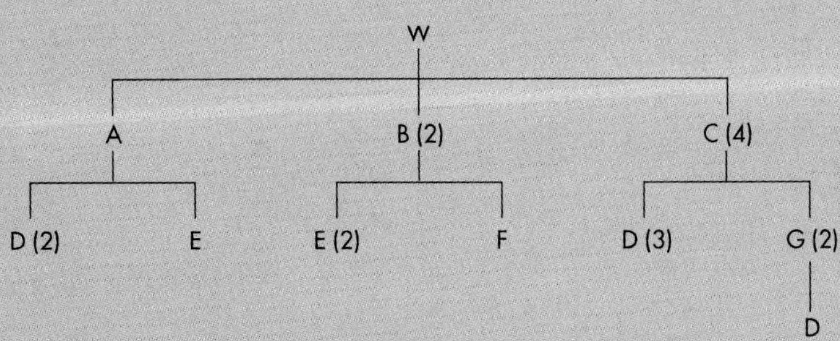

Solution

An easy way to compute and keep track of component requirements is to do it right on the tree, as shown in the following figure. Using low-level coding, it is then relatively simple to group like items in order to determine overall requirements.

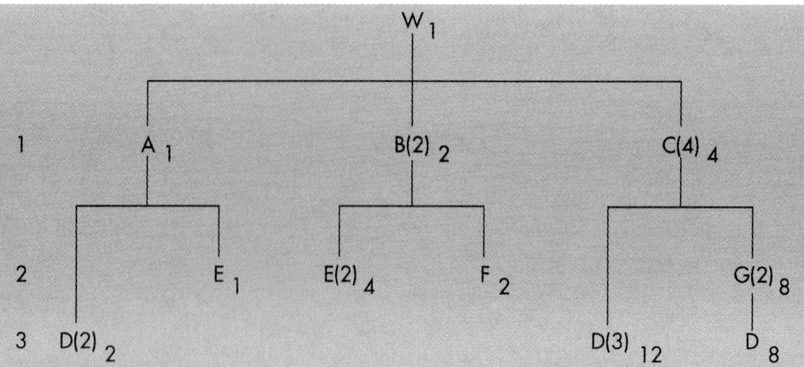

Summary:

Level	Item	(1 W) Quantity	(100 W) Quantity
0	W	1	100
1	A	1	100
	B	2	200
	C	4	400
2	E	5	500
	F	2	200
	G	8	800
3	D	22	2,200

Problem 2

The product structure tree for end item E follows. The manager wants to know the material requirements for ordered part R that will be needed to complete 120 units of E by the start of week 5. Lead times for items are one week for level 0 items, one week for level 1 items, and two weeks for level 2 items. There is a scheduled receipt of 60 units of M at the *end* of week 1 and 100 units of R at the *start* of week 1. Lot-for-lot ordering is used.

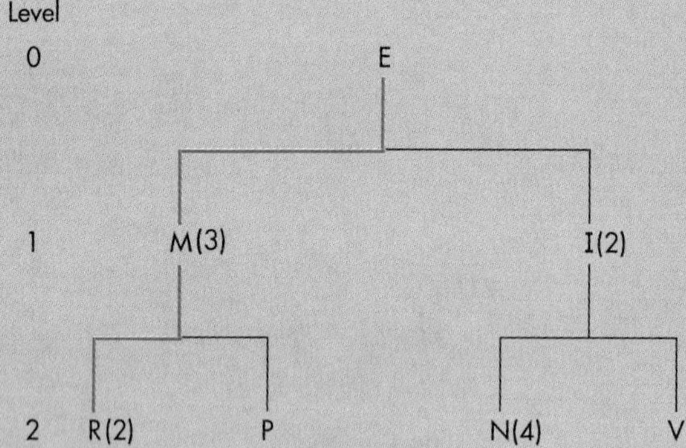

Solution

A partial assembly-time chart that includes R and leads to completion of E by the start of week 5 looks like this:

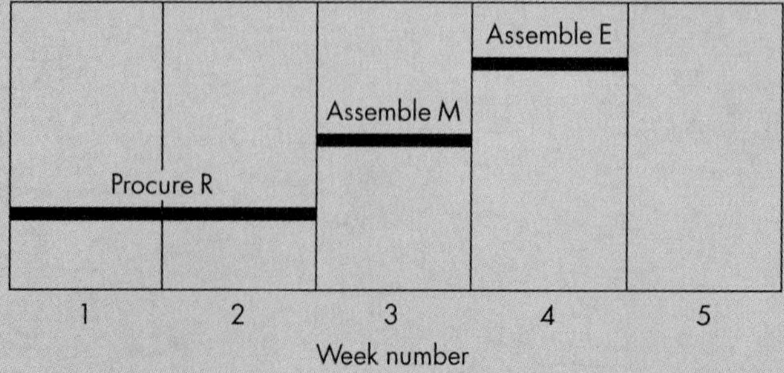

The table entries are arrived at as follows:

Master schedule: 120 units of E to be available at the start of week 5.

Item E: Gross requirements equal the quantity specified in the master production schedule. Since there is no on-hand inventory, net requirements also equal 120 units. Using lot-for-lot ordering, 120 units must be scheduled to be available at the start of week 5. Because there is a one-week lead time for assembly of Es, an order will need to be released (i.e., work started) at the beginning of week 4.

Item M: The *gross* requirements for M are three times the *net* requirements for E, because each E requires three Ms. These must be available at the start of week 4. The net requirements are 60 units less due to the 60 units expected to be on hand at that time. Hence, 300 additional units of M must be available at the start of week 4. With the one-week lead time, there must be an order release at the start of week 3.

Item R: Because each M requires two units of R, 600 Rs will be needed to assemble 300 units of M. However, 100 units will be on hand, so only 500 need to be ordered. Because there is a lead time of two weeks, the 500 Rs must be ordered at the start of week 1.

The master schedule for E and requirements plans for E, M, and R follow.

Master schedule for E

Week number	1	2	3	4	5
Quantity					120

Item: E LT = 1 week					
Gross requirements					120
Scheduled receipts					
Projected on hand					
Net requirements					120
Planned-order receipts					120
Planned-order releases				120	

Multiplied by 3
(see product tree)

Item: M LT = 1 week					
Gross requirements				360	
Scheduled receipts		60			
Projected on hand		60	60	60	
Net requirements				300	
Planned-order receipts				300	
Planned-order releases			300		

Multiplied by 2
(see product tree)

Item: R LT = 2 weeks					
Gross requirements			600		
Scheduled receipts	100				
Projected on hand	100	100	100		
Net requirements			500		
Planned-order receipts			500		
Planned-order releases	500				

Problem 3

Capacity requirements planning. Given the following production schedule in units and the production standards for labor and machine time for this product, determine the labor and machine capacity requirements for each week. Then compute the percent utilization of labor and machines in each week if labor capacity is 200 hours per week and machine capacity is 250 hours per week.

Production Schedule:

Week	1	2	3	4
Quantity	200	300	100	150

Standard Times:

Labor	.5 hours/unit
Machine	1.0 hours/unit

Solution

Convert the quantity requirements into labor and machine requirements by multiplying the quantity requirements by the respective standard times (i.e., multiply each quantity by .5 to obtain the labor hours and multiply each quantity by 1.0 to obtain the machine hours):

Week	1	2	3	4
Quantity	200	300	100	150
Labor hours	100	150	50	75
Machine hours	200	300	100	150

To compute utilization, divide the capacity requirements by the available capacity and multiply by 100. The results are:

Week	1	2	3	4
Labor	50%	75%	25%	37.5%
Machine	80%	120%	40%	60%

Note that machine capacity in week 2 is overutilized (i.e., capacity is insufficient) because the utilization exceeds 100 percent.

Discussion and Review Questions

1. Contrast independent and dependent demand.
2. When is MRP appropriate?
3. Briefly define or explain each of these terms.
   a. Master schedule
   b. Bill of materials
   c. Inventory records file
   d. Gross Requirements
   e. Net requirements
   f. Time-phased plan
   g. Low-level coding
4. How is safety stock included in a material requirements plan?
5. What factors can create safety stock requirements in an MRP system?
6. What is meant by the term *safety time*?
7. Contrast *net-change* systems and *regenerative* systems for MRP.
8. Briefly discuss the requirements for effective MRP.

9. What are some of the main advantages and limitations of MRP?
10. How can the use of MRP contribute to productivity?
11. Briefly describe MRP II and indicate how it relates to MRP.
12. What is lot sizing, what is its goal, and why is it an issue with lumpy demand?
13. Contrast planned-order receipts and scheduled receipts.
14. If seasonal variations are present, is their incorporation into MRP fairly simple or fairly difficult? Explain briefly.

Memo Writing Exercises

1. Your manager, Felix Young, wants to know why your estimate of the time it will take to get an MRP system up and running is so long. Young noted that you included an allowance for "potentially inaccurate bill-of-materials and inventory records," and he wants to know how the company could have operated as well as it has "if the records weren't pretty accurate right now." Write a memo in response to Young's concerns.
2. Suppose you work for a furniture manufacturer, one of whose products is the chair depicted in Figure 14–5. Finished goods inventory is held in a central warehouse in anticipation of customer orders. Finished goods are controlled using EOQ/ROP methods. The warehouse manager, Juan Villa, has suggested using the same methods for controlling component inventory. Write him a brief memo outlining your opinion on doing that.

Problems

1. Given the following diagram for a product, determine the quantity of each component required to assemble one unit of the finished product.

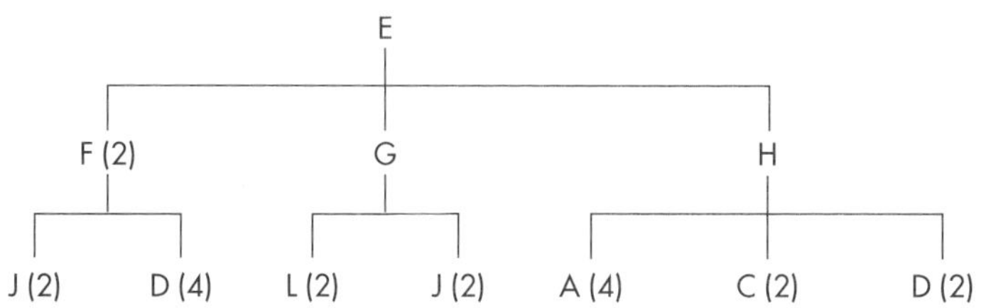

E = 1 H = 1 A = 4
F = 2 J = 6 C = 2
G = 1 L = 2 D = 10

2. The following table lists the components needed to assemble an end item, lead times, and quantities on hand.

a. 138
b. The beginning of week 5.

Item	End	B	C	D	E	F	G	H
LT (wk)	1	2	3	3	1	2	1	2
Amount on hand	0	10	10	25	12	30	5	0

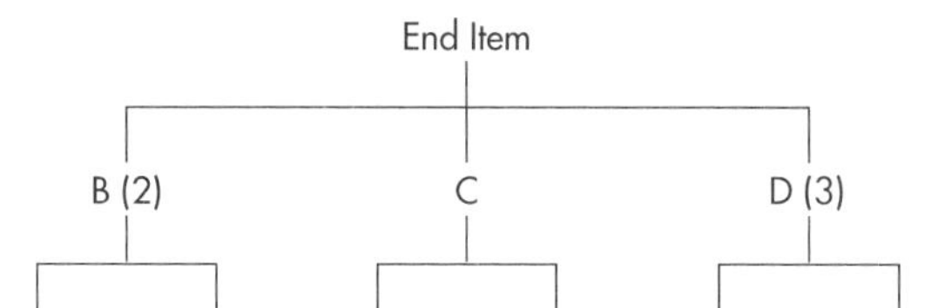

a. If 20 units of the end item are to be assembled, how many additional units of E are needed? (*Hint:* You don't need to develop an MRP plan to determine this.)

b. An order for the end item is scheduled to be shipped at the start of week 11. What is the latest week that the order can be started and still be ready to ship on time? (*Hint:* You don't need to develop an MRP plan for this part either.)

3. The following table lists the components needed to assemble an end item, lead times (in weeks), and quantities on hand.

a. 360
b. The beginning of week 1. (Now!)

Item	Lead Time	Amount on Hand	Direct Components
End	1	—	L(2), C(1), K(3)
L	2	10	B(2), J(3)
C	3	15	G(2), B(2)
K	3	20	H(4), B(2)
B	2	30	
J	3	30	
G	3	5	
H	2	—	

a. If 40 units of the end item are to be assembled, how many additional units of B are needed? (*Hint:* You don't need to develop an MRP plan.)

b. An order for the end item is scheduled to be shipped at the start of week 8. What is the latest week that the order can be started and still be ready to ship on time? (*Hint:* You don't need to develop an MRP plan.)

See IM

4. A very simple product consists of two components, A and B. Two As and four Bs are used to make this product. The department wants to ship 100 units of the product at the beginning of week 6. There are 50 As and 100 Bs on hand. In addition, two scheduled receipts of 100 Bs each are due, one at the start of week 4 and the other at the start of week 5. Lead time is two weeks for the end item, and one week each for A and B. Prepare an MRP plan for this product using lot-for-lot ordering.

See IM

5. End item P is composed of three subassemblies: K, L, and W. K is assembled using 3 Gs and 4 Hs; L is made of 2 Ms and 2 Ns; and W is made of 3 Zs. On-hand inventories are 20 Ls, 40 Gs, and 200 Hs. Scheduled receipts are 10 Ks at the start of week 3, 30 Ks at the start of week 6, and 200 Ws at the start of week 3.

   One hundred Ps will be shipped at the start of week 6, and another 100 at the start of week 7. Lead times are two weeks for subassemblies and one week for components G, H, and M. Final assembly of P requires one week. Include an extra 10 percent scrap allowance in each planned order of G. The minimum order size for H is 200 units. Develop each of the following:

   a. A product structure tree.

   b. An assembly time chart.

   c. A master schedule for P.

   d. A material requirements plan for K, G, and H using lot-for-lot ordering.

See IM

6. A table is assembled using three components, as shown in the accompanying product structure tree. The company that makes the table wants to ship 100 units at the beginning of day 4, 150 units at the beginning of day 5, and 200 units at the beginning of day 7. Receipts of 100 wood sections are scheduled at the beginning of day 2. There are 120 legs on hand. An additional 10 percent of the order size on legs is added for safety stock. There are 60 braces on hand with no safety stock requirement. Lead times (in days) for all items are shown in the following table. Prepare a material requirements plan using lot-for-lot ordering.

Quantity	Lead Time
1–200	1
201–550	2
551–999	3

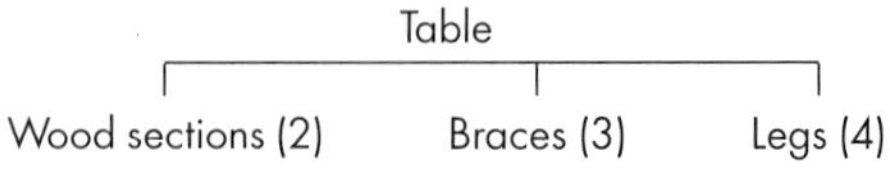

180 units in week 6.

7. The MRP Department has a problem. Its computer "died" just as it spit out the following information: Planned order release for item J27 = 640 units in week 2. The firm has been able to reconstruct all the information they lost except the Master Schedule for end item

565. The firm is fortunate because J27 is used only in 565s. Given the following product structure tree and associated inventory status record information, determine what master schedule entry for 565 was exploded into the material requirements plan that killed the computer.

Part Number	On Hand	Lot Size	Lead Time
565	0	Lot-for-lot	1 week
X43	60	120	1 week
N78	0	Lot-for-lot	2 weeks
Y36	200	Lot-for-lot	1 week
J27	0	Lot-for-lot	2 weeks

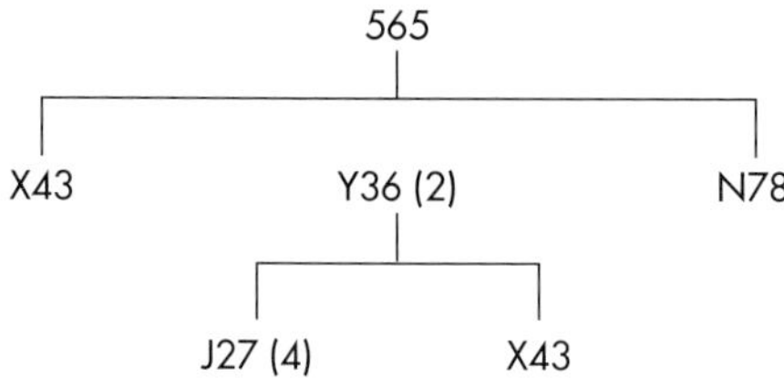

8. The Bloomington Bike Co. produces two models: the Basic and the Supra. Herb Hoosier, the owner, plans to assemble 15 Basics and 10 Supras each week during periods 4–8. Product structure trees for each bike are shown below. MRP in IM.

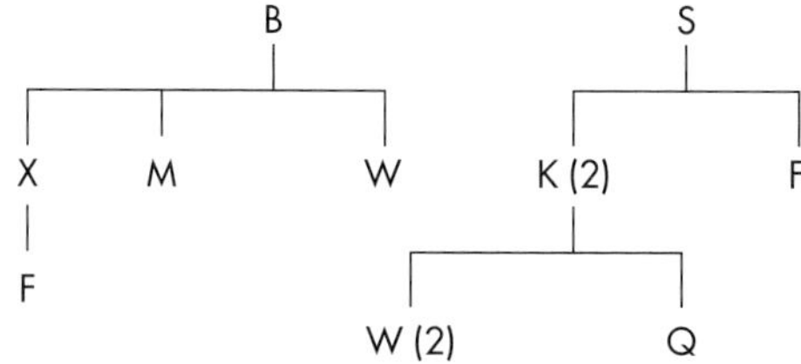

Develop a material requirements plan for component parts K and W over the next eight weeks. Data on all lead times, current inventory, and ordering rules are given below.

Item	Lead Time (weeks)	On Hand	Lot-Sizing Rule
B	2	5	Lot-for-lot
S	2	2	Lot-for-lot
X	1	5	Q = 25
W	2*	2	Multiples of 12
F	1	10	Q = 30
K	1	3	Lot-for-lot
Q	1	15	Q = 30
M	1	0	Lot-for-lot

*LT = 3 weeks for orders of 36 or more units on this item.

Scheduled receipts are:

Period 1: 20 Bs: 18 Ws
Period 2: 20 Ss: 15 Fs

9. Assume that you are the manager of a shop that assembles power tools. You have just received an order for 50 chain saws, which are to be shipped at the start of week 8. Pertinent information on the saws is: Tree, chart, schedule, and MRP in IM.

Item	Lead Time (weeks)	On Hand	Components
Saw.	2	15	A(2), B(1), C(3)
A.	1	10	E(3), D(1)
B.	2	5	D(2), F(3)
C.	2	30	E(2), D(2)
D.	1	20	
E.	1	10	
F.	2	30	

Develop a product structure tree, an assembly time chart, and a master schedule. In addition, develop the material requirements plan for component E using lot-for-lot ordering.

Schedule in IM.
Order 160 units in week 2.

10. Assume that you are the manager of Assembly, Inc. You have just received an order for 40 units of an industrial robot, which is to be delivered at the start of week 7 of your schedule. Using the following information, determine how many units of subassembly G to order and the timing of those orders, given that subassembly G must be ordered in multiples of 80 units and all other components are ordered lot-for-lot. Assume that the components are used only for this particular robot.

Item	Lead Time (weeks)	On Hand	Components
Robot	2	10	B, G, C(3)
B	1	5	E, F
C.	1	20	G(2), H
E	2	4	—
F	3	8	—
G.	2	15	—
H.	1	10	—

a. See IM.
b. See IM.
c. See IM.

11. Determine material requirements plans for parts N and V and subassembly I as described in Solved Problem 2.

    a. Assume that there are currently 100 Ns on hand and scheduled receipts of 40 Is and 10 Vs at the beginning of week 3. No Es are on hand; 120 Es are needed at the start of week 5.

    b. Assume on-hand and scheduled receipts as in Part (*a*). Now suppose that 100 Es are needed at the start of week 5 and 50 at the start of week 7. Also, use multiples of these order sizes: N, 800; V, 200. Use lot-for-lot ordering for I.

    c. Using your answer to Part (*b*), update the MRP for V, using the following additional information for each of these cases: (1) one week has elapsed (making it the start of week 2), and (2) three weeks have elapsed (making it the start of week 4). *Note:* start your revised plans so that the updated time in each case is designated as week 1.

    The updated master schedule now has an order for 100 units of E in week 8 of case 1 (i.e., week 9 under the former master schedule). Assume all orders are released and received as planned.

a. Tree in IM.
b. Chart in IM.
c. Schedule in IM.

12. A firm that produces electric golf carts has just received an order for 200 carts, which must be ready for delivery at the start of week 8. Information concerning the product structure, lead times, and quantities on hand is shown in the following table. Use this information to do each of the following:

    a. Construct a product tree.

    b. Construct an assembly time chart.

    c. Develop a material requirements plan that will provide 200 golf carts by week 8 assuming lot-for-lot ordering.

Parts List for Electric Golf Cart	Lead Time	Quantity on Hand
Electric golf cart	1	0
Top	1	40
Base	1	20
Top		
Supports (4)	1	200
Cover	1	0
Base		
Motor	2	300
Body	3	50
Seats (2)	2	120
Body		
Frame	1	35
Controls	1	0
Wheel assemblies (4)	1	240

13. Refer to Problem 12. Assume that unusually mild weather has caused a change in the quantity and timing of orders for golf carts. The revised plan calls for 100 golf carts at the start of week 6, 100 at the start of week 8, and 100 at the start of week 9.

    a. Develop a master schedule for this revised plan.

    b. Determine the timing and quantities for orders for tops and bases.

    c. Assume that equipment problems reduce the firm's capacity for assembling bases to 50 units per week. Revise your material plan for bases to reflect this, but still meet delivery dates.

a. Schedule in IM.
b. Schedule in IM.
c. Schedule in IM.

14. A manufacturing firm buys a certain part in varying quantities throughout the year. Ordering cost is $11 per order, and carrying cost is $0.14 per piece per month. Given the following demand schedule for the part for the next eight months,

    a. What order sizes would be indicated using an economic part period approach? When should each order be received?

    b. What order sizes and times would be used for EOQ sizing? Assume lead time is zero.

    c. Compare the total cost of fixed internal ordering with a 2-month interval, the EPP approach, and the EOQ approach.

EPP = 78.57
Order 120 units in period 2, 60 units in period 6.

Month	Demand
1	—
2	80
3	10
4	30
5	—
6	30
7	—
8	30

15. A firm periodically produces a part that is a basic component of an assembled product it makes. Each time the part is run, a fixed cost of $125 is incurred. The cost to carry one unit for a week is estimated to be $1.65. For the demand schedule shown, determine the quantity and timing of runs that would be consistent with an economic part period approach. Assume a one-week lead time for each run.

Period	Produce
1	60
3	140
7	80

Week	Demand
1	40
2	20
3	100
4	20
5	—
6	20
7	80

a. Capacity Utilization

Week	Labor	Machine
1	53.3%	60%
2	106.7%	120%
3	80%	90%
4	93.3%	105%

b. Capacity utilization exceeds 100% for both labor and machine in week 2, and for machine in week 4.

16. A company that manufactures paving material for driveways and parking lots expects the following demand for its product for the next four weeks:

Week	1	2	3	4
Material (tons)	40	80	60	70

The company's labor and machine standards and available capacities are:

	Labor	Machine
Production standard (hours per ton)	4	3
Weekly production capacity (hours)	300	200

a. Determine the capacity utilization for labor and machine for each of the four weeks.

b. In which weeks do you foresee a problem? What options would you suggest to resolve any problems? What costs are relevant in making a decision on choosing an option?

a. Schedule in IM.

b. Labor capacity of 700 hours is exceeded on Tuesday in Fabrication and Assembly. Machine capacity of 500 hours also exceeded on Tuesday in those departments. Friday demand exceeded capacity.

17. A company produces two very similar products that go through a three-step sequence of fabrication, assembly, and packaging. Each step requires one day for a lot to be completely processed and moved to the next department. Processing requirements for the departments (hours per unit) are:

	FABRICATION		ASSEMBLY		PACKAGING	
Product	Labor	Machine	Labor	Machine	Labor	Machine
A	2	1	1.5	1	1	.5
B	1	1	1	1	1.5	.5

Department capacities are all 700 hours of labor and 500 hours of machine time, except Friday, when capacities are 200 hours for both labor and machine time. The following production schedule is for next week:

Day	Mon	Tues	Wed	Thurs	Fri
A	200	400	100	300	100
B	300	200	200	200	200

a. Develop a production schedule for each department that shows the capacity requirements for each product and the total load for each day. Ignore changeover time.

b. Evaluate the projected loading for the first three days of the week. Is the schedule feasible? What do you suggest for balancing the load?

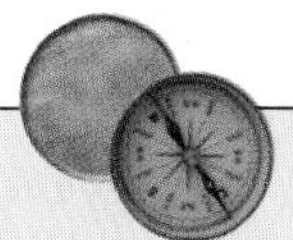

OPERATIONS TOUR
STICKLEY FURNITURE

Introduction

L.&J.G. Stickley was founded in 1900 by brothers Leopold and George Stickley. Located just outside of Syracuse, New York, the company is a producer of fine cherry, white oak, and mahogany furniture. In the late 1980s, the company reintroduced the company's original line of mission oak furniture, which now accounts for over 50 percent of the company's sales.

Over the years, the company experienced both good and bad times, and at one point, it employed over 200 people. But by the early 1970s, the business was in disarray; there were only about 20 full-time employees, and the company was on the brink of bankruptcy. The present owners bought the ailing firm in 1974, and under their leadership, the company has prospered and grown. It now has 900 employees. Stickley has five retail showrooms in New York State, and its furniture is sold nationally by some 100 dealers. In 1994, sales were about $65 million.

Production

The production facility is a large, rectangular building with a 30 foot ceiling. Furniture making is labor intensive, although saws, sanders, and other equipment are very much a part of the process. In fact, electric costs average $40,000 to $50,000 a month. The company has its own tool room where cutting tools are sharpened, and replacement parts are produced as needed.

Worker skills range from low-skilled material handlers to highly skilled craftsmen. For example, three master cabinet makers handle customized orders.

The process (see figure) begins with various sawing operations where large boards received from the lumber mills are cut into smaller sizes. The company recently purchased a computer-controlled "optimizer" saw that greatly improves sawing productivity, and eliminates some waste. Workers inspect and mark knot locations and other defects they find on each piece of lumber before feeding it into the saw. The computer then determines the optimal set of cuttings, given the location of knots and other defects, and standard lengths needed for subsequent operations. Approximately 29,000 board feet are cut each day. Subsequent sawing operations provide additional cuts for specific jobs.

Workers then glue some of the pieces together; they will end up as tops of tables, desks, dressers, or a similar item. Large presses hold 20 to 30 glued sections at a time. Other pieces that will become table or chair legs, chair backs or other items go through various

Optimizer Saw

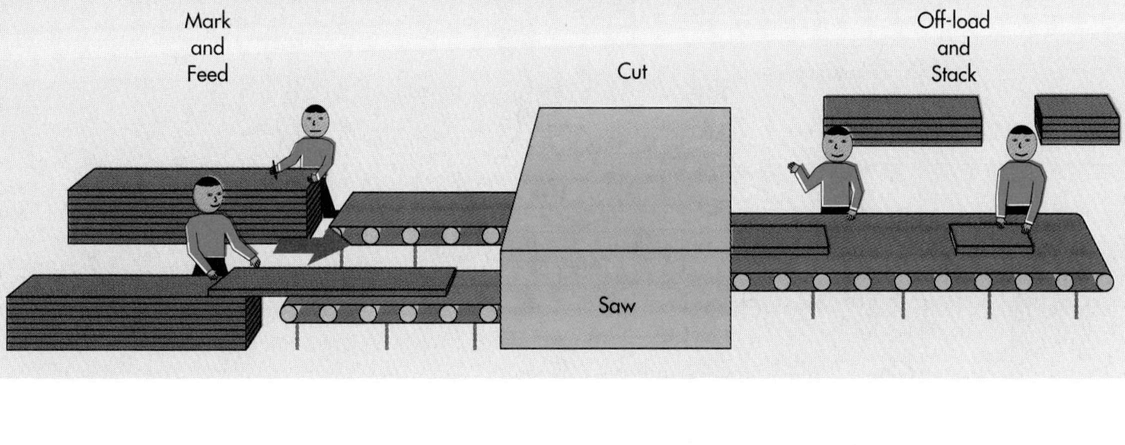

Mark and Feed

Cut

Off-load and Stack

Saw

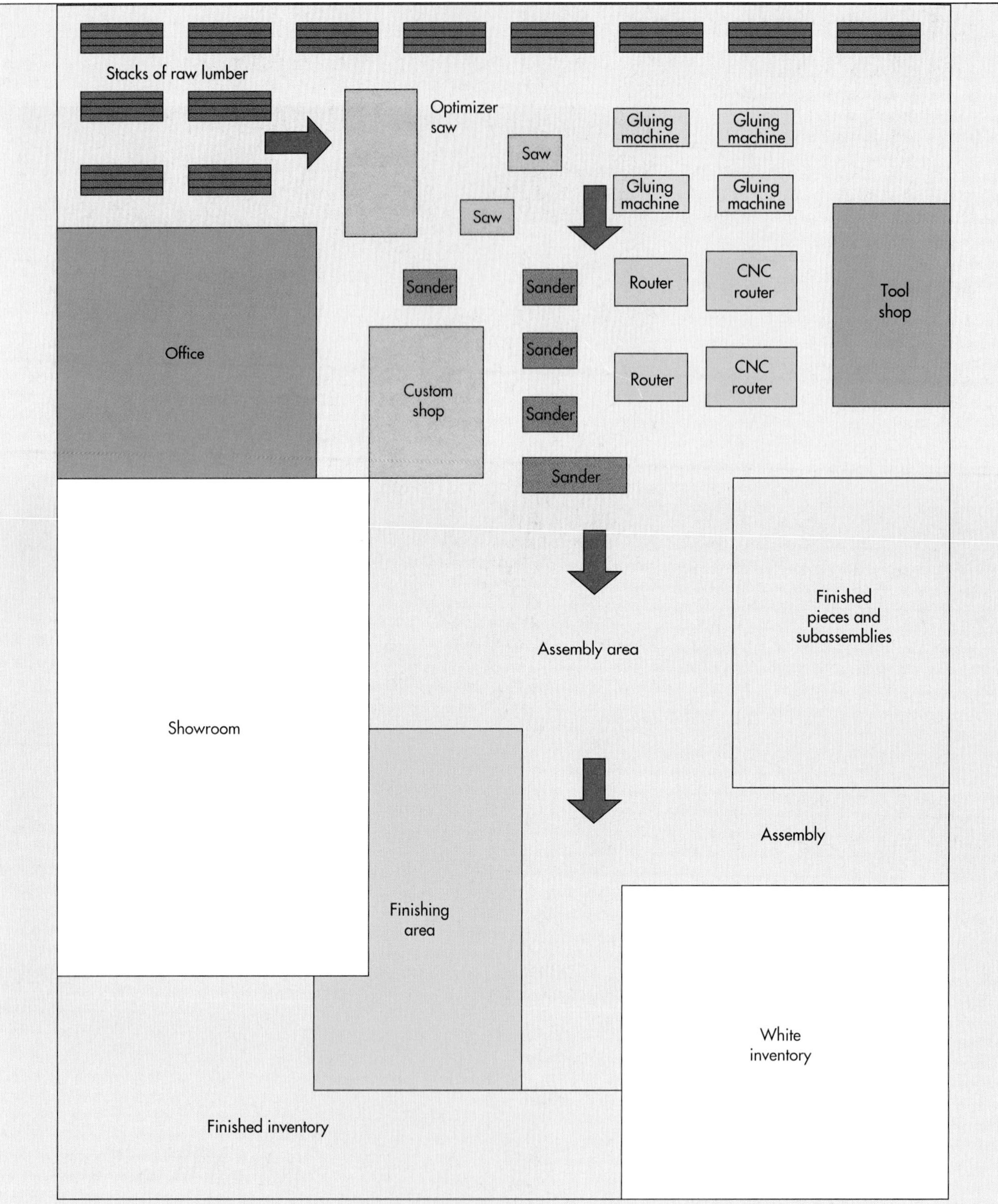

shaping operations. Next comes a series of sanding operations, which remove excess glue from the glued sections, and smooth the surface of both glued pieces and other pieces.

Some of the pieces may require drilling or mortising, an operation in which rectangular holes and other shapes are cut into the wood. The company has a CNC (numerically controlled) router that can be programmed to make grooves and other specialty cuts. Some items require carving, which involves highly skilled workers.

Next, workers assemble the various components, either into subassemblies, or sometimes directly to other components to obtain completed pieces. Each item is stamped with the date of production, and components such as dresser drawers, cabinet doors, and expansion leaves of tables also are stamped to identify their location (e.g., top drawer, left door). Careful records are kept so that if a piece of furniture is ever returned for repairs, complete instructions are available (type of wood, finish, etc.) to enable repair people to closely match the original piece.

The furniture items then usually move to the "white inventory" section, and eventually to the finishing department where workers apply linseed oil or another finish before the items are moved to the finished goods inventory to await shipment to stores or customers.

Aggregate Planning

The company uses a level production plan (maintain steady output and steady labor force). Demand is seasonal; it is highest in the first and third quarters. During the second and fourth quarters, excess output goes into inventory; during the first and third quarters, excess demand is met using inventory. The production scheduler uses a schedule that is set for the next 8 to 10 weeks.

Production Control

Job sequence is determined by the amount of remaining inventory (days supply on hand), and processing time. Lot sizes are determined by factoring in demand, setup costs, and carrying costs. Typical lot sizes are 25 to 60 pieces. There are many jobs being done concurrently. Each job is accompanied by a set of bar codes that identify the job and the operation. As each operation is completed, the operator removes a bar code sticker and delivers it to the scheduling office where it is scanned into the computer, thereby enabling production control to keep track of progress on a job, and to know its location in the shop.

The company's policy of level output coupled with seasonal demand patterns means that prior to peak demand periods, excess output is used to build up inventories, which is then drawn down when demand exceeds production capacity during periods of peak production.

Inventory

In addition to the "white" inventory and a small finished goods inventory, the company maintains an inventory of furniture pieces (e.g., table and chair legs) and partially assembled items. This inventory serves two important functions. One is to reduce the amount of time needed to respond to customer orders rather than having to go through the entire production process to obtain needed items, and secondly, it helps to smooth production and utilize idle machinery/workers. Because of unequal job times on successive operations, some workstations invariably have slack time while others work at capacity. This is used to build an inventory of commonly used pieces and subassemblies. Moreover, because pieces are being made for inventory, there is flexibility in sequencing. This permits jobs that have similar setups to be produced in sequence, thereby reducing setup time and cost.

Quality

Each worker is responsible for checking his or her quality, as well as the quality of materials received from preceding operations, and to report any deficiencies. In addition, on several difficult operations quality control people handle inspections, and work with operators to correct any deficiencies. The company is considering a TQM approach, but has not yet made a decision on whether to go in that direction.

Questions

1. Which type of production processing—job shop, batch, repetitive, or continuous—is the primary mode of operation at Stickley Furniture? Why? What other type of processing is used to a lesser extent? Explain.
2. How does management keep track of job status and location during production?
3. Suppose the company has just received an order for forty mission oak dining room sets. Briefly list the kinds of information the company will need to plan, schedule, and process this job.
4. What benefits, and what problems, would you expect, given the company's level production policy?
5. Can you suggest any changes that might be beneficial to the company? What are they?

Selected Bibliography

Blumberg, D. F. "Factors Affecting the Design of a Successful MRP System." *Production and Inventory Management* 21, no. 4 (4th quarter 1980), pp. 50–62.

Hopp, Wallace, and Mark L. Spearman. *Factory Physics*. Burr Ridge, IL: Richard D. Irwin, 1996.

LaForge, R. C. "MRP and the Part-Period Algorithm." *Journal of Purchasing Management* (Winter 1982), pp. 21–26.

Love, Stephen. *Inventory Control*. New York: McGraw-Hill, 1979.

Orlicky, Joseph. *Material Requirements Planning*. New York: McGraw-Hill, 1975.

Sridharan, V., and R. Lawrence LaForge. "Freezing the Master Production Schedule: Implications for Customer Service." *Decision Sciences* (May–June 1994), pp. 461–69.

Vollmann, Thomas E.; William L. Berry; and D. Clay Whybark. *Manufacturing, Planning and Control Systems*. 4th ed. Burr Ridge, IL: Richard D. Irwin, 1997.

Wight, Oliver W. *The Executive's Guide to Successful MRP II*. Williston, VT: Oliver Wight Limited Publications, 1982.

LEARNING OBJECTIVES

After completing this chapter, you should be able to:

1 Explain what is meant by the term *just-in-time (JIT) production system.*

2 List each of the goals of JIT and explain its importance.

3 List and briefly describe the building blocks of JIT.

4 List the benefits of the JIT systems.

5 Outline the considerations important in converting a traditional mode of production to a JIT system.

6 List some of the obstacles that might be encountered when converting to a JIT system.

CHAPTER FIFTEEN
JUST-IN-TIME SYSTEMS

CHAPTER OUTLINE

just-in-time Repetitive production system in which processing and movement of materials and goods occur just as they are needed, usually in small batches.

The term **just-in-time (JIT)** is used to refer to a production system in which both the movement of goods during production and deliveries from suppliers are carefully timed so that at each step of the process the next (usually small) batch arrives for processing just as the preceding batch is completed—thus the name, *just-in-time.* The result is a system with no idle items waiting to be processed, and no idle workers or equipment waiting for items to process.

The just-in-time phenomenon is characteristic of *lean production systems,* which operate with very little "fat" (e.g. excess inventory, extra workers, wasted space). JIT pertains to the timing of the flow of parts and material through the system, and the timing of services. Companies that employ the JIT/lean production approach typically enjoy a competitive advantage over companies that use a more traditional approach: They have lower processing costs, fewer defectives, greater flexibility, and are able to bring new or improved products to the market more quickly.

One aspect of JIT involves production planning and control, which makes JIT one of the two basic approaches to manufacturing planning and control; the other is material requirements planning. JIT is sometimes viewed as a system for repetitive production operations, while MRP is considered a system for batch production. However, they are sometimes applied to similar situations, but the two systems function somewhat differently. MRP systems are fairly complex, requiring extensive and detailed shop floor controls. JIT systems are much simpler, involving only minimal shop floor controls. Moreover, MRP relies on a computer-based component-*scheduling* system to trigger production and deliveries, whereas JIT relies on visual or audible signals to trigger production and deliveries. The two approaches are compared in more detail later in the chapter.

This chapter describes the JIT/lean production approach, including the basic elements of these systems, and what it takes to make them work effectively. It also points out the benefits of these systems and the potential obstacles that companies may encounter when they attempt to convert from a traditional system to a JIT/lean production system.

INTRODUCTION

www.toyota.com.jp

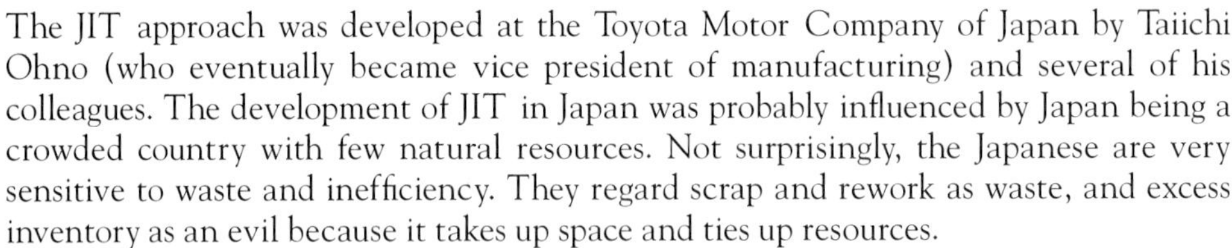

The JIT approach was developed at the Toyota Motor Company of Japan by Taiichi Ohno (who eventually became vice president of manufacturing) and several of his colleagues. The development of JIT in Japan was probably influenced by Japan being a crowded country with few natural resources. Not surprisingly, the Japanese are very sensitive to waste and inefficiency. They regard scrap and rework as waste, and excess inventory as an evil because it takes up space and ties up resources.

READING

THE NUTS AND BOLTS OF JAPAN'S FACTORIES **Urban C. Lehner**

"*If American automobile king Henry Ford I were alive today, I am positive he would have done what we did with our Toyota production system.*"

—Taiichi Ohno

TOYOTA CITY, Japan—Groping to explain "how Japan does it," experts have made much of the close ties between business and government and of the loyalty of Japan's highly skilled workers to their employers. They've noted the fierce competitiveness of Japanese companies in their home market, the nation's high savings rate, even the relative scarcity of lawyers.

Doubtless these are among the pieces needed to solve the puzzle. But some management consultants who've studied how Japan makes such high-quality, competitively priced products say there's another piece often overlooked. The Japanese, they say, have proved themselves increasingly adroit at organizing and running manufacturing operations. Japanese managers may lack the MBAs or the ability to plot big-picture business strategy of their American counterparts. But they know how to run factories.

"There's a growing acceptance that Japanese success is based at least in part on the development of manufacturing techniques that often tend to outrun our own," says management consultant Rex Reid, head of A.T. Kearney's Toyota office.

One of the most interesting examples of Japanese production management skills is a concern quite familiar to Americans: Toyota Motor Co., the largest-selling foreign automaker in the U.S.

Believe in Their System

Toyota officials resist claiming that their way of building autos is better than anyone else's. They're somewhat embarrassed by the exuberant projections of Henry Ford's behavior essayed by their former chief production executive, Taiichi Ohno, in his 1978 book. But Toyota men clearly remain believers in what Mr. Ohno called "the Toyota production system."

For a first-hand look at the system, take a walk through the Tsutsumi plant here in Toyota City, a town of 280,000 in central Japan that's the site of 8 of Toyota's 10 factories. Over here, Muneo Nakahara, 26 years old and an 8-year Toyota veteran, is doing his job. With the help of an overhead crane that Mr. Nakahara controls from a hand-held device, he hoists auto engines onto a conveyor belt that will take them to be matched up with auto bodies.

Mr. Nakahara is lifting the engines onto the conveyor from a small flat-bed truck that has brought them from the engine plant. Only two trucks carrying just 12 engines apiece park at Mr. Nakahara's post at any given time, so every few minutes an empty truck drives back to the engine plant and a new one takes its place.

That's the first feature of the Toyota system: no inventories. Toyota's factories keep on hand only that amount of parts needed for immediate production, from a few minutes' worth up to a few hours' worth, depending on the part. When fresh parts are needed—and only when they're needed—they're delivered from other Toyota plants or from outside suppliers directly to the production line.

Outsiders who've seen Toyota in action often call this the "kanban system," *kanban* being the Japanese word for the piece of paper enclosed in clear plastic that accompanies each bin of parts. (When a worker on the line begins drawing from a new bin, he removes the kanban and routes it back to the supplier, for whom it serves as an order for a new bin of parts.) But Toyota officials say the pieces of paper are just tools. They call this inventory control aspects of their broader system the "just-in-time" system.

The same philosophy guides the meshing of operations within each plant. An assembly line that is building subcomponents makes just that number of subcomponents immediately needed at the next stage of production. When it's made enough, it's changed over to build some other kind of subcomponent for a while. Likewise, the final assembly line first builds one kind of car, then another, in small lots—only as much as called for in actual orders from Toyota's sales unit. Toyota engineers "average" and "level" production among the lines to coordinate output without building inventories. They compare auto assembly to rowing a boat: Everybody has to be pulling on the oars at the same rate.

"They concentrate very heavily on avoiding end-item and intermediate-item storage," says a Ford official in Detroit who's seen the system at work. "They throw out the whole concept of mass production."

The benefits are substantial. Toyota doesn't need space for inventory, people to handle and control inventory, or borrowed money to finance inventory. "It cuts costs in a lot of ways," says an official of Nissan Motor Co., Japan's second-largest automaker, which has adopted an inventory control system similar to its rival's in some plants.

Then there are the side benefits. Because Toyota is constantly changing over its machines to make new things, its workers have become fast at repair and changeover. In his book, Mr. Ohno cites a mold on a press that took two to three hours to change in the 1940s. Today "it takes only three minutes to change the mold," Mr. Ohno says.

Aside from its emphasis on holding down inventories, Toyota's system stresses quality controls. Throughout the Tsutsumi plant are boards with electric lights to indicate conditions on each assembly line. A red light means a line has been stopped because of a problem. Every worker has a button or cord with which he can stop the line, and he's told to use it whenever he thinks something's wrong with the operation of the line or when he spots defects in the product.

"We lose production temporarily this way," concedes Fujio Cho, manager of the production control department at Toyota's headquarters. "But in our experience stopping lines helps us detect problems early and avoid bad practices."

Another feature that becomes clear is the company's penchant for training workers to do more than one job. The man who runs one machine switches off every few moments to run another. The man who feeds rear windows to a robot also "tags" car shells with instructions telling workers farther down the line what to install in them. This versatility allows Toyota to realign its work force more efficiently when business is bad.

Indeed, "recession" thinking underlies a big part of Toyota's system. Much of the system originated in the late 1940s and early 1950s, when Toyota was producing exclusively for a domestic market that wasn't very strong. The company has been operating on the conventional assumption that it's most efficient to produce in large lots, "but that kind of thinking has pushed us close to bankruptcy, because the large lots we were producing couldn't be sold," says Mr. Cho. Toyota couldn't lay off workers—Japan's a "life-time" employment system—so Toyota executives hit upon the simple yet radical idea that still pervades its operations: Overproduction is waste.

Special Relationship

It is, of course, open to question whether Toyota's [system] is the best way to make cars, and Toyota officials themselves doubt whether other automakers could adopt it readily. They note among other problems that it takes a special kind of relationship with suppliers to make the system work.

Fully 50 of Toyota's 250 suppliers are headquartered in Toyota City, and almost all have plants here. They have to be close to make all those deliveries every day. It still shocks Toyota officials to be told that American automakers buy part from suppliers all over the United States and even from suppliers in Europe and Japan. Toyota's most distant supplier is a five-hour drive away.

Then, too, suppliers must have close working relations with Toyota to adjust to Toyota's peculiar needs. It isn't surprising that many of Toyota's suppliers do all or most of their business with Toyota, and that Toyota owns large blocks of the stock of some of its most important suppliers. Many suppliers, even those Toyota doesn't own, have adopted Toyota's production system for their own operations. It improves coordination with Toyota and helps them avoid getting stuck with the inventory buildup that Toyota refuses to get stuck with.

The point isn't whether Toyota's system is best. The point is that it's very good and that Toyota is in many ways typical of Japanese manufacturers in its continual striving to improve production techniques. When experts talk about the competitiveness of Japanese products in international markets, that's something that shouldn't be forgotten.

Vol. 2, seg. 4
Toyota

"The Nuts and Bolts of Japan's Factories" by Urban Lehner provides a number of important insights on the just-in-time approach to manufacturing, and some reasons for the successes Japanese manufacturers have achieved. In some respects, the just-in-time concept was operational over a half century ago at Henry Ford's great industrial complex in River Rouge, Michigan:

> Nothing in the period that followed was too good for the Rouge; it had the best blast furnaces, the best machine tools, the best metal labs, the best electrical systems, the most efficient efficiency experts. At its maturity in the mid-twenties, the Rouge dwarfed all other industrial complexes. It was a mile and a half long and three quarters of a mile wide. Its eleven hundred acres contained ninety-three buildings, twenty-three of them major. There were ninety-three miles of railroad track on it and twenty-seven miles of conveyor belts. Some seventy-five thousand men worked there, five thousand of them doing nothing but keeping it clean, using eighty-six tons of soap and wearing out five thousand mops each month. By the standards of the day the Rouge was, in fact, clean and quiet. Little was wasted. A British historian of the time, J. A. Spender, wrote of its systems: "If absolute completeness and perfect adaptation of means to end justify the word, they are in their own way works of art." Dissatisfied with the supply and quality of the steel he was getting from the steel companies, Ford asked how much it would cost to build a steel plant within the Rouge. About $35 million, Sorensen told him. "What are you waiting for?" said Ford. Equally dissatisfied with both the availability and the quality of glass, he built a glass factory at the Rouge as well. The price of glass had been roughly 30 cents a square foot early in the life of the T; soon it had soared to $1.50 a foot. With the glass plant at the Rouge, the price of glass came down to 20 cents a foot. Barges carrying iron ore would steam into the inland docks, and even as they were tying up, huge cranes would be swinging out to start the unloading. The process was revolutionary. On Monday morning a barge bearing ore would arrive in a slip, and the ore would go to the blast furnace. By Tuesday it would be poured into a foundry mold and later that day would become an engine. John DeVenter, a business historian, wrote in awe: "Here is the conversion of raw material to cash in approximately thirty-three hours." Some sixty years later Toyota would be credited for its just-in-time theory of manufacturing, in which parts arrived from suppliers just in time to be part of the final assembly. But in any real sense that process began at the Rouge. Toasting Philip

Caldwell, the head of Ford who in 1982 was visiting Japan, Eiji Toyoda, of the Toyota company, said, "There is no secret to how we learned to do what we do, Mr. Caldwell. We learned it at the Rouge."—*Source: Reprinted with permission of Rawson Associates/Scribner, a division of Simon & Schuster, Inc., from THE MACHINE THAT CHANGED THE WORLD, by James P. Womack, Daniel T. Jones, and Daniel Roos. © 1990 by James P. Womack, Daniel T. Jones, Daniel Roos, and Donna Sammons Carpenter.*

Evidently, some aspects of JIT are not new at all.

A widely held view of JIT is that it is simply a system for scheduling production that results in low levels of work-in-process and inventory. But in its truest sense, JIT represents a *philosophy* that encompasses every aspect of the process, from design to after the sale of a product. The philosophy is to pursue a system that functions well with minimal levels of inventories, minimal waste, minimal space, and minimal transactions. It must be a system that is not prone to disruptions and is flexible in terms of the product variety and range of volume that it can handle. The ultimate goal is to achieve a balanced system that permits a smooth, rapid flow of materials through the system.

READING

ROMANTIC JIT AND PRAGMATIC JIT

There are two views of JIT presented in the literature. Paul Zipkin refers to these views as *romantic* JIT and *pragmatic* JIT. Romantic JIT consists of various slogans and idealistic goals such as lot sizes of one, zero inventories, and zero defects. JIT is seen as a "revolution" for manufacturing, one that is relatively simple to install and maintain, and one that can lead to dramatic reductions in work-in-process (WIP) inventory and competitive advantage. Pragmatic JIT, on the other hand, consists of a collection of techniques, some fairly technical, that relate to machine changeovers, layout design, product simplification, quality training, equipment maintenance, and so on.

One problem with romantic JIT is that it is apt to appeal to senior managers, who are not much involved in the day-to-day operations on the shop floor level. They are attracted by the promise of inventory reductions and higher quality. They envision JIT as a quick fix to their problems, and look forward to lower WIP levels and increased inventory turnovers, which will lead to decreased inventory investment and improved cash flow. Thinking that JIT is a relatively simple concept, they expect that lower-level managers and workers will quickly and easily convert to this new way of thinking and soon achieve results, not comprehending the complexity that is involved, or the time they will need to achieve meaningful results. For example, it took Toyota over 20 years to perfect its JIT system!

Adopting the romantic view of JIT can lead to much frustration and disappointment as people on the shop floor struggle to achieve what senior management perceives to be relatively easy. Moreover, cutting back WIP inventories without dealing with the reasons for WIP can quickly lead to chaos on the shop floor, with delays and missed deliveries.

Source: Paul Zipkin, "Does Manufacturing Need a JIT Revolution?" *Harvard Business Review,* January–February 1991, pp. 40–50.

In JIT systems, quality is ingrained in both the product and the process. Companies that use JIT have achieved a level of quality that enables them to function with small batch sizes and tight schedules. JIT systems have high reliability; major sources of inefficiency and disruption have been eliminated, and workers have been trained not only to function in the system but also to continuously improve it.

Comment. JIT is not for everyone. Manufacturers often use it for production of items such as automobiles, where it is possible to produce goods in response to customer demand, and thereby avoid large buildups of finished-goods inventory. It is less likely to be used for production of standardized consumer goods. These manufacturers, rather than using the make-to-order (MTO) mode, generally operate in a make-to-stock (MTS) mode, producing items that will be stored in inventory until they are needed. They can do this because they produce standard items on the basis of a marketing forecast. However, the forecast is subject to error due to demand variability. Finished-goods inventories buffer these manufacturers from disruptions caused by differences between the forecast and actual demand.

The following section describes the goals of JIT systems.

JIT GOALS

The ultimate goal of JIT is a balanced system, that is, one that achieves a smooth, rapid flow of materials through the system. The idea is to make the process time as short as possible by using resources in the best possible way. The degree to which the overall goal is achieved depends on how well certain supporting goals are achieved. Those goals are:

1. Eliminate disruptions.
2. Make the system flexible.
3. Reduce setup times and lead times.
4. Minimize inventory.
5. Eliminate waste.

Disruptions have a negative influence on the system by upsetting the smooth flow of products through the system, and they should be eliminated. Disruptions are caused by a variety of factors, such as poor quality, equipment breakdowns, changes to the schedule, and late deliveries. All these should be eliminated as much as possible. This will reduce the uncertainty that the system must deal with.

A *flexible system* is one that is robust enough to handle a mix of products, often on a daily basis, and to handle changes in the level of output while still maintaining balance and throughput speed. This enables the system to deal with some uncertainty.

Setup times and *delivery lead times* prolong a process without adding any value to the product. Moreover, long setup times and long lead times negatively impact the flexibility of the system. Hence, reduction of setup and lead times is very important, and is one objective of continuous improvement.

Inventory is an idle resource, taking up space and adding cost to the system. It should be minimized as much as possible.

Waste represents unproductive resources; eliminating waste can free up resources and enhance production. In the JIT philosophy, wastes include

Vol. 4, tape 2, seg. 2
Federal Signal

a. *Overproduction*—involves excessive use of manufacturing resources.

b. *Waiting time*—requires space, adds no value.

c. *Unnecessary transporting*—increases handling, increases work-in-process inventory.

d. *Inventory*—causes idle resources, hides quality problems and production inefficiencies.

e. *Processing waste*—makes unnecessary production steps, scrap.

f. *Inefficient work methods*—indicates poor layout and material movement patterns, increases work-in-process inventory.

g. *Product defects*—requires rework costs and possible sales loss due to customer dissatisfaction.

The existence of these wastes is an indication that improvement is possible. The list of wastes can also identify potential targets for continuous improvement efforts.

BUILDING BLOCKS

The design and operation of a JIT system provide the foundation for accomplishing the aforementioned goals. The foundation is made up of four building blocks:

1. Product design
2. Process design

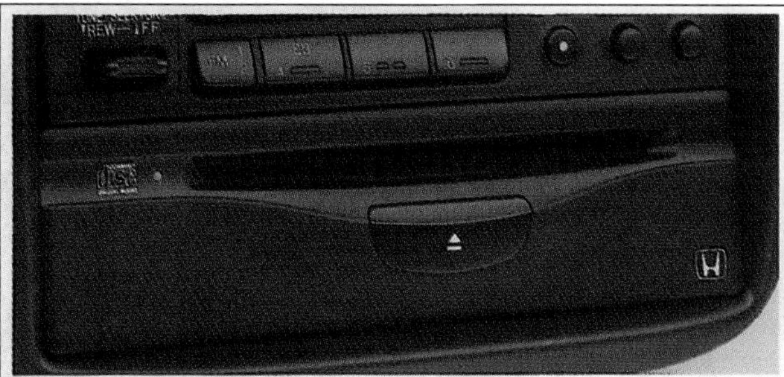

Civic and Odyssey CD Player...$349.00 SUGG. RETAIL. *Faceplate subject to change.*

CD PLAYER

Allows you to drive to crisp, distortion-free digital sounds. This sculpted unit is specifically designed to work in conjunction with any Honda audio system equipped with CD Compact Disc Controller™ technology so you can operate it through the same buttons that control your stereo, eliminating dashboard clutter and simplifying operation.

Prices shown are suggested retail prices only. Dealers are independent and are free to set their own selling prices. Prices shown do not include installation, labor or taxes.

Passport CD Player...$349.00 SUGG. RETAIL.

Accord CD Player...$349.00 SUGG. RETAIL.

CASSETTE PLAYER

is specifically designed to work in conjunction with the Accord, CR-V, Prelude or Civic's audio system. You can operate it through the stereo controls, eliminating dashboard clutter and simplifying operation.

Cassette Player offers the features you'd expect. Powered tape loading and eject for smooth operation. Auto reverse that allows uninterrupted listening to both sides of the tape. Single tracks can be repeated if desired. Dolby® B noise reduction can be selected to reduce tape hiss. Music search allows the listener to skip forward or reverse to find a preferred track. The pinch roller releases to protect against damage to the tape and the cassette transport mechanism when the power is shut off. The tape head is made of a hard permalloy head for long term durability. Even the tape door is illuminated so it is easy to load a tape at night.

CONSOLE ASHTRAY with LIGHTER

Convenience and style punctuate this duo. The ashtray with cigarette lighter is a replacement for the factory installed center instrument cover. Color coordinated to vehicle interior.

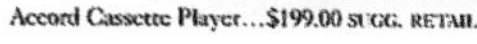

Civic Cassette Player...$199.00 SUGG. RETAIL.
Ashtray with Cigarette Lighter...TBD SUGG. RETAIL.

Honda designs its product options into its vehicles. Various radio/CD/cassette options are interchangeable among vehicle models.

3. Personnel/organizational elements
4. Manufacturing planning and control

www.honda.com/

Speed and simplicity are two common threads that run through these building blocks.

Product Design

Three elements of product design are key to JIT systems:

a. Standard parts

b. Modular design

c. Quality

The first two elements relate to speed and simplicity.

The use of *standard parts* means that workers have fewer parts to deal with, and training times and costs are reduced. Purchasing, handling, and checking quality are more routine and lend themselves to continual improvement. Another important benefit is the ability to use standard processing.

Modular design is an extension of standard parts. Modules are clusters of parts treated as a single unit. This greatly reduces the number of parts to deal with, simplifying assembly, purchasing, handling, training, and so on. Standardization has the added benefit of reducing the number of different parts contained in the bill of materials for various products, thereby simplifying the bill of materials.

Disadvantages of standardization are less product variety and resistance to change in a standard design. These disadvantages are partly offset where different products have some common parts or modules. Using a tactic that is sometimes referred to as **delayed differentiation,** a manager can delay a decision concerning which end-items will be produced while the standard portions are produced. When it becomes apparent exactly which products are needed, the system can quickly respond by producing the remaining unique portions of those products. An example of this occurs in new car sales. One key option when purchasing a new car is the type of radio. By making the different options so they all fit the same space and so that they can be easily installed at the dealership, companies can service a variety of customer wants quickly, instead of making customers custom order (and wait) for a car with the particular option they want.

delayed differentiation A delay in the decision about the manufacture of a certain product while the standard parts or modules are being produced.

Quality is the *sine qua non* ("without which not") of JIT. It is crucial to JIT systems because poor quality can create major disruptions. The systems are geared to a smooth flow of work; the occurrence of problems due to poor quality creates disruption in this flow. Because of small lot sizes and the absence of buffer stock, production must cease when problems occur, and it cannot resume until the problems have been resolved. Obviously, shutting down an entire process is costly and cuts into planned output levels, so it becomes imperative to try to avoid shutdowns, and to quickly resolve problems when they do appear.

JIT systems use a three-part approach to quality. One part is to design quality into the product and the production process. High quality levels can occur because JIT systems produce standardized products that lead to standardized job methods, workers who are very familiar with their jobs, and the use of standardized equipment. Moreover, the cost of product design quality (i.e., building quality in at the *design* stage) can be spread over many units, yielding a low cost per unit. It is also important to choose appropriate quality levels in terms of the final customer and of manufacturing capability. Thus, product design and process design must go hand in hand.

Process Design

Seven aspects of process design are particularly important for JIT systems:

1. Small lot sizes
2. Setup time reduction
3. Manufacturing cells
4. Limited work in process
5. Quality improvement
6. Production flexibility
7. Little inventory storage

Vol. 5, seg. 1, 2
TriState

Small Lot Sizes. In the JIT philosophy, the ideal lot size is one unit, a quantity that may not always be realistic owing to practical considerations requiring minimum lot sizes (e.g., machines that process multiple items simultaneously, heat-treating equipment that processes multiple items simultaneously, and machines with very long setup times). Nevertheless, the goal is still to reduce the lot size as much as possible. *Small lot sizes* in both the production process and deliveries from suppliers yield a number of benefits that enable JIT systems to operate effectively. First, with small lots moving through the system, in-process inventory is considerably less than it is with large lots. This reduces carrying costs, space requirements, and clutter in the workplace. Second, inspection and rework costs are less when problems with quality occur, because there are fewer items in a lot to inspect and rework. Table 15–1 gives a summary of the benefits of small lot sizes.

Small lots also permit greater flexibility in scheduling. Repetitive systems typically produce a small variety of products. In traditional systems, this usually means long production runs of each product, one after the other. Although this spreads the setup

TABLE 15–1

Benefits of small lot sizes

Reduces inventory, lowers carrying costs
Less space required to store inventory
Less rework if defects occur
Less inventory to "work off" before implementing product improvements
Problems are more apparent
Increases production flexibility
Easier to balance operations

cost for a run over many items, it also results in long cycles over the entire range of products. For instance, suppose a firm has three product versions, A, B, and C. In a traditional system, there would be a long run of version A (e.g., covering two or three days or more), then a long run of version B, followed by a long run of version C before the sequence would repeat. In contrast, a JIT system, using small lots, would frequently shift from producing A to producing B and C. This flexibility enables JIT systems to respond more quickly to changing customer demands for output: JIT systems can produce just what is needed, when it is needed. The contrast between small and large lot sizes is illustrated in Figure 15–1.

Small lots and changing product mixes require frequent setups. Unless these are quick and relatively inexpensive, the time and cost to accomplish them can be prohibitive. Often, a system trains workers to do their own setups. Moreover, the system uses programs to reduce setup time and cost to achieve the desired results; a deliberate effort is required, and workers are usually a valuable part of the process.

Setup Time Reduction. Setup tools and equipment and setup procedures must be simple and standardized. Multipurpose equipment or attachments can help to reduce setup time. For instance, a machine with multiple spindles that can easily be rotated into place for different job requirements can drastically reduce job changeover time. Moreover, *group technology* may be used to reduce setup cost and time by capitalizing on similarities in recurring operations. For instance, parts that are similar in shape, materials, and so on may require very similar setups. Processing them in sequence on the same equipment can reduce the need to completely change a setup; only minor adjustment may be necessary.

www.ford.com

More than sixty years have passed since the introduction of Henry Ford's Model A with its all-steel body. Yet, across the world, nearly all motor-vehicle bodies are still produced by welding together about 300 metal parts stamped from sheet steel.

Auto makers have produced these "stampings" by employing one of two different methods. A few tiny craft producers, such as Aston Martin, cut sheets of metal—usually aluminum—to a gross shape, then beat these blanks by hand on a die to their final shape. (A die is simply a hard piece of metal in the precise shape the sheet metal should assume under pounding.)

Any producer making more than a few hundred cars a year—a category that includes auto makers ranging from Porsche to General Motors—starts with a large roll of sheet steel. They run this sheet through an automated "blanking" press to produce a stack of flat blanks slightly larger than the final part they want. They then insert the blanks in massive stamping presses containing matched upper and lower dies. When these dies are pushed together under thousands of pounds of pressure, the two-dimensional blank takes the three-dimensional shape of a car fender or a truck door as it moves through a series of presses.

The problem with this second method, from Ohno's perspective, was the minimum scale required for economical operation. The massive and expensive Western press lines were designed to operate at about twelve strokes per minute, three shifts a day, to make a million or more of a given part in a year. Yet, in the early days, Toyota's entire production was a few thousand vehicles a year.

FIGURE 15-1

JIT versus large-lot run sizes

A = units of product A
B = units of product B
C = units of product C

JIT approach

AAA BBBBBBB CC AAA BBBBBBB CC AAA BBBBBBB CC AAA BBBBBBB CC

Time →

Large-lot approach

AAAAAAAAAAAA BBBBBBBBBBBBBBBBBBBBBBBB CCCCCCCCC AAAAAAAAAAA

Time →

The dies could be changed so that the same press line could make many parts, but doing so presented major difficulties. The dies weighed many tons each, and workers had to align them in the press with absolute precision. A slight misalignment produced wrinkled parts. A more serious misalignment could produce a nightmare in which the sheet metal melted in the die, necessitating extremely expensive and time-consuming repairs.

To avoid these problems, [Western manufacturers] assigned die changes to specialists. Die changes were undertaken methodically and typically required a full day to go from the last part with the old dies to the first acceptable part from the new dies. As volume in the Western industry soared after World War II, the industry found an even better solution to the die-change problem. Manufacturers found they often could "dedicate" a set of presses to a specific part and stamp these parts for months, or even years, without changing dies.

To Ohno, however, this solution was no solution at all. The dominant Western practice required hundreds of stamping presses to make all the parts in car and truck bodies, while Ohno's capital budget dictated that practically the entire car be stamped from a few press lines.

His idea was to develop simple die-change techniques and to change dies frequently—every two to three hours versus two to three months—using rollers to move dies in and out of position and simple adjustment mechanisms. Because the new techniques were easy to master and production workers were idle during the die changes, Ohno hit upon the idea of letting the production workers perform the die changes as well.

By purchasing a few used American presses and endlessly experimenting from the late 1940s onward, Ohno eventually perfected his technique for quick changes. By the late 1950s, he had reduced the time required to change dies from a day to an astonishing three minutes and eliminated the need for die-change specialists. In the process, he made an unexpected discovery—it actually cost less per part to make small batches of stampings than to run off enormous lots.

There were two reasons for this phenomenon. Making small batches eliminated the carrying cost of the huge inventories of finished parts that mass-production systems required. Even more important, making only a few parts before assembling them into a car caused stamping mistakes to show up almost instantly.

The consequences of this latter discovery were enormous. It made those in the stamping shop much more concerned about quality, and it eliminated the waste of large numbers of defective parts—which had to be repaired at great expense, or even discarded—that were discovered only long after manufacture. But to make this system work at all—a system that ideally produced two hours or less of inventory—Ohno needed both an extremely skilled and a highly motivated work force.

If workers failed to anticipate problems before they occurred and didn't take the initiative to devise solutions, the work of the whole factory could easily come to a halt. Holding back knowledge and effort—repeatedly noted by industrial sociologists as a salient feature of all mass-production systems—would swiftly lead to disaster in Ohno's factory.—*Source: Reprinted with permission of Rawson Associates/Scribner, a division of Simon & Schuster, Inc., from THE MACHINE THAT CHANGED THE WORLD, by James P. Womack, Daniel T. Jones, and Daniel Roos. © 1990 by James P. Womack, Daniel T. Jones, Daniel Roos, and Donna Sammons Carpenter, pp. 51–53.*

TABLE 15–2

Guidelines for increasing production flexibility

1. Reduce downtime due to changeovers by reducing changeover time.
2. Use preventive maintenance on key equipment to reduce breakdowns and downtime.
3. Cross-train workers so they can help when bottlenecks occur or other workers are absent. Train workers to handle equipment adjustments and minor repairs.
4. Use many small units of capacity; many small cells make it easier to shift capacity temporarily and to add or subtract capacity than a few units of large capacity.
5. Use off-line buffers. Store infrequently used safety stock away from the production area to decrease congestion and to avoid continually turning it over.
6. Reserve capacity for important customers.

Source: Adapted from Richard J. Schonberger and Edward M. Knod, Jr. *Operations Management: Improving Customer Service,* 4th ed. (Burr Ridge, IL: Richard D. Irwin, 1991), p. 343.

Manufacturing Cells. One characteristic of many JIT systems is multiple *manufacturing cells.* The cells contain the machines and tools needed to process families of parts having similar processing requirements. In essence, the cells are highly specialized and efficient production centers. Among the important benefits of manufacturing cells are reduced changeover times, high utilization of equipment, and ease of cross-training operators. The combination of high cell efficiency and small lot sizes results in very *little work-in-process inventory.*

Quality Improvement. The occurrence of quality defects during the process can disrupt the orderly flow of work. Consequently, problem solving is important when defects occur. Moreover, there is a never-ending quest for *quality improvement,* which often focuses on finding and eliminating the causes of problems so they do not continually crop up.

JIT systems sometimes minimize defects through the use of **autonomation** (note the extra syllable *on* in the middle of the word). This refers to the automatic detection of defects during production. It can be used with machines or manual operations. It consists of two mechanisms: one for detecting defects when they occur and another for stopping production to correct the cause of the defects. Thus, the halting of production forces immediate attention to the problem, after which an investigation of the problem is conducted, and corrective action is taken to resolve the problem.

autonomation Automatic detection of defects during production.

Production Flexibility. The overall goal of a JIT system is to achieve the ability to process a mix of products in a smooth flow. One potential obstacle to this goal is bottlenecks that occur when portions of the system become overloaded. The existence of bottlenecks reflect inflexibilities in a system. Process design can increase *production flexibility* and reduce bottlenecks in a variety of ways. Table 15–2 lists some of the techniques used for this purpose.

Because JIT systems have very little in-process inventory, equipment breakdowns can be extremely disruptive. To minimize breakdowns, companies use **preventive maintenance** programs, which emphasize maintaining equipment in good operating condition and replacing parts that have a tendency to fail before they fail. Workers are often responsible for maintaining their own equipment.

preventive maintenance Maintaining equipment in good operating condition and replacing parts which have a tendency to fail before they actually do fail.

Even with preventive maintenance, occasional equipment failures will occur. Companies must be prepared for this, so they can quickly return equipment to working order. This may mean maintaining supplies of critical spare parts and making other provisions for emergency situations, perhaps maintaining a small force of repair people or training workers to do certain repairs themselves. Note that when breakdowns do occur, they indicate potential opportunities to be exploited in a JIT environment.

Inventory Storage. JIT systems are designed to *minimize* inventory storage. Recall that in the JIT philosophy, inventory storage is a waste. Inventories are buffers that tend to

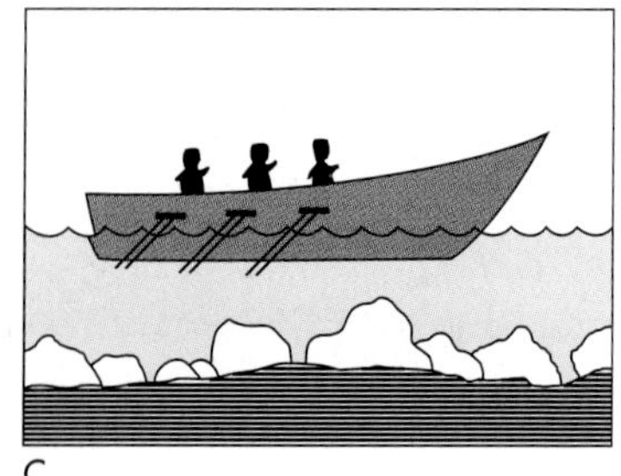

FIGURE 15-2

Large rocks (problems) are hidden by a high water level (inventory) in A. Lower water level (B) reveals rocks. Once the rocks are removed, the water level (inventory) can be lowered (C).

cover up recurring problems that are never resolved, partly because they aren't obvious and partly because the presence of inventory makes them seem less serious. When a machine breaks down, it won't disrupt the system if there is a sufficient inventory of the machine's output to feed into the next workstation. The use of inventory as the "solution" can lead to increasing amounts of inventory if breakdowns increase. A better solution is to investigate the *causes* of machine breakdowns and focus on eliminating them. Similar problems with quality, unreliable vendors, and scheduling can also be solved by having ample inventories to fall back on. However, carrying all that extra inventory creates a tremendous burden in cost and space and allows problems to go unresolved.

The JIT approach is to pare down inventories gradually in order to uncover the problems. Once they are uncovered and solved, the system removes more inventory, finds and solves additional problems, and so on. A useful analogy is a boat on a pond that has large, hidden rocks. (See Figure 15–2.) The rocks represent problems that can hinder production (the boat). The water in the pond that covers the rocks is the inventory in the system. As the water level is slowly lowered, the largest rocks are the first to appear (those problems are the first to be identified). At that point, efforts are undertaken to remove these rocks from the water (resolve these problems). Once that has been accomplished, additional water is removed from the pond, revealing the next layer of rocks, which are then worked on. As more rocks are removed, the need for water to cover them diminishes. Likewise, as more of the major production problems are solved, there is less need to rely on inventory.

Low inventories are the result of a *process* of successful problem solving, one that has occurred over time. Furthermore, because it is unlikely that all problems will be found and resolved, it is necessary to be able to deal quickly with problems when they do occur. Hence, there is a continuing need to identify and solve problems within a short time span to prevent new problems from disrupting the smooth flow of work through the system.

One way to minimize inventory storage in a JIT system is to have deliveries from suppliers go directly to the production floor, which completely eliminates the need to store incoming parts and materials. At the other end of the process, completed units are shipped out as soon as they are ready, which minimizes storage of finished goods. Coupled with low work-in-process inventory, these features result in systems that operate with very little inventory.

Among the advantages of lower inventory are less carrying cost, less space needed, less tendency to rely on buffers, less rework if defects occur, and less need to "work off" current inventory before implementing design improvements. But carrying less inventory also has some risks: The primary one is that if problems arise, there is no safety net. Another is that opportunities may be lost if the system is unable to respond quickly enough.

Personnel Organizational Elements

There are five elements of personnel and organization that are particularly important for JIT systems:

1. Workers as assets
2. Cross-trained workers
3. Continuous improvement
4. Cost accounting
5. Leadership/project management

Workers as Assets. A fundamental tenet of the JIT philosophy is that *workers are assets*. Well-trained and motivated workers are the heart of a JIT system. They are given more authority to make decisions than their counterparts in more traditional systems, but they are also expected to do more.

NEWSCLIP

'PEOPLE' FIRMS BOOST PROFITS, STUDY SHOWS

Companies that treat employees as valuable assets, invest in training programs and use innovative workplace practices are more profitable than those that don't, a study found.

The two-year look at the workplace strategies of American companies was conducted by the management consulting firm Ernst & Young LLP for the Labor Department.

"This is a path-breaking study that shows the surest way to profits and productivity is to treat employees as assets to be developed rather than costs to be cut," Labor Secretary Robert Reich said at a press conference.

For the study, researchers at Harvard and Wharton business schools in partnership with the Ernst & Young Center for Business Innovation, reviewed over 100 papers examining business practices of thousands of U.S. companies.

The report focused on the economic benefits to companies of such Japanese-inspired concepts of labor-management cooperation as Just-In-Time inventory, which moves components to factories only as they are needed.

Among the findings:

- Economic benefits to companies were greatest when they successfully integrated innovations in management and technology with the appropriate employee training and "empowerment" programs.
- Companies investing in employee development enjoy significantly higher market values, on average than their industry peers.
- Companies that were first among their competitors in implementing new management practices reaped the largest rewards.

According to the study, Motorola, Inc. estimates it earns $30 for every $1 invested in employee training, while Xerox Corp. found that in cooperation with its employee union it has reduced manufacturing costs by 30 percent and halved the time needed to develop new products.

Source: Copyright 1995 by the Associated Press.

Cross-Trained Workers. Workers are *cross-trained* to perform several parts of a process and operate a variety of machines. This adds to system flexibility because workers are able to help one another when bottlenecks occur or when a co-worker is absent.

Continuous Improvement. Workers in a JIT system have greater responsibility for quality than workers in traditional systems, and they are expected to be involved in problem solving and *continuous improvement*. JIT workers receive extensive training in statistical process control, quality improvement, and problem solving.

Problem solving is a cornerstone of any JIT system. Of interest are problems that interrupt, or have the potential to interrupt, the smooth flow of work through the system. When such problems surface, it becomes important to resolve them quickly. This may entail increasing inventory levels *temporarily* while the problem is investigated, but the intent of problem solving is to eliminate the problem, or at least greatly reduce the chances of it recurring.

Vol. 4, Tape 1, seg. 3
Detroit Diesel

andon System of lights used at each workstation to signal problems or slowdowns.

Problems that occur during production must be dealt with quickly. Some companies use a light system to signal problems; in Japan, such a system is called **andon.** Each workstation is equipped with a set of three lights. A green light means no problems, an amber light means a worker is falling a little bit behind, and a red light indicates a serious problem. The purpose of the light system is to keep others in the system informed and to enable workers and supervisors to immediately see when and where problems are occurring.

Japanese companies have been very successful in forming teams composed of workers and managers who routinely work on problems. Moreover, workers are encouraged to report problems and potential problems to the teams.

It is important that all levels of management actively support and become involved in problem solving. This includes a willingness to provide financial support and to recognize achievements. It is desirable to formulate goals with the help of workers, publicize the goals, and carefully document accomplishments. Goals give workers something tangible to strive for; recognition can help maintain worker interest and morale.

A central theme of a true just-in-time approach is to work toward continual improvement of the system—reducing inventories, reducing setup cost and time, improving quality, increasing the output rate, and generally cutting waste and inefficiency. Toward that end, problem solving becomes a way of life—a "culture" that must be assimilated into the thinking of management and workers alike. It becomes a never-ending quest for improving operations as all members of the organization strive to improve the system.

Workers in JIT systems have more stress than their counterparts in more traditional systems. Stress comes not only from their added authority and responsibility but also from the high-paced system they work in, where there is little slack and a continual push to improve.

Cost Accounting. Another feature of some JIT systems is the method of allocating overhead. Traditional accounting methods sometimes distort overhead allocation because they allocate it on the basis of direct labor hours. However, that approach does not always accurately reflect the consumption of overhead by different jobs. In addition, the number of direct labor hours in some industries has declined significantly over the years and now frequently accounts for a relatively small portion of the total cost. Conversely, other costs now represent a major portion of the total cost. Therefore, labor-intensive jobs (i.e., those that use relatively large proportions of direct labor) may be assigned a disproportionate share of overhead, one that does not truly reflect actual costs. That in turn can cause managers to make poor decisions. Furthermore, the need to track direct labor hours can itself involve considerable effort.

One alternative method of allocating overhead is *activity-based costing.* This method is designed to more closely reflect the actual amount of overhead consumed by a particular job or activity. Activity-based costing first identifies traceable costs and then assigns those costs to various types of activities such as machine setups, inspection, machine hours, direct labor hours, and movement of materials. Specific jobs are then assigned overhead based on the percentage of activities they consume.

Leadership/Project Management. Another feature of JIT relates to *leadership.* Managers are expected to be leaders and facilitators, not order givers. JIT encourages two-way communication between workers and managers.

Project managers often have full authority over *all* phases of a project; they remain with the project from beginning to end. In the more traditional forms of project management, the project manager often has to rely on the cooperation of other managers to accomplish project goals.

Manufacturing Planning and Control

Five elements of manufacturing planning and control are particularly important for JIT systems:

1. Level loading
2. Pull systems
3. Visual systems
4. Close vendor relationships
5. Reduced transaction processing

Level Loading. JIT systems place a strong emphasis on achieving stable, level daily mix schedules. Toward that end, the master production schedule is developed to provide *level capacity loading.* That may entail a rate-based production schedule instead of the more familiar quantity-based schedule. Moreover, once established, production schedules are relatively fixed over a short time horizon, and this provides certainty to the system. Even so, some adjustments may be needed in day-to-day schedules to achieve level capacity requirements. Suppliers like level loading because it means smooth demand for them.

A level production schedule requires smooth production. When a company produces different products or product models, it is desirable to produce in small lots (to minimize work-in-process inventory and to maintain flexibility), and to spread the production of the different products throughout the day to achieve smooth production. The extreme case would be to produce one unit of one product, then one of another, then one of another, and so on. While this approach would allow for maximum smoothness, it would generally not be practical because it would generate excessive setup costs.

Mixed-model sequencing begins with daily production requirements of each product or model. For instance, suppose a department produces three models, A, B, and C, with these daily requirements:

Model	Daily Quantity
A	10
B	15
C	5

There are three issues that need to be resolved. One is which sequence to use (C-B-A, A-C-B, etc.), another is how many times (i.e., cycles) the sequence should be repeated daily, and the third is how many units of each model to produce in each cycle.

The choice of sequence can depend on several factors, but the key one is usually the setup time or cost, which may vary depending on the sequence used. For instance, if two of the models, say A and C, are quite similar, the sequences A-C and C-A may involve only minimal setup changes, whereas the setup for model B may be more extensive. Choosing a sequence that has A-C or C-A will result in about 20 percent fewer setups over time than having B produced between A and C on every cycle.

The number of cycles per day depends on the daily production quantities. If every model is to be produced in every cycle, which is often the goal, determining the smallest integer that can be evenly divided into each model's daily quantity will indicate the number of cycles. For models A, B, and C shown in the preceding table, there should be five cycles (5 can be evenly divided into each quantity). High setup costs may cause a manager to use fewer cycles, trading off savings in setup costs and level production. If dividing by the smallest daily quantity does not yield an integer value for each model, a manager may opt for using the smallest production quantity to select a number of cycles, but then produce more of some items in some cycles to make up the difference.

Sometimes a manager determines the number of units of each model in each cycle by dividing each model's daily production quantity by the number of cycles. Using five cycles per day would yield the following:

Model	Daily Quantity	Units per Cycle
A	10	10/5 = 2
B	15	15/5 = 3
C	5	5/5 = 1

These quantities may be unworkable due to restrictions on lot sizes. For example, Model B may be packed four to a carton, so producing three units per cycle would mean that at times finished units (inventory) would have to wait until sufficient quantities were available to fill a crate. Similarly, there may be standard production lot sizes for some operations. A heat-treating process might involve a furnace that can handle six units at a time. If the different models require different furnace temperatures, they could not be grouped. What would be necessary here is an analysis of the trade-off between furnace lot size and the advantages of level production.

Example 1

A manager wants to plan production for these three models using the sequence A-B-C:

Model	Daily Quantity
A	7
B	16
C	5

Solution

The smallest daily quantity is 5, but dividing the other two quantities by 5 does not yield integers. The manager might still decide to use five cycles. Producing one unit of Models A and C and three units of Model B in each of the five cycles would leave the manager short two units of Model A and one unit of Model B. The manager might decide to intersperse those units like this to achieve nearly level production:

Cycle	1	2	3	4	5
Pattern	A B(3) C	A(2) B(3) C	A B(4) C	A(2) B(3) C	A B(3) C
Extra unit(s)		A	B	A	

If the requirement for Model A had been 8 units a day instead of 7, the manager might decide to use the following pattern:

Cycle	1	2	3	4	5
Pattern	A(2) B(3) C	A B(3) C	A(2) B(4) C	A B(3) C	A(2) B(3) C
Extra unit(s)	A		A B		A

push system Work is pushed to the next station as it is completed.

pull system A workstation pulls output from the preceding station as it is needed.

Pull Systems. The terms *push* and *pull* are used to describe two different systems for moving work through a production process. In traditional production environments, a **push system** is used: when work is finished at a workstation, the output is *pushed* to the next station; or, in the case of the final operation, it is pushed on to final inventory. Conversely, in a **pull system,** control of moving the work rests with the following operation; each workstation *pulls* the output from the preceding station as it is needed; output of the final operation is pulled by customer demand or the master schedule. Thus, in a pull system, work moves on in response to demand from the next stage in the process, whereas in a push system, work moves on as it is completed, without regard to

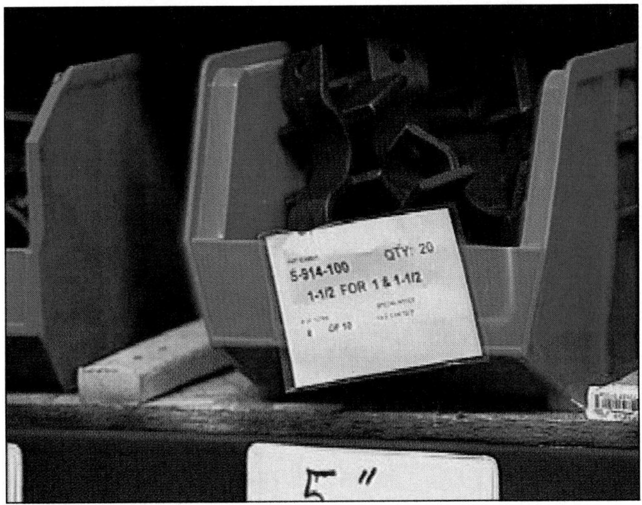

At TriState Industries, a kanban system is in effect to move work through the production system. Shown here, a kanban card provides the authorization to move or work on parts.

Special carts were designed that, when filled, act as a visual signal that they are ready to be moved to the next work cell. An empty cart indicates that it is time to produce, to refill the cart.

the next station's readiness for the work. Consequently, work may pile up at workstations that fall behind schedule because of equipment failure or the detection of a problem with quality.

JIT system communication moves backward through the system from station to station. Each workstation (i.e., customer) communicates its need for more work to the preceding workstation (i.e., supplier), thereby assuring that supply equals demand. Work moves "just in time" for the next operation; the flow of work is thereby coordinated, and the accumulation of excessive inventories between operations is avoided. Of course, some inventory is usually present because operations are not instantaneous. If a workstation waited until it received a request from the next workstation before starting its work, the next station would have to wait for the preceding station to perform its work. Therefore, by design, each workstation produces just enough output to meet the (anticipated) demand of the next station. This can be accomplished by having the succeeding workstation communicate its need for input sufficiently ahead of time to allow the preceding station to do the work. Or there can be a small buffer of stock between stations; when the buffer decreases to a certain level, this signals the preceding station to produce enough output to replenish the buffer supply. The size of the buffer supply depends on the cycle time at the preceding workstation. If the cycle time is short, the station will need little or no buffer; if the cycle time is long, it will need a considerable amount of buffer. However, production occurs only in response to *usage* of the succeeding station; work is *pulled* by the demand generated by the next operation.

Visual Systems. In a pull system work flow is dictated by "next-step demand." A system can communicate such demand in a variety of ways, including a shout or a wave, but by far the most commonly used device is the **kanban** card. *Kanban* is a Japanese word meaning "signal" or "visible record." When a worker needs materials or work from the preceding station, he or she uses a kanban card. In effect, the kanban card is the *authorization* to move or work on parts. In kanban systems, no part or lot can be moved or worked on without one of these cards.

kanban Card or other device that communicates demand for work or materials from the preceding station.

The system works this way: A kanban card is affixed to each container. When a workstation needs to replenish its supply of parts, a worker goes to the area where these parts are stored and withdraws one container of parts. Each container holds a predetermined quantity. The worker removes the kanban card from the container and posts it in a designated spot where it will be clearly visible, and the worker moves the container to the workstation. The posted kanban is then picked up by a stock person who replenishes the stock with another container, and so on down the line. Demand for parts triggers a replenishment, and parts are supplied as usage dictates. Similar withdrawals and replenishments—all controlled by kanbans—occur all the way up and down the line from vendors to finished-goods inventories. If supervisors decide the system is too loose because inventories are building up, they may decide to tighten the system and withdraw some kanbans. Conversely, if the system seems too tight, they may introduce additional kanbans to bring the system into balance.

It is apparent that the number of kanban cards in use is an important variable. One can compute the ideal number of kanban cards using this formula:

$$N = \frac{DT(1 + X)}{C} \qquad (15\text{–}1)$$

where

N = Total number of containers (1 card per container)
D = Planned usage rate of using work center
T = Average waiting time for replenishment of parts plus average production time for a container of parts
X = Policy variable set by management that reflects possible inefficiency in the system (the closer to 0, the more efficient the system)
C = Capacity of a standard container (should be no more than 10 percent of daily usage of the part)

Note that D and T must use the same units (e.g., minutes, days).

Example 2

Usage at a work center is 300 parts per day, and a standard container holds 25 parts. It takes an average of .12 day for a container to complete a circuit from the time a kanban card is received until the container is returned empty. Compute the number of kanban cards (containers) needed if $X = .20$.

Solution

$N = ?$

$D = 300$ parts per day

$T = .12$ day

$C = 25$ parts per container

$X = .20$

$$N = \frac{300(.12)(1 + .20)}{25} = 1.728 \text{ (round to two containers)}$$

Note: Rounding up will cause the system to be looser, and rounding down will cause it to be tighter. Usually, rounding up is used.

Although the goals of MRP II and kanban are essentially the same (i.e., to improve customer service, reduce inventories, and increase productivity), their approaches are different.[1] Neither MRP II nor kanban is a stand-alone system—each exists within a larger framework. MRP II is basically a computerized system; kanban is a manual system.

[1]Based on an article by Walter E. Goddard, "Kanban versus MRP II—Which Is Best for You?" *Modern Materials Handling,* November 5, 1982.

TABLE 15–3

Comparison of kanban and MRP II

Functions	Categories	Kanban System	MRP II
Rates of output	Families of products	Leveling	Production plan
Products to be built	Finished goods for make-to-stock, customer orders for make-to-order	Master production schedule	Master production schedule
Materials required	Components—manufactured and purchased	Kanban cards	Material requirements planning (MRP)
Capacity required	Output for key work centers and vendors	Visual	Capacity requirements planning (CRP)
Executing capacity plans	Producing enough output to satisfy plans	Visual	Input/Output controls (I/O)
Executing material plans—manufactured items	Working on right priorities in factory	Kanban cards	Dispatching reports
Executing material plan—purchased items	Bringing in right items from vendors	Kanban cards and unofficial orders	Purchasing reports
Feedback information	What cannot be executed due to problems	Andon	Anticipated delay reports

Note: The same functions are performed by every manufacturing company; however, the tools used by kanban differ greatly from the MRP II tools. Under kanban, the tools are manual—kanban cards, andon lights, visual checks, and oral orders. Under MRP II, the most important tool is the computer.

Source: Reprinted with permission of *Modern Materials Handling,* June, 1982. © 1982 by Cahners Publishing Company.

Kanban systems usually have very small lot sizes, short lead times, and high-quality output, and they exemplify teamwork. Kanban is essentially a two-bin type of inventory: supplies are replenished semiautomatically when they reach a predetermined level. MRP II is more concerned with projecting requirements and with planning and leveling capacity by means of the computer.

A major benefit of the kanban system is its simplicity; a major benefit of MRP II is its ability to handle complex planning and scheduling quickly and efficiently. It also handles changes fairly easily. In addition, MRP II is capable of simulation (i.e., it enables management to answer "what if" questions) for capacity planning.

Table 15–3 presents a comparison of how eight manufacturing functions are controlled by kanban and MRP II. Note that the two are the same in terms of employing a master production schedule but differ in terms of every other function. Note, too, that the kanban approach is considerably less formal than MRP II.

The philosophies that underlie JIT systems are quite different from those traditionally held by manufacturers. Nonetheless, both approaches have their merits, so it probably would not make sense in most instances to switch from one method of operation to the other. Moreover, to do so would require a tremendous effort. It is noteworthy that at the same time that Western manufacturers are studying kanban systems, some Japanese manufacturers are studying MRP II systems. This suggests the possibility that either system could be improved by incorporating selected elements of the other. That would take careful analysis to determine which elements to incorporate as well as carefully implementing selected elements, and close monitoring to assure that intended results were achieved.

Whether manufacturers should adopt the kanban method is debatable. Some form of it may be useful, but kanban is merely an information system; by itself it offers little in terms of helping manufacturers become more competitive or productive. By the same token, MRP alone will not achieve those results either. Instead, it is the overall approach to manufacturing that is crucial; it is the commitment and support of top management and the continuing efforts of all levels of management to find new ways to improve their manufacturing planning and control techniques, and to adapt those techniques to fit their particular circumstances, that will determine the degree of success.

Comment. The use of either JIT or MRP II does not preclude the use of the other. In fact, it is not unusual to find the two systems used in the same production facility. Some Japanese manufacturers, for example, are turning to MRP systems to help them plan production. Both approaches have their advantages and limitations. MRP systems provide the capability to explode the bill of materials to project timing and material requirements that can then be used to plan production. But the MRP assumption of fixed lead times and infinite capacity can often result in significant problems. At the shop floor level, the discipline of a JIT system, with materials pull, can be very effective. But JIT works best when there is a uniform flow through the shop; a variable flow requires buffers and this reduces the advantage of a pull system.

In effect, some situations are more conducive to a JIT approach, others to an MRP approach. Still others can benefit from a hybrid of the two.

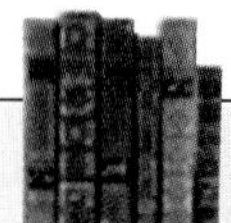

READING

DEVELOPING THE JIT PHILOSOPHY

www.avon.com

Successful implementation of JIT requires involvement from all levels of management in all disciplines

"Our improvement in just-in-time has to be credited to the involvement of higher management and management across many disciplines," says Tom Jensch, Avon's director of materials management.

Every two weeks, at Avon's headquarters in New York City, senior management holds an inventory meeting to discuss progress in JIT. Included in that meeting are top-level managers from finance, marketing, merchandising, purchasing and strategic planning.

"It's a working meeting," says Jensch. "You don't just go in and listen to how the inventory numbers came in at the end of the month. We talk about specific marketing plans for items. How risky are they from an inventory standpoint? What can we do to minimize the risk? We can end up affecting designs of new products by getting away from unique containers and using stock containers if that's important to reduce the inventory risks. We talk about marketing flow and what we are going to do if a product oversells or undersells."

Management involvement doesn't begin and end with that one meeting. "We hold a monthly meeting with lower levels of management and nonmanagement people where we process ideas that have come out of the senior management meeting," says Jensch. "We are working to come up with new ideas. In this meeting are the appropriate people from marketing, finance, purchasing and materials control who are actually the hands-on inventory people. They are much closer to the front lines. It's a communication meeting, to share information and then follow up on various inventory projects we have initiated."

In addition, management from New York meets with inventory control people from Avon's three manufacturing facilities every six weeks. "The purpose of these meetings is to review inventory progress with the manufacturing facilities, to bring to them some of the new ideas that have been stimulated here in New York, to hear from them the new ideas and opportunities that they've identified and to see how we can work together to best process those ideas," says Jensch.

One of the ideas that went through this process and was implemented was the Vendor Proximity Program—the development of suppliers closer to Avon's manufacturing facilities. "Personnel at Avon's manufacturing plant in Suffern, N.Y., were having great success with JIT and we realized that all of the plant's suppliers were within a one-day transit time," says Jensch. After discussing this at inventory meetings, the company decided to seek vendors closer to its other two manufacturing sites in Morton Grove, Ill., and Springdale, Ohio.

Source: Reprinted with permission of *Purchasing*, September 1990. © 1990 by Cahners Publishing Company.

Close Vendor Relationships. JIT systems typically have *close relationships with vendors,* who are expected to provide frequent small deliveries of high-quality goods. Traditionally, buyers have assumed the role of monitoring the quality of purchased goods, inspecting shipments for quality and quantity, and returning poor-quality goods to the vendor for rework. JIT systems have little slack, so poor-quality goods cause a disruption in the smooth flow of work. Moreover, the inspection of incoming goods is viewed as inefficient because it does not add value to the product. For these reasons, the burden of ensuring quality shifts to the vendor. Buyers work with vendors to help them achieve the desired quality levels and to impress upon them the importance of consistent, high-quality goods. The ultimate goal of the buyer is to be able to *certify* a vendor as a producer of high-quality goods. The implication of certification is that a vendor can be relied on to deliver high-quality goods without the need for buyer inspection.

Suppliers must also be willing and able to ship in small lots on a regular basis. Ideally, suppliers themselves will be operating under JIT systems. Buyers can often help suppliers convert to JIT production based on their own experiences. In effect, the supplier becomes part of an extended JIT system that integrates the facilities of buyer and supplier. Integration is easier when a supplier is dedicated to only one or a few buyers. In practice, a supplier is likely to have many different buyers, some using traditional systems and others using JIT. Consequently, compromises may have to be made by both buyers and suppliers.

Traditionally, a spirit of cooperation between buyer and seller has not been present; buyers and vendors have had a somewhat adversarial relationship. Buyers have generally regarded price as a major determinant in sourcing, and they have typically used *multiple-source* purchasing, which means having a list of potential vendors and buying from several to avoid getting locked into a sole source. In this way, buyers play vendors off against each other to get better pricing arrangements or other concessions. The downside is that vendors cannot rely on a long-term relationship with a buyer, and they feel no loyalty to a particular buyer. Furthermore, vendors have often sought to protect themselves from losing a buyer by increasing the number of buyers they supply.

Under JIT purchasing, good vendor relationships are very important. Buyers take measures to reduce their lists of suppliers, concentrating on maintaining close working relationships with a few good ones. Because of the need for frequent, small deliveries, many buyers attempt to find local vendors to shorten the lead time for deliveries and to reduce lead time variability. An added advantage of having vendors nearby is quick response when problems arise.

JIT purchasing is enhanced by long-term relationships between buyers and vendors. Vendors are more willing to commit resources to the job of shipping according to a buyer's JIT system given a long-term relationship. Moreover, price often becomes secondary to other aspects of the relationship (e.g., consistent high quality, flexibility, frequent small deliveries, and quick response to problems).

Suppliers. A key feature of many lean production systems is the relatively small number of suppliers used. In traditional production, companies often deal with hundreds or even thousands of suppliers in a highly centralized arrangement not unlike a giant wheel with many spokes. The company is at the hub of the wheel, and the spokes radiate out to suppliers, each of whom must deal directly with the company. In traditional systems, a supplier does not know the other suppliers or what they are doing. Each supplier works to specifications provided by the buyer. Suppliers have very little basis (or motivation) for suggesting improvements. Moreover, as companies play one supplier off against others, the sharing of information is more risky than rewarding. In contrast, lean production companies may employ a tiered approach for suppliers: They use relatively few first-tier suppliers who work directly with the company or who supply major subassemblies. The first-tier suppliers are responsible for dealing with second-tier suppliers who provide components for the subassemblies, thereby relieving the final buyer from dealing with large numbers of suppliers.

The automotive industry provides a good example of this situation. Suppose a certain car model has an electric seat. The seat and motor together might entail 250 separate parts. A traditional producer might use more than 50 suppliers for the electric seat, but a lean producer might use a single (first-tier) supplier who has the responsibility for the entire seat unit. The company would provide specifications for the overall unit, but leave to the supplier the details of the motor, springs, and so on. The first-tier supplier, in turn, might subcontract the motor to a second-tier supplier, the track to another second-tier supplier, and the cushions and fabric to still another. The second-tier suppliers might subcontract some of their work to third-tier suppliers, and so on. Each tier has only to deal with those just above it or just below it. Suppliers on each level are encouraged to work with each other, and they are motivated to do so because that increases the probability that the resulting item (the seat) will meet or exceed the final buyer's expectations. In this "team of suppliers" approach, all suppliers benefit from a successful product, and each supplier bears full responsibility for the quality of its portion of the product. Figure 15–3 illustrates the difference between the traditional approach and the tiered approach.

Reduced Transaction Processing. Traditional manufacturing systems often have many built-in transactions that do not add value. In their well-known article, "The Hidden Factory,"[2] Jeffrey G. Miller and Thomas Vollmann identify a laundry list of transaction processing that comprises a "hidden factory" in traditional manufacturing planning and control systems, and point out the tremendous cost burden that results. The transactions can be classified as logistical, balancing, quality, or change transactions.

Logistical transactions include ordering, execution, and confirmation of materials transported from one location to another. Related costs cover shipping and receiving personnel, expediting orders, data entry, and data processing.

Balancing transactions include forecasting, production planning, production control, procurement, scheduling, and order processing. Associated costs relate to the personnel involved in these and supporting activities.

Quality transactions include determining and communicating specifications, monitoring, recording, and follow-up activities. Costs relate to appraisal, prevention, internal failures (e.g., scrap, rework, retesting, delays, administration activities) and external failures (e.g., warranty costs, product liability, returns, potential loss of future business).

Change transactions primarily involve engineering changes and the ensuing changes generated in specifications, bills of material, scheduling, processing instructions, and so on. Engineering changes are among the most costly of all transactions.

JIT systems cut transaction costs by reducing the number and frequency of transactions. For example, suppliers deliver goods directly to the production floor, bypassing the storeroom entirely, thereby avoiding the transactions related to receiving the shipment into inventory storage and later moving the materials to the production floor. In addition, vendors are certified for quality, eliminating the need to inspect incoming shipments for quality. The unending quest for quality improvement that pervades JIT systems eliminates many of the above-mentioned quality transactions and their related costs. The use of bar coding (not exclusive to JIT systems) can reduce data entry transactions and increase data accuracy.

The goals and building blocks of JIT are summarized in Figure 15–4 on page 680.

JIT systems have been described and compared with traditional manufacturing systems in the preceding pages. Table 15–4 on page 681 provides a brief overview of those comparisons.

CONVERTING TO A JIT SYSTEM

The success of JIT systems in Japan has attracted keen interest among U.S. manufacturers. A number of well-known firms have converted a portion of their operations to JIT systems.

---

[2]Jeffrey G. Miller and Thomas Vollmann, "The Hidden Factory," *Harvard Business Review,* September–October 1985, pp. 141–50.

a. **Traditional**

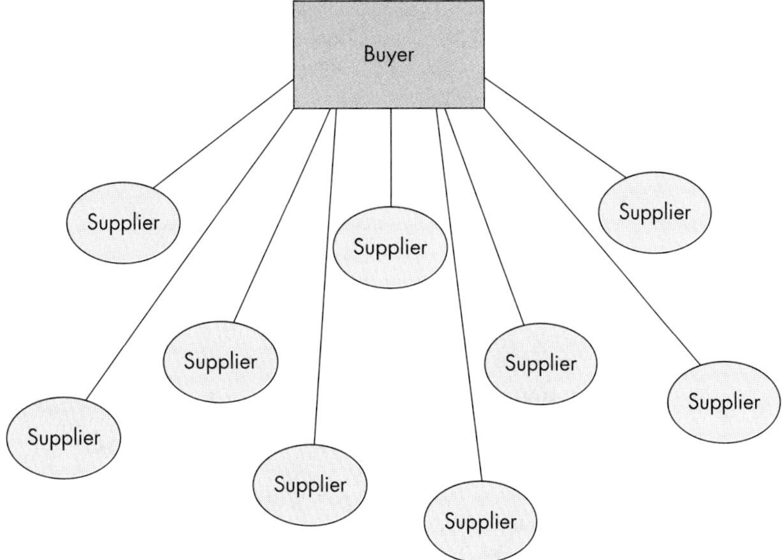

b. **Tiered**

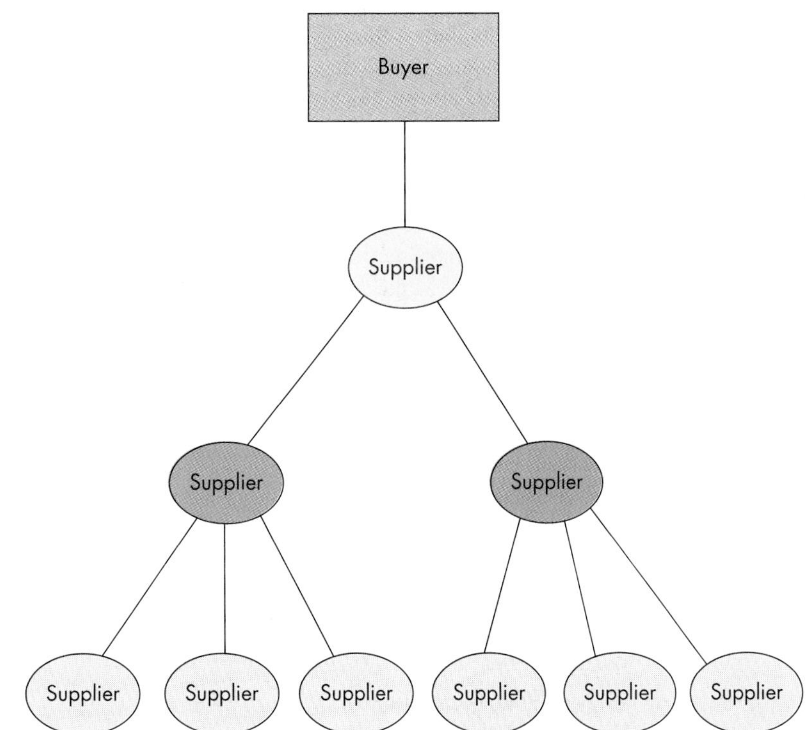

FIGURE 15-3

Traditional supplier network compared to supplier tiers

FIGURE 15-4

Summary of JIT goals and building blocks

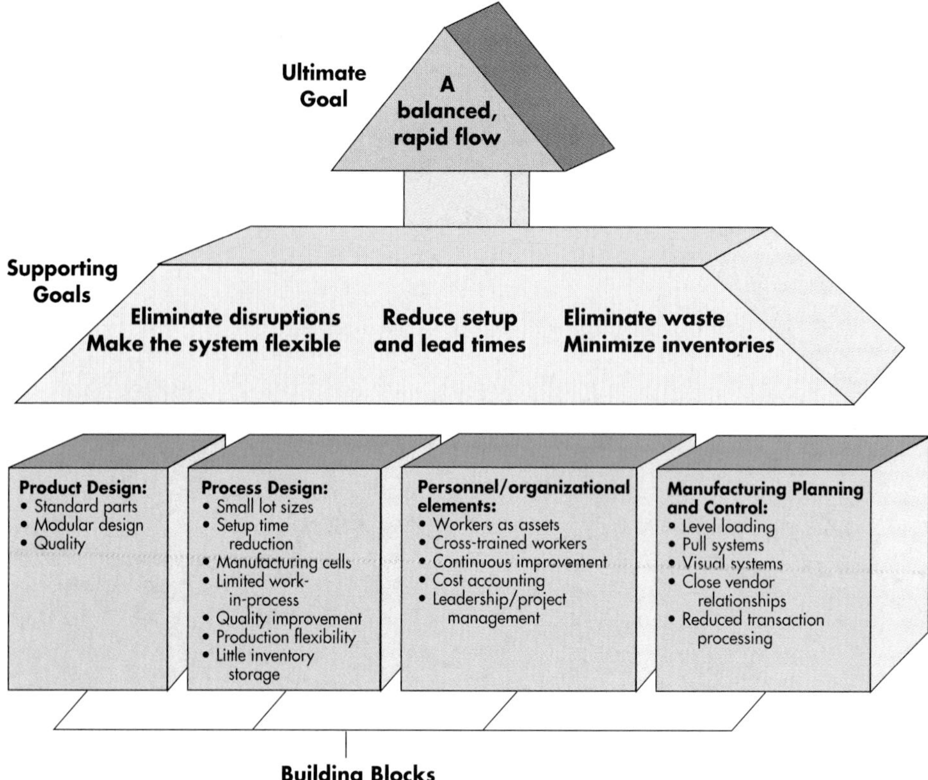

Source: Adapted from Thomas E. Vollmann, William L. Berry, and D. Clay Whybark, *Manufacturing Planning and Control Systems,* 3rd ed. (Burr Ridge, IL: Richard D. Irwin, 1992), p. 76.

Planning a Successful Conversion

To increase the probability of successful conversion, companies should adopt a carefully planned approach that includes the following elements:

1. Make sure top management is committed to the conversion and that they know what will be required. Make sure that management is involved in the process and knows what it will cost, how long it will take to complete the conversion, and what results can be expected.
2. Study the operations carefully; decide which parts will need the most effort to convert.
3. Obtain the support and cooperation of workers. Prepare training programs that include sessions in setups, maintenance of equipment, cross-training for multiple tasks, cooperation, and problem solving. Make sure workers are fully informed about what JIT is and why it is desirable. Reassure workers that their jobs are secure.
4. Begin by trying to reduce setup times while maintaining the current system. Enlist the aid of workers in identifying and eliminating existing problems (e.g., bottlenecks, poor quality).
5. Gradually convert operations, beginning at the *end* of the process and working *backward.* At each stage, make sure the conversion has been relatively successful before moving on. Do not begin to reduce inventories until major problems have been resolved.
6. As one of the last steps, convert suppliers to JIT and be prepared to work closely with them. Start by narrowing the list of vendors, identifying those who are willing to embrace the JIT philosophy. Give preference to vendors who have long-term track records of reliability. Use vendors located nearby if quick response time is important.

Factors	JIT	Traditional Philosophy
Inventory	A liability. Every effort must be expended to do away with it.	An asset. It protects against forecast errors, machine problems, late vendor deliveries. More inventory is "safer."
Lot size	Immediate needs only. A minimum replenishment quantity is desired for both manufactured and purchased parts.	Formulas.
Setups	Make them insignificant. This requires either extremely rapid changeover to minimize the impact on production or the availability of extra machines already set up. Fast changeover makes small lot sizes practical and allows the frequent manufacture of a wide variety of parts.	Low priority issue. Maximum output is the usual goal. Rarely does similar thought and effort go into achieving quick changeover.
Queues	Eliminate them. When problems occur, identify the causes and correct them. The correction process is aided when queues are small.	Necessary investment. Queues permit following operations to continue in the event of a problem with the feeding operation. By providing a selection of jobs, factory management has a greater opportunity to match varying operator skills and machine capabilities, combine setups, and contribute to the efficiency of the operation.
Vendors	Co-workers; part of the team. Multiple deliveries for all active items are expected daily. The vendor takes care of the needs of the customer, and the customer treats the vendor as an extension of the factory.	Adversaries. Multiple sources are the rule, and it's typical to play them off against each other.
Quality	Zero defects. If quality is not 100 percent, production is in jeopardy.	Tolerate some scrap.
Equipment maintenance	Constant and effective. Machine breakdowns must be minimal.	As required. Not critical because of queues.
Lead times	Keep them short. This simplifies the jobs of marketing, purchasing, and manufacturing, as it reduces the need for expediting.	The longer the better. Most foremen and purchasing agents want more lead time, not less.
Workers	Management by consensus. Changes are not made until consensus is reached, with or without a bit of arm twisting. The vital ingredient of "ownership" is achieved.	Management by edict. New systems are installed in spite of the workers, not thanks to the workers. Then measurements are used to determine whether or not they're doing their jobs.

Source: Reprinted with permission of *Modern Materials Handling,* June 1982. © 1982 by Cahners Publishing Company.

TABLE 15–4

Comparison of JIT and traditional manufacturing philosophies

Establish long-term commitments with vendors. Insist on high standards of quality and adherence to strict delivery schedules.

7. Be prepared to encounter obstacles to conversion, which are discussed next.

Obstacles to Conversion

There are numerous obstacles to conversion. Among the most crucial are:

1. Management may not be totally committed or may be unwilling to devote the necessary resources to conversion. This is perhaps the most serious impediment because the conversion is probably doomed without serious commitment.
2. Workers and/or management may not display a cooperative spirit. The system is predicated on cooperation. Managers may resist because JIT shifts some of the responsibility from management to workers and gives workers more control over the work. Workers may resist because of the increased responsibility and stress.
3. Suppliers may resist for several reasons:
   a. Buyers may not be willing to commit the resources necessary to help them adapt to the JIT systems.
   b. They may be uneasy about long-term commitments to a buyer.
   c. Frequent, small deliveries may be difficult, especially if the supplier has other buyers who use traditional systems.

d. The burden of quality control will shift to the supplier.

e. Frequent engineering changes may result from continuing JIT improvements by the buyer.

Comment. JIT systems require a cooperative spirit among workers, management, and vendors. Unless that is present, it is doubtful that a truly effective JIT system can be achieved. The Japanese have been very successful in this regard, partly because respect and cooperation are ingrained in the Japanese culture. In Western cultures, workers, managers, and vendors have historically been strongly at odds with each other. Consequently, a major consideration in converting to a JIT system is whether a spirit of mutual respect and cooperation can be achieved. This requires an appreciation of the importance of cooperation and a tenacious effort by management to instill and maintain that spirit.

The Downside of Conversion. Despite the many advantages of JIT production systems, an organization must take into account a number of other considerations when planning a conversion.

The key considerations are the time and cost requirements for successful conversion, which can be substantial. But it is absolutely essential to eliminate the major sources of disruption in the system. Management must be prepared to commit the resources necessary to achieve a high level of quality and to function on a tight schedule. That means attention to even the smallest of details during the design phase and substantial efforts to debug the system to the point where it runs smoothly. Beyond that, management must be capable of responding quickly when problems arise, and both management and workers must be committed to the continuous improvement of the system. Although each case is different, a general estimate of the time required for conversion is one to three years.

Another consideration pertains to small lot sizes. While small lot sizes allow flexibility in changing the product mix and reduced carrying costs and space requirements, they typically result in (1) increased transportation costs and (2) traffic congestion due to frequent deliveries.

Also, converting to a JIT system can sometimes be a formidable undertaking, as illustrated by the following reading. Hence, firms must decide whether a JIT approach will be advantageous.

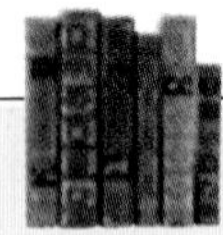

READING

JIT AND QUALITY: A PERFECT FIT

www.dupont.com
www.monsanto.com

A company that plans to implement just-in-time manufacturing must have an established quality program to use as a foundation, according to several quality leaders in the chemical industry.

"Just-in-time is an integral part of the quality process," says Theodore C. Kuchler, quality manager, Chemicals and Pigments Dept., Du Pont Co. "It is a tool that can be used to accomplish continuous improvement."

From New York to Japan the words "continuous improvement" can be found in corporate literature describing quality programs. But where does just-in-time, or JIT, come into play, particularly in the chemical process industry?

Loosely defined, JIT is a manufacturing concept which eliminates anything that does not add value to the product. Successful quality programs revolve around strong supplier relationships—with purchasing playing a critical role—and top management commitment. And JIT is no different.

"There is a misconception of what just-in-time truly is," says Thomas Cummins, manager, total quality, Monsanto Chemical Co. "The heart of just-in-time is process capability, and the very first need is a stable, capable and predictable process effort."

In order to achieve a capable and predictable process, most companies feel an emphasis on quality is necessary. "Quality and just-in-time are interrelated," says David E. Winkler, vice president, manufacturing convenience meals, Campbell Soup Co.

"Companies can say they have just-in-time, but if they don't have a reliable system in place that is compatible with the just-in-time concept, then they are not going to be very successful," he says.

JIT and quality work well together because both strive to improve the way companies operate. JIT, unlike the quality process, cannot stand alone over the long haul. It is considered by many to be part of the quality process, rather than vice versa.

"Just-in-time without quality?" asks Neal Kane, quality manager, specialty chemicals, Du Pont Co. "It may be possible, but it is very risky. There would be limits as to how far a company could proceed without a quality process."

"You could say you have just-in-time without the quality process," says Gary Childers, manager of quality, planning and administration, Union Carbide Corp. "But eventually the system would break down because you would be placing too much of a demand on it. You need to have the quality process."

Excerpt from "JIT and Quality: A Perfect Fit." Reprinted with permission of *CPI Purchasing*, September 1990.

JIT IN SERVICES

The discussion of just-in-time has focused on manufacturing simply because that is where it was developed, and where it has been used most often. Nonetheless, services can and do benefit from many JIT concepts. When just-in-time is used in the context of services, the focus is often on the time needed to provide a service—because speed is often an important order winner for services. Some services do have inventories of some sort, so inventory reduction is another aspect of JIT that can apply to services.

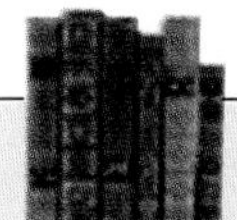

READING

BOX MAKER KEEPS THE LID ON LEAD TIMES

Al Johnson of Davidson Exterior Trim in Americas, Ga., thinks Georgia Box is the best in the business.

"It's our number-one supplier as far as JIT delivery is concerned; it's never early or late," says Johnson. "Moreover, its leadtime on any size box is three or four days; other box makers want at least two weeks. And in a pinch we can call in the morning and get boxes by the afternoon."

Davidson (a division of Textron) makes bumpers and fascias for GM and Chrysler—as many as 14,000 bumpers a day when auto business is going strong. That means Davidson needs a lot of boxes, in different sizes but mostly big boxes. A two-week supply could easily fill 50,000 square feet of warehouse space. But thanks to JIT delivery from Georgia Box, only half that space is utilized today, even though the daily production rate is twice what it used to be. In a good year, Davidson would spend as much as $2 million on corrugated boxes.

How is Georgia Box able to offer shorter lead times than other box makers? Short setup times. According to Georgia Box's president, Stanley Walker, setup times can vary from 30 seconds to 15 minutes tops. The industry norm is anywhere from a half-hour to an hour.

Setups are all computerized, says Walker. "For repeat orders, all one has to do is punch in three numbers; for new orders, box dimensions need to be keyed into the computer. With a traditional manual system, you had to go in with wrenches, adjust the scoring, and set ink registrations.

"Very few sheet plants have this type of computerized equipment," adds Walker. "We can produce 50 million square feet of box board per month on one shift." Even the big-time corrugators would be impressed.

Sheet plants, corrugators—that's how the box business is divided up. The latter group comprises the large integrated paper mills that make their own linerboard as well as corrugated boxes, while the much smaller sheet plants buy their board from the corrugators. Sheet plants handle low-volume orders, while the corrugators prefer big production runs.

With short setup times, Georgia Box is able to handle both types of businesses. Since putting in the computerized equipment three years ago, its sales have doubled, from $18 million to $36 million.

Reprinted with permission of *Purchasing*, September 1990.

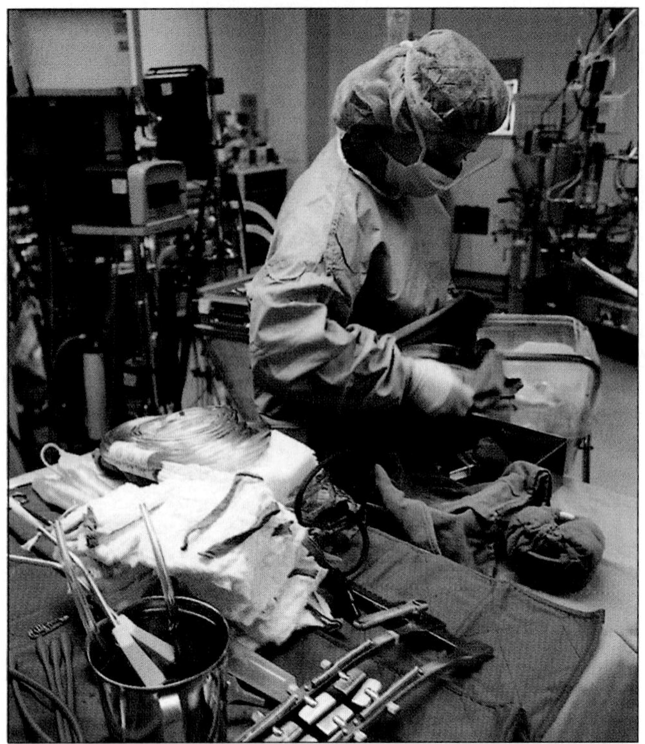

At Duke University Medical Center, the supplies for a bypass operation from Baxter International arrive packaged in the order in which they will be used during surgery. Baxter provides this computer based service, along with the supplies, giving it a competitive advantage in the hospital supply industry.

www.baxter.com

Examples of speedy delivery ("available when requested") are Domino's Pizza, Federal Express and Express Mail, fast-food restaurants, and emergency services through 911. The previous reading describes another example of JIT delivery.

www.dominos.com
www.fedex.com

Process improvement and problem solving can contribute to streamlining a system, resulting in increased customer satisfaction and higher productivity. Some of the ways JIT benefits can be achieved in services are:

- Eliminate disruptions. For example, try to avoid having workers who are servicing customers also answer telephones.
- Make the system flexible. This can cause problems unless approached carefully. Often, it is desirable to standardize work because that can yield high productivity. On the other hand, being able to deal with variety in task requirements can be a competitive advantage. One approach might be to train workers so that they can handle more variety. Another might be to assign work according to specialties, with certain workers handling different types of work according to their specialty.
- Reduce setup times and processing times. Have frequently used tools and spare parts readily available. Additionally, for service calls, try to estimate what parts and supplies might be needed so they will be on hand, and avoid carrying huge inventories.
- Eliminate waste. This includes errors and duplicate work. Keep the emphasis on quality and uniform service.
- Minimize work-in-process. Examples include orders waiting to be processed, calls waiting to be answered, packages waiting to be delivered, trucks waiting to be unloaded or loaded, applications waiting to be processed.
- Simplify the process, especially when customers are part of the system (self-service systems including retail operations, ATM and vending machines, service stations, etc.).

JIT service can be a major competitive advantage for companies that can achieve it. An important key to JIT service is the ability to provide service when it is needed. That requires flexibility on the part of the provider, which generally means short setup times, and it requires clear communication on the part of the requester. If a requester can determine when it will need a particular service, a JIT server can schedule deliveries to correspond to those needs, eliminating the need for continual requests, and reducing the need for provider flexibility—and therefore probably reducing the cost of the JIT service.

Operations Strategy

The JIT philosophy of production offers new perspectives on manufacturing that must be given serious consideration by managers in repetitive manufacturing who wish to be competitive.

Potential adopters should carefully study the requirements and benefits of lean production systems, as well as the difficulties and strengths of their current systems, before making a decision on whether to convert to JIT. Careful estimates of time and cost to convert, and an assessment of how likely workers, managers, and suppliers are to cooperate in such an approach, are essential.

The decision to convert can be sequential, giving management an opportunity to gain firsthand experience with portions of JIT without wholly committing themselves. For instance, improving vendor relations, reducing setup times, improving quality, and reducing waste and inefficiency are desirable goals in themselves. Moreover, a level production schedule is a necessary element of a JIT system, and achieving that will also be useful under a traditional system of operation.

The important thing is to not dismiss the concept of lean production without serious consideration, because it could turn out to give a formidable competitive advantage to one's company—or one's competitors.

Summary

Just-in-time (JIT) is a system of lean production used mainly in repetitive manufacturing, in which goods move through the system and tasks are completed just in time to maintain the schedule. JIT systems require very little inventory because successive operations are closely coordinated.

The ultimate goal of a JIT system is to achieve a balanced, smooth flow of production. Supporting goals include eliminating disruptions to the system, making the system flexible, reducing setup and lead times, eliminating waste, and minimizing inventories. The building blocks of a JIT system are product design, process design, personnel and organization, and manufacturing planning and control.

Lean systems require the elimination of sources of potential disruption to the even flow of work. High quality is essential because problems with quality can disrupt the process. Quick, low-cost setups, special layouts, allowing work to be pulled through the system rather than pushed through, and a spirit of cooperation are important features of lean systems. So, too, are problem solving aimed at reducing disruptions and making the system more efficient, and an attitude of working toward continual improvement.

Key benefits of JIT/lean systems are reduced inventory levels, high quality, flexibility, reduced lead times, increased productivity and equipment utilization, reduced amounts of scrap and rework, and reduced space requirements.

JIT differs in many ways from traditional systems used by Western companies, but its benefits have attracted the attention of U.S. manufacturers, causing many to consider converting their operations to JIT. In doing so, careful attention must be given to obtaining the support of top management, achieving a cooperative spirit throughout the organization, and establishing good relationships with a small number of vendors.

Key Terms

andon 670
autonomation 667
delayed differentiation 664
just-in-time (JIT) 658
kanban 673
preventive maintenance 667
pull system 672
push system 672

Solved Problems

Problem 1

Determine the number of containers needed for a workstation that uses 100 parts per hour if the time for a container to complete a cycle (move, wait, empty, return, fill) is 90 minutes and a standard container holds 84 parts. An efficiency factor of 0.10 is currently being used.

Solution

$N = ?$
$D = 100$ parts per hour
$T = 90$ minutes (1.5 hours)
$C = 84$ parts
$X = 0.10$

$$N = \frac{D(T)(1 + X)}{C} = \frac{100(1.5)(1 + .10)}{84} = 1.96 \text{ (round to 2) containers}$$

Problem 2

Determine the number of cycles per day and the production quantity per cycle for this set of products. The department operates five days a week. Assume the sequence A-B-C-D will be used.

Product	Weekly Quantity
A	20
B	40
C	30
D	15

Solution

Convert weekly quantities to daily quantities, and note the smallest *daily* quantity is 3 units.

Product	Daily Quantity 5 Weekly Quantity 4 5	Units Short Using 3 Cycles
A	20 ÷ 5 = 4	1
B	40 ÷ 5 = 8	2
C	30 ÷ 5 = 6	–
D	15 ÷ 5 = 3	–

Use three cycles, producing all four products in every cycle. Produce units that are short by adding units to some cycles. Disperse the additional units as evenly as possible. There are several possibilities. One is

Cycle	1	2	3
Pattern	A B(3) C(2) D	A B(3) C(2) D	A(2) B(2) C(2) D
Extra unit(s)	B	B	A

Discussion and Review Questions

1. Some key elements of production systems are listed in Table 15–3. Explain briefly how JIT systems differ from traditional production systems for each of those elements.
2. What is the ultimate goal of a JIT system? What are the supporting goals? What are the building blocks?
3. Describe the philosophy that underlies JIT (i.e., what is JIT intended to accomplish?).
4. What are some of the main obstacles that must be overcome in converting from a traditional system to JIT?
5. Briefly discuss vendor relations in JIT systems in terms of the following issues:
   a. Why are they important?
   b. How do they tend to differ from the more adversarial relations of the past?
   c. Why might suppliers be hesitant about JIT purchasing?
6. Certain Japanese have claimed that Henry Ford's assembly line provided some of the rationale for JIT. What features of assembly lines are common to JIT systems?
7. What is the kanban aspect of a JIT system?
8. Contrast push and pull methods of moving goods and materials through production systems.
9. What are the main benefits of a JIT system?
10. What is the hidden factory, and how does JIT eliminate most of it?
11. What are the benefits of small lot sizes?

Memo Writing Exercises

1. Suppose you have been hired as a consultant for a supermarket chain to advise them on a plan to prepare ahead of time and cook complete "package" meals to be sold as convenience items for customers. The menu will vary from day to day, but a typical menu might look like this:

♣♣♣♣♣	♣♣♣♣♣	♣♣♣♣♣
Roast leg of lamb	Filet mignon	Fried chicken
Rosemary potatoes	Pan roast potatoes	Mashed potatoes
Garden peas	Garden salad	Fruit salad
Roll, butter	French bread, butter	Biscuit, honey

   Write a one-page memo to Brad Marlow, Manager of Special Services, proposing the use of delayed differentiation for this concept. Explain how it would work and outline the potential benefits as well as the potential disadvantages of your idea.

2. The manager of production planning and control, Daryll Jones, wants an explanation of the term *hidden factory.* Write a memo to Jones in which you provide some potential benefits from examining the hidden factors at your company.

Problems

1. A manager wants to determine the number of containers to use for a kanban system to be installed next month. The process will have a usage rate of 80 pieces per hour. Because the process is new, the manager has assigned an efficiency factor of .35. Each container holds 45 pieces and takes an average of 75 minutes to complete a cycle. How many containers should be used? As the system improves, will more or fewer containers be required? Why?

   $$N = \frac{80(1.25)(1.35)}{45} = 3$$

2. A JIT system uses kanban cards to authorize production and movement of materials. In one portion of the system, a work center uses an average of 100 pieces per hour while running. The manager has assigned an efficiency factor of .20 to the center. Standard containers are designed to hold six dozen parts each. The cycle time for parts containers is about 105 minutes. How many containers are needed?

   $$N = \frac{100(1.75)(1.20)}{72} = 2.917$$
   (round to 3)

3. A machine cell uses 200 pounds of a certain material each day. Material is transported in vats that hold 120 pounds each. Cycle time for the vats is about two hours. The manager has assigned an efficiency factor of .08 to the cell. The plant operates on an eight-hour day. How many vats will be used?

   $$N = \frac{200(.250)(1.08)}{120} = .45$$
   (round to 1)

3 cycles

4. Determine the number of cycles per day and the production quantity per cycle for this set of products:

Product	Daily Quantity
A	21
B	12
C	3
D	15

4 cycles, plus extra units in each cycle

5. Determine the number of cycles per day and a production quantity per cycle for this set of products that achieves fairly level production:

Product	Daily Quantity
A	22
B	12
C	4
D	15
E	9

Assume the production sequence will be A-B-C-D-E.

5 cycles, plus extra units in each cycle

6. Determine the number of cycles per day and a production quantity per cycle for this set of products that achieves fairly level production:

Product	Daily Quantity
F	9
G	8
H	5
K	6

Assume the production sequence will be F-G-H-K.

CASE

LEVEL OPERATIONS

Level Operations is a small company located in eastern Pennsylvania. It produces a variety of security devices and safes. The safes come in several different designs. Recently, a number of new customers have placed orders, and the production facility has been enlarged to accommodate increased demand for safes. Production manager Stephanie Coles is currently working on a production plan for the safes. She needs a plan for each day of the week. She has obtained the following information from the marketing department on projected demand for the next five weeks:

Model	Weekly Quantity
S1	120
S2	102
S7	48
S8	90
S9	25

The department operates five days a week. One complexity is that partially completed safes are not permitted; each cycle must turn out finished units.

After discussions with engineering, Stephanie determined that the best production sequence for each cycle is S7-S8-S9-S01-S02.

Question

1. What might Stephanie determine as the best production quantity per cycle for each day of the week?

www.boeing.com

OPERATIONS TOUR
BOEING

The Boeing Company, headquartered in Everett, Washington, is one of the two major producers of aircraft in the global market. The other major producer is European Airbus.

Boeing produces three models of aircraft in Everett, 747s, 767s, and 777s. The planes are all produced in the same building. At any one time, there may be as many as six planes in various stages of production. Obviously the building has to be fairly large to accommodate such a huge undertaking. In fact, the building is so large that it covers over 98 acres and it is four stories high, making it the largest building by volume in the world. It is so large that all of Disneyland would fit inside, and still leave about 15 acres for indoor parking! The windowless building has six huge doors along one side, each about 100 yards wide and 40 yards high (the size of a football field)—large enough to allow a completed airplane to pass through.

Wing and a portion of the fuselage being lowered into plane.

Boeing sells airplanes to airlines and countries around the globe. There isn't a set price for the planes; the actual price depends on what features the customer wants. Once the details have been settled and an order submitted, the customer requirements are sent to the design department.

Design

Designers formerly had to construct a mockup to determine the exact dimensions of the plane and to identify any assembly problems that might occur. That required time, materials, labor, time and space. Now they use computers (CAD) to design airplanes, avoiding the cost of the mockups and shortening the development time.

The Production Process

Once designs have been completed and approved by the customer, production of the plane is scheduled, and parts and materials are ordered. Parts come to the plant by rail, airplane, and truck, and are delivered to the major assembly area of the plane they will be used for. The parts are scheduled so they arrive at the plant just prior to when they will be used in assembly, and immediately moved to storage areas close to where they will be used. Time phasing shipments to arrive as parts are needed helps to keep inventory investment low and avoids having to devote space to store parts that won't be used immediately. There is a trade-off, though, because if any parts are missing or damaged and have to be reordered, that could cause production delays. When missing or defective parts are discovered, they are assigned priorities according to how critical the part is in terms of disruption of the flow of work. The parts with the highest priorities are assigned to expediters who determine the best way to replace the part. The expediters keep track of the progress of the parts and deliver them to the appropriate location as soon as they arrive. In the meantime, a portion of the work remains unfinished, awaiting the replacement parts, and workers complete other portions of the assembly. If the supplier is unable to replace the part in a time frame that will not seriously delay assembly, as a last resort, Boeing has a machine shop that can make the necessary part.

The partially assembled portions of the plane, and in later stages, the plane itself, move from station to station as the work progresses, staying about five days at each station. Giant overhead cranes are used to move large sections from one station to the next, although once the wheel assemblies have been installed, the plane is towed to the remaining stations.

Finished planes are painted in one of two separate buildings. Painting usually adds 400 to 600 pounds to the weight of a plane. The painting process involves giving the airplane a negative charge and the paint a positive charge so that the paint will be attracted to the airplane.

Testing and Quality Control

Boeing has extensive quality control measures in place throughout the entire design and production process. Not only are there quality inspectors, individual employees inspect their own work and the work previously done by others on the plane. Buyers' inspectors also check on the quality of the work.

There are 60 test pilots who fly the planes. Formerly planes were tested to evaluate their flight worthiness in a wind tunnel, which required expensive testing and added considerably to product development time. Now new designs are tested using a computerized wind tunnel before production even begins, greatly reducing both time and cost. And in case you're wondering, the wings are fairly flexible; a typical wing can flap by as much as 22 feet before it will fracture.

Re-engineering

Boeing is re-engineering its business systems. A top priority is to upgrade its computer systems. This will provide better links to suppliers, provide more up-to-date information for materials management, and enable company representatives who are at customer sites to create a customized aircraft design on their laptop computer.

Another aspect of the re-engineering involves a shift to lean production. Key goals are to reduce production time and reduce inventory.

Boeing wants to reduce the time that a plane spends at each work station from 5 days to 3 days, a reduction of 40%. Not only will that mean that customers can get their planes much sooner, it will also reduce labor costs, inventory costs, and improve cash flow. One part of this will be accomplished by moving toward late stage customization, or delayed differentiation. That would mean standardizing the assembly of planes as long as possible before adding custom features. This, and other time saving steps will speed up production considerably, giving it a major competitive advantage. It also wants to reduce the tremendous amount of inventory it carries (a 747 jumbo jet has about 6 million parts, including 3 million rivets). One part of the plan is to have suppliers do more predelivery work by assembling the parts into kits that are delivered directly to the staging area where they will be installed on the aircraft instead of delivering separate parts to inventory. That would cut down on inventory carrying costs and save time.

Boeing is also hoping to reduce the number of suppliers it has, and to establish better links and cooperation from suppliers. Currently Boeing has about 3,500 suppliers. Compare that with GM's roughly 2,500 suppliers, and you get an idea of how large this number is.

Engineers setting up wind-tunnel testing of physical scale models.

Employee ceremony celebrates completion of first model 777.

Selected Bibliography

Alster, Norm. "What Flexible Workers Can Do." *Fortune,* February 13, 1989, pp. 62–66.

Blackburn, Joseph D. *Time-Based Competition.* Burr Ridge, IL: Business One Irwin, 1991.

Burton, Terence T. "JIT/Repetitive Sourcing Strategies: 'Tying the Knot' with Your Suppliers." *Production and Inventory Management,* Fourth quarter 1988, pp. 38–41.

Goddard, Walter E. "Kanban versus MRP II—Which Is Best for You?" *Modern Materials Handling,* November 5, 1982, pp. 40–48.

Hall, Robert W. *Attaining Manufacturing Excellence.* Burr Ridge, IL: Richard D. Irwin, 1987.

Hannah, Kimball H. "Just-in-Time: Meeting the Competitive Challenge." *Production and Inventory Management,* Third quarter 1987, pp. 1–3.

Hopp, Wallace J., and Mark L. Spearman. *Factory Physics: Foundations of Manufacturing Management.* Burr Ridge, IL: Richard D. Irwin, 1996.

Karmarkar, Uday. "Getting Control of Just-in-Time." *Harvard Business Review,* Sept.-Oct. 1989, pp. 122–133.

Monden, Yasuhiro. "What Makes the Toyota Production System Really Tick?" *Industrial Engineering* 13, no. 1 (January 1981), pp. 38–46.

Schonberger, Richard J. *Japanese Manufacturing Techniques: Nine Hidden Lessons in Simplicity.* New York: Free Press, 1982.

———. *World Class Manufacturing.* New York: Free Press, 1986.

Shores, A. Richard. *Reengineering the Factory: A Primer for World-Class Manufacturing.* Milwaukee: ASQC Quality Press, 1994.

Stalk, George, Jr. "Time—The Next Source of Competitive Advantage." *Harvard Business Review,* July–August 1988, pp. 41–51.

Vollmann, Thomas E.; William L. Berry; and D. Clay Whybark. *Manufacturing Planning and Control Systems,* 4th ed. Burr Ridge, IL: Richard D. Irwin, 1997.

Zipkin, Paul H. "Does Manufacturing Need a JIT Revolution?" *Harvard Business Review,* January-February 1991, pp. 40–50.

36
35
34
33
31
30
29
28

LEARNING OBJECTIVES

After completing this chapter, you should be able to:

1 Describe the elements of a supply chain.

2 Describe the purchasing cycle.

3 Explain what value analysis is.

4 Discuss the impact of JIT on purchasing.

5 List the key characteristics of good suppliers.

6 Describe supplier audits and supplier certification.

7 Discuss the benefits and risks of supplier partnering.

8 Describe the logistics function.

9 Discuss the impact of JIT on logistics and potential tradeoff decisions.

CHAPTER SIXTEEN
SUPPLY CHAIN MANAGEMENT

CHAPTER OUTLINE

Logistics has become a hot competitive advantage as companies struggle to get the right stuff to the right place at the right time.

It comes down to this: All your TQM and reengineering and teamwork and delighting the customers are riding on the back of a double-clutching, diesel-guzzling, steel-girded mastodon rumbling down the highway while the rest of the economy sleeps. A 23-ton rig laden with odometers and speedometers for General Motors' Saturn assembly plant in Tennessee is just one link in a supply chain that covers more than 99,000 miles every day. That's 36 million miles a year, not long enough to reach the planet that shares the name of the popular GM car, but plenty long enough to command attention.

www.gm.com

Call it distribution or logistics or supply-chain management. By whatever name, it is the sinuous, gritty, and cumbersome process by which companies move materials, parts, and products to customers. In industry after industry, from cars and clothing to computers and chemicals, executives have plucked this once dismal discipline off the loading dock and placed it near the top of the corporate agenda. Hard-pressed to knock out competitors on quality or price, companies are trying to gain an edge through their ability to deliver the right stuff in the right amount at the right time. Says Robert Sabath, a vice president with Mercer Management Consulting in Chicago: "Logistics, long an unsung, operations-intensive area, has suddenly become very strategic."

www.compaq.com

Just how strategic? Compaq Computer, the white-hot company that recently became the world's No. 1 producer of PCs, estimates it has lost $500 million to $1 billion in sales recently because its laptops and desktops weren't available when and where customers were ready to buy them. Says chief financial officer Daryl White: "We've done most of what we need to do to be more competitive. We've changed the way we develop products, manufacture, market, and advertise. The one piece of the puzzle we haven't addressed is logistics. It's the next source of competitive advantage. The possibilities are just astounding." Excerpted from "Delivering the Goods," FORTUNE, November 28, 1994 © 1994, Time Inc.

INTRODUCTION

supply chain Sequence of suppliers, warehouses, processing, and distribution.

A **supply chain** is a sequence of suppliers, warehouses, operations, and retail outlets. Different companies may not have identical supply chains due to the nature of their operations, and whether they are primarily a manufacturing operation or a service operation. Figure 16–1a illustrates a typical supply chain for a manufacturer, and Figure 16–1b illustrates a typical supply chain for a service organization. Note that both

FIGURE 16-1

Overview of the supply chain

a.

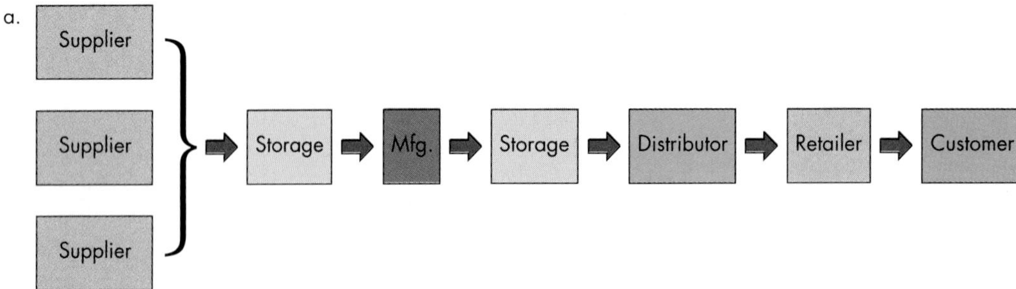

b.

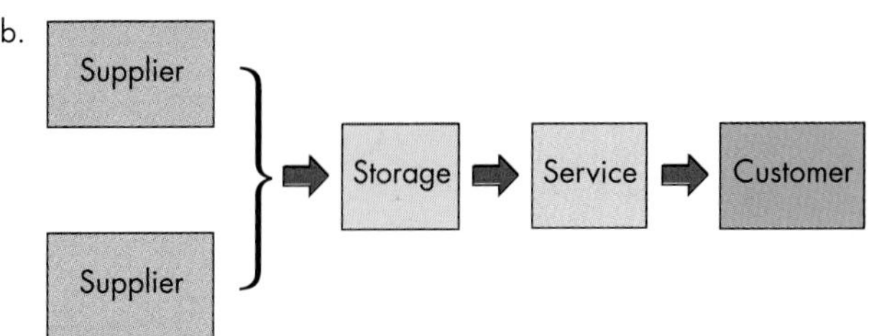

typically require some storage of shipments from suppliers. This can be minimal for service operations, consisting of such items as forms, copier paper and other office suppliers, perhaps spare parts for equipment, printer cartridges, and so on. Manufacturers' inventories would also include parts and raw materials, probably in substantial quantities. Moreover, manufacturers generally require storage of goods during production as well as storage of finished goods, perhaps in a progression of warehouses (e.g., manufacturer, distributor, and retailer).

S

PURCHASING

Purchasing is responsible for obtaining the materials, parts, and supplies needed to produce a product or provide a service. You can get some idea of the importance of purchasing when you consider that in manufacturing upwards of 60 percent of the cost of finished goods comes from purchased parts and materials. Furthermore, the percentages for purchased inventories are even higher for retail and wholesale companies, sometimes exceeding 90 percent. Nonetheless, the importance of purchasing is more than just the cost of goods purchased; other important factors include the *quality* of goods and services and the *timing* of deliveries of goods or services, both of which can have a significant impact on operations.

This section describes the purchasing function as it operates in most organizations. It begins by discussing how purchasing interacts with other functional areas of the organization and with outside suppliers. We will examine the determination of sources of supply and price, and the concept of value analysis.

Purchasing Interfaces

As a service function, purchasing has interfaces with a number of other functional areas, as well as with outside suppliers. Purchasing is the connecting link between the organization and its suppliers. In this capacity, it exchanges information with suppliers and functional areas. The interactions between purchasing and these other areas are briefly summarized in the following paragraphs.

S

Operating units constitute the main source of requests for purchased materials, and close cooperation between these units and the purchasing department is vital if quality, quantity, and delivery goals are to be met. Cancellations, changes in specifications, or changes in quantity or delivery times must be communicated immediately for purchasing to be effective.

The purchasing department may require the assistance of the *legal* department in contract negotiations, in drawing up bid specifications for nonroutine purchases, and to help interpret legislation on pricing, product liability, and contracts with suppliers.

Accounting is responsible for handling payments to suppliers and must be notified promptly when goods are received in order to take advantage of possible discounts. In many firms, *data processing* is handled by the accounting department, which keeps inventory records, checks invoices, and monitors vendor performance.

Design and *engineering* usually prepare material specifications, which must be communicated to purchasing. Because of its contacts with suppliers, purchasing is often in a position to pass information about new products and materials improvements on to design personnel. Also, design and purchasing people may work closely to determine whether changes in specifications, design, or materials can reduce the cost of purchased items (see the following section on value analysis).

Receiving checks incoming shipments of purchased items to determine whether quality, quantity, and timing objectives have been met, and moves the goods to temporary storage. Purchasing must be notified when shipments are late; accounting must be notified when shipments are received so that payments can be made; and both purchasing and accounting must be apprised of current information on continuing vendor evaluation.

FIGURE 16–2

Purchasing interfaces with functional areas of the firm and with suppliers

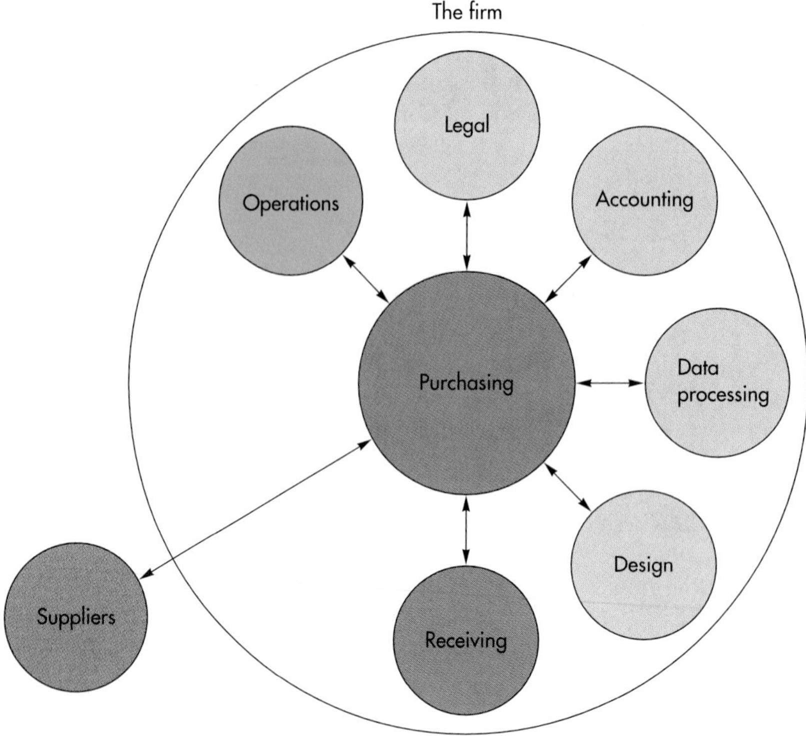

Suppliers or vendors work closely with purchasing to learn what materials will be purchased and what kinds of specifications will be required in terms of quality, quantity, and deliveries. Purchasing must rate vendors on cost, reliability, and so on (see the later section on vendor analysis). Good supplier relations can pay dividends on rush orders and changes, and vendors provide a good source of information on product and material improvements.

Figure 16–2 summarizes the purchasing interfaces.

The Purchasing Cycle

purchasing cycle Series of steps that begin with a request for purchase, and end with notification of shipment received in satisfactory condition.

The **purchasing cycle** begins with a request from within the organization to purchase material, equipment, supplies, or other items from outside the organization, and the cycle ends when the purchasing department is notified that a shipment has been received in satisfactory condition. The main steps in the cycle are these:

1. *Purchasing receives the requisition*. The requisition includes (*a*) a description of the item or material desired, (*b*) the quantity and quality necessary, (*c*) desired delivery dates, and (*d*) who is requesting the purchase.
2. *Purchasing selects a supplier.* The purchasing department must identify suppliers who have the capability of supplying the desired goods. If no suppliers are currently listed in the files, new ones must be sought. Vendor ratings may be referred to in choosing among vendors, or perhaps rating information can be relayed to the vendor with the thought of upgrading future performance.
3. *Purchasing places the order with a vendor.* If the order involves a large expenditure, particularly for a one-time purchase of equipment, for example, vendors will usually be asked to bid on the job, and operating and design personnel may be asked to assist in negotiations with a vendor. Large-volume, continuous-usage items may be covered by blanket purchase orders, which often involve annual negotiation of prices with deliveries subject to request throughout the year. Moderate-volume items may also have blanket purchase orders, or they may be handled on an individual basis. Small

TABLE 16–1
Value analysis overview

1. Select an item that has a high annual dollar volume. This can be a purchased part or material or a service.
2. Identify the function of the item.
3. Obtain answers to questions such as these:
   a. Is the item necessary, does it add value, or can it be eliminated?
   b. Are there alternative sources for the item?
   c. Can the item be provided internally?
   d. What are the advantages of the present arrangement?
   e. What are the disadvantages of the present arrangement?
   f. Could another part, material, or service be used instead?
   g. Can specifications be made less stringent to save cost or time?
   h. Can two or more parts be combined?
   i. Can more (less) processing be done on the item to save on cost?
   j. Do suppliers/providers have suggestions for improvements?
   k. Do employees have suggestions for improvements?
   l. Can packaging be improved or made less costly?
4. Analyze the answers obtained plus answers to other questions that arise, and make recommendations.

purchases may be handled directly between the operating unit requesting the item and the supplier, although some control must be exercised over those purchases or else they could get out of hand.

4. *Monitoring orders.* Routine follow-up on orders, particularly large orders or those with lengthy delivery schedules, allows the purchasing department to foresee delays and relay this information to the appropriate operating units. Likewise, the department must relay changes in quantity and delivery needs of the operating units to suppliers so they have time to adjust their plans.
5. *Receiving orders.* Receiving must check incoming shipments from vendors for quality and quantity. It must notify purchasing, accounting, and the operating unit that requested the goods. If the goods are not received in satisfactory condition, they may have to be returned to the supplier for credit or replacement, or subjected to detailed inspection. Again, purchasing, accounting, and the operating unit must be notified. In either case, vendor evaluation records must be updated.

Value Analysis

value analysis Examination of the function of purchased parts and materials in an effort to reduce cost and/or improve performance.

Value analysis refers to an examination of the *function* of purchased parts and materials in an effort to reduce the cost and/or improve the performance of those items. Typical questions that would be asked as part of the analysis include: Could a cheaper part or material be used? Is the function necessary? Can the function of two or more parts or components be performed by a single part for a lower cost? Can a part be simplified? Could product specifications be relaxed, and would this result in a lower price? Could standard parts be substituted for nonstandard parts? Table 16–1 provides a checklist of questions that can guide a value analysis.

Naturally, purchasing cannot perform an investigation each time materials are ordered. However, it should conduct value analysis periodically on large dollar-volume items because of the potential savings.

Although purchasing does not ordinarily have the authority to implement changes on the basis of a value analysis, it can make suggestions to operating units, designers, and suppliers, which may lead to improved performance of purchased goods and/or reduction of the cost of those goods. Purchasing can offer a different perspective to the analysis, and purchasing people, because of their association with suppliers, possess information not known to others within the organization. If a fair amount of technical knowledge is required to review a part or product, a team can be formed with representatives from design and/or operations to work with purchasing in conducting the analysis.

Outsourcing

outsourcing Buying goods or services from outside sources rather than making or providing them in-house.

Outsourcing refers to buying goods or services from outside sources instead of making the goods or providing the services within house. Some companies do little outsourcing, preferring to do almost everything themselves, while others engage in extensive outsourcing. For example, some personal computer companies buy most or all of the parts for the computers they sell from suppliers, and merely assemble the computers. Services can also be outsourced, and it is not unusual for companies to outsource data processing, payroll and benefits, maintenance, field service and repair, food services, and more.

Companies may outsource for a variety of reasons. Chief among them are that the outside source can provide materials, parts, or services better, cheaper, or more efficiently. Outside suppliers, for example, might be large-scale producers of a certain part or service who—because of economies of scale—can provide the part or service at a lower cost than the company can achieve. Expertise and knowledge are other key reasons for outsources. Or a supplier may hold a patent on a necessary part. Outsourcing also gives a company added flexibility. Outsourcing often increases when companies downsize, as they narrow their focus to core activities.

Outsourcing carries risks as well as benefits. Among the risks are loss of control, greater dependency on suppliers, and not possessing the ability to perform in-house.

Generally, a company takes into account the following factors in deciding whether to outsource:

1. Cost to do it in-house versus cost to buy, including start-up costs, versus cost to outsource.
2. Stability of demand and possible seasonality.
3. Quality available from suppliers compared with a firm's own quality capabilities.
4. The desire to maintain close control over operations.
5. Idle capacity available within the organization.
6. Lead times for each alternative.
7. Who has patents, expertise, and so on, if these are factors.
8. Stability of technology (if a technology is changing, it may be better to use a supplier).
9. The degree to which the necessary operations are consistent with, or in conflict with, current operations.

JIT Purchasing

The adoption of just-in-time manufacturing techniques by some companies has created new challenges for purchasing. It has also made some aspects of purchasing easier. The easy parts include having to deal with fewer suppliers and forming long-term relationships with suppliers that emphasize cooperative spirit rather than low price. Some companies prefer to work with local suppliers when possible, and that can make purchasing activities somewhat easier. The challenges are in having to change from traditional purchasing thinking and practice, and in the need to assure that suppliers will provide quality and timely delivery of goods. On-time delivery is usually the primary need of JIT operations, followed by small lot sizes.

Determining Prices

Companies determine prices in essentially three ways: published price lists, competitive bidding, and negotiation.

In many instances, organizations buy products and services that have fixed or *predetermined prices*. This is generally the case for standard items that are bought infrequently and/or in small quantities.

For large orders of standard products and services, *competitive bidding* is common. The purchasing department sends request for bids to potential suppliers, asking vendors to quote a price for a specified quantity and quality of items or for a specified service to be performed. Government purchases of standard goods or services are usually made through competitive bidding.

Negotiated purchasing is used for special purchasing situations—when specifications are vague, when one or a few customized products or services are involved (e.g., space exploration), and when few potential sources exist. Several myths concerning negotiated purchasing should be recognized:

1. Negotiation is a win–lose confrontation.
2. The main goal is to obtain the lowest possible price.
3. Each negotiation is an isolated transaction.[1]

No one likes to be taken advantage of. Furthermore, contractors and suppliers need a reasonable profit to survive. Therefore, a take-it-or-leave-it approach or one that capitalizes on the weaknesses of the other party will serve no useful purpose and may have detrimental effects that surface later. The most reasonable approach is one of give and take, with each side giving and receiving some concessions.

Centralized versus Decentralized Purchasing

Purchasing can be centralized or decentralized. **Centralized purchasing** means that purchasing is handled by one special department. **Decentralized purchasing** means that individual departments or separate locations handle their own purchasing requirements.

centralized purchasing Purchasing is handled by one special department.

decentralized purchasing Individual departments or separate locations handle their own purchasing requirements.

Centralized purchasing may be able to obtain lower prices than decentralized units if the higher volume created by combining orders enables it to take advantage of quantity discounts offered on large orders. Centralized purchasing may also be able to obtain better service and closer attention from suppliers. In addition, centralized purchasing often enables companies to assign certain categories of items to specialists, who tend to be more efficient because they are able to concentrate their efforts on relatively few items instead of spreading themselves across many items.

Decentralized purchasing has the advantage of awareness of differing "local" needs and being better able to respond to those needs. Decentralized purchasing usually can offer quicker response than centralized purchasing. Where locations are widely scattered, decentralized purchasing may be able to save on transportation costs by buying locally, which has the added attraction of creating goodwill in the community.

Some organizations manage to take advantage of both centralization and decentralization by permitting individual units to handle certain items while centralizing purchases of other items. For example, small orders and rush orders may be handled locally or by departments, while centralized purchases would be used for high-volume, high-value items for which discounts are applicable or specialists can provide better service than local buyers or departments.

The Global Supply Chain

As international trade barriers fall, more companies are expanding global operations. This is presenting tremendous opportunities and opening up previously untapped markets for products and services. It has also increased the number of competitors, and even companies that only operate within a single country are faced with increased foreign competition.

---

[1]Richard J. Tersine, *Production/Operations Management: Concepts, Structure, and Analysis*, 2nd ed. (New York: Elsevier North-Holland Publishing, 1985), p. 598.

Managing a global supply chain that may have far flung customers and/or suppliers magnifies some of the challenges compared to managing a domestic supply chain. Obviously, distances and lead times become more critical as the supply chain becomes more stretched out. So, too, does the possibility of having to deal with (perhaps many) language and cultural differences. Currency differences and monetary fluctuations are other factors that must be dealt with, and possibly with additional modes of transportation.

The following reading "Worldly Desires" describes some of the opportunities related to an expanding worldwide middle class.

READING

WORLDLY DESIRES **Kathleen Martin**

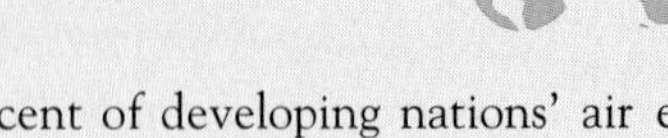

Capitalizing on Global Changes in Transportation to Supply the Worldwide Middle Class

The global economy will experience major changes in the 21st century. Everyone involved in international trade—shippers, carriers, and forwarders—should stay abreast of and understand this coming upheaval. This economic revision promises to transform billions of lives, profoundly affect the revenues of companies engaged in overseas trade, and establish new patterns in international shipping.

A fundamental reason is the emergence and growth of a worldwide middle class. Higher wages are putting disposable income into the pockets of millions of people who previously earned little more than subsistence wages. Workers in Asia, Latin America, and parts of Africa have begun to cultivate healthy appetites for goods once considered out of their reach.

Their desires are spurred by four factors.

1. The telecommunications revolution. Asian farmers, for example, now see beyond their little patch of land. MTV is watched by more people internationally than any other television program. When a Thai farmer and his family see how well other people live, Thailand's current economic woes and currency devaluation notwithstanding, they want to enjoy the same amenities as well, and will work very hard to acquire them.

2. National boundaries no longer stop money from flowing speedily around the world in search of new investment opportunities. The globalization of finance makes billions of dollars available to emerging markets and local enterprises.

3. The information revolution allows factories, warehouses, transportation vendors, and their customers to exchange production details through a web of manufacturing sites that may span the globe. Many of these manufacturing sites are located in developing nations.

4. Free-market mania has transformed the governments of many countries. Free enterprise has become the rule rather than the exception in once staunchly socialist nations like India, China, and Russia. This change in governmental philosophy has unleashed entrepreneurial passions in individuals all over the world, stoking the fires of world trade.

Consumer demand in emerging nations is surging more quickly and powerfully than many American executives realize. Emerging nations used to grow their own foodstuffs and make inexpensive clothes. Thanks to the "green revolution," many of these nations produce food for export; many of these countries also export relatively sophisticated manufactured goods to the developed nations. Manufactured products now account for almost 60 percent of developing nations' air exports, up from 5 percent in the 1960s.

Emerging nations are expected to boost their share of world imports by about two-thirds during the next 25 years. With a more equitable balance of trade, they are now in a better position to pay for their imports.

Taken together, these shifts in buying and selling patterns are modifying trade routes that have been in place since the Industrial Revolution. International transportation is changing more now than it ever has before, with a wider range of products moving in all directions. Astute transportation organizations are fully aware of these changes and are helping their international customers respond to these changes.

Corporate executives should draw a new world map. The important trend to watch is the wholesale geographic shift of commerce into new markets. Latin America is a choice example. South America's business climate has improved immensely after trade flows liberalized, private ventures started earning greater profits, and the flow of investment began to increase.

To handle all of these new trade flow patterns, freight forwarders such as Ideal are striking new partnership arrangements with established local companies in countries like Argentina, Brazil, and Chile; expanding US gateways like the port of Los Angeles for easier transshipments to Latin America from Asia; and increasing computer horsepower to link US operations with those in South America. Managers have wrung most of the possible savings out of the manufacturing, marketing, and finance processes. Transportation is now under scrutiny by corporate America. The transportation industry has responded by changing, changing to a greater degree in the past 5 years than in the previous 50.

New logistics strategies allow manufacturers and distributors to ship their products more fequently and in smaller quantities. Firms have shortened their inventory and distribution cycles. They have increased their use of third-party warehousing and inventory-control companies to store and assemble partially finished goods and send them out on demand as finished products. Sophisticated information systems enable them to track with great confidence the complicated journey of parts from order through delivery.

Advanced information networks will permit freight forwarders to scan the horizon for cargo in both directions so that costs and rates are equalized for both "front" and "back" haul trips. These systems will maximize space utilization and minimize cost, benefiting the producer, the carrier, the forwarder, and ultimately the consumer.

Forward-seeing companies, aided by their transportation providers, will correctly anticipate these changes in global traffic flow. The old days when manufactured items moved from the industrialized northern region of the world to the agricultural south are gone forever. The new economic model will see a far more equal trade balance among the different regions of the world.

Source: Reprinted with permission from *International Business*, January/February 1998.

SUPPLIERS

Good suppliers are a vital link in the supply chain. Late deliveries of parts or materials, or missing or defective items, can wreak havoc for manufacturers by disrupting production schedules, increasing inventory costs, and causing late deliveries of end products. Substandard supplier services can have similar results.

Choosing Suppliers

When choosing suppliers, there are a number of key considerations. Obviously, quality of the products or services and on-time delivery are important, but there are many other factors to take into account. Table 16–2 lists some of those factors and questions that need to be addressed. The list is meant to be representative rather than exhaustive. Moreover, the importance of each factor often differs from company to company, and even with a company from one product or service to another. Consequently, a manager must decide what weight to assign to each factor for each product or service, and use those weights in selecting a supplier.

Evaluating Sources of Supply (Vendor Analysis)

In many respects, choosing a vendor involves taking into account many of the same factors associated with making a major purchase (e.g., a car or stereo system). A company considers price, quality, the supplier's reputation, past experience with the supplier, and service after the sale; this process is called **vendor analysis.** The main difference is that a company, because of the quantities it orders and production requirements, often provides suppliers with detailed specifications of the materials or parts it wants instead of buying items off the shelf, although even large companies buy standard items that way.

vendor analysis Evaluating the sources of supply in terms of price, quality, reputation, and service.

The main factors a company takes into account when it selects a vendor are:

1. *Price.* This is the most obvious factor, along with any discounts offered, although it may not be the most important.
2. *Quality.* A company may be willing to spend more money to obtain high quality.
3. *Services.* Special services can sometimes be very important in choosing a supplier. Replacement of defective items, instruction in the use of equipment, repair of equipment, and similar services can be key in selecting one supplier over another.

TABLE 16-2

Choosing a supplier

Lead Times and On-Time Delivery	• What lead times can the supplier provide? • What procedures does the supplier have for assuring on-time deliveries? • What procedures does the supplier have for documenting and correcting delivery problems? • Are there procedures?
Quality and Quality Assurance	• What procedures does the supplier have for quality control and quality assurance? • Are problems and corrective actions for quality documented? • Are investigations conducted to determine and correct the causes of nonconforming materials?
Flexibility	• How flexible is the supplier in handling changes in quantity, delivery schedules, and product or service changes?
Location	• Is the supplier located nearby?
Price	• Are prices reasonable given the entire package the supplier will provide? • Is the supplier willing to negotiate prices? • Is the supplier willing to engage in a joint effort to reduce costs (and prices)?
Product or Service Changes	• How much advance notification does the supplier give when changes are made in products or services? • To what extent does the buyer have inputs regarding changes?
Reputation and Financial Stability	• What is the reputation of the supplier? • How financially stable is the supplier?
Other Accounts	• Is the supplier heavily dependent on other buyer(s), causing the risk of giving priority to another buyer's needs over ours?

4. *Location.* Location of a supplier can have impact on shipping time, transportation costs, and response time for rush orders or emergency service. Local buying can create goodwill in the community by helping the local economy.
5. *Inventory policy of supplier.* If a supplier maintains an inventory policy of keeping spare parts on hand, this could be helpful in case of an emergency equipment breakdown.
6. *Flexibility.* The willingness and ability of a supplier to respond to changes in demand and to accept design changes could be important considerations.

Supplier Audits

Periodic audits of suppliers are a means of keeping current on suppliers' production (or service) capabilities, quality and delivery problems and resolutions, as well as suppliers' performance on other buyer criteria. If an audit reveals problem areas, a buyer can address these, hopefully before they result in serious problems. Among the factors typically covered by a supplier audit are management style, quality assurance, materials management, the design process used, process improvement policies, and procedures for corrective action and follow-up.

Supplier audits are also an important first step in supplier certification programs.

TABLE 16–3

The supplier as partner versus adversary

Aspect	Partner	Adversary
Number of suppliers	One or a few	Many; play off one against the others
Length of relationship	Long-term	May be brief
Low price	Moderately important	Major consideration
Reliability	High	May not be high
Openness	High	Low
Quality	Insured at the source; vendor-certified	Buyer inspects; may be unreliable
Volume of business	High	May be low due to many suppliers
Location	Proximity may be stressed for short lead times and service	Widely dispersed
Flexibility	Relatively high	Relatively low

Supplier Certification

Supplier certification is a detailed examination of the policies and capabilities of a supplier. The certification process verifies that a supplier meets or exceeds the requirements of a buyer. This is generally important in supplier relationships, but it is particularly important when buyers are seeking to establish long-term relationship with suppliers. Certified suppliers are sometimes referred to as *world class* suppliers. One advantage of using certified suppliers is that the buyer can eliminate much or all of the inspection and testing of delivered goods. And although problems with supplier goods or services might not be totally eliminated, there is much less risk than with noncertified suppliers.

Rather than develop their own certification program, some companies rely on standard industry certifications such as ISO 9000, perhaps the most widely used international certification.

Supplier Partnerships

American firms have become increasingly aware of the importance of building good relations with their suppliers. In the past, too many firms regarded their suppliers as adversaries and dealt with them on that basis. One lesson learned from the Japanese is that numerous benefits derive from good supplier relations, including supplier flexibility in terms of accepting changes in delivery schedules, quality, and quantities. Moreover, suppliers can often help identify problems and offer suggestions for solving them. Thus, simply choosing and switching suppliers on the basis of price is a very shortsighted approach to handling an ongoing need.

Many Japanese firms rely on one or a few suppliers to handle their needs. In contrast, many U.S. firms deal with numerous suppliers. Perhaps they want to remain flexible, and possibly there are advantages in playing off one against the others. Although reducing the number of suppliers to one or a few, as the Japanese have done, may be too extreme, a more realistic approach for U.S. firms is to move in the direction of reducing the number of suppliers.

Keeping good relations with suppliers is increasingly recognized as an important factor in maintaining a competitive edge. Many companies are adopting a view of suppliers as partners. This viewpoint stresses a stable relationship with relatively few reliable suppliers who can provide high-quality supplies, maintain precise delivery schedules, and remain flexible relative to changes in productive specifications and delivery schedules. A comparison of the contrasting views of suppliers is provided in Table 16–3.

An article in *Purchasing* pointed out why it pays for an organization to listen to its suppliers: Suppliers can be a source of ideas that contribute to the competitiveness of an

Ryder company delivers (in trucks that say 'Saturn' on them) bins of speedometers and odometers to Saturn's Spring Hill, Tennessee, plant. Thursday, 9 A.M.: A Ryder truck arrives at a Saturn supplier in Winchester, Virginia.

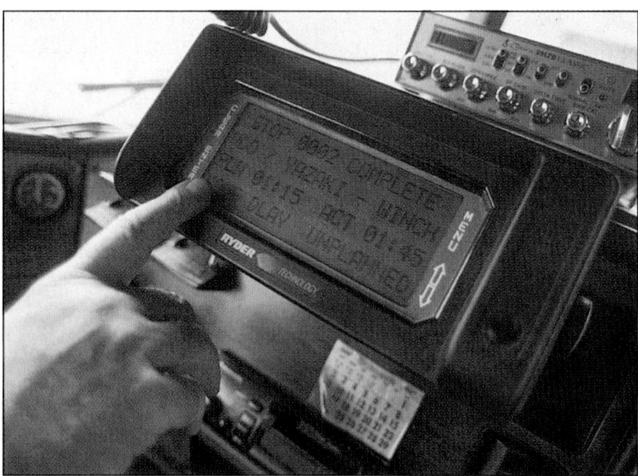

Drivers check the on-board computer, which provides destination and routing information, along with estimated travel times.

Spring Hill, Tennessee, Friday, 3 A.M.: After parking the trailer in a computer-assigned spot in Ryder's switching yard two miles from the Saturn plant, the driver downloads a key-shaped floppy disk

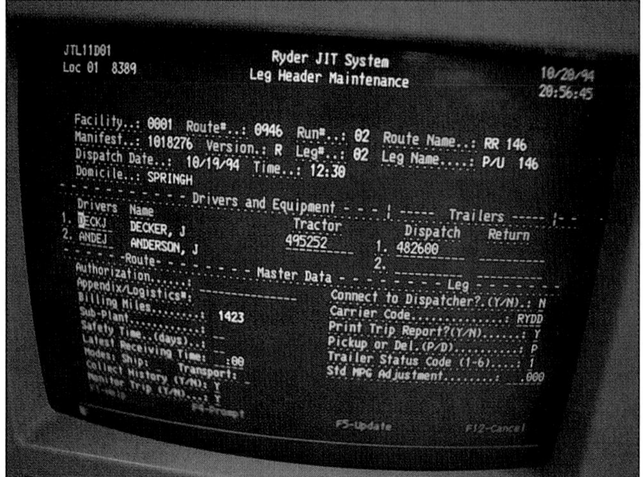

from the on-board computer into Ryder's mainframe, which generates performance reports for Saturn.

12:50 P.M.: The trailer approaches the Saturn plant.

12:53 P.M.: The trailer arrives at one of Saturn's 56 receiving docks just in time for Saturn workers to unload the bins and unwrap the pre-inspected instruments to ready them for the production line.

organization.[2] A poll of 1,000 purchasing/supply operations identified nine areas in which potential ideas from suppliers could lead to improved competitiveness.

www.saturn.com

1. Reduce the cost of making the purchase.
2. Reduce transportation costs.
3. Reduce production costs.
4. Improve product quality.
5. Improve product design.
6. Reduce the time it takes to get the product to market.
7. Improve customer satisfaction.
8. Reduce inventory costs.
9. Introduce new products or processes.

The organizations that stand the best chance of tapping these ideas have good supplier relations and are genuinely open to supplier input.

There are a number of obstacles to supplier partnerships, not the least of which is that, because many of the benefits go to the buyer, suppliers may be hesitant to enter into such relationships. Suppliers may have to increase their investment in equipment, which might put a strain on cash flow. Another possibility is that the cultures of the buyer and supplier might be quite different and not lend themselves to such an arrangement.

Supplier partnerships work best when they lead to savings in inventories and the ability to plan and stabilize operations. Often, the buyer and/or the supplier is a large company, either because large companies are more likely to reap benefits from the relationship, or because their size enables them to force this type of relationship.

www.bose.com

READING

JIT II

JIT II is a philosophy that involves empowering suppliers to assume some of the responsibilities ordinarily done by a company's own buyers. Those responsibilities might include planning, transaction processing, and dealing with quality and delivery on problems that relate to a supplier's products. In some instances, that means having supplier personnel on-site while in others those responsibilities are managed through electronic links from the supplier's site.

BOSE Corporation, a producer of premium audio products, began empowering suppliers back in the mid-1980s. Lance Dixon, director of purchasing and logistics, dubbed it JIT II. To him, JIT II was a natural exten-

Bose Corporation has worked to shorten and improve its supply chain both by empowering its suppliers and by using direct marketing, including significant focus on the design and quality of the product, to distribute directly to the final customer.

[2] Jim Morgan, "Nine Ways Suppliers Can Improve Competitiveness," *Purchasing,* November 24, 1994, pp. 7–9.

sion of the philosophies of agile manufacturing, namely JIT, partnering, and concurrent engineering. Under JIT II, suppliers have access to design engineers; sometimes, supplier personnel are even permitted to pursue their own engineering change orders.

The advantages of empowering suppliers include enhanced communications, elimination of redundant steps, improved planning and fulfillment of materials requirements, early supplier involvement in new product planning and design, less inventory, and cost reductions for both supplier and customer. Some of the risks include possible exposure of proprietary technology, reliance on a supplier's personnel to fill gaps in technical knowledge, and supplier complacency.

The list of companies that have embraced this philosophy in at least some portion of their operations is impressive. In addition to BOSE, it includes IBM, Honeywell, Intel, and Ingersoll-Rand.

Source: Based on "JIT II Comes of Age," pp. 41–46. Reprinted with permission of *Purchasing*, October 1994, © 1994 by Cahners Publishing Company.

LOGISTICS

logistics Movement of materials and supplies.

Vol. 5, seg. 3
American President Lines

Logistics refers to the movement of materials within a production facility and to incoming and outgoing shipments of goods and materials. Materials include all of the physical items used in the production process. In addition to raw materials, there are support items such as fuels, equipment, parts, tools, lubricants, office supplies, and more.

Movement within a Facility

Figure 16–3 shows the many steps where materials move within a production facility:

1. From incoming vehicles to receiving.
2. From receiving to storage.
3. From storage to the point of use (e.g., work center).
4. From one work center to the next, or to temporary storage.
5. From the last operation to final storage.
6. From storage to packaging/shipping.
7. From shipping to outgoing vehicles.

In some instances, the goods being moved are supplies; in other instances, the goods are actual products or partially completed products; and in still other instances, the goods are raw materials or purchased parts.

Movement of materials must be coordinated to arrive at the appropriate destinations at appropriate times. Workers and supervisors must take care so that items are not lost, stolen, or damaged during movement.

Incoming and Outgoing Shipments

traffic management Overseeing the shipment of incoming and outgoing goods.

Overseeing the shipment of incoming and outgoing goods comes under the heading of **traffic management.** This function handles schedules and decisions on shipping methods and times, taking into account costs of various alternatives, government regulations, the needs of the organization relative to quantities and timing, and external factors such as potential shipping delays or disruptions (e.g., highway construction, truckers' strikes).

Computer tracking of shipments often helps to maintain knowledge of the current status of shipments as well as to provide other up-to-date information on costs and schedules.

Evaluating Shipping Alternatives

A situation that arises frequently in some businesses is making a choice between quicker (but more expensive) shipping alternatives such as overnight or 2-day air, and slower, but cheaper alternatives.

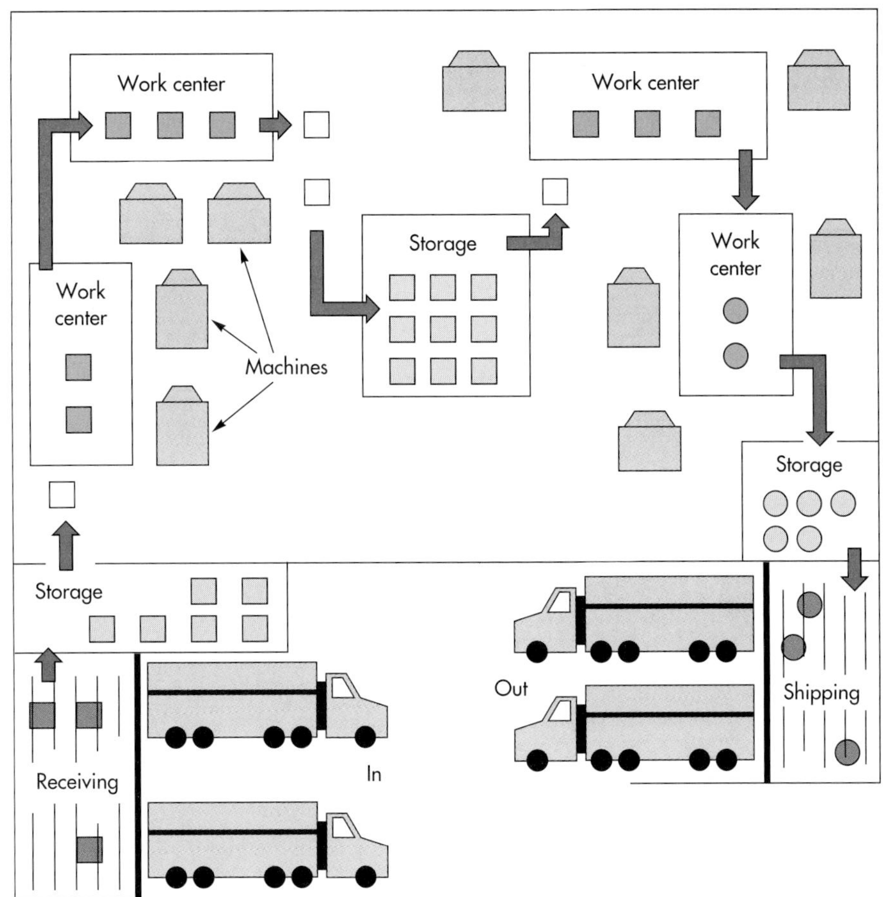

FIGURE 16-3

Overview of materials movement

www.apl.com

American President Lines (APL), one of the world's top five shipping lines, loads truck trailers filled with American products and ships them from Port of Los Angeles to Singapore for distribution to overseas destinations. Each ship can hold 4800 20 foot trailers and can be unloaded in 72 hours. State-of-the-art container location techniques speed local truckers in and out of the terminal in 30 minutes or less. The progress of each shipment is reported to APL's worldwide communications network and directly to customers.

In some instances, there is an overriding factor present that justifies sending a shipment by the quickest means possible, so there is little or no choice involved. However, in many instances, urgency is not the primary consideration, so there is a choice. The decision in such cases often focuses on the cost savings of slower alternatives versus the increased holding cost that would result from using the slower alternative. An important assumption is that the seller gets paid upon receipt of the goods by the buyer (e.g., through EDI).

The incremental holding cost incurred by using the slower alternative is computed as:

$$\text{Incremental Holding cost} = H\left(\frac{d}{365}\right) \tag{16.1}$$

where

H = annual holding cost for the item
d = time saving in days
$d/365$ = fraction of year saved

Example 1

Determine which shipping alternative, 1 day or 3 days is best when the holding cost of an item is $1,000 per year and the 1-day shipping cost is $40, and the 3-day shipping cost is

a. $35
b. $30

Solution

H = $1,000 per year
Time savings = 2 days using 1-day shipping
Holding cost for additional 2 days = $1,000 × (2/365) = $5.48

a. Cost savings = $5. Because the actual savings of $5 is less than the holding cost of $5.48, use the 1-day alternative.

b. Cost savings = $10. Because the actual savings of $10 exceeds the savings in holding cost of $5.48, use the 3-day alternative.

Bar Coding

bar coding Patterns of black lines and white spaces that can be read by scanning devices, containing a variety of information.

Bar codes are the patterns of alternating wide and narrow black lines and white spaces, and numbers and symbols, that you see on everyday products at supermarkets and other retail stores. The bar codes (or universal product codes) are read by scanning devices that use the information for a variety of purposes, such as recording prices and quantities, printing sales receipts, and updating inventory records.

Bar codes are also useful in manufacturing and distribution. In distribution, they enable companies to keep track of items in warehouses and en route to customers; managers can instantaneously determine the location of any item in the system and its status (e.g., en route to customer, delivered to customer). In manufacturing, bar codes track the progress of jobs as they move through the production process. In addition, by scanning bar codes of jobs just before each operation, specific processing instructions for the job at each operation can be provided to operators. They are also used to update inventory records, monitor quality losses, and monitor productivity. Employee attendance and location can also be tracked with employee bar codes. And bar codes can be used for automatic sorting and packing of items as they move down a conveyor line.

Bar codes provide companies with a tremendous ability to monitor and control items in manufacturing and distribution by providing accurate, up-to-date information on quantity, quality, location, and other data. This enables managers to realize significant improvements in productivity and effectiveness, and to provide outstanding levels of customer service.

Electronic Data Interchange

Electronic data interchange (EDI) is the direct transmission of interorganization transactions, computer to computer, including purchase orders, shipping notices, debit or credit memos, and more. There are many benefits to such systems, such as:

electronic data interchange The direct transmission of interorganization transactions, computer to computer.

- Increased productivity.
- Elimination of paperwork.
- Lead time and inventory reduction.
- Facilitation of just-in-time systems.
- Electronic transfer of funds.
- Improved control of operations.
- Reduction in clerical labor.
- Increased accuracy.

The use of EDI linkages with other organizations can be part of a strategy to achieve a competitive advantage by leveraging logistics performance. In addition, in some JIT environments, EDI serves as the "kanban" between the manufacturer and the supplier.

There are many applications of EDI in the retail industries involving electronic communication between retailers and vendors. Dubbed quick response, the approach is based on scanning bar codes and transmitting that information to vendors. The purpose is to create a just-in-time replenishment system that is keyed to customer buying patterns. Retailers use Universal Product Code (UPC) scanning or Point-of-Sale (POS) scanning at the registers which uses price-look-up (PLU) to track customer buying. Efficient consumer response (ECR) is a variation of quick response used by the supermarket industry to provide supermarkets, distributors, and suppliers with key data on buying patterns so that they can make better decisions on replenishment.

Quick response approaches have several benefits. Among them are reducing the dependency on forecasts and the ability to achieve a closer match between supply and demand. In addition, there is the strong possibility of saving on inventory carrying costs.

Wal-Mart has a satellite network for electronic data interchange that allows vendors to directly access point-of-sale data in real time, enabling them to improve their forecasting and inventory management. Wal-Mart also uses the system for issuing purchase orders and receiving invoices from its vendors.

www.wal-mart.com

The following newsclip illustrates how some companies are taking advantage of this.

NEWSCLIP

EFFICIENT CONSUMER RESPONSE

At first glance, your average supermarket looks like a marvel of supply-chain efficiency. Think of the wide aisles, the fully stocked shelves, the wondrous variety, the endless checkout lines (OK, you can't have everything). But pull back the curtain, Toto, and the wizardry fades fast. Behind the scenes is an industry that actually rewards people for building unnecessary inventory. All along the supply chain, sellers push discounted products onto merchandise buyers, who in turn are compensated for purchasing things cheaply even though they may languish in a warehouse for months on end. So addled are the industry's practices that a whole class of characters known as diverters thrives by buying caseloads of discounted food in one state and transporting them to another where the discount isn't being offered.

But the grocery business has been shocked out of its torpor by some bigfooted players with very efficient distribution systems. These "alternative format" retailers, of which Wal-Mart is the largest and most menacing, have been stealing market share from supermarkets.

To parry the threat, food manufacturers, wholesalers, and retailers have rallied around an industrywide initiative known as efficient consumer response (ECR). A study by consultants from Kurt Salmon Associates in Atlanta concluded that the industry could save $24 billion in operating costs and another $6 billion in interest expenses by streamlining its logistics and changing its behavior.

Spartan Stores, a Grand Rapids food distributor and a leader in the ECR movement, was recently able to shut down a 300,000-square-foot warehouse, which cost $1 million a year to operate, when it stopped stockpiling discounted products. Says Keith Wagar, vice president of procurement and inbound logistics: "We've got to eliminate all the activities that don't add value to the consumer. Our only chance is to look at the transaction from manufacturer to consumer as a single process. It's a strategy for survival, a strategy for growth."

Source: Excerpted from FORTUNE, November 28, 1994, © 1994 Time, Inc.

www.spartanstores.com

distribution requirements planning A system for inventory management and distribution planning.

Distribution Requirements Planning

Distribution requirements planning (DRP) is a system for inventory management and distribution planning. It is especially useful in multi-echelon warehousing systems (factory and regional warehouses). It extends the concepts of material requirements planning (MRPII) to multi-echelon warehouse inventories, starting with demand at the end of the channel and working that back through the warehouse system to obtain time-phased replenishment schedules for moving inventories through the warehouse network. In effect, management uses DRP to plan and coordinate transportation, warehousing, workers, equipment, and financial flows.

JIT Deliveries

www.ups.com

JIT systems often require frequent deliveries of small shipments. That can place a tremendous burden on the delivery system in several respects. One is the increased traffic that results. Instead of one large delivery per week, a company switching to JIT may require many smaller loads every day. Multiply that by the number of parts obtained from suppliers, and you can begin to appreciate the potential traffic nightmare, resulting delays in receiving shipments, and disruption in production. Also, there is the likely increase in transportation cost per unit. There is often a fixed cost per delivery (e.g., per truck), and the fact that only a partial load is delivered doesn't change that. Smaller trucks are one possibility, but that, too, will generate a cost.

Another aspect of JIT deliveries became very apparent during a strike of United Parcel Service (UPS) employees in mid-1997. The strike was a prolonged one, and many companies that used UPS for JIT deliveries were forced to scramble to find alternate means of supply. That severely taxed some of the other rapid-delivery organizations, and deliveries fell behind schedule, causing disruptions to JIT-based companies.

Because of these factors, it is necessary to carefully weigh the costs and benefits of using frequent, small deliveries, and select a lot size that *balances* all relevant costs.

Operations Strategy

Supply chain management creates value through changes in time, location, and quantity. Moreover, this area holds great potential for competitive advantage. To realize this potential, management must *integrate* and *streamline* a diverse range of activities that involve purchasing, internal inventory transfers, and physical distribution. The reading, "Delivering the Goods" on page 712 illustrates how industry is accomplishing this.

Summary

Supply chain management is the management of the flow of materials and information over the entire range of the production process, beginning with suppliers and ending with distribution. It covers movement of materials within the organization as well as storage.

Purchasing is responsible for obtaining material inputs for the organization. This includes the actual purchasing of materials, supplier selection, and conducting value analyses on purchased items. The basic design question is whether to centralize or decentralize purchasing. Centralization provides for closer control and may produce certain economies, whereas decentralized purchasing tends to produce quick response and may better serve local needs. Value analysis may be performed periodically to assure that purchased item cost-benefit is being optimized. Vendors are evaluated for cost, service after the sale, reliability, and quality.

Logistics involves movement of materials within the organization, incoming and outgoing shipments, and distribution. Electronic data linkages have increased productivity and accuracy in supply chain management and improved control of operations.

Key Terms

Solved Problem

Problem 1

Determine which shipping alternative is best if the annual holding cost of an item is 25% of unit price, and a single unit with a price of $6,000 is to be shipped, either by 2-day freight at a cost of $400 or 5-day freight at a cost of $350.

Solution

H = .25($6,000) = $1,500 per year
d = 5 days − 2 days = additional 3 days
Cost savings = $400 − $350 = $50

$$\begin{aligned}\text{Incremental Holding cost} &= H\left(\frac{d}{365}\right)\\ &= \$1{,}500\left(\frac{3}{365}\right)\\ &= \$12.33\end{aligned}$$

The added cost to hold the item is less than the cost savings of $50. Hence, use the 5-day alternative.

Discussion and Review Questions

1. Briefly describe supply chain management.
2. Briefly describe how purchasing interacts with other functional areas of the firm such as accounting and design.
3. Describe value analysis. Why is purchasing a good location for this task?
4. Should the supplier with the highest quality–lowest price combination always be selected over others? Explain.
5. Discuss the issue of centralization versus decentralization of the purchasing function.

6. Describe vendor analysis and supplier certification.
7. Discuss the importance of good supplier relations. Compare the viewpoints of supplier-as-partner with supplier-as-adversary.
8. In what ways does just-in-time manufacturing affect purchasing? How does it affect traffic management?
9. Explain what EDI is and the benefits it can provide.

Memo Writing Exercises

1. Write a one-page memo to your boss, Jake Marley, on the trade-offs that relate to continuing to make a certain part in-house versus shutting down the operation and buying the part from a supplier.
2. Assume the role of vice president of operations at a large commercial bank. Write a one-page memo to your boss, CEO Michael Numbers, on why good forecasts are important for decisions concerning logistics planning as well as decisions on possible outsourcing and supplier partnerships.

Problems

Incremental holding cost = $160.44.
Use 2-day freight.

1. A manager at Strateline Manufacturing must choose between two shipping alternatives—2-day freight and 5-day freight. Using 5-day freight would cost $135 less than using 2-day freight. The primary consideration is holding cost, which is $10 per item per year. Two thousand items are to be shipped. Which alternative would you recommend? Explain.
2. Determine which shipping alternative would be most economical to ship 80 boxes of parts when each box has a price of $200 and holding cost is 30% of price, given these shipping alternatives:

Net savings: 2-day = $26.85
6-day = $54.25
Use 6-day.

Alternative	Shipping Cost
Overnight	$300
Two day	260
Six day	180

Ship 2-day using A; all other options result in higher cost.

3. A manager must choose a shipping alternative. There are two possible shippers, A and B. Both offer a two-day rate. In addition, A offers a 3-day rate and a 9-day rate, and B offers a 4-day rate and a 7-day rate. Annual holding cost is 35% of price. Three hundred boxes are to be shipped, and each box is worth $140. Which shipping alternative would you recommend? Explain.

A		B	
Options	Cost	Options	Cost
2 day	$500	2 day	$525
3 days	460	4 days	450
9 days	400	7 days	410

READING

DELIVERING THE GOODS Ronald Henkoff

Last year American companies spent $670 billion—a gaping 10.5 percent of GDP—to wrap, bundle, load, unload, sort, reload, and transport goods. So clogged is the gross national pipeline with unnecessary steps and redundant stockpiles that the grocery industry alone believes it can wash $30 billion, or nearly 10 percent of its annual operating costs, out of the system. A typical box of breakfast cereal spends a stunning 104 days getting from factory to supermarket, haltingly progressing through a concatenation of wholesalers, distributors, brokers, di-

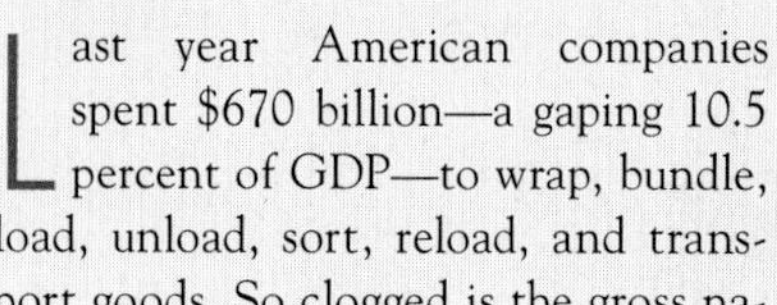

verters, and consolidators, each of which has a warehouse. Says James Alampi, president of Van Waters & Rogers, the largest U.S. chemical distributor: "Frankly, the cost of making a product is almost irrelevant. You have far more opportunity to get cost out of the supply chain than you do out of manufacturing. There's so much duplication and inefficiency."

Think of supply-chain management as nothing less than the reengineering of the entire economy—one convoluted chain at a time. Once you start reconceptualizing your company as a collection of business processes, it becomes dauntingly clear that those processes extend beyond the portals of any one building, the boundaries of any one corporation, and the borders of any one country. Says Harold Sirkin, a vice president of the Boston Consulting Group: "As the economy changes, as competition becomes more global, it's no longer company vs. company but supply chain vs. supply chain."

Streamlining processes that traverse companies and continents is about as easy as reforming the health care system. But the payback can be enormous. In two years National Semiconductor has cut its standard delivery time 47 percent, reduced distribution costs 2.5 percent and increased sales 34 percent. How? By shutting six warehouses around the globe and air-freighting its microchips to customers worldwide from a new, 125,000-square-foot distribution center in Singapore.

What works with chips can work with chintz. Laura Ashley, British purveyor of clothing and home furnishings, turns inventory roughly ten times a year, five times faster than three years ago, after overhauling its information systems and consolidating warehouses. Saturn, whose world-class logistics system links suppliers, factories, and dealers, turns its parts inventory so fast—300 times a year—that it's barely there. Even Compaq, which says it has just begun to fight, has quintupled production since 1991 without expanding factory space, partly by working with suppliers to reduce stockpiles of unassembled components.

Companies that don't import high-priced talent often export their logistics systems—lock, stockpile, and barrel—to outside contractors. Ryder System, best known for renting trucks, manages Saturn's far-flung supply chain from supplier to factory to dealer showroom. Federal Express, the pioneer in overnight delivery of documents, has developed a thriving sideline running transportation and distribution services for companies like National Semiconductor and Laura Ashley. Cass Information Services, a transportation accounting firm in St. Louis, expects the contract logistics business to expand from $16 billion in revenues today to as much as $50 billion by the turn of the century.

Delve into supply-chain management and you quickly confront a litany of recondite concepts, including *heijunka* (Japanese for "even flow"), cross-docking, usable cube, electronic data interchange, efficient consumer response, and activity-based costing. Companies that revamp their logistics operations often wind up adopting solutions that fly in the face of popular management nostrums. Why? In an age of delegation and empowerment, logistics demands centralized control. And sometimes the best way to get closer to your customer is actually to fold your tent—or your warehouse—and move that process to a more efficient location farther away.

National Semiconductor, the world's 13th-largest chipmaker, has gleaned those insights—and many more—on its way to overhauling a logistics network that circles the globe. The Santa Clara, California, company, whose products end up inside cars, computers, and telecommunications gear, began looking at distribution in the early 1990s as part of a broad effort to reverse a spate of corporate losses.

The first step was to figure out how much the company was spending to move and store parts and products along its supply chain. Says Dennis Samaritoni, National's vice president of corporate services: "We had no idea what this process was costing us."

National begins producing chips at six high-tech silicon wafer fabrication plants (known as fabs): four in the U.S., one in Britain, and another in Israel. Next the company ships the wafers to seven assembly plants, most in Southeast Asia. Then—and here's where the real fun begins—the company must get finished products to an array of heavy-hitting customers, including IBM, Toshiba, Compaq, Ford, and Siemens, each with its own network of factories scattered around the world.

Everywhere National's logistics team looked—just collecting and collating the data took a year—it found costly stockpiles. They were sitting at assembly plants, at customer warehouses, and at all points in between: consolidators, forwarders, customs clearers, and distributors. To reach the customers, the chips traveled 20,000 different routes, mostly in the bellies of planes flown by 12 airlines, stopping along the way at 10 warehouses.

National delivered 95 percent of its products to those customers within 45 days of the time they were ordered, leaving considerable room for improvement. But the real trouble was with the other 5 percent, which required as long as 90 days for delivery. Since customers never knew which 5 percent would be late, they demanded 90 days' worth of everything. Says Patrick Brockett, president of National's international business group: "We had buffer stocks all along the line. The whole system was awash in inventory."

Using activity-based costing, a method of measuring the expense of everything done to a part as it meanders along the supply chain, National reached an eye-popping conclusion. Explains Brockett, a bearded Brit who once ran the company's European operations: "We found that most of the stuff we were moving around didn't generate a damn in revenues or profits." As a result, National has slashed the number of products it sells by 45 percent.

To speed its remaining products to market, National set out to craft a drastically simplified system. All finished chips would be transported to a central facility in Asia, where they would be stored and air-freighted to customers. A big problem remained. Says Brockett: "To do that, we would have had to make our company into Federal Express." Instead, National hired Federal Express. As part of a deal that took 16 months to negotiate and involved the chief executives of both companies, FedEx now runs all National's storage, sorting, and shipping activities out of a distribution center in Singapore.

As a result, National can move products from factory to customer in four days or less and is well on its way to its goal of 72 hours. National has increased sales by $584 million in the past two years. But distribution costs have fallen from 2.6 percent of revenues to 1.9 percent. So pleased are National's managers that they want to share their logistics insights with competitors. Call it self-serving altruism: National believes it and the whole electronics industry would benefit if companies could take money now wasted on distribution and redirect it into what they do best—inventing new and better products.

Says Gregory Petsch, senior vice president of corporate operations, who appears mildly bemused at his own choice of metaphors: "What we've done to improve manufacturing is not a home run. It's a base hit. We won't widen our lead unless we send the logistics batter up."

Petsch, a thin, deceptively soft-spoken man with a pronounced Texas drawl, has already enlisted—some might say dragooned—his suppliers into smoothing the flow of parts into Compaq's inordinately clean Houston factories. Picking up an idea from a newspaper article on Wal-Mart, Petsch cleared most of the parts inventories off the factory floor. Those pieces are now stored 12 miles away in a warehouse leased by 35 major Compaq suppliers, all of which moved their inventories to Houston at Compaq's behest. In addition, Compaq convinced seven sheet-metal suppliers to relocate to Texas. All these suppliers now truck the parts of Compaq exactly when and where the computer maker needs them.

Petsch secured the suppliers' cooperation not by promising them an uninterrupted flow of business but by agreeing to keep them better informed about Compaq's production plant, thereby helping them keep their own parts inventories under control. Petsch predicts that the new system, combined with the ever-falling prices of electronic components, will help knock more than $600 million off Compaq's supply costs this year.

Compaq's biggest challenge looms much further along in the supply chain. The company wants to drastically improve the way it gets computers from its factories to distributors, wholesalers, and retailers, whose ranks have swelled tenfold to 31,000 in the past three years. When it comes to deliveries, Compaq, like most computer companies, has an on-time arrival record that would land it in bankruptcy court if it were an airline. On average, only 40 percent of Compaq's computers reach the customer on time.

Compaq cut from four weeks to two the average amount of time a boxed computer spends in its 350,000-square-foot Houston distribution center. Sophisticated software matches customer orders to inventory on hand every day. Radio signals then tell the drivers of Aisle Rangers, the 30-foot-high trucks that run on tracks between the shelves, which items need to be picked up. An increasing number of Compaq's finished products buzz through the distribution center without ever tarrying on a shelf. Workers sort them onto pallets and then load them into waiting trucks, a practice known as cross-docking.

As part of its $100 million corporate reengineering effort, Compaq is testing an automatic replenishment system in its Erskine, Scotland, plant that electronically links the factory to some customers. Compaq will no longer make computers on speculation, hoping that supply will match demand somewhere down the road. Instead, in this build-to-order program, every unit that rolls off the line will be destined for a particular customer.

None of these concepts—cross-docking, build-to-order, or electronically swapping information with customers and suppliers are novel. Compaq logistics czar Duane Weeks helped forge one of the first electronic data interchange (EDI) links between Procter & Gamble and some of its customers way back in 1984. Automakers have long been building cars with specific dealers in mind. Says Weeks: "I've learned that there is nothing new under the sun. But while you could see any one of these practices at other companies, it's unusual to see them all in one place."

A great many of these practices converge at the Saturn operation in Spring Hill, Tennessee. So adroitly does the automaker manage its supply chain that in four years it has had to halt production just once—for only 18 minutes—because the right part was not delivered at the right time. Saturn maintains almost no inventory of components. Instead, a central computer directs trucks to deliver preinspected and presorted parts at precise times to the factory's 56 receiving docks, 21 hours a day, six days a week. Especially striking about this just-in-time system is that most of Saturn's 339 suppliers aren't located anywhere near the plant. They are in 39 states and an average of 550 miles away from Spring Hill.

Charged with making the Saturn network run on schedule is Ryder System, the Miami transportation service company that has become the biggest logistics management firm in nearly constant motion. Tractors, pulling trailers that are, on average, 90 percent full (an exceptionally high ratio), arrive daily at Ryder's command post some two miles from the Saturn factory. There the drivers uncouple the trailers. Specially designed shuttle tractors then take the trailers, which contain bar-coded, reusable plastic bins full of parts, and deliver them to the plant.

The long-haul drivers, meanwhile, hitch their tractors, the most expensive part of Ryder's $32 million of assets, onto other waiting trailers, which are stocked with empty bins to be hauled back to suppliers. In a first for the auto industry, the trucks also carry boxes of service parts to be delivered to Saturn dealers. Drivers plug a plastic key, loaded with electronic data, into an onboard computer. The screen then tells them exactly where to go, which route to take, and how much time to spend getting there.

Yet another of its graceful reinventions is the way Saturn replaces parts. Linked electronically to all its suppliers, the company automatically reorders parts each time a car rolls off the line. This so-called pull system of replenishment replaces the old, endlessly intricate computer programs used by General Motors to forecast future needs. Says Curt Gibbs, Saturn's director of materials flow and transportation: "We had systems so complex they bordered on artificial intelligence. Now, when we use a part, we pay for it, and then we order another one."

Logistics. It may not be cool. But it's flush with potential. Says the Boston Consulting Group's Hal Sirkin: "We're miles ahead of where we were five years ago, but we've only just begun to tap the savings we can get out of the system." May your trucks run full, your warehouses stay empty, and your supply chains come alive with the sound of cooperative suppliers and truly delighted customers.

Source: Excerpted from FORTUNE, November 28, 1994, © 1994 Time, Inc.

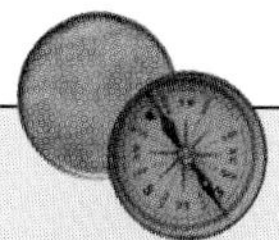

OPERATIONS TOUR

www.pscnet.com

PSC, INC.

PSC designs and produces a variety of laser bar code scanning devices. The products include hand-held, bar code readers, high-speed fixed position industrial scanners, and retail checkout scanners as well as a full line of accessories, software, and supplies to support its products. Headquartered in Webster, N.Y., the company has manufacturing facilities in Webster, Eugene, Oregon, and Paris, France, with roughly 1,200 employees worldwide.

Products

Bar code scanners are designed for a variety of situations that can involve long range scanning, reading small bar codes, and performing in harsh industrial environments. They are used extensively in industry, business, and government to manage and control the entire supply chain, which includes suppliers, production, warehousing, distribution, retail sales, and service. Examples of bar code readers include the familiar point-of-sale scanners encountered at supermarkets and other retail stores. They come in a variety of forms, ranging from hand-held to built-in models. High-speed, unattended scanners are used for automated material handling and sorting. Typical installations include high-volume distribution centers such as JC Penney's catalog operation and airport baggage handling systems. The company also produces "reader engines" that it supplies to other companies for use in their products. These may be as small as 1.2 cubic inches. One application for an "engine product" is found in lottery ticket validation machines. Use of bar code readers has greatly increased the speed and accuracy of data collection, resulting in increased productivity, improved production and inventory tracking and control, and improved market information.

Operations

Forecasting. Forecasting is not a significant activity at PSC due to several factors. There is high standardization of scanner components, which creates stability in usage requirements. Supplier lead times are relatively short, often only a few days. Orders are typically small; seventy percent of all orders are for 10 units or less. There is a fair degree of production flexibility, particularly in terms of product customization. As a result of these factors, the company relies mainly on short-term, moving average forecasts.

Product design. PSC has adopted a "Taguchi" approach" in the design of many of its products, designing them to perform effectively under a broad range of operating conditions. For example, many of its hand-held scanners can operate at temperatures ranging from −22° F to 120° F, and can withstand drops onto concrete surfaces from heights up to six feet and still function. This has enabled the company to offer warranties ranging from 24 to 36 months, far exceeding the industry standard of 3 to 12 months.

Layout. PSC has developed an efficient production layout that consists of assembly lines and work centers. The assembly lines handle standardized production and subassemblies and the work centers handle final assembly and customization of products. Assembly lines are U-shaped to facilitate communication among workers. The work centers are designed for production flexibility; they can be reconfigured in about 4 hours. Work centers are staffed by teams of three to six cross-trained workers who are responsible for an order from start to finish.

The production process. Production involves a combination of assembly line and batch processing that provides high volume as well as flexibility to customized individual orders. Because of the high standardization among the internal components of different scanners, many of the subassemblies can be produced on assembly lines. Customization is done primarily on the external portion of various products according to customer specification.

The production process for scanner engines is depicted in the process flowchart shown in the Figure. The process begins when an order is received from a customer. The order is then configured according to customer specifications. Next it is entered into the computer to obtain a bill of materials (BOM), and the order is transmitted to production control so that it can be scheduled for production. A "traveler" packet containing product specifications and the BOM is created. It will accompany the order throughout the process.

The traveler is sent to the "kitting" area where standard parts and any customized parts are obtained and placed into a bin ("kit"), and placed in a flow rack until the assigned work center is ready for the job (i.e., a pull system).

The next phase of the process transforms unprogrammed, panelized circuit boards into programmed boards. The boards first pass through a screen printer which uses a stencil to coat the boards with a solder paste. Next the boards pass through a chip mounter which enters values for the smaller, passive components of the circuit board at a rate of 25,000 parts per hour. A second mounter enters values for the larger, programmable components at a rate of 7,000 parts per hour. The slower rate for the larger components is offset by the fact that there are fewer of those components. The process ends up being balanced, and no bottlenecks occur.

The programmed boards move by conveyor to a station for visual inspection. Rejects are returned to the chip mounter area, and boards that pass are sent through an oven to solidify the solder, making the programming permanent. The circuit boards are then removed from the panels and placed into the kit. The kits are then taken to designated work centers for customization and placement in scanner engines.

Work centers typically have builders, computer operators, and a tester. A builder mounts the laser diodes on the circuit board and passes it to a computer operator who downloads the customer specifications into the microprocessor of the scan engine. The operator also mounts the optical components and adjusts them for the design of the scanner (e.g., long range scanning). Next, the engine goes to the tester who checks to make sure that the scanner is capable of reading bar codes, and laser characteristics. Engines that fail are sent for repair, and later retested. If the engine fails a second time, it is either returned for further repair or scrapped. Engines which pass are placed in an electrostatic bag which protects them from static electricity that could damage the programming.

Engines are then sent to Audit for another check for performance quality.

Engines that pass are incorporated into the final product, a serial number is added, along with a label, and the product is sent to the packing area, and then shipped to the customer.

Scheduling. Scheduling is done using spreadsheets to generate material requirements plans. An updated master schedule is distributed daily to management.

Inventory. The company uses aspects of both MRPII and JIT for inventory management, and attempts to minimize the amount of inventory it maintains. A computer determines component requirements and generates purchase orders for the components for each order, and then appropriate orders for various components from vendors are prepared. However, the company maintains a stock of standard components which are replenished using a reorderpoint system. Kanbans are used on the production floor to control the movement of work. The company has adopted point-of-use replenishment for some areas of operations, having deliveries come directly to the production floor. Finished products are immediately shipped to the customer, which enhances the company's delivery performance and avoids finished goods inventory.

Suppliers. Approximately 40 vendors supply parts and materials to PSC, each of which has been subjected to a multiple-step supplier certification program that includes the supplier completing a self-evaluation questionnaire; an on-site visit of supplier facilities by a team from PSC made up of people from engineering, purchasing, and operations; a probation period; and rating of products using mil standard 105 specifications. Vendor performance is tracked on product quality, delivery, and service.

When an item is removed from inventory, it is scanned into the computer, and then information on some components is transmitted directly to suppliers, along with purchase orders to restock the components.

Quality. Quality is strongly emphasized at PSC. Employees are trained in quality concepts and the use of quality tools. Training is incorporated on-the-job so that employees can see the practical applications of what they are learning. Employees are responsible for

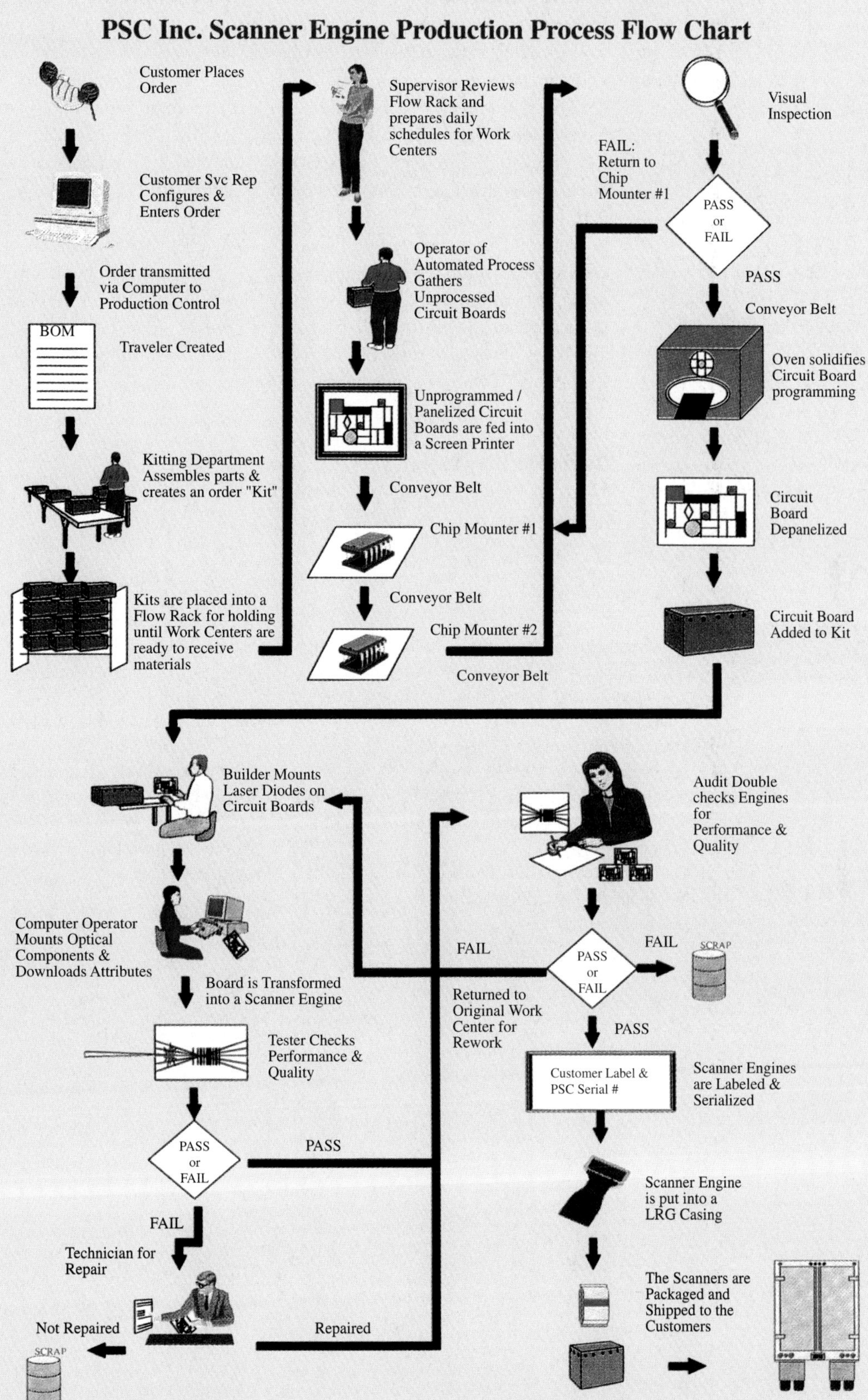
PSC Inc. Scanner Engine Production Process Flow Chart
Customer Places Order
Customer Svc Rep Configures & Enters Order
Order transmitted via Computer to Production Control
BOM
Traveler Created
Kitting Department Assembles parts & creates an order "Kit"
Kits are placed into a Flow Rack for holding until Work Centers are ready to receive materials
Supervisor Reviews Flow Rack and prepares daily schedules for Work Centers
Operator of Automated Process Gathers Unprocessed Circuit Boards
Unprogrammed / Panelized Circuit Boards are fed into a Screen Printer
Conveyor Belt
Chip Mounter #1
Conveyor Belt
Chip Mounter #2
Conveyor Belt
Visual Inspection
FAIL: Return to Chip Mounter #1
PASS or FAIL
PASS
Conveyor Belt
Oven solidifies Circuit Board programming
Circuit Board Depanelized
Circuit Board Added to Kit
Builder Mounts Laser Diodes on Circuit Boards
Computer Operator Mounts Optical Components & Downloads Attributes
Board is Transformed into a Scanner Engine
Tester Checks Performance & Quality
PASS or FAIL
PASS
FAIL
Technician for Repair
Not Repaired
SCRAP
Repaired
FAIL
Returned to Original Work Center for Rework
Audit Double checks Engines for Performance & Quality
PASS or FAIL
FAIL
SCRAP
PASS
Customer Label & PSC Serial #
Scanner Engines are Labeled & Serialized
Scanner Engine is put into a LRG Casing
The Scanners are Packaged and Shipped to the Customers

performing in-process quality checks (quality at the source), and to report any defects they discover to their supervisor. Defects are assigned to one of three categories for problem solving:

- Operator/training error. The supervisor notifies a trainer who then provides appropriate retraining.
- Process/equipment problem. The supervisor notifies the manufacturing engineer who is then responsible for diagnosing the cause and correcting the problem.
- Parts/material problem. The supervisor notifies quality assurance, who then notifies the vendor to correct the problem. Defective parts are either scrapped or returned to the vendor.

Lean Production

PSC strives to operate on lean production principles. In addition to emphasizing high levels of quality, production flexibility, low levels of inventories, and having some deliveries come right to the production floor, its organization structure is fairly flat, and it uses a team approach. Still another feature of lean production is that many of PSC's workers are multi-skilled. The company encourages employees to master new skills through a pay-for-skill program, and bases hourly pay rates on the number of skills a worker can perform.

Business Strategy

The company has developed what it believes is a strong strategy for success. Strategic initiatives include anticipating customer demand for miniaturization and the ability to customize products; expanding its proprietary technology; and expanding internationally into Western Europe (now accounts for about 35% of sales) and the Pacific rim (now accounts for about 10% of sales). Several plants or groups are ISO certified, which has been important for European sales. The company intends to continue to expand its product lines through acquisition of other companies.

Selected Bibliography

Burt, David N., and Michael F. Doyle. *The American Keiretsu: A Strategic Weapon for Global Competitiveness.* Burr Ridge, IL: Business One Irwin, 1993.

Carter, Joseph. *Purchasing.* Burr Ridge, IL: Business One Irwin, 1993.

Copacino, William C. *Supply Chain Management: The Basics and Beyond.* Boca Raton, FL: St. Lucie Press, 1997.

Fernandez, Ricardo R. *Total Quality in Purchasing & Supplier Management.* Delray Beach, FL: St. Lucie Press, 1995.

Leenders, Michael R., and Harold E. Fearon. *Purchasing and Supply Management.* Burr Ridge, IL: Richard D. Irwin, 1997.

Pooler, Victor H., and David J. Pooler. *Purchasing and Supply Chain Management: Creating the Vision.* New York: Chapman & Hall, 1997.

Schorr, John E. *Purchasing in the 21st Century: A Guide to State-of-the-Art Techniques and Strategies.* Essex Junction, VT: Oliver Wright Companies, 1992.

AA
AA
AA
AA
AA
American
American
AA
American
AA
N6803
American
American

LEARNING OBJECTIVES

After completing this chapter, you should be able to:

1. Explain what scheduling involves and the importance of good scheduling.
2. Discuss scheduling needs in high-volume and intermediate-volume systems.
3. Discuss scheduling needs in job shops.
4. Use and interpret Gantt charts, and use the assignment method for loading.
5. Discuss and give examples of commonly used priority rules.
6. Describe some of the unique problems encountered in service systems, and describe some of the approaches used for scheduling service systems.

CHAPTER SEVENTEEN SCHEDULING

CHAPTER OUTLINE

scheduling Establishing the timing of the use of equipment, facilities, and human activities in an organization.

Within an organization, **scheduling** pertains to establishing the timing of the use of specific resources of that organization. It relates to the use of equipment, facilities, and human activities. Scheduling occurs in every organization, regardless of the nature of its activities. For example, manufacturers must schedule production, which means developing schedules for workers, equipment, purchases, maintenance, and so on. Hospitals must schedule admissions, surgery, nursing assignments, and support services such as meal preparation, security, maintenance, and cleaning. Educational institutions must schedule classrooms, instruction, and students. And lawyers, doctors, dentists, hairdressers, and auto repair shops must schedule appointments.

In the decision-making hierarchy, scheduling decisions are the final step in the transformation process before actual output occurs. Many decisions about system design and operation have to be made long before scheduling decisions. They include the capacity of the system, equipment selection, selection and training of workers, and design of products and services. Consequently, scheduling decisions must be made within the constraints established by many other decisions, making them fairly narrow in scope and latitude.

Generally, the objectives of scheduling are to achieve trade-offs among conflicting goals, which include efficient utilization of staff, equipment, and facilities, and minimization of customer waiting time, inventories, and process times.

This chapter covers scheduling in both manufacturing and service environments. Although the two environments have many similarities, some basic differences are important. We begin with manufacturing environments.

Operations Strategy

Scheduling can either help or hinder operations strategy. If a manager does scheduling well, products or services can be made or delivered in a timely manner. Resources can be used to best advantage and customers will be satisfied. Scheduling not performed well will result in inefficient use of resources and possibly dissatisfied customers.

The implication is clear: Management should not overlook the important role that scheduling plays in the success of an organization, giving a competitive advantage if handled well or disadvantage if done poorly. Time-based competition depends on good scheduling. Coordination of materials, equipment use, and employee time is an important function of operations management. It is not enough to have good design, superior quality, and the other elements of a well-run organization if scheduling is done poorly—just as it is not enough to own a well-designed and well-made car, with all the latest features for comfort and safety, if the owner doesn't know how to drive it!

SCHEDULING MANUFACTURING OPERATIONS

Scheduling tasks are largely a function of the volume of system output. High-volume systems require approaches substantially different from those required by job shops, and project scheduling requires still different approaches. In this chapter, we will consider scheduling for high-volume systems, intermediate-volume systems, and low-volume (job shop) scheduling. Project scheduling is discussed in Chapter 18.

Scheduling in High-Volume Systems

Scheduling encompasses allocating workloads to specific work centers and determining the sequence in which operations are to be performed. High-volume systems are characterized by standardized equipment and activities that provide identical or highly similar operations on customers or products as they pass through the system. The goal is to obtain a smooth rate of flow of goods or customers through the system in order to get a high utilization of labor and equipment. High-volume systems are often referred to as **flow systems**; scheduling in these systems is referred to as **flow-shop scheduling.** Examples of high-volume products include autos, personal computers, radios and televisions, stereo equipment, toys, and appliances. In process industries, examples include petroleum refining, sugar refining, mining, waste treatment, and the manufacturing of fertilizers. Examples of services include cafeteria lines, news broadcasts, and mass inoculations. Because of the highly repetitive nature of these systems, many of the loading and sequence decisions are determined during the design of the system. The use of highly specialized tools and equipment, the arrangement of equipment, the use of specialized material-handling equipment, and the division of labor are all designed to enhance the flow of work through the system, since all items follow virtually the same sequence of operations.

flow system High-volume system with standardized equipment and activities.

flow-shop scheduling Scheduling for the high-volume flow system.

A major aspect in the design of flow systems is *line balancing,* which concerns allocating the required tasks to workstations so that they satisfy technical (sequencing) constraints and are balanced with respect to equal work times among stations. Highly balanced systems result in the maximum utilization of equipment and personnel as well as the highest possible rate of output. Line balancing is discussed in Chapter 6.

In setting up flow systems, designers must consider the potential discontent of workers in connection with the specialization of job tasks in these systems; high work rates are often achieved by dividing the work into a series of relatively simple tasks assigned to different workers. The resulting jobs tend to be boring and monotonous and may give rise to fatigue, absenteeism, turnover, and other problems, all of which tend to reduce productivity and disrupt the smooth flow of work. These problems and potential solutions are elaborated on in Chapter 7, which deals with the design of work systems.

In spite of the built-in attributes of flow systems related to scheduling, a number of scheduling problems remain. One stems from the fact that few flow systems are *completely* devoted to a single product or service; most must handle a variety of sizes and models. Thus, an automobile manufacturer will assemble many different combinations of cars—two-door and four-door models, some with air-conditioning and some not, some with deluxe trim and others with standard trim, some with tape decks, some with tinted glass, and so on. The same can be said for producers of appliances, electronic equipment, and toys. Each change involves slightly different inputs of parts, materials, and processing requirements that must be scheduled into the line. If the line is to operate smoothly, a supervisor must coordinate the flow of materials and the work. This requires scheduling the inputs, the processing, and the outputs, as well as scheduling purchases. In addition to achieving a smooth flow, it is important to avoid excessive buildup of inventories. Again, each variation in size or model will tend to have somewhat different inventory requirements, so that additional scheduling efforts will be needed.

One source of scheduling concern is possible disruptions in the system that result in less than the desired output. These can be caused by equipment failures, material shortages, accidents, and absences. In practice, it is usually impossible to increase the rate of output to compensate for these factors, mainly because flow systems are designed to operate at a given rate. Instead, strategies involving subcontracting or overtime are often required, although subcontracting on short notice is not always feasible. Sometimes work that is partly completed can be made up off the line.

The reverse situation can also impose scheduling problems although these are less severe. This happens when the desired output is less than the usual rate. However, instead of slowing the ensuing rate of output, it is usually necessary to operate the system at the usual rate, but for fewer hours. For instance, a production line might operate temporarily for seven hours a day instead of eight.

High-volume systems usually require automated or specialized equipment for processing and handling. Moreover, they perform best with a high, uniform output. Consequently, the following factors often determine the success of such a system:

1. *Process and product design.* Here, cost and manufacturability are important, as is achieving a smooth flow through the system.
2. *Preventive maintenance.* Keeping equipment in good operating order can minimize breakdowns that would disrupt the flow of work.
3. *Rapid repair when breakdowns occur.* This can require specialists as well as stocks of critical spare parts.
4. *Optimal product mixes.* Techniques such as linear programming can be used to determine optimal blends of inputs to achieve desired outputs at minimal costs. This is particularly true in the manufacture of fertilizers, animal feeds, and diet foods.
5. *Minimization of quality problems.* Quality problems can be extremely disruptive, requiring shutdowns while problems are resolved. Moreover, when output fails to meet quality standards, not only is there the loss of output but also a waste of the labor, material, time, and other resources that went into it.
6. *Reliability and timing of supplies.* Shortages of supplies is an obvious source of disruption and must be avoided. On the other hand, if the solution is to stockpile supplies, that can lead to high carrying costs. Shortening supply lead times, developing reliable supply schedules, and carefully projecting needs are all useful.

Scheduling in Intermediate-Volume Systems

Intermediate-volume system outputs fall between the standardized type of output of the high-volume systems and made-to-order output of job shops. Like the high-volume systems, intermediate-volume systems typically produce standard outputs. If manufacturing is involved, the products may be for stock rather than for special order. However, the volume of output in such cases is not large enough to justify continuous production. Instead, it is more economical to process these items *intermittently.* Thus, intermediate-volume work centers periodically shift from one job to another. In contrast to a job shop, the run sizes are relatively large. Examples of products made in these systems include canned foods, baked goods, paint, and cosmetics.

The three basic issues in these systems are the *run size* of jobs, the *timing* of jobs, and the *sequence* in which jobs should be processed.

Sometimes, the issue of run size can be determined by using a model such as the economic run size model discussed in Chapter 13 on inventory management. The run size that would minimize setup and inventory costs is

$$Q_0 = \sqrt{\frac{2DS}{H}}\sqrt{\frac{p}{p-u}} \qquad (17\text{–}1)$$

This approach may require multiple setup costs if setup costs depend on the order of processing: some products will have similar setups, so it may be less expensive to process them in sequence to take advantage of that. Another difficulty arises because usage is not always as smooth as assumed in the model. Some products will tend to be used up faster than expected and have to be replenished sooner. Also, because multiple products are to be processed, it is not always possible to schedule production to correspond with optimum run intervals.

Another approach frequently used is to base production on a master schedule developed from customer orders and forecasts of demand. Companies engaged in assembly operations would then use an MRP approach (described in Chapter 14) to determine the quantity and projected timing of jobs for components. The manager would then compare projected requirements with projected capacity, and develop a feasible schedule from that information. Companies engaged in producing processed rather than assembled goods (e.g., food products, such as canned goods and beverages; publishing, such as magazines; paints and cleaning supplies) would use a somewhat different approach; the *time-phasing* information provided by MRP would not be an important factor.

Solectron Corporation, a provider of customized electronics manufacturing services and winner of the 1997 Baldrige award, provides printed circuit assembly for its customers. Because it customizes assemblies, its products are made to order, yet produced in high volume for its large customers.

www.solectron.com

SCHEDULING IN LOW-VOLUME SYSTEMS

The characteristics of low-volume systems (job shops) are considerably different from those of high- and intermediate-volume systems. Products are made to order, and orders usually differ considerably in terms of processing requirements, materials needed, processing time, and processing sequence and setups. Because of these circumstances, **job-shop scheduling** is usually fairly complex. This is compounded by the impossibility of establishing firm schedules prior to receiving the actual job orders.

job-shop scheduling Scheduling for low-volume systems with many variations in requirements.

Vol. 2, seg. 2
Bernard Welding

Job-shop processing gives rise to two basic issues for schedulers: how to distribute the workload among work centers and what job processing sequence to use.

Loading

Loading refers to the assignment of jobs to processing (work) centers. Loading decisions involve assigning specific jobs to work centers and to various machines in the work centers. In cases where a job can be processed only by a specific center, loading presents little difficulty. However, problems arise when two or more jobs are to be processed and there are a number of work centers capable of performing the required work. In such cases, the operations manager needs some way of assigning jobs to the centers.

loading The assignment of jobs to processing centers.

When making assignments, managers often seek an arrangement that will minimize processing and setup costs, minimize idle time among work centers, or minimize job completion time, depending on the situation.

Gantt Charts. Visual aids called **Gantt charts** are used for a variety of purposes related to loading and scheduling. They derive their name from Henry Gantt, who pioneered the use of charts for industrial scheduling in the early 1900s. Gantt charts can be used in a number of different ways, two of which are illustrated in Figure 17–1, which shows scheduling classrooms for a university and scheduling hospital operating rooms for a day.

Gantt chart Chart used as visual aid for loading and scheduling purposes.

The purpose of Gantt charts is to organize and clarify the actual or intended use of resources in a *time framework*. In most cases, a time scale is represented horizontally, and resources to be scheduled are listed vertically. The use of the resources is reflected in the body of the chart.

Managers may use the charts for trial-and-error schedule development to get an idea of what different arrangements would involve. Thus, a tentative surgery schedule might reveal insufficient allowance for surgery that takes longer than expected and can be revised accordingly. Use of the chart for classroom scheduling would help avoid assigning two different classes to the same room at the same time.

FIGURE 17–1

Examples of charts used for scheduling

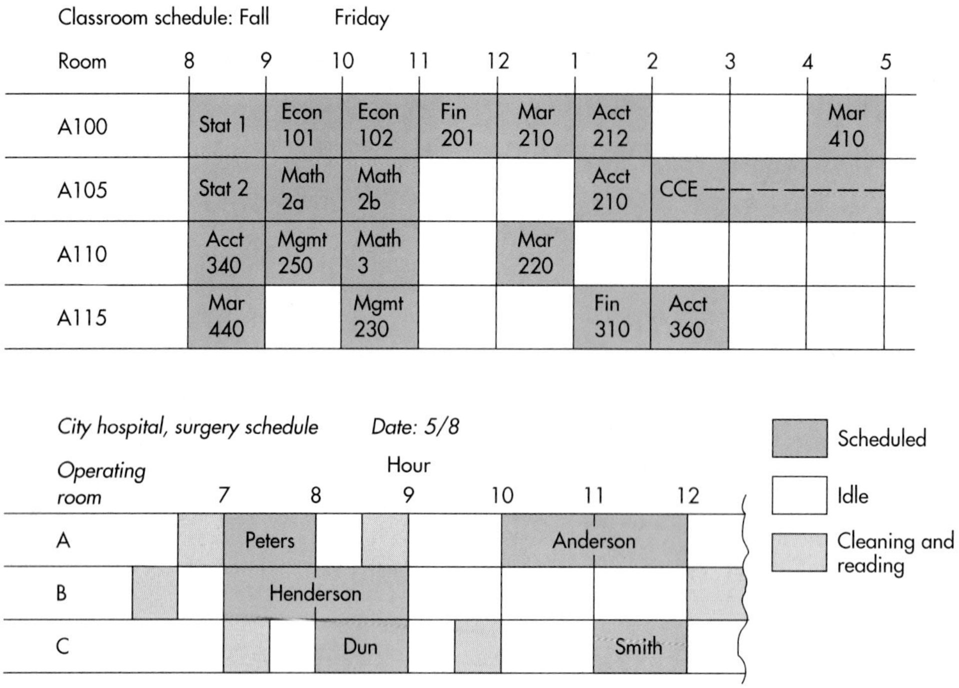

Classroom schedule: Fall — Friday

Room	8	9	10	11	12	1	2	3	4
A100	Stat 1	Econ 101	Econ 102	Fin 201	Mar 210	Acct 212			Mar 410
A105	Stat 2	Math 2a	Math 2b			Acct 210	CCE —	— — —	— — —
A110	Acct 340	Mgmt 250	Math 3		Mar 220				
A115	Mar 440		Mgmt 230			Fin 310	Acct 360		

City hospital, surgery schedule — *Date: 5/8*

Operating room (Hour)	7	8	9	10	11	12
A	Peters			Anderson		
B	Henderson					
C		Dun			Smith	

Scheduled

Idle

Cleaning and reading

FIGURE 17–2

A Gantt load chart

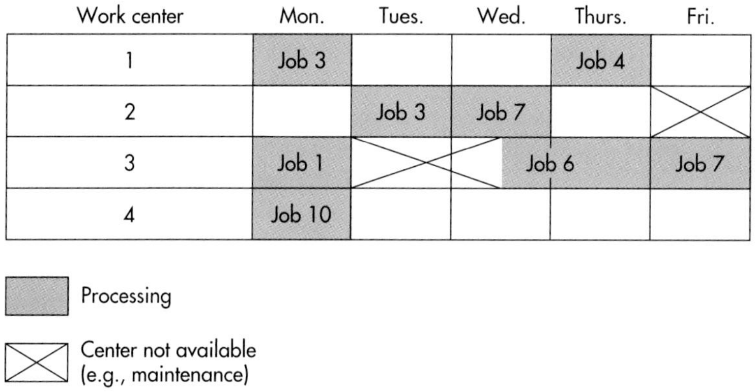

Work center	Mon.	Tues.	Wed.	Thurs.	Fri.
1	Job 3			Job 4	
2		Job 3	Job 7		X
3	Job 1	X	X / Job 6	Job 6	Job 7
4	Job 10				

Processing

Center not available (e.g., maintenance)

There are a number of different types of Gantt charts. Two of the most commonly used are the *load chart* and the *schedule chart.*

load chart A Gantt chart that shows the loading and idle times for a group of machines or list of departments.

A **load chart** depicts the loading and idle times for a group of machines or a list of departments. Figure 17–2 illustrates a typical load chart. This chart indicates that work center 3 is completely loaded for the entire week, center 4 will be available after noon on Tuesday, and the other two centers have idle time scattered throughout the week. This information can help a manager rework loading assignments to better utilize the centers. For instance, if all centers perform the same kind of work, the manager might want to free one center for a long job or a rush order. The chart also shows when certain jobs are scheduled to start and finish, and where to expect idle time.

infinite loading Jobs are assigned to work centers without regard to the capacity of the work center.

finite loading Jobs are assigned to work centers taking into account the work center capacity and job processing times.

Two different approaches are used to load work centers, *infinite* loading and *finite* loading. **Infinite loading** assigns jobs to work centers without regard to the capacity of the work center. The priority sequencing rules described in this chapter use infinite loading. One possible result of infinite loading is the formation of queues in some (or all) work centers. **Finite loading** projects actual job starting and stopping times at each work center, taking into account the capacities of each work center and the processing times of jobs, so that capacity is not exceeded. One output of finite loading is a detailed projection of hours each work center will operate. Schedules based on finite loading may have to be updated often, perhaps daily, due to processing delays at work centers

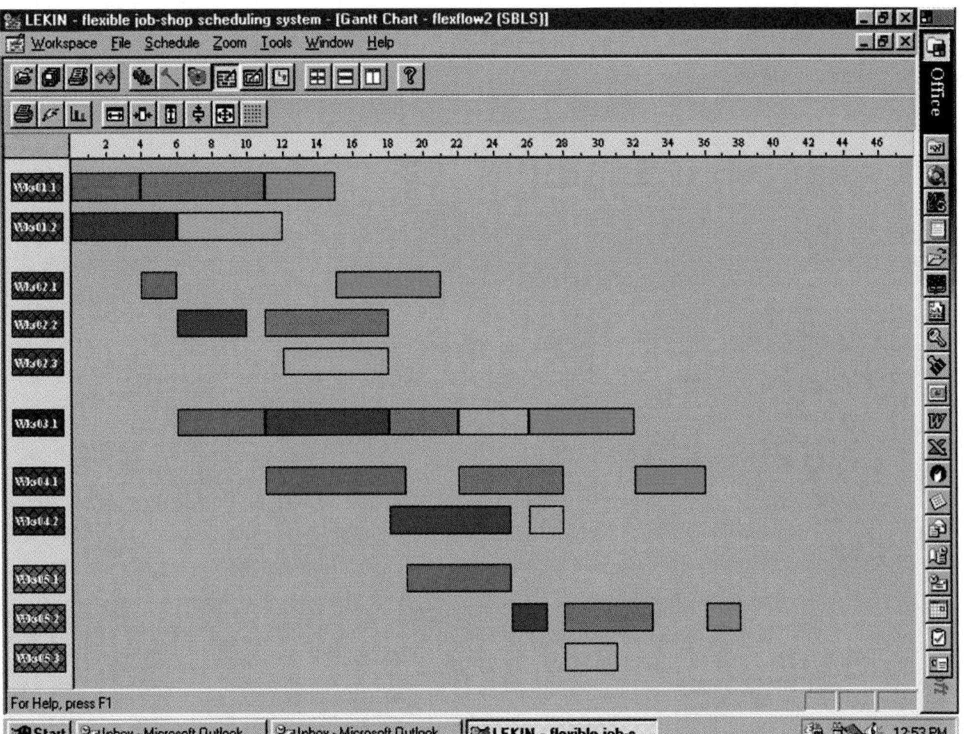

Software such as "LEKIN" developed at NYU includes multiple scheduling routines along with graphics such as this interactive GANTT chart for scheduling a wide variety of manufacturing systems.

and the addition of new jobs or cancellation of current jobs. The following diagram illustrates these two approaches.

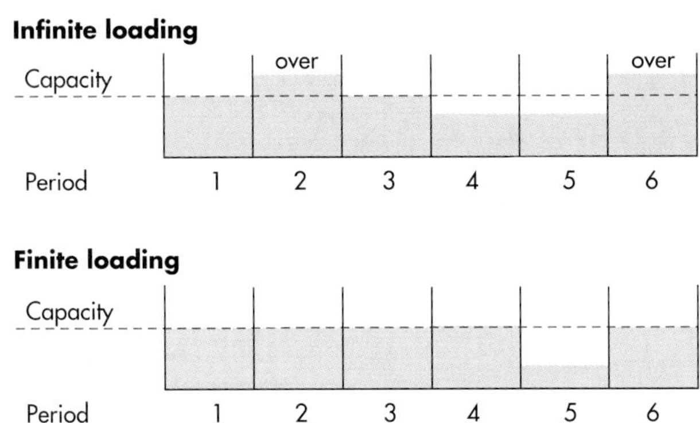

Loading can be done in various ways. **Vertical loading** refers to loading jobs at a work center job by job, usually according to some priority criterion. Vertical loading does not consider the work center's capacity (i.e., infinite loading). In contrast, **horizontal loading** involves loading the job that has the highest priority on all work centers it will require, then the job with the next highest priority, and so on. Horizontal loading is based on finite loading. One possible result of horizontal loading is keeping jobs waiting at a work center even though that center is idle, so the center will be ready to process a higher priority job that is expected to arrive shortly. That would not happen with vertical loading; the work center would be fully loaded, although a higher priority job would have to wait if it arrived while the work center was busy. So horizontal loading takes a more *global* approach to scheduling, while vertical loading uses a *local* approach. Which approach is better? That depends on various factors: the relative costs of keeping

vertical loading Loading jobs at a work center, job by job, usually according to some priority criterion, using infinite loading.

horizontal loading Loading each job on all work centers it will require, then the next job on all work centers, according to some priority, using finite loading.

FIGURE 17–3

Progress chart for landscaping job

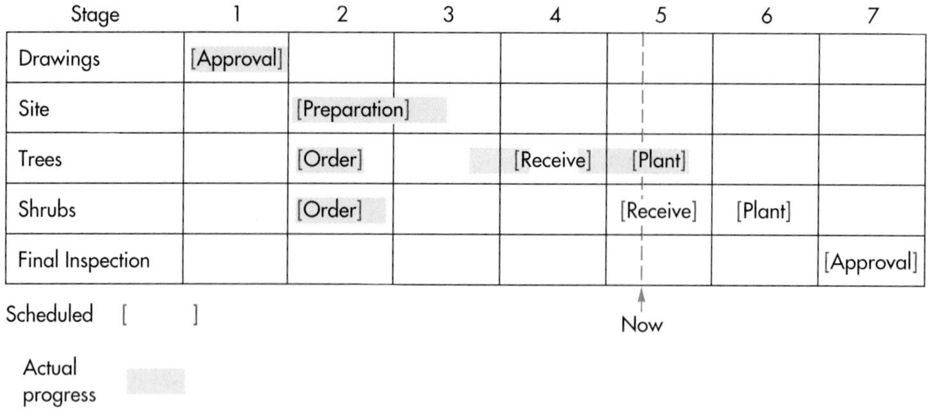

Stage	1	2	3	4	5	6	7
Drawings	[Approval]						
Site		[Preparation]					
Trees		[Order]		[Receive]	[Plant]		
Shrubs		[Order]			[Receive]	[Plant]	
Final Inspection							[Approval]

Scheduled [ ]

Actual progress

Now

higher priority jobs waiting, the cost of having work centers idle, the number of jobs, the number of work centers, the potential for processing disruptions, the potential for new jobs and job cancellations, and so on. As you can see, the circumstances can be quite complex.

With infinite loading, a manager may need to make some response to overloaded work centers. Among the possible responses are shifting work to other periods or other centers, working overtime, or contracting out a portion of the work. Note that the last two options in effect increase capacity to meet the work load.

Finite loading may reflect a fixed upper limit on capacity. For example, a bus line will have only so many buses. Hence, the decision to place into service a particular number of buses fixes capacity. Similarly, a manufacturer might have one specialized machine that it operates around the clock. Thus, it is operated at the upper limit of its capacity, so finite loading would be called for.

forward scheduling Scheduling ahead, from some point in time.

backward scheduling Scheduling by working backwards from the due date(s).

schedule chart A Gantt chart that shows the orders or jobs in progress and whether they are on schedule.

There are two general approaches to scheduling: forward scheduling and backward scheduling. **Forward scheduling** means scheduling ahead from a point in time; **backward scheduling** means scheduling backward from a due date. Forward scheduling is used if the issue is "How long will it take to complete this job?" Backward scheduling would be used if the issue is "When is the latest the job can be started and still be completed by the due date?"

A manager often uses a **schedule chart** to monitor the progress of jobs. The vertical axis on this type of Gantt chart shows the orders or jobs in progress, and the horizontal axis shows time. The chart indicates which jobs are on schedule and which are behind or ahead.

A typical schedule chart is illustrated in Figure 17–3. It shows the current status of a landscaping job with planned and actual starting and finishing times for the five stages of the job. The chart indicates that approval and the ordering of trees and shrubs was on schedule. The site preparation was a bit behind schedule. The trees were received earlier than expected, and planting is ahead of schedule. However, the shrubs have not yet been received. The chart indicates some slack between scheduled receipt of shrubs and shrub planting, so if the shrubs arrive by the end of the week, it appears the schedule can still be met.

Despite the obvious benefits of Gantt charts, they possess certain limitations, the chief one being the need to repeatedly update a chart to keep it current. In addition, a chart does not directly reveal costs associated with alternative loadings. Finally, a job's processing time may vary depending on the work center; certain stations or work centers may be capable of processing some jobs faster than other stations. Again, that situation would increase the complexity of evaluating alternative schedules. Nonetheless, Gantt charts are the most widely used scheduling tools.

		Period 1	2	3	4	5	6
Input	Planned	100	100	90	90	90	90
	Actual	120	95	80	88	93	94
	Deviation	+20	−5	−10	−2	+3	+4
	Cum. dev.	+20	+15	+5	+3	+6	+10
Output	Planned	110	110	100	100	100	95
	Actual	110	105	95	101	103	96
	Deviation	0	−5	−5	+1	+3	+1
	Cum. dev.	0	−5	−10	−9	−6	−5

Backlog	40	50	40	25	12	2	0

Note: Figures represent standard hours of processing time.

FIGURE 17–4

A sample input/output report for a work center

Input/Output Control. **Input/output (I/O) control** refers to monitoring the work flow and queue lengths at work centers. The purpose of I/O control is to manage work flow so that queues and waiting times are kept under control. Without I/O control, demand may exceed processing capacity, causing an overload at the center. Conversely, work may arrive slower than the rate a work center can handle, leaving the work center underutilized. Ideally, a balance can be struck between the input and output rates, thereby achieving effective use of work center capacities without experiencing excessive queues at the work centers. A simple example of I/O control is the use of stop lights on some expressway onramps. These regulate the flow of entering traffic according to the current volume of expressway traffic.

input/output control Managing work flow and queues at work centers.

Figure 17–4 illustrates an input/output report for a work center. A key portion of the report is the backlog of work waiting to be processed. The report reveals deviations-from-planned for both inputs and outputs, thereby enabling a manager to determine possible sources of problems.

The deviations in each period are determined by subtracting "planned" from "actual." For example, in the first period, subtracting the planned input of 100 hours from the actual input of 120 hours produces a deviation of +20 hours. Similarly, in the first period, the planned and actual outputs are equal, producing a deviation of 0 hours.

The backlog for each period is determined by subtracting the "actual output" from the "actual input" and adjusting the backlog from the previous period by that amount. For example, in the first period actual output exceeds actual input by 10 hours. Hence, the previous backlog of 50 hours is reduced by 10 hours to 40 hours.

Assignment Method of Linear Programming. The **assignment model** is a special-purpose linear-programming model that is useful in situations that call for assigning tasks or other work requirements to resources. Typical examples include assigning jobs to machines or workers, territories to salespeople, and telephone line repair jobs to repair crews. The idea is to obtain an optimum *matching* of tasks and resources. Commonly used criteria include costs, profits, efficiency, and performance.

assignment model A linear-programming model for optimal assignment of tasks and resources.

Table 17–1 illustrates a typical problem, where four jobs are to be assigned to four machines. The problem is arranged in a format that facilitates evaluation of assignments. The numbers in the body of the table represent the value or cost associated with each job-machine combination. In this case, the numbers represent costs. Thus, it would cost

TABLE 17–1

A typical assignment problem

		MACHINE			
		A	B	C	D
Job	1	8	6	2	4
	2	6	7	11	10
	3	3	5	7	6
	4	5	10	12	9

$8 to do job 1 on machine A, $6 to do job 1 on machine B, and so on. If the problem involved minimizing the cost for job 1 alone, it would clearly be assigned to machine C, since that combination has the lowest cost. However, that assignment does not take into account the other jobs and their costs, which is important since the lowest-cost assignment for any one job may not be consistent with a minimum-cost assignment when all jobs are considered.

If there are to be n matches, there are $n!$ different possibilities. In this case, there are $4! = 24$ different matches. One approach is to investigate each match and select the one with the lowest cost. However, a much simpler approach is to use a procedure called the **Hungarian method** to identify the lowest-cost solution.

Hungarian method Method of assigning jobs by a one-for-one matching to identify the lowest-cost solution.

To be able to use the Hungarian method, a one-for-one matching is required. Each job, for example, must be assigned to only one machine. It is also assumed that every machine is capable of handling every job, and that the costs or values associated with each assignment combination are known and fixed (i.e., not subject to variation). The number of rows and columns must be the same. Solved Problem 1 at the end of the chapter shows how to adjust the table if they aren't the same.

Once the relevant cost information has been acquired and arranged in tabular form, the basic procedure of the Hungarian method is:

1. Subtract the smallest number in each row from every number in the row. This is called a *row reduction*. Enter the results in a new table.
2. Subtract the smallest number in each column of the new table from every number in the column. This is called a *column reduction*. Enter the results in another table.
3. Test whether an optimum assignment can be made. You do this by determining the *minimum* number of lines needed to cover (i.e., cross out) all zeros. If the number of lines equals the number of rows, an optimum assignment is possible. In that case, go to step 6. Otherwise go on to step 4.
4. If the number of lines is less than the number of rows, modify the table in this way:
   a. Subtract the smallest uncovered number from every uncovered number in the table.
   b. Add the smallest uncovered number to the numbers at *intersections* of covering lines.
   c. Numbers crossed out but not at intersections of crossout lines carry over unchanged to the next table.
5. Repeat steps 3 and 4 until an optimal table is obtained.
6. Make the assignments. Begin with rows or columns with only one zero. Match items that have zeros, using only one match for each row and each column. Cross out both the row and the column after the match.

Example 1

Determine the optimum assignment of jobs to machines for the following data (from Table 17–1).

		MACHINE				Row Minimum
		A	**B**	**C**	**D**	
	1	8	6	2	4	2
Job	**2**	6	7	11	10	6
	3	3	5	7	6	3
	4	5	10	12	9	5

Solution

a. Subtract the smallest number in each row from every number in the row, and enter the results in a new table. The result of this row reduction is

		MACHINE			
		A	**B**	**C**	**D**
	1	6	4	0	2
Job	**2**	0	1	5	4
	3	0	2	4	3
	4	0	5	7	4
Column Minimum		0	1	0	2

b. Subtract the smallest number in each column from every number in the column, and enter the results in a new table. The result of this column reduction is

		MACHINE			
		A	**B**	**C**	**D**
	1	6	3	0	0
Job	**2**	0	0	5	2
	3	0	1	4	1
	4	0	4	7	2

c. Determine the *minimum* number of lines needed to cross out all zeros. (Try to cross out as many zeros as possible when drawing lines.)

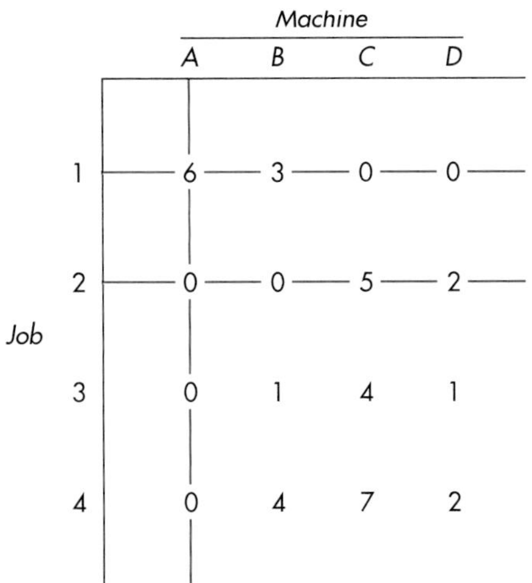

d. Since only three lines are needed to cross out all zeros, this is not the optimum.

e. Subtract the smallest value that hasn't been crossed out (in this case, 1) from every number that hasn't been crossed out, and add it to numbers that are at the intersections of covering lines. The results are

		MACHINE			
		A	**B**	**C**	**D**
Job	**1**	7	3	0	0
	2	1	0	5	2
	3	0	0	3	0
	4	0	3	6	1

f. Determine the minimum number of lines needed to cross out all zeros (four). Since this equals the number of rows, you can make the optimum assignment.

		Machine			
		A	*B*	*C*	*D*
Job	1	7	3	0	0
	2	1	0	5	2
	3	0	0	3	0
	4	0	3	6	1

g. Make assignments: Start with rows and columns with only one zero. Match jobs with machines that have a zero cost.

		Machine			
		A	*B*	*C*	*D*
Job	1	7	3	0	0
	2	1	0	5	2
	3	0	0	3	0
	4	0	3	6	1

As you can see, the progress is relatively simple compared with some of the other linear-programming techniques. It is worth knowing that one extension of this technique can be used to prevent undesirable assignments. For example, union rules may prohibit one person's assignment to a particular job, or a manager might wish to avoid assigning an unpleasant job to the person who did it last. Whatever the reason, specific combinations can be avoided by assigning a relatively high cost to that combination. In the previous example, if we wish to avoid combination 1–A, assigning a cost of $50 to that combination will achieve the desired effect, because $50 is much greater than the other costs. When profits instead of costs are involved, the profits can be converted to *relative costs* by subtracting every number in the table from the largest number and then proceeding as in a minimization problem.

TABLE 17–2

Possible priority rules

FCFS (first come, first served): Jobs are processed in the order in which they arrive at a machine or work center.

SPT (shortest processing time): Jobs are processed according to processing time at a machine or work center, shortest job first.

DD (due date): Jobs are processed according to due date, earliest due date first.

CR (critical ratio): Jobs are processed according to smallest ratio of time remaining until due date to processing time remaining.

S/O (slack per operation): Jobs are processed according to average slack time (time until due date minus remaining time to process). Compute by dividing slack time by number of remaining operations, including the current one.

Rush: Emergency or preferred customers first.

The simplicity of the Hungarian method belies its usefulness when the assumptions are met. Not only does it provide a rational method for making assignments, it guarantees an optimal solution, often without the use of a computer, which is necessary only for fairly large problems.

Sequencing

Although loading decisions determine the machines or work centers that will be used to process specific jobs, they do not indicate the *order* in which the jobs waiting at a given work center are to be processed. **Sequencing** is concerned with determining job processing order. Sequencing decisions determine both the order in which jobs are processed at various work centers, and the order in which jobs are processed at individual **workstations** *within* the work centers.

sequencing Determining the order in which jobs at a work center will be processed.

workstation An area where one person works, usually with special equipment, on a specialized job.

If work centers are lightly loaded and if jobs all require the same amount of processing time, sequencing presents no particular difficulties. However, for heavily loaded work centers, especially in situations where relatively lengthy jobs are involved, the order of processing can be very important in terms of costs associated with jobs waiting for processing and in terms of idle time at the work centers. In this section, we will examine some of the ways in which jobs are sequenced.

Typically, a number of jobs will be waiting for processing. **Priority rules** are simple heuristics used to select the order in which the jobs will be processed. Some of the most common are listed in Table 17–2. The rules generally rest on the assumption that job setup cost and time are *independent* of processing sequence. In using these rules, job processing times and due dates are important pieces of information. **Job time** usually includes setup and processing times. Jobs that require similar setups can lead to reduced setup times if the sequencing rule takes this into account (the rules described here do not). Due dates may be the result of delivery times promised to customers, MRP processing, or managerial decisions. They are subject to revision and must be kept current to give meaning to sequencing choices. Also, it should be noted that due dates associated with all rules except S/O and CR are for the operation about to be performed; due dates for S/O and CR are typically final due dates for orders rather than intermediate, departmental deadlines.

priority rules Simple heuristics used to select the order in which jobs will be processed.

job time Time needed for setup and processing of a job.

Vol. 3, seg. 2
Washburn Guitar

The priority rules can be classified as either *local* or *global*. Local rules take into account information pertaining only to a single workstation; global rules take into account information pertaining to multiple workstations. FCFS, SPT, and DD are local rules; CR and S/O are global rules. Rush can be either local or global. As you might imagine, global rules require more effort than local rules. A major complication in global sequencing is that not all jobs require the same processing or even the same order of processing. As a result, the set of jobs is different for different workstations. Local rules are particularly useful for bottleneck operations, but they are not limited to those situations.

TABLE 17–3

Assumptions of priority rules

The set of jobs is known; no new jobs arrive after processing begins; and no jobs are canceled.
Setup time is independent of processing sequence.
Setup time is deterministic.
Processing times are deterministic rather than variable.
There will be no interruptions in processing such as machine breakdowns, accidents, or worker illness.

A number of assumptions apply when using the priority rules; Table 17–3 lists them. In effect, the priority rules pertain to *static* sequencing: for simplicity it is assumed that there is no variability in either setup or processing times, or in the set of jobs. The assumptions make the scheduling problem manageable. In practice, jobs may be delayed or cancelled, and new jobs may arrive, requiring schedule revisions.

The effectiveness of any given sequence is frequently judged in terms of one or more *performance measures*. The most frequently used performance measures are:

Job Flow Time. This is the length of time a job is in the shop at a particular workstation or work center. It includes not only actual processing time but also any time waiting to be processed, transportation time between operations, and any waiting time related to equipment breakdowns, unavailable parts, quality problems, and so on. Job flow time is the length of time that begins when a job arrives at the shop, workstation, or work center, and ends when it leaves the shop, workstation, or work center. The average flow time for a group of jobs is equal to the total flow time for the jobs divided by the number of jobs.

Job Lateness. This is the length of time the job completion date is expected to exceed the date the job was due or promised to a customer. It is the difference between the actual completion time and the due date. If we only record differences for jobs with completion times that exceed due dates, and assign zeros to jobs that are early, the term we use to refer to that is job *tardiness*.

makespan Total time needed to complete a group of jobs from the beginning of the first job to the completion of the last job.

Makespan. This is the total time needed to complete a *group* of jobs. It is the length of time between the start of the first job in the group and the completion of the last job in the group.

Average Number of Jobs. Jobs that are in a shop are considered to be work-in-process inventory. The average work-in-process for a group of jobs can be computed using the following formula:

$$\text{Average number of jobs} = \frac{\text{Total flow time}}{\text{Makespan}}$$

If the jobs represent equal amounts of inventory, the average number of jobs will also reflect the average work-in-process inventory.

Of these rules, rush scheduling is quite simple and needs no explanation. The other rules and performance measures are illustrated in the following two examples.

Example 2

Processing times (including setup times) and due dates for six jobs waiting to be processed at a work center are given in the following table. Determine the sequence of jobs, the average flow time, average days late, and average number of jobs at the work center, for each of these rules:

a. FCFS

b. SPT

c. DD

d. CR

Job	Processing Time (days)	Due Date (days)
A	2	7
B	8	16
C	4	4
D	10	17
E	5	15
F	12	18

Assume jobs arrived in the order shown.

Solution

a. The FCFS sequence is simple A-B-C-D-E-F. The measures of effectiveness are (see table):

(1) *Average flow time:* 120/6 = 20 days.

(2) *Average tardiness:* 54/6 = 9 days.

(3) The *makespan* is 41 days. *Average number of jobs at the work center:* 120/41 = 2.93.

Job Sequence	(1) Processing Time	(2) Flow Time	(3) Due Date	(2) − (3) Days Late [0 if negative]
A	2	2	7	0
B	8	10	16	0
C	4	14	4	10
D	10	24	17	7
E	5	29	15	14
F	12	41	18	23
	41	120		54

The flow time column indicates *cumulative* processing time, so summing these times and dividing by the total number of jobs processed indicates the average time each job spends at the work center. Similarly, find the average number of jobs at the center by summing the flow times and dividing by the total processing time.

b. Using the SPT rule, the job sequence is A-C-E-B-D-F (see the following table). The resulting values for the three measures of effectiveness are

(1) *Average flow time:* 108/6 = 18 days.

(2) *Average tardiness:* 40/6 = 6.67 days.

(3) *Average number of jobs at the work center:* 108/41 = 2.63.

Sequence	(1) Processing Time	(2) Flow Time	(3) Due Date	(2) − (3) Days Late [0 if negative]
A	2	2	7	0
C	4	6	4	2
E	5	11	15	0
B	8	19	16	3
D	10	29	17	12
F	12	41	18	23
	41	108		40

c. Using earliest due date as the selection criterion, the job sequence is C-A-E-B-D-F. The measures of effectiveness are (see table):

(1) *Average flow time:* 110/6 = 18.33 days.

(2) *Average tardiness:* 38/6 = 6.33 days.

(3) *Average number of jobs at the work center:* 110/41 = 2.68.

Sequence	(1) Processing Time	(2) Flow Time	(3) Due Date	(2) − (3) Days Late [0 if negative]
C	4	4	4	0
A	2	6	7	0
E	5	11	15	0
B	8	19	16	3
D	10	29	17	12
F	12	41	18	23
	41	110		38

d. Using the critical ratio (Due date − Current date) ÷ Processing time, the job sequence is C-F-D-B-E-A, and the resulting values for the measures of effectiveness are:

(1) *Average flow time:* 160/6 = 26.67 days.
(2) *Average tardiness:* 85/6 = 14.17 days.
(3) *Average number of jobs at the work center:* 160/41 = 3.90.

Sequence	(1) Critical Ratio	(2) Processing Time	(3) Flow Time	(4) Due Date	(3) − (4) Days Late
C	1.0	4	4	4	0
F	1.5	12	16	18	0
D	1.7	10	26	17	9
B	2.0	8	34	16	18
E	3.0	5	39	15	24
A	3.5	2	41	7	34
		41	160		85

The results of these four rules are summarized in Table 17–4.

In this example, the SPT rule was the best according to two of the measures of effectiveness and a little worse than the DD rule on average lateness. The CR rule was the worst in every case. For a different set of numbers, the DD rule (or perhaps another rule not mentioned here) might prove superior to SPT in terms of average job lateness or some other measure of effectiveness. However, SPT is always superior in terms of minimizing flow time and, hence, in terms of minimizing the average number of jobs at the work center and completion time. It is best for each shop or organization to consider carefully its own circumstances and the measures of effectiveness it feels are important, when selecting a rule to use.

Generally speaking, the FCFS rule and the CR rule turn out to be the least effective of the rules.

The primary limitation of the FCFS rule is that long jobs will tend to delay other jobs. If a process consists of work on a number of machines, machine idle time for downstream workstations will increase. However, for service systems in which customers are directly involved, the FCFS rule is by far the dominant priority rule, mainly because of the inherent fairness but also because of the inability to obtain realistic estimates of processing time for individual jobs. The FCFS rule also has the advantage of simplicity. If other measures are important when there is high customer contact, companies may adopt the strategy of moving processing to the "backroom" so they don't necessarily have to follow FCFS.

Because the SPT rule always results in the lowest average completion (flow) time, it can result in lower in-process inventories. And because it often provides the lowest average lateness, it can result in better customer service levels. Finally, since it always

Rule	Average Flow Time (days)	Average Lateness (days)	Average Number of Jobs at the Work Center
FCFS	20.00	9.00	2.93
SPT	18.00	6.67	2.63
DD	18.33	6.33	2.68
CR	26.67	14.17	3.90

TABLE 17-4

Comparison of the four rules for Example 2

involves a lower average number of jobs at the work center, there tends to be less congestion in the work area. SPT also minimizes downstream idle time. However, due dates are often uppermost in managers' minds, so they may not use SPT because it doesn't incorporate due dates.

Example 3

Use the S/O rule to schedule the following jobs. Note that processing time includes the time remaining for the current and subsequent operations. In addition, you will need to know the number of operations remaining, including the current one.

Job	Remaining Processing Time	Due Date	Remaining Number of Operations
A	4	14	3
B	16	32	6
C	8	8	5
D	20	34	2
E	10	30	4
F	18	30	2

Solution

Determine the difference between the due date and the processing time for each operation. Divide the amount by the number of remaining operations, and rank them from low to high. This yields the sequence of jobs:

Job	(1) Remaining Processing Time	(2) Due Date	(3) (2) − (1) Slack	(4) Remaining Number of Operations	(5) (3) ÷ (4) Ratio	(6) Rank
A	4	14	10	3	3.33	3
B	16	32	16	6	2.67	2
C	8	8	0	5	0	1
D	20	34	14	2	7.00	6
E	10	30	20	4	5.00	4
F	18	30	12	2	6.00	5

The indicated sequence (see column 6) is C-B-A-E-F-D.

The major disadvantage of the SPT rule is that it tends to make long jobs wait, perhaps for rather long times (especially if new, shorter jobs are continually added to the system). Various modifications may be used in an effort to avoid this. For example, after waiting for a given time period, any remaining jobs are automatically moved to the head of the line. This is known as the *truncated* SPT rule.

The DD rule directly addresses due dates and usually minimizes lateness. Although it has intuitive appeal, its main limitation is that it does not take processing time into account. One possible consequence is that it can result in some jobs waiting a long time, which adds to both in-process inventories and shop congestion.

The CR rule is easy to use and has intuitive appeal. Although it had the poorest showing in Example 2 for all three measures, it usually does quite well in terms of minimizing job tardiness. Therefore, if job tardiness is important, the CR rule might be the best choice among the rules.

Let's take a look now at the S/O (slack per operation) rule.

Using the S/O rule, the designated job sequence may change after any given operation, so it is important to reevaluate the sequence after each operation. Note that any of the previously mentioned priority rules could be used on a station-by-station basis for this situation; the only difference is that the S/O approach incorporates downstream information in arriving at a job sequence.

In reality, many priority rules are available to sequence jobs, and some other rule might provide superior results for a given set of circumstances. Our purpose in examining these few rules is to provide insight into the nature of sequencing rules. The following section describes a special-purpose algorithm that can be used to sequence a set of jobs that must all be processed at the same two machines or work centers.

Sequencing Jobs through Two Work Centers[1]

Johnson's rule Technique for minimizing makespan for a group of jobs to be processed on two machines or at two work centers.

Johnson's rule is a technique that managers can use to minimize the makespan for a group of jobs to be processed on two machines or at two successive work centers.[2] It also minimizes the total idle time at the work centers. For the technique to work, several conditions must be satisfied:

1. Job time (including setup and processing) must be known and constant for each job at each work center.
2. Job times must be independent of the job sequence.
3. All jobs must follow the same two-step work sequence.
4. Job priorities cannot be used.
5. All units in a job must be completed at the first work center before the job moves on to the second work center.

Determination of the optimum sequence involves these steps:

1. List the jobs and their times at each work center.
2. Select the job with the shortest time. If the shortest time is at the first work center, schedule that job first; if the time is at the second work center, schedule the job last. Break ties arbitrarily.
3. Eliminate the job and its time from further consideration.
4. Repeat steps 2 and 3, working toward the center of the sequence, until all jobs have been scheduled.

When significant idle time at the second work center occurs, job splitting at the first center just prior to the occurrence of idle time may alleviate some of it and also shorten throughput time. In Example 4, this was not a concern. The last solved problem at the end of this chapter illustrates the use of job splitting.

[1]For description of a heuristic that can be used for the case where a set of jobs is to be processed through more than two work centers, see Vollmann, Thomas, et al., *Manufacturing Planning and Control Systems,* 4th ed. (Burr Ridge, IL: Richard D. Irwin, 1997), p. 532.

[2]S. M. Johnson, "Optimal Two- and Three-Stage Production with Setup Times Included," *Naval Research Quarterly* 1 (March 1954), pp. 61–68.

Example 4

A group of six jobs is to be processed through a two-step operation. The first operation involves cleaning and the second involves painting. Determine a sequence that will minimize the total completion time for this group of jobs. Processing times are as follows:

	PROCESSING TIME (HOURS)	
Job	**Work Center 1**	**Work Center 2**
A	5	5
B	4	3
C	8	9
D	2	7
E	6	8
F	12	15

Solution

a. Select the job with the shortest processing time. It is job D, with a time of two hours.

b. Since the time is at the first center, schedule job D first. Eliminate job D from further consideration.

c. Job B has the next shortest time. Since it is at the second work center, schedule it last and eliminate job B from further consideration. We now have

1st	2d	3d	4th	5th	6th
D					B

d. The remaining jobs and their times are

Job	**1**	**2**
A	5	5
C	8	9
E	6	8
F	12	15

Note that there is a tie for the shortest remaining time: job A has the same time at each work center. It makes no difference, then, whether we place it toward the beginning or the end of the sequence. Suppose it is placed arbitrarily toward the end. We now have

1st	2d	3d	4th	5th	6th
D				A	B

e. The shortest remaining time is six hours for job E at work center 1. Thus, schedule that job toward the beginning of the sequence (after job D). Thus,

1st	2d	3d	4th	5th	6th
D	E			A	B

f. Job C has the shortest time of the remaining two jobs. Since it is for the first work center, place it third in the sequence. Finally, assign the remaining job (F) to the fourth position and the result is

1st	2d	3d	4th	5th	6th
D	E	C	F	A	B

g. The easiest way to determine the throughput time and idle times at the work centers is to construct a chart:

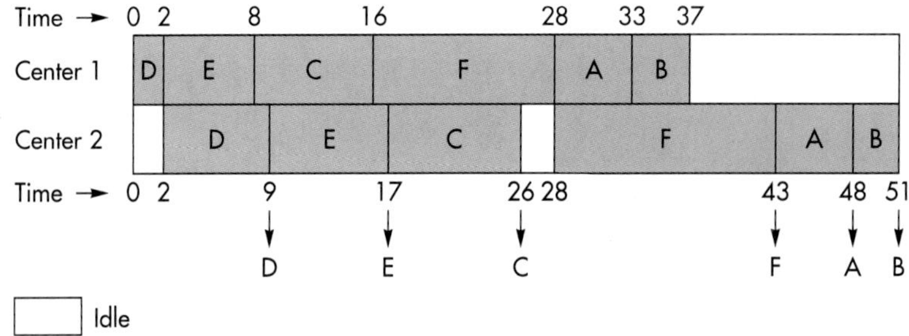

Thus, the group of jobs will take 51 hours to complete. The second work center will wait two hours for its first job and also wait two hours after finishing job C. Center 1 will be finished in 37 hours. Of course, idle periods at the beginning or end of the sequence could be used to do other jobs or for maintenance or setup/teardown activities.

Sequencing Jobs When Setup Times Are Sequence-Dependent

The preceding discussion and examples assumed that machine setup times are independent of processing order, but in many instances that assumption is not true. Consequently, a manager may want to schedule jobs at a workstation taking those dependencies into account. The goal is to minimize total setup time.

Consider the following table, which shows workstation machine setup times based on job processing order. For example, if job A is followed by job B, the setup time for B will be six hours. Furthermore, if job A is completed first, followed by job B, job C will then follow job B, and have a setup time of four hours. If a job is done first, its setup time will be the amount shown in the setup time column to the right of the job. Thus, if job A is done first, its setup time will be three hours.

			Resulting following job setup time (hrs.) is		
		Setup time (hrs.)	**A**	**B**	**C**
If the preceding job is	**A**	**3**	–	6	2
	B	**2**	1	–	4
	C	**2**	5	3	–

The simplest way to determine which sequence will result in the lowest total setup time is to list each possible sequence and determine its total setup time. In general, the number of different alternatives is equal to $n!$, where n is the number of jobs. Here, n is 3, so $n! = 3 \times 2 \times 1 = 6$. The six alternatives and their total setup times are:

Sequence	Setup Times	Total
A-B-C	3 + 6 + 4 =	13
A-C-B	3 + 2 + 3 =	8
B-A-C	2 + 1 + 2 =	5 (best)
B-C-A	2 + 4 + 5 =	11
C-A-B	2 + 5 + 6 =	13
C-B-A	2 + 3 + 1 =	6

Hence, to minimize total setup time, the manager would select sequence B-A-C.

This procedure is relatively simple to do manually when the number of jobs is two or three. However, as the number of jobs increases, the list of alternatives quickly becomes larger. For example, six jobs would have 720 alternatives. In such instances, a manager would employ a computer to generate the list and identify the best alternative(s). (Note that more than one alternative may be tied for the lowest setup time.)

Why Scheduling Is Difficult

Scheduling is difficult for a number of reasons. One is that in reality, an operation must deal with variability in setup times, processing times, interruptions, and changes in the set of jobs. Another major reason is that except for very small problems, there is no method for identifying the optimal schedule, and it would be virtually impossible to sort through the vast number of possible alternatives to obtain the best schedule. As a result, scheduling is far from an exact science and, in many instances, is an ongoing task for a manager.

Things a Manager Can Do to Achieve Good Scheduling Results

There are a number of actions that managers can consider to minimize scheduling problems, such as:

- Setting realistic due dates.
- Focusing on bottleneck operations: First, try to increase the capacity of the operations. If that is not possible or feasible, schedule the bottleneck operations first, and then schedule the nonbottleneck operations around the bottleneck operations.
- Considering lot splitting for large jobs. This probably works best when there are relatively large differences in job times.

SCHEDULING SERVICE OPERATIONS

Scheduling service systems presents certain problems not generally encountered in manufacturing systems. This is due primarily to (1) the inability to store or inventory services, and (2) the random nature of customer requests for service. In some situations, the second difficulty can be moderated by using appointment or reservation systems, but the inability to store services in most cases is a fact of life that managers must contend with.

An important goal in service systems is to match the flow of customers and service capabilities. An ideal situation is one that has a smooth flow of customers through the system. This would occur if each new customer arrives at the precise instant that the preceding customer's service is completed, as in a physician's office, or in air travel where the demand just equals the number of available seats. In each of these situations customer waiting time is minimized, and the service system staff and equipment would be fully utilized. Unfortunately, the random nature of customer requests for service that generally prevails in service systems makes it nearly impossible to provide service capability that matches demand. Moreover, if service times are subject to variability—say, because of differing processing requirements—the inefficiency of the system is compounded. The inefficiencies can be reduced if arrivals can be scheduled (e.g., appointments), as in the case of doctors and dentists. However, in many situations appointments are not possible (supermarkets, gas stations, theaters, hospital emergency rooms, repair of equipment breakdowns). Chapter 19, on waiting lines, focuses on those kinds of situations. There, the emphasis is on intermediate-term decisions related to service capacity. In this section, we will concern ourselves with short-term *scheduling,* in which much of the capacity of a system is essentially fixed, and the goal is to achieve a certain degree of customer service by efficient utilization of that capacity.

Vol. 1, Tape 2, seg. 3
First National Bank of Chicago

Scheduling in service systems may involve scheduling (1) customers, (2) the workforce, and (3) equipment. Scheduling customers often takes the form of appointment systems or reservation systems.

Appointment Systems

Appointment systems are intended to control the timing of customer arrivals in order to minimize customer waiting while achieving a high degree of capacity utilization. A doctor can use an appointment system to schedule office patients during the afternoon, leaving the mornings free for hospital duties. Similarly, an attorney can schedule clients around court appearances. Even with appointments, however, problems can still arise due to lack of punctuality on the part of patients or clients, no-shows, and the inability to completely control the length of contact time (e.g., a dentist might run into complications in filling a tooth and have to spend additional time with a patient, thus backing up later appointments). Some of this can be avoided by trying to match the time reserved for a patient or client with the specific needs of that case rather than setting appointments at regular intervals. Even with the problems of late arrivals and no-shows, the appointment system is a tremendous improvement over random arrivals.

Reservation Systems

Reservation systems are designed to enable service systems to formulate a fairly accurate estimate of the demand on the system for a given time period and to minimize customer disappointment generated by excessive waiting or inability to obtain service. Reservation systems are widely used by resorts, hotels and motels, restaurants, and some modes of transportation (e.g., airlines, car rentals). In the case of restaurants, reservations enable management to spread out or group customers so that demand matches service capabilities. Late arrivals and no-shows can disrupt the system. One approach to the no-show problem is to use decision theory (described in the supplement to Chapter 2). The problem can also be viewed as a single-period inventory problem, as described in Chapter 13.

Scheduling the Workforce

Scheduling customers is demand management. Scheduling the workforce is capacity management. This approach works best when demand can be predicted with reasonable accuracy. This is often true for restaurants, theaters, rush-hour traffic, and similar instances that have repeating patterns of intensity of customer arrivals. Scheduling hospital personnel, police, and telephone operators for catalog sales, credit card companies, and mutual fund companies also comes under this heading. An additional consideration is the extent to which variations in customer demands can be met with workforce flexibility. Thus, capacity can be adjusted by having cross-trained workers who can be temporarily assigned to help out on bottleneck operations during periods of peak demand.

Various constraints can affect workforce scheduling flexibility, including legal, behavioral, technical—such as workers' qualifications to perform certain operations—and budget constraints. Union contracts may provide still more constraints.

Scheduling Multiple Resources

In some situations, it is necessary to coordinate the use of more than one resource. For example, hospitals must schedule surgeons, operating room staffs, recovery room staffs, admissions, special equipment, nursing staffs, and so on. Educational institutions must schedule faculty, classrooms, audiovisual equipment, and students. As you might guess, the greater the number of resources to be scheduled, the greater the complexity of the problem and the less likely that an optimum schedule can be achieved. The problem is further complicated by the variable nature of such systems. For example, educational institutions frequently change their course offerings, student enrollments change, and students exhibit different course-selection patterns.

Some schools and hospitals are using computer programs to assist them in devising acceptable schedules, although many appear to be using intuitive approaches with varying degrees of success.

Airlines are another example of service systems that require the scheduling of multiple resources. Flight crews, aircraft, baggage handling equipment, ticket counters, gate personnel, boarding ramps, and maintenance personnel all have to be coordinated. Furthermore, government regulations on the number of hours a pilot can spend flying place an additional restriction on the system. Another interesting variable is that, unlike most systems, the flight crews and the equipment do not remain in one location. Moreover, the crew and the equipment are not usually scheduled as a single unit. Flight crews are often scheduled so that they return to their base city every two days or more often, and rest breaks must be considered. On the other hand, the aircraft may be in almost continuous use except for periodic maintenance and repairs. Consequently, flight crews commonly follow different trip patterns than that of the aircraft.

Vol. 4, seg. 2
United Airlines

There are also other activities that must be scheduled, some of which are described by American Airlines Chairman Robert L. Crandall in the reading.

www.americanair.com

READING

SERVICING PASSENGER PLANES **Robert L. Crandall**

Before takeoff and after landing, you've probably noticed lots of people and various types of vehicles and equipment busily moving around on the ground outside with no apparent game plan, particularly at larger airports. Actually, all that activity is carefully orchestrated, and because the people involved are an important part of our team, I'd like to tell you what the hustle and bustle is all about.

When a flight lands and approaches the terminal, ramp personnel guide the aircraft to its parking position and after it comes to a stop, put chocks under its wheels. As soon as that's been done, other workers hook up ground-based power and air conditioning. Electric power comes from what looks like a large, industrial-strength extension cord plugged into the lower portion of the nose section. Heated or cooled air goes into the cabin via the big yellow hoses that run from the airplane either to the terminal, in cities where we have central air-handling facilities or, in others, to mobile air-handling units.

On the airplane, meanwhile, flight attendants open the door and as passengers begin deplaning, a mechanic squeezes past them to get a debriefing from the cockpit and check to see if any maintenance work must be done.

Once all deplaning passengers are off, the cabin cleaners go into high gear on most flights, cleaning our seatback pockets, tidying up the cabin, cleaning the lavatories, doing a light vacuuming, repositioning safety belts for each seat's next occupant, and so forth. (A more thorough cleaning is done each night.)

Simultaneously, out on the ramp, our people are unloading baggage, freight, and mail from the airplane's belly compartments and beginning the process of sorting by various categories and destinations. In addition to the bags and cargo that have reached their destination, which must be delivered promptly to passengers and shippers, some must be transferred to other American or American Eagle flights, and still others must be transferred to other carriers. Complicating matters further, baggage, freight, and mail are often handled in different facilities on the airport property.

If a meal has been served or is planned for the outbound flight, catering trucks pull up to service the First Class and main cabin galleys. Another special truck services the lavatory

Meal supply truck arrives to load its supplies onto American Airlines plane before its scheduled departure.

holding tanks, and in the midst of all this, mechanics are dealing with any problems reported by the crew and doing their own walk-around inspections.

Once all that is complete, customers start to board the aircraft for its next flight and everything happens in reverse. Ground workers start loading baggage in the forward belly and freight and mail in the rear. Fuel trucks pull up to refuel most flights. The airplanes must be "watered" as well; fresh water is pumped aboard from either a water truck or servicing equipment built into the gate itself. During cold-weather months, de-icing trucks add another element of ramp activity as they spray fluid on the airplane's wings and fuselage.

Most of the work on the ramp is done by our own people, principally fleet-service personnel, but there are also various services, like catering, which are performed by contractors.

All this simultaneous activity can create conflicts. Baggage carts unloading the forward hold can encroach on the ground space needed by the trucks catering the First Class cabin, for example, and fuelers perform their function in the space needed by those unloading freight and mail. The ramp crew chiefs are the conductors, orchestrating the entire production, overseeing the detail, and seeing to it that all gets done without mishap—and on time.

It's a delicate balancing act, and although our customers rarely come in contact with the folks working on the ramp, they are an important part of a team that aims to serve every customer well and keep our operation running safely and on time.

Source: *American Way,* March 15, 1995.

Summary

Scheduling involves the timing and coordination of operations. Such activities are fundamental to virtually every organization. Scheduling problems differ according to whether a system is designed for high volume, intermediate volume, or low volume. Scheduling problems are particularly complex for job shops (low volume) because of the variety of jobs these systems are required to process.

The two major problems in job-shop scheduling are assigning jobs to machines or work centers, and designating the sequence of job processing at a given machine or work center. Gantt load charts are frequently employed to help managers visualize workloads, and they are useful for describing and analyzing sequencing alternatives. In addition, both heuristic and optimizing methods are used to develop loading and sequencing plans. For the most part, the optimization techniques can be used only if certain assumptions can be made.

Customer requirements in service systems generally present very different circumstances from those encountered in manufacturing systems. Some companies can use appointment and reservation systems for scheduling purposes, although not all systems are amenable to this. When multiple resources are involved, the task of balancing the system can be fairly complex.

Key Terms

assignment model, 729
backward scheduling, 728
finite loading, 726
flow-shop scheduling, 723
flow system, 723
forward scheduling, 728
Gantt chart, 726
horizontal loading, 727
Hungarian method, 730
infinite loading, 726
input/output control, 729
job-shop scheduling, 725
job time, 733
Johnson's rule, 738
load chart, 726
loading, 725
makespan, 734
priority rules, 733
schedule chart, 728
scheduling, 722
sequencing, 733
vertical loading, 727
workstation, 733

Solved Problems

Problem 1

The assignment method. The following table contains information on the cost to run three jobs on four available machines. Determine an assignment plan that will minimize costs.

	MACHINE			
	A	B	C	D
Job 1	12	16	14	10
Job 2	9	8	13	7
Job 3	15	12	9	11

Solution

In order for us to be able to use the assignment method, the numbers of jobs and machines must be equal. To remedy this situation, add a *dummy* job with costs of 0, and then solve as usual.

	MACHINE			
	A	B	C	D
Job 1	12	16	14	10
Job 2	9	8	13	7
Job 3	15	12	9	11
(dummy) 4	0	0	0	0

a. Subtract the smallest number from each row. The results are

	MACHINE			
	A	B	C	D
Job 1	2	6	4	0
Job 2	2	1	6	0
Job 3	6	3	0	2
Job 4	0	0	0	0

b. Subtract the smallest number in each column. (Because of the dummy zeros in each column, the resulting table will be unchanged.)

c. Determine the minimum number of lines needed to cross out the zeros. One possible way is

	Machine			
	A	*B*	*C*	*D*
Job 1	2	6	4	0
Job 2	2	1	6	0
Job 3	6	3	0	2
Job 4	0	0	0	0

d. Since the number of lines is less than the number of rows, modify the numbers.

(1) Subtract the smallest uncovered number (1) from each uncovered number.

(2) Add the smallest uncovered number to numbers at line intersections.

The result is

		MACHINE			
		A	B	C	D
Job	1	1	5	4	0
	2	1	0	6	0
	3	5	2	0	2
	4	0	0	1	1

e. Test for optimality:

		Machine			
		A	*B*	*C*	*D*
Job	1	1	5	4	0
	2	1	0	6	0
	3	5	2	0	2
	4	0	0	1	1

Since the minimum number of lines equals the number of rows, an optimum assignment can be made.

f. Assign jobs to machines. Start with rows 1 and 3, since they each have one zero, and columns A and C, also with one zero each. After each assignment, cross out all the numbers in that row *and* column. The result is

		Machine			
		A	*B*	*C*	*D*
Job	1	1	5	4	[0]
	2	1	[0]	6	0
	3	5	2	[0]	2
	4	[0]	0	1	1

Notice that there is only one assignment in each row, and only one assignment in each column.

g. Compute total costs, referring to the original table:

1-D . . .	$10
2-B	8
3-C . . .	9
4-A . . .	0
	$27

h. The implication of assignment 4-A is that machine A will not be assigned a job. It may remain idle or be used for another job.

Problem 2

Priority rules. Job times (including processing and setup) are shown in the following table for five jobs waiting to be processed at a work center:

Job	Job Time (hours)	Due Date (hours)
a	12	15
b	 6	24
c	14	20
d	 3	8
e	 7	6

Determine the processing sequence that would result from each of these priority rules:

a. SPT

b. DD

c. CR

Solution

Assume job times are independent of processing sequence.

Note: CR = Hour due ÷ Job time

	a. SPT		b. DD		c. CR	
Job	Job Time	Processing Order	Hour Due	Processing Order	CR	Processing Order
a	12	4	15	3	1.25	2
b	6	2	24	5	4.00	5
c	14	5	20	4	1.43	3
d	3	1	8	2	2.67	4
e	7	3	6	1	0.86	1

Problem 3

Priority rules. Using the job times and due dates from Solved Problem 2, determine each of the following performance measures for first-come, first-served processing order:

a. Makespan

b. Average flow time

c. Average tardiness

d. Average number of jobs at the workstation.

Solution

Job	Job Time	Flow Time	Hour Due	Hours Late
a	12	12	15	0
b	6	18	24	0
c	14	32	20	12
d	3	35	8	27
e	7	42	6	36
Total		139		75

a. Makespan = 42 hours

b. $\text{Average flow time} = \dfrac{\text{Total flow time}}{\text{Number of jobs}} = \dfrac{139}{5} = 27.80 \text{ hours}$

c. $\text{Average tardiness} = \dfrac{\text{Total hours late}}{\text{Number of jobs}} = \dfrac{75}{5} = 15 \text{ hours}$

d. $\text{Average number of jobs at workstation} = \dfrac{\text{Total flow time}}{\text{Makespan}} = \dfrac{139}{42} = 3.31$

Problem 4

S/O rule. Using the following information, determine an order processing sequence using the S/O priority rule.

Order	Processing Time Remaining (days)	Due Date (days)	Number of Operations Remaining
A	20	30	2
B	11	18	5
C	10	6	2
D	16	23	4

Solution

Assume times are independent of processing sequence.

Order	(1) Remaining Processing Time	(2) Due Date	(3) (2) − (1) Slack	(4) Number of Operations	(5) Ratio	(6) Rank (sequence)
A	20	30	10	2	5.00	4
B	11	18	7	5	1.40	2
C	10	6	−4	2	−2.00	1
D	16	23	7	4	1.75	3

(Note that one ratio is negative. When negatives occur, assign the *lowest* rank to the *most negative* number.)

Problem 5

Sequencing jobs through two work centers. Use Johnson's rule to obtain the optimum sequence for processing the jobs shown through work centers A and B.

	JOB TIMES (HOURS)	
Job	Work Center A	Work Center B
a	2.50	4.20
b	3.80	1.50
c	2.20	3.00
d	5.80	4.00
e	4.50	2.00

Solution

a. Identify the smallest time: job b (1.50 hours at work center B). Since the time is for B, schedule this job last.

b. The next smallest time is job e (2.00 hours at B). Schedule job e next to last.

c. Identify the smallest remaining job time: job c (2.20 hours at center A). Since the time is in the A column, schedule job c first. At this point, we have:

c, ___, ___, e, b

d. The smallest time for remaining jobs is 2.50 hours for job a at center A. Schedule this job after job c. The one remaining job (job d) fills the remaining slot. Thus, we have:

c-a-d-e-b

Problem 6

For Solved Problem 5, determine what effect splitting jobs c, d, e, and b in work center A would have on the idle time of work center B and on the throughput time. Assume that each job can be split into two equal parts.

Solution

We assume that the processing sequence remains unchanged and proceed on that basis. The solution from the previous problem is shown in the following chart. The next chart shows reduced idle time at center B when splitting is used.

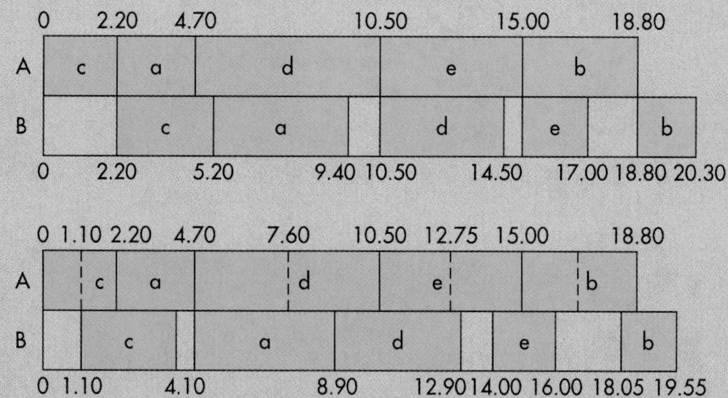

An inspection of these two figures reveals that throughput time has decreased from 20.30 hours to 19.55 hours. In addition, the original idle time was 5.6 hours. After splitting certain jobs, it was reduced to 4.85 hours, so some improvement was achieved. Note that processing times at B are generally less than at A for jobs toward the end of the sequence. As a result, jobs such as e and b at B were scheduled so that they were *centered* around the finishing times of e and b, respectively, at A, to avoid having to break the jobs due to waiting for the remainder of the split job from A. Thus, the greatest advantage from job splitting generally comes from splitting earlier jobs when Johnson's rule is used for sequencing.

Discussion and Review Questions

1. Why is scheduling fairly simple for repetitive systems but fairly complex for job shops?
2. What are the main decision areas of job-shop scheduling?
3. What are Gantt charts? How are they used in scheduling? What are the advantages of using Gantt charts?
4. What are the basic assumptions of the assignment method of linear programming?
5. Briefly describe each of these priority rules:
   a. FCFS
   b. SPT
   c. DD
   d. S/O
   e. Rush
6. Why are priority rules needed?
7. What problems not generally found in manufacturing systems do service systems present in terms of scheduling the use of resources?
8. Doctors' and dentists' offices frequently schedule patients at regularly spaced intervals. What problems can this create? Can you suggest an alternative approach to reduce these problems? Under what circumstances would regularly spaced appointments constitute a reasonable approach to patient scheduling?
9. How are scheduling and productivity related?
10. What factors would you take into account in deciding whether or not to split a job?
11. Explain the term *makespan*.

Memo Writing Exercises

1. The manufacturing company you work for is about to open a call-in service line that will allow customers to ask technical questions of manufacturing people about their orders. Your manager, Veronica Jones, will be responsible for setting up a schedule for staffing the center. She has asked for your opinion on whether that scheduling would be any different than the job-shop scheduling in current use. Write a one-page memo to her on these scheduling differences.
2. Write a short memo to your new employee, Chuck West, contrasting local and global priority rules. Provide him with the names of a few rules of each type.

Problems

1. Use the assignment method to determine the best way to assign workers to jobs, given the following cost information. Compute total cost for your assignment plan.

		JOB		
		A	B	C
Worker	1	5	8	6
	2	6	7	9
	3	4	5	3

	A	B	C
1	[0]	2	1
2	0	[0]	3
3	1	1	[0]

TC = $15

2. Rework problem 1 treating the numbers in the table as profits instead of costs. Compute the total profit.

Optimum: 1-B, 2-C, 3-A
TC = $21

3. Assign trucks to delivery routes so that total costs are minimized, given the cost data shown. What is the total cost?

		ROUTE				
		A	B	C	D	E
Truck	1	4	5	9	8	7
	2	6	4	8	3	5
	3	7	3	10	4	6
	4	5	2	5	5	8
	5	6	5	3	4	9

	A	B	C	D	E
1	[0]	2	6	4	1
2	3	2	6	[0]	[0]
3	3	0	7	[0]	[0]
4	2	[0]	3	2	3
5	2	2	[0]	0	3

Optimum:
1-A, 2-E, 3-D, 4-B, 5-C or
1-A, 2-D, 3-E, 4-B, 5-C
TC = $18

4. Develop an assignment plan that will minimize processing costs, given the information shown, and interpret your answer.

		MACHINE		
		A	B	C
Job	1	12	8	11
	2	13	10	8
	3	14	9	14
	4	10	7	12

	A	B	C	D
1	1	[0]	2	0
2	3	3	[0]	1
3	3	1	5	[0]
4	[0]	0	4	1

Optimum:
1-B, 2-C, 3-D, 4-A

5. Use the assignment method to obtain a plan that will minimize the processing costs in the following table under these conditions:

   a. The combination 2-D is undesirable.

   b. The combinations 1-A and 2-D are undesirable.

a. Optimum:
1-A, 2-B, 3-C, 4-D, 5-E
b. Optimum:
1-E, 2-B, 3-C, 4-D, 5-A

		MACHINE				
		A	**B**	**C**	**D**	**E**
	1	14	18	20	17	18
	2	14	15	19	16	17
Job	**3**	12	16	15	14	17
	4	11	13	14	12	14
	5	10	16	15	14	13

6. The following table contains information concerning four jobs that are awaiting processing at a work center.

Job	Job Time (days)	Due Date (days)
A	14	20
B	10	16
C	7	15
D	6	17

   a. Sequence the jobs using (1) FCFS, (2) SPT, (3) DD, and (4) CR. Assume the list is by order of arrival.

   b. For each of the methods in part *a,* determine (1) the average job flow time, (2) the average lateness, and (3) the average number of jobs at the work center.

   c. Is one method superior to the others? Explain.

a. FCFS: A-B-C-D
SPT: D-C-B-A
DD: C-B-D-A
CR: A-B-C-D

b.

	FCFS	SPT	DD	CR
(1)	26.50	19.75	21.00	26.50
(2)	11.00	6.00	6.00	11.00
(3)	2.86	2.14	2.27	2.86

c. SPT is superior, but ties with DD in average job lateness.

7. Using the information presented in the following table, identify the processing sequence that would result using (1) FCFS, (2) SPT, (3) DD, and (4) CR. For each method, determine (1) average job completion time, (2) average job lateness, and (3) average number of jobs in the system. Jobs are listed in order of arrival. (*Hint:* First determine the total job time for each job by computing the total processing time for the job and then adding in the setup time. All times and due dates are in hours.)

Job	Processing Time per Unit	Units per Job	Setup Time	Due Date
a	0.14	45	0.7	4
b	0.25	14	0.5	10
c	0.10	18	0.2	12
d	0.25	40	1.0	20
e	0.10	75	0.5	15

(FCFS): a-b-c-d-e
(SPT): c-b-a-e-d
(DD): a-b-c-e-d
(CR): a-d-e-b-c

FCFS	SPT	DD	CR
17.40	14.80	16.80	22.60
5.20	5.40	4.60	10.80
2.72	2.31	2.63	3.53

8. The following table shows orders to be processed at a machine shop as of 8 A.M. Monday. The jobs have different operations they must go through. Processing times are in days. Jobs are listed in order of arrival.

   a. Determine the processing sequence at the first work center using each of these rules: (1) FCFS, (2) S/O.

   b. Compute the effectiveness of each rule using each of these measures: (1) average completion time, (2) average number of jobs at the work center.

a. FCFS: A-B-C-D-E
S/O: B-D-C-A-E or
D-B-C-A-E

b. FCFS:
(1) 25.20
(2) 2.93
S/O: [Assume B-D-C-A-E]
(1) 26.80
(2) 3.12

Job	Processing Time (days)	Due Date (days)	Remaining Number of Operations
A	8	20	2
B	10	18	4
C	5	25	5
D	11	17	3
E	9	35	4

B-A-G-E-F-D-C

9. A wholesale grocery distribution center uses a two-step process to fill orders. Tomorrow's work will consist of filling the seven orders shown. Determine a job sequence that will minimize the time required to fill the orders.

	TIME (HOURS)	
Order	Step 1	Step 2
A......	1.20	1.40
B......	0.90	1.30
C	2.00	0.80
D......	1.70	1.50
E......	1.60	1.80
F......	2.20	1.75
G	1.30	1.40

a. e-b-g-h-d-c-a-f
b. See Chart in IM. 18 hrs.
c. New idle time: 15 hrs. Savings of 3 hrs.

10. The times required to complete each of eight jobs on two machines are shown in the table that follows. Each job must follow the same sequence, beginning with machine A and moving to machine B.

a. Determine a sequence that will minimize makespan time.

b. Construct a chart of the resulting sequence, and find machine B's idle time.

c. For the sequence determined in part *a,* how much would machine B's idle time be reduced by splitting the last two jobs in half?

	TIME (HOURS)	
Job	Machine A	Machine B
a.........	16	5
b.........	3	13
c.........	9	6
d.........	8	7
e.........	2	14
f.........	12	4
g.........	18	14
h.........	20	11

a. B-A-C-E-F-D
b. See Chart in IM.

11. Given the operation times provided:

a. Develop a job sequence that minimizes idle time at the two work centers.

b. Construct a chart of the activities at the two centers, and determine each one's idle time, assuming no other activities are involved.

	JOB TIMES (IN MINUTES)					
	A	B	C	D	E	F
Center 1	20	16	43	60	35	42
Center 2	27	30	51	12	28	24

12. A shoe repair operation uses a two-step sequence that all jobs in a certain category follow. For the group of jobs listed,

    a. Find the sequence that will minimize total completion time.

    b. Determine the amount of idle time for workstation B.

    c. What jobs are candidates for splitting? Why? If they were split, how much would idle time and makespan time be reduced?

	JOB TIMES (IN MINUTES)				
	a	**b**	**c**	**d**	**e**
Workstation A	27	18	70	26	15
Workstation B	45	33	30	24	10

a. b-a-c-d-e
b. 18 + 19 = 37 min.
c. Jobs B, A, C, D, E could be split to reduce makespan and idle time. Makespan reduced by 15 minutes to 164 minutes. Idle time reduced by 15 minutes to 22 minutes.

	FCFS	SPT	DD	CR
(1)	17.40	14.80	16.80	22.60
(2)	5.20	5.40	4.60	10.80
(3)	2.72	2.31	2.63	3.53

13. The following schedule was prepared by the production manager of Marymount Metal Shop.

	CUTTING		POLISHING	
Job	**Start**	**Finish**	**Start**	**Finish**
A	0	2	2	5
B	2	6	6	9
C	6	11	11	13
D	11	15	15	20
E	15	17	20	23
F	17	20	23	24
G	20	21	24	28

Determine a schedule that will result in earlier completion of all jobs on this list.

G-A-E-D-B-C-F

14. The production manager must determine the processing sequence for seven jobs through the grinding and deburring departments. The same sequence will be followed in both departments. The manager's goal is to move the jobs through the two departments as quickly as possible. The foreman of the deburring department wants the SPT rule to be used to minimize the work-in-process inventory in his department.

	PROCESSING TIME (HOURS)	
Job	**Grinding**	**Deburring**
A	3	6
B	2	4
C	1	5
D	4	3
E	9	4
F	8	7
G	6	2

a. Prepare a schedule using SPT for the grinding department.

b. What is the flow time in the grinding department for the SPT sequence? What is the total time needed to process the seven jobs in both the grinding and deburring departments?

c. Determine a sequence that will minimize the total time needed to process the jobs in both departments. What flow time will result for the grinding departments?

d. Discuss the trade-offs between the two alternative sequencing arrangements. At what point would the production manager be indifferent concerning the choice of sequences?

a. See schedule in IM. Sequence is C-B-A-D-G-F-E.
b. Grinding flow time is 93 hrs. Total time is 37 hrs.
c. C-B-A-F-E-D-G. Grinding flow time is 108 hrs. Total time is 35 hrs.
d. Discussion in IM.

15. A foreman has determined processing times at a work center for a set of jobs and now wants to sequence them. Given the information shown, do the following:

a.

	Avg. Late.	Avg. Comp. Time	Avg. # of Jobs
(1)	1.7	15.25	3.91
(2)	2.33	11.117	2.85
(3)	0.55	15.60	4.00
(4)	0.63	15.68	4.02

b. Ratio of avg. comp. time to avg. # of jobs all 3.90.

a. Determine the processing sequence using (1) FCFS, (2) SPT, (3) DD, and (4) CR. For each sequence, compute the average job lateness, the average completion time, and the average number of jobs at the work center. The list is in FCFS order.

b. Using the results of your calculations in part *a*, show that the average completion time and the average number of jobs measures are equivalent for all four sequencing rules.

Job	Job Time (days)	Due Date	Operations Remaining
a	4.5	10	2
b	6.0	17	4
c	5.2	12	3
d	1.6	27	5
e	2.8	18	3
f	3.3	19	1

B-C-E-A-D

Job	Slack ÷ # of Remaining Operations
a	1.50
b	−0.25
c	0.25
d	1.67
e	1.00

16. Given the information in the following table, determine the processing sequence that would result using the S/O rule.

Job	Remaining Processing Time (days)	Due Date	Remaining Number of Operations
a	5	8	2
b	6	5	4
c	9	10	4
d	7	12	3
e	8	10	2

	Avg. Comp. Time	Avg. Job Late.
FCFS	12.5	2.83
SPT	11.33	0.83
DD	11.58	0.42
CR	11.75	0.58

17. Given the following information on job times and due dates, determine the optimal processing sequence using (1) FCFS, (2) SPT, (3) DD, and (4) CR. For each method, find the average job completion time and the average job lateness.

Job	Job Time (hours)	Due Date (hours)
a	3.5	7
b	2.0	6
c	4.5	18
d	5.0	22
e	2.5	4
f	6.0	20

Order	Job Time
A	64
B	72
C	30
D	80
E	4

a. A-C-D-B-E
b. 15.2 min.
c. 3.31
d. New avg. job lateness 12 min.

18. The Budd Gear Co. specializes in heat-treating gears for many automobile companies. At 8 A.M., when Budd's shop opened today, five orders (listed in order of arrival) were waiting to be processed.

Order	Order Size (units)	Per Unit Time in Heat Treatment (minutes/unit)	Due Date (min. from now)
A	16	4	160
B	6	12	200
C	10	3	180
D	8	10	190
E	4	1	220

a. If the due date rule is used, what sequence should be used?

b. What will be the average job lateness?

c. What will be the average number of jobs in the system?

d. Would the SPT rule produce better results in terms of lateness?

19. The following table contains order-dependent setup times for three jobs. Which processing sequence will minimize the total setup time? A-B-C

			Following job's setup time (hrs.)		
		Setup time (hrs.)	**A**	**B**	**C**
Preceding Job	**A**	**2**	–	3	5
	B	**3**	8	–	2
	C	**2**	4	3	–

20. The following table contains order-dependent setup times for three jobs. Which processing sequence will minimize the total setup time? C-B-A

			Following job's setup time (hrs.)		
		Setup time (hrs.)	**A**	**B**	**C**
Preceding Job	**A**	**2.4**	–	1.8	2.2
	B	**3.2**	0.8	–	1.4
	C	**2.0**	2.6	1.3	–

21. The following table contains order-dependent setup times for four jobs. For safety reasons, job C cannot follow job A, nor can job A follow job C. Determine the processing sequence that will minimize the total setup time. (*Hint:* There are 12 alternatives.) B-C-D-A

			Following job's setup time (hrs.)			
		Setup time (hrs.)	**A**	**B**	**C**	**D**
Preceding Job	**A**	**2**	–	5	x	4
	B	**1**	7	–	3	2
	C	**3**	x	2	–	2
	D	**2**	4	3	6	–

22. Given this information on planned and actual inputs and outputs for a service center, determine the work backlog for each period. The beginning backlog is 12 hours of work. The figures shown are standard hours of work. See IM

		Period				
		1	2	3	4	5
Input	Planned	24	24	24	24	20
	Actual	25	27	20	22	24
Output	Planned	24	24	24	24	23
	Actual	24	22	23	24	24

Selected Bibliography

Fogarty, Donald W., and Thomas R. Hoffmann. *Production and Inventory Management*. Cincinnati: South-Western Publishing, 1983.

Hopp, Wallace J., and Mark L. Spearman. *Factory Physics: Foundations of Manufacturing Management*. Burr Ridge, IL: Richard D. Irwin, 1996.

Schonberger, Richard J. *Operations Management: Improving Customer Service*. 5th ed. Burr Ridge, IL: Richard D. Irwin, 1994.

Vollmann, Thomas E.; William L. Berry; and D. Clay Whybark. *Manufacturing Planning and Control Systems*. 4th ed. Burr Ridge, IL: Richard D. Irwin, 1997.

SUPPLEMENT TO CHAPTER SEVENTEEN

MAINTENANCE

LEARNING OBJECTIVES

After completing this supplement, you should be able to:

1. Explain the importance of maintenance in production systems.
2. Describe the range of maintenance activities.
3. Discuss preventive maintenance and the key issues associated with it.
4. Discuss breakdown maintenance and the key issues associated with it.
5. State how the Pareto phenomenon pertains to maintenance decisions.

SUPPLEMENT OUTLINE

maintenance All activities that maintain facilities and equipment in good working order so that a system can perform as intended.

Maintaining the production capability of an organization is an important function in any production system. **Maintenance** encompasses all those activities that relate to keeping facilities and equipment in good working order and making necessary repairs when breakdowns occur, so that the system can perform as intended.

Maintenance activities are often organized into two categories: (1) buildings and grounds, and (2) equipment maintenance. Buildings and grounds is responsible for the appearance and functioning of buildings, parking lots, lawns, fences, and the like. Equipment maintenance is responsible for maintaining machinery and equipment in good working condition and making all necessary repairs.

INTRODUCTION

The goal of maintenance is to keep the production system in good working order at minimal cost. There are several reasons for wanting to keep equipment and machines in good operating condition, such as to:

1. avoid production disruptions
2. not add to production costs
3. maintain high quality
4. avoid missed delivery dates

When breakdowns occur, there are a number of adverse consequences:

1. Production capacity is reduced, and orders are delayed.
2. There is no production, but overhead continues, increasing the cost per unit.
3. There are quality issues; product may be damaged.
4. There are safety issues; employees or customers may be injured.

breakdown maintenance Reactive approach; dealing with breakdowns or problems when they occur.

preventive maintenance Proactive approach; reducing breakdowns through a program of lubrication, adjustment, cleaning, inspection, and replacement of worn parts.

Decision makers have two basic options with respect to maintenance. One option is *reactive:* it is to deal with breakdowns or other problems when they occur. This is referred to as **breakdown maintenance.** The other option is *proactive:* it is to reduce breakdowns through a program of lubrication, adjustment, cleaning, inspection, and replacement of worn parts. This is referred to as **preventive maintenance.**

Decision makers try to make a trade-off between these two basic options that will minimize their combined cost. With no preventive maintenance, breakdown and repair costs would be tremendous. Furthermore, hidden costs, such as lost production and the cost of wages while equipment is not in service, must be factored in. So must the cost of injuries or damage to other equipment and facilities or to other units in production. However, beyond a certain point, the cost of preventive maintenance activities exceeds the benefit.

As an example, if a person never had the oil changed in his or her car, never had it lubricated, and never had the brakes or tires inspected, but simply had repairs done when absolutely necessary, preventive costs would be negligible but repair costs would be quite high, considering the wide range of parts (engine, steering, transmission, tires, brakes, etc.) that could fail. In addition, property damage and injury costs might be incurred, plus there would be the uncertainty of when failure might occur (e.g., on the expressway during rush hour, or late at night). On the other hand, having the oil changed and the car lubricated every morning would obviously be excessive because automobiles are designed to perform for much longer periods without oil changes and lubrications. The best approach is to seek a balance between preventive maintenance and breakdown maintenance. The same concept applies to maintaining production systems: Strike a balance between prevention costs and breakdown costs. This concept is illustrated in Figure 17S–1.

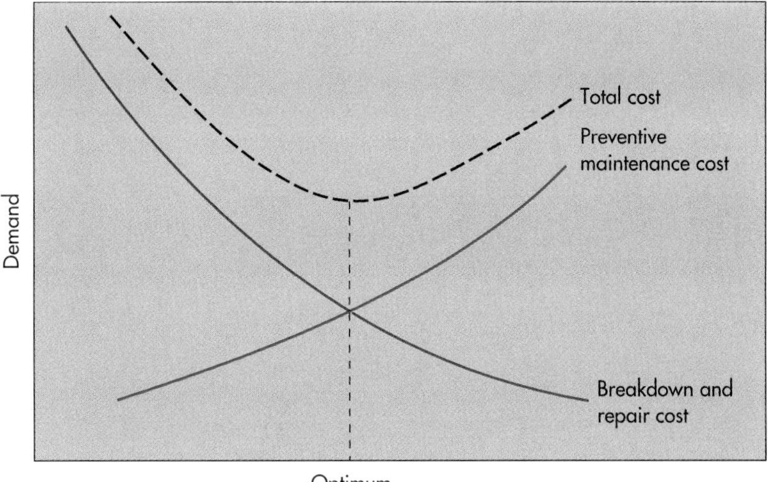

FIGURE 17S-1

Total maintenance cost as a function of preventive maintenance effort

The age and condition of facilities and equipment, the degree of technology involved, the type of production process, and similar factors enter into the decision of how much preventive maintenance is desirable. Thus, in the example of a new automobile, little preventive maintenance may be needed since there is slight risk of breakdowns. As the car ages and becomes worn through use, the desirability of preventive maintenance increases because the risk of breakdown increases. Thus, when tires and brakes begin to show signs of wear, they should be replaced before they fail; dents and scratches should be periodically taken care of before they begin to rust; and the car should be lubricated and have its oil changed after exposure to high levels of dust and dirt. Also, inspection and replacement of critical parts that tend to fail suddenly should be performed before a road trip to avoid disruption of the trip and costly emergency repair bills.

PREVENTIVE MAINTENANCE

The goal of preventive maintenance is to reduce the incidence of breakdowns or failures in the plant or equipment to avoid the associated costs. Those costs can include loss of output; idle workers; schedule disruptions; injuries; damage to other equipment, products, or facilities; and repairs, which may involve maintaining inventories of spare parts, repair tools and equipment, and repair specialists.

Preventive maintenance is *periodic.* It can be scheduled according to the availability of maintenance personnel and to avoid interference with operating schedules. Managers usually schedule preventive maintenance using some combination of the following:

1. The result of planned inspections that reveal a need for maintenance.
2. According to the calendar (passage of time).
3. After a predetermined number of operating hours.

An important issue in preventive maintenance is the frequency of preventive maintenance. As the time between periodic maintenance increases, the cost of preventive maintenance decreases while the risk (and cost) of breakdowns increase. The goal is to strike a balance between the two costs (i.e., to minimize total cost).

Determining the amount of preventive maintenance to use is a function of the expected frequency of breakdown, the cost of a breakdown (including actual repair costs as well as potential damage or injury, lost production, and so on). The following two examples illustrate this.

Example 1

The frequency of breakdown of a machine per month is shown in the table. The cost of a breakdown is $1,000 and the cost of preventive maintenance is $1,250 per month. If preventive maintenance is performed, the probability of a machine breakdown is negligible. Should the manager use preventive maintenance, or would it be cheaper to repair the machine when it breaks down?

Number of breakdowns	0	1	2	3
Frequency of occurrence	.20	.30	.40	.10

Solution

The expected number of breakdowns without preventive maintenance is 1.40:

Number of Breakdowns	Frequency of Occurrence	Expected Number of Breakdowns
0	.20	0
1	.30	.30
2	.40	.80
3	.10	.30
	1.00	1.40

Expected cost using repair policy is: 1.40 breakdowns/month × $1,000/breakdown = $1,400. Preventive maintenance would cost $1,250.

Therefore, preventive maintenance would yield a savings of $150/month.

Example 2

Another approach that might be used relates to the time before a breakdown occurs. Suppose that the average time before breakdown is normally distributed and has a mean of 3 weeks and a standard deviation of .60 weeks. If breakdown cost averages $1,000 and preventive maintenance costs $250, what is the optimal maintenance interval?

Solution

Begin by computing the ratio of preventive cost to the sum of preventive and breakdown cost:

$$\frac{\text{preventive cost}}{\text{preventive cost} + \text{breakdown cost}} = \frac{\$250}{\$250 + \$1{,}000} = .20.$$

Find the number of standard deviations from the mean represented by an area under the normal curve of .60 using Appendix Table B. It is −0.84. Use this value of z to compute the maintenance interval:

mean + z standard deviations = 3 −.84(.60) = 2.496 weeks, or approximately every two and one-half weeks.

predictive maintenance An attempt to determine when best to perform preventive maintenance activities.

Ideally, preventive maintenance will be performed just prior to a breakdown or failure because this will result in the longest possible use of facilities or equipment without a breakdown. **Predictive maintenance** is an attempt to determine when to perform preventive maintenance activities. It is based on historical records and analysis of technical data to predict when a piece of equipment or part is about to fail. The better the predictions of failures are, the more effective preventive maintenance will be. A good preventive maintenance effort relies on complete records for each piece of equipment. Records must include information such as data of installation, operating hours, dates and types of insurance, and dates and types of repairs.

Some Japanese companies have workers perform preventive maintenance on the machines they operate, rather than use separate maintenance personnel for that task. Called **total preventive maintenance,** this approach is consistent with JIT systems and lean production, where employees are given greater responsibility for quality, productivity, and the general functioning of the system.

total preventive maintenance JIT approach where workers perform preventive maintenance on the machines they operate.

In the broadest sense, preventive maintenance extends back to the design and selection stage of equipment and facilities. Maintenance problems are sometimes *designed into* a system. For example, equipment may be designed in such a way that it needs frequent maintenance, or maintenance may be difficult to perform (e.g., the equipment has to be partially dismantled in order to perform routine maintenance). An extreme example of this was a certain car model that required the engine block to be lifted slightly in order to change the spark plugs! In such cases, it is very likely that maintenance will be performed less often than if its performance was less demanding. In other instances, poor design can cause equipment to wear out at an early age or experience a much higher than expected breakdown rate. *Consumer Reports,* for example, publishes annual breakdown data on automobiles. The data indicate that some models tend to break down with a much higher frequency than other models.

www.consumerreports.org/

One possible reason for maintenance problems being designed into a product is that designers have accorded other aspects of design greater importance. Cost is one such aspect. Another is appearance; an attractive design may be chosen over a less attractive one even though it will be more demanding to maintain. Customers may contribute to this situation; the buying public probably has a greater tendency to select an attractive design over one that offers ease of maintenance.

Obviously, durability and ease of maintenance can have long-term implications for preventive maintenance programs. Training of employees in proper operating procedures and in how to keep equipment in good operating order—and providing the incentive to do so—are also important. More and more, U.S. organizations are taking a cue from the Japanese and transferring routine maintenance (e.g., cleaning, adjusting, inspecting) to the users of equipment, in an effort to give them a sense of responsibility and awareness of the equipment they use and to cut down on abuse and misuse of the equipment.

BREAKDOWN PROGRAMS

The risk of a breakdown can be greatly reduced by an effective preventive maintenance program. Nonetheless, occasional breakdowns still occur. Even firms with good preventive practices have some need for breakdown programs. Of course, organizations that rely less on preventive maintenance have an even greater need for effective ways of dealing with breakdowns.

Unlike preventive maintenance, management cannot schedule breakdowns but must deal with them on an irregular basis (i.e., as they occur). Among the major approaches used to deal with breakdowns are the following:

1. *Standby or backup equipment* that can be quickly pressed into service.
2. *Inventories of spare parts* that can be installed as needed, thereby avoiding lead times involved in ordering parts, and buffer inventories, so that other equipment will be less likely to be affected by short-term downtime of a particular piece of equipment.
3. *Operators* who are able to perform at least minor repairs on their equipment.
4. *Repair people* who are well trained and readily available to diagnose and correct problems with equipment.

The degree to which an organization pursues any or all of these approaches depends on how important a particular piece of equipment is to the overall production system. At one extreme is equipment that is the focal point of a system (e.g., printing presses for a newspaper, or vital operating parts of a car, such as brakes, steering, transmission, ignition, and engine). At the other extreme is equipment that is seldom used because it does not perform an important function in the system, and equipment for which substitutes are readily available.

The implication is clear: Breakdown programs are most effective when they take into account the degree of importance a piece of equipment has in the production system, and the ability of the system to do without it for a period of time. The Pareto phenomenon exists in such situations: A relatively few pieces of equipment will be extremely important to the functioning of the system, thereby justifying considerable effort and/or expense; some will require moderate effort or expense; and many will justify little effort or expense.

REPLACEMENT

When breakdowns become frequent and/or costly, the manager is faced with a trade-off decision in which costs are an important consideration: What is the cost of replacement compared with the cost of continued maintenance? This question is sometimes difficult to resolve, especially if future breakdowns cannot be readily predicted. Historical records may help to project future experience. Another factor is technological change; newer equipment may have features that favor replacement over either preventive or breakdown maintenance. On the other hand, the removal of old equipment and the installation of new equipment may cause disruptions to the system, perhaps greater than the disruptions caused by breakdowns. Also, employees may have to be trained to operate the new equipment. Finally, forecasts of future demand for the use of the present or new equipment must be taken into account. The demand for the replacement equipment might differ because of the different features it has. For instance, demand for output of the current equipment might be two years, while demand for output of the replacement equipment might be much longer.

These decisions can be fairly complex, involving a number of different factors. Nevertheless, most of us are faced with a similar decision with our personal automobiles: When is it time for a replacement?

Summary

Maintaining the productive capability of an organization is an important function. Maintenance includes all of the activities related to keeping facilities and equipment in good operating order and maintaining the appearance of buildings and grounds.

The goal of maintenance is to minimize the total cost of keeping the facilities and equipment in good working order. Maintenance decisions typically reflect a trade-off between preventive maintenance, which seeks to reduce the incidence of breakdowns and failures, and breakdown maintenance, which seeks to reduce the impact of breakdowns when they do occur.

Key Terms

breakdown maintenance, 758
maintenance, 758
predictive maintenance, 760
preventive maintenance, 758
total preventive maintenance, 761

Discussion and Review Questions

1. What is the goal of a maintenance program?
2. List the costs associated with equipment breakdown.
3. What are three different ways preventive maintenance is scheduled?
4. Explain the term *predictive maintenance* and the importance of good records.
5. List the major approaches organizations use to deal with breakdowns.

6. Explain how the Pareto phenomenon applies to:
   a. Preventive maintenance.
   b. Breakdown maintenance.
7. Discuss the key points of this supplement as they relate to maintenance of an automobile.
8. What advantages does preventive maintenance have over breakdown maintenance?
9. Explain why having a good preventive maintenance program in place is necessary prior to implementing a JIT system.
10. Discuss the relationship between preventive maintenance and quality.

Problems

1. The probability that equipment used in a hospital lab will need recalibration is given in the following table. A service firm is willing to provide maintenance and provide any necessary calibrations for free for a fee of \$650 per month. Recalibration costs \$500 per time. Which approach would be most cost effective, recalibration as needed or the service contract?

Number of recalibrations	0	1	2	3	4
Probability of occurrence	.15	.25	.30	.20	.10

Expected recalibration cost: \$925/month
Use the service contract

2. The frequency of breakdown of a machine that issues lottery tickets is given in the following table. Repairs cost an average of \$240. A service firm is willing to provide preventive maintenance under either of two options. #1 is \$500 and covers all necessary repairs for free, and #2 is \$350 and covers any repairs after the first one, for free, provide any necessary repairs. Which option would have the lowest expected cost, pay for all repairs, service option #1, or service option #2?

Number of breakdowns/month	0	1	2	3	4
Frequency of occurrence	.10	.30	.30	.20	.10

Expected repair cost: \$456/month
Option #1: \$500
Option #2: \$422

3. Determine the optimum preventive maintenance frequency for each of the pieces of equipment if breakdown time is normally distributed:

Equipment	Average Time (days) between Breakdowns	Standard Deviation
A201	20	2
B400	30	3
C850	40	4

Equipment	Preventive Maintenance Cost	Breakdown Cost
A201	\$300	\$2,300
B400	\$200	\$3,500
C850	\$530	\$4,800

Equipment	Ratio	z	Interval (days)
A201	.115	−1.20	17.60
B400	.054	−1.61	25.17
C850	.099	−1.28	34.88

Selected Bibliography

Hall, Robert W. "Total Productive Maintenance—Essential to Maintain Progress." *Target* 3, no. 3 (Fall 1987), pp. 4–11.

Hora, Michael E. "The Unglamorous Game of Managing Maintenance." *Business Horizons* 30, no. 3 (May–June 1987), pp. 67–75.

Mann, Lawrence, Jr. *Maintenance Management.* Rev. ed. Lexington, MA: Lexington Books, 1983.

Nolden, Carol. "Predictive Maintenance: Route to Zero Unplanned Downtime." *Plant Engineering* 41, no. 4 (February 1987), pp. 38–43.

Wireman, Terry. *Preventive Maintenance,* Reston, VA: Reston Publishing, 1984.

LEARNING OBJECTIVES

After completing this chapter, you should be able to:

1. Discuss the behavioral aspects of projects in terms of project personnel and the project manager.
2. Discuss the nature and importance of a work breakdown structure in project management.
3. Give a general description of PERT/CPM techniques.
4. Construct simple network diagrams.
5. List the kinds of information that a PERT or CPM analysis can provide.
6. Analyze networks with deterministic times.
7. Analyze networks with probabilistic times.
8. Describe activity "crashing" and solve typical problems.

CHAPTER EIGHTEEN PROJECT MANAGEMENT

CHAPTER OUTLINE

The Olympic Games—summer games, winter games, you name it, have much in common besides the festivities, the excitement, national pride, and competition among athletes. They all involve a tremendous amount of planning, preparation, and coordinating work that needs to get done before and during the games. Athletes' living quarters and training facilities must be provided, competition schedules must be developed, arrangements for televising events must be made, equipment and crews must be coordinated, transportation and hotel accommodations must be made, as well as many other activities that go on behind the scenes so that everything goes off smoothly.

www.microsoft.com

The Microsoft Corporation periodically releases new or updated software. Each release is the result of many people working countless hours writing code, testing programs, and revising code. Design, production, and marketing efforts also have to be coordinated. The reputation and profits of the company are closely related to successful software development.

Construction of a brand new supermarket that will replace the old store across the street is nearing completion, and the contents of the old store will be moved to the new store.

www.foxintltv.com

A few years ago major portions of the Santa Ana freeway in California were destroyed by an earthquake and had to be rebuilt. Thousands of activities had to be planned and coordinated. Thousands of pieces of construction equipment and tons of material had to be coordinated. The rebuilding was successful, and actually completed 70 days ahead of schedule.

The movie *Titanic,* released in late 1997, was the leading box office money maker of all time. And a good thing too, because over $250 million was spent on making the movie, also a record. Imagine the immense job of planning and coordinating the script writers, set designers, set builders, camera crews, directors, actors and actresses, makeup, costumes, advertising, and more that went into making that movie.

All of these feats have something very important in common: They were all *projects,* not unlike projects that are going on all around us, every day. Some are large, some are small. Some are successful, and some aren't. But projects are an important way that certain kinds of work gets done.

INTRODUCTION

projects Unique, one-time operations designed to accomplish a specific set of objectives in a limited time frame.

Managers typically oversee a variety of operations. Some of these involve routine, repetitive activities, but others involve nonroutine activities. Under the latter heading are **projects:** unique, one-time operations designed to accomplish a set of objectives in a limited time frame. Other examples of projects include constructing a shopping complex, installing a new computer system, launching a space shuttle, making a movie, putting on a play, and designing and running a political campaign. Examples of projects within business organizations include designing new products or services, designing advertising campaigns, designing information systems, designing databases, and designing web pages. Still other examples include quality improvement, process improvement, re-engineering, moving an organization from one location to another, and designing and implementing a new payroll system.

Projects may involve considerable cost. Some have a long time horizon, and some involve a large number of activities that must be carefully planned and coordinated. Most are expected to be completed within time, cost, and performance guidelines. To accomplish these, goals must be established and priorities set. Tasks must be identified and time estimates made. Resource requirements also must be projected and budgets prepared. Once underway, progress must be monitored to assure that project goals and objectives will be achieved.

The project approach enables an organization to focus attention and concentrate efforts on accomplishing a narrow set of objectives within a limited time and budget framework. This can produce significant benefits compared with other approaches that might be considered. Even so, projects present managers with a host of problems that differ in many respects from those encountered with more routine activities. The problems of planning and coordinating project activities can be formidable for large projects, which typically have hundreds or even thousands of activities that must be carefully planned and monitored if the project is to proceed according to schedule and at a reasonable cost.

The chapter introduces the basic concepts of project management. It begins with a brief discussion of some behavioral aspects of project management, along with some of the difficulties project managers are apt to encounter. The main portion of the chapter is devoted to a description of graphical and computational methods that are used for planning and scheduling projects.

BEHAVIORAL ASPECTS OF PROJECT MANAGEMENT

Project management differs from management of more traditional activities mainly because of its limited time framework and the unique set of activities involved, which gives rise to a host of rather unique problems. This section describes more fully the nature of projects and their behavioral implications. We will pay special attention to the role of the project manager.

The Nature of Projects

Projects go through a series of stages—a life cycle—which include project planning, execution of major activities, and project phaseout. During this life cycle, a variety of skill requirements are involved. The circumstances are analogous to constructing a house. Initially a site must be found, and plans must be drawn up and approved by the owner and possibly a town building commission or other regulatory agency. Then a succession of activities occurs, each with its own skill requirements, starting with the site preparation, then laying the foundation, erecting the frame, roofing, constructing exterior walls, wiring and plumbing, installing kitchen and bathroom fixtures and appliances, interior finishing work, and painting and carpeting work. Similar sequences occur on large construction projects, in R&D work, in the aerospace industry, and in virtually every other instance where projects are being carried out.

Projects typically bring together personnel with diverse knowledge and skills, most of whom remain associated with the project for less than its full life. Some personnel go from project to project as their contributions become needed, and others are "on loan," either on a full-time or part-time basis, from their regular jobs. The latter is usually the case when a special project exists within the framework of a more traditional organization. Certain kinds of organizations are involved with projects on a regular basis; examples include consulting firms, architects, writers and publishers, and construction firms. In those organizations, it is not uncommon for some individuals to spend virtually all of their time with projects.

Some organizations use a *matrix organization* that allows them to integrate the activities of a variety of specialists within a functional framework. For instance, they have certain people who prepare proposals, others who concentrate exclusively on engineering, others who devote their efforts to marketing, and so on.

Key Decisions in Project Management

Much of the success of projects depends on certain key managerial decisions, such as:

- Deciding which projects to implement.
- Selecting the project manager.
- Selecting the project team.
- Planning and designing the project.
- Managing and controlling project resources.
- Deciding if and when a project should be terminated.

Deciding Which Projects to Implement. This consists of determining the criteria that will be used to decide which projects to pursue. Typical factors include budget, availability of appropriate knowledge and skill personnel, and cost–benefit considerations. Of course, other factors may override these criteria, factors such as safety issues, government-mandated actions, and so on.

Selecting the Project Manager. The project manager is the central person in the project. The following section on project managers discusses this topic.

Selecting the Project Team. The team can greatly influence the ultimate success or failure of a project. Important considerations include not only a person's knowledge and skill base, but also how well the person works with others (particularly those who have already been chosen for the project), enthusiasm for the project, other projects the person is involved in, and how likely those other projects might be to interfere with work on this project.

Planning and Designing the Project. Project planning and design require decisions on project goals, a timetable for project completion, what work needs to be done, how it will be done, if some portions will be outsourced, what resources will be needed, and when and how long they will be needed.

Managing and Controlling Project Resources. This involves managing personnel, equipment, and the budget, monitoring progress, and taking corrective action when needed. Also necessary is designing an information system and deciding what project information will be needed, when and by whom it will be needed, and how often it should be updated.

Deciding If and When a Project Should Be Terminated. Sometimes it is better to terminate a project than to invest any more resources. Important considerations here are the likelihood of success, termination costs, and whether resources could be better used elsewhere.

The Project Manager

The project manager bears the ultimate responsibility for the success or failure of the project. The role of the project manager is one of an organizer—a person who is capable of working through others to accomplish the objectives of the project.

Once the project is underway, the project manager is responsible for effectively managing each of the following:

1. The *work,* so that all of the necessary activities are accomplished in the desired sequence.
2. The *human resources,* so that those working on the project have direction and motivation.
3. *Communications,* so that everybody has the information they need to do their work.
4. *Quality,* so that performance objectives are realized.
5. *Time,* so that the project is completed on time.
6. *Costs,* so that the project is completed within budget.

The job of project manager can be both difficult and rewarding. The manager must coordinate and motivate people who sometimes owe their allegiance to other managers in their functional areas. In addition, the people who work on a project frequently possess specialized knowledge and skills that the project manager lacks. Nevertheless, the manager is expected to guide and evaluate their efforts. Project managers must often function in an environment that is beset with uncertainties. Even so, budgets and time constraints are usually imposed, which can create additional pressures on project personnel. Finally, the project manager may not have the authority needed to accomplish all the objectives of the project. Instead, the manager must sometimes rely on persuasion and the cooperation of others to realize project goals.

The position of project manager has high visibility. The rewards of the job of project manager come from the creative challenges of the job, the benefits of being associated with a successful project (including promotion and monetary compensation), and the personal satisfaction of seeing it through to its conclusion.

The Pros and Cons of Working on Projects

People are selected to work on special projects because the knowledge or abilities they possess are needed. In some instances, however, their supervisors may be reluctant to allow them to interrupt their regular jobs, even on a part-time basis, because it may require training a new person to do a job that will be temporary. Moreover, managers don't want to lose the output of good workers. The workers themselves are not always eager to participate in projects because it may mean working for two bosses who impose differing demands, and may cause disruption of friendships and daily routines, and the risk of being replaced on the current job. Furthermore, there may be fear of being associated with an unsuccessful project because of the adverse effect it might have on career advancement. In too many instances, when a major project is phased out and the project team disbanded, team members tend to drift away from the organization for lack of a new project and the difficulty of returning to former jobs. This tendency is more pronounced after lengthy projects and is less likely to occur when a team member works on a part-time basis.

In spite of the potential risks, people are attracted by the potential rewards of being involved in a project. One is the dynamic environment that surrounds a project, often a marked contrast to the staid environment of a routine in which some may feel trapped. Some individuals seem to thrive in more dynamic environments; they welcome the challenge of working under pressure and solving new problems. Then, too, projects may present opportunities to meet new people and to increase future job opportunities, especially if the project is successful. In addition, association with a project can be a source of status among fellow workers. Finally, working on projects frequently inspires a team spirit, increasing morale and motivation to achieve successful completion of project goals.

PROJECT LIFE CYCLE

The size, length, and scope of projects varies widely according to the nature and purpose of the project. Nevertheless, all projects have something in common: They go through a life cycle, which typically consists of five phases.

1. *Concept,* at which point the organization recognizes the need for a project or responds to a request for a proposal from a potential customer or client.

2. *Feasibility analysis,* which examines the expected costs, benefits, and risks of undertaking the project.
3. *Planning,* which spells out the details of the work and provides estimates of the necessary human resources, time, and cost.
4. *Execution* during which the project itself is done. This phase often accounts for the majority of time and resources consumed by a project.
5. *Termination,* during which closure is achieved. Termination can involve reassigning personnel and dealing with any leftover materials, equipment (e.g., selling or transferring equipment), and any other resources associated with the project.

It should be noted that the phases can overlap, so that one phase may not be fully complete before the next phase begins. This can reduce the time necessary to move

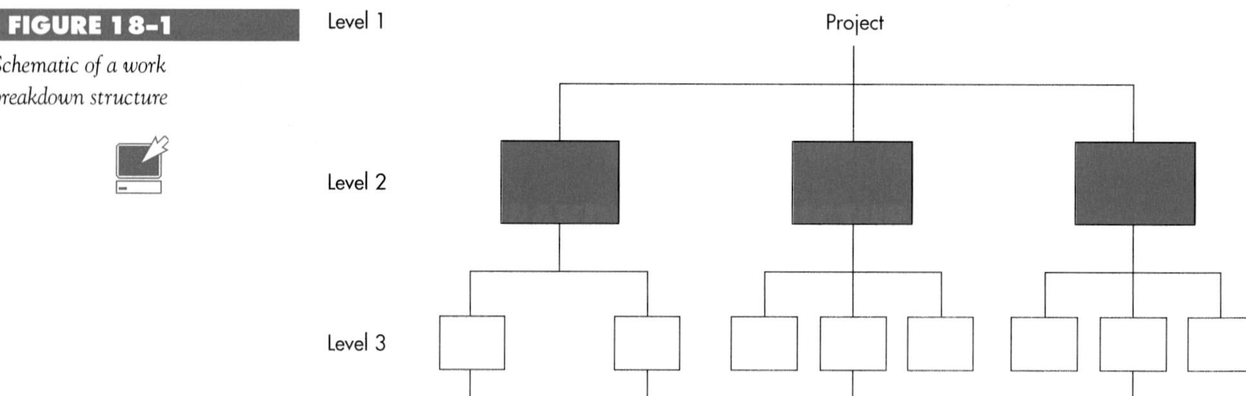

FIGURE 18–1

Schematic of a work breakdown structure

through the life cycle, perhaps generating some competitive advantage and cost saving. Although subsequent decisions in an earlier phase may result in waste for some portion of the activity in a following phase, careful coordination of activities can minimize that risk.

WORK BREAKDOWN STRUCTURE

work breakdown structure A hierarchical listing of what must be done during a project.

Because large projects usually involve a very large number of activities, planners need some way to determine exactly what will need to be done so that they can realistically estimate how long it will take to complete the various elements of the project and how much it will cost. They often accomplish this by developing a **work breakdown structure (WBS),** which is a hierarchical listing of what must be done during the project. This methodology establishes a logical framework for identifying the required activities for the project (see Figure 18–1). The first step in developing the work breakdown structure is to identify the major elements of the project. These are the Level 2 boxes in Figure 18–1. The next step is to identify the major supporting activities for each of the major elements—the Level 3 boxes. Then, each major supporting activity is broken down into a list of the activities that will be needed to accomplish it—the Level 4 boxes. (For purposes of illustration, only a portion of the Level 4 boxes are shown.) Usually there are many activities in the Level 4 lists. Some large projects may involve additional levels, but Figure 18–1 gives you some idea of the concept of the work breakdown structure.

Developing a good work breakdown structure can require substantial time and effort due to the uncertainties associated with a project and/or the size of the project. Typically the portion of time spent on developing the work breakdown structure greatly exceeds the time spent on actually developing a project schedule. The importance of a work breakdown structure is underscored by the fact that it serves as the focal point for planning and doing the project.

PLANNING AND SCHEDULING WITH GANTT CHARTS

The Gantt chart is a popular tool for planning and scheduling *simple* projects. It enables a manager to initially schedule project activities and then to monitor progress over time by comparing planned progress to actual progress. Figure 18–2 illustrates a Gantt chart for a bank's plan to establish a new direct marketing department. To prepare the chart, the vice president in charge of the project had to first identify the major activities that would be required. Next, time estimates for each activity were made, and the sequence

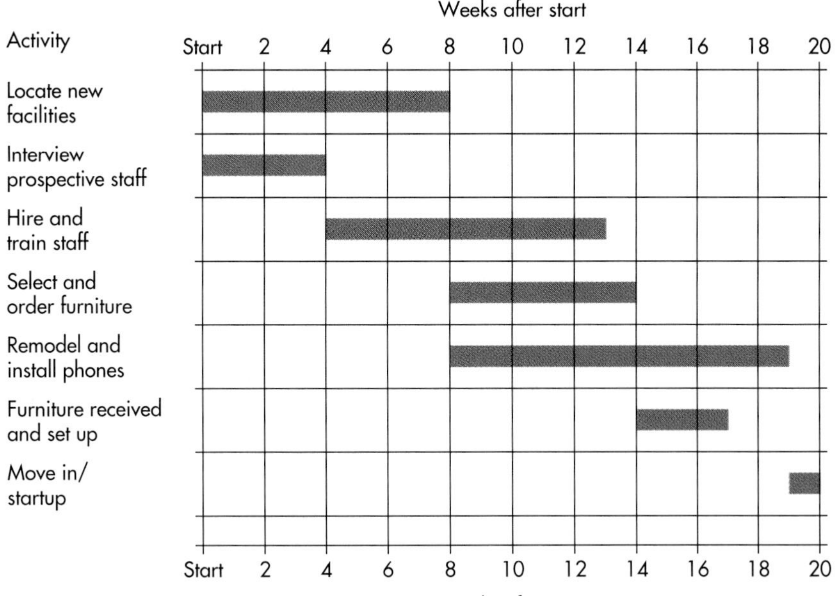

FIGURE 18-2
Gantt chart for bank example

of activities was determined. Once completed, the chart indicated which activities were to occur, their planned duration, and when they were to occur. Then, as the project progressed, the manager was able to see which activities were ahead of schedule and which were delaying the project. This enabled the manager to direct attention where it was needed most to speed up the project in order to finish on schedule.

The obvious advantage of a Gantt chart is its simplicity, and this accounts for its popularity. However, Gantt charts fail to reveal certain relationships among activities that can be crucial to effective project management. For instance, if one of the early activities in a project suffers a delay, it would be important for the manager to be able to easily determine which later activities would result in a delay. Conversely, some activities may safely be delayed without affecting the overall project schedule. The Gantt chart does not directly reveal this because it is most useful for simple projects or the initial project planning on more complex projects, which then gives way to the use of *networks*, the subject of the following section.

PERT AND CPM

PERT (program evaluation and review technique) and **CPM** (critical path method) are two of the most widely used techniques for planning and coordinating large-scale projects. By using PERT or CPM, managers are able to obtain:

PERT Program evaluation and review technique, for planning and coordinating large projects.

CPM Critical path method, for planning and coordinating large projects.

1. A graphical display of project activities.
2. An estimate of how long the project will take.
3. An indication of which activities are the most critical to timely project completion.
4. An indication of how long any activity can be delayed without lengthening the project.

PERT and CPM were developed independently during the late 1950s. PERT evolved through the joint efforts of Lockheed Aircraft, the U.S. Navy Special Projects Office, and the consulting firm of Booz, Allen & Hamilton in an effort to speed up the Polaris missile project. At the time, the U.S. government was concerned that the Soviet Union might be gaining nuclear superiority over the United States, and it gave top priority for the early completion of the project by the Department of Defense. The project was a

In this project within a project, participants raise the Onbashira pillars during the opening ceremony of the Winter Olympics in Nagano, Japan, on February 7, 1998. This is only one of many sites for the various events. Planning and managing the execution of the Olympic Games is a large project that includes long term construction and improvement of sites as well as finely timed activities, such as the opening ceremony.

huge one, with more than 3,000 contractors and thousands of activities. PERT was quite successful and was generally credited with shaving two years off the length of the project. Partly for that reason, PERT or some similar technique is now required on all large government projects.

www.dupont.com

CPM was developed by J. E. Kelly of the Remington Rand Corporation and M. R. Walker of DuPont to plan and coordinate maintenance projects in chemical plants.

Although PERT and CPM were developed independently, they have a great deal in common. Moreover, many of the initial differences between them have disappeared as users borrowed certain features from one technique for use with the other. For example, PERT originally stressed probabilistic activity time estimates, because the environment in which it developed was typified by high uncertainty. In contrast, the tasks for which CPM was developed were much less certain, so CPM originally made no provision for variable time estimates. At present, either technique can be used with deterministic or probabilistic times. Other initial differences concerned the mechanical aspects of developing project networks. However, from a conceptual standpoint, most of these differences were relatively minor. To avoid confusion, we will not delve into the differences. For practical purposes, the two techniques are the same; the comments and procedures described will apply to CPM analysis as well as to PERT analysis of projects.

The Network Diagram

network (precedence) diagram Diagram of project activities that shows sequential relationships by use of arrows and nodes.

One of the main features of PERT and related techniques is their use of a **network** or **precedence diagram** to depict major project activities and their sequential relationships. Recall the bank example that used a Gantt chart (see Figure 18–2). A network diagram for that same problem is shown in Figure 18–3. The diagram is composed of a number of arrows and nodes. The arrows represent the project activities. Note how much clearer the sequential relationship of activities is with a network chart than with a Gantt chart. For instance, it is apparent that ordering the furniture and remodeling both require that

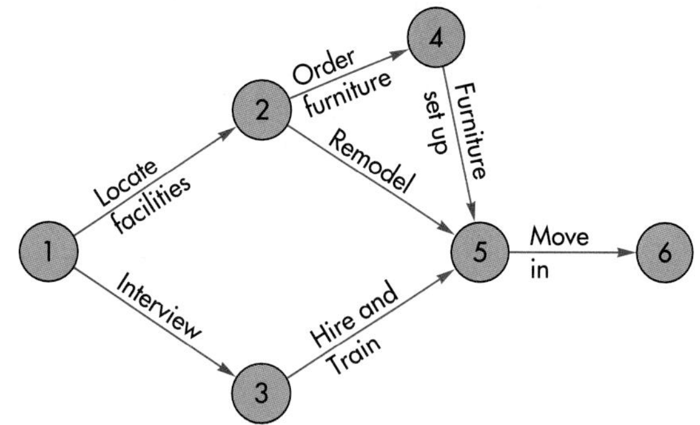

FIGURE 18-3

A simple project network diagram

a location for the office has been identified. Likewise, interviewing must precede training. However, interviewing and training can take place independently of activities associated with locating a facility, remodeling, and so on. Hence, a network diagram is generally the preferred approach for visual portrayal of project activities.

There are two slightly different conventions for constructing these network diagrams. Under one convention, the *arrows* designate activities; under the other convention, the *nodes* designate activities. These conventions are referred to as **activity-on-arrow (A-O-A)** and **activity-on-node (A-O-N).** To avoid confusion, the discussion here will focus primarily on the activity-on-arrow convention, with a comparison of the two conventions provided later in the chapter. For now, we shall use the arrows for activities. **Activities** consume resources and/or *time.* The nodes in the A-O-A approach represent the activities' starting and finishing points, which are called **events.** Events are points in time. Unlike activities, they consume neither resources nor time.

activity-on-arrow Network diagram convention in which arrows designate activities.

activity-on-node Network diagram convention in which nodes designate activities.

Activities can be referred to in either of two ways. One is by their endpoints (e.g., activity 2-4) and the other is by a letter assigned to an arrow (e.g., activity *c*). Both methods are illustrated in this chapter.

activities Project steps that consume resources and/or time.

The network diagram describes sequential relationships among major activities on a project. For instance, activity 2-4 cannot be started, according to the network, until activity 1-2 has been completed. A **path** is a sequence of activities that leads from the starting node to the finishing node. Thus, the sequence 1-2-4-5-6 is a path. There are two other paths in this network: 1-2-5-6 and 1-3-5-6. The length (of time) of any path can be determined by summing the expected times of the activities on that path. The path with the longest time is of particular interest because it governs project completion time. In other words, expected project duration equals the expected time of the longest path. Moreover, if there are any delays along the longest path, there will be corresponding delays in project completion time. Attempts to shorten project completion must focus on the longest sequence of activities. Because of its influence on project completion time, the longest path is referred to as the **critical path,** and its activities are referred to as **critical activities.**

events The starting and finishing of activities, designated by nodes in the A-O-A convention.

path A sequence of activities that leads from the starting node to the finishing node.

critical path The longest path; determines expected project duration.

Paths that are shorter than the critical path can experience some delays and still not affect the overall project completion time as long as the ultimate path time does not exceed the length of the critical path. The allowable slippage for any path is called **slack,** and it reflects the difference between the length of a given path and the length of the critical path. The critical path, then, has zero slack time.

critical activities Activities on the critical path.

slack Allowable slippage for a path; the difference between the length of a path and the length of the critical path.

Network Conventions

Developing and interpreting network diagrams requires some familiarity with networking conventions. Although many could be mentioned, the discussion will concentrate on some of the most basic and most common features of network diagrams. This will provide sufficient background for understanding the basic concepts associated with precedence diagrams and allow you to solve typical problems.

One of the main features of a precedence diagram is that it reveals which activities must be performed in *sequence* (i.e., there is a precedence requirement) and which can be performed independently of each other. For example, in the following diagram, activity *a* must be completed before activity *b* can begin, and activity *b* must be completed before activity *c* can begin.

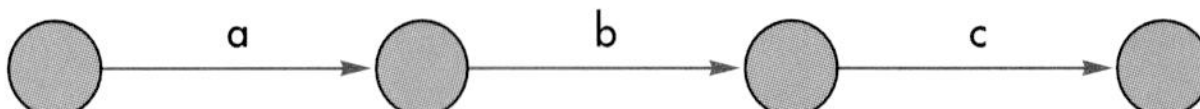

If the diagram had looked like the one following, both activities *a* and *b* would have to be completed before activity *c* could begin, but *a* and *b* could be performed at the same time; performance of *a* is independent of performance of *b*.

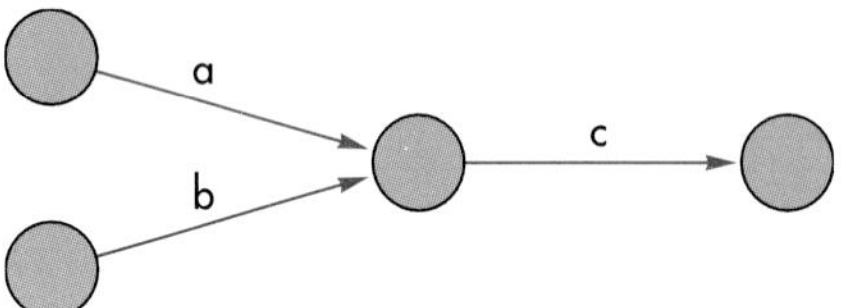

If activity *a* must precede *b* and *c*, the appropriate network would look like this:

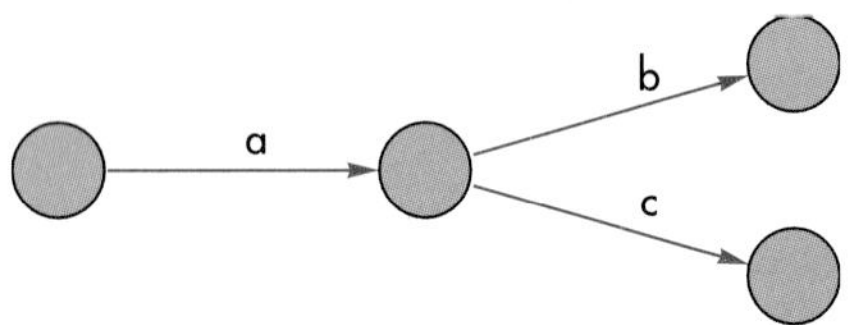

When multiple activities enter a node, this implies that all those activities must be completed before any activities that are to begin at that node can start. Hence, in this next diagram, activities *a* and *b* must both be finished before either activity *c* or activity *d* can start.

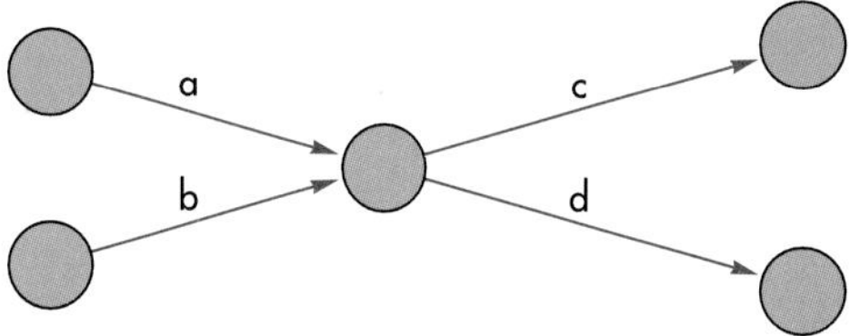

When two activities both have the same beginning and ending nodes, a *dummy* node and activity is used to preserve the separate identity of each activity. In the diagram below, activities *a* and *b* must be completed before activity *c* can be started.

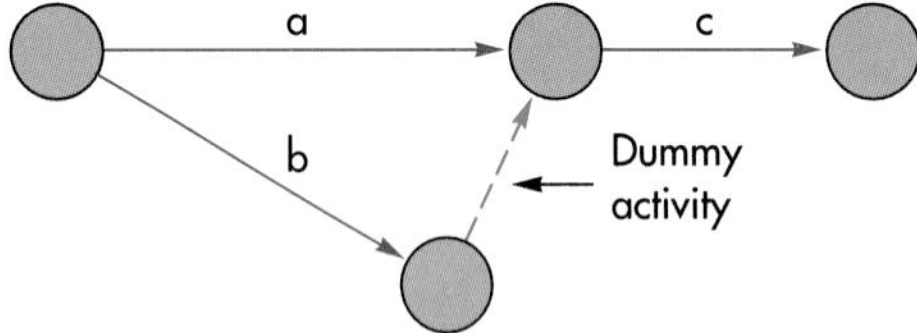

Separate identities are particularly important for computer analysis, because most computer programs identify activities by their endpoints; activities with the same endpoints could not be distinguished from each other, although they might have quite different expected times.

Dummy activities have a number of different uses. Here is another common use:

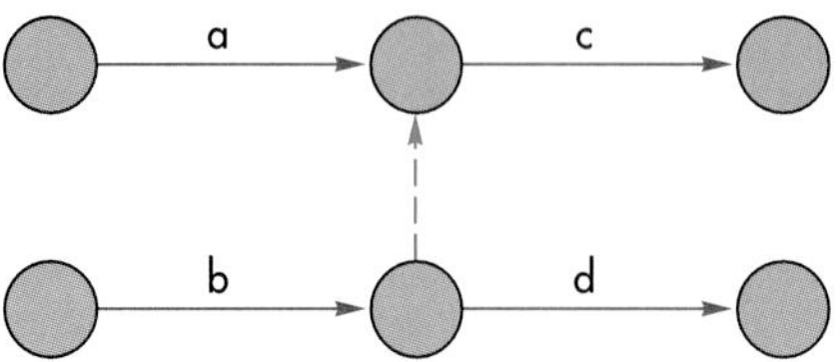

In this situation, activities *a* and *b* must both precede activity *c*. However, the start of activity *d* is dependent only on completion of activity *b*, not on the completion of activity *a*.

The primary function of dummy activities is to clarify relationships. As far as time is concerned, a dummy activity has an activity time equal to zero.

For reference purposes, nodes are numbered typically from left to right:

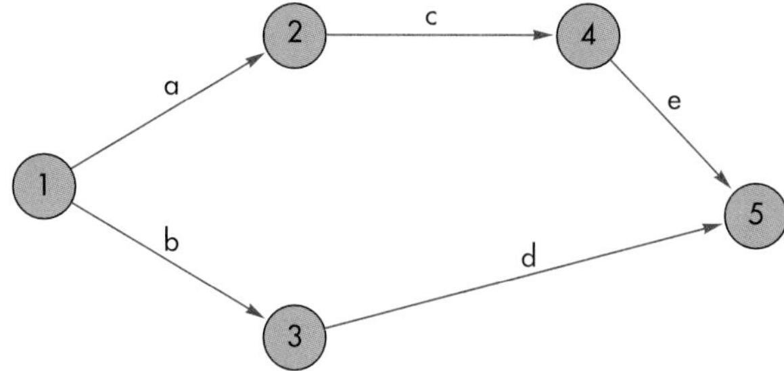

Starting and ending arrows are sometimes used during development of a network for increased clarity.

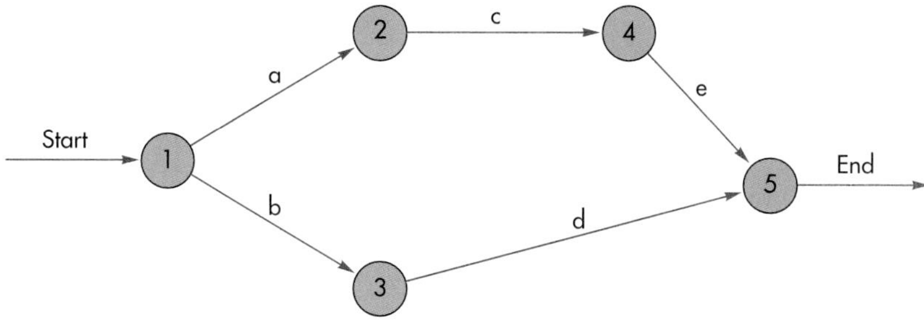

DETERMINISTIC TIME ESTIMATES

The main determinant of the way PERT and CPM networks are analyzed and interpreted is whether activity time estimates are *probabilistic* or *deterministic*. If time estimates can be made with a high degree of confidence that actual times will not differ significantly, we say the estimates are **deterministic.** If estimated times are subject to variation, we say the estimates are **probabilistic.** Probabilistic time estimates must include an indication of the extent of probable variation.

deterministic Time estimates that are fairly certain.

probabilistic Estimates of times that allow for variation.

This section describes analysis of networks with deterministic time estimates. A later section deals with probabilistic times.

One of the best ways to gain an understanding of the nature of network analysis is to consider a simple example.

Example 1

Given the following information:

FIGURE 18-4

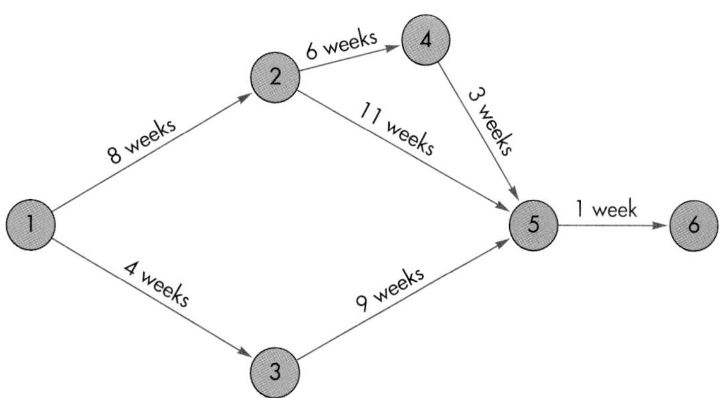

Determine:

a. The length of each path.
b. The critical path.
c. The expected length of the project.
d. Amount of slack time for each path.

Solution

a. As shown in the following table, the path lengths are 18 weeks, 20 weeks, and 14 weeks.
b. The longest path (20 weeks) is 1-2-5-6, so it is the critical path.
c. The expected length of the project is equal to the length of the critical path (i.e., 20 weeks).
d. We find the slack for each path by subtracting its length from the length of the critical path, as shown in the last column of the table. (*Note:* It is sometimes desirable to know the slack time associated with activities. The next section describes a method for obtaining those slack times.)

Path	Length (weeks)	Slack
1-2-4-5-6	8 + 6 + 3 + 1 = 18	20 − 18 = 2
1-2-5-6	8 + 11 + 1 = 20*	20 − 20 = 0
1-3-5-6	4 + 9 + 1 = 14	20 − 14 = 6

*Critical path length.

A COMPUTING ALGORITHM

Many real-life project networks are much larger than the simple network illustrated in the preceding example; they often contain hundreds or even thousands of activities. Because the necessary computations can become exceedingly complex and time-consuming, large networks are generally analyzed by computer programs instead of manually. The intuitive approach just demonstrated does not lend itself to computerization because, in many instances, path sequences are not readily apparent. Instead, planners use an algorithm to develop four pieces of information about the network activities:

ES, the earliest time activity can start, assuming all preceding activities start as early as possible.

EF, the earliest time the activity can finish.

LS, the latest time the activity can start and not delay the project.

LF, the latest time the activity can finish and not delay the project.

Once these values have been determined, they can be used to find:

1. Expected project duration.
2. Slack time.
3. The critical path.

The three examples that follow illustrate how to compute those values using the precedence diagram of Example 1.

Example 2

Compute the earliest starting time and earliest finishing time for each activity in the diagram shown in Figure 18–4.

Solution

Begin by placing brackets at the two ends of each starting activity:

We determine and place in the brackets for each activity, the earliest starting time, ES, and the earliest finishing time, EF, as follows:

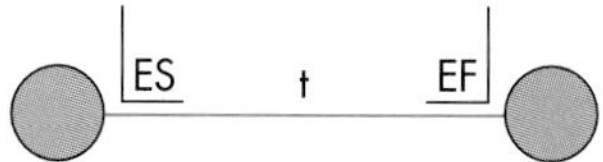

Do this for all activities, beginning at the left side of the precedence diagram and moving to the right side.

Once ES has been determined for each activity, EF can be found by adding the activity time, t, to ES: ES + t = EF.

Use an ES of 0 for all starting activities. Thus, activities 1-2 and 1-3 are assigned ES values of 0. This permits computation of the EF for each of these activities:

$$EF_{1\text{-}2} = 0 + 8 = 8 \qquad \text{and} \qquad EF_{1\text{-}3} = 0 + 4 = 4$$

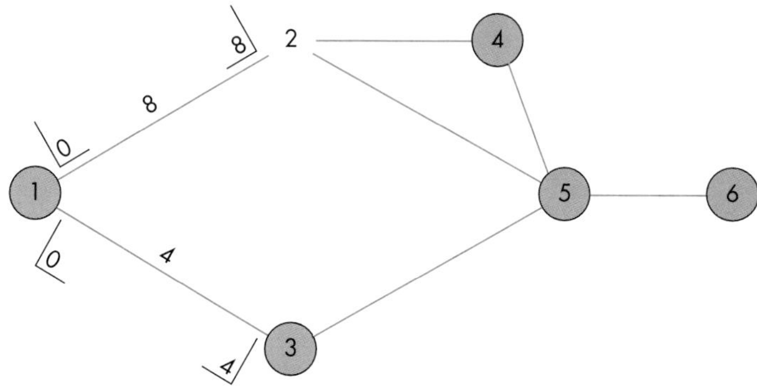

The EF time for an activity becomes the ES time for the next activity to follow it in the diagram. Hence, because activity 1-2 has an EF time of 8, both activities 2-4 and 2-5 have ES times of 8. Similarly, activity 3-5 has an ES time of 4.

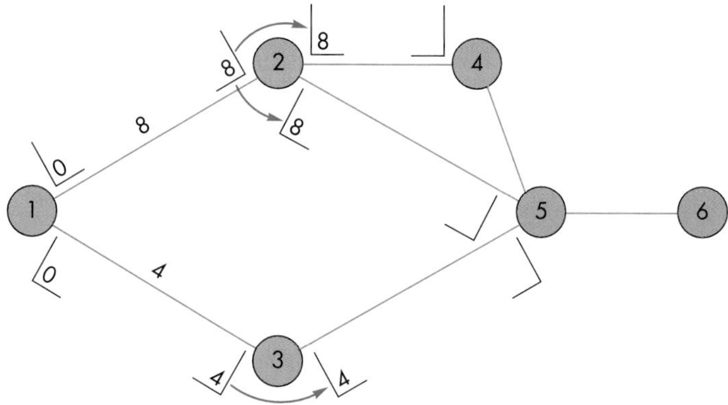

This permits calculation of the EF times for these activities: $EF_{2\text{-}4} = 8 + 6 = 14$; $EF_{2\text{-}5} = 8 + 11 = 19$; and $EF_{3\text{-}5} = 4 + 9 = 13$.

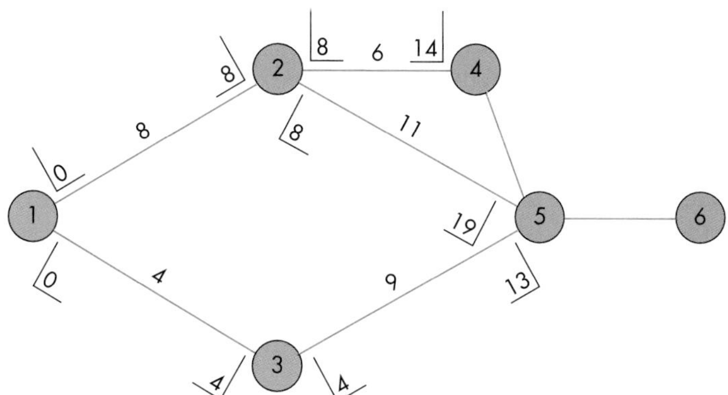

The ES for activity 4-5 is the EF time of activity 2-4, which is 14. Using this value, we find the EF for activity 4-5 is 17; $EF_{4\text{-}5} = 14 + 3 = 17$.

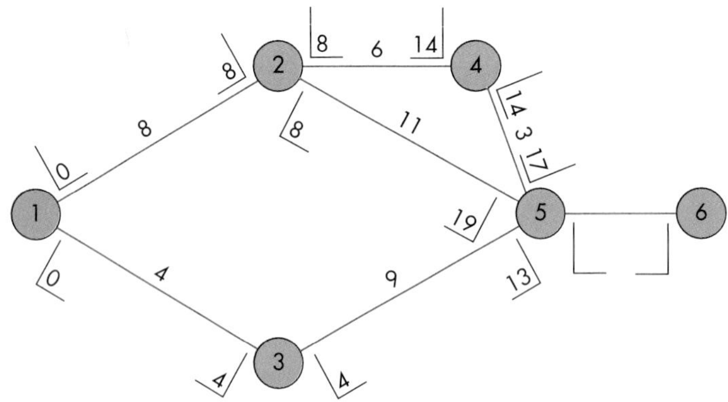

In order to determine the ES for activity 5-6, we must realize that activity 5-6 cannot start until *every* activity that precedes it is finished. Therefore, the *largest* of the EF times for the three activities that precede activity 5-6 determines $ES_{5\text{-}6}$. Hence, the ES for activity 5-6 is 19.

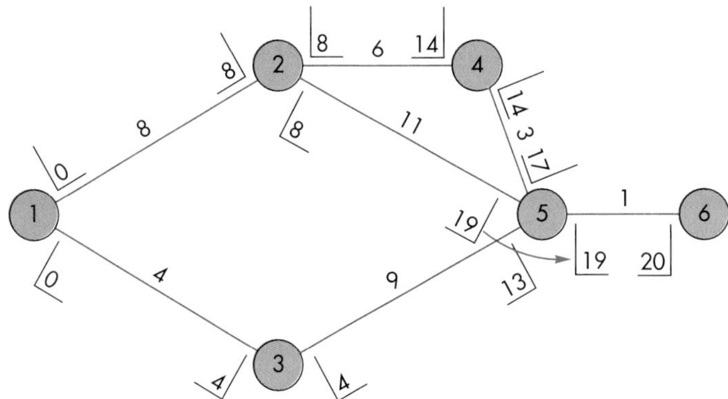

Then the EF for the last activity, 5-6, is 20; $EF_{5\text{-}6} = 19 + 1 = 20$. Note that the latest EF is the project duration. Thus, the expected length of the project is 20 weeks.

Computing ES and EF Times

Computation of earliest starting and finishing times is aided by two simple rules:

1. The earliest finish time for any activity is equal to its earliest start time plus its expected duration, t:

   $$EF = ES + t \tag{18–1}$$

2. ES for activities at nodes with one entering arrow is equal to EF of the entering arrow. ES for activities leaving nodes with multiple entering arrows is equal to the largest EF of the entering arrow.

Computing LS and LF Times

Computation of the latest starting and finishing times is aided by the use of two rules:

1. The latest starting time for each activity is equal to its latest finishing time minus its expected duration:

   $$LS = LF - t \tag{18–2}$$

2. For nodes with one leaving arrow, LF for arrows entering that node equals the LS of the leaving arrow. For nodes with multiple leaving arrows, LF for arrows entering that node equals the smallest LS of leaving arrows.

Finding ES and EF times involves a "forward pass" through the network; finding LS and LF times involves a "backward pass" through the network. Hence, we must begin with the EF of the last activity and use that time as the LF for the last activity. Then we obtain the LS for the last activity by subtracting its expected duration from its LF.

Example 3

Compute the latest finishing and starting times for the precedence diagram developed in Example 2.

Solution

We must add the LS and LF times to the brackets on the diagram.

Begin by setting the LF time of the last activity equal to the EF of that activity. Thus,

$LF_{5\text{-}6} = EF_{5\text{-}6} = 20$ weeks

Obtain the LS time for activity 5-6 by subtracting the activity time, t, from the LF time:

$LS_{5\text{-}6} = LF_{5\text{-}6} - t = 20 - 1 = 19.$

Mark these values on the diagram:

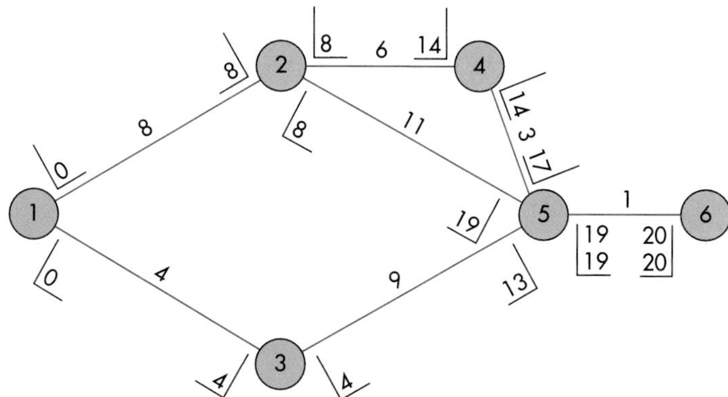

The LS time of 19 for activity 5-6 now becomes the LF time for each of the activities that precede activity 5-6. This permits determination of the LS times for each of those activities: Subtract the activity time from the LF to obtain the LS time for the activity. The LS time for activity 3-5 is 19 − 9 = 10.

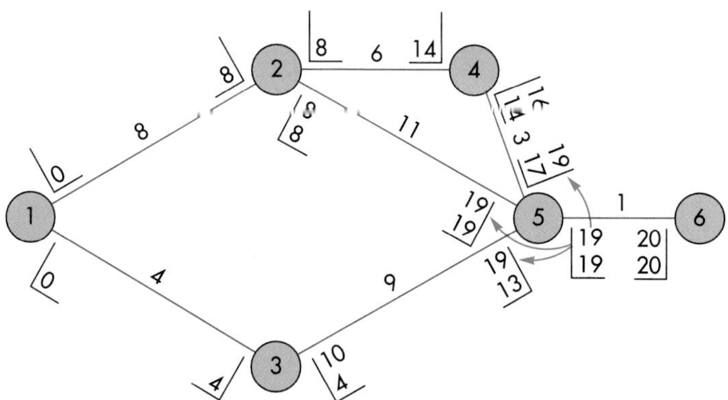

Next, the LS for activity 4-5, which is 16, becomes the LF for activity 2-4, and the LS for activity 3-5, which is 10, becomes the LF for activity 1-3. Using these values, you find the LS for each of these activities by subtracting the activity time from the LF time.

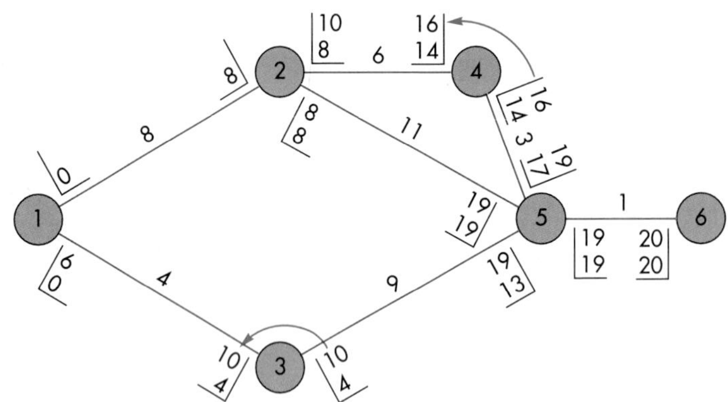

The LF for activity 1-2 is the *smallest* of the two LS times of the activities that 1-2 precedes. Hence, the LF time for activity 1-2 is 8. The reason you use the smallest time is that activity 1-2 must finish at a time that permits all following activities to start no later than their LS times.

Once you have determined the LF time of activity 1-2, find its LS time by subtracting the activity time of 8 from the LF time of 8. Hence, the LS time is 0.

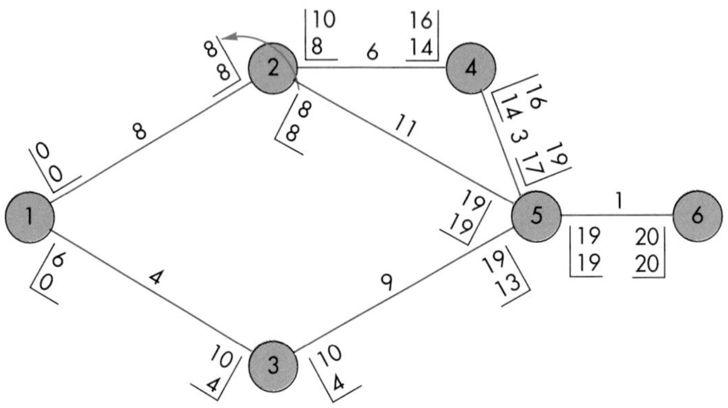

Computing Slack Times

The slack time can be computed in either of two ways:

$$\text{Slack} = \text{LS} - \text{ES} \quad \text{or} \quad \text{LF} - \text{EF} \tag{18–3}$$

The critical path using this computing algorithm is denoted by activities with zero slack time. Thus, the table in Example 4 indicates that activities 1-2, 2-5, and 5-6 are all critical activities, which agrees with the results of the intuitive approach demonstrated in Example 1.

Knowledge of slack times provides managers with information for planning allocation of scarce resources and for directing control efforts toward those activities that might be most susceptible to delaying the project. In this regard, it is important to recognize that the activity slack times are based on the assumption that all of the activities on the same path will be started as early as possible and not exceed their expected times. Furthermore, if two activities are both on the same path (e.g., activities 2-4 and 4-5 in the preceding example) and have the same slack (e.g., two weeks), this will be the *total* slack available to both. In essence, the activities have *shared slack.* Hence, if the first activity uses all the slack, there will be zero slack for the other activity and that much less slack for all following activities on that same path.

TABLE 18–1

Computer printout

		SCHEDULE				
		EARLY		LATE		
ACTIVITY	TIME	ES	EF	LS	LF	SLACK
1-2	8.00	0.00	8.00	0.00	8.00	0.00
1-3	4.00	0.00	4.00	6.00	10.00	6.00
2-4	6.00	8.00	14.00	10.00	16.00	2.00
2-5	11.00	8.00	19.00	8.00	19.00	0.00
3-5	9.00	4.00	13.00	10.00	19.00	6.00
4-5	3.00	14.00	17.00	16.00	19.00	2.00
5-6	1.00	19.00	20.00	19.00	20.00	0.00

THE CRITICAL PATH SEQUENCE IS:

SNODE	FNODE	TIME
1	2	8.00
2	5	11.00
5	6	1.00
		20.00

Example 4

Compute slack times for the precedence diagram of Figure 8–4.

Solution

Either the starting times or the finishing times can be used. Suppose we choose the starting times. Using ES times computed in Example 2 and LS times computed in Example 3, slack times are:

Activity	LS	ES	(LS – ES) Slack
1-2	0	0	0
1-3	6	0	6
2-4	10	8	2
2-5	8	8	0
3-5	10	4	6
4-5	16	14	2
5-6	19	19	0

As noted earlier, this algorithm lends itself to computerization. A computer printout for this problem would appear something like the one shown in Table 18–1.

PROBABILISTIC TIME ESTIMATES

The preceding discussion assumed that activity times were known and not subject to variation. While that assumption is appropriate in some situations, there are many others where it is not. Consequently, those situations require a probabilistic approach.

The probabilistic approach involves *three* time estimates for each activity instead of one:

optimistic time The length of time required under optimal conditions.

pessimistic time The length of time required under the worst conditions.

most likely time The most probable length of time that will be required.

1. **Optimistic time:** The length of time required under optimum conditions; represented by t_o.
2. **Pessimistic time:** The length of time required under the worst conditions; represented by t_p.
3. **Most likely time:** The most probable amount of time required; represented by t_m.

Managers or others with knowledge about the project can make these time estimates.

FIGURE 18–5

A beta distribution is used to describe probabilistic time estimates

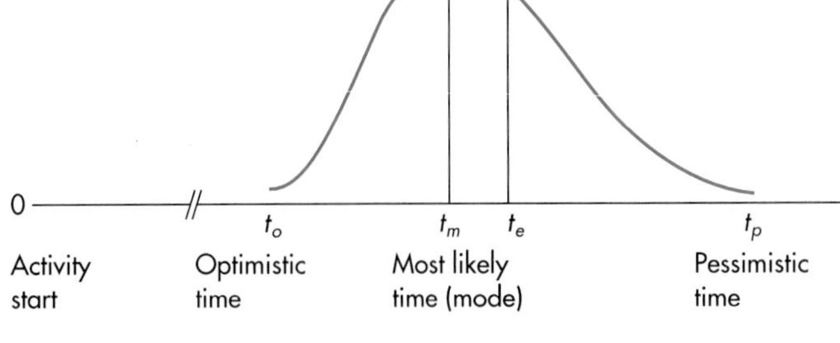

beta distribution Used to describe the inherent variability in activity time estimates.

The **beta distribution** is generally used to describe the inherent variability in time estimates (see Figure 18–5). Although there is no real theoretical justification for using the beta distribution, it has certain features that make it attractive in practice: The distribution can be symmetrical or skewed to either the right or the left according to the nature of a particular activity; the mean and variance of the distribution can be readily obtained from the three time estimates listed above; and the distribution is unimodal with a high concentration of probability surrounding the most likely time estimate.

Of special interest in network analysis are the average or expected time for each activity t_e, and the variance of each activity time σ_i^2. The expected time of an activity, t_e, is a weighted average of the three time estimates:

$$t_e = \frac{t_o + 4t_m + t_p}{6} \tag{18–4}$$

The standard deviation of each activity's time is estimated as one-sixth of the difference between the pessimistic and optimistic time estimates. (Analogously, essentially all of the area under a normal distribution lies within three standard deviations of the mean, which is a range of six standard deviations.) We find the variance by squaring the standard deviation. Thus,

$$\sigma^2 = \left[\frac{(t_p - t_o)}{6}\right]^2 \quad \text{or} \quad \frac{(t_p - t_o)^2}{36} \tag{18–5}$$

The size of the variance reflects the degree of uncertainty associated with an activity's time: the larger the variance, the greater the uncertainty.

It is also desirable to compute the standard deviation of the expected time for *each path.* We can do this by summing the variances of the activities on a path and then taking the square root of that number; that is,

$$\sigma_{\text{path}} = \sqrt{\Sigma(\text{variances of activities on path})} \tag{18–6}$$

The expected time of a path is found by summing the expected times of the activities that are on the path.

Example 5 illustrates these computations.

Example 5

The network diagram for a project is shown in the accompanying figure, with three time estimates for each activity. Activity times are in weeks. Do the following:

a. Compute the expected time for each activity and the expected duration for each path.

b. Identify the critical path.

c. Compute the variance of each activity and the variance of each path.

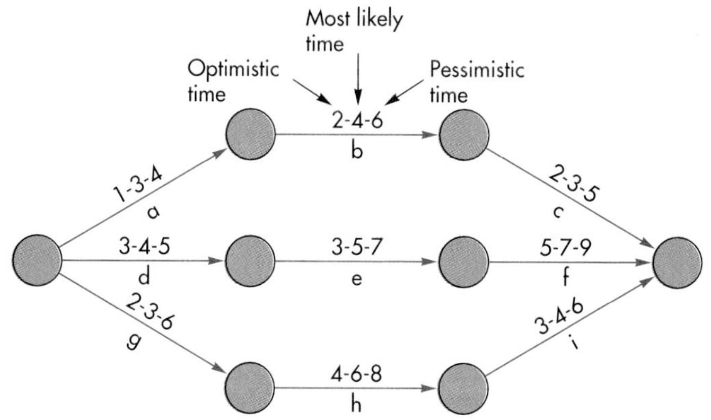

Solution

a.

Path	Activity	TIMES t_o	t_m	t_p	$t_e = \frac{t_o + 4t_m + t_p}{6}$	Path Total
	a	1	3	4	2.83	
a-b-c	b	2	4	6	4.00	10.00
	c	2	3	5	3.17	
	d	3	4	5	4.00	
d-e-f	e	3	5	7	5.00	16.00
	f	5	7	9	7.00	
	g	2	3	6	3.33	
g-h-i	h	4	6	8	6.00	13.50
	i	3	4	6	4.17	

b. The path that has the longest expected duration is the critical path. Because path d-e-f has the largest path total, it is the critical path.

c.

Path	Activity	TIMES t_o	t_m	t_p	$\sigma^2_{act} = \frac{(t_p - t_o)^2}{36}$	σ^2_{path}	σ_{path}
	a	1	3	4	$(4-1)^2/36 = 9/36$		
a-b-c	b	2	4	6	$(6-2)^2/36 = 16/36$	34/36 = 0.944	0.97
	c	2	3	5	$(5-2)^2/36 = 9/36$		
	d	3	4	5	$(5-3)^2/36 = 4/36$		
d-e-f	e	3	5	7	$(7-3)^2/36 = 16/36$	36/36 = 1.00	1.00
	f	5	7	9	$(9-5)^2/36 = 16/36$		
	g	2	3	6	$(6-2)^2/36 = 16/36$		
g-h-i	h	4	6	8	$(8-4)^2/36 = 16/36$	41/36 = 1.139	1.07
	i	3	4	6	$(6-3)^2/36 = 9/36$		

Knowledge of the expected path times and their standard deviations enables a manager to compute probabilistic estimates of the project completion time, such as these:

The probability that the project will be completed by a specified time.

The probability that the project will take longer than its scheduled completion time.

These estimates can be derived from the probability that various paths will be completed by the specified time.

FIGURE 18–6

The path probability is the area under a normal curve to the left of z

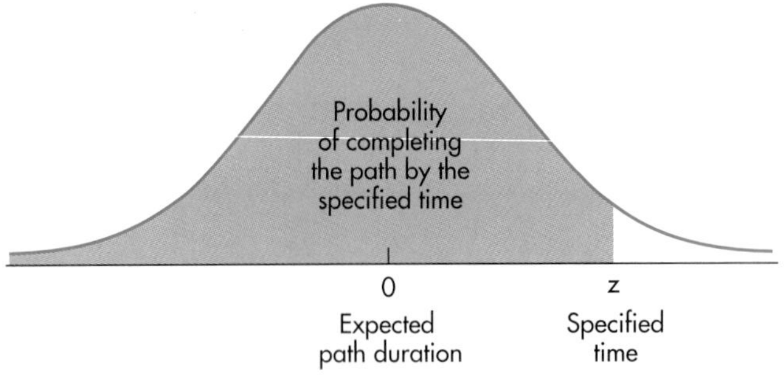

Determining Path Probabilities

The probability that a given path will be completed in a specified length of time can be determined using the following formula:

$$z = \frac{\text{Specified time} - \text{Path mean}}{\text{Path standard deviation}} \tag{18–7}$$

The resulting value of z indicates how many standard deviations of the path the specified time is beyond the expected path duration. (A negative value of z indicates that the specified time is *earlier* than the expected path duration.) Once the value of z has been determined, it can be used to obtain the probability that the path will be completed by the specified time from Appendix Table B. Note that the probability is equal to the area under the normal curve to the left of z, as illustrated in Figure 18–6.

If the value of z is $+2.50$ or more, the path probability is close to 100 percent (for $z = +2.50$ it is .9938). Hence, it is very likely the activities that make up the path will be completed by the specified time. For that reason, a useful rule of thumb is to treat the path probability as being equal to 100 percent if the value of z is $+2.50$ or more.

Rule of thumb: If the value of z is $+2.50$ or more, treat the probability of path completion by the specified time as 100 percent.

The rationale for using a normal distribution is that sums of random variables (activity times) will tend to be normally distributed, irrespective of the distributions of the variables. The normal tendency improves as the number of random variables increases. However, even when the number of items being summed is fairly small, the normal approximation provides a reasonable approximation to the actual distribution.

A project is not completed until *all* of its activities have been completed, not only those on the critical path. It sometimes happens that another path ends up taking more time to complete than the critical path, in which case the project runs longer than expected. Hence, it can be risky to focus exclusively on the critical path. Instead, one must consider the possibility that at least one other path will delay timely project completion. This requires determining the probability that *all* paths will finish by a specified time. To do that, find the probability that each path will finish by the specified time, and then multiply those probabilities. The result is the probability that the *project* will be completed by the specified time.

independence Assumption that path duration times are independent of each other; requiring that activity times be independent, and that each activity is on only one path.

It is important to note the assumption of **independence.** It is assumed that path duration times are independent of each other. In essence, this requires two things: Activity times are independent of each other, and each activity is only on one path. For activity times to be independent, the time for one must not be a function of the time of another; if two activities were always early or late together, they would not be considered independent. The assumption of independent *paths* is usually considered to be met if only a *few* activities in a large project are on multiple paths. Even then, common sense should govern the decision of whether the independence assumption is justified.

Example 6

Using the information from Example 5, answer the following questions:

a. Can the paths be considered independent? Why?

b. What is the probability that the project can be completed within 17 weeks of its start?

c. What is the probability that the project will be completed within 15 weeks of its start?

d. What is the probability that the project will *not* be completed within 15 weeks of its start?

Solution

a. Yes, the paths can be considered independent, since no activity is on more than one path and you have no information suggesting that any activity times are interrelated.

b. To answer questions of this nature, you must take into account the degree to which the path distributions "overlap" the specified completion time. This overlap concept is illustrated in the accompanying figure, which shows the three path distributions, each centered on that path's expected duration, and the specified completion time of 17 weeks.

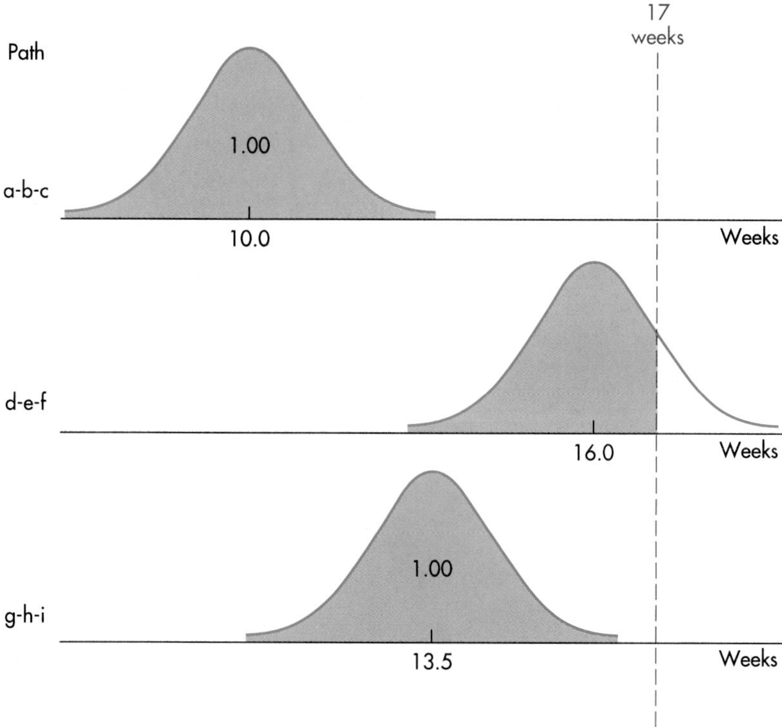

The shaded portion of each distribution corresponds to the probability that the part will be completed within the specified time. Observe that paths a-b-c and g-h-i are well enough to the left of the specified time, so that it is highly likely that both will be finished by week 17, but the critical path overlaps the specified completion time. In such cases, you need consider only the distribution of path d-e-f in assessing the probability of completion by week 17. To do so, you must first compute the value of z using Formula 18–7:

$$z = \frac{17 - 16}{1.00} = +1.00$$

Turning to Appendix Table B with $z = +1.00$, you will find that the area under the curve to the left of z is .8413. Hence, the probability of the project finishing within 17 weeks of its start is .8413.

In part *c* you will learn how to determine if a path's distribution overlaps the specified completion time.

c. Although the figure is useful in illustrating the concept of overlapping paths, you need a more rigorous approach to determine which paths to consider and what the probability of completion is for each path. This requires computing a value of z using Formula 18–7 for each path. Then, any path with a z of more than +2.50 is treated as having a completion probability of 100 percent (i.e., it does *not* overlap the specified time); the others will be used to determine the *joint* probability of all finishing by the specified time. For this problem, with a specified time of 15 weeks, the z values are:

Path	$z = \frac{15 - \text{Expected path duration}}{\text{Path standard deviation}}$	Probability of Completion in 15 Weeks
a-b-c	$\frac{15 - 10.00}{0.97} = +5.15$	1.0000
d-e-f	$\frac{15 - 16.00}{1.00} = -1.00$	.1587
g-h-i	$\frac{15 - 13.50}{1.07} = +1.40$	.9192

Paths d-e-f and g-h-i overlap because their z values are less than +2.50.

From Appendix Table B, the area to the *left* of $z = -1.00$ is .1587, and the area to the *left* of $z = +1.40$ is .9192. The path distributions are illustrated in the figure. The joint probability of all finishing before week 15 is the product of their probabilities: 1.00(.1587)(.9192) = .1459.

d. The probability of not finishing before week 15 is the complement of the probability obtained in part *c*: 1 − .1459 = .8541.

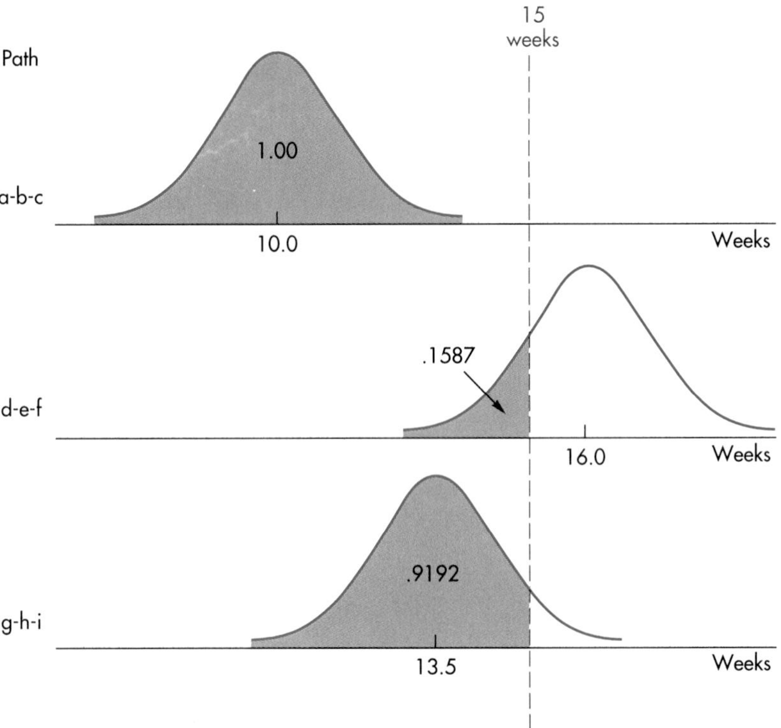

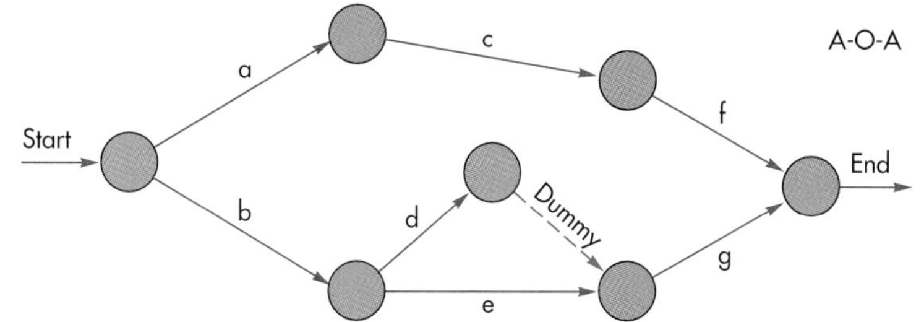

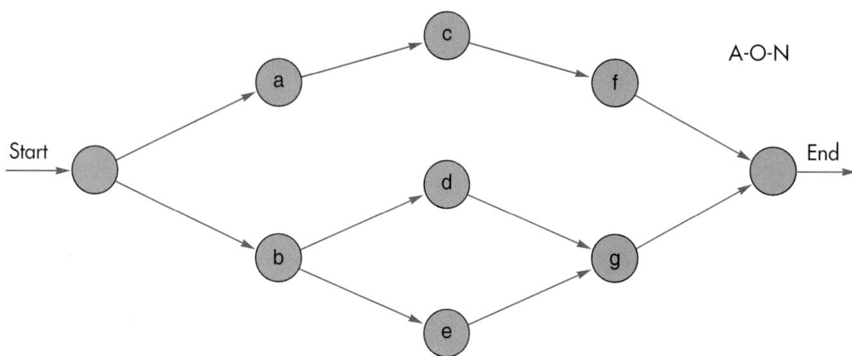

FIGURE 18–7

A comparison of the A-O-A and A-O-N methods of network representation

ACTIVITY-ON-NODE DIAGRAM

An alternate method for drawing network diagrams involves placing the activities on nodes (A-O-N) rather than on arrows. The result is a slightly different diagram. We can begin to get an appreciation for the differences by comparing the two approaches to a given problem. Consider this set of project activities:

Activity	Precedes
a	c
b	d, e
c	f
d	g
e	g
f	End
g	End

Figure 18–7 illustrates both types of network.

One obvious difference in the two approaches is that the A-O-N approach has more nodes. Generally, this will be the case. A second difference is that the A-O-A network has a dummy arrow that is necessary to correctly show the precedence relationship that exists. A-O-N networks eliminate the need for dummy activities, which is an advantage of having activities on nodes.

In practice, both approaches are used. Most PERT/CPM computer programs can handle either convention, but some do require one or the other. Often the choice of method depends more on personal preference or established procedures.

PROJECT MANAGEMENT SOFTWARE

The methods described in this chapter are sufficient for reasonably small projects. However, as the number of activities increases, computations and updating can become time-consuming, making a computerized approach highly desirable, even essential. There are many software packages to choose from, ranging from mainframe versions to minicomputer and PC versions. Because the market is changing rapidly, we will not discuss specific software here. However, for a discussion of how to choose project software, see D. S. Kezsbom, D. L. Schilling, and K. A. Edward, *Dynamic Project Management: A Practical Guide for Managers and Engineers,* published by John Wiley & Sons, New York, 1989.

The reality is that there is more to successfully managing a project than choosing computer software. The project manager must make countless decisions, including the key decisions noted earlier in the chapter.

SIMULATION

We have examined a method for computing the probability that a project would be completed in a specified length of time. That discussion assumed that the paths of the project were *independent*; that is, the same activities are not on more than one path. If an activity were on more than one path and it happened that the completion time for that activity far exceeded its expected time, all paths that included that activity would be affected and, hence, their times would not be independent. Where activities are on multiple paths, one must consider if the preceding approach can be used. For instance, if only a few activities are on multiple paths, particularly if the paths are *much* shorter than the critical path, that approach may still be reasonable. Moreover, for purposes of illustration, as in the text problems and examples, we treat the paths as being independent when, in fact, they may not be.

In practice, when *dependent* cases occur, project planners often use an approach called *simulation*. It amounts to a form of repeated sampling wherein many passes are made through the project network. In each pass, a randomly selected value for each activity time is made based on the characteristics of the activity's probability distribution (e.g., its mean, standard deviation, and distribution type). After each pass, the expected project duration is determined by adding the times along each path and designating the time of the longest path as the project duration. After a large number of such passes (e.g., several hundred), there is enough information to prepare a frequency distribution of the project duration times. Planners can use this distribution to make a probabilistic assessment of the actual project duration, allowing for some activities that are on more than one path. Problem 19 in the supplement to Chapter 19 illustrates this.

TIME–COST TRADE-OFFS: CRASHING

Estimates of activity times for projects usually are made for some given level of resources. In many situations, it is possible to reduce the length of a project by injecting additional resources. The impetus to shorten projects may reflect efforts to avoid late penalties, to take advantage of monetary incentives for timely or early competition of a project, or to free resources for use on other projects. In new product development, shortening may lead to a strategic benefit: beating the competition to the market. In some cases, however, the desire to shorten the length of a project merely reflects an attempt to reduce the indirect costs associated with running the project, such as facilities and equipment costs, supervision, and labor and personnel costs. Managers often have certain options at their disposal that will allow them to shorten, or **crash,** certain activities. Among the most obvious options are the use of additional funds to support additional personnel or more efficient equipment, and the relaxing of some work specifications. Hence, a project manager may be able to shorten a project by increasing *direct* expenses to speed up the project, thereby realizing savings on indirect project costs. The goal in evaluating time–cost trade-offs is to identify activities that will reduce the sum of the indirect and direct project costs.

crash Shortening activity durations.

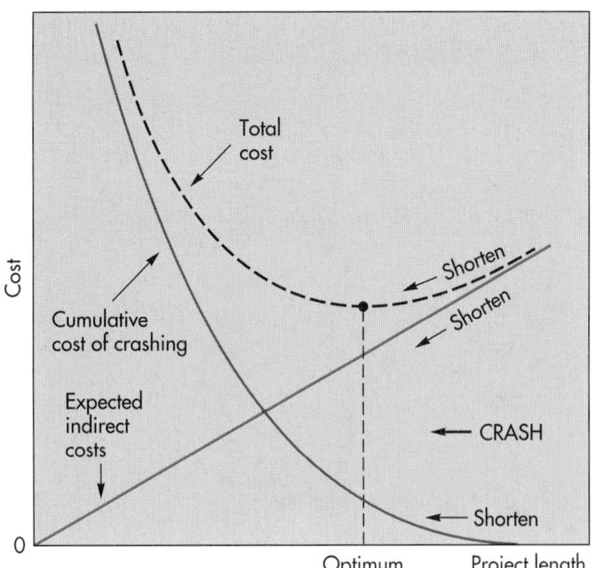

★ Crashing activities reduce indirect project costs and increase direct costs; the optimum amount of crashing results in minimizing the sum of these two types of costs.

FIGURE 18–8

*Crashing activities**

In order to make a rational decision on which activities, if any, to crash and on the extent of crashing desirable, a manager needs certain information:

1. Regular time and crash time estimates for each activity.
2. Regular cost and crash cost estimates for each activity.
3. A list of activities that are on the critical path.

Activities on the critical path are potential candidates for crashing, because shortening noncritical activities would not have an impact on total project duration. From an economic standpoint, activities should be crashed according to crashing costs: Crash those with the lowest crash costs first. Moreover, crashing should continue as long as the cost to crash is less than the benefits derived from crashing. These benefits might take the form of incentive payments for early project completion as part of a government contract, or they might reflect savings in the indirect project costs, or both. Figure 18–8 illustrates the basic relationship between indirect, direct, and total project costs due to crashing.

The general procedure for crashing is:

1. Obtain estimates of regular and crash times and costs for each activity.
2. Determine the lengths of all paths and path slack times.
3. Determine which are the critical activities.
4. Crash critical activities, in order of increasing costs, as long as crashing costs do not exceed benefits. Note that two or more paths may become critical as the original critical path becomes shorter, so that subsequent improvements will require simultaneous shortening of two or more paths. In some cases, it will be more economical to shorten an activity that is on two (or more) of the critical paths. This is true whenever the crashing cost of a joint activity is less than the sum of crashing one activity on each separate path.

Example 7

Using the following information, develop the optimal time–cost solution. Indirect project costs are $1,000 per day.

Activity	Normal Time	Crash Time	Cost per Day to Crash
a	6	6	—
b	10	8	$500
c	5	4	300
d	4	1	700
e	9	7	600
f	2	1	800

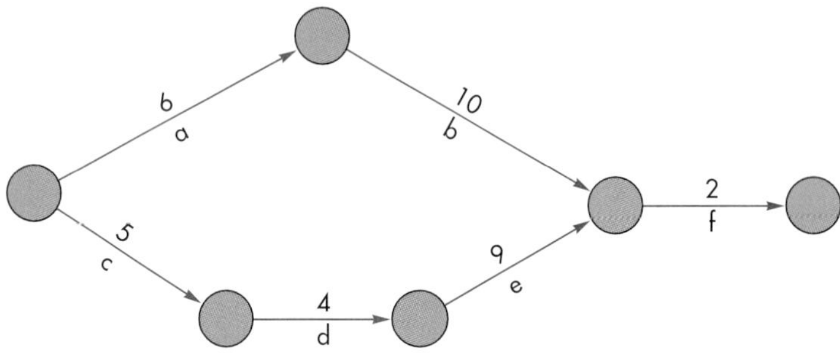

Solution

a. Determine which activities are on the critical path, its length, and the length of the other path:

Path	Length
a-b-f	18
c-d-e-f	20 (critical path)

b. Rank the critical path activities in order of lowest crashing cost, and determine the number of days each can be crashed.

Activity	Cost per Day to Crash	Available Days
c	$300	1
e	600	2
d	700	3
f	800	1

c. Begin shortening the project, one day at a time, and check after each reduction to see which path is critical. (After a certain point, another path may equal the length of the shortened critical path.) Thus:

(1) Shorten activity c one day at a cost of $300. The length of the critical path now becomes 19 days.

(2) Activity c cannot be shortened any more. Shorten activity e one day at a cost of $600. The length of path c-d-e-f now becomes 18 days, which is the same as the length of path a-b-f.

(3) The paths are now both critical, further improvements will necessitate shortening both paths.

The remaining activities for crashing and their costs are:

Path	Activity	Crash Cost (per day)
a-b-f	a	No reduction possible
	b	$500
	f	800
c-d-e-f	c	No further reduction possible
	d	$700
	e	600
	f	800

At first glance, it would seem that crashing f would not be advantageous, because it has the highest crashing cost. However, f is on both paths, so shortening f by one day would shorten *both* paths (and hence, the project) by one day for a cost of $800. The option of shortening the least expensive activity on each path would cost $500 for b and $600 for e, or $1,100. Thus shorten f by one day. The project duration is now 17 days.

(4) At this point, no additional improvement is feasible. The cost to crash b is $500 and the cost to crash e is $600, for a total of $1,100, and that would exceed the indirect project costs of $1,000 per day.

(5) The crashing sequence is summarized below:

	LENGTH AFTER CRASHING *n* DAYS:			
Path	*n* = 0	1	2	3
a-b-f	18	18	18	17
c-d-e-f	20	19	18	17
Activity crashed		c	e	f
Cost		$300	$600	$800

ADVANTAGES OF USING PERT AND POTENTIAL SOURCES OF ERROR

PERT and similar project scheduling techniques can provide important services for the project manager. Among the most useful features are these:

1. Use of these techniques forces the manager to organize and quantify available information and to recognize where additional information is needed.
2. The techniques provide a graphic display of the project and its major activities.
3. They identify (*a*) activities that should be closely watched because of the potential for delaying the project and (*b*) other activities that have slack time and so can be delayed without affecting project completion time. This raises the possibility of reallocating resources to shorten the project.

No analytical technique is without potential errors. Among the more important sources of errors are:

1. When developing the project network, managers may unwittingly omit one or more important activities.
2. Precedence relationships may not all be correct as shown.
3. Time estimates may include a fudge factor; managers may feel uncomfortable about making time estimates because they appear to commit themselves to completion within a certain time period.

Summary

Projects are composed of a unique set of activities established to realize a given set of objectives in a limited time span. The nonroutine nature of project activities places a set of demands on the project manager that are different in many respects from those the manager of more routine operations activities experiences, both in planning and coordinating the work and in the human problems encountered.

PERT and CPM are two commonly used techniques for developing and monitoring projects. Although each technique was developed independently and for expressly different purposes, time and practice have erased most of the original differences, so that now there is little distinction between the two. Either provides the manager with a rational approach to project planning and a graphical display of project activities. Both depict the sequential relationships that exist among activities and reveal to managers which activities must be completed on time to achieve timely project completion. Managers can use that information to direct their attention toward the most critical activities.

Two slightly different conventions can be used for constructing a network diagram. One designates the arrows as activities; the other designates the nodes as activities. To avoid confusion, this chapter emphasizes only the activity-on-arrow model.

The task of developing and updating project networks quickly becomes complex for projects of even moderate size, so computer programs, which involve the use of some computing algorithm, are often used.

A deterministic approach is useful for estimating the duration of a project when activity times can be fairly well established. When activity times are subject to some uncertainty, a probabilistic approach is more realistic, and estimates of the length of such projects should be couched in probabilistic terms.

In some instances, it may be possible to shorten, or crash, the length of a project by shortening one or more of the project activities. Typically, such gains are achieved by the use of additional resources, although in some cases, it may be possible to transfer resources among project activities. Generally, projects are shortened to the point where the cost of additional reduction would exceed the benefit of additional reduction, or to a specified time.

Key Terms

activities, 773
activity-on-arrow, 773
activity-on-node, 773
beta distribution, 782
CPM, 771
crash, 788
critical activities, 773
critical path, 773
deterministic, 775
events, 773
independence, 784
most likely time, 781
network diagram, 772
optimistic time, 781
path, 773
PERT, 771
pessimistic time, 781
precedence diagram, 772
probabilistic, 775
projects, 766
slack, 773
work breakdown structure, 770

Solved Problems

Problem 1

The following table contains information related to the major activities of a research project. Use the information to do the following:

a. Draw a precedence diagram.

b. Find the critical path.

c. Determine the expected length of the project.

Activity	Precedes	Expected Time (days)
a.	c, b	5
c	d	8
d.	i	2
b.	i	7
e.	f	3
f	m	6
i.	m	10
m	End	8
g	h	1
h.	k	2
k	End	17

Solution

a. In constructing networks, these observations can be useful.

(1) Activities with no predecessors are at the beginning (left side) of the network.

(2) Activities with multiple predecessors are located at path intersections.

Complete the diagram in sections. Go down the activity list, in order when possible, to avoid overlooking any activities.

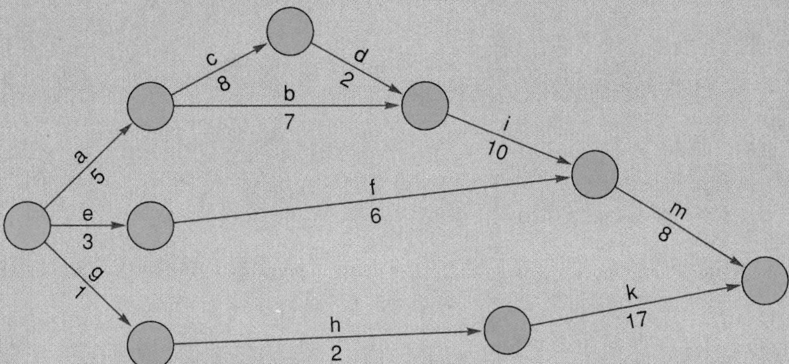

Here are some additional hints for constructing a precedence diagram.

(1) Use pencil.

(2) Start and end with a single node.

(3) Avoid having paths cross each other.

(4) Number nodes from left to right.

(5) Have activities go from left to right.

(6) Use only one arrow between any pair of nodes.

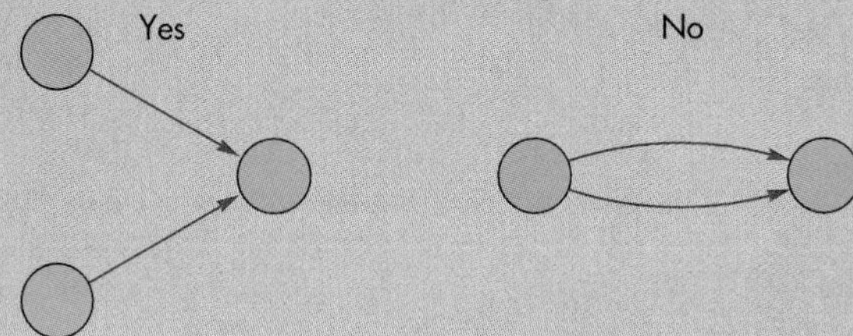

b. and *c.*

Path	Length (days)
a-c-d-i-m*	5 + 8 + 2 + 10 + 8 = 33†
a-b-i-m	5 + 7 + 10 + 8 = 30
e-f-m	3 + 6 + 8 = 17
g-h-k	1 + 2 + 17 = 20

*Critical path.

†Expected project duration.

Problem 2

Using the computing algorithm, determine the slack times for the following diagram. Identify the activities that are on the critical path.

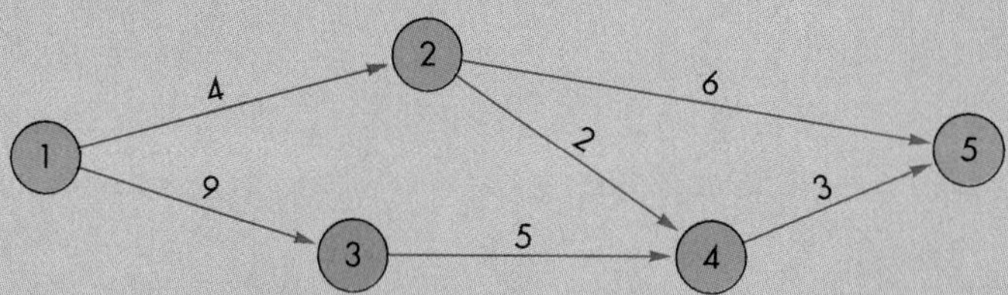

Solution

The task of determining ES, EF, LS, and LF times can be greatly simplified by setting up two brackets for each activity, as illustrated:

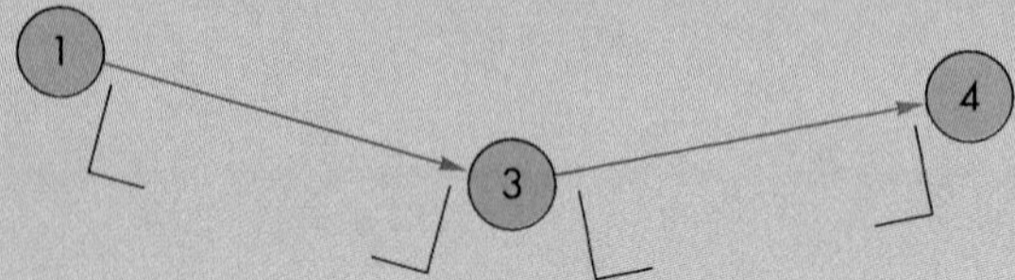

The bracket at the left of each activity will eventually be filled in with the earliest and latest *starting* times, and the bracket at the right end of each activity will be filled in with the earliest and latest *finishing* times:

This is accomplished in a two-step process. First, determine the earliest starting times and earliest finishing times, working from left to right, as shown in the following diagram.

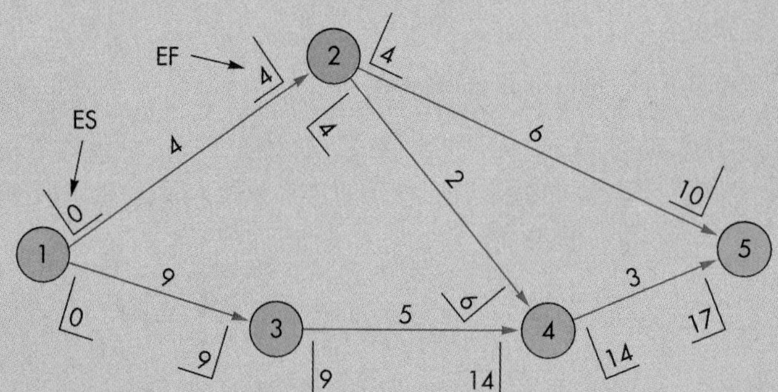

Thus, activity 1-2 can start at 0. With a time 4, it can finish at 0 + 4 = 4. This establishes the earliest start for all activities that begin at node 2. Hence, 2-5 and 2-4 can start no earlier than time 4. Activity 2-5 has an early finish of 4 + 6 = 10, and activity 2-4 has an early finish of 4 + 2 = 6. At this point, it is impossible to say what the earliest start is for 4-5; that will depend on which activity, 3-4 or 2-4, has the latest EF. Consequently, it is necessary to compute ES and EF along the lower path. Assuming an ES of 0 for activity 1-3, its EF will be 9, so activity 3-4 will have an ES of 9 and an EF of 9 + 5 = 14.

Considering that the two activities entering node 4 have EF times of 6 and 14, the earliest that activity 4-5 can start is the *larger* of these, which is 14. Hence, activity 4-5 has an ES of 14 and an EF of 14 + 3 = 17.

Now compare the EFs of the activities entering the final node. The largest of these, 17, is the expected project duration.

The LF and LS times for each activity can now be determined by working backward through the network (from right to left). The LF for the two activities entering node 5 is

17—the project duration. In other words, to finish the project in 17 weeks, these last two activities must both finish by that time.

In the case of activity 4-5, the LS necessary for an LF of 17 is 17 − 3 = 14. This means that both activities 2-4 and 3-4 must finish no later than 14. Hence, their LF times are 14. Activity 3 has an LS time of 14 − 5 = 9, making the LF of activity 1-3 equal to 9, and its LS equal to 9 − 9 = 0.

Activity 2-4, with an LF time of 14, has an LS time of 14 − 2 = 12. Activity 2-5 has an LF of 17 and therefore an LS of 17 − 6 = 11. Thus, the latest activity 2-5 can start is 11, and the latest 2-4 can start is 12 in order to finish by week 17. Since activity 1-2 precedes *both* of these activities, it can finish no later than the *smaller* of these, which is 11. Hence, activity 1-2 has an LF of 11 and an LS of 11 − 4 = 7.

The ES, EF, LF, and LS times are shown on the following network.

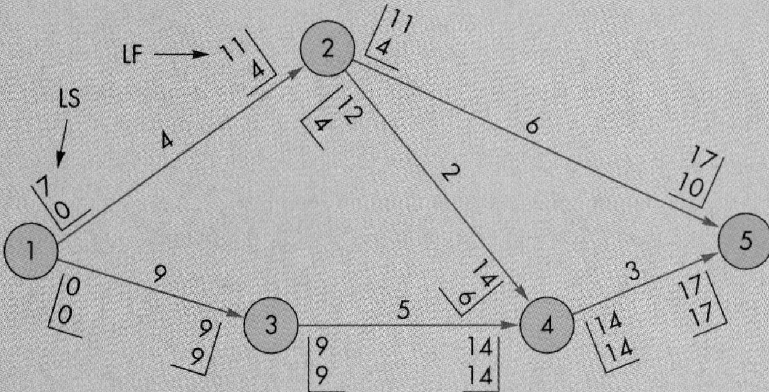

The slack time for any activity is the difference between *either* LF and EF *or* LS and ES. Thus,

Activity	LS	ES	Slack	or	LF	EF	Slack
1-2	7	0	7		11	4	7
2-5	11	4	7		17	10	7
2-4	12	4	8		14	6	8
1-3	0	0	0		9	9	0
3-4	9	9	0		14	14	0
4-5	14	14	0		17	17	0

The activities with zero slack times indicate the critical path. In this case the critical path is 1-3-4-5.

When working problems of this nature, keep in mind the following:

a. The ES time for leaving activities of nodes with multiple entering activities is the *largest* EF of the entering activities.

b. The LF for an entering activity for nodes with multiple leaving activities is the *smallest* LS of the leaving activities.

Problem 3

Expected times and variances for the major activities of an R&D project are depicted in the following PERT chart. Determine the probability that project completion time will be

a. Less than 50 weeks.

b. More than 50 weeks.

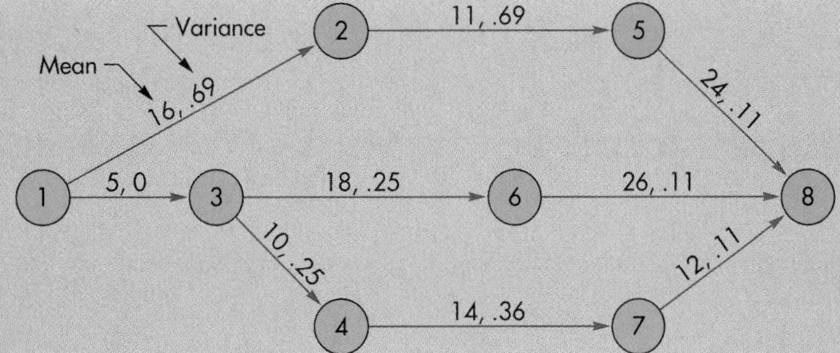

Solution

a. Compute the mean and standard deviation for each path:

Path	Expected Time (weeks)	Standard Deviation (weeks)
1-2-5-8	16 + 11 + 24 = 51	$\sqrt{.69 + .69 + .11} = 1.22$
1-3-6-8	5 + 18 + 26 = 49	$\sqrt{.00 + .25 + .11} = 0.60$
1-3-4-7-8	5 + 10 + 14 + 12 = 41	$\sqrt{.00 + .25 + .36 + .11} = 0.85$

b. Compute the z for each path for the length specified. For any path that has a z of +2.50 or more, treat its probability of completion before the specified time as 1.00. Use

$$z = \frac{50 - t_{\text{path}}}{\sigma_{\text{path}}}$$

The probability that each path will be completed in 50 weeks or less is shown in the corresponding diagram. (Probabilities are from Appendix Table B.) The probability that the project will be completed in 50 weeks or less depends on all three paths being completed in that time. Because z for path 1-3-4-7-8 is greater than +2.50, it is treated as having a probability of completion in 50 weeks of 100 percent. It is less certain that the other two paths will be completed in that time. The probability that *both* will not exceed 50 is the *product* of their individual probabilities of completion. Thus, .2061(.9525) = .1963.

The probability that the project *will* exceed 50 weeks is the complement of this number, which is 1.000 − .1963 = .8037. (Note that it is *not* the product of the path probabilities.)

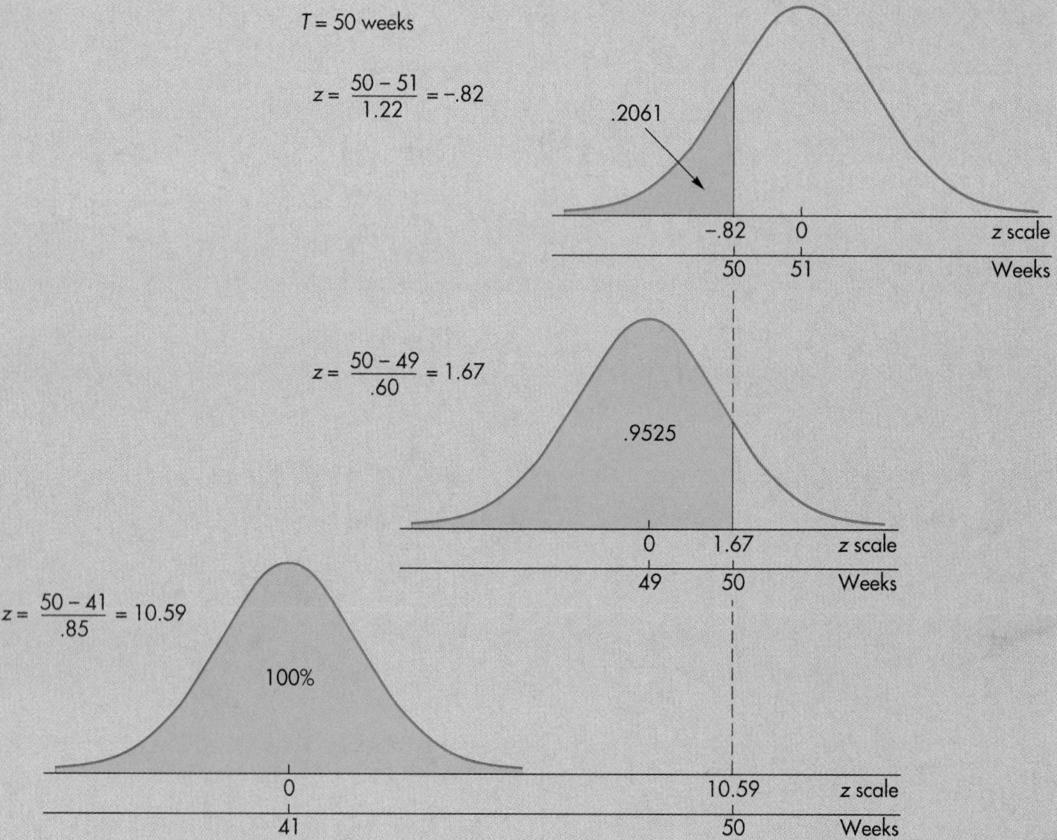

Problem 4

Indirect costs for a project are $12,000 per week for as long as the project lasts. The project manager has supplied the cost and time information shown. Use the information to:

a. Determine an optimum crashing plan.

b. Graph the total costs for the plan.

Activity	Crashing Potential (weeks)	Cost per Week to Crash
a	3	$11,000
b	3	3,000 first week, $4,000 others
c	2	6,000
d	1	1,000
e	3	6,000
f	1	2,000

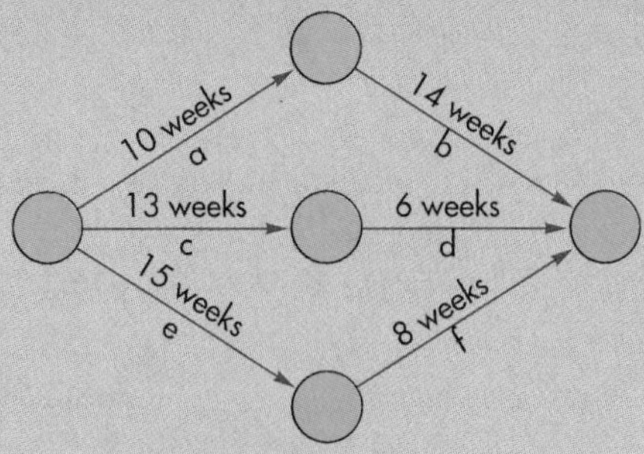

Solution

a. (1) Compute path lengths and identify the critical path:

Path	Duration (weeks)
a-b	24 (critical path)
c-d	19
e-f	23

(2) Rank critical activities according to crash costs:

Activity	Cost per Week to Crash
b	$ 3,000
a	11,000

Activity b should be shortened one week since it has the lower crashing cost. This would reduce indirect costs by $12,000 at a cost of $3,000, for a net savings of $9,000. At this point, paths a-b and e-f would both have a length of 23 weeks, so both would be critical.

(3) Rank activities by crashing costs on the two critical paths:

Path	Activity	Cost per Week to Crash
a-b	b	$ 4,000
	a	11,000
e-f	e	6,000
	f	2,000

Choose one activity (the least costly) on each path to crash: b on a-b and f on e-f, for a total cost of $4,000 + $2,000 = $6,000 and a net savings of $12,000 − $6,000 = $6,000.

(4) Check to see which path(s) might be critical: a-b and e-f would be 22 weeks in length, and c-d would still be 19 weeks.

(5) Rank activities on the critical paths:

Path	Activity	Cost per Week to Crash
a-b	b	$ 4,000
	a	11,000
e-f	e	6,000
	f	(no further crashing possible)

Crash b on path a-b and e on e-f for a cost of $4,000 + $6,000 = $10,000, for a net savings of $12,000 − $10,000 = $2,000.

(6) At this point, no further improvement is possible: paths a-b and e-f would be 21 weeks in length, and one activity from each path would have to be shortened. This would mean a at $11,000 and e at $6,000 for a total of $17,000, which exceeds the $12,000 potential savings in indirect costs.

b. The following table summarizes the results, showing the length of the project after crashing *n* weeks:

Path	$n = 0$	1	2	3
a-b	24	23	22	21
c-d	19	19	19	19
e-f	23	23	22	21
Activity crashed		b	b,f	b,e
Crashing costs ($000)		3	6	10

A summary of costs for the preceding schedule would look like this:

Project Length	Cumulative Weeks Shortened	Cumulative Crashing Costs ($000)	Indirect Costs ($000)	Total Costs ($000)
24	0	0	24(12) = 288	288
23	1	3	23(12) = 276	279
22	2	3 + 6 = 9	22(12) = 264	273
21	3	9 + 10 = 19	21(12) = 252	271
20	4	19 + 17 = 36	20(12) = 240	276

The graph of total costs is as follows.

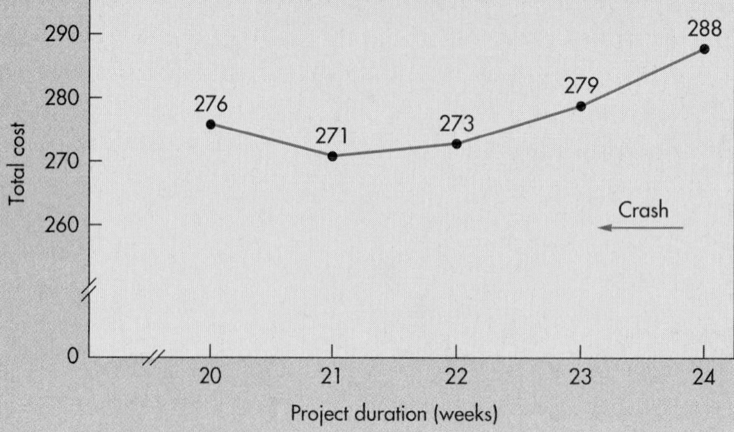

Discussion and Review Questions

1. What is a work breakdown structure, and how is it useful for project planning?
2. "A probabilistic approach to estimating project duration is generally preferred." Comment.
3. Identify the term being described for each of the following:
   a. A sequence of activities in a project.
   b. The longest time sequence of activities in a project.
   c. Used when two activities have the same starting and finishing points.
   d. The difference in time length of any path and the critical path.
   e. Used to denote the beginning or end of an activity.
   f. The statistical distribution used to describe variability of an activity time.
   g. The statistical distribution used to describe project variability.
   h. Shortening an activity by allocating additional resources.
4. Explain the concept of shared slack.
5. In what kind of situation might resources be *subtracted* from a project, in a kind of reverse time–cost trade-off? Explain briefly.
6. List the main advantages of PERT. List the main limitations.
7. What are dummy activities? Why are they used?
8. Why might a probabilistic estimate of a project's completion time based solely on the variance of the critical path be misleading? Under what circumstances would it be acceptable?
9. Define each of these terms, and indicate how each is determined.
   a. Optimistic time estimate.
   b. Most likely time estimate.
   c. Pessimistic time estimate.
   d. Expected activity time.
   e. Variance of an activity time.
   f. Standard deviation of a path's time.
10. Could PERT or CPM be used to schedule nonproject activities? Explain your reasoning briefly.
11. Why might a person wish to be associated with a critical path activity? What are some of the reasons one might have for not wanting this association?
12. What are some of the potential benefits of working on a special project in one's firm? What are some of the risks?
13. What are some aspects of the project manager's job that make it more demanding than the job of a manager working in a more routine organizational framework?
14. What is the main benefit of a project organization over more traditional forms of operations management for project work?

Memo Writing Exercises

1. Write a memo to your new employee, June Farber, in which you briefly describe the nature and importance of a work breakdown structure for project planning.
2. You have been assigned to work on a special project outside your department for the next two years. It has not been determined whether this assignment will be full-time or part-time. Write a memo to your supervisor, Henry Armbruster, outlining your reservations about each type of assignment.

Problems

Path	Time
1-2-4-7-10-12	23
1-2-5-8-10-12	24
1-3-6-9-11-12	31*

Path	Time
1-2-4-6-8-9	37*
1-2-5-6-8-9	30
1-2-5-7-8-9	35
1-3-7-8-9	30

Path	Time
1-2-5-12-16	44*
1-3-6-13-16	39
1-3-7-14-16	41
1-4-8-9-10-11-15-16 1-4-8-10-11-15-16	38

1. For each of the following network diagrams, determine both the critical path and the expected project duration. The quantities on the arrows represent expected activity times.

a.

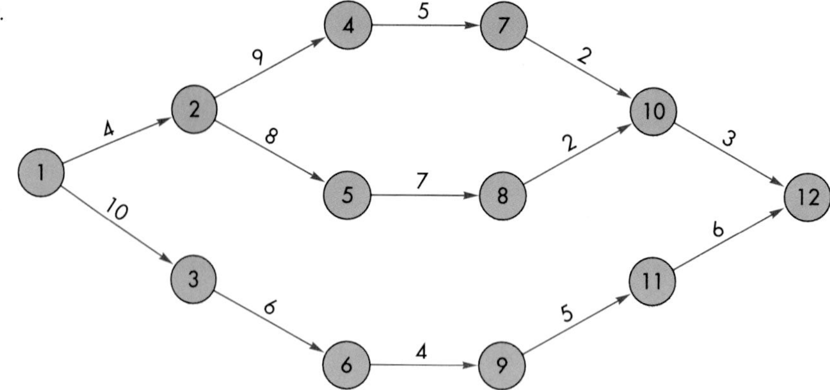

b.

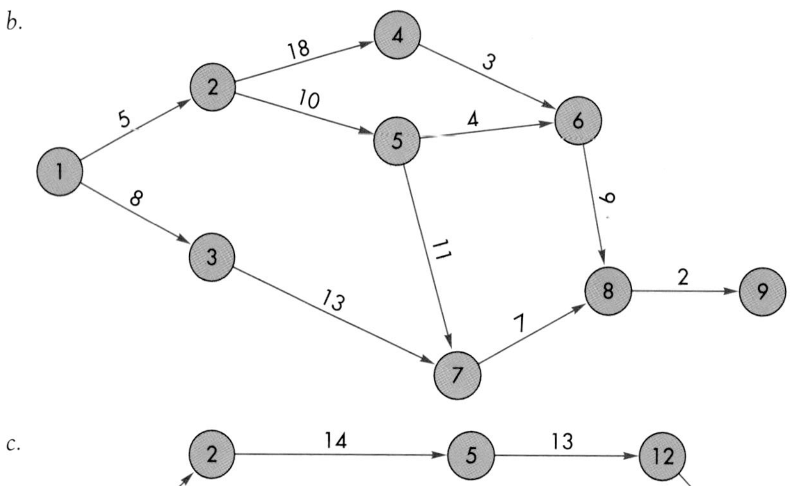

c.

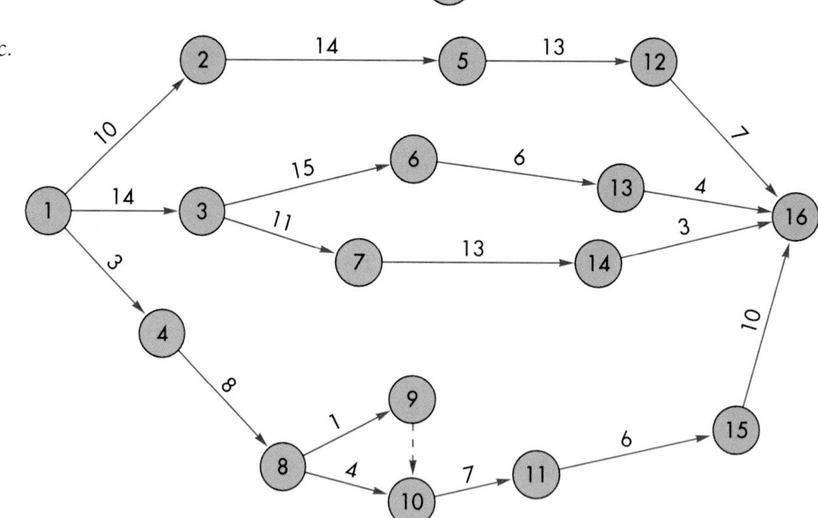

2. Chris received new word processing software for her birthday. She also received a check, with which she intends to purchase a new computer. Chris's college instructor has assigned a paper due next week. Chris has decided that she will prepare the paper on the new computer. She has made a list of the activities she will need to do and their estimated times.
   a. Arrange the activities into two logical sequences.
   b. Construct a network diagram that incorporates all of these activities.
   c. Determine the critical path and the expected duration time.
   d. What are some possible reasons for the project to take longer than the expected duration?

a. paper: Ch-Lib-Out-Write-Ck-Sub
computer: Sh-Sel-Inst

b. See IM.

c. paper; 7.5 hours

d. See IM.

Estimated Time (hrs.)	Activity (abbreviation)
0.8	Install software (Inst)
0.4	Outline the paper (Out)
0.2	Submit paper in to instructor (Sub)
0.6	Choose a topic (Ch)
0.5	Use grammar-checking routine and make corrections (Ck)
3.0	Write the paper using the word-processing software (Write)
2.0	Shop for a new computer (Sh)
1.0	Select and purchase computer (Sel)
2.0	Library research on chosen topic (Lib)

3. The information in the table pertains to a project that is about to commence. As the project manager, which activities would you be concerned with in terms of timely project completion? Explain.

See diagram in IM.

Concerned with activities f, h, j, and k which are on the critical path. Also a, b, c, and e.

Activity	Precedes	Estimated Time (days)
a	b	15
b	c, d	12
c	e	6
d	End	5
e	End	3
f	g, h	8
g	i	8
h	j	9
i	End	7
j	k	14
k	End	6

4. Construct a precedence diagram for each of the following cases. Note that each case requires the use of a dummy activity.

a. See diagram in IM.
b. See diagram in IM.

a. Activity	Precedes Activity	*b.* Activity	Precedes Activity
a	d	j	l, n
b	e, f	k	r
c	g	l	m
d	k	m	end
e	k	n	p
f	h, i	p	end
g	i	q	s, t
h	end	r	v
i	end	s	v
k	end	v	end
		t	v, w
		w	end

5. For each of the problems listed, determine the following quantities for each activity: the earliest start time, latest start time, earliest finish time, latest finish time, and slack. List the critical activities, and determine the expected duration of the project.

See IM.

a. Problem 1*a*.

b. Problem 1*b*.

c. Problem 3.

Project can be completed in a total of 30 weeks.

6. Reconsider the network diagram of Problem 1*a*. Suppose that after 12 weeks, activities 1-2, 1-3, and 2-4 have been finished; activity 2-5 is 75 percent finished; and activity 3-6 is half finished. How many weeks after the original start time should the project be finished?

a. See diagram in IM.
b. 24: .9686
21: .2350
c. 24: .9328
21: .0186

7. Three recent college graduates have formed a partnership and have opened an advertising firm. Their first project consists of activities listed in the following table.

   a. Draw the precedence diagram.

   b. What is the probability that the project can be completed in 24 days or less? In 21 days or less?

   c. Suppose it is now the end of the seventh day and that activities a and b have been completed while activity d is 50 percent completed. Time estimates for the completion of activity d are 5, 6, and 7. Activities c and h are ready to begin. Determine the probability of finishing the project by day 24 and the probability of finishing by day 21.

		TIME IN DAYS		
Activity	**Precedes**	**Optimistic**	**Most Likely**	**Pessimistic**
a	c	5	6	7
b	h	8	8	11
c	e	6	8	11
d	f	9	12	15
e	End	5	6	9
f	g	5	6	7
g	End	2	3	7
h	i	4	4	5
i	End	5	7	8

a. .6881
b. .3984
c. .0204

8. The new director of special events at a large university has decided to completely revamp graduation ceremonies. Toward that end, a PERT chart of the major activities has been developed. The chart has five paths with expected completion times and variances as shown in the table. Graduation day is 16 weeks from now. Assuming the project begins now, what is the probability that the project will be completed before:

   a. Graduation time?

   b. The end of week 15?

   c. The end of week 13?

Path	**Expected Duration (weeks)**	**Variance**
A	10	1.21
B	8	2.00
C	12	1.00
D	15	2.89
E	14	1.44

1 − .7291(.8944) = .3479

9. What is the probability that the following project will take more than 10 weeks to complete if the activity means and standard deviations are as shown below:

Activity	Mean	Standard Deviation
1-2	5	1.3
2-3	4	1.0
1-3	8	1.6

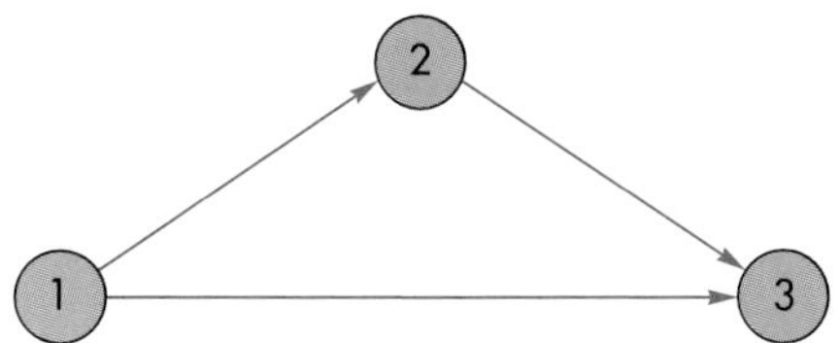

10. The project described in the following table has just begun. It is scheduled to be completed in 11 weeks.

    a. If you were the manager of this project, would you be concerned? Explain.

    b. If there is a penalty of $5,000 a week for each week the project is late, what is the probability of incurring a penalty of at least $5,000?

a. Yes. The probability of finishing on schedule is only about .27.
b. 1 − .9599(.5000) = .52

Activity	Estimated Time (weeks)	Standard Deviation (wks.)
1-2	4	0.70
2-4	6	0.90
1-3	3	0.62
3-4	9	1.90

11. The following precedence diagram reflects three time estimates for each activity. Determine:

    a. The expected completion time for each path and its variance.

    b. The probability that the project will require more than 49 weeks.

    c. The probability that the project can be completed in 46 weeks or less.

a.

Path	Time	Std. Dev.
1-2-3-8-11	37.34	1.03
1-2-4-6-11	43.00	2.26
1-2-4-7-11	40	0.90
1-2-5-9-10-11	46.83	1.12

b. .0262
c. .2085

1-2: 8-8-8
2-3: 11-12-13
2-4: 5-6-7
2-5: 10-12-14
3-8: 9-10-12
4-6: 14-18-26
4-7: 13-13-13
5-9: 7-10-12
6-11: 8-10-14
7-11: 10.5-13-15.5
8-11: 5-7-10
9-10: 10-11-12
10-11: 6-6-6

12. A project manager has compiled a list of major activities that will be required to install a computer information system in her firm. The list includes estimated completion times for activities and precedence relationships.

Path	Mean	Std. Dev.
a-d-e-h	24.34	1.354
a-f-g	15.50	1.258
b-i-j-k	14.83	1.014
c-m-n-o	26.17	1.658

P($1000): .4099
P($500): .6742

Activity	Precedes	Estimated Times (weeks)
a	d, f	2-4-6
d	e	6-8-10
e	h	7-9-12
h	End	2-3-5
f	g	3-4-8
g	End	5-7-9
b	i	2-2-3
i	j	2-3-6
j	k	3-4-5
k	End	4-5-8
c	m	5-8-12
m	n	1-1-1
n	o	6-7-11
o	End	8-9-13

If the project is finished within 26 weeks of its start, the project manager will receive a bonus of $1,000; and if the project is finished within 27 weeks of its start, the bonus will be $500. Find the probability of each bonus.

Diagram in IM.
Diagram in IM.

13. *a.* Construct an activity-on-node diagram for the set of activities listed in Problem 3.

    b. Construct an activity-on-node diagram for the set of activities listed in Problem 12.

Diagram in IM. Crash schedule: (1 week each): c, c, f, f, e, p

14. The project manager of a task force planning the construction of a domed stadium had hoped to be able to complete construction prior to the start of the next college football season. After reviewing construction time estimates, it now appears that a certain amount of crashing will be needed to ensure project completion before the season opener. Given the following time and cost estimates, determine a minimum-cost crashing schedule that will shave five weeks off the project length.

Activity	Precedes	Normal Time (weeks)	CRASHING COSTS	
			First Week	Second Week
a	b	12	$15,000	$20,000
b	k	14	10,000	10,000
c	d, e, f	10	5,000	5,000
d	g	17	20,000	21,000
e	h	18	16,000	18,000
f	i	12	12,000	15,000
g	m	15	24,000	24,000
h	n, p	8	—	—
i	j	7	30,000	—
j	p	12	25,000	25,000
k	End	9	10,000	10,000
m	End	3	—	—
n	End	11	40,000	—
p	End	8	20,000	20,000

15. A construction project has indirect costs totaling $40,000 per week. Major activities in the project and their expected times are shown in this precedence diagram:

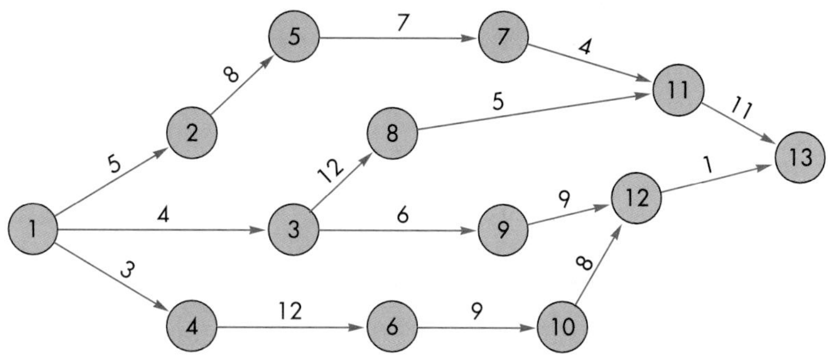

a. Crash 4 weeks.
(1) 7-11, (2) 1-2, (3) 7-11 and 6-10, (4) 11-13 and 4-6, (5) 11-13, (6) 1-3 and 6-10.
b. Plot in IM.

Crashing costs for each activity are:

Activity	CRASHING COSTS ($000) First Week	Second Week	Third Week
1-2	$18	$22	$—
2-5	24	25	25
5-7	30	30	35
7-11	15	20	—
11-13	30	33	36
1-3	12	24	26
3-8	—	—	—
8-11	40	40	40
3-9	3	10	12
9-12	2	7	10
12-13	26	—	—
1-4	10	15	25
4-6	8	13	—
6-10	5	12	—
10-12	14	15	—

a. Determine the optimum time–cost crashing plan.

b. Plot the total-cost curve that describes the least expensive crashing schedule that will reduce the project length by six weeks.

16. Chuck's Custom Boats (CCB) builds luxury yachts to customer order. CCB has landed a contract with a mysterious New York doctor (Mr. T). Relevant data are shown below. The complication is that Mr. T wants delivery in 32 weeks or he will impose a penalty of $375 for each week his yacht is late.

Project Length	Shorten Activity	Crash Cost
39 wk	—	0
38	Z	90
37	N, L	170
36	Q	200
35	Q	225
34	M, N	345

Stop crashing here.

Activity	Precedes	Normal Time (weeks)	CRASHING COSTS 1st Week	2nd Week
K	L, N	9	$410	$415
L	M	7	125	—
N	J	5	45	45
M	Q	4	300	350
J	Q	6	50	—
Q	P, Y	5	200	225
P	Z	8	—	—
Y	End	7	85	90
Z	End	6	90	—

Develop a crashing schedule.

CASE

THE CASE OF THE MEXICAN CRAZY QUILT

"The mission of the project which you will head is to get our new Mexican subsidiary ready for take-over by Mexican managers. My hope is that you will be able to do this in about two years," explained Robert Linderman, president of Linderman Industries, Inc., to Carl Conway, newly appointed manager for "Operation Mexicano." Conway had been hired specifically for this assignment because of his experience in managing large defense projects in the aerospace industry.

"The first thing that I will have to do is put a project team together," said Conway. "I imagine that you have in mind my drawing people from the functional divisions."

"Yes, I have already sent memoranda to the division managers informing them that you will be asking for some of their key people to work under you for about two years," said Linderman. "In addition, I have advised them to be prepared to process work orders from Operation Mexicano with the personnel and equipment of their organizations. Later on in the project's life, you will begin to get Mexican personnel, both managers and technicians, into your organization. These people will have Mexican supervisors, but until the mission is accomplished, they also will report to you. I will have to admit that you are going to have some complex authority relationships, especially as you personally will be responsible to the president of the subsidiary, Felix Delgado, as well as to me."

Conway began to make his plans for the project team. The plant building was available and empty in Mexico City, and it was important to get equipment purchased and installed as soon as possible. A plant layout would have to be prepared, but before that could be done there would have to be a manufacturing plan. Therefore, he needed to recruit an industrial engineer, a production planner, and an equipment buyer. They, in turn, would have to build their own staffs.

He made an appointment with Sam Sargis, corporate manager of industrial engineering. "I have had a preliminary talk with Bob Cates about his joining Operation Mexicano, and he is quite interested," Carl said. "Will you release him to me?"

"Why, I'm grooming Cates to take over my job when I retire," replied Sargis. "He is my best man. Let me pick someone else for you, or better still, you just tell me what industrial engineering work you want done, and I will have it done for you."

"Sorry, I want Cates," said Carl firmly. "And besides, you are not due to retire for five years. This will be good experience for him."

For production planning, Carl had in mind Bert Mill, an older man with extensive experience in managing production operations, but Mill rejected his offer. "I talked it over with my wife," he said, "and we feel that at my age I shouldn't take a chance on not having a job to come back to when Operation Mexicano is finished."

Carl next talked to Emil Banowetz, who was assistant to Jim Burke, the vice president for manufacturing, and Banowetz decided that he would like to join the project team. However, Burke told Conway that if Banowetz were forcibly taken away from him, he would give Mr. Linderman his resignation, so Carl decided to back down. He finally accepted a man that Burke recommended.

Filling the equipment buyer's slot was not easy. The director of procurement phoned Carl and said that a senior buyer, Humberto Guzman, had requested permission to ask for the assignment, and that he strongly recommended him. Guzman had been purchasing agent for a large mining company in Mexico for about 10 years.

Carl had about the same experiences in getting the people he wanted for the functions of engineering, quality control, cost, marketing, and advertising as he did for the first three positions; in other words, he won some confrontations with the division managers and lost some. For personnel, he got Dr. Juan Perez, who was slated to be personnel director of the subsidiary company, to affiliate temporarily with the project team.

The first brush that Project Mexicano had in getting a functional division to do work for it came when Carl's engineering man, Frank Fong, reported to him that the engineering vice president, who was formerly Fong's boss, refused to authorize top priority to the changing of dimensions in the production drawings to the metric system. Carl had to take this issue to Linderman, who ruled in his favor. The defeated vice president, of course, did not take kindly to the decision.

The next incident revolved about Carl's desire to have a pilot run of products made with metric measurements for shipment to Mexico. The purpose was to test the market acceptance of the Linderman articles. Jim Burke stated flatly that there was no way that his production workers could be trained to work with metric drawings. Carl quickly saw that this was an issue that he was not going to win, so he had his buyer, Guzman, work with the newly appointed manufacturing manager for the subsidiary in getting a run of the products subcontracted in Mexico City.

Bob Cates made a special trip from Mexico City to present Carl with an interesting problem. The Mexican industrial engineer, whom Bob was supposed to be training, had his own ideas about plant layout. When they differed from Bob's as they usually did, he would take his complaint directly to Felix Delgado, the president of the Mexican subsidiary. Because Delgado's competence was primarily in finance, he would not know how to decide the argument and would simply table it. Carl took examples of some of the disagreements to Bob's former boss, Sam Sargis, who quite unexpectedly ruled against Bob's proposed methods. Carl saw that there was bad feeling by Sargis against Bob for leaving his department, which boded ill for Bob's return. To solve the immediate problem, however, Carl asked Dr. Perez to try to reconcile the situation in Mexico City.

Despite these problems, and many more of a similar nature, Project Mexicano was successful, and the transition to Mexican management was made in just a little over two years. By a curious twist, through Dr. Perez's intercession Felix Delgado became very impressed by Bob Cates and convinced him to accept the job of director of industrial engineering for the Mexican company. Humberto Guzman also stayed on to head the procurement operation.

Other members of the project team were not so fortunate. Linderman Industries was laying off personnel when the project ended, and only the project production man was able to get a job in the company at as high a level as the one he had when he joined the team. The cost expert elected to leave Linderman because he said the glamour of Project Mexicano had spoiled him for any routine job.

Carl Conway had a difficult decision of his own to make. Robert Linderman said that he was extremely pleased with his performance and that something good would open up in the company for him soon. In the meantime, there was a staff assignment available for him. Carl had seen enough project managers in the aerospace industry who had figuratively rotted on staff assignments when their projects were completed to be somewhat wary.

Questions

1. Was Linderman Industries' adoption of project organization an appropriate one for getting the Mexican subsidiary started?
2. In consideration of Robert Linderman's letting the division managers know that the project manager would be asking for some of their key people, why would Conway have any difficulty in getting the ones he wanted?
3. Would you expect that many people would turn down a chance to join a project organization, as Bert Mill did?
4. Why would Conway take his problem with the engineering vice president to Linderman and have it resolved in his favor, yet back down in two disputes with the manufacturing vice president?
5. What could Linderman Industries have done to assure good jobs for the people coming off Project Mexicano, including Carl Conway, the project manager?

Source: *Management: The Key to Organizational Effectiveness*, rev. ed. by Clayton Reeser and Marvin Loper.

CASE
FANTASY PRODUCTS

Company Background

The Fantasy Product Company is a manufacturer of high-quality small appliances intended for home use. Their current product line includes irons, a small hand-held vacuum, and a number of kitchen appliances such as toasters, blenders, waffle irons, and coffeemakers. Fantasy Products has a strong research and development department that continually searches for ways to improve existing products as well as developing new products.

Currently, the research and development department is working on the development of a new kitchen appliance that will chill foods quickly much as a microwave oven heats them quickly, although the technology involved is quite different. Tentatively named The Big Chill, the product will initially carry a price tag of around $125 and therefore the target market consists of upper income groups. At this price, it is expected to be a very profitable item. R&D engineers now have a working prototype and are satisfied that, with cooperation from the production and marketing people, the product can be ready in time for the all important Christmas buying season. A target date has been set for product introduction that is 24 weeks away.

Current Problem

Fantasy Products Marketing Vice President Vera Sloan has recently learned from reliable sources that a competitor is also in the process of developing a similar product, which it intends to bring out at almost exactly the same time. In addition, her source indicated that the competitor plans to sell its product, which will be smaller than The Big Chill, for $99 in the hope of appealing to upper-middle as well as upper income groups. Vera, with the help of several of her key people who are to be involved in marketing The Big Chill, has decided that in order to compete, the selling price for The Big

Chill will have to be lowered to within $10 of the competitor's price. At this price level, it will still be profitable, although not nearly as profitable as originally anticipated.

However, Vera is wondering whether it would be possible to expedite the usual product introduction process in order to beat the competition to the market. If possible, she would like to get a six-week jump on the competition; this would put the product introduction date only 18 weeks away. During this initial period, Fantasy Products could sell The Big Chill for $125, reducing the selling price to $109 when the competitor's product actually enters the market. Since forecasts based on market research show that sales during the first six weeks will be about 2,000 per week, there is an opportunity for considerable extra profit if the early introduction can be accomplished. In addition, there is a certain amount of prestige involved in being first to the market. This should help enhance The Big Chill's image during the anticipated battle for market share.

Data Collection

Since Fantasy Products has been through the product-introduction process a number of times, the R&D department has developed a list of the tasks that must be accomplished and the order in which they must be completed. Although the times and costs vary depending on the particular product, the basic process does not. The list of activities involved and their precedence relationships are presented in Table 1. Time and cost estimates for the introduction of The Big Chill are presented in Table 2. Note that some of the activities can be completed on a crash basis, with an associated increase in cost.

Questions

Fantasy Products needs to decide whether to bring The Big Chill to market 18 weeks from now as Vera Sloan is recommending. As the project management specialist in the R&D department, you have been asked to answer the following questions:

1. When would the project be completed using normal times?
2. Is it possible to complete the project in 18 weeks? What would the associated additional cost be? Which activities would need to be completed on a crash basis?
3. Is the additional cost justified in terms of the increased profits expected?
4. The estimated demand is very uncertain; how much can this number vary without changing the recommendations you are making?
5. Is there some time frame other than the 18 weeks Vera has recommended that would make more sense in terms of profits?

Source: Used with permission from Robert J. Thieraus, Margaret Cunningham, and Melanie Blackwell, Xavier University, Cincinnati, Ohio.

TABLE 1

List of activities and precedence relationships

Activity	Description	Immediate Predecessor
A	Select and order equipment	—
B	Receive equipment from supplier	A
C	Install and set up equipment	A
D	Finalize bill of materials	B
E	Order component parts	C
F	Receive component parts	E
G	First production run	D, F
H	Finalize marketing plan	—
I	Produce magazine ads	H
J	Script for TV ads	H
K	Produce TV ads	J
L	Begin ad campaign	I, K
M	Ship product to consumers	G, L

Activity	Normal Time	Normal Cost	Crash Time	Crash Cost
A	3	$ 2,000	2	$ 4,500
B	8	9,000	6	12,000
C	4	2,000	2	7,000
D	2	1,000	1	2,000
E	2	2,000	1	3,000
F	5	0	5	0
G	6	12,000	3	24,000
H	4	3,500	2	8,000
I	4	5,000	3	8,000
J	3	8,000	2	15,000
K	4	50,000	3	70,000
L	6	10,000	6	10,000
M	1	5,000	1	5,000

TABLE 2

Time and cost estimates

Selected Bibliography

Bierman, Harold, Jr.; Charles P. Bonini; and Warren H. Hausman. *Quantitative Analysis for Business Decisions.* 8th ed. Burr Ridge, IL: Richard D. Irwin, 1991.

Kerzner, Harold. *Project Management for Executives.* New York: Van Nostrand Reinhold, 1984.

Levin, Richard I.; Charles A. Kirkpatrick; and David S. Rubin. *Quantitative Approaches to Management.* 5th ed. New York: McGraw-Hill, 1982.

Meredith, Jack R., and Samuel J. Mantel, Jr. *Project Management: A Managerial Approach.* New York: John Wiley & Sons, 1985.

Moder, J. E.; E. W. Davis; and C. Phillips. *Project Management with CPM and PERT.* New York: Van Nostrand Reinhold, 1983.

Peterson, P. "Project Management Software Survey." *PMNETwork* 8, no. 5 (May 1994), pp. 33–41.

Rogers, T. "Project Management Emerging as a Requisite for Success." *Industrial Engineering,* June 1993, pp. 42–43.

LEARNING OBJECTIVES

After completing this chapter, you should be able to:

1. Explain why waiting lines form in underloaded systems.
2. Identify the goal of queuing (waiting-line) analysis.
3. List the measures of system performance that are used in queuing.
4. Discuss the assumptions of the basic queuing models presented.
5. Solve typical problems.

CHAPTER NINETEEN
WAITING LINES

CHAPTER OUTLINE

READING

WAITING—A NEW POPULAR PASTIME: MISS MANNERS **Judith Martin**

Many things in life are worth waiting for, but not all that long. Miss Manners would put a time limit on how long one should wait for salespeople to finish their conversations with each other before writing up one's order, or for a spouse who has departed with someone else to realize what a terrible mistake that was.

Nevertheless, waiting is now in a class with working as a popular pastime. A *waitologist* has estimated that the average adult spends one tenth of his or her waking moments waiting, at a minimum. There are waits for buses, banks, stores, theaters, gas stations, court cases, elevators, driver's licenses, and dentist appointments.

One could easily pass one's life enduring just such basic waits. But there are also intermediary waits, such as waiting for the rain to stop, and advanced waits, such as waiting for your ship to come in. Some of these go in fashions. There was a time when all of America was waiting to be discovered by a movie talent scout in a drugstore, and now everyone is waiting for a television camera to come along and ask her to tell the world what she thinks.

It is the elementary and comparatively short-term wait with which Miss Manners is concerned. If you want to hear about the others, you will just have to wait.

It is perfectly correct, although not many people realize it, to refuse to wait on the telephone. When Miss Manners is asked "Can you hold on for a minute?" she often replies, "No," and it is too bad that the person on the other end ties up his own line by putting her on hold anyway, because that person has not waited for Miss Manners' reply.

One should also refuse to wait for inefficient or indefinite service. A restaurant should be able to tell you how long the wait will be, and a service person should not keep you waiting except to attend a previous customer.

It is rude to refuse to wait by announcing that one's needs take precedence over those of other waiting people. Miss Manners can think of no circumstances in which a person transacting the ordinary business of life can plead with legitimacy that it is more outrageous to expect her to wait than to expect it of others. "Let me go through, please—I'm in labor," perhaps, but then what are you doing at the stockings sale, anyway?

The only polite way to wait, if one must do so, is to bring one's own portable work or amusement. An unoccupied person waiting in line is by definition a potential raving maniac. A nice Jane Austen novel ready-to-go has preserved even the naturally tranquil spirits of Miss Manners. Even using conversations as a means to pass the time is dangerous, in Miss Manners' opinion. Two people quietly discussing what a shame it is to have to wait are, by the same definition, a potential mob.

The "Miss Manners" article pokes fun at one of life's realities: having to wait in line. No doubt those waiting in line would all agree that the solution to the problem is obvious: simply add more servers or else do something to speed up service. Although both ideas may be potential solutions, there are certain subtleties that must be dealt with. For one thing, most service systems have the capacity to process more customers over the long run than they are called on to process. Hence, the problem of customers waiting is a short-term phenomenon. The other side of the coin is that at certain times the system is empty, and servers are idle, waiting for customers. Thus, by increasing the service capacity, the server idle time would increase even more. Consequently, in designing service systems, the designer must weigh the cost of providing a given level of service capacity against the potential (implicit) cost of having customers wait for service. This planning and analysis of service capacity frequently lends itself to **queuing theory,** which is a mathematical approach to the analysis of waiting lines.

queuing theory Mathematical approach to the analysis of waiting lines.

The foundation of modern queuing theory is based on studies about automatic dialing equipment made in the early part of this century by Danish telephone engineer A. K. Erlang. Prior to World War II, very few attempts were made to apply queuing theory to business problems. Since that time, queuing theory has been applied to a wide range of problems.

The mathematics of queuing can be complex; for that reason, the emphasis here will not be on the mathematics but the concepts that underlie the use of queuing in analyzing waiting-line problems. We shall rely on the use of formulas and tables for analysis.

Waiting lines are commonly found wherever customers arrive *randomly* for services. Some examples of waiting lines we encounter in our daily lives include the lines at supermarket checkouts, fast-food restaurants, airport ticket counters, theaters, post offices, and toll booths. In many situations, the "customers" are not people but orders waiting to be filled, trucks waiting to be unloaded, jobs waiting to be processed, or equipment awaiting repairs. Still other examples include ships waiting to dock, planes waiting to land, hospital patients waiting for a nurse, and cars waiting at a stop sign.

One reason that queuing analysis is important is that customers regard waiting as a non-value-added activity. Customers may tend to associate this with poor service quality, especially if the wait is long. Similarly, in an organizational setting, having work or employees wait is non-value-added—the sort of waste that workers in JIT systems strive to reduce.

Our discussion of queuing begins with an examination of what is perhaps the most fundamental issue in waiting-line theory: Why is there waiting?

WHY IS THERE WAITING?

Many people are surprised to learn that waiting lines tend to form even though a system is basically underloaded. For example, a fast-food restaurant may have the capacity to handle an average of 200 orders per hour and yet experience waiting lines even though the average number of orders is only 150 per hour. The key word is *average*. In reality, customers arrive at random intervals rather than at evenly spaced intervals, and some orders take longer to fill than others. In other words, both arrivals and service times exhibit a high degree of variability. As a result, the system at times becomes temporarily overloaded, giving rise to waiting lines; at other times, the system is idle because there are no customers. Thus, although a system may be *underloaded* from a *macro* standpoint, variabilities in arrivals and service mean that at times the system is *overloaded* from a *micro* standpoint. It follows that in systems where variability is minimal or nonexistent (e.g., because arrivals can be scheduled and service time is constant), waiting lines do not ordinarily form.

GOAL OF QUEUING ANALYSIS

The goal of queuing is essentially to minimize total costs. There are two basic categories of cost in a queuing situation: those associated with customers waiting for service and those associated with capacity. Capacity costs are the costs of maintaining the ability to provide service. Examples include the number of bays at a car wash, the number of checkouts at a supermarket, the number of repair people to handle equipment breakdowns, and the number of lanes on a highway. When a service facility is idle, capacity is lost since it cannot be stored. The costs of customer waiting include the salaries paid to employees while they wait for service (mechanics waiting for tools, the drivers of trucks waiting to unload), the cost of the space for waiting (size of doctor's waiting room, length of driveway at a car wash, fuel consumed by planes waiting to land), and any loss of business due to customers refusing to wait and possibly going elsewhere in the future.

A practical difficulty frequently encountered is pinning down the cost of customer waiting time, especially since major portions of that cost are not a part of accounting data. One approach often used is to treat waiting times or line lengths as a policy variable: A manager simply specifies an acceptable level of waiting and directs that capacity be established to achieve that level.

The traditional goal of queuing analysis is to balance the cost of providing a level of service capacity with the cost of customers waiting for service. Figure 19–1 illustrates this concept. Note that as capacity increases, its cost increases. For simplicity, the increase is shown as a linear relationship. Although a step function is often more appropriate, use of a straight line does not significantly distort the picture. As capacity increases, the number of customers waiting and the time they wait tend to decrease,

Visitors wait to ride the Big Thunder Mountain Railroad at Walt Disney World in Florida. Disney has studied queuing and can predict which rides will draw how many people. To make the wait more palatable, waiting times are often posted, and many of the more popular rides include entertainment and snack vendors along the way.

www.disney.com/

FIGURE 19-1

The goal of queuing analysis is to minimize the sum of two costs: customer waiting costs and service capacity costs

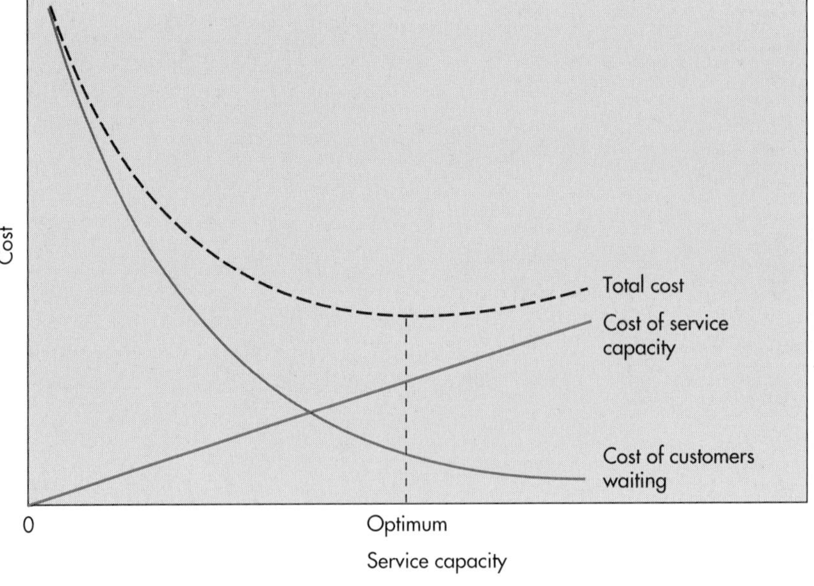

thereby decreasing waiting costs. As is typical in trade-off relationships, total costs can be represented as a U-shaped curve. The goal of analysis is to identify a level of service capacity that will minimize total cost. (Unlike the situation in the inventory EOQ model, the minimum point on the total cost curve is *not* usually where the two cost lines intersect.)

In situations where those waiting in line are *external* customers (as opposed to employees), the existence of waiting lines can reflect negatively on an organization's *quality* image. Consequently, some organizations are focusing their attention on providing faster service—speeding up the rate at which service is delivered rather than merely increasing the number of servers. The effect of this is to shift the total cost curve downward, because the cost of customer waiting decreases.

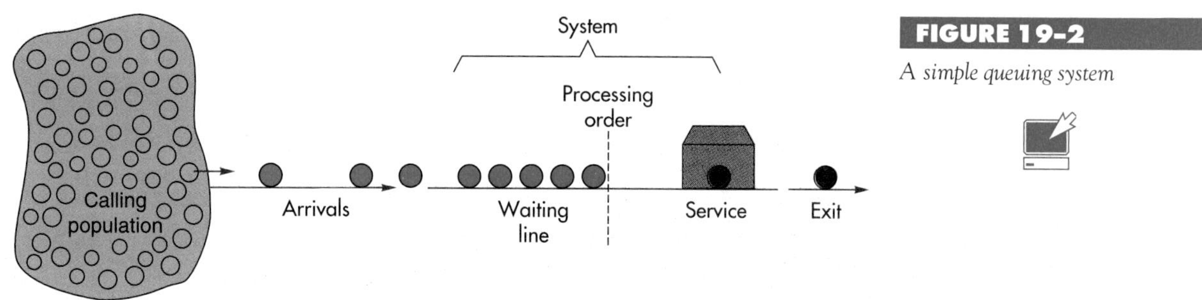

FIGURE 19-2
A simple queuing system

SYSTEM CHARACTERISTICS

There are numerous queuing models from which an analyst can choose. Naturally, much of the success of the analysis will depend on choosing an appropriate model. Model choice is affected by the characteristics of the system under investigation. The main characteristics are:

1. Population source.
2. Number of servers (channels).
3. Arrival and service patterns.
4. Queue discipline (order of service).

Figure 19–2 depicts a simple queuing system.

Population Source

The approach to use in analyzing a queuing problem depends on whether the potential number of customers is limited. There are two possibilities: *infinite-source* and *finite-source* populations. In an **infinite-source** situation, the potential number of customers greatly exceeds system capacity. Infinite-source situations exist whenever service is *unrestricted.* Examples are supermarkets, drugstores, banks, restaurants, theaters, amusement centers, and toll bridges. Theoretically, large numbers of customers from the "calling population" can request service at any time. When the potential number of customers is limited, a **finite-source** situation exists. An example is the repairman responsible for a certain number of machines in a company. The potential number of machines that might need repairs at any one time cannot exceed the number of machines assigned to the repairer. Similarly, an operator may be responsible for loading and unloading a bank of four machines, a nurse may be responsible for answering patient calls for a 10-bed ward, a secretary may be responsible for taking dictation from three executives, and a company shop may perform repairs as needed on the firm's 20 trucks.

infinite source Customer arrivals are unrestricted.

finite source The number of potential customers is limited.

Number of Servers (Channels)

The capacity of queuing systems is a function of the capacity of each server and the number of servers being used. The terms *server* and **channel** are synonymous, and it is generally assumed that each channel can handle one customer at a time. Systems can be either *single-* or *multiple-channel.* (A group of servers working together as a team, such as a surgical team, is treated as a single-channel system.) Examples of single-channel systems are small grocery stores with one checkout counter, some theaters, single-bay car washes, and drive-in banks with one teller. Multiple-channel systems (those with more than one server) are commonly found in banks, at airline ticket counters, at auto service centers, and at gas stations.

channel A server in a service system.

A related distinction is the number of steps or *phases* in a queuing system. For example, at theme parks, people go from one attraction to another. Each attraction constitutes a separate phase where queues can (and usually do) form.

Figure 19–3 illustrates some of the most common queuing systems. Because it would not be possible to cover all of these cases in sufficient detail in the limited amount of space available here, our discussion will focus on *single-phase* systems.

Arrival and Service Patterns

Waiting lines are a direct result of arrival and service variability. They occur because random, highly variable arrival and service patterns cause systems to be temporarily overloaded. In many instances, the variabilities can be described by theoretical distributions. In fact, the most commonly used models assume that the customer arrival *rate* can be described by a Poisson distribution and that the service *time* can be described by a negative exponential distribution. Figure 19–4 illustrates these distributions.

The Poisson distribution often provides a reasonably good description of customer arrivals per unit of time (e.g., per hour). Figure 19–5A illustrates how Poisson-distributed arrivals (e.g., accidents) might occur during a three-day period. In some hours, there are three or four arrivals, in other hours one or two arrivals, and in some hours no arrivals.

The negative exponential distribution often provides a reasonably good description of customer service times (e.g., first aid care for accident victims). Figure 19–5B illustrates how exponential service times might appear for the customers whose arrivals are illustrated in Figure 19–5A. Note that most service times are very short—some are close to zero—but a few require a relatively long service time. That is typical of a negative exponential distribution.

Waiting lines are most likely to occur when arrivals are bunched or when service times are particularly lengthy, and they are very likely to occur when both factors are present. For instance, note the long service time of customer 7 on day 1, in Figure 19–5B. In Figure 19–5A, the seventh customer arrived just after 10 o'clock and the next two customers arrived shortly after that, making it very likely that a waiting line formed. A similar situation occurred on day 3 with the last three customers: The relatively long service time for customer 13 (Figure 19–5B), and the short time before the next two arrivals (Figure 19–5A, day 3) would create (or increase the length of) a waiting line.

It is interesting to note that the Poisson and negative exponential distributions are alternate ways of presenting the same basic information. That is, if service time is exponential, then the service rate is Poisson. Similarly, if the customer arrival rate is Poisson, then the interarrival time (i.e., the time between arrivals) is exponential. For example, if a service facility can process 12 customers per hour (rate), average service time is five minutes. And if the arrival rate is 10 per hour, then the average time between arrivals is six minutes. Hence, the models described here generally require that arrival and service rates lend themselves to description using a Poisson distribution, or equivalently, that interarrival and service times lend themselves to description using a negative exponential distribution. In practice, it is necessary to verify that these assumptions are met. Sometimes this is done by collecting data and plotting them, although the preferred approach is to use a chi-square goodness-of-fit test for that purpose. A discussion of the chi-square test is beyond the scope of this text, but most basic statistics textbooks cover the topic.

Research has shown that these assumptions are often appropriate for customer arrivals but less likely to be appropriate for service. In situations where the assumptions are not reasonably satisfied, the alternatives would be to (1) develop a more suitable model, (2) search for a better (and usually more complex) existing model, or (3) resort to computer simulation. Each of these alternatives requires more effort or cost than the ones presented here.

Queue Discipline

queue discipline The order in which customers are processed.

Queue discipline refers to the order in which customers are processed. All but one of the models to be described shortly assume that service is provided on a *first-come, first-served* basis. This is perhaps the most commonly encountered rule. We find first-come

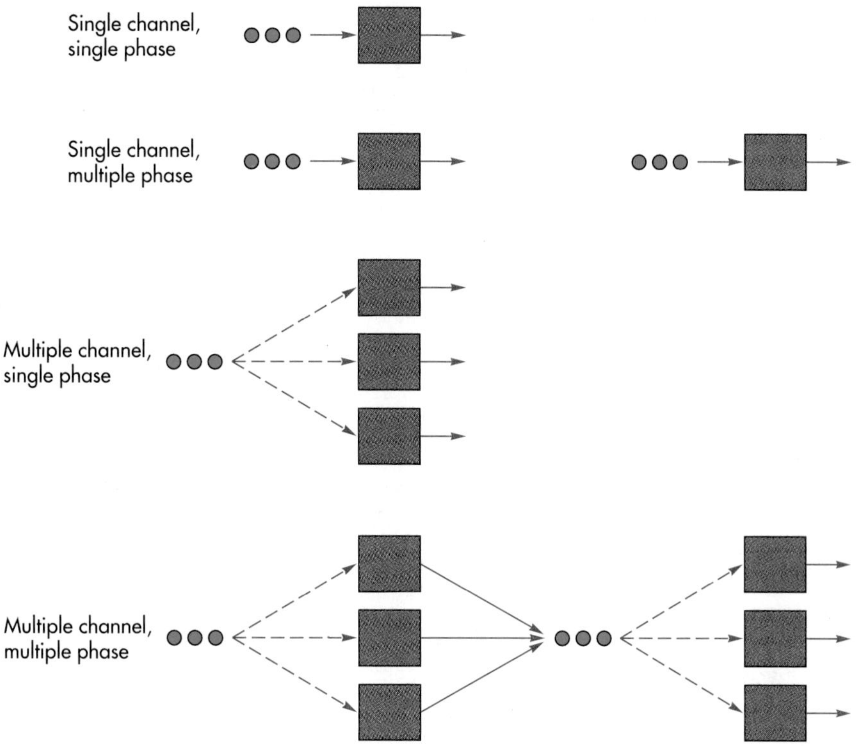

FIGURE 19-3

Four common variations of queuing systems

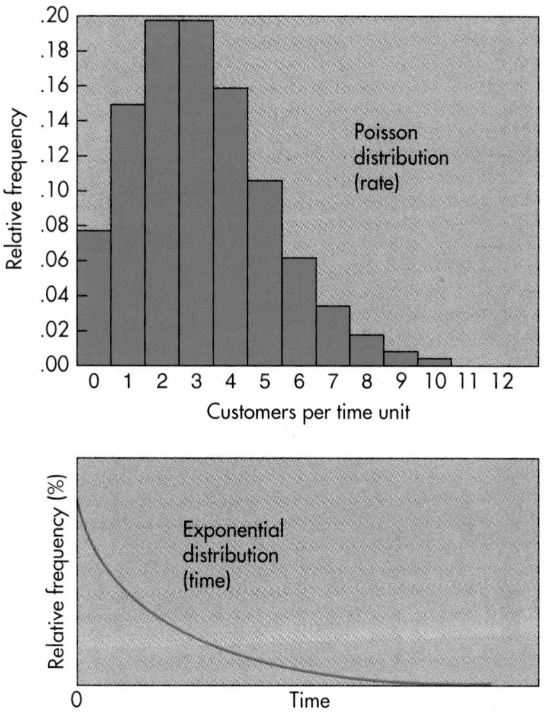

FIGURE 19-4

Poisson and negative exponential distributions

service at banks, stores, theaters, restaurants, four-way stop signs, registration lines, and so on. Examples of systems that do not serve on a first-come basis include hospital emergency rooms, rush orders in a factory, and main frame computer processing of jobs. In these and similar situations, customers do not all represent the same waiting costs; those with the highest costs (e.g., the most seriously ill) are processed first, even though other customers may have arrived earlier.

FIGURE 19-5

Poisson arrivals and exponential service times

A. Customer Arrivals

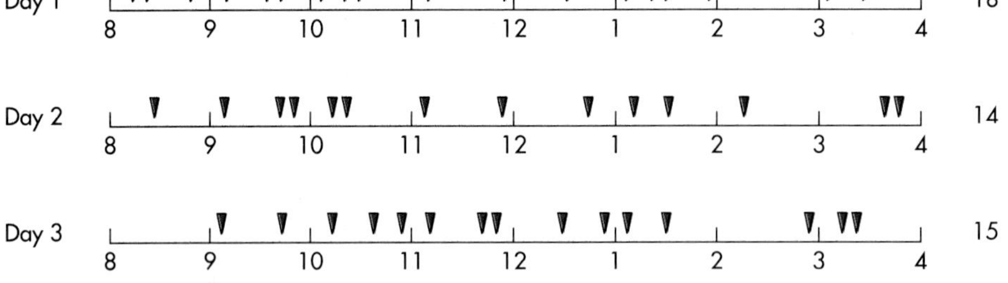

B. Service Times

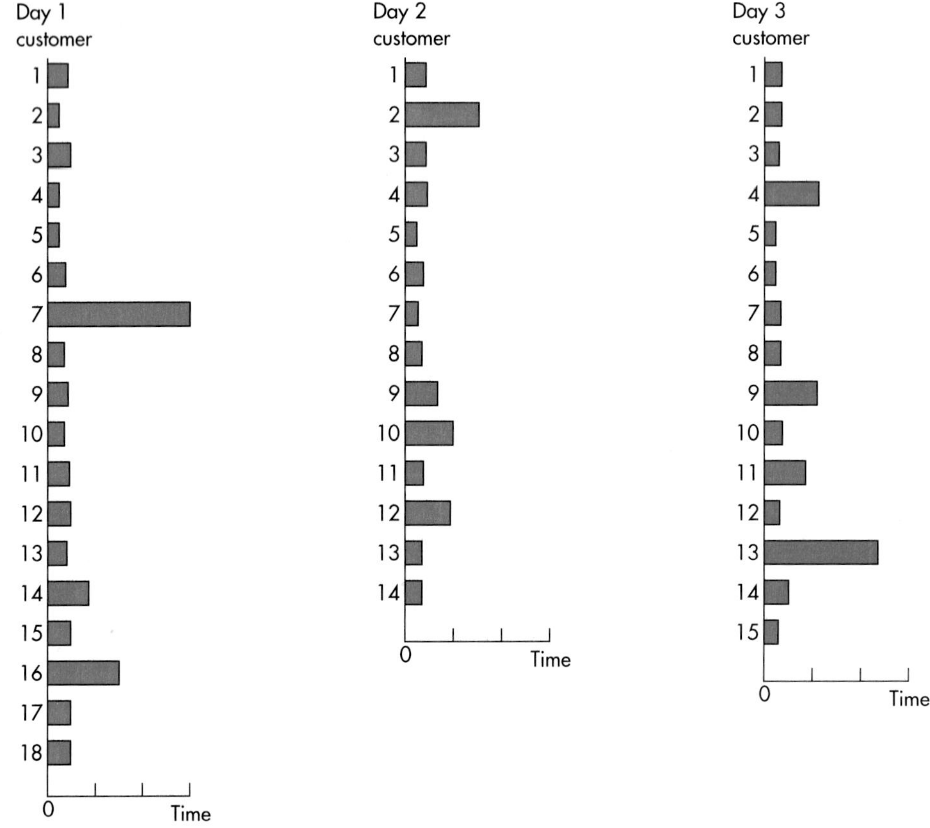

MEASURES OF SYSTEM PERFORMANCE

The operations manager typically looks at five measures when evaluating existing or proposed service systems. Those measures are:

1. The average number of customers waiting, either in line or in the system.
2. The average time customers wait, either in line or in the system.
3. System utilization, which refers to the percentage of capacity utilized.
4. The implied cost of a given level of capacity and its related waiting line.
5. The probability that an arrival will have to wait for service.

Of these measures, system utilization bears some elaboration. It reflects the extent to which the servers are busy rather than idle. On the surface, it might seem that the operations manager would want to seek 100 percent utilization. However, as Figure

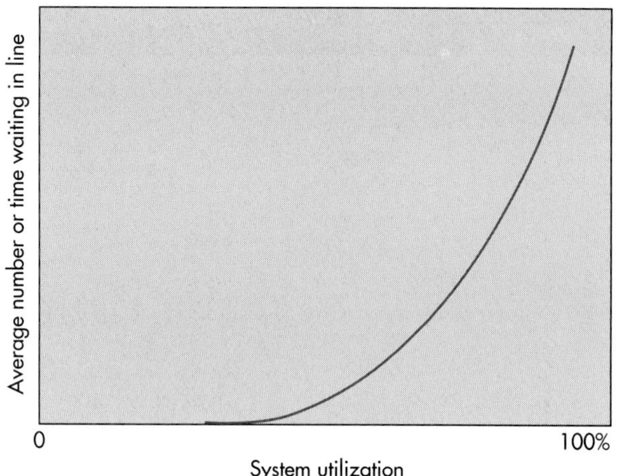

FIGURE 19–6

The average number waiting in line and the average time customers wait in line increase exponentially as the system utilization increases

19–6 illustrates, increases in system utilization are achieved at the expense of increases in both the length of the waiting line and the average waiting time. In fact, these values become exceedingly large as utilization approaches 100 percent. The implication is that under normal circumstances, 100 percent utilization is not a realistic goal. Instead, the operations manager should try to achieve a system that minimizes the sum of waiting costs and capacity costs.

QUEUING MODELS: INFINITE-SOURCE

Many queuing models are available for a manager or analyst to choose from. The discussion here includes four of the most basic and most widely used models. The purpose is to provide an exposure to a range of models rather than an extensive coverage of the field. All assume a Poisson arrival rate. Moreover, the models pertain to a system operating under *steady state* conditions; that is, they assume the average arrival and service rates are stable. The four models described are:

1. Single channel, exponential service time.
2. Single channel, constant service time.
3. Multiple channel, exponential service time.
4. Multiple priority service, exponential service time.

To facilitate your use of queuing models, Table 19–1 provides a list of the symbols used for the infinite-source models.

Basic Relationships

There are certain basic relationships that hold for all infinite-source models. Knowledge of these can be very helpful in deriving desired performance measures, given a few key values. Here are the basic relationships:

> Note: The arrival and service rates, represented by λ and μ, must be in the same units (e.g., customers per hour, customers per minute).

System utilization: This reflects the ratio of demand (as measured by the arrival rate) to supply or capacity (as measured by the product of the number of servers, M, and the service rate, μ).

$$\rho = \frac{\lambda}{M\mu} \qquad (19\text{–}1)$$

TABLE 19-1
Infinite-source symbols

Symbol	Represents
λ	Customer arrival rate
μ	Service rate
L_q	The average number of customers waiting for service
L_s	The average number of customers in the system (waiting and/or being served)
ρ	The system utilization
W_q	The average time customers wait in line
W_s	The average time customers spend in the system (waiting in line and service time)
$1/\mu$	Service time
P_0	The probability of zero units in the system
P_n	The probability of n units in the system
M	The number of servers (channels)
L_{max}	The maximum expected number waiting in line

The average number of customers being served:

$$r = \frac{\lambda}{\mu} \tag{19–2}$$

The average *number* of customers:

waiting in line for service $\quad L_q$ [Model dependent. Obtain using a table or formula.]

in the system (line plus being served) $\quad L_s = L_q + r \quad$ (19–3)

The average *time* customers are:

waiting in line $\quad W_q = \dfrac{L_q}{\lambda} \quad$ (19–4)

in the system $\quad W_s = W_q + \dfrac{1}{\mu} = \dfrac{L_s}{\lambda} \quad$ (19–5)

All infinite-source models require that system utilization be less than 1.0; the models apply only to underloaded systems.

The average number waiting in line, L_q, is a key value because it is a determinant of some of the other measures of system performance, such as the average number in the system, the average time in line, and the average time in the system. Hence, L_q will usually be one of the first values you will want to determine in problem solving.

Example 1

Customers arrive at a bakery at an average rate of 18 per hour on weekday mornings. The arrival distribution can be described by a Poisson distribution with a mean of 18. Each clerk can serve a customer in an average of four minutes; this time can be described by an exponential distribution with a mean of 4.0 minutes.

a. What are the arrival and service *rates*?

b. Compute the average number of customers being served at any time.

c. Suppose it has been determined that the average number of customers waiting in line is 3.6. Compute the average number of customers in the system (i.e., waiting in line or being served), the average time customers wait in line, and the average time in the system.

d. Determine the system utilization for M = 2, 3, and 4 servers.

Solution

a. The arrival rate is given in the problem: $\lambda = 18$ customers per hour. Change the service time to a comparable *hourly* rate by first restating the time in hours and then taking its reciprocal. Thus, (4 minutes per customer)/(60 minutes per hour) = $1/15 = 1/\mu$. Its reciprocal is $\mu = 15$ customers per hour.

b. $r = \dfrac{\lambda}{\mu} = \dfrac{18}{15} = 1.2$ customers

c. Given: $L_q = 3.6$ customers.

$$L_s = L_q + r = 3.6 + 1.2 = 4.8 \text{ customers}$$

$$W_q = \frac{L_q}{\lambda} = \frac{3.6}{18} = 0.20 \text{ hours per customer, or } 0.20 \text{ hours} \times 60 \text{ minutes/hour} = 12 \text{ minutes}$$

W_s = Waiting in line plus service

$$= W_q + \frac{1}{\mu} = 0.20 + \frac{1}{15} = 0.267 \text{ hours, or approximately 16 minutes}$$

d. System utilization is $\rho = \dfrac{\lambda}{M\mu}$

$$\text{For } M = 2, \rho = \frac{18}{2(15)} = .60$$

$$\text{For } M = 3, \rho = \frac{18}{3(15)} = .40$$

$$\text{For } M = 4, \rho = \frac{18}{4(15)} = .30$$

Hence, as the system capacity as measured by $M\mu$ increases, the system utilization for a given arrival rate decreases.

Model 1: Single Channel, Exponential Service Time

The simplest model involves a system that has one server (or a single crew). The queue discipline is first-come, first-served, and it is assumed that the customer arrival rate can be approximated by a Poisson distribution and service time by a negative exponential distribution. There is no limit on length of queue.

Table 19–2 lists the formulas for the single-channel model, which should be used in conjunction with formulas 19–1 through 19–5.

TABLE 19–2

Formulas for basic single-server model

Performance Measure	Equation	
Average number in line	$L_q = \dfrac{\lambda^2}{\mu(\mu - \lambda)}$	(19–6)
Probability of zero units in the system	$P_0 = 1 - \left(\dfrac{\lambda}{\mu}\right)$	(19–7)
Probability of *n* units in the system	$P_n = P_0\left(\dfrac{\lambda}{\mu}\right)^n$	(19–8a)
Probability of less than *n* units in the system	$P_{<n} = 1 - \left(\dfrac{\lambda}{\mu}\right)^n$	(19–8b)

Example 2

An airline is planning to open a satellite ticket desk in a new shopping plaza, staffed by one ticket agent. It is estimated that requests for tickets and information will average 15 per hour, and requests will have a Poisson distribution. Service time is assumed to be exponentially distributed. Previous experience with similar satellite operations suggests that mean service time should average about three minutes per request. Determine each of the following:

a. System utilization.

b. Percent of time the server (agent) will be idle.

c. The expected number of customers waiting to be served.

d. The average time customers will spend in the system.

e. The probability of zero customers in the system and the probability of four customers in the system.

Solution

$\lambda = 15$ per hour

$$\mu = \frac{1}{\text{Service time}} = \frac{1 \text{ customer}}{3 \text{ minutes}} \times 60 \text{ minutes per hour}$$
$$= 20 \text{ customers per hour}$$

a. $\rho = \dfrac{\lambda}{M\mu} = \dfrac{15}{1(20)} = .75$

b. Percentage idle time $= 1 - \rho = 1 - .75 = .25$, or 25 percent

c. $L_q = \dfrac{\lambda^2}{\mu(\mu - \lambda)} = \dfrac{15^2}{20(20-15)} = 2.25$ customers

d. $W_s = \dfrac{L_q}{\lambda} + \dfrac{1}{\mu} = \dfrac{2.25}{15} + \dfrac{1}{20} = 0.20$ hours, or 12 minutes

e. $P_0 = 1 - \dfrac{\lambda}{\mu} = 1 - \dfrac{15}{20} = .25$ and $P_4 = P_0\left(\dfrac{\lambda}{\mu}\right)^4 = .25\left(\dfrac{15}{20}\right)^4 = .079$

Model 2: Single Channel, Constant Service Time

As noted previously, waiting lines are a consequence of random, highly variable arrival and service rates. If a system can reduce or eliminate the variability of either or both, it can shorten waiting lines noticeably. A case in point is a system with constant service time. The effect of a constant service time is to cut in half the average number of customers waiting in line:

$$L_q = \frac{\lambda^2}{2\mu(\mu - \lambda)} \tag{19–9}$$

The average time customers spend waiting in line is also cut in half. Similar improvements can be realized by smoothing arrival times (e.g., by use of appointments).

Example 3

Wanda's Car Wash & Dry is an automatic, five-minute operation with a single bay. On a typical Saturday morning, cars arrive at a mean rate of eight per hour, with arrivals tending to follow a Poisson distribution. Find:

a. The average number of cars in line.

b. The average time cars spend in line and service.

Solution

$\lambda = 8$ cars per hour

$\mu = 1$ per 5 minutes, or 12 per hour

$$a.\ L_q = \frac{\lambda^2}{2\mu(\mu - \lambda)} = \frac{8^2}{2(12)(12 - 8)} = 0.667 \text{ cars}$$

$$b.\ W_s = \frac{L_q}{\lambda} + \frac{1}{\mu} = \frac{0.667}{8} + \frac{1}{12} = 0.167 \text{ hours, or 10 minutes}$$

Model 3: Multiple Channel

A multiple-channel system exists whenever there are two or more servers working *independently* to provide service to customer arrivals. Use of the model involves the following assumptions:

1. A Poisson arrival rate and exponential service time.
2. Servers all work at the same average rate.
3. Customers form a single waiting line (in order to maintain first-come, first-served processing).

Formulas for the multiple-channel model are listed in Table 19–3. Obviously, the multiple-channel formulas are more complex than the single-channel formulas, especially the formulas for L_q and P_0. These formulas are shown primarily for completeness; you can actually determine their values using Table 19–4, which gives values of L_q and P_0 for selected values of λ/μ and M.

To use Table 19–4, compute the value of λ/μ and round according to the number of decimal places given for that ratio in the table. Then simply read the values of L_q and P_0 for the appropriate number of channels, M. For instance, if $\lambda/\mu = 0.50$ and M = 2, the table provides a value of 0.033 for L_q and a value of .600 for P_0. These values can then be used to compute other measures of system performance. Note that the formulas in Table 19–3 and the values in Table 19–4 yield *average* amounts (i.e., expected values). Note also that Table 19–4 can also be used for single-channel problems (i.e., M = 1).

Example 4

Alpha Taxi and Hauling Company has seven cabs stationed at the airport. The company has determined that during the late-evening hours on weeknights, customers request cabs at a rate that follows the Poisson distribution with a mean of 6.6 per hour. Service time is exponential with a mean of 50 minutes per customer. Assume that there is one customer per cab. Find each of the performance measures listed in Table 19–3, and the system utilization.

TABLE 19–3

Multiple-channel queuing formulas

Performance Measure	Equation	
Average number in line	$L_q = \dfrac{\lambda\mu\left(\dfrac{\lambda}{\mu}\right)^M}{(M-1)!(M\mu-\lambda)^2}P_0$	(19–10)
Probability of zero units in the system	$P_0 = \left[\displaystyle\sum_{n=0}^{M-1}\frac{\left(\frac{\lambda}{\mu}\right)^n}{n!} + \frac{\left(\frac{\lambda}{\mu}\right)^M}{M!\left(1-\frac{\lambda}{M\mu}\right)}\right]^{-1}$	(19–11)
Average waiting time for an arrival not immediately served	$W_a = \dfrac{1}{M\mu - \lambda}$	(19–12)
Probability that an arrival will have to wait for service	$P_w = \dfrac{W_q}{W_a}$	(19–13)

TABLE 19-4

Infinite-source values for L_q and P_0 given λ/μ and M

λ/μ	M	L_q	P_0	λ/μ	M	L_q	P_0	λ/μ	M	L_q	P_0
0.15	1	0.026	.850	1.3	3	0.130	.264	2.7	3	7.354	.025
	2	0.001	.860		4	0.023	.271		4	0.811	.057
0.20	1	0.050	.800		5	0.004	.272		5	0.198	.065
	2	0.002	.818	1.4	2	1.345	.176		6	0.053	.067
0.25	1	0.083	.750		3	0.177	.236		7	0.014	.067
	2	0.004	.778		4	0.032	.245	2.8	3	12.273	.016
0.30	1	0.129	.700		5	0.006	.246		4	1.000	.050
	2	0.007	.739	1.5	2	1.929	.143		5	0.241	.058
0.35	1	0.188	.650		3	0.237	.211		6	0.066	.060
	2	0.011	.702		4	0.045	.221		7	0.018	.061
0.40	1	0.267	.600		5	0.009	.223	2.9	3	27.193	.008
	2	0.017	.667	1.6	2	2.844	.111		4	1.234	.044
0.45	1	0.368	.550		3	0.313	.187		5	0.293	.052
	2	0.024	.633		4	0.060	.199		6	0.081	.054
	3	0.002	.637		5	0.012	.201		7	0.023	.055
0.50	1	0.500	.500	1.7	2	4.426	.081	3.0	4	1.528	.038
	2	0.033	.600		3	0.409	.166		5	0.354	.047
	3	0.003	.606		4	0.080	.180		6	0.099	.049
0.55	1	0.672	.450		5	0.017	.182		7	0.028	.050
	2	0.045	.569	1.8	2	7.674	.053		8	0.008	.050
	3	0.004	.576		3	0.532	.146	3.1	4	1.902	.032
0.60	1	0.900	.400		4	0.105	.162		5	0.427	.042
	2	0.059	.538		5	0.023	.165		6	0.120	.044
	3	0.006	.548	1.9	2	17.587	.026		7	0.035	.045
0.65	1	1.207	.350		3	0.688	.128		8	0.010	.045
	2	0.077	.509		4	0.136	.145	3.2	4	2.386	.027
	3	0.008	.521		5	0.030	.149		5	0.513	.037
0.70	1	1.633	.300		6	0.007	.149		6	0.145	.040
	2	0.098	.481	2.0	3	0.889	.111		7	0.043	.040
	3	0.011	.495		4	0.174	.130		8	0.012	.041
0.75	1	2.250	.250		5	0.040	.134	3.3	4	3.027	.023
	2	0.123	.455		6	0.009	.135		5	0.615	.033
	3	0.015	.471	2.1	3	1.149	.096		6	0.174	.036
0.80	1	3.200	.200		4	0.220	.117		7	0.052	.037
	2	0.152	.429		5	0.052	.121		8	0.015	.037
	3	0.019	.447		6	0.012	.122	3.4	4	3.906	.019
0.85	1	4.817	.150	2.2	3	1.491	.081		5	0.737	.029
	2	0.187	.404		4	0.277	.105		6	0.209	.032
	3	0.024	.425		5	0.066	.109		7	0.063	.033
	4	0.003	.427		6	0.016	.111		8	0.019	.033
0.90	1	8.100	.100	2.3	3	1.951	.068	3.5	4	5.165	.015
	2	0.229	.379		4	0.346	.093		5	0.882	.026
	3	0.030	.403		5	0.084	.099		6	0.248	.029
	4	0.004	.406		6	0.021	.100		7	0.076	.030
0.95	1	18.050	.050	2.4	3	2.589	.056		8	0.023	.030
	2	0.277	.356		4	0.431	.083		9	0.007	.030
	3	0.037	.383		5	0.105	.089	3.6	4	7.090	.011
	4	0.005	.386		6	0.027	.090		5	1.055	.023
1.0	2	0.333	.333		7	0.007	.091		6	0.295	.026
	3	0.045	.364	2.5	3	3.511	.045		7	0.019	.027
	4	0.007	.367		4	0.533	.074		8	0.028	.027
1.1	2	0.477	.290		5	0.130	.080		9	0.008	.027
	3	0.066	.327		6	0.034	.082	3.7	4	10.347	.008
	4	0.011	.367		7	0.009	.082		5	1.265	.020
1.2	2	0.675	.250	2.6	3	4.933	.035		6	0.349	.023
	3	0.094	.294		4	0.658	.065		7	0.109	.024
	4	0.016	.300		5	0.161	.072		8	0.034	.025
	5	0.003	.301		6	0.043	.074		9	0.010	.025
1.3	2	0.951	.212		7	0.011	.074	3.8	4	16.937	.005

TABLE 19–4
(Concluded)

λ/μ	M	L_q	P_0	λ/μ	M	L_q	P_0	λ/μ	M	L_q	P_0
3.8	5	1.519	.017	4.6	5	9.289	.004	5.3	8	0.422	.005
	6	0.412	.021		6	1.487	.008		9	0.155	.005
	7	0.129	.022		7	0.453	.009		10	0.057	.005
	8	0.041	.022		8	0.156	.010		11	0.021	.005
	9	0.013	.022		9	0.054	.010		12	0.007	.005
3.9	4	36.859	.002		10	0.018	.010	5.4	6	6.661	.002
	5	1.830	.015	4.7	5	13.382	.003		7	1.444	.004
	6	0.485	.019		6	1.752	.007		8	0.483	.004
	7	0.153	.020		7	0.525	.008		9	0.178	.004
	8	0.050	.020		8	0.181	.008		10	0.066	.004
	9	0.016	.020		9	0.064	.009		11	0.024	.005
4.0	5	2.216	.013		10	0.022	.009		12	0.009	.005
	6	0.570	.017	4.8	5	21.641	.002	5.5	6	8.590	.002
	7	0.180	.018		6	2.071	.006		7	1.674	.003
	8	0.059	.018		7	0.607	.008		8	0.553	.004
	9	0.019	.018		8	0.209	.008		9	0.204	.004
4.1	5	2.703	.011		9	0.074	.008		10	0.077	.004
	6	0.668	.015		10	0.026	.008		11	0.028	.004
	7	0.212	.016	4.9	5	46.566	.001		12	0.010	.004
	8	0.070	.016		6	2.459	.005	5.6	6	11.519	.001
	9	0.023	.017		7	0.702	.007		7	1.944	.003
4.2	5	3.327	.009		8	0.242	.007		8	0.631	.003
	6	0.784	.013		9	0.087	.007		9	0.233	.004
	7	0.248	.014		10	0.031	.007		10	0.088	.004
	8	0.083	.015		11	0.011	.077		11	0.033	.004
	9	0.027	.015	5.0	6	2.938	.005		12	0.012	.004
	10	0.009	.015		7	0.810	.006	5.7	6	16.446	.001
4.3	5	4.149	.008		8	0.279	.006		7	2.264	.002
	6	0.919	.012		9	0.101	.007		8	0.721	.003
	7	0.289	.130		10	0.036	.007		9	0.266	.003
	8	0.097	.013		11	0.013	.007		10	0.102	.003
	9	0.033	014	5.1	6	3.536	.004		11	0.038	.003
	10	0.011	.014		7	0.936	.005		12	0.014	.003
4.4	5	5.268	.006		8	0.321	.006	5.8	6	26.373	.001
	6	1.078	.010		9	0.117	.006		7	2.648	.002
	7	0.337	.012		10	0.042	.006		8	0.823	.003
	8	0.114	.012		11	0.015	.006		9	0.303	.003
	9	0.039	.012	5.2	6	4.301	.003		10	0.116	.003
	10	0.013	.012		7	1.081	.005		11	0.044	.003
4.5	5	6.862	.005		8	0.368	.005		12	0.017	.003
	6	1.265	.009		9	0.135	.005	5.9	6	56.300	.000
	7	0.391	.010		10	0.049	.005		7	3.113	.002
	8	0.133	.011		11	0.017	.006		8	0.939	.002
	9	0.046	.011	5.3	6	5.303	.003		9	0.345	.003
	10	0.015	.011		7	1.249	.004		10	0.133	.003

Solution

$\lambda = 6.6$ per hour $\qquad$ M = 7 cabs (servers)

$$\mu = \frac{1 \text{ customer per trip}}{50 \text{ minutes per trip} \div 60 \text{ minutes per hour}}$$

$= 1.2$ customers per hour per cab

$\lambda/\mu = 5.5$. From Table 19–4, with M = 7; $L_q = 1.674$ and $P_0 = .003$

a. $L_q = 1.674$ customers

b. $P_0 = .003$

c. $W_a = \frac{1}{M\mu - \lambda} = \frac{1}{7(1.2) - 6.6} = 0.556$ hours or 33.36 minutes

d. $W_q = \frac{L_q}{\lambda} = \frac{1.674}{6.6} = .2536$ hours or 15.22 minutes

$P_W = \frac{W_q}{W_a} = \frac{.2536}{.556} = .456$

e. $\rho = \frac{\lambda}{M\mu} = \frac{6.6}{7(1.2)} = .786$

Using the Excel template, the solution to Example 4 would appear as follows:

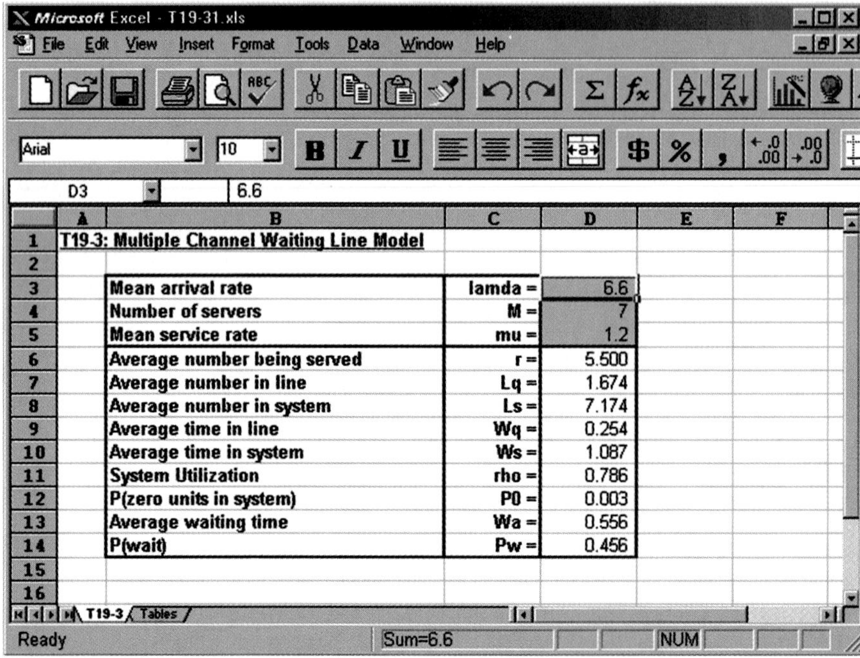

The process can also be worked in reverse; that is, an analyst can determine the capacity needed to achieve specified levels of various performance measures. This approach is illustrated in the following example.

Example 5

Alpha Tax and Hauling also plans to have cabs at a new rail station. The expected arrival rate is 4.8 customers per hour, and the service rate (including return time to the rail station) is expected to be 1.5 per hour. How many cabs will be needed to achieve an average time in line of 20 minutes or less?

Solution

$\lambda = 4.8$ customers per hour

$\mu = 1.5$ customers per hour

$M = ?$

$r = \frac{\lambda}{\mu} = \frac{4.8}{1.5} = 3.2$

W_q(desired) = 20 minutes, or 0.333 hour

Using $L_q = \lambda \times W_q$, you can solve for L_q: 4.8/hour (0.333 hour) = 1.6 units. Thus, the average number waiting should not exceed 1.6 customers. Referring to Table 19–4, with $r = 3.2$, $L_q = 2.386$ for $M = 4$ and 0.513 for $M = 5$. Hence, five cabs will be needed.

Cost Analysis

The design of a service system often reflects the desire of management to balance the cost of capacity with the expected cost of customers waiting in the system. (Note that customer waiting cost refers to the costs incurred by the organization due to customer waiting.) For example, in designing loading docks for a warehouse, the cost of docks plus loading crews must be balanced against the cost of trucks and drivers that will be in the system, both while waiting to be unloaded and while actually being unloaded. Similarly, the cost of having a mechanic wait for tools at a tool crib must be balanced against the cost of servers at the crib. In cases where the customers are not employees (e.g., retail sales), the costs can include lost sales when customers refuse to wait, the cost of providing waiting space, and the cost of added congestion (lost business, shoplifting).

The optimal capacity (usually in terms of number of channels) is one that minimizes the sum of customer waiting costs and capacity or server costs. Thus, the goal is

$$\text{Minimize} \quad \begin{matrix}\text{Total}\\ \text{cost}\end{matrix} = \begin{matrix}\text{Customer}\\ \text{waiting cost}\end{matrix} + \begin{matrix}\text{Capacity}\\ \text{cost}\end{matrix}$$

The simplest approach to a cost analysis involves computing *system* costs; that is, computing the costs for customers in the system and total capacity cost.

An iterative process is used to identify the capacity size that will minimize total costs. Capacity is incremented one unit at a time (e.g., increase the number of channels by one) and the total cost is computed at each increment. Because the total cost curve is U-shaped, usually the total cost will initially decrease as capacity is increased and then it will eventually begin to increase. Once it begins to increase, additional increases in capacity will cause it to continue to increase. Hence, once that occurs, the optimal capacity size can be readily identified.

The computation of customer waiting costs is based on the average *number* of customers in the *system*. This is perhaps not intuitively obvious; instead, it might seem that customer waiting *time* in the system would be more appropriate. However, that approach would pertain to only *one* customer—it would not convey information concerning *how many* customers would wait that long. Obviously, an average of five customers waiting would involve a lower waiting cost than an average of nine. Therefore, it is necessary to focus on the *number* waiting. Moreover, if an average of two customers are in the system, this is equivalent to having *exactly* two customers in the system at all times, even though in reality there will be times when zero, one, two, three, or more customers are in the system.

Example 6

Trucks arrive at a warehouse at a rate of 15 per hour during business hours. Crews can unload the trucks at a rate of five per hour. The high unloading rate is due to cargo being containerized. Recent changes in wage rates have caused the warehouse manager to reexamine the question of how many crews to use. The new rates are: crew and dock cost is $100 per hour; truck and driver cost is $120 per hour.

Solution

(L_q values are from Table 19–4 using $\frac{\lambda}{\mu} = \frac{15}{5} = 3.0$.)

Crew Size	Crew/Dock Cost	$\left[L_s = L_q + \frac{\lambda}{\mu}\right]$ Average Number in System	$[L_s \times \$120]$ Driver/Truck Cost	Total Cost
4	$400	1.528 + 3.0 = 4.528	$543.36	$ 943.36
5	500	.354 + 3.0 = 3.354	402.48	902.48 (minimum)
6	600	.099 + 3.0 = 3.099	371.88	971.88
7	700	.028 + 3.0 = 3.028	363.36	1,063.36

Five crews will minimize the total cost. Because the total cost will continue to increase once the minimum is reached, it is not really necessary to compute total costs for crew sizes larger than six, because total cost increased as the crew size was increased from five to six, indicating that a crew of five is optimal.

One additional point should be made concerning cost analysis. Because both customer waiting costs and capacity costs often reflect estimated amounts, the apparent optimal solution may not represent the true optimum. One ramification of this is that when computations are shown to the nearest penny, or even the nearest dollar, the total cost figures may seem to imply a higher degree of precision than is really justified by the cost estimates. This is compounded by the fact that arrival and service rates either may be approximations or not be exactly represented by the Poisson/exponential distributions. Another ramification is that if cost estimates can be obtained as *ranges* (e.g., customer waiting cost is estimated to range between \$40 and \$50 per hour), total costs should be computed using both ends of the range to see whether the optimal solution is affected. If it is, management must decide whether to expend additional effort to obtain more precise cost estimates or choose one of the two indicated optimal solutions. Management would most likely choose to employ the latter strategy if there were little disparity between total costs of various capacity levels close to the indicated optimal solutions.

Maximum Line Length

Another question that often comes up in capacity planning is the amount of space to allocate for waiting lines. Theoretically, with an infinite population source, the waiting line can become infinitely long. This implies that no matter how much space is allocated for a waiting line, one can never be completely sure that the space requirements won't exceed that amount. Nonetheless, as a practical matter, one can determine a line length that will not be exceeded a specified proportion of the time. For instance, an analyst may wish to know the length of line that will probably not be exceeded 98 percent of the time, or perhaps 99 percent of the time, and use that number as a planning value.

The approximate line length that will satisfy a specified percentage can be determined by solving the following equation for n:

$$n = \frac{\log K}{\log \rho} \text{ or } \frac{\ln K}{\ln \rho} \quad \text{where } K = \frac{1 - \begin{array}{c}\text{Specified}\\ \text{percentage}\end{array}}{L_q(1-\rho)} \tag{19–14}$$

The resulting value of n will not usually be an integer. Generally, round *up* to the next integer and treat the value as n. However, as a practical matter, if the computed value of n is less than 0.10 above the next lower integer, round down. Thus, 15.2 would be rounded to 16, but 15.06 would be rounded to 15.

Example 7

Determine the maximum length of a waiting line for specified probabilities of 95 percent and 98 percent, for a system in which $M = 2$, $\lambda = 8$ per hour, and $\mu = 5$ per hour.

Solution

$$r = \frac{8}{5} = 1.6 \quad \text{and} \quad \rho = \frac{8}{2(5)} = .80$$

From Table 19–4, $L_q = 2.844$ customers. For 95 percent, using Formula 19–14.

$$K = \frac{1 - .95}{2.844(1 - .80)} = .088$$

$$n = \frac{\ln .088}{\ln .80} = \frac{-2.4304}{-.2231} = 10.89, \text{ which rounds to 11.}$$

For 98 percent

$$K = \frac{1 - .98}{2.844(1 - .80)} = .035$$

$$n = \frac{\ln .035}{\ln .80} = \frac{-3.352}{-.2231} = 15.02, \text{ which rounds to 15.}$$

Model 4: Multiple Priorities

In many queuing systems, processing occurs on a first-come, first-served basis. However, there are situations in which that rule is inappropriate. The reason is that the cost or penalty incurred is not the same for all customers. In a hospital emergency waiting room, a wide variety of injuries and illnesses need treatment. Some may be minor (e.g., sliver in finger) and others may be much more serious, even life-threatening. It is more reasonable to treat the most serious cases first, letting the nonserious cases wait until all serious cases have been treated. Similarly, computer processing of jobs often follows rules other than first-come, first-served (e.g., shortest job first). In such cases, a **multiple-priority model** is useful for describing customer waiting times.

multiple-priority model Customers are processed according to some measure of importance.

In these systems, arriving customers are assigned to one of several *priority classes,* or categories, according to a predetermined assignment method (e.g., in a hospital emergency room, heart attacks, serious injuries, and unconscious persons are assigned to the highest priority class; sprains, minor cuts, bruises, and rashes are assigned to the lowest class; and other problems are assigned to one or more intermediate classes). Customers are then processed by class, highest class first. Within each class, processing is first-come, first-served. Thus, all customers in the highest class would be processed before those in the next lower class, then processing would move to that class, and then to the next lower class. Exceptions would occur only if a higher-priority customer arrived; that customer would be processed *after* the customer currently being processed (i.e., service would not be *preemptive*).

This model incorporates all of the assumptions of the basic multiple-server model except that it uses priority serving instead of first-come, first-served. Arrivals to the system are assigned a priority as they arrive (e.g., highest priority = 1, next priority class = 2, next priority class = 3, and so on). An existing queue might look something like this:

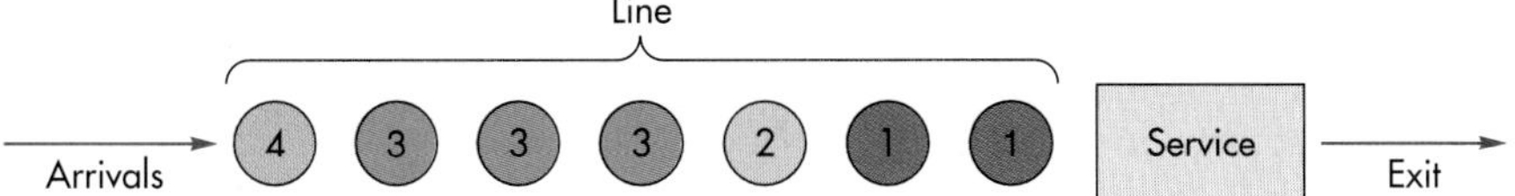

Within each class, waiting units are processed in the order they arrived (i.e., first-come, first-served). Thus, in this sequence, the first 1 would be processed as soon as a server was available. The second 1 would be processed when that server or another one became available. If, in the interim, another 1 arrived, it would be next in line *ahead of the first 2.* If there were no new arrivals, the only 2 would be processed by the next available server. At that point, if a new 1 or 2 arrived, it would be processed ahead of the 3s and the 4. Conversely, if a new 4 arrived, it would take its place at the end of the line.

Obviously, a unit with a low priority could conceivably wait a rather long time for processing. In some cases, units that have waited more than some specified time are reassigned a higher priority.

Table 19–5 gives the appropriate formulas for this multiple-channel priority service model.

Example 8

A machine shop handles tool repairs in a large company. As each job arrives in the shop, it is assigned a priority based on urgency of the need for that tool. Requests for repair can be described by a Poisson distribution. Arrival rates are: $\lambda_1 = 2$ per hour, $\lambda_2 = 2$ per

TABLE 19-5

Multiple-server priority service model

Performance Measure	Formula	Formula Number
System utilization	$\rho = \frac{\lambda}{M\mu}$	(19–15)
Intermediate values (L_q from Table 19–4)	$A = \frac{\lambda}{(1-\rho)L_q}$	(19–16)
	$B_k = 1 - \sum_{c=1}^{k} \frac{\lambda}{M\mu}$	(19–17)
	$(B_0 = 1)$	
Average waiting time in line for units in *k*th priority class	$W_k = \frac{1}{A \cdot B_{k-1} \cdot B_k}$	(19–18)
Average time in the system for units in the *k*th priority class	$W = W_k + \frac{1}{\mu}$	(19–19)
Average number waiting in line for units in the *k*th priority class	$L_k = \lambda_k \times W_k$	(19–20)

hour, and $\lambda_3 = 1$ per hour. The service rate is one tool per hour for each server, and there are six servers in the shop. Determine the following information.

a. The system utilization.

b. The average time a tool in each of the priority classes will wait for service.

c. The average time a tool spends in the system for each priority class.

d. The average number of tools waiting for repair in each class.

Solution

$\lambda = \Sigma\lambda_k = 2 + 2 + 1 = 5$ per hour

M = 6 servers

$\mu = 1$ customer per hour

a. $\rho = \frac{\lambda}{M\mu} = \frac{5}{6(1)} = .833$

b. Intermediate values. For $\lambda/\mu = 5/1 = 5$; from Table 19–4, $L_q = 2.938$

$$A = \frac{5}{(1 - .833)2.938} = 10.19$$

$$B_0 = 1$$

$$B_1 = 1 - \frac{2}{6(1)} = \frac{2}{3} = .667$$

$$B_2 = 1 - \frac{2+2}{6(1)} = \frac{1}{3} = .333$$

$$B_3 = 1 - \frac{2+2+1}{6(1)} = \frac{1}{6} = .167$$

$$W_1 = \frac{1}{A \cdot B_0 \cdot B_1} = \frac{1}{10.19(1)(.667)} = 0.147 \text{ hours}$$

$$W_2 = \frac{1}{A \cdot B_1 \cdot B_2} = \frac{1}{10.19(.667)(.333)} = 0.442 \text{ hours}$$

$$W_3 = \frac{1}{A \cdot B_2 \cdot B_3} = \frac{1}{10.19(.333)(.167)} = 1.765 \text{ hours}$$

c. Average time in system = $W_k + 1/\mu$. In this case, $1/\mu = 1/1 = 1$. Thus, you have:

Class	$W_k + 1 = W$ Hours
1	0.147 + 1 = 1.147
2	0.442 + 1 = 1.442
3	1.765 + 1 = 2.765

d. The average number of units waiting in each class is $L_k = \lambda_k \cdot W_k$. Thus, you have:

Class	$\lambda_k \cdot W_k = L_k$ Units
1	2(0.147) = 0.294
2	2(0.442) = 0.884
3	1(1.765) = 1.765

Using the Excel template, the solution to Example 8 would appear as follows.

Microsoft Excel - T19-41.xls

File Edit View Insert Format Tools Data Window Help

Arial 10

B16 Average number in line

	A	B	C	D	E	F	G
1	T19-4: Multiple Priorities Waiting Line Model						
2							
3			Class ->		1	2	3
4		Mean arrival rate	lamda =	5	2	2	1
5		Number of servers	M =	6			
6		Mean service rate	mu =	1			
7		Average number being served	r =	5.000			
8		Average number in line	Lq =	2.938			
9		Average number in system	Ls =	7.938			
10		System Utilization	rho =	0.833			
11		P(zero units in system)	P0 =	0.005			
12			A =	10.212			
13			Bk =		0.667	0.333	0.167
14		Average time in line	Wk =		0.147	0.441	1.763
15		Average time in the system	W =		1.147	1.441	2.763
16		Average number in line	Lk =		0.294	0.881	1.763

T19-4 / Tables /

Ready Sum=0 NUM

Revising Priorities. If any of the waiting times computed in Example 8 are deemed too long by management (e.g., a waiting time of 0.147 hours for tools in the first class might be too long), there are several options. One is to increase the number of servers. Another is to attempt to increase the service rate, say, by introducing new methods. If such options are not feasible, another approach is to reexamine the membership of each of the priority classifications because, if some repair requests in the first priority class, for example, can be reassigned to the second priority class, this will tend to decrease the average waiting times for repair jobs in the highest priority classification, simply because the arrival rate of those items will be lower.

Example 9 offers several interesting results. One is that through reduction of the arrival rate of the highest priority class, the average waiting time for units in that class has decreased. In other words, removing some members of the highest class and placing them into the next lower class reduced the average waiting time for units that remained in the highest class. Note that the average waiting time for the second priority class also was reduced, even though units were added to that class. Although this may appear counterintuitive, it is necessary to recognize that the *total* waiting time (when all arrivals are taken into account) will remain unchanged. We can see this by noticing that the average *number* waiting (see Example 8, part *d*) is .294 + .884 + 1.765 = 2.943. In Example 9, using the average waiting times just computed, the average number waiting in all three classes is

$$\sum_{k=1}^{3} \lambda_k W_k = 1.5(.131) + 2.5(.393) + 1.0(1.765) = 2.944$$

Aside from a slight difference due to rounding, the totals are the same.

Another interesting observation is that the average waiting time for customers in the third priority class did not change from the preceding example. The reason for this is that the *total* arrival rate for the two higher-priority classes did not change, and the average arrival rate for this class did not change. Hence, units assigned to the lowest class must still contend with a combined arrival rate of 4 for the two higher-priority classes.

Example 9

The manager of the repair shop, after consulting with the managers of the departments that use the shop's services, has revised the list of tools that are given the highest priorities. This is reflected by revised arrival rates. Suppose that the revised rates are: $\lambda_1 = 1.5$, $\lambda_2 = 2.5$, and λ_3 remains unchanged at 1.0. Determine the following information:

a. The system utilization.

b. The average waiting time for units in each priority class.

Solution

$$\lambda = \Sigma\lambda_k = 1.5 + 2.5 + 1.0 = 5.0$$

$$M = 6$$

$$\mu = 1$$

Note that these values are the same as in the previous example.

a. $\rho = 5.0/6(1) = .833$, the same as in the previous example.

b. The value of A, since it is a function of M, μ, and λ, is the same as in the preceding example because these values are the same. Therefore, A = 10.19 and

$$B_0 = 1 \text{ (always)}$$

$$B_1 = 1 - \frac{1.5}{6(1)} = .75$$

$$B_2 = 1 - \frac{1.5 + 2.5}{6(1)} = .333$$

$$B_3 = 1 - \frac{1.5 + 2.5 + 1.0}{6(1)} = .167$$

Then

$$W_1 = \frac{1}{10.19(1)(.75)} = 0.131 \text{ hours}$$

$$W_2 = \frac{1}{10.19(.75)(.333)} = 0.393 \text{ hours}$$

$$W_3 = \frac{1}{10.19(.333)(.167)} = 1.765 \text{ hours}$$

TABLE 19-6

Finite-source queuing formulas and notation

Formulas

Service factor	$X = \frac{T}{T+U}$	(19–21)
Average number waiting	$L = N(1 - F)$	(19–22)
Average waiting time	$W = \frac{L(T+U)}{N-L} = \frac{T(1-F)}{XF}$	(19–23)
Average number running	$J = NF(1 - X)$	(19–24)
Average number being served	$H = FNX$	(19–25)
Number in population	$N = J + L + H$	(19–26)

Notation†

D = Probability that a customer will have to wait in line
F = Efficiency factor: 1 − percentage waiting in line
H = Average number of customers being served
J = Average number of customers not in line or in service
L = Average number of customers waiting for service
M = Number of service channels
N = Number of potential customers
T = Average service time
U = Average time between customer service requirements per customer
W = Average time customers wait in line
X = Service factor

Cycle

	Not waiting or being served	Waiting	Being served
Average number:	J	L	H
Average time:	U	W	T

$$F = \frac{J+H}{J+L+H}\,^{*}$$

*The purpose of this formula is to provide an understanding of F. Because the value of F is needed to compute J, L, and H, the formulas cannot be used to actually compute F. The finite queuing tables must be used for that purpose.

†Adapted from L. G. Peck and R. N. Hazelwood, *Finite Queuing Tables* (New York: John Wiley & Sons, 1958). Reprinted by permission.

QUEUING MODEL: FINITE-SOURCE

The finite-source model is appropriate for cases in which the calling population is limited to a relatively small number of potential calls. For instance, one person may be responsible for handling breakdowns on 15 machines; thus, the size of the calling population is 15. However, there may be more than one server or channel; for example, due to a backlog of machines awaiting repairs, the manager might authorize an additional person to work on repairs.

As in the infinite-source models, arrival rates are required to be Poisson and service times exponential. A major difference between the finite- and infinite-source models is that the arrival rate of customers in a finite situation is *affected by* the length of the waiting line; the arrival rate decreases as the length of the line increases simply because there is a decreasing proportion of the population left to generate calls for service. The limit occurs when *all* of the population are waiting in line; at that point the arrival rate is zero since no additional units can arrive.

Because the mathematics of the finite-source model can be complex, analysts often use finite-queuing tables in conjunction with simple formulas to analyze these systems. Table 19–6 contains a list of the key formulas and definitions. You will find it helpful to study the diagram of a cycle that is presented in the table.

Table 19–7 is an abbreviated finite-queuing table used to obtain values of D and F. (Most of the formulas require a value for F.) In order to use the finite-queuing table, follow this procedure:

1. List the values for
   a. N, population size.
   b. M, number of servers/channels.
   c. T, average service time.
   d. U, average time between calls for service per customer.
2. Compute the service factor, $X = T/(T + U)$.
3. Locate the section of the finite-queuing tables for N.
4. Using the value of X as the point of entry, find the values of D and F that correspond to M.
5. Use the values of N, M, X, D, and F as needed to determine the values of the desired measures of system performance.

TABLE 19-7

Finite-queuing tables

X	M	D	F	X	M	D	F	X	M	D	F	X	M	D	F
Population 5					1	.386	.950	.200	3	.028	.998		2	.422	.912
.012	1	.048	.999	.105	2	.059	.997		2	.194	.976		1	.892	.597
.019	1	.076	.998		1	.404	.945		1	.689	.801	.330	4	.012	.999
.025	1	.100	.997	.110	2	.065	.996	.210	3	.032	.998		3	.112	.986
.030	1	.120	.996		1	.421	.939		2	.211	.973		2	.442	.904
.034	1	.135	.995	.115	2	.017	.995		1	.713	.783		1	.902	.583
.036	1	.143	.994		1	.439	.933	.220	3	.036	.997	.340	4	.013	.999
.040	1	.159	.993	.120	2	.076	.995		2	.229	.969		3	.121	.985
.042	1	.167	.992		1	.456	.927		1	.735	.765		2	.462	.896
.044	1	.175	.991	.125	2	.082	.994	.230	3	.041	.997		1	.911	.569
.046	1	.183	.990		1	.473	.920		2	.247	.965	.360	4	.017	.998
.050	1	.198	.989	.130	2	.089	.933		1	.756	.747		3	.141	.981
.052	1	.206	.988		1	.489	.914	.240	3	.046	.996		2	.501	.880
.054	1	.214	.987	.135	2	.095	.933		2	.265	.960		1	.927	.542
.056	2	.018	.999		1	.505	.907		1	.775	.730	.380	4	.021	.998
	1	.222	.985	.140	2	.102	.992	.250	3	.052	.995		3	.163	.976
.058	2	.019	.999		1	.521	.900		2	.284	.955		2	.540	.863
	1	.229	.984	.145	3	.011	.999		1	.794	.712		1	.941	.516
.060	2	.020	.999		2	.109	.991	.260	3	.058	.994	.400	4	.026	.997
	1	.237	.983		1	.537	.892		2	.303	.950		3	.186	.972
.062	2	.022	.999	.150	3	.012	.999		1	.811	.695		2	.579	.845
	1	.245	.982		2	.115	.990	.270	3	.064	.994		1	.952	.493
.064	2	.023	.999		1	.553	.885		2	.323	.944	.420	4	.031	.997
	1	.253	.981	.155	3	.013	.999		1	.827	.677		3	.211	.966
.066	2	.024	.999		2	.123	.989	.280	3	.071	.993		2	.616	.826
	1	.260	.979		1	.568	.877		2	.342	.938		1	.961	.471
.068	2	.026	.999	.160	3	.015	.999		1	.842	.661	.440	4	.037	.996
	1	.268	.978		2	.130	.988	.290	4	.007	.999		3	.238	.960
.070	2	.027	.999		1	.582	.869		3	.079	.992		2	.652	.807
	1	.275	.977	.165	3	.016	.999		2	.362	.932		1	.969	.451
.075	2	.031	.999		2	.137	.987		1	.856	.644	.460	4	.045	.995
	1	.294	.973		1	.597	.861	.300	4	.008	.999		3	.266	.953
.080	2	.035	.998	.170	3	.017	.999		3	.086	.990		2	.686	.787
	1	.313	.969		2	.145	.985		2	.382	.926		1	.975	.432
.085	2	.040	.998		1	.611	.853		1	.869	.628	.480	4	.053	.994
	1	.332	.965	.180	3	.021	.999	.310	4	.009	.999		3	.296	.945
.090	2	.044	.998		2	.161	.983		3	.094	.989		2	.719	.767
	1	.350	.960		1	.683	.836		2	.402	.919		1	.980	.415
.095	2	.049	.997	.190	3	.024	.998		1	.881	.613	.500	4	.063	.992
	1	.368	.955		2	.117	.980	.320	4	.010	.999		3	.327	.936
.100	2	.054	.997		1	.665	.819		3	.103	.988		2	.750	.748

TABLE 19-7

(Continued)

X	M	D	F	X	M	D	F	X	M	D	F	X	M	D	F
	1	.985	.399		1	.337	.982	.115	3	.081	.996		3	.300	.968
.520	4	.073	.991	.040	2	.050	.999		2	.324	.971		2	.692	.854
	3	.359	.927		1	.354	.980		1	.843	.776		1	.987	.497
	2	.779	.728	.042	2	.055	.999	.120	4	.016	.999	.210	5	.025	.999
	1	.988	.384		1	.371	.978		3	.090	.995		4	.108	.992
.540	4	.085	.989	.044	2	.060	.998		2	.346	.967		3	.333	.961
	3	.392	.917		1	.388	.975		1	.861	.756		2	.728	.835
	2	.806	.708	.046	2	.065	.998	.125	4	.019	.999		1	.990	.474
	1	.991	.370		1	.404	.973		3	.100	.994	.220	5	.030	.998
.560	4	.098	.986	.048	2	.071	.998		2	.369	.962		4	.124	.990
	3	.426	.906		1	.421	.970		1	.878	.737		3	.366	.954
	2	.831	.689	.050	2	.076	.998	.130	4	.022	.999		2	.761	.815
	1	.993	.357		1	.437	.967		3	.110	.994		1	.993	.453
.580	4	.113	.984	.052	2	.082	.997		2	.392	.958	.230	5	.037	.998
	3	.461	.895		1	.454	.963		1	.893	.718		4	.142	.988
	2	.854	.670	.054	2	.088	.997	.135	4	.025	.999		3	.400	.947
	1	.994	.345		1	.470	.960		3	.121	.993		2	.791	.794
.600	4	.130	.981	.056	2	.094	.997		2	.415	.952		1	.995	.434
	3	.497	.883		1	.486	.956		1	.907	.699	.240	5	.044	.997
	2	.875	.652	.058	2	.100	.996	.140	4	.028	.999		4	.162	.986
	1	.996	.333		1	.501	.953		3	.132	.991		3	.434	.938
.650	4	.179	.972	.060	2	.106	.996		2	.437	.947		2	.819	.774
	3	.588	.850		1	.517	.949		1	.919	.680		1	.996	.416
	2	.918	.608	.062	2	.113	.996	.145	4	.032	.999	.250	6	.010	.999
	1	.998	.308		1	.532	.945		3	.144	.990		5	.052	.997
.700	4	.240	.960	.064	2	.119	.995		2	.460	.941		4	.183	.983
	3	.678	.815		1	.547	.940		1	.929	.662		3	.469	.929
	2	.950	.568	.066	2	.126	.995	.150	4	.036	.998		2	.844	.753
	1	.999	.286		1	.562	.936		3	.156	.989		1	.997	.400
.750	4	.316	.944	.068	3	.020	.999		2	.483	.935	.260	6	.013	.999
	3	.763	.777		2	.133	.994		1	.939	.644		5	.060	.996
	2	.972	.532		1	.577	.931	.155	4	.040	.998		4	.205	.980
.800	4	.410	.924	.070	3	.022	.999		3	.169	.987		3	.503	.919
	3	.841	.739		2	.140	.994		2	.505	.928		2	.866	.732
	2	.987	.500		1	.591	.926		1	.947	.627		1	.998	.384
.850	4	.522	.900	.075	3	.026	.999	.160	4	.044	.998	.270	6	.015	.999
	3	.907	.702		2	.158	.992		3	.182	.986		5	.070	.995
	2	.995	.470		1	.627	.913		2	.528	.921		4	.228	.976
.900	4	.656	.871	.080	3	.031	.999		1	.954	.610		3	.537	.908
	3	.957	.666		2	.177	.990	.165	4	.049	.997		2	.886	.712
	2	.998	.444		1	.660	.899		3	.195	.984		1	.999	.370
.950	4	.815	.838	.085	3	.037	.999		2	.550	.914	.280	6	.018	.999
	3	.989	.631		2	.196	.988		1	.961	.594		5	.081	.994
	Population 10				1	.692	.883	.170	4	.054	.997		4	.252	.972
.016	1	.144	.997	.090	3	.043	.998		3	.209	.982		3	.571	.896
.019	1	.170	.996		2	.216	.986		2	.571	.906		2	.903	.692
.021	1	.188	.995		1	.722	.867		1	.966	.579		1	.999	.357
.023	1	.206	.994	.095	3	.049	.998	.180	5	.013	.999	.290	6	.022	.999
.025	1	.224	.993		2	.237	.984		4	.066	.996		5	.093	.993
.026	1	.232	.992		1	.750	.850		3	.238	.978		4	.278	.968
.028	1	.250	.991	.100	3	.056	.998		2	.614	.890		3	.603	.884
.030	1	.268	.990		2	.258	.981		1	.975	.890		2	.918	.672
.032	2	.033	.999		1	.776	.832	.190	5	.016	.999		1	.999	.345
	1	.285	.988	.105	3	.064	.997		4	.078	.995	.300	6	.026	.998
.034	2	.037	.999		2	.279	.978		3	.269	.973		5	.106	.991
	1	.301	.986		1	.800	.814		2	.654	.873		4	.304	.963
.036	2	.041	.999	.110	3	.072	.997		1	.982	.522		3	.635	.872
	1	.320	.984		2	.301	.974	.200	5	.020	.999		2	.932	.653
.038	2	.046	.999		1	.822	.795		4	.092	.994		1	.999	.333

TABLE 19–7

(Concluded)

X	M	D	F	X	M	D	F	X	M	D	F	X	M	D	F
.310	6	.031	.998		4	.591	.887	.520	8	.026	.998		5	.882	.759
	5	.120	.990		3	.875	.728		7	.115	.989		4	.980	.614
	4	.331	.957		2	.991	.499		6	.316	.958		3	.999	.461
	3	.666	.858	.420	7	.034	.993		5	.606	.884	.700	9	.040	.997
	2	.943	.635		6	.130	.987		4	.864	.752		8	.200	.979
.320	6	.036	.998		5	.341	.954		3	.980	.575		7	.484	.929
	5	.135	.988		4	.646	.866		2	.999	.385		6	.772	.836
	4	.359	.952		3	.905	.700	.540	8	.034	.997		5	.940	.711
	3	.695	.845		2	.994	.476		7	.141	.986		4	.992	.571
	2	.952	.617	.440	7	.045	.997		6	.363	.949	.750	9	.075	.994
.330	6	.042	.997		6	.160	.984		5	.658	.867		8	.307	.965
	5	.151	.986		5	.392	.943		4	.893	.729		7	.626	.897
	4	.387	.945		4	.698	.845		3	.986	.555		6	.870	.792
	3	.723	.831		3	.928	.672	.560	8	.044	.996		5	.975	.666
	2	.961	.600		2	.996	.454		7	.171	.982		4	.998	.533
.340	7	.010	.999	.460	8	.011	.999		6	.413	.939	.800	9	.134	.988
	6	.049	.997		7	.058	.995		5	.707	.848		8	.446	.944
	5	.168	.983		6	.193	.979		4	.917	.706		7	.763	.859
	4	.416	.938		5	.445	.930		3	.991	.535		6	.939	.747
	3	.750	.816		4	.747	.822	.580	8	.057	.995		5	.991	.625
	2	.968	.584		3	.947	.646		7	.204	.977		4	.999	.500
.360	7	.014	.999		2	.998	.435		6	.465	.927	.850	9	.232	.979
	6	.064	.995	.480	8	.015	.999		5	.753	.829		8	.611	.916
	5	.205	.978		7	.074	.994		4	.937	.684		7	.879	.818
	4	.474	.923		6	.230	.973		3	.994	.517		6	.978	.705
	3	.798	.787		5	.499	.916	.600	9	.010	.999		5	.998	.588
	2	.978	.553		4	.791	.799		8	.072	.994	.900	9	.387	.963
.380	7	.019	.999		3	.961	.621		7	.242	.972		8	.785	.881
	6	.083	.993		2	.998	.417		6	.518	.915		7	.957	.777
	5	.247	.971	.500	8	.020	.999		5	.795	.809		6	.995	.667
	4	.533	.906		7	.093	.992		4	.953	.663	.950	9	.630	.938
	3	.840	.758		6	.271	.966		3	.996	.500		8	.934	.841
	2	.986	.525		5	.553	.901	.650	9	.021	.999		7	.994	.737
.400	7	.026	.998		4	.830	.775		8	.123	.988				
	6	.105	.991		3	.972	.598		7	.353	.954				
	5	.292	.963		2	.999	.400		6	.651	.878				

Source: L. G. Peck and R. N. Hazelwood, *Finite Queuing Tables* (New York: John Wiley & Sons, 1958). Reprinted by permission.

Example 10

One operator loads and unloads a group of five machines. Service time is exponentially distributed with a mean of 10 minutes per cycle. Machines run for an average of 70 minutes between loading and unloading, and this time is also exponential. Find:

a. The average number of machines waiting for the operator.

b. The expected number of machines running.

c. Average downtime.

d. The probability that a machine will not have to wait for service.

Solution

$N = 5$

$T = 10$ minutes

$M = 1$

$U = 70$ minutes

$$X = \frac{T}{T + U} = \frac{10}{10 + 70} = .125$$

From Table 19–7, with $N = 5$, $M = 1$, and $X = .125$, $D = .473$ and F .920.

a. Average number waiting, $L = N(1 - F) = 5(1 - .920) = 0.40$ machines.

b. Expected number running, $J = NF(1 - X) = 5(.92)(1 - .125) = 4.025$ machines.

c. Downtime = Waiting time + Service time:

$$\text{Waiting time, } W = \frac{L(T + U)}{N - L} = \frac{.40(10 + 70)}{5 - .40} = 6.957 \text{ minutes}$$

Downtime = 6.957 minutes + 10 minutes = 16.957 minutes

d. Probability of not waiting = 1 − Probability of waiting

$$= 1 - D$$
$$= 1 - .473 = .527$$

Using the Excel template, the solution to Example 10 would appear as follows:

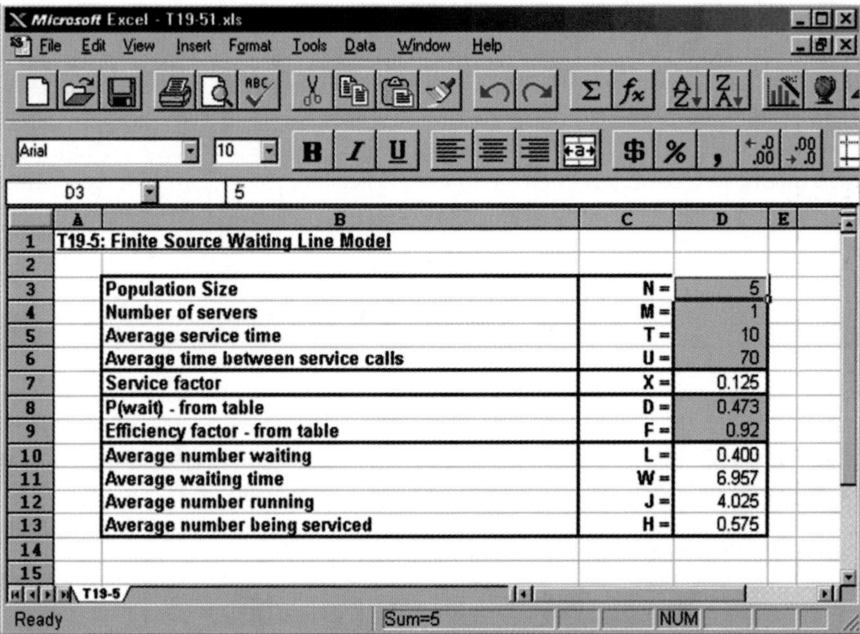

Example 11

Suppose that in Example 10, operators are paid $10 per hour, and machine downtime costs $16 per hour. Should the department add another operator if the goal is cost optimization?

Solution

Compare the total cost of the present system with the expected total cost of the proposed system:

M	Average Number Down, $N - J$	Average Down Cost (per hour), $(N - J)$ $16	Operator Cost (per hour)	Total Cost (per hour)
1	.975	$15.60	$10	$25.60
2	.651	10.42	20	30.42

Hence, the present system is superior because its total cost is less than the expected total cost using two operators.

OTHER APPROACHES

The discussion in this chapter has focused on designing service systems that achieve a balance between service capacity and customer waiting time. The implication is that decision makers can determine an appropriate level of service capacity. In certain instances, such an approach may not be practical for a variety of reasons. One is that the system may be currently in operation and indicated design changes may be too costly, or space restrictions may prevent changes from being made reasonably. One alternative that is particularly suited for queuing systems in which the customers are people rather than inanimate objects is to provide some form of diversion so that the waiting time becomes more tolerable. For example, magazines and newspapers can be placed in waiting rooms, as is usually the case in doctors' and dentists' offices. Auto repair shops sometimes use radio or television, and airlines may provide in-flight movies to help occupy the time. Airlines also serve meals and snacks, which help to make the time spent waiting (flying) more pleasant. Some other measures include placing mirrors where people wait for elevators, and asking people to fill out forms, which makes waiting somewhat constructive.

To carry this concept one step further, it is sometimes possible to derive some benefit from customer waiting. For instance, supermarkets position impulse items near checkout counters and gain additional sales, banks advertise current rates and place brochures describing bank services within easy reach of waiting customers, and restaurants have bars where customers can relax and spend money while waiting for their tables.

The implication in these ideas is that imagination and creativity can often play an important role in system design and that mathematical approaches are not the only ones worth considering.

Summary

Analysis of waiting lines can be an important aspect of the design of service systems. Waiting lines have a tendency to form in such systems even though, in a macro sense, the system is underloaded. The arrival of customers at random times and variability of service times combine to create temporary overloads. When this happens, waiting lines appear. By the same token, at other times the servers are idle.

A major consideration in the analysis of queuing systems is whether the number of potential customers is limited (finite source) or whether entry to the system is unrestricted (infinite source). Five basic queuing models are described, four dealing with infinite-source populations and one dealing with finite-source populations. In general, the models assume that customer arrival rates can be described by a Poisson distribution and that service time can be described by a negative exponential distribution.

Key Terms

channel, 815
finite source, 815
infinite source, 815
multiple-priority model, 829
queue discipline, 816
queuing theory, 812

Solved Problems

Problem 1

Infinite source. One of the features of a new machine shop will be a well-stocked tool crib. The manager of the shop must decide on the number of attendants needed to staff the crib. Attendants will receive $9 per hour in salary and fringe benefits. Mechanics' time will be worth $30 per hour, which includes salary and fringe benefits plus lost work time caused by waiting for parts. Based on previous experience, the manager estimates requests for parts will average 18 per hour with a service capacity of 20 requests per hour per attendant. How many attendants should be on duty if the manager is willing to assume that arrival and service rates will be Poisson-distributed? (Assume the number of mechanics is very large, so an infinite-source model is appropriate.)

Solution

$\lambda = 18$ per hour

$\mu = 20$ per hour

The solution requires a trial-and-error approach that reveals the total cost of feasible alternatives (i.e., a utilization less than 100 percent) and selection of the lowest-cost alternative. Note that the total-cost curve will always be U-shaped; increase the number of servers until the total cost shows an increase over the previous value. The optimum will be the number of servers that produced the previous total cost value. Thus,

Number of Servers, M	L_q*	$L_q + \frac{\lambda}{\mu} = L$	$9 · M: Server Cost (per hour)	$30 · L_s: Mechanic Cost (per hour)	Total Cost (per hour)
1	8.1	8.1 + 0.9 = 9.0	$ 9	$270	$279
2	0.229	0.229 + 0.9 = 1.129	$18	$ 33.87	$ 52†
3	0.03	0.03 + 0.9 = 0.93	$27	$ 27.9	$ 55†

*L_q from Table 19–4, with $r = \lambda/\mu = 18/20 = 0.9$.
†Rounded.

Hence, two servers will produce the lowest total cost.

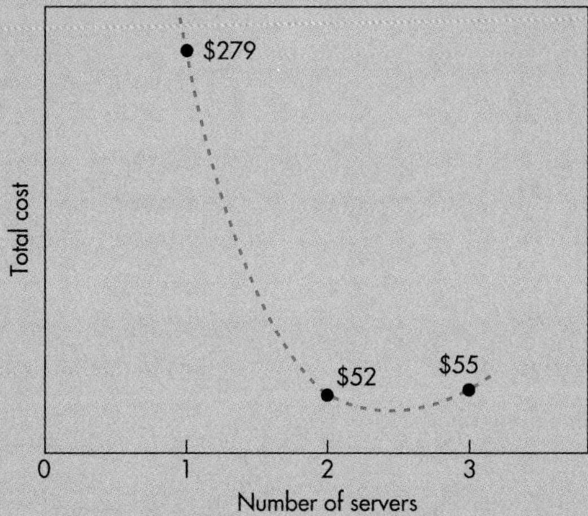

Problem 2

Infinite source. The following is a list of service times for three different operations:

Operation	Service Time
A	8 minutes
B	1.2 hours
C	2 days

a. Determine the service rate for each operation.

b. Would the calculated rates be different if these were interarrival times rather than service times?

Solution

a. The service rate is the reciprocal of service time. Thus, the rates are:
A: 1/8 per minute = .125 per minute, or .125/min. × 60 min./hr. = 7.5 per hour
B: 1/1.2 per hour = .833 per hour
C: 1/2 per day = .50 per day

b. No. In either case, the rate is simply the reciprocal of the time.

Problem 3

Finite source. A group of 10 machines is loaded and unloaded by one of three servers. The machines run for an average of six minutes per cycle, and average time to unload and reload is nine minutes. Each time can be described by an exponential distribution. While running, the machines produce at a rate that would yield 16 units per hour if they did not have to pause to wait for a server and be loaded and unloaded. What is the average hourly output of each machine when waiting and serving are taken into account?

Solution

$T = 9$ minutes

$U = 6$ minutes; $X = \frac{T}{T+U} = \frac{9}{9+6} = .60$

$M = 3$ servers; from Table 19–7, $F = .500$

$N = 10$ machines

a. Compute the average number of machines running:

$J = NF(1 - X) = 10(.500)(.40) = 2$

b. Determine the percentage of machines running, and multiply by output while running:

$\frac{J}{N} \times (16 \text{ per hour}) = \frac{2}{10} \times (16 \text{ per hour}) = 3.2 \text{ per hour}$

Discussion and Review Questions

1. In what kinds of situations is queuing analysis most appropriate?
2. Why do waiting lines form even though a service system is underloaded?
3. What are the most common measures of system performance in a queuing analysis?
4. What effect would decreasing arrival and service variability have on the effective capacity of a system?
5. What approaches do supermarkets use to offset variations in customer traffic intensity?
6. Contrast *finite* and *infinite* population sources.
7. Will doubling the service rate of a single-channel system reduce the average waiting time in line by a factor of one half? Explain briefly.
8. In a multiple-channel system, what is the rationale for having customers wait in a single line, as is now being done in many banks and post offices, rather than multiple lines? (*Hint:* The average waiting time is unaffected.)
9. What happens to the length of a waiting line in a highly variable (queuing) setting if a manager attempts to achieve a high percentage of capacity utilization?

Memo Writing Exercises

1. Write a half-page memo to your manager, Betty Davis-Brown, comparing the costs and benefits of two options being considered for the placement of a tool crib in your plant, which is fairly large. One option is to have a single, centralized tool crib where employees can check out tools they need, and the other is to have two cribs, one at each end of the plant.
2. Write a short memo to your manager, Tom Peters, explaining why you believe that the multiple priority queuing model should be used for customer complaints.

Problems

$M = 1$
$\lambda = 3$/day
$\mu = 4$/day
a. 2.25 customers
b. 75 percent

1. Repair calls are handled by one repairman at a photocopy shop. Repair time, including travel time, is exponentially distributed, with a mean of two hours per call. Requests for copier repairs come in at a mean rate of three per eight-hour day (assume Poisson). Determine:

   a. The average number of customers awaiting repairs.

   b. System utilization.

c. The amount of time during an eight-hour day that the repairman is not out on a call.

d. The probability of two or more customers in the system.

c. 2 hours
d. .5625

2. A vending machine dispenses hot chocolate or coffee. Service time is 30 seconds per cup and is constant. Customers arrive at a mean rate of 80 per hour, and this rate is Poisson-distributed. Determine:

   a. The average number of customers waiting in line.

   b. The average time customers spend in the system.

   c. The average number in the system.

$M = 1$
$\lambda = 80/\text{hr.}$
$\mu = 120/\text{hr.}$
a. .67 customer
b. 1 minute
c. 1.33 customer

3. Many of a bank's customers use its automatic teller machine to transact business after normal banking hours. During the early evening hours in the summer months, customers arrive at a certain location at the rate of one every other minute. This can be modeled using a Poisson distribution. Each customer spends an average of 90 seconds completing his or her transactions. Transaction time is exponentially distributed. Determine:

   a. The average time customers spend at the machine, including waiting in line and completing transactions.

   b. The probability that a customer will not have to wait upon arriving at the automatic teller machine.

   c. The average number waiting to use the machine.

a. 6 minutes
b. 0.25
c. 2.25 customers

4. A small town with one hospital has two ambulances to supply ambulance service. Requests for ambulances during nonholiday weekends average 0.8 per hour and tend to be Poisson-distributed. Travel and assistance time averages one hour per call and follows an exponential distribution. Find:

   a. System utilization.

   b. The average number of customers waiting.

   c. The average time customers wait for an ambulance.

   d. The probability that *both* ambulances will be busy when a call comes in.

$M = 2$
$\lambda = .8/\text{hr.}$
$\mu = 1/\text{hr.}$
a. 40 percent
b. .152
c. .19 hour
d. .2280

5. The following information pertains to telephone calls to a motel switchboard on a typical Tuesday.

Period	Incoming Rate (calls per minute)	Service Rate (calls per minute per operator)	Number of Operators
Morning	1.8	1.5	2
Afternoon	2.2	1.0	3
Evening	1.4	0.7	3

   a. Determine the average time callers wait to have their calls answered for each period and the probability that a caller will have to wait for each period.

   b. For each case in the previous problem, determine the maximum line length for a probability of 96 percent.

a. M: .375 minutes, .45
A: .678 minutes, .54
E: .635 minutes, .44
b. M: 4, A: 8, E: 5

6. Trucks are required to pass through a weighing station so that they can be checked for weight violations. Trucks arrive at the station at the rate of 40 an hour between 7 P.M. and 9 P.M. Currently two inspectors are on duty during those hours, each of whom can inspect 25 trucks an hour.

   a. How many trucks would you expect to see at the weighing station, including those being inspected?

   b. If a truck was just arriving at the station, about how many minutes could the driver expect to be at the station?

   c. What is the probability that both inspectors would be busy at the same time?

a. 4.444 trucks
b. 6.67 minutes
c. .7104
d. 6 minutes
e. The system would be overloaded.
f. 13.186

d. How many minutes, on average, would a truck that is not immediately inspected have to wait?

e. What condition would exist if there was only one inspector?

f. What is the maximum line length for a probability of .97?

a. $\lambda = 4/\text{day}$, $\mu = 5/\text{day}$

Docks	Cost
1	$2,300
2	$2,485.69

Choose 1 dock

b. $\mu = 5.71/\text{day}$

Docks	Cost
1	$1,900
2	$2,904

Choose 1 dock

7. The manager of a regional warehouse must decide on the number of loading docks to request for a new facility in order to minimize the sum of dock costs and driver-truck costs. The manager has learned that each driver-truck combination represents a cost of $300 per day and that each dock plus loading crew represents a cost of $1,100 per day.

   a. How many docks should be requested if trucks arrive at the rate of four per day, each dock can handle five trucks per day, and both rates are Poisson?

   b. An employee has proposed adding new equipment that would speed up the loading rate to 5.71 trucks per day. The equipment would cost $100 per day for each dock. Should the manager invest in the new equipment?

a. 0.952 mechanics
b. 0.229
c. 0.056 hr
d. 0.60
e. Two

8. The parts department of a large automobile dealership has a counter used exclusively for mechanics' requests for parts. The time between requests can be modeled by a negative exponential distribution that has a mean of five minutes. A clerk can handle requests at a rate of 15 per hour, and this can be modeled by a Poisson distribution that has a mean of 15. Suppose there are two clerks at the counter.

   a. On average, how many mechanics would be at the counter, including those being served?

   b. What is the probability that a mechanic would have to wait for service?

   c. If a mechanic has to wait, how long would the average wait be?

   d. What percentage of time are the clerks idle?

   e. If clerks represent a cost of $20 per hour and mechanics a cost of $30 per hour, what number of clerks would be optimal in terms of minimizing total cost?

$M = 1, N = 5$
$U = 4$ days, $T = 1$ day
a. .995 customer
b. 2.24 days
c. 31.1 percent
d. .875 customer

9. One field representative services five customers for a computer manufacturer. Customers request assistance at an average (Poisson-distributed) rate of once every four working days. The field representative can handle an average (Poisson-distributed) of one call per day. Determine:

   a. The expected number of customers waiting.

   b. The average length of time customers must wait from the initial request for service until the service has been completed.

   c. The percentage of time the service rep will be idle.

   d. By how much would your answer to part *a* be reduced if a second field rep were added?

$N = 10, M = 2$
$T = 14$ min., $U = 86$ min.
a. .437
b. 0.53 machine
c. 1.33 machines
d. 40.72 pieces
e. 3 operators

10. Two operators handle adjustments for a group of 10 machines. Adjustment time is exponentially distributed and has a mean of 14 minutes per machine. The machines operate for an average of 86 minutes between adjustments. While running, each machine can turn out 50 pieces per hour. Find:

    a. The probability that a machine will have to wait for an adjustment.

    b. The average number of machines waiting for adjustment.

    c. The average number of machines being serviced.

    d. The expected hourly output of each machine, taking adjustments into account.

    e. Machine downtime represents a cost of $70 per hour; operator cost (including salary and fringe benefits) is $15 per hour. What is the optimum number of operators?

$N = 5, M = 1$
$U = 90$ min., $T = 35$ min.
a. 28.56 pieces
b. 2

11. One operator services a bank of five machines. Machine running time and service time are both exponential. Machines run for an average of 90 minutes between service requirements, and service time averages 35 minutes. The operator receives $20 per hour in salary and fringe benefits, and machine downtime costs $70 per hour per machine.

a. If each machine produces 60 pieces per hour while running, find the average hourly output of each machine, when waiting and service times are taken into account.

b. Determine the optimum number of operators.

12. A milling department has 10 machines. Each operates an average of eight hours before requiring adjustment, which takes an average of two hours. While running, each machine can produce 40 pieces an hour.

    a. With one adjuster, what is the net average hourly output per machine?

    b. If machine downtime cost is $80 per hour and adjuster cost is $30 per hour, how many adjusters would be optimal?

a. 15.9 pieces
b. 3 adjusters

13. Trucks arrive at the loading dock of a wholesale grocer at the rate of 1.2 per hour. A single crew consisting of two workers can load a truck in about 30 minutes. Crew members receive $10 per hour in wages and fringe benefits, and trucks and drivers reflect an hourly cost of $60. The manager is thinking of adding another member to the crew. The service rate would then be 2.4 trucks per hour. Assume rates are Poisson.

    a. Would the third crew member be economical?

    b. Would a fourth member be justifiable if the resulting service capacity were 2.6 trucks per hour?

Crew Size	Cost
2	$110
3	$ 90
4	$ 91.43

3 crew members is optimal

14. Customers arriving at a service center are assigned to one of three categories, with category 1 given the highest priority. Records indicate that an average of nine customers arrive per hour and that one-third are assigned to each category. There are two servers, and each can process customers at the rate of five per hour. Arrival and service rates can be described by Poisson distributions.

    a. What is the utilization rate for this system?

    b. Determine the average waiting time for units in each class.

    c. Find the average number of customers in each class that are waiting for service.

a. .90
b. $W_1 = .12$ hour
$W_2 = .3045$ hour
$W_3 = 2.13$ hours
c. $L_1 = .36$
$L_2 = .91$
$L_3 = 6.39$

15. A manager must determine requirements for waiting space for customers. A priority system is used to process customers, who are assigned to one of two classes when they enter the processing center. The highest-priority class has an arrival rate of four per hour; the other class has an arrival rate of two per hour. Both can be described as Poisson-distributed. There are two servers, and each can process customers in an average of six minutes.

    a. What is the system utilization?

    b. Determine the number of customers of each class that are waiting for service.

    c. Determine the average waiting time for each class.

    d. If the manager could alter the assignment rules so that arrival rates of the two classes were equal, what would be the revised average waiting time for each priority class?

a. 30
b. $L_1 = .034$
$L_2 = .025$
c. $W_1 = .52$ min.
$W_2 = .74$ min.
d. $W_1 = .008$ hr.
$W_2 = .0116$ hr.

16. A priority waiting system assigns arriving customers to one of four classes. Arrival rates (Poisson) of the classes are shown in the following table:

Class	Arrivals per Hour
1	2
2	4
3	3
4	2

Five servers process the customers, and each can handle three customers per hour.

a. What is the system utilization?

a. .733
b. $W_1 = .035$ hr.
$W_2 = .059$ hr.
$W_3 = .128$ hr.
$W_4 = .284$ hr.
$L_1 = .070$
$L_2 = .236$
$L_3 = .384$
$L_4 = .568$

16. *c.* $W_1 = .035$ hr.
$W_2 = .053$ hr.
$W_3 = .115$ hr.
$W_4 = .284$ hr.
$L_1 = .070$
$L_2 = .159$
$L_3 = .458$
$L_4 = .568$
17. *a.* 55
b. $W_1 = .010, L_1 = .020$
$W_2 = .014, L_2 = .056$
$W_3 = .023, L_3 = .069$
$W_4 = .036, L_4 = .072$
c. $W_1 = .010, L_1 = .020$
$W_2 = .013, L_2 = .039$
$W_3 = .022, L_3 = .088$
$W_4 = .036, L_4 = .072$
18. *a.* Approx. 0.011
b. Approx. 0.444

b. What is the average wait for service by customers in the various classes? How many are waiting in each class, on the average?

c. If the arrival rate of the second priority class could be reduced to three units per hour by shifting some arrivals into the third priority class, how would your answers to part *b* change?

d. What observations can you make based on your answers to part *c*?

17. Referring to Problem 16, suppose that each server could handle four customers per hour. Answer the questions posed in the problem. Explain why the impact of reassigning customers is much less than in Problem 16.

18. During the morning hours at a catalog sales department, telephone calls come in at the rate (Poisson) of 40 per hour. Calls that cannot be answered immediately are put on hold. The system can handle eight callers on hold. If additional calls come in, they receive a busy signal. The three customer service representatives who answer the calls spend an average of three minutes with a customer.

    a. What is the probability that a caller will get a busy signal? (*Hint:* Solve for log *K* or ln *K* using trial and error.)

    b. What is the probability that a customer will be put on hold?

Selected Bibliography

Buffa, Elwood. *Operations Management.* 3rd ed. New York: John Wiley & Sons, 1972.

Griffin, W. *Queuing: Basic Theory and Applications.* Columbus, OH: Grid Publishing, 1978.

Hillier, Frederick S., and Gerald J. Lieberman. *Introduction to Operations Research,* 3rd ed. San Francisco: Holden-Day, 1980.

Katz, K. L.; B. M. Larson; and R. C. Larson. "Prescriptions for the Waiting-in-Line Blues: Entertain, Enlighten, and Engage." *Sloan Management Review,* 32, no. 2 (Winter 1991) pp. 44–53.

Stevenson, William J. *Introduction to Management Science.* 2nd ed. Burr Ridge, IL: Richard D. Irwin, 1992.

SUPPLEMENT TO CHAPTER NINETEEN

SIMULATION

LEARNING OBJECTIVES

After completing this supplement, you should be able to:

1. Explain what is meant by the term *simulation*.
2. List some of the reasons for simulation's popularity as a tool for decision making.
3. Explain how and why random numbers are used in simulation.
4. Outline the advantages and limitations of simulation.
5. Describe the alternatives that a manager would reject before choosing simulation as a decision-making tool.
6. Solve typical problems that require the use of simulation.

SUPPLEMENT OUTLINE

simulation A descriptive technique that enables a decision maker to evaluate the behavior of a model under various conditions.

Simulation is a descriptive technique in which a model of a process is developed and then experiments are conducted on the model to evaluate its behavior under various conditions. Unlike many of the other models described in the text, simulation is not an optimizing technique. It does not produce a solution per se. Instead, simulation enables decision makers to test *their* solutions on a model that reasonably duplicates a real process; simulation models enable decision makers to experiment with decision alternatives using a *what if* approach.

The use of simulation as a decision-making tool is fairly widespread, and you are undoubtedly familiar with some of the ways it is used. For instance, space engineers simulate space flight in laboratories to permit future astronauts to become accustomed to working in a weightless environment. Similarly, airline pilots often undergo extensive training with simulated landings and takeoffs before being allowed to try the real thing. Many video games are simulations, and universities use management games as a means of simulating business environments. Tire designers evaluate alternative tread designs using machines that simulate conditions that produce tire wear and handling problems.

INTRODUCTION

Simulation has applications across a broad spectrum of operations management problems. In some instances, the simulations are quite modest, while others are rather complex. Their usefulness in all cases depends on the degree to which decision makers are able to answer their *what if* questions.

A list of operations management topics would reveal that most have simulation applications. For instance, simulation is often helpful in product design and testing, facilities layout, line balancing, job design, aggregate planning, testing alternative inventory policies, scheduling, and project management. Actually, the list is quite long. The books and articles in the bibliography indicate the richness and breadth of those applications and offer the interested reader an opportunity to explore this fascinating and useful subject in more detail.

Generally, analysts use the simulation approach either because optimization techniques are unavailable or because the assumptions required by an optimizing technique are not reasonably satisfied in a given situation. Waiting-line problems are a good example of the latter reason. Although waiting-line problems are pervasive, the rather restrictive assumptions of arrival and service distributions in many cases are simply not met. Very often, analysts will then turn to simulation as a reasonable alternative for obtaining descriptive information about the system in question.

Other reasons for the popularity of simulation include:

1. Many situations are too complex to permit development of a mathematical solution; the degree of simplification needed would seriously affect the results. In contrast, simulation models are often able to capture the richness of a situation without sacrificing simplicity, thereby enhancing the decision process.
2. Simulation models are fairly simple to use and understand.
3. Simulation enables the decision maker to conduct experiments on a model that will help in understanding process behavior while avoiding the risks of conducting tests on the model's real-life counterpart.
4. Extensive computer software packages make it easy to use fairly sophisticated models.
5. Simulation can be used for a wide range of situations.
6. There have been numerous successful applications of these techniques.

STEPS IN THE SIMULATION PROCESS

Regardless of the type of simulation involved, certain basic steps are used for all simulation models:

1. Identify the problem and set objectives.
2. Develop the simulation model.
3. Test the model to be sure that it reflects the system being studied.
4. Develop one or more experiments (conditions under which the model's behavior will be examined).
5. Run the simulation and evaluate the results.
6. Repeat steps 4 and 5 until you are satisfied with the results.

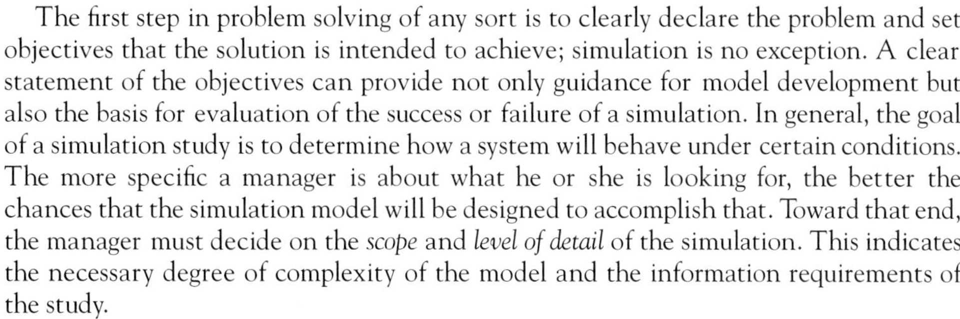

The first step in problem solving of any sort is to clearly declare the problem and set objectives that the solution is intended to achieve; simulation is no exception. A clear statement of the objectives can provide not only guidance for model development but also the basis for evaluation of the success or failure of a simulation. In general, the goal of a simulation study is to determine how a system will behave under certain conditions. The more specific a manager is about what he or she is looking for, the better the chances that the simulation model will be designed to accomplish that. Toward that end, the manager must decide on the *scope* and *level of detail* of the simulation. This indicates the necessary degree of complexity of the model and the information requirements of the study.

The next step is model development. Typically, this involves deciding on the structure of the model and using a computer to carry out the simulations. (For instructional purposes, the examples and problems in this chapter are primarily manual, but in most real-life applications computers are used. This stems from the need for large numbers of runs, the complexity of simulations, and the need for record-keeping of results.) Data gathering is a significant part of model development. The amount and type of data needed are a direct function of the scope and level of detail of the simulation. The data are needed for both model development and evaluation. Naturally, the model must be designed to enable evaluation of key decision alternatives.

The validation phase is closely related to model development. Its main purpose is to determine if the model adequately depicts real system performance. An analyst usually accomplishes this by comparing the results of simulation runs with known performance of the system under the same circumstances. If such a comparison cannot be made because, for example, real-life data are difficult or impossible to obtain, an alternative is to employ a test of reasonableness, in which the judgments and opinions of individuals familiar with the system or similar systems are relied on for confirmation that the results are plausible and acceptable. Still another aspect of validation is careful consideration of the assumptions of the model and the values of parameters used in testing the model. Again, the judgments and opinions of those familiar with the real-life system and those who must use the results are essential. Finally, note that model development and model validation go hand in hand: Model deficiencies uncovered during validation prompt model revisions, which lead to the need for further validation efforts and perhaps further revisions.

The fourth step in simulation is designing experiments. Experiments are the essence of a simulation; they help answer the *what if* questions posed in simulation studies. By going through the process, the manager or analyst learns about system behavior.

The fifth step is to run the simulation model. If a simulation model is deterministic and all parameters are known and constant, only a single run will be needed for each *what if* question. But if the model is probabilistic, with parameters subject to random variability, multiple runs will be needed to obtain a clear picture of the results. In this text, probabilistic simulations are the focal point of the discussion, and the comments are limited to them. Probabilistic simulation is essentially a form of random sampling, with each run representing one observation. Consequently, statistical theory can be used to determine appropriate sample sizes. In effect, the larger the degree of variability inherent in simulation results, the greater the number of simulation runs needed to achieve a reasonable level of confidence in the results as true indicators of model behavior.

The last step in the simulation process is to analyze and interpret the results. Interpretation of the results depends to a large extent on the degree to which the simulation model approximates reality; the closer the approximation, the less need to "adjust" the results. Moreover, the closer the approximation of the model to reality, the less the risk inherent in applying the results.

MONTE CARLO SIMULATION

Monte Carlo method Probabilistic simulation technique used when a process has a random component.

random Chance.

There are many different kinds of simulation techniques. The discussion will focus on probabilistic simulation using the **Monte Carlo method.** The technique gets its name from the famous Mediterranean resort associated with games of chance. The chance element is an important aspect of Monte Carlo simulation, and this approach can be used only when a process has a **random,** or chance, component.

In the Monte Carlo method, a manager identifies a probability distribution that reflects the random component of the system under study. Random samples taken from this probability distribution are analogous to observations made on the system itself. As the number of observations increases, the results of the simulation will more closely approximate the behavior of the real system, provided an appropriate model has been developed. Sampling is accomplished by the use of random numbers.

The basic steps in the process are as follows:

1. Identify a probability distribution for each random component of the system.
2. Work out an assignment so that intervals of random numbers will correspond to the probability distribution.
3. Obtain the random numbers needed for the study.
4. Interpret the results.

The random numbers used in Monte Carlo simulation can come from any source that exhibits the necessary randomness. Typically, they come from one of two sources: Large studies depend on computer-generated random numbers, and small studies commonly make use of numbers from a table of random digits like the one shown in Table 19S–1. The digits are listed in pairs for convenience, but they can be used singly, in pairs, or in whatever grouping a given problem calls for.

Two important features of the sets of random numbers are essential to simulation. One is that the numbers are uniformly distributed. This means that for any size grouping

TABLE 19S–1

Random digits

	1	2	3	4	5	6	7	8	9	10	11	12
1	18	20	84	29	91	73	64	33	15	67	54	07
2	25	19	05	64	26	41	20	09	88	40	73	34
3	73	57	80	35	04	52	81	48	57	61	29	35
4	12	48	37	09	17	63	94	08	28	78	51	23
5	54	92	27	61	58	39	25	16	10	46	87	17
6	96	40	65	75	16	49	03	82	38	33	51	20
7	23	55	93	83	02	19	67	89	80	44	99	72
8	31	96	81	65	60	93	75	64	26	90	18	59
9	45	49	70	10	13	79	32	17	98	63	30	05
10	01	78	32	17	24	54	52	44	28	50	27	68
11	41	62	57	31	90	18	24	15	43	85	31	97
12	22	07	38	72	69	66	14	85	36	71	41	58

of digits (e.g., two-digit numbers), every possible outcome (e.g., 34, 89, 00) has the same probability of appearing. The second feature is that there are no discernible patterns in sequences of numbers to enable one to predict numbers further in the sequence (thus the name *random* digits). This feature holds for any sequence of numbers; the numbers can be read across rows and up or down columns.

When using the table, it is important to avoid always starting in the same spot; that would result in the same sequence of numbers each time. Various methods exist for choosing a random starting point. One can use the serial number of a dollar bill to select the row, column, and direction of number selection. Or use rolls of a die. For our purposes, the starting point will be specified in each manual example or problem so that everyone obtains the same results.

The process of simulation will become clearer as we work through some simple problems.

Example S-1

The manager of a machine shop is concerned about machine breakdowns. He has made a decision to simulate breakdowns for a 10-day period. Historical data on breakdowns over the last 100 days are given in the following table:

Number of Breakdowns	Frequency
0	10
1	30
2	25
3	20
4	10
5	5
	100

Simulate breakdowns for a 10-day period. Read two-digit random numbers from Table 19S–1, starting at the top of column 1 and reading down.

Solution

a. Develop cumulative frequencies for breakdowns:

(1) Convert frequencies into relative frequencies by dividing each frequency by the sum of the frequencies. Thus, 10 becomes 10/100 = .10, 30 becomes 30/100 = .30, and so on.

(2) Develop cumulative frequencies by successive summing. The results are shown in the following table:

Number of Breakdowns	Frequency	Relative Frequency	Cumulative Frequency
0	10	.10	.10
1	30	.30	.40
2	25	.25	.65
3	20	.20	.85
4	10	.10	.95
5	5	.05	1.00
	100	1.00	

b. Assign random-number intervals to correspond with the cumulative frequencies for breakdowns. (*Note:* Use two-digit numbers because the frequencies are given to two decimal places.) You want a 10 percent probability of obtaining the event "0 breakdowns" in our simulation. Therefore, you must designate 10 percent of the possible random numbers as corresponding to that event. There are 100 two-digit numbers, so we can assign the 10 numbers 01 to 10 to that event.

Similarly, assign the numbers 11 to 40 to "one breakdown," 41 to 65 to "two breakdowns," 66 to 85 to "three breakdowns," 86 to 95 to "4 breakdowns" and 96 to 00 to five breakdowns.

Number of Breakdowns	Frequency	Probability	Cumulative Probability	Corresponding Random Numbers
0	10	.10	10	01 to 10
1	30	.30	.40	11 to 40
2	25	.25	.65	41 to 65
3	20	.20	.85	66 to 85
4	10	.10	.95	86 to 95
5	5	.05	1.00	96 to 00
	100	1.00		

c. Obtain the random numbers from Table 19S–1, column 1, as specified in the problem:

18 25 73 12 54 96 23 31 45 01

d. Convert the random numbers into numbers of breakdowns:

18 falls in the interval 11 to 40 and corresponds, therefore, to one breakdown on day 1.
25 falls in the interval 11 to 40, which corresponds to one breakdown on day 2.
73 corresponds to three breakdowns on day 3.
12 corresponds to one breakdown on day 4.
54 corresponds to two breakdowns on day 5.
96 corresponds to five breakdowns on day 6.
23 corresponds to one breakdown on day 7.
31 corresponds to one breakdown on day 8.
45 corresponds to two breakdowns on day 9.
01 corresponds to no breakdowns on day 10.

The following table summarizes these results:

Day	Random Number	Simulated Number of Breakdowns
1	18	1
2	25	1
3	73	3
4	12	1
5	54	2
6	96	5
7	23	1
8	31	1
9	45	2
10	01	0
		17

The mean number of breakdowns for this 10-period simulation is $17/10 = 1.7$ breakdowns per day. Compare this to the *expected* number of breakdowns based on the historical data:

$$0(.10) + 1(.30) + 2(.25) + 3(.20) + 4(.10) + 5(.05) = 2.05 \text{ per day}$$

Several points are worth noting:

1. This simple example is intended to illustrate the basic concept of Monte Carlo simulation. If our only goal were to estimate the average number of breakdowns, we would not have to simulate; we could base the estimate on the historical data alone.
2. The simulation should be viewed as a *sample*; it is quite likely that additional runs of 10 numbers would produce different means.

3. Because of the variability inherent in the results of small samples, it would be unwise to attempt to draw any firm conclusions from them; in an actual study, much larger sample sizes would be used.

In some cases, it is helpful to construct a flowchart that describes a simulation, especially if the simulation will involve periodic updating of system values (e.g., amount of inventory on hand), as illustrated in Example S–2.

The excel spreadsheet formulation for this problem is shown below. Note that the alignment of values in columns B, C, and D must be exactly as shown.

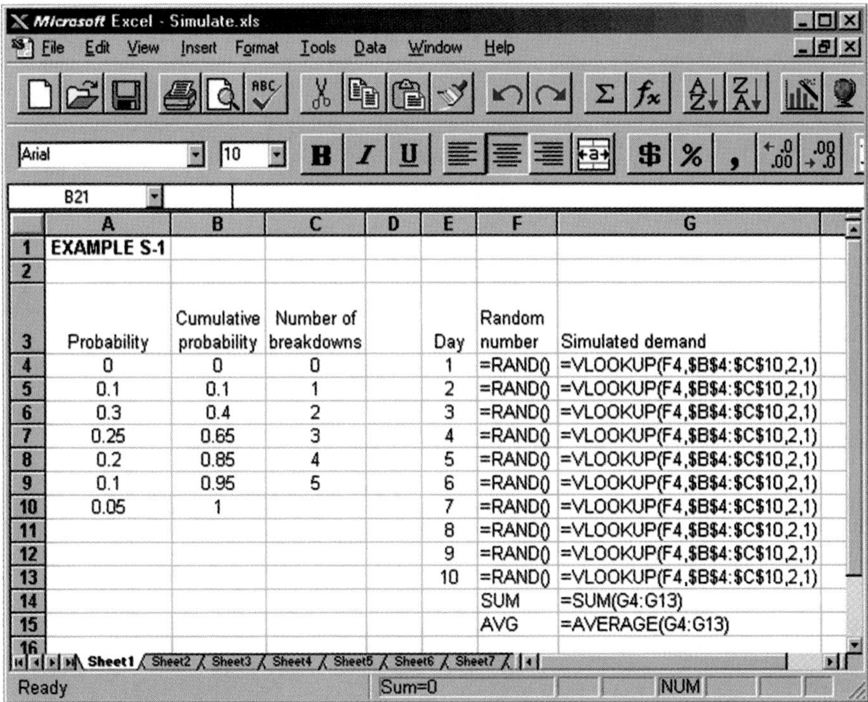

Microsoft Excel - Simulate.xls

	A	B	C	D	E	F	G
1	EXAMPLE S-1						
2							
3	Probability	Cumulative probability	Number of breakdowns		Day	Random number	Simulated demand
4	0	0	0		1	=RAND()	=VLOOKUP(F4,$B$4:$C$10,2,1)
5	0.1	0.1	1		2	=RAND()	=VLOOKUP(F4,$B$4:$C$10,2,1)
6	0.3	0.4	2		3	=RAND()	=VLOOKUP(F4,$B$4:$C$10,2,1)
7	0.25	0.65	3		4	=RAND()	=VLOOKUP(F4,$B$4:$C$10,2,1)
8	0.2	0.85	4		5	=RAND()	=VLOOKUP(F4,$B$4:$C$10,2,1)
9	0.1	0.95	5		6	=RAND()	=VLOOKUP(F4,$B$4:$C$10,2,1)
10	0.05	1			7	=RAND()	=VLOOKUP(F4,$B$4:$C$10,2,1)
11					8	=RAND()	=VLOOKUP(F4,$B$4:$C$10,2,1)
12					9	=RAND()	=VLOOKUP(F4,$B$4:$C$10,2,1)
13					10	=RAND()	=VLOOKUP(F4,$B$4:$C$10,2,1)
14						SUM	=SUM(G4:G13)
15						AVG	=AVERAGE(G4:G13)

The simulation results are shown in the following screen. Use key F4 to do a simulation or another simulation.

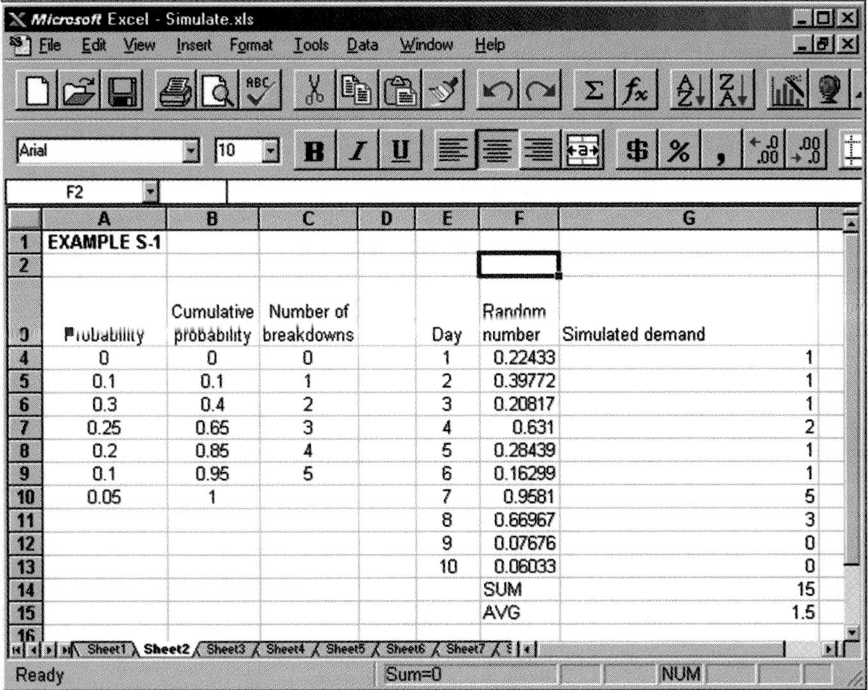

Microsoft Excel - Simulate.xls

	A	B	C	D	E	F	G
1	EXAMPLE S-1						
2							
3	Probability	Cumulative probability	Number of breakdowns		Day	Random number	Simulated demand
4	0	0	0		1	0.22433	1
5	0.1	0.1	1		2	0.39772	1
6	0.3	0.4	2		3	0.20817	1
7	0.25	0.65	3		4	0.631	2
8	0.2	0.85	4		5	0.28439	1
9	0.1	0.95	5		6	0.16299	1
10	0.05	1			7	0.9581	5
11					8	0.66967	3
12					9	0.07676	0
13					10	0.06033	0
14						SUM	15
15						AVG	1.5

Example S-2

The manager of a small truck dealership wants to acquire some insight into how a proposed policy for reordering trucks might affect order frequency. Under the new policy, two trucks are to be ordered whenever the number of trucks on hand is five or fewer. Due to the nearness of the dealer to the home office, orders can be filled overnight. According to the dealer's records, the probability distribution for daily demand is:

Demand, x	P(x)
0	.50
1	.40
2	.10

a. Construct a flowchart that describes a 10-day simulation.

b. Use two-digit random numbers from Table 19S–1, column 11, reading down. Assume a beginning inventory of seven trucks.

Solution

a.

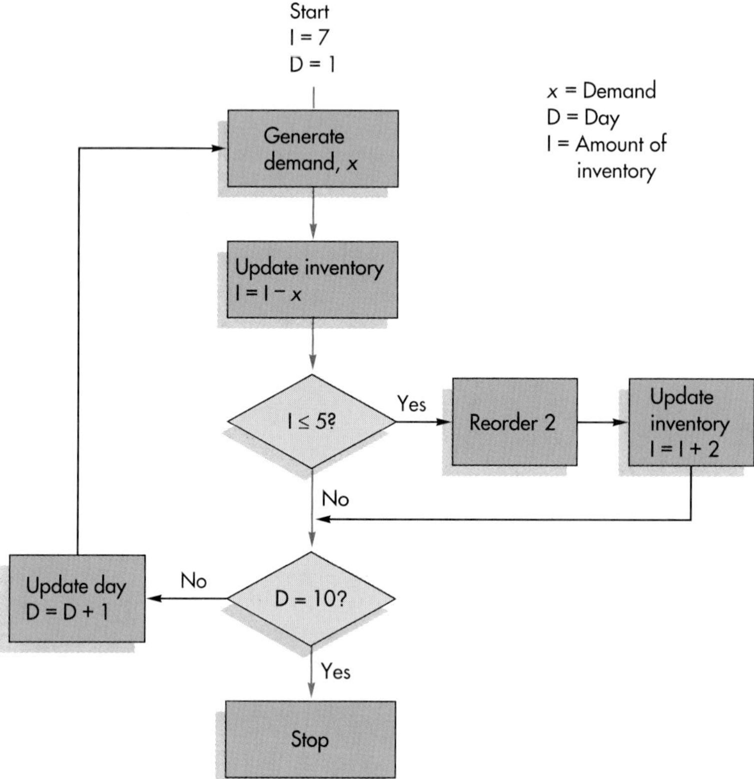

b. (1) Specify random number ranges for demand:

x	P(x)	Cumulative P(x)	Ranges
0	.50	.50	01–50
1	.40	.90	51–90
2	.10	1.00	91–00

(2) Obtain random numbers, convert to demand, update inventory accordingly, and reorder when necessary:

Day	Random Number	Demand, x	Beginning Inventory	Ending Inventory
1	54	1	7	6
2	73	1	6	5 (reorder 2; new beginning inventory = 5 + 2)
3	29	0	7	7
4	51	1	7	6
5	87	1	6	5 (reorder 2; new beginning inventory = 5 + 2)
6	51	1	7	6
7	99	2	6	4 (reorder 2; new beginning inventory = 4 + 2)
8	18	0	6	6
9	30	0	6	6
10	27	0	6	6

Simulating Theoretical Distributions

In many instances, a simulation will involve the use of theoretical distributions. Among the most frequently encountered theoretical distributions are the Poisson, normal, and exponential distributions. Being able to simulate these distributions will greatly enhance your knowledge and appreciation of simulation.

Simulation of a Poisson distribution requires the mean of the distribution. Knowledge of the mean enables one to obtain cumulative probabilities for the distribution from Appendix Table C; these, in turn, provide the basis for random-number assignments. You can use Table 19S–1 to obtain random numbers; you must read three-digit random numbers from Table 19S–1 to achieve correspondence. Example S–3 illustrates these concepts.

The normal distribution can be important in many problems. There are a number of ways to simulate a normal distribution, but perhaps the simplest is to use a table of normally distributed random numbers, such as Table 19S–2. The basis of the table is a normal distribution with a mean of 0 and a standard deviation of 1.00. Like all such tables, the numbers are arranged randomly, so that when they are read in any sequence they exhibit "randomness." In order to use the table, we must have the parameters of a normal distribution (i.e., its mean and standard deviation) in mind. Numbers obtained from the random number table can then be converted to "actual" values by multiplying the standard deviation by the random number and adding this amount to the mean. That is:

$$\text{Simulated value} = \text{Mean} + \text{Random number} \times \text{Standard deviation} \quad (19S\text{–}1)$$

In effect, the random number equates to a normal z value, which indicates how far a particular value is above or below the distribution mean.

Example S-3

The number of lost-time accidents at a large plant has been determined from historical records to be two per day. Moreover, it has been determined that this accident rate can be well approximated by a Poisson distribution that has a mean of 2.0. Simulate five days of accident experience for the plant. Read random numbers from columns 1 and 2 of Table 19S–1.

TABLE 19S–2

Normally distributed random numbers

	1	2	3	4	5	6	7	8	9	10
1	1.46	−0.09	−0.59	0.19	−0.52	−1.82	0.53	−1.12	1.36	−0.44
2	−1.05	0.56	−0.67	−0.16	1.39	−1.21	0.45	−0.62	−0.95	0.27
3	0.15	−0.02	0.41	−0.09	−0.61	−0.18	−0.63	−1.20	0.27	−0.50
4	0.81	1.87	0.51	0.33	−0.32	1.19	2.18	−2.17	1.10	0.70
5	0.74	−0.44	1.53	−1.76	0.01	0.47	0.07	0.22	−0.59	−1.03
6	−0.39	0.35	−0.37	−0.52	−1.14	0.27	−1.78	0.43	1.15	−0.31
7	0.45	0.23	0.26	−0.31	−0.19	−0.03	−0.92	0.38	−0.04	0.16
8	2.40	0.38	−0.15	−1.04	−0.76	1.12	−0.37	−0.71	−1.11	0.25
9	0.59	−0.70	−0.04	0.12	1.60	0.34	−0.05	−0.26	0.41	0.80
10	−0.06	0.83	−1.60	−0.28	0.28	−0.15	0.73	−0.13	−0.75	−1.49

Solution

First obtain the cumulative distribution from Appendix Table C for a mean of 2.0, and make the range assignments:

x	Cumulative Probability	Random Number Ranges
0	.135	001 to 135
1	.406	136 to 406
2	.677	407 to 677
3	.857	678 to 857
4	.947	858 to 947
5	.983	948 to 983
6	.995	984 to 995
7	.999	996 to 999
8	1.000	000

Next obtain three-digit numbers from Table 19S–1. Reading from column 1 and 2 as instructed, you find 182, 251, 735, 124, and 549.

Finally, convert the random numbers into number of lost-time accidents using the established set of ranges. Since 182 falls in the second range, it corresponds to one accident on day 1. The second random number, 251, falls in the same range, indicating one accident on day 2. The number 735 falls between 678 and 857, which corresponds to three accidents on day 3; 124 corresponds to 0 accidents on day 4; and 549 corresponds to two accidents on day 5.

Example S–4

It has been determined that the time required to perform a certain task can be described by a normal distribution that has a mean of 30 minutes and a standard deviation of 4 minutes. Simulate times for three jobs using the first three values in column 1 of Table 19S–2.

Solution

The first three values are: 1.46, −1.05, and 0.15. The simulated values are:

For 1.46, $30 + 1.46(4) = 35.84$ minutes.

For −1.05, $30 - 1.05(4) = 25.80$ minutes.

For 0.15, $30 + 0.15(4) = 30.60$ minutes.

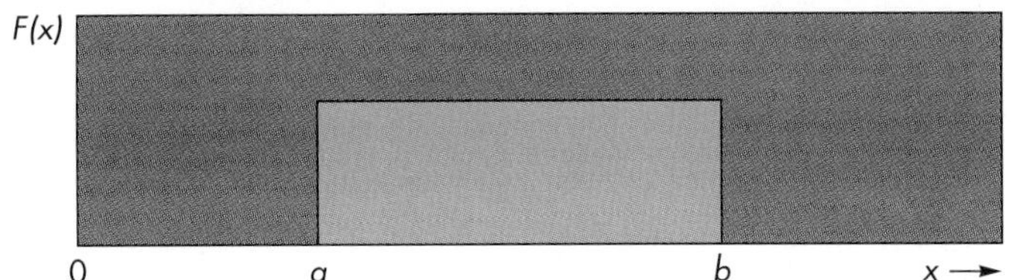

FIGURE 19S–1

A uniform distribution

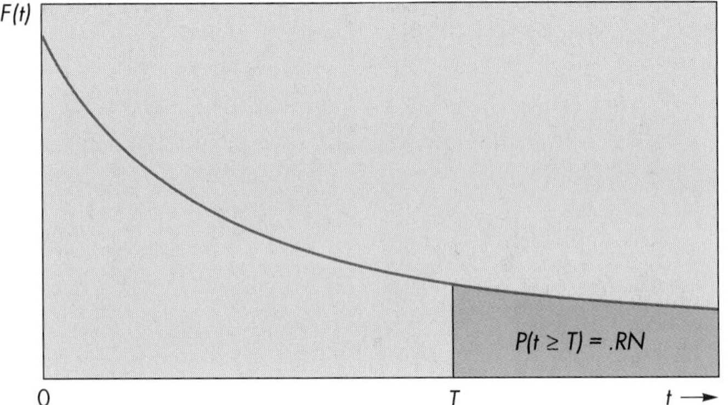

FIGURE 19S–2

A negative exponential distribution

It is important to recognize that Example 4 involves a continuous variable, whereas the previous examples have involved discrete variables. (Remember that discrete variables typically take on only integer values, whereas continuous variables can take on integer and noninteger values.) Whenever possible, a model of a continuous variable should be able to simulate noninteger values as well as integer values.

Another continuous type of distribution we can consider is the uniform distribution, in which values may occur anywhere over a continuous range between two extremes, a and b, as illustrated in Figure 19S–1.

$$\text{Simulated value} = a + (b - a)(\text{Random number as a percentage}) \qquad (19S\text{–}2)$$

Converting the random number to a percentage simply involves placing a decimal point to the left of the number. For example, 77 becomes .77.

A third continuous distribution is the exponential distribution. We will concern ourselves with simulating values of negative exponential distributions, as portrayed in Figure 19S–2.

Example S-5

Job times vary uniformly between 10 and 15 minutes. Use Table 19S–1 to simulate job times for four jobs. Read numbers from column 9, going down.

Solution

$a = 10$ minutes, $\quad b = 15$ minutes, $\quad b - a = 5$ minutes

a. Obtain the random numbers: 15, 88, 57, and 28.

b. Convert to simulated values:

Random Number	Computation		Simulated Value (minutes)
15	10 + 5(.15)	=	10.75
88	10 + 5(.88)	=	14.40
57	10 + 5(.57)	=	12.85
28	10 + 5(.28)	=	11.40

With a negative exponential distribution, the probability is fairly high that the random variable will assume a value close to zero. Moreover, the probability decreases as the specified value of the random variable increases. The probability that a random variable will take on a value greater than some specified value T, given that the variable can be described by an exponential distribution with a mean equal to $1/\lambda$, is given by the equation.

$$P(t \geq T) = e^{-\lambda t} \qquad (19S\text{–}3)$$

To simulate exponential values, we obtain a random number, place a decimal point to the left of it, set this equal to the probability $P(T)$, and solve Formula 19S–3 for t. The result is a simulated value from an exponential distribution with a mean of λ.

We can obtain an expression for t by taking the natural logarithm of both sides of the equation. Thus, with $P(T)$ = .RN (for *random number*), we have

$$\ln(.\text{RN}) = \ln(e^{-\lambda t})$$

The natural logarithm of a power of e is equal to the power itself, so

$$\ln(.\text{RN}) = \ln(e^{-\lambda t}) = -\lambda t$$

Then

$$t = -\frac{1}{\lambda}\ln(.\text{RN}) \qquad (19S\text{–}4)$$

This concept is illustrated in Figure 19S–2.

Values of random numbers can be obtained using Table 19S–1, as demonstrated in the following example.

Example S–6

Times between breakdowns of a certain type of equipment can be described by an exponential distribution with a mean of five hours. Simulate the time between two breakdowns. Read two-digit random numbers from column 3 of Table 19S–1.

Solution

The mean, $1/\lambda$, is 5 hours. The random numbers are 84 and 05. Using Formula 19S–4, the simulated times are:

For 84, $t = -5[\ln(.84)] = -5[-0.1744] = 0.872$ hours.
For 05, $t = -5[\ln(.05)] = -5[-2.9957] = 14.979$ hours.

Note that the smaller the value of the random number, the larger the simulated value of t.

COMPUTER SIMULATION

Although the emphasis in this supplement has been on manual simulation in order to convey the main concepts, most real-life simulations involve the use of a computer. Computers offer relatively quick and easy means of obtaining results. Many real-life applications involve fairly complex simulation models that have substantial record-keeping requirements for which computers offer a practical solution.

Over the years, programmers have developed a number of simulation languages that make the task of writing simulation programs much simpler. Some of the general-purpose languages are SIMSCRIPT II.5, GPSS/H, GPSS/PC, and RESQ. In addition, a number of other simulation packages are available, some of which have very narrow focuses that relate to queuing or network problems. Most of the simulation packages have certain features in common. For example, they generally provide for random number generation from a variety of statistical distributions, as well as collection and tabulation of simulation results and time keeping.

It should be noted that some managers prefer to write their own simulations or have a member of their staff do so, using spreadsheet software such as MSExcel rather than a simulation language, as illustrated on page 851. In cases where simulation is used infrequently, it can be more practical to use that approach rather than go through the time and effort required to use a specialty language. For simple problems, the slight inefficiency related to the use of a standard language is not usually an important consideration.

ADVANTAGES AND LIMITATIONS OF USING SIMULATIONS

Among the main advantages of simulation are these:

1. It lends itself to problems that are difficult or impossible to solve mathematically.
2. It permits an analyst to experiment with system behavior while avoiding possible risks inherent in experimenting with the actual system.
3. It compresses time so that managers can quickly discern long-term effects.
4. It can serve as a valuable tool for training decision makers by building up their experience and understanding of system behavior under a wide range of conditions.

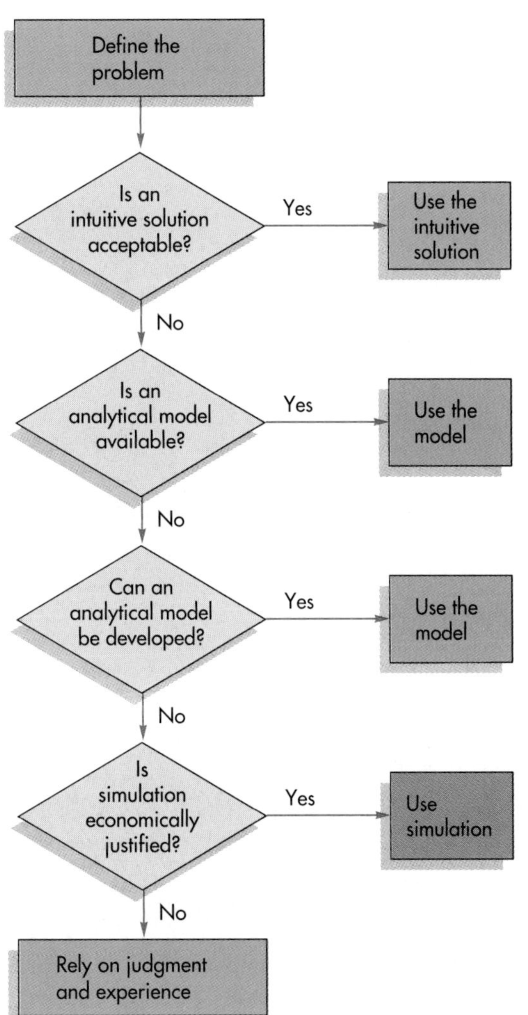

FIGURE 19S-3

Choosing a solution technique

Certain limitations are also associated with simulation. Chief among these are:

1. Simulation does not produce an optimum solution; it merely indicates an *approximate* behavior for a given set of inputs. There are two reasons for this:
   a. By design, there is inherent randomness (i.e., random numbers) in simulation.
   b. Simulations are based on models, and models are only approximations of reality.
2. For large-scale simulation, it can require considerable effort to develop a suitable model as well as considerable computer time to obtain simulations.

Because simulation produces an approximate answer rather than an exact solution, and because of the cost of running a simulation study, simulation is not usually the first choice of a decision maker. Instead, depending on the complexity of the situation, intuitive or analytical methods should first be investigated. In simple cases, an intuitive solution very often is acceptable. In more complex cases, an analytical solution is preferable, assuming an appropriate technique is available. If not, it may be possible to develop an analytical model that could be used to generate a solution. If these measures do not suffice, simulation becomes the next logical possibility. Of course, if that is not economically justifiable, the decision maker will have to rely on judgment and experience; in effect, after reevaluating all of the alternatives, the decision maker may revert to an intuitive solution, even though initially that approach did not appear acceptable. Figure 19S–3 on page 857 outlines this process.

Key Terms

Solved Problems

Problem 1

The number of customers who arrive at a transmission repair shop can be described by a Poisson distribution that has a mean of three per hour. Assuming the distribution holds for an entire eight-hour day, simulate customer arrivals for the first four hours of a day. Read random numbers from Table 19S–1, columns 4 and 5, going down.

Solution

a. Obtain cumulative probabilities of the Poisson distribution from Appendix Table C for the mean specified. (Those values are given below for convenience.) Determine the random number ranges.

x	Cumulative Probability
0	.050
1	.199
2	.423
3	.647
4	.815
5	.916
6	.966
7	.988
8	.996
9	.999
10	1.000

b. Obtain the random numbers: 299, 642, 350, and 091 (Note: Three-digit numbers are needed because the probabilities are given to three decimal places.)

c. Convert the random numbers to numbers of arrivals. Note where each number falls in the random number range list. For instance, 299 falls between 199 and 423. Interpret this to mean two customers arrive in the first hour. Similarly, 642 is interpreted to mean three customers arrive in the second hour, 350 implies two customers in the third hour, and 091 implies one customer in the fourth hour.

In sum, the number of customers per hour for the four-hour simulation is:

Hour	Number of Arrivals
1	2
2	3
3	2
4	1

Problem 2

Jobs arrive at a workstation at fixed intervals of one hour. Processing time is approximately normal and has a mean of 56 minutes per job and a standard deviation of 4 minutes per job. Using the fifth row of the table of normally distributed random numbers (Table 19S–2), simulate the processing times for four jobs, and determine the amount of operator idle time and job waiting time. Assume the first job arrives at time = 0.

Solution

a. Obtain the random numbers from the table: 0.74, −0.44, 1.53, and −1.76.

b. Convert the random numbers to simulated processing times:

Random Number	Computation	Simulated Time
0.74	56 + 4(0.74) =	58.96
−0.44	56 + 4(−0.44) =	54.24
1.53	56 + 4(1.53) =	62.12
−1.76	56 + 4(−1.76) =	48.96

Note that three of the times are less than the interarrival times for the jobs, meaning the operator may be idle after those three jobs. One time exceeds the one-hour interval, so the next job must wait, and possibly the job following it if the waiting plus processing time exceeds 60 minutes.

c. Compute waiting and idle times:

Job Number	Arrives at	Processing Time, t (minutes)	$60 - t$ Operator Idle (minutes)	$t - 60$ Next Job Waits (minutes)
1	0	58.96	1.04	—
2	60	54.24	5.76	—
3	120	62.12	—	2.12
4	180	48.96	8.92*	—
			15.72	2.12

*60 − 2.12 − 48.96 = 8.92

Problem 3

The time between mechanics' requests for tools in a large plant is normally distributed with a mean of 10 minutes and a standard deviation of 1 minute. The time to fill requests is also normal with a mean of 9 minutes per request and a standard deviation of 1 minute. Mechanics' waiting time represents a cost of $2 per minute, and servers represent a cost of $1 per minute. Simulate arrivals for the first nine mechanic requests and their service times, and determine the mechanics' waiting time, assuming one server. Would it be economical to add another server? Explain. Use Table 19S–2, column 8 for requests and column 9 for service.

Solution

a. Obtain random numbers and convert to times [see columns (*a*) and (*b*) in the following table for requests and columns (*f*) and (*g*) for service].

CUSTOMER ARRIVALS				SERVICE			
(*a*) Random → Number	(*b*) Time between Arrivals	(*c*) Cumulative Arrival Time	(*d*) (*e* − *c*) Customer Waiting Time	(*e*) Service Begins	(*f*) Random → Number	(*g*) Service Time	(*h*) (*e* + *g*) Service Ends
−1.12	8.88	8.88	0.00	8.88	1.36	10.36	19.24
−0.62	9.38	18.26	0.98	19.24	−0.95	8.05	27.29
−1.20	8.80	27.06	0.23	27.29	0.27	9.27	36.56
−2.17	7.83	34.89	1.67	36.56	1.10	10.10	46.66
0.22	10.22	45.11	1.55	46.66	−0.59	8.41	55.07
0.43	10.43	55.54	0.00	55.54	1.15	10.15	65.69
0.38	10.38	65.92	0.00	65.92	−0.04	8.96	74.88
−0.71	9.29	75.21	0.00	75.21	−1.11	7.89	83.10
−0.26	9.74	84.95	0.00	84.95	0.41	9.41	94.36
			4.43				

b. Determine arrival times [column (*c*)] by successive adding to times between arrivals in column (*b*).

c. Use arrival times for service start *unless service is still in progress on a previous request*. In that case, determine how long the arrival must wait (*e* − *c*). Column (*e*) values are the sum of starting time and service time [column (*g*)], which is the time service ends [column (*h*)]. Thus, service on each new request begins [column (*e*)] at the same time that service on the previous request ends [column (*h*)].

d. The simulation and resulting waiting times for the first nine arrivals are shown in the table. Total waiting time is 4.43 minutes.

e. The total cost for the 94.36 minutes (end of service on the ninth request) of the simulation is:

Waiting cost	4.43 minutes at $2 per minute =	$ 8.86
Server cost	94.36 minutes at $1 per minutes =	94.36
		$103.22

f. Usually, a second simulation with two servers would be needed (but with the same arrival times so that the results are comparable). However, in this case it is apparent that a second server would increase server cost by about $94 but could not eliminate more than $8.86 of waiting cost. Hence, the second server would not be justified.

Discussion and Review Questions

1. What is a simulation?
2. What are some of the primary reasons for the widespread use of simulation techniques in practice?
3. What are some of the ways managers can use simulation?
4. What role do random numbers play in Monte Carlo simulations?
5. How would you respond to the following comment? "I ran the simulation several times, and each run gave me a different result. Therefore, the technique does not seem to be useful. I need answers!"
6. List the main advantages of simulation.
7. What are some of the limitations of simulation as a tool for decision making?

Problems

1. The number of jobs received by a small shop is to be simulated for an eight-day period. The shop manager has collected the following data:

Number of Jobs	Frequency
2 or less	0
3	10
4	50
5	80
6	40
7	16
8	4
9 or more	0
	200

Use the third column of Table 19S–1 and read two-digit numbers, going down. Determine the average number of jobs per day for the eight-day simulation period.

2. Jack M. sells insurance on a part-time basis. His records on the number of policies sold per week over a 50-week period are:

Number Sold	Frequency
0	8
1	15
2	17
3	7
4	3
	50

Simulate three five-day periods. Use Table 19S–1, column 6 for the first simulation, column 7 for the second, and column 8 for the third. In each case, read two-digit numbers, beginning at the *bottom* of the column and going *up*. For each simulation, determine the percentage of days on which two or more policies are sold.

3. After a careful study of requests for a special tool at a large tool crib, an analyst concluded that demand for the tool can be adequately described by a Poisson distribution with a mean of two requests per day. Simulate demand for a 12-working-day period for this tool using Table 19S–1. Read three-digit numbers from columns 5 and 6 combined, starting at the *top* and reading *down* (e.g., 917, 264, 045).

4. The number of lost-time accidents at a logging firm can be described using a Poisson distribution that has a mean of four accidents per month. Using the last two columns of Table 19S–1 (e.g. 540, 733, 293), simulate accidents for a 12-month period.

5. The time a physician spends with patients can be modeled using a normal distribution that has a mean of 20 minutes and a standard deviation of 2 minutes. Using the table of normally distributed random numbers (Table 19S–2), simulate the times the doctor might spend with the next seven patients. Use column 4 of the table; start at the bottom of the column and read *up*.

6. Jobs are delivered to a workstation at random intervals. The time between job arrivals tends to be normally distributed with a mean of 15 minutes and a standard deviation of 1 minute. Job processing time is also normally distributed with a mean of 14 minutes per job and a standard deviation of 2 minutes.

   a. Using Table 19S–2, simulate the arrival and processing of five jobs. Use column 4 of the table for job arrival times and column 3 for processing times. Start each column at row 4. Find the total times jobs wait for processing.

Avg # of jobs: 5.25

# of Jobs
6
3
6
5
4
5
7
6
42

1: 80 percent
2: 40 percent
3: 40 percent

3.

Day	Demand	Day	Demand
1	4	7	0
2	1	8	2
3	0	9	1
4	1	10	1
5	2	11	4
6	1	12	3

4.

Month	# Acc.	Month	# Acc.
1	4	7	10
2	5	8	2
3	3	9	3
4	4	10	3
5	6	11	3
6	4	12	3

5.

Patient	Time
1	19.44 min.
2	20.24
3	17.92
4	19.38
5	18.96
6	16.48
7	20.66

6*a*.

Job	Waiting
1	0 min.
2	1.78
3	4.36
4	2.93
5	3.49

Total: 12.56 min.

b.

Job	Waiting Time
1	0 min.
2	.27
3	.32
4	0
5	0

Total lost $44.05
Stay with plan
(*a*) cost $37.68

7.

Day	Usage	Day	Usage
1	43.932	6	30.438
2	48.036	7	33.298
3	34.908	8	41.678
4	32.758	9	33.526
5	42.186	10	30.904

8.

Week	Usage	Week	Usage
1	2	6	4
2	1	7	5
3	3	8	3
4	1	9	1
5	1	10	2

a. See IM.
b.

Day	Break	Order
1	2	—
2	0	—
3	0	—
4	0	—
5	1	3
6	0	—
7	0	—
8	1	1
9	0	—
10	0	—
11	2	3
12	1	—

12. *a.*

Cust.	Wait	Process
1	0	12.76
2	0	13.76
3	0	12.6
4	0	10.66
5	0	15.44
6	1.61	15.86
7	0	15.76
8	3.07	13.58
9	0	14.48

b. The company is considering the use of new equipment that would result in processing time that is normal with a mean of 13 minutes and a standard deviation of 1 minute. Job waiting represents a cost of $3 per minute, and the new equipment would represent an additional cost of $.50 per minute. Would the equipment be cost justified? (Note: Use the same arrival times and the same random numbers for processing times.)

7. Daily usage of sugar in a small bakery can be described by a uniform distribution with endpoints of 30 pounds and 50 pounds. Assuming usage is independent of the day of the week, simulate daily usage for a 10-day period. Read four-digit numbers from Table 19S–1, columns 5 and 6, going *up* from the bottom.

8. Weekly usage of spare parts for a specialized machine can be described by a Poisson distribution with a mean of 2.8 parts per week. Lead time to replenish the supply of spare parts is two weeks (constant). Simulate the total usage of parts during lead time 10 times, and then determine the frequency of lead time demands (i.e., what percentage of times was the demand equal to 2, 3, 4, etc.?). Read four-digit numbers from Table 19S–1, columns 8 and 9, going down.

9. (*Computer exercise.*) Repeat Problem 8 for 100 lead time periods.

10. A repair shop breaks an average of 0.6 tool per day. The average number of days required to obtain replacements is six. (Parts are delivered by mail once each day.) Both breakages and delivery times can be described by Poisson distributions. Tools are reordered whenever three or more must be replaced.

    a. Draw a flowchart to describe this process.

    b. Simulate breakage and ordering for a 12-day period. Read three-digit numbers from Table 19S–1, columns 5 and 6, going down (e.g., 917, 264), for tool breakage, and columns 7 and 8, going down (e.g., 643, 200), for delivery time. Assume zero tools in inventory to start.

11. (*Computer exercise.*) Repeat Problem 10 for 150 days.

12. Customers arrive randomly at a catalog department of a large store. The time between arrivals varies uniformly between 10 and 20 minutes. Service time is normal with a mean of 15 minutes and a standard deviation of 2 minutes.

    a. Simulate processing and waiting times for nine customers. Read three-digit numbers going down columns 9 and 10 of Table 19S–1 for arrivals (e.g., 156, 884, 576). Use column 8, Table 19S–2, for processing time.

    b. If management can reduce the range of arrival times to between 13 and 17 minutes, what would the impact be on customer waiting times? (Use the same service times and the same random numbers for arrival times from part *a.*) Round arrival times to two decimal places.

13. Probabilities have been determined for the movements of the ball in a pinball game. These are shown in the accompanying table, along with the points awarded if the ball strikes a given position. Simulate the paths of three balls, and compute the number of points awarded for each ball. Use column 1 of Table 19S–1 for the first ball, column 2 for the second ball, and column 3 for the third ball, reading down the columns.

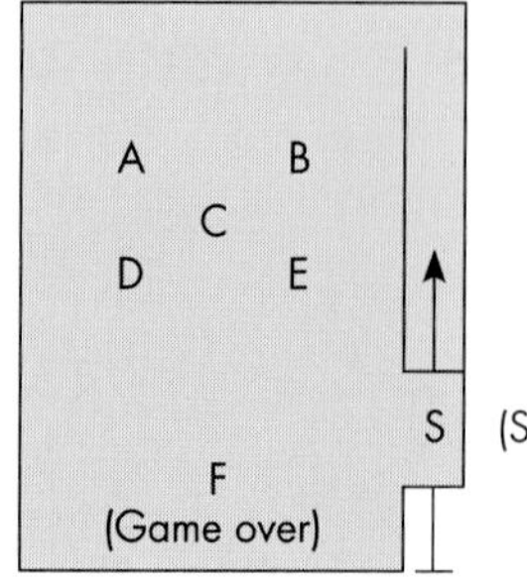

Position	Points
A	400
B	300
C	200
D	100
E	50
F	20

Path	Probability	Path	Probability	Path	Probability
S to A	.30	B to A	.30	D to A	.10
B	.30	B	.20	B	.05
C	.25	C	.20	C	.25
D	.10	D	.15	D	.20
E	.04	E	.10	E	.25
F	.01	F	.05	F	.15
A to A	.25	C to A	.10	E to A	.05
B	.25	B	.15	B	.10
C	.25	C	.20	C	.15
D	.10	D	.15	D	.15
E	.10	E	.25	E	.20
F	.05	F	.15	F	.35

(*Hint:* Construct a cumulative probability distribution for each of the six cases. Each new random number indicates the next position to move to. Each ball must start at S.)

14. Repeat problem 13, using columns 1, 2, and 3 for the next three balls, reading *up* from the bottom.

15. An analyst found that the length of telephone conversations in an office could be described by an exponential distribution with a mean of four minutes. Reading two-digit random numbers from Table 19S–1, column 6, simulate the length of five calls and compute the simulated average time. Why is the simulated average different from the mean of four minutes?

16. The length of time between calls for service of a certain piece of equipment can be described by an exponential distribution with a mean of 40 minutes. Service time can be described by a normal distribution with a mean of eight minutes and a standard deviation of two minutes. Simulate the time until the first breakdown and the times between three breakdowns, as well as the four service times. For breakdowns, read two-digit numbers from Table 19S–1, column 7; for service times, read numbers from Table 19S–2, column 8.

17. The number of jobs per day that a repair shop gets can be described by a Poisson distribution with a mean of 3.0. Repair time per job can be described by an exponential distribution with a mean of six hours. Simulate the number of jobs received for a four-day period and the repair time for each job. What is the simulated total repair time per day? For number of jobs received, read three-digit numbers from Table 19S–1, row 1; for repair times, read two-digit numbers from row 2.

18. A service operation consists of three steps. The first step can be described by a uniform distribution that ranges between five and nine minutes. The second step can be described by a normal distribution with a mean of seven minutes and a standard deviation of one minute, and the third step can be described by an exponential distribution with a mean of five minutes. Simulate three cycles using two-digit numbers from Table 19S–1, row 4 for step 1; Table 19S–2, row 6 for step 2; and two-digit numbers from column 4 of Table 19S–1 for step 3. Determine the simulated time for each of the three cycles.

19. A project consists of five major activities, as illustrated in the accompanying diagram. Activity times are normally distributed with means and standard deviations as shown in the following table. Note that there are two paths through the project: 1-2-3-5 and 1-2-4-5. Project duration is defined as the largest sum of times along a path. Simulate 12 times for each activity. Use columns 1 and 2 of Table 19S–2 for activity 1-2, columns 3 and 4 for activity 2-3, columns 5 and 6 for activity 2-4, columns 7 and 8 for activity 3-5, and columns 9 and 10 for activity 4-5. Determine the project duration for each of the 12 sets, and prepare a frequency distribution of project duration. Use categories of 25 to less than 30, 30 to less than 35, 35 to less than 40, 40 to less than 45, and 45 or more. Determine the proportion of

b.

Cust.	Wait
1	0
2	0
3	0
4	0
5	0
6	.91
7	.55
8	2.24
9	0

13. Ball 1: 1,520 pts.
Ball 2: 1,120 pts.
Ball 3: 1,170 pts.

Ball 1: 2,120 pts.
Ball 2: 720 pts.
Ball 3: 770 pts.

Call	Time
1	1.259
2	3.566
3	2.616
4	1.848
5	3.766

Avg. = 2.611 min/call

Break	Wait	Service
1	0	5.76
2	0	6.76
3	0	5.6
4	3.12	3.66

Day	Jobs	Time	Total Time
1	1	8.32	8.32
2	1	9.96	9.96
3	2	17.97 2.68	20.65
4	1	8.08	8.08

Cycle	T1	T2	T3
1	5.48	6.61	6.19
2	6.92	7.35	2.23
3	6.48	6.63	5.25

Cycle	Total Time
1	18.28
2	16.5
3	18.36

19. See IM.

time that a simulated duration of less than 40 days occurred. How might this information be used?

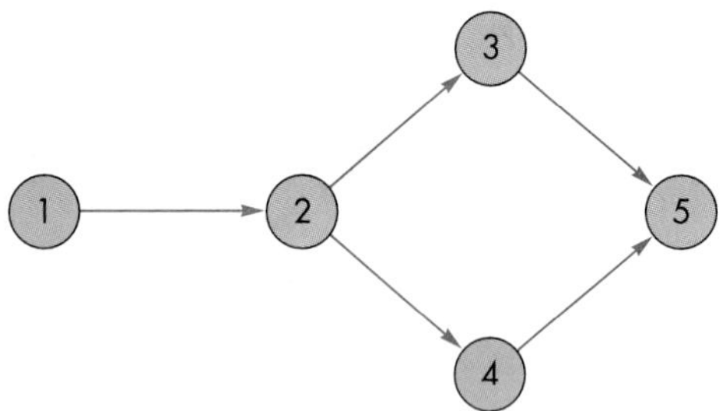

Activity	Mean (days)	Standard Deviation (days)
1-2	10	2
2-3	12	2
2-4	15	3
3-5	14	2
4-5	8	1

CASE
COQUILLE REFINERY CORPORATION

The Coquille Refinery Corporation is contemplating building a crude-oil storage and docking facility on the southern coast of France. They plan to import crude oil by ship from the Middle East and distribute this crude oil by pipeline to refineries in the area.

The construction of this facility represents a substantial capital investment. Furthermore, the cost of such a facility is principally determined by its crude-oil storage capacity. You have been asked to study the problem and recommend an appropriate storage capacity; bear in mind that too large a capacity represents an unnecessary expense, but too small a capacity will result in costly later additions to the facility.

A long-term contract has been made with the Middle East supplier to furnish an average daily supply of 300,000 barrels of crude oil. Because its fleet of ships consists of 200,000-barrel tankers, the supplier expects that the arrival of its tankers will follow the distribution below:

Tanker Arrivals per Day	Probability
0	0.1
1	0.5
2	0.2
3	0.2

A review of past production records of refineries in the area suggests the following distribution of crude-oil demand per day:

Barrels per Day	Probability
100,000	0.1
200,000	0.2
300,000	0.3
400,000	0.4

Questions

1. Consider the following issues before you simulate:
   a. What is the expected daily demand for crude oil? Why must this be so?
   b. What assumptions concerning the timing of crude-oil receipts and deliveries would require the greatest oil storage capacity?
   c. What assumption concerning receipts and deliveries would require the least oil storage capacity?
   d. Give a reason based on systems-analysis considerations why back orders should be filled from the next day's receipts rather than considered as lost sales.
2. Develop a Monte Carlo simulation model for Coquille that will generate information useful for resolving the storage capacity problem and simulate 10 days of capacity.
3. Assume a computer program was written to simulate 10,000 days of activity under the assumptions listed in question 1*b*. From the results of such a simulation the following distribution of oil in storage, after a day's receipts, is determined. (Note: Negative figures represent back orders.)

Oil in Storage, in Thousands	Probability
−300	0.01
−200	0.04
−100	0.06
0	0.07
⋮	⋮
1,000	0.09
1,100	0.08
1,200	0.07
1,300	0.05
1,400	0.03
1,500	0.02
	1.00

   a. What level of crude-oil safety stock should be carried to ensure 95 percent protection against stock-outs (1 stock-out in 20 days)?
   b. If Coquille decides to use the above safety stock, what should be the oil storage capacity to ensure 90 percent protection against overruns (i.e., sufficient

capacity to accommodate completely receipts 9 days out of 10)?

4. Note two ways one might determine if the simulation run length of 10,000 days is adequate.
5. Make a list of cost factors that would influence the final selection of oil storage capacity.

Source: James A. Fitzsimmons and Robert S. Sullivan, *Service Operations Management* (New York: McGraw-Hill, 1982), pp. 78–80. Reprinted by permission.

Selected Bibliography

Graybeal, W., and U. Pooch. *Simulation: Principles and Methods.* Cambridge: Winthrop Publishers, 1980.

Law, A. M. "Introduction to Simulation." *Industrial Engineering,* May 1986, pp. 46–63.

Law, A. M., and W. D. Kelton. *Simulation Modeling and Analysis.* 2nd ed. New York: McGraw-Hill, 1991.

Watson, Hugh J. *Computer Simulation in Business.* New York: John Wiley & Sons, 1981.

APPENDIX A

ANSWERS TO SELECTED PROBLEMS

Chapter 2: Productivity, Competitiveness, and Strategy

1. Last week: Labor productivity = 37.5 bookcases per worker.
   Week before: Labor productivity = 40 bookcases per worker.
2. A crew size of two had the highest productivity (250 yards installed).
3. Week 1: 5.62.
   Week 2: 5.45.
   Week 3: 5.20.
4. *a*. Before Labor productivity = 16 carts per worker per hour.
   After: Labor productivity = 21 carts per worker per hour.
   b. Before: Multifactor productivity = .89 carts per dollar.
   After: Multifactor productivity = .93 carts per dollar.

Chapter 2 Supplement: Decision Making

1. *a*. Expand (80).
   b. Do nothing (50).
   c. Indifferent between do nothing and subcontract (55).
   d. Subcontract (10).
2. *a*. Expand (62).
   c. \$9 (000).
3. Do nothing: $P(\text{high}) < .50$.
   Subcontract: $.50 < P(\text{high}) < .67$.
   Expand: $P(\text{high}) > .67$.
4. *a*.

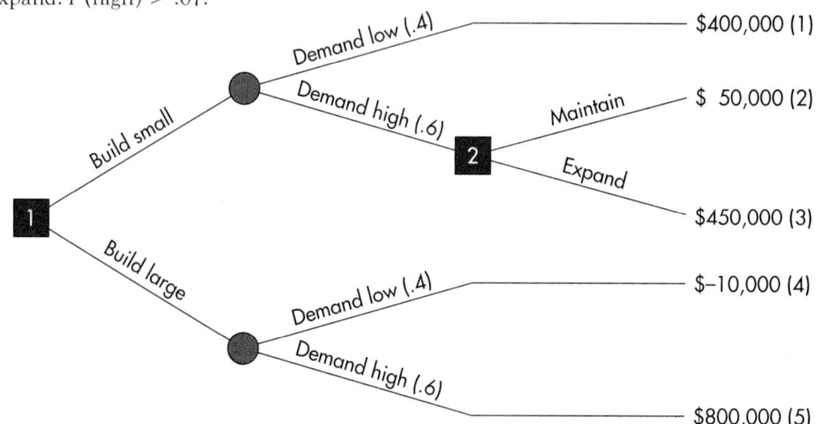

   b. \$164,000.
   c. small 0 to .46.
   large .46 to 1.00.
5. *b*. 4.
   c. 0.
6. *a*. Relocate.
   b. Renew.
   c. Relocate.
   d. Relocate.
7. *a*. Renew.
   b. EVPI = \$1,575,000.
   c. Yes.
9. *b*. Build large: \$53.6 million.
   Build small: \$42 million.
   c. \$12.4.
   d. Build small for $P(\text{high}) < .721$.
   Build large for $P(\text{high}) > .721$.
10. Buy two (\$113.5).
11. A: 22.
    B: 25.
12. \$4.
13. *a*. New staff.
    b. Redesign.
    c. New staff.

d. New staff or redesign.

16. *b.* Alternative C.
 c. $P(2) > .625$.
 d. $P(1) < .375$.

17. *b.* Alternative B.
 c. $P(2) < .444$.
 d. $P(1) > .556$.

Chapter 3: Forecasting

1. *d.* Demand did not exceed supply.
2. *b.* (1) 20.86, (2) 19, (3) 19.26, (4) 20, (5) 20.4.
3. *a.* 88.16 percent.
 b. 88.54 percent.
4. *a.* 22. *b.* 20.75. *c.* 20.72.
5. *a.* Increasing by 15,000 bottles per year.
 b. 275 (i.e., 275,000 bottles).
6. $500 - 20t$.
7. *a.* $Y_t = 204.09 + 19.76t$. *c.* 500.49.
8. Y_{11}: 267.63; Y_{12}: 274.13; Y_{13}: 280.63; Y_{14}: 287.13.
11. Q_1: 157.85; Q_2: 175; Q_3: 126.3; Q_4: 325; Q_1 = 322.85.
12. Fri. = 0.79, Sat. = 1.34, Sun. = 0.87.

15.

Day	Relative
1	.902
2	.836
3	.919
4	1.034
5	1.416
6	1.487
7	.427

17. *b.* Jan. 800
 Feb. 810
 Mar. 900
 Apr. 810
20. *b.* $29,000.
21. *b.* $17.90.
23. *b.* $Y = 66.33 + .584x$. *c.* 90.27.
24. *a.* $r = +.94$. *b.* $Y = .955 + 5.276x$.
 c. 11.51 mowers.
27. *a.* $MAD_5 = 5$ *b.* $TS_5 = 1.40$
 $MAD_6 = 5.9$ $TS_6 = -0.17$
 $MAD_7 = 4.73$ $TS_7 = -0.63$
 $MAD_8 = 3.911$ $TS_8 = -0.26$
 $MAD_9 = 4.238$ $TS_9 = -1.42$
 etc. etc.
28. *a.*

	MSE	MAD
Forecast 1	10.44	2.8
Forecast 2	42.44	3.6
Naive	156	10.7

29. *a.* Initial MAD = 4.727. The tracking signal for month 15 is 4.087, so at that point, the forecast would be suspect.
 b. Σ errors = -1, Σ errors^2 = 345. Control limits: 0 ± 12.38 (in control). Plot reveals cycles in errors.

Chapter 4 Supplement: Reliability

1. *a.* .81. *b.* .9801. *c.* .9783.
2. .9033.
3. .9726.
4. .93.
5. *a.* .9315. *b.* .9953. *c.* .994.
6. *a.* .7876. *b.* .90 component.
7. *a.* Plan 2 (.9934).
8. .996.
9. .995.
10. .006.
11. *a.* (1) .2725. *b.* (1) .6671.
 (2) .2019. (2) .3935.
 (3) .1353. (3) .1813.
 c. (1) 21 months. (3) 90 months.
 (2) 57 months. (4) 138 months.
12. *a.* .6321. *b.* Three months or 90 days.
13. *a.* .3012. *b.* .1813. *c.* .5175.
14. *a.* .2231. *b.* .8647.
 c. .0878. *d.* .0302.
15. *a.* .2266. *b.* .4400. *c.* .3830.
16. *a.* (1) .9772. *b.* Approximately zero.
 (2) .5000.
 (3) .0013.
18. *a.* .93. *b.* .98.
19. .96.

Chapter 5: Process Selection and Capacity Planning

1. *a.* 46,000 units. *b.* (1) $3,000.
 (2) $8,200.
 c. 126,000 units. *d.* 25,556 units.
2. *a.* A: 8,000 units. *b.* 10,000 units.
 B: 7,500 units.
 c. A: $20,000.
 B: $18,000.
3. *a.* 39,683 units. *b.* $1.71 (rounded up).
4. *a.* A: $82.
 B: $92.
 C: $100.
 c. A: 0 to less than 178.
 B: Never.
 C: 178+.
5. ⅓ day, ⅔ evening.
6. Vendor best for $Q < 63{,}333$. For larger quantities, produce in-house at $4 per unit.
7. *a.* Vendor B best for 10,000 and 20,000.
8. 3 cells.
9. Buy 2 Bs.
11. *a.* one: $Q = 80$. two: $Q = 152$.
12. One line.

Chapter 5 Supplement: Linear Programming

1. *a.* (1) $x_1 = 2, x_2 = 9, Z = 35$.
 (2) No.
 (3) No.
 (4) No.
 b. (1) $x_1 = 1.5, x_2 = 6.25, Z = 65.6$.
 (2) No.

(3) Yes, S has surplus of 15.
(4) No.
c. (1) $A = 24$, $B = 20$, $Z = \$204$.
(2) Yes. Labor, 420 hr.
(3) No.
(4) No.
2. a. $S = 8$, $T = 20$, $Z = \$58.40$.
b. (1) $x_1 = 4.2$, $x_2 = 1.6$, $Z = 13.2$.
(2) Yes. $E = 4.6$.
(3) No.
(4) No.
3. $H = 132$ units, $W = 36$ units, Profit = \$6,360.
4. Deluxe = 90 bags, Standard = 60 bags, Profit = \$243.
5. $x = 327.27$, $y = 163.64$, $Z = \$1{,}636.36$.
6. 500 apple, 200 grape, Revenue = \$990. Fifty cups of sugar will be unused.
7. a. $x_1 = 4$, $x_2 = 0$, $x_3 = 18$.
$s_1 = 3$, $s_2 = 0$, $s_3 = 0$.
$Z = 106$.
b. $x_1 = 15$, $x_2 = 10$, $x_3 = 0$.
$s_1 = 0$, $s_2 = 0$, $s_3 = 5$.
$Z = 210$.
9. $A = 0$, $B = 80$, $C = 50$.
$Z = 350$.
C_A (insignificance): \$0 to \$3.04.
C_B (optimality): \$1.95 to \$3.75.
C_C (optimality): \$2.00 to \$5.00.
11. a. board = 0, holder = 50.
b. Cutting = 16 minutes, gluing = 0 minutes, finishing = 210 minutes.
12. a. ham = 37.14, deli = 18, cost = \$165.42.
b. ham = 20, deli = 84, profit = \$376.
14. $Z = \$433$.
15. a. Machine and materials are binding.
b. No change.
c. No change.
d. Only s_2 would change. It would be 46.
e. None.
f. Yes; \$844
16. a. \$1.50; range is 510 to 750.
b. \$1.50/pound.
c. \$0; range 375 to infinity.
d. None.
e. 150 pounds of pine bark.
f. Optimal quantities would not change; Z would increase by \$75.
g. Yes, \$1,155. h. Yes, \$1,110.

Chapter 6: Facilities Layout

1. a. Minimum is 2.4 minutes, maximum is 18 minutes.
b. 25 units to 187.5 units. c. Eight.
d. 3.6 minutes. e. (1) 50 units.
(2) 30 units.

2. a.

Station	Tasks	Time
1	a	1.4
2	b, e	1.3
3	d, c, f	1.8
4	g, h	1.5

3. a.

Station	Tasks	Time
1	f, a	8
2	d, g	13
3	b, c, e	10
4	h, i	14

4. a. (3) 11.54%. (4) 323 units per day.
b. 2.3 minutes. (3) 182.6 units per day.
(4) 91.3 unts per day.
5. b. 2 minutes. c. Three stations.
6. c. (1) 11.1%. (2) 11.1%. (3) 11.1%.
7. b. CT = .84 min. or 50.4 sec.
c. N = 3.83 (round to 4) stations.

10.

	2	3	4	5	6
1	A	X	O	O	O
2		O	O	A	O
3			O	O	O
4				O	O
5					A

11.

1	5	4
3	8	7
6	2	

13.

3	1	8
9	7	4
5	2	6

14.

A #1	B #4	C #2
	D #3	

TC = \$14,150.

16. A: 3; B: 5; C: 1; D: 4; E: 6; F: 2.
17. A: 1; B: 2; C: 5; D: 4; E: 9; F: 8; G: 6; H: 10; I: 7; J: 3.

Chapter 7: Design of Work Systems

1. 15.08 minutes.
2. a. 1.2 minutes.
b. 1.14 minutes.
c. 1.27 minutes.

3.

Element	OT	NT	ST
1 . . .	0.46	0.414	0.476
2 . . .	1.505	1.280	1.472
3 . . .	0.83	0.913	1.050
4 . . .	1.16	1.160	1.334

4.

Element	Average
1	4.1
2	1.5
3	3.3
4	2.7

7. 5.85 minutes.
8. 7.125 minutes.
10. 57 observations.
11. 37 cycles.
12. *a.* 12%. *b.* 163 observations.
13. 377 observations.

Chapter 7 Supplement: Learning Curves

1. *a.* 178.8 hours.
   b. 1,121.4 hours.
   c. 2,914.8 hours.
2. *a.* 41.47 hours.
   b. 60.55 hours.
   c. 72.20 hours.
3. *a.* 56.928 days.
   b. 42.288 days.
   c. 37.512 days.
5. *a.* P = 85 percent.
   b. 26.21 minutes.
6. 87.9 minutes.
7. 201.26 hours.
8. *a.* 11.35 hours.
   b. 13.05 hours.
   c. 13.12 hours.
9. *a.* $80.31.
   b. 10 units.
10. B and C.
12. 30.82 hours.
13. No.
14. 18.76 hours.

Chapter 8: Location Planning and Analysis

1. Kansas City: $256,000.
2. *a.* A: 16,854; B: 17,416; C: 17,753.
   b. C: $14,670.
3. *a.* 120 units. *b.* A: 0 to 119; B: 121+.
4. *a.* B: 0 to 33; C: 34 to 400; A: 400+.
5. C($270,000).
6. Biloxi ($160,000).
7. *a.* (1) outside; (2) city. *b.* 230 cars.
9. A.
10. *a.* B = C > A.
    b. B > C > A.
11. (5,4) is optimal.
12. (6,7).
13. (6,6).
14. (3,2.21).

Chapter 8 Supplement: The Transportation Model

1.

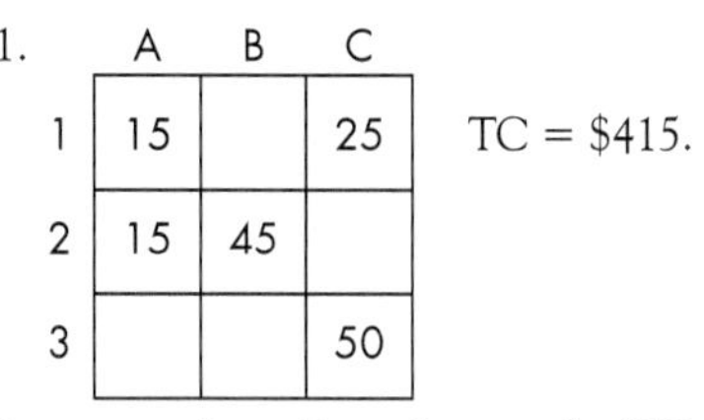

	A	B	C
1	15		25
2	15	45	
3			50

TC = $415.

2. *a.*

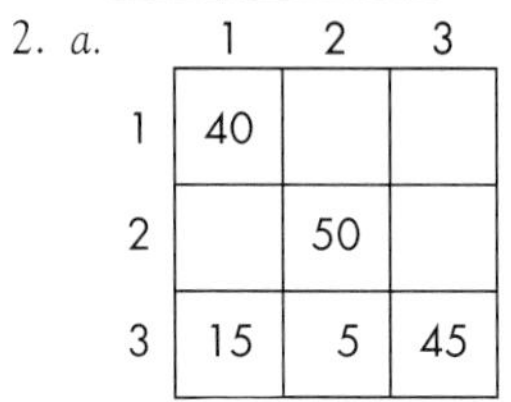

	1	2	3
1	40		
2		50	
3	15	5	45

b. TC = $485.
c. No.

3.

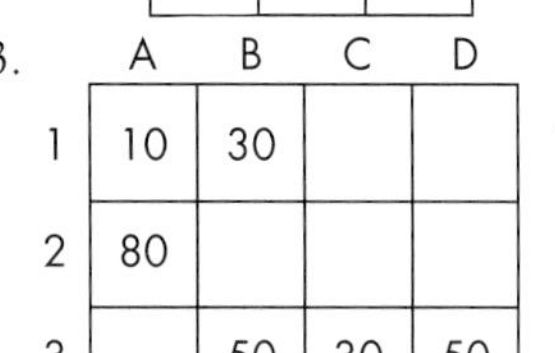

	A	B	C	D
1	10	30		
2	80			
3		50	30	50

TC = $6,310.

4.

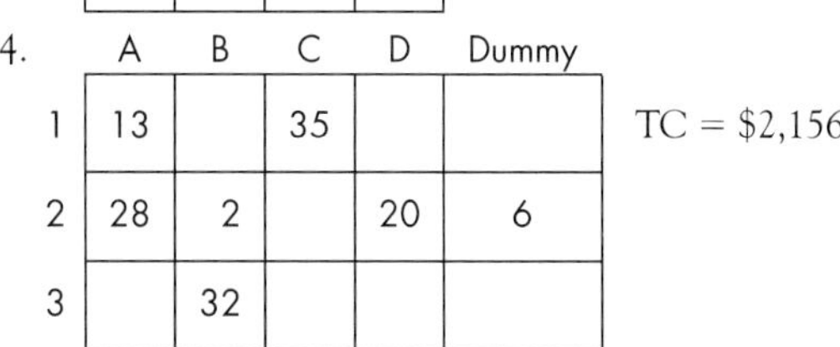

	A	B	C	D	Dummy
1	13		35		
2	28	2		20	6
3		32			

TC = $2,156.

5.

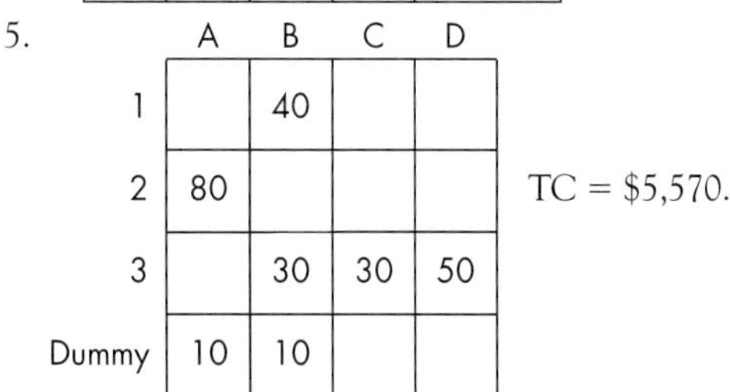

	A	B	C	D
1		40		
2	80			
3		30	30	50
Dummy	10	10		

TC = $5,570.

6.

	A	B	C	D	Dummy
1	41		7		
2				56	
3		32			
Balt.		2	28	4	16

	A	B	C	D	Dummy
1	41		7		
2		2		54	
3		32			
Ph.			28	6	16

7. *b.* $1,310.
   c. No.
8.

	A	B	C	Dummy
1		500		
2	400			
N1		100	350	50

	A	B	C	Dummy
1		500		
2	400			
N2		100	350	50

9.

	A	B	C
1			210
2	140		
3	80	60	10
Tol.		160	

	A	B	C
1			210
2	60	80	
3		140	10
Cin.	160		

Chapter 10: Quality Control

1. *a.* .0124. *b.* 24.35 pounds and 24.65 pounds.
2. *a.* LCL: 0.996 liters. *b.* Not in control.
   UCL: 1.004 liters.
3. *a.* Mean: LCL is 3.019, UCL is 3.181.
   Range: LCL is 0.1845, UCL is 0.7155.
   b. Yes.
4. Mean: LCL is 78.88 cm.
   UCL is 81.04 cm.
   Range: LCL is 0 cm.
   UCL is 4.009 cm.
   Process in control.
5. *a.*

1	2	3	4
.020	.010	.025	.045

   b. 2.5 percent.
   c. Mean = .025, standard deviation = .011.
   d. LCL = .0011, UCL = .0489.
   e. .0456.
   f. Yes.
   g. Mean = .02, standard deviation = .01.
   h. LCL = 0, UCL = .04.
6. LCL: 0.
   UCL: .0234.
   Revised limits based on 12 samples:
   LCL: 0.
   UCL: .0197.
7. Yes, UCL = 16.266, LCL = 0.
8. Yes, UCL = 5.17, LCL = 0.
9. Yes, UCL = .098, LCL = .010.
10. 35 pieces.
11. One in 30 is "out." Tolerances seem to be met. Approximately 97% will be acceptable.
12. *a.* LCL: 3.73.
    UCL: 3.97.
    Out of control.
    b. Random variations.
13.

	Test	NUMBER OF RUNS Observed	Expected	Standard Deviation	z	Conclude
a.	Med	18	14	2.50	1.6	Okay
	U/D	17	17	2.07	0	Okay
b.	Med	 8	14	2.50	−2.40	Nonrandom
	U/D	22	17	2.07	2.41	Nonrandom

14. *a.* Med: $z = -0.52$.
    U/D: $z = -1.47$.
    b. Med: $z = -1.11$.
    U/D: $z = -1.36$.
15. Med: $z = -2.34$.
    U/D: $z = -1.45$.
16. 440 pieces.
18. *a.* 566 units.
    b. 62 units.
    c. $1,160.
19. Med: $z = +0.9177$.
    U/D: $z = +0.5561$.
20. *b.* 4.5, .192.
    c. 4.5, .086.
    d. 4.242 to 4.758.
    f. None.
21. *a.* 1.11. *b.* Yes.
22. Machines 004 and 005 are capable.

Chapter 10 Supplement: Acceptance Sampling

1. *a.* (1) Yes. (2) Yes. *b.* .0067.
2. *b.* .0390.
3. *c.* .0024.
5. *a.* 0.16.
   b. 2.

c. (1) Accept.
(2) .0362.
(3) .9638.
(4) p(Type I) = .0362.
p(Type II) = 0.

Chapter 11: Total Quality Management (TQM)

2.

	Res.	Com.
Noisy	10	3
Failed	7	2
Odor	5	7
Warm	3	4

Chapter 12: Aggregate Planning

1. *b.* $6,350.
2. *a.* $4,670.
b. $4,800.
3. *b.* $4,640.
4. $4,970.
7. *a.* $31,250.
b. $31,520.
8. *a.* $4,600.
b. $4,680.
c. $4,660.
10. B: $14,340.
C: $14,370.
11. $13,475.
12. $13,885.
13. $12,930.
15. $124,960.
16. $126,650; additional cost: $230.

Chapter 13: Inventory Management

1.

Item	Category
4021	A
9402	C
4066	B
6500	C
9280	C
4050	C
6850	B
3010	C
4400	B

2. *a.*

Item	Category
K34	C
K35	A
K36	B
M10	C
M20	C
Z45	A
F14	B
F95	A
F99	C
D45	B
D48	C
D52	C
D57	B
N08	C
P05	B
P09	C

3. *a.* 36 bags.
b. 18 bags.
c. 135.
d. $1,080.
e. Increase by $127.48.
4. *a.* 204 packages.
b. $611.74.
c. Yes.
d. No; TC = $612; only save $.26.
5. *a.* 1,039 pots; $519.62.
b. $44.55.
6. $364.
7. *a.* 1-6: 75 units; 7-12: 91 units.
b. EOQ requirement.
c. 1-6: 50 units; 7-12: 100 units.
8. *a.* $1.32.
b. $24.30.
9. *a.* 4,812.
b. 15.59 (approx. 16).
c. .96.
10. *a.* 10,328 bags.
b. 3,098 bags.
c. 10.33 days.
d. 7.75 (approx. 8).
e. $774.50.
11. *a.* 1,414 units.
b. 7.07 days.
c. 120 units.
d. No.
13. *a.* 10,000 boxes.
b. 1.8 orders.
14. *a.* 600 stones.
b. 600 stones.
c. 150 stones on hand.
15. Indifferent between 495 and 1,000 pulleys.
16. A, 400 units.
17. Buy 300 gallons from vendor X. TC = $6,685.
18. 6,600 feet.
19. *a.* 370 units.
b. 70 units.

   c. Both smaller.
20. *a.* 91 pounds.
   b. ROP = 691 pounds.
21. *a.* 8.39 gallons.
   b. 40 gallons.
   c. .4168.
22. 70.14 gallons.
23. .1093.
24. ROP = 70.14.
25. *a.* 400 gallons.
   b. 45.02 gallons.
26. *a.* 72 boxes.
   b. .0023.
   c. .0228.
27. *a.* 749 pounds. *b.* 4.078 pounds.
28. *a.* 134 rolls.
   b. 36 rolls.
   c. .055 per cycle [$\sigma_{dLT} = \sqrt{LT}\sigma_d$].
   d. .9996.
29. ROP = 77 cases.
30. *a.* −0.40 gal.
   b. −1.98 gal.
31. KO33: 581.
   K144: 458.
   L700: 0.
32. *a.* P34: ROP = 131.1 units.
   P 35: Every 4 weeks.
   b. 306 units.
   c. 334 units.
34. 25 dozen.
35. *a.* Nine spares. *b.* $C_s \leq \$78.03$.
36. 78.9 pounds.
37. $4.89 per quart.
38. Five cakes.
39. 421.5 pounds.
40. *a.* $0.53 to $1.76.
   c. $56.67 to $190.00.

Chapter 14: Material Requirements Planning

1. F = 2, G = 1, H = 1, J = 6, D = 10, L = 2, A = 4, C = 2.
2. *a.* E = 138.
   b. Week 5.
3. *a.* 360.
   b. Day 1 (now).

4.

End Item: A(2), B(4)

Master Schedule	Week	1	2	3	4	5	6
	Quantity						100

End Item	LT = 2	1	2	3	4	5	6
Gross requirements							100
Scheduled receipts							
Projected on hand							100
Net requirements							100
Planned-order receipts							100
Planned-order releases					100		

A (2)	LT = 1	1	2	3	4	5	6
Gross requirements					200		
Scheduled receipts							
Projected on hand	50	50	50	50	50		
Net requirements					150		
Planned-order receipts					150		
Planned-order releases				150			

B (4)	LT = 1	1	2	3	4	5	6
Gross requirements					400		
Scheduled receipts					100	100	
Projected on hand	100	100	100	100	100	100	100
Net requirements					200		
Planned-order receipts					200		
Planned-order releases				200			

5. *c.* Master Schedule

Weeks	1	2	3	4	5	6	7
Quantity						100	100

d.

P LT = 1 wk.	1	2	3	4	5	6	7
Gross requirements						100	100
Scheduled receipts							
Projected on hand							
Net requirements						100	100
Planned-order receipts						100	100
Planned-order releases					100	100	

K LT = 2 wk.	1	2	3	4	5	6	7
Gross requirements					100	100	
Scheduled receipts			10			30	
Projected on hand			10	10	10		
Net requirements					90	70	
Planned-order receipts					90	70	
Planned-order releases			90	70			

G (3) LT = 1 wk.	1	2	3	4	5	6	7
Gross requirements			270	210			
Scheduled receipts							
Projected on hand 40	40	40	40				
Net requirements			230	210			
Planned-order receipts			253	231			
Planned-order releases		253	231				

H (4) LT = 1 wk.	1	2	3	4	5	6	7
Gross requirements			360	280			
Scheduled receipts							
Projected on hand 200	200	200	200	40			
Net requirements			160	240			
Planned-order receipts			200	240			
Planned-order releases		200	240				

6. Master Schedule

Day	1	2	3	4	5	6	7
Quantity				100	150		200

Table	1	2	3	4	5	6	7
Gross requirements				100	150		200
Scheduled receipts							
Projected on hand							
Net requirements				100	150		200
Planned-order receipts				100			200
Planned-order releases			100	150		200	

Wood Sections (2)	1	2	3	4	5	6	7
Gross requirements			200	300		400	
Scheduled receipts		100					
Projected on hand		100	100				
Net requirements			100	300		400	
Planned-order receipts			100	300		400	
Planned-order releases		400		400			

Braces (3)	1	2	3	4	5	6	7
Gross requirements			300	450		600	
Scheduled receipts							
Projected on hand 60	60	60	60				
Net requirements			240	450		600	
Planned-order receipts			240	450			
Planned-order releases	240	450	600				

Legs (4)	1	2	3	4	5	6	7
Gross requirements			400	600		800	
Scheduled receipts							
Projected on hand 120	120	120	120				
Net requirements			280	600		800	
Planned-order receipts			308	660		880	
Planned-order releases	968		880				

10. Order 160 units in week 2.
11. *a.* Master Schedule for E.

Week number	1	2	3	4	5	6	7	8
Quantity					120			

Item: E LT = 1 week								
Gross requirements					120			
Scheduled receipts								
Projected on hand								
Net requirements					120			
Planned-order receipts					120			
Planned-order releases				120				

Item: I (2) LT = 1 week	1	2	3	4	5	6	7	8
Gross requirements				240				
Scheduled receipts			40					
Projected on hand			40	40				
Net requirements				200				
Planned-order receipts				200				
Planned-order releases			200					

Item: N (4) LT = 2 weeks	1	2	3	4	5	6	7	8
Gross requirements			800					
Scheduled receipts								
Projected on hand	100	100	100					
Net requirements			700					
Planned-order receipts			700					
Planned-order releases	700							

Item: V LT = 2 weeks	1	2	3	4	5	6	7	8
Gross requirements			200					
Scheduled receipts			10					
Projected on hand								
Net requirements			190					
Planned-order receipts			190					
Planned-order releases	190							

13. *c.* Master Schedule for golf carts.

Week number	1	2	3	4	5	6	7	8	9
Quantity						100		100	100

Item: Golf cart LT = 1 week									
Gross requirements						100		100	100
Scheduled receipts									
Projected on hand									
Net requirements						100		100	100
Planned-order receipts						100		100	100
Planned-order releases					100		100	100	

Item: Bases LT = 1 week									
Gross requirements					100		100	100	
Scheduled receipts									
Projected on hand 20	20	20	20	50	100	50	100	50	
Net requirements					0		0	50	
Planned-order receipts			30	50	50	50	50	50	
Planned-order releases		30	50	50	50	50	50		

14. *a.* EPP = 78.57. Order 120 units in period 2 and 60 units in period 6.
15. EPP = 75.76.

Chapter 15: Just-in-Time Systems

1. 3.
2. 3.
4. 3 cycles.
5. 4 cycles.

Chapter 16: Supply Chain Management

1. Use 2-day freight.
2. Use 6-day.
3. Ship 2-day using A.

Chapter 17: Scheduling

1. 1-A, 2-B, 3-C, TC = 15.
2. 1-B, 2-C, 3-A, TC = 21.
3. 1-A, 2-E, 3-D, 4-B, 5-C; or 1-A, 2-D, 3-E, 4-B, 5-C.
4. 1-B, 2-C, 4-A, TC = 18.
5. *a.* 1-A, 2-B, 3-C, 4-D, 5-E.
   b. 1-E, 2-B, 3-C, 4-D, 5-A.
6. *b.*

	FCFS	SPT	DD	CR
Av. compl. time	26.5	19.75	21	26.50
Av. job lateness	11	6	6	11
Av. no. of jobs	2.86	2.14	2.27	2.86

7. FCFS: a-b-c-d-e.
   SPT: c-b-a-e-d.
   DD: a-b-c-e-d.
   CR: a-d-e-b-c.

	FCFS	SPT	DD	CR
Av. compl. time	17.40	14.80	16.80	22.60
Av. job lateness	5.20	5.40	4.60	10.80
Av. no. of jobs	2.72	2.31	2.63	3.53

9. B-A-G-E-F-D-C.
10. *a.* e-b-g-h-d-c-a-f.
    c. 2 hours.
11. *a.* B-A-C-E-F-D.
12. *a.* b-a-c-d-e.
    b. 37 minutes.
    c. 15 minutes.
13. G-A-E-D-B-C-F.
14. *a., b.* Grinding flow time is 93 hours. Total time is 37 hours.
    c. Grinding flow time is 107 hours. Total time is 35 hours.

15. *a.*

	FCFS	SPT	DD	CR
Av. compl. time	15.25	11.12	15.60	15.68
Av. job lateness	1.7	2.33	0.55	0.63
Av. no. of jobs	3.91	2.85	4.00	4.02

16. b-c-e-a-d.
19. A-B-C.
20. C-B-A.
21. B-C-D-A.

Chapter 17 Supplement: Maintenance

1. Expected recalibration cost = $925 a month.
   Use the service contract.
2. Expected repair cost = $456 a month.
   Option #1: $500.
   Option #2: $422.
3.

Equipment	Ratio	Interval (days)
A201	.115	17.60
B400	.054	25.17
C850	.099	34.88

Chapter 18: Project Management

1. *a.* 1-3-6-9-11-12: 31.
   b. 1-2-4-6-8-9: 37.
   c. 1-2-5-12-16: 44.
3. Concerned about critical path activities f, h, and j, but also about a, b, c and e.
5. *a.* Summary:

Activity	ES	EF	LF	LS	Slack
1- 2	0	4	11	7	7
2- 4	4	13	21	12	8
4- 7	13	18	26	21	8
7-10	18	20	28	26	8
10-12	21	24	31	28	7
2- 5	4	12	19	11	7
5- 8	12	19	26	19	7
8-10	19	21	28	26	7
1- 3	0	10	10	0	0
3- 6	10	16	16	10	0
6- 9	16	20	20	16	0
9-11	20	25	25	20	0
11-12	25	31	31	25	0

   b. Summary:

Activity	ES	EF	LF	LS	Slack
1-2	0	5	5	0	0
2-4	5	23	23	5	0
4-6	23	26	26	23	0
6-8	26	35	35	26	0
8-9	35	37	37	35	0
2-5	5	15	17	7	2
5-6	15	19	26	22	7
5-7	15	26	28	17	2
7-8	26	33	35	28	2
1-3	0	8	15	7	7
3-7	8	21	28	15	7

6. 30 weeks.
7. *b.* 24 days: .9686; 21 days: .2350.
8. *a.* .6881.
   b. .3984.
   c. .0204.
9. .3479.
10. *b.* .52.
11. *b.* .0262.
    c. .2085.
12.

Path	Mean	Standard Deviation
a-d-e-h	24.34	1.354
a-f-g	15.50	1.258
b-i-j-k	14.83	1.014
c-m-n-o	26.17	1.658

    27 weeks: .6742; 26 weeks: .4099.
14. Crash schedule (1 week each): c, c, f, f, e, p.
15. *a.* Crash four weeks:
    (1) 7-11, (2) 1-2, (3) 7-11 and 6-10, (4) 11-13 and 4-6.

Chapter 19: Waiting Lines

1. *a.* 2.25 customers.
   b. 75 percent.
   c. Two hours.
   d. .5625.
2. *a.* 0.67 customers.
   b. One minute.
   c. 1.33 customers.
3. *a.* 6 minutes.
   b. 0.25.
   c. 2.25 customers.
4. *a.* 40 percent.
   b. 0.1523.
   c. 0.19 hour.
   d. .2288.
5. *a.* Morning: 0.375 minute; .45.
   Afternoon: 0.677 minute; .54.
   Evening: 0.635 minute; .44.
   b. M: 4; A: 8; E: 5.
6. *a.* 4.444 trucks.
   b. 6.67 minutes.
   c. .7104.
   d. 6 minutes.

e. The system would be overloaded.
f. 13.186.

7. *a.* One dock.
   b. One dock.
8. *a.* 0.952 mechanics.
   b. 0.229.
   c. 0.056 hr.
   d. 0.60.
   e. Two.
9. *a.* 0.995 customer.
   b. 2.24 days.
   c. 31.1 percent.
   d. 0.875 customer.
10. *a.* .437
    b. 0.53 machine.
    c. 1.33 machines.
    d. 40.72 pieces.
    e. Three.
11. *a.* 28.56 pieces.
    b. Two.
12. *a.* 15.9 pieces.
    b. Three.
13. Three.
14. *a.* .90.
    b. $W_1 = .12$ hour.
    $W_2 = .3045$ hour.
    $W_3 = 2.13$ hours.
    c. $L_1 = .36$.
    $L_2 = .91$.
    $L_3 = 6.39$.
15. *a.* 30.
    b. $L_1 = .034$.
    $L_2 = .025$.
18. *a.* approx. 0.011.
    b. approx. 0.444.

Chapter 19 Supplement: Simulation

1. 5.25 jobs.
2. 1: 80 percent, 2: 40 percent, 3: 40 percent.

5.

Patient	Time
1	19.44
2	20.24
3	17.92
4	19.38
5	18.96
6	16.48
7	20.66

7.

Day	Usage
1	43.932
2	48.036
3	34.908
4	32.758
5	42.186
6	30.438
7	33.298
8	41.678
9	33.526
10	30.904

8.

Period	Usage
1	2
2	1
3	3
4	1
5	1
6	4
7	5
8	3
9	1
10	2

10. *b.*

Day	Order Size
1	–
2	–
3	–
4	–
5	3
6	–
7	–
8	1
9	–
10	–
11	3
12	–

13. Ball 1: 1,520 points; ball 2: 1,120 points; ball 3: 1,170 points.
14. Ball 1: 2,120 points; ball 2: 720 points; ball 3: 770 points.

APPENDIX B

TABLES

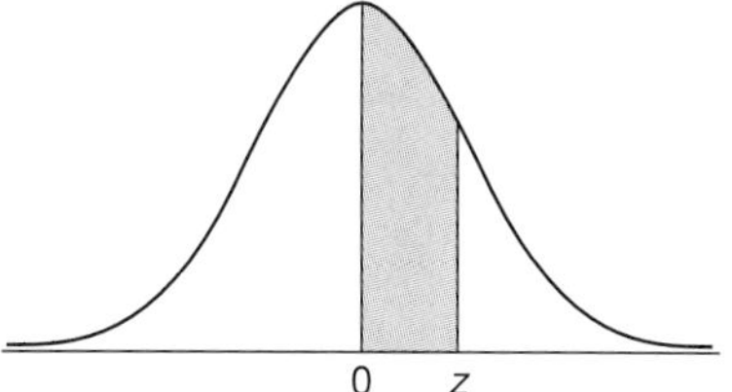

TABLE A

Areas under the normal curve, 0 to z

z	.00	.01	.02	.03	.04	.05	.06	.07	.08	.09
0.0	.0000	.0040	.0080	.0120	.0160	.0199	.0239	.0279	.0319	.0359
0.1	.0398	.0438	.0478	.0517	.0557	.0596	.0636	.0675	.0714	.0753
0.2	.0793	.0832	.0871	.0910	.0948	.0987	.1026	.1064	.1103	.1141
0.3	.1179	.1217	.1255	.1293	.1331	.1368	.1406	.1443	.1480	.1517
0.4	.1554	.1591	.1628	.1664	.1700	.1736	.1772	.1808	.1844	.1879
0.5	.1915	.1950	.1985	.2019	.2054	.2088	.2123	.2157	.2190	.2224
0.6	.2257	.2291	.2324	.2357	.2389	.2422	.2454	.2486	.2517	.2549
0.7	.2580	.2611	.2642	.2673	.2703	.2734	.2764	.2794	.2823	.2852
0.8	.2881	.2910	.2939	.2967	.2995	.3023	.3051	.3078	.3106	.3133
0.9	.3159	.3186	.3212	.3238	.3264	.3289	.3315	.3340	.3365	.3389
1.0	.3413	.3438	.3461	.3485	.3508	.3531	.3554	.3577	.3599	.3621
1.1	.3643	.3665	.3686	.3708	.3729	.3749	.3770	.3790	.3810	.3830
1.2	.3849	.3869	.3888	.3907	.3925	.3944	.3962	.3980	.3997	.4015
1.3	.4032	.4049	.4066	.4082	.4099	.4115	.4131	.4147	.4162	.4177
1.4	.4192	.4207	.4222	.4236	.4251	.4265	.4279	.4292	.4306	.4319
1.5	.4332	.4345	.4357	.4370	.4382	.4394	.4406	.4418	.4429	.4441
1.6	.4452	.4463	.4474	.4484	.4495	.4505	.4515	.4525	.4535	.4545
1.7	.4554	.4564	.4573	.4582	.4591	.4599	.4608	.4616	.4625	.4633
1.8	.4641	.4649	.4656	.4664	.4671	.4678	.4686	.4693	.4699	.4706
1.9	.4713	.4719	.4726	.4732	.4738	.4744	.4750	.4756	.4761	.4767
2.0	.4772	.4778	.4783	.4788	.4793	.4798	.4803	.4808	.4812	.4817
2.1	.4821	.4826	.4830	.4834	.4838	.4842	.4846	.4850	.4854	.4857
2.2	.4861	.4864	.4868	.4871	.4875	.4878	.4881	.4884	.4887	.4890
2.3	.4893	.4896	.4898	.4901	.4904	.4906	.4909	.4911	.4913	.4916
2.4	.4918	.4920	.4922	.4925	.4927	.4929	.4931	.4932	.4934	.4936
2.5	.4938	.4940	.4941	.4943	.4945	.4946	.4948	.4949	.4951	.4952
2.6	.4953	.4955	.4956	.4957	.4959	.4960	.4961	.4962	.4963	.4964
2.7	.4965	.4966	.4967	.4968	.4969	.4970	.4971	.4972	.4973	.4974
2.8	.4974	.4975	.4976	.4977	.4977	.4978	.4979	.4979	.4980	.4981
2.9	.4981	.4982	.4982	.4983	.4984	.4984	.4985	.4985	.4986	.4986
3.0	.4987	.4987	.4987	.4988	.4988	.4989	.4989	.4989	.4990	.4990

TABLE B

1. Areas under the standardized normal curve, from $-\infty$ to $-z$

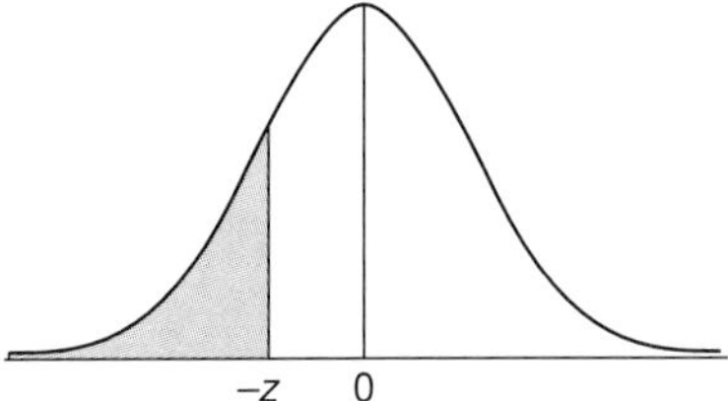

.09	.08	.07	.06	.05	.04	.03	.02	.01	.00	z
.0002	.0003	.0003	.0003	.0003	.0003	.0003	.0003	.0003	.0003	−3.4
.0003	.0004	.0004	.0004	.0004	.0004	.0004	.0005	.0005	.0005	−3.3
.0005	.0005	.0005	.0006	.0006	.0006	.0006	.0006	.0007	.0007	−3.2
.0007	.0007	.0008	.0008	.0008	.0008	.0009	.0009	.0009	.0010	−3.1
.0010	.0010	.0011	.0011	.0011	.0012	.0012	.0013	.0013	.0013	−3.0
.0014	.0014	.0015	.0015	.0016	.0016	.0017	.0018	.0018	.0019	−2.9
.0019	.0020	.0021	.0021	.0022	.0023	.0023	.0024	.0025	.0026	−2.8
.0026	.0027	.0028	.0029	.0030	.0031	.0032	.0033	.0034	.0035	−2.7
.0036	.0037	.0038	.0039	.0040	.0041	.0043	.0044	.0045	.0047	−2.6
.0048	.0049	.0051	.0052	.0054	.0055	.0057	.0059	.0060	.0062	−2.5
.0064	.0066	.0068	.0069	.0071	.0073	.0075	.0078	.0080	.0082	−2.4
.0084	.0087	.0089	.0091	.0094	.0096	.0099	.0102	.0104	.0107	−2.3
.0110	.0113	.0116	.0119	.0122	.0125	.0129	.0132	.0136	.0139	−2.2
.0143	.0146	.0150	.0154	.0158	.0162	.0166	.0170	.0174	.0179	−2.1
.0183	.0188	.0192	.0197	.0202	.0207	.0212	.0217	.0222	.0228	−2.0
.0233	.0239	.0244	.0250	.0256	.0262	.0268	.0274	.0281	.0287	−1.9
.0294	.0301	.0307	.0314	.0322	.0329	.0336	.0344	.0351	.0359	−1.8
.0367	.0375	.0384	.0392	.0401	.0409	.0418	.0427	.0436	.0446	−1.7
.0455	.0465	.0475	.0485	.0495	.0505	.0516	.0526	.0537	.0548	−1.6
.0559	.0571	.0582	.0594	.0606	.0618	.0630	.0643	.0655	.0668	−1.5
.0681	.0694	.0708	.0721	.0735	.0749	.0764	.0778	.0793	.0808	−1.4
.0823	.0838	.0853	.0869	.0885	.0901	.0918	.0934	.0951	.0968	−1.3
.0985	.1003	.1020	.1038	.1056	.1075	.1093	.1112	.1131	.1151	−1.2
.1170	.1190	.1210	.1230	.1251	.1271	.1292	.1314	.1335	.1357	−1.1
.1379	.1401	.1423	.1446	.1469	.1492	.1515	.1539	.1562	.1587	−1.0
.1611	.1635	.1660	.1685	.1711	.1736	.1762	.1788	.1814	.1841	−0.9
.1867	.1894	.1922	.1949	.1977	.2005	.2033	.2061	.2090	.2119	−0.8
.2148	.2177	.2206	.2236	.2266	.2296	.2327	.2358	.2389	.2420	−0.7
.2451	.2483	.2514	.2546	.2578	.2611	.2643	.2676	.2709	.2743	−0.6
.2776	.2810	.2843	.2877	.2912	.2946	.2981	.3015	.3050	.3085	−0.5
.3121	.3156	.3192	.3228	.3264	.3300	.3336	.3372	.3409	.3446	−0.4
.3483	.3520	.3557	.3594	.3632	.3669	.3707	.3745	.3783	.3821	−0.3
.3859	.3897	.3936	.3974	.4013	.4052	.4090	.4129	.4168	.4207	−0.2
.4247	.4286	.4325	.4364	.4404	.4443	.4483	.4522	.4562	.4602	−0.1
.4641	.4681	.4721	.4761	.4801	.4840	.4880	.4920	.4960	.5000	−0.0

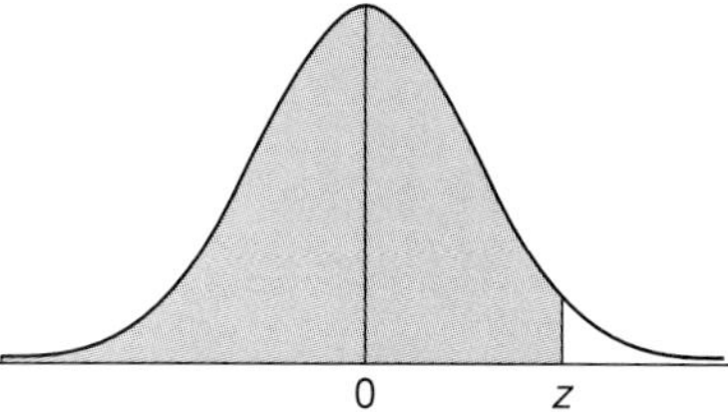

TABLE B

(concluded)

2. *Areas under the standardized normal curve, from* $-\infty$ *to* $+z$

z	.00	.01	.02	.03	.04	.05	.06	.07	.08	.09
.0	.5000	.5040	.5080	.5120	.5160	.5199	.5239	.5279	.5319	.5359
.1	.5398	.5438	.5478	.5517	.5557	.5596	.5636	.5675	.5714	.5753
.2	.5793	.5832	.5871	.5910	.5948	.5987	.6026	.6064	.6103	.6141
.3	.6179	.6217	.6255	.6293	.6331	.6368	.6406	.6443	.6480	.6517
.4	.6554	.6591	.6628	.6664	.6700	.6736	.6772	.6808	.6844	.6879
.5	.6915	.6950	.6985	.7019	.7054	.7088	.7123	.7157	.7190	.7224
.6	.7257	.7291	.7324	.7357	.7389	.7422	.7454	.7486	.7517	.7549
.7	.7580	.7611	.7642	.7673	.7703	.7734	.7764	.7794	.7823	.7852
.8	.7881	.7910	.7939	.7967	.7995	.8023	.8051	.8078	.8106	.8133
.9	.8159	.8186	.8212	.8238	.8264	.8289	.8315	.8340	.8365	.8389
1.0	.8413	.8438	.8461	.8485	.8508	.8531	.8554	.8577	.8599	.8621
1.1	.8643	.8665	.8686	.8708	.8729	.8749	.8770	.8790	.8810	.8830
1.2	.8849	.8869	.8888	.8907	.8925	.8944	.8962	.8980	.8997	.9015
1.3	.9032	.9049	.9066	.9082	.9099	.9115	.9131	.9147	.9162	.9177
1.4	.9192	.9207	.9222	.9236	.9251	.9265	.9279	.9292	.9306	.9319
1.5	.9332	.9345	.9357	.9370	.9382	.9394	.9406	.9418	.9429	.9441
1.6	.9452	.9463	.9474	.9484	.9495	.9505	.9515	.9525	.9535	.9545
1.7	.9554	.9564	.9573	.9582	.9591	.9599	.9608	.9616	.9625	.9633
1.8	.9641	.9649	.9656	.9664	.9671	.9678	.9686	.9693	.9699	.9706
1.9	.9713	.9719	.9726	.9732	.9738	.9744	.9750	.9756	.9761	.9767
2.0	.9772	.9778	.9783	.9788	.9793	.9798	.9803	.9808	.9812	.9817
2.1	.9821	.9826	.9830	.9834	.9838	.9842	.9846	.9850	.9854	.9857
2.2	.9861	.9864	.9868	.9871	.9875	.9878	.9881	.9884	.9887	.9890
2.3	.9893	.9896	.9898	.9901	.9904	.9906	.9909	.9911	.9913	.9916
2.4	.9918	.9920	.9922	.9925	.9927	.9929	.9931	.9932	.9934	.9936
2.5	.9938	.9940	.9941	.9943	.9945	.9946	.9948	.9949	.9951	.9952
2.6	.9953	.9955	.9956	.9957	.9959	.9960	.9961	.9962	.9963	.9964
2.7	.9965	.9966	.9967	.9968	.9969	.9970	.9971	.9972	.9973	.9974
2.8	.9974	.9975	.9976	.9977	.9977	.9978	.9979	.9979	.9980	.9981
2.9	.9981	.9982	.9982	.9983	.9984	.9984	.9985	.9985	.9986	.9986
3.0	.9987	.9987	.9987	.9988	.9988	.9989	.9989	.9989	.9990	.9990
3.1	.9990	.9991	.9991	.9991	.9991	.9992	.9992	.9992	.9993	.9993
3.2	.9993	.9993	.9994	.9994	.9994	.9994	.9994	.9995	.9995	.9995
3.3	.9995	.9995	.9995	.9996	.9996	.9996	.9996	.9996	.9996	.9997
3.4	.9997	.9997	.9997	.9997	.9997	.9997	.9997	.9997	.9997	.9998

TABLE C

Cumulative Poisson probabilities

$$P(x \le c) = \sum_{x=0}^{x=c} \frac{\mu^x \cdot e^{-\mu}}{x!}$$

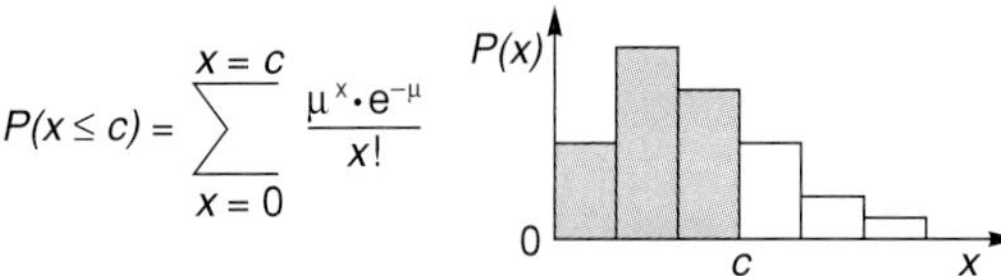

μ\x	0	1	2	3	4	5	6	7	8	9
0.05	.951	.999	1.000							
0.10	.905	.995	1.000							
0.15	.861	.990	.999	1.000						
0.20	.819	.982	.999	1.000						
0.25	.779	.974	.998	1.000						
0.30	.741	.963	.996	1.000						
0.35	.705	.951	.994	1.000						
0.40	.670	.938	.992	.999	1.000					
0.45	.638	.925	.989	.999	1.000					
0.50	.607	.910	.986	.998	1.000					
0.55	.577	.894	.982	.998	1.000					
0.60	.549	.878	.977	.997	1.000					
0.65	.522	.861	.972	.996	.999	1.000				
0.70	.497	.844	.966	.994	.999	1.000				
0.75	.472	.827	.960	.993	.999	1.000				
0.80	.449	.809	.953	.991	.999	1.000				
0.85	.427	.791	.945	.989	.998	1.000				
0.90	.407	.772	.937	.987	.998	1.000				
0.95	.387	.754	.929	.984	.997	1.000				
1.0	.368	.736	.920	.981	.996	.999	1.000			
1.1	.333	.699	.900	.974	.995	.999	1.000			
1.2	.301	.663	.880	.966	.992	.998	1.000			
1.3	.273	.627	.857	.957	.989	.998	1.000			
1.4	.247	.592	.833	.946	.986	.997	.999	1.000		
1.5	.223	.558	.809	.934	.981	.996	.999	1.000		
1.6	.202	.525	.783	.921	.976	.994	.999	1.000		
1.7	.183	.493	.757	.907	.970	.992	.998	1.000		
1.8	.165	.463	.731	.891	.964	.990	.997	.999	1.000	
1.9	.150	.434	.704	.875	.956	.987	.997	.999	1.000	
2.0	.135	.406	.677	.857	.947	.983	.995	.999	1.000	
2.2	.111	.355	.623	.819	.928	.975	.993	.998	1.000	
2.4	.091	.308	.570	.779	.904	.964	.988	.997	.999	1.000
2.6	.074	.267	.518	.736	.877	.951	.983	.995	.999	1.000
2.8	.061	.231	.470	.692	.848	.935	.976	.992	.998	.999

TABLE C *(concluded)*

μ\x	0	1	2	3	4	5	6	7	8	9	10	11	12	13	14	15	16	17	18	19	20
3.0	.050	.199	.423	.647	.815	.916	.966	.988	.996	.999	1.000										
3.2	.041	.171	.380	.603	.781	.895	.955	.983	.994	.998	1.000										
3.4	.033	.147	.340	.558	.744	.871	.942	.977	.992	.997	.999	1.000									
3.6	.027	.126	.303	.515	.706	.844	.927	.969	.988	.996	.999	1.000									
3.8	.022	.107	.269	.474	.668	.816	.909	.960	.984	.994	.998	.999	1.000								
4.0	.018	.092	.238	.433	.629	.785	.889	.949	.979	.992	.997	.999	1.000								
4.2	.015	.078	.210	.395	.590	.753	.868	.936	.972	.989	.996	.999	1.000								
4.4	.012	.066	.185	.359	.551	.720	.844	.921	.964	.985	.994	.998	.999	1.000							
4.6	.010	.056	.163	.326	.513	.686	.818	.905	.955	.980	.992	.997	.999	1.000							
4.8	.008	.048	.143	.294	.476	.651	.791	.887	.944	.975	.990	.996	.999	1.000							
5.0	.007	.040	.125	.265	.441	.616	.762	.867	.932	.968	.986	.995	.998	.999	1.000						
5.2	.006	.034	.109	.238	.406	.581	.732	.845	.918	.960	.982	.993	.997	.999	1.000						
5.4	.005	.029	.095	.213	.373	.546	.702	.822	.903	.951	.978	.990	.996	.999	1.000						
5.6	.004	.024	.082	.191	.342	.512	.670	.797	.886	.941	.972	.988	.995	.998	.999	1.000					
5.8	.003	.021	.072	.170	.313	.478	.638	.771	.867	.929	.965	.984	.993	.997	.999	1.000					
6.0	.003	.017	.062	.151	.285	.446	.606	.744	.847	.916	.957	.980	.991	.996	.999	.999	1.000				
6.2	.002	.015	.054	.134	.259	.414	.574	.716	.826	.902	.949	.975	.989	.995	.998	.999	1.000				
6.4	.002	.012	.046	.119	.235	.384	.542	.687	.803	.886	.939	.969	.986	.994	.997	.999	1.000				
6.6	.001	.010	.040	.105	.213	.355	.511	.658	.780	.869	.927	.963	.982	.992	.997	.999	.999	1.000			
6.8	.001	.007	.030	.082	.173	.301	.450	.599	.729	.830	.915	.955	.978	.990	.996	.998	.999	1.000			
7.0	.001	.007	.030	.082	.173	.301	.450	.599	.729	.830	.901	.947	.973	.987	.994	.998	.999	1.000			
7.2	.001	.006	.025	.072	.156	.276	.420	.569	.703	.810	.887	.937	.967	.984	.993	.997	.999	.999	1.000		
7.4	.001	.005	.022	.063	.140	.253	.392	.539	.676	.788	.871	.926	.961	.980	.991	.996	.998	.999	1.000		
7.6	.001	.004	.019	.055	.125	.231	.365	.510	.648	.765	.854	.915	.954	.976	.989	.995	.998	.999	1.000		
7.8	.000	.004	.016	.048	.112	.210	.338	.481	.620	.741	.835	.902	.945	.971	.986	.993	.997	.999	1.000		
8.0	.000	.003	.014	.042	.100	.191	.313	.453	.593	.717	.816	.888	.936	.966	.983	.992	.996	.998	.999	1.000	
8.2	.000	.003	.012	.037	.089	.174	.290	.425	.566	.692	.796	.873	.926	.960	.979	.990	.995	.998	.999	1.000	
8.4	.000	.002	.010	.032	.079	.157	.267	.400	.537	.666	.774	.857	.915	.952	.975	.987	.994	.997	.999	1.000	
8.6	.000	.002	.009	.030	.074	.150	.256	.386	.523	.653	.763	.849	.909	.949	.973	.986	.993	.997	.999	1.000	
8.8	.000	.002	.007	.024	.062	.128	.226	.348	.482	.614	.729	.822	.889	.935	.964	.981	.990	.995	.998	.999	1.000
9.0	.000	.001	.006	.021	.055	.116	.207	.324	.456	.587	.706	.803	.876	.926	.959	.978	.989	.995	.998	.999	1.000
9.5	.000	.001	.004	.015	.040	.089	.165	.269	.392	.522	.645	.752	.836	.898	.940	.967	.982	.991	.996	.998	.999

TABLE D

Cumulative binomial probabilities

$$P(x \le c) = \sum_{x=0}^{c} \binom{n}{x} p^x (1-p)^{n-x}$$

n	x	.05	.10	.15	.20	.25	.30	.35	.40	.45	.50	.55	.60	.65	.70	.75	.80	.85	.90
1	0	.9500	.9000	.8500	.8000	.7500	.7000	.6500	.6000	.5500	.5000	.4500	.4000	.3500	.3000	.2500	.2000	.1500	.1000
	1	1.0000	1.0000	1.0000	1.0000	1.0000	1.0000	1.0000	1.0000	1.0000	1.0000	1.0000	1.0000	1.0000	1.0000	1.0000	1.0000	1.0000	1.0000
2	0	.9025	.8100	.7225	.6400	.5625	.4900	.4225	.3600	.3025	.2500	.2025	.1600	.1225	.0900	.0625	.0400	.0225	.0100
	1	.9975	.9900	.9775	.9600	.9375	.9100	.8775	.8400	.7975	.7500	.6975	.6400	.5775	.5100	.4375	.3600	.2775	.1900
	2	1.0000	1.0000	1.0000	1.0000	1.0000	1.0000	1.0000	1.0000	1.0000	1.0000	1.0000	1.0000	1.0000	1.0000	1.0000	1.0000	1.0000	1.0000
3	0	.8574	.7290	.6141	.5120	.4219	.3430	.2746	.2160	.1664	.1250	.0911	.0640	.0429	.0270	.0156	.0080	.0034	.0010
	1	.9928	.9720	.9393	.8960	.8438	.7840	.7183	.6480	.5748	.5000	.4253	.3520	.2818	.2160	.1563	.1040	.0608	.0280
	2	.9999	.9990	.9966	.9920	.9844	.9730	.9571	.9360	.9089	.8750	.8336	.7840	.7254	.6570	.5781	.4880	.3859	.2710
	3	1.0000	1.0000	1.0000	1.0000	1.0000	1.0000	1.0000	1.0000	1.0000	1.0000	1.0000	1.0000	1.0000	1.0000	1.0000	1.0000	1.0000	1.0000
4	0	.8145	.6561	.5220	.4096	.3164	.2401	.1785	.1296	.0915	.0625	.0410	.0256	.0150	.0081	.0039	.0016	.0005	.0001
	1	.9860	.9477	.8905	.8192	.7383	.6517	.5630	.4752	.3910	.3125	.2415	.1792	.1265	.0837	.0508	.0272	.0120	.0037
	2	.9995	.9963	.9880	.9728	.9492	.9163	.8735	.8208	.7585	.6875	.6090	.5248	.4370	.3483	.2617	.1808	.1095	.0523
	3	1.0000	.9999	.9995	.9984	.9961	.9919	.9850	.9744	.9590	.9375	.9085	.8704	.8215	.7599	.6836	.5904	.4780	.3439
	4	1.0000	1.0000	1.0000	1.0000	1.0000	1.0000	1.0000	1.0000	1.0000	1.0000	1.0000	1.0000	1.0000	1.0000	1.0000	1.0000	1.0000	1.0000
5	0	.7738	.5905	.4437	.3277	.2373	.1681	.1160	.0778	.0503	.0313	.0185	.0102	.0053	.0024	.0010	.0003	.0001	.0000
	1	.9974	.9185	.8352	.7373	.6328	.5282	.4284	.3370	.2562	.1875	.1312	.0870	.0540	.0308	.0156	.0067	.0022	.0005
	2	.9988	.9914	.9734	.9421	.8965	.8369	.7648	.6826	.5931	.5000	.4069	.3174	.2352	.1631	.1035	.0579	.0266	.0086
	3	1.0000	.9995	.9978	.9933	.9844	.9692	.9460	.9130	.8688	.8125	.7438	.6630	.5716	.4718	.3672	.2627	.1648	.0815
	4	1.0000	1.0000	.9999	.9997	.9990	.9976	.9947	.9898	.9815	.9688	.9497	.9222	.8840	.8319	.7627	.6723	.5563	.4095
	5	1.0000	1.0000	1.0000	1.0000	1.0000	1.0000	1.0000	1.0000	1.0000	1.0000	1.0000	1.0000	1.0000	1.0000	1.0000	1.0000	1.0000	1.0000
6	0	.7351	.5314	.3771	.2621	.1780	.1176	.0754	.0467	.0277	.0156	.0083	.0041	.0018	.0007	.0002	.0001	.0000	.0000
	1	.9672	.8857	.7765	.6554	.5339	.4202	.3191	.2333	.1636	.1094	.0692	.0410	.0223	.0109	.0046	.0016	.0004	.0001
	2	.9978	.9842	.9527	.9011	.8306	.7443	.6471	.5443	.4415	.3438	.2553	.1792	.1174	.0705	.0376	.0170	.0059	.0013
	3	.9999	.9987	.9941	.9830	.9624	.9295	.8826	.8208	.7447	.6563	.5585	.4557	.3529	.2557	.1694	.0989	.0473	.0159
	4	1.0000	.9999	.9996	.9984	.9954	.9891	.9777	.9590	.9308	.8906	.8364	.7667	.6809	.5798	.4661	.3446	.2235	.1143
	5	1.0000	1.0000	1.0000	.9999	.9998	.9993	.9982	.9959	.9917	.9844	.9723	.9533	.9246	.8824	.8220	.7379	.6229	.4686
	6	1.0000	1.0000	1.0000	1.0000	1.0000	1.0000	1.0000	1.0000	1.0000	1.0000	1.0000	1.0000	1.0000	1.0000	1.0000	1.0000	1.0000	1.0000

TABLE D *(continued)*

n	x	.05	.10	.15	.20	.25	.30	.35	.40	.45	.50	.55	.60	.65	.70	.75	.80	.85	.90
7	0	.6983	.4783	.3206	.2097	.1335	.0824	.0490	.0280	.0152	.0078	.0037	.0016	.0006	.0002	.0001	.0000	.0000	.0000
	1	.9556	.8503	.7166	.5767	.4449	.3294	.2338	.1586	.1024	.0625	.0357	.0188	.0090	.0038	.0013	.0004	.0001	.0000
	2	.9962	.9743	.9262	.8520	.7564	.6471	.5323	.4199	.3164	.2266	.1529	.0963	.0556	.0288	.0129	.0047	.0012	.0002
	3	.9998	.9973	.9879	.9667	.9294	.8740	.8002	.7102	.6083	.5000	.3917	.2898	.1998	.1260	.0706	.0333	.0121	.0027
	4	1.0000	.9998	.9988	.9953	.9871	.9712	.9444	.9037	.8471	.7734	.6836	.5801	.4677	.3529	.2436	.1480	.0738	.0257
	5	1.0000	1.0000	.9999	.9996	.9987	.9962	.9910	.9812	.9643	.9375	.8976	.8414	.7662	.6706	.5551	.4233	.2834	.1497
	6	1.0000	1.0000	1.0000	1.0000	.9999	.9998	.9994	.9984	.9963	.9922	.9848	.9720	.9510	.9176	.8665	.7903	.6794	.5217
	7	1.0000	1.0000	1.0000	1.0000	1.0000	1.0000	1.0000	1.0000	1.0000	1.0000	1.0000	1.0000	1.0000	1.0000	1.0000	1.0000	1.0000	1.0000
8	0	.6634	.4305	.2725	.1678	.1001	.0576	.0319	.0168	.0084	.0039	.0017	.0007	.0002	.0001	.0000	.0000	.0000	.0000
	1	.9428	.8131	.6572	.5033	.3671	.2553	.1691	.1064	.0632	.0352	.0181	.0085	.0036	.0013	.0004	.0001	.0000	.0000
	2	.9942	.9619	.8948	.7969	.6785	.5518	.4278	.3154	.2201	.1445	.0885	.0498	.0253	.0113	.0042	.0012	.0002	.0000
	3	.9996	.9950	.9786	.9437	.8862	.8059	.7064	.5941	.4470	.3633	.2604	.1737	.1061	.0580	.0273	.0104	.0029	.0004
	4	1.0000	.9996	.9971	.9896	.9727	.9420	.8939	.8263	.7396	.6367	.5230	.4059	.2936	.1941	.1138	.0563	.0214	.0050
	5	1.0000	1.0000	.9998	.9988	.9958	.9887	.9747	.9502	.9115	.8555	.7799	.6848	.5722	.4482	.3215	.2031	.1052	.0381
	6	1.0000	1.0000	1.0000	.9999	.9996	.9987	.9964	.9915	.9819	.9648	.9368	.8936	.8309	.7447	.6329	.4967	.3428	.1869
	7	1.0000	1.0000	1.0000	1.0000	1.0000	.9999	.9998	.9993	.9983	.9961	.9916	.9832	.9681	.9424	.8999	.8322	.7275	.5695
	8	1.0000	1.0000	1.0000	1.0000	1.0000	1.0000	1.0000	1.0000	1.0000	1.0000	1.0000	1.0000	1.0000	1.0000	1.0000	1.0000	1.0000	1.0000
9	0	.6302	.3874	.2316	.1342	.0751	.0404	.0207	.0101	.0046	.0020	.0008	.0003	.0001	.0000	.0000	.0000	.0000	.0000
	1	.9288	.7748	.5995	.4362	.3003	.1960	.1211	.0705	.0385	.0195	.0091	.0038	.0014	.0004	.0001	.0000	.0000	.0000
	2	.9916	.9470	.8591	.7382	.6007	.4628	.3373	.2318	.1495	.0898	.0498	.0250	.0112	.0043	.0013	.0003	.0000	.0000
	3	.9994	.9917	.9661	.9144	.8343	.7297	.6089	.4826	.3614	.2539	.1658	.0994	.0536	.0253	.0100	.0031	.0006	.0001
	4	1.0000	.9991	.9944	.9804	.9511	.9012	.8283	.7334	.6214	.5000	.3786	.2666	.1717	.0988	.0489	.0196	.0056	.0009
	5	1.0000	.9999	.9994	.9969	.9900	.9747	.9496	.9006	.8342	.7461	.6386	.5174	.3911	.2703	.1657	.0856	.0339	.0083
	6	1.0000	1.0000	1.0000	.9997	.9987	.9957	.9888	.9750	.9502	.9102	.8505	.7682	.6627	.5372	.3993	.2618	.1409	.0530
	7	1.0000	1.0000	1.0000	1.0000	.9999	.9996	.9986	.9962	.9909	.9805	.9615	.9295	.8789	.8040	.6997	.5638	.4005	.2252
	8	1.0000	1.0000	1.0000	1.0000	1.0000	1.0000	.9999	.9997	.9992	.9980	.9954	.9899	.9793	.9596	.9249	.8658	.7684	.6126
	9	1.0000	1.0000	1.0000	1.0000	1.0000	1.0000	1.0000	1.0000	1.0000	1.0000	1.0000	1.0000	1.0000	1.0000	1.0000	1.0000	1.0000	1.0000
10	0	.5987	.3487	.1969	.1074	.0563	.0282	.0135	.0060	.0025	.0010	.0003	.0001	.0000	.0000	.0000	.0000	.0000	.0000
	1	.9139	.7361	.5443	.3758	.2440	.1493	.0860	.0464	.0233	.0107	.0045	.0017	.0005	.0001	.0000	.0000	.0000	.0000
	2	.9885	.9298	.8202	.6778	.5256	.3828	.2616	.1673	.0996	.0547	.0274	.0123	.0048	.0016	.0004	.0001	.0000	.0000
	3	.9990	.9872	.9500	.8791	.7759	.6496	.5138	.3823	.2660	.1719	.1020	.0548	.0260	.0106	.0035	.0009	.0001	.0000
	4	.9999	.9984	.9901	.9672	.9219	.8497	.7515	.6331	.5044	.3770	.2616	.1662	.0949	.0473	.0197	.0064	.0014	.0001
	5	1.0000	.9999	.9986	.9936	.9803	.9527	.9051	.8338	.7384	.6230	.4956	.3669	.2485	.1503	.0781	.0328	.0099	.0016
	6	1.0000	1.0000	.9999	.9991	.9965	.9894	.9740	.9452	.8980	.8281	.7340	.6177	.4862	.3504	.2241	.1209	.0500	.0128
	7	1.0000	1.0000	1.0000	.9999	.9996	.9984	.9952	.9877	.9726	.9453	.9004	.8327	.7384	.6172	.4744	.3222	.1798	.0702
	8	1.0000	1.0000	1.0000	1.0000	1.0000	.9999	.9995	.9983	.9955	.9893	.9767	.9536	.9140	.8507	.7560	.6242	.4557	.2639
	9	1,0000	1.0000	1.0000	1.0000	1.0000	1.0000	1.0000	.9999	.9997	.9990	.9975	.9940	.9865	.9718	.9437	.8926	.8031	.6513
	10	1.0000	1.0000	1.0000	1.0000	1.0000	1.0000	1.0000	1.0000	1.0000	1.0000	1.0000	1.0000	1.0000	1.0000	1.0000	1.0000	1.0000	1.0000

TABLE D *(concluded)*

n	x	.05	.10	.15	.20	.25	.30	.35	.40	.45	.50	.55	.60	.65	.70	.75	.80	.85	.90
15	0	.4633	.2059	.0874	.0352	.0134	.0047	.0016	.0005	.0001	.0000	.0000	.0000	.0000	.0000	.0000	.0000	.0000	.0000
	1	.8290	.5490	.3186	.1671	.0802	.0353	.0142	.0052	.0017	.0005	.0001	.0000	.0000	.0000	.0000	.0000	.0000	.0000
	2	.9638	.8159	.6042	.3980	.2361	.1268	.0617	.0271	.0107	.0037	.0011	.0003	.0001	.0000	.0000	.0000	.0000	.0000
	3	.9945	.9444	.8227	.6482	.4613	.2969	.1727	.0905	.0424	.0176	.0063	.0019	.0005	.0001	.0000	.0000	.0000	.0000
	4	.9994	.9873	.9383	.8358	.6865	.5155	.3519	.2173	.1204	.0592	.0255	.0093	.0028	.0007	.0001	.0000	.0000	.0000
	5	.9999	.9978	.9832	.9389	.8516	.7216	.5643	.4032	.2608	.1509	.0769	.0338	.0124	.0037	.0008	.0001	.0000	.0000
	6	1.0000	.9997	.9964	.9819	.9434	.8689	.7548	.6098	.4522	.3036	.1818	.0950	.0422	.0152	.0042	.0008	.0001	.0000
	7	1.0000	1.0000	.9994	.9958	.9827	.9500	.8868	.7869	.6535	.5000	.3465	.2131	.1132	.0500	.0173	.0042	.0006	.0000
	8	1.0000	1.0000	.9999	.9992	.9958	.9848	.9578	.9050	.8182	.6964	.5478	.3902	.2452	.1311	.0566	.0181	.0036	.0003
	9	1.0000	1.0000	1.0000	.9999	.9992	.9963	.9876	.9662	.9231	.8491	.7392	.5968	.4357	.2784	.1484	.0611	.0168	.0022
	10	1.0000	1.0000	1.0000	1.0000	.9999	.9993	.9972	.9907	.9745	.9408	.8796	.7827	.6481	.4845	.3135	.1642	.0617	.0127
	11	1.0000	1.0000	1.0000	1.0000	1.0000	.9999	.9995	.9981	.9937	.9824	.9576	.9095	.8273	.7031	.5387	.3518	.1773	.0556
	12	1.0000	1.0000	1.0000	1.0000	1.0000	1.0000	.9999	.9997	.9989	.9963	.9893	.9729	.9383	.8732	.7639	.6020	.3958	.1841
	13	1.0000	1.0000	1.0000	1.0000	1.0000	1.0000	1.0000	1.0000	.9999	.9995	.9983	.9948	.9858	.9647	.9198	.8329	.6814	.4510
	14	1.0000	1.0000	1.0000	1.0000	1.0000	1.0000	1.0000	1.0000	1.0000	1.0000	.9999	.9995	.9984	.9953	.9866	.9648	.9126	.7941
	15	1.0000	1.0000	1.0000	1.0000	1.0000	1.0000	1.0000	1.0000	1.0000	1.0000	1.0000	1.0000	1.0000	1.0000	1.0000	1.0000	1.0000	1.0000
20	0	.3585	.1216	.0388	.0115	.0032	.0008	.0002	.0000	.0000	.0000	.0000	.0000	.0000	.0000	.0000	.0000	.0000	.0000
	1	.7358	.3917	.1756	.0692	.0243	.0076	.0021	.0005	.0001	.0000	.0000	.0000	.0000	.0000	.0000	.0000	.0000	.0000
	2	.9245	.6769	.4049	.2061	.0913	.0355	.0121	.0036	.0009	.0002	.0000	.0000	.0000	.0000	.0000	.0000	.0000	.0000
	3	.9841	.8670	.6477	.4114	.2252	.1071	.0444	.0160	.0049	.0013	.0003	.0000	.0000	.0000	.0000	.0000	.0000	.0000
	4	.9974	.9568	.8298	.6296	.4148	.2375	.1182	.0510	.0189	.0059	.0015	.0003	.0000	.0000	.0000	.0000	.0000	.0000
	5	.9997	.9887	.9327	.8042	.6172	.4164	.2454	.1256	.0553	.0207	.0064	.0016	.0003	.0000	.0000	.0000	.0000	.0000
	6	1.0000	.9976	.9781	.9133	.7858	.6080	.4166	.2500	.1299	.0577	.0214	.0065	.0015	.0003	.0000	.0000	.0000	.0000
	7	1.0000	.9996	.9941	.9679	.8982	.7723	.6010	.4159	.2520	.1316	.0580	.0210	.0060	.0013	.0002	.0000	.0000	.0000
	8	1.0000	.9999	.9987	.9900	.9591	.8867	.7624	.5956	.4143	.2517	.1308	.0565	.0196	.0051	.0009	.0001	.0000	.0000
	9	1.0000	1.0000	.9998	.9974	.9861	.9520	.8782	.7553	.5914	.4119	.2493	.1275	.0532	.0171	.0039	.0006	.0000	.0000
	10	1.0000	1.0000	1.0000	.9994	.9961	.9829	.9468	.8725	.7507	.5881	.4086	.2447	.1218	.0480	.0139	.0026	.0002	.0000
	11	1.0000	1.0000	1.0000	.9999	.9991	.9949	.9804	.9435	.8692	.7483	.5857	.4044	.2376	.1133	.0409	.0100	.0013	.0001
	12	1.0000	1.0000	1.0000	1.0000	.9998	.9987	.9940	.9790	.9420	.8684	.7480	.5841	.3990	.2277	.1018	.0321	.0059	.0004
	13	1.0000	1.0000	1.0000	1.0000	1.0000	.9997	.9985	.9935	.9786	.9423	.8701	.7500	.5834	.3920	.2142	.0867	.0219	.0024
	14	1.0000	1.0000	1.0000	1.0000	1.0000	1.0000	.9997	.9984	.9936	.9793	.9447	.8744	.7546	.5836	.3828	.1958	.0673	.0113
	15	1.0000	1.0000	1.0000	1.0000	1.0000	1.0000	1.0000	.9997	.9985	.9941	.9811	.9490	.8818	.7625	.5852	.3704	.1702	.0432
	16	1.0000	1.0000	1.0000	1.0000	1.0000	1.0000	1.0000	1.0000	.9997	.9987	.9951	.9840	.9556	.8929	.7748	.5886	.3523	.1330
	17	1.0000	1.0000	1.0000	1.0000	1.0000	1.0000	1.0000	1.0000	1.0000	.9998	.9991	.9964	.9879	.9645	.9087	.7939	.5951	.3231
	18	1.0000	1.0000	1.0000	1.0000	1.0000	1.0000	1.0000	1.0000	1.0000	1.0000	.9999	.9995	.9979	.9924	.9757	.9308	.8244	.6083
	19	1.0000	1.0000	1.0000	1.0000	1.0000	1.0000	1.0000	1.0000	1.0000	1.0000	1.0000	1.0000	.9998	.9992	.9968	.9885	.9612	.8784
	20	1.0000	1.0000	1.0000	1.0000	1.0000	1.0000	1.0000	1.0000	1.0000	1.0000	1.0000	1.0000	1.0000	1.0000	1.0000	1.0000	1.0000	1.0000

PHOTO CREDITS

Chapter 1
Page 2: © John Abbott; **page 4:** Courtesy of Southwest Airlines; **page 24:** © Brownie Harris; **page 33:** Courtesy of Wegmans Food Markets, Inc.; **page 33:** Courtesy of Wegmans Food Markets, Inc.; **page 34:** Courtesy of Wegmans Food Markets, Inc.; **page 34:** Courtesy of Wegmans Food Markets, Inc.

Chapter 2
Page 36: © Spencer Grant: PhotoEdit; **page 43:** Courtesy of Texas Instruments; **page 50:** Courtesy of Lexus; **page 50:** Courtesy of Federal Express Corporation; **page 50:** Cathlyn Melloan: Tony Stone Images; **page 51:** Courtesy of Ford PhotoMedia.

Chapter 3
Page 86: © Photodisc; **page 106:** Courtesy of Toro; **page 106:** Courtesy of Toro.

Chapter 4
Page 148: Courtesy of Ford Photomedia Services; **page 152:** © Kevin Horan: Tony Stone Images; **page 152:** Courtesy of Chrysler Corporation; **page 152:** Courtesy of Ford Photomedia Services; **page 156:** Courtesy of The Burdick Group; **page 157:** Courtesy of Images Group, Inc. and Imperial Schrade Corporation; **page 157:** Courtesy of Toshiba America Information Systems, Inc.; **page 160:** Art reprinted from October 25, 1991, issue of *Business Week* by special permission, copyright 1991 by McGraw-Hill, Inc.; **page 160:** Courtesy of Eastman Kodak; **page 164:** Courtesy of Ford Photomedia Services; **page 167:** Courtesy of Toyota Motor Sales, U.S.A, Inc.; **page 173:** Caroline Parsons.

Chapter 5
Page 194: Courtesy of USX Corporation; **page 200:** © David Joel: Tony Stone Images; **page 200:** Courtesy of Ben & Jerry's Homemade; **page 200:** Courtesy of Saturn Corporation; **page 200:** © Tom Tracy; **page 203:** Courtesy of ABB Robotics Inc.; **page 205:** © Michael Rosenfeld: Tony Stone Images; **page 212:** © Alan Levin: Tony Stone Images; **page 212:** © Phil Borden: PhotoEdit; **page 231:** Courtesy of High Acres Landfill and Recycling Center/Fairport, New York..

Chapter 6
Page 266: © 1984 Walt Disney Productions: Disney Enterprises, Inc.; **page 269:** © Jim Howard: Tony Stone Images; **page 273:** Courtesy of Electric Boat Corporation: a General Dynamics Company; **page 289:** © Tom and Dee Ann McCarthy: The Stock Market; **page 289:** © Tom and Dee Ann McCarthy: The Stock Market.

Chapter 7
Page 308: Courtesy of Bridgestone Corporation; **page 313:** © Michael H. Dunn: The Stock Market; **page 320:** © Michael Newman: PhotoEdit; **page 321:** Courtesy of Mercedes-Benz of North America, Inc.; **page 323:** © Greg Girard: The Stock Market; **page 326:** © Alan Levenson: Tony Stone Images; **page 314:** © Gerald Scott; **page 314:** © Gerald Scott; **page 315:** © Gerald Scott.

Chapter 8
Page 360: Courtesy of Kraft Foods International; **page 366:** Courtesy of Tropicana Products, Inc.; **page 371:** Courtesy of Mall of America; **page 371:** Courtesy of Mall of America.

Chapter 9
Page 418: Courtesy of Saatchi & Saatchi; **page 423:** Courtesy of Mitutoya Corporation; **page 427:** © Aldo Mauro; **page 427:** © Blaine Harrington; **page 427:** Courtesy of Career IV; **page 433:** Courtesy of Solectron Corporation.

Chapter 10
Page 438: © Sylvain Coffie: Tony Stone Images; **page 444:** © José Azel: Aurora; **page 452:** Courtesy of The Hertz Corporation; **page 456:** © Michael L. Abramson; **page 463:** Courtesy of Motorola Corporation; **page 476:** © Rich Hein: Chicago Sun-Times.

Chapter 11
Page 490: Courtesy of Trek Bicycle Corporation; **page 496:** Courtesy of L. L. Bean Inc.; **page 509:** Courtesy of Xerox Corporation.

Chapter 12
Page 522: © Rex Rystedt; **page 526:** Courtesy of Rubbermaid Incorporated; **page 529:** © John Neubauer: PhotoEdit; **page 533:** Courtesy of International Business Machines Company; **page 543:** © Larry Edwards: The Stock Market.

Chapter 13
Page 556: The Stock Market; **page 562:** © Tom Tracey; **page 565:** © Matthew McVay: SABA; **page 587:** FPG International; **page 587:** © Ed Wheeler: The Stock Market.

Chapter 14
Page 616: Courtesy of International Business Machines Company; **page 623:** Courtesy of International Business Machines Company; **page 627:** Courtesy of Kirby U.S.A. Inc.; **page 653:** Courtesy of L. & J. G. Stickley, Inc.

Chapter 15
Page 656: Courtesy of Ford Motor Company; **page 663:** Courtesy of Honda Automotive; **page 673:** Tri-State Industries; footage courtesy of MacLean Media, Inc. Chicago, IL; **page 684:** Ron Havis: SABA; **page 689:** Courtesy of the Boeing Company; **page 690:** Courtesy of the Boeing Company.

Chapter 16
Page 692: © Bruce Hands: Tony Stone Images; **page 704:** © JB Pictures, Ltd. (7 photos); **page 705:** © 1998 Bose Corporation; **page 707:** Courtesy of American President Line.

Chapter 17
Page 720: Courtesy of American Airlines; **page 725:** Courtesy of Solectron; **page 743:** © Tony Freeman: PhotoEdit.

Chapter 18
Page 764: © Bill Pogue: Tony Stone Images; **page 772:** © AP/Wide World Photos.

Chapter 19
Page 811: Courtesy of American Airlines; **page 841:** © Jeff Greenberg: PhotoEdit.

COMPANY INDEX

SUBJECT INDEX

N

Z

Areas under the standardized normal curve, from $-\infty$ to $+z$

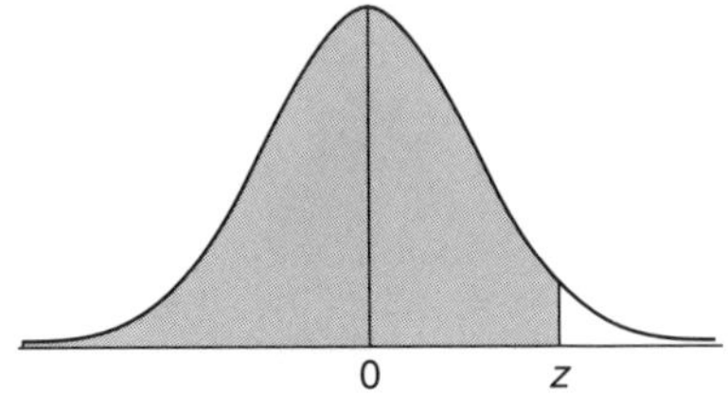

z	.00	.01	.02	.03	.04	.05	.06	.07	.08	.09
.0	.5000	.5040	.5080	.5120	.5160	.5199	.5239	.5279	.5319	.5359
.1	.5398	.5438	.5478	.5517	.5557	.5596	.5636	.5675	.5714	.5753
.2	.5793	.5832	.5871	.5910	.5948	.5987	.6026	.6064	.6103	.6141
.3	.6179	.6217	.6255	.6293	.6331	.6368	.6406	.6443	.6480	.6517
.4	.6554	.6591	.6628	.6664	.6700	.6736	.6772	.6808	.6844	.6879
.5	.6915	.6950	.6985	.7019	.7054	.7088	.7123	.7157	.7190	.7224
.6	.7257	.7291	.7324	.7357	.7389	.7422	.7454	.7486	.7517	.7549
.7	.7580	.7611	.7642	.7673	.7703	.7734	.7764	.7794	.7823	.7852
.8	.7881	.7910	.7939	.7967	.7995	.8023	.8051	.8078	.8106	.8133
.9	.8159	.8186	.8212	.8238	.8264	.8289	.8315	.8340	.8365	.8389
1.0	.8413	.8438	.8461	.8485	.8508	.8531	.8554	.8577	.8599	.8621
1.1	.8643	.8665	.8686	.8708	.8729	.8749	.8770	.8790	.8810	.8830
1.2	.8849	.8869	.8888	.8907	.8925	.8944	.8962	.8980	.8997	.9015
1.3	.9032	.9049	.9066	.9082	.9099	.9115	.9131	.9147	.9162	.9177
1.4	.9192	.9207	.9222	.9236	.9251	.9265	.9279	.9292	.9306	.9319
1.5	.9332	.9345	.9357	.9370	.9382	.9394	.9406	.9418	.9429	.9441
1.6	.9452	.9463	.9474	.9484	.9495	.9505	.9515	.9525	.9535	.9545
1.7	.9554	.9564	.9573	.9582	.9591	.9599	.9608	.9616	.9625	.9633
1.8	.9641	.9649	.9656	.9664	.9671	.9678	.9686	.9693	.9699	.9706
1.9	.9713	.9719	.9726	.9732	.9738	.9744	.9750	.9756	.9761	.9767
2.0	.9772	.9778	.9783	.9788	.9793	.9798	.9803	.9808	.9812	.9817
2.1	.9821	.9826	.9830	.9834	.9838	.9842	.9846	.9850	.9854	.9857
2.2	.9861	.9864	.9868	.9871	.9875	.9878	.9881	.9884	.9887	.9890
2.3	.9893	.9896	.9898	.9901	.9904	.9906	.9909	.9911	.9913	.9916
2.4	.9918	.9920	.9922	.9925	.9927	.9929	.9931	.9932	.9934	.9936
2.5	.9938	.9940	.9941	.9943	.9945	.9946	.9948	.9949	.9951	.9952
2.6	.9953	.9955	.9956	.9957	.9959	.9960	.9961	.9962	.9963	.9964
2.7	.9965	.9966	.9967	.9968	.9969	.9970	.9971	.9972	.9973	.9974
2.8	.9974	.9975	.9976	.9977	.9977	.9978	.9979	.9979	.9980	.9981
2.9	.9981	.9982	.9982	.9983	.9984	.9984	.9985	.9985	.9986	.9986
3.0	.9987	.9987	.9987	.9988	.9988	.9989	.9989	.9989	.9990	.9990
3.1	.9990	.9991	.9991	.9991	.9991	.9992	.9992	.9992	.9993	.9993
3.2	.9993	.9993	.9994	.9994	.9994	.9994	.9994	.9995	.9995	.9995
3.3	.9995	.9995	.9995	.9996	.9996	.9996	.9996	.9996	.9996	.9997
3.4	.9997	.9997	.9997	.9997	.9997	.9997	.9997	.9997	.9997	.9998

Areas under the normal curve, 0 to z

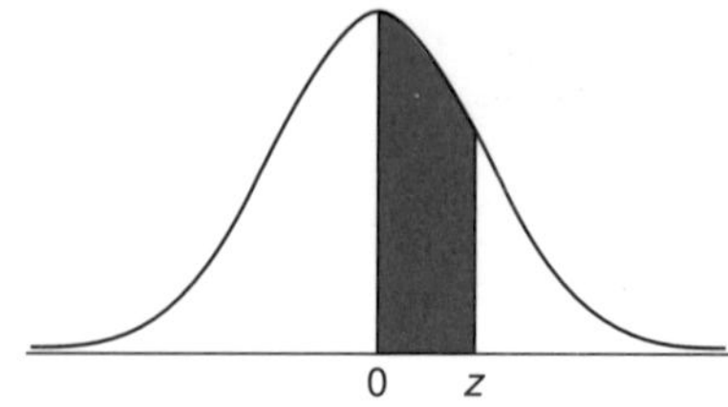

z	.00	.01	.02	.03	.04	.05	.06	.07	.08	.09
0.0	.0000	.0040	.0080	.0120	.0160	.0199	.0239	.0279	.0319	.0359
0.1	.0398	.0438	.0478	.0517	.0557	.0596	.0636	.0675	.0714	.0753
0.2	.0793	.0832	.0871	.0910	.0948	.0987	.1026	.1064	.1103	.1141
0.3	.1179	.1217	.1255	.1293	.1331	.1368	.1406	.1443	.1480	.1517
0.4	.1554	.1591	.1628	.1664	.1700	.1736	.1772	.1808	.1844	.1879
0.5	.1915	.1950	.1985	.2019	.2054	.2088	.2123	.2157	.2190	.2224
0.6	.2257	.2291	.2324	.2357	.2389	.2422	.2454	.2486	.2517	.2549
0.7	.2580	.2611	.2642	.2673	.2703	.2734	.2764	.2794	.2823	.2852
0.8	.2881	.2910	.2939	.2967	.2995	.3023	.3051	.3078	.3106	.3133
0.9	.3159	.3186	.3212	.3238	.3264	.3289	.3315	.3340	.3365	.3389
1.0	.3413	.3438	.3461	.3485	.3508	.3531	.3554	.3577	.3599	.3621
1.1	.3643	.3665	.3686	.3708	.3729	.3749	.3770	.3790	.3810	.3830
1.2	.3849	.3869	.3888	.3907	.3925	.3944	.3962	.3980	.3997	.4015
1.3	.4032	.4049	.4066	.4082	.4099	.4115	.4131	.4147	.4162	.4177
1.4	.4192	.4207	.4222	.4236	.4251	.4265	.4279	.4292	.4306	.4319
1.5	.4332	.4345	.4357	.4370	.4382	.4394	.4406	.4418	.4429	.4441
1.6	.4452	.4463	.4474	.4484	.4495	.4505	.4515	.4525	.4535	.4545
1.7	.4554	.4564	.4573	.4582	.4591	.4599	.4608	.4616	.4625	.4633
1.8	.4641	.4649	.4656	.4664	.4671	.4678	.4686	.4693	.4699	.4706
1.9	.4713	.4719	.4726	.4732	.4738	.4744	.4750	.4756	.4761	.4767
2.0	.4772	.4778	.4783	.4788	.4793	.4798	.4803	.4808	.4812	.4817
2.1	.4821	.4826	.4830	.4834	.4838	.4842	.4846	.4850	.4854	.4857
2.2	.4861	.4864	.4868	.4871	.4875	.4878	.4881	.4884	.4887	.4890
2.3	.4893	.4896	.4898	.4901	.4904	.4906	.4909	.4911	.4913	.4916
2.4	.4918	.4920	.4922	.4925	.4927	.4929	.4931	.4932	.4934	.4936
2.5	.4938	.4940	.4941	.4943	.4945	.4946	.4948	.4949	.4951	.4952
2.6	.4953	.4955	.4956	.4957	.4959	.4960	.4961	.4962	.4963	.4964
2.7	.4965	.4966	.4967	.4968	.4969	.4970	.4971	.4972	.4973	.4974
2.8	.4974	.4975	.4976	.4977	.4977	.4978	.4979	.4979	.4980	.4981
2.9	.4981	.4982	.4982	.4983	.4984	.4984	.4985	.4985	.4986	.4986
3.0	.4987	.4987	.4987	.4988	.4988	.4989	.4989	.4989	.4990	.4990